AF598555

northumbria
UNIVERSITY
LIBRARY
REFERENCE
ONLY
08.03 _ ERDF

ERRATA

VOLUME 13

Page	Line	For:	Read:
63	10	di-*o*-tolyguanidine	di-*o*-tolylguanidine
85	42	17.6–81.6	7270–7720
	43	8.82	3650
362	Table 6	structure of dicumyl peroxide (see text)	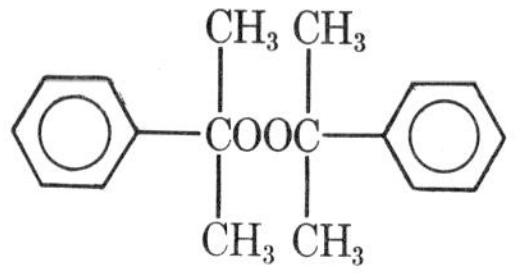 $C_6H_5-C(CH_3)_2-OO-C(CH_3)_2-C_6H_5$
		structure for 2,5-di (*tert*-butylperoxy)-2,5-dimethylhex-3-yne	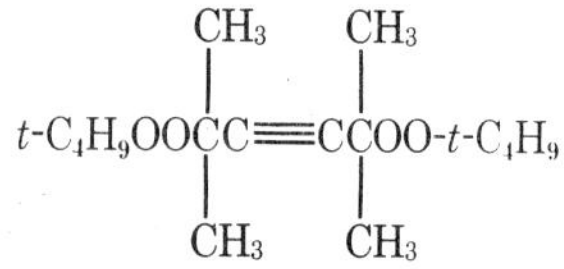 $t\text{-}C_4H_9OOC(CH_3)_2C{\equiv}CC(CH_3)_2OO\text{-}t\text{-}C_4H_9$
374	17	$\times 10^6$	$\times 10^9$
424	structure	see text	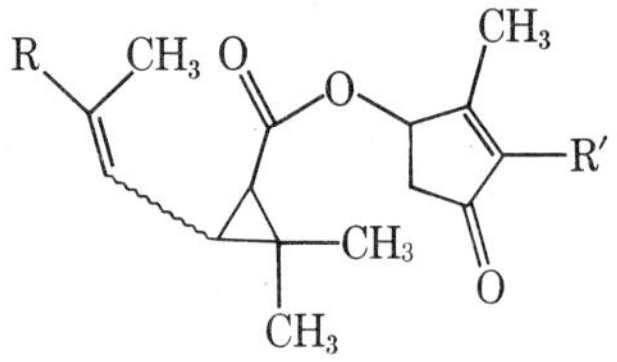 R, CH_3, O, O, CH_3, R′, CH_3, O, CH_3
455	9	3-dimethylamino-ethyleneiminophenyl	3-dimethylaminomethyleneimino-phenyl
967	cross refs. LAC to LACTOSE		transpose to p. 941 right after the letter L

KIRK-OTHMER

ENCYCLOPEDIA OF CHEMICAL TECHNOLOGY

Third Edition

VOLUME 13

Hydrogen-Ion Activity
to
Laminated Materials, Glass

KIRK-OTHMER

ENCYCLOPEDIA OF CHEMICAL TECHNOLOGY

THIRD EDITION

VOLUME 13

HYDROGEN-ION ACTIVITY
TO
LAMINATED MATERIALS, GLASS

A WILEY-INTERSCIENCE PUBLICATION
John Wiley & Sons
NEW YORK • CHICHESTER • BRISBANE • TORONTO

Library of Congress Cataloging in Publication Data:

Main entry under title:
Encyclopedia of chemical technology.

At head of title: Kirk-Othmer.
"A Wiley-Interscience publication."
Includes bibliographies.
1. Chemistry, Technical—Dictionaries. I. Kirk, Raymond Eller, 1890–1957. II. Othmer, Donald Frederick, 1904–
III. Grayson, Martin. IV. Eckroth, David. V. Title: Kirk-Othmer encyclopedia of chemical technology.

TP9.E685 1978 660′.03 77-15820
ISBN 0-471-02066-4

Printed in the United States of America

CONTENTS

EDITORIAL STAFF FOR VOLUME 13

CONTRIBUTORS TO VOLUME 13

Richard C. Bailie, *West Virginia University, Morgantown, West Virginia,* Incinerators

D. W. Bannister, *Toms River Chemical Corp., Toms River, New Jersey,* Indole

Roger G. Bates, *University of Florida, Gainesville, Florida,* Hydrogen-ion activity

Gregory B. Bennett, *Sandoz Inc., East Hanover, New Jersey,* Hypnotics, sedatives, anticonvulsants

S. Edmund Berger, *Allied Chemical Corporation, Buffalo, New York,* Hydroxy dicarboxylic acids

Bohdan V. Burachinsky, *Inmont Corporation, Clifton, New Jersey,* Inks

James W. Butler, *Naval Research Laboratory, Washington, D. C.,* Ion-implantation

D. H. Chadwick, *Mobay Chemical Corporation, New Martinsville, West Virginia,* Isocyanates, organic

C. C. Chamis, *National Aeronautics and Space Administration, Cleveland, Ohio,* Laminated and reinforced metals

George D. Clayton, *Clayton Environmental Consultants, Inc., San Luis Rey, California,* Industrial hygiene and toxicology

J. T. Clemens, *Bell Telephone Laboratories, Incorporated, Allentown, Pennsylvania,* Integrated circuits

T. H. Cleveland, *Mobay Chemical Corporation, New Martinsville, West Virginia,* Isocyanates, organic

Burton B. Crocker, *Monsanto Co., St. Louis, Missouri,* Incinerators

Hugh Dunn, *Inmont Corporation, Clifton, New Jersey,* Inks

Richard A. Durst, *National Bureau of Standards, Washington, D. C.,* Hydrogen-ion activity

James K. Ely, *Inmont Corporation, Clifton, New Jersey,* Inks

J. Feinman, *United States Steel Corporation, Monroeville, Pennsylvania,* Iron by direct reduction

Martin S. Frant, *Foxboro Analytical, a division of the Foxboro Corporation, Burlington, Massachusetts,* Ion-selective electrodes

Margaret H. Graham, *Exxon Research and Engineering Company, Linden, New Jersey,* Information retrieval

G. E. Ham, *Dow Chemical U. S. A., Freeport, Texas,* Imines, cyclic

Robert H. Hasek, *Tennessee Eastman Company, Kingsport, Tennessee,* Ketenes and related substances

John E. Hogan, *The Okonite Co., Ramsey, New Jersey,* Wire and cable coverings under Insulation, electric

William J. Houlihan, *Sandoz Inc., East Hanover, New Jersey,* Hypnotics, sedatives, anticonvulsants

Charles R. Jokel, *Bolt, Beranek & Newman, Cambridge, Massachusetts,* Insulation, acoustic

Vasanth Kamath, *Pennwalt Corporation, Buffalo, New York,* Initiators

J. R. Kircher, *E. I. du Pont de Nemours & Co., Inc., Wilmington, Delaware,* Hydrogen peroxide

W. A. Knepper, *United States Steel Corp., Pittsburgh, Pennsylvania,* Iron

E. F. Labuda, *Bell Telephone Laboratories, Incorporated, Allentown, Pennsylvania,* Integrated circuits

Alexis B. Lamy, *Essochem Europe, Inc. Diegem, Belgium,* Information retrieval

Barbara Lawrence, *Exxon Corporation, Linden, New Jersey,* Information retrieval

C. Michael Lederer, *Lawrence Berkeley Laboratory, University of California, Berkeley, California,* Isotopes

L. J. Lefevre, *Dow Chemical U. S. A., Walnut Creek, California,* Ion exchange

Charles J. Mazac, *PPG Industries, Corpus Christi, Texas,* Iodine and iodine compounds

James V. McArdle, *University of Maryland, College Park, Maryland,* Iron compounds

Frederick J. McGarry, *Massachusetts Institute of Technology, Cambridge, Massachusetts,* Laminated and reinforced plastics

Robert L. Metcalf, *University of Illinois at Urbana-Champaign, Urbana, Illinois,* Insect control technology

E. F. Milner, *Cominco Ltd., Trail British Columbia, Canada,* Indium and indium compounds

Richard M. Mullins, *The Dow Chemical Company, Midland, Michigan,* Hydroxybenzaldehydes

R. H. Neisel, *Johns-Manville Sales Corporation, Denver, Colorado,* Insulation, thermal

Anthony J. Papa, *Union Carbide Corporation, South Charleston, West Virginia,* Ketones

Edward N. Peters, *General Electric, Pittsfield, Massachusetts,* Inorganic high polymers

James J. Pitts, *Olin Corporation, New Haven, Connecticut,* Industrial antimicrobial agents

William M. Saltman, *The Goodyear Tire & Rubber Company, Akron, Ohio,* Isoprene

R. N. Sampson, *Westinghouse Electric Corporation, Pittsburgh, Pennsylvania,* Properties and materials under Insulation, electric

Chester S. Sheppard, *Pennwalt Corporation, Buffalo, New York,* Initiators

Paul D. Sherman, Jr., *Union Carbide Corporation, South Charleston, West Virginia,* Ketones

Robert M. Sowers, *Ford Motor Company, Lincoln Park, Missouri,* Laminated materials, glass

Lorraine Y. Stroumtsos, *Exxon Research and Engineering Company, Linden, New Jersey,* Information retrieval

Bryce E. Tate, *Pfizer Inc., Groton, Connecticut,* Itaconic acid and derivatives

Samuel I. Trotz, *Olin Corporation, New Haven, Connecticut,* Industrial antimicrobial agents

G. F. Tutwiler, *McNeil Laboratories, Fort Washington, Pennsylvania,* Insulin and other antidiabetic agents

J. H. Van Ness, *Monsanto Industrial Chemicals Co., St. Louis, Missouri,* Hydroxy carboxylic acids

J. Varagnat, *Rhône-Poulenc, Lyon, Cedex, France,* Hydroquinone, resorcinol, and catechol

J. D. Verschoor, *Johns-Manville Sales Corporation, Denver, Colorado,* Insulation, thermal

Clifton Warren, *AGA Infrared System AB, Danderyd, Sweden,* Infrared technology

R. M. Wheaton, *Dow Chemical U. S. A., Walnut Creek, California,* Ion exchange

C. E. T. White, *Indium Corporation of America, Utica, New York,* Indium and indium compounds

Theodore J. Williams, *Purdue University, West Lafayette, Indiana,* Instrumentation and control

Stewart Wong, *Boehringer-Ingelheim Ltd., Ridgefield, Connecticut,* Immunotherapeutic agents

NOTE ON CHEMICAL ABSTRACTS SERVICE REGISTRY NUMBERS AND NOMENCLATURE

Chemical Abstracts Service (CAS) Registry Numbers are unique numerical identifiers assigned to substances recorded in the CAS Registry System. They appear in brackets in the *Chemical Abstracts* (CA) substance and formula indexes following the names of compounds. A single compound may have many synonyms in the chemical literature. A simple compound like phenethylamine can be named β-phenylethylamine or, as in *Chemical Abstracts,* benzeneethanamine. The usefulness of the Encyclopedia depends on accessibility through the most common correct name of a substance. Because of this diversity in nomenclature careful attention has been given the problem in order to assist the reader as much as possible, especially in locating the systematic CA index name by means of the Registry Number. For this purpose, the reader may refer to the CAS Registry Handbook-Number Section which lists in numerical order the Registry Number with the *Chemical Abstracts* index name and the molecular formula; eg, **458-88-8,** Piperidine, 2-propyl-, (*S*)-, $C_8H_{17}N$; in the Encyclopedia this compound would be found under its common name, coniine [*458-88-8*]. The Registry Number is a valuable link for the reader in retrieving additional published information on substances and also as a point of access for such on-line data bases as Chemline, Medline, and Toxline.

In all cases, the CAS Registry Numbers have been given for title compounds in articles and for all compounds in the index. All specific substances indexed in *Chemical Abstracts* since 1965 are included in the CAS Registry System as are a large number of substances derived from a variety of reference works. The CAS Registry System identifies a substance on the basis of an unambiguous computer-language description of its molecular structure including stereochemical detail. The Registry Number is a machine-checkable number (like a Social Security number) assigned in sequential order to each substance as it enters the registry system. The value of the number lies in the fact that it is a concise and unique means of substance identification, which is independent of, and therefore bridges, many systems of chemical nomenclature. For polymers, one Registry Number is used for the entire family; eg, polyoxyethylene (20) sorbitan monolaurate has the same number as all of its polyoxyethylene homologues.

Registry numbers for each substance will be provided in the third edition cumulative index and appear as well in the annual indexes (eg, Alkaloids shows the Registry Number of all alkaloids (title compounds) in a table in the article as well, but the intermediates will have their Registry Numbers shown only in the index). Articles such as Analytical methods, Batteries and electric cells, Chemurgy, Distillation, Economic evaluation, and Fluid mechanics have no Registry Numbers in the text.

Cross-references are inserted in the index for many common names and for some systematic names. Trademark names appear in the index. Names that are incorrect, misleading or ambiguous are avoided. Formulas are given very frequently in the text to help in identifying compounds. The spelling and form used, even for industrial names, follow American chemical usage, but not always the usage of *Chemical Abstracts* (eg, *coniine* is used instead of *(S)-2-propylpiperidine, aniline* instead of *benzenamine,* and *acrylic acid* instead of *2-propenoic acid*).

There are variations in representation of rings in different disciplines. The dye industry does not designate aromaticity or double bonds in rings. All double bonds and aromaticity are shown in the *Encyclopedia* as a matter of course. For example, tetralin has an aromatic ring and a saturated ring and its structure appears in the

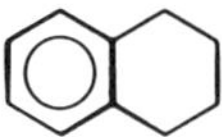

Encyclopedia with its common name, Registry Number enclosed in brackets, and parenthetical CA index name, ie, tetralin, [*119-64-2*] (1,2,3,4-tetrahydronaphthalene). With names and structural formulas, and especially with CAS Registry Numbers the aim is to help the reader have a concise means of substance identification.

CONVERSION FACTORS, ABBREVIATIONS, AND UNIT SYMBOLS

SI Units (Adopted 1960)

A new system of measurement, the International System of Units (abbreviated SI), is being implemented throughout the world. This system is a modernized version of the MKSA (meter, kilogram, second, ampere) system, and its details are published and controlled by an international treaty organization (The International Bureau of Weights and Measures) (1).

SI units are divided into three classes:

BASE UNITS

length	meter[†] (m)
mass[‡]	kilogram (kg)
time	second (s)
electric current	ampere (A)
thermodynamic temperature[§]	kelvin (K)
amount of substance	mole (mol)
luminous intensity	candela (cd)

[†] The spellings "metre" and "litre" are preferred by ASTM; however "-er" are used in the Encyclopedia.

[‡] "Weight" is the commonly used term for "mass."

[§] Wide use is made of "Celsius temperature" (t) defined by

$$t = T - T_0$$

where T is the thermodynamic temperature, expressed in kelvins, and T_0 = 273.15 K by definition. A temperature interval may be expressed in degrees Celsius as well as in kelvins.

SUPPLEMENTARY UNITS

plane angle	radian (rad)
solid angle	steradian (sr)

DERIVED UNITS AND OTHER ACCEPTABLE UNITS

These units are formed by combining base units, supplementary units, and other derived units (2–4). Those derived units having special names and symbols are marked with an asterisk in the list below:

Quantity	*Unit*	*Symbol*	*Acceptable equivalent*
*absorbed dose	gray	Gy	J/kg
acceleration	meter per second squared	m/s^2	
*activity (of ionizing radiation source)	becquerel	Bq	1/s
area	square kilometer	km^2	
	square hectometer	hm^2	ha (hectare)
	square meter	m^2	
*capacitance	farad	F	C/V
concentration (of amount of substance)	mole per cubic meter	mol/m^3	
*conductance	siemens	S	A/V
current density	ampere per square meter	A/m^2	
density, mass density	kilogram per cubic meter	kg/m^3	g/L; mg/cm^3
dipole moment (quantity)	coulomb meter	C·m	
*electric charge, quantity of electricity	coulomb	C	A·s
electric charge density	coulomb per cubic meter	C/m^3	
electric field strength	volt per meter	V/m	
electric flux density	coulomb per square meter	C/m^2	
*electric potential, potential difference, electromotive force	volt	V	W/A
*electric resistance	ohm	Ω	V/A
*energy, work, quantity of heat	megajoule	MJ	
	kilojoule	kJ	
	joule	J	N·m
	electron volt†	eV†	
	kilowatt-hour†	kW·h†	

† This non-SI unit is recognized by the CIPM as having to be retained because of practical importance or use in specialized fields (1).

Quantity	Unit	Symbol	Acceptable equivalent
energy density	joule per cubic meter	J/m^3	
*force	kilonewton	kN	
	newton	N	$kg{\cdot}m/s^2$
*frequency	megahertz	MHz	
	hertz	Hz	1/s
heat capacity, entropy	joule per kelvin	J/K	
heat capacity (specific), specific entropy	joule per kilogram kelvin	J/(kg·K)	
heat transfer coefficient	watt per square meter kelvin	$W/(m^2{\cdot}K)$	
*illuminance	lux	lx	lm/m^2
*inductance	henry	H	Wb/A
linear density	kilogram per meter	kg/m	
luminance	candela per square meter	cd/m^2	
*luminous flux	lumen	lm	cd·sr
magnetic field strength	ampere per meter	A/m	
*magnetic flux	weber	Wb	V·s
*magnetic flux density	tesla	T	Wb/m^2
molar energy	joule per mole	J/mol	
molar entropy, molar heat capacity	joule per mole kelvin	J/(mol·K)	
moment of force, torque	newton meter	N·m	
momentum	kilogram meter per second	kg·m/s	
permeability	henry per meter	H/m	
permittivity	farad per meter	F/m	
*power, heat flow rate, radiant flux	kilowatt	kW	
	watt	W	J/s
power density, heat flux density, irradiance	watt per square meter	W/m^2	
*pressure, stress	megapascal	MPa	
	kilopascal	kPa	
	pascal	Pa	N/m^2
sound level	decibel	dB	
specific energy	joule per kilogram	J/kg	
specific volume	cubic meter per kilogram	m^3/kg	
surface tension	newton per meter	N/m	
thermal conductivity	watt per meter kelvin	W/(m·K)	
velocity	meter per second	m/s	
	kilometer per hour	km/h	
viscosity, dynamic	pascal second	Pa·s	
	millipascal second	mPa·s	
viscosity, kinematic	square meter per second	m^2/s	

Quantity	*Unit*	*Symbol*	*Acceptable equivalent*
	square millimeter per second	mm^2/s	
volume	cubic meter	m^3	
	cubic decimeter	dm^3	L(liter) (5)
	cubic centimeter	cm^3	mL
wave number	1 per meter	m^{-1}	
	1 per centimeter	cm^{-1}	

In addition, there are 16 prefixes used to indicate order of magnitude, as follows:

Multiplication factor	*Prefix*	*Symbol*	*Note*
10^{18}	exa	E	
10^{15}	peta	P	
10^{12}	tera	T	
10^{9}	giga	G	
10^{6}	mega	M	
10^{3}	kilo	k	
10^{2}	hecto	h[a]	[a] Although hecto, deka, deci, and centi are SI prefixes, their use should be avoided except for SI unit-multiples for area and volume and nontechnical use of centimeter, as for body and clothing measurement.
10	deka	da[a]	
10^{-1}	deci	d[a]	
10^{-2}	centi	c[a]	
10^{-3}	milli	m	
10^{-6}	micro	μ	
10^{-9}	nano	n	
10^{-12}	pico	p	
10^{-15}	femto	f	
10^{-18}	atto	a	

For a complete description of SI and its use the reader is referred to ASTM E 380 (4) and the article Units and Conversion Factors which will appear in a later volume of the *Encyclopedia.*

A representative list of conversion factors from non-SI to SI units is presented herewith. Factors are given to four significant figures. Exact relationships are followed by a dagger. A more complete list is given in ASTM E 380-76(4) and ANSI Z210.1-1976 (6).

Conversion Factors to SI Units

To convert from	*To*	*Multiply by*
acre	square meter (m^2)	4.047×10^3
angstrom	meter (m)	1.0×10^{-10}†
are	square meter (m^2)	1.0×10^{2}†
astronomical unit	meter (m)	1.496×10^{11}
atmosphere	pascal (Pa)	1.013×10^{5}
bar	pascal (Pa)	1.0×10^{5}†
barn	square meter (m^2)	1.0×10^{-28}†

† Exact.

To convert from	*To*	*Multiply by*
barrel (42 U.S. liquid gallons)	cubic meter (m^3)	0.1590
Bohr magneton μ_β	J/T	9.274×10^{-24}
Btu (International Table)	joule (J)	1.055×10^3
Btu (mean)	joule (J)	1.056×10^3
Btu (thermochemical)	joule (J)	1.054×10^3
bushel	cubic meter (m^3)	3.524×10^{-2}
calorie (International Table)	joule (J)	4.187
calorie (mean)	joule (J)	4.190
calorie (thermochemical)	joule (J)	4.184†
centipoise	pascal second (Pa·s)	1.0×10^{-3}†
centistoke	square millimeter per second (mm^2/s)	1.0†
cfm (cubic foot per minute)	cubic meter per second (m^3/s)	4.72×10^{-4}
cubic inch	cubic meter (m^3)	1.639×10^{-5}
cubic foot	cubic meter (m^3)	2.832×10^{-2}
cubic yard	cubic meter (m^3)	0.7646
curie	becquerel (Bq)	3.70×10^{10}†
debye	coulomb·meter (C·m)	3.336×10^{-30}
degree (angle)	radian (rad)	1.745×10^{-2}
denier (international)	kilogram per meter (kg/m)	1.111×10^{-7}
	tex‡	0.1111
dram (apothecaries')	kilogram (kg)	3.888×10^{-3}
dram (avoirdupois)	kilogram (kg)	1.772×10^{-3}
dram (U.S. fluid)	cubic meter (m^3)	3.697×10^{-6}
dyne	newton (N)	1.0×10^{-5}†
dyne/cm	newton per meter (N/m)	1.0×10^{-3}†
electron volt	joule (J)	1.602×10^{-19}
erg	joule (J)	1.0×10^{-7}†
fathom	meter (m)	1.829
fluid ounce (U.S.)	cubic meter (m^3)	2.957×10^{-5}
foot	meter (m)	0.3048†
footcandle	lux (lx)	10.76
furlong	meter (m)	2.012×10^{-2}
gal	meter per second squared (m/s^2)	1.0×10^{-2}†
gallon (U.S. dry)	cubic meter (m^3)	4.405×10^{-3}
gallon (U.S. liquid)	cubic meter (m^3)	3.785×10^{-3}
gallon per minute (gpm)	cubic meter per second (m^3/s)	6.308×10^{-5}
	cubic meter per hour (m^3/h)	0.2271
gauss	tesla (T)	1.0×10^{-4}
gilbert	ampere (A)	0.7958
gill (U.S.)	cubic meter (m^3)	1.183×10^{-4}
grad	radian	1.571×10^{-2}
grain	kilogram (kg)	6.480×10^{-5}
gram force per denier	newton per tex (N/tex)	8.826×10^{-2}

† Exact.

‡ See footnote on p. xiv.

To convert from	*To*	*Multiply by*
hectare	square meter (m^2)	$1.0 \times 10^{4\dagger}$
horsepower (550 ft·lbf/s)	watt (W)	7.457×10^2
horsepower (boiler)	watt (W)	9.810×10^3
horsepower (electric)	watt (W)	$7.46 \times 10^{2\dagger}$
hundredweight (long)	kilogram (kg)	50.80
hundredweight (short)	kilogram (kg)	45.36
inch	meter (m)	$2.54 \times 10^{-2\dagger}$
inch of mercury (32°F)	pascal (Pa)	3.386×10^3
inch of water (39.2°F)	pascal (Pa)	2.491×10^2
kilogram force	newton (N)	9.807
kilowatt hour	megajoule (MJ)	$3.6^\dagger$
kip	newton (N)	4.48×10^3
knot (international)	meter per second (m/s)	0.5144
lambert	candela per square meter (cd/m^2)	3.183×10^3
league (British nautical)	meter (m)	5.559×10^3
league (statute)	meter (m)	4.828×10^3
light year	meter (m)	9.461×10^{15}
liter (for fluids only)	cubic meter (m^3)	$1.0 \times 10^{-3\dagger}$
maxwell	weber (Wb)	$1.0 \times 10^{-8\dagger}$
micron	meter (m)	$1.0 \times 10^{-6\dagger}$
mil	meter (m)	$2.54 \times 10^{-5\dagger}$
mile (statute)	meter (m)	1.609×10^3
mile (U.S. nautical)	meter (m)	$1.852 \times 10^{3\dagger}$
mile per hour	meter per second (m/s)	0.4470
millibar	pascal (Pa)	1.0×10^2
millimeter of mercury (0°C)	pascal (Pa)	$1.333 \times 10^{2\dagger}$
minute (angular)	radian	2.909×10^{-4}
myriagram	kilogram (kg)	10
myriameter	kilometer (km)	10
oersted	ampere per meter (A/m)	79.58
ounce (avoirdupois)	kilogram (kg)	2.835×10^{-2}
ounce (troy)	kilogram (kg)	3.110×10^{-2}
ounce (U.S. fluid)	cubic meter (m^3)	2.957×10^{-5}
ounce-force	newton (N)	0.2780
peck (U.S.)	cubic meter (m^3)	8.810×10^{-3}
pennyweight	kilogram (kg)	1.555×10^{-3}
pint (U.S. dry)	cubic meter (m^3)	5.506×10^{-4}
pint (U.S. liquid)	cubic meter (m^3)	4.732×10^{-4}
poise (absolute viscosity)	pascal second (Pa·s)	$0.10^\dagger$
pound (avoirdupois)	kilogram (kg)	0.4536
pound (troy)	kilogram (kg)	0.3732
poundal	newton (N)	0.1383
pound-force	newton (N)	4.448
pound per square inch (psi)	pascal (Pa)	6.895×10^3
quart (U.S. dry)	cubic meter (m^3)	1.101×10^{-3}

† Exact.

To convert from	*To*	*Multiply by*
quart (U.S. liquid)	cubic meter (m^3)	9.464×10^{-4}
quintal	kilogram (kg)	$1.0 \times 10^{2\dagger}$
rad	gray (Gy)	$1.0 \times 10^{-2\dagger}$
rod	meter (m)	5.029
roentgen	coulomb per kilogram (C/kg)	2.58×10^{-4}
second (angle)	radian (rad)	4.848×10^{-6}
section	square meter (m^2)	2.590×10^{6}
slug	kilogram (kg)	14.59
spherical candle power	lumen (lm)	12.57
square inch	square meter (m^2)	6.452×10^{-4}
square foot	square meter (m^2)	9.290×10^{-2}
square mile	square meter (m^2)	2.590×10^{6}
square yard	square meter (m^2)	0.8361
stere	cubic meter (m^3)	$1.0^\dagger$
stokes (kinematic viscosity)	square meter per second (m^2/s)	$1.0 \times 10^{-4\dagger}$
tex	kilogram per meter (kg/m)	$1.0 \times 10^{-6\dagger}$
ton (long, 2240 pounds)	kilogram (kg)	1.016×10^{3}
ton (metric)	kilogram (kg)	$1.0 \times 10^{3\dagger}$
ton (short, 2000 pounds)	kilogram (kg)	9.072×10^{2}
torr	pascal (Pa)	1.333×10^{2}
unit pole	weber (Wb)	1.257×10^{-7}
yard	meter (m)	$0.9144^\dagger$

Abbreviations and Unit Symbols

Following is a list of commonly used abbreviations and unit symbols appropriate for use in the *Encyclopedia*. In general they agree with those listed in *American National Standard Abbreviations for Use on Drawings and in Text (ANSI Y1.1)* (6) and *American National Standard Letter Symbols for Units in Science and Technology (ANSI Y10)* (6). Also included is a list of acronyms for a number of private and government organizations as well as common industrial solvents, polymers, and other chemicals.

Rules for Writing Unit Symbols (4):

1. Unit symbols should be printed in upright letters (roman) regardless of the type style used in the surrounding text.
2. Unit symbols are unaltered in the plural.
3. Unit symbols are not followed by a period except when used as the end of a sentence.
4. Letter unit symbols are generally written in lower-case (eg, cd for candela) unless the unit name has been derived from a proper name, in which case the first letter of the symbol is capitalized (W,Pa). Prefix and unit symbols retain their prescribed form regardless of the surrounding typography.
5. In the complete expression for a quantity, a space should be left between the numerical value and the unit symbol. For example, write 2.37 lm, *not* 2.37lm, and 35 mm, *not* 35mm. When the quantity is used in an adjectival sense, a hyphen is often used, for example, 35-mm film. *Exception:* No space is left between the numerical value and the symbols for degree, minute, and second of plane angle, and degree Celsius.

6. No space is used between the prefix and unit symbols (eg, kg).

7. Symbols, not abbreviations, should be used for units. For example, use "A," not "amp," for ampere.

8. When multiplying unit symbols, use a raised dot:

$$\mathrm{N{\cdot}m} \text{ for newton meter}$$

In the case of W·h, the dot may be omitted, thus:

$$\mathrm{Wh}$$

An exception to this practice is made for computer printouts, automatic typewriter work, etc, where the raised dot is not possible, and a dot on the line may be used.

9. When dividing unit symbols use one of the following forms:

$$\mathrm{m/s} \textit{ or } \mathrm{m{\cdot}s^{-1}} \text{ or } \frac{\mathrm{m}}{\mathrm{s}}$$

In no case should more than one slash be used in the same expression unless parentheses are inserted to avoid ambiguity. For example, write:

$$\mathrm{J/(mol{\cdot}K)} \textit{ or } \mathrm{J{\cdot}mol^{-1}\cdot K^{-1}} \textit{ or } \mathrm{(J/mol)/K}$$

but *not*

$$\mathrm{J/mol/K}$$

10. Do not mix symbols and unit names in the same expression. Write:

joules per kilogram *or* J/kg *or* $\mathrm{J{\cdot}kg^{-1}}$

but *not*

joules/kilogram *nor* joules/kg *nor* $\mathrm{joules{\cdot}kg^{-1}}$

ABBREVIATIONS AND UNITS

A	ampere
A	anion (eg, H*A*)
a	atto (prefix for 10^{-18})
AATCC	American Association of Textile Chemists and Colorists
ABS	acrylonitrile–butadiene–styrene
abs	absolute
ac	alternating current, *n.*
a-c	alternating current, *adj.*
ac-	alicyclic
ACGIH	American Conference of Governmental Industrial Hygienists
ACS	American Chemical Society
AGA	American Gas Association
Ah	ampere hour
AIChE	American Institute of Chemical Engineers
AIP	American Institute of Physics
alc	alcohol(ic)
Alk	alkyl
alk	alkaline (not alkali)
amt	amount
amu	atomic mass unit
ANSI	American National Standards Institute
AOAC	Association of Official Analytical Chemists
AOCS	American Oil Chemist's Society
AP	atomic orbital
APHA	American Public Health Association

API	American Petroleum Institute
aq	aqueous
Ar	aryl
ar-	aromatic
as-	asymmetric(al)
ASHRAE	American Society of Heating, Refrigerating, and Air Conditioning Engineers
ASM	American Society for Metals
ASME	American Society of Mechanical Engineers
ASTM	American Society for Testing and Materials
at no.	atomic number
at wt	atomic weight
av(g)	average
bbl	barrel
bcc	body-centered cubic
Bé	Baumé
BET	Brunauer-Emmett-Teller (adsorption equation)
bid	twice daily
Boc	*t*-butyloxycarbonyl
BOD	biochemical (biological) oxygen demand
bp	boiling point
Bq	becquerel
C	coulomb
°C	degree Celsius
C-	denoting attachment to carbon
c	centi (prefix for 10^{-2})
c	critical
ca	circa (approximately)
cd	candela; current density; circular dichroism
CFR	Code of Federal Regulations
cgs	centimeter–gram–second
CI	Color Index
cis-	isomer in which substituted groups are on same side of double bond between C atoms
cl	carload
cm	centimeter
CMA	Chemical Manufacturers' Association
cmil	circular mil
cmpd	compound
CNS	cerebral nervous system
COA	coenzyme A
COD	chemical oxygen demand
coml	commercial(ly)
cp	chemically pure
cph	close-packed hexagonal
CPSC	Consumer Product Safety Commission
cryst	crystalline
cub	cubic
D-	denoting configurational relationship
d	differential operator
d-	dextro-, dextrorotatory
da	deka (prefix for 10^{1})
dB	decibel
dc	direct current, *n.*
d-c	direct current, *adj.*
dec	decompose
detd	determined
detn	determination
dia	diameter
dil	dilute
dl-; DL-	racemic
DMF	dimethylformamide
DMG	dimethyl glyoxime
DMSO	dimethyl sulfoxide
DOD	Department of Defense
DOE	Department of Energy
DOT	Department of Transportation
dp	dew point; degree of polymerization
DPH	diamond pyramid hardness
dstl(d)	distill(ed)
dta	differential thermal analysis
(*E*)-	entgegen; opposed
ϵ	dielectric constant (unitless number)
e	electron
ECU	electrochemical unit
ed.	edited, edition, editor
ED	effective dose
EDTA	ethylenediaminetetraacetic acid
emf	electromotive force
emu	electromagnetic unit
eng	engineering

EPA	Environmental Protection Agency
epr	electron paramagnetic resonance
eq.	equation
esp	especially
esr	electron-spin resonance
est(d)	estimate(d)
estn	estimation
esu	electrostatic unit
exp	experiment, experimental
ext(d)	extract(ed)
F	farad (capacitance)
F	faraday (96,487 C)
f	femto (prefix for 10^{-15})
FAO	Food and Agriculture Organization (United Nations)
fcc	face-centered cubic
FDA	Food and Drug Administration
FEA	Federal Energy Administration
fob	free on board
fp	freezing point
FPC	Federal Power Commission
FRB	Federal Reserve Board
frz	freezing
G	giga (prefix for 10^9)
G	gravitational constant = 6.67×10^{-11} N·m^2/kg^2
g	gram
(g)	gas, only as in $H_2O(g)$
g	gravitational acceleration
gem-	geminal
glc	gas-liquid chromatography
g-mol wt; gmw	gram-molecular weight
GNP	gross national product
grd	ground
Gy	gray
H	henry
h	hour; hecto (prefix for 10^2)
ha	hectare
HB	Brinell hardness number
Hb	hemoglobin
hcp	hexagonal close-packed
hex	hexagonal
HK	Knoop hardness number
HRC	Rockwell hardness (C scale)
HV	Vickers hardness number
hyd	hydrated, hydrous
hyg	hygroscopic
Hz	hertz
^{i}(eg, Pr^i)	iso (eg, isopropyl)
i-	inactive (eg, *i*-methionine)
IACS	International Annealed Copper Standard
ibp	initial boiling point
ICC	Interstate Commerce Commission
ICT	International Critical Table
ID	inside diameter; infective dose
IPS	iron pipe size
IPT	Institute of Petroleum Technologists
IPTS	International Practical Temperature Scale (NBS)
ir	infrared
IRLG	Interagency Regulatory Liaison Group
ISO	International Organization for Standardization
IU	International Unit
IUPAC	International Union of Pure and Applied Chemistry
IV	iodine value
J	joule
K	kelvin
k	kilo (prefix for 10^3)
kg	kilogram
L	denoting configurational relationship
L	liter (for fluids only)(5)
l-	*levo*-, levorotatory
(l)	liquid, only as in $NH_3(l)$
LC_{50}	conc lethal to 50% of the animals tested
LCAO	linear combination of atomic orbitals
LCD	liquid crystal display
lcl	less than carload lots
LD_{50}	dose lethal to 50% of the animals tested
LED	light-emitting diode
liq	liquid
lm	lumen
ln	logarithm (natural)
LNG	liquefied natural gas

log	logarithm (common)
LPG	liquefied petroleum gas
ltl	less than truckload lots
lx	lux
M	mega (prefix for 10^6); metal (as in *MA*)
M	molar
m	meter; milli (prefix for 10^{-3})
m	molal
m-	meta
max	maximum
MEK	methyl ethyl ketone
meq	milliequivalent
mfd	manufactured
mfg	manufacturing
mfr	manufacturer
MIBC	methyl isobutyl carbinol
MIBK	methyl isobutyl ketone
MIC	minimum inhibiting concentration
min	minute; minimum
mL	milliliter
MLD	minimum lethal dose
MO	molecular orbital
mo	month
mol	mole
mol wt	molecular weight
mom	momentum
mp	melting point
MR	molar refraction
ms	mass spectrum
mxt	mixture
μ	micro (prefix for 10^{-6})
N	newton (force)
N	normal (concentration)
N-	denoting attachment to nitrogen
n (as n_D^{20}	index of refraction (for 20°C and sodium light)
^{n} (as Bu^n), *n*-	normal (straight-chain structure)
n	neutron
n	nano (prefix for 10^{-9})
na	not available
NAS	National Academy of Sciences
NASA	National Aeronautics and Space Administration
nat	natural
NBS	National Bureau of Standards
neg	negative
NF	*National Formulary*
NIH	National Institutes of Health
NIOSH	National Institute of Occupational Safety and Health
nmr	nuclear magnetic resonance
NND	New and Nonofficial Drugs (AMA)
no.	number
NOI-(BN)	not otherwise indexed (by name)
NOS	not otherwise specified
nqr	nuclear quadruple resonance
NRC	Nuclear Regulatory Commission; National Research Council
NRI	New Ring Index
NSF	National Science Foundation
NTA	nitrilotriacetic acid
NTP	normal temperature and pressure (25°C and 101.3 kPa or 1 atm)
NTSB	National Transportation Safety Board
O-	denoting attachment to oxygen
o-	ortho
OD	outside diameter
OPEC	Organization of Petroleum Exporting Countries
OSHA	Occupational Safety and Health Administration
owf	on weight of fiber
Ω	ohm
P	peta (prefix for 10^{15})
p	pico (prefix for 10^{-12})
p-	para
p	proton
p.	page
Pa	pascal (pressure)
pd	potential difference

pH	negative logarithm of the effective hydrogen ion concentration
pmr	proton magnetic resonance
POP	polyoxypropylene
pos	positive
pp.	pages
ppb	parts per billion
ppm	parts per million
PPO	poly(phenyl oxide)
ppt(d)	precipitate(d)
pptn	precipitation
Pr (no.)	foreign prototype (number)
pt	point; part
PVC	poly(vinyl chloride)
pwd	powder
qv	quod vide (which see)
R	univalent hydrocarbon radical
(*R*)-	rectus (clockwise configuration)
r	precision of data
rad	radian; radius
rds	rate determining step
ref.	reference
rf	radio frequency, *n.*
r-f	radio frequency, *adj.*
rh	relative humidity
RI	Ring Index
rps	revolutions per second
RT	room temperature
^{s} (eg, Bu^{s}); *sec*-	secondary (eg, secondary butyl)
S	siemens
(*S*)-	sinister (counterclockwise configuration)
S-	denoting attachment to sulfur
s-	symmetric(al)
s	second
(s)	solid, only as in $H_2O(s)$
SAE	Society of Automotive Engineers
SAN	styrene–acrylonitrile
sat(d)	saturate(d)
satn	saturation
SBS	styrene–butadiene–styrene
SCF	self-consistent field; standard cubic feet
Sch	Schultz number
SFs	Saybolt Furol seconds
SI	Le Système International d'Unités (International System of Units)
sl sol	slightly soluble
sol	soluble
soln	solution
soly	solubility
sp	specific; species
sp gr	specific gravity
sr	steradian
std	standard
STP	standard temperature and pressure (0°C and 101.3 kPa)
SUs	Saybolt Universal seconds
syn	synthetic
^{t} (eg, Bu^{t}), *t*-, *tert*-	tertiary (eg, tertiary butyl)
T	tera (prefix for 10^{12}); tesla (magnetic flux density)
t	metric ton (tonne) temperature
TAPPI	Technical Association of the Pulp and Paper Industry
tex	tex (linear density)
TGA	thermogravimetric analysis
THF	tetrahydrofuran
tlc	thin layer chromatography
TLV	threshold limit value
trans-	isomer in which substituted groups are on opposite sides of double bond between C atoms
TSCA	Toxic Substance Control Act
TWA	time-weighted average
Twad	Twaddell
UL	Underwriters' Laboratory
USDA	United States Department of Agriculture
USP	*United States Pharmacopeia*
uv	ultraviolet
V	volt (emf)

var	variable	Wh	watt hour
vic-	vicinal	WHO	World Health Organization (United Nations)
vol	volume (not volatile)	wk	week
vs	versus	yr	year
v sol	very soluble	(*Z*)-	zusammen; together
W	watt		
Wb	Weber		

Non-SI (Unacceptable and Obsolete) Units		*Use*
Å	angstrom	nm
at	atmosphere, technical	Pa
atm	atmosphere, standard	Pa
b	barn	cm^2
bar†	bar	Pa
bbl	barrel	m^3
bhp	brake horsepower	W
Btu	British thermal unit	J
bu	bushel	m^3; L
cal	calorie	J
cfm	cubic foot per minute	m^3/s
Ci	curie	Bq
cSt	centistokes	mm^2/s
c/s	cycle per second	Hz
cu	cubic	exponential form
D	debye	C·m
den	denier	tex
dr	dram	kg
dyn	dyne	N
dyn/cm	dyne per centimeter	mN/m
erg	erg	J
eu	entropy unit	J/K
°F	degree Fahrenheit	°C; K
fc	footcandle	lx
fl	footlambert	lx
fl oz	fluid ounce	m^3; L
ft	foot	m
ft·lbf	foot pound-force	J
gf den	gram-force per denier	N/tex
G	gauss	T
Gal	gal	m/s^2
gal	gallon	m^3; L
Gb	gilbert	A
gpm	gallon per minute	(m^3/s); (m^3/h)
gr	grain	kg
hp	horsepower	W
ihp	indicated horsepower	W
in.	inch	m
in. Hg	inch of mercury	Pa
in. H_2O	inch of water	Pa
in.-lbf	inch pound-force	J
kcal	kilogram-calorie	J
kgf	kilogram-force	N
kilo	for kilogram	kg

† Do not use bar (10^5Pa) or millibar (10^2Pa) because they are not SI units, and are accepted internationally only for a limited time in special fields because of existing usage.

Non-SI (Unacceptable and Obsolete) Units		*Use*
L	lambert	lx
lb	pound	kg
lbf	pound-force	N
mho	mho	S
mi	mile	m
MM	million	M
mm Hg	millimeter of mercury	Pa
mμ	millimicron	nm
mph	miles per hour	km/h
μ	micron	μm
Oe	oersted	A/m
oz	ounce	kg
ozf	ounce-force	N
η	poise	Pa·s
P	poise	Pa·s
ph	phot	lx
psi	pounds-force per square inch	Pa
psia	pounds-force per square inch absolute	Pa
psig	pounds-force per square inch gauge	Pa
qt	quart	m^3; L
°R	degree Rankine	K
rd	rad	Gy
sb	stilb	lx
SCF	standard cubic foot	m^3
sq	square	exponential form
thm	therm	J
yd	yard	m

BIBLIOGRAPHY

1. The International Bureau of Weights and Measures, BIPM, (Parc de Saint-Cloud, France) is described on page 22 of Ref. 4. This bureau operates under the exclusive supervision of the International Committee of Weights and Measures (CIPM).
2. *Metric Editorial Guide* (*ANMC-78-1*) 3rd ed., American National Metric Council, 1625 Massachusetts Ave. N.W., Washington, D.C. 20036, 1978.
3. *SI Units and Recommendations for the Use of Their Multiples and of Certain Other Units* (*ISO 1000-1973*), American National Standards Institute, 1430 Broadway, New York, N. Y. 10018, 1973.
4. Based on *ASTM E 380-79* (*Standard for Metric Practice*), American Society for Testing and Materials, 1916 Race Street, Philadelphia, Pa. 19103, 1979.
5. *Fed. Regist.*, Dec. 10, 1976 (41 FR 36414).
6. For ANSI address, see Ref. 3.

R. P. LUKENS
American Society for Testing and Materials

HYDROGEN-ION ACTIVITY

The effective concentration of hydrogen ion in solution is expressed in terms of pH, which is the negative logarithm of the hydrogen-ion activity:

$$\mathrm{pH} = -\log_{10} a_{\mathrm{H}^+} \tag{1}$$

The relationship between activity and concentration is

$$a = \gamma c \tag{2}$$

where the activity coefficient γ is a function of the ionic strength of the solution and approaches unity as the ionic strength decreases; ie, the difference between the activity and the concentration of hydrogen ion diminishes as the solution becomes more dilute. The pH of a solution may have little relationship to the titratable acidity of a solution that contains weak acids or buffering substances; the pH of a solution indicates only the free hydrogen-ion activity. If total acid concentration is to be determined, an acid–base titration must be performed.

Thermodynamically, the activity of a single ionic species (and, eg, the pH) is an inexact quantity, and a conventional pH scale has been adopted which is defined by reference solutions with assigned pH values. These reference solutions, in conjunction with equation 3, define the pH.

$$\mathrm{pH}(X) = \mathrm{pH}(S) - \frac{(E_X - E_S)F}{2.303\,RT} \tag{3}$$

E_S is the electromotive force (emf) of the cell: reference electrode|KCl ($\geq 3.5\,M$)‖solution S|H_2(g),Pt, and E_X is the emf of the same cell when the reference buffer solution S is replaced by the sample solution X. The quantities R, T, and F are the gas constant, the thermodynamic temperature, and the Faraday constant (96,487 C), respectively. For routine pH measurements, the hydrogen gas electrode [H_2(g),Pt] usually is replaced by a glass membrane electrode.

The availability of multiple pH-reference solutions makes possible an alternative definition of pH:

$$\text{pH}(X) = \text{pH}(S_1) + [\text{pH}(S_2) - \text{pH}(S_1)]\,\frac{(E_X - E_{S_1})}{(E_{S_2} - E_{S_1})} \tag{4}$$

where E_{S_1} and E_{S_2} are the measured cell potentials when the sample solution X is replaced in the cell by the two reference solutions S_1 and S_2 such that the values E_{S_1} and E_{S_2} are on either side of and as near as possible to E_X. Equation 4 assumes linearity of the pH vs E response between the two reference solutions, whereas equation 3 assumes both linearity and ideal Nernstian response of the pH electrode. The two-point calibration procedure is recommended if the electrode that is responsive to hydrogen ion is a glass pH electrode.

pH Determination

There are two methods used to measure pH (1–6). The most common utilizes the commercial pH meter with a glass electrode. This procedure is a determination of the difference between the pH of an unknown or test solution and that of a standard solution. The instrument measures the emf developed between the glass electrode and a reference electrode of constant potential. The difference in emf when the electrodes are removed from the standard solution and placed in the test solution is converted to a difference in pH. The second method, which has more limited applications, is the indicator method. The success of this procedure depends upon matching the color that is produced by the addition of a suitable indicator dye to a portion of the unknown solution with the color produced by adding the same quantity of the same dye to a series of standard solutions of known pH. The results obtained by the indicator method are less accurate relative to those obtained using a pH meter. The former method, however, is simple to apply and inexpensive.

Reference Buffer Solutions. The uncertainties of the salt bridge and liquid junction which exist in conventional electrochemical cells can be avoided by using a cell without transference, eg:

$$\text{Ag, AgCl}|\text{MCl}\ (m),\ \text{solution}\ S|\text{H}_2(\text{g}),\text{Pt}$$

A soluble chloride salt, MCl, of molality m is added to each reference solution. If the standard potential, $E°$, of the cell and the molality of the chloride ion, m_{Cl}, are known, emf measurements yield values of the acidity function $\text{p}(a_{\text{H}}\gamma_{\text{Cl}})$, as shown by the following equation:

$$\text{p}(a_{\text{H}}\gamma_{\text{Cl}}) \equiv -\log(m_{\text{H}}\gamma_{\text{H}}\gamma_{\text{Cl}}) = \frac{(E° - E)F}{2.303\,RT} + \log m_{\text{Cl}} \tag{5}$$

In order to eliminate the effect of the added MCl on the acidity of the buffer solution, $\text{p}(a_{\text{H}}\gamma_{\text{Cl}})$ is determined for three or more portions of the buffer solution that contain different amounts of added chloride. The limiting value of the acidity function is obtained by extrapolation to zero molality of chloride ion. If the activity coefficient of chloride ion in the buffer solution could be obtained, the activity of hydrogen ion would be readily accessible.

$$pa_H \equiv -\log(m_H\gamma_H) = p(a_H\gamma_{Cl}) + \log \gamma_{Cl} \qquad (6)$$

In order to establish a conventional scale of hydrogen-ion activity, it has been suggested that the activity coefficient of chloride ion in selected reference buffer solutions having ionic strengths of ≤ 0.1 be defined as (7):

$$-\log \gamma_{Cl} = \frac{A\sqrt{I}}{1 + 1.5\sqrt{I}} \qquad (7)$$

where A is a constant of the Debye-Hückel theory and I is the ionic strength. pa_H for these selected reference solutions are identified with pH(S) in the operational definition:

$$pa_H \equiv \text{pH}(S) \qquad (8)$$

The pH(S) values at 25°C of the NBS primary and secondary reference buffer solutions are listed in Table 1 (2).

Accuracy and Interpretation of Measured pH Values. The acidity function $p(a_H\gamma_{Cl})$, which is the experimental basis for the assignment of pH(S), is reproducible within about 0.003 pH unit from 10 to 40°C. If the ionic strength is known, the assignment of numerical values to the activity coefficient of chloride ion does not add to the uncertainty. However, errors in the standard potential of the cell, in the composition of the buffer materials, and in the preparation of the solutions, may raise the uncertainty to 0.005 pH unit.

The reproducibility of the practical scale that has been defined using the seven primary standards includes the possible inconsistencies introduced in the standardization of the instrument with seven different standards of different composition and concentration. These inconsistencies are the result of variations in the liquid-junction potential when one solution is replaced by another and are unavoidable. The accuracy of the practical scale from 10 to 40°C therefore appears to be from 0.008 to 0.01 pH unit.

Table 1. pH Standards, Molality Scale[a]

Solution composition	pH(S) at 25°C
Primary standards	
potassium hydrogen tartrate (saturated at 25°C)	3.557
0.05 m potassium dihydrogen citrate	3.776
0.05 m potassium hydrogen phthalate	4.004
0.025 m KH_2PO_4 + 0.025 m Na_2HPO_4	6.863
0.008695 m KH_2PO_4 + 0.03043 m Na_2HPO_4	7.415
0.01 m $Na_2B_4O_7.10H_2O$	9.183
0.025 m $NaHCO_3$ + 0.025 m Na_2CO_3	10.014
Secondary standards	
0.05 m potassium oxalate.$2H_2O$	1.679
0.01667 m tris(hydroxymethyl)aminomethane + 0.05 m tris(hydroxymethyl)aminomethane.HCl	7.699
$Ca(OH)_2$ (saturated at 25°C)	12.454

[a] Ref. 2.

Variations in the liquid-junction potential may be increased when the standard solutions are replaced by test solutions that do not clearly match the standards with respect to the types and concentrations of solutes, or to the composition of the solvent. Under these circumstances, the pH remains a reproducible number, but it may have little or no meaning in terms of the conventional hydrogen-ion activity of the medium. It is impossible to use the experimental pH numbers as a measure of the extent of acid–base reactions or to obtain thermodynamic equilibrium constants. Such interpretations are justified when the pH of the medium is between 2.5 and 11.5 and when the mixture is an aqueous solution of simple solutes in total concentration of ca ≤ 0.2 M.

When these restricted experimental conditions are present, the measured pH is regarded as an experimental value of pa_H, that is, of $-\log m_H\gamma_H$. In order to utilize this pa_H, it is necessary to evaluate one or more activity coefficients, either γ_H or the corresponding activity coefficients for the acidic and basic species of a conjugate pair. For most dilute solutions the activity coefficients of all univalent ions are considered equal to that of chloride ion as defined by equation 7, and the activity coefficients of ions (subscript i) of other charges (z_i) are related to that for chloride ion through the z^2 relationship of the Debye-Hückel law. Thus, the activity coefficient γ_i of any ion i is given by

$$\ln \gamma_i = z_i^2 \ln \gamma_{Cl} \tag{9}$$

Sources of Error. Although subject to fewer interferences and other types of error than most potentiometric ionic-activity sensors, ie, ion-selective electrodes (qv), pH electrodes must be used with an awareness of their particular response characteristics as well as the potential sources of error which may affect the other components of the measurement system, especially the reference electrode (see also pH Measurement System Electrodes). Several common causes of measurement problems are electrode interferences and/or fouling of the pH sensor, sample matrix effects, reference electrode instability, and improper calibration of the measurement system (8).

In general, the potential of an electrochemical cell, E_{cell}, is the sum of three potential terms:

$$E_{cell} = E_{pH} - E_{ref} + E_{lj} \tag{10}$$

where E_{pH} and E_{ref} are the potentials of the pH and reference electrodes, respectively, and E_{lj} is the ubiquitous liquid-junction potential. After substitution of the Nernst equation for the pH electrode potential term in equation 10,

$$E_{cell} = E^{\circ}_{pH} - \frac{RT}{F} \ln a_{H+} - E_{ref} + E_{lj} \tag{11}$$

it can be calculated that a 1 mV error in any of the potential terms corresponds to an error of ca 4% in the hydrogen-ion activity. Under carefully controlled experimental conditions, the potential of a pH cell can be measured with an uncertainty as small as 0.3 mV, which corresponds to a ± 0.005 uncertainty in pH.

pH Measurement System Electrodes

Glass Electrodes. The glass electrode is the hydrogen-ion sensor in most pH-measurement systems. The pH-responsive surface of the glass electrode consists of a thin membrane formed from a special glass which, after suitable conditioning, develops a surface potential that is an accurate index of the acidity of the solution in which the electrode is immersed. To permit changes in the potential of the active surface of the glass membrane to be measured, an inner reference cell of constant potential is placed on the opposite side of the glass membrane. The inner reference cell consists of a solution that has a stable hydrogen-ion concentration and that contains counterions to which the inner electrode is reversible. The ionic concentrations are sufficient to guarantee constancy of the inner cell over extended periods of time. The choice of an inner cell has a bearing on the temperature coefficient of the emf of the pH assembly. The inner cell commonly consists of a silver–silver chloride electrode or calomel electrode in a buffered chloride solution.

Immersion electrodes are the most common glass electrodes. These are roughly cylindrical and consist of a barrel or stem of inert glass which is sealed at the lower end to a tip, which is often hemispherical, of special pH-responsive glass. The tip is completely immersed in the solution during measurements. Miniature and subminiature electrodes are used widely, particularly in physiological studies. Capillary electrodes permit the use of small samples and protect them from exposure to air during the measurements. They are advantageous in the determination of blood pH. Sometimes the pH-sensitive glass membrane is fabricated in the form of a tube surrounded by the inner solution; this type of electrode may be provided with a water jacket for temperature control.

The membrane of pH-responsive glass usually is made as thin as is consistent with adequate mechanical strength; nevertheless, its electrical resistance is high, eg, from 10–250 MΩ. Therefore, electronic amplifiers must be used to obtain adequate accuracy in the measurement of the surface potentials of glass electrodes. The versatility of the glass electrode results from its mechanism of operation, which is one of proton transfer rather than electron transfer; hence, oxidizing and reducing agents in the solution do not affect the pH response.

Most modern electrode glasses contain silicon dioxide, either sodium or lithium oxide, and either calcium, barium, cesium, or lanthanum oxide. The latter oxides are added to reduce spurious response to alkali metal ions in high pH solutions. The composition of the glass has a profound effect upon the electrical resistance, the chemical durability of the pH-sensitive surface, and the accuracy of the pH response in alkaline solutions. Both the electrical and the chemical resistance of the electrode glasses decrease rapidly with a rise in temperature. Therefore, it is difficult to design an electrode that is sufficiently durable for extended use at high temperatures, and yet, when used at room temperature, free from the sluggish response often characteristic of pH cells of excessively high resistance. Some manufacturers use different glass compositions for electrodes depending upon whether the intended use is at low, intermediate, or high temperatures.

The mechanism of the glass electrode response is uncertain. It is clear, however, that when a freshly blown membrane of pH-responsive glass is first conditioned in water, the sodium or lithium ions that occupy the interstices of the silicon–oxygen network in the glass surface are exchanged for protons from the water. Although glass electrodes usually are conditioned in water, it is probable that the conditioning process proceeds in the same way when other amphiprotic solvents are used.

The protons find stable sites in the conditioned gel layer of the glass surface. Exchange of the labile protons between these sites and the solution phase appears to be the mechanism by which the surface potential reflects changes in the acidity of the external solution. When the glass electrode and the hydrogen electrode are immersed in the same solution, their potentials usually differ by a constant amount, even though the pH of the medium is raised from 1 to 10 or greater. In this range, the potential E_g of a glass electrode may be written

$$E_g = E_g^\circ + \frac{RT}{F} \ln a_{\mathrm{H}^+} \qquad (12)$$

where E_g° is the standard potential of that particular glass electrode on the hydrogen scale.

Departures from the ideal behavior expressed by equation 12 usually are found in alkaline solutions containing alkali metal ions in appreciable concentration, and often in solutions of strong acids. The supposition that the alkaline error is associated with the development of an imperfect response to alkali metal ions is substantiated by the successful design of cation-sensitive electrodes that are used to determine sodium, silver, and other monovalent cations (3).

The advantage of the lithium glasses over the sodium glasses in the reduction of alkaline error is attributed to the smaller size of the proton sites remaining after elution of the lithium ions from the glass surface. This view is consistent with the relative magnitudes of the alkaline errors for various cations: these errors decrease rapidly as the diameter of the cation becomes larger. The error observed in concentrated solutions of the strong acids is characterized by a marked drift of potential with time, which is thought to result from the penetration of acid anions as well as of protons into the glass surface (9).

The immersion of glass electrodes in strongly dehydrating media should be avoided. If the electrode is used in solvents of low water activity, frequent conditioning in water is advisable, as dehydration of the gel layer of the surface causes a progressive alteration in the electrode potential with a consequent drift of the measured pH. Slow dissolution of the pH-sensitive membrane is unavoidable and it leads to mechanical failure. Standardization of the electrode with two buffer solutions is the best means of early detection of electrode failure.

Fouling of the pH sensor may occur in solutions containing surface-active constituents that coat the electrode surface and result in sluggish response and drift of the pH reading. Prolonged measurements in blood, sludges, and various industrial process materials and wastes can cause such drift; therefore, it is necessary to mechanically or chemically clean the membrane at intervals that are consistent with the magnitude of the effect and the precision of the results required. These cleaning procedures can be performed manually or can be automated.

Reference Electrodes and Liquid Junctions. The electrical circuit of the pH cell is completed through a salt bridge that usually consists of a concentrated solution of potassium chloride. The solution makes contact at one end with the test solution and at the other with a reference electrode of constant potential. The liquid junction is formed at the area of contact between the salt bridge and the test solution. The mercury–mercurous chloride electrode—the calomel electrode—provides a highly reproducible potential in the potassium chloride bridge solution and is the most widely used reference electrode. However, mercurous chloride is converted readily into mercuric ion and mercury when in contact with concentrated potassium chloride solutions above 80°C. This disproportionation reaction causes an unstable potential with calomel electrodes. Therefore, the silver–silver chloride electrode and the thallium amalgam–thallous chloride electrode often are used for measurements above 80°C. However, because silver chloride is highly soluble in concentrated solutions of potassium chloride, the solution in the electrode chamber must be saturated with silver chloride.

Like the glass electrode, the commercially used reference electrode-salt bridge combination usually is of the immersion type. The salt bridge chamber may surround the electrode element. Some provision is made to allow a slow leakage of the bridge solution out of the tip of the electrode to establish the liquid junction with the standard solution or test solution in the pH cell. An opening is also provided through which the chamber may be refilled. Various devices are used to impede the outflow of bridge solution, eg, fibers, porous disks, capillaries, ground-glass joints, and controlled cracks such as those formed around a piece of palladium that has been sealed into the glass tip of the salt bridge tube. Such commercial electrodes normally give very satisfactory results, but there is some evidence that the type and structure of the junction may affect the reference potential when measurements are made at very low pH and, possibly, at high alkalinities.

Combination electrodes have increased in use and are a consolidation of the glass and reference electrodes in a single probe, usually in a concentric arrangement, with the reference electrode compartment surrounding the pH sensor. The advantages of combination electrodes include the convenience of using a single probe and the ability to measure small volumes of sample solution or in restricted-access containers, eg, test tubes and narrow-mouth flasks. In addition, the surrounding electrolyte solution in the reference electrode compartment provides excellent electrical shielding of the pH sensor, which greatly reduces noise and susceptibility to polarization.

Theoretical considerations favor junctions by which cylindrical symmetry and a steady state of diffusion are achieved. For the most accurate results, therefore, a junction established by the leakage of the potassium chloride solution into the less dense test solution should be avoided. Special cells in which a stable junction can be achieved are not difficult to construct and are available commercially.

A solution of potassium chloride that is saturated at room temperature usually is used for the salt bridge. It has been shown that the higher the concentration of the solution of potassium chloride, the more effective the bridge solution is in reducing the liquid-junction potential (10). Also, the saturated calomel reference electrode is stable, reproducible, and easy to prepare. However, the saturated electrode is not without its disadvantages. For example, it shows a marked hysteresis with changes of temperature. After long periods and upon temperature-lowering, the salt bridge chamber may become filled with crystals of potassium chloride which block the flow of bridge solution and thereby impair the reproducibility of the junction potential and raise the resistance of the cell. A slightly undersaturated (eg, 3.5 M) solution of potassium chloride is preferred.

Samples that contain suspended matter are one of the most difficult types from which to obtain accurate pH readings because of the suspension effect, ie, the suspended particles produce abnormal liquid-junction potentials at the reference electrode (11). This effect is especially noticeable with soil slurries, pastes, and other types of colloidal suspensions. In the case of a slurry that separates into two layers, pH differences of several units may result, depending on the placement of the electrodes in the layers. Although several theories have been proposed to explain this effect, none has been confirmed, and the accuracy of the pH values obtained for these samples often are questionable. Internal consistency is achieved by pH measurement using carefully prescribed measurement protocols, as has been used in the determination of soil pH (12).

Another effect that may result in spurious pH readings is caused by streaming potentials. Presumably, these are attributable to changes in the reference electrode liquid junction that are caused by variations in the flow rate of the sample solution. Factors that affect the observed pH include the magnitude of the flow rate changes, the geometry of the electrode system, and the concentration of the salt-bridge electrolyte; therefore, this problem may be avoided by maintaining constant flow and geometry characteristics and calibrating the system under operating conditions that are identical to those of the sample measurement.

pH Instrumentation

The pH meter is an electronic voltmeter that provides a direct conversion of voltage differences to differences of pH at the measurement temperature. One class of instruments is the direct-reading analogue which has a deflection meter with a large scale calibrated in mV and pH units. Several versions of the direct-reading meter make possible scale expansion; eg, a selected pH range that is one or two pH units wide can be expanded to full scale to provide increased accuracy in the reading. Many direct-reading meters have digital displays of the emf or pH. The types range from very inexpensive meters that read to the nearest 0.1 pH unit to the research models capable of measuring pH with a precision of 0.001 pH unit and drifting less than 0.003 pH unit over 24 h; however, the fundamental meaning of these measured values is considerably less certain than the precision of the measurement.

Because of the very large resistance of the glass membrane in a conventional pH electrode, an input amplifier of high impedance (usually 10^{12}–10^{14} Ω) is required to avoid errors in the pH (or mV) readings. Most pH meters have field-effect transistor (FET) amplifiers which typically exhibit bias currents of only a picoampere (10^{-12} ampere) which, for an electrode resistance of 100 MΩ, results in an emf error of only 0.1 mV (0.002 pH unit) (see Integrated circuits).

In addition, most of these devices provide operator control of settings for temperature and/or response slope, isopotential point, zero or standardization, and function (pH, mV, or monovalent/bivalent cation/anion). Microprocessors are incorporated in some meters to facilitate calibration and calculation of measurement parameters (see Computers). Furthermore, pH meters are provided with output connectors for continuous readout via a strip-chart recorder and often with BCD (binary-coded decimal) output for computer interconnections or connection to a printer. Although the accuracy of the measurement is not increased by the use of a recorder, the readability of the displayed pH (on analogue models) can be expanded, and recording provides a permanent record and information on response and equilibration times during measurement (5).

Temperature Effects

The emf E of a pH cell may be written

$$E = E° + k\text{pH} \tag{13}$$

where k is $(2.303\,RT)/F$, and $E°$, the standard potential, includes the liquid-junction potential and the emf of the cell on the reference side of the glass membrane. Changes of temperature alter the scale length because k is proportional to T. The scale position also is changed because the standard potential is temperature dependent: $E°$ is usually a quadratic function of the temperature.

The objective of temperature compensation in a pH meter is to nullify changes in emf from any source except changes in the true pH of the test solution. Nearly all pH meters provide automatic or manual adjustment for the change of k with T. If correction is not made for the change of standard potential, however, the instrument must always be standardized at the temperature at which the pH is to be determined. In industrial pH control, standardization of the assembly at the temperature of the measurements is not always possible, and compensation for shift of the scale position, though imperfect, is useful. If the value of $E°$ were a linear function of T, it would be easy to show that the straight lines representing the variation of E and pH at different temperatures would intersect at a point, the isopotential point or pH_i. Even though $E°$ does not usually vary linearly with T, these plots intersect at about pH_i when the range of temperatures is narrow. By providing a temperature-dependent bias potential of $k\text{pH}_i$, which is measured in volts, an approximate correction for the change of the standard potential with temperature can be supplied automatically (1,5).

Nonaqueous Solvents

The activity of the hydrogen ion is affected by the properties of the solvent in which it is measured. Scales of pH only apply to the medium (single solvent or mixed solvents, eg, water–alcohol) for which they are developed. The comparison of the pH values of a buffer in aqueous solution to one in a nonaqueous solvent has neither direct quantitative nor thermodynamic significance. Consequently, operational pH scales must be developed for the individual solvent systems. In certain cases, correlation to the aqueous pH scale can be made but, in others, only relative pH values are used as indicators of the hydrogen-ion activity.

Other difficulties of measuring pH in nonaqueous solvents are the complications that result from dehydration of the glass pH membrane, increased sample resistance, and large liquid-junction potentials. These effects are complex and are highly dependent on the type of solvent or mixture used (1,5).

Indicator pH Measurements

The indicator method is especially convenient when the pH must be measured with an accuracy no greater than 0.5 pH unit of a well-buffered colorless solution at room temperature. Under optimum conditions an accuracy of 0.2 pH is obtainable. A list of acid–base indicators is given in Table 2 with their corresponding transformation ranges. The theory of the indicator color change and of the salt effect is discussed in refs. 1 and 13.

Table 2. Acid–Base Indicators

Indicator	pH range	Color change
crystal violet	0.0–1.8	yellow–blue
acid cresol red	0.2–1.8	red–yellow
quinaldine red	1.0–2.2	colorless–red
acid metacresol purple	1.2–2.8	red–yellow
acid thymol blue	1.2–2.8	red–yellow
benzopurpurine 4B	2.2–4.2	violet–red
bromophenol blue	3.0–4.6	yellow–blue
methyl orange	3.2–4.4	red–yellow
bromocresol green	3.8–5.4	yellow–blue
ethyl red	4.0–5.8	colorless–red
methyl red	4.4–6.0	red–yellow
propyl red	4.8–6.6	red–yellow
chlorophenol red	5.2–6.8	yellow–red
bromocresol purple	5.2–6.8	yellow–purple
alizarin	5.6–7.2	yellow–red
bromothymol blue	6.0–7.6	yellow–blue
brilliant yellow	6.6–7.8	yellow–orange
phenol red	6.8–8.4	yellow–red
cresol red	7.2–8.8	yellow–red
metacresol purple	7.6–9.2	yellow–purple
thymol blue	8.0–9.6	yellow–blue
phenolphthalein	8.2–10.0	colorless–pink
thymolphthalein	9.4–10.6	colorless–blue
tolyl red	10.0–11.6	red–yellow
alizarin	11.0–12.4	red–purple
parazo orange	11.4–12.6	yellow–orange
2,4,6-trinitrotoluene	11.5–13.0	colorless–orange
1,3,5-trinitrobenzene	12.0–14.0	colorless–orange

Because they are weak acids or bases, the indicators may affect the pH of the sample, especially in the case of a poorly buffered solution. Variations in the ionic strength or solvent composition, or both, also can produce large uncertainties in pH measurements, presumably caused by changes in the equilibria of the indicator species. Specific chemical reactions also may occur between solutes in the sample and the indicator to produce appreciable pH errors. Examples of such interferences include binding of the indicator forms by proteins and colloidal substances and direct reaction with sample components, eg, bleaches and heavy metal ions.

Industrial Process Control

Specialized equipment for industrial measurements and automatic control have been developed (14) (see Instrumentation and control). In general, the pH of an industrial process need not be controlled with great precision. Consequently, frequent standardization of the cell assembly may be unnecessary. On the other hand, the temperature and humidity conditions under which the industrial control measurements are made may be such that the pH meter must be much more rugged than those intended for laboratory use. In order to avoid costly downtime for repairs, pH instruments may be constructed of modular units, permitting rapid removal and replacement of a defective subassembly.

The pH meter usually is coupled to a recorder and often to a pneumatic or electric controller. The controller governs the addition of reagent so that the pH of the process stream is maintained at the desired level.

Immersion-cell assemblies are designed for continuous pH measurement in tanks, troughs, or other deep vessels containing process solutions at different levels under various conditions of agitation and pressure. The electrodes are guarded from mechanical damage and sometimes are provided with devices to remove surface deposits as they accumulate. Process flow chambers are designed to introduce the pH electrodes directly into piped sample streams or bypass sample loops which may be pressurized. Electrode chambers of both types usually contain a temperature-sensing element that controls the temperature-compensating circuits of the measuring instrument.

Glass electrodes for process control do not differ materially from those used for pH measurements in the laboratory, but the emphasis in industrial applications is on rugged construction to withstand both mechanical stresses and high pressures. Pressurized salt bridges, which ensure slow leakage of bridge solution into the process stream even under very high pressures, have been developed. For less severe process monitoring conditions, reference electrodes are available with no-flow polymeric or gel-filled junctions that can be used without external pressurization.

BIBLIOGRAPHY

"Hydrogen-Ion Concentration" in *ECT* 1st ed., Vol. 7, pp. 711–726, by R. G. Bates and E. R. Smith, National Bureau of Standards; "Hydrogen-Ion Concentration" in *ECT* 2nd ed., Vol. 11, pp. 380–390, by Roger G. Bates, National Bureau of Standards.

1. R. G. Bates, *Determination of pH,* 2nd ed., Wiley-Interscience, New York, 1973.
2. R. A. Durst, *Standardization of pH Measurements,* NBS Spec. Publ. 260-53, U.S. Govt. Printing Office, Washington, D.C., 1975.
3. G. Eisenman, ed., *Glass Electrodes for Hydrogen and Other Cations,* Marcel Dekker, Inc., New York, 1967.
4. G. Mattock, *pH Measurement and Titration,* The Macmillan Co., New York, 1961.
5. C. C. Westcott, *pH Measurements,* Academic Press, Inc., New York, 1978.
6. "pH of Aqueous Solutions with the Glass Electrode," *ASTM Method E 70-77,* ASTM, Philadelphia, Pa., 1977.
7. R. G. Bates and E. A. Guggenheim, *Pure Appl. Chem.* **1,** 163 (1960).
8. R. A. Durst, "Sources of Error in Ion-Selective Electrode Potentiometry" in H. Freiser, ed., *Ion-Selective Electrodes in Analytical Chemistry,* Plenum Publishing Corp., New York, 1978, Chapt. 5.
9. K. Schwabe, "pH Measurements and Their Applications" in H. W. Nürnberg, ed., *Electroanalytical Chemistry,* John Wiley & Sons, Inc., New York, 1974, Chapt. 7.
10. E. A. Guggenheim, *J. Am. Chem. Soc.* **52,** 1315 (1930).
11. H. Jenny, T. R. Nielson, N. T. Coleman, and D. E. Williams, *Science* **112,** 164 (1950).
12. A. M. Pommer, "Glass Electrodes for Soil Waters and Soil Suspensions" in G. Eisenman, ed., *Glass Electrodes for Hydrogen and Other Cations,* Marcel Dekker, Inc., New York, 1967, Chapt. 14.
13. I. M. Kolthoff and C. Rosenblum, *Acid-Base Indicators,* The Macmillan Co., New York, 1937.
14. F. G. Shinskey, *pH and pIon: Control in Process and Waste Streams,* John Wiley & Sons, Inc., New York, 1973.

General References

Refs. 1–6 are also general references.

RICHARD A. DURST
National Bureau of Standards

ROGER G. BATES
University of Florida

HYDROGEN PEROXIDE

Hydrogen peroxide [*7722-84-1*] (H_2O_2), mol wt 34.016, is a weakly acidic, clear colorless liquid, miscible with water in all proportions. The four atoms are covalently bound in a nonpolar H—O—O—H structure (1–2). It is commercially available in aqueous solution over a wide concentration range. Discovered by Thenard in 1818, it has been an article of commerce since the mid-19th century. Its scale of manufacture and use have increased markedly since about 1925 when electrolytic processes were introduced to the United States and industrial bleach applications were developed. Now prepared primarily by anthraquinone autoxidation processes, hydrogen peroxide is used widely to prepare other peroxygen compounds and as a nonpolluting oxidizing agent.

Physical Properties

Properties of pure hydrogen peroxide are listed in Table 1. Physical constants have been determined or calculated for its aqueous solutions, the only form in which hydrogen peroxide is commercially available (see Table 2). Numerous other physical property data appear in the literature, including approximation coefficients for free energy function calculations (7), coefficients of diffusion (8), distribution coefficients (9), spectroscopic studies (10–17), thermodynamic properties (18), and third-law entropy (19). Ref. 20 contains mathematical correlations for vapor pressure, surface tension, heat of vaporization and liquid density, heat capacity, thermal conductivity, and viscosity values over 0–450°C; heat and free energy of formation, vapor heat capacity, thermal conductivity, and viscosity values over 0–1200°C.

Table 1. Properties of Hydrogen Peroxide

Property	Value	Ref.
mp[a], °C	−0.41	3
bp, °C	150.2	4
density at 25°C, g/mL	1.4425	4
viscosity at 20°C, mPa·s (= cP)	1.245	5
surface tension at 20°C, mN/m (= dyn/cm)	80.4	4
specific conductance at 25°C, per Ω·cm	4×10^{-7}	4
heat of fusion, J/g[b]	367.52	4
specific heat at 25°C, J/g[b]	2.628	4
heat of vaporization at 25°C, kJ/g[b]	1.517	6
dissociation constant[c] at 20°C	1.78×10^{-12}	4
heat of dissociation, kJ/mol[b]	34.3	4

[a] Tends to supercool.
[b] To convert J to cal, divide by 4.184.
[c] At zero ionic strength.

Table 2. Physical Properties of Aqueous Hydrogen Peroxide

Liquid composition wt % H_2O_2	Freezing point[a], °C	Boiling[b] point, °C	Vapor composition[b], wt % H_2O_2	Density[c] at 25°C, g/mL	Heat of vaporization[d] at 25°C, kJ/g[e]
10	− 6.4	101.7	0.9	1.0324	2.357
20	−14.6	103.6	2.1	1.0694	2.274
30	−25.7	106.2	4.2	1.1081	2.192
40	−41.4	109.6	7.6	1.1487	2.105
50	−52.2	113.8	13.0	1.1914	2.017
60	−55.5	119.0	20.8	1.2364	1.926
70	−40.3	125.5	33.4	1.2839	1.832
80	−24.8	132.9	51.5	1.3339	1.733
90	−11.5	141.3	75.0	1.3867	1.627

[a] Ref. 5.
[b] At 101.3 kPa (1 atm).
[c] Ref. 4.
[d] Ref. 6.
[e] To convert J to cal, divide by 4.184.

Chemical Properties

The reactions of hydrogen peroxide are:

$$\textit{decomposition},\ 2\,H_2O_2 \rightarrow 2\,H_2O + O_2 \tag{1}$$

$$\textit{molecular additions},\ H_2O_2 + Y \rightarrow Y.H_2O_2 \tag{2}$$

$$\textit{substitutions},\ H_2O_2 + RX \rightarrow ROOH + HX \tag{3}$$

$$H_2O_2 + 2\,RX \rightarrow ROOR + 2\,HX \tag{4}$$

$$\textit{oxidations},\ H_2O_2 + W \rightarrow WO + H_2O \tag{5}$$

$$\textit{or reductions}\ H_2O_2 + Z \rightarrow ZH_2 + O_2 \tag{6}$$

Hydrogen peroxide may react directly or after it has first ionized or dissociated into free radicals. In many cases, the reaction mechanism is extremely complex and may involve catalysis or be dependent upon the reaction environment.

Ionization. Hydrogen peroxide is a weak acid. The pH of its aqueous solutions can be measured reproducibly with a glass electrode, but a correction factor dependent on the concentration must be added to obtain the true pH value. Correction values for the most common commercial solutions are listed in Table 3.

Free-Radical Formation. Hydrogen peroxide can form free radicals by homolytic cleavage of either an O—H bond or the O—O bond.

$$HOOH \rightarrow H\cdot + \cdot OOH \quad \Delta H = 380\ \text{kJ/mol (90 kcal/mol)} \tag{7}$$

$$HOOH \rightarrow 2\,\cdot OH \quad \Delta H = 210\ \text{kJ/mol (50 kcal/mol)} \tag{8}$$

Equation 8 predominates in uncatalyzed vapor-phase decomposition and photochemically initiated reactions. In catalytic reactions, especially in solution, the nature of the reactants determines which reaction is predominant.

Table 3. Apparent and True pH of Aqueous Hydrogen Peroxide

H_2O_2 conc, wt %	Equivalence point[a]	True pH	Correction factor
35	3.9	4.6	0.7
50	2.8	4.3	1.5
70	1.6	4.4	2.8
90	0.2	5.1	4.9

[a] Measured with glass electrode.

Decomposition. Decomposition of hydrogen peroxide must be controlled at all times, because the resultant generation of gas and heat (about 100.4 kJ/mol or 24 kcal/mol) may cause safety problems (see under Health and Safety Factors). It is readily detected by the evolution of gas and an increase in temperature.

The mechanism of decomposition depends on many factors, including temperature, pH, or the presence of a catalyst (21–27), such as metal ions or metal oxides and hydroxides.

The stability of pure hydrogen peroxide solutions increases with increasing concentration and is at a maximum between pH 3.5–4.5. The temperature coefficient for the decomposition of ultrapure or stabilized hydrogen peroxide solutions is about 2.2 for a 10 degree rise; values as low as 1.6 have been noted for impure, unstabilized solutions.

Decomposition of hydrogen peroxide solutions is minimized by purification during manufacture, control of contaminants, and the addition of stabilizers.

Stabilization. Pure hydrogen peroxide solutions stored in clean passive containers are relatively stable. Commercial solutions, however, invariably contain and are exposed to varying amounts of catalytic impurities and must be stabilized with reagents that deactivate catalytic impurities, either through complex formation or by adsorption. Sodium pyrophosphate, for example, can be added to acidic solutions as a complexing agent, whereas sodium stannate forms a protective colloid. Alkali metal silicates stabilize alkaline solutions.

Various stabilizing systems have been devised, particularly combinations of tin salts and phosphates. Recent patents cover modifications of the stannate–phosphate systems (28–31) and a variety of organic chelating agents, with or without stannate and phosphate (32–44). Stabilizers have been developed for such specific applications as metal and mineral treatment (45–56), chemiluminescent systems (57), and alkaline textile treatment baths (58), or for protection against chloride ion (59) or aluminum-induced decomposition (60). Some stabilizers include organic compounds that cannot be used in concentrated solutions but are effective in dilute solutions. The efficiency of various stabilizers against iron- and zinc-promoted decomposition has been studied (61). An electrolytic procedure has been patented for the removal of inorganic stabilizers from hydrogen peroxide used for fuel cells (62) (see Batteries).

The quantity of stabilizer required decreases with increasing concentration. The quantity and type in commercial solutions are adjusted to meet specific use requirements. Dilution-grade 70% H_2O_2 solutions contain enough special stabilizers to allow dilution at time of delivery to 35 or 50% concentration.

Container corrosion is a source of catalytic impurities. Aluminum is commonly used for shipping and storing hydrogen peroxide and nitrate salts are frequently added as corrosion inhibitors (see Corrosion and corrosion inhibitors).

Molecular Addition. Oxyacid salts, metal peroxides, nitrogen compounds, zirconyl acetate (63), and 1,4-diazabicyclo[2.2.2]octane (64) form crystalline peroxyhydrates with hydrogen peroxide. When dissolved in water, these peroxyhydrates react as solutions of their components. The peroxyhydrates formed by sodium carbonate and urea are commercially available.

Sodium perborate, formed from hydrogen peroxide and sodium borate and frequently cited as an example of a peroxyhydrate, has been shown to be a true peroxy compound (65). It properly should be called sodium peroxyborate trihydrate.

Substitution. A variety of peroxygen compounds are prepared by substitution reactions of hydrogen peroxide with organic reagents. These compounds are commercially useful polymerization catalysts and oxidizing agents for a number of special reactions. For example, with alkylating agents they form alkylhydroperoxides, with carboxylic acids, peroxy acids, with acid chlorides, diacyl peroxides, and with ketones, ketone peroxides. These derivatives are considerably more hazardous than hydrogen peroxide. Information on their safe handling should be obtained from the manufacturer.

Inorganic peroxygen compounds can be prepared by similar reactions with inorganic reagents. Alkaline earth metal peroxides are prepared from hydrogen peroxide and the corresponding hydroxide, and monoperoxysulfuric acid by reaction of hydrogen peroxide with sulfur trioxide, or sulfuric acid.

Oxidation. Hydrogen peroxide is a strong oxidant and most of its uses and those of its derivatives depend on this property. It oxidizes a wide variety of organic and inorganic compounds, ranging from iodide ions to color bodies of unknown structure in cellulosic fibers. The mechanisms of these reactions are varied and dependent on the reductive substrate, the reaction environment, and catalysis. Specific reactions are discussed in a number of general and other references (66–67).

Reduction. Hydrogen peroxide reduces stronger oxidizing agents such as chlorine, sodium hypochlorite, potassium permanganate, and ceric sulfate; the last two are used for the volumetric determination of hydrogen peroxide. Reductive destruction of residual chlorine and hypochlorite values in industrial waste streams is used for pollution abatement.

Manufacture

Hydrogen peroxide can be formed from compounds that contain the peroxy group, from hydrogen–oxygen mixtures or water and oxygen by thermal, photochemical, electrochemical or similar processes, and by the uncatalyzed reaction of molecular oxygen with an appropriate hydrogen-containing species. It has been manufactured commercially by processes based on the reaction of barium or sodium peroxide with an acid, the electrolysis of sulfuric acid solutions, and the autoxidation of anthraquinones, isopropyl alcohol, and hydrazobenzene. All manufacturing facilities constructed in the United States since 1957 have been based on the autoxidation of an anthraquinone.

AUTOXIDATION METHODS

Anthraquinone Autoxidation. The first commercial anthraquinone autoxidation process (30 t H_2O_2 per month) was operated by I. G. Farbenindustrie in Germany during World War II. All subsequent anthraquinone autoxidation processes developed worldwide retain the basic features of the Riedl-Pfleiderer process shown in Figure 1. A 2-alkylanthraquinone dissolved in a suitable solvent or solvent mixture is reduced catalytically to the corresponding anthraquinol (or anthrahydroquinone, eq. 9). The anthraquinone is commonly called the reaction carrier or working material, whereas the anthraquinone–solvent mixture is called the working solution. The working solution containing the anthrahydroquinone is separated from the hydrogenation catalyst and aerated with an oxygen-containing gas, usually air, to reform the anthra-

$$\text{(O, O, R)} + H_2 \xrightarrow{\text{catalyst}} \text{(OH, OH, R)} \qquad (9)$$

$$\text{(OH, OH, R)} + O_2 \longrightarrow \text{(O, O, R)} + H_2O_2 \qquad (10)$$

quinone and simultaneously produce hydrogen peroxide (eq. 10). The hydrogen peroxide is extracted with water and the aqueous solution purified and concentrated to the degree required. The extracted solution is recycled.

The alkylanthrahydroquinone is subject to a number of secondary reactions during each process cycle (68). Catalytic reduction of the unsubstituted aromatic ring yields the 5,6,7,8-tetrahydroanthrahydroquinone (**1**); further ring reduction generates

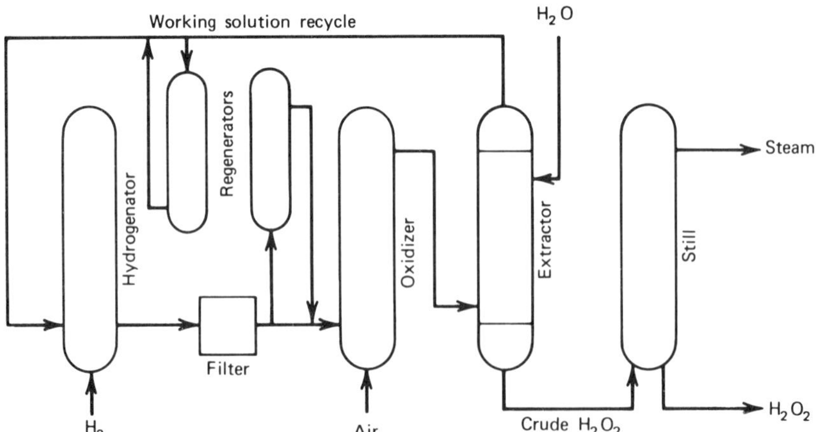

Figure 1. Anthraquinone autoxidation, Riedl-Pfleiderer process.

the 1,2,3,4,5,6,7,8-octahydroanthrahydroquinone (**2**). Tautomerization of the anthrahydroquinone can yield the hydroxyanthrones (**3**) and (**4**) which can be reduced to anthrones (**5**) and (**6**). The tetrahydroanthrahydroquinone (**1**), once formed, is an active participant in the process cycle. It is oxidized to the corresponding tetrahydroanthraquinone and hydrogen peroxide in the oxidizer, is reformed by catalytic reduction or reaction of the tetrahydroanthraquinone with an anthrahydroquinone in the hydrogenator (eq. 11), and is the apparent precursor to a tetrahydroanthraquinone epoxide (**7**). With the exception of the tetrahydroanthrahydroquinone, the first and second generation anthraquinone derivatives do not produce hydrogen peroxide, but they are susceptible to further reaction by oxidative or reductive mechanisms (see also Anthraquinone derivatives).

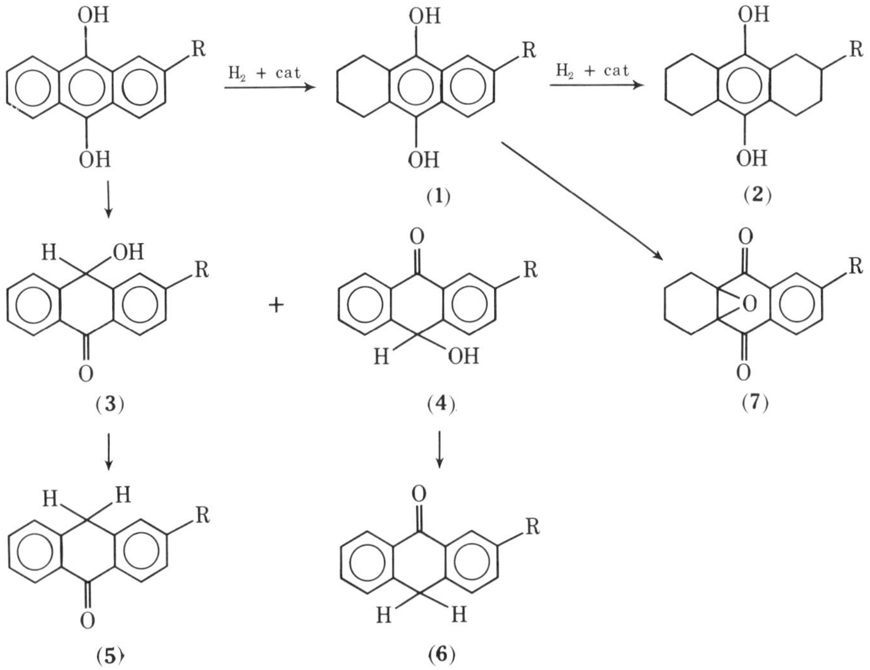

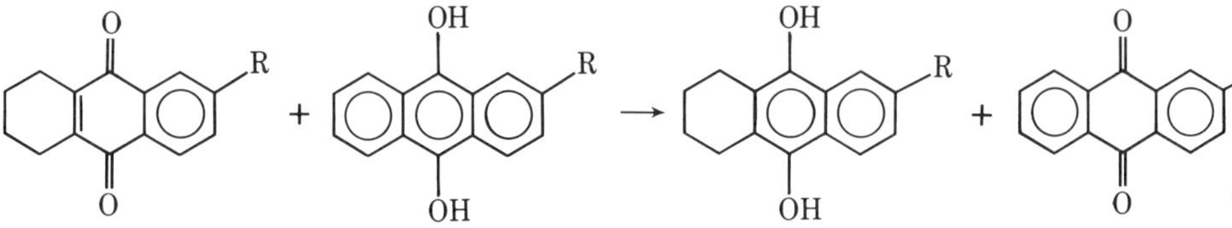

H_2 + cat → (**1**) H_2 + cat → (**2**)

(**3**) + (**4**) (**7**)

(**5**) (**6**)

(11)

Although the chemical yield of hydrogen peroxide per process cycle is very high, secondary reactions necessitate regeneration of the solution and the hydrogenation catalyst and removal of organic material from the extracted hydrogen peroxide.

The first large-scale anthraquinone autoxidation process was put into operation in the United States by E. I. du Pont de Nemours & Co., Inc., in 1953, followed by FMC Corporation (United States), LaPorte Chemicals, Ltd. (UK), Degussa (FRG), Mitsubishi-Gas Chemical Co. (Japan), and others (69–70).

Working Solution Composition. The working solution in an anthraquinone process is composed of anthraquinone, the by-products from the reduction and oxidation steps, and solvent. Once the working solution has been defined for a process, its composition and properties must be maintained within prescribed limits for optimum process operation.

The theoretical production capability of an anthraquinone process is the product of the working solution flow rate and the capacity where capacity is defined as the maximum quantity of hydrogen peroxide (in g H_2O_2/L of working solution) that can be produced per process cycle. The maximum capacity of the working solution for a continuous flow process is limited by the solubility of the working material in both the reduced and oxidized state. The capacity of the working solution can be increased by modifying the substituent group on the anthraquinone to alter its solubility characteristics and by more powerful solvents or solvent combinations. Although only 2-ethyl, 2-*tert*-butyl-, and 2-amylanthraquinones are used commercially, a number of others have been patented for use in autoxidation processes. In most cases the substituent of monosubstituted anthraquinones is located at the 2-position. Disubstituted naphthoquinones (71) and anthraquinones have also been claimed, including dialkylanthraquinones (72), where the two substituents contain a total of three to fifteen carbon atoms. Improved solvency is also claimed for mixtures of alkylanthraquinones, tetrahydroanthraquinones (73) or tetrahydroanthraquinones with anthraquinones (74–76), and for 1-alkyl substituted naphthoquinones (77–79).

The working material should be chemically stable throughout the process cycle. The chemical stability and solubility characteristics of alkylanthraquinones depend on the substituent and can vary significantly.

A high partition coefficient for hydrogen peroxide between water and the working solution maximizes process efficiency. The solvent itself should be chemically stable, poorly water-soluble, and for safety reasons, should have a high flash point and low volatility and be nontoxic. The solvent system used in the Riedl-Pfleiderer process was a 50:50 mixture of benzene and a mixture of C_7–C_9 secondary alcohols. Solvent systems have been patented; the most recent contain trialkylphosphates (80–81), tetraalkyl-substituted ureas (82–83), dialkylcarboxylic acid amides (84), a 1,3,5-triazene (85), a 2,6-dialkycyclohexane (86), an acetate (87) or pivalate (88) ester or a mono- or diacetylbenzoquinone (89) or diacetylbenzene (90).

Process modifications to avoid working material solubility limitations include use of anthraquinone (91) or anthrahydroquinone slurries, water-soluble anthraquinones (92), high molecular weight ion-exchange resins, or polymeric anthraquinones.

Hydrogenation and Catalyst. The working material is catalytically reduced to the hydroquinone form in the hydrogenator. The reduction is carried out at slightly elevated hydrogen partial pressure and below 100°C, with most patent examples citing temperatures in the 40–50°C range. The degree of hydrogenation (conversion of quinone to hydroquinone) is ordinarily limited to about 50% to minimize secondary reactions, but conversions greater than 80% have been cited (93).

The hydrogenation rate is maintained constant by the periodic addition or exchange of the hydrogenation catalyst, or by controlled variation of the hydrogen partial pressure (94). The rate, however, is not affected by hydrogen pressures greater than 405 kPa (4 atm). Reduced hydrogen partial pressures limit ring reduction of the anthraquinone. The hydrogenation rate is increased by the addition of ammonia, water-soluble amines, or ammonium salts to the working solution; degradation of the working material is decreased by addition of insoluble particulate inorganic alkaline reagents (95). The agglomeration of slurry catalysts caused by water separation in the hydrogenator can be prevented by operating the hydrogenator at a temperature higher than that in the extractor.

Tubular (96), draft-tube (97), and fixed-bed (98) hydrogenators have been patented as has a carbon filter for use with a palladium black catalyst (99). Alternatively, hydrogen sulfide can be used with an amine catalyst (100) or sodium amalgam (101–102) to form the anthraquinol. The photochemical reduction of a water-soluble anthraquinone has been studied (103) and a mathematical model derived to determine the optimum process parameters for a large-scale anthraquinone autoxidation plant (104).

The catalyst should be active, selective, stable, and have a long life. The Raney nickel catalyst used in the original Riedl-Pfleiderer process was active, but was also pyrophoric, deactivated by hydrogen peroxide, and caused excessive ring hydrogenation.

Residual hydrogen peroxide can be destroyed catalytically or by reaction with stannous ion before the oxidized and extracted working solution is recycled to the hydrogenator. The catalyst selectivity can be improved either by pretreatment or by the addition of moderators to the working solution (105–106). It can be regenerated (107) and its concentration continuously monitored when used as a finely dispersed slurry (108). Supported Raney nickel catalysts are claimed to have better selectivity (109). Nickel boride and molybdenum-promoted nickel boride catalysts, however, promoted the unselective reduction of 2-ethylanthraquinone (110).

A palladium catalyst avoids the problems associated with Raney nickel. It can be used as palladium black, as wire screen, or supported on a carrier for slurry or fixed-bed application. The activity and selectivity of the supported catalysts are influenced by both the chemical structure and physical construction of the support (111). Pretreatment of alumina supports with alkali (112) or a strong inorganic oxidizing agent have been patented, as have support pretreatments to improve the abrasion resistance of the palladium (113–114). The selectivity of supported palladium catalysts reportedly is improved by treatment with hydrogen at high temperature (105) or by the addition of platinum-group-metal modifiers (115–117). Metal-loaded ion-exchange resins (118), a magnetic metal catalyst and numerous fixed-bed hydrogenation catalysts have been patented (119–120) (see also Catalysis).

Oxidation. The hydrogenated working solution, free of catalyst, is oxidized with an oxygen-containing gas in an uncatalyzed reaction whose overall rate is dependent on the oxygen partial pressure, the temperature, and the phase-boundary surface area. Either co- or countercurrent flow of gas and liquid in a single- or multiple-stage system with or without packing can be employed (121–123).

Side reactions and solvent losses can be minimized by limiting the contact time of the working solution with air in a countercurrent, packed-tower oxidizer. The heat of reaction can be controlled by cooling part of the oxidized working solution and recycling it with the hydrogenated feed solution. Gas diffusers maximize the oxidation rate. An alkylanthrahydroquinone-based working solution reduces by-product formation (124). The addition of a small quantity of an organic base to the oxidizer is claimed to increase the oxidation rate of the hydrogenated working solution (125).

Hydrogen Peroxide Recovery. The dilute solution of hydrogen peroxide obtained in the oxidation step can be treated with an aqueous solution of sodium metaborate to give sodium perborate (126–127) or with an aliphatic carboxylic acid to give a peroxycarboxylic acid (128). Usually the hydrogen peroxide is recovered by extraction with water, which can be carried out with mechanical extractors, in packed, sieve-plate extraction columns, spray columns (129), or unfilled towers (130).

Polyethylene, polypropylene, polytetrafluoroethylene, or poly(vinyl chloride) are used as column-packing materials or to coagulate working solution emulsions (131). An inorganic stabilizer, such as a phosphate or sodium stannate, maintains the strength of the hydrogen peroxide extract.

Depending on the composition of the working solution, aqueous hydrogen peroxide concentrations of 20% and higher can be obtained by extraction. The hydrogen peroxide also can be recovered directly from the working solution by reduced pressure distillation or by stripping with organic solvents (132). The organic solutions obtained by solvent stripping can be used to prepare peroxycarboxylic acids or it can be extracted with water to obtain 60% H_2O_2. Other patents describe the transfer of hydrogen peroxide from one solvent, including water, to another (133–135).

Working Solution Regeneration and Purification. Degradation products formed during the cyclic reduction and oxidation of the working solution must be removed or regenerated. Treatment of the working solution or isolated quinone mixtures with a dehydrogenation catalyst converts tetrahydroanthraquinones to the corresponding anthraquinone. Treatment with an olefin in the presence of a hydrogenation catalyst converts hydroxyanthrones into the corresponding anthraquinone (136), and passage of hydrogenated working solution through sodium aluminum silicate converts tetrahydroanthraquinone epoxides into the corresponding tetrahydroanthraquinone (137–138).

The working solution is regenerated by a large variety of procedures, such as treatment with alkali (139–140), acids (141), or metal oxides, chlorides, silicates, or aluminosilicates (142), and dithionates, with oxygen (143) or ozone (144), or purification by washing with water (145).

Recovery of working materials includes recrystallization (146–148), extraction of crude mixtures with lower alcohols (149), procedures based on distillation of the working solution (150), use of an anion-exchange resin (151), or isolation as the hydroquinone (152). The residue obtained after removal of at least 90% of the working solution solvents is heated and recycled to avoid the accumulation of inert by-products (153).

The alumina or sodium aluminosilicate catalysts used to regenerate degraded working solution lose activity on use, but can be regenerated (154–157).

Hydrogenation Catalyst Regeneration. Procedures for the recovery of palladium from spent hydrogenation catalyst have been developed (150–159) as well as methods for *in situ* or external regeneration. The latter include treatment with wet steam, strong oxidants, liquid ammonia, or concentrated ammonium hydroxide, highly alkaline solutions, carboxylic acids, hydrogen peroxide (160–161), organic solvents, or heating in air at 250°C. Multiple-step regeneration procedures include sequential treatment with a polar organic solvent, aqueous ammonium hydroxide, and steam and an oxygen-containing gas (162), and multiple-wash treatments to remove adsorbed organic materials before dissolution and redeposition of the palladium on the support (163). Periodic reduction of the hydrogen partial pressure and treatment with an inert gas during the hydrogenation step prolongs catalyst life. Regeneration never fully restores the catalyst activity to its original level, except in treatments where the catalyst is essentially reconstituted, but it does permit longer use periods for more economical operation.

Alcohol Autoxidation. The partial oxidation of primary or secondary alcohols in either the liquid or vapor phase gives hydrogen peroxide and an aldehyde or ketone in high yield (164–166). A liquid-phase process based on the autoxidation of isopropyl alcohol has been installed at Norco, Louisiana, by Shell Chemical Company.

$$(CH_3)_2CHOH + O_2 \rightarrow (CH_3)_2C{=}O + H_2O_2 \qquad (12)$$

Liquid-phase alcohol autoxidation is carried out at temperatures of 70–160°C and pressures of 10–20 atm. The process does not require special catalysts although they can be added. Aqueous isopropyl alcohol (2-propanol–water azeotrope) containing hydrogen peroxide is treated with an oxygen-containing gas at a temperature and pressure combination that allows practical hydrogen peroxide production rates and maintains the liquid phase. Acids, nitriles (167), stabilizers, and trace quantities of sequestered transition metal oxides improve process economics. The product mixture, containing hydrogen peroxide, acetone, and unreacted isopropyl alcohol is separated in a falling-film evaporator. Organic materials and water are taken overhead and further distilled to recover by-product acetone and isopropyl alcohol for recycle. The hydrogen peroxide, which concentrates at the base of the evaporator, is continuously diluted with water to keep its concentration below 50% and minimize safety problems associated with heating and concentrating hydrogen peroxide in the presence of organic materials. The recovered hydrogen peroxide is purified by countercurrent solvent extraction, and treatment with anion- and cation-exchange resins, followed by stabilization. It may also be recovered as a urea complex (168). Propyl acetate has been used as an oxidation medium (169).

Acidic degradation products affect the oxidation rate and may cause reaction control problems at high conversion levels. A multistage continuous process (171–175) alleviates this problem and improves the yield. Aluminum equipment washed with hydrogen peroxide stabilizers before being placed in operation has been proposed to minimize hydrogen peroxide decomposition (176). Hydrogen peroxide yields, disregarding purification loss, of 97% have been reported (174), although yields of about 87% are probably more typical (177). Process modifications and the use of other secondary alcohols (170,178) have been described.

Other patented alcohol oxidation processes include the vapor-phase oxidation of primary or secondary alcohols and their photochemical oxidation in the presence of titanium dioxide.

Hydrocarbon Autoxidation. Hydrogen peroxide can be prepared directly by the vapor-phase partial oxidation of hydrocarbons and indirectly from several hydrocarbon hydroperoxides. Although numerous process patents have been granted for producing hydrogen peroxide from hydrocarbon feedstocks, none are known to be practiced commercially (see Hydrocarbon oxidation).

The vapor-phase oxidation processes generate a great number of by-products which complicate recovery, purification, and concentration of the hydrogen peroxide. Several recovery methods have been patented (179–182), eg, extractive distillation of the crude reaction mixture with acetic acid (183).

Organic hydroperoxides can be prepared in high yield from selected hydrocarbons by liquid-phase oxidation. Cyclic processes for hydrogen peroxide based on dihydroanthracene hydroperoxide (184) and *tert*-butyl hydroperoxide (185) have been patented. Other patents (186–188) describe the reaction of *tert*-butyl hydroperoxide with dilute to 55% sulfuric acid to obtain hydrogen peroxide and *tert*-butyl alcohol or *tert*-butyl peroxide (di-*tert*-butyl peroxide).

Nitrogen Compound Autoxidation. Hydrogen peroxide can be prepared commercially by the oxidation and reduction of hydrazobenzene in a cyclic process. Such

$$C_6H_5\text{—NH—NH—}C_6H_5 + O_2 \xrightarrow{\text{solvent}} C_6H_5\text{—N=N—}C_6H_5 + H_2O_2 \quad (13)$$

$$C_6H_5\text{—N=N—}C_6H_5 + H_2 \xrightarrow{\text{catalyst}} C_6H_5\text{—NH—NH—}C_6H_5 \quad (14)$$

processes were developed in the United States (Mathieson Alkali Corp.) and Germany (BASF) during World War II. However, they could not compete economically with anthraquinone autoxidation processes.

Cyclic processes based on phenazine compounds (eqs. 15 and 16) have been patented (189).

$$\text{dihydrophenazine (N—H, N—H)} + O_2 \xrightarrow{\text{solvent}} \text{phenazine} + H_2O_2 \quad (15)$$

$$\text{phenazine} + H_2 \xrightarrow{Pd/Al_2O_3} \text{dihydrophenazine (N—H, N—H)} \quad (16)$$

HYDROLYSIS OF PEROXYCARBOXYLIC ACIDS

Peroxyacetic acid is produced commercially by the controlled autoxidation of acetaldehyde. Under hydrolytic conditions, it forms an equilibrium mixture with acetic acid and hydrogen peroxide. Recovery of the hydrogen peroxide from the equilibrium mixture by extractive distillation (190) or by precipitating the peroxide as its calcium salt followed by carbonation with carbon dioxide have been patented but not commercialized. Alternatively, peroxycarboxylic acids can be treated with alcohols in the presence of an esterification catalyst to form hydrogen peroxide and the corresponding ester (191–192) (see also Peroxides and peroxy compounds).

ELECTROLYTIC METHODS

Electrolytic processes for commercial production are based on the hydrolysis of peroxydisulfates obtained by the anodic oxidation of acidic, sulfate-containing solutions, as shown below for the Weissenstein (peroxydisulfuric acid) process (4).

$$2\ H_2SO_4 + \text{electrical energy} \rightarrow H_2S_2O_8 + H_2 \quad (17)$$

$$H_2S_2O_8 + H_2O \rightarrow H_2SO_5 + H_2SO_4 \quad (18)$$

$$H_2SO_5 + H_2O \rightarrow H_2O_2 + H_2SO_4 \quad (19)$$

Electrolytic processes require extensive and continuous purification of the electrolyte and because of their high capital and power requirements, cannot compete economically with anthraquinone autoxidation processes.

The cathodic reduction of oxygen using an alkaline electrolyte yields hydrogen peroxide directly. It is recovered by distillation after neutralizing the electrolyte or by forming an insoluble metal peroxide from which it can be subsequently regenerated (193–194). Attempts to develop the method commercially using high oxygen pressures and metal cathodes have been unsuccessful. Activated carbon cathodes operate at lower oxygen pressures and current process development efforts are concentrated in this area (see Electrochemical processing).

DIRECT COMBINATION OF HYDROGEN AND OXYGEN

Hydrogen peroxide can be formed directly by the thermal, electric-discharge, or metal-activated reaction of hydrogen and oxygen. Silent electric discharge processes have been patented and a process piloted in Germany, but their power requirements are too high for commercial use. Recent patent activity (195–199) and a laboratory study (200) have focused on metal-catalyzed processes where hydrogen and oxygen react in an acidic, aqueous medium in the presence of a platinum-group metal, and additives. Hydrogen peroxide reportedly has been obtained in 12.8% concentration at 88% yield (201).

INORGANIC METHODS

Before the development of electrolytic processes, hydrogen peroxide was manufactured solely from metal peroxides. Early methods based on barium peroxide, obtained by air-roasting barium oxide, used sulfuric and/or phosphoric acid to form hydrogen peroxide in 3–8% concentration and the insoluble barium salt. Recent patents propose acidification of the barium peroxide slurry with carbon dioxide and calcination of the by-product barium carbonate to the oxide for recycle.

Acid ion-exchange resins have been proposed for sodium peroxide-based processes. Alternatively, the use of an acid and a water-insoluble organic solvent has been suggested to avoid the hydrogen peroxide recovery problems otherwise encountered with the more soluble sodium salts (202–203). A cyclic synthesis of hydrogen peroxide and hydrazine has been proposed, using an alkali metal–alkali metal peroxide as the common electron donor–acceptor pair (204).

PURIFICATION AND CONCENTRATION

The crude product from any hydrogen peroxide process can be used as such, but commercial grades are further purified, concentrated, and stabilized. Crude products from organic-based processes contain organic impurities which affect their color, odor, surface tension, and stability, and are normally pretreated to reduce their carbon content before final purification and concentration by total distillation. Procedures patented include solvent extraction followed optionally by air blowing to remove residual solvent and treatment with synthetic resins (205), polyethylene, waxes, carbon (206), and aluminum and magnesium hydroxides and alumina (207–209). Active ion-exchange resins are used to remove both metallic (201–212) and acidic (210–216) impurities. The amount of nonvolatile organic impurities present in the crude extracted hydrogen peroxide from anthraquinone autoxidation processes may be reduced by washing the working solution with water and discarding the washings.

Concentration of hydrogen peroxide prepared by autoxidation processes can be carried out safely and conveniently by distillation at reduced pressures. Procedures analogous to those developed for electrolytic process products are used to assure safe operations and to alleviate yield problems associated with heating and concentrating hydrogen peroxide with organic materials (217).

The hydrogen peroxide is concentrated during distillation and is usually marketed as a 50–70 wt % product. It is concentrated to higher strengths by redistillation (217) or freezing methods (218). A procedure for concentrating very dilute solutions via formation of organic nitrogen-compound peroxyhydrates has been patented (219).

Economic Aspects

U.S. production of hydrogen peroxide, as reported by the U.S. Bureau of the Census (220), is shown in Table 4. A substantial increase is expected in 1980 owing to new plants in Texas (FMC at Bayport and Interox at Deer Park). Total U.S. consumption in 1979 was 108,800 t. U.S. manufacturers (221) are shown in Table 5. The January, 1980 prices for various commercial grades are shown in Table 6. Manufacturing costs have been rising at 12–13% yr, which will be reflected in future price increases.

Table 4. U.S. Production of Hydrogen Peroxide[a]

Year	Thousands of metric tons
1960	26.0
1970	55.7
1972	66.5
1974	85.2
1976	85.0
1978	100.2

[a] 100% basis.

Table 5. U.S. Manufacturers of Hydrogen Peroxide

Manufacturer	Site	Process	Capacity[a], thousand metric tons
DuPont	Memphis, Tenn.	anthraquinone	56.7
FMC	S. Charleston, W. Va.	anthraquinone	36.2
	Bayport, Texas	anthraquinone	27.2
	Vancouver, Wash.	electrolytic	8.1
Interox	Deer Park, Texas	anthraquinone	39.9
Shell	Norco, La.	2-propanol	16.3
PPG	Barberton, Ohio	anthraquinone	4.5

[a] 100% Basis.

Table 6. Hydrogen Peroxide, January 1980 Price

Concentration, %	Price, ¢/kg fob producer
30[a]	51[b]
35	57[b]
35	40[c]
50	56[c]
70	79[c]

[a] Reagent grade.
[b] Drums.
[c] Tank cars and trucks.

Specifications and Standards

Aqueous hydrogen peroxide is sold in grades ranging from 3 to 98%, mainly containing 35, 50, 70, or 90% H_2O_2.

The 3–6% H_2O_2 solutions for cosmetic and medicinal use are obtained by diluting a more concentrated grade, usually with the addition of extra stabilizer. There is a USP specification for 3% H_2O_2 (222).

The 30% H_2O_2 concentration is used as a laboratory reagent and in some special use areas, such as electronics. Several grades recommended for use in the electronics industry are marketed with very low impurity and stabilizer concentrations. There is an ACS reagent specification (see Fine chemicals).

The 35 and 50% H_2O_2 concentrations are used for most industrial applications. Standard grades contain sufficient stabilizers to ensure safety. Grades with lower stabilizer concentrations are available for food and other special use areas.

The 70% H_2O_2 concentration is used for certain organic oxidations. Dilution-grade 70% H_2O_2 is stabilized to allow for later dilution to 35 or 50% concentration.

The 90% H_2O_2 concentration is used as a propellant in military and space programs and for a few industrial applications where minimum water content is required. Concentrations above 90% are special products of interest in military applications and have not been sold in large quantities (see Explosives and propellants).

The ICC classifies solutions containing over 8% H_2O_2 by weight as corrosive liquids. Special containers and labels are required (223–224). Aluminum or polyethylene-lined drums and aluminum tank cars and trucks are used to ship grades containing up to 52% H_2O_2. Grades containing over 52% H_2O_2 are shipped in special double-headed aluminum drums, or aluminum tank trucks or tank cars.

Analytical and Test Methods

Analytical methods for the qualitative or quantitative determination of hydrogen peroxide are based on its redox or physical properties. Aqueous hydrogen peroxide solutions are usually titrated with potassium permanganate. Colorimetric and specific procedures are compiled in refs. 225–226. For other test procedures (impurities and stability) see USP (222), *Food Chemicals Codex* (227), and refs. 224 and 228.

High pressure liquid chromatography can be used for the detection of anthraquinone derivatives in a hydrogen peroxide process working solution (229).

Health and Safety Factors

Hydrogen peroxide, especially in high concentrations, is a high energy material and a strong oxidant. It can be handled safely provided proper precautions are observed (224).

Physiological Effects. Hydrogen peroxide is irritating to the skin, eyes, and mucous membranes. However, low concentrations (3–6%) are used in medicinal and cosmetic applications. Precautions should be taken with higher strengths to prevent vapor inhalation and ingestion of or other direct contact with solutions. In case of accidental contact with hydrogen peroxide, the affected area should be washed immediately with excess water.

Decomposition and Explosive Hazards. Hydrogen peroxide decomposes with the generation of heat and oxygen. Decomposition is promoted by catalytic impurities and proceeds at a rate that increases approximately 2.2 times for each 10 degree rise in temperature over the range of 20–100°C (224). The decomposition can become self-accelerating and must be controlled for the safe storage and handling of hydrogen peroxide. Containers and equipment should be constructed of suitable materials, be adequately vented, and maintained free of contaminants (224). Decomposition hazards increase with increasing concentration.

It reportedly is impossible to obtain a propagating detonation of commercial hydrogen peroxide at ambient temperature under normal conditions of storage (230). Concentrations of 86% and above detonate, but only with a high energy ignition source. In the vapor phase, explosions occur under certain conditions. The lower explosive limit at atmospheric pressure is about 26 mol % H_2O_2 which is equal to the equilibrium vapor concentration over a boiling solution of ca 75% H_2O_2. Explosive vapor concentrations with 90% H_2O_2 can be avoided by maintaining temperatures below ca 115°C.

The hazard of an explosion is usually present in organic oxidations with high-strength hydrogen peroxide. The concentration range for explosive mixtures increases with decreasing water content. Data have been published on the explosive limits for a number of hydrogen peroxide–organic compound mixtures. The most sensitive compositions are those that contain the stoichiometric quantities of hydrogen peroxide required to oxidize the organic compound to water and carbon dioxide. Detonable compositions can be obtained with mixtures of organic materials and relatively dilute hydrogen peroxide solutions if excess unreacted hydrogen peroxide is concentrated by evaporation or distillation of lower boiling components, including water.

Contact of hydrogen peroxide with some inorganic reagents should be avoided; eg, mercurous oxide and hydrogen peroxide react with explosive violence (231).

Practically all solid combustible materials contain sufficient catalytic impurities to rapidly decompose hydrogen peroxide, especially at 70–90% concentrations.

Uses

Hydrogen peroxide use areas are given in Table 7. Most applications are based on the oxidizing properties of hydrogen peroxide; some are derived from substitution or decomposition or the formation of perhydrates.

Bleaching. The largest single use for hydrogen peroxide in the U.S. is the bleaching of cotton textiles (see Bleaching agents); lesser quantities are used to bleach wool, silk, and some vegetable and animal fibers (see also Textiles). Today, most cotton fabrics are bleached with stabilized alkaline hydrogen peroxide solutions in continuous processes. The most effective stabilizers are alkaline earth silicates, although phosphates and organic chelating agents also have been used. Alkaline hydrogen peroxide bleach systems can also be used for cotton–synthetic fiber blends and do not affect many dyes. It has been estimated that 85% of all cotton fabrics are bleached with hydrogen peroxide (233). In addition to its other advantages, hydrogen peroxide is a nonpolluting oxidant, a factor that is becoming increasingly important.

Alkaline hydrogen peroxide solutions are also used to bleach groundwood, and chemimechanical, and chemical pulps (see Pulp). Hydrogen peroxide is used in the final stage in the bleaching sequence of chemical pulps to obtain a very white product with maximum color stability. This use is expected to increase since it avoids the environmentally undesirable chlorine compounds formed by Cl_2, HOCl, or ClO_2 bleaching (234–236).

Hydrogen peroxide is used in the deinking process in the recycle of waste paper (237) (see Recycling), to bleach solid surfaces such as wood or linoleum, and to improve the color of oils and waxes.

Environmental Applications. Hydrogen peroxide is an ecologically desirable pollution control agent since it yields only water and/or oxygen upon decomposition. It has been used in increasingly greater quantities to convert domestic and industrial effluents to an environmentally compatible state (see Water; Wastes).

Table 7. Estimates of Hydrogen Peroxide Use Areas, %

Use area	1978[a]	1979
chemicals	28[b]	33
textiles	25	17
pulp and paper	10	13
pollution control	10	19
metallurgy	7	6
others	20	12

[a] Ref. 232.
[b] Including plasticizers.

Hydrogen peroxide is used to treat wastewaters (238–243) and sewage effluents, and to control hydrogen sulfide generated by the anaerobic reaction of raw sewage in sewer lines or collection points (244). It has been proposed as a supplemental oxygen source for overloaded activated sludge plants (245–251). It reportedly controls denitrification in secondary clarifiers (246) and improves bulking conditions (252–254). It has been used as a flotation assistant (255–256). Activated sludge-treated wastewater COD values have been decreased using hydrogen peroxide with an iron salt catalyst (257–258). It has been generated in a wastewater reservoir by the cathodic reduction of oxygen (259).

Industrial liquid and gaseous effluent detoxification systems based on hydrogen peroxide have been developed (260–266). Hydrogen peroxide is reported to be especially suitable for cyanide-containing effluents with high free-cyanide concentration or containing organic impurities (267–268). Hydrogen peroxide systems with formaldehyde (269), copper salts (270–272), water glass (273), or iodide and silver ion (274) added to increase reaction rates and efficiency have been described or patented. A method and apparatus for cyanide destruction by monoperoxysulfuric acid, generated when and as necessary by the reaction of hydrogen peroxide and sulfuric acid has been claimed (275). Procedures have been described for the removal of nitrite ion (276) from waste streams and for the recovery of arsenic acid or arsenates from arsenic-containing wastewaters (277). Similar procedures using hydrogen peroxide have been developed for detoxifying organic pollutants (278–281), including formaldehyde, phenol, and phenol–formaldehyde mixtures (282–284), acetic acid (285), lignin sugars (286), surfactants, amines and/or glycol ethers (287–288), and sulfur derivatives (289–294) in wastewater streams.

Toxic or malodorous pollutants can be removed from industrial gas streams by reaction with hydrogen peroxide (295). Numerous liquid-phase methods have been patented for the removal of NO_x gases (296–316), sulfur dioxide, reduced sulfur compounds, and amines (317–327) and phenols (328). Other treatments aimed at effluent improvement include reduction of BOD and COD (329), color (330), odor (331–333), and chlorine (334) concentration.

Chemical. Hydrogen peroxide is employed in the manufacture of numerous organic and inorganic chemicals, either directly or after conversion to a peroxycarboxylic acid (335). The electrophilic epoxidation (qv) of soybean oil, linseed oil, and related unsaturated esters with peroxyformic or peroxyacetic acid formed *in situ* from aqueous hydrogen peroxide has been used to prepare poly(vinyl chloride) plasticizers and stabilizers. Processes have been developed recently for the continuous epoxidation of low molecular weight olefins, for example, propylene (336–339), using water-immiscible organic solutions of a peroxycarboxylic acid, and for the continuous preparation of the peroxy acid solutions from aqueous hydrogen peroxide (340–347). Additional patents describe the epoxidation of water-insoluble olefins with hydrogen peroxide with a metal or fluoroketone catalyst (348–350).

Hydroxylation of olefinic compounds is an important use for hydrogen peroxide, especially in the synthesis of glycerol (qv) from propylene. The catalyzed hydroxylation of phenol yields catechol and hydroquinone (335,351–355). Other significant applications are in the manufacture of long-chain *tert*-amine oxides for use in liquid detergents and thiourea dioxide for textile processing. Hydrogen peroxide is used extensively in the latter area to oxidize sulfur and vat dyes.

Interest in hydrogen peroxide as an active chemical reagent for the preparation of other large volume chemicals is growing. Processes have been developed for the preparation of hydrazine (qv) (335,356) and caprolactam (335,357–361) (see Polyamides). Hydrogen peroxide can also be used to prepare cyanogen (362), cyanogen chloride (335,363), bromine (364), and iodic acid (365).

Derivative Formation. Hydrogen peroxide is an important reagent in the manufacture of organic peroxides (366), including *tert*-butyl hydroperoxide, benzoyl peroxide, peroxyacetic acid, esters such as *tert*-butyl peroxyacetate, and ketone derivatives such as methyl ethyl ketone peroxide. These derivatives are used as polymerization catalysts, cross-linking agents, and oxidants (see Peroxides and peroxy compounds). Zinc, calcium, and magnesium peroxides prepared from hydrogen peroxide are used as specialized oxidants where a slow release of hydrogen peroxide is desired.

Hydrogen peroxide is used, particularly in Europe, to manufacture sodium peroxyborate for household and detergent bleach applications and for use in a number of organic reactions as a solid source of hydrogen peroxide. Sodium carbonate perhydrate and urea peroxyhydrate are prepared for similar uses. Potassium monoperoxysulfate is manufactured for uses requiring a highly effective bleaching agent.

Mining. Use of hydrogen peroxide as an oxidant for the in-place solution mining of low-grade uranium ores with ammonium bicarbonate (367–368), an alkaline earth carbonate or bicarbonate (369), or carbonic acid (370) is a relatively new technique. Hydrogen peroxide can also be used to precipitate uranium from ion-exchange eluate and solvent-extraction strip liquor, and gives better results than ammonia (371). Hydrogen peroxide has been proposed for use in the *in situ* mining of low copper-containing formations as a fracturing agent (372). Procedures for leaching copper (373–375), manganese (376), zinc (377), and gold and silver (378–379) from milled ore and for ruthenium (380–381) and uranium (382–383) recovery using hydrogen peroxide have been patented. Hydrogen peroxide can be a heat source for the secondary recovery of oil (384).

Propellant. High strength (70% and above) hydrogen peroxide can be used as a propellant (see Explosives and propellants). It is catalytically decomposed to steam and oxygen and the hot gases are used as the energy source or it is used as an oxidant with a variety of fuels to obtain a greater amount of energy then is available from hydrogen peroxide alone.

Other Uses. A great number of smaller, special uses are based on the ability of hydrogen peroxide to oxidize metals or metal ions to higher valence states. Thus, metal salts have been freed of iron by oxidation and precipitation of ferric hydroxide. Hydrogen peroxide is used in the chemical polishing of copper, iron, and steel and as an oxidant in a number of other metal surface treatments, including phosphatizing, enameling, and etching (see Metal surface treatments).

The catalytic decomposition of hydrogen peroxide yields oxygen which can be used as a blowing agent. Certain types of foam rubber have been prepared by this method (see Foamed plastics). The mild bactericidal properties of hydrogen peroxide have led to medicinal and disinfectant applications, whereas cosmetic uses are based on its bleaching and oxidizing properties (see Disinfectants and antiseptics). A new use is for the preparation of chemiluminescent compositions (385–387) (see Chemiluminescence).

BIBLIOGRAPHY

"Hydrogen Peroxide" in *ECT* 1st ed., Vol. 7, pp. 727–741, by N. A. Milas, Massachusetts Institute of Technology; Suppl. 1, pp. 418–429, by H. A. Bewick and J. K. Farrell, Solvay Process Division, Allied Chemical & Dye Corporation; "Hydrogen Peroxide" in *ECT* 2nd ed., Vol. 11, pp. 391–417, by A. F. Chadwick and G. L. K. Hoh, E. I. du Pont de Nemours & Company, Inc.

1. W. R. Busing and H. A. Levy, *J. Chem. Phys.* **42,** 3054 (1965).
2. G. A. Khachkuruzov and I. N. Przhevolskii, *Opt. Spektrosk.* **35,** 374 (1973); *Opt. Spektrosk.* **36,** 299 (1974); *Opt. Spektrosk.* **36,** 304 (1974); *Opt. Spektrosk.* **44**(1), 194 (1978).
3. P. A. Giguere, *J. Chem. Ed.* **51,** 470 (1974).
4. W. C. Schumb, C. N. Satterfield, and R. L. Wentworth, *Hydrogen Peroxide,* Reinhold Publishing Corp., New York, 1955.
5. P. A. Giguere, "Peroxyde d'Hydrogene et Polyoxydes d'Hydrogene" in P. Pascal, ed., *Complements au Nouveau Traite de Chemie,* Vol. 4, Masson et cie, Paris, 1975.
6. P. A. Giguere and J. L. Carmichael, *J. Chem. Eng. Data* **7,** 526 (1962).
7. I. B. Rozhdestvenskii, V. N. Gutov, and N. A. Zhigulskaya, *Sb. Tr-Glavniiprockt Energ. Inst.* **7,** 88 (1973); *Chem. Abstr.* **85,** 83876u (1976).
8. O. K. Borggaard, *Acta Chem. Scand.* **26,** 3393 (1973).
9. H. Nakamura, T. Hatamoto, and I. Nakamori, *Kagaku Kogaku Rombunshu* **2,** 606 (1976).
10. J. L. Arnau, P. A. Giguere, A. Motoko, and R. C. Taylor, *Spectrochim. Acta* **30A,** 777 (1974).
11. P. A. Giguere and T. K. K. Srinivasan, *J. Raman Spectrosc.* **2**(2), 225 (1974).
12. J. L. Arnau and P. A. Giguere, *Phys. Chem. Ice Pap. Symp. 1972,* 66 (1973).
13. M. G. Faulkner, *Diss. Abstr. Int. B* **35,** 4096 (1975).
14. J. L. Arnau and P. A. Giguere, *Can. J. Spectrosc.* **17**(4), 121 (1972).
15. H. Chen and P. A. Giguere, *Spectrochim. Acta* **29A,** 1611 (1973).
16. K. Osafune and K. Kimura, *Chem. Phys. Lett.* **25**(1), 47 (1974).
17. A. W. Ellenbroek and A. Dymanus, *Chem. Phys.* **31**(1), 107 (1978).
18. L. I. Nekrasov, *Zh. Fiz. Khim.* **46**(11), 2743 (1972).
19. P. A. Giguere, *J. Chem. Thermodyn.* **6,** 1013 (1974).
20. C. L. Yaws and H. S. N. Setty, *Chem. Eng.* **81,** 67 (1974).
21. H. F. Bell, *Diss. Abstr. B* **29,** 1948 (1968).
22. N. G. Klynchnikov and V. S. Dukhanin, *Uch. Zap. Mosk. Gos. Pedagog. Inst.* **340,** 353 (1971); *Chem. Abstr.* **72,** 156937m (1972).
23. A. Krause, *Sci. Pharm.* **40,** 211 (1972).
24. A. McAuley, *Inorg. React. Mech.* **2,** 88 (1972).
25. V. K. Aleksondrov, A. I. Gorbunov, and K. N. Fastova, *Tezisy Dokl. Vses. Soveshch. Khim. Neorg. Perekisnykh Svedin,* 153 (1973); *Chem. Abstr.* **82,** 175790C (1975).
26. I. I. Vasilenko, *Sb. Tr. Belgorod. Tekhnol. Inst. Stroit. Materialov* **15,** 87 (1975); *Chem. Abstr.* **88,** 111135j (1978).
27. A. McAuley, *Inorg. React. Mech.* **5,** 107 (1977).
28. U.S. Pat. 3,356,457 (Dec. 5, 1967), and U.S. Pat. 3,429,666 (Feb. 25, 1969), G. V. Morris and P. B. Weill (to U.S. Dept. of the Navy).
29. U.S. Pat. 3,333,925 (Aug. 1, 1967), J. H. Young (to DuPont).
30. Jpn. Pat. 68 11,656 (May 16, 1968), J. Koryo, H. Yoshida, and A. Hirata.
31. U.S. Pat. 3,591,341 (July 6, 1971), and U.S. Pat. 3,607,053 (Sept. 21, 1971), V. J. Reilly (to DuPont).
32. U.S. Pat. 3,208,825 (Sept. 28, 1965), R. E. Meeker (to Shell Oil Co.).
33. Fr. Pat. 1,473,779 (Mar. 17, 1967), G. T. Carnine and L. R. Darbee (to FMC Corp.).
34. U.S. Pat. 3,387,939 (June 11, 1968), V. J. Reilly and N. G. Stalter (to DuPont).
35. U.S. Pat. 3,681,022 (Aug. 1, 1972), W. H. Kibbel Jr. (to FMC Corp.).
36. U.S. Pat. 3,687,627 (Aug. 29, 1972), N. J. Stalter (to DuPont).
37. Ger. Offen. 2,152,741 (Apr. 27, 1972), K. J. Radimer and T. F. Munday (to FMC Corp.).
38. Ger. Offen. 1,817,812 (Feb. 24, 1972), R. H. Carlson (to Hooker Chemical Corp.).
39. Jpn. Kokai 74 73,398 (July 16, 1974), 74 74,697 (July 18, 1974), 74 74,697 (July 18, 1974), S. Shiga (to Furukawa Electric Co. Ltd.).
40. Jpn. Kokai 75 146,597 (Nov. 25, 1975), N. Saisho, S. Kawahara, H. Kawata and M. Aruga (to Yamamouchi Pharmaceutical Co. Ltd.).
41. U.S. Pat. 4,034,064 (July 5, 1977), J. A. Cook (to PPG Industries, Inc.).

42. U.S. Pats. 4,035,471 and 4,061,721 (Dec. 6, 1977), W. A. Strong (to PPG Industries, Inc.).
43. U.S. Pat. 4,133,869 (Oct. 31, 1977) and U.S. Pat. 4,132,762 (Jan. 2, 1979), L. Kim (to Shell Oil Co.).
44. U.S. Pat. 4,140,772 (Feb. 2, 1979), T. F. Korenowski (to Dart Industries, Inc.).
45. Brit. Pat. 1,157,038 (July 2, 1969), (to Lancy Laboratories, Inc.).
46. Jpn. Kokai 72 33,993 (Nov. 20, 1972), S. Shiga (to Furukawa Electric Co., Ltd.).
47. U.S. Pat. 3,781,409 (Dec. 25, 1973), K. J. Radimer (to FMC Corp.).
48. Ger. Offen. 2,256,579 (May 24, 1973), J. C. Solenberger (to DuPont).
49. Fr. Demande 2,216,222 (Aug. 30, 1974), and U.S. Pat. 3,903,244 (Feb. 2, 1973), D. C. Winkley (to FMC Corp.).
50. Jpn. Kokai 74 109,296 (Oct. 17, 1974), K. Kushibe (to Tokai Electro-Chemical Co., Ltd.).
51. Jpn. Kokai 74 113,799 (Oct. 30, 1974), A. Naito, Y. Oshida, T. Fukano, and I. Ikeya (to Mitsubishi Gas Chemical Co., Inc.).
52. Ger. Offen. 2,441,725 (Mar. 13, 1975), J. C. Watts (to DuPont).
53. U.S. Pat. 3,864,271 (Feb. 4, 1975), N. J. Stalter (to DuPont).
54. Ger. Offen. 2,441,725 (Mar. 13, 1975), J. C. Watts (to DuPont).
55. Fr. Demande 2,343,797 (Oct, 7, 1977), J. P. Zumbrunn (to L'Air Liquide).
56. Jpn. Kokai 78 102,297 (Sept. 6, 1978), S. Shiga and T. Inada (to Furukawa Electric Co., Ltd.).
57. U.S. Pat. 3,974,086 (Aug. 10, 1976) and U.S. Pat. 4,064,064 (Dec. 20, 1977), A. M. Semsel (to American Cyanamid Co.).
58. Ger. Offen. 2,532,866 (July 23, 1975), (to Degussa).
59. U.S. Pat. 3,869,401 (Mar. 4, 1975), R. E. Ernst (to DuPont).
60. U.S. Pat. 3,037,622 (June 5, 1962), R. S. Treseder (to Shell Oil Co.).
61. A. A. Kassem, S. A. Zaki, E. N. El-Din, and A. E. Fouli, *Bull. Fac. Pharm. Cairo Univ.* **10**, 265 (1971).
62. Ger. Offen. 2,109,495 (Sept. 14, 1972), A. Winsel and H. H. Von Doehren (to Varta A.-G.).
63. Ger. Offen. 2,742,907 (Oct. 26, 1978), (to U.S. Secretary of Commerce (USDA)).
64. P. G. Cookson, A. G. Davies, and N. Fazal, *J. Organomet. Chem.* **99**(2), C31 (1975).
65. A. Hansson, *Acta. Chem. Scand.* **15**, 934 (1961).
66. W. A. Waters, *Mechanism of Oxidation of Organic Compounds,* Methuen & Co., Ltd., London, 1974, p. 39.
67. C. A. Bunton in J. O. Edwards, ed., *Peroxide Reaction Mechanisms,* Interscience Publishers, a Division of John Wiley & Sons, Inc., New York, 1962, p. 11.
68. W. M. Weigert, H. Delle, and G. Kaebisch, *Chem. Ztg.* **99**(3), 101 (1975).
69. M. Miwa, *Jpn. Chem. Quart.* **2**(4), 36 (1966).
70. M. Cerar, *Vestn. Slov. Kem. Drus.* **25**(2), 187 (1978); *Chem. Abstr.* **89**, 1317599 (1978).
71. Belg. Pat. 652,956 (Mar. 11, 1965), (to LaPorte Chemicals, Ltd.).
72. Jpn. Pats. 76 20,198 and 76 20,199 (June 23, 1976), (to Mitsubishi Gas-Chemical Co.).
73. Belg. Pat. 769,675 (Jan. 10, 1972), (to Solvay & Cie).
74. U.S.S.R. Pat. 124,420 (July 9, 1966), V. I. Franchuk, M. D. Bobyshev, and T. B. Andryanov; *Chem. Abstr.* **66**, 39475w (1967).
75. U.S.S.R. Pat. 192,762 (Mar. 2, 1967), V. I. Franchuk, L. I. Ovchinnikova, V. F. Kosareva, and T. B. Andrianova; *Chem. Abstr.* **68**, 70768j (1968).
76. Ger. Pat. 1,667,515 (Nov. 20, 1975), V. I. Franchuk, L. I. Ovchinnikova, and V. F. Kosareva.
77. Brit. Pat. 1,425,202 (Feb. 18, 1976), L. G. Vaughan (to DuPont).
78. Brit. Pat. 3,923,967 (Dec. 2, 1975), J. R. Kirchner and L. G. Vaughan (to DuPont).
79. Brit. Pat. 1,425,250 (Feb. 18, 1976), L. G. Vaughan (to DuPont).
80. U.S. Pat. 3,328,128 (June 27, 1967), G. Kabisch (to FMC Corp.).
81. U.S. Pat. 3,761,581 (Sept. 23, 1973), G. Kabisch and R. Trube (to Degussa).
82. Ger. Offen. 2,018,686 (Oct. 21, 1971), G. Schreyer, O. Weiberg, and W. Weigert (to Degussa).
83. Ger. Offen. 2,065,155 (Nov. 2, 1972) and 2,532,819 (Jan. 27, 1977), G. Giesselmann, G. Schreyer, and W. Weigert (to Degussa).
84. Ger. Offen. 2,446,053 (Apr. 3, 1975), L. G. Vaughan (to DuPont).
85. U.S. Pat. 3,789,114 (Jan. 29, 1974), G. Giesselmann, G. Schreyer, and W. Weigert (to Degussa).
86. Jpn. Pat. 73 996 (Jan. 12, 1973), S. Matsumura and co-workers (to Mitsubishi Gas-Chemical Co.).
87. Fr. Pat. 1,516,938 (Mar. 15, 1968), M. Coingt (to Oxysynthese).
88. U.S. Pat. 3,699,217 (Oct. 17, 1972), G. Schreyer, O. Weiberg, and W. Weigert (to Degussa).
89. Jpn. Pat. 74 20,477 (May 24, 1974), S. Matsumura and co-workers (to Mitsubishi Gas-Chemical Co.).

90. Jpn. Pat. 74 18,919 (May 14, 1974), S. Matsumura and co-workers (to Mitsubishi Gas-Chemical Co.).
91. Fr. Pat. 1,428,528 (Feb. 18, 1966), (to L'Air Liquide).
92. Can. Pat. 715,979 (Aug. 17, 1965), G. R. Hoey and D. Hudson (to Canadian Industries Ltd.).
93. Brit. Pat. 1,119,820 (May 6, 1970), (to Oxysynthese).
94. Ger. Pat. 812,426 (Aug. 30, 1951), H. J. Riedl (to BASF).
95. Ger. Pat. 1,244,129 (July 13, 1967), H. Herzog and G. Kabisch (to Degussa); U.S. Pat. 3,307,909 (Mar. 7, 1967), V. J. Reilly (to DuPont).
96. U.S. Pat. 3,423,176 (Jan. 21, 1969), G. Kabisch and H. Herzog (to Degussa).
97. Brit. Pat. 718,307 (Nov. 10, 1954), J. A. Williams and C. W. LeFeuvre (to LaPorte Chemicals, Ltd.).
98. Ger. Offen. 1,925,034 (Dec. 4, 1969) and 2,151,104 (Apr. 27, 1972), (to FMC Corp.).
99. U.S. Pat. 3,433,358 (Mar. 18, 1969), H. Herzog and G. Kabisch (to Degussa).
100. U.S. Pat. 3,923,966 (Dec. 2, 1975), L. G. Vaughan (to DuPont).
101. Fr. Pat. 1,463,808 (Dec. 30, 1966), F. Coussemant and J. Vidal (to Institut Francais du Petrole).
102. U.S. Pat. 3,551,104 (Dec. 29, 1970), S. Lynn and H. H. Paalman (to Dow Chemical Co.).
103. D. Tresselt, *Monatsber. Dent. Akad. Wiss. Berlin* **10**(1), 76 (1968).
104. K. K. Kirdin and I. E. Balabin, *Khim. Prom.* (*Moscow*) **47**(2), 83 (1971).
105. Ger. Pat. 1,297,087 (June 12, 1968), M. Bobyshev and co-workers.
106. Ger. Offen. 1,951,568 (May 21, 1970), S. Matsumura and co-workers (to Mitsubishi Edogawa Chemical Co.).
107. Ger. Pat. 1,120,432 (Dec. 28, 1961), U. Hauschild and H. Nicolaus (to Kali-Chemie A.G.).
108. V. A. Gusev, E. M. Emelyanov, L. A. Frangulyan, and V. I. Franchuk, *Khim. Prom.* (*Moscow*) **48,** 783 (1972); *Chem. Abstr.* **78,** 45511u (1973).
109. Fr. Pat. 1,118,442 (June 6, 1956), (to L'Air Liquide).
110. Z. Csuros, J. Petro, J. Morgos, and B. Losonczi, *Magy. Kem. Lapja* **26**(10), 497 (1971).
111. Ger. Pat. 1,079,604 (Apr. 14, 1969), D. H. Porter (to Degussa).
112. U.S. Pat. 4,061,598 (Dec. 6, 1977), K. M. Makar (to DuPont).
113. Brit. Pat. 776,991 (June 12, 1957), R. Lait and C. W. LeFeuvre (to LaPorte Chemicals, Ltd.).
114. Brit. Pat. 922,022 (Mar. 27, 1963), R. Lait and D. R. Lloyd-Owen (to LaPorte Chemicals, Ltd.).
115. U.S. Pat. 3,488,150 (Jan. 6, 1970), G. Kabisch and H. Wittmann (to Degussa).
116. Fr. Pat. 1,509,429 (Jan. 12, 1968), (to Degussa).
117. Jpn. Pat. 74 5,120 (Feb. 5, 1974), S. Matsumura and co-workers (to Mitsubishi Gas-Chemical Co.).
118. Fr. Pat, 1,502,268 (Nov. 18, 1967), (to LaPorte Chemicals, Ltd.).
119. Ger. Offen. 2,029,394 (Jan. 14, 1971), C. D. Keith, K. W. Cornely, and N. D. Lee (to Engelhard Minerals and Chemicals Corp.).
120. Ger. Offen. 2,029,427 (Feb. 4, 1971), J. D. Lee (to FMC Corp.).
121. U.S. Pat. 3,323,868 (June 6, 1967), A. Ogilvie and B. Laurence (to LaPorte Chemicals, Ltd.).
122. Ger. Offen. 2,003,268 (July 29, 1971), M. Liebert, H. Delle, and G. Kabisch (to Degussa).
123. Ger. Offen. 2,419,534 (Nov. 28, 1974), B. G. Franzen (to Elektrokemiska AB.).
124. Jpn. Pat. 76 39,640 (Oct. 28, 1976), T. Suzuki and co-workers (to Mitsubishi Gas-Chemical Co.).
125. Brit. Pat. 867,679 (May 10, 1961), (to Allied Chemical Corp.).
126. Ital. Pat. 598,191 (Sept. 23, 1959), (to Montecatini).
127. Fr. Pat. 1,504,848 (Dec. 8, 1967), (to Alpine Chemische A.G.).
128. Fr. Pat. 1,331,385 (July 5, 1963), J. C. Courtier (to Usines de Melle).
129. Ger. Offen. 1,945,753 (May 6, 1971), G. Kabisch and H. Wittmann (to Degussa).
130. Ger. Pat. 1,945,752 (Feb. 11, 1971), G. Kabisch and S. Raupach (to Degussa).
131. Fr. Pat. 1,492,339 (Aug. 18, 1967), (to LaPorte Chemicals, Ltd.).
132. Ger. Pat. 1,802,003 (Mar. 25, 1971); Ger. Offen. 1,951,211 (April 22, 1971); Ger. Offen. 2,025,237 (Dec. 9, 1971); Ger. Offen. 2,042,522 (Mar. 2, 1972), (to Degussa).
133. Ger. Offen. 2,018,957 (Nov. 19, 1970), K. Jones, E. H. Joscelyne, and P. H. Harvey (to Petrocarbon Developments, Ltd.).
134. Ger. Offen. 2,038,319 (Feb. 10, 1972) and 2,038,320 (Feb. 17, 1972), H. Waldmann, W. Schwerdtel, and W. Swodenk (to Farbenfabriken Bayer A.G.).
135. Ger. Offen. 2,041,124 (Feb. 24, 1972), and 2,125,159 (Dec. 21, 1972), (to Degussa).
136. U.S. Pat. 3,965,251 (June 22, 1976), H. Shin and co-workers (to Mitsubishi Gas-Chemical Co.).
137. Fr. Pat. 1,468,707 (Feb. 10, 1967), (to Degussa).
138. U.S. Pat. 3,912,766 (Oct. 14, 1975), (to LaPorte Industries, Ltd.).
139. U.S. Pat. 2,901,491 (Aug. 25, 1959), C. W. Eller and J. M. Snyder (to DuPont).

140. Brit. Pat. 1,002,391 (Aug. 25, 1965), A. Skull (to LaPorte Chemicals, Ltd.).
141. Can. Pat. 649,850 (Oct. 2, 1962), W. S. Hinegardner (to DuPont).
142. Fr. Pat. 1,465,543 (Jan. 13, 1967), M. Charret (to Oxysynthese).
143. U.S. Pat. 2,940,987 (June 14, 1960), C. W. Eller and J. M. Snyder (to DuPont).
144. U.S. Pat. 3,432,267 (Mar. 11, 1969), N. D. Lee and N. N. Schwartz (to FMC Corp.).
145. Can. Pat. 664,876 (June 11, 1963), J. G. Schwemberger (to DuPont).
146. Fr. Pat. 1,335,700 (July 15, 1963), D. Schumann, H. Frömbgen, and H. Schell (to Kali-Chemie A.G.).
147. Ger. Pat. 1,667,476 (Oct. 22, 1979), G. Demmering, G. Weber, and S. Geigel (to Elektrochemische Werke Muenchen A.G.).
148. Brit. Pat. 1,252,822 (Nov. 10, 1971), R. J. Sinnott and W. R. Logan (to LaPorte Industries, Ltd.).
149. U.S. Pat. 3,179,672 (Apr. 20, 1965), H. Herzog and G. Kabisch (to FMC Corp.).
150. U.S. Pat. 3,949,063 (Apr. 6, 1976), M. Coingt and P. Thirion (to Oxysynthese).
151. Ger. Offen. 2,012,988 (Sept. 24, 1979), W. R. Logan (to LaPorte Industries, Ltd.).
152. Brit. Pat. 1,228,494 (Apr. 15, 1971), J. Mackenzie (to LaPorte Industries, Ltd.).
153. U.S. Pat. 3,330,625 (July 11, 1967), P. H. Baker and J. A. Cook (to Pittsburgh Plate Glass Co.).
154. Fr. Pat. 1,340,901 (Oct. 25, 1963), M. Charret (to Oxysynthese).
155. Ger. Offen. 2,036,553 (Feb. 3, 1972), G. Kabisch and W. Kunkel (to Degussa).
156. Ger. Offen. 2,340,430 (Feb. 21, 1974), J. N. Browning (to FMC Corp.).
157. Jpn. Kokai 72 23,386 (Oct. 12, 1972), M. Terashi, Y. Miyata, and M. Saito (to Japan Peroxide Co., Ltd.).
158. Brit. Pat. 922,021 (Mar. 27, 1963), R. Lait and D. R. Lloyd-Owen (to LaPorte Chemicals, Ltd.).
159. Belg. Pat. 653,225 (Mar. 18, 1965), (to LaPorte Chemicals, Ltd.).
160. Ger. Offen. 2,042,523 (Mar. 2, 1972), G. Schreyer, F. Theissen, O. Weiberg, and W. Weigert (to Degussa).
161. U.S. Pat. 3,998,936 (Dec. 21, 1976), R. E. Ernst, B. Amini, and W. H. Bareford (to DuPont).
162. U.S. Pat. 3,901,822 (Aug. 26, 1975), J. N. Browning, N. D. Lee, and G. Smee (to FMC Corp.).
163. Ger. Pat. 1,935,478 (May 13, 1971), G. Kabisch and S. Raupach (to Degussa).
164. U.S. Pat. 2,479,111 (Aug. 16, 1949), C. R. Harris (to DuPont).
165. Brit. Pat. 708,339 (May 5, 1954), (to N. V. de Bataafschi Petroleum Maatschappij).
166. Ger. Pat. 935,303 (Nov. 17, 1955), F. R. Rust (to N. V. de Bataafche Petroleum Maatschappij).
167. Jpn. Pat. 69 17,493 (Aug. 1, 1969), G. Inone, H. Tanaka, and T. Kobayashi (to Asahi Chemical Industry Co., Ltd.).
168. Ger. Offen. 1,917,033 (Oct. 23, 1969), (to Petrocarbon Developments, Ltd.).
169. Ger. Offen. 2,051,662 (July 22, 1971), A. W. Dawkins (to Burmah Oil Trading Ltd.).
170. Brit. Pat. 1,113,687 (May 15, 1968), E. G. E. Hawkins (to Distillers Co., Ltd.).
171. Ger. Pat. 1,002,295 (Feb. 14, 1957), J. R. Skinner and S. E. Steinle (to N. V. de Bataafsche Petroleum Maatschappij).
172. Jpn. Pat. 76 14,116 (Feb. 25, 1976), E. Y. Buneva.
173. Ger. Offen. 2,326,804 (Dec. 12, 1974), E. Y. Pneva and co-workers.
174. Can. Pat. 1,003,189 (Jan. 11, 1977), A. A. Shkurkina.
175. Hung. Pat. 12,766 (Jan. 28, 1977), E. Y. Pneva.
176. Ger. Pat. 970,174 (Aug. 31, 1958), F. R. Rust and L. M. Porter (to N. V. de Bataafsche Petroleum Maatschappij).
177. P. W. Sherwood, *Chem. Ing. Tech.* **32,** 459 (1960).
178. L. L. Zalygen and co-workers, *Zh. Prikl. Khim.* (*Leningrad*) **47,** 1599 (1974); *Chem. Abstr.* **81,** 104817k (1974).
179. Neth. Pat. 61,465 (Oct. 16, 1948), J. Overhoff.
180. U.S. Pat. 2,461,988 (Feb. 15, 1949), P. L. Kooijman.
181. U.S. Pat. 2,533,581 (Dec. 12, 1950), C. R. Harris (to DuPont).
182. U.S. Pat. 2,614,907 (Oct. 21, 1952), G. A. Cook.
183. U.S. Pat. 3,398,185 (Aug. 20, 1968), A. F. MacLean and C. C. Hobbs (to Celanese Corp.).
184. U.S. Pat. 4,131,646 (Dec. 26, 1978), L. W. Gosser (to DuPont).
185. U.S. Pat. 3,899,576 (Aug. 12, 1974), R. Rosenthal and J. A. Kieras (to Atlantic Richfield Co.).
186. Ger. Offen. 2,100,784 (July 22, 1971), J. O. Turner (to Sun Oil Co.).
187. U.S. Pat. 3,737,518 (June 5, 1973), G. Bonetti, R. Rosenthal, and J. A. Kieras (to Atlantic Richfield Co.).
188. Neth. Appl. 7,533/74 (Dec. 20, 1974), (to Atlantic Richfield Co.).
189. U.S. Pat. 2,862,794 (Dec. 2, 1958), C. Dufraisse, A. Etienne, and E. Toromanoff (to L'Air Liquide).

190. U.S. Pat. 3,341,297 (Sept. 12, 1967), A. F. MacLean and A. L. Stautzenberger (to Celanese Corp.).
191. U.S. Pat. 3,124,421 (Mar. 10, 1964), W. Lohringer and J. Sixt (to Wacker-Chemie G.m.b.H.).
192. Jpn. Kokai 76 2,700 (Jan. 10, 1976), N. Isogai, T. Ikawo, and T. Tokeda (to Mitsubishi Gas-Chemical Co., Inc.).
193. Czech. Pat. 136,978 (June 15, 1970), J. Vachuda and W. Havlicek.
194. Ger. Offen. 2,331,296 (Jan. 16, 1975) and Ger. Offen. 2,501,342 (July 22, 1976), B. Kastening and W. Paul (to Kernforschungsanlage Juelich G.m.b.H.).
195. Brit. Pats. 1,041,045 and 1,041,046 (Sept. 1, 1966); Brit. Pats. 1,056,121–1,056,126 (Jan. 25, 1967); Brit. Pat. 1,094,804 (Dec. 13, 1967); Fr. Pat. 1,366,253 (June 1, 1964), (to Imperial Chemical Industries, Ltd.).
196. Fr. Pat. 1,214,015 (Apr. 5, 1960), (to Engelhard Industries, Inc.).
197. Jpn. Kokai 75 145,394, 75 145,395, and 75 145,396 (Nov. 21, 1975); Jpn. Kokai 75 146,596 (Nov. 25, 1975); Ger. Offen. 2,655,920 (Aug. 4, 1977); Jpn. Kokai 78 72,799 (June 28, 1978), (to Tokuyama Soda Co., Ltd.).
198. Ger. Offen. 2,615,625 (Oct. 21, 1976), L. Kim and G. W. Schoenthal (to Shell International Research Maatschappij B.V.).
199. U.S. Pat. 4,128,627 (Dec. 5, 1978), P. N. Dyer and F. Moseley (to Air Products and Chemicals, Inc.).
200. M. Equchi and N. Morita, *Kogyo Kagaku Zasshi* **71,** 783 (1968).
201. U.S. Pat. 4,009,252 (Feb. 22, 1977), Y. Izumi, H. Miyazaki, and S. Kawahara (to Tokuyama Soda Co., Ltd.).
202. U.S. Pat. 3,317,280 (May 2, 1967), R. S. Long and A. T. Noma (to Dow Chemical Co.).
203. U.S. Pat. 3,330,626 (July 11, 1967), J. O. Archambault (to Canadian Industries).
204. Ger. Offen. 2,654,514 (June 1, 1978), R. Radebold and W. Seiler.
205. Fr. Pat. 1,539,843 (Sept. 20, 1968), P. Thirion (to Oxysynthese).
206. Fr. Pat. 1,475,926 (Apr. 7, 1967), (to Solvey & Cie).
207. Can. Pats. 591,795 and 591,796 (Feb. 2, 1960), J. H. Young (to DuPont).
208. U.S. Pat. 3,353,913 (Nov. 21, 1967), D. O. Flach and G. W. Siwinski (to FMC Corp.).
209. U.S. Pat. 3,410,659 (Nov. 12, 1968), C. K. Muehlhausser (to FMC Corp.).
210. Pol. Pat. 50,982 (Apr. 15, 1966), H. Eisermann and F. Piecharczek; *Chem. Abstr.* **66,** 97013c (1967).
211. Ger. (GDR) Pat. 51,025 (Oct. 20, 1966), S. Richter.
212. Pol. Pat. 55,378 (June 25, 1968), S. Marzec, J. Czeckowska, J. Krawczyk, and J. Piatek (to Zaklady Elektrochemiczne).
213. U.S. Pat. 3,294,488 (Dec. 27, 1966), A. K. Dunlog, R. E. Meeker, and G. J. Pierotti (to Shell Oil Co.).
214. Span. Pat. 328,719 (Jan. 4, 1967), (to Foret S.A.).
215. U.S. 3,297,404 (Jan. 10, 1967), R. B. Elliott and J. H. Young (to DuPont).
216. Jpn. Pat. 71 26,095 (July 28, 1971), T. Sakamaki, Y. Miyata, and M. Saotome (to Nippon Peroxide Co., Ltd.).
217. Ger. Offen. 1,945,754 (May 19, 1971), G. Kaebisch (to Degussa).
218. Ger. Offen. 2,125,192 (Nov. 23, 1972), G. Kaebisch and R. Truebe (to Degussa).
219. Ger. Offen. 2,233,159 (Jan. 24, 1974), A. Becker and U. Schwenk (to Farbwerke Hoechst A.-G.).
220. *Inorganic Chemicals and Gases,* U.S. Bureau of Census, Current Industrial Reports M28A, U.S. Dept. of Commerce, Washington, D.C.
221. *Chem. Mark. Rep.* **213**(22), 9 (1978); *Chem. Week* **123**(17), 58 (1978); *Chem. Eng. News* **56**(43), 11 (1978).
222. *The United States Pharmacopeia XX (USP XX–NF XV),* The United States Pharmacopeial Convention, Inc., Rockville, Md., 1980, p. 318.
223. *Section 173.266,* ICC Regulations, U.S. Government Printing Office, Washington, D.C., 1979.
224. *Chemical Safety Data Sheet SD-53—Properties and Essential Information for Safe Handling and Use of Hydrogen Peroxide,* Manufacturing Chemists Association, Washington, D.C., 1969.
225. G. A. Parker, *Chem. Anal.* (*N.Y.*) **8,** 253 (1978).
226. J. Gallus-Olender and B. Franc, *Chem. Anal.* (*Warsaw*) **19**(1), 203 (1974); *Chem. Abstr.* **81,** 44942k (1974).
227. *Food Chemicals Codex,* 3rd ed., National Academy of Sciences–National Research Council, Washington, D.C., 1980.
228. N. J. Stalter, *Soap. Chem. Spec.* **45**(6), 62 (1969).

229. D. B. Thorburn and R. K. Harle, *Anal. Chim. Acta* **100,** 563 (1978).
230. *U.S. Bur. Mines Inform. Circ.* (*8387*) (1968).
231. *Manual of Hazardous Chemical Reactions No. 491,* National Fire Protection Association, 1971.
232. *Chem. Mark. Rep.* **213**(22), 9 (1978).
233. T. E. Bell, "Bleaching of Textiles" in N. M. Bikales, ed., *Encyclopedia of Polymer Science and Technology,* Vol. 2, Interscience Publishers, a division of John Wiley & Sons, Inc., New York, 1965, p. 428.
234. S. Rothenburg and co-workers, *Tappi* **58,** 182 (1975).
235. Ger. Pat. 2,327,900 (Aug. 3, 1978), (to Degussa).
236. Fr. Demande 2,367,859 (May 12, 1978), (to Degussa).
237. G. Papageorges and J. Deceuster, *Indian Chem. Age* **27,** 451 (1976).
238. C. A. Geisler, C. Chin, and G. Hicks, *Eng. Bull. Purdue Univ. Eng. Ext. Serv.* **140,** 897 (1973).
239. J. W. Haskins, Jr., *DuPont Innovation* **5**(1), 6 (1973).
240. W. H. Kibbel, Jr., C. W. Raleigh and J. S. Shaperd, *Eng. Bull. Purdue Univ. Eng. Ext. Ser.* **141,** 824 (1972).
241. W. H. Kibbel, Jr., *Ind. Water Eng.* **13**(4), 6 (1976).
242. W. G. Strunk, *Treat. Disposal Ind. Waste Waters Residues Proc. Natl. Conf.,* 119 (1977).
243. J. Eley, *J. Mo. Water Sewerage Conf.* **47,** 25 (1976).
244. U.S. Pat. 2,070,856 (Feb. 16, 1937), E. E. Butterfield; Ger. Offen. 2,207,920 (Aug. 31, 1972), J. S. Shepherd and M. F. Hobbs (to FMC Corp.); S. R. Lindstrom, *Pollut. Eng.* **7**(10), 40 (1975).
245. K. H. Hunken, J. Sekoulov, and D. Bardthe, *Gas Wasserfach Wasser Abwasser* **114**(4), 176 (1973).
246. C. A. Cole, D. L. Ochs, and F. C. Funnell, *J. Water Pollut. Control Fed.* **46,** 2579 (1974).
247. Can. Pat. 966,234 (Apr. 15, 1975), C. A. Cole (to DuPont).
248. C. Chin and G. Hicks, *J. Water Pollut. Control Fed.* **42,** 1327 (1970).
249. C. Chin, G. Hicks, and C. A. Geisler, *J. Water Pollut. Control Fed.* **45,** 283 (1973).
250. J. C. Young and R. E. Baumann, *Water Pollut. Control Fed. Highlights* **10**(1), 4 (1973).
251. K. H. Hunken and D. Bardthe, *Gas Wasserfoch* **114,** 176 (1973).
252. Ger. Offen. 1,926,813 (Dec. 11, 1969), L. D. Friedman (to FMC Corp.).
253. C. A. Cole, J. B. Stamberg, and D. F. Bishop, *J. Water Pollut. Contr. Fed.* **45,** 829 (1973).
254. P. J. Keller and J. A. Cole, *Eng. Bull. Purdue Univ. Eng. Ext. Ser.* **141,** 756 (1972).
255. Ger. Offen. 2,446,511 (Apr. 15, 1976), N. Wolters and U. Loll.
256. Jpn. Kokai 78 128,145 (Nov. 8, 1978), T. Yoshida and T. Iwamoto (to Nippon Peroxide Co.).
257. Ger. Offen. 2,211,890 (Sept. 20, 1973), G. Siegmund.
258. Jpn. Kokai 76 32,057 (Mar. 18, 1976), and 123,556 (Oct. 28, 1978), (to Toa Gosei Chemical Industry Co., Ltd.).
259. U.S. Pat. 3,788,967 (Jan. 29, 1974), M. Kawahata and K. R. Price (to General Electric Co.).
260. J. P. Zumbrunn, *Trib. CEBEDEAU* **26**(357–358), 332 (1973).
261. J. P. Zumbrunn, *Inf. Chim.* **124,** 189 (1973).
262. *Ind. Miljoe* **6**(6), 27, 33 (1975); *Chem. Abstr.* **85,** 11236e (1976).
263. W. H. Kibble Jr., *Ind. Wastes (Chicago)* **24**(3), 26 (1978).
264. P. N. Cheremisinoff, R. M. Bethea, T. M. Hellman, and O. B. Lauren, *Pollut. Eng.* **7**(10), 24 (1975).
265. C. W. Raleigh, *Ind. Odor. Technol. Assess.,* 369 (1975).
266. J. Shapiro and Y. Thiffault, *Eau Que* **11**(1), 113 (1978).
267. Jpn. Kokai 75 8,369 (Jan. 28, 1975), K. Kurosawa and Y. Hirotani (to Fuji Sekiyu K.K.).
268. H. Knorre, *Galvanotechnik* **66**(5), 374 (1975).
269. U.S. Pat. 3,617,582 (Nov. 2, 1971), B. C. Lawes and O. B. Mathre (to DuPont).
270. R. Armand, *Pasteur Lyon* **2**(1), 51 (1969).
271. Fr. Pat. 1,564,915 (Apr. 25, 1969), J. P. Zumbrunn (to Air Liquide).
272. U.S. Pat. 3,617,567 (Nov. 2, 1971), O. B. Mathre (to DuPont).
273. Ger. Offen. 2,219,645 (Oct. 25, 1973), W. Sebb, H. Schwarzer, and J. Schmitz (to Peroxide-Chemie G.m.b.H.).
274. Ger. Offen. 2,352,856 (Apr. 30, 1975), J. Fischer, H. Knorre, and G. Pohl (to Degussa).
275. Ger. Offen. 2,337,733 (May 16, 1974), E. Jourdan-LaPorte (to Air Liquide).
276. Jpn. Kokai 76 88,863 (Aug. 3, 1976), Y. Yamada, T. Sawasugi, T. Madokow, and S. Kimura (to Kubota, Ltd.).
277. Jpn. Kokai 74 47,299 (May 7, 1974), Y. Fujii, K. Masuda, and Y. Toda (to Sumitimo Chemical Co., Ltd.).

278. U.S. Pat. 3,878,208 (Apr. 15, 1975), R. H. Carlson, R. N. Mesiah, and H. R. Chancey (to FMC Corp.).
279. Brit. Pat. 1,526,190 (Sept. 27, 1978), (to Erdolchemie G.m.b.H.).
280. Jpn. Kokai 76 26,757 (Mar. 5, 1976), I. Teraoka (to Kowa Co., Ltd.).
281. Japan. Kokai 75 61,057 (May 26, 1975), K. Mizukami, R. Tsukihashi, Y. Kuriyama, and Y. Ishiuchi (to Mitsubishi Kakoki).
282. Ger. Pat. 2,511,432 (Oct. 30, 1975), H. Jundermann and H. Schwab (to Degussa).
283. Fr. Demande 2,181,551 (Jan. 11, 1974), J. P. Zumbrunn (to Air Liquide).
284. Belg. Pats. 863,321 and 863,322 (July 25, 1978), (to Degussa).
285. Jpn. Kokai 76 35,672 (Mar. 26, 1976), K. Tsukamoto (to Ebara-Infilco Co., Ltd.).
286. Ger. Offen. 2,521,893 (Nov. 20, 1975), A. H. Karlsson (to SCA Development AB).
287. Jpn. Kokai 76 2,252 (Jan. 9, 1976), K. Sonouchi and J. Hanawa (to Nippon Hyukazai Co., Ltd.).
288. U.S. Pat. 3,200,069 (Aug. 10, 1965), R. Eisenhauer (to DuPont).
289. Jpn. Kokai 76 90,163 (Aug. 7, 1976), T. Matsuda, T. Motohashi, and H. Danda (to Sumitomo Chemical Co., Ltd.).
290. Jpn. Kokai 78 95,170 (Aug. 19, 1978), (to Wako Pure Chemical Ind.).
291. Jpn. Kokai 73 53,967 (July 28, 1973), H. Sano and Y. Somo (to Agency of Industrial Sciences and Technology).
292. U.S. Pat. 3,721,624 (Mar. 20, 1973), R. Fisch and N. Newman (to Minnesota Mining and Manufacturing Co.).
293. Jpn. Kokai 78 108,068 (Sept. 20, 1978), S. Ikuta and T. Shimomura (to Mitsubishi Gas-Chemical Co., Inc.).
294. Jpn. Pat. 78 40,591 (Oct. 27, 1978), (to Nippon Kogyo Senjo).
295. Ger. Offen. 2,754,932 (June 15, 1978), S. Azuhata and co-workers (to Hitachi, Ltd.).
296. Jpn. Pat. 1,926/73 (Jan. 16, 1973), K. Yoshioka and M. Matsumoto.
297. Jpn. Kokai 15,766 (Feb. 28, 1973), N. Hagiwara, S. Iwasaki, and K. Matsuzawa (to Kanto Denka Kogyo Co., Ltd.).
298. Jpn. Kokai 73 83,069 (Nov. 6, 1973), T. Ozawa.
299. Jpn. Kokai 74 8,465 (Jan. 25, 1974), K. Yudate, M. Iwasaki, and M. Ohnaka (to Sumitomo Chemical Engineering Co., Ltd.).
300. Jpn. Kokai 74 30,272 (Mar. 18, 1974). Y. Hirose and T. Ueno (to Nippon Muki Kagaku Kogyo Co., Ltd.).
301. Jpn. Kokai 74 74,658 and 74 74,659 (July 18, 1974), H. Kawasaki and co-workers (to Nissan Engineering Co., Ltd.).
302. Jpn. Kokai 74 95,871 (Sept. 11, 1974), A. Tanaka (to Daito Kakoki Co., Ltd.).
303. Jpn. Kokai 74 99,954 (Sept. 20, 1974), K. Azuma (to Mitsubishi Electric Co.).
304. Jpn. Kokai 74 128,867 (Dec. 10, 1974), Y. Numi, S. Inenago, and Y. Kusumoto (to Sumitomo Metal Industries, Ltd.).
305. Jpn. Kokai 75 8,769 (Jan. 29, 1975), Y. Tijika and A. Katsushima.
306. Jpn. Kokai 75 62,892 (May 29, 1975), H. Ito, S. Mitsuda, and N. Niwa (to Kawasaki Heavy Industries, Ltd.).
307. Jpn. Kokai 75 83,261 (July 5, 1975), M. Atsukawa and co-workers (to Mitsubishi Heavy Industries, Ltd.).
308. Jpn. Kokai 76 6,884 (Jan. 20, 1976), T. Yamada and co-workers (to Osaka Soda Co., Ltd.).
309. Jpn. Kokai 78 33,975 (Mar. 30, 1978), H. Hayasaka, Y. Sekiguchi, and N. Okigami (to Hitachi Ship Building and Equipment Co., Ltd.).
310. Ger. Offen. 2,447,202 (April 24, 1975), H. Yoshida and K. Saga (to Hodogaya Chemical Co., Ltd.).
311. Ger. Offen. 2,524,116 (Dec. 4, 1975), M. Atsukawa, N. Shinoda, and N. Ukawa (to Mitsubishi Heavy Industries, Ltd.).
312. Ger. Offen. 2,436,363 (Feb. 27, 1975), R. Depommier and E. Martin (to Products Chemiques Pechiney-Saint-Gobain).
313. Fr. Pat. 2,373,327 (Aug. 11, 1978), (to Hitachi KK).
314. U.S.S.R. Pat. 312,479 (Oct. 7, 1969), I. E. Kuznetsov.
315. Ger. Pat. 2,524,115 (Aug. 10, 1978), (to Mitsubishi Jukogyo).
316. Ger. Pat. 2,654,234 (July 10, 1978), (to Ugine Kuhlmann).
317. U.S.S.R. Pat. 245,012 (June 4, 1969), I. E. Kuznetsov and S. H. Ganz.
318. Ger. Offen. 2,163,631 (June 29, 1972), A. E. Konopik (to DuPont).
319. Ger. Offen. 2,148,244 (Apr. 6, 1972), J. B. Conilland and J. Louise (to Air Liquide).

320. U.S. Pat. 3,760,061 (Sept. 18, 1973), M. G. Hammond (to DuPont).
321. Ger. Offen. 2,445,598 (Mar. 27, 1975), A. E. Konopik and J. D. Kusko (to DuPont).
322. Fr. Demande 2,170,852 (Oct. 26, 1973), P. Thirion (to Oxysynthese).
323. Jpn. Kokai 74 97,790 (Sept. 17, 1974), T. Ohtsu and H. Nakamura (to Japan Gasoline Co., Ltd.).
324. Jpn. Kokai 76 12,376 (Jan. 30, 1976), S. Yamashita (to Kubota, Ltd.).
325. Jpn. Kokai 75 16,664 (Feb. 21, 1975), J. Sugamo, Y. Kuriyama, and Y. Ishiuchi (to Mitsubishi Gas-Chemical Co., Ltd.).
326. Ger. Offen. 2,453,434 (May 22, 1975), O. R. Wallergard (to Elektrokemishka AB).
327. Ger. Offen. 803,148 (June 26, 1969, A. Deschamps and P. Renault (to Institut Francais du Petrole).
328. Ger. Offen. 2,511,581 (Oct. 9, 1975), H. Suzuki and co-workers, (to Sintokigio, Ltd.).
329. U.S. Pat. 3,721,624 (Mar. 20, 1973), R. Fisch and N. Newman (to Minnesota Mining and Manufacturing Co.).
330. Swiss. Pat. 573,874 (Mar. 31, 1976), P. Lang (to Cellulose Allisholz A.-G.).
331. Can. Pat. 960,437 (Jan. 7, 1975), P. B. Lonnes and co-workers (to Environmental Research Corp.).
332. Jpn. Kokai 78 110,961 (Sept. 28, 1978), (to C. Soda).
333. Belg. Pat. 867,389 (Nov. 23, 1978), (to Degussa).
334. Ger. Offen. 2,067,497 (July 8, 1971), F. Bourdin, M. Costanlini, M. Jouffret, and G. Lartigan (to Rhone-Poulenc S.A.).
335. W. M. Weigert and co-workers, *Chem. Ztg.* **99**(3), 106 (1975).
336. Fr. Pats. 2,379,519 and 2,379,520 (Oct. 6, 1978), (to Interox Chemicals).
337. U.S. Pat. 4,137,242 (Jan. 30, 1978), G. Prescher and co-workers (to Degussa).
338. Belg. Pats. 841,208 (Oct. 28, 1976), and 847,664 (Apr. 27, 1977), (to Bayer A.G.).
339. Brit. Pat. 1,520,821 (Aug. 9, 1978), (to Olin Corp.).
340. Ger. Offen. 2,807,344 (Aug. 31, 1978), A. M. Hildon, T. D. Manly, and A. J. Jaggers (to Propylox).
341. Ger. Offen. 2,038,318 (Feb. 10, 1972), H. Waldmann, W. Schwerdtel, and W. Swodenk (to Bayer AG).
342. Ger. Offen. 2,125,160 (Nov. 30, 1972), P. Hoffmann and co-workers (to Degussa).
343. Ger. Offen. 2,141,156 (Mar. 1, 1973), M. Krueger and co-workers (to Degussa).
344. Belg. Pats. 841,197–841,200; 841,206; 841,209–841,211 (Oct. 28, 1976), (to Degussa).
345. U.S. Pat. 4,101,570 (July 18, 1978), M. Krueger, G. Schreyer, and O. Weiberg (to Degussa).
346. U.S.S.R. Pat. 496,716 (May 15, 1976), (to Bayer A.G.).
347. Belg. Pats. 841,203–841,205 (Oct. 28, 1976), (to Bayer A.G.).
348. Fr. Pats. 2,378,773 and 2,378,774 (Sept. 29, 1978); Fr. Pat. 2,372,161 (July 28, 1978); Belg. Pats. 848,522 (May 20, 1977) and 863,237 (July 24, 1978); Ger. Pat. 2,446,830 (Aug. 24, 1978), (to Ugine Kuhlmann).
349. Fr. Pat. 2,082,811 (Jan. 14, 1972), C. Bocard, H. Mimoun, and I. Seree de Roch (to Institut Francais du Petrole).
350. Ger. Offen. 2,239,681 (Feb. 22, 1973), L. Kim (to Shell International Research Maatschappij N.F.).
351. Fr. Pat. 2,318,851 (Mar. 25, 1977), (to Rhone-Poulenc S.A.).
352. Ger. Offen. 2,167,040 (Oct. 20, 1977), (to Brichma S.p.A.).
353. Jpn. Kokai 73 36,130 (May 28, 1973), (to Toa Gosei Chemical Industry Co., Ltd.).
354. Jpn. Kokai 75 130,727 (Oct. 16, 1975), (to Ube Industries, Ltd.).
355. U.S. Pat. 4,053,523 (Oct. 11, 1977), H. Seifert and co-workers (to Bayer A.G.).
356. Ger. Offen. 2,361,932 (July 11, 1974), 2,403,810 (Aug. 29, 1974) and 2,504,924 (Aug. 14, 1975), (to Ugine Kuhlmann).
357. Jpn. Pat. 69 28,701 (Nov. 25, 1969); Brit. Pat. 1,108,826 (Apr. 3, 1968) and 1,092,899 (Nov. 29, 1967); Fr. Pat. 1,468,022 (Feb. 3, 1967), (to Toa Gosei Chemical Co., Ltd.).
358. Brit. Pat. 1,177,495 (Jan. 14, 1979), (to Farbenfabriken Bayer, A.G.).
359. E. G. E. Hawkins, *J. Chem. Soc. C*, 2663 (1969).
360. Ger. Offen. 2,109,923 (Sept. 30, 1971), and 2,116,500 (Nov. 18, 1971), (to BP Chemicals International Ltd.).
361. Ger. Offen. 2,004,440 (Feb. 17, 1972), (to Degussa).
362. Ger. Offen. 2,012,509 (Sept. 30, 1971) and Ger. Offen. 2,022,455 (Nov. 18, 1971), (to Degussa).
363. U.S. Pat. 4,046,862 (Sept. 6, 1977), W. Heimberger and G. Schreyer (to Degussa).
364. Ger. Pat. 2,534,541 (Aug. 17, 1978), (to Ugine Kuhlmann); Ger. Offen. 2,713,345 (Sept. 28, 1978), (to Chemische Fabrik Kalk G.m.b.H.).
365. Jpn. Kokai 78 108,096 (Sept. 20, 1978), (to Ise Chemical Industries Co., Ltd.).
366. D. Swern, *Organic Peroxides*, John Wiley & Sons, Inc., New York, 1970.

367. B. C. Lawes, *In Situ* **2**(2), 75 (1978).
368. Fr. Pat. 2,299,410 (Oct. 1, 1976), (to Wyoming Mineral Corp.).
369. Fr. Pat. 2,376,215 (Sept. 1, 1978), (to Minatonic Corp.).
370. Fr. Pat. 2,380,410 (Oct. 13, 1978), (to Union Oil Co., California).
371. M. Shabbir and K. E. Tame, *Report of Investigations 7931 Hydrogen Peroxide Precipitation of Uranium,* U.S. Dept. of the Interior, Bureau of Mines, 1974.
372. U.S. Pats. 3,865,435 (Feb. 11, 1975) and 3,896,879 (July 29, 1975), S. S. Sareen, L. Girard III, and R. A. Hard (to Kennecott Copper Corp.).
373. Jpn. Pat. 11,534/74 (Mar. 18, 1974), (to Furukawa Electric Co., Ltd.).
374. U.S. Pat. 3,959,436 (May 25, 1976), J. C. Watts (to DuPont).
375. Jpn. Kokai 76 75,616 (June 30, 1976), (to Hachinohe Smelting Co., Ltd.).
376. Jpn. Kokai 72 30,505 and 72 30,515 (Nov. 9, 1972), (to Nisshin Steel Co., Ltd.).
377. Belg. Pat. 866,937 (Nov. 13, 1978), (to Interox Chemicals).
378. U.S. Pat. 3,826,723 (July 30, 1974), J. L. Woods and T. A. Pittman (to Elmet, Inc.).
379. Brit. Pat. 1,534,485 (Dec. 6, 1978), (to Soc. Mines Fond Zinc SA).
380. Brit. Pat. 1,527,758 (Oct. 11, 1978), (to Japan Carlit KK).
381. U.S. Pat. 4,132,569 (Jan. 2, 1979), R. S. DePablo, D. E. Harring, and D. E. Bramstedt (to Diamond Shamrock Corp.).
382. Ger. Pat. 2,623,977 (Aug. 17, 1978), (to Nukim Nuclear-Chem. G.m.b.H.).
383. Jpn. Pat. 78 41,120 (Oct. 31, 1978), (to Nippon Nuclear Fuel Co., Ltd.).
384. U.S.S.R. Pat. 570,700 (Oct. 20, 1977), (to Azerb Petrol Ind.).
385. U.S. Pats. 3,671,450 (June 20, 1972), 3,816,325 (June 11, 1974), 3,781,329 (Dec. 25, 1973), 3,843,549 (Oct. 22, 1974), 3,909,440 (Sept. 30, 1975), and 3,970,660 (July 20, 1976); Ger. Offen. 2,329,779 (Jan. 3, 1974); S. Afr. Pat. 2,681/73 (Feb. 28, 1974), (to American Cyanamid Co.).
386. Jpn. Kokai 74 26,191 (Mar. 8, 1974), T. Tarui.
387. U.S. Pat. 3,850,836 (Nov. 26, 1974), H. P. Richter, C. A. Heller, W. P. Norris, and W. S. McEwan (to U.S. Dept. of the Navy).

General References

Refs. 4 and 5 are also general references.

H. Pistor, "Die Peroxoverbindungen" in K. Winnacker and L. Kuchler, eds., *Chemische Technolgie,* Band 1 Anorganische Technologie 1, Hanser, Munich, 1969 (a review of industrial manufacturing processes).

R. Powell, *Hydrogen Peroxide Manufacture,* Chemical Process Review No. 20, Noyes Development Corporation, Park Ridge, New Jersey, 1968.

J. G. Wallace, *Hydrogen Peroxide in Organic Chemistry,* E. I. du Pont de Nemours & Co., Inc., Wilmington, Del., 1962.

W. M. Weigert, ed., *Wasserstoffperoxid and Seine Derivate: Chemie and Anwendung,* Huethig, Heidelberg, Germany, 1978 (a review of uses).

W. Machu, *Das Wasserstoffperoxyd und die Perverbindungen,* Springer Verlag, Vienna, 1951.

O. Kausch, *Das Wasserstoffsuperoxyd,* Wilhelm Knapp, Halle, 1938 (Photo-lithoprinted, Edwards Bros., Ann Arbor, Mich.).

C. A. Crampton, G. Faber, R. Jones, J. P. Leaver, and S. Schelle, *Chem. Soc. Spec. Publ. 31,* 232 (1977).

W. S. Wood, *Hydrogen Peroxide,* Royal Institute of Chemistry, Monograph No. 2, London, 1954.

J. R. KIRCHNER
E. I. du Pont de Nemours & Co., Inc.

HYDROGEN SULFIDE, H_2S. See Sulfur compounds.

HYDROMETALLURGY. See Extractive metallurgy.

HYDROPROCESSES. See Petroleum refinery processes.

HYDROQUINONE, RESORCINOL, AND CATECHOL

HYDROQUINONE

Hydroquinone (**1**) (1,4-benzenediol, 1-4-dihydroxybenzene, *p*-dihydroxybenzene) is a white crystalline compound that first was obtained in 1820 by dry distillation (qv) of quinic acid. In 1844 Woehler established its structure also after obtaining it by reduction of quinone which had been obtained by oxidation of quinic acid in 1838. Important uses are in photographic developers and include the production of polymerization inhibitors and rubber and food antioxidants (see Photography; Antioxidants and antiozonants; Rubber).

OH

OH

(**1**)

RESORCINOL

Resorcinol (**2**) (1,3-benzenediol, 1,3-dihydroxybenzene, *m*-dihydroxybenzene) is a crystalline compound with a faint aromatic odor, and a sweet and bitter taste. It was first obtained in 1864 by the alkali fusion of galbanum and asafetida resins; its name was derived from resin and orcinol to indicate its origin and analogous structure with orcinol (5-methylresorcinol). Important uses of resorcinol include the production of adhesives for rubber, starch, tires, and wood and for the preparation of dyestuffs, uv absorbers, and pharmaceuticals (see Dyes and dye intermediates).

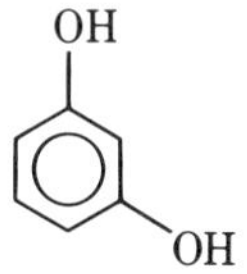

OH

OH

(**2**)

CATECHOL

Catechol (**3**) (1,2-benzenediol, 1,2-dihydroxybenzene, *o*-dihydroxybenzene) is a crystalline compound with a phenolic odor and a sweet and bitter taste. It was first obtained in 1839 by dry distillation of catechin, which is found in the aqueous extract of catechu. Important uses of catechol are in fur dyeing, leather (qv) tanning, and photographic developers and in the fabrication of polymerization inhibitors, perfumes (qv), pharmaceuticals, and pesticides (see Poisons, economic).

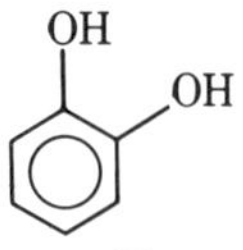

OH
OH

(**3**)

Hydroquinone, resorcinol, catechol, and their derivatives (that have been represented structurally in the text) are listed with their CAS Registry Numbers at the end of this article.

Physical Properties

Hydroquinone has three crystalline modifications. By sublimation, a labile γ form is obtained as monoclinic prisms, whereas the labile β form separates as trigonal prisms on crystallization from methanol. The stable α form is obtained as white trigonal needles or prisms by crystallization from water. Its crystals are triboluminescent. Its solutions are oxidized in air.

Resorcinol crystallizes in the orthorhombic hemimorphic system, and its crystals are colorless, triboluminescent, and piezoelectric. X-ray diffraction measurements indicate that resorcinol crystallizes in the α form which is converted to the β form at 74°C; the latter is more dense (d = 1.33 g/cm^3), as a result of hydrogen bonding between oriented molecules in the crystal lattice. It acquires a pink tint when exposed to light and air.

Catechol crystallizes in the monoclinic system, and the crystals are colorless. It sublimes and is volatile with steam. Air and light cause discoloration.

The ir (1), uv (2), Raman (3), and ^{13}C-nmr (4) spectra of each dihydroxybenzene are known; physical constants and solubility data are listed in Table 1.

Chemical Properties

Dihydrozybenzenes (DHBs) are weak acids with two dissociation constants. They form mono- and di-salts in solutions of alkali hydroxides or carbonates. With its two adjacent hydroxyl groups, catechol is able to complex most metallic salts. Many of the complexes of catechol and catechol derivatives are useful analytical reagents (6) (see Analytical methods); a review of transition metal complexes has been published (7) (see Organometallics).

Dihydroxybenzenes are more easily oxidized than is phenol. However, only catechol and hydroquinone are converted by most oxidizing agents to the corresponding *ortho*- and *para*-benzoquinones. The first step consists of a one-electron transfer with formation of a semibenzoquinone radical in alkaline solution (**4**) as well as in acid solution (**5**). The radical has a sufficiently long lifetime for characterization which explains the antioxidant activity of catechol, hydroquinone, and their derivatives. The quinones that are formed, 2,5-cyclohexadiene-1,4-dione (**6**) and 3,5-cyclohexadiene-1,2-dione (**7**), depending on the starting material, can react further by nucleophilic attack of hydroxide ions to give hydroxybenzoquinone (**8**) (2-hydroxy-2,5-cyclohexadiene-1,4-dione); such results have been obtained with alkaline ferricyanide (8–9),

Table 1. Properties of Hydroquinone, Resorcinol, and Catechol

Properties	Hydroquinone	Resorcinol	Catechol
mp, °C	172	110	105
bp (at 101.3 kPa)[a], °C	287	277	245
flash point (closed cup), °C	165	127	127
autoignition temperature, °C	515	607	510
specific gravity, d_4^{15}, g/cm^3	1.332	1.272	1.344
refractive index, n_D^{20}		1.620	1.615
dissociation constant at 30°C			
K_1	1.22×10^{-10}	7.11×10^{-10}	7.5×10^{-10}
K_2	9.28×10^{-13}	4.78×10^{-12}	8.37×10^{-13}
dipole moment, C·m[b]	4.7×10^{-30}	6.91×10^{-30}	8.74×10^{-30}
molar heat of combustion at constant volume, J/mol[c]	2864	2855	2863
molar heat of sublimation, kJ/mol[c]	99		80.6
solubility at 60°C g/100 g solution[d]			
water	26.0	83.3	80.4
ethyl alcohol	45.7	73.0	74.0
ethyl ether	12.1		65.3
acetone	45.9	75.1	75.2
benzene	0.8	14.1	7.3
chloroform		1.2	0.7
carbon tetrachloride	20.1	0.3	0.8

[a] To convert kPa to mm Hg, multiply by 7.5.
[b] To convert C·m to debye, divide by 3.336×10^{-30}.
[c] To convert J to cal, divide by 4.184.
[d] Ref. 5.

(1)

but with excess sodium periodate, intermediate *o*-benzoquinone is oxidized to *cis,-cis*-muconic acid **(9)** [(*Z*,*Z*)-2,4-hexadienedioic acid] (10).

The redox potential of resorcinol (E_0 in Table 2) cannot be determined experimentally because its oxidation is irreversible, but the polarographic half-wave potential ($E_{1/2}$ in Table 2) is a measure of its oxidizability.

Resorcinol behaves similarly to monophenols; following one-electron oxidation with potassium ferricyanide (8) or ceric sulfate (13), it affords an unstable *meta*-

OH OH (3) $\xrightarrow{K_3Fe(CN)_6}$ O O (7) → O OH O (8) (2)

(7) → HO C O C O OH (9)

Table 2. **Redox and Half-Wave Potentials of Hydroquinone, Resorcinol, and Catechol**

Potential	Hydroquinone	Resorcinol	Catechol
E_0, mV[a]	699	non det.	792
$E_{1/2}$ at pH = 5, 6, mV[b]	234	613	349
$E_{1/2}$ at pH = 0, mV[a]	560	800	600

[a] Ref. 11.
[b] Ref. 12.

semibenzoquinone which produces a mixture of C—C- and C—O-coupled dimeric and polymeric compounds (8). Resorcinol and its derivatives are only slowly attacked by periodate (10), but reaction with hydrogen peroxide in presence of tungstic oxide yields maleic acid (14).

Hydroquinone and *p*-benzoquinone form an equimolecular complex by charge transfer between the π orbitals of the rings. This quinhydrone complex crystallizes to red–brown needles that have a green surface shine (mp = 171°C, d = 1.401 g/cm^3). Quinhydrone is more easily oxidized than hydroquinone. Hydroquinone, when used as a developing agent in photography, reduces exposed silver bromide at a substantially greater rate than the unexposed or underexposed grains. To avoid the formation of quinhydrone, sodium sulfite is added to give sodium hydroquinonesulfonate (**10**) (2,5-dihydroxybenzenesulfonic acid, sodium salt), a weaker developing agent than hydroquinone.

OH OH (1) + 2 AgBr + Na_2SO_3 → OH SO_3Na OH (10) + 2 Ag + HBr + NaBr (3)

DHBs undergo all of the typical reactions of phenols. Only those that are important in the preparation of DHB derivatives are reported here. Electron densities at the nucleus explain the relative reactivity and the regioselectivity in electrophilic substitution (15). The symmetry of hydroquinone (**1**) indicates that all four available

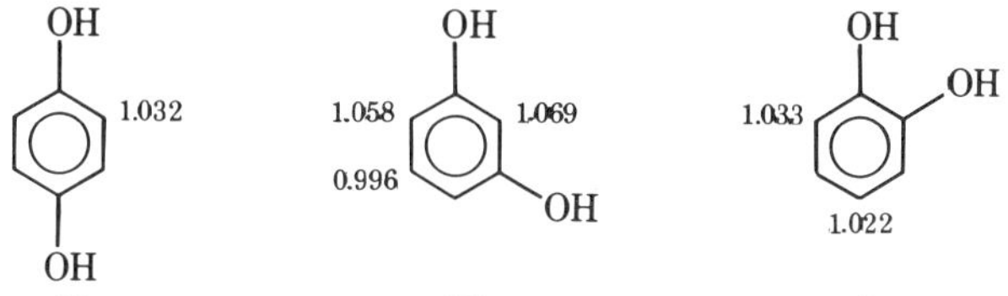

Figure 1. Electron densities of hydroquinone, resorcinol, and catechol.

positions are equivalent in monosubstitution. The 2 position of resorcinol (**2**) has the highest electron density but is hindered sterically by both adjacent hydroxyl groups so that monosubstitution occurs mainly in the 4 or 6 position. Catechol (**3**) is less reactive than resorcinol; monosubstitution occurs in the 3 and 4 positions.

DHBs react with phthalic anhydride to form multiring systems according to the reaction conditions. Hydroquinone yields quinazarin (**11**) (1,4-dihydroxy-9,10-anthracenedione), resorcinol yields fluorescein (**12**), and catechol gives hystazarin (**13**) (2,3-dihydroxy-9,10-anthracenedione) and alizarin (**14**) (1,2-dihydroxy-9,10-anthracenedione).

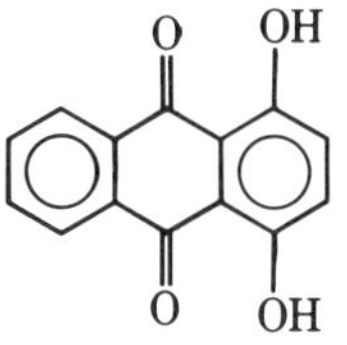

Both mono- and diethers of DHBs can be prepared readily by the usual methods (see Ethers). However it is difficult to confine etherification to the monoether stage. This reaction is used to prepare guaiacol (**15**) (2-methoxyphenol), veratrole (**16**) (1,2-dimethoxybenzene), guethol (**17**) (2-ethoxyphenol), *p*-hydroxyanisole or hydroquinone monomethyl ether (**18**) (4-methoxyphenol), decyloxyresorcinol (**19**) (2-decyloxyphenol), guaiacol glyceryl ether (**20**) [3-(2-methoxyphenoxy)-1,2-propanediol], and 1,4-di(2-hydroxyethoxy)benzene (**21**) (2,2′-[1,4-phenylenebisoxy] bisethanol).

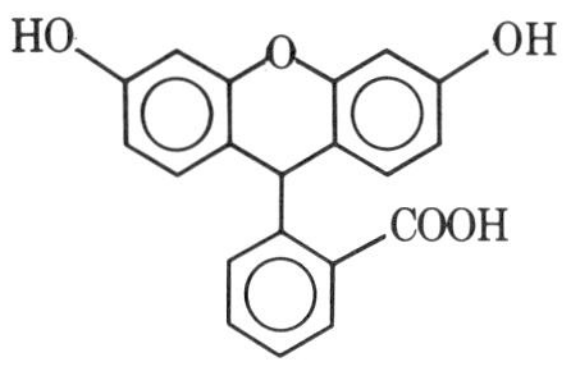

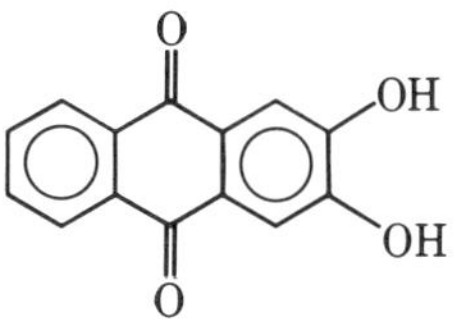

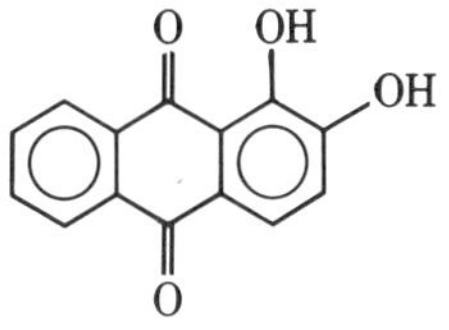

Catechol with its two adjacent hydroxyl groups undergoes a cyclization reaction with methylene chloride to give methylenedioxybenzene (**22**) (1,3-benzodioxole) (16) and, with di(2-chloroethyl) ether, yields dibenzo-18-crown-6-polyether (**23**) (17).

DHBs and their monoethers are *C*-alkylated by all of the common agents that alkylate phenol in the presence of usual Friedel-Crafts catalysts (see Friedel-Crafts reactions; Catalysis). Hydroquinone with *tert*-butanol gives *tert*-butylhydroquinone (**24**) [2-(1,1-dimethylethyl)-1,4-benzenediol] and *p*-hydroxyanisole in the presence of a silica catalyst gives a mixture of 2-*tert*-butyl-4-methoxyphenol (**25**) [2-(1,1-dimethyl)-4-methoxyphenol] and 3-*tert*-butyl-4-methoxyphenol; research has been directed to maximizing the former reaction (18). Catechol is *tert*-butylated with isobutylene in the presence of ion-exchange resins to 4-*tert*-butylcatechol (**26**) [4-(1,1-dimethylethyl)-1,2-benzenediol]; allyl chloride alkylates catechol and guaiacol to give

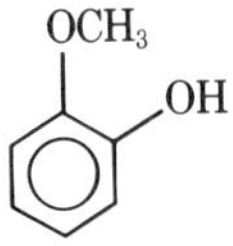

(15) 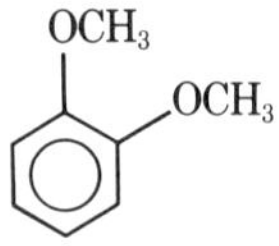 (16) 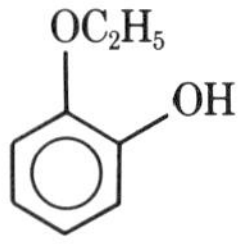 (17) 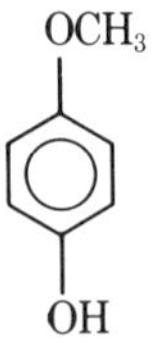 (18) 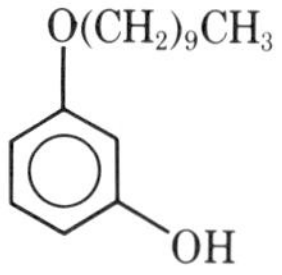 (19) 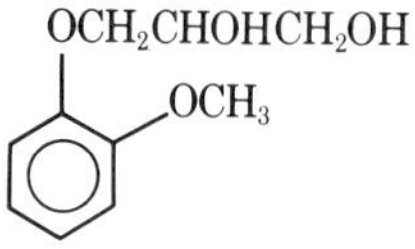 (20)

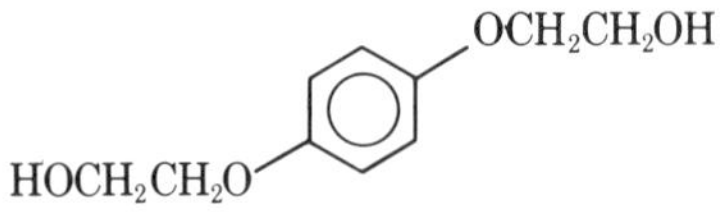

(21)

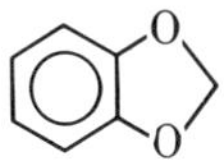

(22) 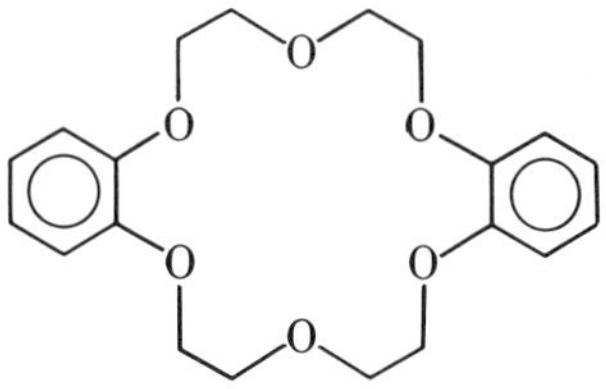 (23)

4-allylcatechol (**27**) [4-(2-propenyl)-1,2-benzenediol] and eugenol (**28**) [2-methoxy-4-(2-propenyl)phenol], respectively (19).

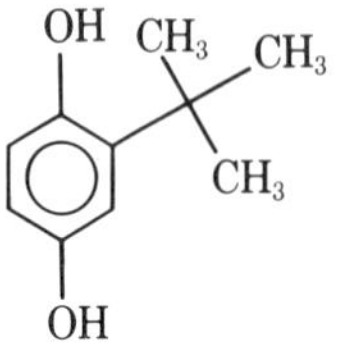

(24) 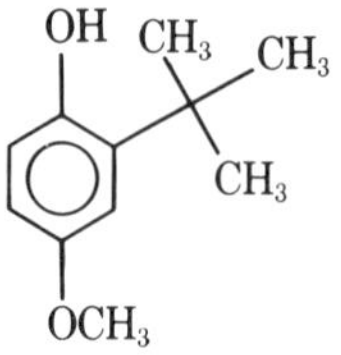 (25) 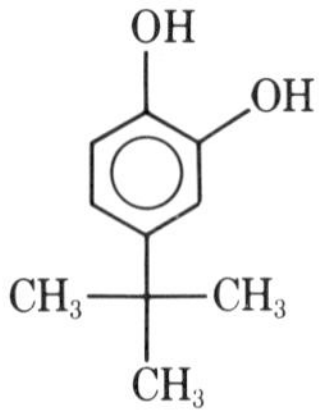 (26) 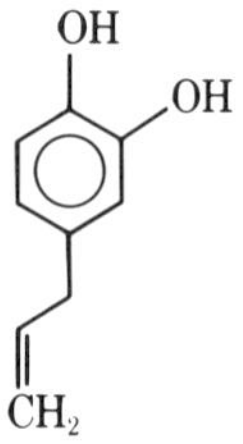 (27) 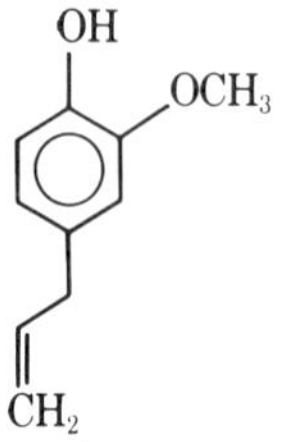 (28)

Acylation of DHBs with carboxylic halides or esters gives corresponding mono- and diesters which can undergo Fries rearrangement (see Esters, organic). Chloroacetyl chloride reacts with catechol to give 4-chloroacetylcatechol (**29**) [2-chloro-1-(3,4-dihydroxyphenyl)ethanone] by Fries reaction (20). In the same way, resorcinol reacts with caproyl chloride to give 4-hexanoylresorcinol (**30**) [1-(2,4-dihydroxyphenyl)-1-hexanone] and with benzoyl chloride to give 2,4-dihydroxybenzophenone (**31**) [(2,4-dihydroxyphenyl)phenylmethanone].

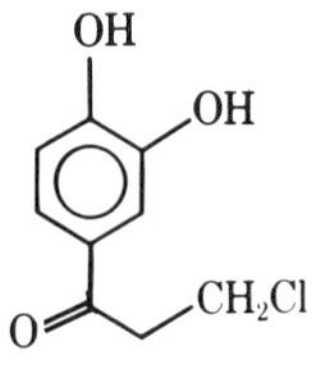

(29) 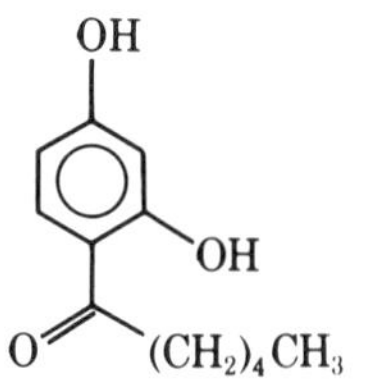 (30) 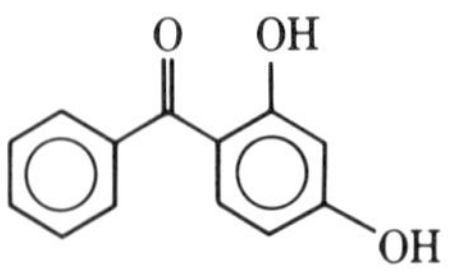 (31)

Hydroquinone, resorcinol, and catechol react with ammonia to give the corresponding *p*-, *m*-, and *o*-aminophenol (**32–34**); alkyl- and arylamines react similarly. Thus hydroquinone with methylamine and aniline gives *p*-methylaminophenol (**35**) and *N*,*N*′-diphenyl-*p*-phenylenediamine (**36**), respectively.

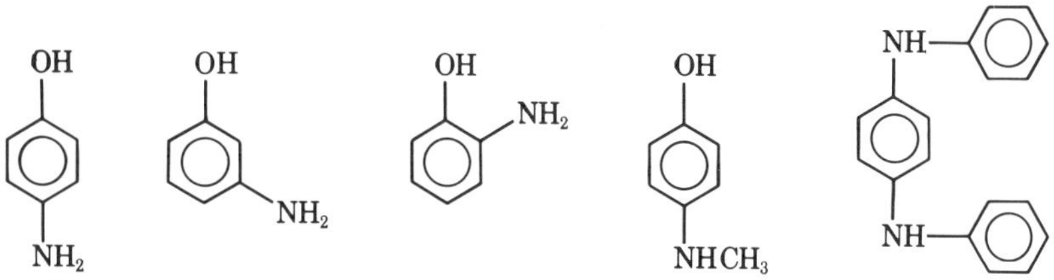

OH, NH_2 (**32**) OH, NH_2 (**33**) OH, NH_2 (**34**) OH, $NHCH_3$ (**35**) NH—, NH— (**36**)

DHBs undergo the Reimer-Tiemann reaction. Catechol reacts with chloroform under alkaline conditions to give a mixture of proto- and catechuic aldehydes: (**37**) (3,4-dihydroxybenzaldehyde) and (**38**) (2,3-dihydroxybenzaldehyde), respectively. Vanillin (qv) (**39**) (4-hydroxy-3-methoxybenzaldehyde) and ethylvanillin (**40**) (3-ethoxy-4-hydroxybenzaldehyde) can be prepared from guaiacol (**15**) and guethol (**17**), respectively, following reaction with glyoxylic acid in an alkaline medium; the substituted mandelic acid that is formed is catalytically oxidized to the corresponding substituted aldehyde (21).

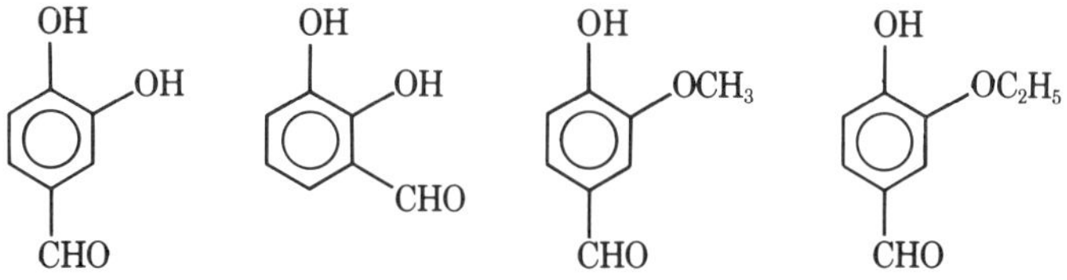

OH, OH, CHO (**37**) OH, OH, CHO (**38**) OH, OCH_3, CHO (**39**) OH, OC_2H_5, CHO (**40**)

DHBs undergo Kolbe-Schmitt carboxylation by reaction of carbon dioxide with their alkali metal salts. Hydroquinone gives gentisic acid (**41**) (2,5-dihydroxybenzoic acid) and 2,5-dihydroxyterephthalic acid (**42**). Resorcinol gives β-resorcylic acid (**43**) (2,4-dihydroxybenzoic acid) and γ-resorcylic acid (**44**) (2,6-dihydroxybenzoic acid).

Hydroquinone, resorcinol, and catechol react with formaldehyde under both acidic and basic conditions to give methylol derivatives which undergo condensation to yield high molecular weight condensation products; resorcinol has the highest reactivity of the three and has its most important use in this reaction. DHBs readily couple with aryldiazonium salts to give azo and bisazo compounds which are used in dyes. Dihydroxybenzenes are hydrogenated catalytically to the corresponding cyclohexanediols.

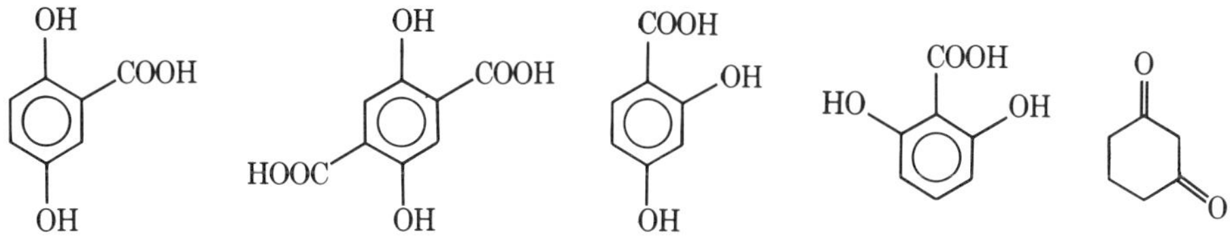

OH, COOH, OH (**41**) OH, COOH, HOOC, OH (**42**) COOH, OH, OH (**43**) COOH, HO, OH (**44**) O, O (**45**)

In the case of resorcinol, hydrogenation can be limited to 1,3-cyclohexanedione (**45**) (22).

Coumarin derivatives are obtained readily by condensation of resorcinol with malic acid to give umbelliferone (**46**) (7-hydroxy-2*H*-1-benzopyran-2-one) and with ethyl acetoacetate to give β-methylumbelliferone (**47**) (7-hydroxy-3-methyl-2*H*-1-benzopyran-2-one).

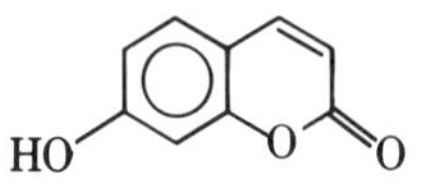

(**46**) 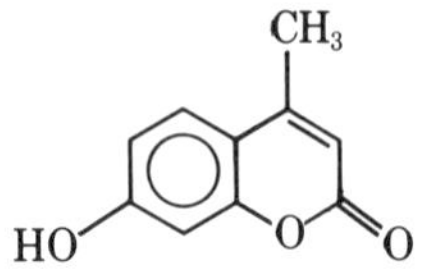 (**47**) 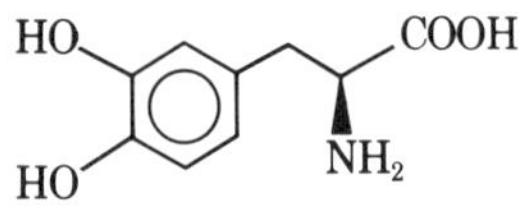 (**48**)

L-Dopa (**48**) (3-hydroxy-L-tyrosine) is produced by culturing whole cells of *Erwinia herbicola* in the presence of catechol and D,L-serin (23).

Occurrence

Hydroquinone has been found as β-D-glucopyranoside (**49**) (4-hydroxyphenyl-β-D-glucopyranoside)—also called arbutin—in the leaves of blueberry, cranberry, cowberry, and bearberry plants. A review of the occurrence of arbutin in plants has been published (24). Although arbutin is easily hydrolyzed in plants, gallotannin prevents enzymes, eg, β-D-glucosidase from splitting it into hydroquinone and glucose.

Resorcinol does not seem to occur in plants or animals. It has been found in cigarette smoke, in the effluents from the production of coal-tar chemicals, and in distillates obtained by delayed coking of higher fractions of oil shale (qv).

Catechol and some of its derivatives are found in higher plants, and they are biosynthetically related to the aromatic aminoacids and are derived from the shikimic acid (**50**) [[3*R*-(3α,4α,5β)] trihydroxy-1-cyclohexene-1-carboxylic acid] pathway, which is closely connected with sugar metabolisms. Catechol has been found in the tannin layer of mycorrhizas of Douglas pine; in leaves and branches of oak and willow; and in apples, onions, and crude beet sugar. A review of catechol and its derivatives that are found in lignin (qv), wood, and other plants has been published (25). 1,2-Catechol sulfate has been found in horse and human urine, and its amount is increased by exposure to benzene or phenol. It also has been found in castoreum and cigarette smoke.

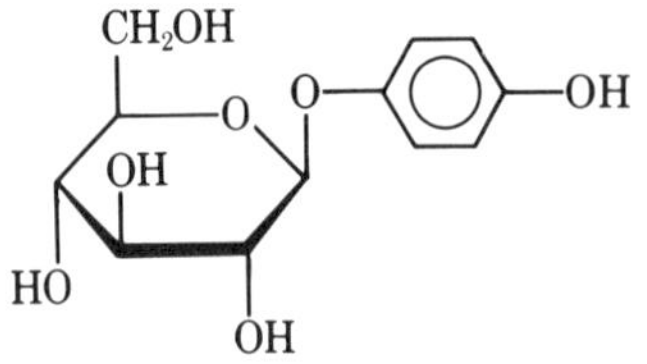

(**49**) 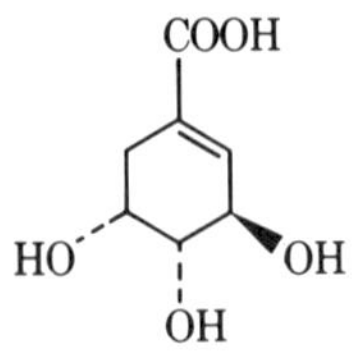 (**50**)

Manufacture and Processing

Manufacture of dihydroxybenzenes is based on the processes listed in Table 3; world production capacities for each also are reported in the table. In Europe, catechol has been recovered industrially from gasworks' ammoniacal liquor and has been obtained from the low temperature carbonization of lignite and brown coal (qv). A review describing the latter process has been published (38); however, the corresponding plants were closed in 1973.

Hydrolysis of *o*-Chlorophenol. During the 1930s catechol was manufactured by hydrolysis of *o*-chlorophenol in aqueous solution of barium and sodium hydroxides with cuprous chloride as catalyst. A review of this process is presented in ref. 39. More recent technology involves a similar process in which sodium hydroxide is used (40–41). The principal steps of this process are: (*1*) alkaline hydrolysis of *o*-chlorophenol in 75% sodium hydroxide aqueous solution with copper sulfate as catalyst at 190°C. A 300% molar excess of sodium hydroxide is used; (*2*) dissolution of the reaction mixture in water, and neutralization with 50% aqueous sulfuric acid; (*3*) extraction of crude catechol with ethyl acetate—Crude catechol contains: 0.8 mol % resorcinol, 1.0 mol % phenol, and 1.5 mol % *o*-chlorophenol, and some high molecular weight compounds (1 wt %); and (*4*) solvent distillation and recycling—the crude catechol is distilled to produce technical-grade catechol.

Table 3. Manufacturing Processes and World Production Capacities for Dihydroxybenzenes

Dihydroxybenzene	World capacity[a], metric tons	Process	Location	Refs.
hydroquinone	>40,000	aniline oxidation	U.S.	
			FRG	
			Japan	
			U.K.	
			COMECON[b]	
			People's Republic of China	
		phenol hydroxylation	France	26
			Italy	27
			Japan	28
		p-diisopropylbenzene oxidation	U.S.	29
			Japan	30–31
resorcinol	30,000–35,000	benzene disulfonation	U.S.	32–33
			Italy	34
			FRG	34
			U.K.	
			Puerto Rico	35
			Japan	36
		m-diisopropylbenzene hydroperoxidation	Japan	
catechol	>20,000	*o*-chlorophenol hydrolysis	none	37
		phenol hydroxylation	France	
			Italy	27
			Japan	28
		coal-tar distillation	U.K.	
			COMECON[b]	

[a] Estimated for 1979.
[b] COMECON = Council for Mutual Economic Assistance (Communist-bloc nations).

The yield of catechol is 87% from *o*-chlorophenol. 2-2′-Dihydroxydiphenyl ether can be extracted from the distillation residue (42). This process produces 1.5 kg of sodium sulfate for 1 kg of catechol. The alkaline hydrolysis step is carried out in a copper vessel. With the same technology but without a copper catalyst, mixtures of *o*- and *p*-chlorophenols give dihydroxybenzenes rich in resorcinol (43).

Hydroxylation of Phenol. Although phenol is less easily oxidized than dihydroxybenzenes are, it can be hydroxylated directly to catechol and hydroquinone. Three plants use this reaction. The hydroxylation agent is hydrogen peroxide and the reaction occurs in the presence of catalytic amounts of strong mineral acids (44) or of ferrous or cobaltous salts (45–46). These two processes have been reviewed (47–49). The overall reaction is:

$$C_6H_5OH + H_2O_2 \xrightarrow[Fe^{2+}]{H^+ \text{ or}} (1 - x)\ o\text{-}C_6H_4(OH)_2 + x\ p\text{-}C_6H_4(OH)_2 + H_2O \quad (4)$$

With a ferrous catalyst, some resorcinol is produced, because of a nonchain free-radical reaction. The principal steps of both processes are the same: (*1*) phenol is hydroxylated at 80°C, with a 70% hydrogen peroxide aqueous solution in the presence of a catalyst solution of perchloric acid or ferrous sulfate; (*2*) the phenols are extracted with a solvent after the oxidation mixture is washed with water; and (*3*) successive distillations separate the solvent and phenol (both of which are recycled); melted technical-grade catechol is fed to a scaling machine and melted technical-grade hydroquinone is solidified in a flaker.

Alkaline Fusion of *m*-Benzenedisulfonic Acid. The manufacture of resorcinol was first described (50) in 1878 as disulfonation of benzene followed by alkaline fusion of the disodium salt of *m*-benzenedisulfonic acid; this process is still used, and three schematic flow charts have been published (51–53). The steps of the process are: (*1*) Benzene is monosulfonated with sulfuric acid at 80°C (54). This reaction has been reviewed (55–56) and used industrially for phenol production. (*2*) *m*-Benzenedisulfonic acid is obtained by adding 30% oleum at 180°C. To prevent formation of sulfones, which yield phenol and heavy phenols after alkaline fusion (57), 5 wt % of sodium sulfate is added to the solution (58). Water that is formed during sulfonation hydrolyzes SO_3 to sulfuric acid. (*3*) Excess sulfuric acid is vaporized in a thin-film evaporator at 240°C and 0.7 kPa (5 mm Hg) (52). Sulfuric acid is recycled to step (*1*). (*4*) Crude *m*-benzenedisulfonic acid, containing sodium sulfate, is neutralized with caustic soda (52). Sodium disulfonate solution is cooled to 50°C; the organic salt is filtered and dried. (*5*) Solid sodium disulfonate and caustic soda in 5% excess react in a kneader reactor (59) at 350°C. The resulting solid is cooled and dissolved in water. (*6*) The solution is neutralized with sulfuric acid and the crude resorcinol is extracted with ethyl ether in a column. (*7*) The ether extract is distilled, the ether is recycled, and the light ends (phenol) and bottoms (heavy phenols) are incinerated. Molten resorcinol is cooled and flaked. (*8*) The sulfite–sulfate solution is transferred to a dryer. The dry products are marketed to kraft mills. The resorcinol yield is 94% (calculated on *m*-

$$m\text{-}C_6H_4(SO_3H)_2 + Na_2SO_3 \longrightarrow m\text{-}C_6H_4(SO_3Na)_2 + SO_2 + H_2O \quad (5)$$

$$m\text{-}C_6H_4(SO_3Na)_2 + 4\ NaOH \longrightarrow m\text{-}C_6H_4(ONa)_2 + 2\ Na_2SO_3 + 2H_2O \quad (6)$$

benzenedisulfonic acid). The process produces 2.3 kg of sodium sulfite per kg of resorcinol according to reaction 6. To minimize this by-product, *m*-benzenedisulfonic acid can be neutralized with the sodium sulfite solution from step 6 (51). However, according to reaction 5, only one half of the sulfite that is produced is used. For economic purposes lime also has been used in the neutralization step (52).

Hydroperoxidation of Diisopropylbenzene. Analogous to the cumene process for phenol production, the preparation of resorcinol from *m*-diisopropyl benzene (*m*-DIPB) has been studied and reviewed (60) (see Phenol; Cumene). From *p*-diisopropyl benzene (*p*-DIPB) and by the same process, hydroquinone is produced according to the dihydroperoxide reaction 7. Hydroquinone and resorcinol can be produced alternatively in the same plant. *o*-Diisopropylbenzene (*o*-DIPB) is an inert compound in radical oxidation, because its two adjacent isopropyl groups are sterically hindered.

A two-step process for producing purified *m*- or *p*-DIPB has been described (61). In one reactor cumene is alkylated with 3 moles of propylene in the liquid phase with silica–alumina catalyst (62) at 200°C. In another reactor, the unwanted di- and tri-isopropylbenzene isomers are transalkylated with benzene at 270°C in the liquid phase with silica–alumina catalyst (63). In the feed, the molar ratio of propyl groups to benzene rings is ca 1. For *p*-DIPB production, the fractionation of the product occurs from effluents from both reactors. For *m*-DIPB production, the product is fractionated only from transalkylation effluents; indeed higher temperature and residence time in the transalkylation reactor minimize *o*-DIPB formation. This is necessary because

dihydroperoxide (DHP) $\xrightarrow{H^+}$ hydroquinone + 2 CH_3COCH_3 (7)

p-DIPB $\xrightarrow{O_2}$ monohydroperoxide (MHP) $\xrightarrow{-[O]}$ $\xrightarrow{O_2}$ hydroxyhydroperoxide (HHP) $\xrightarrow{H^+}$ *p*-isopropenylphenol + CH_3COCH_3 + H_2O (8)

monohydroperoxide (MHP) $\xrightarrow{O_2}$ dihydroperoxide (DHP)

monohydroperoxide (MHP) $\xrightarrow{H^+}$ *p*-isopropylphenol + CH_3COCH_3 (9)

o- and *m*-DIPB cannot be separated because both have ca the same boiling point, 203.8°C and 203.2°C, respectively (64). Such units can be designed for alternatively producing the *m*- and *p*-isomers.

Instead of silica–alumina catalyst, aluminium chloride or boron trifluoride catalysts can be used; the process requires lower temperatures and residence times but 1,1,3-trimethylindan is produced (eq. 10) from *o*-DIPB (65–66).

Diisopropylbenzenes (DIPB) are by-products of cumene production in a phenol plant. Unfortunately the separation of purified *p*- and chiefly *m*-DIPB is not achieved easily from such a feed.

$$\text{C}_6\text{H}_4[\text{CH}(\text{CH}_3)_2]_2 \longrightarrow \text{1,1,3-trimethylindan} + \text{H}_2 \qquad (10)$$

Production of hydroquinine or resorcinol from purified *p*- or *m*-diisopropylbenzene, respectively, includes the following steps: (*1*) Air oxidation of slightly alkaline DIPB to the corresponding dihydroperoxide (DHP reaction (eq. 7)) at 90°C is a two-step reaction. Monohydroperoxide (MHP reaction (eq. 8)) is formed first and then is oxidized to DHP. By-products are cumyl methyl ketone and cumyldimethylcarbinol; the latter is oxidized to hydroxyhydroperoxide (HHP reaction (eq. 8)). To minimize by-product formation, DHP effluent concentration is maintained at 5.5 wt % and that of MHP at 53 wt %. (*2*) DHP is separated from the oxidation products and the rest is recycled to the oxidation reaction. This can be done in two ways: (*a*) DHP is a weak acid and can be extracted with a 4% aqueous sodium hydroxide solution. The caustic raffinate is recycled, DHP is extracted from aqueous alkaline solution with methylisobutyl ketone (67). The alkaline solution also can be neutralized with CO_2; DHP forms an oily layer which is separated, dried in an evaporator, and immediately dissolved in acetone (68–69). (*b*) The alternative route (70) consists of adding *n*-hexane to the oxidate and chilling the mixture to −20°C. DHP crystallizes and is separated by centrifuging which is followed by dissolution in acetone. Pure DHPs have high melting points: 64°C and 147°C for the meta and para isomers, respectively (71). The separation of molten or crystallized DHP requires special safety precautions (72). (*3*) DHP is cleaved with 0.2 wt % sulfuric acid at 80°C in a tubular heat exchanger (73). In the feed, hydrogen peroxide is added to oxidize HHP to DHP; a 25% molar excess is used (74). If no hydrogen peroxide is added, HHP yields *p*-isopropenylphenol (eq. 9) which forms heavy products with dihydroxybenzene (mainly with resorcinol) and they have to be cracked at 250°C (75). (*4*) Cleavage products are neutralized with a lime slurry followed by solid calcium sulfate filtration. (*5*) Ketones are distilled and are stocked for acetone (qv) production or are recycled to step (*2*). (*6*) Dihydroxybenzene solution is purified by toluene extraction. (*7*) Concentration of the aqueous solution and cooling, crystallization centrifuging, and vacuum drying of the technical-grade dihydroxybenzene follows (see Drying). For alternative production of resorcinol and hydroquinone, the water balance of this step is effected by the two dihydroxybenzenes whose solubilities in water are 58 and 66 wt %, respectively, at 20°C (38).

The yield of dihydroxybenzene is 71% (based on DIPB). The high reactivity of resorcinol towards acetone indicates the critical nature of step 3. This multistep process produces 1 kg of acetone for 1 kg of dihydroxybenzene; therefore, its economy depends on acetone demand.

Aniline Oxidation. Aniline oxidation to benzoquinone was the first process used in the United States by DuPont after World War I and originated at Usines du Rhône in France. Sodium bichromate was used as the oxidizing agent and the quinone was reduced to hydroquinone with sulfur dioxide. Oxidation of aniline to quinone is still one of the most widely used processes for hydroquinone production (76). A more current oxidizing agent is manganese dioxide. Reactions 11 and 12 are the overall aniline oxidation and quinone reduction reactions:

$$2\,C_6H_5NH_2 + 4\,MnO_2 + 5\,H_2SO_4 \longrightarrow 2\,C_6H_4O_2 + (NH_4)_2SO_4 + 4\,MnSO_4 + 4\,H_2O \quad (11)$$

$$C_6H_4O_2 + Fe + H_2O \longrightarrow C_6H_4(OH)_2 + FeO \quad (12)$$

The main steps of the process are: (*1*) Oxidation of aniline sulfate, with a 20% excess of manganese dioxide at 5°C in sulfuric acid. Pyrolusite ore, which contains manganese dioxide, generally is used. Oxidation time is a function of the quality of pyrolusite and generally it is 8 h. (*2*) Quinone is steam-stripped from the oxidation solution, whereupon quinone is removed for final processing (77); (*3*) The quinone–steam mixture can be reduced with an aqueous suspension of iron (78); the iron and iron oxide are filtered off. Instead of iron reduction, catalytic hydrogenation can be used (79). (*4*) The filtrate is concentrated by evaporation under reduced pressure at 5°C. Hydroquinone (which crystallizes in the concentrated liquor) is separated by centrifuging and is dried in a vacuum drier (see Centrifugal separation). This process produces technical-grade hydroquinone and technical-grade manganese sulfate. The stripped oxidation mixture from step *2* is neutralized with magnesium hydroxide and is filtered. The filtrate is evaporated and, after crystallization, centrifuging, and drying, technical-grade manganese sulfate is sold as a fertilizer or as an animal feed additive (see Fertilizers; Pet and other livestock feeds). The oxidizers for step *1* are wood-stave vessels or lead-and-brick-lined steel reactors. The yield of hydroquinone is 90% (based on aniline).

Production of photographic-grade hydroquinone includes dissolution of technical-grade hydroquinone in demineralized water, treatment of the solution with decolorizing charcoal, crystallization at 5–10°C, centrifuging, and drying in a nonoxidizing atmosphere.

Miscellaneous Preparations. Many preparations have been described and some of them have been studied on a pilot-plant scale.

Catechol has been prepared from cyclohexane derivatives. The cyclohexanone is chlorinated and subsequent hydrolysis gives 2-hydroxycyclohexanone and 1,2-cyclohexanedione. This mixture is dehydrogenated to catechol (80).

Resorcinol yields of up to 95% have been reported by hydrolysis of *m*-phenylenediamine at 230°C in acidic conditions (81). Resorcinol also has been prepared by dehydrogenation (82) of 1,3-cyclohexanedione which is obtained by alkaline cyclization of ethyl 5-oxohexanoate (83–84) according to reaction 13:

$$CH_3CO(CH_2)_3COOC_2H_5 \xrightarrow{HO^-} \text{1,3-cyclohexanedione} \xrightarrow{-H_2} \text{1,3-}C_6H_4(OH)_2 \qquad (13)$$

Hydroquinone can be formed, based on Reppe's synthesis, by carbonylation of acetylene. The reaction is carried out under pressure (85–87). Hydroquinone also is obtained in high yield from the reaction of p-isopropenylphenol and 30% aqueous hydrogen peroxide in acidic conditions (88) according to reaction 14. p-Isopropenylphenol is obtained by alkaline cracking of bisphenol A (89) (eq. 15); however, care has to be taken because the p-isopropenylphenol polymerizes easily (90). Phenol and acetone, which are by-products, are recycled to synthesize bisphenol A (eq. 16).

$$HO{-}C_6H_4{-}C(=CH_2)CH_3 + H_2O_2 \xrightarrow{H^+} HO{-}C_6H_4{-}OH + CH_3COCH_3 \qquad (14)$$

$$HO{-}C_6H_4{-}C(CH_3)_2{-}C_6H_4{-}OH \xrightarrow{OH^-} HO{-}C_6H_4{-}C(=CH_2)CH_3 + C_6H_5{-}OH \qquad (15)$$

$$2\ HO{-}C_6H_5 + CH_3COCH_3 \xrightarrow{H^+} HO{-}C_6H_4{-}C(CH_3)_2{-}C_6H_4{-}OH + H_2O \qquad (16)$$

Phenol can be oxidized catalytically to p-benzoquinone with a selectivity of 95% at 50% phenol conversion. Catalytic systems are cupric or cuprous chloride in dimethylformamide or acetonitrile. Partial oxygen pressure must be higher than 3.5 MPa (510 psi) at 40–60°C. p-Benzoquinone is reduced easily to hydroquinone (91–92). Hydroquinone also is formed from p-nitrosophenol by catalytic hydrogenation in acidic solution. Yields of 50% are reported (93).

Yields of 50–60% of hydroquinone can be obtained by heating nitrobenzene in aqueous sulfuric or phosphoric acid at 250–260°C (94). Higher yields are obtained (75%) by heating nitrobenzene in aqueous $NaHSO_4$ at 250°C (95). Although the chemical reactions that are involved are not described, they seem related to Bamberger's reaction.

Electrolytic oxidation of benzene or phenol in dilute sulfuric acid, gives benzoquinone which can undergo electrolytic reduction to hydroquinone (96–97). Good reviews on these processes have been published (98–99).

Economic Aspects

During 1977 hydroquinone was produced by two United States companies, and the estimated total production was 15,000 metric tons (100–101). Estimated Western European production was 7000 t. In 1975 two Japanese producers manufactured 2900 t of hydroquinone.

The United States use of hydroquinone in 1977 was 45% in photographic developers, 50% as antioxidants and polymerization inhibitors, 5% for other uses. These figures were 70%, 15%, and 15%, respectively, in Western Europe and 20%, 35%, and 45% in Japan. Prices of hydroquinone, resorcinol, and catechol and their derivatives on January 1st, 1979 are reported in Table 4.

During 1974 resorcinol was produced by Koppers Co., the sole United States resorcinol manufacturer (100). In 1977 United States production was estimated at 16,000 t (103). United States exports (104) totalled 5200 t, 38% of which went to Western Europe, 17% to Eastern Europe, and 15% to Japan. Estimated Western European production was 8500 t. The sole Japanese manufacturer in 1975 produced an estimated 1000 t, of which about 100 t was exported. Japanese imports in that year were 1200–1800 t and were primarily from the FRG and the United States (105).

In 1977 the United States use of resorcinol was 65% in the manufacture of rubber products, 20% in wood adhesives, and 15% for miscellaneous uses. The corresponding figures for Western Europe were 35%, 50%, and 15%, respectively, and for Japan, 55%, 20%, and 25%, respectively.

Table 4. Prices (1979) for Hydroquinone, Resorcinol, Catechol and their Derivatives[a]

Compound	Price, $/kg
hydroquinone, photograde	3.74
technical grade	3.39
butylated hydroxyanisole, food grade	14.0
p-aminophenol	4.29
resorcinol, USP crystal	9.12
USP powder	9.67
technical-grade flakes	3.50
resorcinol, monoacetate	4.63
hexylresorcinol, USP	66.13
catechol, cp crystals	6.50
extra pure	7.60
technical grade	3.00
guaiacol, technical grade	5.75
vanillin USP	11.79
epinephrine base USP	0.58/g
alizarin	7.60
eugenol USP	9.37
glyceryl guaiacolate	14.33
heliotropine	17.08
isoeugenol	15.10
methyleugenol	7.82
papaverine hydrochloride NF	83.77
piperonyl butoxide	8.82

[a] Ref. 102.

Before 1972 catechol was in short supply, was high priced, and was produced mainly from brown coal tar (see Lignite and brown coal) or from lignin (qv) contained in wastes from wood pulping (see Pulp). Many catechol derivatives were not produced from catechol but from other raw materials. In 1973 catechol plants were built and catechol demand increased; however, the market has not stabilized.

In 1974 only one United States company reported commercial production of catechol and, in the same year, United States imports were 502 t (101). United States demand in 1977 has been estimated to have been 650 t. Catechol production in Western Europe is estimated to have been 8000 t in 1977. In 1975 the sole Japanese producer manufactured 600 t (105).

The demand for pyrocatechol in 1977 in the United States was 75% for *tert*-butylcatechol and 25% for specialty products. The breakdown was 75% and 25%, respectively, in Japan.

Specifications and Standards

Hydroquinone is available in the United States, Western Europe, and Japan in photographic and technical grades. Manufacturer's specifications are given in Table 5. *USP XX–NF XV* describes specifications for hydroquinone used in pharmaceuticals (108).

Table 5. Manufacturer's Specifications for Hydroquinone[a]

Property	Specifications	
	Technical	Photographic[b]
physical state	crystals	crystals
color	light tan to pink	white
mp, °C (min)	169	170–174
purity, wt % (min)	98.5	99
heavy metal content (eg, Pb), % (max)		0.001
iron content, wt % (max)		0.001
ash content, wt % (max)	0.07	0.05

[a] Ref. 106.
[b] Ref. 107.

Table 6. Manufacturer's Specifications for Resorcinol[a]

Property	Specifications	
	Technical	USP XX
physical state	flakes	crystal or powder
color	white or slightly colored	white or nearly white
purity, wt % (min)	99	99
mp, °C		109–111
fp, °C (min)	109.1	
phenol content, wt % (max)	1.0	not perceptible
pyrocatechol content, wt % (max)	0.1	no turbidity with Pb(II)
ash content, wt % (max)	0.005	0.05

[a] Ref. 109.

Identification of hydroquinone is by determination of physical constants and by oxidation with ferric chloride, which gives a brown–yellow coloration and the characteristic pungent odor of quinone.

Resorcinol is available in the United States in crystal and powder, USP and technical grades. Manufacturer's specifications are given in Table 6. Technical-grade resorcinol is available in Europe and Japan. Identification of resorcinol is by ir analysis and by the reaction of resorcinol and chloroform in an alkaline solution to produce an intense crimson color which, with addition of hydrochloric acid, turns to pale yellow.

Catechol is available in the United States in technical, chemically pure, and extra pure grades. Manufacturer's specifications are given in Table 7.

In Western Europe and Japan, catechol is commercially available in technical and chemically pure grades. Impurities in catechol are the two other dihydroxybenzenes. Identification of catechol is by determination of the physical constants and by the precipitation of lead salt of catechol at pH 8–9.

Table 7. Manufacturer's Specifications for Catechol [a]

Specifications	Technical	Chemically pure	Extra pure
physical state	flakes	white crystals	white crystals
purity, wt % (min)	98	99	99.5
mp, °C (min)	102		
fp, °C (min)		103.5	103.2
ash content, wt% (max)	0.05	0.05	0.05

[a] Ref. 110.

Analytical and Test Methods

Dihydroxybenzenes can be determined by gas chromatography (111) using a glass capillary column (112) and by liquid chromatography (113). Semiquantitative determination of DHBs by tlc gives detection limits of 0.008–4 μg depending on which reagent spray is used (114). Thin layer chromatography can be used for the separation of DHBs and of their oxidation products in waste waters (115). DHBs can be separated by paper chromatography (116) but only catechol and hydroquinone can be exposed on chromatographic paper if the paper has been treated with 5% sodium periodate; this method is used for microgram quantities (117).

Spectrophotometric determination of DHBs can be carried out with potassium iodate in dilute nitric acid (118) or by determination of products formed after coupling DHB with diazotized *p*-nitroaniline (119). A colorimetric determination of hydroquinone in waste waters containing catechol and resorcinol has been described (120). Titrimetric determinations of DHBs have been made with iodine (121) or with iodine cyanide or bromine cyanide in aqueous or nonaqueous media (122). Titrimetric determination of hydroquinone with bichromate is possible in the presence of resorcinol but not with catechol (123). Stannous chloride allows differentiated complexometric determination of catechol in various media containing DHBs (124). Redox mercurimetric titration of DHBs in nonaqueous solution has a relative error of less 2% (125). Selective potentiometric determination of catechol is possible with Pb(II) salts (126).

Health and Safety

Except for resorcinol, dihydroxybenzenes are more toxic than phenol. Experimental studies on humans and animals have demonstrated low chronic toxicity and rapid excretion of DHBs. Systemic effects from industrial exposure have not been observed. Safety data are given in Table 8.

DHBs are absorbed readily from the gastroenteric tract and, in suitable solution or salves, are absorbed readily through human skin. After absorption of catechol or hydroquinone, one part of the molecule is oxidized to the more toxic compound, quinone, whereas another part of the molecule conjugates with hexuronic, sulfuric, and other acids. Resorcinol is excreted in a free state and is conjugated to acids. The symptoms of intoxication by DHBs resemble those induced by phenol poisoning: nausea, dizziness, a sense of suffocation, an increased rate of respiration, vomiting, pallor, convulsions, headache, dyspnea, cyanose, delerium, and coma. Phenol-like toxicity symptoms are induced in experimental animals that are given toxic or lethal doses of DHBs but the antipyretic action is more marked in them than in humans. Catechol can cause depression of the central nervous system and a rise of blood pressure resulting from peripheral vasoconstriction. Unlike phenol, DHBs do not have a tumor-promoting activity, but catechol is reported to have a co-carcinogenic activity (131).

Table 8. Safety Data for Dihydroxybenzenes[a]

Property	Hydroquinone	Resorcinol	Pyrocatechol
threshold limit value (TLV), mg/m^3 [b]	2	45	20
LD_{50} (oral, rats), mg/kg	370	370	3890
LD_{50} (subcutaneous, mice), mg/kg	160	450	179
LD (*Escherichia coli*), mg/L	50	1000	90
LD (*Daphnia*), mg/L	0.6	0.8	4
LC (goldfish, after 48 h), mg/L	0.28	57.4	14

[a] Refs. 127–129.
[b] Ref. 130.

Cases of dermatitis have resulted from skin contact with hydroquinone. Cases of keratitis and discoloration of the conjunctiva have been reported among men exposed to hydroquinone vapor or dust. Persons with poor visual acuity caused by astigmatism, keratoconus, or preexisting corneal injury should be excluded from repeated unprotected exposure. Hydroquinone induces hyperemia of abdominal organs that are rich in pigments, pathological changes in the liver and kidney, and bronchopneumonia.

Resorcinol may cause irritation to the eyes and induce dermatitis particularly in sensitive individuals. The cutaneous application of solution or salves of resorcinol may result in local hyperemia, itching, dermatis, edema, and the loss of superficial layers of the skin. Resorcinol induces marked siderosis of the spleen, marked tubular injury in the kidney, fatty changes and anemia in the liver, fatty changes in the heart, and edema and emphysema in the lungs.

Catechol contact with the skin can cause an eczematous dermatitis. Catechol induces degenerative changes in the tubuli of the kidney characterized by red blood cells and fibrin clots in the lumina.

Average daily concentration of DHB dust should be maintained below the threshold limit value (130). Operations, eg, sieving, blending, and packaging usually will require enclosure and local controlled ventilation. Good housekeeping practices and personal contamination control are essential. In the case of hydroquinone, complete eye protection may be useful for short exposure where other controls are not feasible.

Uses

Dihydroxybenzenes are high priced chemical specialties. The main outlets of DHB and their derivatives are, in decreasing order of importance, photographic developers, tire adhesives, rubber antioxidants and antiozonants and monomer inhibitors, wood adhesives, ultraviolet absorbers and optical brighteners, dyestuffs, and miscellaneous derivatives (see Photography; Antioxidants and antiozonants; Uv absorbers; Brighteners, fluorescent).

The largest single use of DHB (hydroquinone particularly) is as a developer in black-and-white photography and related graphic arts. Catechol is used much less than hydroquinone. Both products (stabilized by sodium sulfite in buffer solution) reduce the exposed silver halide grains in a photographic emulsion at a faster rate than that of unexposed grains and provide a very fine grain and high contrast images. Hydroquinone is used in conjunction with one of its derivatives: Metol (**35**) (*p*-methylaminophenol) or with Phenetidone (1-phenyl-3-pyrazolidinone). Hydroquinone also is used as a developer in graphic arts, eg, lithography, photoengraving, and rotogravure and for medical and industrial x-ray films. Hydroquinone and its ethers are used in the preparation of intermediates for diazo developers, in dry-processable color photothermographic compositions (132), in photographic color-reversal developers (133), in photographic heat developable compositions (134), in photoresist composition for printed electric circuits (135–137), and as a developer for the rapid processing of reflected holograms (see Holography; Reprography; Color photography) (138).

The second largest use of DHBs is in the rubber industry, as tire and rubber adhesives. Only resorcinol [as a resorcinol–formaldehyde resin (RF)] is used; catechol–formaldehyde resins can be used as well and are available under the tradename Rhodylene. They are used to reinforce tires, hoses, and belts. Generally, no adhesive is required with cotton fibers, but high tenacity rayons or polyamide fibers (nylon-6,6 or -6) and fiberglass require an adhesive treatment of the resin–latex type (see Adhesives). The RF resin is added to 2-vinylpyridine–styrene–butadiene terpolymer latices thereby yielding the adhesive, through which the textile is dipped. Nylon fibers require higher proportions of vinyl–pyridine latex than rayon. The resorcinol content of the dry RF latex is 8–12% (139). Polyester fibers require different adhesive systems. Steel (qv) fibers are brass-plated and are not dip-coated, so that the rapid growth of steel fibers in the tire industry represents a threat to the growth of this resorcinol use. Nevertheless new RF adhesives are being developed, eg, triarylcyanurate–resorcinol–formaldehyde (140) or treatment of steel tire cords (qv) with silane derivatives to improve adhesion of RF resins (141).

An alternative system to develop adhesion is to incorporate a bonding system, which is a mixture of resorcinol, hexamine, and silica, in the rubber compound. Improvements are achieved when this system is used for reinforcement through the incorporation of short fibers in a rubber matrix. Resorcinol-based adhesives also are used for the production of a wide variety of rubber-derived products, eg, hoses, textiles, gloves, and belts. RF resins also have been used as vulcanizing agents for noncrystallizing rubbers, eg, butyl and neoprene rubbers (142–143). A process has been described for the production of nonstaining reclaimed rubber by the addition of resorcinol and Formol (aqueous CH_2O) to a conventional reclaiming process (144). Hydroquinone is used as the vulcanizing agent for fluororubbers (145) (see Elastomers, synthetic).

Another use of DHBs is in the manufacture of rubber antioxidants and antiozonants, monomer inhibitors, and food antioxidants. This is the second largest outlet for hydroquinone and the largest for catechol and its derivatives (see Antioxidants and antiozonants). Hydroquinone derivatives that are used as antioxidants are dialkylated hydroquinone, *N*-alkyl-*p*-aminophenol, and, mainly, diaryl-*p*-phenylenediamines. Hydroquinone derivatives that are used as antiozonants are dialkyl and alkylaryl substituted *p*-phenylenediamines. The second largest antioxidant use of hydroquinone and catechol derivatives is their addition to monomers to inhibit radical polymerization during processing, refining, shipping and storage (see Polymerization mechanisms and processes). The usage level of these derivatives vary from 5 to 300 ppm. In acrylic and methacrylic esters, acrylonitrile, and vinyl acetate, the preferred inhibitors are hydroquinone for in-process stabilization, and hydroxyanisole (**18**) for storage. In styrene and butadiene, *tert*-butylcatechol (**26**) is used during shipment and storage. Hydroquinone and its derivatives stabilize unsaturated polyester resins; derivatives that are used are *p*-benzoquinone, diarylbenzoquinone, and mono- and di-*tert*-butylhydroquinone. The latter two also are used as color stabilizers and for short-stopping of rubber.

One of the approved antioxidants in foods is butylated hydroxyanisole (**25**) which is a mixture of 2-*tert*-butyl- and 3-*tert*-butyl-4-methoxyphenol (see Food additives). The antioxidant activity of the 2 isomer in preventing rancidity is higher than that of the 3 isomer. The most important property of BHA, which accounts for its great popularity as a food antioxidant, is its ability to remain active in baked and fried foods. FDA approval was given in 1972 for use of *tert*-butylhydroquinone in the food industry as a preservative for salad oil and snack foods (146).

DHBs and their derivatives also are used to prevent deterioration in many oxidizable products. Hydroquinone and BHA stabilize vitamin A in fish oil, vitamins D and E, β-carotene, and antibiotics (qv) in feeds. Stabilization of soaps, cosmetics (qv), textile oils, pharmaceuticals, essential oils, and lubricants, by DHB derivatives are reported (147) (see Vitamins; Lubrication and lubricants; Soap; Oils, essential).

Dihydroxybenzenes have an important outlet in wood adhesives, which is the second largest use of resorcinol. The advantages of resorcinol adhesives are room temperature curing, high durability of bond, specific adhesion of the resorcinol resins for one or both adherents, and water resistance (which permits exterior and marine uses) (148). Both resorcinol-formaldehyde (RF) and resorcinol-modified phenol formaldehyde (RPF) resins are used to laminate beams, repair plywood, scarf two beams by a finger-joint technique, and laminate arches when stress is imparted to the adhesive system (149). For the latter application, RF have replaced casein. RF resins cure faster and at lower temperatures than phenolic resins (qv) and have been used for plywood manufacture. RF resins are used in the production of particle board where they compete with urea adhesives which can be used only for boards not intended for outdoor use (see Laminated and reinforced wood). RF resins are used as an additive in starch (qv) adhesives in the manufacture of shipping containers requiring a high degree of moisture resistance (see Phenolic resins).

Resorcinol is the starting material for an important class of ultraviolet absorbers (qv), eg, benzophenones, which are derivatives of 2,4-dihydroxybenzophenone (**31**); β-resorcylic acid (**43**) and resorcinol esters also are used. Other derivatives of resorcinol, eg, umbelliferone (**46**) and β-methylumbelliferone (**47**) and the more substituted derivatives (150–151) are used as fluorescent brighteners (qv) in plastics, textiles, soap, laundry products, paper, and in sunscreen lotions and creams. A derivative of hydroquinone, 2,5-dimethoxyterephthalic acid, is used as a fluorescent brightener in fiber forming of polyester (152) (see Polyesters).

Resorcinol and catechol also are used as raw materials for the preparation of dyestuffs, primarily of xanthene and the azo types. Fluorescein (**12**) is a xanthene dye used as a marine dye marker because of its intense green fluorescence. Resorcinol and catechol are used in the oxidation of fur dyes, eg, for Persian lamb. Eosin, a tetrabromo derivative of fluorescein used for dyeing some drugs and cosmetics. Other small-volume resorcinol derivatives that are used as dye intermediates are α-resorcylic acid (**44**) and β-resorcylic acid (**43**) (153).

Hydroquinone Derivatives. Hydroquinone is used in ointments as a bleaching and depigmenting agent (154). Gentisic acid (**41**) is used in medicinal preparations as an analgesic and antirheumatic. The chlorinated derivative of hydroquinone dimethyl ether: Chloroneb, is a systemic fungicide used in treatment of cotton and bean seeds and of turf (see Fungicides). Hydroquinone is a good antiskimming agent in printing inks (qv). *p*-Benzoquinone is an insolubilizing agent and bacteriostat in the processing of papermakers felt. Hydroquinone is used in highly specialized polyester-imide resins and as a scorch inhibitor in polyether-based flexible polyurethane foams (155). Hydroquinone di-(β-hydroxyethyl) ether (**21**) is used as an extender for molding one-component thermoset urethanes (156) (see Urethane polymers). Nematic liquid crystal compositions contain phenyl benzoate derivatives of hydroquinone and are used in electro-optical devices (157) (see Liquid crystals). Polyester filaments are flame-retarded by incorporating the condensation product of hydroquinone with phenylphosphonic dichloride (157) (see Flame retardants, phosphorus compounds). Polymers of hydroquinone and phosphoric anhydride are electrically insulating varnishes (159) (see Insulation, electric). Redox ion exchangers are based on copolymerization of diphenyloxidehydroquinone and formaldehyde (160). Polycondensation products of hydroquinone and *p*-phenylenediamine have semiconductor properties (161).

Hydroquinone Ethers. Hydroxyanisole (**18**) (162) is used in ointments, as a bleaching and depigmenting agent (163), as heat stabilizer for ir absorbers and methacrylic acid, as a polymerization inhibitor for unsaturated polyesters and methacrylate esters, and as a stabilizer for chloroethylenes, 1,1,1-trichloroethane, and polyethers (see Bleaching agents; Heat stabilizers; Methacrylic acid and derivatives). Hydroxyanisole is *tert*-butylated with *tert*-butyl alcohol in the presence of a silica catalyst to manufacture *tert*-butylhydroxyanisole (**25**) (BHA).

1,4-Dimethoxybenzene is a by-product in the Williamson synthesis; it has antimicrobial activity, and is used as a photoconductor for xerography (see Electrophotography) (164) and as a raw material for the manufacture of Chloroneb (see Fungicides) (165).

Bis-hydroxyethylhydroquinone (**21**) is used as a chain extender in thermosetting urethane polymers (156), as a cross-linking and vulcanizing agent for urethane rubbers (166) and as a curing agent for urethane polymer resins. It is used to improve the dyeability of polyester fibers (167). Polymers with 3,3′-thiodipropionic acid are used as stabilizers for olefinic polymers (168) and, with terephthalic acid, are used as binders for electrophotographic toners (169).

Ring-Substitution Products. Chlorination of hydroquinone yields mixtures of mono-, di-, tri-, and tetrachloro products. 2-Chlorohydroquinone and 2,3,5,6-tetrachlorohydroquinone are used as photographic developers and antioxidants. Tetrachlorohydroquinone is a metabolite of pentachlorophenol and is used as a heat stabilizer for polyester tire cords. Its reaction products with chlorinated phosphorus compounds are fireproofing agents for plastics and polyesters (170).

Hydroquinone is *tert*-butylated with isobutylene or *tert*-butyl alcohol in the presence of a Friedel-Crafts catalyst to produce *tert*-butylhydroquinone (171). The dibutylated derivative in the Friedel-Crafts alkylation of hydroquinone, 2,5-di-*tert*-butylhydroquinone can be used in many instances when hydroquinone is unsuitable; it is an effective polymerization inhibitor and antioxidant.

Kolbe-Schmitt carboxylation of hydroquinone yields gentisic acid (**41**), which is used as an analgesic and an antirheumatic.

Sulfonation of hydroquinone yields hydroquinonesulfonic acid (**10**). Its sodium salt is used in photographic developers for its superadditivity efficiency; its calcium salt is an inflammation inhibitor in circulatory disorder treatment, and its diethylammonium salt (Ethamsylate) is a hemostatic.

Amino Derivatives. The reaction of hydroquinone with amines gives *N*-substituted hydroxyanilines. This process is used in the commercial preparation of *p*-methylaminophenol (**35**) (Metol), which is used as a developer in black-and-white photography, and *N*,*N*′-diphenyl-*p*-phenylenediamine (**36**), which is used as an antioxidant for rubber and food. The condensation product of hydroquinone with aniline has antihemolytic activity, is an inflammation inhibitor, is used for liver protection in carbon tetrachloride poisoning (172) and in teratogenesis prevention (173).

Resorcinol Derivatives. Resorcinol and its esters have antiseptic properties and are used to disinfect solutions for the meat industry and in topically applied medicines for the treatment of seborrheic keratosis, alopecia senilis (174), allergic dermatitis, and dandruff. A carcinogenic or toxic potential that would affect uses of these agents in man has not been detected (175). Hexylresorcinol, which is prepared by reduction of hexanoylresorcinol (**30**), is an anthelmintic and a topical antiseptic. *p*-Aminosalicylic acid, which is prepared by carboxylation of *m*-aminophenol (**33**), is an antitubercular and its sodium and calcium salts and its hydrazide also are used. Mercurochrom, a derivative of fluorescein (**12**), is a well-known antiseptic. Derivatives of *m*-aminophenol are used as the herbicides Karbutilate (165) and Phenmediphan (165) (see Herbicides). Resorcinol is a good coupler for oxidation hair dyes (176) which are neither systemically toxic nor carcinogenic (177). The lead salt of 2,4,6-trinitroresorcinol, styphnic acid, as well as 2-nitroresorcinol are used in heat-sensitive explosive detonators for military purposes (see Explosives). Resorcinol is a corrosion inhibitor for copper, brass, and aluminum-based copper alloys in alkaline media (178) and for aluminium alloys in hydrochloric acid (179). Resorcinol–formaldehyde resins are used to decrease the water permeability of cement and as speciality tanning agents (180). Resorcinol is used in compositions for diazo process multicolor copy (181). Resorcinol and its esters are used to stabilize and plasticize cellulose esters and ethers and as a cross-linking agent for epoxy resins (qv) and unsaturated polyesters. Resins from hydroxyquinoline, ethylenediamine, resorcinol, and formaldehyde are used for tungsten recovery from brine containing as little as 0.07 g/L of WO_3 (182). Heavy metals, such as cadmium, can be collected by RF resins containing nitrohumic acid (183). Polyester filaments are flame-retarded by incorporation of the condensation product of resorcinol with cyclohexylphosphonic dichloride (184) (see Flame retardants).

Ring-Substitution Products. The condensation of resorcinol with hexanoyl chloride yields an ester which undergoes a Fries rearrangement to produce 4-hexanoylresorcinol (**30**) which, by subsequent reduction with zinc amalgam in dilute hydrochloric acid, yields hexylresorcinol (4-hexylresorcinol). The latter compound has anthelminthic activity on ascaris and is used in veterinary compositions (185) (see Veterinary drugs). It has sporicide, bactericide, fungicide, and antiseptic activity and is used in pharmaceutical topical compositions (186). Hexylresorcinol is used with hexachlorophene in bactericidal preparations; hexylresorcinol also has been used as a fungicide for lubricants (187) (see Disinfectants and antiseptics).

2,4-Dihydroxybenzophenone (**31**) (benzoylresorcinol) is manufactured by condensation of benzoyl chloride with resorcinol followed by Fries rearrangement, or by treating resorcinol with benzotrichloride at 50°C (188). Benzoylresorcinol and its alkyl derivatives are used as uv light absorbers in paper, wool (as a yellowing inhibitor), nylon, exposed coatings, paints (qv), vinyl chloride polymers, olefin polymers, and detergents (qv) (see Olefin polymers; Vinyl polymers, vinyl chloride). It is also used as a vulcanization accelerator for urethane rubbers (189) and fluoro rubbers (190), and as a yellow dye-forming coupler for diazo copying compositions (191).

β-Resorcylic acid (**43**) is produced from resorcinol by the Kolbe-Schmitt synthesis; specifications of a commercial product have been published (171). β-Resorcylic acid is a starting material for the production of dyes, photographic chemicals, and cosmetic preparations, and it has antimicrobial and antirheumatic activity (192). Its coupling with diazonium salts results in the formation of azo dyes of deeper and more intense color than those formed from salicylic acid (see Salicylic acid and related compounds). It is used in heat-sensitive copying coatings and in electroplating for suppression of halogen gas evolution.

Styphnic acid (2,4,6-trinitro-1,3-benzenediol) is produced by nitration of resorcinol or by nitrosation and subsequent oxidation by adding nitric acid (193). It is used in propellants (see Explosives and propellants) containing nitrocellulose (194). Its lead salt is used as a nonbrisant primary explosive, as a detonator for explosives and fuses, and in primers for small arms ammunition.

Umbelliferone (**46**) is produced by Pechmann condensation of resorcinol with malic acid. It is used as an optical brightener in plastics, textiles, soap, laundry products, and sunscreen lotions. It also is used as an hypotensive (195) and in brucellosis treatment (196). It exhibits fungistatic, bactericidal, bacteriostatic, and choleretic activity and is used in ointments (197). It forms an exciplex used in laser emission (198) (see Lasers).

β-Methylumbelliferone (**47**) (hymecromone) is produced by Pechmann condensation of resorcinol with ethyl acetoacetate. Specifications for the commercial product have been published (199). β-Methylumbelliferone is useful for masking yellow tints in soaps, detergents, waxes, plastics, fibers, and paper and as a uv screening agent in suntan lotions. It exhibits anticoagulant, anti-inflammatory (200), antimicrobial, and bactericidal activity (201). Like its sodium salt, it shows antispasmodic and choleretic action (202–203). Like umbelliferone, it is used in laser emission to form an exciplex (= excitated complex) (204).

Miscellaneous Derivatives. Resorcinol monoacetate is used in pharmaceutical compositions for its antiseborrheic and keratolytic activity and as an acne remedy. *m*-Aminophenol (**33**) is a dye intermediate (see Aminophenols) and is used in the production of adhesives for bonding nylon, rayon, and polyester cords to rubber. Phenol–formaldehyde adhesives for wood are improved by the addition of *m*-aminophenol (205). It is a corrosion inhibitor for brass and copper and a brightener in the electroplating of brass and lead–tin alloy. It is the starting material for the production of *p*-aminosalicylic acid which is used in tuberculosis treatment and the production of Karbutilate and Phenmediphan.

Catechol Derivatives. Guaiacol (**15**) and guaiacol benzoate, guaiacol carbonate, guaiacol salicylate, potassium guaiacol sulfonate, and guaiacol glyceryl ether (**20**) are used in pharmaceutical preparations as expectorants (see Expectorants and antitussives). Guaiacol (206) is a starting material in the production of vanillin (qv) (**39**) (207). Its homologue, ethylvanillin (**40**), is produced from guethol (**17**). Veratrole (**16**) is used as an antiseptic. Eugenol (**28**), the allyl derivative of guaiacol, is used in perfumery and as a dental analgesic (208). Protocatechuic aldehyde (**37**), which is prepared by the Reimer-Tiemann reaction with catechol, is a raw material for piperonal which finds uses in perfumery and in cherry and vanilla flavors; it also exhibits a pediculicide action (208). Piperonal and its derivatives are used as insecticidal synergists for pyrethrum and rotenone and as an ingredient for insecticidal preparations; the high reactivity of its methylene bridge irreversibly destroys cytochrome C 450 enzyme activity. Safrole is the allyl derivative of methylenedioxybenzene (**22**), is the main constituent of sassafras, and is used in perfumery and as a topical antiseptic. It is a raw material for the manufacture of piperonal.

Two important systemic insecticides can be derived from catechol: Carbofuran (**51**), which is produced in the United States and Propoxur (**52**), which is produced in the FRG (165):

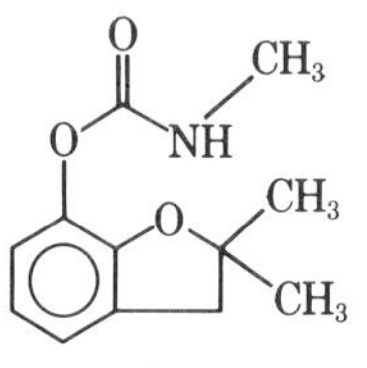

(**51**) 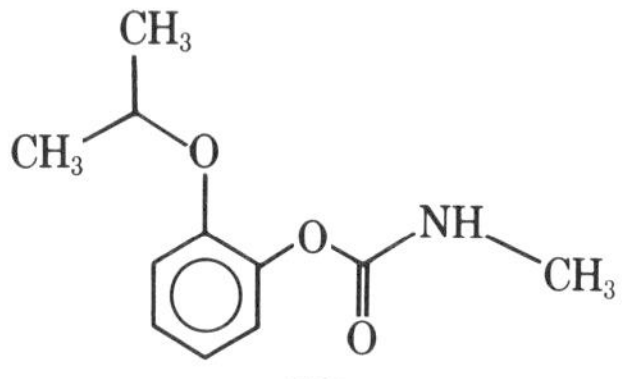 (**52**)

Chloracetylcatechol (**29**) is a starting material in the production of adrenalone (**53**) which is a local anesthetic (208) and is a precursor for adrenaline (**54**), the principal hormone produced by the adrenal medula. L-Adrenaline (209) is a sympathomimetic vasoconstrictor, cardiac stimulant, and bronchodilatator. Its homologues, nordefrine

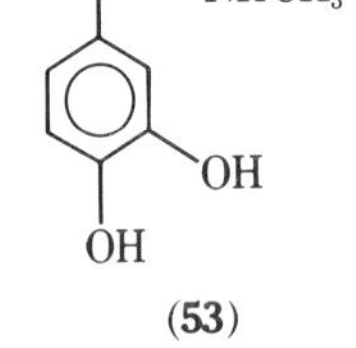

(**53**) 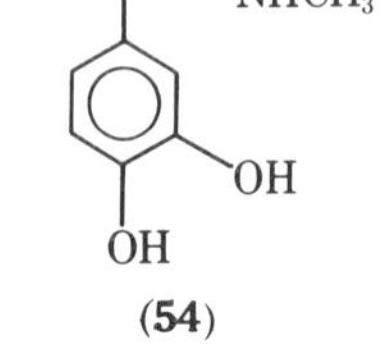 (**54**) 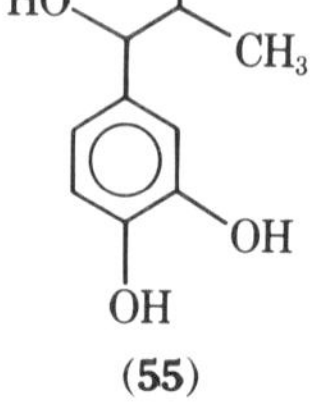 (**55**) 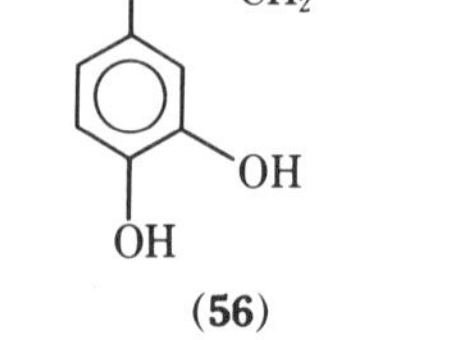 (**56**)

(**55**) (210) and norepinephrine (**56**) (209), are used in medicinal preparations as sympathomimetics and vasoconstrictors (see Epinephrine and norepinephrine). Dopamine (**57**) (211), which is produced from homoveratrylamine, is a derivative of vanillin (**39**) and is used as an adrenergic. Levodopa (**48**), which has anticholinesterase and antiparkinsonian activities, is used in pharmaceutical preparations. Methyldopa (**58**) is an antihypertensive. Papaverine (**59**) (212) and its oxidation product papaveraldine (**60**) are alkaloids that are found in opium and are prepared from veratraldehyde; they are used as smooth-muscle relaxants and cerebral vasodilatators. The ethoxy analogue of papaverine, ethaverine (208), is used as an antispasmodic.

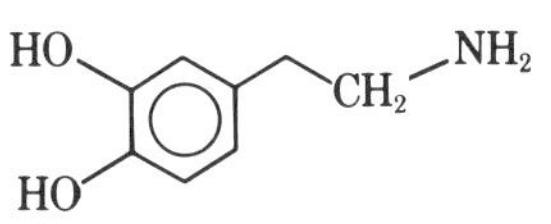

(**57**) 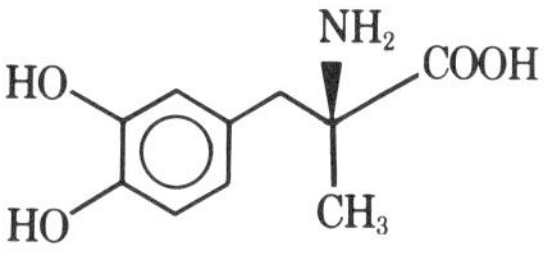 (**58**)

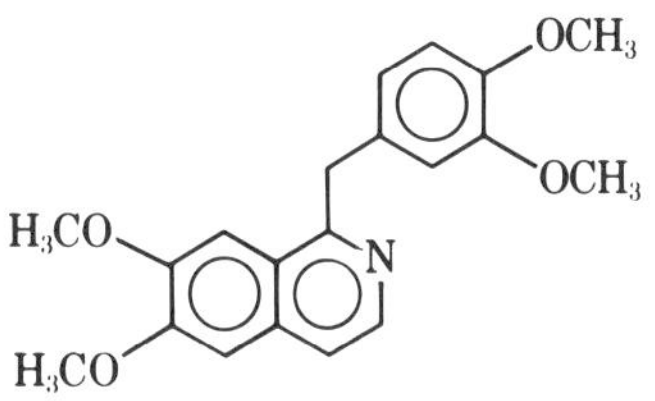

(**59**) 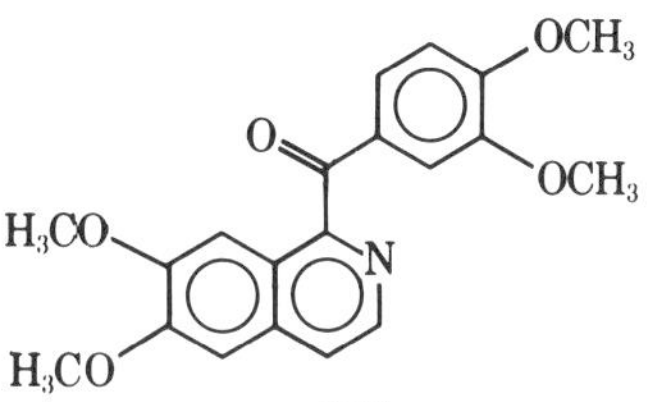 (**60**)

The di-*o*-tolyguanidine salt of dicatechol borate is used as a rubber accelerator and catechol phosphonate is claimed to be a plant growth accelerator (213) (see Plant growth substances). Phosphoric and thiophosphoric esters of catechol are used as fungicides (214) (see Fungicides). Stannous derivatives of catechol are employed as gel catalysts for polyurethane foams (215). Salts of *tert*-butylpyrocatechol (**26**) are used to stabilize poly(vinyl chloride) polymers (216).

Some plant metabolites are allyl or propenyl derivatives of guaiacol (**15**) or methylenedioxybenzene (**22**) (see Allyl compounds). Eugenol is used in perfumery instead of oil of cloves. Isoeugenol occurs in ylang-ylang and other essential oils and is produced by isomerization of eugenol. Safrole and isosafrole occur in oils of sassafras. The trans form of isosafrole has an odor of anise. Other plant metabolites are formyl derivatives of catechol, methylenedioxybenzene, guaiacol, and guethol (**17**). Piperonal (heliotropine) is produced by methylenation of protocatechuic aldehyde. Piperine is an alkaloid that occurs in black pepper (see Alkaloids). Some guaiacol, which is a plant metabolite, also is formulated into proprietary antiskinning and antioxidant additives for paints, varnishes, enamels, and printing inks. However, guaiacol shows the most promise as a vanillin precursor since ligninsulfonate has come under tightened EPA waste-disposal regulations (217). Veratrole (**16**) is used in liquid smoke for meat cold cuts (218) and in electroplating of bright tin (219). After undergoing chloromethylation (220) and subsequent cyanation, it is used as a starting material for the production of homoveratronitrile, homoveratric acid, and homoveratrylamine. Guethol (**17**) is used in liquid smoke for food, and its methylcarbamate has a synergistic action with pyrethroid insecticides (221). It is the principal starting material for ethylvanillin manufacture (222–223).

tert-Butylcatechol (**26**) (TBC), when applied topically, is a depigmenting agent in human and animal skin although the mechanism of this effect remains unclear (224). TBC is used primarily as a stabilizer or polymerization inhibitor for styrene (qv), butadiene (qv), and other reactive monomers. It is an antioxidant for cottonseed and safflower oil,lard, polypropylene glycol polyesters and olefin polymers. It also is a heat stabilizer for polyurethanes.

Structure No.	*Compound*	*CAS Registry No.*
(1)	hydroquinone	*[123-31-9]*
(2)	resorcinol	*[108-46-3]*
(3)	catechol	*[120-80-9]*
(6)	2,5-cyclohexadiene-1,4-dione	*[106-51-4]*
(7)	3,5-cyclohexadiene-1,2-dione	*[583-63-1]*
(8)	hydroxybenzoquinone	*[2474-72-8]*
(9)	*cis,cis*-muconic acid	*[1119-72-8]*
(10)	hydroquinonesulfonic acid	*[88-46-0]*
(11)	quinazarin	*[81-64-1]*
(12)	fluorescein	*[2321-07-5]*
(13)	hystazarin	*[483-35-2]*
(14)	alizarin	*[72-48-0]*
(15)	guaiacol	*[90-05-1]*
(16)	veratrole	*[91-16-7]*
(17)	guethol	*[94-71-3]*
(18)	*p*-hydroxyanisole	*[150-76-5]*
(19)	decyloxyresorcinol	*[25934-35-4]*
(20)	guaiacol glyceryl ether	*[93-14-1]*
(21)	1,4-di(2-hydroxyethoxy)benzene	*[104-38-1]*
(22)	methylenedioxybenzene	*[274-09-9]*
(23)	dibenzo-18-crown-6-polyether	*[14605-55-1]*
(24)	*tert*-butylhydroquinone	*[1948-33-0]*
(25)	2-*tert*-butyl-4-methoxyphenol	*[121-00-6]*
(26)	4-*tert*-butylcatechol	*[98-29-3]*
(27)	4-allylcatechol	*[1126-61-0]*
(28)	eugenol	*[97-53-0]*
(29)	4-chloroacetylcatechol	*[99-40-1]*
(30)	4-hexanoylresorcinol	*[3144-54-5]*
(31)	2,4-dihydroxybenzophenone	*[131-56-6]*
(32)	*p*-aminophenol	*[123-30-8]*
(33)	*m*-aminophenol	*[591-27-5]*
(34)	*o*-aminophenol	*[95-55-6]*
(35)	*p*-methylaminophenol	*[150-75-4]*
(36)	di-*N*-phenyl-*p*-phenylenediamine	*[74-31-7]*
(37)	protocatechuic aldehyde	*[139-85-5]*
(38)	pyrocatechuic aldehyde	*[24677-78-9]*
(39)	vanillin	*[121-33-5]*
(40)	ethylvanillin	*[121-32-4]*
(41)	gentisic acid	*[490-79-9]*
(42)	2,5-dihydroxyterephthalic acid	*[610-92-4]*
(43)	β-resorcylic acid	*[89-86-1]*
(44)	γ-resorcylic acid	*[303-07-1]*

(45)	1,3-cyclohexanedione	[*504-02-9*]
(46)	umbelliferone	[*93-35-6*]
(47)	β-methylumbelliferone	[*62252-23-7*]
(48)	L-dopa	[*59-92-7*]
(49)	β-D-glucopyranoside	[*497-76-7*]
(50)	shikimic acid	[*138-59-0*]
(51)	carbofuran	[*1563-66-2*]
(52)	propoxur	[*114-26-1*]
(53)	adrenalone	[*99-45-6*]
(54)	L-adrenaline	[*51-43-4*]
(55)	nordefrine	[*61-96-1*]
(56)	norepinephrine	[*586-17-4*]
(57)	dopamine	[*51-61-6*]
(58)	methyldopa	[*555-30-6*]
(59)	papaverine	[*58-74-2*]
(60)	veraldine	[*522-57-6*]

BIBLIOGRAPHY

"Hydroquinone" in *ECT* 1st ed., Vol. 7, pp. 755–762, by P. W. Vittum, Eastman Kodak Co.; "Resorcinol" in *ECT* 1st ed., Vol. 11, pp. 711–720, by R. A. V. Raff, Koppers Company, Inc.; "Pyrocatechol" in *ECT* 1st ed., Vol. 11, pp. 307–314, by R. A. V. Raff, Koppers Company, Inc.; "Hydroquinone, Resorcinol, Pyrocatechol" in *ECT* 2nd ed., Vol. 11, pp. 462–492, by R. Raff and B. V. Ettling, Washington State University.

1. A. Hidalgo and C. Otero, *Spectrochim. Acta* **16,** 528 (1960).
2. A. I. Kiss and J. Molnar, *Bull. Soc. Chim. Fr.,* 275 (1949).
3. K. W. F. Kohlrausch and A. Pongratz, *Monatsh. Chem.* **65,** 6 (1935).
4. Y. Mukoyama and T. Tanno, *J. Polym. Sci. Polym. Chem. Ed.* **11,** 3193 (1973).
5. W. H. Walker, A. R. Collett, and C. L. Lazzell, *J. Phys. Chem.* **35,** 3259 (1931).
6. C. N. Reylley and A. J. Barnard, L. Meytes, ed., *Handbook of Analytical Chemistry,* McGraw-Hill, New York, 1963.
7. Kh. R. Rakhimov, *Nauch. Tr. Tashkent Gos. Univ.* **399,** 177 (1970).
8. H. Musso, *Angew. Chem.* **75,** 965 (1963); *Angew. Chem. Int. Ed.* **2,** 723 (1963).
9. T. J. Stone and W. A. Waters, *J. Chem. Soc.,* 1488 (1965).
10. E. Adler, L. Junghahn and U. Lindberg, *Acta Chem. Scand.* **14,** 1261 (1960); E. Adler, *Angew. Chem.* **71,** 580 (1959).
11. H. Musso and H. Doepp, *Chem. Ber.* **100,** 3267 (1967).
12. P. J. Elving and A. F. Krivis, *Anal. Chem.* **30,** 1645 (1958).
13. G. E. Panketh, *J. Appl. Chem.* (*London*) **7,** 512 (1957).
14. N. R. Kar, *J. Indian Chem. Soc.* **12,** 470 (1935).
15. H. Hagemann and K. Imkampe, *Z. Naturforsch.* **27B,** 705 (1972).
16. W. Bonthrone and J. W. Cornforth, *J. Chem. Soc. C,* 1202 (1969).
17. C. J. Pedersen, *Org. Synth.* **52,** 66 (1972).
18. U.S. Pat. 2,616,931 (Nov. 4, 1952), (to Union Oil Products).
19. S. G. Mel'Kanoviiskaya, *Zh. Org. Khim.* **1,** 325 (1965).
20. E. Miller, W. H. Hartung, H. T. Rock, and F. S. Crossley, *J. Am. Chem. Soc.* **60,** 7 (1930).
21. Jpn. Kokai 51 128,934 (Apr. 30, 1975); 53 065,840 (Nov. 22, 1976), (to UBE Industries).
22. Fr. Pat. 767,619 (July 21, 1934), (to Hoffmann La Roche).
23. Jpn. Kokai 76 09,394 (Mar. 26, 1976), (to Ajinomoto Co., Inc.).
24. G. Klein, *Handbuch der Pflanzenanalyse,* Vol. 3, 2nd part, Wien, Austria, 1932.
25. C. J. West, *Tappi Monogr.* (6), (1948).
26. *Eur. Chem. News* **30,** 32 (Mar. 10, 1978).
27. *Chem. Mark. Rep.,* 3, 31 (May 3, 1976).
28. *Chem. Age* **108,** 13 (Apr. 26, 1974).

29. *Chem. Eng.*, 50 (June 9, 1975).
30. *Jpn. Chem.* **18,** 2 (Oct. 13, 1977).
31. *Jpn. Chem. Rev.* **19,** 42 (Mar. 1978).
32. *Chem. Mark. Rep.*, 9 (Apr. 10, 1972).
33. *Chem. Mark. Rep.*, 11 (Oct. 30, 1978).
34. *Jpn. Chem.* **13,** 2 (May 18, 1972).
35. *Eur. Chem. News* **30,** 36 (Jan. 20, 1978).
36. *Chem. Week* **118,** 32 (June 16, 1976).
37. *Chem. Mark. Rep.*, 331 (Apr. 21, 1975).
38. W. Lowenstein-Lom, *Petroleum (London)* **13,** 61 (1950).
39. J. G. Werk, *PB 63959,* U.S. Dept. of Commerce Office of Technical Service, Washington, D.C.
40. Belg. Pat. 765,949 (July 22, 1979), (to Chemopetrol).
41. Ger. Pat. 2,237,808 (Aug. 1, 1972), (to Bayer A.G.).
42. Jpn. Pat. 0,111,030 (Dec. 4, 1974), (to Ube Industries).
43. Brit. Pat. 1,131,530 (May 8, 1965), (to The Dow Chemical Co.).
44. Fr. Pat. 2,071,464 (Dec. 30, 1969); Belg. Pat. 798,995 (May 3, 1972); U.S. Pat. 3,943,179 (Sept. 27, 1973), (to Rhône-Poulenc).
45. U.S. Pat. 3,914,323 (Aug. 18, 1972), P. Maggioni (to Brichima).
46. U.S. Pat. 3,920,756 (Mar. 23, 1973), S. Tahara, S. Nagai, Y. Hayashi, K. Hoshide, S. Kawazoe, K. Harada, and K. Yamoto (to Ube Industry).
47. J. Varagnat, *Ind. Eng. Chem. Product Res. Dev.* **15,** 212 (1976).
48. P. Maggioni and F. Minisci, *Chim. Ind. (Milan)* **59,** 239 (1977).
49. Belg. Pat. 849,860 (Dec. 24, 1975), (to Rhône-Poulenc).
50. H. Bindschedler and co-workers, *Chem. Ind.* **I,** 371 (1878).
51. S. Billbrough, *PB63,* U.S. Dept. of Commerce Office of Technical Service, Washington, D.C., p. 627.
52. E. E. Gilbert, *Ind. Eng. Chem.* **43,** 2022 (1951).
53. F. L. Allen, *Chem. Eng.* **74,** 118 (1967).
54. U.S. Pat. 2,578,823 (Dec. 18, 1951), V. Molinari and H. G. Affholter (to Union Carbide and Carbon Corp.).
55. P. H. Groggins, *Unit Processes in Organic Chemistry,* 5th ed., McGraw-Hill, New York, 1958.
56. E. E. Gilbert, *Sulfonation and Related Reactions,* Wiley-Interscience, New York, 1965.
57. A. P. Shestov and co-workers, *J. Gen. Chem. USSR* **26,** 2235 (1956).
58. Brit. Pat. 679,827 (Jan. 30, 1948), (to Monsanto Chemical Corp.).
59. U.S. Pat. 2,736,754 (Feb. 28, 1956), G. A. Webb (to Koppers Co., Inc.).
60. A. R. Graham, *World Petroleum Congress, Proc. 7th, 1968,* p. 29.
61. J. P. Fortuin and co-workers, *Pet. Ref.* **38,** 189 (June 1959); *World Petroleum Congress, Proc. 5th, New York, 1960,* p. 205.
62. Brit. Pat. 778,014 (June 30, 1954), (to Bataafsche Petroleum Mij.).
63. Brit. Pat. 749,187 (Mar. 6, 1953), (to Bataafsche Petroleum Mij.).
64. F. W. Melpolder and co-workers, *J. Am. Chem. Soc.* **70,** 935 (1948).
65. Brit. Pat. 785,607 (Oct. 30, 1957), (to Distillers Co., Ltd.).
66. J. M. Delderik and co-workers, *Brennstoff Chem.* **40,** 13 (1959).
67. Brit. Pat. 921,557 (Mar. 21, 1963), (to Phenolchemie).
68. Brit. Pat. 727,498 (Apr. 6, 1955), (to Distillers Co., Ltd.).
69. U.S. Pat. 3,360,570 (Dec. 26, 1967), T. Bewley and W. Webster (to Hercules, Inc.).
70. Brit. Pat. 799,963 (Aug. 13, 1958), (to Bataafsche Petroleum Mij.).
71. Brit. Pat. 1,050,093 (Dec. 14, 1966), (to Nauchno-Issledovatelsky Institute).
72. *Chem. Eng.*, 75 (May 16, 1970).
73. Ger. Pat. 2,053,196 (Oct. 29, 1969), (to Signal Chemical Co.).
74. Brit. Pat. 910,735 (Nov. 21, 1962), (to Distillers Co., Ltd.).
75. Brit. Pat. 775,813 (Jan. 12, 1956), (to Distillers Co., Ltd.).
76. W. H. Shearo, L. G. Davy, and H. von Bramer, *Ind. Eng. Chem.* **44,** 1730 (1952).
77. U.S. Pat. 2,148,669 (Feb. 28, 1939), (to Eastman Kodak).
78. U.S. Pat. 1,880,534 (Oct. 4, 1932), (to Eastman Kodak).
79. U.S. Pat. 2,495,521 (Jan. 24, 1950), (to Rhône-Poulenc).
80. Belg. Pat. 770,112 (Jan. 17, 1972), (to Chem Systems).

81. U.S. Pat. 3,462,497 (Aug. 19, 1969); Ger. Pat. 1,808,389 (Oct. 16, 1969), N. P. Greco (to Koppers Co., Inc.).
82. Brit. Pat. 1,188,387 (Apr. 15, 1979), (to British Oxygen).
83. Ger. Pat. 2,749,437 (Nov. 5, 1976), (to Stamicarbon).
84. Belg. Pats. 844,736; 844,739 (July 30, 1975), (to Hoechst).
85. U.S. Pat. 3,420,895 (Jan. 7, 1969), (to Ajinomoto).
86. Ger. Pat. 1,232,974 (Aug. 3, 1967), (to Badische Anilin-und Soda Fabrik).
87. Brit. Pat. 1,215,568 (Dec. 9, 1979), (to Lonza).
88. Ger. Pat. 2,214,971 (Nov. 16, 1972), (to Upjohn Co.).
89. Ger. Pat. 1,235,894 (Apr. 9, 1967), (to Farbenfabriken Bayer).
90. Brit. Pat. 1,436,647 (Aug. 10, 1973), (to Mitsui Toatsu Chemicals).
91. U.S. Pats. 3,859,317 (Jan. 7, 1978); 3,870,731 (Mar. 11, 1975), (to Goodyear).
93. U.S. Pat. 3,676,503 (July 11, 1972), (to Upjohn Co.).
94. St. Penchev, *God. Nauchnoizsled. Inst. Koksoklim. Neftsprecab.* **7,** 261 (1969).
95. Jpn. Kokai 77 102,235 (Feb. 23, 1976), (to Mitsui Toatsu).
96. Ger. Pats. 1,102,171 (June 19, 1956); 1,101,436 (Jan. 21, 1956), K. Duerkes.
97. Belg, Pat. 765,476 (Apr. 9, 1979), (to Union Carbide Corp.).
98. M. Fremery, H. Hoever, and G. Schwarzlose, *Chem. Ing. Technol.* **46,** 635 (1974).
99. M. Ya. Fioshin, *Electrokhimiya* **13,** 381 (1977).
100. *Synthetic Organic Chemicals U.S. Production and Sales, 1974, ITC Publication 776,* U.S. International Trade Commission, Washington, D.C., pp. 44, 103, 208.
101. *Imports of Benzenoid Chemicals and Products, 1974, ITC Publication 762,* U.S. International Trade Commission, Washington, D.C., pp. 20, 26, 85.
102. *Chem. Mark. Rep.,* (Jan. 1, 1979).
103. *U.S. Import for Consumption, IM 146,* U.S. Department of Commerce, Bureau of the Census, Washington, D.C., 1977.
104. *U.S. Exports, FT 410,* U.S. Department of Commerce, Bureau of the Census, Washington, D.C., 1977.
105. *I.A.R.C. Monographs, Evaluation of the Carcinogenic Risk of Chemicals to Man,* Vol. 15, International Agency for Research on Cancer, Lyon, Fr., Aug. 1977, pp. 155–175.
106. *Tecquinol, Hydroquinone Technical Grade, D-104 B,* Eastman Chemical Products, Inc., Kingsport, Tenn.
107. *U.S. Standard, USAS PH4 126,* American National Standard, Washington, D.C., 1962.
108. *The United States Pharmacopeia XX (USP XX–NF XV),* The United States Pharmacopeial Convention, Inc., Rockville, Md., 1980.
109. *Resorcinol, Technical Bulletin CD-2-424,* Koppers Co., Inc., Pittsburgh, Pa.
110. *Catechol, Technical Bulletin CD-0-527,* Koppers Co., Inc., Pittsburgh, Pa.
111. P. Buryan and J. Macak, *J. Chromatogr.* **150,** 246 (1978).
112. B. E. Nadin, V. V. Malynov and M. M. Zelenkov, *Vestsi Akad. Navuk. BSSR* **4,** 110 (1978).
113. T. Seki and V. Kusy, *J. Chromatogr.* **115,** 262 (1975); **57,** 132 (1971).
114. H. Thielemann, *Z. Chem.* **15,** 231 (1975).
115. S. S. Tmofeeva and D. I. Stom, *Zh. Anal. Khim.* **31,** 198 (1976).
116. H. Thielemann, *Mikrochim. Acta* **4,** 696 (1971).
117. M. N. Clifford and J. Wight, *J. Chromatogr.* **86,** 222 (1973).
118. Q. M. Usmani and I. C. Shukla, *Analyst (London)* **102,** 306 (1977).
119. W. Schwartz and C. Jaehrling, *Wiss. Z. Tech. Hochsch. Magdebourg* **17,** 585 (1973).
120. Yu. Yu. Lur'e and Z. V. Vikolaeva, *Zavodsk. Lab.* **31,** 802 (1965).
121. R. Parkash, *Mikrochim. Acta* **2,** 201 (1975); *Microchem. J.* **20,** 193 (1975).
122. M. R. Ashworth and G. Lazik, *Anal. Chim. Anal.* **60,** 464 (1972).
123. J. Mlodecka, *Chem. Anal. (Warsaw)* **10,** 861 (1965).
124. E. N. Zelenina, *Zh. Prikl. Khim.* **44,** 2588 (1971).
125. L. N. Balyatinskaya and T. V. Kurchenko, *Zh. Anal. Khim.* **31,** 957 (1976).
126. E. G. Chikryzova and V. A. Khomenko, *Zh. Anal. Khim.* **30,** 993 (1975).
127. K. Verschueren, *Handbook of Environmental Data on Organic Chemicals,* van Nostrand Reinhold Co., New York, 1977, pp. 168, 383, 562.
128. Nisa, *Dangerous Properties of Industrial Material,* 4th ed., van Nostrand Reinhold Co., 1975, pp. 820, 1070, 1078.

129. F. A. Patty, *Industrial Hygiene and Toxicology,* 2nd ed., Vol. 2, Wiley-Interscience, New York, 1963, p. 1375.
130. *Toxic and Hazardous Substances, U.S. Code of Federal Regulations, Title 29,* U.S. Occupational Safety and Health Administration, Washington, D.C., 1976, Part 1910.1000, p. 29.
131. C. E. Searle, *Chemical Carcinogens ACS Monograph 173,* ACS, Washington, D.C., 1976, pp. 34, 354.
132. Fr. Pat. 2,173,929 (Jan. 3, 1972), S. P. Birkeland.
133. Ger. Pat. 2,264,986 (Oct. 12, 1971), (to Minnesota Mining and Mfg. Co.).
134. Ger. Pat. 2,162,714 (Dec. 21, 1970), (to Agfa Gevaert A.G.).
135. S. Afr. Pat. 75 07,987 (Dec. 23, 1975), (to Dynachem Corp.).
136. U.S. Pat. 4,017,371 (May 5, 1972), C. R. Morgan (to W. R. Grace and Co.).
137. S. Afr. Pat. 75 07,984 (Dec. 23, 1975), (to Dynachem Corp.).
138. V. D. Petrov, *Zh. Nauchn. Prikl. Fotogr. Kinematogr.* **21,** 214 (1976); *Chem. Abstr.* **86,** 36,269j (1977).
139. W. C. Wake, *Developments in Adhesives,* 1st ed., Applied Science Publisher, Ltd., London, Eng., 1978, p. 181.
140. Jpn. Kokai 77 28,579 (Aug. 29, 1975), (to Yokohama Rubber Co., Ltd.).
141. Ger. Pat. 2,636,611 (Aug. 15, 1975), (to Monsanto Co.).
142. J. A. Brydson, *Rubber Chemistry,* Applied Science Publisher, Ltd., London, Eng., 1977, p. 255.
143. U.S. Pat. 3,988,306 (Apr. 28, 1975), N. L. Turner (to Petro Tec Chem. Corp.).
144. U.S. Pat. 3,657,160 (Apr. 18, 1972), A. E. Creplau (to Uniroyal Inc.).
145. U.S. Pat. 3,988,502 (Aug. 19, 1968), K. U. Patel and J. E. Maier (to Minnesota Mining and Mfg. Co.).
146. *Chem. Mark. Rep.* 11 (Nov. 25, 1974).
147. W. O. Lundberg, *Antioxidation and Antioxidants,* Vol. 2, Wiley-Interscience, New York, 1962, p. 985.
148. C. B. Hemming in I. Skeist, ed., *Handbook of Adhesives,* Reinhold Publishing Co., New York, 1962, pp. 306–309, 505–511.
149. R. H. Moult, *Handbook Adhesives,* 2nd ed., Van Nostrand Rheinhold, New York, 1977, p. 417.
150. U.S. Pat. 3,351,482 (Nov. 7, 1967), (to Farben Fabrik Bayer).
151. U.S. Pat. 3,525,753 (Aug. 25, 1970), (to Hickson and Welch).
152. U.S. Pat. 3,344,115 (Sept. 26, 1967), (to Vereinigte Glanzstoff Fabriken).
153. *Colour Index,* Vol. 4, Society of Dyes and Colourists, London, Eng., 1973.
154. B. N. Hemswerth, *J. Soc. Cosmet. Chem.* **24,** 727 (1973).
155. *Plast. Technol.,* 113 (July 1977).
156. *Prod. Eng. (New York),* 35 (June 1976).
157. Jpn. Kokai 75 133,185 (Apr. 9, 1974), (to Matsushita Electric Industrial Co.).
158. Jpn. Kokai 77 68,293 (Dec. 2, 1975), (to Kanebo, Ltd.).
159. Jpn. Kokai 77 03,836 (Apr. 18, 1964), (to Hitachi, Ltd.).
160. E. E. Ergozhin, A. R. Agibaeva, and K. D. Dzhandosova, *Izv. Akad. Nauk, Kaz. SSR Ser. Khim.* **27,** 65 (1977).
161. V. M. Vozzhennikov, V. V. Kopylov, and A. N. Pravednikov, *Vysokomol. Soed. Ser. B* **19,** 417 (1977).
162. Jpn. Kokai 75 47,935 (Aug. 28, 1973), (to Mitsubishi Petrochemical Co.).
163. T. Fitzvatrick, Y. Hori, K. Toda, and S. Kinebuchi, *Biol. Norm. Abnorm. Melanocytes U.S. Japan Sem. 1970; Chem. Abstr.* **76,** 30997 (1972).
164. Fr. Pat. 1,477,139 (Apr. 28, 1965), (to Rank Xerox Ltd.).
165. M. Sittig, *Pesticids Processes Encyclopedia,* Noyes Data Corp., Park Ridge, N.J., 1977.
166. U.S. Pat. 3,892,713 (July 1, 1975), J. Burkus, R G. Lederc, and L. V. Eposito (to Uniroyal Inc.).
167. Jpn. Kokai 70 32,955 (Nov. 18, 1967), (to Toray Industries Inc.).
168. Czech. Pat, 120,854 (July 4, 1963), L. Balaban, D. Risauy, and M. Uhlir.
169. Ger. Pat. 2,253,401 (Dec. 22, 1971), (to Xerox Corp.).
170. Jpn. Kokai 73 43,373 (July 20, 1970), (to Toyobo Co., Ltd.).
171. *Bull. C-8-131 β-Resorcylic Acid,* Koppers Co., Inc., Pittsburgh, Pa.
172. C. G. Crafton and N. R. Di Luzio, *Proc. Soc. Exp. Biol. Med.* **124,** 1321 (1967).
173. D. J. Pizzarello and J. C. Kloss, *Experientia* **23,** 589 (1967).
174. Ger. Pat. 2,620,849 (May 11, 1975), (to DSO "Farmakhim").
175. F. Stenback and P. Shibik, *Toxic. Appl. Pharmacol.* **30,** 7 (1974).

176. Ger. Pat. 2,357,215 (Nov. 16, 1973), (to Henkel Co.).
177. C. Burnett and co-workers, *Food Cosmet. Toxicol.* **13,** 353 (1975).
178. M. N. Desai and V. K. Shah, *Labdev Part B* **8,** 64 (1970); *Chem. Abstr.* **73,** 69246j (1970).
179. N. K. Patel, S. C. Makwana, and K. C. Patel, *J. Indian Chem. Soc.* **51,** 971 (1974).
180. H. W. Jones and W. Windus, *J. Am. Leather Chem. Assoc.* **63,** 452 (1968).
181. Can. Pat. 992,373 (Oct. 26, 1970), (to Mita Industrial Co., Ltd.).
182. *Chem. Eng. (New York),* 71 (Feb. 27, 1978).
183. Jpn. Kokai 77 15,486 (July 29, 1975), (to Kohkoku Chemical Industries Co., Ltd.).
184. U.S. Pat. 4,035,442 (July 12, 1977), W. P. Dunworth (to E. I. du Pont de Nemours & Co., Inc.).
185. R. Hermoso and M. Monteoliva, *Rev. Iber. Parasitol.* **30,** 567 (1970).
186. Fr. Pat. M. 4334 (Mar. 26, 1965), (to Maternite Heureuse).
187. M. S. Rodionova and A. Baigozhin, *Probl. Biol. Povsezhdenii Oheastanii Nater.,* 148, 1972.
188. Brit. Pat. 1,246,958 (Jan. 29, 1970), (to Furukawa Electric Co., Ltd.).
189. Ger. Pat. 2,314,147 (Apr. 11, 1972), (to Goodyear Tire and Rubber Co.).
190. Ger. Pat. 2,433,265 (July 11, 1974), (to Dynamit Nobel A.G.).
191. Czech. Pat. 159,444 (Mar. 16, 1972), V. Chmatal and J. Kroupa.
192. J. Catalan and J. I. Fernandez-Alonso, *Experientia Suppl.* **23,** 177 (1976).
193. Ger. Pat. 1,959,930 (Nov. 29, 1968), (to Ministery of Technology, London).
194. U.S. Pat. 3,335,185 (Aug. 8, 1967), H. W. H. Dykes (to U.S. Dept. of the Army).
195. S. Afr. Pat. 68 01,383 (Nov. 20, 1967), (to Lipha).
196. Fr. Pat. 2,150,208 (Aug. 20, 1971), E. Greib.
197. C. Popescu, Cl. Braileanu, and C. Fica, *Farmacia (Bucharest)* **14,** 711 (1966).
198. N. A. Borisevich, V. V. Grozinskii, N. R. Paltarak. and P. I. Petrovich, *Zh. Prikl. Spektr.* **12,** 926 (1970).
199. *Bull. C-9-111, β-methylumbelliferone,* Koppers Co., Inc.
200. L. Fontaine, M. Grand, D. Molho, and E. Boschetti, *Med. Pharmacol. Exp.* **17,** 497 (1967).
201. S. A. Vichkonova, M. A. Rubinchick, and V. Adgina, *Rastit. Resur.* **9,** 370 (1973).
202. S. Bonfils and G. Madexclaire, *Therapie* **22,** 521 (1967).
203. F. Grossi and S. Spada, *Clin. Ter.* **51,** 41 (1969).
204. U.S. Pat. 3,906,399 (Sept. 16, 1975), A. M. Trozzolo (to Bell Telephone Laboratories, Inc.).
205. R. E. Kreibich, *Adhes. Age* **17,** 26 (1974).
206. *Chem. Mark. Rep.,* 11 (Apr. 9, 1973).
207. *Eur. Chem.,* 297 (July 18, 1978).
208. M. Windholz, ed., *Merck Index,* 9th ed., Merck and Co. Inc., Rahway, N.J., 1976.
209. H. Loewe, *Arzneimittel Forschung* **4,** 583 (1954).
210. Ger. Pat. 639,126 (Nov. 28, 1936), (to I. G. Farben).
211. C. Schoepp and H. Bayeler, *Annalen* **513,** 196 (1934).
212. A. A. Goldberg, *Chem. Prod. Chem. News* **17,** 371 (1954).
213. Fr. Pat. 1,556,007 (Feb. 23, 1967), (to General Anilin and Film Corp.).
214. Fr. Pat. 1,527,848 (Mar. 4, 1966), (to Battelle Development Corp.).
215. U.S. Pat. 4,011,181 (Mar. 4, 1977), B. G. van Leuwen Trumbull and J. J. Pitts (to Olin Corporation).
216. Ger. Pat. 1,261,315 (June 30, 1960), (to Bayer).
217. *Chem. Week,* 33, 34 (Aug. 17, 1977).
218. M. A. Gabriel'yants and L. G. Il'ina, *Sb. Nauch Tr. Zasch. Inst. Sor. Iorg. RSFR* 3, 123, 1969.
219. Fr. Pat. 2,190,940 (June 28, 1972), (to Rhône-Poulenc).
220. Ger. Pat. 2,223,957 (May 31, 1971), J. Dory, L. Feuer, B. Javor, and G. Matolcsy.
221. Jpn. Kokai 73 61,630 (Dec. 6, 1971), (to Dainippon Jochugiku Co., Ltd.).
222. Czech. Pat. 172,112 (Nov. 14, 1974), A. Hora.
223. Jpn. Kokai 78 65,840 (Nov. 22, 1976); Ger. Pat. 2,804,063 (Feb. 7, 1977), (to Ube Industries Ltd.).
224. J. D. Mansu, K. Fukuyama, and G. A. Gellin, *J. Invest. Derm.* **70,** 275 (1978).

J. VARAGNAT
Rhône-Poulenc

HYDROXYBENZALDEHYDES

Hydroxybenzaldehydes are organic compounds of the general formula

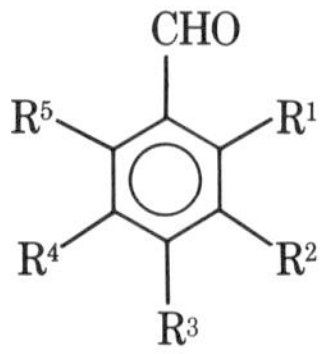

CHO

R^5 R^1

R^4 R^2

R^3

where R^1 through R^5 = H or OH; but at least one R group is OH. All of the nonidentical isomeric mono-, di-, and trihydroxybenzaldehydes are known (see Table 1). Of the higher polyhydroxybenzaldehydes, only 2,3,4,5-tetrahydroxybenzaldehyde has been reported (1).

There are two commercially important hydroxybenzaldehydes, the *o*- and *p*-hydroxy isomers. The more important isomer is the ortho isomer, which is more commonly known as salicylaldehyde. Salicylaldehyde (salicylic aldehyde, *o*-hydroxybenzaldehyde, salicylal) is a clear, straw-colored liquid with a strong almondlike odor. It occurs naturally in oils of spirea plants, bird cherries, and cassia oil (see Benzaldehyde).

Salicylaldehyde and its derivatives are utilized as ingredients in agricultural chemicals, electroplating, perfumes, petroleum chemicals, polymers, and fibers.

Table 1. Physical Properties of Hydroxybenzaldehydes

Chemical Abstracts name	CAS Reg. No.	Common name	mp, °C	bp, °C
2-hydroxybenzaldehyde	[*90-02-8*]	salicylaldehyde	−7	197
3-hydroxybenzaldehyde	[*100-83-4*]	*m*-hydroxybenzaldehyde	108	240
4-hydroxybenzaldehyde	[*123-08-0*]	*p*-hydroxybenzaldehyde	117	sublimes
2,3-dihydroxybenzaldehyde	[*24677-78-9*]	*o*-pyrocatechualdehyde	108	
2,4-dihydroxybenzaldehyde	[*95-01-2*]	β-resorcylic aldehyde	201	220–228 (2.9 kPa)[a]
2,5-dihydroxybenzaldehyde	[*1194-98-5*]	gentisaldehyde	99	
2,6-dihydroxybenzaldehyde	[*387-46-2*]	γ-resorcylic aldehyde	156	
3,4-dihydroxybenzaldehyde	[*139-85-5*]	protocatechualdehyde	154 (dec)	
3,5-dihydroxybenzaldehyde	[*26153-38-8*]	α-resorcylic aldehyde	157	
2,3,4-trihydroxybenzaldehyde	[*2144-08-3*]		162	
2,3,5-trihydroxybenzaldehyde	[*74186-01-9*]		152	
2,3,6-trihydroxybenzaldehyde	[*64168-39-4*]			
2,4,5-trihydroxybenzaldehyde	[*35094-87-2*]		223	
2,4,6-trihydroxybenzaldehyde	[*487-70-7*]	phloroglucinaldehyde		
3,4,5-trihydroxybenzaldehyde	[*13677-79-7*]	gallaldehyde	212	
2,3,4,5-tetrahydroxybenzaldehyde	[*74186-02-0*]			

[a] To convert kPa to mm Hg, multiply by 7.5.

p-Hydroxybenzaldehyde (4-formylphenol) is a colorless–tan solid with a slight, agreeable, aromatic odor. It occurs naturally in some plants in small amounts. *p*-Hydroxybenzaldehyde and its derivatives have commercial application in agricultural chemicals, electroplating, flavors and fragrances, pharmaceuticals, and polymers.

This article deals primarily with salicylaldehyde and *p*-hydroxybenzaldehyde, which represent more than 99% of the hydroxybenzaldehydes market.

Physical Properties

With the exception of salicylaldehyde, all of the hydroxybenzaldehydes are solids at room temperature. The location of the hydroxyl and aldehyde groups ortho to one another in salicylaldehyde results in intramolecular hydrogen bonding, and is the reason for the lower melting point. Solubility in water and other polar solvents generally increases as the number of hydroxyl groups increases. Conversely, solubility in nonpolar solvents generally decreases as the number of hydroxyl groups increases. A summary of a few of the physical properties of the more common hydroxybenzaldehydes is given in Table 1.

Chemical Properties

The effect of the aldehyde group on the phenolic hydroxyl is primarily an increase in its acidity; all three monohydroxybenzaldehydes are stronger acids than phenol. The aldehyde group, however, has little affect on the reactions of the hydroxyl group.

The deactivating effect of the phenolic hydroxyl on the aldehyde group is more pronounced. For example, the monohydroxybenzaldehydes undergo the Cannizzaro reaction much less readily than benzaldehydes, and in the case of the ortho and para isomers, the reaction requires catalysis. The hydroxybenzaldehydes, however, still undergo most of the normal aldehyde reactions such as bisulfite addition, and oxime and hydrazone formation. The reactions of the *o*- and *p*-hydroxybenzaldehydes are outlined below.

Reactions of the Aldehyde Group. ***Oxidation and Reduction.*** Oxidation of hydroxybenzaldehydes can result in the formation of a variety of different compounds, depending on the reagents and conditions used. For example, salicylaldehyde is readily oxidized to salicylic acid by reaction with solutions of permanganate or aqueous silver oxide suspension. However gentisaldehyde (2,5-dihydroxybenzaldehyde) is the product of the oxidation of salicylaldehyde with alkaline potassium persulfate (2). Replacement of the aldehyde function by a hydroxyl group results when *p*-hydroxybenzaldehyde is treated with hydrogen peroxide and sodium hydroxide at 45–50°C (3).

Both salicylaldehyde and *p*-hydroxybenzaldehyde can be reduced by catalytic hydrogenation over palladium or platinum to yield the corresponding hydroxybenzyl alcohols (see Manufacture).

Salicylaldehyde and *p*-hydroxybenzaldehyde undergo a self-oxidation reduction reaction, the Canizzaro reaction, when in the presence of metals such as nickel, cobalt, and silver to yield the corresponding hydroxybenzoic acids and hydroxybenzyl alcohols (4–5).

Reaction with Amines and Amides. Hydroxybenzaldehydes undergo the normal reactions with aliphatic and aromatic primary amines to form imines and Schiff bases, ie, reaction with hydroxylamine gives an oxime, reaction with hydrazines gives hydrazones, and reaction with semicarbazide gives a semicarbazone. The reaction of *p*-hydroxybenzaldehyde with hydroxylamine hydrochloride is a convenient method for the preparation of *p*-cyanophenol (**1**) (6).

$$HO-C_6H_4-CHO + NH_2OH \cdot HCl \xrightarrow{HCOOH} HO-C_6H_4-C\equiv N$$

(**1**)

The reaction product of salicylaldehyde with 1,2-propylenediamine is an excellent chelating agent (**2**), and has wide commercial use in the petroleum industry (see Uses, Petroleum Products) (see Chelating agents).

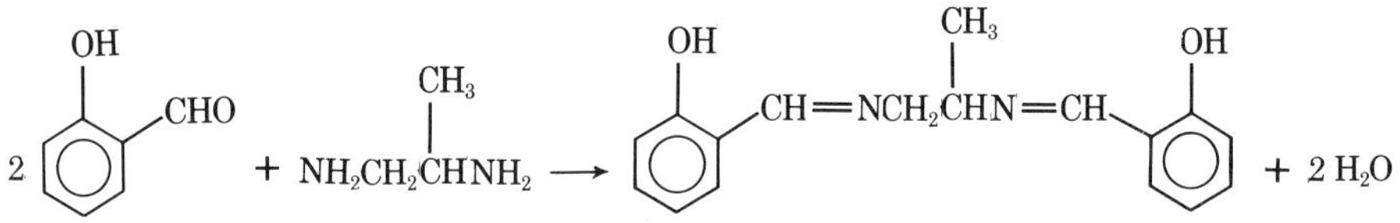

$$2\ (2\text{-}HOC_6H_4CHO) + NH_2CH_2\underset{CH_3}{CH}NH_2 \longrightarrow 2\text{-}HOC_6H_4CH{=}NCH_2\overset{CH_3}{CH}N{=}CHC_6H_4OH\text{-}2 + 2\ H_2O$$

(**2**)

Amides condense with hydroxybenzaldehydes in a manner similar to amines. This reaction is often conducted in the presence of sodium acetate or an organic base such as pyridine. For example, the reaction of salicylaldehyde and propionamide produces salicylidenepropionamide (**3**) (7).

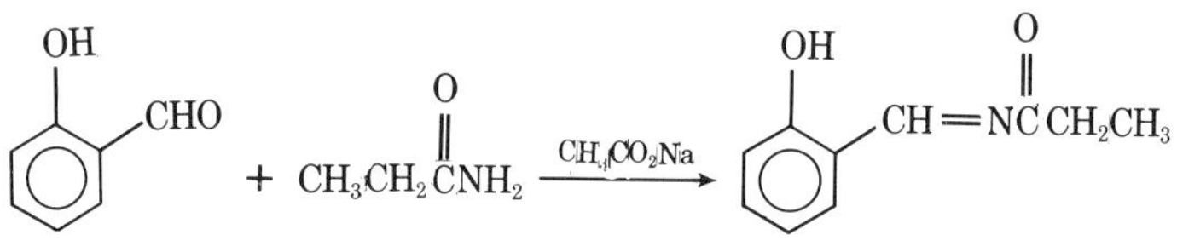

$$2\text{-}HOC_6H_4CHO + CH_3CH_2\overset{O}{\overset{\|}{C}}NH_2 \xrightarrow{CH_3CO_2Na} 2\text{-}HOC_6H_4CH{=}N\overset{O}{\overset{\|}{C}}CH_2CH_3$$

(**3**)

Other Reactions. The reaction of *p*-hydroxybenzaldehyde with sodium cyanide and ammonium chloride, a Strecker synthesis, yields *p*-hydroxyphenylglycine (**4**), a key intermediate in the manufacture of semisynthetic penicillins and cephalosporins (see Uses, Pharmaceuticals).

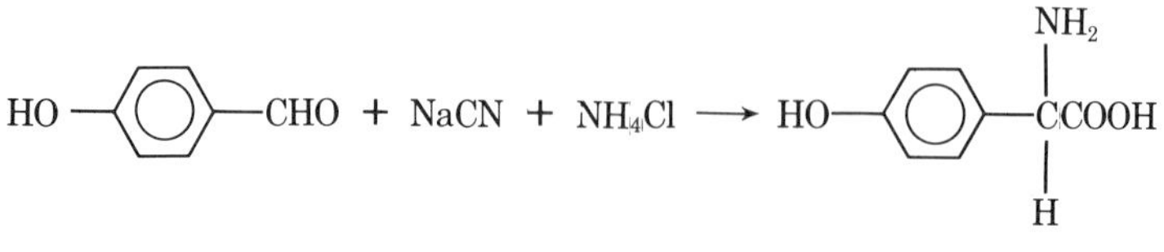

$$HO-C_6H_4-CHO + NaCN + NH_4Cl \longrightarrow HO-C_6H_4-\underset{H}{\overset{NH_2}{C}}COOH$$

(**4**)

Hydroxybenzaldehydes readily react with compounds containing methyl or methylene groups adjacent to one or two carboxyl, carbonyl, nitro-, or similar strong electron-withdrawing groups. The products are usually α-unsaturated compounds. For example, reaction of malonic acid (qv) with salicylaldehyde in the presence of sodium acetate yields 2-hydroxycinnamic acid (**5**).

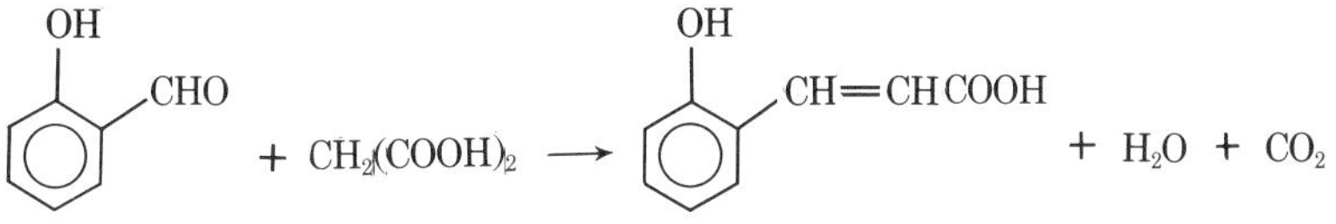

$$2\text{-HOC}_6\text{H}_4\text{CHO} + CH_2(COOH)_2 \longrightarrow 2\text{-HOC}_6\text{H}_4\text{CH{=}CHCOOH} + H_2O + CO_2$$

(**5**)

A product of significant commercial importance, coumarin (qv) (**6**), is made by the reaction of salicylaldehyde with acetic anhydride and sodium acetate—a Perkin reaction .

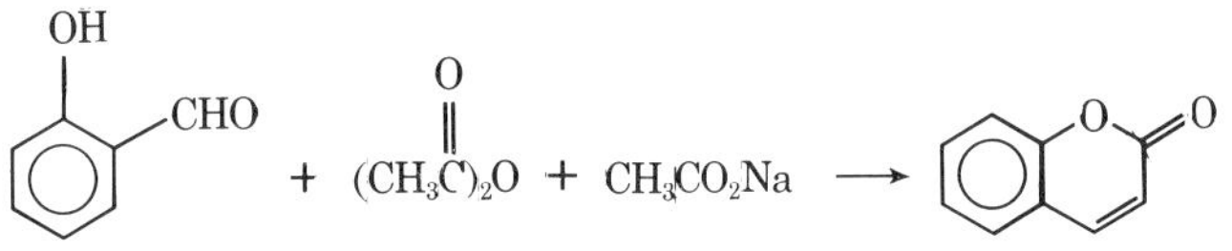

$$2\text{-HOC}_6\text{H}_4\text{CHO} + (CH_3\overset{O}{\overset{\|}{C}})_2O + CH_3CO_2Na \longrightarrow \text{coumarin}$$

(**6**)

Hydroxybenzaldehydes react readily with aldehydes and ketones to form α-unsaturated carbonyl compounds—the Claisen-Schmidt or crossed-aldol condensation.

Reactions of the Hydroxyl Group and the Aromatic Ring. The hydroxyl proton of hydroxybenzaldehydes is acidic and reacts with alkalies to form salts. The sodium, potassium, lithium, and copper salts of salicylaldehyde exist as chelates. The stability constants of numerous salicylaldehyde–metal ion coordination compounds have been measured (8).

Both salicylaldehyde and *p*-hydroxybenzaldehyde are readily converted to the corresponding anisaldehyde (**7**) by reaction with a metal halide or methyl sulfate (9). Other ethers can be made as well by the use of the appropriate halide.

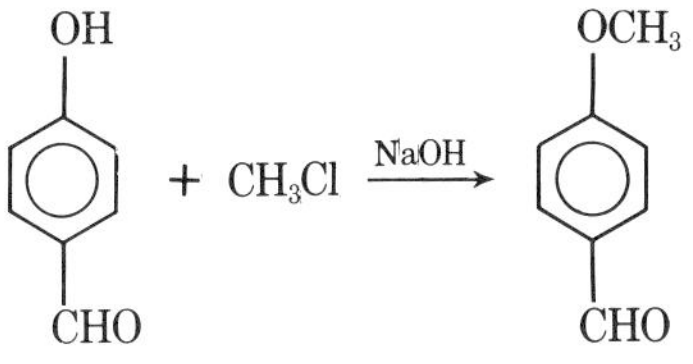

$$4\text{-HOC}_6\text{H}_4\text{CHO} + CH_3Cl \xrightarrow{NaOH} 4\text{-CH}_3\text{OC}_6\text{H}_4\text{CHO}$$

(**7**)

The aromatic ring of hydroxybenzaldehydes participates in several typical aromatic electrophilic substitution reactions. Chlorination and bromination yield mono- and di-halo derivatives depending on reaction conditions.

Manufacture

The two main processes for the manufacture of hydroxybenzaldehydes are both based on phenol. The most widely used process is the saligenin process. Saligenin (*o*-hydroxybenzyl alcohol) (8) and *p*-hydroxybenzyl alcohol are produced from based-catalyzed reaction of formaldehyde with phenol (9). Air oxidation of saligenin over a suitable catalyst such as platinum or palladium produces salicylaldehyde (10).

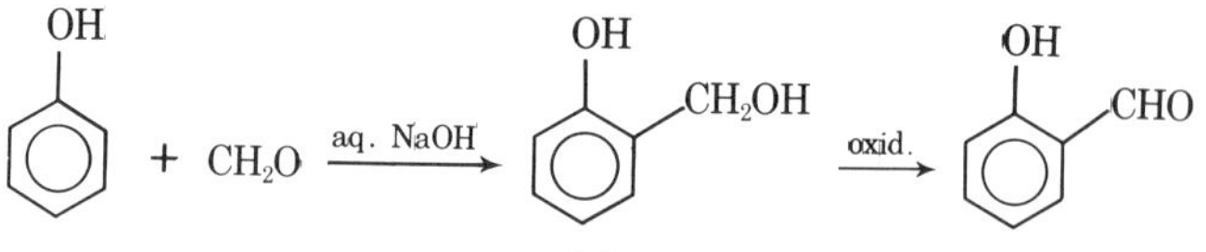

OH; + CH_2O —aq. NaOH→ OH, CH_2OH (8) —oxid.→ OH, CHO

In a refinement of the process, the reaction of phenyl metaborate with formaldehyde followed by catalytic oxidation under atmospheric pressure has been reported to give salicylaldehyde directly from phenol without isolation of any intermediate products (11).

Although *p*-hydroxybenzaldehyde can be made by the saligenin route, it has been made historically by a different route. The Reimer-Tiemann process produces both salicylaldehyde and *p*-hydroxybenzaldehyde in the same reaction (12). Treatment of phenol with aqueous chloroform and aqueous sodium hydroxide results in the formation of benzal chlorides which are rapidly hydrolyzed by the alkaline medium to aldehydes. Acidification of the phenoxide by hydrochloric acid results in the formation of the final products, salicylaldehyde and *p*-hydroxybenzaldehyde. The ratio of ortho to para is flexible and can be controlled within certain limits. The overall reaction scheme is shown below:

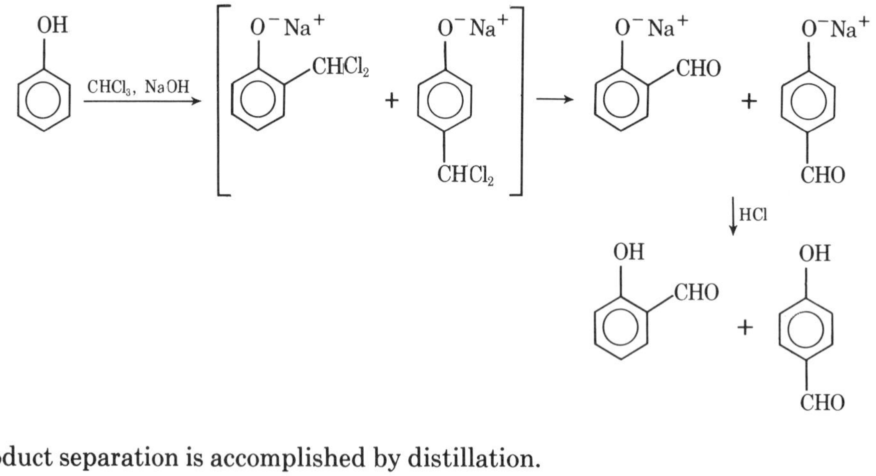

OH —$CHCl_3$, NaOH→ [O^-Na^+, $CHCl_2$ + O^-Na^+, $CHCl_2$] → O^-Na^+, CHO + O^-Na^+, CHO —HCl→ OH, CHO + OH, CHO

Product separation is accomplished by distillation.

Two other routes that have been practiced industrially on a smaller scale are the electrolytic reduction of salicylic acid (13), and the chlorination of esters of *o*-cresol (the carbonate or phosphate) at high temperatures (14) (see Salicylic acid and related compounds).

Salicylaldehyde is available in drums and in bulk quantities. The normal specification is a freezing point minimum of 1.4°C. *p*-Hydroxybenzaldehyde is available in fiber drums, and has a normal specification requirement of a 114°C initial melting point. More refined analytical methods are used where certain uses require more stringent specifications.

Economic Aspects

At present there are two producers of hydroxybenzaldehydes in the United States, Dow Chemical U.S.A. manufactures salicylaldehyde and *p*-hydroxybenzaldehyde for merchant sales. Rhone-Poulenc manufacturers only salicylaldehyde, of which a large portion is used captively in the manufacture of coumarin. The remainder is available for the merchant market. Rhone-Poulenc is also the largest European salicylaldehyde producer.

Worldwide capacity figures for salicylaldehyde are not published; however, the estimated capacity at present is approximately 4000–6000 metric tons per year. The supply–demand picture for salicylaldehyde has been reasonably well balanced recently, as producers have expanded capacity to meet the growing market need. The price of salicylaldehyde was fairly stable at approximately $2.20/kg during the late 1960s and early 1970s; however, prices rose rapidly in the mid-1970s. The current price (1980) of salicylaldehyde is in the $6.00/kg range.

Since at present, Dow Chemical U.S.A. is the only domestic commercial producer of *p*-hydroxybenzaldehyde, published capacity figures are not available, but it is known to be less than 1000 t/yr. Current pricing is in the range of $9.00/kg.

Health and Safety Factors

Salicylaldehyde has a moderate acute oral toxicity; the LD_{50} for rats is 0.3–2.0 g/kg of body weight. *p*-Hydroxybenzaldehyde has a low acute oral toxicity; the LD_{50} for rats is 4.0 g/kg of body weight. Neither material is likely to present a problem from ingestion incidental to its handling and industrial use. It should be recognized, however, that serious effects may result if substantial amounts are swallowed.

Tests performed on rabbits indicate that neither material is absorbed through the skin in toxic amounts. Skin contact with *p*-hydroxybenzaldehyde is essentially nonirritating; however, contact with salicylaldehyde is capable of causing a severe burn, especially when confined to the skin for prolonged or repeated contact. Hence, such contact should be avoided.

p-Hydroxybenzaldehyde is slightly irritating to the eyes and can cause slight transient irritation and slight transient corneal injury. Salicylaldehyde is appreciably irritating to the eyes and may cause pain, irritation, and some corneal injury.

If large amounts of either material are swallowed, vomiting should be induced by tickling the back of the throat with a finger, or by an emetic such as two tablespoonfuls (ca 30 cm^3) of table salt in a glass of water. Medical attention should be obtained promptly.

Protective clothing such as impervious gloves, apron, and shoes should be worn as needed to prevent exposure. If contact with the skin should occur, the material should be removed by washing the exposed area with soap and water. Contaminated clothing and shoes should be removed promptly and not worn again until free of the material. Goggles or safety glasses with side shields should be used to protect the eyes when contact with either material is likely. If the eyes are accidentally contaminated, they should be washed promptly with plenty of flowing water for at least 15 min, and medical attention should be obtained.

Uses

The hydroxybenzaldehydes are primarily used as chemical intermediates to make a variety of products. The largest single use of salicylaldehyde is in the manufacture of coumarin (**6**) (see Chemical Properties). Coumarin, itself, is an important commercial chemical used in soaps, flavors and fragrances, and electroplating (see Coumarin). The other major uses of both salicylaldehyde and *p*-hydroxybenzaldehyde are summarized below.

Agricultural Chemicals. Salicylaldehyde is a valuable intermediate in the manufacture of herbicides (qv) and pesticides. The phenylhydrazones of salicylaldehyde are used to inhibit cereal rusts (15), and as herbicides for a variety of weed species (16), eg, *amaranthus retroflexus*. In addition, both the hydrazone and phenylhydrazone of salicylaldehyde have been found to be very effective antimicrobials (17) (see Disinfectants; Industrial antimicrobial agents). The *N*-methyl- and *N,N*-dimethylcarbamates of salicylaldehyde acetals and mercaptals are very effective insecticides. In fact, the *N*-methylcarbamate of the cyclic acetal of salicylaldehyde (from salicylaldehyde and ethylene glycol) is an important commercial insecticide (18). This product, dioxacarb (**9**), [2(1,3-dioxolan-2-yl)phenyl-*N*-methylcarbamate], is widely used in European and African countries for the protection of potatoes and cocoa (19) (see Insect control technology).

Another important use of salicylaldehyde is in the synthesis of micronutrients. In particular, ferric ion chelates have been used on alkaline or calcerous soils which can occur in citrus or olive groves (see Fertilizers; Mineral nutrients).

p-Hydroxybenzaldehyde has extensive use as an intermediate in the synthesis of a variety of agricultural chemicals. Halogenation of *p*-hydroxybenzaldehyde, followed by conversion to the oxime, and subsequent dehydration results in the formation of a 3,5-dihalo-4-hydroxybenzonitrile. Both the dibromo- and diiodo- compounds are commercially important contact herbicides, bromoxynil (**10**) and ioxynil (**11**), respectively (20). Several hydrazone derivatives have also been shown to be active herbicides (21).

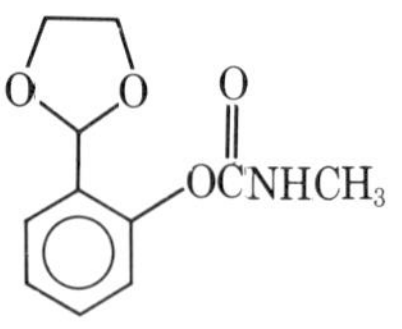

O O O ‖ $OCNHCH_3$

(**9**)

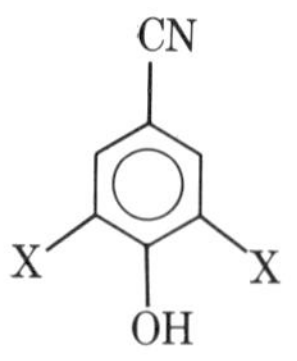

CN X X OH

(**10**) X = Br
(**11**) X = I

Electroplating. Salicylaldehyde is a starting material in the synthesis of coumarin (**6**), which is widely used by the electroplating (qv) industry as a brightener and leveling agent in nickel plating (see Metal surface treatments). The imine resulting from the reaction of salicylaldehyde with an alkanolamine (containing a primary amine group) is an effective brightening agent in the electroplating of zinc on iron and steel (22). In another zinc electroplating process in a noncyanide alkaline bath containing a polyaminesulfone, salicylaldehyde itself was found to aid in producing a bright zinc electroplate on steel (23).

Both *p*-hydroxybenzaldehyde and its methyl ether, *p*-methoxybenzaldehyde (**7**) (*p*-anisaldehyde) have found extensive use in electroplating. The most widespread application has been in alkaline bright zinc plating, both in noncyanide (24) and in cyanide-containing (25) baths. The aldehydes act as both brightening and leveling agents.

Flavors and Fragrances. As mentioned earlier, salicylaldehyde is a starting material in the synthesis of coumarin, which finds extensive use in the soap (qv) and perfume (qv) industries (see Coumarin).

Salicylaldehyde can be used itself as a preservative in essential oils and perfumes (see Oils, essential). The antibacterial activity of salicylaldehyde is strong enough to allow its use at very low concentrations (26).

p-Hydroxybenzaldehyde has an agreeable aromatic odor, but it is not itself a fragrance. It is, however, a very useful intermediate in the synthesis of fragrances. The methyl ether of *p*-hydroxybenzaldehyde, ie, *p*-anisaldehyde, is a commercially important fragrance. *p*-Anisaldehyde can be made in a simple one-step synthesis from *p*-hydroxybenzaldehyde and methyl chloride.

Another important fragrance, 4-(*p*-hydroxyphenyl)butanone, commonly referred to as raspberry ketone, can be prepared from the reaction of *p*-hydroxybenzaldehyde and acetone, followed by reduction (see Flavors and spices).

Petroleum Products. Condensation products of salicylaldehyde and amines are used in various forms for the removal or neutralization of the metallic ions that cause oxidative degradation in petroleum products (qv). The product formed from propylenediamine and salicylaldehyde, ie, *N,N′*-disalicylidene-1,2-propanediamine (**2**), has proven itself to be a commercially important chelating agent, primarily because of its combination of outstanding chelation properties plus its oil solubility (27). Other adducts are used as detergents in lubricating oils and gasoline (28); sludge inhibitors in fuel oils and gasoline (29); and antioxidants to improve the high temperature stability of polyester lubricants (30), gasoline (31), and petroleum oils (32).

Pharmaceuticals. *p*-Hydroxybenzaldehyde is often a convenient intermediate in the manufacture of pharmaceuticals (qv). For example, 2-(*p*-hydroxyphenyl)glycine can be prepared in a two-step synthesis starting with *p*-hydroxybenzaldehyde (33). This amino acid is an important commercial intermediate in the preparation of the semisynthetic penicillin, amoxicillin (**12**) (see Antibiotics, β-lactams). Many cephalosporin-type antibiotics can be made by this route as well (34).

The antiemetic, trimethobenzamide (**13**), is conveniently prepared from *p*-hydroxybenzaldehyde (35) (see Gastrointestinal agents).

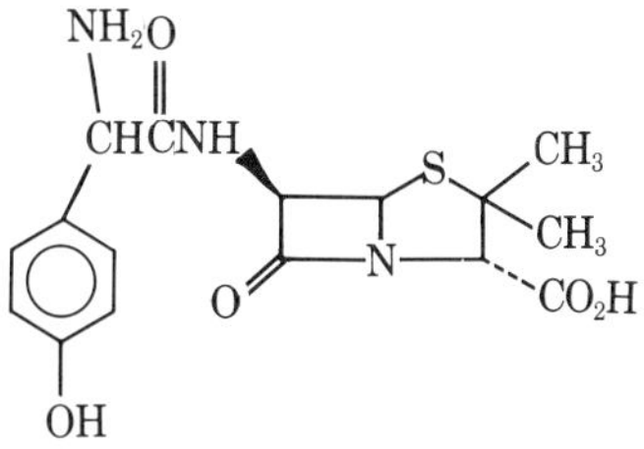

(**12**)

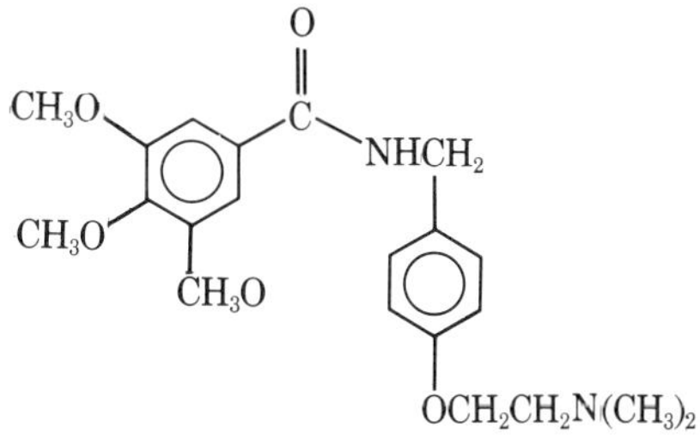

(**13**)

Polymer Applications. The reaction of salicylaldehyde with poly(vinyl alcohol) to form an acetal has been used to provide dye receptor sites on poly(vinyl alcohol) fibers (36) and to improve the light stability of blend fibers from vinyl chloride resin and poly(vinyl alcohol) (37) (see Vinyl polymers).

The metal coordination complexes of both salicylaldehyde phenylhydrazone (38) and salicylaldoxime provide antioxidant (39) protection and uv stability to polyolefins (see Antioxidants and antiozonants; Uv absorbers). In addition, the imines resulting from the reaction of salicylaldehyde and aromatic amines, eg, *p*-aminophenol or α-naphthylamine, can be used at very low levels as heat stabilizers (qv) in polyolefins (40).

The resiliency and dyeability of poly(vinyl alcohol) fibers is improved by a process incorporating *p*-hydroxybenzaldehyde to provide a site for the formation of a stable Mannich base. Hydroxyl groups on the fiber are converted to acetal groups by *p*-hydroxybenzaldehyde. Subsequent reaction with formaldehyde and ammonia or an alkylamine is rapid and forms a very stable Mannich base that is attached to the polymer backbone (41).

Miscellaneous. The reaction products of salicylaldehyde with certain compounds containing active methylene groups, eg, acetylacetone, are excellent uv absorbers. Films containing these compounds can be used as uv filters to protect light-sensitive foods, wood products, paper, dyes, fibers, and plastics (42).

The reaction product of salicylaldehyde and hydroxylamine, i.e., salicylaldoxime, has been found to be effective in photography in the prevention of fogging of silver halide emulsions on copper supports (43). It also forms the basis for an electrolytic facsimile-recording paper (44) and in combination with a cationic polymer, is used in another electrolytic dry-recording process (45) (see Electrophotography; Reprography).

The copper-chelating ability of salicylaldoxime has been used to remove copper from brine in a seawater desalination plant effluent. A carbon-sorbate bed produced by sorption of the oxime on carbon proved to be extremely effective in the continuous process (46). In another application, the chelating ability of salicylaldoxime with iron and copper was used to stabilize bleaching powders containing inorganic peroxide salts (47).

BIBLIOGRAPHY

"Phenolic Aldehydes" in *ECT* 1st ed., Vol. 10, pp. 320–325, by W. R. Brookes, Chemical Division, General Electric Company; "Phenolic Aldehydes" in *ECT* 2nd ed., Vol. 15, pp. 160–165, by Donald B. G. Jaquiss, General Electric Company.

1. F. Wessely and F. Lechner, *Monatsh.* **60,** 159 (1932).
2. W. Baker and N. C. Brown, *J. Chem. Soc.* **151,** 2303 (1948).
3. H. D. Dakin, *Organic Synthesis Col. Vol. I,* Wiley-Interscience, New York, 1941, pp. 149–153.
4. I. A. Pearl, *J. Org. Chem.* **12,** 85 (1947).
5. G. I. Kudryaustiv and E. I. Shilov, *Dokl. Akad. Nauk SSSR* **64,** 73 (1949).
6. T. Van Es, *J. Chem. Soc.* 1564 (1965).
7. K. C. Padya and T. S. Sohdi, *Proc. Ind. Acad. Sci.* **7A,** 361 (1938).
8. D. P. Meller and L. Maley, *Nature* **159,** 370 (1947).
9. K. C. Eapen and L. M. Yeddanapalli, *Makromol. Chem.* **119,** 4 (1968).
10. Ger. Offen. 2,612,844 (Oct. 7, 1976), J. LeLudec (to Rhone-Poulenc S.A.).
11. U.S. Pat. 3,321,526 (May 23, 1967), P. A. R. Marchard and J. B. Grenet (to Rhone-Poulenc S.A.).
12. H. Wynberg, *Chem. Rev.* **60,** 169 (1960).

13. K. S. Udupa, G. S. Subramanian, and H. V. K. Udupa, *Ind. Chemist* **39,** 238 (1963).
14. U.S. Pat. 3,641,158 (March 10, 1969), A. J. Deinet and D. X. Klein (to Tenneco Chemicals, Inc.).
15. U.S. Pat. 2,818,367 (Dec. 31, 1957), E. G. Jaworski and V. R. Gaertner (to Monsanto Chemical Co.).
16. M. Mazza, L. Montanari, and F. Pavanetto, *Farmaco, Ed. Sci.* **31**(5), 334 (1976).
17. M. N. Rotmistrov, G. V. Kulik, E. M. Skrnik, and A. N. Bredikhina, *Mikrobiol. Zh. (Kiev)* **36**(2), 244 (1974).
18. E. F. Nikles, *J. Agr. Food Chem.* **17**(5), 939 (1969).
19. R. P. Ouellette and J. A. King, *Chemical Week Pesticides Register,* McGraw Hill, New York, 1976.
20. U.S. Pat. 3,397,054 (Aug. 13, 1968), R. D. Hart and H. E. Harris (to Schering Corporation).
21. M. Mazza, L. Montanari, and F. Pavanetto, *Farmaco, Ed. Sci.* **31**(5), 334 (1976).
22. Ger. Offen. 1,961,812 (Oct. 1, 1970), R. P. Cope and J. A. Von Pless (to Stauffer Chemical Co.).
23. Ger. Offen. 2,608,644 (Sept. 9, 1976), S. Fujita, K. Murayama, and T. Kaneda (to Japan Metal Finishing Co.).
24. U.S. Pat. 3,871,974 (March 18, 1975), J. R. Duchene and P. J. DeChristopher (to Richardson Chemical Company).
25. Ger. Offen. 1,919,665 (Oct. 15, 1970), S. Acimovic (to Riedel and Company).
26. M. G. deNavarre, ed., *The Chemistry and Manufacture of Cosmetics,* 2nd ed., Vol. III, pp. 85–100, 1975.
27. P. Polss, *Hydrocarbon Process.* 61, (Feb. 1973).
28. U.S. Pat. 3,919,094 (Nov. 11, 1978), S. Schiff (to Phillips Petroleum Co.).
29. Brit. Pat. 1,077,760 (Aug. 2, 1967), (to Mobil Oil Corp.).
30. U.S. Pat. 3,634,248 (Jan. 11, 1972), H. J. Andress (to Mobil Oil Corp.).
31. U.S. 3,399,041 (Aug. 27, 1968), L. J. McCabe (to Mobil Oil Corp.).
32. Ger. (DDR) Pat. 60,835 (March 20, 1968), D. Hoerding.
33. Neth. Appl. 6,607,754 (Dec. 5, 1966), (to Rohm & Haas G.m.b.H.).
34. U.S. Pat. 3,946,003 (March 23, 1976), R. D. Cooper (to Eli Lilly and Company).
35. U.S. Pat. 2,879,293 (March 24, 1959), M. W. Goldberg (to Hoffmann-La Roche, Inc.).
36. Jpn. Pat. 5561 (Aug. 9, 1955), T. Kenichi and co-workers (to Kurashiki Rayon Co.).
37. Jpn. Kokai 73 87,198 (Nov. 16, 1973), M. Furuno and S. Hoshino (to Asahi Dow Ltd.).
38. U.S. Pat. 3,208,968 (Sept. 28, 1965), H. A. Cyba and A. K. Sparks (to Universal Oil Products and Sun Oil Co.).
39. Neth. Appl. 6,614,765 (April 24, 1967), (to Imperial Chemical Industries Ltd.).
40. U.S.S.R. Pat. 253,349 (Sept. 30, 1969) E. N. Matveeva and co-workers (to State Scientific Research Institute of Polymerized Plastics).
41. K. Matsubayashi and K. Tanabe, *Kogyo Kagaku Zasshi* **62,** 1753 (1959).
42. Ger. Pat. 1,087,902 (Aug. 25, 1960), D. Lauerer and M. Pestemer (to Farbenfabriken Bayer Akt.-Ges.).
43. Fr. Demande 2,003,606 (Nov. 7, 1969), T. I. Abbott (to Eastman Kodak Co.).
44. U.S. Pat. 2,864,748 (Dec. 16, 1958), A. H. Mones (to Faximile, Inc.).
45. Jpn. Kokai 73 45,243 (June 28, 1973), Y. Sekine and W. Shimotsuma (to Matsushita Electric Industrial Co., Ltd.).
46. R. H. Moore, *U.S. Office Saline Water, Research and Development Progress Report, 651,* 87 pp., 1971.
47. Ger. Offen. 2,420,009 (Nov. 7, 1974), T. Fujino, M. Yamanaka, and K. Deguchi (to Kao Soap. Co., Ltd.).

RICHARD M. MULLINS
The Dow Chemical Company

HYDROXY CARBOXYLIC ACIDS

Lactic Acid

Lactic acid [*50-21-5*] (2-hydroxypropanoic acid, 2-hydroxypropionic acid), $CH_3CHOHCOOH$, is a naturally occurring organic acid. It is present in many foodstuffs, eg, as a primary acid component in sour milk and a constituent in animal blood and muscle tissue. It is the simplest hydroxy acid and is optically active. The active forms are water-soluble, colorless liquids at ordinary temperatures or are low melting solids in the pure state (1–3). Since they are extremely soluble in both water and water-miscible organic solvents but insoluble in most other organic solvents, the optical isomers of lactic acid are obtained with considerable difficulty, and only as low melting, hygroscopic, and ill-defined solids.

Reference 4 is an excellent treatise on the function of lactic acid in human metabolism. The energy that is necessary for muscular action is normally supplied by the conversion of glycogen, through a complicated series of reactions, into lactic acid. Abnormally high lactic acid content has been observed in human blood in cases of pneumonia, tuberculosis, and heart failure. Lactic acid has been identified in yeast fermentation (qv) (see Yeasts), and is a major component of corn steep liquor which is a by-product of the corn wet-milling industry. It also is found in food products, eg, sauerkraut, pickles, beer, buttermilk, and cheese (3,5). The manufacture and chief uses for lactic acid were developed by United States chemists. All United States and European plants were based on fermentation processes until 1963 when Monsanto synthesized lactic acid from lactonitrile. Two Japanese companies (Musashino and Daicel) synthesize lactic acid by a similar process.

Physical Properties. The properties and chemistry of lactic acid and its derivatives have been reviewed (6). Lactic acid originally was crystallized by fractional distillation of a racemic solution of lactic acid at ca 70–130 Pa (0.5–1 mm Hg) (7). After repeated distillations, pure lactic acid was obtained (mp 18°C). Lactic acid in aqueous solutions readily forms intermolecular esters which complicate the preparation of crystalline lactic acid. Vacuum distillation can remove the excess water at low temperatures and the lactic acid then is distilled from the intermolecular esters. Lactic acid can be crystallized only when it is substantially free of lactoyllactic acid [*26811-96-1*]. Melting point values are found in refs. 8–9.

Both optical isomers of lactic acid occur in nature, but the commercial acid is the optically inactive form (3). Optically active acids have been prepared by direct fermentation under controlled conditions and by the resolution of the racemic mixture. The resolution of lactic acid is unnecessary because microorganisms are available for specific production of the optically pure compound (3). Ref. 3 includes a laboratory method for the preparation of the optical isomers by fermentation with strains of *Lactobacillus delbrueckii* and *Lactobacillus leichmannii*. L(+) lactic acid [*79-33-4*] can be obtained commercially in almost an optically pure condition.

One method used for the resolution of lactic acid depends upon the good solubilities of zinc lactate and zinc ammonium lactate, and the ability of these salts to form supersaturated solutions (10). Resolution also can be achieved with optically active organic bases of which the most effective is morphine, however, strychnine, brucine, quinine, and α-methylbenzylamine also can be used.

Lactic acid has an asymmetric carbon. The following formulas were proposed for the stereo isomers in 1891:

$$\begin{array}{c} COOH \\ | \\ HO—C—H \\ | \\ CH_3 \end{array} \qquad \begin{array}{c} COOH \\ | \\ H—C—OH \\ | \\ CH_3 \end{array}$$

L(+) lactic acid D(−) lactic acid

The acid, frequently known as sacrolactic or paralactic acid, which occurs in the blood, has a (+) rotation but the L configuration and its enantiomorph is D(−) lactic acid [*10326-41-7*]. Optically active lactic acid shows only slight rotation of polarized light. L(+) lactic acid is dextrorotatory, whereas the salts and esters of this isomeric acid generally are levorotatory.

The composition of dilute solutions containing less than about 20% lactic acid corresponds essentially to monomer lactic acid and water. Solutions of higher concentration, however, are more complex because of the self-esterification of the polylactic acids of various chain lengths. The equilibrium composition of aqueous lactic acid, therefore, depends upon the concentration (2,11–12) as shown in Figure 1 (13). The thermodynamic properties of lactic acid are listed in Table 1.

Chemical Properties. Because lactic acid has both hydroxyl and and carboxyl functional groups, it undergoes self-esterification (when it is concentrated by evaporation) and forms linear polyesters (qv). The first esterification product is the dimer, lactoyllactic acid.

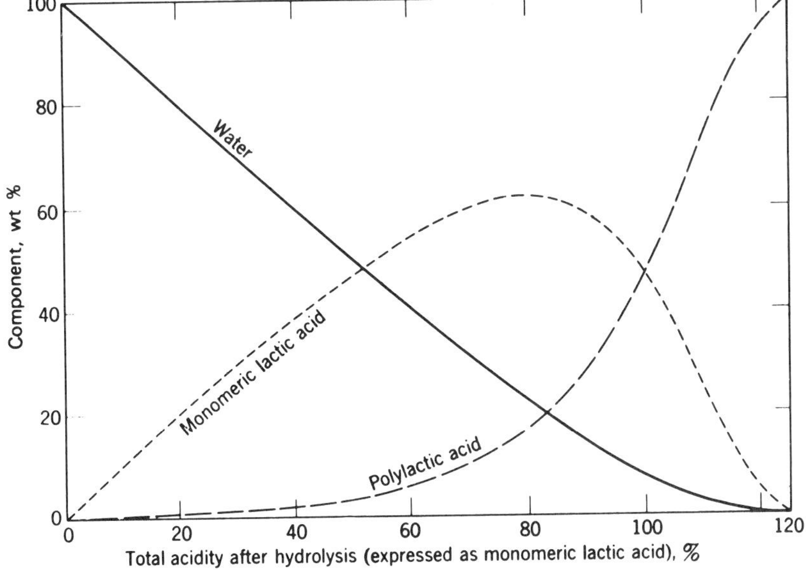

Figure 1. Composition of lactic acid–water systems (13).

Table 1. Thermodynamic Properties of Lactic Acid[a]

Property	Value
dissociation constants[b]	
pK_a (at 25°C)	3.862
K_a (at 25°C)	1.37×10^{-4}
heat of dissociation (ΔH, at 25°C), J/mol[c]	263
free energy of dissociation (ΔF), kJ/mol[c]	20.9
heat of solution (ΔH) at 25°C for L(+), kJ/mol[c]	7.79
heat of dilution (ΔH), kJ/mol[c]	−4.19
heat of fusion (ΔH), kJ/mol[c]	
racemic	11.35
L(+) lactic acid	16.87
free energy (ΔF) of solution, kJ/mol[c]	0
free energy (ΔF) of dilution, kJ/mol[c]	0
free energy (ΔF) of fusion, kJ/mol[c]	0
entropy (ΔS) of solution, J/(mol·K)[c]	26
entropy (ΔS) of dilution, J/(mol·K)[c]	15
entropy (ΔS) of fusion, J/(mol·K)[c] racemic	39

[a] Ref. 6 (most values are estimated).
[b] The acid probably contained lactoyllactic acid.
[c] To convert J to cal, divide by 4.184.

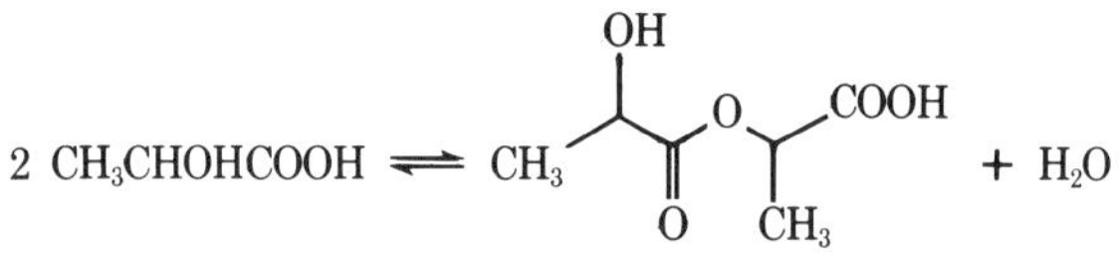

$$2\ CH_3CHOHCOOH \rightleftharpoons CH_3CH(OH)C(=O)OCH(CH_3)COOH + H_2O$$

However, as the concentration increases, higher linear esters, ie, polylactic acids (anhydrides) or trimeric, tetrameric, and polymeric lactic acids, are formed (11–13). The following formula shows the general structure of the polyacids:

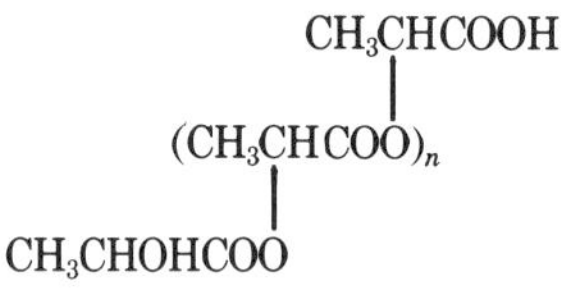

$$CH_3CHOHCOO\text{—}(CH_3CHCOO)_n\text{—}CH_3CHCOOH$$

Equilibrium among the several components is attained slowly and may take several weeks at room temperature. The linear polymers predominate but cyclic polymers also may form. The cyclic dimer, lactide [*95-96-5*] (3,6-dimethyl-*p*-dioxane-2,5-dione), can be formed in good yield:

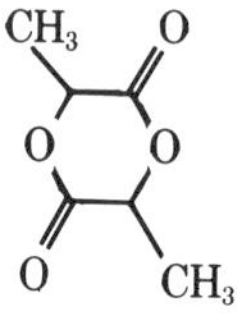

[Structure: lactide, six-membered ring with two ring O, two C=O, and two CH_3 substituents]

When lactic acid is pyrolyzed it decomposes to acetaldehyde (qv), carbon monoxide, and water. At low temperatures, and upon heating of lactic acid in the presence of catalytic amounts of mineral acids, polylactic acid is formed; whereas, with larger amounts of mineral acid, decomposition into acetaldehyde, carbon monoxide, formic acid, and water occurs (see Formic acid and derivatives). Pyrolysis of the esters or acyl derivatives of lactic acid results in the formation of acetaldehyde and carbon monoxide:

$$CH_3CHOHCOOR \rightarrow CH_3CHO + CO + ROH$$

$$CH_3CH(OOCCH_3)COOH \rightarrow CH_3CHO + CO + CH_3COOH$$

Under mild conditions and, particularly in the presence of a mineral acid catalyst, the lactic esters undergo self-alcoholysis to produce esters of polylactic acid [*26100-51-6*] (14):

$$n\ CH_3CHOHCOOR \xrightleftharpoons{H^+} HO[CH(CH_3)COO]_nR + (n-1)ROH$$

In these reactions the lactic monoesters exhibit a behavior similar to that of lactic acid (see Esters, organic).

The diesters of lactic acid, however, show a completely different behavior when subjected to pyrolysis; acrylic esters and acetic acid are formed from acetylated lactic esters (15–16):

$$CH_3CH(OOCCH_3)COOR \xrightarrow{550°C} CH_2{=}CHCOOR + CH_3COOH$$

The yield of acrylic ester is markedly dependent on the structure of the alcohol radical R. When R is a thermally stable group that is not capable of forming an olefin, the yield of acrylic ester is high (see Acrylic acid and derivatives; Acrylic ester polymers).

Lactic acid shows the typical reactions of organic acids; many salts have been reported and generally they are water-soluble. Lactic acid readily undergoes esterification with many alcohols, and numerous lactic esters have been prepared by acid-catalyzed esterification. Esterification usually takes place under anhydrous conditions but produces both lactic and polylactic esters. The production of the latter can be minimized by using an excess of alcohol. The acid catalyst has to be neutralized with a base, eg, sodium acetate, and preferably under reduced pressure, in order to prevent self-alcoholysis of the lactic ester during distillation. The by-products are esters of polylactic acid which may be recycled to produce lactic acid; therefore, the final yield of the lactic ester usually is high. The equilibrium constant for the esterification of lactic acid with ethyl alcohol is 2.71 ± 0.06.

Lactic esters hydrolyze more readily than esters of fatty acids, such as acetic acid, but less readily than esters of glycolic acid. The lactates of secondary alcohols are considerably more resistant to hydrolysis than the esters of corresponding primary alcohols (2). Lactic esters and lactide readily undergo alcoholysis with various alcohols. Alcoholysis of lactide in the presence of a trace of catalyst produces esters of lactoyllactic acid (19), whereas larger amounts of catalyst yield the monomeric lactic ester. An excellent method for preparing lactic esters is the alcoholysis of methyl lactate with a higher alcohol.

Lactamide [*2043-43-8*], $CH_3CHOHCONH_2$, and substituted lactamides can be prepared readily by ammonolysis or aminolysis of methyl lactate or by the dehydration of amine or ammonia salts of lactic acid.

Lactic acid also shows many of the reactions characteristic of alcohols. Thus, it can be esterified with organic acids, anhydrides, and acid chlorides. It can be alkylated with alkylating agents such as diazomethane and dimethyl sulfate. Lactic acid and its esters can be converted into chloroformates with phosgene (qv), into carbamates with cyanic acid, and into allophanates and urethanes (see Urethane polymers) with isocyanates (see Isocyanates, organic). Lactic esters undergo oxidation to the corresponding pyruvic esters and can be converted into esters of inorganic acids by treatment with phosphorus oxychloride or thionyl chloride.

In addition to self-esterification, lactic acid undergoes reactions involving simultaneous participation of the hydroxyl and carboxyl groups. When treated with aldehydes and ketones, a cyclic acetal, a 2-substituted-5-methyl-1,3-dioxol-4-one is formed. Alkyl lactates react with urea to give 5-methyl-oxazolidine-2,4-dione, which also is prepared by the interaction of lactamide and methyl carbonate.

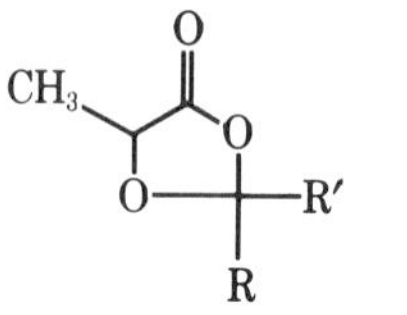

2-substituted-5-methyl-1,3-dioxol-4-one, R′ = H, alkyl, or aryl

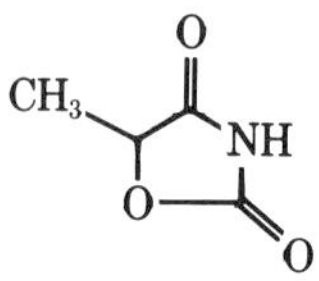

5-methyoxazolidine-2,4-dione

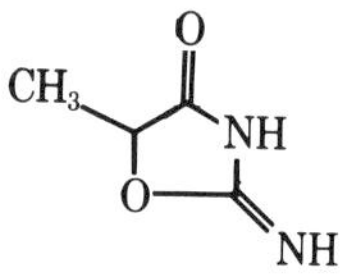

2-imino-5-methyloxazolidin-4-one

When an alkyl lactate is treated with guanidine, 2-imino-5-methyl-oxazolidin-4-one is formed.

Manufacture. Lactic acid can be manufactured either by fermentation or by synthesis; both methods are used commercially. In the United States, more than 85% of the lactic acid that is sold is made by a synthetic route, and all of the lactic acid that is produced in Japan is synthetic. Lactic acid from European manufacturers, whose combined capacity is approximately half of the estimated world capacity, is produced by fermentation processes. The world capacity is estimated at 28,000–30,000 metric tons per year.

Fermentation. There are two types of microorganisms that produce lactic acid. Heterolactic fermentation organisms produce some lactic acid as well as other fermentation products, ie, carbon dioxide, ethyl alcohol, and acetic acid. These organisms are of little use for making lactic acid industrially. Commercial lactic acid production is by homolactic fermentation organisms which form lactic acid exclusively or predominantely from carbohydrates (qv). Homolactic bacteria, such as *Lactobacillus delbrueckii*, *L. bulgarcius*, and *L. leichmanii* are used. The following materials have been used as substrates in this fermentation: products containing starch (qv), ie, corn starch, potato starch, or rice starch; whey; cane sugar, beet sugar, or molasses of beet sugar (see Sugar); and sulfite lyes. Ammonium salts serve as nitrogen sources. The operating margin in fermentation is small, and the cost of a relatively pure substrate versus an impure substrate must be weighed against the cost of refining the acid to market specifications.

The procedure involves fermentation of a carbohydrate with suitable mineral and proteinaceous nutrients in the presence of an excess of calcium carbonate. A thermophilic strain of the *L. delbrueckii* type, which exhibits its optimum activity at about 50°C and at a pH of 5.0–5.5, is preferred. Such a fermentation system eliminates most contamination problems and permits the use of a growing medium which only needs to be pasturized (not autoclaved) under pressure for complete sterilization, as is normally required for most mesophilic fermentations. This procedure converts all of the lactic acid to calcium lactate. Magnesium hydroxide, limestone, and other nutrients and ingredients that may have been in the water are removed by filtration. The filtrate is acidified with sulfuric acid to regenerate the lactic acid and to precipitate the calcium as calcium sulfate. The resulting filtrate consists of about a 10% solution of crude lactic acid, which is concentrated to >50% and then further refined. The refining processes include treatment with activated vegetable carbon to remove organic impurities, treatment with sodium ferrocyanide to remove heavy metals, and filtration to remove impurities that have coagulated during the concentration. The solution is filtered and is passed through ion-exchange resins to remove the last traces of contamination (see Ion exchange; Resins, water-soluble).

Synthesis. The synthesis of lactic acid is based on lactonitrile, which is a by-product from the acrylonitrile (qv) synthesis. The lactic acid synthesis reaction was discovered in 1863 by Wislicenus who prepared lactonitrile from acetaldehyde and hydrogen cyanide and hydrolyzed it to lactic acid:

$$CH_3CHO + HCN \rightarrow CH_3CHOHCN$$

$$CH_3CHOHCN + 2\ H_2O + HCl \rightarrow CH_3CHOHCOOH + NH_4Cl$$

The same reactions are used today: the lactic acid is isolated and purified by esterification with methyl alcohol, and the resulting methyl lactate is purified by distillation (18–19). Once the methyl ester has been produced and distilled, it can be hydrolyzed with a strong acid catalyst to produce a semirefined lactic acid. Purification is achieved by a combination of steaming, carbon treatment, and ion exchange. The synthetic acid is produced in three grades: technical, food, and USP, and in two concentrations: 50 and 88% (see Fine chemicals). Synthetic lactic acid is water-white and has excellent heat stability.

Other possible chemical syntheses of lactic acid include the degradation of sugars (preferably sucrose with sodium hydroxide); interaction of acetaldehyde, carbon monoxide, and water at elevated temperatures and pressures (3,20); hydroformylation of vinyl acetate, oxidation, and hydrolysis (21); and hydrolysis of a chloropropionic acid (prepared by the chlorination of propionic acid). A lactic acid synthesis, based on the nitric acid oxidation of propylene and which yields a by-product of oxalic acid, has been realized but has not been developed on an industrial scale (22).

Another method for synthesizing lactic acid is based on the continuous hydrolysis of lactonitrile in an aqueous sulfuric acid solution (22). The resulting lactic acid is extracted with isopropyl ether and is recovered by reextraction with water. The aqueous extract is concentrated under reduced pressure 33.3 kPa (250 mm Hg) to the desired concentration.

Economic Aspects. Since 1963 there have been only two producers of lactic acid: Clinton Corn Products (a division of Standard Brands) and Monsanto. The production of lactic acid has gradually increased to a level of 17.6–81.7 metric tons in 1977 from 8.82 t in 1963. The reason for this upsurge may be accounted for by three new applications for lactic acid: the use of stearoyl lactylates as emulsifiers and dough conditioners in the baking industry (24–30) (see Bakery processes and leavening agents), the use of sodium lactate for intravenous feeding, and the use of synthetically produced in place of natural fermentation lactic acid for the souring and curdling of milk or cream in the production of cheese and other dairy products (31) (see Milk products).

Lactic acid is imported from the UK, The Netherlands, and the FRG into the United States because fermentation lactic acid is inexpensive relative to synthetic lactic acid. The import duty on lactic acid prior to 1968 was 16% *ad valorem* and the duty declined to 8% in 1972. The Commerce Commission is reviewing the duty and may reduce it further. Therefore, imports have increased because they have gained sales where customer specifications are flexible.

The price of lactic acid remained relatively stable from 1950 to the 1973–1974 energy crisis. Since 1973 the price has inflated just as feedstocks (qv) and energy. The average price of 88 wt % lactic acid from 1963 to 1979 is shown in Table 2.

Analysis. The standard method for the determination of lactic acid is described in ref. 32. Lactic acid is oxidized, under carefully controlled conditions, with potassium permanganate to acetaldehyde, which is absorbed in sodium bisulfite and titrated iodometrically. A quantitative determination can be made by acid–base titration.

Gas and liquid chromatographic methods have been developed to analyze aqueous lactic acid solutions; Porapak N and Tenax GC are excellent column packings for gas chromatography. Because of the self-esterification of lactic acid with resultant formation of the polylactic acids, these methods are unsatisfactory for quantitative analysis. However, they are very useful in detecting levels of impurities in lactic acid solutions.

Uses. In the United States, the primary uses of lactic acid are in foodstuffs and in pharmaceutical products; a relatively small volume is used in industrial applications (see Food additives; Pharmaceuticals). The main use for food-grade synthetic lactic acid is in the manufacture of calcium and sodium stearoyl-2-lactylates for the baking industry. Only the synthetically produced lactic acid has the water-white clarity and high heat stability needed to make these and other food emulsifiers. Calcium stearoyl-2-lactylate (Verv) increases the mixing tolerance of doughs (33) and is approved in the Bread Standard of Identity. It simplifies processing and improves the quality of the yeast-leavened bakery goods. This calcium salt has very little emulsifying ability in water-oil systems. However there is little need for emulsifying properties for bread and the calcium salt exhibits good dough conditioning which is not observed with other acyl lactylates. However, for a high fat, yeast-leavened baked product, the sodium stearoyl-2-lactylate (Emplex) is used because it is both a dough conditioner and an emulsifier.

The lactylated fatty acid esters of mono- and diglycerides are used in prepared cake mixes and other bakery products and in liquid shortenings (34–38). In prepared cake mixes, glyceryl lactopalmitate improve cake texture, and glyceryl lactostearate increases cake volume and permit mixing tolerances. Studies on the metabolism (39) and composition (40) of glyceryl lactopalmitate have been published.

As a food acidulant, lactic acid has a mild acidic taste in contrast with the sharp taste of some of the other food acids. It is relatively nonvolatile and practically odorless. It is a relatively strong acid, however, and is a good preservative (when used in a brine

Table 2. Average Prices of 88 Wt % Lactic Acid, ¢/kg

Grade	1963	1965	1970	1975	1977	1978	1979
food	73.4	73.7	73.7	134.2	134.2	145.2	157.3
technical	63.8	60.5	60.5	128.7	128.7	139.7	151.8

with salt and water) or pickling agent (when used with vinegar) for sauerkraut, pickles, olives, and similar acid-preserved foods. It also is distributed in liquid form and, thus, is ready for use. One or more of these properties have contributed to the increasing use of food-grade lactic acid in animal food and remedies (see Pet and other livestock feeds), bakery products, candy, flavoring extracts, liquid pectin, mincemeat, soft drinks, soups, and sherberts. Lactic acid is used widely to adjust pH in diary products (including pan- or spray-dried egg whites), beer (qv), jams, and jellies. In mayonnaise lactic acid can be used to arrest bacterial spoilage (41). Poultry that is treated with lactic acid shows less tendency to dehydrate in storage (42). Lactic acid also is used in combination with other food acids to improve flavor in carbonated fruit-juice drinks (43) (see Fruit juices). It also is reported to be excellent in acidifying fruit juice for the production of wine (qv).

Lactic acid occurs naturally in many food ingredients. Its salts are soluble and, in many cases, they partly can replace the acid. Lactic acid and its sodium and calcium salts are completely nontoxic and are classified GRAS (generally recognized as safe) for general-purpose food additives by the FDA; similar decisions have been made in other countries.

Technical-grade lactic acid has long been in use in the leather (qv) tanning industry as an acidulant for deliming hides, and in vegetable tanning. It also is the raw material used in the preparation of esters such as methyl, ethyl, and *n*-butyl lactates. Technical lactic acid is used in acid dyeing of wool (qv) and other textiles (see Textiles). Other nonfood uses include applications in adhesives (qv), cleaning and polishing formulations (see Polishes), electroplating and electropolishing, insecticides (see Insect control technology) and fungicides (qv), lithographic developers (see Printing processes), plastics and resins, special inks, textiles, and the treatment of oil and water wells. In the manufacture of phenol–formaldehyde resins, lactic acid neutralizes the alkali catalyst; a noncrystallizable salt is formed which does not impair the clarity or strength of the resin. A high-purity lactic acid, essentially free from iron, is required in this application.

Applications of lactic acid as an extender for glycerol in alkyd resins (qv) is described in the literature (44). The preparation of coatings by the reaction of polylactic acid with drying oils or with small amounts (1–6%) of salts of polyvalent metals, such as aluminum, chromium, cobalt, copper, iron, lead, manganese, thorium, tin, titanium, and zinc, has been reported (45) (see Driers). The addition of dicyclopentadiene to lactic acid (46) results in a dihydrodicyclopentadienyl ether–ester of lactic acid. This ester rapidly absorbs oxygen from the air, especially in the presence of driers, to give an insoluble varnishlike film. This ether–ester also can be polymerized by heating with peroxides to a viscous autoxidizable oil (see Cyclopentadiene and dicyclopentadiene; Peroxides and peroxy compounds, organic).

Many high boiling lactic esters have been evaluated as plasticizers for resins (8,47–49). Mixed esters of lactic and adipic acids of the type $ROOC(CH_2)_4COO$-$CH(CH_3)COOR$ and $(ROCOCH(CH_3)OCOCH_2CH_2)_2$ are efficient plastizers for vinyl chloride copolymers (48) (see Vinyl polymers, poly(vinyl chloride)). Lactic acid esters also are applied as lacquer solvents.

Lactic acid is preferred as the pH controller during cellophane-coating processes. In textiles the finishing or brightening of silk (qv) and rayon (qv) involves passing the yarn through a weakly acidic bath. Lactic acid is widely used to control pH because it helps improve yarn brightening without harming the fiber strength and elasticity (see Hydrogen-ion activity). Lactic acid is used in the chrome–mordant process of wool dyeing because it reduces chromate content in the wool and prevents oxidation of the fibers (see Fibers, vegetable; Dyes, application and evaluation).

Lactic acid has been used as a flux promoter in the preparation of cellulose acetate reverse-osmosis (qv) membranes (50), and it is used in wood treatments (51), waterproofing cement, and electroplating (qv). It reacts with maleic anhydride to form carboxymethoxysuccinic acid-type detergent builders (see Maleic anhydride, maleic acid, and fumaric acid.) Lactic acid is used in an acid formulation to make an electrolyte for the preparation of surfaces for offset printing plates. The formulation yields a metal surface that has the desired degree of graininess and good printing properties (52). It also is used in a formulation to produce a water-soluble antacid of magnesium and aluminum salts (53).

Derivatives. ***Calcium and Sodium Lactates.*** Calcium and sodium lactates are the most important and the most widely used lactic acid salts. They can be prepared as the primary fermentation products by controlling the pH during fermentation with either calcium or sodium carbonates. The salts also can be prepared by the reaction of the free acid with metal carbonates or hydroxides.

Calcium lactate, $(CH_3CHOHCOO)_2Ca.5H_2O$, crystallizes as the colorless pentahydrate, which is only slightly efflorescent and loses water of hydration at 100–120°C. The hydrate has a solubility in 100 g H_2O of 3.1 g at 0°C, 5.4 g at 15°C, and 7.9 g at 30°C, but it is very soluble in hot water. Calcium lactate is used in calcium therapy and has been employed as a blood coagulant in the treatment of hemorrhages and is administered prior to dental operations to inhibit bleeding (see Blood). It has numerous food applications, eg, to firm apples during processing and to inhibit or suppress discoloration of many fruits and vegetables. It is an ingredient in baking powder and improves properties of dry milk powders, sweetened condensed milk, and baked food products. It also is used to dry cheese curd for Swiss cheese. Calcium lactate is used as a gelatinizing agent in alginate-based puddings and pie fillings and is used to accelerate curing of meat products.

Sodium lactate, $CH_3CHOHCOONa$, is very hygroscopic and is sold in aqueous solution as 50% technical grade or 50 and 60% food grade. The physical properties of aqueous sodium lactate solutions have been reported (53). It is difficult to obtain the crystalline salt because it is very hygroscopic and extremely soluble in water (32) and alcohol. Its hygroscopicity makes it highly adaptable as a glycerol (qv) substitute, as a plasticizer, as a humectant in paper (qv) and textiles, and in tobacco curing. However the largest use of sodium lactate is for intravenous feeding.

Optically pure sodium lactate crystals can be prepared by spray drying a solution of sodium lactate of high optical purity and low lactide content. Sodium lactate contents can be determined by titration with acetone or dioxane that contains 0.1N HCl (90% acetone) using 0.1% bromophenol blue indicator.

Other Salts. Aluminum lactate, $(CH_3CHOHCOO)_3Al$, a granular solid that is freely water-soluble, and zirconium lactate are used as antiperspirants. Antimony lactate, $(CH_3CHOHCOO)_3Sb$, is used in the mordant dyeing of textiles. Copper lactate, (cupric lactate), $(CH_3CHOHCOO)_2Cu.3H_2O$, which is soluble in water, can be used in electroplating baths. Iron lactate (ferrous lactate), $(CH_3CHOHCOO)_2Fe.3H_2O$, which also is soluble in water, is used for treatment of anemia, as an ingredient of tonics, and in infant food. Titanium lactate is applied as a glass coating to prevent scratches or abrasion during mechanical handling on production lines (see Abrasives).

Esters. Conventional esterification processes are involved in the commercial production (with high yields) of methyl, ethyl, and *n*-butyl lactates. Lactic acid usually is the starting material; however, salts of lactic acid also may be used. In the case of methyl lactate, polylactic acid is a more desirable intermediate than lactic acid, because the ester is formed by esterification and by alcoholysis reactions. Thus, the problem of separating methyl lactate from water practically can be eliminated by using polylactic acid that has a high degree of polymerization. Ethyl lactate is prepared synthetically from fermentation lactic acid. The former process has not been published but it is understood to utilize acetaldehyde cyanohydrin (lactonitrile) which, upon interaction with ethyl alcohol, yields ethyl lactate. Likewise, methyl lactate can be produced from the reaction of methanol with an aqueous solution of lactonitrile using an acid catalyst, ie, transesterification of one ester into another by reaction of one with a higher alcohol. Thus, butyl lactate can be made by transesterifying methyl lactate with butanol. Lactic acid can be converted into methyl acrylate by esterification to methyl lactate, acetylation, and pyrolysis at 550°C.

Many esters of lactic acid have been reported in the literature. Some of these are high boiling liquids with properties characteristic of plasticizers for cellulose plastics and vinyl resins. The azeotropes of methyl and ethyl lactate also have been reported (54). In general, the rate of hydrolysis of organic acid esters is roughly proportional to the strength of the organic acid. Lactic esters are hydrolyzed more rapidly than the corresponding propionic esters. The rate of saponification of *n*-alkyl lactates decreases as the size of the alkyl group increases; the most pronounced difference in rate is between methyl and ethyl groups. The rate of acid-catalyzed hydrolysis of lower *n*-alkyl lactates to butyl lactates is independent of the alkyl group, and the same is true in uncatalyzed hydrolysis. Lactic esters of secondary alcohols are considerably more resistant to hydrolysis than those of the corresponding primary alcohols (2).

Methyl, ethyl, and butyl lactates are acetylated easily with acetic anhydride to produce the acetyl derivatives which are slightly soluble in water and which boil at about 25°C above the corresponding lactic ester. Some of the properties of methyl lactate, $CH_3CHOHCOOCH_3$, are mp −66°C; $bp_{101\ kPa\ (1\ atm)}$ 145°C; $bp_{12\ kPa\ (90\ mm\ Hg)}$ 42°C; d_4^{20} 1.0939 g/cm^3; n_D^{20} 1.4139; viscosity (at 20°C), 2.94 mPa·s (= cP); miscible with water, distills as an azeotrope (about 25% ester, bp 99°C). The recovery of lactic acid as the methyl ester from a crude fermentation beer may be a practical method for purifying lactic acid. Methyl lactate can be acetylated with acetic acid or acetic anhydride to produce either methyl α-acetoxypropionate or α-acetoxypropionic acid [*2382-59-4*] as the predominant product. In ammonolysis, methyl lactate is considerably more reactive than the higher alkyl lactates.

Ethyl lactate, $CH_3CHOHCOOCH_2CH_3$, mp −25°C; $bp_{101\ kPa\ (1\ atm)}$ 154°C; $bp_{(9.3\ kPa\ or\ 70\ mm\ Hg)}$ 51°C; d_4^{20} 1.034 g/cm^3; n_D^{20} 1.4132; viscosity (at 20°C), 2.61 mPa·s (= cP) is miscible with water. The lactic esters of higher alcohols can be prepared readily by transesterification with the appropriate alcohol. Ethyl lactate also may serve as a volatile ester in the purification of lactic acid. It is used in solvents for lacquers and sometimes as a lubricant in the manufacture of medicinal tablets and tablet casings.

Butyl lactate, $CH_3CHOHCOO(CH_2)_3CH_3$, mp −43°C, bp 187°C is used as a high boiling solvent in lacquer formulations. Lactide is the cyclic dimeric ester of lactic acid (11–13,55). Because of its low solubility and slow hydrolysis rate in water, lactide can be used as a mild continuous acidifier in aqueous systems. In anhydrous systems, lactide is neutral and relatively nonreactive; however, it reacts with and neutralizes alkaline materials that are present or formed in such systems. Lactide is used, therefore, as a neutral component in anhydrous systems where small amounts of alkali may form during reaction or storage and which must be neutralized immediately.

Amides. Lactamides are prepared readily by the interaction of lactic esters with ammonia and amines:

$$CH_3CHOHCOOR + NH_3 \rightarrow CH_3CHOHCONH_2 + ROH$$

$$CH_3CHOHCOOR + R'NH_2 \rightarrow CH_3CHOHCONHR' + ROH$$

The reactivity of lactic esters in ammonolysis is considerably greater than that of esters of unsubstituted aliphatic acids except of formic acid. Methyl lactate is the most reactive lactic ester toward ammonolysis: the yield is almost quantitative. Many substituted lactamides have been prepared by aminolysis of methyl lactate. The primary alkylamines react readily and almost quantitatively at room temperature. The secondary amines, with a few exceptions, are not suitable for this reaction. The notable exceptions are dimethylamine, morpholine, piperidine, pyrrolidine, and diethanolamine, which give very high yields. However, lactamide derivatives of unreactive secondary amines are easily obtained by dehydration of the lactic acid amine salt by co-distillation with a high boiling solvent such as xylene.

The *N*-alkyl-substituted lactamides generally are low melting solids having high boiling points. Methyl-, ethyl-, *n*-propyl-, and *n*-butyllactamide, are very soluble in water. The *N*-hydroxyalkyl-substituted lactamides, such as *N*-2-hydroxyethyl-, *N*-bis(2-hydroxyethyl)-, and *N*-2 hydroxypropyllactamide, are viscous water-soluble liquids with hygroscopic properties.

Lactamide is a white crystalline compound. Racemic lactamide melts at 76.4–77.6°C but the melting point is severely influenced by impurities. Lactamide is very soluble in water and ethanol and is slightly soluble in ethyl ether and benzene. The density of a 50% aqueous solution of lactamide at 25°C is 1.093 g/cm^3. The solid lactamide has a density of 1.138 g/cm^3 at 25°C.

Hydroxyacetic Acid

Hydroxyacetic acid [*79-14-1*] (glycolic acid), $HOCH_2COOH$, is the first and simplest member of the series of hydroxy carboxylic acids. It is a colorless, translucent, and odorless solid. The acid was made first in 1848 by treating glycine with nitrous acid, and it was characterized in 1851. Hydroxyacetic acid is used in textile and leather (qv) processing, detergents, metal cleaning and plating, and dairy sanitation (see Metal surface treatment, cleaning, pickling, and related processes).

Properties. The following physical properties of glycolic acid are known: mp 79–80°C; bp 100°C (dec); d_4^{25} 1.49; K_a at 25°C, 1.5×10^{-4}; pH of 1*M* solution, 2.4; heat of combustion, 697.1 kJ/mol (166.6 kcal/mol); heat of solution (infinite dilution), −154 J/g (−36.8 cal/g); flash point, above 300°C (dec). Glycolic acid crystallizes in space group P2./c and has unit space cell dimensions of a = 0.8965 nm, b = 1.0563 nm, c = 0.7826 nm, β = 115.083° at 24 ± 1°C with Z = 8. Analyses have been made by x-ray diffraction (56) and single crystal neutron diffraction (57). The structure consists of a loose three-dimensional H-bonded network of two closely similar but crystallographically distinct types of molecules.

Glycolic acid is slightly volatile in steam but it cannot be distilled because it readily loses water by self-esterification to form poly(hydroxyacetic acid) [*26124-68-5*] [poly(glycolic acid)]. Glycolic acid is very soluble in water, methanol, ethanol, acetone, and ethyl acetate; slightly soluble in ethyl ether; and only sparingly soluble in hydrocarbon solvents.

Reactions. Hydroxyacetic acid contains both a carboxyl group and a primary hydroxyl group. Thus, it can react as an acid or as an alcohol or as both. As an acid it forms salts, esters, amides, etc; as an alcohol, it forms esters with other acids and with itself, acetals, ethers, etc. When reacting as both an alcohol and an acid, it forms derivatives, eg, alkyl acyl oxyacetates and polyhydroxyacetic acid.

Aqueous solutions of the acid contain, in addition to the free acid, soluble polyesters, eg, the corresponding trimer ($\overline{OCH_2CO_2CH_2CO_2CH_2CO}$), and, possibly the dimer glycolide [*502-97-6*] (diglycolide). The relative concentration of free acid and polymers is a function of the total concentration of acid in the solution; for every concentration of hydroxyacetic acid, there is an equilibrium concentration of free acid and of polymers. The formation of polymers in the equilibrium solution may be represented by the following reaction:

$$n\ HOCH_2COOH \rightleftharpoons H(OCH_2CO)_nOH + (n-1)\ H_2O$$

Solutions of hydroxyacetic acid can be brought to equilibrium by refluxing for a maximum of about two hours. Although the position of the equilibrium is changed only slightly by a change in temperature, its rate of establishment is affected markedly by temperature change and is very slow at or below room temperature.

On oxidation with hydrogen peroxide in the presence of a ferrous salt, hydroxyacetic acid first gives glyoxylic acid [*298-12-4*] (HCOCOOH), and then formaldehyde, formic acid, carbon dioxide, and water. With concentrated nitric acid or lead oxide and water at 160°C, it forms oxalic acid (qv). On treatment with zinc and sulfuric acid, hydroxyacetic acid is completely reduced to acetic acid.

Glycolic acid reacts with maleic anhydride under basic conditions in the presence of zinc or calcium ion to form a carboxymethyl oxysuccinic acid (38) by a Michael addition in accordance with the following equation:

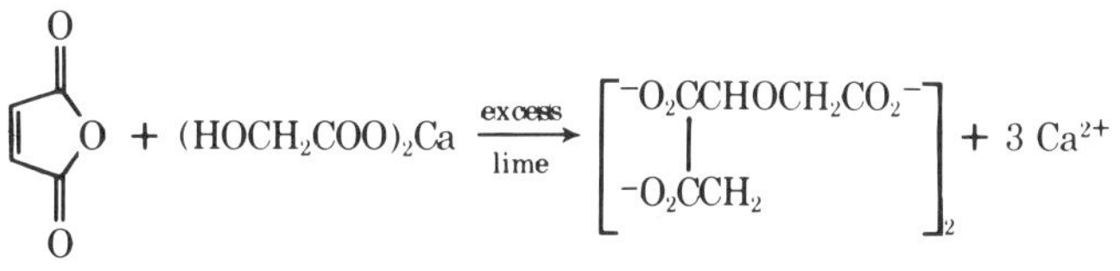

$$\text{(maleic anhydride)} + (HOCH_2COO)_2Ca \xrightarrow[\text{lime}]{\text{excess}} \left[\begin{array}{l} ^{-}O_2CCHOCH_2CO_2^{-} \\ \quad | \\ ^{-}O_2CCH_2 \end{array}\right]_2 + 3\ Ca^{2+}$$

N-Hydroxyacetamide may be prepared by the reaction of ammonia and the polyacid or by treating an ester of hydroxyacetic acid with aqueous ammonium hydroxide. Substituted amides are prepared by the reaction of an amine with hydroxyacetic acid.

Manufacture. Hydroxyacetic acid occurs in nature in sugar beets, unripe grapes, and spent sulfite liquor from pulp processing. However, it has never been produced commercially from these sources. The acid was available before 1940 only in limited quantities and was prepared from the hydrolysis of monohalogenated acetic acids. It may have been produced in Germany at one time by the electrolytic reduction of oxalic acid.

Hydroxyacetic acid is produced commercially in the United States as an intermediate in the manufacture of ethylene glycol (see Glycols). Under high pressure,

30.4–91.2 MPa (300–900 atm) and from 160–200°C (59–61), formaldehyde reacts with carbon monoxide and water in the presence of an acidic catalyst, ie, sulfuric, hydrochloric, or phosphoric acids, to form hydroxyacetic acid. In the presence of HF catalyst, this reaction rapidly produces hydroxyacetic acid even at 20–60°C (62); more moderate pressures are required for the reaction. The advantage of this catalyst is less reactor corrosion at the lower temperatures and easy separation of the catalyst from the reaction mass by distillation. The reaction that is catalyzed by sulfuric acid, on the other hand, is purified by treating the reaction mass with granulated carbon for decolorization, then by treating it in a weak anion-resin column to remove the sulfuric acid, followed by steam stripping to remove low boiling impurities, and finally by treatment with a cation-exchange resin to remove metallic impurities (63).

Another method for producing hydroxyacetic acid is by the hydrolysis of glyconitrile with an acid (eg, H_3PO_3 or H_2SO_3) having a pK_a of about 1.5–2.5 and at 100–150°C. The glyconitrile is produced by reaction of formaldehyde with hydrogen cyanide (64). Preparation of glycolic acid, based on the oxidation of hexoses, isoamyl benzoate, and 5-oxodigluconic acid also has been studied (65–67).

Specifications. Hydroxyacetic acid has been offered commercially since 1950 and is available only as a 70% aqueous solution; however, the quality of the acid has been improved significantly. Specifications for the commercial technical-grade solution are reported in ref. 68.

Uses. Hydroxyacetic acid is produced in large volume. Nevertheless, it has found uses in a number of areas, eg, adhesives, metal cleaning, electroplating (qv), dairy cleaning, biodegradable polymers, dyeing, water-well cleaning, masonry, textiles, and detergents. Numerous properties of hydroxyacetic acid contribute to its excellence for use in cleaning dairy and food processing equipment. One of the novel soap (qv) builders consists of salts of α-substituted β-sulfosuccinic acids that have the following general formula:

$$\mathrm{MO\overset{\overset{\displaystyle O}{\|}}{C}\underset{\underset{\displaystyle SO_3M}{|}}{C}H\overset{\overset{\displaystyle ZR}{|}}{C}H\overset{\overset{\displaystyle O}{\|}}{C}OM}$$

where R is hydrogen or an organic moiety; Z is either O, S, SO, SO_2, N, or NO; and M is an alkali metal. One soap builder is prepared from the reaction of sulfomaleic anhydride and ethyl glycolate; the product is a tetrasodium salt of α-carboxymethyl β-sulfosuccinic acid (69).

A similar but simpler alkali metal carboxyalkoxysuccinate that also is a useful detergent component, particularly in detergent formulations, has the general formula (70–71):

$$\mathrm{HO\overset{\overset{\displaystyle O}{\|}}{C}CH_2\overset{\overset{\displaystyle R}{|}}{C}H\overset{\overset{\displaystyle O}{\|}}{C}OH}$$

where R is a carboxyalkoxy radical (—OR′COOH) in which R′ is 1–6 methylenes. A process used to make the simplest compound in this series is based on the reaction (in a lime system) of an alkaline earth metal glycolate and maleate. The product can be recovered from the mother liquor by decanting or filtering. Another detergent builder may be produced by the reaction of glycolic acid and silicon tetrachloride followed by neutralization to form the sodium salt (72) (see Surfactants and detersive systems).

Glycolic acid has been used with a cross-linking agent in cellulosic fabrics to create durable-press quality and to render the fabric responsive to basic dyes. Glycolic acid serves as a catalyst and a reactive additive producing a cross-linked cellulosic fabric with pendant carboxylic groups. The carboxylic groups permit selective absorption of basic dyes by the chemically modified cellulosic textile (73–74). Glycolic acid also has been used to cure flame-retardant finishes for cellulose textiles (75).

Hydroxyacetic acid is used extensively for dying wool (qv). It readily exhausts level-dyeing and milling-type acid colors with minimum damage to wool fibers, and its mild reducing properties make it useful in setting mordant dyes on the wool in the bottom-chrome dying process (see Dyes and dye intermediates). It also is used in nylon dyeing. Hydroxyacetic acid has been used to obtain low pH in dye baths. It has the advantage of having a mild odor and of being nonvolatile; therefore, it does not evaporate from dye baths at elevated temperatures. Hydroxyacetic acid complexes with metal ions present in the bath water, rinse, etc. It is used for the dyeing of chromic colors but is not recommended for colors that are easily reduced.

Hydroxyacetic acid also is used in a variety of metal-treating applications, eg, to impart luster to copper and copper alloys. This luster is retained longer than when strong oxidizing acids are used alone. Likewise, iron or iron alloy surfaces can be brightened when treated with an aqueous solution containing hydrogen peroxide (qv), ammonium bifluoride, and hydroxyacetic acid (76). Hydroxyacetic acid and either phosphoric or sulfuric acid are used in electropolishing stainless steel. Aluminum alloys can be anodized for subsequent electroplating with a solution containing phosphoric, sulfuric, and hydroxyacetic acids. Also, sodium and potassium salts are excellent substitutes for Rochelle salts which are used as bath additives. Inasmuch as hydroxyacetic acid forms complexes with virtually all multivalent metals, its salts are used in many electroplating baths, eg, for chrome, lead, cobalt, tin, and nickel. In the electroless plating (qv) of nickel or copper, sodium salts of hydroxyacetic acid have been used. They act as catalysts or activators for electroless deposition of copper or nickel on glass, ceramics, and organic polymers (77). Hydroxyacetic acid has been utilized in the etching of lithograph plates. It also is useful in the prevention of unwarranted precipitates by forming soluble metallic complexes in both alkaline and acid environments, eg, in metal-cleaning formulations. It gives a low rate of corrosion of nonferrous metals and is effective in dissolving deposits, eg, smelt on aluminum. These cleaning formulations can be rinsed easily and they reduce spotting caused by trace-metal residues. These properties make is especially applicable for cleaning electronic printed circuits and metal-plating operations.

Mixtures of hydroxyacetic acid and phosphoric acid can be used as a masonry cleaner both in the manufacture of bricks and in cleaning freshly laid bricks or tiles. Certain clays used in making white bricks contain traces of metals which will stain the brick during firing and hydroxyacetic acid is effective in removing these stains (see Clays, survey). The addition of a salt of hydroxyacetic acid to cement (qv) improves its setting and strength (78). Hydroxyacetic acid also is useful as a retarding admixture for concrete products. Its use permits reduction of cement content without sacrificing compressive strength or improvement of the compressive strength when used with higher proportions of cement.

The iron salts resulting from mixed acid cleaning solutions (formic and hydroxyacetic acids) exhibit higher solubility and less tendency to precipitate than when any of the acids is used alone. Therefore, longer contact time and more deposit removal with less danger of cement plugging is possible (78).

Hydroxyacetic acid removes hard-water scale from all types of heat-exchange equipment (see Heat exchange technology), as well as calcium carbonate, mill scale, and magnetic iron oxide. It has been considered for pH adjustment of cooling water to prevent scale buildup. The corrosion rates of hydroxyacetic acid solutions on stainless steel are low. A solution containing 2% hydroxyacetic acid and 1% formic acid has been developed for cleaning high pressure stainless steel steam generators. Ammoniated hydroxyacetic acid (79) is effective in removing deposits of gypsum. It as well as mixtures of hydroxyacetic and citric acid (qv) that are ammoniated to pH 4 show excellent dissolution for oxides and salts. Ammoniated hydroxyacetic acid gives negligible corrosion on stainless steel, zinc alloys, and other resistant metals.

Hydroxyacetic acid exhibits good carbonate and iron scale removal and is used for cleaning water wells. It complexes ferric iron thereby preventing reprecipitation when the material is diluted. Corrosion by hydroxyacetic acid on construction materials that are used in wells is much below that of mineral acids. Its use in well cleaning, therefore, eliminates the need for separate addition of chelating agents (qv) and bactericides. Hydroxyacetic acid, however, does not replace the need for phosphate treatment. It may be used in an admixture with other acids such as hydrochloric acid.

Hydroxyacetic acid is used to replace mineral acids where high-quality leather (qv) is desired, and it is used in the deliming operation and in pH adjustment during tanning and dyeing. It also prevents staining of the leather caused by formation of colored metal-tanning salts.

Recent applications of hydroxyacetic acid include: preparation of reverse-osmosis membranes (see Reverse osmosis) (80); reduction of polyether discoloration during chlorination (81); preparation of a stabilizing solution for fluorescent whiteness (82); and protection against oxidation of silver images in photography.

Derivatives. ***Alcohol Anhydride (Ether).*** Diglycolic acid [*110-99-6*] (oxydiacetic acid), $HOOCCH_2OCH_2COOH$, is the alcohol anhydride of glycolic acid. It is a white solid which is crystallized from water in monoclinic prisms (mp 148°C). Its ionization constants at 25°C are $K_1 = 1.1 \times 10^{-3}$ and $K_2 = 3.7 \times 10^{-5}$. Its reactions are typical of a dibasic acid.

Diglycolic acid is formed with glycolic acid by boiling monochloroacetic acid with calcium hydroxide, barium hydroxide, or magnesium hydroxide, as well as by oxidation of diethylene glycol, $O(CH_2CH_2OH)_2$. It is used in the manufacture of resins and plasticizers and in organic synthesis (80).

Salts. Sodium hydroxyacetate (glycolate), $HOCH_2COONa$, has been used in the dyeing of unwashed carbonized wool, as a masking agent in zirconium leather tanning, and in a bath used for the electroless plating of nickel (81). Zinc glycolate and zinc naphthenate are used in automobile-primer coatings (qv). Butyl glycolate is added to these primers to prevent zinc glycolate plugging of spray nozzles.

Esters. Esters of hydroxyacetic acid may be obtained by direct esterification of the free acid or by alcoholysis of the methyl ester. Distillation of hydroxyacetic esters in the presence of either acidic or basic catalysts is accompanied by decomposition (self-alcoholysis) to the alcohol and ester of polyhydroxyacetic acid:

$$2\ HOCH_2COOR \rightarrow HOCH_2COOCH_2COOR + ROH, \text{etc}$$

Therefore, it is advisable to neutralize an esterification product before attempting to distill the ester. The distillation should be carried out at reduced pressure.

Hydroxyacetic esters of the common alcohols are colorless liquids with pleasant characteristic odors. They have considerably higher boiling points than the corresponding esters of acetic acid. The methyl and isobutyl esters readily dissolve cellulose nitrate, cellulose acetate, cellulose acetate propionate, and poly(vinyl acetate) (see Cellulose acetate and triacetate fibers; Vinyl polymers, poly(vinyl acetate). The isobutyl ester also dissolves ethyl cellulose, poly(vinyl butyral), ester gum, and a number of other synthetic and natural resins (see Resins, natural; Resins, water-soluble).

Other Hydroxy Acids

Preparation. α-Hydroxy acids may be prepared by the hydrolysis of an α-halo acid or by the acid hydrolysis of the cyanohydrins (qv) of an aldehyde or ketone:

$$RCHO \xrightarrow{HCN} \underset{\displaystyle OH}{\underset{|}{RCHCN}} \xrightarrow[HCl]{H_2O} \underset{\displaystyle OH}{\underset{|}{RCHCOOH}} + NH_4Cl$$

Aliphatic α-hydroxy acids that do not have side chains can be prepared in good yields by the hydrolysis of α-nitrato acids with aqueous sulfite solutions. The α-nitrato acids are obtained by the reaction of olefins (qv) and N_2O_4 in the presence of oxygen. The α-hydroxy acids that are obtained can be esterified or acylated directly to yield anhydro ester acids which, in turn, give the α-hydroxy acid on saponification. For example, 2-nitrooxyoctanoic acid, $C_6H_{13}CH(ONO_2)CO_2H$ (from 1-octene), is added to a solution of sodium sulfite (2.5 mol/mol acid) at 60–90°C; it then is cooled and acidified. The organic layer yields 95% $C_6H_{13}CHOHCOOH$, 2-hydroxyoctanoic acid [*617-73-2*] (mp 55–65°C, saponification no. 354, and acid no. 320) (83).

β-Hydroxy acids may be made by catalytic reduction of β-keto esters followed by hydrolysis. β-Hydroxy esters also can be obtained by the Reformatsky reaction.

The Reformatsky reaction is analogous to the Grignard reaction (qv). It permits the formation of β-hydroxy esters, or α,β-unsaturated esters and is characterized by the reaction of an aldehyde or a ketone and an α-haloester in the presence of activated zinc in anhydrous ethyl ether (84). Ether is the solvent that usually is used but benzene, toluene, xylene, or mixtures of these with ether, depending upon the temperature desired, also may be used. The reaction product is treated with an aqueous acid to hydrolyze the organo zinc complex to produce the β-hydroxy ester.

Unlike the Grignard reaction, the success of the Reformatsky reaction depends upon the fact that organozinc compounds are considerably less reactive towards carbonyl groups than Grignard reagents. These organozinc compounds react with the carbonyl function rather than with the less reactive ester function. The carbonyl compounds that may be used in the Reformatsky reaction include saturated and unsaturated aliphatic aldehydes, aromatic aldehydes, and aliphatic, aromatic, alicyclic, saturated and unsaturated ketones. The α-bromo esters generally give the best results in the Reformatsky reaction; α-chloro esters react slowly or not at all and the α-iodo esters are not commonly available.

β-Hydroxypropionic acid can be formed by the reaction of formaldehyde and lead acetate trihydrate with catalytic amounts of pyridine and hydroquinone under pressure at 175°C for 2 h. After acidification with excess HCl, the reaction yields lead chloride and a red solution. When the latter is extracted with ether, β-hydroxypropionic acid is obtained. In a similar manner, formaldehyde, lead acetate, and acetic acid give β-acetoxypropionic acid (85).

Reactions. α-Hydroxy acids undergo bimolecular esterification with the formation of a six-membered ring, a lactide:

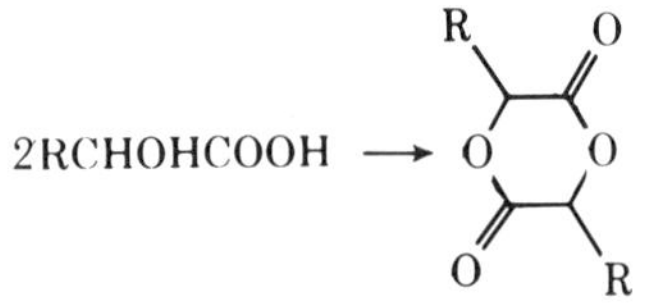

$$2RCHOHCOOH \longrightarrow \text{lactide}$$

lactide

The ease with which α-hydroxy acids self-esterify makes it almost impossible to keep α-hydroxy acids in their monomolecular state except when in the form of their sodium salts.

When an α-hydroxy acid is boiled with dilute sulfuric acid, decarbonylation occurs, and water is lost, and an aldehyde is formed. The carbon cation that is formed is the conjugate acid of the aldehyde:

$$RCHOHCOOH \xrightarrow{H^+} RCHOH\overset{+}{C}(OH)_2 \xrightarrow[-H_2O]{} RCHOH\overset{+}{C}O \xrightarrow[-CO]{} \{R\overset{+}{C}HOH \leftrightarrow RCH{=}\overset{+}{O}H\} \xrightarrow[-H^+]{} RCHO$$

This reaction is valuable for the synthesis of higher aldehydes. Since one carbon atom is lost in the process, the series of reactions may be used for stepwise degradation of a carbon chain.

β-Hydroxy acids lose water (especially when heated with sulfuric acid) to give α,β-unsaturated acids and, frequently, β,γ-unsaturated acids. β-Hydroxy acids do not form β-lactones readily because of the difficulty of four-membered ring formation. The simplest β-lactone, β-propiolactone, is prepared by the reaction of ketene (qv) and formaldehyde in the presence of methyl borate.

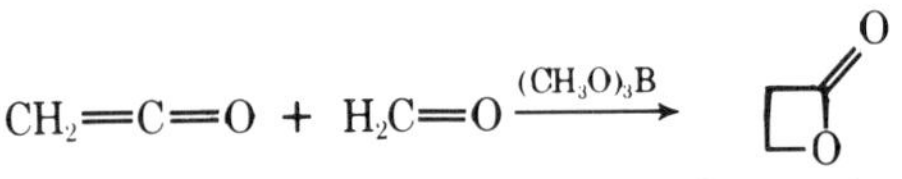

$$CH_2{=}C{=}O + H_2C{=}O \xrightarrow{(CH_3O)_3B} \beta\text{-propiolactone}$$

β-propiolactone

The reaction also may be carried out in a 20% $AlCl_3$ solution in methanol ≤20°C. Other catalysts, eg, $ZnCl_2$, may be present in the substantially anhydrous solution. With the zinc chloride catalyst system, β-propiolactone can be made from ketene and formaldehyde in excellent yields (86).

The four-membered ring of the β-lactone is opened easily by a variety of nucleophiles to give β-substituted propionic acids in high yields, which suggests that the ring is opened by S_N2 attack on the β-carbon rather than by an initial attack on the carbonyl carbon. The former mechanism involves opening of the ring and relieves the strain at all four atoms in the ring. By contrast, the latter mechanism would involve first cleaving the carbon–oxygen double bond. The position of the ring cleavage is dependent on the acidity of the medium, eg, aqueous ammonia gives β-hydroxypropionamide but ammonia dissolved in acetonitrile gives β-aminopropionic acid (β-alanine). The base-catalyzed reaction with methanol opens the lactone ring to form the ester, whereas the product in the reaction with methanol without the catalyst and the primary product of the acid-catalyzed reaction, the ether acid, is formed by the cleavage of the ether rather than of the ester linkage of the lactone. The diversity of these reactions indicate that β-propiolactone is a useful intermediate for organic synthesis. However, caution should be exercised when handling the β-lactone because it is a carcinogen.

γ-Hydroxy acids seldom are obtained in the free state because of the ease with which they form monomeric inner esters as a result of their having five-membered rings (see Furan and furan derivatives):

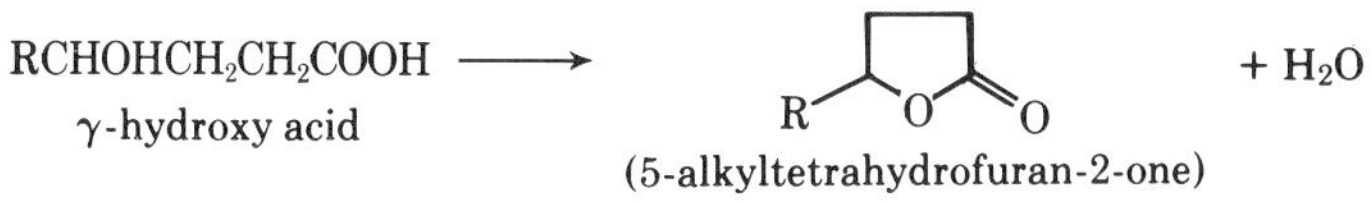

$$RCHOHCH_2CH_2COOH \longrightarrow \text{(5-alkyltetrahydrofuran-2-one)} + H_2O$$

γ-hydroxy acid

(5-alkyltetrahydrofuran-2-one)

Cyclic esters that are derived from γ-hydroxy acids are γ-lactones. γ-Lactones are very stable, neutral substances. However, heating the γ-lactone in alkali breaks the ring and results in the formation of the corresponding acid salt.

γ-Butyrolactone (tetrahydrofuran-2-one) is made commercially by the catalytic dehydrogenation of tetramethylene glycol:

$$HOCH_2(CH_2)_2CH_2OH \xrightarrow[200°C]{Cu\text{–}SiO_2} \gamma\text{-butyrolactone} + 2\,H_2$$

tetramethylene glycol

γ-butyrolactone

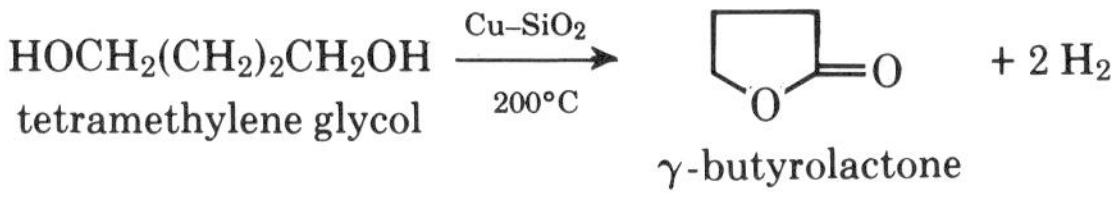

γ-Butyrolactone is an intermediate in the preparation of pyrrolidine (see Acetylene-derived chemicals). Both γ-hydroxybutyrate ion and γ-butyrolactone, when administered orally or intravenously, produce sleep. The anion is the active compound.

δ-Lactones can be prepared but with difficulty and they are less stable than γ-lactones since the δ-lactone ring opens more easily. The δ-lactone often changes spontaneously into linear polyesters. δ-Valerolactone, when heated, becomes thick and more viscous on standing and finally becomes a solid polymer with about seven recurring units. The melting points of some hydroxy acids are listed in Table 3.

Table 3. Melting Points of Some Hydroxy Acids

Acid	CAS Registry No.	mp, °C
3-hydroxypropionic acid[a]	*[503-66-2]*	143
3-hydroxybutyric acid[a]	*[300-85-6]*	
2-hydroxybutyric acid	*[565-70-8]*	49–50
4-hydroxybutyric acid	*[591-81-1]*	−17

[a] Exists as a syrup.

Occurrence and Uses

Almost all of the normal hydroxy aliphatic acids occur in nature. The literature is filled with biochemical studies of these acids, particularly 3-hydroxy- and 4-hydroxybutyric acids. The investigation of the latter compounds include studies on the determination of the effect and the level and/or actions that these acids have in the brain, central nervous system, blood, kidneys, urine, muscles, and stomach. In addition to their importance as cellular constituents of plant and animal tissues, they are produced by microorganisms during the various processes of metabolism. γ-Butyrolactone and its hydrolysis product, γ-hydroxybutyric acid [*591-81-1*], exert a variety of pharmacological effects when administered systematically to humans and laboratory animals (see Psychopharmacological agents). Bacteria appear to use γ-hydroxybutyric acid as a carbon source after converting it to succinic acid. *In vitro,* γ-hydroxybutyric acid is utilized as a mitochondrial substrate and for investigating the properties of the enzymes of body metabolism (87). *In vivo,* it is used to investigate the metabolism of ketone bodies and their effects in test animals; γ-hydroxybutyric acid is the preferred isomer to use because it is metabolized physiologically. However, the DL form also frequently is used. Source of the D(−) isomer are urine or bacterial poly D(−)-β-hydroxybutyrate (88).

γ-Hydroxybutyric acid (GHB) and its derivatives, particularly its sodium salt, have been studied as anesthetics (qv), tranquilizers, sedatives, and hypnotics (see Hypnotics, sedatives, anticonvulsants). GHB is used as an anesthetic for surgery and in general obstetrics. When administered intraperitoneally into mice, it produces a dose-dependent increase in the concentration of dopamine in the mouse brain thereby causing a sleeplike state. GHB restores the normal ratio between sleep phases and increases duration of sleep. Electroencephalography has been used to measure the effects of GHB in relation to dopaminergic neuron function. The physiochemical properties, metabolism, toxicity, pharmacological and clinical aspects, and uses of GHB are reviewed in refs. 89–91.

Mevalonic Acid

The racemic form of mevalonic acid [*150-97-0*] is synthesized from 4-hydroxy-2-butanone, which is the aldol addition product of acetone and formaldehyde. The proton-catalyzed reaction with ketene yields the acetate. The acetate in turn reacts with another molecule of ketene in the presence of boron trifluoride to give the β-lactone of the mevalonic acid:

$$CH_3C(=O)CH_2CH_2OH + CH_2{=}C{=}O \xrightarrow{H^+} CH_3C(=O)CH_2CH_2OC(=O)CH_3$$

$$CH_3C(=O)CH_2CH_2OC(=O)CH_3 + CH_2{=}C{=}O \xrightarrow{BF_3} \beta\text{-lactone (2-oxetanone bearing } CH_3 \text{ and } CH_2CH_2OC(=O)CH_3)$$

$$\xrightarrow{2\,H_2O} CH_3C(OH)(CH_2COOH)CH_2CH_2OH + HOC(=O)CH_3$$

mevalonic acid

Mevalonic acid was discovered as a growth factor for lactobacilli in 1956 (see Plant growth substances). It is the key intermediate in the formation of squalene which is an intermediate to cholesterol formation. Mevalonic acid, also is an intermediate in the biosynthesis of terpenes (see Terpenoids) and steroids (qv).

Aldonic Acids

Aldonic acids are poly(hydroxy)monocarboxylic aliphatic acids, and they can be prepared by mild oxidation of aldoses:

$$\underset{\text{aldose}}{\begin{array}{c}CHO\\|\\(CHOH)_n\\|\\CH_2OH\end{array}} \xrightarrow{Br_2 + H_2O} \underset{\text{aldonic acids}}{\begin{array}{c}COOH\\|\\(CHOH)_n\\|\\CH_2OH\end{array}} \xrightarrow{HNO_3} \underset{\text{aldaric acids}}{\begin{array}{c}COOH\\|\\(CHOH)_n\\|\\COOH\end{array}}$$

where, in general, $n = 2, 3$, or 4. Aldonic acids usually are isolated as the metallic salts or lactones but they readily form amides, hydrazides, and other derivatives. They are helpful in the characterization of the large number of sugar configurations (see Sugar). The preparation of an aldonic acid having the same number of carbon atoms as the aldose it was prepared from has been regarded as proof of the aldehyde structure. By contrast, ketoses undergo chain splitting to form aldonic acids that have fewer carbons. Aldonic acids also are precursors in the preparation of aldoses that have fewer carbons. Hydrogen peroxide and iron salts oxidize an aldonic acid to produce an aldose containing one less carbon than the acid. For example D-gluconic acid is converted to D-arabinose and D-galactonic acid is converted to D-lyxose. Methods used to lengthen the carbon chain of sugars may involve the formation of aldonic acids as intermediates. The Kiliani cyanohydrin synthesis produces two aldonic acids with one more carbon than the original aldose.

Properties. Free aldonic acids seldom exist as such in aqueous solutions; instead they form lactones by the elimination of water. Either of the hydroxyls in the γ- and δ-positions can take part in the formation of lactones. δ-Lactones usually hydrolyze easily and mutarotate rapidly in aqueous solutions. However, γ-lactones are more stable and hydrolyze very slowly in water to form the equilibrium mixture of the free acids and lactones. This equilibrium mixture is attained only after many days at room temperature but it can be accelerated by the presence of strong acids and can be detected by paper chromotography. Equilibrium solutions of mannose contain large proportions of the γ-lactones whereas those of glucose contain a large proportion of the δ-lactone and free acids. Solvents affect the equilibrium composition, eg, mannonic acid dissolved in acetic acid with 16% water shows a higher positive rotation than in water alone. The mutarotation is slower in the acid solution but there is a greater conversion to the δ-lactone than to the γ-lactone. The pH, temperature, and concentration also have an effect on the final equilibrium composition.

The positions of the hydroxyl groups on carbons 4 and 5 influence the rotations of the lactones. A lactone is more dextrorotatory than the free acid if the hydroxyl group involved in the lactone formation lies on the right side in the Fischer projectional formula and vice versa. Most aldonic acids have only small rotations, whereas the lactones possess fairly strong rotations. Therefore, the lactones can be divided into levo- and dextrorotatory groups. Both γ- and δ-lactones of gluconic and mannonic acids are dextrorotatory; gluconic and galactonic acids form levorotatory γ-lactones and dextrorotatory δ-lactones, respectively. D-Allonic-γ-lactone is an exception to this rule since it has a small negative rotation (D-α −6.8°) instead of the predicted positive rotation.

Aqueous solutions of aldonic acids have a pH of 2–3. The free acids are soluble in water and slightly soluble in ethanol. They are more soluble in nonpolar solvents than sugars but are less soluble in nonpolar solvents than the lactones. Salts can be formed readily but some are unstable.

Esters of aldonic acids are prepared from δ-lactones and (slowly) from the γ-lactones by their reaction with an alcohol in the presence of hydrogen chloride or the free aldonic acid. Aldonic acid can be recrystallized from hot methanol without much consequent esterification. Ethyl D-mannonate decomposes when heated to its mp into the γ-lactone and ethanol.

Lactones give a positive ester test when treated with alkaline hydroxylamine by forming a hydroxamic acid. Acids give an intense color with ferric chloride. This test can be used to detect lactones on paper chromatograms. Diazomethane converts acids to esters; therefore, by treating chromatogram paper first with this reagent, the acids and lactones can be detected. Lactones are less reactive to alkali than to acids. A solution of free acids can be neutralized with calcium carbonate. Sodium carbonate reacts with the δ-lactones and an excess of sodium hydroxide reacts with the γ-lactones.

Lactones also react with liquid ammonia to form amides of aldonic acids. These derivatives often can be crystallized and are useful for the characterization of the acid. Phenylhydrazine reacts with either the acid or the lactones to form phenylhydrazides. Hydrolysis of the latter by alkalies is slow and usually is incomplete. The melting points of some aldonic acids and their derivatives are listed in Table 4.

Table 4. Melting Points of Hydroxy Acids and Their Derivatives, °C

Acid	CAS Registry No.	Amide	Ester	Lactone
erythronic	[*13752-84-6*]	D(−), 91–92	DL(−)(butyl), 62–64	DL(−) γ, 52.5–53
		L(−), 91–92		D(−) γ, 104–105
				D(−) γ, 105
threonic, 98–99	[*3909-12-4*]	DL(−), 116		D(−) γ, 75–77
		L(−), 105.5–107		L(−) γ, 65–68
L(−) arabinonic, 118–119	[*608-53-7*]	L(−), 136		DL(−) γ, 115–116
				D(−) γ, 98–99
D(−) lyxonic, 144	[*526-92-1*]			L(−) γ, 98–100
				D(−) γ, 110
xylonic	[*526-91-0*]	D(−), 81–82	tetraacetate, 134.5	D(−) γ, 99–103
galactonic	[*576-36-3*]	D(−), 172–172.5		DL(−) γ, 122–5
				D(−) γ, 170
D(−) gluconic, 131	[*526-95-4*]	D(−), 143–144	D(−), ethyl 62–63	D(− γ, 134–136
				L(−) γ, 134–135
mannonic	[*20248-27-5*]	D(−), 176	D(−), ethyl 164	DL(−) γ, 155
		L(−), 171–172		D(−) γ, 151
				D(−) δ, 235
				L(−) γ, 150.5–151
				L(−) δ, 96.7

Preparation. Bromine or nitric acid are the oxidizing agents that usually are used in the preparation of aldonic acids. Nitric acid is used under very mild conditions; the best results are obtained with bromine in a slightly acid buffered solution (pH 5–6). The product generally is isolated as the salt by direct crystallization from the reaction solution or by precipitation from ethanol. Commercially, the indirect use of bromine is employed in the electrolytic oxidation process that involves calcium bromide as a catalyst. Gluconic acid soluble salts are used as cation sequestering agents to introduce appropriate metallic ions, such as iron, bismuth, and particularly calcium, into the human body in a neutral and easily assimilable form. Gluconic acid and its salts are not metabolized but are excreted in the urine. When the acid or salt is given orally, only a small portion is absorbed. In proper dosages, gluconic acid and its salts produce a decrease in urine acidity.

BIBLIOGRAPHY

"Lactic Acid" in *ECT* 1st ed., Vol. 8, pp. 167–180, by E. M. Filachione, Eastern Regional Research Laboratory, Bureau of Agriculture and Industrial Chemistry, U.S. Department of Agriculture; "Lactic Acid" in *ECT* 2nd ed., Vol. 12, pp. 170–188, by G. T. Peckham, Jr., Clinton Corn Processing Co., and E. M. Filachione, Agricultural Research Service, U.S. Department of Agriculture, Wyndmoor, Pa.

1. H. Borsook, H. M. Huffman, and Y. P. Liu, *J. Biol. Chem.* **102,** 449 (1933).
2. C. H. Fisher and E. M. Filachione, *U.S. Dept. Agric. Bur. Agric. Ind. Chem. Mimeo. Circ. Ser.* **AIC-279** (Oct. 1956).
3. G. T. Peckham, Jr., *Chem. Eng. News* **22,** 440 (1944).
4. *Symp. N.Y. Acad. Sci. Ann. N.Y. Acad. Sci.* **119,** 851 (1965).
5. C. F. Woodward, *Chemurgic Dig.* **9**(6), 9 (1950).
6. C. H. Holten, A. Muller, and D. Rehbinder, *Lactic Acid,* Verlag Chemie, International Research Association, Copenhagen, Den., 1971.
7. F. Krafft and W. A. Dyes, *Ber. Dtsch. Chem. Ges.* **28,** 2589 (1895).
8. Lockwood, L. B., Yoder, D. E., Zienty, and M. Ann, *N.Y. Acad. Sci.* **119,** 854 (1965).
9. H. Borsook, H. M. Huffman, and Y. P. Liu, *J. Biol. Chem.* **102,** 449 (1933).
10. T. Puride and J. W. Walker, *J. Chem. Soc.* **67,** 616 (1895).
11. C. A. Crutchfield, Jr., W. M. McNabb, and J. F. Hazel, *J. Inorg. Nucl. Chem.* **24,** 291 (1962).
12. R. Dietzel and E. Rosenbaum, *Z. Electrochem.* **33,** 196 (1927).
13. E. M. Filachione and C. H. Fisher, *Ind. Eng. Chem.* **36,** 223 (1944).
14. F. M. Filachione and co-workers, *U.S. Dept. Agric. Bur. Agric. Ind. Chem. Mimeo. Arc. Ser.* **AIC-295,** 11 (Feb. 1951).
15. R. Burns and co-workers, *J. Chem. Soc.,* 400 (1935).
16. L. T. Smith and co-workers, *Ind. Eng. Chem.* **34,** 473 (1942).
17. U.S. Pat. 2,371,281 (Mar. 13, 1945), H. V. Claborn (to the U.S. Government).
18. A. Sepitka, *Prum. Potravin* **14,** 606 (1963).
19. *Ibid.,* **15,** 193 (1964).
20. S. K. Bhatlacharyya and co-workers, *Ind. Eng. Chem. Prod. Res. Div.* **9**(1), 92 (1970).
21. Ger. Offen. 2,623,763 (Dec. 16, 1976), H. B. Tinker (to Monsanto Corp.).
22. Fr. Pat. 1,465,640 (Jan. 13, 1967), J. Boichand and co-workers (to Rhone Poulenc).
23. H. D. Hoperman and A. F. D'Adamo, Jr., *J. Org. Chem.* **25,** 30 (1960).
24. U.S. Pat. 2,973,270 (Feb. 28, 1961), J. B. Thompson and B. D. Buddenmeyer (to Paniplus).
25. U.S. Pat. 3,068,103 (Dec. 11, 1962), N. H. Kuhrt and L. J. Swickrick (to Eastman).
26. U.S. Pat. 3,144,341 (Aug. 11, 1964), S. W. Thompson (to Lever).
27. U.S. Pat. 3,145,107 (Aug. 18, 1964), N. B. Howard (to Proctor & Gamble).
28. U.S. Pat. 3,173,796 (Mar. 16, 1965), M. Pader (to Lever).
29. U.S. Pat. 3,180,736 (Apr. 27, 1965), B. W. Lanfried (to Top Scor).
30. U.S. Pat. 2,982,654 (May 2, 1961), E. G. Hammond and D. D. Deane (to Iowa St. College).
31. Can. Pat. 635,192 (Sept. 16, 1962), H. G. Foster, H. Crest, and E. H. Cornwell, (to Swift and Co.).
32. T. E. Freidemann and J. B. Graeser, *J. Biol. Chem.* **100,** 291 (1933).

33. J. B. Thompson and B. D. Buddemeyer, *Cereal Chem.* **31,** 296 (1954).
34. *Fed. Reg.* **121,** 1047 (Apr. 10, 1961).
35. U.S. Pats. 2,690,971 (Oct. 5, 1954); 2,752,376 (June 26, 1956); 2,957,932 (Oct. 25, 1960), H. T. Iverson and co-workers (to Glidden).
36. U.S. Pats. 2,864,703; 2,864,705 (Dec. 16, 1958), G. Scheilman (to Glen Labs).
37. U.S. Pats. 2,968,562; 2,968,963 (Jan. 17, 1961), C. J. Houser (to National Dairy Products).
38. U.S. Pat. 3,004,853 (Oct. 17, 1961), I. J. Perry and co-workers (to Glidden).
39. J. F. Treon and co-workers, *Agric. Food Chem.* **10,** 111 (1962).
40. H. M. Fett, *J. Am. Oil Chem.* **30,** 81 (1963).
41. M. S. Iszard, *Canning Age* **4,** 434 (1927).
42. U.S. Pat. 3,025,170 (Mar. 13, 1962), J. F. Murphy and R. E. Murphy (to Swift & Co.).
43. I. E. Althof, *Soft Drinks Trade J.,* (Mar. 1961).
44. J. T. Stearns, B. Makower, and P. T. Groggins, *Ind. Eng. Chem.* **32,** 1335 (1940).
45. P. D. Watson, *Ind. Eng. Chem.* **40,** 19393 (1948).
46. U.S. Pat. 2,400,873 (May 28, 1946), H. A. Bruson.
47. M. L. Fein and C. H. J. Fisher, *Org. Chem.* **15,** 530 (1950).
48. C. E. Rekberg and co-workers, *Ind. Eng. Chem.* **44,** 2191 (1952).
49. *Ibid.,* **42,** 2374 (1950).
50. U.S. Pat. 3,607,329 (Sept. 28, 1971), S. Manjekian.
51. U.S. Pat. 3,622,380 (Nov. 23, 1971), F. D. Williams (to UOP).
52. Czech. Pat. 145,868 (Oct. 15, 1972), J. Papajanovsky.
53. U.S. Pat. 3,655,833 (Apr. 11, 1972), H. Eggensperger, V. Franzen, and H. Stephen (to Deutsche Advance Produktion).
54. T. S. Rumsey and C. H. Noller, *J. Chromatogr.* **24,** 375 (1966).
55. R. A. Troupe and co-workers, *Ind. Eng. Chem.* **43,** 1143 (1951).
56. W. P. Pijper, *Acta Crystallogr. Sec. B* **27**(2), 344 (1971).
57. R. D. Ellison and co-workers, *Acta Crystallogr. Sec. B* **27**(2), 333 (1971).
58. U.S. Pat. 3,821,295 (June 28, 1974), J. H. Blumberg (to FMC).
59. U.S. Pat. 2,152,852 (Apr. 4, 1939), D. J. Loder (to E. I. du Pont de Nemours & Co., Inc.).
60. U.S. Pat. 2,153,064 (Apr. 4, 1939), D. J. Loder (to E. I. du Pont de Nemours & Co., Inc.).
61. U.S. Pat. 2,211,624 (Aug. 13, 1940), D. J. Loder (to E. I. du Pont de Nemours & Co., Inc.).
62. U.S. Pat. 3,911,003 (Oct. 7, 1975), S. Suzuki (to Chevron).
63. U.S. Pat. 3,859,349 (Jan. 7, 1975), N. F. Cody (to E. I. du Pont de Nemours & Co., Inc.).
64. U.S. Pat. 3,867,440 (Feb. 18, 1975), P. Kobetz and K. L. Lindsay (to Ethyl Corp.).
65. Indian Pat. 96,663 (Feb. 4, 1967), H. G. Vartak and S. G. Patel.
66. I. I. Korol'kov and V. I. Krupenskii, *Zh. Prikl. Khim.* (*Moscow*) (12), 2741 (1974).
67. USSR Pat. 271,506 (May 26, 1970), L. A. Merozov and co-workers.
68. *Hydroxyacetic Acid, Technical Bull. E12172,* E. I. du Pont de Nemours & Co., Inc., Wilmington, Del., 1976.
69. U.S. Pats. 3,912,663 (Oct. 14, 1975); 3,914,297 (Oct. 21, 1975); 3,917,601 (Nov. 4, 1975); 3,935,206 (Jan. 27, 1976), V. Lamberti (to Lever).
70. U.S. Pat. 3,862,219 (Jan. 21, 1975), V. Lindsay and co-workers (to Ethyl Corp.).
71. Ger. Offen. 2,209,132 (Oct. 5, 1972), Stukks (to Unilever).
72. U.S. Pat. 3,806,533 (Apr. 23, 1974), M. M. Tessler (to National Starch).
73. U.S. Pat. 3,778,804 (Jan. 29, 1974), R. J. Harper and co-workers (to U.S. Dept. of Agriculture).
74. U.S. Pat. 3,868,216 (Feb. 25, 1975), W. E. Franklin and co-workers (U.S. Dept. of Agriculture).
75. U.S. Pat. 3,796,596 (Mar. 12, 1974), Dargle and co-workers (to U.S. Dept. of Agriculture).
76. U.S. Pat. 3,537,926 (Nov. 3, 1970), G. Fisher (to Laney Lab.).
77. U.S. Pat. 3,704,156 (Nov. 28, 1972), E. F. Foley, Jr., and co-workers (to E. I. du Pont de Nemours & Co., Inc.).
78. Jpn. Kokai 72 13,680 (Apr. 24, 1972), K. Horazawa (to Honny Chemical Co.).
79. U.S. Pat. 3,808,143 (Apr. 30, 1974), T. R. Gardner and co-workers (to Halliburton Co.).
80. W. M. Bruner and L. T. Sherwood, Jr., *Ind. Eng. Chem.* **41,** 1653 (1949).
81. Ger. Offen. 2,156,133 (May 17, 1973), G. Kuenstle (to Wacker Chemie Co.).
82. A. Brenner and G. J. Riddell, *Res. Natl. Bur. Stand.* **39,** 385 (1947).
83. Ger. Pat. 1,257,766 (Jan. 4, 1968), W. Mueller and co-workers (to Lentia G.m.b.H. Chemie U. Pharm. Erzeugnisse-Ind.).

84. C. R. Noller, *Chemistry of Organic Compounds,* 3rd ed., W. B. Saunders Co., Philadelphia, Pa., 1965.
85. U.S. Pat. 3,202,702 (Aug. 24, 1965), K. L. Olivier (to Union Oil).
86. U.S. Pat. 3,069,433 (Dec. 18, 1962), K. A. Dunn (to Celanese Corp.).
87. V. G. Yanovich and co-workers, *Fiziol. Biokhim. Sil's'kogospod. Tvarin.* **13,** 109 (1970).
88. B. J. Passingham and R. N. Barton, *Biochemistry,* **62**(2), 418 (1975).
89. R. Chevais, *Ann. Anesthisiol Fr.* **16**(1), 1 (1975).
90. A. Pesce and G. Rugerio, *Minerva Anestesiol.* **42**(1), 88 (1976).
91. V. Andreoli and co-workers, *Frascastro* **69**(1–2), 87 (1976).

J. H. VAN NESS
Monsanto Industrial Chemicals Co.

HYDROXY DICARBOXYLIC ACIDS

Many natural and synthetic organic compounds are hydroxy dicarboxylic acids (see also Hydroxy carboxylic acids). This section deals with malic, thiomalic, tartaric, tartronic, and phloionic acids; thiomalic acid is included because of its structural similarity to malic acid.

Malic Acid

Malic acid (hydroxysuccinic acid, hydroxybutanedioic acid, 1-hydroxy-1,2-ethanedicarboxylic acid), $C_4H_6O_5$, is a white, crystalline material. The levorotatory isomer, *S*(−)-malic acid [*97-67-6*] (L-malic acid), is a natural constituent and common metabolite of plants and animals. It is the principal acid found in apples. The racemic compound, *R,S*-malic acid [*617-48-1*] (DL-malic acid), is a widely used organic food acidulant. This material also is used in some industrial applications to sequester ions, neutralize bases, and as a buffer for pH control. *R*(+)-Malic acid [*636-61-3*] (D-malic acid) is available as a laboratory chemical. In the United States, *R,S*-malic acid first was produced synthetically in 1923. Until the early 1960s, it was produced batchwise on a small scale and had limited industrial application. Following the introduction of a modern, continuous manufacturing process in the early 1960s, malic acid gradually became a large-volume industrial organic acid.

Physical Properties. Malic acid crystallizes from aqueous solutions as white, translucent, anhydrous crystals. The *S*(−) isomer melts at 100–103°C (1) and the *R*(+) isomer at 98–99°C (2). On heating, malic acid decomposes at ca 180°C. Under normal conditions, malic acid is stable; under conditions of high humidity, it is hygroscopic.

Malic acid is a relatively strong organic acid. Its dissociation constants and other constants are given in Table 1. The pH of a 0.001% aqueous solution is 3.80, that of a 0.1% solution is 2.80, and that of a 1.0% solution is 2.35. Many of its physical properties are similar to those of citric acid (qv) (4). Solubility characteristics are shown in Table 2 and densities of aqueous solutions are listed in Table 3.

Table 1. Physical Properties of *R,S*-Malic Acid[a]

Property	Value
mol wt	134.09
appearance	white, crystalline
crystal system	triclinic
melting point, °C	ca 130
d_4^{20}	1.601
dissociation constant	
K_1	4×10^{-4}
K_2	9×10^{-6}
heat of combustion (at 20°C), MJ/mol (kcal/mol)	1.340 (320.1)
heat of solution, kJ/mol (kcal/mol) solute	−20.515 (4.903)
viscosity (50% aqueous solution at 25°C), mPa·s (= cP)	6.5

[a] Ref. 3.

Table 2. Solubility of *R,S*(±)-Malic Acid[a]

Solvent	Temperature, °C	Value
water	5	48 g/100 g solution
	25	58 g/100 g solution
	75	80 g/100 g solution
ethanol (qv)	25	39.15 g/100 mL solvent
ethyl ether	25	1.41 g/100 mL solvent

[a] Ref. 3.

Table 3. Density of Aqueous Malic Acid Solutions at 15°C

Concentration, g/L	d_{15}^{15}, g/mL	(°Bé)	Concentration, g/L	d_{15}^{15}, g/mL	(°Bé)
30	1.115	(14.9)	46	1.169	(21.0)
32	1.121	(15.7)	48	1.179	(22.0)
34	1.129	(16.6)	50	1.186	(22.7)
36	1.138	(17.6)	52	1.192	(23.4)
38	1.146	(18.5)	54	1.199	(24.1)
40	1.151	(19.0)	56	1.208	(25.0)
42	1.158	(19.8)	58	1.212	(25.4)
44	1.165	(20.5)	60	1.220	(26.1)

Chemical Properties. ***Configuration.*** Malic acid is optically active because of its chiral center. The levorotatory enantiomer was confirmed as having the spatial configuration (**1**) (5) when tartaric acid was first reduced to malic acid (6).

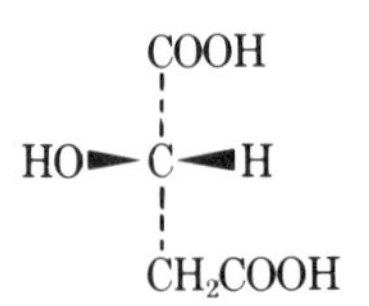

COOH | HO▶C◀H | CH_2COOH

(**1**) *S*(−)-malic acid

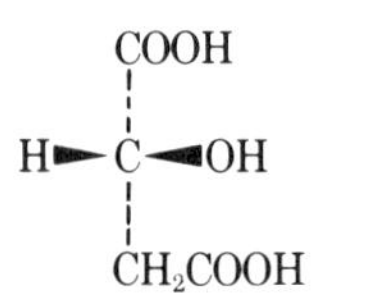

COOH | H▶C◀OH | CH_2COOH

(**2**) *R*(+)-malic acid

The other enantiomer (2) is assigned the R configuration. The optically inactive compound has the R,S symbol. A detailed discussion of configuration assignment by the sequence rule or the R and S system is given in ref. 7.

The optical activity of malic acid changes with dilution (8). The naturally occurring, levorotatory acid shows a most peculiar behavior in this respect; a 34% solution at 20°C is optically inactive. Dilution results in increasing levo rotation, whereas more concentrated solutions show dextro rotation. The effects of dilution are explained by the postulation that an additional form, the epoxide

$$HOOCCH_2CH\overset{O}{\frown}C(OH)_2,$$

occurs in solution and that the direction of rotation of the normal (linear) and epoxide forms is reversed (8). Synthetic or R,S-malic acid can be resolved into the two enantiomers by crystallization of its cinchonine salts.

Reactions. Malic acid undergoes many of the characteristic reactions of dibasic acids, monohydric alcohols, and α-hydroxy carboxylic acids. When heated to 170–180°C, it decomposes to fumaric acid and maleic anhydride which sublimes on further heating (see Maleic anhydride, maleic acid, and fumaric acid). Malic acid forms two types of condensation products: linear malomalic acids and the cyclic dilactone or malide; it does not form an anhydride.

As a dibasic acid, malic acid forms the usual salts, esters, amides, and acyl chlorides (see Esters, organic). Thus, malic acid yields the usual diesters with an alcohol in the presence of an esterification catalyst (see Esterification). Monoesters can be prepared easily by refluxing malic acid, an alcohol, and boron trifluoride as a catalyst (9). With polyhydric alcohols and polycarboxylic aromatic acids, malic acid yields alkyd polyester resins (10) (see Alcohols, polyhydric; Alkyd resins). Complete esterification results from the reaction of the diester of malic acid with an acid chloride, eg, acetyl or stearoyl chloride (11).

Alkyl halides in the presence of silver oxide react with alkyl malates to yield alkoxy derivatives of succinic acid, eg, 2-ethoxysuccinic acid, $HOOCCH_2CH(OC_2H_5)COOH$ (12–13). Another synthetic approach to ethers of malic acid is the reaction of malic esters and sodium alkoxides which affords 2-alkoxysuccinic esters (14).

The expected amides are obtained when alkyl esters of malic acid are treated with ammonia in alcoholic solution. Hydrazine reacts in a similar manner to yield malic dihydrazide (15) (see Hydrazine and its derivatives). Depending on the proportions of water that are present in the reaction, aniline and malic acid form N,N'-diphenylmalamide or the cyclic compound, N-phenylmalimide (16). When monoanilinium malate is distilled under reduced pressure, a mixture of C-anilino-N-phenylsuccinimide, N-phenylmalanil, and N-phenylsuccinimide is obtained (17–18).

Malic acid yields coumalic acid when treated with fuming sulfuric acid (19). Similar treatment of malic acid in the presence of phenol and substituted phenols is a facile method of synthesizing coumarins that are substituted in the aromatic nucleus (20–21) (see Coumarin). Similar reactions take place with thiophenol and substituted thiophenols, yielding, among other compounds, a red dye (22) (see Dyes and dye intermediates). Oxidation of an aqueous solution of malic acid with hydrogen peroxide (qv) catalyzed by ferrous ions yields oxalacetic acid (23). If this oxidation is performed in the presence of chromium, ferric, or titanium ions, or mixtures of these, the product is tartaric acid (24). Chlorals react with malic acid in the presence of sulfuric acid or other acidic catalysts to produce 4-ketodioxolones (25–26).

In aqueous solution, malic acid can be mildly corrosive toward carbon steels. Under normal conditions, it is not corrosive to stainless steels, which usually are the construction materials for processes involving malic acid. Malic acid also is virtually noncorrosive to tinplate and other materials used to package acidulated foods and beverages.

At proper pH in aqueous media, malic acid forms complexes or chelates with metal ions (see Chelating agents). These chelating reactions are useful in industrial processes requiring elimination or control of metal-ion catalysis (eg, of oxidation), removal of corrosion products (eg, rust), lowering of metal oxidation potentials (electroplating), etc. The chelating properties of malic acid vary with different metal cations, ionic strength, pH, etc and, in many cases, approximate those of other hydroxy carboxylic acids.

Occurrence. *S*(−)-Malic acid occurs widely in biological systems. It is the predominant acid in many fruits (Table 4); it is present with tartaric acid in grapes. However, malic acid occurs in relatively low concentrations, thus making its isolation from natural sources expensive and impractical. An ion-exchange process has been published (30) for the isolation of this acid from apple juice, which may contain from 0.4–0.7% malic acid (31) (see Fruit juices).

In addition to its presence in fruits, *S*(−)-malic acid has been found in cultures of a variety of microorganisms including *Aspergilli*, yeasts, *Sclerotinias,* and *Penicillium brevi-compactum.* Yields of levorotatory malic acid as high as 55 g/100 g (of D-glucose amounting to 74% of theory) have been reported for *Aspergillus flavus* and *Aspergillus parasiticus.* Iron, manganese, chromium, or aluminum ions reportedly enhance malic acid production. *S*(−)-Malic acid is involved in two respiratory metabolic cycles: the Krebs tricarboxylic acid cycle (see Citric acid) and the glyoxylic acid cycle. These metabolic cycles appear to play two essential roles in cellular metabolism: they account for the terminal oxidation system which supplies energy and provide the carbon skeletons from which many of the amino acids of proteins are derived.

Table 4. Malic Acid in Fruits[a]

Fruit	% of total acid	Fruit	% of total acid
apple	97.2	orange pulp	trace
apricot	23.7–69.8	peach	50.0–96.2
banana	53.7–92.3	pear	33.0–86.6
blueberry	6.0	persimmon	100.0
cherry	94.2	pineapple	12.5
cranberry	19.1–23.5	plum	98.5
gooseberry	46.2	quince	100.0
grape (Concord)	60.0	rhubarb	77.0
grapefruit	5.6	strawberry	9.9–11.0
lemon	4.5	watermelon	100.0
orange peel	59.6–80.0		

[a] Refs. 27–29.

Manufacture. In the United States, Canada, and Europe, only the synthetic, *R,S*-malic acid is produced commercially, whereas both the *S* and the *R,S* forms are produced in Japan.

Biosynthesis of S(−)-Malic Acid. Aqueous fumaric acid is converted to levorotatory malic acid by the intracellular enzyme, fumarase, which is produced by various microorganisms. A Japanese process for continuous commercial production of *S*(−)-malic acid from fumaric acid is based on the use of immobilized *Brevibacterium ammoniagenes* cells (32). The yield of pyrogen-free *S*(−)-malic acid that is suitable for pharmaceutical use is ca 70% of the theoretical. In this process, the *B. ammoniagenes* cells are immobilized in a polyacrylamide gel (see Enzymes, immobilized). A suspension of cells in saline solution is mixed with acrylamide and a suitable cross-linking agent and polymerization promoter. The resulting gel is formed into particles that are 3 mm in diameter, washed, and immersed in a 1 *M* sodium fumarate solution containing ca 0.3% bile extract. The solution is held at 37°C for 20 h. The treated cells then serve as packing in a reactor column. The feedstream is 1 *M* sodium fumarate and is passed through the system at 37°C. *S*(−)-Malic acid is separated from the effluent by conventional methods. The bile extract is used to enhance fumarate activity and depress the formation of succinic acid by-product. The process is carried out in a column reactor at 37°C and a space velocity of 0.23 h^{-1}. An equilibrium yield of *S*(−)-malic acid is produced, and the half life of fumarase activity in the immobilized cells is about two months. With a 100-L column, a 200-L/h feed flow would produce a theoretical yield of 15.4 t/mo of *S*(−)-malic acid. In practice, the process gives ca 70% of the theoretical yield.

Commercial Synthesis of R,S-Malic Acid. The commercial synthesis of *R,S*-malic acid involves hydration of maleic or fumaric acid at elevated temperature and pressure. A Japanese patent (33), describing a manufacturing procedure for malic acid, claims the direct hydration of maleic acid at 180°C and 1.03–1.21 MPa (150–175 psi). These workers suggest that in the hydration of maleic acid, fumaric acid is formed as a by-product and is hydrated slowly under the conditions of the reaction. If an amount of fumaric acid equivalent to that formed at equilibrium is charged with the maleic acid, it is possible to hydrate maleic acid to malic acid in a short time without the formation of additional fumaric acid.

The conventional commercial processes are commonly carried out in aqueous solution at elevated temperatures above 150°C, at pressures above 1.4 MPa (200 psi), and for a residence time of 3–5 h (34). The resulting mixture contains, primarily, malic acid and fumaric acid in equilibrium with a small percentage of maleic acid. The purification of the malic acid that is obtained can be accomplished by a two-stage crystallization process (35) involving the following steps: (*1*) adjusting the aqueous solution to a malic acid concentration of 40% by weight at 40°C; (*2*) cooling the solution to ca 15°C until equilibrium is reached; (*3*) separating solid fumaric acid from the slurry, preferably by filtration; (*4*) concentrating the mother liquor from the previous step to a malic acid concentration of at least 62% by weight at a temperature of at least 40°C to effect the crystallization of malic acid; (*5*) separating solid malic acid from the resulting slurry at about 40°C; (*6*) washing the malic acid with an aqueous solution that is substantially free from maleic and fumaric acids; (*7*) redissolving the washed malic acid crystals in water; (*8*) removing insolubles, principally fumaric acid, by filtration; (*9*) passing the mother liquor through a carbon column to remove colored contaminants; (*10*) adjusting the resulting solution to a malic acid concentration of 62% by weight at about 40°C and maintaining this temperature to effect the crystallization of malic acid; (*11*) separating solid malic acid at ca 40°C; and (*12*) washing the solid malic acid with an aqueous solution that is substantially free of maleic and fumaric acids. The purified malic acid contains less than 0.05% maleic acid and less than 1% fumaric acid. Additional purification can be achieved by use of cation- and anion-exchange columns after the carbon treatment and before the second-stage crystallization (36) (see Ion exchange). The cation exchange removes heavy-metal ions that were introduced into the system and the anion exchange removes unsaturated organic acids, primarily maleic and fumaric acids.

The purified malic acid crystals that are obtained after the final washing step are dried and classified before packaging. Figure 1 is a flow diagram of the process.

The initial step of production is carried out in a titanium reactor (34) because of the high corrosivity of maleic acid to most metals under the drastic reaction conditions used. The other steps are performed in stainless steel equipment. Improved purification processes for malic acid have been patented (37–38).

Energy Requirements. Despite the relatively high temperatures that are used, the energy requirements of the maleic acid hydration step are relatively low, because a substantial part of the heat that is supplied to the reaction is recovered. Thus, most of the 2.4–5 kg of steam required to produce 1 kg of malic acid is used for purification.

Shipment and Storage. Malic acid is shipped in 23-kg, multiwall, paper bags and in 114 kg fiber drums. A technical-grade, 50% solution of malic acid may be shipped in tank cars or tank trucks. Malic acid can be stored in dry form without difficulty; however, conditions of high humidity and elevated temperatures should be avoided to prevent caking.

Economic Aspects. U.S. and Canadian annual production capacity in 1978 was estimated to be 16,800 and 3,200 metric tons, respectively (39), and (1980) U.S. production is 7,000 t (40). The price (1980) of U.S. malic acid is 1.52/kg (41).

Specifications. *R,S*-Malic acid that is sold in the U.S. meets the specifications of the *Food Chemicals Codex* which are listed in Table 5 (42). Malic acid is available in the following U.S. standard sieve sizes:

granular:	min	100 wt % through 2.00 mm (10 mesh) sieve
	max	10 wt % through 0.30 mm (50 mesh) sieve
fine granular:	min	99 wt % through 0.71 mm (25 mesh) sieve
	max	5 wt % through 0.15 mm (100 mesh) sieve

Analytical and Test Methods. Aqueous titration with 1 *N* sodium hydroxide is the usual malic acid assay. Maleic and fumaric acid are determined by a polarographic method. Analytical methods are described in ref. 42.

Health and Safety. The FDA has affirmed *R,S*- and *S*(−)-malic acids as substances that are generally-recognized-as-safe (GRAS) as a flavor enhancer, flavoring agent and adjuvant, and a pH control agent, at levels varying from 6.9% for hard candy to 0.7% for miscellaneous food uses (43). *R,S*- and *S*(−)-malic acid may not be used in baby foods.

Uses. *R,S*-Malic acid has shown increasing use in food and nonfood applications because of its pleasant tartness, flavor-retention characteristics, high water solubility, chelating and buffering properties, and frequently, lower effective cost. Malic acid also is a reactive intermediate in chemical synthesis.

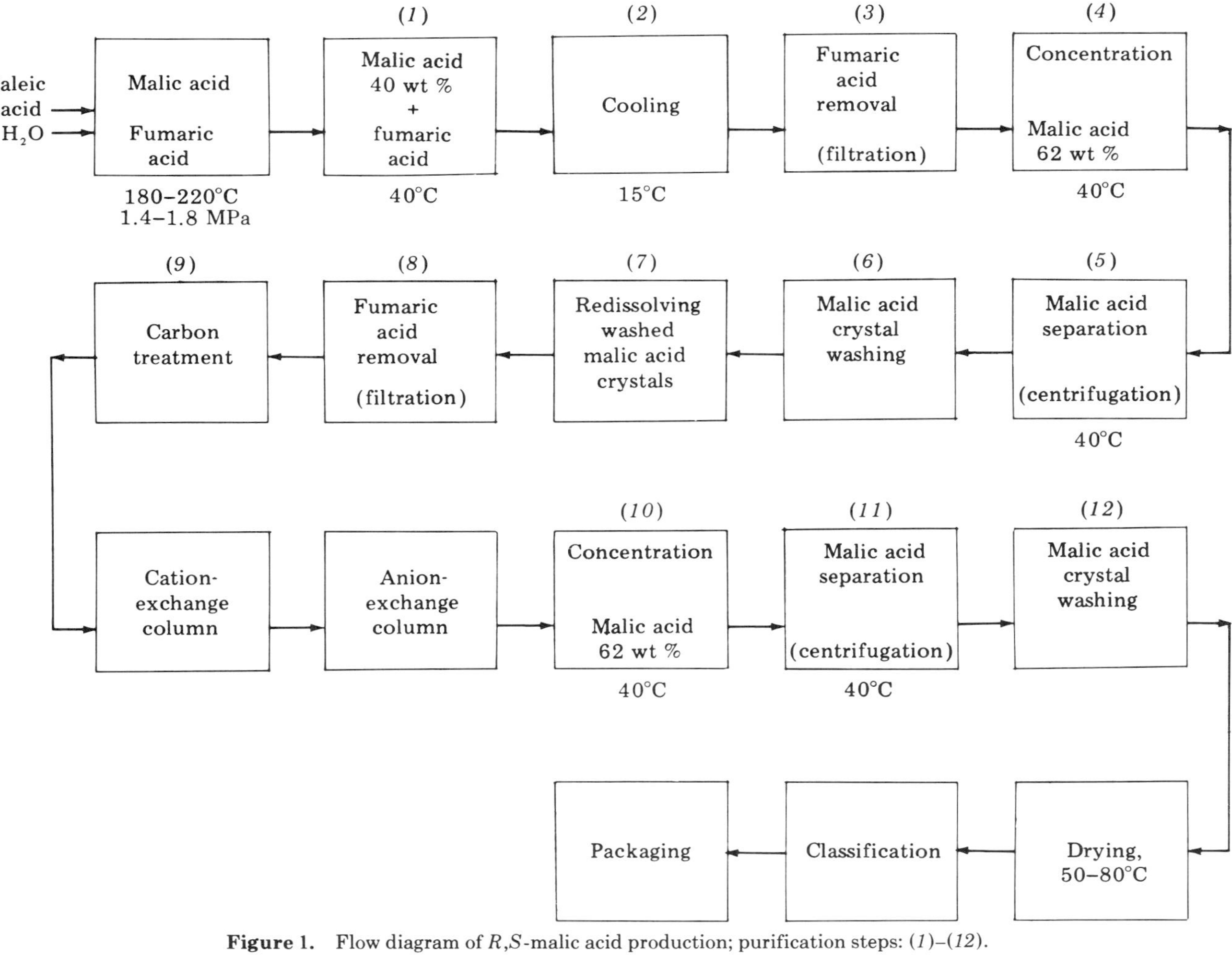

Figure 1. Flow diagram of *R*,*S*-malic acid production; purification steps: (*1*)–(*12*).

Table 5. ***Food Chemicals Codex*** **Specifications for *R,S*-Malic Acid**[a]

Property	Value
appearance	white or nearly white powder or granules
solubility	1.0 g dissolves in 0.8 mL of water
	1.0 g dissolves in 1.4 mL of ethanol
optical activity	solutions are optically inactive
assay	not less than 99.0% of $C_4H_6O_5$[a]
melting point	ca 130°C[a]
arsenic (as As)	3 ppm max
heavy metals (as Pb)	20 ppm max
lead	10 ppm max
maleic acid	0.05% max
residue on ignition	0.1% max
fumaric acid	1.0% max[a]
water-insoluble matter	0.1% max

[a] Ref. 42.

Food. Malic acid is a general-purpose soft drink acidulant. It is used widely in carbonated and still beverages and powdered drink mixes to provide tartness and the low pH required for the effective functioning of antimicrobial preservatives (eg, sodium benzoate). To a significant degree its acceptance as an additive that imparts tartness and flavor results from its advantageous economics (44). Malic acid is also used in candy, particularly hard candy where its low melting point and sugar inversion properties are of interest (45–46). Because of its ability to chelate trace metals (47), malic acid is used to protect vegetable oils (qv) during processing. Malic acid also is used in jellies, jams, preserves, desserts, canned fruits, vegetables, salad dressings, and in many other foods to provide tartness and pH control. A recent study shows (48) that R,S(±)-malic acid is a very valuable wine (qv) acidulant and, in some cases, is up to 30% more economical to use than citric acid (see Food additives).

Other. Malic acid is used in pharmaceuticals (qv), cosmetics (qv), dentifrices (qv) (49), metal cleaning, electroless plating (qv) (50), wash-and-wear textile finishing (51–53) (see Textiles, crease-resistant and durable-press finishes), for stabilization of heat-sensitive copying paper (54) (see Reprography), as an inhibitor of gelation, livering, and agglomeration in cellulose nitrate lacquers, and in many other applications.

Thiomalic Acid

Thiomalic acid [*70-49-5*] (mercaptosuccinic acid), $C_4H_6O_4S$, mol wt 150.2, is a sulfur analogue of malic acid. The properties of the crystalline, solid thiomalic acids are given in Table 6.

R,S-Thiomalic acid can be prepared from bromosuccinic acid by reaction with K_2S (55–56). The two enantiomers can be obtained from the corresponding optically active potassium bromosuccinates (56–57). Salts or amides of maleic acid react with NaSH to give the thiomalic derivatives (58). In the presence of a ferric salt in aqueous ammoniacal solution, thiomalic acid is oxidized to the disulfide (59). This oxidation also is catalyzed by other metal ions. Oxidation with diluted HNO_3 gives sulfosuccinic acid (60). The thiomalic acids form salts, chelates, and esters. *R,S*-Thiomalic acid gives a red color with $FeCl_3 + NH_4OH$ (61). The reaction with aqueous AuCN affords aurothiomalic acid (62).

Table 6. Properties of Thiomalic Acids[a]

Acid	CAS Registry No.	Mp, °C	Solubility, Water	Solubility, Ethanol	$[\alpha]_D^{17}$ (5% in ethanol)	Dissociation at 25°C, pK_{a1}	Dissociation at 25°C, pK_{a2}
R,S-	[*644-87-1*]	151	very sol	very sol		3.30	4.94
R	[*20182-99-4*]	154	sol	sol	+64.4°		
S	[*74708-34-2*]	152–153	sol	slightly sol	−64.8°		

Thiomalic acid is a skin sensitizer (63) and an antidote in heavy-metal poisoning (64). It is a component of cold permanent hair-waving solutions (see Hair preparations) (65–66) and of rust-removing (67) and corrosion-inhibiting compositions (see Corrosion and corrosion inhibitors) (68). Sodium aurothiomalate (Myochrisin) and other gold thiomalate complexes have antiarthritic properties (69) (see Gold and gold compounds). The well known insecticide, malathion, is the thiomalate *S*-ester of *O,O*-dimethylphosphonodithioic acid (see Insect control technology).

Tartaric Acid

Tartaric acid [*526-83-0*] (2,3-dihydroxybutanedioic acid, 2,3-dihydroxysuccinic acid), $C_4H_6O_6$, is a dihydroxy dicarboxylic acid with two chiral centers. It exists as the dextro- and levorotatory acid, the meso-form (which is inactive owing to internal compensation), and the racemic mixture (which commonly is known as racemic acid). The commercial product in the U.S. is the natural, dextrorotatory form, (*R*-*R**,*R**)-tartaric acid [*87-69-4*], (L(+)-tartaric acid). This enantiomer occurs in grapes as its acid potassium salt (cream of tartar). In the fermentation (qv) of wine, this salt deposits in the vats; free crystallized tartaric acid was first obtained from such fermentation residues by Scheele in 1769.

In Europe, South Africa, and Japan, racemic (*R**,*R**)-tartaric acid [*133-37-9*] (DL-tartaric acid) also is produced commercially.

Physical Properties. When crystallized from aqueous solutions above 5°C, natural, (*R*-*R**,*R**)-tartaric acid is obtained in the anhydrous form. Below 5°C, tartaric acid forms a monohydrate which is unstable at room temperature. The optical rotation of an aqueous solution varies with the concentration. It is stable in air, and racemizes with great ease on heating. Some of the physical properties of (*R*-*R**,*R**)-tartaric acid are listed in Table 7.

The solubility of (*R*-*R**,*R**)-tartaric acid in water is given in Table 8. One hundred grams of absolute ethanol dissolves 20.4 g of tartaric acid at 18°C and 100 g of ethyl ether dissolves 0.3 g at 18°C. Densities of (*R*-*R**,*R**)-tartaric acid solutions are listed in Table 9.

Some physical properties of the four enantiomeric tartaric acids are compared in Table 10.

Chemical Properties. ***Configuration.*** The assignment of systematic names that would unambiguously reflect the configurations of the four tartaric acids is difficult. The notation used by *Chemical Abstracts* has been adopted as follows:

```
      COOH                 COOH                 COOH
       |                    |                    |
   H—C—OH             HO—C—H              H—C—OH
       |                    |                    |
  HO—C—H               H—C—OH             H—C—OH
       |                    |                    |
      COOH                 COOH                 COOH
```

(*R*-*R**,*R**)-tartaric acid [*87-69-4*] (*S*-*R**,*R**)-tartaric acid [*147-71-7*] *meso*-tartaric acid [*147-73-9*]

Table 7. Physical Properties of (*R*-*R,*R**)-Tartaric Acid**

Property	Value
mol wt	150.086
appearance	colorless or translucent crystals
crystal system	monoclinic
mp, °C (anhydrous)	169–170
d^{20}, g/cm^3	1.76
$[\alpha]_D^{20}$ (for concentration, c, from 20–50%)	15.050–0.1535 c
pK_1 (at 25°C)	1.04×10^{-3}
pK_2 (at 25°C)	4.55×10^{-5}
heat of combustion, MJ/mol (kcal/mol)	1.08 (257)[a]
heat of solution, kJ/mol (kcal/mol)	−13.8 (−3.3)[b]

[a] Ref. 70.
[b] Ref. 3.

Table 8. Solubility of (*R*-*R,*R**)-Tartaric Acid in Water**

Temp, °C	Soly, g/100 g H_2O	Temp, °C	Soly, g/100 g H_2O	Temp, °C	Soly, g/100 g H_2O
0	115	30	156	70	244
5	120	40	176	80	273
10	125	50	195	90	307
20	139	60	218	100	343

Table 9. Density of (*R*-*R,*R**)-Tartaric Acid Solutions at 15°C[a]**

wt % (at 15°C)	d_4^{15}
1	1.0045
10	1.0469
20	1.0969
30	1.1505
40	1.2078
50	1.2696

[a] Ref. 71.

Racemic acid is an equimolar mixture of the two optically active enantiomers and, hence, like the meso acid, is optically inactive.

Reactions. When free (*R*-*R**,*R**)-tartaric acid is heated above its melting point, amorphous anhydrides are formed which, on boiling with water, regenerate the acid. Further heating causes simultaneous formation of pyruvic acid ($CH_3COCOOH$), pyrotartaric acid ($HOOCCH_2CH(CH_3)COOH$), and (finally) a black, charred residue. In the presence of ferrous salt and with hydrogen peroxide, dihydroxymaleic acid, $HOOCCOH{=}COHCOOH$, is formed. Nitrating the acid yields a dinitro ester which, on hydrolysis, is converted to dihydroxytartaric acid, $HOOCC(OH)_2C(OH)_2COOH$. The latter forms tartronic acid, $HOCH(COOH)_2$, on oxidation with nitric acid. Acetyl chloride or acetic anhydride form diacetyltartaric anhydride, which is readily hydrolyzable to diacetyltartaric acid.

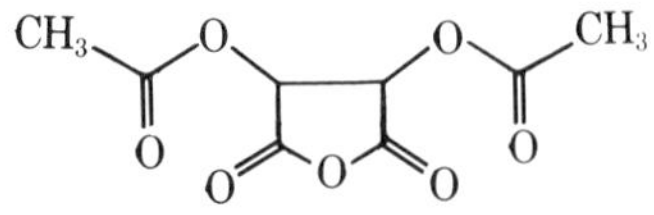

CH_3 O O CH_3 O O O O O

Table 10. Physical Properties of Tartaric Acids

Properties	Natural	Levorotatory	Racemic	Meso
mp, °C (anhydrous)	169–170	169–170	205–206	159–160
soly, g/100 g H_2O (in water, at 20°C)	139	139	20.6	125
soly of acid potassium salt, g/100 g H_2O (at 25°C)	0.84	0.84	0.72	16.7
soly of Ca salt, g/100 g	0.020[a]	0.025[b]	0.004[a]	0.034[b]
moles of water in hydrate of calcium salt	4	4	8	3

[a] At 25°C.
[b] At 20°C.

Tartaric acid is reduced stepwise with concentrated hydriodic acid, first to *R*(+)-malic acid and then to succinic acid. Ammoniacal silver solution is reduced with the formation of a silver mirror. In common with some other hydroxy organic acids, tartaric acid complexes many metal ions; thus, tartaric acid and its salts are useful as sequestering agents. In aqueous solution, tartaric acid can be mildly corrosive toward carbon steels. Under normal conditions, it is not corrosive to stainless steels.

Occurrence. (*R*-*R**,*R**)-Tartaric acid occurs in the juice of the grape and in a few other fruits and plants. It is not as widely distributed as citric or *S*(−)-malic acid. The only commercial source is the residues from the wine industry.

(*S*-*R**,*R**)-Tartaric acid has been found in the fruit and leaves of *Bauhinia reticulata*, a tree native to Mali (western Africa). Like the dextrorotatory acid, it forms anhydrous monoclinic crystals.

The racemic acid is not a primary product of plant processes but is formed readily from the dextrorotatory acid by heating alone or with strong alkali or strong acid. As the classical example of an optically inactive compound originating from the combinations of molar proportions of the dextro and levo enantiomers, it is of considerable historical interest. The methods by which such racemic compounds can be separated into the optically active modifications were devised by Pasteur and were applied first to the racemic acid. Racemic acid crystallizes as the dihydrate $(C_4H_6O_6)_2.2H_2O$ in triclinic prisms. It becomes anhydrous on drying at 110°C and melts incongruently at 205°C. Calcium racemate, $(C_4H_4O_6Ca)_2$, is even less soluble in water than calcium tartrate; thus, a dilute racemic acid solution is precipitated by a saturated calcium sulfate solution, whereas active tartaric acid is not.

meso-Tartaric acid is not found in nature. It is obtained from the other isomers by prolonged boiling with caustic alkali. The free acid crystallizes as a monohydrate, $C_4H_6O_6.H_2O$, in monoclinic prisms. On drying at 110°C, it becomes anhydrous and melts at 159–160°C.

Synthesis. Racemic acid is obtained synthetically by treatment of maleic acid with hydrogen peroxide in the presence of a catalyst, eg, tungstic acid (72) (see also Manufacture). Other synthetic routes that have been explored include the production of (*R*-*R**,*R**)-tartaric acid by bacterial fermentation of glucose (73) or 5-keto-D-gluconic acid (74), catalytic oxidation of 5-keto-D-gluconic acid with gaseous oxygen (75–76), and nitric acid oxidation of carbohydrates (qv), eg, glucose (77). Production of (*R**,*R**)-tartaric acid by catalytic chlorate oxidation of fumaric or maleic acid also has been described (78).

meso-Tartaric acid can be prepared by microbiological conversion of *trans*-

epoxysuccinic acid (readily obtainable by fermentation of an aqueous medium containing glucose or ethanol, eg, with *Aspergillus fumigatus* No. 12) (79). *cis*-Epoxysuccinic acid does not undergo this conversion. With exception of the hydrogen peroxide oxidation of maleic acid, none of the foregoing processes have achieved commercial status.

Manufacture. ***(R-R*,R*)-Tartaric Acid and Its Salts.*** The raw materials available for the manufacture of tartaric acid and tartrates are by-products of wine making. Crude tartars are recoverable from the following sources:

(*1*) The press cakes from grape juice (ie, unfermented (marcs) or partly fermented (pomace)), are boiled with water, and alcohol, if present, is distilled. The hot mash is settled, decanted, and the clear liquor is cooled to crystallize. The recovered high-test crude cream of tartar (vinaccia) has an 85–90% cream of tartar content.

(*2*) Lees, which are the dried slimy sediments in the wine fermentation vats, consist of yeast cells, pectinous substances, and tartars. Their content of total tartaric acid equivalent ranges from 16–40%.

(*3*) The crystalline crusts that form in the vats in the secondary fermentation period (argols) contain more than 40% tartaric acid; they are high in potassium hydrogen tartrate and low in the calcium salt.

It usually is advantageous to combine the manufacture of tartaric acid, cream of tartar, and Rochelle salt in one plant. This permits the most favorable disposition of the mother liquors from the three processes. The chemical reactions involved are: formation of calcium tartrate from crude potassium acid tartrate,

$$2\ KHC_4H_4O_6 + Ca(OH)_2 + CaSO_4 \rightarrow 2\ CaC_4H_4O_6 + K_2SO_4 + 2\ H_2O$$

formation of tartaric acid from calcium tartrate,

$$CaC_4H_4O_6 + H_2SO_4 \rightarrow H_2C_4H_4O_6 + CaSO_4$$

formation of Rochelle salt from argols,

$$2\ KHC_4H_4O_6 + Na_2CO_3 \rightarrow 2\ KNaC_4H_4O_6 + CO_2 + H_2O$$

and formation of cream of tartar from tartaric acid and Rochelle salt liquors,

$$2\ H_6C_4O_6 + 2\ KNaC_4H_4O_6 + K_2SO_4 \rightarrow 4\ KHC_4H_4O_6 + Na_2SO_4$$

Calcium Tartrate. The oldest calcium tartrate preparation procedure is the decantation process, in which fresh, wet lees or finely ground dry lees are treated with sufficient hydrochloric acid to dissolve all tartrates. The acidified magma is diluted with sufficient water to obtain at least 50% supernatant solution after settling. The acidic liquor is drawn off and the insolubles are washed by repeated suspension in water and decantation. The combined extracts are neutralized with ground limestone or hydrated lime which leaves the end reaction slightly acidic (see Lime and Limestone). Tartar recovery from winery still residues may be accomplished by exchange adsorption on an ion exchanger in the chloride form. Normal salt solution serves as the regenerant. Concentration of the tartrate in solution is increased 15–18 times more than the original still residue. High purity calcium tartrate is the final product.

In the neutral pressure process, the ground tartars are suspended in water, heated to boiling with direct steam, and then nearly neutralized with a slurry of hydrated lime. The magma is autoclaved at ca 35 kPa (50 psi) for 2–3 h. After cooling to 30°C in an open tank, the batch is heated with calcium chloride in slight excess, is cooled further, filtered, and washed. It is then promptly acidified with sulfuric acid to prevent bacterial activity.

In the roasting process, the crude material is ground so that it passes through a 0.59–0.84-mm (20–30 mesh) screen. Roasting time is from 2–6 h at 115–165°C. The roasted material is neutralized with hydrated lime slurry in a wooden precipitation tank that is provided with a cooling coil. A slurry of precipitated calcium sulfate is added until a 20% excess is present. The reaction mixture is cooled to 30°C, filtered, and washed with cold water. The filtrate is saved for recovery of potassium sulfate. Potassium sulfate recovery is effected by treatment with sodium carbonate which precipitates the calcium salt. The slurry is filtered and concentrated in a triple-effect evaporator. Crystallized potassium sulfate is collected on drain boxes.

In the decomposition of calcium tartrate, the washed press cake is slurried with wash water from a previous acid charge and is acidified with cold 70% sulfuric acid to a pH of 0.8 at sp gr of 1.12. The batch is filtered and the filtrate is evaporated under vacuum until a heavy magma of crystals is obtained. The charge is passed through lead-lined, stirred granulators and is centrifuged to remove the mother liquor. The mother liquor is boiled again in vacuum to yield a second crop of crystals. The filtrate from this crop is treated to yield a third crop, although this is of poorer quality and often is used to make calcium tartrate or cream of tartar.

A concentrated solution of the crystals is made in wooden tanks. Sulfate is removed by adding barium carbonate, and iron is removed with addition of calcium ferrocyanide. The solution is filtered and evaporated to crystallization in vacuum tanks. The extraction of tartaric acid from by-products of winemaking by percolation with a mixture of nitric and sulfuric acids (80) and a modified method for isolating calcium tartrate (81) have been patented.

Biochemical methods of production of (*R*-*R**,*R**)-tartaric acid include enzymatic hydrolysis of synthetic *cis*-epoxysuccinic acid (82–83) and the fermentation of carbohydrates (84).

Production of the Rochelle salt involves roasting of the blended argols at 160–165°C, mixing with wash liquor from a previous charge, and treatment with sodium carbonate and afterwards with potassium oxalate. Evaporation of the resulting filtrate followed by centrifugation, washing, and drying yields the finished product.

Cream of tartar can be produced (*1*) directly from the argol–sodium carbonate cook; (*2*) by combining tartaric acid solutions (which are available from the manufacture of tartaric acid) with Rochelle salt solution derived from suitable crude potassium bitartrate, eg, high-grade argols or recovered cream of tartar; and (*3*) by saturating a water suspension of argols with sulfur dioxide. An accelerated process for removing potassium bitartrate from crude tartaric acid solutions has been described (85).

For the production of tartar emetic, potassium bitartrate and antimony oxide, Sb_2O_3, are added simultaneously to water in a stainless-steel reactor. The reaction mixture is diluted, filtered, and collected in jacketed granulators where crystallization takes place after cooling. Centrifuging, washing, and drying completes the process.

(*R**,*R**)-Tartaric (racemic) acid is obtained synthetically by epoxidation (qv) of maleic acid with hydrogen peroxide in the presence of a catalyst followed by hydrolysis of the resulting *cis*-epoxysuccinic acid (72,86–92). This commercial process is used in South Africa (91), and it involves the addition of 60% H_2O_2 to 40% aqueous maleic acid. The addition of a molybdenum catalyst permits the reaction to take place at 70°C in 20 h. The resulting *cis*-epoxysuccinic acid is hydrolyzed by boiling, and (*R**,*R**)-tartaric acid is isolated by cooling, centrifuging, washing, and drying. A 20%

Table 11. Imports Into the U.S. of Tartaric Acid[a], kg

Country	1976	1977	1978
Argentina	108,226	176,419	226,465
Austria	na	na	1002
Brazil	na	na	18,038
Canada	36,075	na	[b]
France	35,041	34,551	187,810
FRG	25,884	11,139	na
India	50	na	na
Italy	335,309	365,485	302,643
Netherlands	8217	na	4545
Spain	620,124	1,013,597	832,106
Switzerland	425	224	364
Total	*1,159,353*	*1,601,415*	*1,572,973*
Value, $	1,212,015	1,599,393	1,621,964

[a] Ref. 94.
[b] Tartaric acid was not imported from Canada in 1978.

Table 12. Imports Into the U.S. of Cream of Tartar[a], kg

Country	1976	1977	1978
FRG	[b]	2505	[b]
Italy	423,012	402,856	501,638
Spain	238,945	164,856	277,833
UK	[b]	[b]	1165
Total	*661,957*	*570,276*	*780,636*
Value, $	646,422	479,371	673,542

[a] Ref. 94.
[b] Data not available.

excess of maleic acid is used and the yield (based on H_2O_2) is 84%. Unreacted maleic acid is converted either to fumaric acid or malic acid. The tartaric acid in the mother liquor is recovered.

(R^*,R^*)-Tartaric acid also can be produced by racemization of (R-R^*,R^*)-tartaric acid in the presence of *meso*-tartaric acid (93). In this process, formation of *meso*-tartaric acid during racemization does not occur.

Economic Aspects. In 1978, the United States imported 1600 metric tons of tartaric acid and 700 t of cream of tartar. These materials came from wine-producing countries, eg, Spain, Italy, France. Some crude tartars also were imported and refined to tartaric acid and tartrates by the U.S. chemical industry. Approximately 30,000 t of tartaric acid is used annually worldwide (91). U.S. imports of the finished products for 1976–1978 are shown in Tables 11 and 12. At the end of 1978, U.S. pricing was as follows (95): tartaric acid, NF: $1.87–2.09/kg; potassium bitartrate, NF: $1.50–1.87/kg; Rochelle salt: $1.23–1.65/kg; and tartar emetic: $3.85–4.07/kg.

Specifications. (R-R^*,R^*)-Tartaric acid that is sold in the United States meets the specifications of the *Food Chemicals Codex* (42) (Table 13) and the USP (96).

Tartaric acid is supplied as Fine-Granular and Powder in 45-kg bags. It should be stored in tightly closed containers.

Table 13. *Food Chemicals Codex* Description and Specifications for (*R*-*R**,*R**)-Tartaric Acid

appearance	colorless or translucent crystals or a white, fine-to-granular, crystalline powder
solubility	1.0 g dissolves in 0.8 mL of water at 25°C, in ca 0.5 mL of boiling water, and in ca 3 mL of ethanol
specific rotation	$[\alpha]_D^{25°}$ + 12.0–13.0°, determined in a solution containing 2 g in each 10 mL sample[a]
assay	not less than 99.7% of $C_4H_6O_6$ after drying
arsenic (as As)	3 ppm max
heavy metals (as Pb)	10 ppm max
loss on drying	0.5% max
oxalate	passes test
residue on ignition	0.05% max
sulfate	passes test

[a] Ref. 42.

Analytical and Test Methods. Test methods for tartaric acid and some tartrates are described in refs. 42 and 97.

Health and Safety (Toxicology). In the past, the FDA considered (*R*-*R**,*R**)-tartaric acid as a generally-recognized-as-safe (GRAS) food substance (98). Recently, the FDA also proposed the affirmation of this GRAS status (99).

Tartaric acid and tartrates are poorly absorbed from the intestine. Their metabolism is different from that of citric acid in that tartaric acid is only slightly oxidized. The acid that is absorbed is excreted unchanged in the urine. Information regarding the physiological properties of tartaric acid is available in the literature (99). So far as is known, all nutritional and physiological investigations have been made with the dextrorotatory enantiomer.

Uses. (*R*-*R**,*R**)-Tartaric acid is used as an acidulant in carbonated and still beverages, including beverage powders, as well as in a number of other acidulated food products, (eg, fruit jellies, hard candy, starch jelly candy).

The complexing ability of tartaric acid and its salts has been employed to advantage in metal cleaning and finishing (see Metal surface treatments). These chemicals can be used as paste and powder cleaners and as bath components in electrolytic polishing of copper and its alloys, aluminum, and ferrous metals. They also are employed in baths for removal of rust, scale, and oxide and in baths for electroless plating (qv) of metals. Tartrates are valuable sequestrants in aluminum etching. The iron–tartaric acid complex is an oxidation-stable solvent for cellulose (100).

Rochelle salt is well known as an important bath component in electroplating of many metals and alloys. It is used in the silvering of mirrors, in crystal-controlled electronic oscillators, and as a component of mild saline cathartic preparations. It can be used as an emulsifying agent in the manufacture of processed cheese.

Cream of tartar is used in baking powder and prepared baking mixes (see Bakery processes and leavening agents). It also is used in metal cleaners and in gold and silver coating of various metals. Solutions of cream of tartar can be used for rinsing and storing metal components between acid plating and pickling. This use is important in plating titanium and other metals that tend to form an oxide coating after abrasive or chemical etching. A rinse with a 5% tartrate solution retards oxide formation and improves plate adhesion.

Tartar emetic is used in small doses as an expectorant in cough syrups (see Expectorants and antitussives). It is one of the best drugs available for treating infections caused by *Schistosoma japonium*. It is no longer used as an emetic.

Diacetyltartaric acid and its mono- and diglycerides are used in the baking industry as dough emulsifiers and conditioners.

Derivatives. Tartaric acid forms stable salts and esters. The salts of commercial importance are: acid potassium tartrate (potassium hydrogen tartrate, cream of tartar, potassium bitartrate), $KHC_4H_4O_6$; soly, g/100 mL water: 0.6 at 20°C, and 6.1 at 100°C. Potassium sodium tartrate tetrahydrate (Rochelle salt), $KNaC_4H_4O_6.4H_2O$, soly, g/100 mL water: 26 at 26°C and 66 at 100°C. Potassium antimony tartrate (tartar emetic), $KSbC_4H_2O_6.1\frac{1}{2}H_2O$; soly, g/100 mL water: 8.7 at 25°C and 35.7 at 100°C. Calcium tartrate, $CaC_4H_4O_6.4H_2O$, is the usual intermediate in the commercial production of tartaric acid. The neutral esters of the lower aliphatic alcohols, formed by the usual esterification method are liquids or low-melting solids.

Fehling's solution (alkaline cupric tartrate) is used widely for the determination of reducing sugars. It consists of two equal volumetric amounts of solutions: solution A contains 173 g of crystallized potassium sodium tartrate tetrahydrate and 50 g of sodium hydroxide in a 500-mL solution with water; solution B contains 34.66 g of cupric sulfate pentahydrate (showing no trace of efflorescence) in a 500-mL solution of water. On reaction with a reducing sugar, the reagent yields cuprous oxide (Cu_2O) and gives a dark red color.

Tartronic Acid

Tartronic acid [*80-69-3*] (hydroxypropanedioic acid, hydroxymalonic acid), $C_3H_4O_5$, mol wt 120.06, is a crystalline substance, mp 156–158°C (decomp), is soluble in water and ethanol and slightly soluble in ether. It forms salts, chelates, and esters. It is formed in the ozonization of malonic acid (101) and in the oxidation of glucose (102). It can be synthesized from tartaric acid (103) and from diethyl benzyloxymalonate.

Tartronic acid occurs in various plants. In the United States, it is available only as a laboratory chemical; it is not manufactured on a large scale.

Tartronic acid is not metabolized by the rat. It causes the contraction of coronary blood vessels (105) and exerts a beneficial effect in the topical treatment of keratosis (106).

Phloionic Acid

Phloionic acid [*137-21-3*] (9,10-dihydroxyoctadecanedioic acid), $C_{18}H_{34}O_6$, mol wt 346.45, in esterified form is a constituent of cork (qv) from which it can be obtained by alkaline hydrolysis (107–108). The compound that is isolated is optically inactive, has the (R^*,R^*)-(±)phloionic acid [*23843-52-9*] configuration, and melts at 121°C. The two enantiomers can be obtained from the racemic acid by resolution with brucine (109). Phloionic acid is insoluble in cold water but slightly soluble in boiling water; it is soluble in ethanol, acetone, and acetic acid. Its stereochemistry is discussed in ref. 109.

Phloionic acid has been synthesized in several ways (110–113). It forms salts, esters, and amides and can be used to make polymers, eg, glycol polyesters which, in turn, react with diisocyanates to give polyurethanes (114).

BIBLIOGRAPHY

"Tartaric Acid" in *ECT* 1st ed., Vol. 13, pp. 645–656, by Richard Pasternack, Chas. Pfizer & Co., Inc.; "Tartaric Acid" in *ECT* 2nd ed., Vol. 19, pp. 723–732, by D. F. Chichester, Chas. Pfizer & Co., Inc.; "Malic Acid" in *ECT* 2nd ed., Vol. 12, pp. 837–839, by W. E. Irwin, L. B. Lockwood, and M. F. Zienty, Miles Laboratories, Inc., Chemicals Division.

1. T. S. Patterson and C. Buchanan, *J. Chem. Soc.*, 3006 (1928).
2. A. E. Dunstan and F. B. Thole, *J. Chem. Soc.* **93,** 1815 (1908).
3. W. H. Gardner, *Food Acidulants,* Allied Chemical Corp., 1966, p. 7.
4. *Malic Acid, Tech. Bull. TS-17,* National Aniline Div., Allied Chemical Corp., 1964.
5. E. Fischer, *Ber.* **29**(2), 1377 (1896).
6. M. G. J. W. Bremer, *Bull. Soc. Chim. Fr. Ser. 2* **25,** 6 (1876).
7. J. F. Stoddart in J. F. Stoddart, ed., *Comprehensive Organic Chemistry,* Vol. 1, *Stereochemistry, Hydrocarbons, Halo Compounds, Oxygen Compounds,* Pergamon Press, Oxford, 1979, pp. 3–33.
8. W. D. Bancroft and H. L. Davis, *J. Phys. Chem.* **34,** 897 (1931).
9. J. A. Niewland, R. R. Vogt, and W. L. Foohey, *J. Am. Chem. Soc.* **52,** 1018 (1930).
10. U.S. Pat. 1,489,744 (Apr. 8, 1924), G. R. Downs and L. Weisberg (to Barrett Co.).
11. K. Freudenberg and A. Lux, *Ber.* **61B,** 1083 (1928).
12. T. Purdie and G. B. Neave, *J. Chem. Soc.* **97,** 1517 (1910).
13. T. Purdie and Pitkeathly, *J. Chem. Soc.* **75,** 157 (1899).
14. L. H. Flett and W. H. Gardner, *Maleic Anhydride Derivatives,* John Wiley & Sons, Inc., New York, 1952, p. 64.
15. T. Curtius and C. von Hofe, *J. Prakt. Chem.* **95**(2), 210 (1917).
16. A. E. Arppe, *Ann. Chem.* **96,** 106 (1855).
17. R. Anschütz and Q. Wirtz, *Ann. Chem.* **239,** 137 (1887).
18. J. B. Tingle and S. J. Bates, *J. Am. Chem. Soc.* **32,** 1233 (1910).
19. H. von Pechmann, *Ann. Chem.* **264,** 272 (1891).
20. V. V. K. Sastry, *J. Indian Chem. Soc.* **19,** 403 (1942).
21. R. C. Shah and co-workers, *J. Indian Chem. Soc.* **14,** 717 (1937).
22. J. Smiles and E. W. McClelland, *J. Chem. Soc.* **119,** 1810 (1921).
23. H. J. H. Fenton and H. O. Jones, *J. Chem. Soc.* **77,** 77 (1900).
24. Fr. Pat. 849,852 (Dec. 4, 1939), H. Goldstein and A. Bonn.
25. F. H. Yorston, *Rec. Trav. Chim.* **46,** 711 (1927).
26. N. M. Shah, *J. Indian Chem. Soc.* **16,** 285 (1939).
27. T. J. Sausville, *Food Technol.* **19,** 67 (1965).
28. T. J. Sausville, *Glass Packer* **44,** 27 (1965).
29. L. H. Meyer, *Food Chemistry,* Reinhold Publishing Corp., New York, 1960, pp. 275–277.
30. R. E. Buck and H. H. Mottern, *Ind. Eng. Chem.* **37,** 635 (1945).
31. R. E. Buck and H. H. Mottern, *Ind. Eng. Chem.* **39,** 1087 (1947).
32. M. K. McAbee, *Chem. Eng. News* **53,** 32 (Aug. 25, 1975).
33. Jpn. Pat. 4360 (Dec. 16, 1950), K. Saito, Y. Ono, and Y. Mikawa.
34. U.S. Pat. 3,379,756 (Apr. 23, 1968), C. R. Ahlgren (to Allied Chemical Corp.).
35. U.S. Pat. 3,391,187 (July 2, 1968), M. A. Cullen, Jr. and M. R. Ingleman (to Allied Chemical Corp.).
36. U.S. Pat. 3,371,112 (Feb. 27, 1968), L. O. Winstrom and M. R. Ingleman (to Allied Chemical Corp.).
37. U.S. Pat. 3,983,170 (Sept. 28, 1976), S. Sumikawa and R. Maida (to International Organics, Inc., and Kyowa Hakko Kogyo Co., Ltd.).
38. U.S. Pat. 4,035,419 (July 12, 1977), S. Sumikawa, S. Sakaguchi, and T. Okiura.
39. L. Marion, *Chem. Eng.* **85**(25), 82 (1978).
40. C. Andres, *Food Process.* **41,** 72 (Mar. 1980).
41. *Chem. Mark. Rep.* 23 (June 23, 1980).

42. *Food Chemicals Codex,* 3rd ed., National Academy of Sciences–National Research Council, Washington, D.C., 1980.
43. *Fed. Regist.* **44,** 20,655 (Apr. 6, 1979).
44. S. E. Berger, *Beverage World* **96,** 52 (1977).
45. M. B. Sherman, *Manuf. Confect.* **45**(4), 39 (1965).
46. S. E. Berger and D. R. Carr, *Food Technol.* **20,** 1477–8 (1966).
47. S. E. Berger, C. V. Krolewski and L. C. Wizermann, *J. Am. Oil Chem. Soc.* **47**(5), 168 (1970).
48. J. Buechsenstein and C. S. Ough, *Am. J. Enol. Vitic.* **30**(2), 93 (1979).
49. Ref. 3, pp. 175–179.
50. U.S. Pat. 2,935,425 (May 3, 1960), G. Gutzeit, P. Talmey, and W. G. Lee (to General American Transportation Corp.).
51. Brit. Pat. 1,375,537 (Nov. 27, 1974), H. M. Ferrarini (to Sun Chemical Corp.).
52. U.S. Pat. 3,788,804 (Jan. 29, 1974), R. J. Harper and co-workers (to United States Dept. of Agriculture).
53. A. G. Pierce, Jr., E. A. Boudreaux, and J. D. Reid, *Am. Dyestuff Rep.* **59**(5), 50 (1970).
54. U.S. Pat. 3,074,809 (Jan. 22, 1963), R. Owen (to Minnesota Mining and Manufacturing Co.).
55. L. Carius, *Ann. Chem.* **129,** 6 (1864).
56. B. Holmberg, *Ark. Kem. Min. Och Geol.* **6**(1), 4 (1915); *Chem. Zentr.* **I,** 968 (1916).
57. P. A. Levene and L. A. Mikeska, *J. Biol. Chem.* **60,** 687 (1924).
58. Brit. Pat. 670,702 (Apr. 23, 1952), R. L. Evans.
59. E. Biilmann, *Ann. Chem.* **348,** 132 (1906).
60. *Beil.* **3,** 439 (1921).
61. R. Andreasch, *Monatsh.* **49,** 131 (1918).
62. U.S. Pat. 2,370,593 (Feb. 27, 1945), N. R. Trenner and F. A. Bacher (to Merck & Co.).
63. J. G. Voss, *J. Invest. Dermatol.* **31,** 273 (1958).
64. F. Meidinger, *Arch. Intern. Pharmacodyn.* **76,** 351 (1948).
65. Ger. Pat. 1,035,856 (Aug. 7, 1958), J. H. Brant (to Gillette Co.).
66. A. Shansky, *Soap Cosmet. Chem. Spec.* **52**(9), 32 (1976).
67. U.S. Pat. 3,277,012 (Oct. 4, 1966), E. W. Krockow (to Dr. Spiess GmbH).
68. N. K. Patel and J. Franco, *Indian J. Technol.* **13,** 239 (1975).
69. G. Jasmin, *J. Pharmacol. Exptl. Therap.* **120,** 349 (1957).
70. M. S. Kharasch, *Bur. Std. J. Res.* **2,** 359 (1929).
71. H. W. Ockerman, *Source Book for Food Scientists,* The Avi Publishing Co., Inc., Westport, Conn., 1978, p. 276.
72. J. M. Church and R. Blumberg, *Ind. Eng. Chem.* **43,** 1780 (1951).
73. U.S. Pat. 2,314,831 (Mar. 24, 1943), J. Kamlet (to Miles Laboratories).
74. U.S. Pat. 2,559,650 (July 10, 1951), L. B. Lockwood and G. E. N. Nelson (to the United States of America, as represented by the Secretary of Agriculture).
75. U.S. Pat. 2,197,021 (Apr. 16, 1940), R. Pasternack and E. V. Brown (to Chas. Pfizer & Co., Inc.).
76. U.S. Pat. 2,417,230 (Mar. 11, 1947), W. E. Barch (to Standard Brands).
77. U.S. Pats. 2,419,019 and 2,419,020 (Apr. 15, 1947), R. A. Hales (to Atlas Powder Co.).
78. U.S. Pat. 2,000,213 (May 7, 1935), G. Braun (to Standard Brands).
79. U.S. Pat. 2,947,665 (Aug. 2, 1960), J. W. Foster.
80. U.S. Pat. 3,069,230 (Dec. 18, 1962), B. Pescarolo and V. Bianchi (to Montecatini).
81. U.S. Pat. 3,114,770 (Dec. 17, 1963), I. Dabul (to Orandi & Massera).
82. Ger. Offen. 2,605,921 (Oct. 21, 1976), Y. Miura and co-workers (to Tokuyama Soda Co., Ltd.).
83. Ger. Offen. 2,619,311 (Nov. 18, 1976), Y. Kamatani and co-workers (to Takeda Chemical Industries, Ltd.).
84. U.S. Pat. 3,585,109 (June 15, 1971), K. Yamada and co-workers (to the United States of America, as represented by the Secretary of Agriculture).
85. N. Ya. Novotel'nova, R. A. Yurchenko, and L. F. Petrova, *Khlebopek. Konditer. Prom-st.* (1), 30 (1978); *Chem. Abstr.* **88,** 465 (1978).
86. Brit. Pat. 1,442,748 (July 14, 1976), D. F. Lewis and A. B. Rodriguez (to Imperial Chemical Industries Ltd.).
87. Jpn. Kokai 77 85,119 (July 15, 1977), N. Kazutani and co-workers (to Nippon Peroxide Co., Ltd.).
88. Jpn. Kokai 76,113,822 (Oct. 7, 1976), M. Kataoka, K. Hosoi, and M. Ono (to Toray Industries, Inc.).

89. Ger. Offen. 2,555,699 (July 1, 1976), K. Petritsch, P. Korl, and F. Pogoriach (to Oesterreichische Chemische Werke GmbH).
90. Ger. Offen. 2,543,333 (May 26, 1977), G. Prescher and G. Schreyer (to Deutsche Gold-und Silber-Scheideanstalt vorm. Roessler).
91. J. A. Bewsey, *Chem. Ind. (London)* (3), 119 (1977).
92. Fr. Demande 2,285,370 (Apr. 16, 1976), (to Etablissements Louis Francois).
93. Fr. Demande 2,303,778 (Sept. 8, 1976), M. Saotome and co-workers (to Nippon Peroxide Co., Ltd.; Showa Chemical Co., Ltd.).
94. *IM 146, U.S. Imports for Consumption,* Bureau of the Census, U.S. Department of Commerce.
95. *Chem. Mark. Rep.,* (Jan. 1, 1979).
96. *The United States Pharmacopeia XX, (USP XX–NFXV),* The United States Pharmacopeial Convention, Inc., Rockville, Md., 1980.
97. W. Horwitz, ed., *Official Methods of Analysis of the Association of Official Analytical Chemists,* Washington, D.C., 1975, p. 1090.
98. *Code of Federal Regulations,* Title 22, 182.1099.
99. *Fed. Regist.* **44,** 2687 (Jan. 12, 1979).
100. A. H. Nissan, G. K. Hunger, and S. S. Sternstein, "Cellulose" in N. M. Bikales, ed., *Encyclopedia of Polymer Science and Technology,* Vol. 3, John Wiley & Sons, Inc., New York, 1965, pp. 166–167.
101. J. Cartlidge and C. F. H. Tipper, *Chem. Ind. (London)* **37,** 853 (1959).
102. W. Poethke, *Pharmazie* **4,** 214 (1949).
103. B. Bak, *Ann.* **537,** 286 (1939).
104. P. Chang and S. L. Wu, *Hua Hsüeh Hsüeh Pao* **23,** 76 (1957); *Chem. Abstr.* **52,** 14567 (1958).
105. J. Tripod and co-workers, *Arch. Intern. Pharmacodyn.* **104,** 121 (1955).
106. U.S. Pat. 3,988,470 (Oct. 26, 1976), E. J. Van Scott and R. J. Yu.
107. Fr. Pat. 998,071 (Jan. 14, 1952), A. Guillemonat.
108. U.S. Pat. 2,872,464 (Feb. 3, 1959), R. Z. Brown and B. Rosen (to Crown Cork and Seal Co., Inc.).
109. W. J. Gensler and H. N. Schlein, *J. Am. Chem. Soc.* **78,** 169 (1956).
110. L. Ruzicka, Pl. A. Plattner, and W. Widmer, *Helv. Chim. Acta* **25,** 1086 (1942).
111. W. J. Gensler and H. N. Schlein, *J. Am. Chem. Soc.* **77,** 4846 (1955).
112. J. F. McGhie and co-workers, *J. Chem. Soc.* (1), 1 (1968).
113. J. F. McGhie and co-workers, *Chem. Ind. (London)* (13), 536 (1972).
114. J. Bontan Yanes, and O. Gonzales-Babe, *Rev. Plast. Mod.* **20,** 691, 712 (1969).

S. Edmund Berger
Allied Chemical Corporation

HYGROMETRY. See Drying.

HYPNOTICS, SEDATIVES, ANTICONVULSANTS

Hypnotics are central nervous system (CNS) depressants that induce sleep when given in appropriate doses (see Neuroregulators). At lower doses these substances frequently exhibit a calming or sedative action and at higher doses they may produce anesthesia, coma, or possibly death. Anticonvulsants are agents that are used primarily to prevent epileptic seizures and to control convulsions produced by seizures. Because their main use is in epilepsy, they frequently are referred to as antiepileptics.

Hypnotics and Sedatives

Insomnia is the inability to sleep properly. The cause for the sleep disturbance frequently is related to chronic anxiety, psychiatric disorders, stress, noises, and overwork (see Noise pollution). Most insomniacs have either difficulty in falling asleep, do not sleep as well as they would like to, or do not feel refreshed on awakening. The ideal hypnotic agent should alleviate these problems without affecting the ability of the patient to function normally after sleep.

Ethyl alcohol in the form of wine (qv) or beer (qv) has been and still is used since the earliest days of history to induce sleep. The more specific hypnotics were introduced to medical practice in the mid-19th century. Chloral hydrate, first prepared by Liebig in 1832, was employed as a hypnotic by Liebreich in 1869 (1). Adolf Baeyer's discovery of barbituric acid [*67-52-7*] in 1864, the parent compound of the barbiturate hypnotics was not capitalized on until diethyl barbituric acid [*57-44-3*] was introduced as a hypnotic in 1903 (2). Progress in the development of hypnotics and sedatives proceeded slowly until 1960 when the first 1,4-benzodiazepine was introduced (3). This class of compounds has led to the development of a number of safe and effective hypnotics, and sedatives, eg, diazepam and flurazepam, that dominate the field. The properties of the hypnotics and sedatives, that are discussed below, are listed according to their chemical classes in Table 1.

Total sales of hypnotics and sedatives in the United States for 1978 was \$84,332,000 with the nonbarbiturates representing \$61,837,000 of the total. The sales trend in recent years has shown an increase in nonbarbiturate and a decrease in barbiturate sales (6). The unit price of the common dosage form for sedative–hypnotics that are marketed in the United States is given in Table 2.

Alcohols. Although ethyl alcohol probably is the most widely used sedative or hypnotic that is available without a physician's prescription, its danger for abuse prohibits it as a useful therapeutic agent (9). To overcome this problem, a number of chlorinated alcohols with greater potency than ethyl alcohol have been developed.

Ethchlorvynol (1), C_7H_9ClO (1-chloro-3-ethyl-1-penten-4-yn-3-ol) has a camphoracious odor.

$$HC{\equiv}C\overset{\overset{\displaystyle OH}{|}}{\underset{\underset{\displaystyle C_2H_5}{|}}{C}}CH{=}CHCl$$

(1)

Table 1. Properties of Hypnotics and Sedatives

Class	Structure no.	Compound[a]	CAS Registry No.	Appearance	mp, °C
alcohols					
	(1)	ethchlorvynol	[113-18-8]	liquid	
	(2)	chloral hydrate	[302-17-0]	white solid	55
cyclic amides					
	(4)	methyprylon	[125-64-4]	white solid	74–77
	(7)	glutethimide	[77-21-4]	white crystalline powder	86–89
	(10)	methaqualone	[72-44-6]	white solid	114–117
acylureas and urethanes					
	(12)	carbromal	[77-65-6]	white powder	115–117
	(14)	bromisovalum	[496-67-3]	white solid	145–151
	(15)	ethinamate	[126-52-3]	solid	95–98
	(17)	meprobamate	[57-53-4]	white powder	103–107
benzodiazepines					
	(21)	diazepam	[439-14-5]	off-white crystals	125–126
	(25)	flurazepam hydrochloride-	[1172-18-5]	pale yellow crystals	215.5–217.5
	(27)	flunitrazepam	[1622-62-4]	pale yellow	166–167
	(29)	lorazepam	[846-49-1]	nearly white powder	166–168
	(31)	nitrazepam	[145-22-5]		224–226
	(33)	quazepam	[36735-22-5]		138–139
	(36)	temazepam	[846-50-4]	white crystalline powder	156–162
	(38)	clobazam	[22316-47-8]		180–182
	(41)	estazolam	[29975-16-4]		228–229
	(44)	triazolam	[28911-01-5]		223–224
	(47)	ketazolam	[27223-35-4]		182–183.5
	(48)	chlorazepate dipotassium	[57109-90-7]	pale yellow powder	
phenothiazines					
	(50)	mesoridazine besylate	[5588-33-0]	white powder	178 (dec)
	(52)	1-methotrimetrazine maleate	[60-99-1]	(light-sensitive) solid	190
	(54)	promethazine hydrochloride-	[58-33-3]	white (air-sensitive) powder	230–232

[a] Ref. 4.
[b] Ref. 5.

Ethchlorvynol is useful for short-term hypnotic therapy and for management of insomnia. It does not induce addiction upon continued administration (10).

Chloral hydrate (**2**), $C_2H_3Cl_3O_2$ (2,2,2-trichloro-1,1-ethanediol), has a penetrating, acrid odor, and a slightly bitter caustic taste. It is prepared by chlorination of ethanol followed by treatment with water.

Table 2. Sedative-Hypnotics Marketed in the United States

Generic name	Structure no.	Control class[a]	Trade name, supplier[b]	Common form, dosage, mg[b]	Price per 100[c], $
ethchlorvynol	(1)	C IV	Placidyl, Abbott	capsule, 500	10.18
chloral hydrate	(2)	C IV	many proprietary and generic	capsule, 500	1.80–3.84
methyprylon	(4)	C III	Noludar, Roche	capsule, 250	8.81
glutethimide	(7)	C III	Doriden, USV	capsule, 250	8.81
methaqualone	(10)	C II	Quaalude, Lemmon; Sopor, Arnar-Stone	tablet, 150	6.95, 10.02
carbromal	(12)	C III	Carbrital, Parke–Davis	capsule, 4 grains	6.41
meprobamate	(17)	C IV	many proprietary and generic names	tablet, 400	
diazepam	(21)	C IV	Valium, Roche	tablet, 5	10.48
flurazepam	(25)	C IV	Dalmane, Roche	capsule, 30	11.72
mesoridazine	(50)	℞	Serentil, Boehringer–Ingelheim	tablet, 25	10.74
methotrimetrazine	(52)	℞	Levoprome, Lederle	ampul, 20 mg/cm³	18.95 per 25
promethazine	(54)	℞	Phenergan, Wyeth; Promethazine, Purepac	tablet, 25	1.64, 7.91

[a] *Controlled Substance Act of 1970* schedule classification.
[b] Ref. 7.
[c] Ref. 8.

Chloral hydrate is useful as a hypnotic at doses of 500–1000 mg. When effective, it will maintain sleep for about 6–8 h and rarely causes morning hangover. The drug commonly causes gastric irritation and continued use may lead to toleration and habituation. The hypnotic action of chloral hydrate is believed to result from trichloroethanol [*115-20-8*] (**3**), a metabolite that is formed in the tissues (11). The activity and toxicity of chloral hydrate is potentiated by the simultaneous consumption of ethyl alcohol and this dangerous combination is commonly known as knockout drops.

$$\underset{(2)}{Cl_3CCH(OH)_2} \qquad \underset{(3)}{Cl_3CCH_2OH}$$

Cyclic Amides. *Methyprylon* (**4**), $C_{10}H_{17}NO_2$ (3,3-diethyl-5-methyl-2,4-piperidinedione), can be prepared (12) by treating the tetrahydropyridinedione (**5**), with formaldehyde and hydrogenating the resulting 5-hydroxymethyl derivative (**6**).

$$(5) \xrightarrow{CH_2O} (6) \xrightarrow{H_2} (4)$$

Methyprylon is used as a hypnotic in 300–400 mg doses; it is effective for at least seven consecutive nights of medication. It causes withdrawal symptoms that are similar to the barbiturates (13).

Glutethimide (**7**), $C_{13}H_{15}NO_2$ (3-methyl-3-phenyl-2,6-piperidinedione), is synthesized (14) by the hydrolysis and ring closure of the dinitrile (**9**), which is obtained by Michael addition of acrylonitrile (qv) to 2-phenylbutanonitrile (**8**).

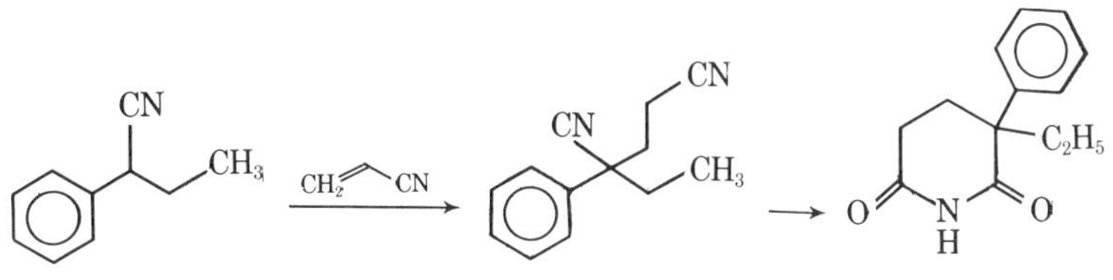

(**8**) $\xrightarrow{CH_2{=}CHCN}$ (**9**) $\longrightarrow$ (**7**)

Glutethimide at doses of 250–500 mg is useful for all types of insomnia: on preoperative nights and as a sedative in the first stage of labor. It induces a rapid onset of sleep that lasts for 4–8 h and seldom causes hangover the morning after. Both physical and psychological dependence have been reported (15).

Methaqualone (**10**), $C_{16}H_{14}N_2O$ (2-methyl-3-*o*-tolyl-3*H*-4-quinazolinone, Quaalude), can be prepared by the reaction of *N*-acetylanthranilic acid (**11**), with *o*-tolidine in the presence of phosphorus chlorides (16).

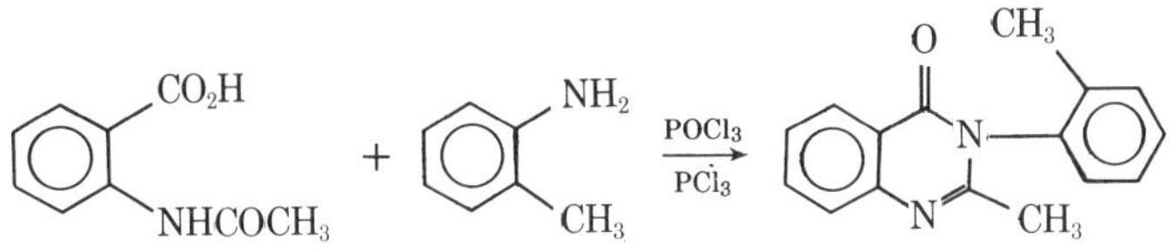

(**11**) + *o*-toluidine $\xrightarrow[PCl_3]{POCl_3}$ (**10**)

Methaqualone produces sleep and daytime sedation. The usual dosage for sleep is 150–300 mg. Psychological dependence has been reported with methaqualone although physical dependence is rare (17).

Acylureas and Urethanes. *Carbromal* (**12**), $C_7H_{13}BrN_2O_2$, is prepared by brominating the acylurea (**13**), in the alpha position (18).

$$\underset{(13)}{(C_2H_5)_2CHCONHCONH_2} \xrightarrow{Br_2} \underset{(12)}{(C_2H_5)_2CBrCONHCONH_2}$$

The combination, carbromal–pentobarbital (4:1.5 mixture) has been rated as possibly effective as a sedative and a hypnotic at a dosage of 0.36–0.71 g (5.5–11.0 grains). Long, continued use could result in bromism (19).

Bromisovalum (**14**), $C_6H_{11}BrN_2O_2$ ((2-bromo-3-methylbutyryl)urea), is no longer available on the United States market. It is prepared by treating 2-bromo-3-methylbutyryl chloride with urea (20) (see Urea and urea derivatives). Bromisovalum can be used as a sedative at 300 mg or as a hypnotic at 600 mg.

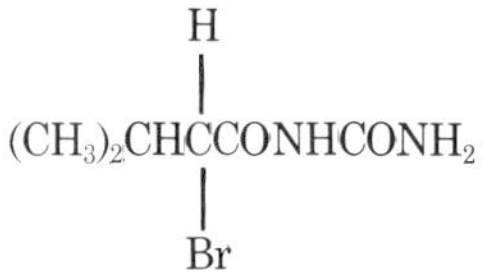

$$(CH_3)_2CH\overset{H}{\underset{Br}{C}}CONHCONH_2$$

(**14**)

Ethinamate (**15**), $C_9H_{13}NO_2$ (1-ethynylcyclohexyl carbamate), is no longer available on the United States market. It is prepared by treating cyclohexanone with acetylene in the presence of sodium amide and treating the resulting 1-ethynylcyclohexanol (**16**), with carbamoyl chloride (21).

$$\text{cyclohexanone} \xrightarrow[NaNH_2]{HC{\equiv}CH} \underset{(\mathbf{16})}{\text{1-(OH)-1-}(C{\equiv}CH)\text{cyclohexane}} \xrightarrow{ClCONH_2} \underset{(\mathbf{15})}{\text{1-}(OCONH_2)\text{-1-}(C{\equiv}CH)\text{cyclohexane}}$$

Ethinamate is useful as a daytime sedative when barbiturates are not desirable and as a hypnotic in doses of 500–1000 mg.

Meprobamate (**17**), $C_9H_{18}N_2O_4$ (2-methyl-2-*n*-propyl-1,3-propanediol dicarbamate), is prepared by an aldol reaction of the aldehyde (**18**) with formaldehyde and the diol (**19**), followed by treatment with phosgene and then ammonia (22).

$$\underset{(\mathbf{18})}{HC(CH_3)(C_3H_7)CHO} \xrightarrow{CH_2O} \underset{(\mathbf{19})}{HOCH_2C(CH_3)(C_3H_7)CH_2OH} \xrightarrow[(2)\ NH_3]{(1)\ COCl_2} \underset{(\mathbf{17})}{H_2NC(=O)OCH_2C(CH_3)(C_3H_7)CH_2OC(=O)NH_2}$$

Meprobamate is used to promote sleep in anxious, tense patients at 400–800 mg (23).

Benzodiazepines. Between 1955 and 1960 a new class of compounds, the 1,4-benzodiazepines, entered the therapeutic armamentarium of physicians (24) (see Psychopharmacological agents). A number of these 1,4-benzodiazepines possess pharmacological properties in those tests generally used in the preliminary screening of tranquilizers and sedatives, eg, muscle relaxant, sedative, and anticonvulsant properties, and have found their way to use in human medicine.

The more active members of this series have a 1,3-dihydro-5-aryl-2*H*-1,4-benzodiazepin-2-one framework (**20**), with a Cl or NO_2 at position 7, a CH_3 or $CH_2CH_2N(C_2H_5)_2$ group at position 1, and a phenyl or *o*-fluorophenyl as the aryl group at position 5. Recently, an additional hetero ring has been fused to positions 1–2 (estrazolam and triazolam) or at positions 4–5 (ketazolam) to give more potent compounds.

The central nervous system pharmacological activities of the benzodiazepines correlate with the net charge on the carbonyl oxygen atom and the total dipole moment of the molecule (25).

The benzodiazepines are metabolized in humans according to three main pathways (26): *N*-demethylation, hydroxylation at position 3 or *N*-demethylation and 3-hydroxylation. In the case of diazepam (**21**), these metabolites contribute to the therapeutic effect of the parent drug and have led to the synthesis of the clinically useful temazepam and oxazepam (27).

Benzodiazepines tend to cause hypnotic hangover and, at high doses, cause REM (rapid eye movement) and slow-wave sleep deprivation, thereby adversely affecting

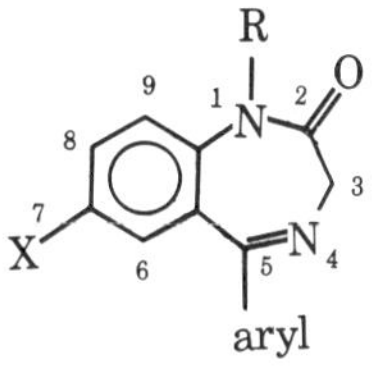

(20)

	R	aryl	X
[2011-67-8]	CH_3	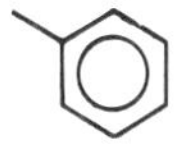 phenyl	NO_2
[3900-31-0]	CH_3	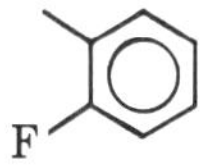 2-fluorophenyl	Cl
[1101-71-9]	$(CH_2)_2N(C_2H_5)_2 \cdot 2HCl$	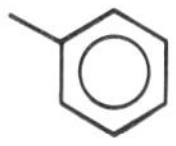 phenyl	Cl
[30195-52-9]	$(CH_2)_2N(C_2H_5)_2$	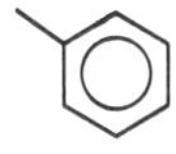 phenyl	NO_2
[17617-23-1]	$(CH_2)_2N(C_2H_5)_2$	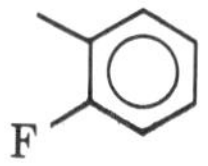 2-fluorophenyl	Cl
diazepam	CH_3	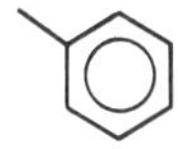 phenyl	Cl
flunitrazepam	CH_3	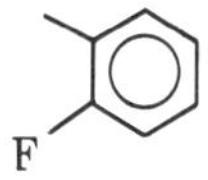 2-fluorophenyl	NO_2

the quality of sleep they produce. In addition, there are rebound abnormalities that occur on withdrawal. These side effects are of marginal importance, however, when considered in light of the short- to intermediate-term efficacy and exceptionally high safety margins displayed by the benzodiazepines. As such they are the safest drugs presently available for the induction and maintenance of sleep (28).

Diazepam (**21**), Valium, $C_{16}H_{13}ClN_2O$ (7-chloro-1,3-dihydro-1-methyl-5-phenyl-2*H*-1,4-benzodiazepin-2-one), is insoluble in water. Diazepam (**21**) can be prepared by several routes, eg, treatment of the amino oxime (**22**) first with chloroacetyl chloride, then with base, and finally by reduction of the *N*-oxide (**23**) with hydrogen (29).

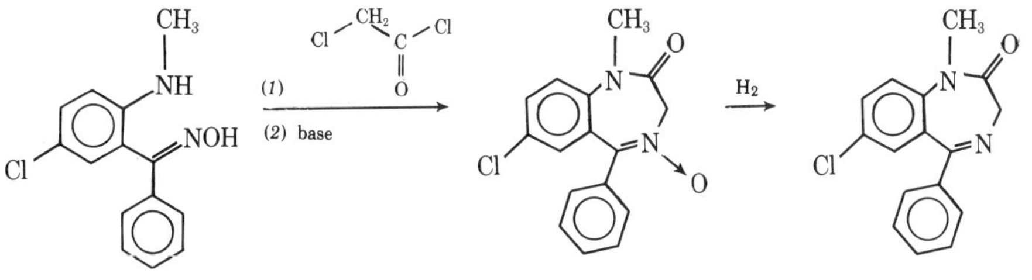

(1) $ClCH_2COCl$, (2) base; H_2

(**22**) (**23**) (**21**)

In animals, diazepam appears to act on parts of the limbic system, the thalamus, and the hypothalamus. It has much stronger sedative, muscle relaxant, taming, and anticonvulsant properties than chlordiazepoxide [*58-25-3*] (**24**), which was the first of the marketed benzodiazepines.

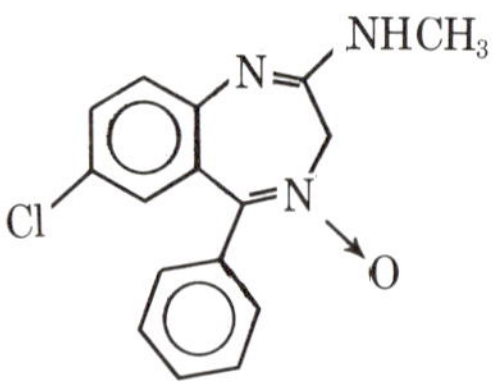

(**24**)

At a dose of 5–15 mg, diazepam is effective in anxious insomniacs, aiding both sleep induction and maintenance and decreasing slow wave and REM (rapid eye movement) sleep. It is particularly recommended in patients already taking diazepam in the daytime for anxiety; however, it is indicated for tension and anxiety and not as a hypnotic (see Psychopharmacological agents) (30).

Flurazepam hydrochloride (**25**), $C_{21}H_{23}ClFN_3O.HCl$ (7-chloro-1-(2-(dimethylaminoethyl))-5-(*o*-fluorophenyl)-1,3-dihydro-2*H*-1,4-benzodiazepin-2-one hydrochloride), is freely soluble in ethanol and very soluble in water. It can be prepared (31) by the oxidative ring expansion of the diaminoindole (**26**).

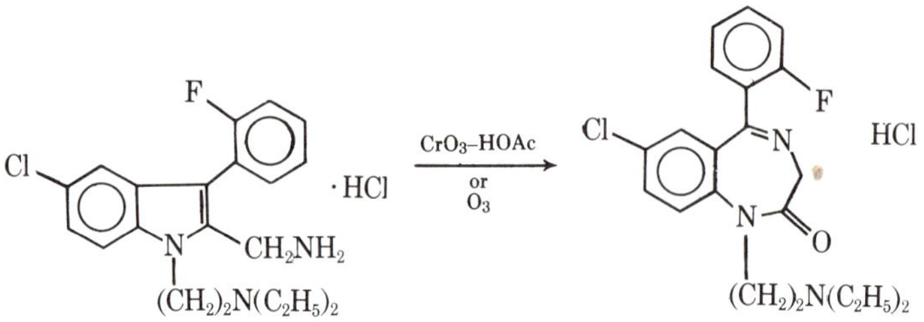

(**26**) (**25**)

Introduced into the United States in 1970, it is the most widely studied of the benzodiazepine sedative–hypnotics. Like the other benzodiazepines, it is primarily a central-muscle relaxant that exerts its action by intraneuronal blockade at the supraspinal level (32). Following oral administration, it is rapidly absorbed from the gastrointestinal tract and is distributed widely throughout body tissues. Its hypnotic effect begins within 45–50 min after oral administration and continues for 4–8 h. Thus it is useful clinically as both a sleep inducer and a sleep maintainer by reducing sleep latency and increasing total sleep time (33). Morning drowsiness or hangover is the most common unwanted side effect, with claims of significantly reduced REM and slow-wave sleep time; however, no REM rebound occurs when the drug is withdrawn (34). The usual dose is 15 or 30 mg.

Flunitrazepam (**27**), $C_{16}H_{12}FN_3O_3$ (5-(2-fluorophenyl)-1,3-dihydro-1-methyl-7-nitro-2*H*-1,4-benzodiazepin-2-one), is available in Europe as Rohypnol (Roche). It is insoluble in water and is prepared (35) in a manner similar to diazepam (**21**) except that the starting material is amino oxime (**28**).

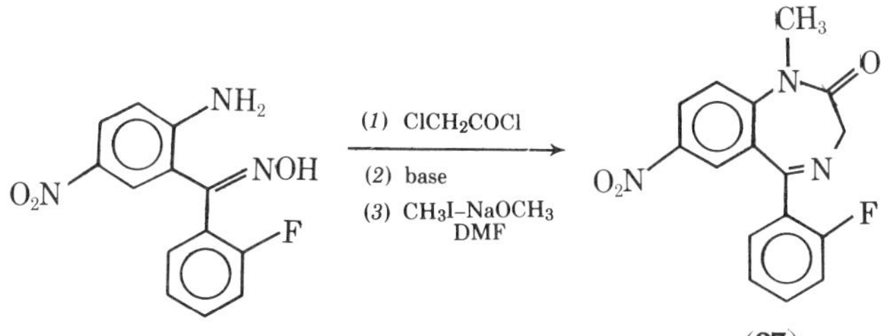

(1) $ClCH_2COCl$
(2) base
(3) CH_3I–$NaOCH_3$ DMF

(28) → (27)

Clinically, 2 mg appears effective in improving sleep induction and maintenance on short-term use (36). Both REM and slow-wave sleep were significantly reduced with a rebound upon withdrawal (37). There was a shift to faster EEG (electroencephalogram) frequencies. Compared with nitrazepam, there is a much more rapid onset and longer duration of sleep.

Lorazepam (**29**), $C_{15}H_{10}Cl_2N_2O$ (7-chloro-5-(*o*-chlorophenyl)-1,3-dihydro-3-hydroxy-2*H*-1,4-benzodiazepine-3-one), is almost insoluble in water. It is prepared by a Polonovski-type rearrangement in acetic anhydride of the benzodiazepine-*N*-oxide (**30**), followed by mild basic hydrolysis of the intermediate acetate ester (38).

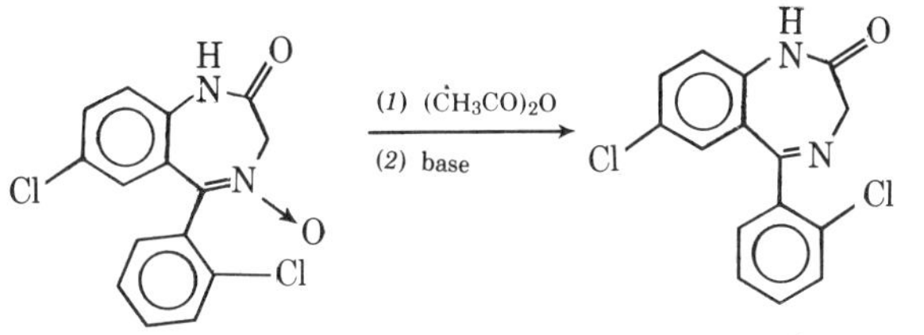

(1) $(CH_3CO)_2O$
(2) base

(30) → (29)

In monkeys, it produces a sedation of long duration. It suppresses the polysynaptic linguomandibular reflex without suppressing the monosynaptic patellar reflex (39). After oral dosing, it reaches a peak level in the plasma within 3 h. Clinically, it has anxiolytic activity that is roughly equivalent to diazepam (**21**) and muscle relaxant activity that is between diazepam and oxazepam. It is used for insomnia that results from anxiety or transient situational stress at a dose of 2–4 mg at bedtime. Lorazepam is another example of the versatility of the benzodiazepines: they generally can be interchanged as anxiolytics or hypnotics provided that the dosage is appropriately adjusted (39).

Nitrazepam (**31**), $C_{15}H_{11}N_3O_3$ (1,3-dihydro-7-nitro-5-phenyl-2*H*-1,4-benzodiazepine-2-one), is soluble in alcohol but practically insoluble in water and ether. It is synthesized (35) in a manner similar to that of diazepam, but using a different amino oxime precursor (**32**).

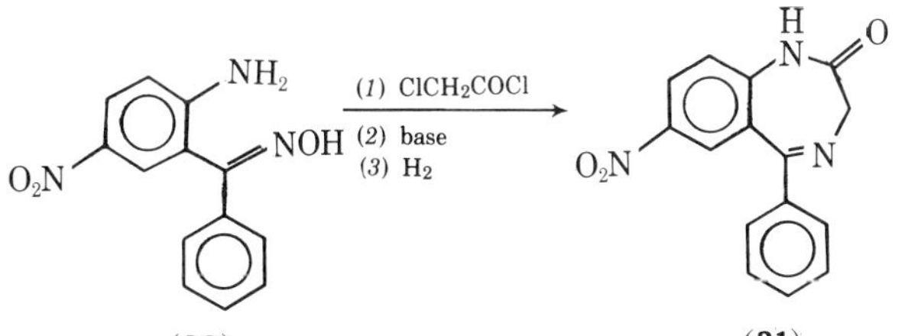

(1) $ClCH_2COCl$
(2) base
(3) H_2

(32) → (31)

Peak plasma levels of unchanged drug were reached in about 2 h following oral administration in man with only 65–75% of the drug undergoing renal excretion during the first 120 h. The slowness of elimination is one of the major disadvantages of the drug (28). Clinical doses up to 20 mg cause drug-induced EEG fast-wave abnormalities including a reduction of the proportion of sleep spent in REM, hangover effects, impairment of psychomotor performance, and a liability to fall asleep during daytime (28).

Quazepam (**33**), $C_{17}H_{11}ClF_4N_2S$ (7-chloro-1,3-dihydro-5-(2-fluorophenyl)-3-hydroxy-1-(2,2,2-trifluoroethyl)-2*H*-1,4-benzodiazepin-2-thione), is synthesized by treatment of lactam (**35**) with P_2S_5 (21). The lactam compound is obtained from the aminobenzophenone (**34**), bromoacetyl chloride, and ammonia treatment (40).

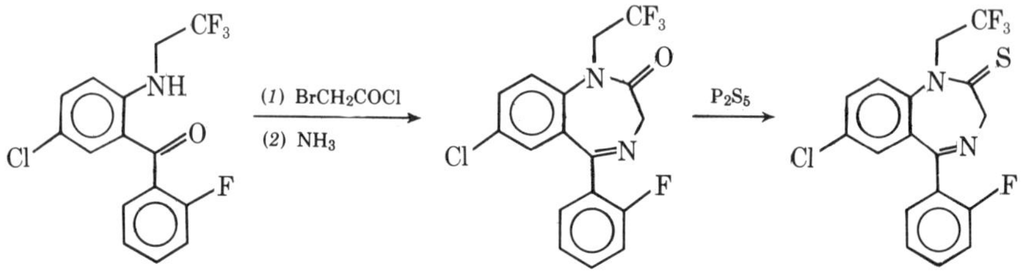

(*1*) $BrCH_2COCl$ (*2*) NH_3 ; P_2S_5

(**34**) (**35**) (**33**)

Clinically, in healthy volunteers, doses of 10.25–50 mg decreased the time necessary to fall asleep, suppressed REM but not slow-wave sleep. At the highest dose, a hangover effect was noticed (41).

Temazepam (**36**), $C_{16}H_{13}ClN_2O_2$ (7-chloro-1,3-dihydro-3-hydroxy-1-methyl-5-phenyl-2*H*-1,4-benzodiazepin-2-one), has little or no odor. It is prepared in a manner similar to lorazepam (**30**), ie, from diazepam-*N*-oxide [*2888-64-4*] (**23**) (42), or by the *N*-methylation of oxazepam [*604-75-1*] (**37**) (43).

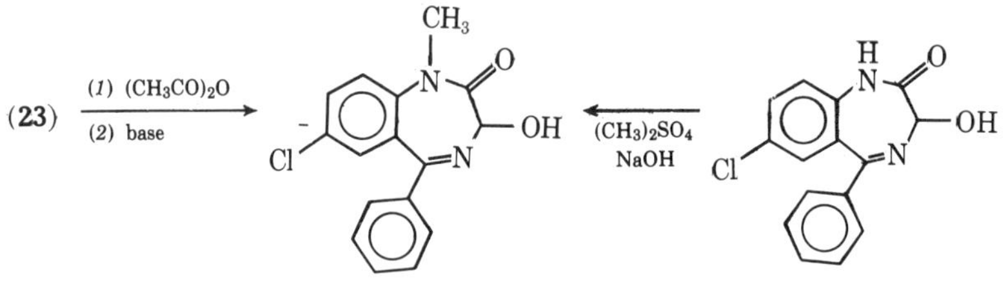

(**23**) (*1*) $(CH_3CO)_2O$ (*2*) base ; $(CH_3)_2SO_4$ NaOH

(**36**) (**37**)

It is particularly effective in tests that reflect sleep enhancement (both induction and maintenance) in Cebus monkeys and rats. Results from oral administration of radioactive temazepam disclosed that the onset of absorption is rapid with peak concentrations 1 h after dosing, and clearance is rapid with the elimination of all radioactive material within 72 h. In human blood, the main temazepam-derived compounds were the free and conjugated parent drug. Like other benzodiazepines, it is a relatively nontoxic substance (44). It possesses a beneficial therapeutic effect on anxiety and insomnia at daily doses of 10–100 mg. Side effects are few and relatively minor. At doses of 15–30 mg, it reduces early morning awakenings, reduces sleep induction time, and increases the duration of sleep. It does not appear to cause any residual effects or an effect in total REM sleep. Furthermore, there appears to be a lack of tolerance and treatment withdrawal effects following long-term administration (45).

Clobazam (**38**), $C_{16}H_{13}ClN_2O_2$ (7-chloro-1-methyl-5-phenyl-1*H*-1,5-benzodiazepine-2,4(3*H*,5*H*)dione), a benzodiazepine analogue in which the nitrogens of the heterocyclic ring are in the 1,5 rather than in the more common 1,4 position, is prepared from the nitroaniline (**39**) by acylation, nitro reduction, and lactam formation to (**40**) followed by *N*-alkylation with NaH and methyl iodide (46).

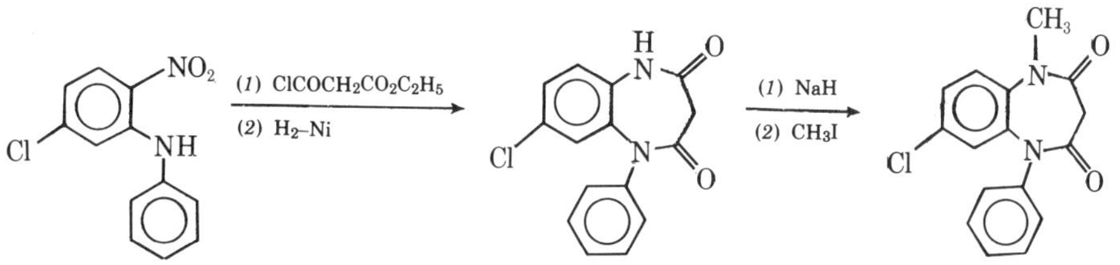

(1) $ClCOCH_2CO_2C_2H_5$ (2) H_2–Ni; (1) NaH (2) CH_3I

(**39**) (**40**) (**38**)

Following oral administration in man, it has a half-life of 5 h (47). In mice, it is a sedative at a dose that is much lower than that which produces ataxia (48). Clinically, it is effective at 20–40-mg daily doses in relieving anxiety (47) but produces side effects lasting up to 2 wk (49). It is available in Europe as Frisium (Hoechst).

Estazolam (**41**), $C_{16}H_{11}ClN_4$ (8-chloro-6-phenyl-4*H*-*s*-triazolo[4,3-*a*][1,4]benzodiazepine), is prepared by treating the thiolactam (**42**) with hydrazine and the resulting amidrazine (**43**) with ethyl formate (50).

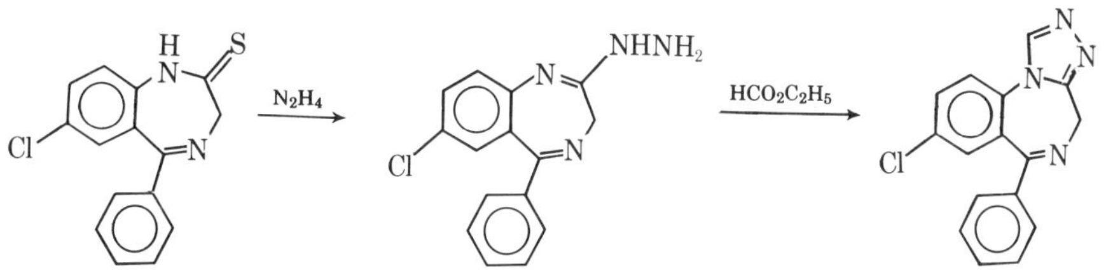

N_2H_4; $HCO_2C_2H_5$

(**42**) (**43**) (**41**)

In mice, estazolam (**41**) is about twice as potent as diazepam (**21**) in affecting overt behavior and is equivalent to the effect of pentobarbital, but is considerably poorer than pentobarbital on blocking electrically or chemically induced convulsions.

In moderately anxious middle-aged men at doses of 2, 4, and 6 mg over 3 wk, estazolam did not appear to have any residual effects after 8 h of sleep on short-term visual memory, recall, perception, or reaction times, which indicates a lack of drug-induced residual or hangover effects (51).

Triazolam (**44**) $C_{17}H_{12}Cl_2N_4$ (8-chloro-6-(*o*-chlorophenyl)-1-methyl-4*H*-*s*-triazolo[4,3-*a*][1,4]benzodiazepine), is synthesized from the lactam (**45**) in a sequence similar to that used for estazolam (**41**) with treatment of the amidrazine (**46**) with acetic anhydride in sulfuric acid (52).

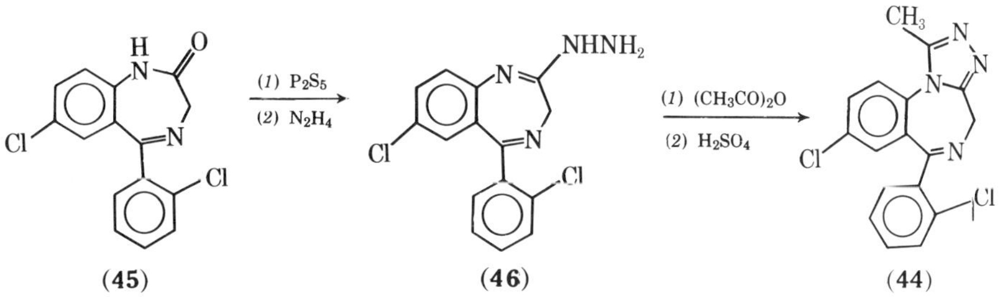

(1) P_2S_5 (2) N_2H_4; (1) $(CH_3CO)_2O$ (2) H_2SO_4

(**45**) (**46**) (**44**)

In mice, it is ten times as potent as diazepam (**21**); in overt behavior, two to three times as potent in potentiating the effects of pentobarbital; and about twenty times as potent in protecting mice against chemically induced convulsions (53). In insomniacs, triazolam (on a weight basis) is thirty to sixty times more potent as a hypnotic agent than flurazepam (**25**) with a recommended dose of 0.5–1 mg. There appears to be a good degree of patient satisfaction and a low incidence of side effects (54) along with its effectiveness as both a sleep inducer and a sleep maintainer. Upon intermediate-term use it appears to lose effectiveness (based on sleep-laboratory studies), with sleep worsening beyond baseline after drug withdrawal (55); there was relatively little impairment of performance on the day following therapy (56).

Ketazolam (**47**), $C_{20}H_{17}ClN_2O_3$ (11-chloro-8-12b-dihydro-2,8-dimethyl-12b-phenyl-4*H*[1,3]oxazino[3,2-*d*][1,4]-benzodiazepine-4,7(6*H*)dione), is prepared by the condensation of diazepam (**21**) and diketene (57).

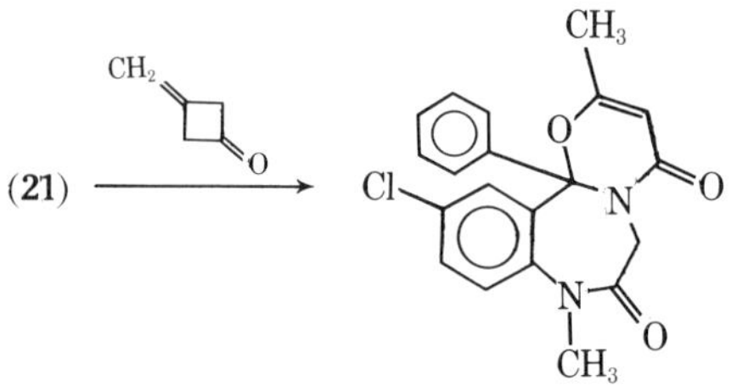

(**21**) → (**47**)

In laboratory animals, ketazolam acts similarly to other benzodiazepines in antagonizing chemically induced convulsions (58). The identified metabolites of ketazolam include oxazepam (**38**), diazepam (**21**), *N*-desmethyldiazepam, and several hydroxylated derivatives of diazepam (59). At 45 mg, it has been shown to be clinically effective in the treatment of anxiety (60) and as a hypnotic (58), indicating a fairly rapid onset of action. The major clinical effects were psychosedation and muscle relaxation (61).

Chlorazepate dipotassium (**48**), $C_{16}H_{11}ClN_2O_4K_2$ (7-chloro-2,3-dihydro-2,2-dihydroxy-5-phenyl-1*H*-1,4-benzodiazepine-3-carboxylic acid, dipotassium salt), is very water soluble. It is prepared (55) by the pH-sensitive lactonization of the diacidimine (**49**).

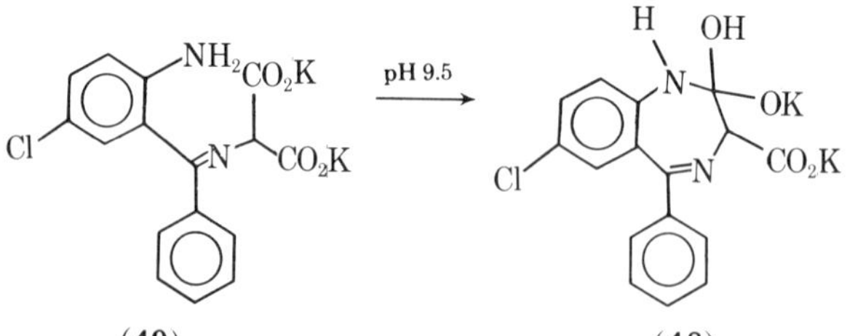

(**49**) —pH 9.5→ (**48**)

In animals, it displays muscle relaxant, ataractic, and anticonvulsant activity (62) and is especially effective against aggressive behavior (63). Orally, in man, the drug is rapidly and completely converted to *N*-desmethyldiazepam by spontaneous dehydration and decarboxylation in the stomach. This metabolites' half-life is about 50 h (64). It is used primarily for the symptomatic relief of anxiety (65). In the treatment of certain epileptic patients (66), 15 mg increases sleep time, shortens sleep latency, and reduces awakenings with a depression of stage-4 sleep (62). Also, a daily dose of 22.5 mg of (**48**) decreased baseline arousal but facilitated central nervous system response to stimulation, whereas diazepam, at 15 mg daily, depressed baseline and stimulation arousal (67).

Phenothiazines. The main clinical use of phenothiazine derivatives is as antipsychotic and antihistamine agents. A number of these substances possess a sedative–hypnotic component and have been used to relieve anxiety or to induce sleep.

Mesoridazine besylate (**50**), $C_{21}H_{26}N_2OS_2.C_6H_5SO_3H$ (10-(2-(1-methyl-2-piperidinyl)ethyl)-2-(methylsulfinyl)-10*H*-phenothiazone), is freely soluble in water and methanol. It is prepared by the alkylation of the sulfoxide (**51**) (68).

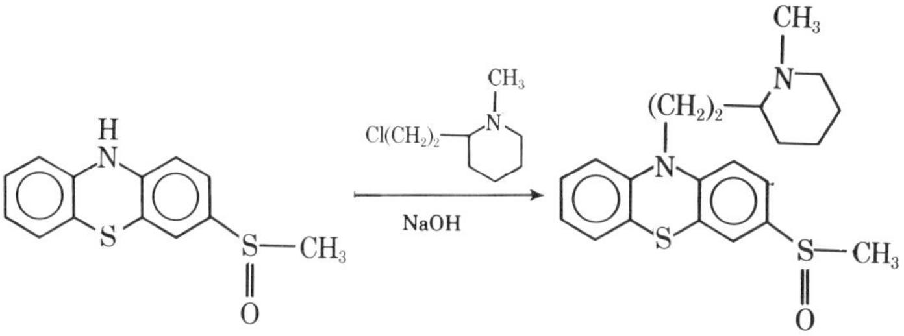

(**51**) → $Cl(CH_2)_2$–(1-methyl-2-piperidinyl), NaOH → (**50**)

In animals, it has a spectrum of pharmacodynamic actions that are typical of a major tranquilizer, including inhibition of spontaneous motor activity and reinduction of hexobarbital-induced sleeping. It is indicated for schizophrenia, behavioral problems in mental deficiency and chronic brain syndrome, alcoholism, and psychoneurotic manifestations. It works by reducing the symptoms of anxiety and tension, thus aiding sleep. Although not marketed as a sedative–hypnotic, it has been claimed to induce (in man and monkeys) a sleep that is undistinguishable from natural sleep using subjective, behavioral, and EEG criteria.

1-Methotrimetrazine maleate (**52**), $C_{19}H_{24}N_2OS.C_4H_4O_4$ (2-methoxy-*N*,*N*,*β*-trimethyl-10-*H*-phenothiazine-10-propaneamine), is sparingly soluble in water. It is prepared by alkylation of methoxy phenothiazine (**53**) (69).

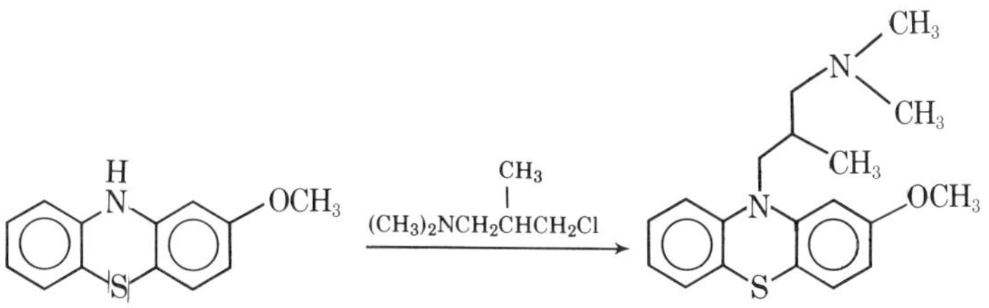

(**53**) → $(CH_3)_2NCH_2CH(CH_3)CH_2Cl$ → (**52**)

Methotrimetrazine maleate is a potent central nervous system depressant producing suppression of sensory impulses, reduction of motor activity, sedation, and tranquilization and has antihistamine, anticholinergic, and antiadrenergic effects. It is absorbed rapidly with 20–40 min to maximal effect, and is actively metabolized. Elimination usually continues for several days after dosing. It produces an analgesic effect and sedation in man, and is indicated in cases where respiratory depression is to be avoided. It is used as a preanesthetic medication for producing sedation, somnolence, and relief of apprehension and anxiety. The most significant side effect is its ability to produce orthostatic hypotension, usually within 10–20 min following intramuscular injection, which lasts up to 6 h. For preoperative sedation, it is given at a dose of 2–20 mg 45 min to 3 h before surgery.

Promethazine hydrochloride (**54**), $C_{17}H_{20}N_2S.HCl$ (N,N,α-trimethyl-10H-phenothiazine-10-ethanamine), is odorless and is prepared by the alkylation of phenothiazine (70).

$$\text{phenothiazine} \xrightarrow[\text{base}]{(CH_3)_2NC(CH_3)HCH_2Cl} \text{promethazine}$$

(**54**)

It possesses antihistiminic, sedative, antimotion sickness, antiemetic, and anticholinergic effects. The duration of action generally is 4–6 h. Although it does show a dose-related suppression of REM sleep, the REM rebound following withdrawal is weaker than that occurring with pentobarbital withdrawal (71). In humans, it relieves apprehension and induces a quiet sleep from which the patient can be aroused easily. It is recommended at a dose of 25–50 mg for nighttime, presurgical, or obstetrical sedation.

Anticonvulsants

The main application of anticonvulsant drugs is in the control and prevention of seizures associated with epilepsy. A recent international classification (72) of epileptic seizures recognizes partial, generalized, unilateral, and unclassified seizures. A common form of partial seizures is the Jacksonian seizure, a type of focal epilepsy with localized convulsions and retention of consciousness. The two main generalized seizures are petit mal, which is associated with myoclonic jerks, akinetic seizures, transient loss of consciousness, but without convulsions and grand mal which manifests itself in a continuous series of seizures and convulsions, and loss of consciousness.

The symptoms of the disease were mentioned as early as 2000 BC in the Hammurabi Code of Babylonian law. Early treatment involved such agents as dried figs, fleabane, mustard, and weasel's blood. In the Middle Ages, when the disease was associated with the idea of possession by demons, treatment was associated with the occult power, and frog's liver, mistletoe, human urine, and other substances associated with witchcraft were used (73).

In the 19th century, bromides were introduced for epilepsy and, in the early 20th century, the barbiturates became available. A major advance in drug therapy of epilepsy was the introduction of phenytoin (**56**) in 1938. Until the early 1960s the drugs were structurally related to phenytoin. In 1963, the first of several benzodiazepines appeared and in 1974 valproic acid (**71**) was introduced. Most antiepileptic drugs in use are weak acids that presumably exert their action on neurons or glial cells, or both, of the central nervous system (74). The majority of these compounds are characterized by the presence of at least one amide unit and one or more benzene rings that are present as a phenyl group or part of a cyclic system. A notable exception to this is valproic acid which lacks both of these features. Most antiepileptic drugs are inactivated by the mixed-function oxidase system (cytochrome P-450 system) located in hepatic microsomes that produce an oxidized metabolite which is excreted free or in conjugated form (74).

Despite the availability of a variety of drugs (75), many epileptics fail to experience seizure control and others do so at the expense of significant toxic side effects (76). The need for more selective and less toxic drugs has caused the NIH to initiate a large screening program for new antiepileptic drugs (77). The anticonvulsants that are discussed below are listed in Table 3, according to their chemical classes.

Total sales (6) of anticonvulsants in the United States for 1978 was $61,952,000. The unit price of the common dosage forms of anticonvulsants that are marketed in the United States is given in Table 4.

Hydantoins. The hydantoins are useful in the control of generalized convulsive seizures and all forms of partial seizures.

Phenytoin (**56**), $C_{15}H_{12}N_2O_2$ (5,5-diphenyl-2,4-imidazolidinedione), when in the form of its sodium salt, can be formulated into a propylene glycol–alcohol–water solution that is used for parenteral administration. The more common method of synthesis (78) involves condensation of benzil with urea to give (**57**) and rearrangement accompanied by loss of water to give phenytoin.

(**57**) (**56**)

Table 3. Properties of Anticonvulsants[a]

Class	Structure no.	Compound	CAS Registry No.	Appearance	mp, °C
hydantoins					
	(**56**)	phenytoin	[*57-41-0*]	white solid	295–298
	(**57**)	mephenytoin	[*50-12-4*]	white crystalline powder	136–137
oxazolidine-diones					
	(**59**)	trimethadione	[*127-48-0*]	white crystalline solid	45–47
	(**61**)	paramethadione	[*115-67-3*]	colorless liquid	
succinimides					
	(**62**)	phensuximide	[*86-34-0*]	crystalline powder	68–74
	(**63**)	methsuximide	[*77-41-8*]	white crystalline powder	50–56
	(**64**)	ethosuximide	[*77-67-8*]	white solid	47–52
other					
	(**65**)	primidone	[*125-33-7*]	white powder	279–284
	(**67**)	clonazepam	[*1622-61-3*]	white crystalline solid	236.5–238.5
	(**69**)	carbamazipine	[*294-46-4*]	white or yellowish white solid	189–193
	(**70**)	phenacemide	[*63-98-9*]	white powder	212–216
	(**71**)	valproic acid	[*99-66-1*]	colorless liquid	
	(**72**)	sulthiame	[*61-56-3*]	white crystalline powder	185–187
	(**21**)	diazepam (see Table 1)			

[a] Ref. 5.

Table 4. Anticonvulsants Marketed in the United States

Generic name	Structure no.	Trade name, supplier[a]	Common form, dosage, mg[a]	Price per 100 units[b], $
phenytoin	(**56**)	Dilantin, Parke-Davis; Dihycon, Consolidated Midland	capsule, 100	4.22
mephenytoin	(**57**)	Mesantoin, Sandoz	tablet, 100	4.22
trimethadione	(**59**)	Tridione, Abbott	tablet, 300	5.96
paramethadione	(**61**)	Paradione, Abbott	tablet, 300	9.81
phensuximide	(**62**)	Milontin, Parke-Davis	tablet, 500	10.93
methsuximide	(**63**)	Celontin, Parke-Davis	tablet, 300	11.63
ethosuximide	(**64**)	Zarontin, Parke-Davis	tablet, 250	11.40
primidone	(**65**)	Mysoline, Ayerst	tablet, 250	5.24
clonazepam	(**67**)	Clonopin, Roche	tablet, 1	7.86
carbamazepine	(**69**)	Tegretol, CIBA-GEIGY	tablet, 200	11.37
phenacemide	(**70**)	Phenurone, Abbott	tablet, 500	7.36
valproic acid	(**71**)	Depakene, Abbott	capsule, 250	17.34

[a] Ref. 7.
[b] Ref. 8.

Phenytoin is the antiepileptic of choice for most patients with tonic–clonic status epilepticus of the grand mal type with focal and psychomotor seizures (74). The dosage should be individualized, but most adults require 100 mg three times a day orally for satisfactory maintenance. Phenytoin also can be used to control seizures occurring during neurosurgery when given at 100–200 mg intramuscularly at 4-h intervals. The drug should not be used during pregnancy since there is evidence that it may cause birth defects.

Mephenytoin (**57**), $C_{12}H_{14}N_2O_2$ (5-ethyl-3-methyl-5-phenyl-2,4-imidazolinedione), can be prepared by the method of Bucherer from propiophenone, potassium cyanide, and ammonium carbonate to give the hydantoin (**58**), followed by methylation (79).

KCN, $(NH_4)_2CO_3$ → (**58**); $(CH_3)_2SO_4$ → (**57**)

Mephenytoin can be used for the control of grand mal, local, Jacksonian, and psychomotor seizures in patients who are refractory to less toxic anticonvulsants (74). The average daily adult oral dosage for seizure control is 2–6 100 mg tablets.

Oxazolidinediones. The oxazolidinediones are useful in the control of generalized nonconvulsive seizures.

Trimethadione (**59**), $C_6H_9NO_3$ (3,5,5-trimethyl-1,3-oxazolidine-2,4-dione), is prepared by condensing ethyl dimethylglycolate with urea followed by methylation of (**60**) with dimethyl sulfate in the 3 position (80).

$(CH_3)_2C(OH)CO_2C_2H_5$ —NH_2CONH_2, $NaOC_2H_5$→ (**60**) —$(CH_3)_2SO_4$→ (**59**)

Trimethadione is used to treat petit mal seizures. The usual oral daily dose range for adults is 900–2400 mg. The drug should only be used when less toxic drugs are ineffective since it has potential to produce fetal malformations and other serious side effects (81).

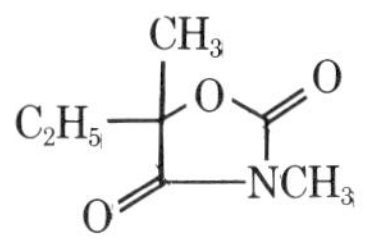

CH_3 C_2H_5 O O NCH_3 O

(**61**)

Paramethadione (**61**), $C_7H_{11}NO_3$ (3,5-dimethyl-5-ethyl-1,3-oxazolidine-2,4-dione), has an aromatic odor. It is prepared in a similar manner as trimethadione, is used clinically in the same dosages for the same indications, and suffers the same toxic effects as trimethadione (81).

Succinimides. Like their structural analogues, the oxazolidinediones, the succinimides are useful in generalized nonconvulsive seizures.

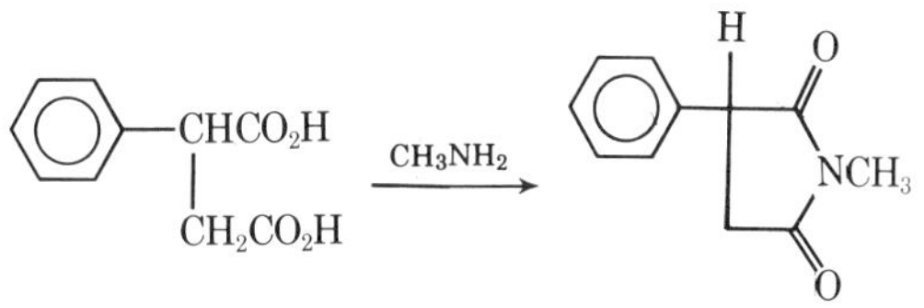

$C_6H_5CHCO_2H$ / CH_2CO_2H $\xrightarrow{CH_3NH_2}$ H O NCH_3 O

(**62**)

Phensuximide (**62**), $C_{11}H_{11}NO_2$ (1-methyl-3-phenyl-2,5-pyrrolidinedione), is prepared by heating 2-phenylsuccinic acid with methylamine in a variety of solvents.

Phensuximide primarily is used for the control of absence (petit mal) seizures in the dose range of 500–1000 mg two to three times daily (81).

Methsuximide (**63**), $C_8H_{13}NO_2$ (1,3-dimethyl-3-ethyl-2,5-pyrrolidinedione) is prepared in the same manner as phensuximide.

CH_3 O C_2H_5 NCH_3 O

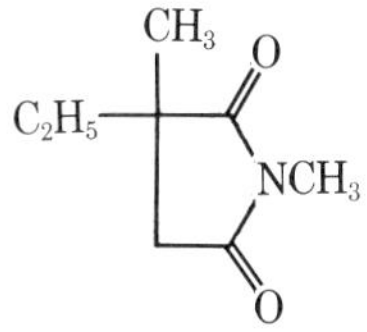

(**63**)

Methsuximide is used for the control of absence (petit mal) seizures that are refractory to other drugs in the daily dosage range of 300–1200 mg (81).

Ethosuximide (**64**), $C_7H_{11}NO_2$ (3-ethyl-3-methyl-2,5-pyrrolidinedione) is prepared in the same manner as phensuximide and is used for the control of absence (petit mal) and minor motor–myoclonic seizures in the daily dosage range of 500–1500 mg.

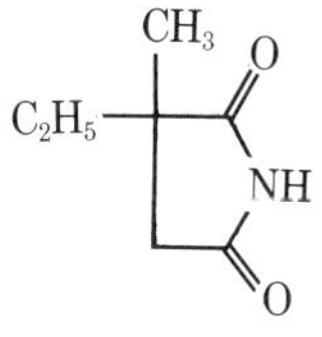

CH_3 O C_2H_5 NH O

(**64**)

Side effects include sedation and ataxic and central nervous system stimulation (see Stimulants) (81).

Others. *Primidone* (**65**), $C_{12}H_{14}N_2O_2$ (5-ethyl-5-phenylhexahydropyrimidine-4,6-dione), is prepared by reductive desulfurization of 5-ethyl-5-phenyl-2-thiobarbituric acid (**66**) in the presence of a hydrogenation catalyst.

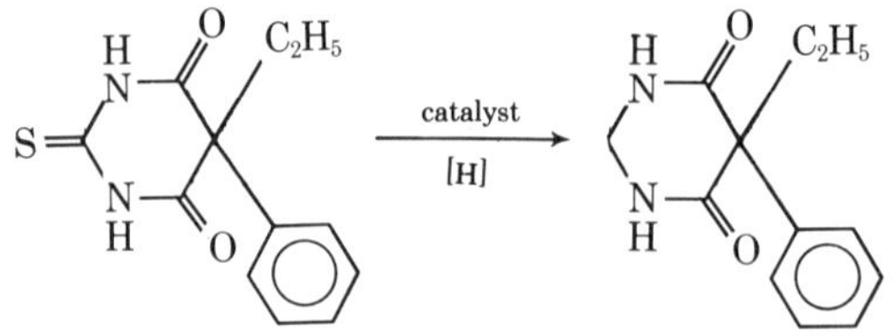

catalyst
[H]

(**66**) (**65**)

Primidone is used in the control of grand mal, psychomotor, and focal epileptic seizures in dose ranges of 250–2000 mg daily (81). It also may control grand mal seizures that are refractory to other anticonvulsant therapy. Its main side effect is sedation.

Clonazepam (**67**), $C_{15}H_{10}ClN_3O_3$ (5-(*o*-chlorophenyl)-1,3-dihydro-7-nitro-2*H*-1,4-benzodiazepin-2-one) is prepared (35) in a manner similar to diazepam (**21**), starting with the aminooxime (**68**). It is available in Europe as Clonopin (Roche).

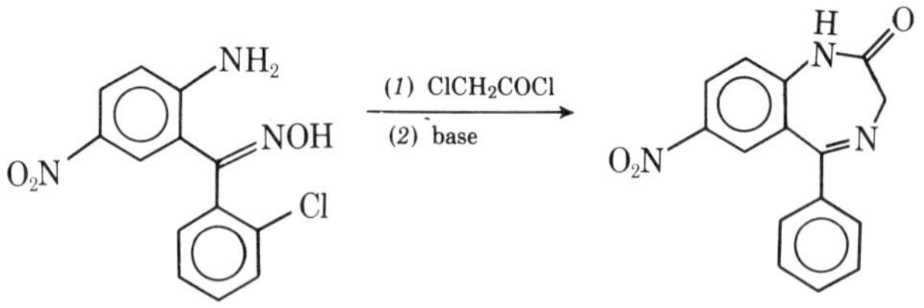

(*1*) $ClCH_2COCl$
(*2*) base

(**68**) (**67**)

Pharmacological results indicate good antiepileptic action on primary and secondary epileptic activity, good effect on propagation of focal epileptic discharges to distant brain regions, and no or very poor effect on focal epileptic activity (82). Of the benzodiazepines, clonazepam appears to have one of the largest safety margins in mice, rats, cats, and monkeys.

After oral administration, it reaches maximal blood levels within 1–2 h with a biological half-life of 18–50 h. When administered orally, the most beneficial use of the agent may be as an adjuvant with existing medication for refractory petit mal epilepsy. This would allow for lower doses and, thereby, avoid the side effects of drowsiness, ataxia, and some bothersome behavioral changes (83). It is useful in the management of minor motor seizures, petit mal seizures, and in generalized seizures of the akinetic, atonic, and myoclonic variety (84). It is not recommended for grand mal seizures. The recommended daily dose is 1.5 mg three times a day with increases of 0.5–1 mg/d until seizures are adequately controlled. The results obtained with protracted oral treatment in epileptic patients are controversial. A decrease in the drug's antiepileptic potency over time has been noticed (85).

Diazepam (**21**) (see Hypnotics and Sedatives). Intravenous (iv) diazepam is indicated for the immediate management of tonic–clonic status epilepticus in the convulsing patient. The initial adult dosage is 5–10 mg (iv) and the injection may be repeated if necessary at 10–15-min intervals up to a maximum dose of 30 mg. The most worrisome complications of iv diazepam are respiratory depression and hypotension (74).

Carbamazepine (**69**), $C_{15}H_{12}N_2O$ (5*H*-dibenz[*b*,*f*]azepine-5-carboxamide), is prepared by treatment of 5*H*-dibenz[*b*,*f*]azepine with phosgene followed by ammonia gas (86). Marketed in the United States for treatment of pain associated with trigeminal and glassopharyngeal analgesia, it is most useful in the treatment of partial seizures with complex symptomalogy and generalized tonic–clonic seizures (grand mal) in dose ranges of 400–1200 mg/d (81). It is not recommended as first-choice therapy since serious and sometimes fatal abnormalities of blood cells have been reported.

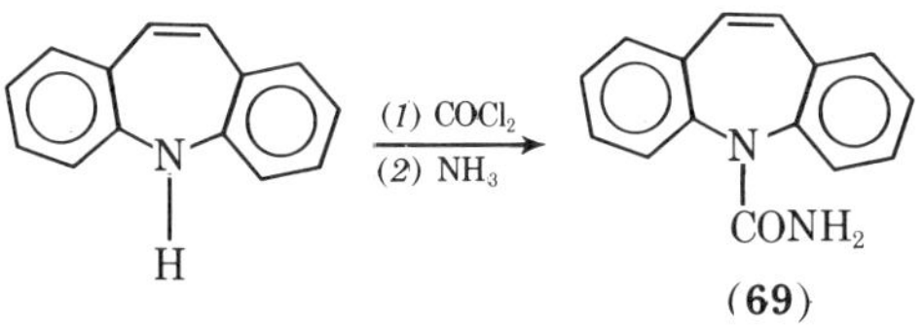

N–H —(*1*) $COCl_2$, (*2*) NH_3→ N–$CONH_2$

(**69**)

Phenacemide (**70**), $C_9H_{10}N_2O_2$ (phenacetylurea), is prepared by reaction of phenylacetyl chloride and urea.

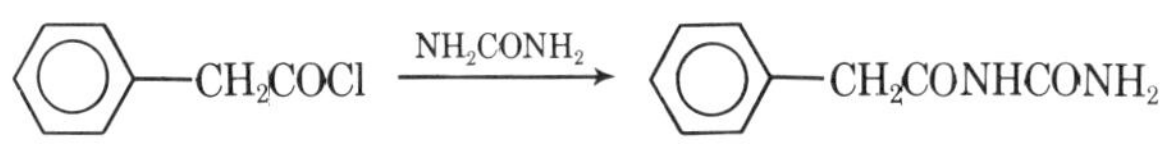

$C_6H_5CH_2COCl \xrightarrow{NH_2CONH_2} C_6H_5CH_2CONHCONH_2$

(**70**)

Valproic acid (**71**), $C_8H_{16}O_2$ (2-propylpentanoic acid), has a characteristic odor.

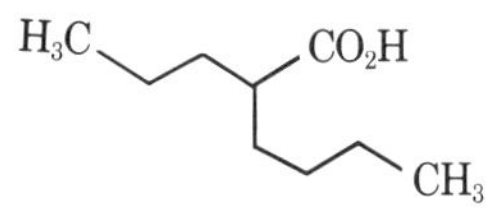

H_3C, CO_2H, CH_3

(**71**)

Valproic acid was first synthesized in 1881 and was used as a solvent for animal studies until 1963 when it was noted that various compounds that were dissolved in valproic acid protected mice and rabbits from pentylenetetrazole-induced seizures (87). Valproic acid as well as the sodium and magnesium salts have been marketed in many countries for the treatment of various epilepsies (87). It is useful as sole or adjunctive therapy in simple and complex absence seizures, including petit mal, and adjunctively for multiple seizure types which include absence (petit mal) seizures. Adequate control of seizures usually is achieved with doses of 1000–1600 mg daily. Valproic acid is thought to control seizures by increasing brain levels of γ-aminobutyric acid (GABA) in the appropriate regions of the brain. Compared with other anticonvulsant drugs, valproic acid is remarkably free of side effects; gastrointestinal problems, eg, nausea, vomiting, abdominal cramps, and diarrhea, are the usual complaints (88).

Sulthiame (**72**), $C_{10}H_{14}N_2O_4S_2$ (4-(tetrahydro-2*H*-1,2-thiazin-2-yl)-benzenesulfonamide-*S*,*S*-dioxide) is not available on the United States market. It is prepared (89) by alkali cyclization of disulfonamide (**73**).

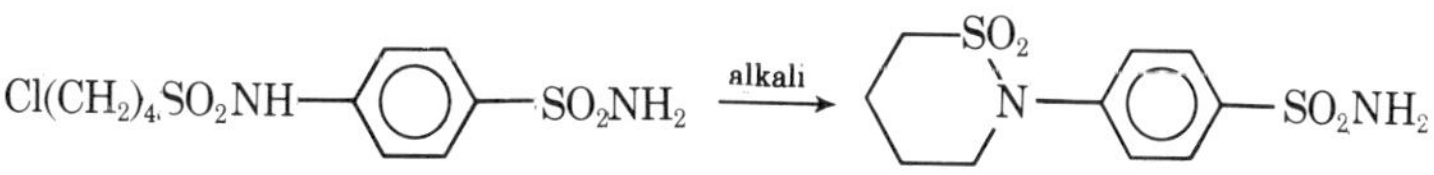

$Cl(CH_2)_4SO_2NH-C_6H_4-SO_2NH_2 \xrightarrow{alkali}$ SO_2 / N$-C_6H_4-SO_2NH_2$

(**73**) (**72**)

Sulthiame has been used widely in Europe since the 1960s under the trade name Ospolot (Bayer) for the treatment of seizures caused by focal or psychomotor epilepsy at doses of about 1200 mg daily (90). Its values as a primary anticonvulsant agent is limited because it has been associated with possible renal failures and has caused status epilepticus and hyperpnea during treatment.

BIBLIOGRAPHY

"Antispasmodics" in *ECT* 1st ed., Vol. 2, pp. 105–107, by J. C. Krantz, Jr., University of Maryland; "Hypnotics and Sedatives" in *ECT* 1st ed., Vol. 7, pp. 771–779, by Koert Gerzon, Eli Lilly and Company; "Hypnotics, Sedatives, Anticonvulsants" in *ECT* 2nd ed., Vol. 11, pp. 508–525, by Paul Turi, Sandoz, Inc.

1. O. Liebreich, *Wien. Med. Wochschr.*, 1087 (1869).
2. E. Fischer and J. von Mering, *Ther. Ggw.* **44,** 97 (1903).
3. S. Garattini, E. Mussini, and L. O. Randall, eds., *The Benzodiazepines,* Raven Press, New York, 1973.
4. *USAN and the USP Dictionary of Drug Names,* The United States Pharmacopeial Convention, Inc., Rockville, Md., 1975.
5. *The United States Pharmacopeia XX* (*USP XX–NF XV*), The United States Pharmacopeial Convention, Inc., Rockville, Md., 1980.
6. *Drugstore and Hospital Sales, 1977–1978,* IMS America, Ambler, Pa., 1978.
7. *Physicians Desk Reference,* 33rd ed., Medical Economics Company, Oradell, N.J., 1979.
8. *Redbook 1979,* Medical Economics Company, Oradell, N.J., 1979.
9. A. Burger, *Symposium on Sedative and Hypnotic Drugs, Elkhart, Ind., 1954.*
10. L. Lasagna, *Med. Clin. N. Am.* **41,** 359 (1957).
11. T. C. Butler, *J. Pharmacol. Exp. Ther.* **95,** 360 (1948).
12. K. Vogler and M. Kofler, *Helv. Chim. Acta* **39,** 1387 (1956).
13. E. H. Loughlin and co-workers, *Int. Rec. Med.* **168,** 52 (1955).
14. E. Tagmann and co-workers, *Helv. Chim. Acta* **35,** 1541 (1952).
15. Ref. 7, pp. 1729–1730.
16. Brit. Pat. 843,073 (Aug. 4, 1960), (to Laboratories Toraude).
17. A. H. Amin, D. R. Mehta, and S. S. Samarth in E. Jücker, ed., *Progress in Drug Research,* Vol. 14, Birkhäuser, Verlag, Basel, Switz., 1970, pp. 218–268.
18. Brit. Pat. 15,933 (1913), (to Beckmann Chem. Fabrik Ges.).
19. Ref. 7, p. 1276.
20. Ger. Pat. 185,962 (Apr. 17, 1906), (to Knoll and Co.).
21. D. Papa and co-workers, *J. Am. Chem. Soc.* **76,** 4446 (1954).
22. B. J. Ludwig and E. C. Piech, *J. Am. Chem. Soc.* **73,** 5779 (1951).
23. F. M. Berger, *J. Pharmacol. Exp. Ther.* **104,** 229 (1952).
24. P. Tyrer, *Lancet,* 709 (1974).
25. T. Blair and G. A. Webb, *J. Med. Chem.* **20,** 1206 (1977).
26. M. A. Schwartz, B. A. Koechlin, E. Postma, S. Palmer, and G. Krol, *J. Pharmacol. Exp. Ther.* **149,** 423 (1965).
27. G. K. Woo, S. J. Kolii, and M. A. Schwartz, *J. Pharmacol. Exp. Ther.* **149,** 783 (1965).
28. I. Oswald, S. A. Lewis, J. Tagney, H. Firth, and I. Haider in ref. 3, pp. 613–625.
29. L. H. Sternbach and E. Reeder, *J. Org. Chem.* **26,** 1111 (1961).
30. A. Kales and M. B. Scharf in ref. 3, pp. 577–598.
31. L. H. Sternach, G. A. Archer, and J. V. Earley, *J. Med. Chem.* **8,** 815 (1965).
32. D. J. Greenblat, R. I. Shader, and J. Koch-Weser, *Clin. Pharmacol. Ther.* **17,** 1 (1975).
33. *Psychopharmacologic Drugs: A Pocket Reference,* G. F. Stickley Co., Philadelphia, Pa., 1978.
34. I. MacPhail, A. Ogilvie, and C. R. Purvis, *J. Int. Med. Res.* **3,** 49 (1975).
35. L. H. Sternbach, R. I. Fryer, O. Keller, W. Metlesics, G. Sach, and N. Steiger, *J. Med. Chem.* **6,** 261 (1963).

36. E. O. Bixler, A. Kales, C. R. Soldatos, and J. D. Kales, *J. Clin. Pharmacol.* **17,** 569 (1977).
37. J. M. Monti and H. Altier, *Psychopharm.* **32,** 343 (1973).
38. S. J. Childress and M. I. Gluckman, *J. Pharm. Sci.* **53,** 577 (1964).
39. M. I. Gluckman, *Arzneim. Forsch.* **21,** 1049 (1971).
40. J. Castaner and P. Thorpe, *Drugs of the Future* **3,** 139 (1978).
41. F. R. Freeman, *J. Clin. Pharmacol.* **17,** 398 (1977).
42. M. A. Schwartz in ref. 3, pp. 53–74.
43. S. C. Bell and S. J. Childress, *J. Org. Chem.* **27,** 1691 (1962).
44. A. Longinil, V. Mandelli and I. Pessotti in ref. 3, pp. 347–354.
45. L. K. Fowler, *J. Int. Med. Res.* **52,** 295, 297 (1977).
46. K.-H. Weber, A. Bauer, and K.-H. Hauptmann, *Ann. Chem.* **756,** 128 (1972).
47. S. Deventhan and S. Channabasaranna, *Curr. Ther. Res.* **21,** 361 (1977).
48. F. Barzaghi, R. Fournes, and P. Mantegazza, *Arzneim. Forsch.* **23,** 683 (1973).
49. J. Hussar, J. Seffen, and G. K. Wolf, *Arzneim. Forsch.* **25,** 1650 (1975).
50. K. Meguro and Y. Kuwada, *Tetrahedron Lett.,* 4039 (1970).
51. J. Licé, R. M. Rees, T. L. Tyler, and J. D. Arnold, *Pharmacologist* **20,** 17 (1978).
52. K. Meguro, H. Tawada, H. Miyano, Y. Sato, and Y. Kuwada, *Chem. Pharm. Bull.* **21,** 2382 (1973).
53. J. B. Hester, Jr., A. D. Rudzik, and B. V. Kamdar, *J. Med. Chem.* **14,** 1078 (1971).
54. R. I. Wang and S. L. Stockdale, *J. Int. Med. Res.* **1,** 600 (1973).
55. A. Kales, E. O. Bixler, J. D. Kales, and M. B. Scharf, *J. Clin. Pharmacol.* **19,** 207 (1977).
56. M. Leibowitz and A. Sunshine, *J. Clin. Pharmacol.* **20,** 302 (1978).
57. J. Szmuskovicz, C. G. Chidester, D. J. Duchamp, F. A. MacKellar, and G. Slomp, *Tetrahedron Lett.,* 3665 (1971).
58. L. F. Fabre, Jr., and R. T. Harris, *Curr. Therp. Clin. Exp.* **16,** 848 (1974).
59. F. S. Eberts, Jr., and L. M. Reineke, *Pharmacologist* **18,** 153 (1976).
60. L. F. Fabre, Jr., R. T. Harris, and D. F. Stubbs, *J. Int. Med. Res.* **4,** 50 (1976).
61. S. S. Chatterjee, *Drugs of the Future* **1,** 293 (1976).
62. A. N. Nicholson, B. M. Stone, C. H. Clarke, and H. M. Ferris, *Br. J. Clin. Pharmacol.* **3,** 429 (1976).
63. M. Misic, J. Kuftinec, V. Sunjic, F. Kajbez, and N. Blazevik, *Abstr. 064, Fifth International Symposium on Medicinal Chemistry, July 19–22, 1976, Paris, Fr.*
64. C. Post, S. Lindgren, A. Bertless, and H. Malmgren, *Psychopharmacol.* **53,** 105 (1977).
65. N. P. Plotnikoff, *Res. Commun. Chem. Pathol. Pharmacol.* **5,** 128 (1973).
66. H. E. Booker, *J. Am. Med. Assoc.* **229,** 552 (1974).
67. Y. D. Lapiene, *J. Clin. Pharmacol.* **11,** 315 (1975).
68. U.S. Pat. 3,084,161 (Apr. 2, 1963), J. Renz, J. P. Bourquin, and G. Schwab (Sandoz Ltd.).
69. U.S. Pat. 2,837,518 (June 3, 1958), R. M. Jacob and J. G. Robert (to Société des Usines chimiques Rhône-Poulenc).
70. U.S. Pat. 2,607,773 (Aug. 19, 1952), S. S. Berg and J. N. Ashley (to Société des Usines chimiques Rhône-Poulenc).
71. A. M. Risberg, J. Risberg, and D. H. Ingnar, *Psychopharmacol.* **43,** 279 (1975).
72. H. W. Baird and co-workers, *Patient Care,* 2 (1977).
73. W. H. Hartung, ed., *Medicinal Chemistry,* Vol. 5, John Wiley & Sons, Inc., New York, 1961.
74. T. R. Browne, *Am. J. Hosp. Pharm.* **35,** 915, 1048 (1978).
75. J. A. Vida, ed., *Anticonvulsants,* Wiley-Interscience, New York, 1979.
76. M. J. Eadie, *Drugs* **17,** 213 (1979); T. R. Browne, *Am. J. Hosp. Pharm.* **35,** 915 (1978).
77. R. L. Krall, J. K. Penry, H. J. Kupferberg, and E. A. Swinyard, *Epilepsia* **19,** 393, 409 (1978).
78. H. Blitz and K. Seydel, *Ber.* **44,** 411 (1911).
79. Swiss Pat. 166,004 (Feb. 16, 1934), (to Chem. Fabrik v. Sandoz).
80. M. A. Spielman, *J. Am. Chem. Soc.* **66,** 1244 (1944).
81. J. K. Penry and M. E. Newmark, *Ann. Int. Med.* **90,** 207 (1979).
82. G. F. Rossi, C. DiRocco, G. Maira, and M. Maglio in ref. 3, pp. 461–488.
83. P. F. Bladin, *Med. J. Aust.* **683** (1973).
84. R. M. Pinder, R. N. Brogden, T. M. Speight, and G. S. Avery, *Drugs* **12,** 321 (1976).
85. T. R. Browne, *Arch. Neurol.* **33,** 326 (1976).
86. U.S. Pat. 2,948,718 (Aug. 9, 1960), W. Schindler.
87. D. Simon and J. K. Penry, *Epilepsia* **16,** 549 (1975).

88. *Drug Intell. Clin. Pharm.* **13,** 90 (1979).
89. Ger. Pat. 1,111,191 (July 20, 1961), B. Helferich, R. Behnisch, and W. Wirth (to Farbenfabriken Bayer A.G.).
90. M. Tchicaloff, *Schweiz Apotheker-Zeit.* **103,** 562 (1965).

WILLIAM J. HOULIHAN
GREGORY B. BENNETT
Sandoz Inc.

HYPOGLYCEMIC AGENTS.

See Insulin and other anti-diabetic agents.

I

IMINES, CYCLIC

Imines are those compounds containing the bivalent —NH— (imine) group. This term, when used to name cyclic imines such as ethylenimine, denotes a combination of the cyclic group (ethylene, —CH_2CH_2—) and the imine group. The term is also used as a class name for compounds containing the imine group attached with a double bond to a single carbon atom, such as acetalimine ($CH_3CH{=}NH$).

Cyclic imines are classified according to the number of atoms in the alkylenimine ring. Each ring size is individually named as a different heterocyclic series. Thus, derivatives of ethylenimine are named as substituted aziridines, trimethylenimines as azetidines, tetramethylenimines as pyrrolidines, and pentamethylenimines as piperidines. This article is concerned only with the simplest series of cyclic imines, the aziridines.

Aziridine(ethylenimine) is the most important member of this group, followed by 2-methylaziridine(propylenimine) and various aziridine derivatives. Aziridine, first prepared in 1888 by Gabriel, was correctly identified in 1899 by Marckwald, and was first manufactured in 1938 by I. G. Fabenindustrie A.G. Following World War II, Badische Anilin-und Soda-Fabrik in Germany and Chemirad Corporation in the United States continued its production. In the early 1960s, Interchemical Corporation made 2-methylaziridine available and The Dow Chemical Company began production of aziridine by a new process. In the mid-to-late-1960s, Union Carbide Corporation offered 1-(2-hydroxyethyl)aziridine and other derivatives were made available by The Dow Chemical Company. Throughout the 1970s Cordova Chemical Company supplied various aziridine derivatives and in 1979 announced the completion of facilities to produce aziridine commercially (1).

Physical Properties

The low molecular weight aziridines are colorless mobile liquids. Aziridine, 2-methylaziridine, and 1-(2-hydroxyethyl)aziridine are miscible in all proportions with water and most organic solvents. Aziridine and 2-methylaziridine have an odor very similar to that of ammonia. Table 1 lists physical properties of some aziridines that are, or have been, produced commercially or in developmental quantities (2). Table 2 gives the vapor pressure of aziridine and 2-methylaziridine at various temperatures (2), Tables 3 and 4 give thermal and other properties.

Chemical Properties

Aziridines tend to undergo ring-opening reactions because of the strained three-membered ring. Except under unusual conditions, the aziridine nitrogen atom must be present as an ammonium (aziridinium) nitrogen atom (or be substituted with an electron-withdrawing or conjugating group) for the ring to be opened. Thus, most ring-opening reactions are catalyzed by acid and may be formulated as nucleophilic substitutions.

Aziridines that are unsubstituted on the aziridine nitrogen atom undergo reactions typical of secondary amines. However, to avoid ring opening, temperatures should be kept low and the pH neutral or basic.

Table 1. Physical Properties of Aziridines

Name	CAS Registry No.	Formula	Freezing point, °C	Boiling point, °C	Refractive index, n_D^{25}	Density, d^{25}, g/mL	Basicity constant, K_B at 25°
aziridine	[*151-56-4*]	$HN<(CH_2)(CH_2)$ (ring)	−73.96	56.72	1.4123	0.831	7.92×10^{-7}
2-methylaziridine	[*75-55-8*]	$HN<(CH(CH_3))(CH_2)$ (ring)	−65.00	66.0	1.4084	0.802	
2-ethylaziridine	[*2549-67-9*]	$HN<(CH(C_2H_5))(CH_2)$ (ring)	−80.00	89.0	1.4165	0.834	2.0×10^{-6}
2,2-dimethylaziridine	[*2658-24-4*]	$HN<(C(CH_3)_2)(CH_2)$ (ring)	−47.00	70.0	1.4052	0.784	4.3×10^{-6}
1-(2-hydroxyethyl)aziridine	[*1072-52-2*]	$HO(CH_2)_2—N<(CH_2)(CH_2)$ (ring)		155.0	1.453		1.96×10^{-7}

Table 2. Vapor Pressure of Aziridine and 2-Methylaziridine, kPa[a]

Temperature, °C	Aziridine	2-Methylaziridine
−60.0	0.08	
−50.4	0.19	
−40.7	0.45	
−30.9	0.99	
−21.2	2.01	
−10.0	4.93	
0	7.8	
9.7	13.3	
20.0	22.7	14.9
30.0		23.9
39.0	52.0	35.9
51.0		58.1
56.72	101.3	
66.0		101.3

[a] To convert kPa to mm Hg, multiply by 7.5.

Table 3. Properties of Aziridine and 2-Methylaziridine

Property	Value	Reference
Aziridine		
flash point, °C	−11.1	
autoignition temperature, °C	323	
viscosity at 25°C, mPa·s (= cP)	0.418	
surface tension of liq at 25°C, mN/m (= dyn/cm)	32.8	
heat of vaporization, kJ/mol[a]	34.11	3
heat of combustion of liq at 25°C, MJ/mol[a]	1.59	3
heat of formation at 25°C, kJ/mol[a]		
gas	127.2	
liq	93.3	
heat capacity of liq, J/(g·K)[a]		4
at 0°C	2.39	
at 25°C	2.51	
at 42°C	2.59	
heat of mixing to produce 20% by wt soln in H_2O, kJ/mol[a]	13.8	4
dielectric constant at 25°C	18.3	
dipole moment, C·m[b]	$(7.74–7.84) \times 10^{-30}$	
specific conductance $(\Omega \cdot cm)^{-1}$	8×10^{-6}	2
2-Methylaziridine		
heat of vaporization, kJ/mol[a]	33.23	5
heat of mixing to produce 5% by wt soln in H_2O, kJ/mol[a]	18.8	5
viscosity at 25°C, mPa·s (= cP)		

[a] To convert J to cal, divide by 4.184.
[b] To convert C·m to debye, divide by 3.336×10^{-30}.

Reactions. The reactions discussed here refer to aziridines unsubstituted on the nitrogen atom, unless otherwise stated.

With Acids. Aziridines react with protonic acids to form salts. These salts are usually unstable if the anion of the acid is nucleophilic and there are no substituents on the aziridine (6–7).

Table 4. Thermal Properties of Aziridine, Ideal Gas State[a]

T, K	$(H° - H_0°)/T$, J/K[b]	Entropy $S°$, J/K[b]	Specific heat, $Cp°$, J/K[b]
200	33.9135	232.5261	37.4758
298.15	37.1540	250.1910	51.2394
400	42.9887	267.7161	68.8595
600	56.6418	301.3643	97.2493
800	69.3680	332.1697	116.7008
1000	80.310	359.7986	130.7506
1200	89.6330	384.6107	141.2483
1400	97.5996	407.0084	148.3117
1500	101.1493	417.4144	152.4223

[a] Calculated values.
[b] To convert J to cal, divide by 4.184.

$$2\ \underset{CH_2}{\overset{CH_2}{|}}\!\!>NH + H_2SO_4 \longrightarrow \left(\underset{CH_2}{\overset{CH_2}{|}}\!\!>NH_2^+\right)_2 SO_4^{2-}$$

[29086-57-5]
decomposes in less than 2 h at room temperature

$$\underset{CH_2}{\overset{CH_2}{|}}\!\!>NH + HBF_4 \longrightarrow \underset{CH_2}{\overset{CH_2}{|}}\!\!>NH_2^+\ BF_4^-$$

In solution, a ring-opening reaction occurs involving substitution of the acid anion at an aziridine carbon atom, and two moles of acid are used per mol of aziridine. Yields with acids that have highly nucleophilic anions are excellent.

$$\underset{CH_2}{\overset{CH_2}{|}}\!\!>NH + 2\ HCl \longrightarrow ClCH_2CH_2NH_2 \cdot HCl$$

Carboxylic acids react to form 2-aminoethyl esters (8). The corresponding 2-hydroxyethyl amide is a by-product unless an excess of the acid is used.

$$\underset{CH_2}{\overset{C_2H_5\,CH}{|}}\!\!>NH + 2\ C_6H_5CO_2H \longrightarrow C_6H_5CO_2CH_2\overset{C_2H_5}{\overset{|}{C}}HNH_2 \cdot HO_2CC_6H_5$$

With Acyl Halides. Aziridines react readily with acyl (or other acid) halides at room temperature or below (9). Acyl aziridines are obtained in good yields in an inert solvent in the presence of a base, such as triethylamine. Polyfunctional derivatives also are prepared readily (9).

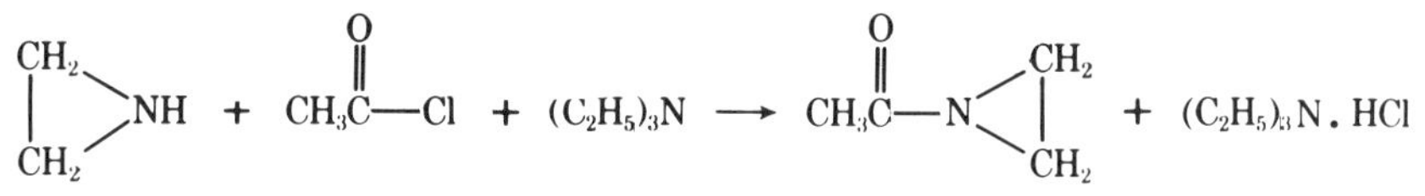

$$(CH_2)_2NH + CH_3C(=O)Cl + (C_2H_5)_3N \longrightarrow CH_3C(=O)N(CH_2)_2 + (C_2H_5)_3N \cdot HCl$$

[*460-07-1*]

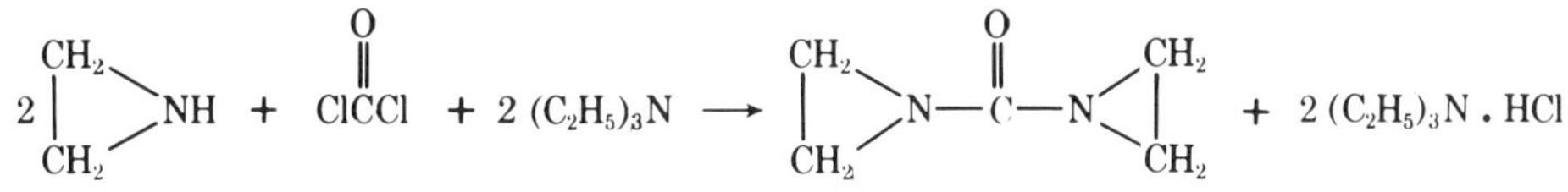

$$2\,(CH_2)_2NH + ClC(=O)Cl + 2\,(C_2H_5)_3N \longrightarrow (CH_2)_2N{-}C(=O){-}N(CH_2)_2 + 2\,(C_2H_5)_3N \cdot HCl$$

[*1192-75-2*]

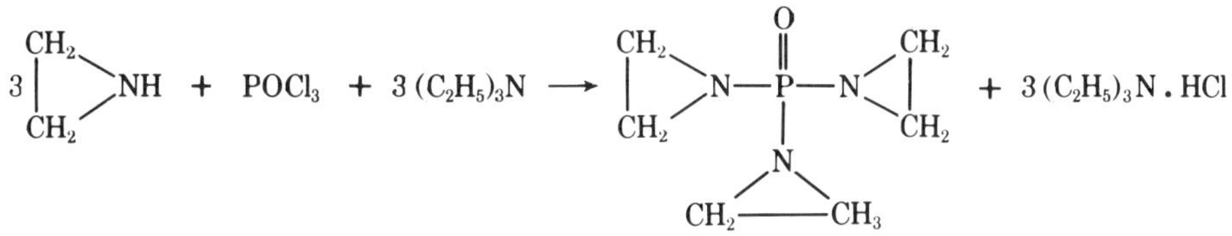

$$3\,(CH_2)_2NH + POCl_3 + 3\,(C_2H_5)_3N \longrightarrow [(CH_2)_2N]_2P(=O){-}N(CH_2{-}CH_3) + 3\,(C_2H_5)_3N \cdot HCl$$

[*545-55-1*]

$$3\,(CH_2)_2NH + C_6H_3(COCl)_3 + 3\,(C_2H_5)_3N \longrightarrow C_6H_3[C(=O)N(CH_2)_2]_3 + 3\,C_2H_5N \cdot HCl$$

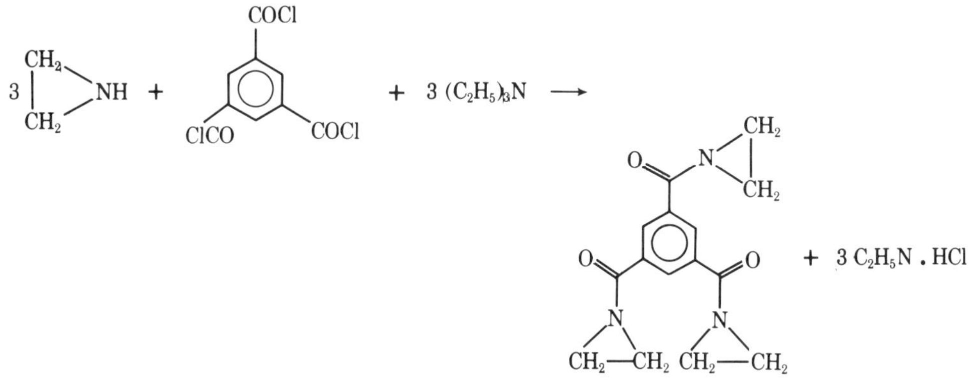

[*16044-74-9*]

If the base in these reactions is omitted, the ring is opened and *N*-2-chloroethyl amides are obtained in good yield (9). This ring opening also occurs readily with 1-alkyl aziridines.

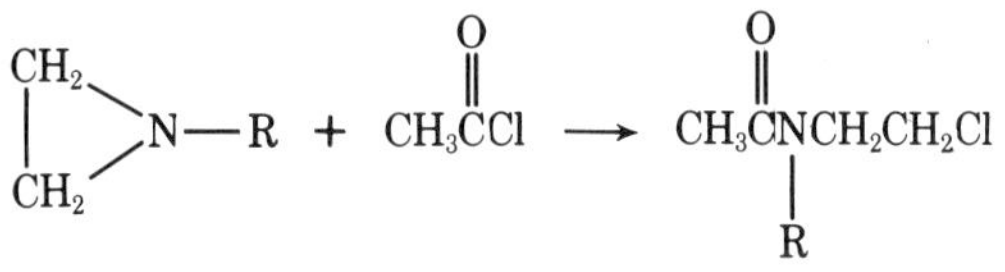

$$(CH_2)_2N{-}R + CH_3C(=O)Cl \longrightarrow CH_3C(=O)N(R)CH_2CH_2Cl$$

$R = C_2H_5$ [*1072-45-3*] $R = C_2H_5$

$$(CH_2)_2NH + C_6H_5SO_2Cl \longrightarrow C_6H_5SO_2NHCH_2CH_2Cl$$

$$2\,(CH_2)_2NH + ClC(=O)OCH_2CH_2OC(=O)Cl \longrightarrow ClCH_2CH_2NHC(=O)OCH_2CH_2OC(=O)NHCH_2CH_2Cl$$

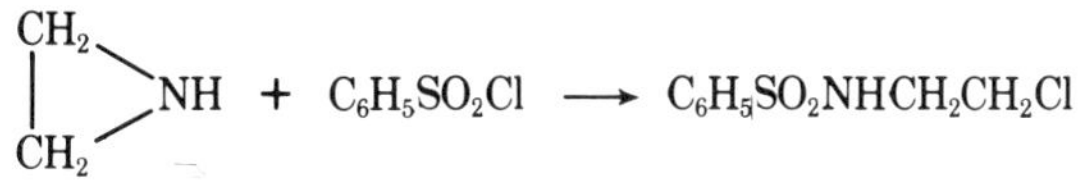

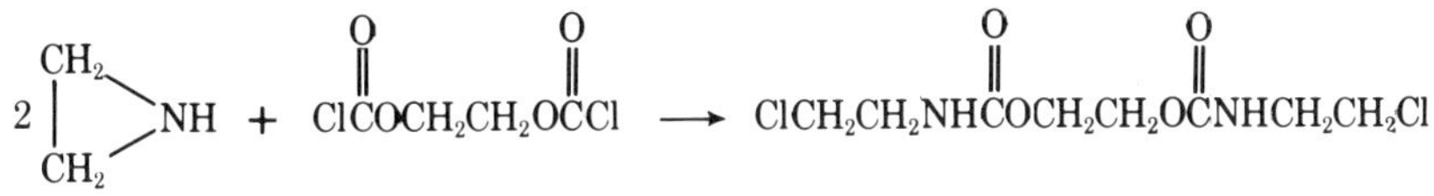

Acid anhydrides react similarly to acyl halides in both types of reactions.

With Alkyl, Substituted Alkyl, and Aryl Halides. Aziridines may be converted to 1-alkyl or 1-aryl aziridines by treatment with the alkyl or aryl halides. The reaction with reactive chlorides or bromides is usually performed at room temperature or below in the presence of a base in an inert solvent (9–10).

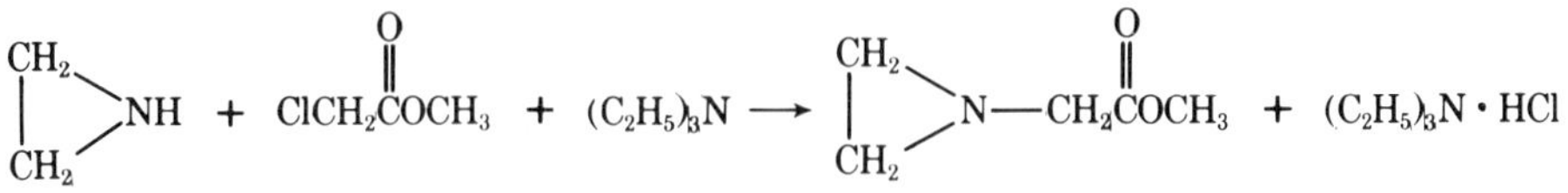

$$(CH_2)_2NH + ClCH_2COCH_3 + (C_2H_5)_3N \longrightarrow (CH_2)_2N{-}CH_2COCH_3 + (C_2H_5)_3N \cdot HCl$$

[*14745-52-9*]

$$(CH_2)_2NH + C_6H_5CHBrCOCH_3 + (C_2H_5)_3N \longrightarrow (C_2H_5)_3N \cdot HBr + C_6H_5CH(COCH_3){-}N(CH_2)_2$$

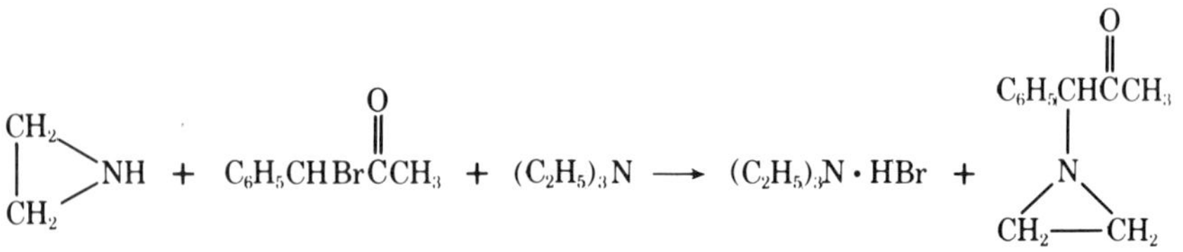

[*6713-46-8*]

With less reactive halides, excess aziridine and inorganic base improves the yield (11).

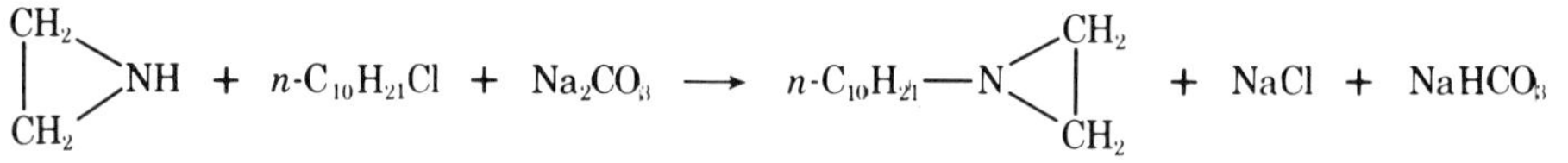

$$(CH_2)_2NH + n\text{-}C_{10}H_{21}Cl + Na_2CO_3 \longrightarrow n\text{-}C_{10}H_{21}{-}N(CH_2)_2 + NaCl + NaHCO_3$$

[*46237-34-7*]

Aromatic halides that contain electronegative groups on the aryl ring react under similar conditions (12).

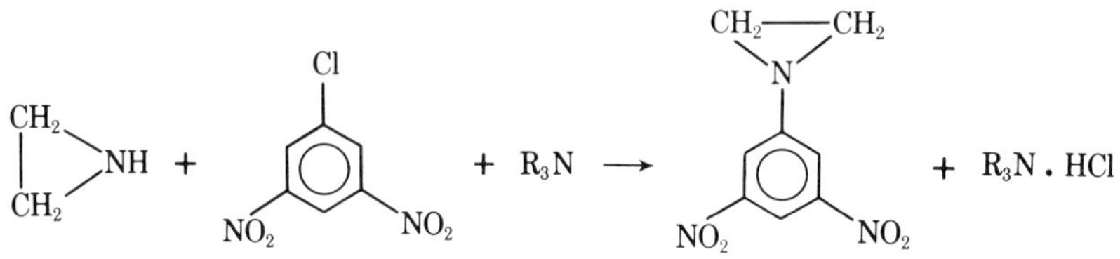

$$(CH_2)_2NH + 1\text{-}Cl\text{-}3,5\text{-}(NO_2)_2C_6H_3 + R_3N \longrightarrow 1\text{-}[(CH_2)_2N]\text{-}3,5\text{-}(NO_2)_2C_6H_3 + R_3N \cdot HCl$$

[*27141-65-7*]

With unreactive aryl halides, such as halobenzenes, the aziridine is first converted to the alkali metal salt by treatment with alkali metal, alkali metal alloys, alkali metal amides, or compounds such as organolithium compounds (13).

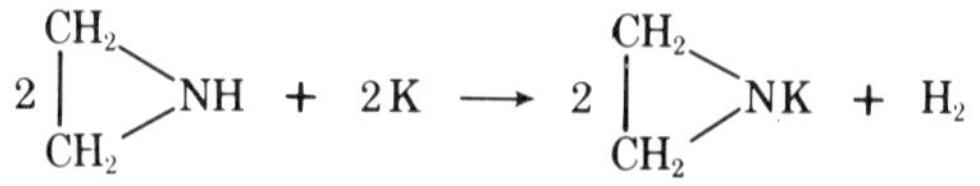

$$2\,(CH_2)_2NH + 2K \longrightarrow 2\,(CH_2)_2NK + H_2$$

[*29521-21-9*]

$$(CH_2)_2NK + C_6H_5Br \longrightarrow C_6H_5{-}N(CH_2)_2 + KBr$$

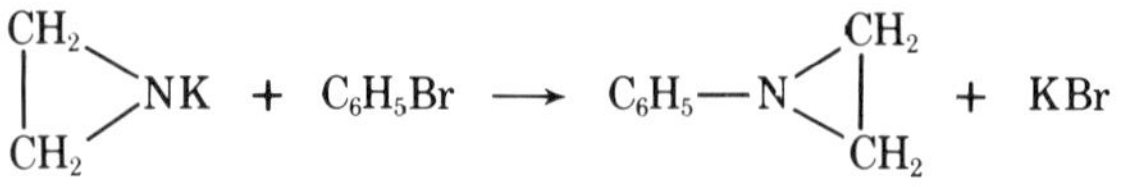

[*696-18-4*]

This same procedure, ie, via the aziridine alkali metal salt, may be used with unreactive alkyl halides.

Treatment of aziridines with a large excess of reactive alkyl halide opens the ring and gives 2-haloethyltrialkylammonium halides.

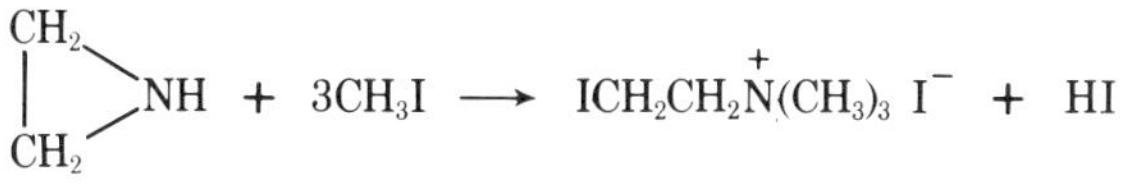

$$\overset{CH_2}{\underset{CH_2}{|}}\!\!>NH + 3CH_3I \longrightarrow ICH_2CH_2\overset{+}{N}(CH_3)_3\ I^- + HI$$

With Other Halogen-Containing Compounds. A variety of other halogen compounds, wherein the halogen atom is attached to an atom other than carbon, react readily with aziridines to produce the corresponding aziridine derivative (14–19). The reactions are performed on alkali metal salts of the aziridine or on the aziridine in the presence of base or a large excess of the aziridine, or both. Some derivatives, eg, the 1-arylazoaziridines, are unstable and may decompose at room temperature with explosive violence. The yields from such reactions vary from ca 10 to ca 90%.

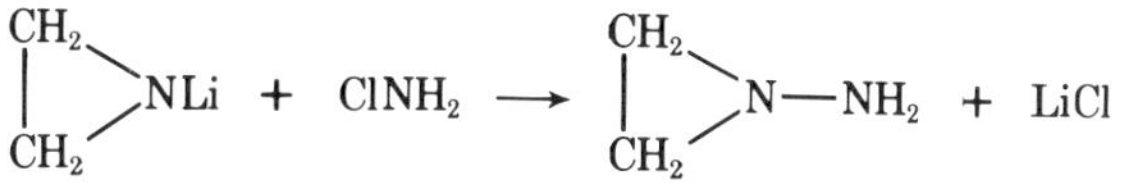

$$\overset{CH_2}{\underset{CH_2}{|}}\!\!>NLi + ClNH_2 \longrightarrow \overset{CH_2}{\underset{CH_2}{|}}\!\!>N{-}NH_2 + LiCl$$

[33734-99-5] [1721-30-8]

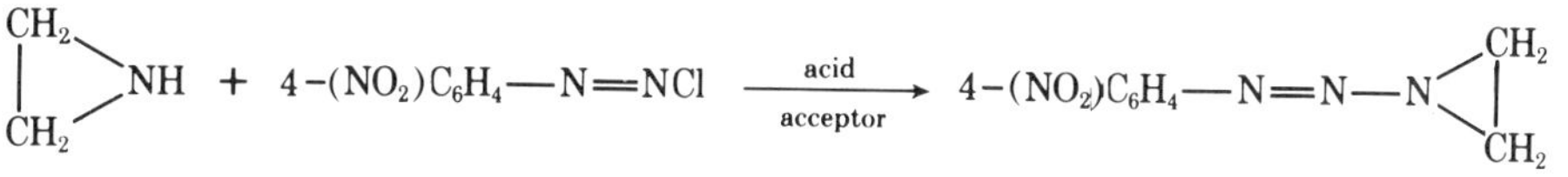

$$\overset{CH_2}{\underset{CH_2}{|}}\!\!>NH + 4\text{-}(NO_2)C_6H_4{-}N{=}NCl \xrightarrow[\text{acceptor}]{\text{acid}} 4\text{-}(NO_2)C_6H_4{-}N{=}N{-}N\!\!<\overset{CH_2}{\underset{CH_2}{|}}$$

[26439-16-7]

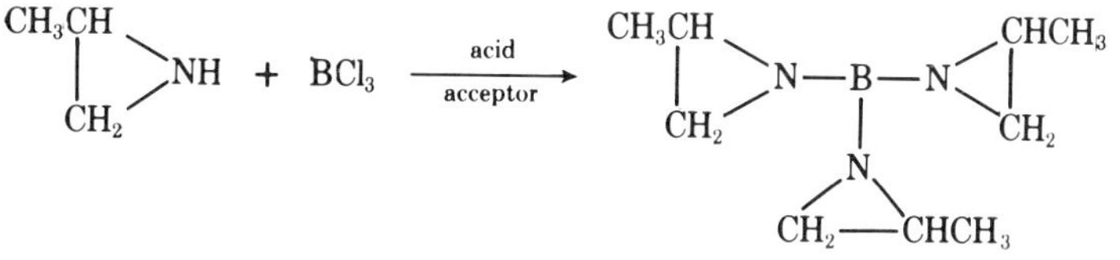

$$\overset{CH_3CH}{\underset{CH_2}{|}}\!\!>NH + BCl_3 \xrightarrow[\text{acceptor}]{\text{acid}} \overset{CH_3CH}{\underset{CH_2}{|}}\!\!>N{-}\underset{\underset{CH_2{-}CHCH_3}{N}}{\underset{|}{B}}{-}N\!\!<\overset{CHCH_3}{\underset{CH_2}{|}}$$

[17862-61-2]

$$2\overset{CH_2}{\underset{CH_2}{|}}\!\!>NH + C_6H_5AsCl_2 \xrightarrow[\text{acceptor}]{\text{acid}} C_6H_5{-}\overset{\overset{CH_2{-}CH_2}{N}}{\overset{|}{As}}{-}N\!\!<\overset{CH_2}{\underset{CH_2}{|}}$$

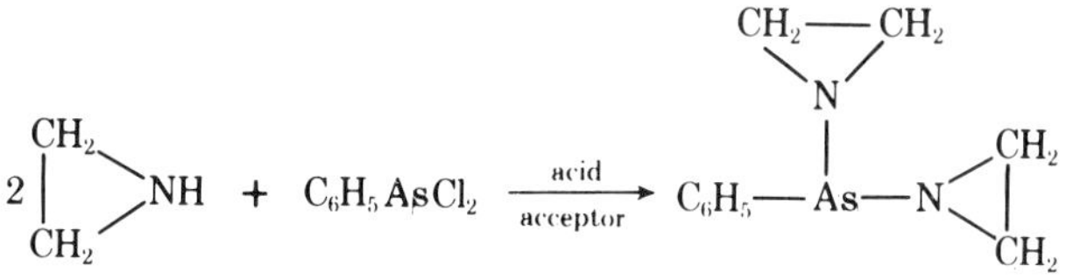

[20825-66-5]

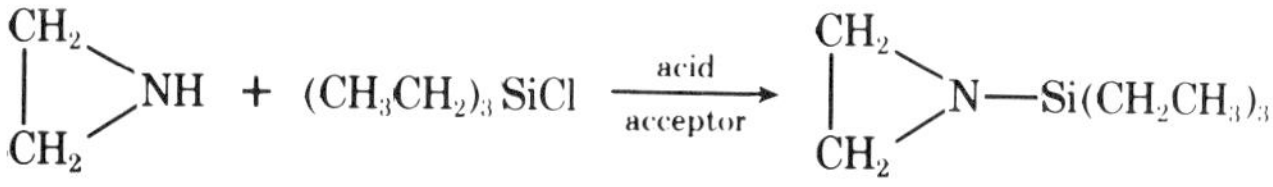

$$\overset{CH_2}{\underset{CH_2}{|}}\!\!>NH + (CH_3CH_2)_3SiCl \xrightarrow[\text{acceptor}]{\text{acid}} \overset{CH_2}{\underset{CH_2}{|}}\!\!>N{-}Si(CH_2CH_3)_3$$

[15000-97-2]

$$2\overset{CH_2}{\underset{CH_2}{|}}\!\!>NH + SCl_2 \xrightarrow[\text{acceptor}]{\text{acid}} \overset{CH_2}{\underset{CH_2}{|}}\!\!>N{-}S{-}N\!\!<\overset{CH_2}{\underset{CH_2}{|}}$$

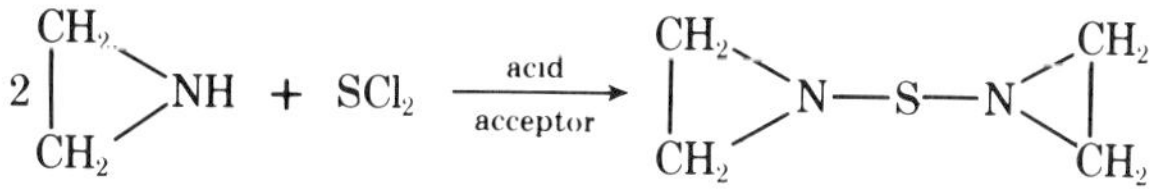

[2881-79-0]

Nitrosyl chloride and aziridines in the presence of a base at room temperature give olefins and nitrous oxide (20).

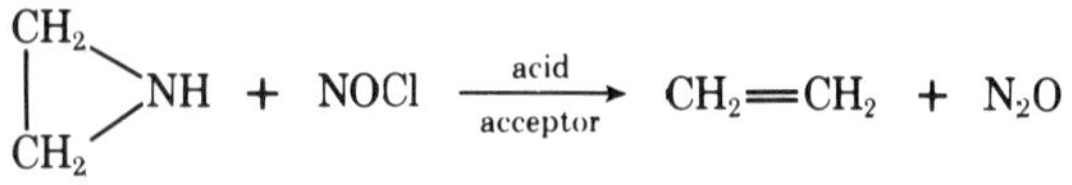

$$\text{(CH}_2)_2\text{NH} + \text{NOCl} \xrightarrow[\text{acceptor}]{\text{acid}} \text{CH}_2{=}\text{CH}_2 + \text{N}_2\text{O}$$

This ring rupture is caused by any reagent able to react with an aziridine to form the 1-nitroso derivative, such as nitrous acid. The 1-nitroso derivative is stable at very low temperatures, but decomposes above ca $-20°C$.

With Aldehydes and Ketones. Aziridines form 1:1 adducts with aldehydes and ketones that are relatively stable when compared to such adducts formed from other secondary amines (21–22).

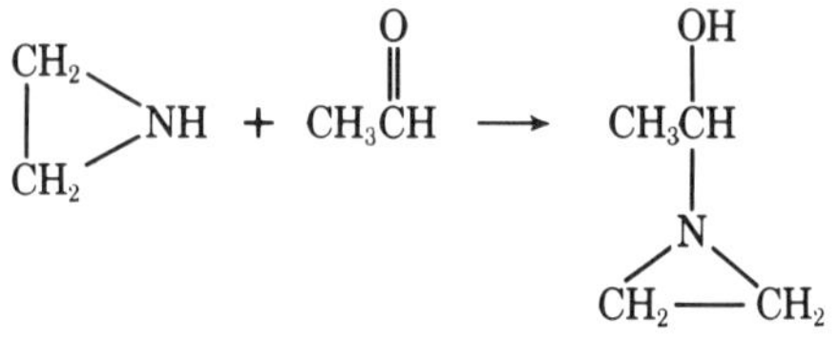

$$\text{(CH}_2)_2\text{NH} + \text{CH}_3\text{CHO} \longrightarrow \text{CH}_3\text{CH(OH)}\text{—N(CH}_2)_2$$

[*13148-34-0*]

$$\text{(CH}_2)_2\text{NH} + \text{CF}_3\text{COCF}_3 \longrightarrow \text{CF}_3\text{—C(OH)(CF}_3)\text{—N(CH}_2)_2$$

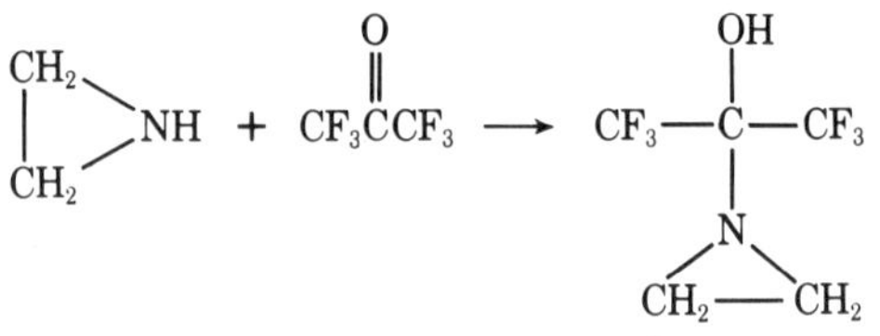

[*773-23-9*]

With aromatic aldehydes, eg, benzaldehyde, a ring-opening reaction occurs (23):

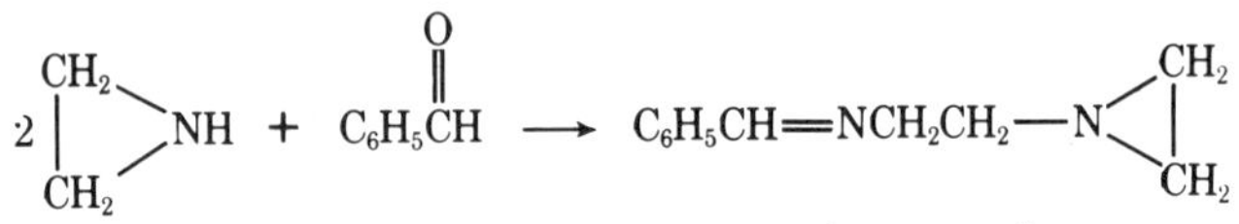

$$2\,\text{(CH}_2)_2\text{NH} + \text{C}_6\text{H}_5\text{CHO} \longrightarrow \text{C}_6\text{H}_5\text{CH}{=}\text{NCH}_2\text{CH}_2\text{—N(CH}_2)_2$$

[*4025-38-1*]

Aziridines react very readily with H_2S and aldehydes or ketones. This route offers a very convenient method to prepare certain thiazolidines (9,24–25).

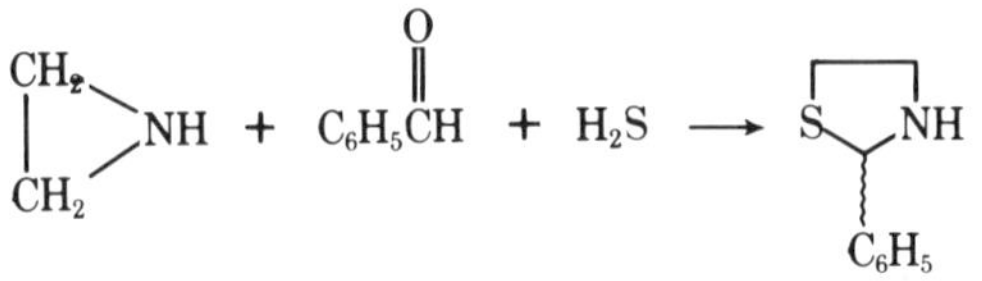

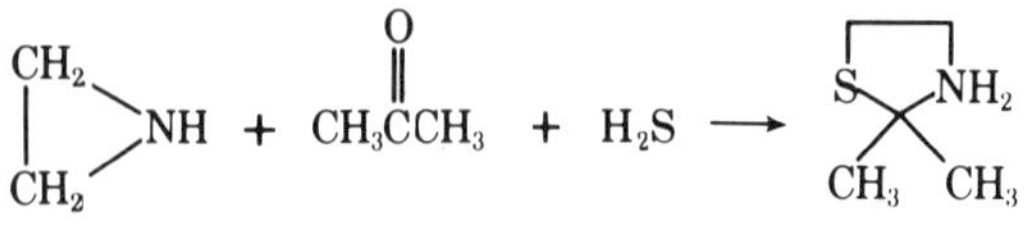

$$\text{(CH}_2)_2\text{NH} + \text{C}_6\text{H}_5\text{CHO} + \text{H}_2\text{S} \longrightarrow \text{2-C}_6\text{H}_5\text{-thiazolidine (S, NH ring)}$$

$$\text{(CH}_2)_2\text{NH} + \text{CH}_3\text{COCH}_3 + \text{H}_2\text{S} \longrightarrow \text{2,2-(CH}_3)_2\text{-thiazolidine (S, NH}_2\text{ ring)}$$

With Amines. Ammonia and primary and secondary amines react with aziridines in the presence of catalytic amounts of acid to give ethylenediamines. A large excess of the amine is generally used (26).

$$\text{(CH}_2\text{)}_2\text{NH} + R_2NH \xrightarrow{H^+} R_2N\text{—}CH_2CH_2\text{—}NH_2$$

$$\text{(CH}_2\text{)}_2\text{N—R} + NH_3 \xrightarrow{H^+} H_2N\text{—}CH_2CH_2\text{—}NHR$$

Reaction with amines is also effected in the presence of equimolar amounts of aluminum chloride. Yields are high and only a moderate excess of the amine is required. Aluminum chloride is generally used with aromatic amines (27).

$$\text{(CH}_2\text{)}_2\text{NH} + ArNH_2 \xrightarrow{AlCl_3} ArNH\text{—}CH_2CH_2\text{—}NH_2$$

With Anionic Reagents. Aziridines or 1-alkylaziridines do not react with most anionic reagents. If, however, the anion is highly nucleophilic such as a carbanion, ring opening occurs at elevated temperatures (28–29).

$$\text{(CH}_2\text{)}_2\text{N—R} + CH_2(\overset{O}{\overset{\|}{C}}OC_2H_5)_2 \xrightarrow{LiOC_2H_5} \text{1-R-3-(CO}_2\text{C}_2\text{H}_5\text{)-2-pyrrolidinone} + C_2H_5OH$$

$$\text{(CH}_2\text{)}_2\text{NH} + (C_6H_5)_2PK \longrightarrow (C_6H_5)_2PCH_2CH_2NHK$$

The Grignard reagent, 3-indolemagnesium bromide, reacts similarly with aziridine to give tryptamine.

With Aromatic Hydrocarbons. Aziridines, in the form of aluminum chloride complexes, undergo ring opening when heated with excess aromatic hydrocarbon at 120–180°C for 7–8 h (30).

$$\text{(CH}_2\text{)}_2\text{NH.AlCl}_3 + C_6H_6 \longrightarrow C_6H_5CH_2CH_2NH_2\text{.}AlCl_3$$

With Epoxides and Episulfides. A variety of epoxides and diepoxides react with 1-unsubstituted aziridines to give 1-(2-hydroxyalkyl) aziridines. An excess of the aziridine is generally used (31).

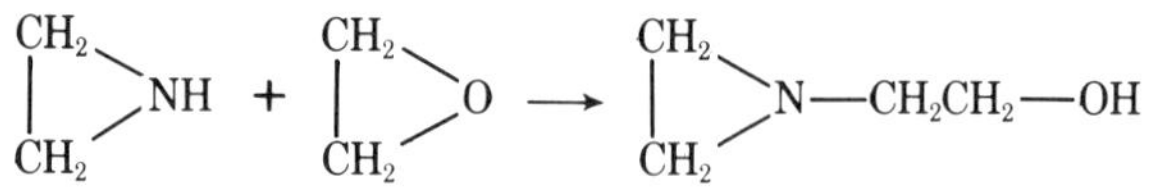

$$\underset{CH_2}{\overset{CH_2}{|}}\!\!>NH + \underset{CH_2}{\overset{CH_2}{|}}\!\!>O \longrightarrow \underset{CH_2}{\overset{CH_2}{|}}\!\!>N{-}CH_2CH_2{-}OH$$

[*1072-52-2*]

With episulfides, use of a large excess of either reactant is not possible since the product will undergo reaction with either reactant. Comparable amounts of each product shown may be formed in the following reaction (32).

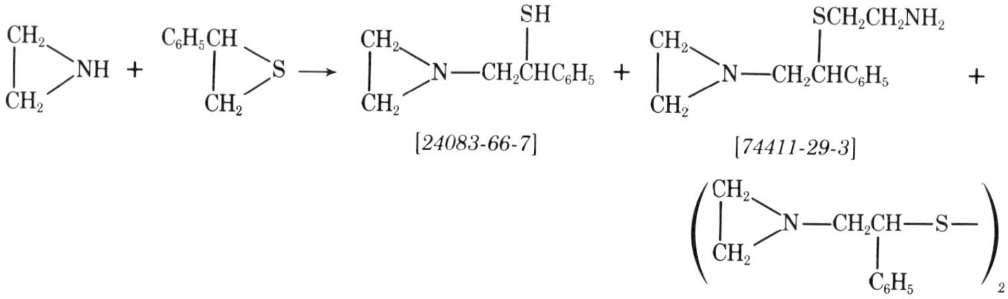

$$\underset{CH_2}{\overset{CH_2}{|}}\!\!>NH + \underset{CH_2}{\overset{C_6H_5CH}{|}}\!\!>S \longrightarrow \underset{CH_2}{\overset{CH_2}{|}}\!\!>N{-}CH_2\overset{SH}{\overset{|}{C}}HC_6H_5 + \underset{CH_2}{\overset{CH_2}{|}}\!\!>N{-}CH_2\overset{SCH_2CH_2NH_2}{\overset{|}{C}}HC_6H_5 +$$

[*24083-66-7*] [*74411-29-3*]

$$\left(\underset{CH_2}{\overset{CH_2}{|}}\!\!>N{-}CH_2\underset{C_6H_5}{\underset{|}{C}}H{-}S{-}\right)_2$$

[*75422-15-0*]

With Hydrogen. Aziridines are hydrogenated in the presence of typical hydrogenation catalysts to give alkyl amines (33).

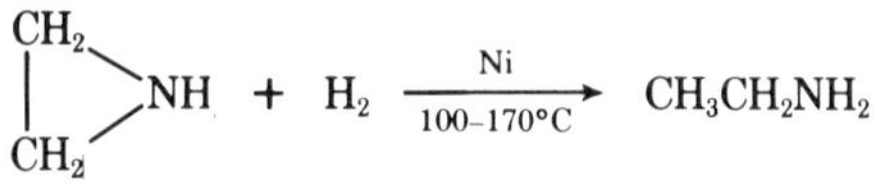

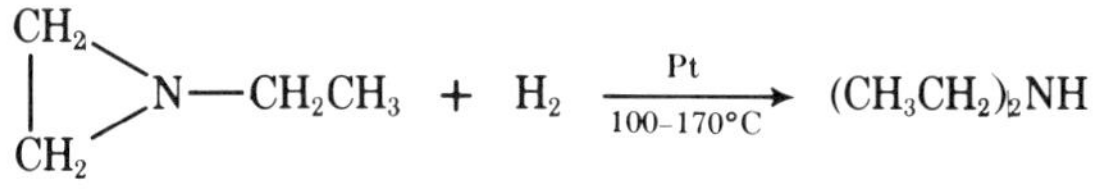

$$\underset{CH_2}{\overset{CH_2}{|}}\!\!>NH + H_2 \xrightarrow[100\text{–}170°C]{Ni} CH_3CH_2NH_2$$

$$\underset{CH_2}{\overset{CH_2}{|}}\!\!>N{-}CH_2CH_3 + H_2 \xrightarrow[100\text{–}170°C]{Pt} (CH_3CH_2)_2NH$$

With Hydroxy Compounds. Hydroxy compounds cause ring opening (34–35).

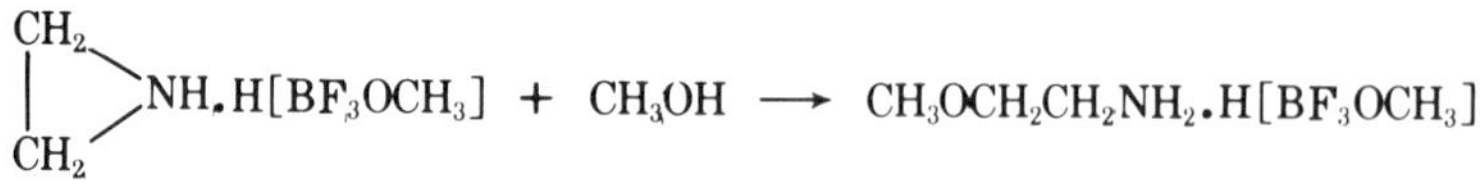

$$\underset{CH_2}{\overset{CH_2}{|}}\!\!>NH.H[BF_3OCH_3] + CH_3OH \longrightarrow CH_3OCH_2CH_2NH_2.H[BF_3OCH_3]$$

[*74411-32-8*]

$$\underset{CH_2}{\overset{CH_2}{|}}\!\!>NH.HBF_4 + C_6H_5OH \longrightarrow C_6H_5OCH_2CH_2NH_2.HBF_4$$

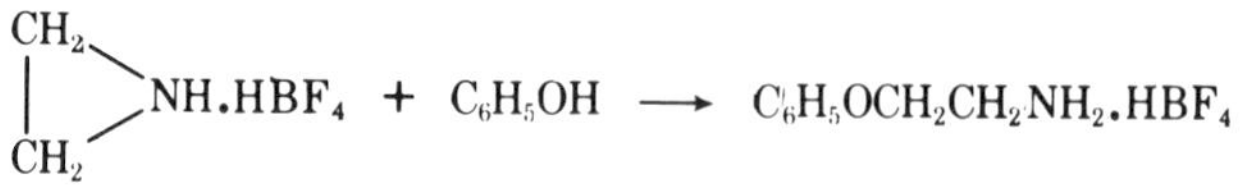

[*17409-60-8*]

With Isocyanates, Isothiocyanates, Ketenes, and Carbodimides. Addition of aziridines to compounds containing cumulated double bonds occurs very readily at room temperature or below. Quantitative yields are frequently obtained (36–38).

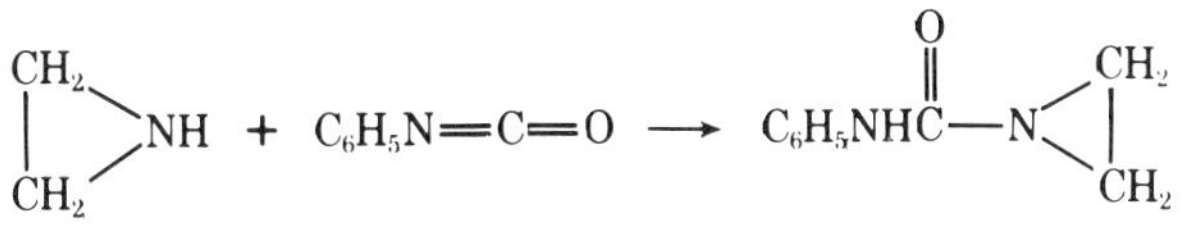

$$\underset{CH_2}{\overset{CH_2}{|}}\!\!>NH + C_6H_5N{=}C{=}O \longrightarrow C_6H_5NH\overset{O}{\overset{\|}{C}}{-}N\!<\!\underset{CH_2}{\overset{CH_2}{|}}$$

[13279-22-6]

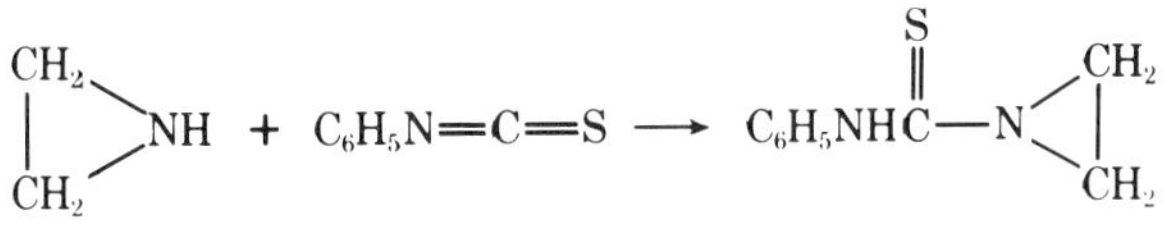

$$\underset{CH_2}{\overset{CH_2}{|}}\!\!>NH + C_6H_5N{=}C{=}S \longrightarrow C_6H_5NH\overset{S}{\overset{\|}{C}}{-}N\!<\!\underset{CH_2}{\overset{CH_2}{|}}$$

[19116-37-1]

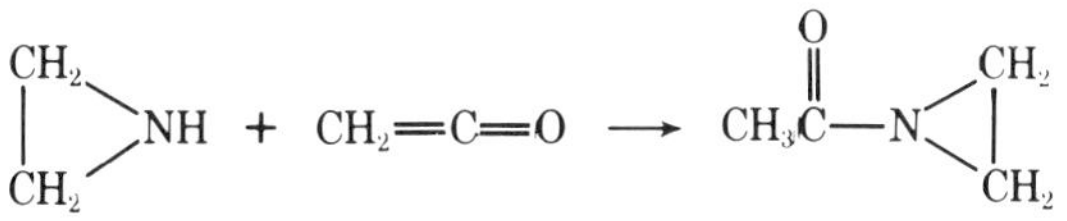

$$\underset{CH_2}{\overset{CH_2}{|}}\!\!>NH + CH_2{=}C{=}O \longrightarrow CH_3\overset{O}{\overset{\|}{C}}{-}N\!<\!\underset{CH_2}{\overset{CH_2}{|}}$$

[460-07-1]

$$\underset{CH_2}{\overset{CH_2}{|}}\!\!>NH + C_6H_5N{=}C{=}NC_6H_5 \longrightarrow C_6H_5NH\overset{N-C_6H_5}{\overset{\|}{C}}{-}N\!<\!\underset{CH_2}{\overset{CH_2}{|}}$$

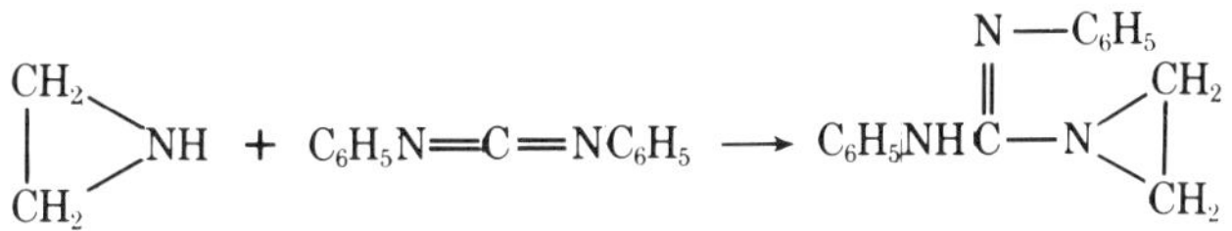

[74411-26-0]

If the aziridine is first converted to an aziridinium salt with an acid containing a nonnucleophilic anion, ring expansion occurs (39–40).

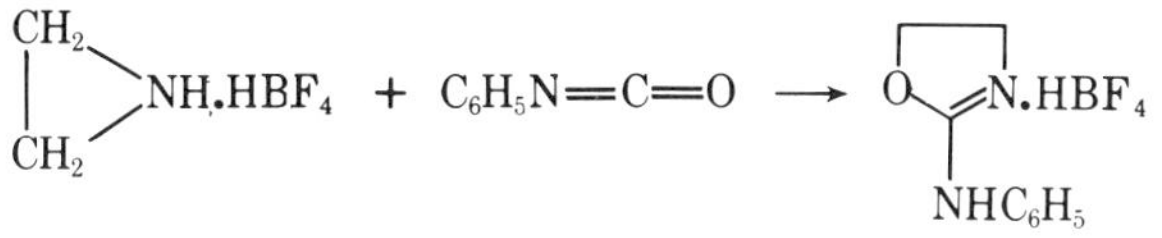

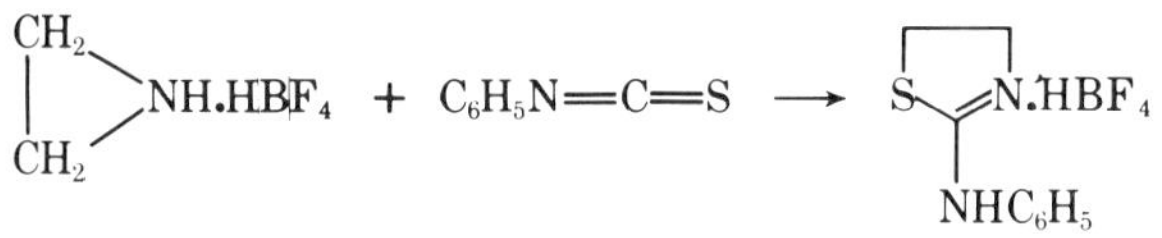

$$\underset{CH_2}{\overset{CH_2}{|}}\!\!>NH.HBF_4 + C_6H_5N{=}C{=}O \longrightarrow \text{(oxazoline ring: O, =N.HBF}_4\text{, with NHC}_6H_5\text{ on C-2)}$$

$$\underset{CH_2}{\overset{CH_2}{|}}\!\!>NH.HBF_4 + C_6H_5N{=}C{=}S \longrightarrow \text{(thiazoline ring: S, =N.HBF}_4\text{, with NHC}_6H_5\text{ on C-2)}$$

This type of ring expansion occurs also with alkylidenimines to form imidazolidines and with nitriles to form imidazolines.

With Metal Salts. Aziridines readily form complexes with metal salts which typically form amine complexes (41).

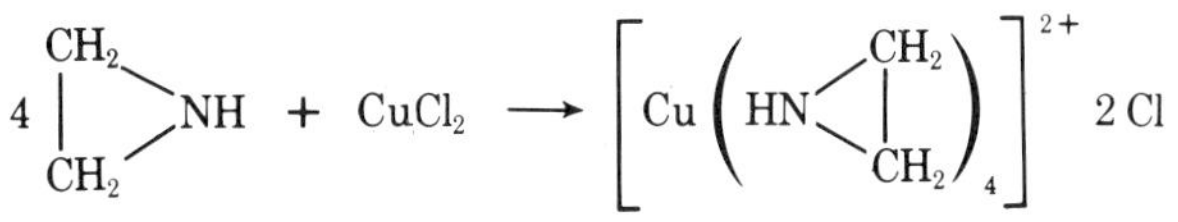

$$4\ \underset{CH_2}{\overset{CH_2}{|}}\!\!>NH + CuCl_2 \longrightarrow \left[Cu\left(HN\!<\!\underset{CH_2}{\overset{CH_2}{|}}\right)_4\right]^{2+} 2\,Cl$$

[28174-42-7]

Heating such complexes in solution results in a ring-opening reaction of the aziridine with itself and eventually in polymerization (42).

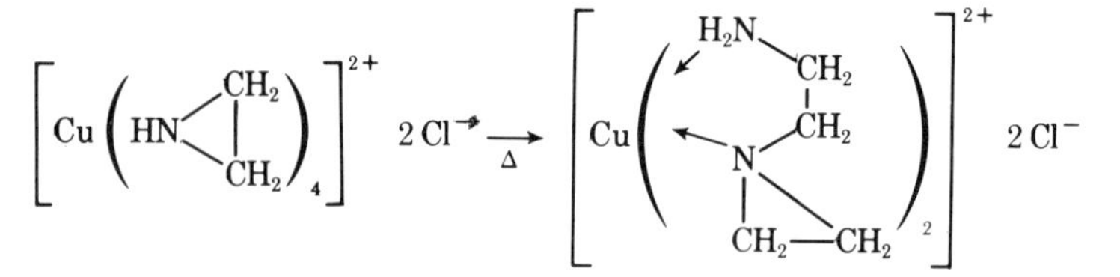

$$\left[\mathrm{Cu}\left(\mathrm{HN}\begin{matrix}CH_2\\|\\CH_2\end{matrix}\right)_4\right]^{2+} 2\,Cl^- \xrightarrow{\Delta} \left[\mathrm{Cu}\left(\begin{matrix}H_2N-CH_2\\ \quad\ \ |\\ \ \ N-CH_2\\ \ \ |\qquad\\ CH_2-CH_2\end{matrix}\right)_2\right]^{2+} 2\,Cl^-$$

[73214-90-1]

With Olefinic and Acetylenic Compounds. Aziridines add to a wide variety of compounds containing carbon–carbon double bonds; yields are frequently very good (43–44).

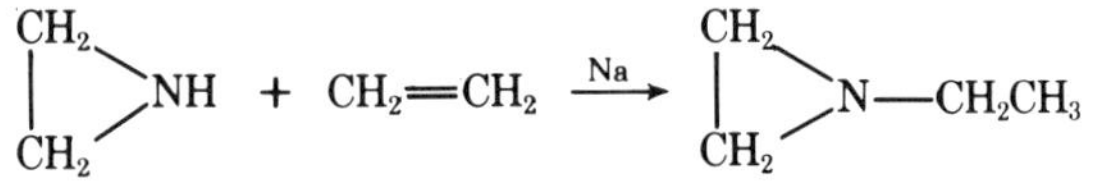

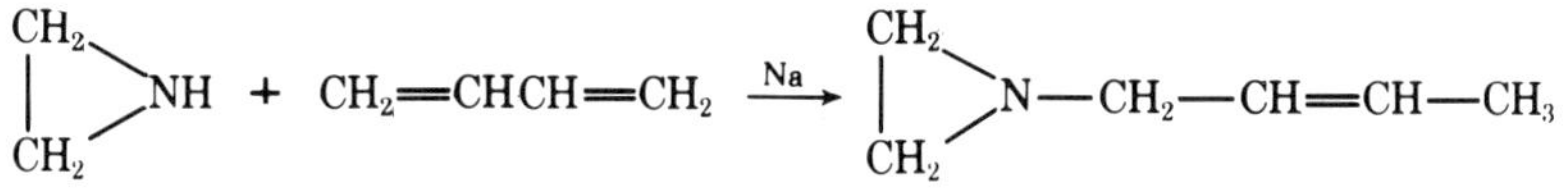

$$\begin{matrix}CH_2\\|\\CH_2\end{matrix}\!\!>NH + CH_2{=}CH_2 \xrightarrow{Na} \begin{matrix}CH_2\\|\\CH_2\end{matrix}\!\!>N{-}CH_2CH_3$$

$$\begin{matrix}CH_2\\|\\CH_2\end{matrix}\!\!>NH + CH_2{=}CHCH{=}CH_2 \xrightarrow{Na} \begin{matrix}CH_2\\|\\CH_2\end{matrix}\!\!>N{-}CH_2{-}CH{=}CH{-}CH_3$$

[41104-25-0]

The reactions with nonconjugated olefins require a catalyst such as an alkali metal and elevated temperatures. Michael-type additions to α,β-unsaturated esters, nitriles, ketones, and the like may be performed at room temperature in the absence of a catalyst (45–47).

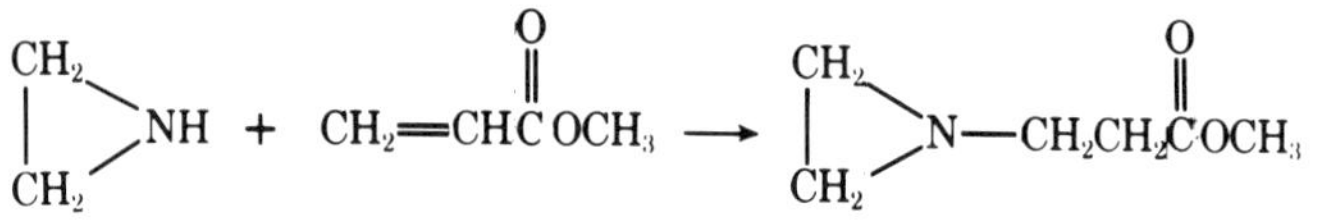

$$\begin{matrix}CH_2\\|\\CH_2\end{matrix}\!\!>NH + CH_2{=}CH\overset{O}{\overset{\|}{C}}OCH_3 \longrightarrow \begin{matrix}CH_2\\|\\CH_2\end{matrix}\!\!>N{-}CH_2CH_2\overset{O}{\overset{\|}{C}}OCH_3$$

[1073-77-4]

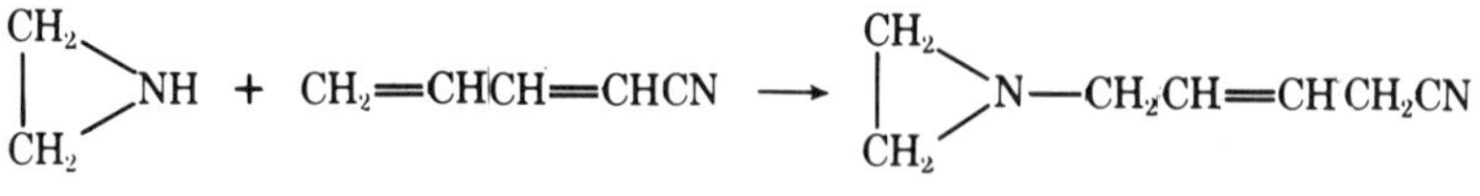

$$\begin{matrix}CH_2\\|\\CH_2\end{matrix}\!\!>NH + CH_2{=}CHCH{=}CHCN \longrightarrow \begin{matrix}CH_2\\|\\CH_2\end{matrix}\!\!>N{-}CH_2CH{=}CHCH_2CN$$

[74411-28-2]

$$\begin{matrix}CH_2\\|\\CH_2\end{matrix}\!\!>NH + HC{\equiv}C\overset{O}{\overset{\|}{C}}OCH_2CH_3 \longrightarrow \begin{matrix}CH_2\\|\\CH_2\end{matrix}\!\!>N{-}CH{=}CH\overset{O}{\overset{\|}{C}}OCH_2CH_3$$

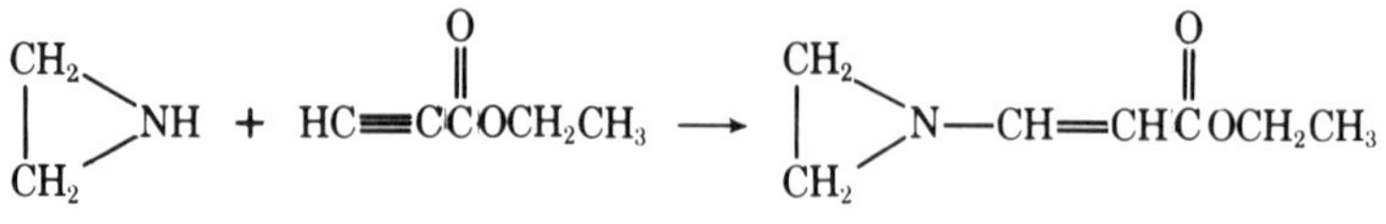

[50868-08-1]

With compounds containing a halogen atom on the β-olefinic carbon, a product is obtained in the presence of base wherein the halogen atom is replaced with a 1-aziridinyl group. The reaction involves addition of the aziridine followed by elimination of the halogen acid (48).

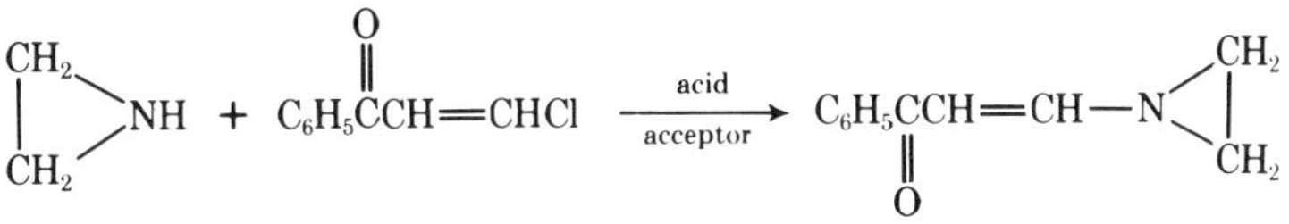

$$\text{(CH}_2\text{CH}_2\text{)NH} + C_6H_5\overset{\displaystyle O}{\overset{\|}{C}}CH{=}CHCl \xrightarrow[\text{acceptor}]{\text{acid}} C_6H_5\underset{\displaystyle O}{\underset{\|}{C}}CH{=}CH{-}N(CH_2CH_2)$$

[22048-28-8]

These reactions occur readily with quinones and substituted quinones to give 1-aziridinyl-substituted hydroquinones and quinones.

Vinyl ethers and vinyl amines, including 1-vinylaziridine, readily add to 1-unsubstituted aziridines (49).

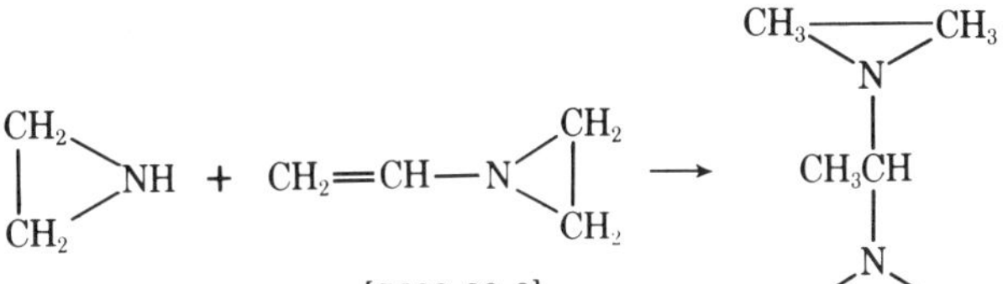

$$\text{(CH}_2\text{CH}_2\text{)NH} + CH_2{=}CH{-}N(CH_2CH_2) \longrightarrow (CH_3CH_3)N{-}CH_3CH{-}N(CH_2CH_2)$$

[5628-99-9] [5498-98-6]

A ring-opening reaction between aziridines and olefins at high temperatures (350°C) gives 3-substituted pyrrolidines. The reaction works best with olefins containing electronegative groups and also occurs with alkynes to produce pyrrolines (50–51).

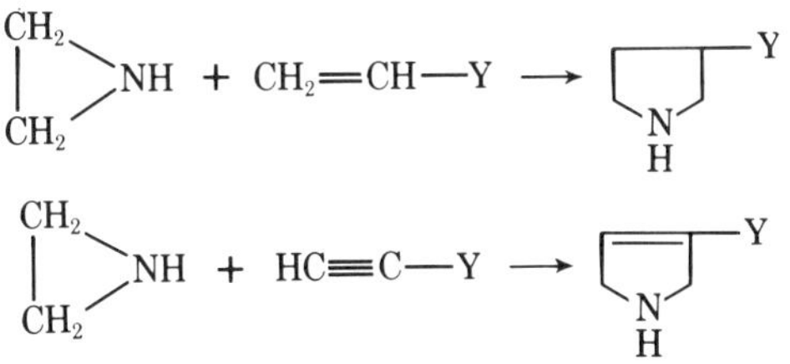

$$\text{(CH}_2\text{CH}_2\text{)NH} + CH_2{=}CH{-}Y \longrightarrow \text{3-Y-pyrrolidine (NH)}$$

$$\text{(CH}_2\text{CH}_2\text{)NH} + HC{\equiv}C{-}Y \longrightarrow \text{3-Y-pyrroline (NH)}$$

Y = electronegative group

With Thiocarbonyl Compounds. A ring-opening reaction occurs with many compounds containing the $>C{=}S$ group (52).

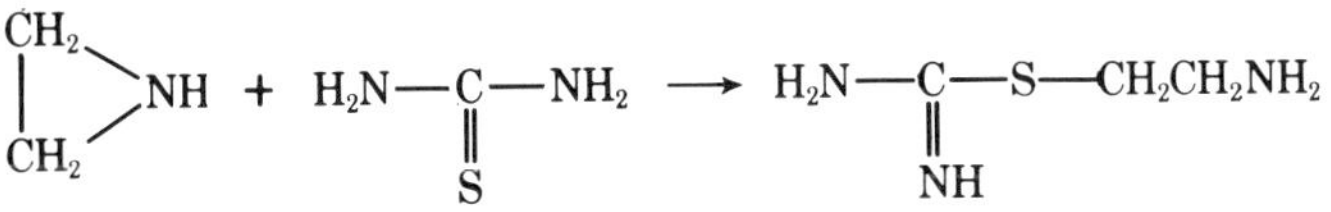

$$\text{(CH}_2\text{CH}_2\text{)NH} + H_2N{-}\underset{\displaystyle S}{\underset{\|}{C}}{-}NH_2 \longrightarrow H_2N{-}\underset{\displaystyle NH}{\underset{\|}{C}}{-}S{-}CH_2CH_2NH_2$$

With Thiols. Hydrogen sulfide and thiols very readily open the aziridine ring (9,53).

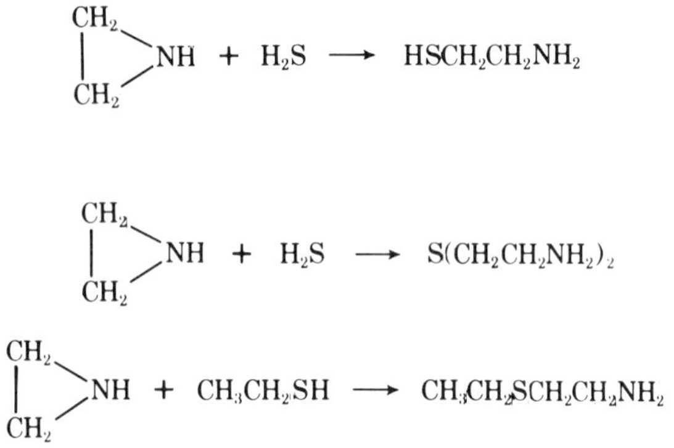

$$\underset{CH_2}{\overset{CH_2}{|}}\!\!>NH + H_2S \longrightarrow HSCH_2CH_2NH_2$$

$$\underset{CH_2}{\overset{CH_2}{|}}\!\!>NH + H_2S \longrightarrow S(CH_2CH_2NH_2)_2$$

$$\underset{CH_2}{\overset{CH_2}{|}}\!\!>NH + CH_3CH_2SH \longrightarrow CH_3CH_2SCH_2CH_2NH_2$$

Reaction of carbon disulfide with aziridines gives 2-thiothiazolines (54).

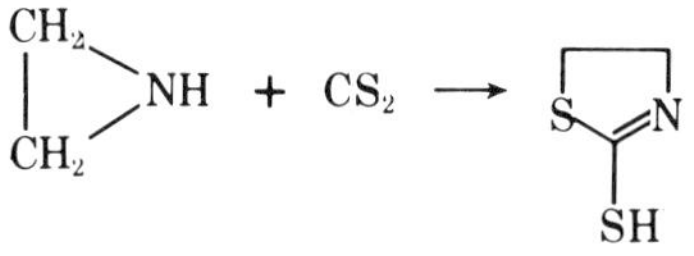

$$\underset{CH_2}{\overset{CH_2}{|}}\!\!>NH + CS_2 \longrightarrow \text{(thiazoline ring S–C(SH)=N)}$$

SH

Polymerization. Aziridines polymerize at elevated temperatures in the presence of catalytic amounts of acid. The reaction is highly exothermic and can be violent with concentrated solutions. After polymerization has been initiated with acid and heat, cooling is frequently necessary. A proposed reaction scheme is shown below.

Initiation

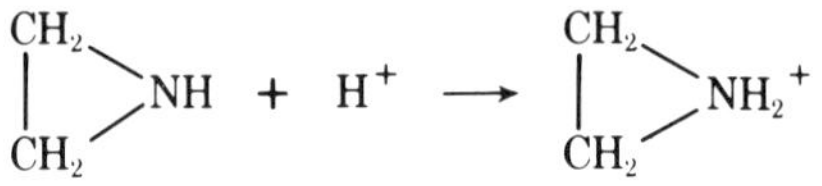

$$\underset{CH_2}{\overset{CH_2}{|}}\!\!>NH + H^+ \longrightarrow \underset{CH_2}{\overset{CH_2}{|}}\!\!>NH_2^+$$

First polymerization step:

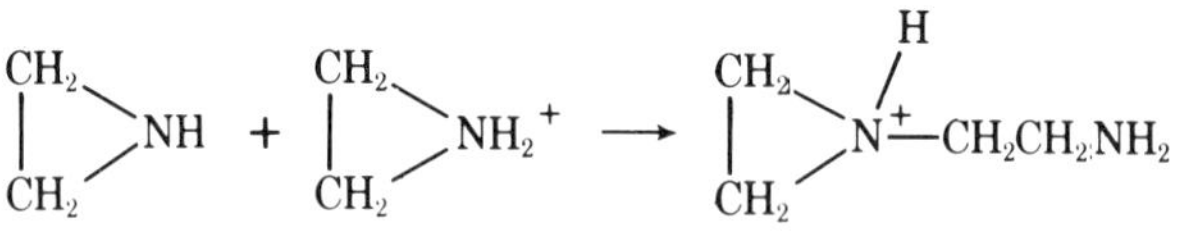

$$\underset{CH_2}{\overset{CH_2}{|}}\!\!>NH + \underset{CH_2}{\overset{CH_2}{|}}\!\!>NH_2^+ \longrightarrow \underset{CH_2}{\overset{CH_2}{|}}\!\!>\overset{H}{N^+}\!-CH_2CH_2NH_2$$

Second polymerization step:

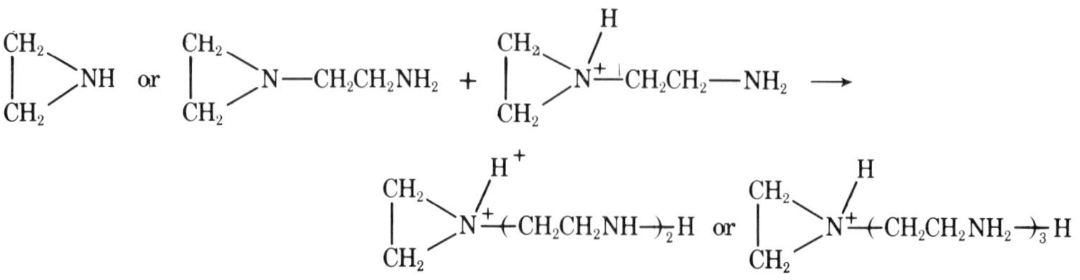

$$\underset{CH_2}{\overset{CH_2}{|}}\!\!>NH \text{ or } \underset{CH_2}{\overset{CH_2}{|}}\!\!>N-CH_2CH_2NH_2 + \underset{CH_2}{\overset{CH_2}{|}}\!\!>\overset{H}{N^+}\!-CH_2CH_2-NH_2 \longrightarrow$$

$$\underset{CH_2}{\overset{CH_2}{|}}\!\!>\overset{H^+}{N^+}\!\!\left(CH_2CH_2NH\right)_2H \text{ or } \underset{CH_2}{\overset{CH_2}{|}}\!\!>\overset{H}{N^+}\!\!\left(CH_2CH_2NH_2\right)_3H$$

The secondary amine groups in the growing polymer react as well as the primary amine groups shown above. Thus, the polymer is highly branched, containing an average of about one tertiary amine group for every three-chain nitrogen atoms (55).

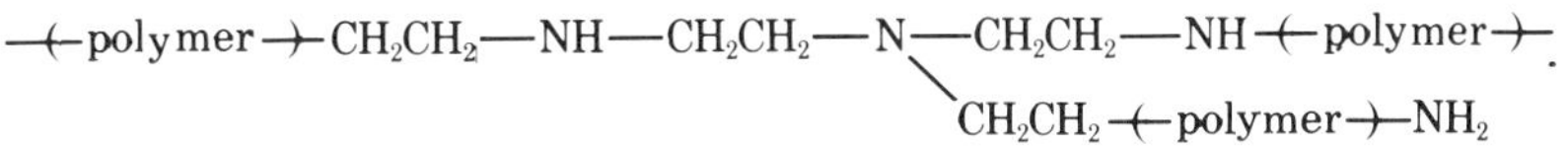

$$\text{-(polymer)-}CH_2CH_2\text{—}NH\text{—}CH_2CH_2\text{—}N(CH_2CH_2\text{-(polymer)-}NH_2)\text{—}CH_2CH_2\text{—}NH\text{-(polymer)-}$$

The polymerization is usually conducted in a solvent, eg, water, at concentrations low enough to provide an adequate rate of heat removal.

The molecular weight of the polymer produced using only acid catalyst, such as hydrochloric acid, corresponds to a degree of polymerization of about 100 or less. With an ethylene dihalide catalyst, a degree of polymerization of about 700–1000 is obtained. The ethylene dihalide generates hydrohalic acid for polymerization and, in effect, links the lower molecular weight polymers.

Aziridine–alkylamine mixtures may be polymerized in the absence of solvent to give a polymer wherein the molecular weight is determined by the ratio of amine to aziridine used (56).

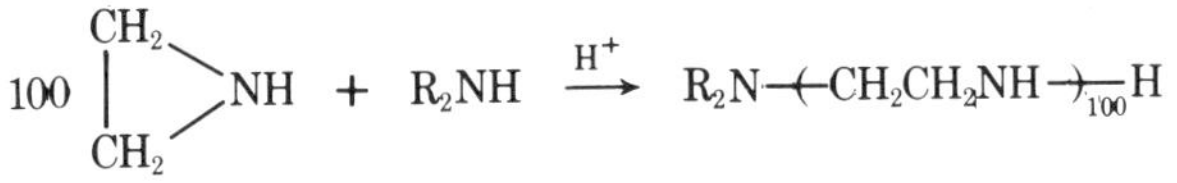

$$100\ \begin{matrix}CH_2\\ |\\ CH_2\end{matrix}\!\!>\!NH + R_2NH \xrightarrow{H^+} R_2N\text{-(}CH_2CH_2NH\text{)}_{100}H$$

Polymerization in the presence of an equal amount of water below ca 40°C, gives a linear polyaziridine [*9002-98-6*] as the water-insoluble crystalline hydrate (57).

Oligomerization. Oligomers of aziridines containing the 1-aziridinyl end-group are obtained from the acid-catalyzed polymerization if the acid catalyst is neutralized before the aziridine has been completely converted. This occurs with solvents less polar than water, eg, ethanol or methanol.

Heating aziridine in a nonpolar solvent, eg, benzene, in the presence of trialkylaluminum catalyst gives oligomers containing the 1-aziridinyl end group (58).

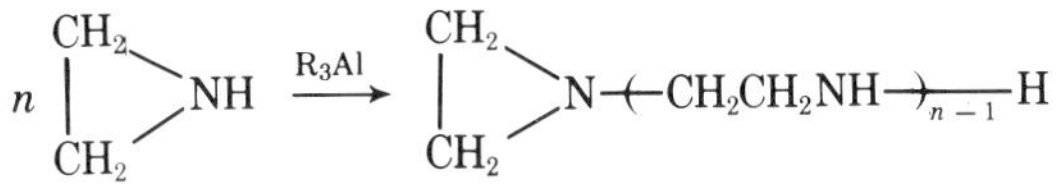

$$n\ \begin{matrix}CH_2\\ |\\ CH_2\end{matrix}\!\!>\!NH \xrightarrow{R_3Al} \begin{matrix}CH_2\\ |\\ CH_2\end{matrix}\!\!>\!N\text{-(}CH_2CH_2NH\text{)}_{n-1}H$$

Refluxing 1-alkylaziridines in ethanol in the presence of acid catalyst gives cyclic oligomers, predominantly with twelve members (59).

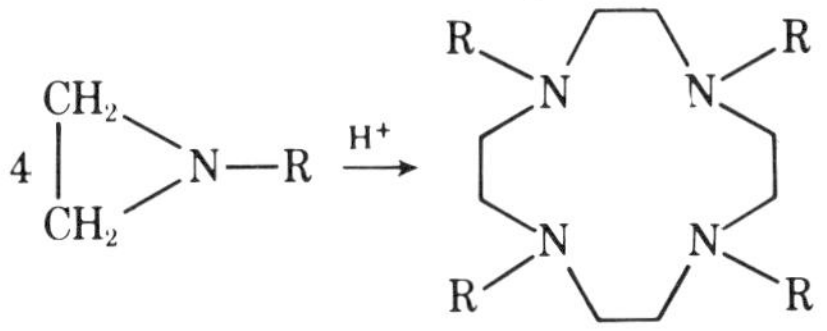

$$4\ \begin{matrix}CH_2\\ |\\ CH_2\end{matrix}\!\!>\!N\text{—}R \xrightarrow{H^+} \text{1,4,7,10-tetra-R-1,4,7,10-tetraazacyclododecane (ring: N(R) at four positions linked by }CH_2CH_2\text{)}$$

Manufacture

The first commercial production of aziridine was based on the following reactions:

$$H_2NCH_2CH_2OH + H_2SO_4 \longrightarrow H_3\overset{+}{N}CH_2CH_2OSO_3^- + H_2O$$

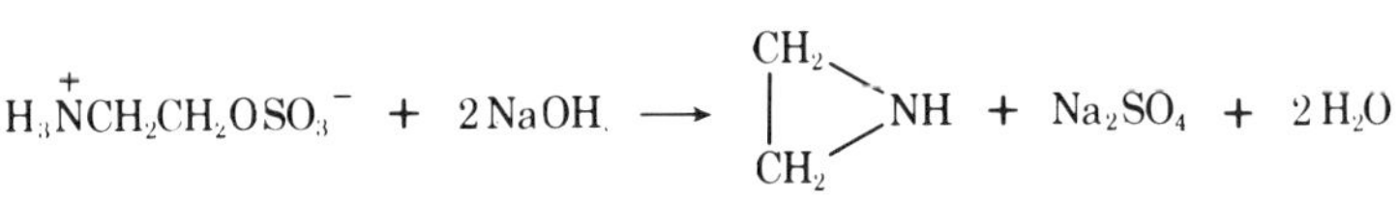

$$H_3\overset{+}{N}CH_2CH_2OSO_3^- + 2\,NaOH \longrightarrow \begin{matrix}CH_2\\ |\\ CH_2\end{matrix}\!\!>\!NH + Na_2SO_4 + 2\,H_2O$$

2-Chloroethylamine hydrochloride can be substituted for 2-aminoethyl hydrogen sulfate.

A number of modifications have improved yields by more rapid removal of the product from the reaction mixture.

A process based on ethylene dichloride and excess ammonia or an inorganic acid acceptor, such as calcium oxide, was patented in 1967 (60).

$$ClCH_2CH_2Cl + NH_3 + CaO \longrightarrow \begin{matrix} CH_2 \\ | \\ CH_2 \end{matrix} \!\!>\! NH + CaCl_2 + H_2O$$

2-Methylaziridine may be produced by either of the above processes. The process based on a sulfuric acid ester gives better yields.

The process based on ethylene dichloride may be used to produce 1-(2-hydroxyethyl)aziridine from monoethanolamine. However, this derivative is more readily produced from aziridine and ethylene oxide.

The catalytic dehydration of monoethanolamine (61) is particularly suitable for commercial production of aziridine.

$$H_2NCH_2CH_2OH \xrightarrow[\Delta]{cat} \begin{matrix} CH_2 \\ | \\ CH_2 \end{matrix} \!\!>\! NH + H_2O$$

Although yields of aziridine are good at low conversions, no commercial process using this route is known at this time.

Economic Aspects

In 1964 the U.S. production of aziridine was about 750 metric tons. By 1978 this annual production rate had more than doubled. Current world capacity for production is ca 12,500 t.

The price of aziridine in tank-car quantities increased at a moderate rate from \$2.09/kg in 1966 to \$2.76/kg in 1972, and then more rapidly to over \$6.00/kg in 1979. Considerably higher prices have been quoted for 2-methylaziridine and other aziridines that are available in small quantities.

Specifications

Typical specifications for aziridine are given in Table 5; they vary for different manufacturers.

Table 5. Typical Specifications for Aziridine

Specification	Value
purity, min mol %	99.0
APHA color, max	15
sodium, ppm	50–150
water, wt %, max	0.2
nonvolatile material, wt %, max	0.2

Analytical and Test Methods

Trace quantities of aziridines in aqueous solution are determined colorimetrically by reaction with 4-(*p*-nitrobenzyl)pyridine, followed by alkalinization. The color intensities, with absorbance maximum at 575–580 nm, conform to Beer's Law in the concentration range of 0.1 to 10 mg/L for aziridine; alkylating agents, eg, alkyl halides, interfere.

Trace quantities (<5 ppm) of volatile aziridines in air are determined by vapor-phase chromatography (62).

The purity of most aziridines can be determined by a time–temperature melting curve (ASTM D 1015-55). The method cannot be used for aziridine when water is present in greater than 5 mol % concentration.

Storage and Handling

Undiluted aziridines, particularly aziridine itself, tend to polymerize with explosive violence in the presence of acids and acid-forming and oxidizing materials. Therefore, contact of such materials must be avoided.

Storage tanks should be provided with a water-spray system to combat fires, spills, or other accidents. A high temperature alarm should also be provided as well as safety relief valves or rupture disks that are vented to an acid scrubber. Storage tanks should be padded with pure nitrogen at a pressure of 136 kPa (5 psig) to prevent contamination.

Mild steel, stainless steels 304 or 316, and glass are satisfactory construction materials for storage and handling. Copper, copper-bearing alloys, and silver are attacked by aziridine and should never be used. Plastics and elastomers are generally unsatisfactory because of excessive swelling. However, ethylene–propylene elastomers show good resistance and may be used for gasketing. Polytetrafluoroethylene is satisfactory for gasket material and is recommended in tape form as the only pipe dope for transfer lines. Canned or sealless pumps should be used for transfer.

Aziridine is shipped in mild-steel tank cars and cylinders. The cylinders should be stored in a well-ventilated location away from acidic materials, flame sources, and highly flammable substances.

Health and Safety Factors

Aziridines are hazardous chemicals because they are generally very toxic, can polymerize violently if undiluted, and the lower molecular weight analogues, such as aziridine itself, are extremely flammable.

The vapor pressures of aziridine and the lower alkyl substituted aziridines are high enough to provide an inhalation hazard. The vapors may cause severe irritation of the eyes and throat and can produce inflammation of the upper and lower respiratory tract. Serious illness may result following inhalation and initial symptoms may not be evident until several hours after exposure. The odor threshold for aziridine is 2 ppm (63). A TLV of 0.5 ppm in air for normal exposure has been established by the ACGIH (64).

Liquid aziridine is readily absorbed through most living tissue (eye, skin) and causes serious burns from contacts of even a few seconds. The appearance of irritation and burns may be delayed.

Precautions must be taken to prevent all contact when handling aziridines. Any skin or eye contact should be followed immediately by washing with copious amounts of water for at least 30 minutes. If inhalation has occurred, the victim should be removed to fresh air, kept warm, and artificial respiration applied if breathing stops.

The acute toxicity of aziridine to test animals (mouse, rat, guinea pig, rabbit) is shown by following LD_{50} values:

After	LD_{50}	*Ref.*
skin absorption, mg/kg	14–15	65
injection, mg/kg	3.8	66
ingestion, mg/kg	14–15	67
10 min inhalation, mg/L air	3.9	68

Although aziridines have been studied as cancer chemotherapeutic drugs (see under Uses) and one such derivative has been approved for human use (69) (see Chemotherapeutics, antimitotic), the regulations issued by OSHA to limit exposure of workers to carcinogens include aziridine (70).

Polymerization of highly concentrated solutions of undiluted aziridines may be inhibited by the presence of sodium hydroxide which neutralizes small amounts of acid contaminants, such as carbon dioxide in the air.

The lower molecular weight aziridines are extremely flammable (see Table 3). The explosive limits for mixtures of aziridine vapor and air are 3.3% to 54.8%. The vapors may ignite or explode if exposed to electric sparks, static electricity, excess heat, or open flame.

Uses

Aziridines, particularly aziridine itself and derivatives of aziridine, are used in a wide variety of industrial applications. The single most important derivative is polyaziridine which is cationic and substantive to many naturally occurring anionic materials, such as cellulose. This derivative can thus be used in small amounts to promote binding effects such as adhesion, improvement of paper strength, or flocculation of colloids (see Flocculating agents). Aziridine or its nitrogen-substituted derivatives are used to modify the properties of various polymeric materials, both naturally occurring and synthetic, that contain functional groups that cause a ring-opening reaction. Thus, a water-soluble polymer containing carboxylic acid groups may react with aziridine to form a water-soluble cationic flocculant, a latex with such groups present may react similarly to form a paint with improved adhesive properties, or a cotton cloth may react with a polyfunctional aziridine to render the cloth crease resistant (see Textiles). The biological effects of the aziridines make these compounds useful agents for producing mutations in plant breeding and for controlling insect pests by sexual sterilization (see Insect control technology).

Paper Industry. Polyaziridine with various modifications imparts wet strength to paper (71–74). Polyaziridine added as a wet-end additive in papermaking, increases the drainage or dewatering rate of the paper stock on the machine wire and enhances the retention of dyes, pigments, fillers, and paper fines in the paper sheet (75–78) (see Papermaking additives; Paper). A vinyl polymer containing pendant carboxylic acid groups that have reacted with aziridine (79) is also useful in this latter application. Modifications are also utilized in the preparation of sizing agents for paper and electroconductive papers, agents for obtaining color uniformity of paper surfaces, and grease-proofing agents (80–83).

The use of aziridine in the paper industry accounts for a major portion of its production and is one of its oldest commercial applications.

Adhesives. Polyaziridine is used to laminate plastic films to paper, other cellulose materials, or metal foils (84). The polymer meets the requirements for food packaging as listed in the *Federal Register* (80), and is used in such applications as milk cartons from cardboard laminates with polyethylene (see Adhesives).

Pressure-sensitive adhesives may be formulated from mixtures of poly(vinyl alcohol) or poly(*N*-vinylpyrrolidinone) and polyaziridine. Mixtures with acrylic latexes are claimed as excellent adhesives (85).

Textiles. Aziridines are used in the textile industry to attach various modifying chemicals to fabrics or fibers and thus to impart specific durable properties to the product. Although cotton products are most often modified in this manner, wool and synthetics also are treated. Aziridines used include mono- and polyfunctional derivatives containing the modifying groups as well as polymers of substituted aziridines and polyaziridine. Dyeing and printing, antistatic properties, creaseproofing, shrinkproofing, waterproofing, softening, and flame resistance are improved (86–90).

An abrasion-resistant cotton fabric may be obtained by the use of poly[1-(2-cyanoethyl)aziridine] [*9016-05-1*] (91) whereas poly(fluoroalkylaziridines) impart oil repellency (92).

Mixtures of polyaziridine and finishing compounds, such as organosilicon compounds for hydrophobic wool finishes (93), are among the more recent techniques for textile treatments. Similarly, prepolymers of chloroalkyl phosphonates and polyaziridine are claimed to improve flame resistance (94) (see Flame retardants for textiles).

Wastewater Treatment and Flocculation. Polyaziridine and various modifications (95–96) are very effective at low concentrations in flocculating anionic colloids suspended in aqueous media. This use represents one of the major applications for aziridine. In addition to the purification of varied industrial wastewaters (97), petroleum-containing wastewaters (98), and municipal wastewaters (99), the polymer is also useful for clarifying fruit juices (80), and coal concentration (100) (see Wastes).

Coatings and Plastics. In both coatings (qv) and plastics, aziridines are used primarily to improve adhesion or effect a cross-linking (curing) reaction. Both of these functions are utilized in the water-thinnable epoxy coating compositions of a polymer containing carboxylic acid groups that have reacted with aziridine (101).

Acrylic latex paints that contain carboxylic acid groups exhibit much improved adhesion properties after reaction of the acid groups with aziridine.

Aziridines and multifunctional aziridines are used as curing agents for epoxy resins (qv) and carboxyl-terminated polybutadiene (102–103).

Ion-Exchange and Metal-Ion Complexing. Polyaziridines (particularly polyaziridine itself) are used to prepare weakly basic ion-exchange resins by reaction with a cross-linking agent. Because of the wide variety of cross-linking agents available, resins with different physical properties can be obtained. These include alkylene dihalides, difunctional epoxides, diisocyanates, formaldehyde, and epichlorohydrin (see Ion exchange).

The vicinal diamine structure of polyaziridine makes such resins suitable for removing metal ions from aqueous wastes (80). Chelating polymers may also be prepared from aziridine-grafted polystyrene (104).

Closely related to this application, anion-exchange membranes and monovalent-cation permselective-exchange membranes may be prepared with polyaziridine (105–107). Ion-exchange fibers are prepared from polyacrylonitrile and polyaziridine (108).

Biological Applications. 2,4,6-Tris(1-aziridinyl)-*s*-triazine [*51-18-3*] (triethylene melamine) is useful as a cancer chemotherapy drug and many other aziridine derivatives have been studied in this application (109–111). The mechanism accounting for the biological activity of aziridines remains a subject of continued study (112–114).

Aziridines are used to promote mutagenesis in plant life, such as wheat or barley, and microorganisms. This method of influencing plant yields through induced mutation has been much studied, particularly in the USSR (115–124).

Various aziridines have been successfully employed in the control of insect pests by chemosterilization (125).

Although polyaziridine does not possess the extreme biological activity shown by compounds containing the aziridine ring, medicinal uses, eg, to reduce blood cholesterol concentration and as cancer chemotherapeutic agents are cited (126–128).

Other Uses. Among the varied applications for aziridine, the compound is used as an intermediate for the preparation of other compounds containing the aminoethyl group. 1-Alkylaziridines are among the best catalysts for preparation of a polyurethane from a diol and a diisocyanate (129). Polyaziridine and its modifications exhibit such catalytic activity in certain reactions as to suggest them as models for synthetic enzymes (130–131) (see Biopolymers).

Modified polyaziridine has been incorporated into germicidal detergent compositions in shampoos for dandruff control (132). Other applications for polyaziridines or poly(substituted aziridines) include use in the cement industry (133–134), in electroplating (135), as an antioxidant for styrene–butadiene rubber (136), as a rocket-propellant binder (137–138), as a lubricating-oil dispersant (139), and in thin-layer chromatography (140).

Polyfunctional aziridines are useful as hardening agents in the preparation of photographic films (141).

BIBLIOGRAPHY

"Imines, Cyclic" in *ECT* 2nd ed., Vol. 11, pp. 526–548, by Otis C. Dermer, Oklahoma State University, and Andrew W. Hart, The Dow Chemical Company.

1. *Chem. Week* **125,** 11 (July 4, 1979).
2. O. C. Dermer and G. E. Ham, *Ethylenimine and Other Aziridines,* Academic Press, Inc., New York, 1969, pp. 87–105.
3. W. D. Good and co-workers, *Proc. JANAF Thermochemical Working Group Symposium Douglas Advan. Res. Lab.,* Chem. Propulsion Inform. Agency, Huntington Beach, Calif., 1968.
4. *Ethylenimine,* The Dow Chemical Co., 1965.
5. *Propylene Imine,* Interchemical Corp., 1966.
6. U.S. Pat. 3,248,385 (Apr. 26, 1966), D. N. Kinsey, F. R. Jensen, and J. H. Hennes (to The Dow Chemical Co.).
7. U. Harder, E. Pfeil, and K. F. Zenner, *Chem. Ber.* **97,** 510 (1964).
8. D. H. Powers, Jr., V. B. Schatz, and L. B. Clapp, *J. Am. Chem. Soc.* **78,** 907 (1956).
9. H. Bestian, *Ann. Chem.* **566,** 210 (1950).

10. F. Fischer and H. Roensch, *Chem. Ber.* **94,** 901 (1961).
11. U.S. Pat. 3,197,463 (July 27, 1965), G. C. Tesoro and S. B. Sello (to J. P. Stevens and Co.).
12. W. D. Stephens and co-workers, *J. Chem. Eng. Data* **8,** 625 (1963).
13. R. G. Kostyanovskii and O. A. Pan'shin, *Izv. Akad. Nauk SSSR Ser. Khim.,* 1554 (1964).
14. U.S. Pat. 3,173,910 (Mar. 16, 1965), A. F. Graefe (to Aerojet-General Corp.).
15. H. W. Heine and D. L. Tomalia, *J. Am. Chem. Soc.* **84,** 993 (1962).
16. U.S. Pat. 3,393,184 (July 16, 1968), J. A. Hoffman (to American Cyanamid Co.).
17. G. Kamai and Z. L. Khisamova, *Dokl. Akad. Nauk SSSR* **156,** 365 (1964).
18. S. Z. Kvin, V. K. Promonenko, and G. V. Konopatova, *Zh. Obshch. Khim.* **37,** 1681 (1967); *J. Gen. Chem. USSR (Eng.)* **37,** 1600 (1967).
19. M. G. Voronkov, L. A. Fedotova and D. Rinkis, *Khim. Geterotsikl. Soedin. Akad.,* 488 (1965).
20. E. Bertele and co-workers, *Angew. Chem.* **76,** 393 (1964).
21. W. J. Rabourn and W. L. Howard, *J. Org. Chem.* **27,** 1039 (1962).
22. R. G. Kostyanovskii, *Dokl. Akad. Nauk SSSR* **139,** 877 (1961).
23. Y. Oshiro, K. Yamamoto, and S. Komori, *Yuki Gosei Kagako Kyokai Shi* **24,** 945 (1966).
24. G. Drefahl and M. Heubner, *J. Prakt. Chem.* **23**(4), 149 (1964).
25. J. L. Larice, J. Roggero, and J. Metzger, *Bull. Soc. Chim. Fr.,* 3637 (1967).
26. L. B. Clapp, *J. Am. Chem. Soc.* **70,** 184 (1948).
27. G. H. Coleman and J. E. Callen, *J. Am. Chem. Soc.* **68,** 2006 (1946).
28. H. Stamm, *Chem. Ber.* **99,** 2556 (1966).
29. K. Issleib and D. Haferburg, *Z. Naturforsch.* **20b,** 916 (1965).
30. G. I. Braz, *Dokl. Akad. Nauk SSSR* **87,** 589 (1952); *Dokl. Akad. Nauk. SSSR* **87,** 747 (1952).
31. A. Funke and G. Benoit, *Bull. Soc. Chim. Fr.,* 1021 (1953).
32. Fr. Pat. 1,504,888 (Dec. 8, 1967), (to Eastman Kodak Co.).
33. U.S. Pat. 3,242,166 (Mar. 22, 1966), G. E. Ham and P. M. Phillips (to The Dow Chemical Co.).
34. U. Harder, E. Pfeil, and K. F. Zenner, *Chem. Ber.* **97,** 510 (1964).
35. L. B. Clapp, *J. Am. Chem. Soc.* **73,** 2584 (1951).
36. H. Najer and co-workers, *Bull. Soc. Chim. Fr.,* 323 (1963).
37. A. S. Deutsch and P. E. Fanta, *J. Org. Chem.* **21,** 892 (1956).
38. B. Adcock and A. Lawson, *J. Chem. Soc.,* 474 (1965).
39. E. Pfeil and K. Milzner, *Angew. Chem.* **78,** 677 (1966).
40. A. P. Sineakov, V. N. Galdysheva, and V. S. Etlis, *Khim. Geterotsikl. Soedin. Akad. Nauk Latv. SSR,* 567 (1968); *Chem. Abstr.* **69,** 96538 (1968).
41. T. B. Jackson and J. O. Edwards, *J. Am. Chem. Soc.* **83,** 355 (1961).
42. U.S. Pat. 4,170,591 (Oct. 9, 1979), J. M. Motes (to The Dow Chemical Co.).
43. U.S. Pat. 3,520,875 (July 21, 1970), R. E. Lane and T. L. Ashby (to The Dow Chemical Co.).
44. U.S. Pat. 2,654,737 (Oct. 6, 1953), M. Erlenbach and A. Sieglitz (to Farbwerke Hoechst).
45. D. Rosenthal and co-workers, *J. Org. Chem.* **30,** 3689 (1965).
46. J. M. Stewart, *J. Am. Chem. Soc.* **76,** 3228 (1954).
47. J. E. Dolfini, *J. Org. Chem.* **30,** 1298 (1965).
48. Ger. Pat. 1,111,637 (Feb. 15, 1962), (to Badische Anilin-und Soda-Fabrik, A.-G.).
49. U.S. Pat. 3,236,835 (Feb. 22, 1966), W. J. Rabourn (to The Dow Chemical Co.).
50. U.S. Pat. 3,342,833 (Sept. 19, 1967), M. I. Fremery (to Shell Oil Co.).
51. U.S. Pat. 3,259,632 (July 5, 1966), M. I. Fremery (to Shell Oil Co.).
52. L. D. Spicer and co-workers, *J. Org. Chem.* **33,** 1350 (1968).
53. G. Tsatsas, C. Sandris, and D. Kontonassios, *Bull. Soc. Chim. Fr.,* 3100 (1964).
54. C. S. Dewey and R. A. Bafford, *J. Org. Chem.* **30,** 491 (1965).
55. C. R. Dick and G. E. Ham, *J. Macromol. Sci. Chem.* **A4,** 1301 (1970).
56. U.S. Pat. 3,519,687 (July 7, 1970), J. G. Schneider, C. R. Dick, and G. E. Ham (to The Dow Chemical Co.).
57. P. A. Gembitskii and co-workers, *Vysokomol. Soedin* **A20,** 1505 (1978).
58. U.S. Pat. 3,492,289 (Jan. 27, 1970), R. H. Symm, G. E. Ham, and C. R. Dick (to The Dow Chemical Co.).
59. U.S. Pat. 4,093,615 (June 6, 1978), G. E. Ham and R. L. Krause (to The Dow Chemical Co.).
60. U.S. Pat. 3,336,294 (Aug. 15, 1967), G. R. Miller, G. E. Ham, J. E. Cobb, and F. R. Jensen (to The Dow Chemical Co.).
61. Jpn. Pat. 75 10,593 (Apr. 22, 1975), T. Agawa, Y. Ohki, and K. Hotta.
62. *Ethylenimine,* The Dow Chemical Co., Midland, Mich., 1976.

63. *Code of Federal Regulations,* Title 46, Parts 146, 149, Coast Guard, 1967, pp. 31, 232; *U.S. Coast Guard Manual CG388, Chemical Data Guide for Bulk Shipment by Water,* 1966, p. 65.
64. The American Conference of Governmental Industrial Hygienists, *Documentation of Threshold Limit Values,* ACGIH, 1966, p. 84.
65. C. P. Carpenter, H. F. Smyth, Jr., and C. B. Shaffer, *J. Ind. Hyg. Toxicol.* **30,** 2 (1948).
66. R. M. V. James, *Biochem. Pharmacol.* **14,** 915 (1965).
67. H. F. Smyth, Jr., J. Seaton, and L. Fischer, *J. Ind. Hyg. Toxicol.* **23,** 259 (1941).
68. S. D. Silver and R. P. McGrath, *J. Ind. Hyg. Toxicol.* **30,** 7 (1948).
69. *Fed. Regist.* **30,** 6497 (1965).
70. *Fed. Regist.* **39,** 3756 (1974); *Fed. Regist.* **38**(85), (1973).
71. U.S. Pat. 3,520,774 (July 14, 1970), H. H. Roth (to The Dow Chemical Co.).
72. G. G. Maher and co-workers, *Tappi* **55**(9), 1938 (1972).
73. W. M. Reif, *TAPPI Papermakers Conf.,* 47 (1972).
74. G. Troemel, *Farben Rev.* **18,** 55 (1970).
75. U.S. Pat. 3,658,641 (Apr. 25, 1972), K. T. Sten (to Ciba Corp.).
76. B. S. Das and H. Lamas, *Pulp Pap. Mag. Can.* **74,** T281 (1973).
77. H. Pummer and A. G. Sandoz, *Papier Darmstadt* **27,** 417 (1973).
78. F. Mueller and U. Beck, *Papier Darmstadt* **32**(10A), 25 (1978).
79. U.S. Pat. 3,372,149 (Mar. 5, 1968), J. Fertig, E. D. Mazzarella, and M. Skoultch (to National Starch and Chemical Corp.).
80. *PEI Polymers,* The Dow Chemical Co., Midland, Mich., 1974.
81. Ger. Offen. 2,012,217 (Sept. 30, 1971), E. Luecke (to Badische Anilin-und Soda-Fabrik A.-G.).
82. K. W. Britt, A. G. Dillon, and L. A. Evans, *TAPPI Papermakers Conf.,* 39 (1977).
83. H. Palonen and P. Stenius, *Kerm. Kemi* **5**(4), 113 (1978); *Chem. Abstr.* **89,** 45263S (1978).
84. A. M. DeRoo in I. Skeist, ed., *Handbook of Adhesives,* 2nd ed., Van Nostrand-Reinhold, New York, 1977, p. 592.
85. U.S. Pat. 3,639,327 (Feb. 1, 1972), A. H. Drelich and P. M. Condon (to Johnson and Johnson).
86. T. Shinozaki, *Sen'i Kako* **27,** 351 (1975); *Sen'i Kako* **27,** 415 (1975).
87. F. Friedrich, *Melliand Textilber. Int.* **52,** 1326 (1971); *Melliand Textilber. Int.* **52,** 1455 (1971).
88. H. E. Bille and R. R. Fikentscher, *Chem. Aftertreat. Text.,* 383 (1971).
89. R. R. Fikentscher and H. E. Bille, *Chem. Aftertreat. Text.,* 357 (1971).
90. K. Hoshino, *Chem. Aftertreat. Text.,* 235 (1971).
91. U.S. Pat. 3,617,346 (Nov. 2, 1971), L. Chance and G. L. Drake (to The United States Department of Agriculture).
92. U.S. Pat. 3,575,961 (Apr. 20, 1971), G. C. Tesoro and R. Ring (to J. P. Stevens and Co., Inc.).
93. G. V. Ivanova, E. Y. Danilova, I. M. Strukova, L. V. Emets, and L. A. Vol'f, *Issled Obl. Khim. Polietilenimina Ego Primenenie Prom-sti,* 151 (1977); *Chem. Abstr.* **89,** 181151F (1978).
94. N. R. Bertoniere and S. P. Rowland, *Text. Res. J.* **48**(7), 385 (1978).
95. N. Ise and T. Okubo, *Polymer* **13,** 552 (1972).
96. K. Akagae and G. G. Allan, *Skikizai Kyokaish* **46,** 239 (1973).
97. G. N. Naletskaya, *Issled. Obl. Khim. Polietilenimina Ego Primenenie Prom-sti,* 106 (1977); *Chem. Abstr.* **89,** 117201V (1978).
98. L. N. Butseva and co-workers, *Issled. Obl. Khim. Polietilenimina Ego Primenenie Prom-sti,* 120 (1977); *Chem. Abstr.* **89,** 117200U (1978).
99. Y. I. Veitser, G. N. Lutsenko, A. I. Tsvetkova, and N. Y. Tugusheva, *Issled. Obl. Khim. Poletilenimina Ego Primenenie Prom-sti,* 126 (1977); *Chem. Abstr.* **89,** 17199A (1978).
100. L. D. Aleksandrova, M. A. Borts, and D. I. Stepanova, *Issled. Obl. Khim. Polietilenimina Ego Primenenie Prom-sti,* 95 (1977); *Chem. Abstr.* **89,** 200108R (1978).
101. U.S. Pat. 3,719,629 (Mar. 6, 1973), P. H. Martin and R. T. McFadden (to The Dow Chemical Co.).
102. T. Endo and T. Tomisawa, *Sekiyu To Sekiyu Kagaku* **15**(5), 79; *Sekiyu To Sekiyu Kagaku* **15**(7), 118 (1971); *Sekiyu To Sekiyu Kagaku* **15**(8), 80 (1971).
103. K. Matsumaga, *Kogyo Kayaku* **38**(4), 171 (1977).
104. T. Saegusa, S. Kobayashi, K. Hayashi, and A. Yamada, *Polym. J.* **10,** 403 (1978).
105. U.S. Pat. 3,714,010 (Jan. 30, 1973), A. Suszer (to The United States Department of Interior).
106. T. Sata, *Kolloid Z. Z. Polym.* **250,** 980 (1972).
107. T. Sata and R. Izuo, *Colloid Polym. Sci.* **256,** 757 (1978).
108. I. B. Ratushnyak and co-workers, *Khim. Volokna* **2,** 17 (1978).
109. V. Bicker, *Fortschr. Med.* **96,** 661 (1978).

110. T. S. Safonova and V. A. Chernov, *Puti Sin. Izyskaniya Protivoopukholevykh Prepr.* **3,** 55 (1970); *Chem. Abstr.* **75,** 35967W (1970).
111. L. D. Protsenko and P. V. Rodionov, *Puti Sin. Izyskaniya Protivoopukholevykh Prepr.* **3,** 69 (1970); *Chem. Abstr.* **75,** 35824X (1970).
112. L. D. Protsenko, *Fiziol. Akt. Veshhestva* **10,** 3 (1978); *Chem. Abstr.* **90,** 33620P (1979).
113. T. A. Connors, *Handb. Exp. Pharmakol.* **38**(Pt. 2), 18 (1975).
114. T. Watabe and K. Akamatsu, *Proc. Symp. Drug Metab. Action 5th,* 219 (1974).
115. L. I. Grinikh, *Chuvstvitel'nost Organizmov Mutagennym Fabtoram Vozniknovenie Mutatsii,* 78 (1973); *Chem. Abstr.* **81,** 10076R (1974).
116. S. P. Kovalenko, *Tsitol. Genet.* **7,** 371 (1973); *Chem. Abstr.* **79,** 111927H (1973).
117. K. Nikolov, *Eksp. Mutagenez Selek.,* 83 (1972); *Chem. Abstr.* **79,** 74098R (1973).
118. S. I. Alikhanyan, *Radiat. Radioisotop. Ind. Microorgan. Proc.* 251 (1971).
119. P. Y. Golodriga and co-workers, *Eksp. Mutatsii Selek. Rosl.,* 227 (1971); *Chem. Abstr.* **77,** 97248N (1972).
120. G. Obe, *Umschau* **72,** 359 (1972).
121. N. N. Zoz, T. V. Sal'nikova, and A. V. Pukhal'skii, *Prakt. Khim. Mutageneza Sb. Tr. Vses. Soveschch.,* 213 (1971); *Chem. Abstr.* **77,** 14708P (1972).
122. N. N. Zoz, *Prakt. Khim. Mutageneza Sb. Tr. Vses. Soveshch.,* 7 (1971); *Chem. Abstr.* **77,** 14706M (1972).
123. M. Denic and J. Dumanovic, *Nucl. Sci. Abstr.* **25**(22), 5510 (1971).
124. A. G. Pavlova and N. N. Zoz, *Teor. Khim. Mutageneza Mater. Vses. Soveshch 4th,* 216 (1971); *Chem. Abstr.* **76,** 68586V (1971).
125. H. Koike, *Noyaku Seisan Gijutsu* **26,** 1 (1971).
126. U.S. Pat. 3,308,020 (Mar. 7, 1967), F. J. Wolfe and D. M. Tennent (to Merck and Co.).
127. H. Moroson, *Chemother. Cancer Dissem. Metastasis,* 245 (1973).
128. I. Iliev, M. Georgieva, and V. Kabaivanor, *Vsp. Khim.* **43**(1), 134 (1974); *Chem. Abstr.* **81,** 58011T (1975).
129. J. W. Britain and P. G. Gemeinhardt, *J. Appl. Polym. Sci.* **4,** 207 (1960).
130. R. S. Johnson, J. A. Walder, and I. M. Klotz, *J. Am. Chem. Soc.* **100**(16), 5159 (1978).
131. Y. Okahata and T. Kunitake, *J. Polym. Sci. Polym. Chem. Ed.* **16,** 1865 (1978).
132. U.S. Pat. 3,769,398 (Oct. 30, 1973), G. T. Hewitt (to Colgate-Palmolive Co.).
133. U.S. Pat. 3,491,049 (Jan. 20, 1970), D. L. Gibson and C. H. Kucera (to The Dow Chemical Co.).
134. A. A. Danilova and G. M. Tarnarutskii, *Issled. Obl. Khim. Polietilenimina Ego Primenenie Prom-sti,* 162 (1977); *Chem. Abstr.* **89,** 151471T (1977).
135. U.S. Pat. 3,767,540 (Oct. 23, 1973), W. E. Rosenberg (to R. O. Hull and Co., Inc.).
136. Ger. Pat. 1,942,697 (Feb. 26, 1970), A. M. DeRoo (to The Dow Chemical Co.).
137. U.S. Pat. 3,725,152 (Apr. 3, 1973), E. T. Niles and D. L. Stevens (to The Dow Chemical Co.).
138. U.S. Pat. 3,798,086 (Mar. 19, 1974), A. H. Muenker, L. K. Beach, and H. T. White (to Esso Research and Engineering Co.).
139. Brit. Pat. 851,615 (Oct. 19, 1960), (to Shell Internationale Research Maatschappi, B.V.).
140. J. S. Miller and R. R. Burgess, *Biochemistry* **17,** 2059 (1978).
141. Jpn. Kokai 77 148,124 (Dec. 9, 1977), S. Miyazawa and F. Veda (to Mitsubishi Paper Mills, Ltd.).

General References

Ref. 2 is the most comprehensive review of the early literature dealing with aziridines through 1968.

P. A. Gembitskii, D. S. Zhuk, and V. A. Kargin, *Polyethylenimine,* Nauka, Moscow, USSR, 1971.

G. E. Ham "Alkylene Imines" in K. C. Frisch, ed., *High Polymers, Vol. 26: Cyclic Monomers,* John Wiley & Sons, Inc., New York, 1972, pp. 313–339.

IARC Monographs on the Evaluation of the Carcinogenic Risk of Chemicals to Man, Vol. 9: Some Aziridines, N-, S- and O-Mustards and Selenium, International Agency for Research on Cancer, Geneva, Switz., 1975.

G. E. Ham "Alkylenimine Polymers" in N. M. Bikales, ed., *Encyclopedia of Polymer Science and Technology,* Suppl. Vol. 1, John Wiley & Sons, Inc., New York, 1976, pp. 25–51.

D. S. Zhuk, ed., *Methods of Synthesizing and Ways of Using Polyethylenimine in the National Economy,* Nauka, Moscow, USSR, 1976.

D. R. Crist and N. J. Leonard, *Angew. Chem. Int. Ed. Engl.* **8,** 962 (1969).

D. Seyferth and W. Tronich, *J. Organometal. Chem.* **21**(1), P3 (1970).

M. Miocque and J. P. Duclos, *Chim. Ther.* **4,** 363 (1969).
D. Swern, *Ann. N.Y. Acad. Sci.* **163**(Art. 2), 601 (1969).
F. N. Gladysheva, A. P. Sineakov, and V. S. Etlis, *Vsp. Khim.* **39**(2), 235 (1970).
S. Hirai and W. Nagata, *Kagaku Nc Ryoiki Zokan* **87,** 213 (1969).
J. F. Bieron and F. J. Dinan, *Chem. Amides,* 245 (1970).
H. W. Heine, *Mech. Mol. Migr.* **3,** 145 (1971).
J. W. Lown and K. Matsumoto, *Yuki Gosei Kagaku Kyokai Shi* **29,** 760 (1971).
T. Ento and A. Tomizawa, *Sekiyu To Sekiyu Kagaku* **15**(7), 118 (1971).
J. W. Lown, *Rec. Chem. Progr.* **32**(2), 51 (1971).
G. Maerker, *Top. Lipid Chem.* **2,** 159 (1971).
M. Veyama and K. Tori, *Nippon Kagaku Zasshi* **92,** 741 (1971).
K. Ichimura, *Sen'i Kobunshi Zairyo Keňkyusho Kenkyo Hokoku* **19**(94), 1 (1971).
G. J. Matthews and A. Hassner, *Org. React. Steroid Chem.* **2,** 1 (1972).
M. Polievka and M. Paviovic, *Petrochemia* **10,** 207 (1970).
N. P. Grechkin and I. A. Noretdinov, *Khim. Primen. Fosfororg. Soedin. Tr. Vses. Konf. 3rd,* 338 (1972).
H. Distler, *Mech. React. Sulfur Compounds* **4,** 7 (1969).
T. Saegusa and S. Kobayashi, *Kobunshi* **22**(3), 159 (1973).
N. R. Bertoniere and G. W. Griffin, *Org. Photochem.* **3,** 115 (1973).
K. Matsumoto, *Kagaku No Ryoiki* **27**(2), 148 (1973).
D. Donescu, *Stud. Cercet. Chim.* **21,** 1189 (1973).
G. Subrahmanyam and A. Adhya, *J. Sci. Ind. Res.* **33,** 308 (1974).
V. Bicker, *Arch. Geschwulstforsch.* **44,** 312 (1975).
E. Boehme and G. Schultz, *Methods Enzymol.* **38**(Pt. C), 27 (1974).
P. A. Gembitskii, V. A. Andronov, and D. S. Zhuk, *Sintez Svoistva Prakti Ispol'z. Polietilenimina,* 7 (1974).
G. L'Abbe, *Bull. Soc. Chim. Fr.* **5–6**(Pt. 2), 1127 (1975).
E. V. Koroleva, *Sovrem. Probl. Org. Khim.* **4,** 124 (1975).
A. N. Studenikov, *Sorrem. Probl. Org. Khim.* **4,** 73 (1975).
O. Cervinka, *Method Chim.* **6,** 591 (1975).
W. Nagata, *Lect. Heterocycl. Chem.* **1,** 29 (1972).
R. C. Gupta, E. Randerath, and K. Randerath, *Nucleic Acid Res.* **3,** 2915 (1976).
E. J. Goethals, *J. Polym. Sci. Polym. Symp.* **56,** 271 (1976).
E. Liepins, *Latv. PSR Zinat. Akad. Vestis* **9,** 32 (1976).
E. J. Goethals, E. H. Schacht, P. Bruggeman, and P. Bossaer, *ACS Symp. Ser.* **59,** 1 (1977).
E. J. Goethals, *Adv. Polym. Sci.* **23,** 103 (1977).
A. I. Nekhaev, *Khim. Zhizn.* **7,** 24 (1977).
S. V. Ris and R. S. Storr, *Vspekhi Khim. Geterotsiklov,* 108 (1976).
J. L. Pierre, P. Baret, and E. M. Rivoirard, *J. Heterocycl. Chem.* **15,** 817 (1978).

G. E. HAM
Dow Chemical U.S.A.

ICE. See Refrigeration; Water.

ICE COLORS. See Azo dyes.

ICE CREAM. See Milk products.

IDITOL, $CH_2OH(CHOH)_4CH_2OH$. See Alcohols, polyhydric.

IDOSE, $CH_2OH(CHOH)_4CH_2O$. See Carbohydrates; Sugars.

ILANG-ILANG OIL. See Oils, essential.

ILMENITE, $FeTiO_3$. See Titanium.

IMMUNOSUPPRESSANTS. See Analgesics, antipyretics, and anti-inflammatory agents; Chemotherapeutics, antimitotic; Immunotherapeutic agents.

IMMUNOTHERAPEUTIC AGENTS

Immunotherapeutic agents are substances that, because of their effects on the immune system, can be used in the treatment of various diseases or conditions with immunological disorders (1–3). Much progress in immunology has been made regarding the cellular and molecular basis of the immune system. Intricate immunologic mechanisms are being defined, and abnormal, exaggerated, or inappropriate immunological responses lead to disease states, eg, autoimmunity, immediate and delayed hypersensitivities, allograft reactions, and neoplasia. The new knowledge will influence the development of better and safer immunotherapeutic agents that will normalize, modulate, or correct the aberrant immune system.

The application of immunologic methods for the prevention and treatment of cancer (4) was made in the early twentieth century. Cancer patients were treated with vaccines and antisera that were prepared with cancer tissues (see Vaccine technology). Lack of consistent clinical benefit and deficient knowledge of cancer immunology led to failure and abandonment of cancer immunotherapy; however, the medical community recognizes and hopes that immunology can offer solutions to a number of other diseases as well as neoplasia. Also, new disciplines, eg, immunopharmacology (5) and immunopathology (6), are being formed.

The Immune System

The immune system constantly is exposed to antigens from many sources. These include, exogenously, multicellular parasites and organ transplants. At the unicellular level, the body often is infested with protozoa and a host of microorganisms (see Chemotherapeutics, antiprotozoal). Red and white blood cells containing antigens are introduced during transfusions. Viruses, food, and blood substitutes are subcellular sources of antigens.

Endogenously, physically or chemically induced tissue injury releases antigenic material. Cell turnover, which may be accelerated under various conditions, leads to increased antigen load. Many of the human biopolymers (qv), eg, proteins (qv), nucleic acids, polysaccharides, and immunoglobulins, can be antigenic especially when denatured or partially metabolized. As the body's anabolic and catabolic processes increase or decrease, the availability of these antigens fluctuate with the state of health or disease. During aging, active diabetes, starvation, or a drastic crash-diet program, deterioration of tissue components may present a stream of antigens. The ability of the immune system to cope with the sum total of these exogenous and endogenous antigenic materials determines the difference between normality and health or pathology and disease.

The immune system is composed of two major components, the two limbs: the humoral immune system is primarily the domain of the B-lymphocytes, and the cell-mediated immune system primarily is the domain of the T-lymphocytes. In a normal healthy state, regulatory mechanisms control the two limbs so that they function in proper balance. The humoral immune system produces antibodies that react specifically with the antigen, and the cell-mediated immune system mobilizes the phagocytic leucocytes to ingest and destroy the invading organisms that contain the antigen. In the healthy state, the two limbs are regulated by complex feedbacks through numerous mediators and by cell-to-cell cooperation. Any sudden increases in antigen supply can immediately elicit amplification of the immune system through clonal proliferation, an expansion process by which a few antigen-sensitive memory cells can generate a large number of responding cells (7). The degree and rate of the amplification at which components will be amplified depends to a great degree on the specific memory cells available at the moment when there is an antigenic challenge. Regardless of how it can be achieved, the antigen must be controlled appropriately in order that the host organism survive without deleterious effects.

If deficiency occurs in one component of the immune system, the other component must adjust accordingly with appropriate amplification to meet the challenge and maintain homeostasis. Therefore, regulation, especially appropriate regulation of immune amplification, is the key to efficient function of the immune system.

The mechanisms involved in immune regulation are depicted in Figure 1. Precursor stem cells from the bone marrow are processed by the thymus to become T-lymphocytes. B-lymphocytes become mature B-cells under the influence of the Bursa equivalent. Lymphocytes may become activated either directly by the primary antigen (Ag_p) or indirectly by macrophages (M) which have been exposed to the specific antigen. Most of the antigen (ca 90%) is destroyed by the macrophage through engulfment and degradation, but some of it is retained, processed, and presented to the appropriate lymphocytes with their resultant activation. Amplification of the immune system occurs with proliferation of both T- and B-lymphocytes. Most of these newly formed cells participate in the development of an immune response. A few remain dormant for long periods of time as memory cells, which can become activated on reexposure to the specific antigen; consequently, a more intense immune response occurs.

Certain subpopulations of T-lymphocytes are known as helper T-cells and sup-

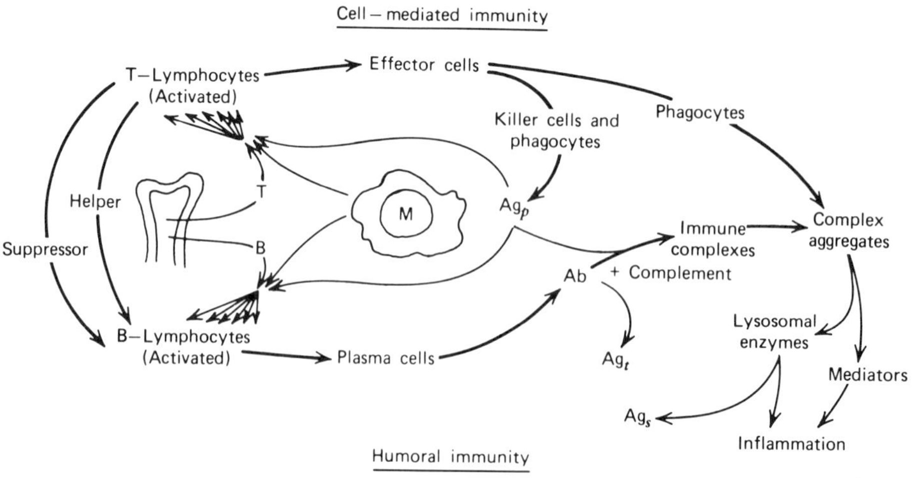

Figure 1. Mechanisms involved in immune regulation.

pressor T-cells. These cells regulate or influence the B-lymphocyte to differentiate to plasma cells which then produce antibodies (Ab). The antibody produced reacts specifically with the antigen which started the reaction to form immune complexes with or without complement activation. The antigen becomes neutralized and the threat to the host could be completely terminated if the antibody concentration is equal to or greater than the antigen concentration. However, when the antigen is in excess of its specific antibody, the immune complex that is produced is more soluble and can circulate through the body systemically and gain access to the synovium of joints. Lymphoid cells of the synovial membrane thus are exposed to primary antigens (Ag_p), and consequently produce antibodies that are specific for that antigen. These immune complexes also may adhere to the endothelial lining of blood vessels inducing vasculitis or are caught between the endothelium and the basement membrane of glomeruli which causes inflammation. In rheumatoid arthritis (RA), these immune complexes are responsible for synovitis (inflammation of joints or the synovial membrane) by activation of the complement system—a key mediator of inflammation and tissue damage.

After the antigen–antibody complex gains access to the joint, the lymphoid cells in the synovial membrane become activated, produce antibodies that shift the antigen–antibody ratio toward unity, and the resulting immune complex precipitates and forms aggregates. These aggregates are too large to be cleared by the lymphatic system alone. Phagocytic effector cells of the cell-mediated immune system that are attracted by chemotactic factors and mediators that are generated through the complement system migrate into the synovial space and participate in the removal of these immune-complex aggregates. If the quantity of aggregates is not excessive, their removal leads to a clean, normal, and healthy functional joint. However, if the amount of aggregates is excessive and lysosomal enzymes are accidently released during phagocytosis, tissue injury and an inflammatory condition are generated through the mediators of inflammation. The formation of secondary antigens (Ag_s) occurs as biopolymers are denatured or partially metabolized by the lysosomal enzymes. The immune system must handle all the secondary antigens by the same mechanisms used against the primary antigens, and the inflammation intensifies.

In the presence of excessive antigen load, suppressor T-cells are deficient and, under the influence of helper T-cells, production of antibody becomes exaggerated. The excessive quantities of antibody become another source of antigen, the tertiary antigen (Ag_t), which results in the formation of rheumatoid factor (RF), an antibody to antibodies or to immunoglobulins. The inflammation becomes more intense and tissue destruction is perpetuated by the growing granulation tissue that forms in synovitis.

The effector cells function as a surveillance system in the healthy state to kill and remove alien cells that do not belong in the system or have been misplaced. In neoplasia, these effector cells, known as killer cells, fail to recognize the alien cell as foreign or nonself and allow the alien cell to proliferate to massive proportions. When tissues and organs are transplanted from one animal to another, the effector cells of the recipient recognize the allogenic graft as alien or histo-incompatible. The graft is destroyed or rejected. Histocompatible or syngenic grafts are accepted because they are not considered foreign by the phagocytic effector cells. Histocompatibility of tissues is governed by genetic mechanisms. Tissues exchanged between close relatives, eg, identical twins, are readily accepted; those exchanged between nonrelatives usually are rejected. However, appropriate immunotherapeutic agents can be administered to alter the immune system so that the effector cells are not able to recognize and destroy the alien graft.

There is a gradual decline in effective immunologic function during aging in humans and animals. Though a decline in antibody production capacity is observed, this primarily results from a decrease in T-helper cell function and not from a decrease in B cell or macrophage function. A corresponding decline in T-suppressor cell activity leads somehow to the production of antibodies that are reactive to self-antigens, whereas the percentage of peripheral B-cells remains fairly constant throughout adult life. Such immune changes appear to be correlated with the incidence of immunologic diseases that are associated with a deficiency in T-cell function.

It is difficult to establish the cause and effect between T-cell deficiency and disease states, eg, rheumatoid arthritis and cancer. However, evidence strongly suggests that the disease states associated with aging result from immunological deficiency. The human thymus undergoes progressive atrophy during adulthood (8), which leads to a decrease in immunocompetent T-lymphocytes and a gradual impairment of the thymus-dependent immune responses. It is possible that the decreased production of thymic hormones is responsible for the lack of competent T-lymphocytes during the aging process.

Immunodeficiency also may occur as a result of exposure to certain toxic chemicals (Fig. 2) that are released accidentally into the environment. Organometallic compounds (9), eg, aryl- and alkyltin and methylmercury; inorganic metals, eg, lead and cadmium (10); and halogenated aromatics (10–11), eg, polychlorinated biphenyls and tetrachlorodibenzo-*p*-dioxin (TCDD) (**1**) affect the immune system adversely. Polybrominated biphenyl compounds (PBB) contaminating dairy farm products were consumed by Michigan residents (11). Immunologic alterations were suspected. Preliminary studies in rodents showed thymic atrophy and compromised thymic-dependent immune functions with little or no effect on humoral immune functions. TCDD selectively inhibits suppressor T-cell with little or no effect on helper T-cell function (12).

Complexities of the immune system are evident when multiple antigens are considered. Bacterial infections commonly accompany viral influenza. Treatment

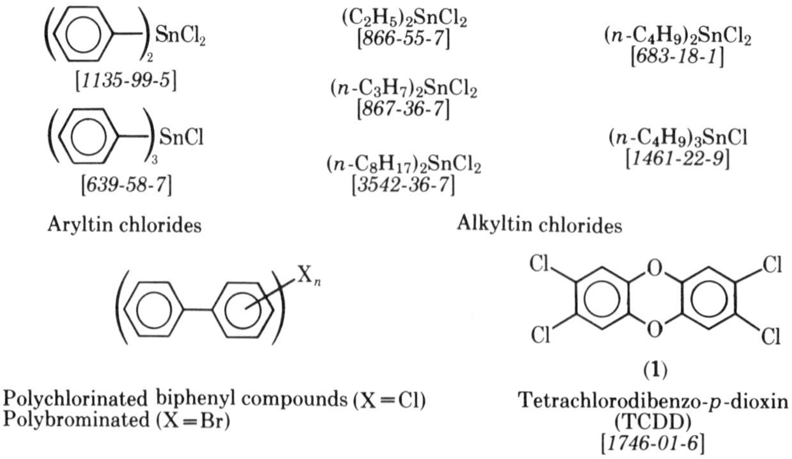

$(C_6H_5)_2SnCl_2$
[1135-99-5]

$(C_6H_5)_3SnCl$
[639-58-7]

Aryltin chlorides

$(C_2H_5)_2SnCl_2$
[866-55-7]

$(n\text{-}C_3H_7)_2SnCl_2$
[867-36-7]

$(n\text{-}C_8H_{17})_2SnCl_2$
[3542-36-7]

$(n\text{-}C_4H_9)_2SnCl_2$
[683-18-1]

$(n\text{-}C_4H_9)_3SnCl$
[1461-22-9]

Alkyltin chlorides

X_n

Polychlorinated biphenyl compounds (X = Cl)
Polybrominated (X = Br)

Cl Cl O O Cl Cl

(**1**)
Tetrachlorodibenzo-*p*-dioxin
(TCDD)
[1746-01-6]

Figure 2. Some chemicals that modify the immune system.

of such conditions with antibiotic therapy is essential and effective, especially when the immune system that is being considered is in an RA patient taking steroids (see Hormones, adrenal-cortical) or in a cancer patient receiving immunosuppressive therapy (see Chemotherapeutics, antimitotic).

Immunotherapeutic Agents

Immunosuppressive agents are used to alter normal immune function to prevent destruction of transplanted tissues (1); however, there are undesirable side effects and narrow safety margins of known agents. It is the nonspecific character of immunosuppressive agents that leads to the undesirable effects. However, the immune system has tremendous adaptability because of the intricate regulatory mechanisms involved. It may be possible to achieve beneficial results by the administration of either immunosuppressants or immunostimulants. Doses of antigen that are optimal for antibody production may suppress cell-mediated immunity (2,13). Such reciprocal relationships between humoral immunity and cell-mediated immunity are noted frequently. Cyclophosphamide (**2**), in low doses, can augment cell-mediated responses without affecting antibody responses (14). On the other hand, T-cell-mediated immunity can be potentiated by selective suppression of antibody formation with cyclophosphamide (15–16). Duality of immunotherapeutic agents can exist; the duality is a property of the immune system and not of the agent employed. An understanding of such factors permits explanation of the mechanism of action of these agents. Therefore, a definitive line cannot be drawn to separate these agents clearly into suppressants and stimulants.

It is possible to modulate the immune system chemically at many different points because of the various immunological pathways and mechanisms (17). This magnifies the possibilities for the discovery of better and perhaps more appropriate immunotherapeutic agents.

Immunosuppressants. A number of drugs with immunosuppressive activity have been used in the treatment of a variety of chronic inflammatory diseases with immunological features. In general, immunosuppressants may have a mild favorable effect on the clinical course of a disease, but their undesirable side effects can be fatal or chronically disabling (18). This situation presents an opportunity for the medicinal chemist to synthesize compounds that are specifically and/or selectively immunosuppressive, and for the immunopharmacologist to select relevant biological experiments that will identify the specific activity of such compounds (19).

The volume of literature describing the chemistry and pharmacology of immunosuppressants is massive. A brief review is given below. Some of these agents may serve as potential prototypes for the development of future specific immunosuppressive agents.

Chemotherapeutic Agents. Major developments in molecular and cellular biology have contributed to understanding the mechanisms by which chemotherapeutic agents (1,20–23) affect cell growth. Eradication of the entire population of neoplastic cells, ie, total killing of cells, is required for a cure. A single clonogenic malignant cell, which can survive after chemotherapy, could produce sufficient progeny to kill the host (see Chemotherapeutics, antimitotic).

Many chemical compounds have been synthesized and tested in experimental animals, but only a few were found to be useful as chemotherapeutic agents with acceptable levels of toxicity. Those that generally are available may be classified as alkylating agents, antimetabolites, and natural products.

The alkylating agents are subdivided into five types according to their basic chemical structures: nitrogen mustards, aziridines (ethylenimines), alkylsulfonates, nitrosoureas, and triazines (see Table 1). The antimetabolites are subdivided into three types and are listed in Table 2: folic acid analogues, pyrimidine analogues, and purine analogues. The natural products are Vinca alkaloids (qv), antibiotics (qv), and enzymes (qv). The chemotherapeutic agents that are presented in the tables have a marked cytotoxic effect on lymphoid tissues by disturbing the fundamental mechanisms involved in cell growth, mitotic activity, differentiation, and function. The ability of these drugs to interfere with mitosis and cell division in rapidly proliferating tissues provides the basis for their therapeutic use in oncology as well as their toxic and immunosuppressive activity. These compounds may affect resting and nondividing cells, but the proliferating cells are the most susceptible cells to these agents.

When the administered dose is great enough possibly to achieve eradication of a malignancy, the immunological defense mechanisms, based on amplification by clonal proliferation of lymphocytes, have been devastated. The patient's susceptibility to infections caused by pathogenetic bacteria or opportunistic microorganisms is markedly increased. Therefore, the use of these potent nonspecific and dangerous agents is not indicated for chronic inflammatory diseases, unless there is no further option.

The folic acid antagonists are potent inhibitors of certain immune reactions and have been used as immunosuppressive agents in organ transplantation to inhibit tissue rejection. Methotrexate (**3**) can inhibit the establishment of delayed hypersensitivity. It is effective against the graft-versus-host response and autoimmune conditions. Cyclophosphamide (**2**) and azathioprine (**4**) have been used in the treatment of chronic inflammatory disease, eg, rheumatoid arthritis and systemic lupus erythematosus,

Table 1. Chemotherapeutic Alkylating Agents[a]

Agent (trade names)	CAS Registry No.	Structure	Neoplastic diseases
Nitrogen mustard[a]			
cyclophosphamide (**2**) (Cytoxan, Endoxan)	[*6055-19-2*]	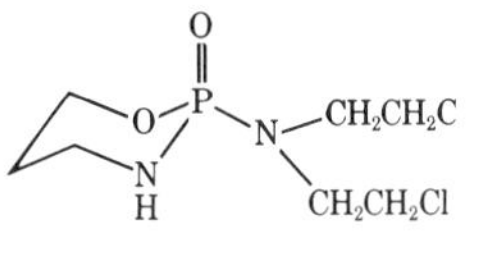 O=P(N(CH$_2$CH$_2$Cl)$_2$) in ring O–P–NH (oxazaphosphorinane)	acute and chronic lymphocytic leukemias, Hodgkin's disease, non-Hodgkin's lymphomas, multiple myeloma, neuroblastoma, breast, ovary, lung, Wilm's tumor, rhabdomyosarcoma
Aziridines (ethylenimines)			
triethylenemelamine (TEM) (**13**)	[*51-18-3*]	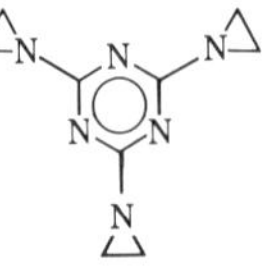 triazine ring with three aziridinyl N groups	Hodgkin's disease, non-Hodgkin's lymphomas, retinoblastoma, breast, ovary, chronic leukemias
triethylene-phosphoramide (**14**)	[*545-55-1*]	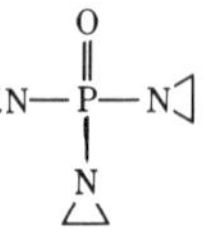 O=P(N-aziridinyl)$_3$	Hodgkin's disease, non-Hodgkin's lymphomas, retinoblastoma, breast, ovary

[a] Other chemotherapeutic alkylating agents are described in Table 1 of the article Chemotherapeutics, antimitotic.

Table 2. Chemotherapeutic Antimetabolites[a]

Antimetabolite (trade name)	CAS Registry No.	Structure	Neoplastic diseases
Folic acid analogue			
methotrexate (3 (Amethopterin)	*[59-05-2]*		acute lymphocytic leukemia, choriocarcinoma, mycosis fungoides, breast, testis, oropharyngeal, lung, osteogenic sarcoma
Pyrimidine analogue			
azauridine (**15**) (Triazure)	*[39455-15-7]*		mycosis fungoides, polycythemia vera
Purine analogue			
azathioprine (**4**)	*[446-86-6]*		

[a] Other chemotherapeutic antimetabolites are described in Table 2 of the article Chemotherapeutics, antimitotic.

with some degree of success. The serious side effects limit their use generally. It is difficult to envision how general suppression of the immune system can lead to beneficial effects in chronic inflammation or oncologic diseases.

Steroids and Nonsteroids. The adrenocorticosteroids have numerous and diversified physiologic functions and pharmacologic effects (24–28). The adrenal cortex is essential for the maintenance of homeostasis, which is mediated through mineralcorticoid and glucocorticoid activity. Since the discovery and identification of the naturally occurring glucocorticosteroids, hydrocortisone (**5**) and corticosterone (**6**) (Fig. 3), a bewildering number of synthetic analogues have been synthesized (Fig. 4). Synthetically induced changes in structure have resulted in greater separation between anti-inflammatory activity and sodium retention, but many share the same side effects and differ only with respect to potency. There is no member of this group of compounds that is significantly safer than another. Other adrenocorticoids are described in the article Hormones, adrenal-cortical (see Steroids).

The adrenocorticosteroids have a striking effect on lymphoid tissue (28). There is an increase in lymphoid tissue mass and lymphocytosis associated with Addison's disease. On the other hand, in Cushing's syndrome (hypersecretion of corticosteroids), lymphocytopenia is accompanied by a decrease in the mass of lymphoid tissue. These observations may represent the prolonged effect of hyposecretion and hypersecretion of adrenocorticoids in humans and animals that are somewhat resistant to steroids. In steroid-sensitive animals, eg, the rat and the mouse, glucocorticoids cause a rapid lysis of lymphatic tissues and lymphoid cells (29).

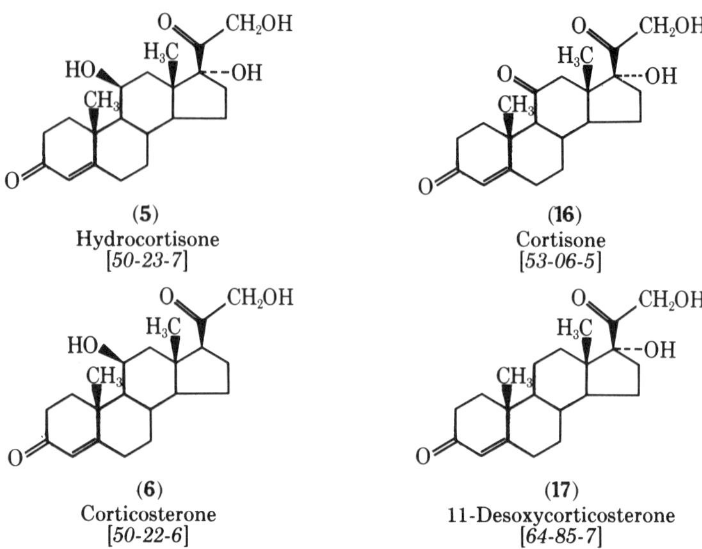

Figure 3. Natural adrenocorticosteroids.

The clinical course of a variety of diseases, in which immune mechanisms play a significant role, appears to be modified, often dramatically, by corticosteroids. However, the titer of circulating immunoglobulin IgG and IgE antibodies are not influenced by these agents. The steroids do not interfere with the normal processes in the development of cell-mediated immunity, but they do inhibit the inflammatory responses resulting from hypersensitivity reactions. Though the mechanism of action is not well understood, such beneficial effects of steroids, both natural or synthetic, may be accomplished through the blocking of the action of the migration inhibitory factor (MIF) which is secreted by activated lymphocytes (30). The movement of macrophages, then, is not inhibited by MIF and the cells do not accumulate at the site of inflammation. As a consequence, the phagocytosis, lysosomal enzyme release, and tissue damage sequence by macrophages do not ensue; however, tissue damage in joints does occur. Bone and cartilage changes continue in rheumatoid arthritic patients who are treated with steroids on a long-term basis.

The effectiveness of the steroids in suppressing inflammation is impressive, but this capacity has an additional hazardous potential in the corticosteroid-treated patient. Infection may continue without monitorable signs of inflammation or a peptic ulcer may perforate without any clinical symptoms. Such dangers, as well as retardation of growth and development in children, have directed research away from steroidal agents.

Nonsteroidal anti-inflammatory agents (NSAIAS) (26–27) such as phenylbutazone and indomethacin were developed subsequent to the steroid era. The second generation NSAIAs (Fig. 5), ibuprofen, naproxen, fenoprofen, and tolmetin, became available in 1975–1976 in the United States. In 1979, sulindac was marketed as a therapeutic agent for a variety of inflammatory conditions (see Analgesics, antipyretics, and anti-inflammatory agents).

The immunological profiles for the NSAIAs are unknown. Preliminary studies indicate that some of these agents have immunosuppressive properties. It has been

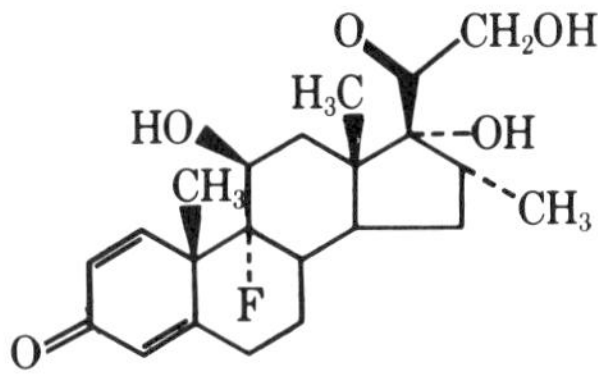

(18)
Dexamethasone
[*50-02-2*]

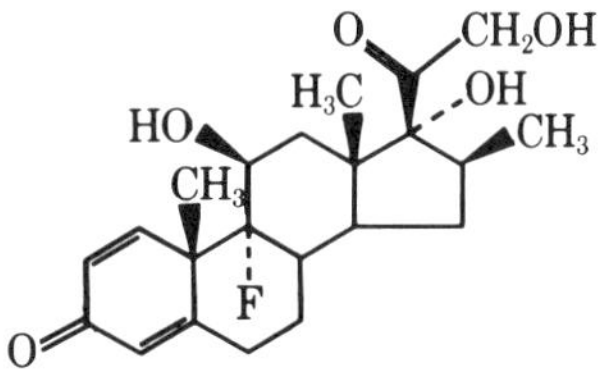

(19)
Betamethasone
[*378-44-9*]

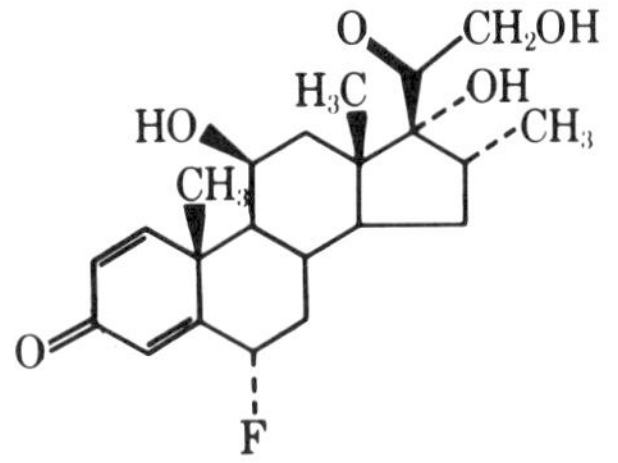

(20)
Paramethasone
[*53-33-8*]

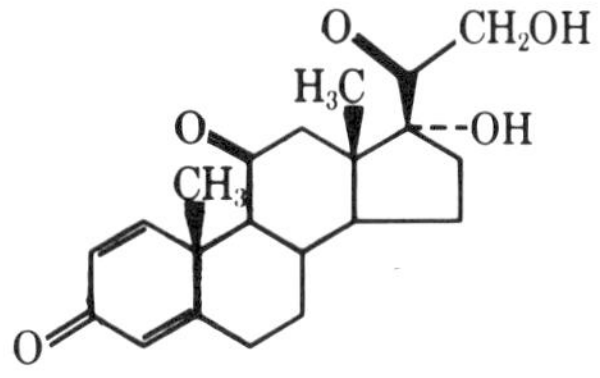

(21)
Prednisone
[*53-03-2*]

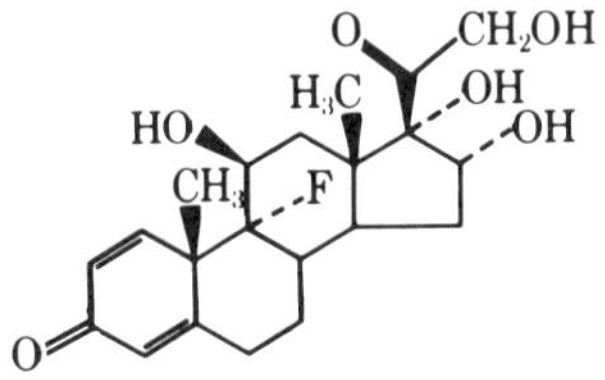

(22)
Triamcinolone
[*124-94-7*]

Figure 4. Synthetic steroids.

shown that aspirin (**7**), phenylbutazone (**8**), indomethacin (**9**), naproxen (**10**), and acetaminophen (**11**) inhibit the lymphocyte-blast transformation induced by phytohemagglutinin (PHA), concanavalin A (Con A), and pokeweed mitogen (PWM) (31). The inhibitory activity of acetaminophen was achieved at high concentrations. A low concentration of acetaminophen could enhance these mitogen responses. The biphasic property of acetaminophen is similar to that of levamisole.

Various anti-inflammatory agents are shown in Figure 5 (see Immunostimulants below). D-Penicillamine (32) produces suppression of clinical and laboratory manifestations of rheumatoid arthritis. It is not generally immunosuppressive nor anti-inflammatory, but the normalization of immunological and serological abnormalities suggests an immunosuppressive effect. Significant improvement of symptoms occurs only after a prolonged period of treatment with penicillamine. Its mechanism of action is unknown. Gold compounds (33) (Fig. 5) are used primarily in rheumatoid arthritis, particularly in early, active disease that progresses despite standard treatment with salicylates (see Gold and gold compounds). Gold salts often may arrest the joint degeneration, prevent involvement of unaffected joints, improve grip strength, and decrease the erythrocyte sedimentation rate. Abnormal plasma glycoprotein and fibrinogen levels also may be lowered.

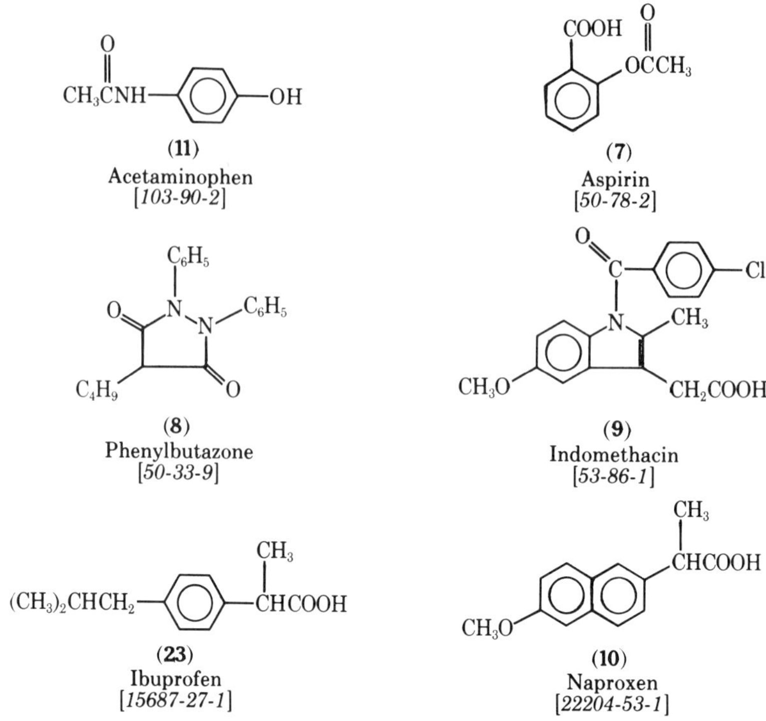

Figure 5. Nonsteroidal anti-inflammatory agents.

Immunostimulants. Emphasis has been primarily on immunosuppression, but more recently, research is directed toward treatment of various conditions by immunostimulation (34–37). Immunostimulants include a spectrum of substances that enhance or augment immunologic function. Immunomodulators, agents that enhance one immunologic component or function and elicit a decrease in a corresponding but opposite component, are included as well as immunoadjuvants which have little or no effect without the presence of an antigen. Therefore, immunostimulant describes the direct activity of the agent.

The use of mycobacteria (complete Freund's adjuvant) in combination with an antigen for the augmentation of the immune response to the antigen is well known. Both humoral and cell-mediated immune responses to an antigen can be enhanced by combination with *bacillus Calmette-Guerin* (BCG). Strains of other mycobacteria and certain fractions of BCG have similar activity. BCG has been used as an immunotherapeutic agent for the treatment of neoplasia during the past decade (37). Prolongation of chemotherapeutic remissions were reported for acute myeloblastic leukemia, metastatic myeloma, and metastatic breast cancer. Postoperative remission of malignant melanoma and colon cancer was delayed by the administration of BCG. A methanol-insoluble fraction of BCG has been studied extensively. The immunostimulatory and antitumor effects are similar to those of BCG (35). Other microorganisms, eg, *Corynebacterium parvum* and *C. granulosum* also were reported to enhance the effects of chemotherapy in cancer (37).

Lipopolysaccharide (LPS), the active principle of Gram-negative bacterial en-

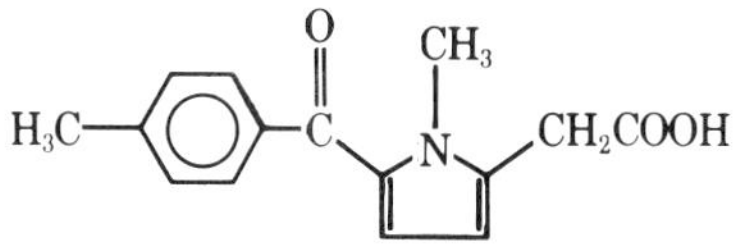

(**24**)
Tolmetin
[*26171-23-3*]

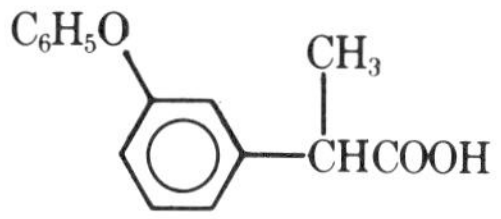

(**25**)
Fenoprofen
[*31879-05-7*]

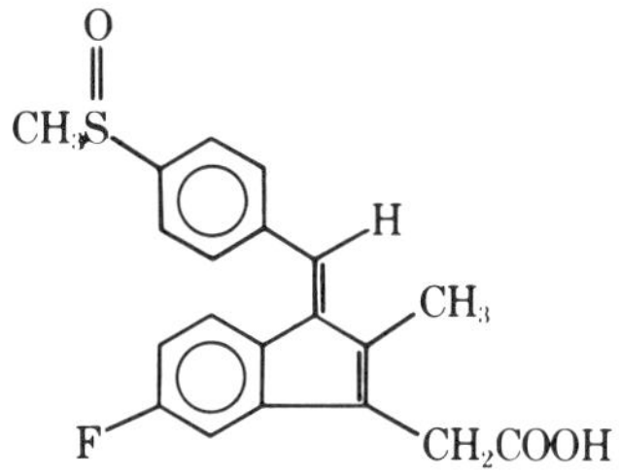

(**26**)
Sulindac
[*38194-50-2*]

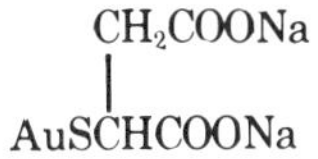

(**27**)
Gold thiomalate
[*24145-43-5*]

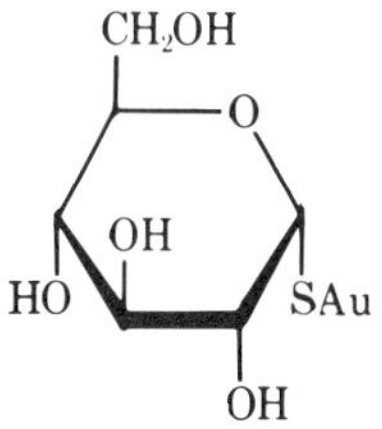

(**28**)
Aurothioglucose
[*12192-57-3*]

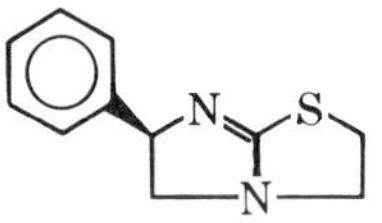

(**12**)
Levamisole
[*14769-73-4, 53631-68-8*]

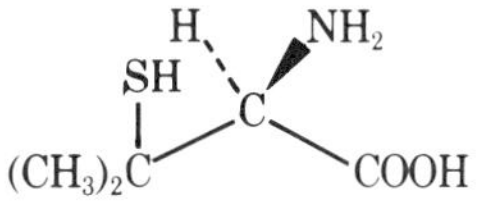

(**29**)
D-Penicillamine
[*52-67-5*]

Figure 5. (*continued*)

dotoxins, can enhance or suppress immune reactions under different experimental conditions. When LPS is given before the antigen, T-cell responses are inhibited and antibody production is enhanced (38–39). If LPS is administered after antigen, the T-cell responses are enhanced with a corresponding decrease in antibody production. The immunological activity of LPS has been reviewed in detail and the mechanism of LPS action has been postulated (17).

Polysaccharides have been extracted from a number of natural sources and have potent immunostimulant and antitumor activity (37). Lentinan [*9051-97-2*], an adjuvant derived from mushrooms, enhances helper T-cell response and stimulates antibody-dependent cell-mediated cytotoxic activity with no apparent affect on other T-cell responses. Glucan [*9037-91-6*], obtained from yeast, enhances the T-cell-dependent antigenic response to sheep red-blood cells. The antitumor activity of this substance may be mediated by the nonspecific activation of macrophages.

Levamisole. The anthelmintic agent, levamisole (**12**), (Fig. 5) has been shown to modulate the immune system in animals and humans (40–41). Its immunopharmacologic profile suggests that levamisole may be an important prototype for the research and development of future immunologic agents to be used in chronic inflammatory diseases and neoplasia.

Levamisole has been shown to influence virtually all functions involved in cell-mediated immune reactions and to restore these functions in compromised hosts, but it had little or no effect in the normal animal. Responses to levamisole depend not only on the initial immune status of the host, but also on the concentration of antigen, dose of levamisole, timing, and duration of treatment. Experiments on host-defense mechanisms in animals show that levamisole increases the number of neutrophils and the number of phagocytized staphylococci at the site of injection. No significant effect could be observed in immunologically competent normal animals, but a definite and significant effect could be elicited in animals with immature or impaired immunity. Antigen-stimulated phagocytosis and intracellular killing by macrophages also are significantly stimulated by levamisole. Lymphocyte cytotoxicity against tumor cells induced by tumor cell vaccine was increased and prolonged in melanoma patients, but levamisole had little effect on primary tumors. However, if the tumor load was decreased, levamisole had a good effect.

Using the PHA stimulation of lymphocyte transformation (*in vitro*) test, levamisole (**12**) has been shown to restore deficient T-cell responses of cancer and tuberculosis (TB) patients to normal; only in cases involving hyporesponsive cells were increases above normal. Levamisole restores E-rosette-forming cells toward normalcy but seldom above normal. Skin-delayed hypersensitivity responses were boosted by levamisole in anergic patients with malignant and nonmalignant diseases. Studies on humoral immunity showed that levamisole had little affect on antibody production in healthy subjects. However, in rheumatoid arthritis or chronic Aleutian disease of mink, where there is an elevation of immunoglobulin levels, levamisole produced a gradual decrease with general improvement of the disease. Following levamisole treatment, a fall in circulating immune complexes occurred in RA patients. These humoral effects appear to result from T-cell or macrophage activation rather than from a direct B-cell response.

Since 1971, levamisole has been used in human diseases where immune imbalance has been either known or postulated. The duration, severity, and frequency of disease episodes in chronic and recurrent infections were markedly reduced with levamisole treatment. In rheumatoid arthritis, preliminary results indicated significant improvement in 50–70% of the patients.

It had been assumed that inflammation and hyperactive immunological events were responsible for autoimmunity and joint destruction in rheumatoid arthritis. Treatment, initially, was aimed at suppressing these processes. However, some investigators have demonstrated that autoimmunity primarily results from immune deficiency and not immune hyperfunction. This concept led to the use of BCG and levamisole in patients with rheumatoid arthritis. Initial reports on the results of six open clinical studies showed that levamisole produced beneficial effects in RA patients. Levamisole also was shown to be as effective as D-penicillamine and both were more effective than a placebo (42). However, one report denied that levamisole had any therapeutic effect in rheumatoid arthritis. Since these preliminary results were reported, intensive investigations have been pursued (41).

Levamisole also has been shown to normalize the RA-deficient T-lymphocyte response to PHA *in vitro* and *in vivo* (43). Patients showing good clinical response showed a greater enhancement of T-cell function as compared to those patients showing a poor clinical response to levamisole. B-cell function decreased and immunoglobulin and autoantibody levels also decreased with levamisole treatment. These studies suggest that the mechanism of levamisole action is normalization of T-cell function with an indirect effect on B-cell autoantibody production.

In rheumatic diseases other than rheumatoid arthritis, it is difficult to make a reliable evaluation of the effect of levamisole. Possible improvement in some of these diseases, eg, systemic lupus, scleroderma, ankylosing spondylitis, and Reiter's syndrome, may be obtained with levamisole. The clinical effects of levamisole treatment in cancer patients have been reviewed (41). Good results were observed when levamisole was used to prevent relapse after tumor burden had been decreased or removed. A beneficial overall trend with respect to survival was associated with the group responding to primary cytoreductive therapy. There was no benefit from levamisole in those patients whose tumor grew despite the chemotherapy. The addition of levamisole to cyclic chemotherapy has resulted, in some instances, in an increased remission rate.

Toxicity

The LD_{50}s of a few immunotherapeutic agents and anti-inflammatory agents are shown in Table 3. Toxicities of immunotherapeutic corticosterioids are described in references 49 and 50. Eventually, tumor immunology will make use of antibodies from cloned cancer-specific immune cells, eg, leukemias and lymphomas (51) (see Genetic engineering).

Table 3. Toxicities of Selected Immunotherapeutic Agents and Nonsteroidal Anti-inflammatory Agents

Compound	LD_{50}, mg/kg		Animal	Reference
	Oral	Intraperitoneal		
Immunotherapeutic agents				
azathioprine (**4**)	2500	650	mouse	44
	400	310	rat	44
levamisole (**12**)	[20 mg/(kg·d) for 2 mo]		dog	45
D-penicillamine (**29**)	(no LD_{50} recorded)			46
Nonsteroidal anti-inflammatory agents				
phenylbutazone (**8**)	700		rat	47
indomethacin (**9**)	19		rat	47
naproxen (**10**)	543		rat	47
ibuprofen (**23**)	800		mouse	48
	1600		rat	48

Nomenclature

Ab	= antibody
Ag_p	= primary antigen
Ag_s	= secondary antigen
Ag_t	= tertiary antigen
B-cell	= bone marrow derived cells (maturation is influenced by Bursa of Fabricus or equivalent)
BCC	= *bacillus Calmette-Guerin*
Con A	= concanavalin A
IgE	= E-type immunoglobin
IgG	= G-type immunoglobin
LPS	= lipopolysaccharide
M	= macrophages
MIF	= migration inhibitory factor
NSAIA	= nonsteroidal anti-inflammatory agent
PBB	= polybrominate biphenyl
PHA	= phytohemagglutinin
PWM	= pokeweed mitogen
RA	= rheumatoid arthritis
RF	= rheumatoid factor
T-cell	= thymus derived cells (maturation is influenced by thymus and/or thymic hormones)
TB	= tuberculosis
TCDD	= tetrachlorodibenzo-*p*-dioxin

BIBLIOGRAPHY

1. J. E. Harris and R. C. Bagai in *Antineoplastic and Immunosuppressive Agents,* Vol. I, 1974, pp. 618–641.
2. A. D. Steinberg and G. W. Williams in D. W. Talmage, B. Rose, K. F. Austen, and J. H. Vaughan, eds., *Immunological Diseases,* 3rd ed., Vol. II, Little, Brown and Company, Boston, Mass., 1978, pp. 1522–1539.
3. F. K. Hess and K. R. Freter in M. E. Wolff, ed., *Burger's Medicinal Chemistry,* Part III, John Wiley & Sons, Inc., New York, 1979, pp. 671–704.
4. C. M. Southam, *Ann. N.Y. Acad. Sci.* **277,** 1 (1976).
5. R. A. Good in J. W. Hadden, R. G. Coffrey, and F. Spreafico, eds., *Immunopharmacology,* Plenum Medical Book Co., New York, 1977, pp. ix–xi.
6. S. Cohen, P. A. Ward, and R. T. McCluskey, eds., *Mechanisms of Immunopathology,* John Wiley & Sons, Inc., New York, pp. vii–viii, 1979.
7. K. B. von Eschen and J. A. Rudbach, *J. Exp. Med.* **140,** 1604 (1974).
8. J. Singh and A. K. Singh, *Clin. Exp. Immunol.* **37,** 507 (1979).
9. W. Seinen and A. Penninks, *Ann. N.Y. Acad. Sci.* **320,** 499, 517 (1979).
10. J. G. Vos, *Crit. Rev. Toxicol.* **5,** 67 (1977).
11. R. P. Sharma and P. J. Gehring, *Ann. N.Y. Acad. Sci.* **320,** 487 (1979).
12. M. I. Luster, R. E. Faith, and G. Clark, *Ann. N.Y. Acad. Sci.* **320,** 473 (1979).
13. A. L. DeWeck and J. R. Frey, *Antibiot. Chemother.* **15,** 110 (1969).
14. P. W. Askenase, B. J. Hayden, and R. K. Gershon, *J. Exp. Med.* **141,** 697 (1975).
15. P. H. Lagrange, G. B. Mackaness, and T. E. Miller, *J. Exp. Med.* **139,** 1529 (1974).
16. J. L. Turk and D. Parker, *J. Immunol.* **1**(2), 127 (1979).
17. J. W. Hadden, *Springer Semin. Immunopathol.* **2,** 35 (1979).
18. E. M. Tan and N. F. Rothfield in D. W. Talmage, B. Rose, K. F. Austen, and J. H. Vaughan, eds., *Immunological Diseases,* 3rd ed., Little, Brown and Company, Boston, Mass., 1978, pp. 1038–1060.
19. P. Davies and R. J. Bonney, *Ann. Rep. Med. Chem.* **121,** 152 (1977).
20. A. R. Kraska, *Ann. Rep. Med. Chem.* **14,** 132 (1979).
21. A. R. Kraska and J. S. Wolff, *Ann. Rep. Med. Chem.* **13,** 120 (1978).
22. J. S. Driscoll and J. A. Beisler, *Ann. Rep. Med. Chem.* **12,** 120 (1977).
23. P. Calabresi and R. E. Parks, Jr. in L. S. Goodman and A. Gilman, eds., *The Pharmacological Basis of Therapeutics,* Section XV, MacMillan Publishing Co., Inc., New York, 1975, pp. 1248–1307.

24. A. S. Fauci, *Ann. Rep. Med. Chem.* **13,** 179 (1978).
25. M. J. Green and B. N. Lutsky, *Ann. Rep. Med. Chem.* **11,** 149 (1976).
26. S. Wong, *Ann. Rep. Med. Chem.* **10,** 172 (1975).
27. M. E. Rosenthale, *Ann. Rep. Med. Chem.* **9,** 193 (1974).
28. R. C. Haynes, Jr. and J. Larner in Ref. 23, pp. 1472–1506.
29. H. N. Claman, *New Eng. J. Med.* **287,** 388 (1972).
30. F. Spreafico and A. Anaclerio in Ref. 5, pp. 245–278.
31. R. S. Panush, *Arth. Rheum.* **19,** 907 (1976).
32. I. A. Jaffe in V. R. Ott and K. L. Schmidt, eds., *Die Behandlung der Rheumatoiden Arthritis mit D-Penicillamine,* Dr. Dietrich Steinkopf Verlag, Darmstadt, 1974, pp. 84–94.
33. S. C. Harvey, in Ref. 23, pp. 931–935.
34. G. Renoux and M. Renoux in M. A. Chirigos ed., *Progress in Cancer Research and Therapy,* Raven Press, New York, 1977, pp. 235–237.
35. P. Dukor, L. Tarcsay, and G. Baschang, *Ann. Rep. Med. Chem.* **14,** 146 (1979).
36. J. G. Lombardino, *Ann. Rep. Med. Chem.* **13,** 167 (1978).
37. J. W. Hadden, L. Delmonte, and H. F. Oettgen in Ref. 5, pp. 279–313.
38. P. H. Lagrange, G. B. Mackaness, T. E. Miller, and P. Pardon, *J. Immunol.* **114,** 442 (1975).
39. P. H. Lagrange and G. B. Mackaness, *Ann. Rep. Med. Chem.* **10,** 447 (1975).
40. J. Symoens and M. Rosenthal, *J. Reticulo. Soc.* **21,** 175 (1977).
41. *Proceedings International Symposium, Levamisole in Rheumatoid Arthritis, held at the University of Antwerp, Belgium, June 23, 1978, under the auspices of the European League Against Rheumatism (EULAR).*
42. E. E. Huskisson, *Drugs Exptl. Clin. Res.* **5,** 115 (1978).
43. J. Levy, *J. Rheumatol.* **5,** 63 (1978).
44. "Imuran" Azothioprine, Burroughs Wellcome and Co., Tuckahoe, New York, 1968.
45. J. Symoens and Y. Schuermans, "Levamisole," in E. C. Huskisson, ed., *Clinics in Rheumatic Diseases,* W. B. Saunders Ltd., London, 1979, No. 2, p. 603.
46. W. H. Lyle, "Penicillamine," in Ref. 45, p. 569.
47. M. Sondervorst, "Azopropazone," in Ref. 45, p. 465.
48. S. S. Adams and J. W. Buckler, "Ibuprofen and Flurbiprofin," in Ref. 45, p. 359.
49. *Evaluations of Drug Interactions,* 2nd ed., American Pharmaceutical Association, Washington, D.C., 1976, pp. 411–413.
50. F. D. Hart, ed., *Drug Treatment of the Rheumatic Diseases,* University Park Press, Baltimore, Md., 1978, pp. 63–74.
51. L. H. Schloen, *The Sciences,* 15 (July/Aug. 1980).

General References

"Mechanism of Tissue Injury," *Ann. N.Y. Acad. Sci.* **256,** (1975).

S. Sell, ed., *Immunology, Immunopathology and Immunity,* 2nd ed., Harper & Row Publishers, Inc., New York, 1975.

R. T. McCluskey and S. Cohen, eds., *Mechanisms of Cell-Mediated Immunity,* John Wiley & Sons, Inc., New York, 1974.

"Present Concepts as a Basis for the Development of Immunopharmacological Agents" in G. H. Werner and F. Floch eds., *The Pharmacology of Immunoregulation,* Academic Press, Inc., New York, 1978.

M. E. Rosenthale and H. C. Mansmann, Jr., eds., *Immunopharmacology,* Spectrum Publications, Inc., New York, 1975.

"Diseases with Immunological Features" in M. Sampter, ed., *Immunological Diseases,* Little, Brown and Company, Boston, Mass., (1978).

J. E. Goodnight, Jr. and D. L. Morton, "Immunotherapy for Malignant Disease," *Ann. Rev. Med.* **29,** 231 (1978).

H. Friedman and C. Southam, "International Conference on Immunobiology of Cancer," *Ann. N.Y. Acad. Sci.* **276–277,** (1976).

H. Friedman, "Thymus Factors in Immunity," *Ann. N.Y. Acad. Sci.* **249,** (1975).

F. H. Bach and R. A. Good, eds., *Clinical Immunobiology,* Vol. 1, Academic Press, Inc., New York, 1972.

H. L. F. Currey "Immune Suppression and Immune Enhancement" in J. L. Gordon and B. L. Hazelman eds., *Rheumatoid Arthritis,* Elsevier/North Holland Biomedical Press, Amsterdam, 1977, pp. 157–163.

T. Y. Shen, "Mechanism of Action Considerations in the Discovery of New Antiarthritic Drugs," *Drugs Exptl. Clin. Res.* **5**, 127 (1979).

STEWART WONG
Boehringer-Ingelheim Ltd.

INCENDIARIES. See Chemical warfare; Pyrotechnics.

INCINERATORS

The concept of a furnace operated specifically to burn refuse originated in England around 100 years ago. These furnaces were originally called destructors. The term, incinerator, did not come into usage until the early twentieth century.

The first U.S. incinerator was built on Governor's Island, New York, in 1885, and the first municipal incinerator was one of 27 metric tons per day capacity constructed at Allegheny City, Pa., in the same year. By 1921 more than 200 municipal incinerators were in operation. These early incinerators were little better than an enclosed bonfire. Today, however, it is possible to build large, advanced central plants which are virtually nuisance-free and esthetically acceptable (1–2).

Small, individual, backyard incinerators were a popular means of disposal of waste in the 1930–1940s in communities without waste-collection facilities. They were, however, sources of air pollution and are now generally prohibited. Larger incinerators are used at locations developing significant amounts of solid waste such as shopping centers and large apartment complexes.

Without recovery of energy, incineration of municipal refuse is not the cheapest method for the disposal of municipal waste. Landfill is generally cheaper, but the growing shortage of disposal sites near population centers, the increasing cost of transportation, and the growing reluctance of smaller communities and rural areas to accept waste from other localities eliminates it as a future disposal method. Composting and biodegradation of organic wastes has not been successful on a large scale. Thus, incineration and related energy-recovery combustion techniques appear to be the most suitable means of disposal of municipal wastes from large communities in the near future.

Combining recovery and utilization of heat with incineration in a well-designed and operated plant may be justified. The economics of installing heat-recovery equipment is based on the income from the sale of steam or power to offset the additional cost of utilizing the excess heat. The two basic designs used for heat recovery are: (*1*) A combustor combined with a waste-heat boiler. The fuel is burned in a refractory-lined vessel and the hot gases are sent to a waste-heat boiler to produce steam; (*2*) A waterwall combustion chamber where heat is removed from the combustion chamber to produce steam.

Refractory-lined chambers require 150–200% excess air and waterwall chambers 50–100%. Low excess air reduces the volume of flue gas, increases the recoverable heat, and reduces the size of the necessary air pollution control equipment. Because of the variations in energy available from the solid waste, production ranges from 1 to 3.5 kg steam per kg solid waste burned. In heat-recovery operations provisions must be made for cleaning the boiler tubes.

In Europe and Japan heat recovery is preferred. In Europe about 250 units are in operation compared to fewer than 20 waste-to-energy systems in the United States. The first waterwalled incinerator in the United States was constructed in Chicago in 1970. Today the recovery of the energy value of municipal waste is an item of national priority. Utilization of solid wastes to produce liquid, gaseous, or solid fuels by pyrolysis processes and by beneficiation of municipal waste to provide fuel for power boilers is receiving extensive examination (see Fuels from waste).

Industrial wastes cover an even greater diversity of substances than municipal wastes. In addition to dry solid combustibles, wet sludges and combustible and hazardous liquids must be incinerated. A classification of industrial wastes is given below (3).

Waste type:

(*1*) Mixed solid combustible materials (paper, wood).

(*2*) Pumpable, high heating value, moderately low ash, high moisture (skimmer emulsions, tank bottoms, heavy ends).

(*3*) Wet, semisolids (refuse and water-treatment sludge).

(*4*) Uniform, solid burnables (waste polymers).

(*5*) Pumpable, high ash, low heating value materials (acid or caustic sludges or sulfonates).

(*6*) Difficult or hazardous materials (toxic compounds, hydrofluoric acid or pesticide residues).

(*7*) Other materials.

Waste form:

(*1*) Solid, dry materials suitable for handling by front-end loader or buckets.

(*2*) Semisolid, wet material suitable for handling by front-end loader or buckets.

(*3*) Semisolid, sticky or tarry substances not suitable for handling by front-end loader or buckets.

(*4*) Mixtures of solids and liquids not suitable for handling by front-end loader.

(*5*) Viscous liquids pumpable when hot or with special pumps.

(*6*) Liquids suitable for normal pumping equipment.

(*7*) Materials to be handled in combustible boxes or cartons.

(*8*) Materials handled in 208-L (55-gal) steel drums.

Different classes of wastes require incinerators of different design. In general, industrial wastes can be handled by providing a special incinerator for each type of waste product. This unit may be more closely designed to meet the particular wastes. Examples are fume incinerators for paint solvents and sewage sludge incinerators. An alternative method is a central facility that can accept a wide spectrum of waste products. For example, the rotary kiln system used by Dow Chemical among others.

It may be desirable to destroy liquid wastes or sludges with high water content by incineration and evaporation because of their acidic, basic, toxic, or biologically active nature. The installation for such materials must include special equipment for pollution control such as scrubbers for containment of acidic gases resulting from the combustion. Special attention must be given to corrosion problems and in some instances, incinerators have been constructed of high alloy materials (stainless steel, Incoloy, Hastelloy). Special incinerators for combustion of toxic and waste materials of all types are being developed as an environmental control measure. With increasing emphasis on adequate control of hazardous materials and chemicals under the Toxic Substances Control Act (TSCA) and the Resource Conservation and Recovery Act (RCRA), even greater utilization of specialized industrial incinerators can be expected.

New methods include slagging kilns, fluidized-bed combustion, suspension burning, pyrolysis, and the use of waste as an auxiliary fuel in power boiler installations.

Classification of Municipal Waste

Refuse characteristics, particularly chemical composition, heating value, and volatile and ash content affect the design and capacity of the incinerator and the type of pollutants that must be controlled. Refuse composition varies rather widely between communities (2) and is sometimes subject to local regulations. In most of the world outside of the United States, the percentage of paper in the waste is much smaller, whereas in Japan, moisture content is greatly increased by the presence of large quantities of fish scraps and vegetable waste. Table 1 shows the range of refuse composition for 23 U.S. communities and the calculated average. Table 2 shows typical composition of American waste and heating values (2,4). An empirical chemical formula for a typical municipal waste stream is given by $C_{30}H_{48}O_{19}N_{0.5}S_{0.05}$.

The Incinerator Institute of America provided a classification system for solid waste (see Table 3). Most of the wastes classified contain polymeric material (largely cellulose) composed largely of carbon, hydrogen, and oxygen. Table 4 contains heating values of some representative materials found in a municipal waste stream.

Sludge from a municipal sewage treatment plant is preferably disposed of by incineration. To save energy, solid waste can be incinerated at the same time.

Table 1. Typical Municipal Refuse for 23 U.S. Communities, %[a]

Classification	Range	Calculated average
food wastes	2.3–34.6	16.2
glass, ceramics	3.3–17.9	8.3
metal	4.6–14.5	8.5
paper products	17.5–61.8	42.8
plastics	0.2–3.4	1.3
leather, rubber	0.8–4.7	1.7
textiles	0.4–4.8	2.3
wood	0.4–22.4	2.6
yard wastes	1.6–23.8	9.6
miscellaneous	0.2–23.6	6.7
Total		*100.0*

[a] Ref. 2.

Table 2. Estimated Proximate and Ultimate Analysis of U.S. Urban Refuse[a]

Proximate analysis, %	
moisture	28.0
volatile matter	43.4
fixed carbon	6.6
glass, ash, metal	22.0
Ultimate analysis, %	
moisture	28.0
carbon	25.0
hydrogen, net[b]	3.3
oxygen	21.1
nitrogen	0.5
sulfur	0.1
glass, ceramics, stones	9.3
metals	7.2
ash, inerts	5.5
Total	*100.0*
higher heating value, MJ/kg[c]	10.5
stoichiometric combustion air, kg/kg refuse	3.2

[a] Refs. 2, 4.
[b] Net hydrogen is hydrogen in excess of oxygen equivalent to H_2O.
[c] To convert MJ/kg to Btu/lb, multiply by 430.

Incinerator Pollutants

Atmospheric emissions from incinerators are classified into particulates (fly ash, smoke, and condensation products), combustible gases (carbon monoxide, hydrocarbons, and partially oxidized hydrocarbons), and noncombustible gases (nitrogen oxides and hydrogen halides). Heavy metals in the feed present a problem (5). Entrained ash can arise from incinerator feed and combustion turbulence. Soot is the result of incompletely burned organic volatile matter that condenses as a fine aerosol and is then heated to cracking temperature. Soot, once formed, requires better mixing, higher temperatures and longer residence times for its destruction than other products of incomplete combustion. Smoke can be composed of condensed materials such as distilled organic matter and volatile heavy metals such as lead and zinc, often present in the effluent as the chloride or sulfate. Combustible gases can contain carbon monoxide, hydrogen, a variety of hydrocarbons, and polynuclear aromatic compounds which may be carcinogens. They arise from poor incinerator design or operations creating pockets of poorly mixed gas, exposure to low temperatures in the oxidation zone, insufficient residence time at temperature, and inadequate combustion turbulence and oxygen.

The presence of halogenated and sulfur-bearing wastes results in the formation of HCl, HF, H_2S, and SO_2 in the incinerator gases. These must be removed with suitable scrubbing equipment before discharge to the atmosphere. Nitrogen oxides are produced during high temperature combustion by reaction between nitrogen and oxygen in the air. Their formation is reduced by lowering the combustion temperature or excess air, but such action may cause the formation of other pollutants. Nitrogen content of waste material is generally low, but the presence of nitrogen-containing materials (nitrates, ammonium compounds, etc) greatly increases NO_x emissions.

Atmospheric emissions are generally controlled by careful incinerator design,

Table 3. Classification of Municipal Wastes, Density, and Heating Value

Type	Description	Density, kg/m³	Heating value, MJ/kg[a]
0	trash, highly combustible paper, cardboard, wood boxes, sweepings; up to 10% plastics and rubber	128–160	19.7
1	rubbish; combustible paper, cardboard, wood, foliage, sweepings; up to 20% food waste, no plastics or rubber	128–160	15.1
2	refuse; approximately even mixture of rubbish and garbage	241–320	10.0
3	garbage; animal and vegetable wastes	481–562	5.8
4	pathological; human and animal remains	722–883	2.3
5	gaseous, liquid or semiliquid by-product waste[b]		
6	solid, by-product waste[b]; mixtures containing over 10% plastics by weight		
7	compact, solid waste (mostly paper)	562–803	

[a] To convert MJ/kg to Btu/lb, multiply by 430.
[b] Frequently found in the chemical process industry.

Table 4. Approximate Heating Values of Wastes

Material	Density, kg/m³	Heating value, MJ/kg[a]
animal fats	963	39.4
brown paper	112	16.8
citrus rinds	642	3.9
coated milk cartons	80	26.3
corn cobs	241	18.6
corrugated boxes	112	16.3
coffee grounds	482	23.2
cotton seed hulls	482	20.0
latex	722	23.2
leather	321	16.8
linoleum	1445	25.5
magazines	562	12.2
newspapers	112	18.5
polyethylene	803	67.3
polyurethane (foamed)	32	30.2
rags (silk and wool)	241	19.7
rags (cotton and linen)	241	16.7
rubber waste	1605	23.2
wood sawdust	193	19.7

[a] To convert MJ/kg to Btu/lb, multiply by 430.

and, where necessary, the installation of control equipment. Particulate material may be removed by cyclones and electrostatic precipitators. Scrubbers are generally required for gaseous pollutants and can remove particulates as well. Baffle towers with overflow weirs and impingement baffle screens with flushing sprays are frequently used. Where fine particulates must be caught, high energy venturi scrubbers may be used. Scrubbers are generally avoided where possible because they create corrosion problems and problems of treating the effluent water. Condensation of the hot saturated effluent results in an unsightly plume and corrosion of nearby structures. Corrosion-resistant wet electrostatic precipitators constructed of plastic parts may control some industrial incinerator effluents. Regulations developed by the EPA under RCRA set specific requirements that should have considerable impact on future incinerator designs.

Solid Waste Incineration

Figure 1 is a diagram of a typical modern municipal waste incinerator with heat recovery in the form of steam. The most recent facility in the United States uses many of the features shown in the figure (6). The system is composed of three main areas. The front-end system is composed of units 1 to 4 and provides a feed for the thermal part of the system composed of units 5 to 9. Units 10 to 13 are associated with the proper disposal of the incinerator discharge, both solid and gaseous.

Waste Destruction. Polymeric or carbonaceous solids are degraded by high temperature. In the presence of oxygen, carbon, hydrogen, and sulfur present are oxidized to CO_2, H_2O, and SO_2, respectively. In any study of incineration or thermal destruction, the laws of conservation of mass and energy must be considered in the process analysis. The theoretical air requirements may be estimated as 0.32 kg air (ca 0.28 m^3 or 10 ft^3 at STP) for every 1054 kJ (1000 Btu) of heat released. The rate of incineration increases rapidly with temperature. A range of 700–760°C is generally required for combustion, and most general purpose incinerators operate between 760 and 1100°C. For an incinerator to operate without auxiliary fuel or air preheating, the refuse must not contain over 50% moisture or 60% ash and have more than 25% combustibles.

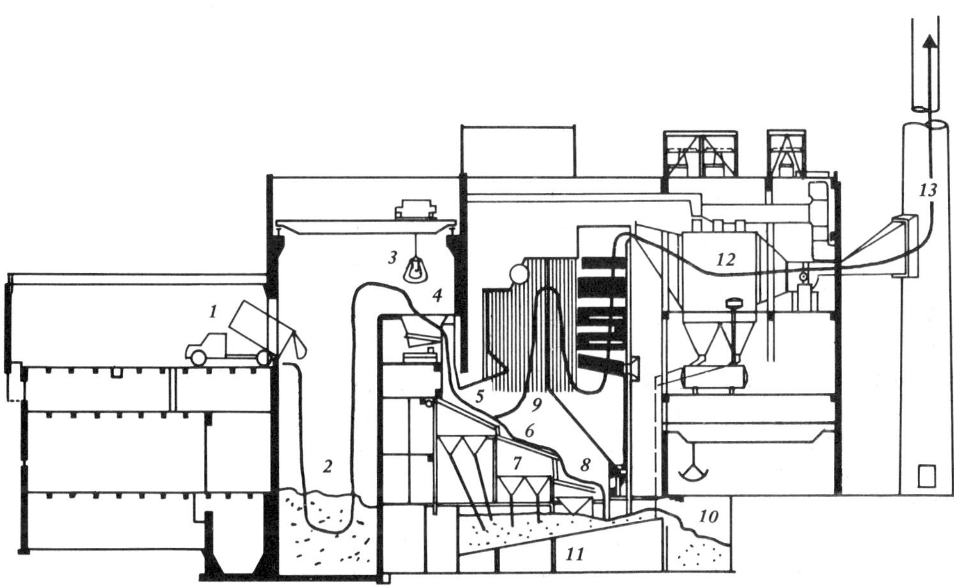

Figure 1. Typical waterwall furnace incinerator (6). RESCO Facility, Saugus, Mass. *1*. Unloading shed; *2*. refuse pit; *3*. loading crane; *4*. vibrating feeder; *5*. drying grate; *6*. combustion grate; *7*. grate movement; *8*. burnout grate; *9*. boiler section; *10*. residue bunker; *11*. residue channel; *12*. electrostatic precipitator; *13*. stack.

In addition to complete combustion, wastes may be destroyed by treatment at high temperatures either without oxygen (pyrolysis), limited oxygen (partial combustion), or reactive atmospheres (gasification), such as steam, hydrogen, or carbon dioxide. These and other thermal processes are largely under development.

Pyrolysis. If the solid is heated in the absence of oxygen (other than that contained in the feed) the organic material breaks down and combustible gas, organic liquids, and char are produced. The yield depends upon the properties of the organic solid, the temperature, and heating rate. Most of the chemical energy (as measured by the heat of combustion) that was originally in the solid appears as chemical energy in the pyrolysis products. To obtain the elevated temperatures required for the reaction, heat must be added to the reaction zone.

Starved-Air Combustion (Partial Combustion). To obtain the temperatures needed for the pyrolysis reaction to occur, a limited amount of oxygen is allowed to enter the reaction zone. This oxygen reacts with the feed or pyrolysis product and releases the needed energy within the reactor. Both pyrolysis and combustion products are obtained. The products leaving the system will contain a large amount of chemical energy. By proper monitoring of oxygen into the system, the temperature may be controlled without adding or removing heat

Stoichiometric Air. If the stoichiometric amount of air is added to the reactor, the temperature is controlled by removing heat from the system. No energy leaves the system as chemical energy. The heat is removed by generating steam.

Excess Air. Rather than remove the heat by generating steam, excess air is added to the system. The energy is carried from the reactor as sensible heat of this excess air and the products of combustion.

Moving-Grate Incinerators. Figure 2 shows the passage of the waste through the furnace on a moving grate. It is apparent that a grate must provide support for the refuse; admit the underfire air through openings; transport the refuse from the feed chute to the ash quench; and agitate the bed to bring fresh charge to the surface.

The refuse fed to the furnace is first dried and preheated by radiation from the hot combustion gases and refractory furnace lining . The refuse, as it is heated further, first pyrolyzes and then ignites. Combustion takes place both in the solid to burn out the residue and in the gas space to burn out the pyrolysis products. Overfire air jets greatly assist the mixing and combustion in the overfire air space.

In U.S. practice, the trend has been from stationary hearths to traveling, rocking, and reciprocating grates (7). Stokers that provide high agitation are preferred even though the agitation increases the particulate loading from the furnace chamber, but not necessarily from the stack since the air pollution control (APC) system can be designed to handle whatever loading is encountered in the furnace gases. Agitation increases the rate of heat transfer and, hence, combustion in the bed and in covering blow holes on the grate, ie, the small areas where complete burn-out or bulky objects leave spaces through which the underfire air may flow preferentially.

The underfire air rate must be selected to balance the conflicting demands of high air rates for high burning rates and of low air rates to minimize the particulate loading on the APC system. In many incinerators 60–100% of the stoichiometric air requirements are supplied under the grates. The balance of the air is fed through overfire jets. The residence time for solids in a municipal incinerator is 15–70 minutes, with grate loadings typically about 27 kg/m^2 (60 lb/ft^2) and burning rates of 0.068–0.136 $kg/(m^2 \cdot s)$ (50–100 $lb/(ft^2 \cdot h)$) or 475–950 $kJ/(m^2 \cdot s)$ (150,000 to 300,000 $Btu/(ft^2 \cdot h)$).

The most frequently encountered systems are:

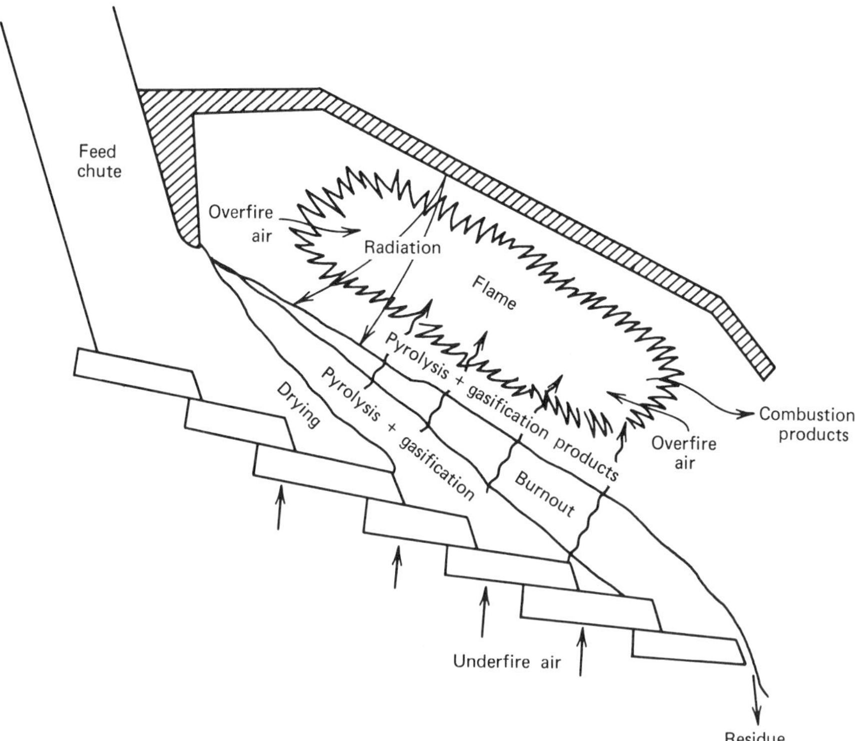

Figure 2. Schematic diagram of processes occurring on grate and in combustion chamber.

Traveling Grates. Traveling grates are continuous belt-like conveyors, ie, two or more grates are positioned at different elevations with the solids tumbling from one to another to provide agitation.

Reciprocating Grates. In a reciprocating grate system, movable and fixed sections alternate. The refuse is pushed forward by the fixed sections as the movable sections slide across them. This grate produces more agitation than the traveling grate, but slag formation on the surface interferes with the reciprocating action.

Rocking Grates. In rocking grates, each section is pivoted. Alternate rows are mechanically rocked to produce an upward and forward motion, thus advancing and agitating the solid waste. This type of grate provides more agitation than the reciprocating grate. Here again, slagging is likely, but generally maintenance difficulties are minimal.

Rotating Drum Grates. Sequential rotating drum grates provide good tumbling, rolling, and burning action.

Circular Grates. The circular grate is used in vertical, circular furnaces. It is commonly used with a central rotating cone grate with extended rabble arms that agitate the fuel bed.

Most modern incineration technology is based on European systems. They are divided evenly between five manufacturers: Martin (FRG), Vereinigte Kessel Werke (FRG), Von Roll (Switzerland), Bruin and Sorensen (Denmark), and Volund (Denmark).

For a fuller description of moving grates, see ref. 7. European and Japanese technology is reviewed in references 7–13. Additional information on incinerators is given in some general references (14–17); articles containing extensive bibliographies (16–17); and cost comparisons (19–21). Reference 17 discusses the use of refuse as a low sulfur fuel.

Multi-Chamber Incinerators. The multi-chamber incinerator is primarily used for commercial and industrial installations (1). These on-site incinerators with capacities up to a few hundred kilograms per hour handle small volumes of materials. The multi-chamber system provides maximum reduction of waste with minimum air contamination over a wide range of operating conditions. They are best used for Type 1 and Type 2 refuse (Table 3).

The combustion process proceeds in two stages: in the primary section the solid phase burns and volatile gases are driven off, and in the secondary section, these volatile gases are burned.

Grate loadings range from 2×10^{-2} to 3.4×10^{-2} kg/(m^2·s) (15–25 lb/(ft^2·h)), with burning rates from 1×10^{-3} to 1×10^{-2} kg/s (75–750 lb/h). The combustion of Type 2 waste often requires an auxiliary burner to maintain sufficient temperature for complete combustion. Large amounts of excess air as high as 300% are frequently used.

Nonconventional Incinerators. ***Suspension-Fired Units.*** The suspension incinerator requires refuse to be shredded to accelerate combustion to ca 2 s. The prepared waste is fed into the furnace by specially designed feeders that distribute it throughout the furnace. This waste is carried in suspension in the combustion air. As a result of the high surface area per unit mass, combustion is rapid. Large particles that do not burn fall on a moving grate where sufficient time is allowed for combustion. The Hamilton Ontario incinerator (22–23) uses suspension firing. The large pieces fall to a receiving grate on the bottom where they form a thin layer and burn.

Total Incineration. If maximum volume reduction is desired, the residue remaining after incineration may be liquefied to a slag at temperatures approaching 1650°C. Slagging incineration reduces volume up to 97.5% compared to 50–85% for conventional incineration. However, nitrogen oxide formation increases as a result of high temperature, auxiliary fuel may be required, and the operation is likely to be more complex. Furthermore, materials of construction are more expensive. Several such systems are under development (24).

Starved-Air Incinerators. In order to avoid the entrainment of excessive amounts of particulates from the burning refuse bed, air can be passed more slowly through the grate, producing a so-called starved-air incineration. In these units the combustion of the pyrolysis products leaving the fuel bed is not completed above the burning bed but in a secondary combustion chamber. Sometimes a secondary fuel supply is provided in the secondary combustion chamber for periods when wastes with low heating values are burned and during start-up and shut-down. The starved-air incinerator has been successfully employed in commercial operations.

Vortex Incinerator. The U.S. Bureau of Mines has developed a vertical incinerator for the burning of special wastes (25). The combustion air is injected tangentially above the burning bed, spirals down through the outside of the bed, and up through the inside. The burning rates reported for this design are only slightly lower, per unit cross-section of the incinerator, than the rates encountered in conventional grate units. The incinerator was originally developed for the treatment of paper wastes; with this or any waste yielding a finely subdivided ash, all inert material is carried over with the combustion products to the APC device. Such units therefore, can be operated continuously without provision for residue removal from the combustion chamber.

Refuse Beneficiation. It is extremely difficult to burn and recover useful energy from unsorted municipal waste, because of its heterogeneity in size, shape, chemical composition, and heating value. However, preparation of the waste before thermal treatment facilitates burning. Such pretreatment contributes to the front-end costs but reduces the furnace costs. The waste is upgraded by separation of nonorganic fraction and drying, shredding, and finally, densifying the solid. The fuels thus prepared are referred to as RDF (refuse-derived fuels) and are described in a number of publications (26–27).

Industrial Waste Incineration

In many medium-sized or large industrial plants, the costs of waste collection and disposal may warrant an on-site operation. Industrial waste is usually more homogeneous waste than municipal waste. It includes packaging, office, and process wastes and scrap lumber. For the chemical industry process wastes are often classified as hazardous, and a specially designed system may be required. The industrial system is usually smaller than the municipal incinerator. As a general rule, any system below 150 kg/h is uneconomical except for disposal of extremely hazardous wastes. The moving grate as well as other systems described for municipal waste may be employed. Many systems including rotary kilns, and multi-hearth and fluidized beds have been used for incineration of industrial waste. The multi-chamber incinerator described above is often used for treatment of certain industrial waste.

Rotary Kiln. The rotary kiln has been used to incinerate a large variety of liquid and industrial wastes (28–29). Any liquid capable of being atomized by steam or air can be incinerated, as well as heavy tars, sludges, pallets, and filter cakes. This ability to accept diverse feeds is the outstanding feature of the rotary kiln and, therefore, this type is often selected by the chemical industry.

The rotary kiln has no moving parts in the high temperature region. As a high capital cost installation, it is not practical for low feed rates. Since the motion of the kiln precludes the use of suspended brick, the refractory is susceptible to thermal shock damage and continuous operation should be maintained. Even under careful operation, rebricking large portions of the kiln is required annually. The system must be supported carefully and aligned to assure that there is no flexing of the unit as it rotates.

Heat recovery from the exhaust gas is a common practice. The rotary kiln is the most common unit for the destruction of chemical wastes on a commercial scale and is used by Dow (30), which built the first such system, 3M (31–32), Union Carbide (33), Eastman Kodak (34), as well as the principal contractor for destruction of toxic wastes, Rollins Environmental Services (35).

Multi-Hearth Furnace. Multi-hearth furnaces (28,36–38) are most often used for the incineration of municipal and industrial sludge and for the generation and reactivation of activated char (see Furnaces).

The main components of the multi-hearth are: a refractory-lined shell; a central rotating shaft; a series of solid flat hearths; a series of rabble arms with teeth for each hearth; an afterburner (possibly above the top hearth); an exhaust blower; fuel burners; an ash-removal system; and a feeding system.

The feed is normally introduced to the top hearth where the rabble arms and teeth attached to the central shaft rotate and spiral the solid across the hearth to the center where an opening is provided and the solid drops to the next hearth. The teeth of the rabble arms on this hearth spiral the solid toward the outside to ports that let the solid drop down to the next hearth. The solid continues downward, traversing each hearth until it reaches the bottom and the ash is discharged. The primary advantage of this system is the long residence time in the furnace which is controlled by the speed of the central shaft and the pitch of the teeth.

Burners and combustion air ports are located in the walls of the furnace to introduce either heat or air where it is needed. The air path is countercurrent to the solid, flowing up from the bottom and across each hearth. The top hearth operates at 310–540°C and dries the feed material. The middle hearths, at 760–980°C, provide the combustion of the waste, whereas the bottom hearth cools the ash and preheats the air. If the gas leaving the top hearth is odorous and detrimental to the environment, afterburning is essential.

The moving parts in such a system are exposed to high temperatures. The hollow central shaft is cooled by passing the combustion air through it.

Fluidized-Bed Incinerator. Fluidized-bed incinerators have been in commercial operation for less than thirty-five years. They are employed in the paper and petroleum industries, in the processing of nuclear wastes, and the disposal of sewage sludge. They are quite versatile and can be used for the disposal of solids, liquids, and gaseous combustible wastes (39–43). Several hundred fluid-bed installations are operating throughout the world for a variety of thermal processes. The world's largest sludge incineration of chemical waste is in Germany and consists of three fluid-bed reactors (44).

The basic fluid-bed unit consists of: a refractory-lined vessel; a perforated plate that supports a bed of granular material and distributes air; a section above the distributor containing granular solids referred to as the fluid bed; a space above the fluid bed referred to as the freeboard; an air blower to move the air through the unit; a cyclone to remove all but the smallest particulates and return to the fluid bed; an air preheater for thermal economy; an auxiliary heater for start-up; and a system to move and distribute the feed in the bed.

The air is distributed across the cross-section of the bed by the distributor and fluidizes the granular solids. Over a proper range of air-flow rates (usually 0.8 to 3.0 m/s), the solid becomes suspended in the air and moves freely through the bed. The fluidized bed has many desirable characteristics. Because of the movement of the particles, the bed operates isothermally and minimizes hot or cold regions. Large fluctuations in fuel quality are damped out as a result of this thermal capacity. The solid particles are reduced in the bed until they become small and light enough to be carried out of the bed (see Fluidization).

The diameter is limited to about 15 m and the depth ranges from 0.5 to 3 m. The bed material may be chosen to react with some impurity in the waste to remove it from the gas stream, eg, the removal of SO_2 from coal-combustion units. As a result of the excellent air-to-solid contact, the fluid bed may be operated at low excess air rates. The high heat-transfer rates allow large quantities of heat to be removed by a small heat-transfer area in the bed. Because of the isothermal operation (between 760 and 980°C) only small amounts of nitrogen oxides are formed.

The fluid bed is not effective in handling materials that contain components with a low ash-melting or softening temperature, since it is often difficult to distribute such materials over the bed cross section.

Table 5 compares several popular incineration processes for use with various wastes. If the waste has a high heating value, it may be desirable to remove heat from the bed. This reduces the need to cool with excess air. Extremely high temperatures are not desirable in any incinerator, because of nitrogen oxide formation and the effect on the materials of construction. When the waste contains large amounts of nitrogen, it is desirable to pyrolyze it at a low temperature to release the nitrogen and then complete the combustion. Several papers describe and compare the basic schemes discussed (45–49).

Thermal Conversion Processes

An alternative method of energy recovery from organic wastes by incineration is the conversion of organic waste into fuel forms that are acceptable to conventional combustion equipment. Gas of medium heating value may be substituted for natural gas in package boilers and pyrolytic liquids for fuel oil in utility boilers. A wide variety of thermal process systems are being developed that convert the waste to a more useful fuel form (50–51). These systems involve pyrolysis, gasification, or liquefaction, and have the advantages of: production of storable fuel as opposed to steam; recovery of char that may be used as fuel converted to activated carbon or synthetic gas; cost reduction of air pollution control because of lower gas volumes compared to incinerators; and utilization of existing boiler facilities by providing a substitute fuel.

Most North American systems use direct heat transfer, with the shaft kiln as the

Table 5. Matching Waste Type to Incineration Processes[a]

Waste type	Rotary kiln	Multi-hearth	Fluidized bed	Multi-chamber incinerator
Solids				
granular homogeneous	X	X	X	
irregular bulky (pallets, etc)	X			X
low melting (tars, etc)	X		X	
organic compounds with fusible ash constituents	X	X		
Liquids				
high organic strength aqueous wastes, often toxic	X[b]			
organic liquids	X[b]		X	
Solids/liquids				
waste contains halogenated aromatic compounds	X		X	
aqueous organic sludges	X[c]	X	X	

[a] Ref. 45.
[b] If equipped with auxiliary liquid-injection nozzles.
[c] Provided waste does not become sticky upon drying.

dominant reactor type. Oxygen-blown systems are rare. European systems more frequently use indirect heating methods, and the shaft kiln dominates. Japanese developers strongly favor fluidized beds with emphasis on indirect heating.

Occidental Petroleum Flash Pyrolysis. The Occidental Petroleum Company (52) has developed the flash pyrolysis system, originally the Garret Research and Development flash pyrolysis system (53). This process represents a truly pyrolytic system and produces a petroleumlike product. The heart of the unit is an entrained bed where the waste and recycled char are carried up through the system with a gas stream. Reactor temperature is held at 480°C. The feed is crushed to provide small-size particles which heat up quickly when introduced to the reactor. Control of gas velocity and reactor height regulates the residence time of both gas and solid. By limiting the contact time, the primary decomposition products that enter the gas are not subjected to further cracking. The exit gas is rapidly quenched, depositing liquid organic products. The gas is recycled to the process.

The lower heating value, highly corrosive behavior, instability during storage, and poor pumping characteristics of the liquid product make it a poor substitute for petroleum fuel.

A 200 t/d facility was built in El Cajon, California in 1976. The plant has never been operated as it was designed and is currently inactive.

Union Carbide Purox. The Purox system (54–55) was developed by Union Carbide Corporation and piloted in a 200 t/d facility at South Charleston, West Virginia. The key element is a vertical kiln, which is blown with pure oxygen at a rate of 0.2 t oxygen per ton waste. The resulting combustion temperatures are so high that they melt the noncombustible portion of the waste and the molten slag is continually tapped from the bottom.

The gas resulting from pyrolysis has a heating value of 11.2 MJ/m^3 (300 Btu/ft^3) and is not diluted with nitrogen. It can be burned in most combustion equipment with little or no retrofit. It may be used as a chemical feedstock and converted to methane and methanol if desired. Overall, 80% by weight of the solid waste is converted to gas and 20% to an inert residue. The volume of the residue is 2 to 3% of the input waste.

Landgard System. The Landgard pyrolysis system (56–57) was piloted by Monsanto Company and a full-scale unit was built to handle the waste for the city of Baltimore. The plant capacity was designed at 1000 t/d (10.5 kg/s). The heart of the unit is a rotary kiln pyrolysis reactor 5.5 m in diameter, 30.5 m in length, and rotating at 2 rpm. A mixture of air and oil is introduced into the kiln countercurrent to the solids flow. The oil burned together with a portion of the solid provides heat for the pyrolysis of the remaining waste. The gas produced has a heating value of 2.8–3.7 MJ/m^3 (75–100 Btu/ft^3). It enters an afterburner where additional air is added to complete combustion. The combustion products are used by a waste-heat boiler to produce steam that is utilized on site. Many scale-up problems were encountered. After extensive modification, the system continues to be operated by the city of Baltimore at reduced capacity. The steam generated is utilized by the Baltimore Gas and Electric Company.

Dual-Bed Processes. In Japan, two fluidized reactors are used by Tsukishima Kikai (58) in its 450 t/d plant that handles municipal waste near Tokyo, and Ebara Corp. is operating a 100 t/d plant in Yokohama City. The two-reactor system produces a fuel gas similar to the gas obtained from the Purox system but does not require an oxygen plant. The basic diagram is shown in Figure 3; one unit operates as a combustion unit and burns char formed during pyrolysis, whereas the other unit operates as a pyrolyzer unit. The temperature of the combustion unit is typically controlled at 927–977°C with the pyrolysis unit 100 to 150°C lower. The fluidized beds are made up of sand that circulates between the two beds. The sand transports the energy required for pyrolysis from the combustion unit to the pyrolysis unit. It does not need a heat-transfer surface as required by other indirectly heated systems.

Andco-Torrax. The Andco-Torrax system (59–60) uses high temperature slagging pyrolysis. The reactor unit is basically a vertical shaft furnace. The refuse is fed to the top without pretreatment except for shearing the oversize items to about 1 m. The ascending hot gases become an effective countercurrent heat exchanger. The refuse descends through three distinct zones: drying, pyrolysis, and combustion. The gas leaves the top at about 400–500°C and contains about 90% of the energy content of the refuse in the form of combustible gases, entrained particles, and sensible and latent heat. This gas passes immediately into a secondary combustion chamber where the gases are burned in a small amount of excess air. They pass then into a waste-heat boiler and generate steam. To reach the high temperatures required for slagging in the bottom of the gasifier (1650°C), the air is preheated to over 1000°C by passing a portion of the combustion products through regenerating towers. The ash leaves the gasifier as a black, glassy slag.

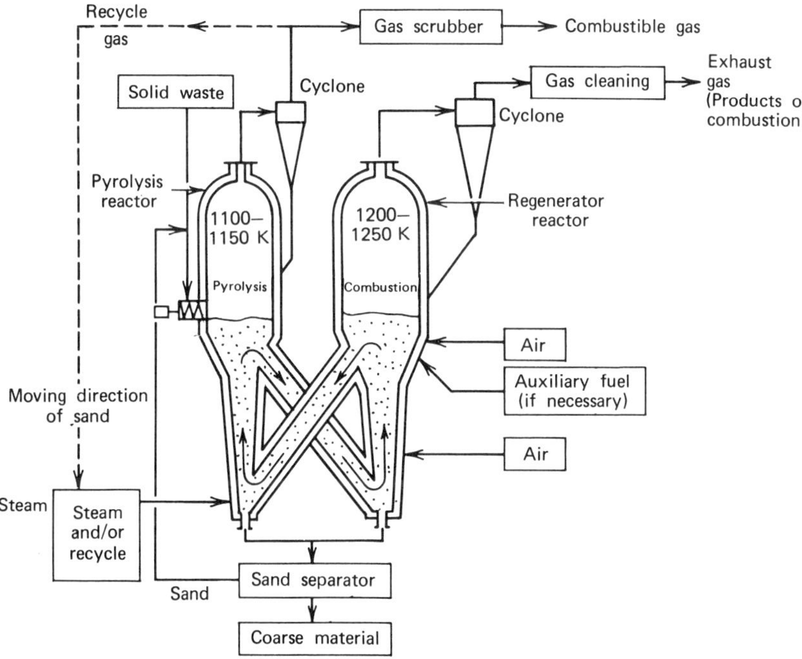

Figure 3. Two-fluidized-bed reactor system.

Incinerators for Liquid Wastes

Liquid waste requiring disposal by incineration seldom results from municipal waste collection and treatment systems. Such wastes generally originate from process operations, although spent lubricating oils may be occasionally disposed of in this fashion. In the 1940s, aqueous wastes, either contaminated or containing dissolved salts, were generally dumped into streams, impounded in ponds, or more recently injected into deep wells. Pollution regulations today generally prohibit the disposal of aqueous wastes containing toxic or hazardous compounds by any of these methods and evaporation of the water and destruction of the combustible compounds by incineration is becoming a costly necessity. Even aqueous wastes containing soluble, nontoxic inorganic compounds or nonbiodegradable organic compounds may require such treatment under increasingly stringent regulations.

Originally, only waste organic products, which were difficult to dispose of in other fashions, such as tarry still bottoms and corrosive liquids (chlorinated compounds) were disposed of in crude open-pit incinerators.

If pumpable and atomizable, the liquid would be sprayed into the refractory-lined pit and burned, often with the evolution of smoke and other products of incomplete combustion.

Liquid, not easily atomized, might be allowed to flow over the wall and cascade down over a series of refractory steps to aid in breaking up the liquid and increase its surface exposed to air. A sump at the bottom would collect the remaining unburned oil until it was all consumed. Burners fueled with gas or oil and firing down over the wall might be used either initially or continuously to provide ignition for the waste. Overfiring air jets were pointed downward to provide combustion air and to increase completeness of the combustion. With viscous feeds, a burner might be introduced through the side wall to provide additional heat for vaporization of the liquid and its ignition. Although much used in the past, such simple incinerators are becoming less acceptable today because of incomplete combustion and the release of substituted hydrocarbons and HCl, SO_2, and NO_x to the atmosphere without treatment. Where only unsubstituted hydrocarbons are to be burned and atomizing burners such as vortex burners can be used, such an incinerator still may have occasional application. In open-pit incinerators with viscous liquid bottom feed, combustion occurs by pool burning (61).

Atomizing Burners and Chambers. With the need for better control of the combustion process and the need to collect, quench, and scrub the combustion products, more complex furnaces and burners are being designed. Most liquids burn only after vaporization and the use of atomizing burners that break the liquid into 40 μm droplets or smaller are generally preferred. Since liquids burn slower than gases, longer flames usually result and larger combustion chambers, maintained above ignition temperatures, and longer residence times are required for complete oxidation (see also Burner technology). Pumpable liquids generally must have viscosities below 22 cm^2/s (22 St). For atomization, however, viscosities should be below 1.65 cm^2/s (1.65 St). Viscosity of the liquid is usually controlled by preheating, either in storage or with heat exchangers in the line to the burner. Where heating is impractical, admixture with a liquid of lower viscosity may be advisable. When heating is considered, possible undesirable chemical reactions should be investigated, including polymerization, oxidation, nitration, rapid decomposition, etc. Other pretreatment may include filtration, degassing, or neutralization, to prevent problems with pumping, burner pluggage, or corrosion.

Mechanical atomization includes pressure atomization and rotary methods such as the rotary cup. In the latter method, liquid is admitted through a hollow shaft to a rapidly spinning rotary cup. A thin film of liquid is centrifugally torn from the lip of the cup and droplets are formed. To achieve a conical flame, primary air at low pressure but high velocity in an annulus is directed axially around the cup. The ratio of air to liquid controls the spread of the flame and is adjusted to prevent impingement of the flame on the walls of the incinerator. This method requires little liquid pressurization and is ideal for atomizing liquids with high solids content. Burner turndown is about 5:1 and capacities from 1–280 cm^3/s (1–265 gal/h) are available. In single-fluid atomizing nozzle burners, the liquid is given a swirl as it passes through an orifice with internal tangential guide slots. Moderate liquid pressures of 0.7–1.03 MPa (100–150 psi) provide good atomization with low to moderate liquid viscosity. In the simplest form, the waste is fed directly to the nozzle but turndown is limited to 2.5 to 3:1 since the degree of atomization drops rapidly with decrease in pressure. In a modified form, involving a return flow of liquid, turndown up to 10:1 can be achieved. Secondary combustion air is generally introduced around the conical spray of droplets. Flames tend to be short, bushy, and of low velocity. Combustion tends to be slower as only secondary air is supplied and a larger combustion chamber is usually required. Typical burner capacities are in the range of 11–1100 cm^3/s (0.17–17.5 gal/min). Disadvantages of pressure atomization are erosion of the burner orifice and a tendency toward pluggage with solids or liquid pyrolysis products, particularly in smaller sizes.

The two-fluid atomizing nozzles may be of the low or high pressure nozzle variety, the latter being more common with high viscosity materials. In low pressure atomizers, air from blowers at pressures from 105–136 kPa (0.5–5 psig) is used to aid atomization of the liquid. A viscous tar, heated to a viscosity of 15–18 mm^2/s (15–18 cSt) requires air at a pressure of somewhat more than 112 kPa (1.5 psig) whereas a low viscosity or aqueous waste can be atomized with 105 kPa (0.5 psig) air. The waste liquid is supplied at a pressure of 132–222 kPa (4.5–17.5 psig). Burner turn-down ranges from 3:1 up to 6:1. Atomization air required ranges from 2.8 to 7.4 m^3/L (375–1000 ft^3/gal) waste liquid. Less air is required as atomizing pressure is increased. The flame is relatively short as up to 40% of the stoichiometric air may be admixed with the liquid in atomization. High pressure burners require compressed air or steam at pressures from 0.3–1.1 MPa (30–150 psig). Air consumption is from 0.6–1.6 m^3/L (80–215 ft^3/gal) waste and steam requirements may be 0.25–0.5 kg/L (2–4 lb/gal) with careful control of the operation. Turndown is relatively poor (3:1 or 4:1) and considerable energy is employed for atomization. Since only a small fraction of stoichiometric air is used for atomization, flames tend to be relatively long. The principal advantage of such burners is the ability to burn pumpable liquids without further viscosity reduction. Steam atomization tends to reduce soot formation of wastes that would normally burn with a smoky flame.

The burner, regardless of type, is frequently mounted in a refractory block or ignition tile. Its purpose serves to maintain ignition, confine the primary air, and ensure proper mixing with the atomized waste. The shape of the ignition tile cavity affects the shape of the flame and the quantity of primary air that must be introduced at the burner. When the waste contains a fusible ash content, considerable care must be given to design of the tiles and selection of the refractory material to prevent rapid fluxing of the tile and slag buildup. Some burners and tiles are arranged to aspirate hot combustion gases back into the tile which aids in vaporizing the liquid and increasing flame temperature more rapidly.

The burners must be placed in such a way that the flame does not impinge on refractory walls and that multiple burners, if provided, do not impact on each other. Secondary air, admitted through separate inlets, should be added to increase turbulence and speed combustion. Radiant refractory walls should be placed to aid maintenance of proper flame temperatures and fuel vaporization. Heat recovery or interchange surfaces, if provided, should not be located so as to cool the flame until after combustion is completed. Choice of materials for heat-recuperative devices is extremely critical when corrosive gases are generated as in the combustion of chlorinated organic compounds. Incinerator chambers for liquids are generally simple refractory-lined cylinders. In axial or side-fired nonswirling units, the burner is mounted either on the end firing down the length of the chamber or in a sidewall firing along a radius. Such designs, although simple and easy to construct, are relatively inefficient in their use of combustion volume. Table 6 lists typical heat-release rates per combustion volume (62). The lower range of these values applies to nonswirling designs. Improved utilization of combustion space and higher heat-release rates can be achieved with the utilization of swirl, vortex burners, or designs involving tangential entry. Vortex burners equipped for high pressure two-fluid atomization are suitable for burning organic liquids with up to 15% water without auxiliary fuel. When large quantities of aqueous waste are to be evaporated and incinerated, a high velocity gas or oil burner for supplementary fuel is mounted tangentially on the side of the combustion chamber. The combination jet from such a burner may issue at velocities of 90–150 m/s and temperatures of 1650°C. Aqueous wastes atomized with compressed gas are injected into the hottest part of this cyclonic combustion zone. Contact between the high velocity–high temperature combustion gases and the liquid waste provides instant mixing, very high heat transfer rates, and extremely rapid vaporization and ignition of the wastes. Such a design is suitable for handling wastes with more than 75% water content. In other commercial designs, such as Figure 4, separate fuel and waste guns are introduced through the side walls of the combustion chamber and secondary air is introduced tangentially. Designs with high turbulence can approach the higher heat-release rates given in Table 6. Such designs may be both horizontal and vertical (63).

Other incinerator applications are discussed in refs. 64–65. Liquid wastes that contain fusible ash may best be burned in horizontal combustion chambers designed for continuous or intermittent slag removal similar to a cyclone coal-combustion furnace or a vertical downflow-combustion chamber in which the gases make a 180° turn into a quench tower.

Control of Pollutant Combustion Products. An acceptable incinerator provides for the collection and removal of objectionable combustion products such as HCl, Cl_2, SO_2, SO_3, NO_x, and inorganic particulate fumes. Usually the hot gases are introduced into a quench chamber where the gases are cooled adiabatically by the evaporation of water. Typical designs utilize spray towers and water-cascading baffle towers. Under some circumstances, sufficient contact and gas absorption may occur that such gases can be released to the atmosphere after quenching. More frequently, additional control equipment may be required for efficient collection such as a packed section for gas absorption following cooling. Where in addition to pollutant gases, particulates or tars are present, a high energy venturi scrubber can be employed. For effluents containing

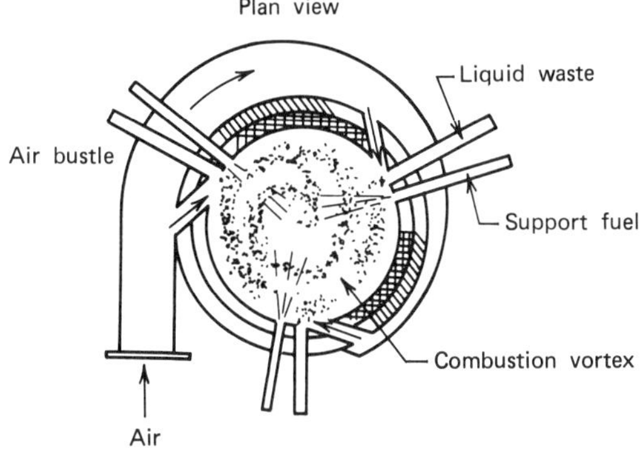

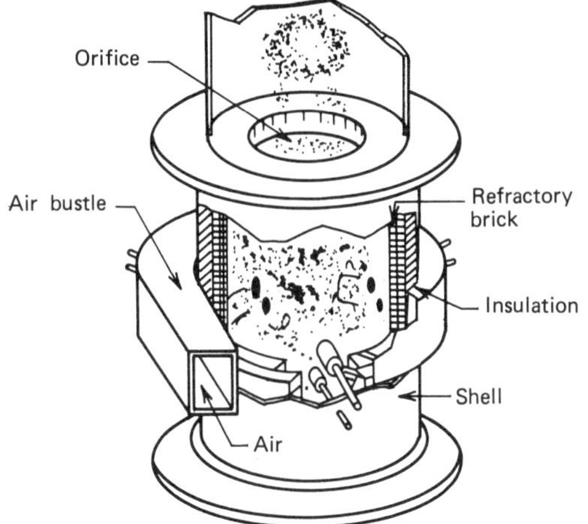

Figure 4. Commercial liquid incinerator with separate atomizing guns for fuel and waste liquid. Secondary air is admitted tangentially which creates high shear, turbulence, rapid mixing, combustion, and heat transfer.

Table 6. Typical Heat Release for Liquid Incinerator Combustion Chambers, per Unit Combustion Volume[a]

Temperature, K	kW/m^3 [b]
575–1075	35–150
1075–1375	150–410
1375–1675	410–580
1675–1925+	580–1050

[a] Recalculated from ref. 62.
[b] To convert kW/m^3 to $Btu/(ft^3 \cdot h)$ multiply by 96.7.

fine inorganic fumes, irrigated fiber beds, wet electrostatic precipitators, and other electrically augmented scrubbers are more energy efficient than venturi scrubbers (see also Air pollution control methods). Unless the waste contains nitrogen compounds, NO_x is obtained only from atmospheric fixation by combustion. The usual steps to reduce NO_x formation during combustion such as lower temperatures, less excess air, and two-stage combustion may not be acceptable for liquid wastes. A given temperature may be required for complete destruction. Utilization of less than the stoichiometric amount of air can result in soot formation which is very difficult to oxidize in a second combustion step (66).

Removal of many pollutants by scrubbing can be facilitated by neutralization. However, many of the resulting compounds are water soluble, creating a water pollution or disposal problem. Therefore, a number of specialized liquid-waste incinerators and treatment systems have been devised in which the recovered pollutant is obtained in a usable or salable form or concentration. In the Nittetu process for chlorinated waste liquid incineration, the waste is burned in a submerged combustion burner (67). Depending on the system, 15–20, 35, or 100% HCl can be recovered. A French system for burning waste with up to 70% chlorine content recovers 95% of the chlorine as HCl (68). An incineration system is available in which waste liquids are atomized horizontally through the wall into a downwardly directed auxiliary fuel-fired chamber (69). Organic compounds are cracked and oxidized, inorganic salts are quickly dried and decomposed to oxides, carbonates, sulfates, etc, which can then be recovered.

Auxiliary Facilities and Alternative Incineration Methods. Since liquid-waste incinerators may be operated for only a few hours per week, suitable tankage and piping facilities for storage, mixing, and blending are required. It is desirable to maintain the composition of the waste and its heating value more or less constant to minimize the need for changes in incinerator operating conditions and techniques. Therefore, various liquid wastes are frequently blended to some approximately constant composition. Liquids with high freezing points may need heating coils, insulation, and steam-traced or jacketed lines. Other liquids may undergo phase separation or solids sedimentation on standing and agitation or recirculation may be required.

Some wastes may be too viscous to pump even with heating, or too unstable chemically for flame burning. Others may have an ash or solids content too high for a liquid-waste incinerator. Frequently, such liquid wastes are disposed of in devices normally used for solid wastes such as fluidized beds and rotary kilns. For example, a waste containing 80% chlorine has been disposed of in a fluid-bed incinerator without the need for auxiliary fuel, very difficult to achieve in an incinerator for liquids (70). Incinerators are often needed for the disposal of other wastes such as solids, sludges, etc, and it is desirable to treat all wastes in one integrated operation. Liquids may be sprayed into solid-waste incinerators, burned in the incinerator as auxiliary fuel, sprayed on a solid-waste bed, or admixed with the solid-waste feed. Viscous wastes, and hazardous and corrosive materials may be placed in drums that are punctured just as they are charged to a solid-waste incinerator capable of handling bulky materials such as a rotary kiln. An integrated waste-disposal system has been described (71) for handling organic, halogenated, metallic, and aqueous wastes.

Incinerators for Vapors and Gases

Undesirable combustible gases and vapors can be destroyed by heating them to their autoignition temperature in the presence of sufficient oxygen to ensure complete oxidation to CO_2 and H_2O. Typical examples are hydrocarbon and organic vapors (particularly unsaturated compounds that are photochemically reactive in the atmosphere), CO, mercaptans, H_2S, H_2, NH_3, and long-lived chlorinated organic vapors which might build up in the environment.

Selection. Gas incinerators are applied to streams too dilute to support combustion. The gas composition is limited typically to 25% or less of the lower explosive limit. Gases that are sufficiently concentrated to support combustion are burned in flares, waste-heat boilers, in conjunction with other fuels in boilers and kilns, or used as process fuel. Occasionally, such gases may be burned in specially designed furnaces incorporating heat recovery. Protection against flame flashback must be provided between the incinerator and the process or source of the waste gas. Such protection is necessary to guard against unusual gas compositions even when the waste gas stream does not support combustion. Design of the flashback-prevention device depends on stream composition. Typical parallel-plate or multiple-screen flame arrestors, or a water seal are usually adequate but special precautions must be taken when the gas contains appreciable quantities of hydrogen. Knockout devices should be considered to prevent carrying slugs of combustible entrained liquid into the incinerator. Safety controls are desirable to protect against overheating caused by erratic flows and sudden composition changes. Noncombustible particles can present problems. They may be melted and collected as a corrosive liquid pool. Removal of particulates prior to incineration may be desirable, but if they are too fine (especially submicrometer in size), efficient removal may not be economical. In such cases, direct-flame incinerators specifically designed for slag removal may be preferred.

Before selecting an incinerator for gases, recovery of valuable organic solvents and hydrocarbons should be considered. Even if not reusable, recovered combustible liquids generally have fuel value. Existing combustion equipment such as a coal-fired boiler can be an excellent incinerator when passing the combustible gases, contained in excess air, through an incandescent bed of carbon. If the waste effluent is high in oxygen, it may be used in a combustion operation as primary or secondary air.

Temperature, time, and turbulence have been referred to as the three Ts of incinerator design (72). Increasing these variables (assuming adequate stoichiometric quantities of oxygen) can make combustion more complete. Data on incinerator kinetics as a function of these variables are needed for design. Reaction kinetics (73–74) have been reported for destruction of vinyl chloride, acrolein, ethyl acrylate, and benzene as a function of temperature and time (see Fig. 5) in a laboratory oven incinerator. Rate equations are also reported, eg, for benzene:

$$k = 2.51 \times 10^{22}\, e^{-99{,}650/RT}$$

where R = 8.314 J/mol or 1.987 cal/mol, and T = K. For maximum reliability this equation should be used for temperatures above 965 K. However, the degree of turbulence and speed of mixing could vary considerably with equipment size and design.

The degree of turbulence is related to time and temperature. For instance, the EPA generally requires direct-flame incineration at 980 K and 0.3 s residence time for nonhazardous pollutants. Proposed RCRA regulations specify direct-flame incineration at 1280–1480 K and 2-s residence time for hazardous materials, chlorinated organics, and possible cancer-suspect materials. Although increased turbulence can permit reduced temperature or time of incineration, this factor has been neglected in environmental regulations. Conversely, incinerator scale-up without adequate attention to turbulence can result in poor destruction even at desirable temperatures and residence times.

Two types of gas incinerators are in use: direct flame and catalytic. An indirectly

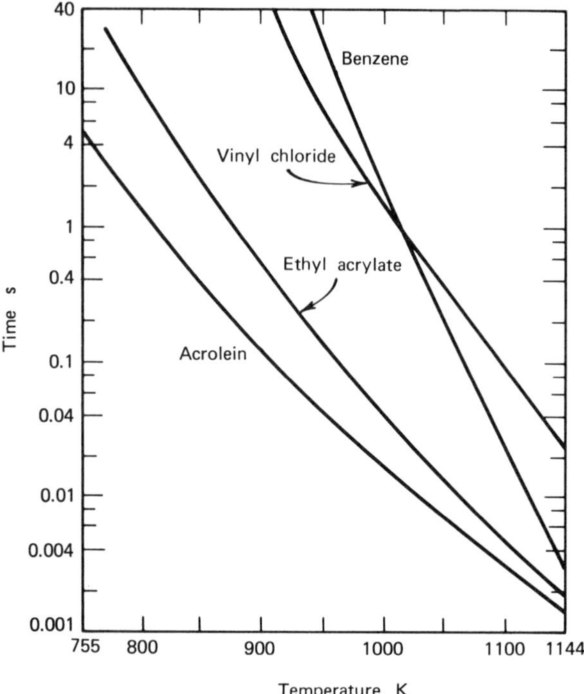

Figure 5. Time–temperature relations for 99.9% destruction of specific organic vapors in laboratory incinerator tests (73).

heated incinerator is feasible, but requires a higher temperature to secure ignition and complete oxidation, and thus offers little advantage. Catalytic incinerators are more compact than direct-flame incinerators, operate at lower temperatures, often require little fuel, and produce little or no NO_x from atmospheric fixation. However, the catalytic bed must be preheated and carefully temperature controlled. Thus, they are generally unsuited to intermittent and highly variable gas flows.

Direct-Flame Incinerators. In direct-flame incineration, the waste gases are heated in a fuel-fired refractory chamber to their autoignition temperature where oxidation occurs with or without a visible flame. The autoignition temperatures for common organic compounds are tabulated in ref. 75. A fuel flame aids mixing and ignition. Excess oxygen is required, since incomplete oxidation produces aldehydes, organic acids, carbon monoxide, carbon soot, and other undesirable materials. Naphtha vapor may be incinerated at 775–785 K whereas 1145 K may be required for certain aromatic compounds or streams containing a high percentage of inert material. The EPA has specified maintaining gases at 980 K for 0.3 s as generally adequate for gases that are not rated as hazardous or highly toxic. The effluent concentration of unoxidized organic compounds is often in the range of 40–50 ppm which would be unsatisfactory for materials requiring essentially total destruction. Increasing the temperature to 1200–1255 K and residence time to 0.7–1.0 s should result in efficient destruction of most pollutants. Operation at 1200 K and 0.7 s has been found adequate for benzene and HCN (76).

A simple direct-fired incinerator may be a refractory-lined furnace arranged for good mixing and fitted with a burner. Such an incinerator is low in capital cost and suitable for infrequent burning of a process purge gas during plant shutdown. However, such an incinerator in continuous service on other than a very small vent would have extremely high operating cost for fuel. Sufficient fuel must be provided to completely heat the incinerator gas to the ignition temperature and this heat is not further utilized. Some type of heat recovery is usually included. For waste-gas flows above 2.35 m^3/s (5000 ft^3/min), steam generation is the lowest cost heat-recovery method if there is a need for process steam. If steam is not needed, moderate to large size incinerators have a heat interchanger between the incoming gases and the hot combustion products. Because of the temperatures involved, heat-transfer surfaces are generally of alloy construction or ceramic materials. The latter can be damaged by thermal shock if sudden step changes in operation occur, whereas alloys may be attacked by corrosive gases (such as halogens and sulfates), if present. Typical heat-recuperation devices are finned gas exchangers, ceramic heat wheels, and Ljungstrom air preheaters. The cost of incinerators with heat interchange is appreciably greater than those designed for steam generation.

Turbulence is enhanced in direct-fired incinerators by imparting shear from gas streams directed in opposing directions or having differing velocities (such as jets introduced with cyclonic flow). Baffles (disk, doughnut, or checker grills) have the same effect, whereas radiant refractory surfaces in contact with the gas speed combustion through surface catalysis. For incinerator design, the composition of the waste gas must be known, and the heat of combustion and gas temperature rise must be calculated in order to choose refractories with adequate service temperature. Safety devices should be provided to shut off flows in the event of flame failure, to provide purging prior to restarting and ignition, and to limit peak temperatures.

Catalytic Incineration. Catalytic oxidation offers appreciable savings in fuel over direct-flame incinerators. However, catalytic incinerators are less suited to intermittent and highly variable flow rates. The catalyst can be deactivated or poisoned by certain gaseous compounds (see Exhaust control, industrial).

Adsorption Concentration for Incineration. The energy cost of heating large volumes of gas containing trace quantities of combustible organic material can be very high even with heat recovery. In many cases, operating costs can be reduced appreciably by concentrating the pollutants, especially when in the range of 20–100 ppmv, by carbon adsorption prior to incineration. The adsorbed organic compounds are stripped from the adsorbent with hot air under conditions that can effect up to a 40-fold increase in concentration. Comparative economics for flow streams of 3.8–28.3 m^3/s (8,000–60,000 ft^3/min) are presented in ref. 77. Adsorption concentration is efficient at hydrocarbon concentrations of 10–300 ppmv. Efficiency increases rapidly with increase in total gas volume

Flares. Flares are used for burning concentrated gases that support combustion such as hydrocarbon blow-down gases, tank venting, and emergency releases. They are located well above other structures. Pilot flames must be arranged to ensure ignition of all combustibles, especially when the flare handles only emergency releases. Design details that should be considered are: (*1*) pilots that stay lit and can be relit even in hurricane winds; (*2*) flare height and location selected to protect both personnel and the surroundings from the combustion heat release. Radiant heat can raise the temperature of combustible materials such as frame buildings and dry vegetation as well as burn personnel on elevated platforms. The flare flame can burn almost horizontally in strong winds; (*3*) Protection against flashback to the process, and production of explosive mixtures in the flare pipe from intermittent flows. Water seals at the base of the flare and molecular seals serve both purposes. A continuous flow of purge gas (inert or combustible materials) can prevent air backflow into the flare pipe; and (*4*) removal of entrained combustible liquids from the flare gas with centrifugal knockout drums to prevent the fall of burning liquid droplets from the flare.

For environmental reasons, burning should be smokeless. Long-chain and unsaturated hydrocarbons crack in the flame producing soot. Steam injection produces clean burning by eliminating carbon through the water–gas reaction. The quantity of steam required can be as high as 0.05–0.3 kg water per kg of gas burned. Since this quantity of steam is not always available, combustion air is injected into the flare flame. A multijet flare can also be used in which the gas burns from a number of small nozzles parallel to radiant refractory rods which provide a hot surface catalytic effect to aid combustion.

When burning hazardous vapors requiring destruction of the molecular species to less than 5–10 ppmv, an open flare may not maintain the combustible gases at a sufficiently high temperature for a sufficient period of time to meet such requirements. Either pretesting of flares should be made, or the flare should be enclosed in an open-ended refractory chamber to maintain combustion temperatures. For occasional emergency releases, an enclosure built of typical refractory brick still gives inadequate destruction until the refractory is heated to high temperatures. Maintaining such a refractory lining at operating temperature with an auxiliary fuel over a long period of time can be very expensive. Lightweight designs lined with high temperature refractory-fiber blankets which heat very rapidly are being investigated for such problems.

Particulate Aerosols. An incinerator to burn airborne suspensions of combustible particles such as coal and coke dust is a distinct possibility, but little research along such lines has been reported. Such materials could well be destroyed in a furnace of suitable size using a burner much like a pulverized-coal burner. The aerosol must be heated in the flame to the ignition temperature of the particles and sufficient residence time provided for their complete combustion. Since solids burn slower than gases, fairly long residence times would be required. If the suspended particles have an ash content, some control methods for ash recovery should be provided.

BIBLIOGRAPHY

1. R. C. Corey, ed., *Principles and Practices of Incineration,* Wiley-Interscience, New York, 1969.
2. W. R. Niessen, *Combustion and Incineration Processes,* Marcel Dekker, New York, 1978.
3. J. A. Sher, *Chem. Eng. Prog.* **67,** 81 (March 1971).
4. E. R. Kaiser, C. D. Zeit, and J. B. McCaffery, *Proceedings of National Incinerator Conference,* ASME, New York, p. 142, 1968.
5. Cudahy, *Proceedings Mid-Atlantic State Section Air Pollution Control Assoc. Conf. Newark, N.J., April 27, 1979.*
6. *Environ. Sci. and Technol.* **8,** 692 (Aug. 1974).
7. J. De Marco and co-workers, *Incineration Guidelines,* PHS Publication 2012, Cincinnati, Ohio, 1969.
8. D. J. De Renzo, ed., *European Technology for Obtaining Energy from Solid Waste,* Noyes Data Corporation, Park Ridge, N.J., 1978.
9. C. A. Rogus, *Public Works* **97,** 100 (May 1966).

10. H. Eberhardt, *Combustion* **38,** 8 (1966).
11. G. Stabenow, *Proceedings of National Incinerator Conference, New York,* pp. 278–286.
12. K. S. Feindler and K. H. Thoeman, *Proceedings of 1978 National Waste Processing Conference, New York, May 7–10, 1978,* pp. 117–157.
13. K. Matsumoto, R. Asukata, and T. Kawishima, *Proceedings of 1968 National Incinerator Conference, New York, May 5–8, 1968,* pp. 180–197.
14. B. Baum and co-workers, *Solid Waste Disposal,* Ann Arbor Science, Ann Arbor, Mich., 1973.
15. A. F. Sarafim in D. Wilson, ed., *Handbook on Solid Waste Management Section 6, Thermal Processing and Pyrolysis,* Van Nostrand Reinhold Co., New York, 1977, pp. 166–196.
16. National Center for Resource Recovery, *Incineration—A State of the Art,* Lexington Books, Lexington, Mass., 1974.
17. Energetics Company, *Systems Evaluation of Refuse as a Low Sulfur Fuel—Vol. II, PB 209 277,* Available through NTIS, 5285 Port Royal Road, Springfield, Va. 22151, Nov. 1971.
18. A. T. Warner, C. H. Parker, and B. Baum, *Plastics Solid Waste Disposal by Incineration or Landfill,* Manufacturers Chemist Association, 1825 Connecticut Ave., Washington, D.C., Dec. 1971.
19. Midwest Research Institute, *Resource Recovery—A Catalog of Processes, PB 214 148,* Available through NTIS, 5285 Port Royal Road, Springfield, Va. 22151, Feb. 1973.
20. W. C. Achinger and L. E. Daniels, *Proceedings of 1970 National Incinerator Conference, Cincinnati, Ohio, May 1970.*
21. Day and Zimmerman Associates, *Special Studies for Incineration in Washington, D.C., PHS No. 1748,* 1968.
22. R. C. Schwieger, *Power* S-1 (Feb. 1975).
23. R. S. Rochford and S. J. Witkowski, *Proceedings of National Waste Processing Conference, Chicago, Ill., June 1978,* pp. 45–59.
24. R. Zinn, C. LaMantla, and W. Niessen, *Ind. Water Eng.* **7,** 29 (July, 1970).
25. C. H. Schwartz, *Proceedings of National Incinerator Conference, New York, June 1972,* pp. 145–154.
26. "Refuse Derived Fuels—Energy for Industry," *National Center for Resource Recovery Bulletin,* Winter 1974, Spring 1975.
27. H. Hollander, *Proceedings of National Waste Processing Conference, Boston, Mass., May 1976,* pp. 67–79.
28. D. A. Hitchcock, *Chem. Eng.* **89,** 185 (May 21, 1979).
29. D. Ackerman and co-workers, *Destroying Chemical Wastes in Commercial Scale Incinerators, Facility Report No. 6, PB-270-897/2GA,* Rollins Environmental Service, Inc., Deer Park, Texas, 1978 (Available through NTIS, Springfield, Va.).
30. R. G. Novak, *Chem. Eng.* **77,** 78 (Oct. 5, 1970).
31. C. R. Lewis, R. E. Edwards, and M. A. Santoro, *Chem. Eng. Deskbook Issue,* 115 (Oct. 18, 1976).
32. C. R. Lewis, R. E. Edwards, and M. A. Santoro, *Proceedings of National Waste Processing Conference, Boston, Mass., May 23–26, 1976,* pp. 291–300.
33. J. M. Rhodes, *Proceedings of National Waste Processing Conference, Boston, Mass., May 23–26, 1976,* pp. 285–290.
34. R. E. Bastian and W. R. Seeman, *Proceedings of National Waste Processing Conference, Chicago, Ill., May 7–10, 1978,* pp. 557–565.
35. H. A. Alsentzer, *Chem. Eng. Prog.* **68,** 73 (Aug. 1972).
36. R. C. Novak and co-workers, *Chem. Eng.* **84,** 131 (May 9, 1977).
37. R. C. Detora, *Proceedings of National Waste Processing Conference, Boston, Mass., May 23–26, 1976,* pp. 313–328.
38. P. J. Cardinal and F. P. Sebastian, *Proceedings of National Incinerator Conference, New York, June 4–6, 1972,* pp. 290–299.
39. K. D. Becker and C. J. Wall, *Hydrocarbon Process.* **152,** 88 (Oct. 1973).
40. K. D. Becker and C. J. Wall, *Chem. Eng. Progr.* **72,** 61 (Oct. 1976).
41. R. C. Bailie and R. S. Burton, *Chemical Engineering Symposium Series* **68**(122), 140 (1972).
42. H. S. Kwon, *Proceedings of National Waste Processing Conference, Chicago, Ill., May 1978,* pp. 1–13.
43. Battelle Laboratories, "Fluidized Bed Incineration of Selected Carbonaceous Industrial Wastes," *State of Ohio, Dept. of Natural Resources and U. S. Environmental Protection Agency, Grant 12120 FYF,* March 1972.
44. *Chem. Week,* 39 (April 2, 1975).

45. *Pollut. Equip. News* (Oct. 1978).
46. C. A. Hescheles and S. L. Zeid, *Proceedings of National Incinerator Conference, New York, May 1972,* pp. 265–280.
47. T. T. Shen, M. Chen, and J. Lauber, *Pollut. Eng.* 45 (Oct. 1978).
48. P. N. Cheremisinoff and R. A. Young, *Pollut. Eng.* 20 (June, 1975).
49. P. N. Cheremisinoff and R. A. Young, *Pollut. Eng.* 29 (June, 1974).
50. J. Jones, *Chem. Eng.* **85,** 87 (Jan. 1978).
51. J. Jones, *Proceedings of National Waste Processing Conference, Chicago, Ill., May 1978,* pp. 387–396.
52. E. Cone in J. Jones and S. Radding, eds., *Am. Chem. Soc. Symp. Ser.* **76,** 287 (1978).
53. C. S. Finney and D. E. Garret, paper presented at *66th Annual Meeting of AIChE, Philadelphia, Pa., Nov. 1973,* Garrett Research Report No. 73-021.
54. J. E. Anderson, *Proceedings of National Incinerator Conference, Miami, Fl., May 1974,* pp. 337–346.
55. J. T. Fisher, M. L. Kasbohm, and J. R. Rivero, *Proceedings of National Incinerator Conference, Boston, Mass., May 1976,* pp. 125–132.
56. T. Bielski and A. C. J. Ellenberger, *Proceedings of National Incinerator Conference, Miami, Fla., May 1974,* pp. 331–346.
57. A. J. Helmstetier and D. B. Sussman, *Proceedings of National Waste Processing Conference, Chicago, Ill., May 1978,* pp. 465–475.
58. M. Hasegawa, J. Fukuda, and D. Kunii, *2nd Annual Pacific Chemical Engineering Congress, Denver, Col., August 28–31, 1977.*
59. P. E. Davidson and T. W. Lucas, Jr., in ref. 52, pp. 45–62.
60. E. Legille, F. A. Berczynski, and K. Guenther Heiss, *Conversion of Refuse to Energy, First International Conference and Technical Exhibition, Montreux-Switzerland, Nov. 3–5, 1975.*
61. W. R. Niessen, *Combustion and Incineration Processes,* Marcel Dekker, New York, 1978, p. 60.
62. Ref. 1, p. 285.
63. E. Skinner, *Australian Waste Conference Proceedings: Waste Management, Control, Recovery, and Reuse,* Ann Arbor Science Publ., Mich., 1975, pp. 113–116.
64. J. J. Santoleri, *Proceedings of the Second National Conference on Complete Water Reuse: Water's Interface with Energy, Air and Solids,* AIChE, New York, 1975, pp 457–466.
65. J. J. Santoleri, *Proceedings of the 22nd Annual Meeting of the Institute of Environmental Science,* Inst. of Environmental Sci., Mount Prospect, Ill., 1976, pp. 520–527.
66. G. S. Coulson and R. Harrison, *Recycling Waste Disposal* **2,** 203, 205 (Oct. 1977).
67. J. J. Santoleri, *CEP* **69,** 68 (Jan. 1973).
68. J. C. Zimmer and R. Guaitella, *Hydrocarbon Process.* **55,** 117 (Aug. 8, 1976).
69. S. W. Sim, *Chem. Process* **20,** 32 (May 5, 1974).
70. W. M. Walker, *Amdel Bull* (16), 41 (Oct. 16, 1973).
71. Y.-H. Kiang, *CEP* **72,** 71 (Jan. 1976).
72. *Air Pollution Manual. Part II: Control Equipment,* American Industrial Hygiene Assn., Detroit, Mich., 1968.
73. K. Lee, H. J. Jahnes, and D. C. Macauley, *Preprint 78-58.6, APCA 71st Annual Meeting, Houston, Tex., June 25–30, 1978.*
74. K. Lee, J. L. Hansen, and D. C. Macauley, *Preprint 79-10.1, APCA 72nd Annual Meeting, Cincinnati, Ohio, June 24–29, 1979.*
75. R. D. Ross, *CEP* **68,** 59 (Aug. 1972).
76. B. B. Crocker, *Reprint 360-376, Proceedings of the Conference on Control of Specific (Toxic) Pollutants, Feb. 1979,* APCA, Pittsburgh, Pa., 1979.
77. B. Grandjacques, *Pollution Eng.* **9,** 28 (Aug. 1977).

BURTON B. CROCKER
Monsanto Co.

RICHARD C. BAILIE
West Virginia University

INCLUSION COMPOUNDS. See Clathration.

INDANTHRENE, INDANTHRENE DYES. See Anthraquinone dyes.

INDENE. See Coumarone–indene resins under Hydrocarbon resins.

INDICATORS. See Hydrogen-ion concentration.

INDIGOID DYES. See Dyes, natural.

INDIUM AND INDIUM COMPOUNDS

Indium

Indium [*7440-74-6*], In, is an element of Subgroup IIIA in the periodic table between gallium and thallium. It is a soft, lustrous, silver-white metal, highly malleable and ductile with a face-centered tetragonal crystalline structure (a = 0.4583 nm, c = 0.4936 nm). Indium was discovered in 1863 in sphaleritic ore by Reich and Richter at the Freiburg School of Mines in Germany, and named for the indigo blue spectral lines that led to its identification.

The abundance of indium in the earth's crust is probably ca 0.1 ppm. It is found in trace amounts in many minerals, particularly in the sulfide ores of zinc and to a lesser extent in association with sulfides of copper and iron. Its content in these ores is often related to that of tin. Indium follows zinc through flotation (qv) concentration, and commercial recovery of the metal is achieved by treating residues, flue dusts, slags and metallic intermediates in zinc smelting and associated lead smelting (see Extractive metallurgy).

Properties. Table 1 lists many of the physical, thermal, mechanical, and electrical properties of indium. The highly plastic properties which are indium's most notable feature are due to deformation from mechanical twinning. It retains these plastic properties at cryogenic temperatures. Indium does not work-harden, can endure considerable deformation through compression, cold-welds easily, and has a distinctive "cry" like tin on bending.

Indium metal is not oxidized by air at ordinary temperature, but at red heat it burns to form the trioxide In_2O_3. On heating, indium also reacts directly with metalloids (arsenic, antimony, selenium, tellurium) and with halogens, sulfur, and phosphorus. It dissolves in mineral acids and mercury but is not affected by alkalis, boiling water, and most organic acids. The metal surface passivates easily. In general, the chemistry of indium in its trivalent compounds stems from nonionic or covalent bonding characteristics. Indium is electroplated easily from a variety of baths including cyanide (1), sulfate (2), fluoborate (3), and sulfamate (4) (see Electroplating).

Production. Indium usually is recovered from fumes, dusts, slags (5), residues (6) and alloys from zinc or lead–zinc smelting. The source material itself, a reduction bullion or electrolytic-slime intermediate, is leached with sulfuric or hydrochloric acid, the solutions are concentrated, if necessary, and crude indium is sponged on zinc or

Table 1. Physical Properties of Indium

Property	Value
atomic weight	114.82
atomic number	49
melting point, °C	156.6
boiling point, °C	2080
latent heat of fusion, kJ/mol[a]	3.27
latent heat of vaporization at bp, kJ/mol[a]	55.57
specific heat, J/(mol·K)[a] at 20°C	27.4
coefficient of linear expansion at 0–100°C, $\times 10^6$ per °C	25
electrical resistivity, μΩ·cm	
at 0°C	8.4
at 22°C	8.8
at 156.6°C	29.0
at 300°C	36
at 600°C	44
at 3.38 K	superconducting
electrode oxidation–reduction potential, V (H_2 = 0.0 V)	0.34
density, kg/m^3	
at 20°C	7.31
at 156.6°C	7.03
volume change in melting, % increase	2.5
thermal conductivity at 0–100°C, W/(m·K)	71.1
surface tension at temperature T, K between mp and bp, mN/m (= dyn/cm)	$602 - 0.1\ T$
stable isotopes, wt % (relative abundance)	
^{113}In	4.23
^{115}In	95.77
vapor pressure p in kPa[b] at temperature T K, between mp and bp $\log_{10} p = 9.835 - 12{,}860/T - 0.7 \log_{10} T$	
valence	usually 3, but also 2 and 1
thermal neutron cross-section at 2200 m/s, m^2/atom[c]	
absorption	$(190 \pm 10) \times 10^{-28}$
scattering	$(2.2 \pm 0.5) \times 10^{-28}$
Brinell hardness number, HB	0.9
tensile strength, MPa[b]	2.645
elongation, %	22
modulus of elasticity, GPa[b]	10.8

[a] To convert J to cal, divide by 4.184.
[b] To convert kPa to atm, divide by 101.3; to convert MPa to atm, divide by 0.101; and to convert GPa to atm, divide by 1.01×10^{-4}.
[c] To convert m^2 to barn, multiply by 10^{28}.

aluminum. Solvent extraction with di(2-ethylhexyl) phosphoric acid (7–9) or tributyl phosphate (10) is effective in removing indium from dilute solution. Another recovery procedure is precipitation of indium phosphate [*14693-82-4*] selectively from slightly acidic solution, conversion to oxide by leaching in strong caustic soda solution, and then reduction of oxide to the metal (11). Indium, distilling with zinc in zinc retort smelting processes, concentrates in the zinc–lead bottom metal of the first stage evaporation–reflux purification of zinc, and can be separated by chlorination under cover of a suitable molten salt mixture (12). Procedures are reported for separation of indium from lead–tin alloys (13–14).

Sponge indium is refined to standard grade and thereafter to higher purity by soluble-anode electrolysis in aqueous halide systems. The anode indium may be as shot or in amalgam. The refining may be supplemented by heating under vacuum for removal of volatile impurities. Much of the recovery and refining technology used by producers is unpublished and unpatented.

Economic Aspects. Production of indium has been reported from the following countries (15): Belgium, Canada, FRG, Italy, Japan, Netherlands, Peru, UK, United States, and USSR.

Production reported from Belgium, the Netherlands, and the UK is thought to be from imported concentrates, residues, and scraps. Reliable statistics on indium cannot be obtained since many producers do not report their production. The U.S. Bureau of Mines estimate of world production in 1977 was 40 metric tons (1,300,000 troy oz), a drop of 40% from 1970.

Major shifts in production during recent years have been the decline in Asarco's (U.S.) and Cominco's (Canada) production, due presumably to changes in their ore supplies. This shortfall has to a large extent been offset by increased production from Hoboken-Overpelt (Belgium) and the joint venture of the Indium Corporation of America and the New Jersey Zinc Company (U.S. and France).

Indium pricing has been based on the so-called standard grade material (99.97% pure) with higher purity grades (99.99%, 99.999%, and 99.9999%) available at a premium. The first quotation in September, 1930, was $15.00/g. As supplies became available the price decreased, holding in the range of $48–64/kg until 1972 when it began a steady rise to a peak of $337/kg in January, 1977. After a modest falloff in 1977 and early 1978, the price of indium began to firm again in late 1978 and has risen to $620/kg (March, 1980).

Analysis. Indium can be detected to 0.01 ppm in spectroscopic analysis by its characteristic lines in the indigo blue region of wavelengths 4511.36, 4101.76, 3256.09, and 3039.36 nm. Procedures for the quantitative determination of indium in ores, compounds, alloys, and for the analysis of impurities in indium metal are covered thoroughly in ref. 16.

Safety and Handling. Physiologically indium is a nonessential element (see Mineral nutrients). It is classified as toxic but there have been no reported cases of systemic effects in human exposure to indium. The threshold limit value of the ACGIH is 0.1 mg/m^3. The primary toxic effects of ionic indium are on the kidneys. Indium has no demonstrated irritating effect on skin but there may be effects on the respiratory system. Ingestion through digestion is ca 0.5% of intake and through respiration about 5%. There have been no reports of accidental or industrial poisoning, indicating that normal safety and containment measures are adequate.

Alloys. Indium alloys with a wide range of metals, and many binary and ternary systems have been studied extensively (17). Indium generally increases the strength, corrosion resistance, and hardness of a system to which it is added. Low-melting-point alloys of indium with lead, tin, bismuth, and cadmium with melting points as low as 47°C are used in surgical casts, patternmaking, lens blocking, and electrical fuses. The alloy In 24%–Ga 76% is molten at 16°C and is used in nuclear reactors to circulate gamma activity, and In 5%–Ag 15%–Cd 80% is used in control rods. Tin–indium and lead–tin–indium solder alloys have melting points in the range 100–300°C and offer good corrosion resistance to alkalies. The 50–50 tin–indium alloy wets glass, quartz, and many ceramics. Indium is also used in specialty brazing alloys with copper, silver, and gold, and in dental alloys.

Uses. Indium's first major use, and still an important outlet for standard-grade metal, was in the production of bearings for heavy duty and high speed service (see Bearing materials). The addition of indium increases strength and hardness, improves resistance to corrosion, and results in improved antiseizure properties. Indium-containing bearings are used in piston-type aircraft engines, high-performance automobile engines (particularly in Europe), and diesel engines.

The solder and alloy market, including the low-melting-point or fusible alloys, is the largest single use of indium (see Solders and brazing alloys). Indium's unique properties, including improved corrosion resistance, increased hardness and fatigue resistance, and general improvement in soldering and braze bonding have promoted its use in this field.

The supplanting of germanium-based semiconductor (qv) devices by silicon devices has almost eliminated the use of indium in the related alloy junctions. However, the loss of this semiconductor market has been offset to some extent by increasing use of indium-containing solders in silicon-device fabrication, where their improved fatigue resistance and reduced gold-leaching, as compared to lead–tin solders, offer advantages.

Applications for electroplated indium coatings include corrosion protection and production of low resistance contacts on, in particular, aluminum electric wiring.

Low pressure sodium-vapor lamps employ an interior indium oxide coating on the lamp envelope for improved efficiency and luminosity.

Indium is used in gaskets for cryogenic applications and for glass-to-metal seals. Indium oxide and tin-doped indium oxide films are used as conductive coatings on glass and ceramics and are a key ingredient in a broad range of electrooptical devices.

In the nuclear field, indium is used in neutron-monitoring badges and in reactor control-rod alloys.

Other minor uses include temperature standards and specialized electroluminescent panels.

A substantial amount of indium continues to be used in research and development, indicating continued investigation of the unique characteristics of the metal and its semiconducting compounds.

Indium Compounds

The usual valence of indium is three, although monovalent and bivalent compounds of indium with oxygen, halogens and Group VI elements are well known. The lower valence compounds tend to disproportionate into the trivalent compound and indium metal; the trivalent compounds are stable.

Halides. Indium trichloride [*10025-82-8*], $InCl_3$, can be made by heating indium in excess chlorine or by chlorinating lower chlorides. It is a white, crystalline solid, very deliquescent, soluble in water, and has a high vapor pressure. $InCl_3$ also forms chloroindates, double salts with chlorides of alkali metals and organic bases.

Indium dichloride [*13465-11-7*], $InCl_2$, is made by heating indium to 200°C in hydrogen chloride or by reduction of $InCl_3$ in H_2/HCl below 600°C. It forms colorless crystals. Indium monochloride [*13465-10-6*], InCl, is formed by passing $InCl_3$ vapor over heated indium. Both the lower chlorides decompose in water to indium and $InCl_3$.

Mono-, di-, and trivalent bromides [*14280-53-6; 21264-43-7; 13465-09-3*] and iodides [*13966-94-4; 13779-78-7; 13510-35-5*] may be made by methods similar to the chlorides. The lower-valence salts also disproportionate in water. Indium trifluoride [*7783-52-0*], InF_3, is sparingly soluble in water. It forms an ammonium double salt, $3NH_4F.InF_3$ [*15273-84-4*], which decomposes on heating to indium nitride [*25617-98-5*], InN.

Oxides. Four oxides of indium are known, the trioxide [*1312-43-2*], In_2O_3; suboxide [*12030-22-7*], In_2O; monoxide [*12136-26-4*], InO; and the sesquioxide [*66525-54-0*], In_3O_4, which is probably a mixture of In_2O_3 and InO. There are a great many compound oxides of indium with other metals. In_2O_3 is formed by burning indium in air or by calcining the carbonate, nitrate or sulfate. It is pale yellow when cold, red-brown when hot. The yellow form dissolves easily in acids but the red-brown form is relatively insoluble. In_2O_3 is easily reduced to metal by hydrogen and other reducing gases at temperatures in the 400–500°C range. Reduction under carefully controlled conditions gives the lower oxides. In_2O_3 has been used to color glass yellow. A hydrated oxide, often referred to as the hydroxide and written $In(OH)_3$ [*56108-30-6*], is precipitated as a gelatinous compound by raising the pH of indium salt solutions. It calcines to In_2O_3.

Sulfates. Indium metal and its oxides dissolve in warm sulfuric acid to give a solution of the trisulfate [*13464-82-9*], $In_2(SO_4)_3$. It is a white, crystalline, deliquescent solid, readily soluble in water that forms double salts with alkali sulfates and some organic substituted ammonium bases. Concentration of the acidified trisulfate solution precipitates the acid sulfate [*57344-73-7*], $In(HSO_4)_2$, and other reaction conditions give basic sulfates.

Sulfides. The main sulfide of indium is In_2S_3 [*12030-24-9*], which can be prepared by heating the metal with sulfur or by precipitation from weakly acid solutions of indium salts by H_2S. Precipitated In_2S_3 varies in color from yellow to red-brown and in crystal size depending on formation conditions. It dissolves in acids and sodium sulfide solution. Other reported sulfides of indium are InS [*12030-14-7*], a red-brown solid, In_2S [*12196-52-0*], and In_4S_5 [*12142-00-6*].

Other Salts. Indium nitrate [*13770-61-1*], $In(NO_3)_3.3H_2O$, is a very soluble salt prepared by dissolution of the metal or oxide in nitric acid. Dehydration at 100°C gives the monohydrate [*13770-61-1*]. Indium phosphate [*14693-82-4*], $InPO_4$, is precipitated by adding phosphate ions to a solution of an indium salt. It is very insoluble in water.

Organic Compounds. Best known are the indium alkyls, trimethyl [*3385-78-2*], $In(CH_3)_3$; triethyl [*923-34-2*], $In(C_2H_5)_3$; and triphenyl [*3958-47-2*], $In(C_6H_5)_3$ which may be made by treating indium with the corresponding mercury alkyls, or by the Grignard reaction. The first two when pure ignite spontaneously in air. The oxalate [*38051-36-4*], $In_2(C_2O_4)_3$, and methoxide [*40521-21-9*], $In(OCH)_3$, also are known (see Organometallics).

Intermetallic and Semiconducting Compounds. Indium forms intermetallic compounds with a great many metals and combinations of metals including alkali metals, magnesium, the iron group, rare earths and precious metals.

Indium also combines with nonmetallic elements and with metalloids such as N, P, Sb, As, Te, Se. Many of the latter compounds are semiconducting as are the oxide and sulfide. These are discussed in more depth under Semiconductors (qv). Indium antimonide [*1312-41-0*], InSb, indium arsenide [*1303-11-3*], InAs, and indium phosphide [*22398-80-7*], InP, are the principal semiconducting compounds. They are all prepared by direct combination of the highly purified elements at elevated temperature under controlled conditions.

BIBLIOGRAPHY

"Indium" in *ECT* 1st ed., Vol. 7, pp. 834–839, by A. A. Smith, Jr., American Smelting and Refining Co.; "Indium and Indium Compounds" in *ECT* 2nd ed., Vol. 11, pp. 581–584, by A. A. Smith, Jr., American Smelting and Refining Co.

1. U.S. Pat. 1,965,251 (July 3, 1934), W. S. Murray and D. Gray (to Oneida Community Ltd.).
2. H. B. Linford, *Trans. Electrochem. Soc.* **79,** 443 (1941).
3. U.S. Pat. 2,409,983 (Oct. 22, 1946), W. M. Martz (to General Motors Corp.).
4. U.S. Pat. 2,458,839 (Jan. 11, 1949), J. R. Dyer, Jr., and T. J. Rowan (to Indium Corporation of America.).
5. B. G. Hunt, C. E. T. White, and R. A. King, *Can. Min. Met. Bull.* **52,** 359 (1959).
6. U.S.S.R. Pat. 501,094 (Jan. 30, 1976), L. S. Gelsin and co-workers (to Almalyk Mining-Metallurgical Combine).
7. J. M. Auchapt and J. Tostain, *Commis. Energ. At. (France) Report,* CEA, 1975, Bib 218(1) and 218(2).
8. A. Cornea and T. Segarceanu, *Rev. Chim. Bucharest* **23,** 673 (1972).
9. Sato and Taichi, *J. Inorg. Nucl. Chem.* **37,** 1485 (1975).
10. M. Golinski in J. G. Gregory, ed., *Proceedings International Solvent Extraction Conference,* Vol. 1, pp. 603–615, 1971.
11. U.S. Pat. 2,241,438 (May 13, 1941), C. Zischkau (to American Smelting and Refining Co.).
12. U.S. Pat. 2,433,770 (Dec. 30, 1947), Y. E. Lebedeff (to American Smelting and Refining Co.).
13. U.S.S.R. Pat. 537,521 (July 5, 1978), V. E. D'Yakov, T. I. Tumanova, V. D. Nikitina, and I. I. Andreev (to Novosibirsk Tin Combine).
14. Ger. Offen. 2,711,508 (Sept. 21, 1978), V. E. D'Yakov and co-workers (to Novosibirsk Tin Combine).
15. E. A. Peretti in M. T. Ludwick, ed., *Indium,* 2nd ed., Indium Corp. of America, Utica, NY, 1959, pp. 24–112.
16. C. E. T. White, *Minor Metals Survey, Metal. Bull. (London)* 47 (1977).
17. J. M. Ramaradhya, R. C. Bell, S. Brownlow, and C. J. Mitchell in S. D. Snell and L. S. Eltre eds., *Encyclopedia of Industrial Chemical Analysis,* John Wiley & Sons, Inc., New York, Vol. 19, 1971, p. 518.

General References

M. T. Ludwick, *Indium,* 2nd ed., Indium Corporation of America, Utica, N.Y., 1959 (a comprehensive bibliography).

"Indium" in C. A. Hampel, ed., *Encyclopedia of the Chemical Elements,* Reinhold Book Corporation, New York, 1968.

E. F. MILNER
Cominco Ltd.

C. E. T. WHITE
Indium Corporation of America

INDOLE

Indole [*120-72-9*] (**1**) (1-*H*-indole) is a benzopyrrole in which the benzene and pyrrole rings are united through the 2,3-positions of the pyrrole. The indole nucleus is found in a large number of naturally occurring compounds; the compound constitutes about 2.5% of jasmine oil and 0.1% of orange-blossom oil, and in both cases, it contributes to their fragrances (see Oils, essential). It is of commercial importance as a component of perfumes (qv). Although synthetic methods have been described for the manufacture of indole (**1**), extraction from the 240–260°C coal-tar distillate fraction is the only commercial source.

Isoindole [*270-68-8*] (**2**) (1-*H*-isoindole), the isomer in which the benzene and pyrrole rings are fused through the 3- and 4-positions of the pyrrole, is not stable. A few of its derivatives are known, the simplest being *N*-methylisoindole.

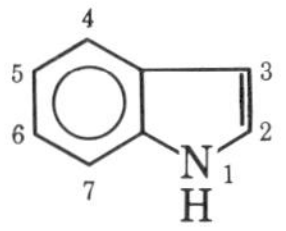

(**1**) 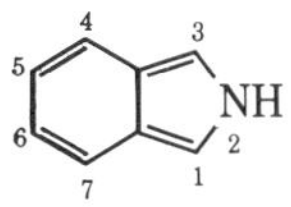 (**2**)

Indole was first obtained (and its structure elucidated) in 1866 by Adolf von Baeyer. The original preparation was by zinc-dust pyrolysis of oxindole which had been obtained by the reduction of isatin, an oxidation product of the natural dyestuff, indigo. Interest in indole chemistry revived about 1930 when it was discovered that the essential amino acid, tryptophan, the plant growth hormone, heteroauxin, and several groups of important alkaloids are indole derivatives (see Amino acids; Plant growth substances; Alkaloids). It was shown that 3-methylindole (skatole) is produced with indole during pancreatic digestion or putrefactive decomposition of proteins and, hence, both are found in the intestines and feces. Interest has centered on medicinal and biochemical aspects of indole chemistry. Serotonin, which has been identified as a metabolite in brain chemistry; the psychotomimetic indoles, psilocin and psilocybin from mushrooms; the tranquilizer reserpine; and the melanin pigments are a few of the compounds that have been studied (see Hormones, brain oligopeptides; Psychopharmacological agents).

Properties

Indole is a colorless crystalline solid (mp 52–54°C, bp (dec) 254°C). The heat of combustion at constant volume is 4.268 MJ/mol (1020 kcal/mol). The molecule is planar and has only moderate polarity. Indole has good solubility in a wide range of solvents including petroleum ether, benzene, chloroform, and hot water. The solubility in cold water is only 1:540 at 25°C; thus, water is a good solvent for purification by recrystallization.

Indole forms salts with high concentrations of both strong bases and strong acids. With strong bases, the anion is stabilized by delocalization of the electron over the aromatic system and indole, with a pK for deprotonation of 16.97, is a stronger acid than simple amines (2). Conversely, because the lone pair of electrons on the nitrogen atom is an integral part of the aromatic system, indole is a much weaker base when combined with strong acids than the alkylamines or pyridine and gives a pK value of −3.5 for protonation (3). The protonation occurs at the N-1- or C-3-positions and gives the indolium ion (**3**) or indoleninium ion (**4**).

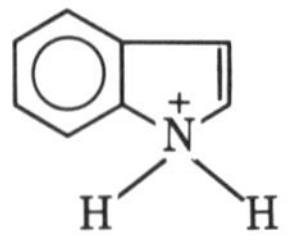

(**3**) 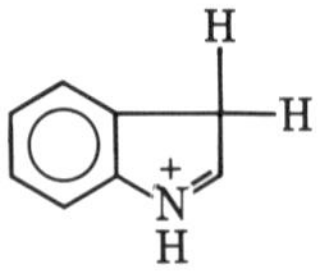 (**4**)

Spectroscopic evidence indicates that in strong sulfuric or perchloric acid, protonation occurs at the C-3 position (4). The formation of dimers (**5**) and trimers (**6**) complicates the situation and, except for 2-substituted indoles (which, for steric reasons, do not dimerize), an equilibrium exists between the indole, the dimer, the trimer, and their salts (5).

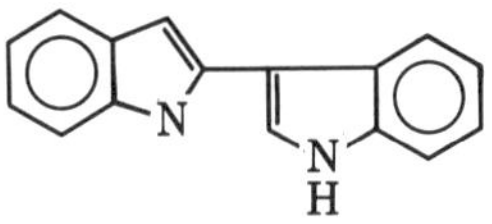

(**5**) 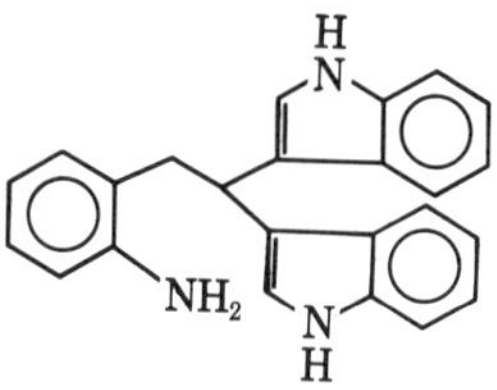 (**6**)

This oligomerization under acid conditions accompanies many reactions of indole, especially those involving electrophilic substitution.

The most important resonance structures for indole are (**7**) and (**8**):

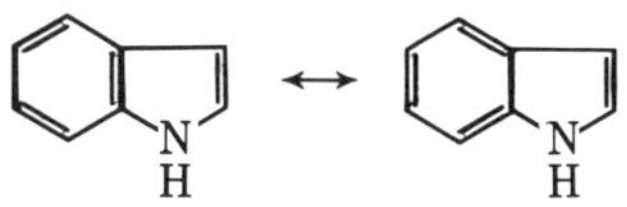

(**7**) (**8**)

The 2,3 double bond is intact and thereby preserves the aromaticity of both the benzene and the pyrrole systems.

Structures (**9**) and (**10**), in which the resonance of the benzene ring is intact, are the most important of the charge-separated contributors.

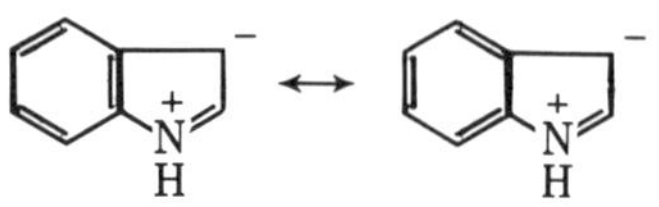

(**9**) (**10**)

If the C-3 atom is unsubstituted, C-3 is the preferred site for the attack of an electrophilic reagent. If C-3 is substituted, attack occurs at the N-1-position. However, many of the reactions on this nitrogen are reversible and products of substitution at C-2 are found. If both the C-2- and C-3-positions are blocked, or if the C-2 is substituted (preventing dimerization) and the C-3 is protonated, the substituents may enter the benzene ring.

Reactions. ***Halogenation.*** Direct bromination or chlorination of indole with bromine or chlorine is not possible because of the dimerization reaction that occurs under acid conditions. The method most frequently used for the preparation of 3-bromoindole was the bromination of N-benzoylindole with bromine or N-bromosuccinimide followed by hydrolysis of the benzoyl group (6). Newer methods are based on the use of mild brominating agents, eg, dioxane dibromide or pyridinium bromide perbromide with an acid scavenger, eg, pyridine (7).

Nitration. Indole may be nitrated in the 3-position by the action of benzoyl nitrate in acetonitrile with a consequent yield of 35% (8). If the indole carries a substituent on the 2-position, direct nitration in sulfuric acid is possible because the dimerization reaction is suppressed. However, as a result of the protonation of the C-3 atom which deactivates the pyrrole system, the nitro group enters the benzene ring in the 5-position (9).

Sulfonation. Sulfonation of indole with pyridine–sulfur trioxide complex below 50°C gives indole-1-sulfonic acid. At higher temperatures indole-2-sulfonic acid is formed by intramolecular rearrangement (10).

Diazo Coupling. Diazonium compounds couple with indole in the 3-position. For example, 4-nitrobenzenediazonium chloride couples with indole to form 3-(4′-nitrophenylazo)indole at pH 4–6 in almost quantitative yield.

Alkylation. The products of alkylation reactions of indole are complex because they are subject to both kinetic and equilibrium control. With methyl iodide, eg, initial attack is on the 3-position to give first 3-methylindole then 3,3-dimethylindolenine. This is followed by equilibration between the 2- and 3-positions to form 2,3-dimethylindole which, in turn, is attacked to give 2,3,3-trimethylindolenine and, finally, 1,2,3,3-tetramethylindoleninium iodide (11). The latter compound loses HI, when treated with aqueous alkali, to form 1,3,3-trimethyl-2-methyleneindoline, a widely used intermediate for the manufacture of cationic dyestuffs (see Dyes and dye intermediates).

Acylation. Acetylation of indole with acetic anhydride gives a mixture of 1-acetylindole and 1,3-diacetylindole (12). The reaction of oxalyl chloride and indoles to form 3-glyoxalyl chloride derivatives (11) is important in the synthesis of tryptamines (13).

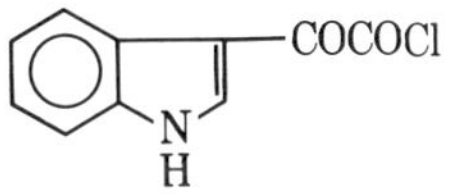

COCOCl
N
H

(11)

Acyl halides react with the alkali salts of indole to form mainly the 1-acyl derivative; although these may rearrange to the 3-substituted compounds at high temperatures. The action of acylating agents on indoylmagnesium bromide in ether to form the 3-alkanoyl derivatives is indicative of the covalent nature of the N—Mg bond in the solvent (14). The products of acylation with ethyl oxalyl chloride and chloroacetyl chloride also are intermediates for the synthesis of tryptamines.

Mannich Reaction. The high electron density at the 3-position enables indole to take part in Mannich reactions with an aldehyde and a secondary amine (15). Thus, with formaldehyde and dimethylamine, gramine (**12**) is formed:

$$\text{indole (N–H)} + HCHO + HN(CH_3)_2 \longrightarrow \text{3-}CH_2N(CH_3)_2\text{-indole (N–H)}$$

(**12**)

Mild conditions, eg, warming the reactants in ethanol, usually are sufficient.

Michael Reactions. Compounds having an olefinic linkage in conjugation with an electron withdrawing group, eg, a carbonyl, cyano, or nitro group, react with indole under neutral conditions in the 3-position if that position is free and in the 2-position if the 3-position is blocked (16). Thus, methyl vinyl ketone (**13**) gives 1-(3′-indoyl)-butan-3-one (**14**):

$$\text{indole (N–H)} + H_2C{=}CHCOCH_3 \xrightarrow[CH_3CO_2H]{(CH_3CO)_2O} \text{3-}CH_2CH_2COCH_3\text{-indole (N–H)}$$

(**13**) (**14**)

Under alkaline conditions, attack is at least partially on the *N*-position; although, at higher temperatures, the *C*-3-derivative alone is formed (17).

Vilsmeier–Haack Formylation. Indole-3-carboxyaldehydes and 3-acylindoles have been prepared by the action of the complex formed between phosphorus oxychloride or phosgene and a tertiary amide upon the indole (18). With dimethylformamide and indole, indole-3-carboxyaldehyde (**15**) is formed:

$$\text{indole (N–H)} + \begin{matrix} POCl_3 \\ HC({=}O)N(CH_3)_2 \end{matrix} \longrightarrow \text{3-}CHO\text{-indole (N–H)}$$

(**15**)

Gattermann and Houben-Hoesch Reactions. These hydrochloric acid-catalyzed reactions for the preparation of indole-3-carboxyaldehydes and ketones by reaction with zinc cyanide and organic nitriles, respectively, are limited in application to those indoles having substituents in the 2-position because they do not dimerize under acid conditions.

Reactions with Carbenes. When indole is treated with chloroform and potassium hydroxide or potassium ethoxide (Reimer-Tiemann reaction), indole-3-carboxaldehyde and some 3-chloroquinoline are formed (19). It generally is accepted that this reaction involves a chlorocarbene intermediate. Reactions with diazoesters and diazoketones in the presence of copper powder are believed to proceed through the formation of a carbene. For example, when heated with ethyl diazoacetate, indole forms ethyl indole-3-acetate (**16**):

$$\text{indole (N–H)} + {:}CHCO_2C_2H_5 \longrightarrow \text{3-}CH_2CO_2C_2H_5\text{-indole (N–H)}$$

$$N_2CHCO_2C_2H_5 \xrightarrow{Cu} {:}CHCO_2C_2H_5$$

(**16**)

Oxidation. Autoxidation of indole by air and light yields first indoxyl (**17**) which can react further to give indigo or leucoindoxyl red (**18**) or indoxyl red (**19**) and finally the trimer (**20**) (20).

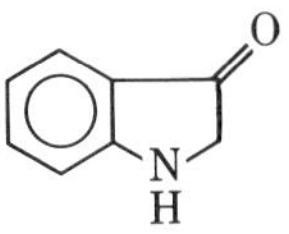

(**17**) 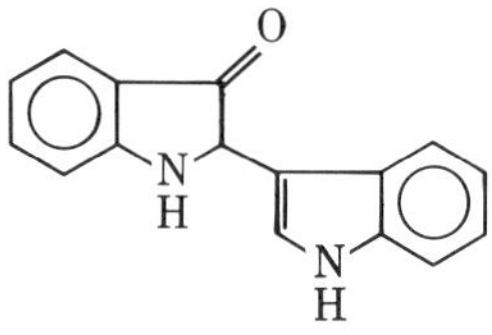 (**18**) 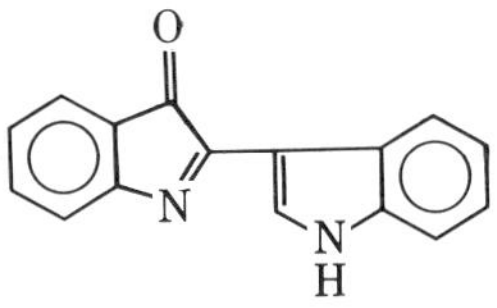 (**19**) 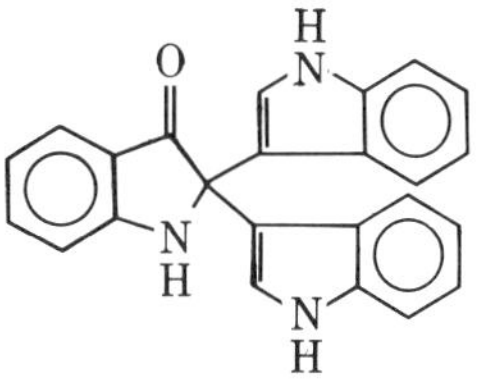 (**20**)

Chemical oxidation of indole with relatively mild reagents gives indoxyl (**17**), indigo, and some indirubin (**21**), which presumably is formed by the reaction of indoxyl with the by-product isatin (**22**) (21).

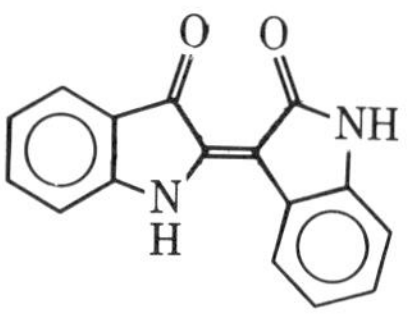

(**21**) 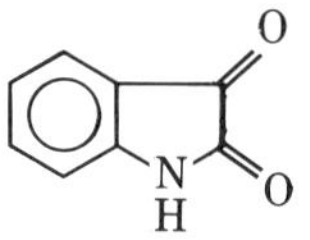 (**22**)

Stronger oxidizing agents cleave the pyrrole ring and form 2-formamidobenzaldehyde (**23**) and 2-formamidobenzoic acid (**24**) (22):

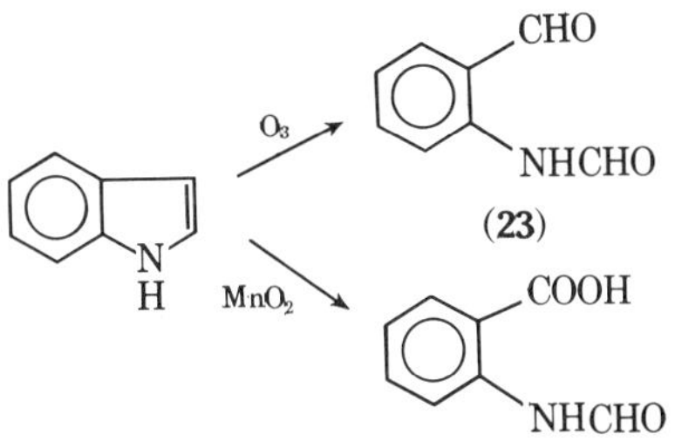

O_3 → (**23**) CHO, NHCHO

MnO_2 → (**24**) COOH, NHCHO

Reduction. The high resonance energy and abundance of electrons make indoles resistant to reduction under neutral conditions. Catalytic reduction at 8.6 MPa (85 atm) and 90–100°C with Raney nickel gives indoline (**25**), whereas, at 25.3 MPa (250 atm) and 250°C, the product is octahydroindole (**26**) (23). Under acid conditions, protonation gives the indolinium cation which is easily reduced but which also is susceptible to polymerization. For reduction to be successful, the faster reaction must occur. This is the case with indole which yields indoline (**25**) in good yield at RT and atmospheric pressure in a mixture of ethanol and aqueous fluoroboric acid with platinum as the catalyst (24). Chemical reducing agents are not effective in neutral solution but under acid conditions (eg, zinc in phosphoric acid) indolines are formed (25). Alkali metals in liquid ammonia, which are very powerful reducing systems, attack the benzene but not the pyrrole ring to form 4,7-dihydroindole (**27**) and 4,5,6,7-tetrahydroindole (**28**) (26).

(**25**) (**26**) (**27**) (**28**)

Preparation

The methods used for the synthesis of indoles involve the formation of the pyrrole ring by the cyclization of an appropriately substituted aniline derivative. In the Fischer indole synthesis, the phenylhydrazone of an aldehyde, ketone, or ketonic acid is converted to a substituted indole by heating with an acid catalyst, such as zinc chloride, sulfuric acid, formic acid, acetic acid, or boron trifluoride. For example, 2-methylindole (**30**) (1980 price: ca $100/kg) is obtained from acetone phenylhydrazone (**29**) (27):

$$(\mathbf{29}) \xrightarrow[\text{catalyst}]{\text{acid}} (\mathbf{30}) + NH_3$$

This synthesis is widely applicable to the preparation of indole derivatives that are substituted on the benzene or pyrrole rings, but fails for indole itself.

The Bischler indole synthesis involves the reaction of an arylamine with an α-haloketone, α-hydroxyketone, or α-anilinoketone (28):

$$RC_6H_4NHR^1 + XCH(R^2)COCH_2R^3 \longrightarrow \text{(indole, 2-}CH_2R^3\text{, 3-}R^2\text{)} + \text{(indole, 2-}R^2\text{, 3-}CH_2R^3\text{)}$$

where R = R^1 = H or alkyl; R^2 = R^3 = H, (substituted) alkyl or (substituted) aryl; and X = halogen, OH, or NHAr. This reaction has been used for the preparation of a number of tetrahydrocarbazoles from anilines and 2-chlorocyclohexanones, and *N*-substituted indole-3-acetic acids and acetamides which are useful for the preparation of tryptamines.

In the Madelung indole synthesis, *N*-acylated *o*-alkylanilines are cyclized in the presence of a strong base such as sodium ethoxide or sodamide (29). Thus, *o*-acetotoluidide (**31**), when heated with sodamide, forms 2-methylindole (**32**) in good yield:

$$(\mathbf{31}) \xrightarrow[200\,^{\circ}C]{\text{base}} (\mathbf{32})$$

The Reissert indole synthesis is a convenient method for the synthesis of indole and indoles substituted in the benzene ring (30). In this method, *o*-nitrophenylpyruvic acid (**35**) [prepared from *o*-nitrotoluene (**33**) and diethyl oxalate (**34**)] is reductively cyclized to indole-2-carboxylic acid (**36**), which readily decarboxylates:

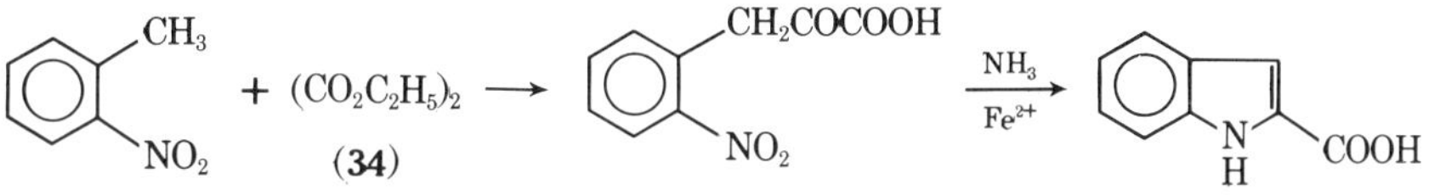

CH_3, NO_2 (**33**) + $(CO_2C_2H_5)_2$ (**34**) → $CH_2COCOOH$, NO_2 (**35**) $\xrightarrow[Fe^{2+}]{NH_3}$ N H COOH (**36**)

Many other syntheses of indole compounds have been reported. An efficient synthesis of indole (**1**) and its derivative beginning with lithiation at the methyl group of *o*-tolyl isocyanide (**37**) follows (31):

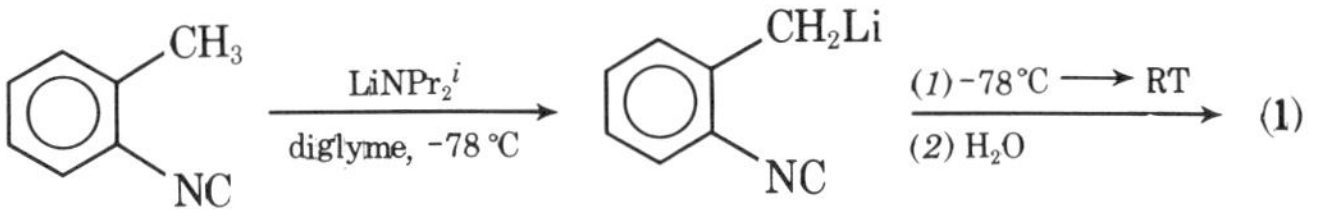

CH_3, NC (**37**) $\xrightarrow[\text{diglyme, -78 °C}]{LiNPr_2^i}$ CH_2Li, NC $\xrightarrow[(2)\ H_2O]{(1)\ -78\,°C \rightarrow RT}$ (**1**)

Indole can be prepared from reduction of *o*-nitro-β-dimethyl aminostyrene (**38**) (32). Compound (**38**) is a condensation product of *o*-nitrotoluene (**33**) and a DMF acetal. Derivatives of indole substituted at *C*-2 can be made by the procedure (33).

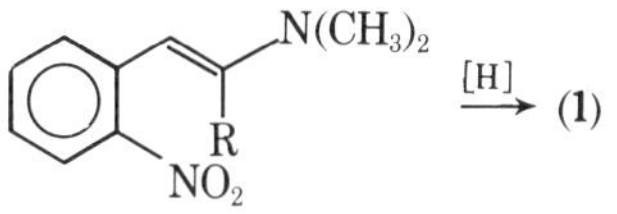

$N(CH_3)_2$, R, NO_2 $\xrightarrow{[H]}$ (**1**)

(**38**)
R = H

The Nenitzescu method (34) involves the reaction of *p*-benzoquinone with ethyl β-aminocrotonate to give 5-hydroxy-2-methylindole-3-carboxylic acid. Other methods are based on the oxidation of β-(hydroxyphenyl)ethylamines (35). Such specialized syntheses are valuable in the oxidation of biologically important compounds, eg, tyrosine and adrenaline, and in the formation of the melanin pigments (36).

Indolines have been used as a source for indoles (37). The introduction of substituents into the indole nucleus by direct reaction is limited to a few cases. Indolines may be regarded as ortho-alkylated arylamines which undergo the usual electrophilic substitution reactions of the latter compounds. Subsequent dehydrogenation of the indolines gives indoles.

The 1980 price of indole is \$46/kg.

Derivatives

Indole is the parent substance of a large number of important compounds that occur in nature. Tryptophan (2-amino-3-(3′-indolyl)propionic acid) (**39**) (1980 price of DL-tryptophan is ca \$200/kg) is one of the naturally occurring essential amino acids. Higher plants degrade tryptophan to heteroauxin (indole-3-acetic acid) (**40**) (1980 price: ca \$300/kg), a plant hormone (see Plant growth substances).

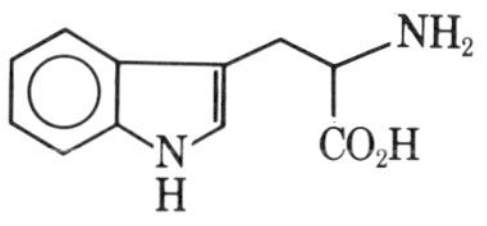

NH_2, N H, CO_2H (**39**)   N H, CO_2H (**40**)

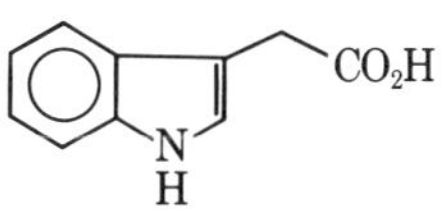

The compounds 3-(3′-indoyl)propionic acid (**41**), indole-3-pyruvic acid (**42**), and the 1-, 2-, and 5-methylindole-3-acetic acids possess similar activity.

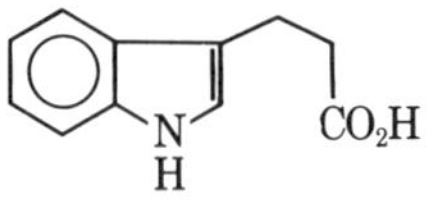

(**41**) 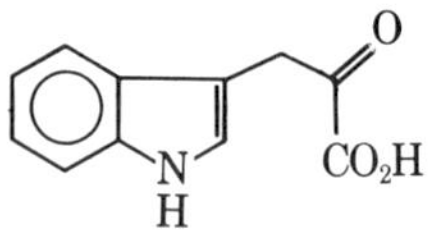 (**42**)

Bacteria degrade tryptophan to tryptamine (2-(3′-indoyl)ethylamine) (**43**), which is the basis for some of the condensed ring alkaloids.

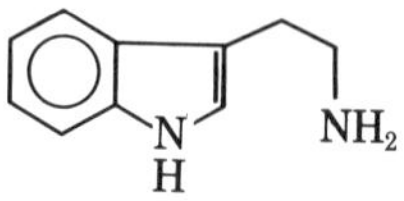

(**43**)

Gramine (3-(dimethylaminomethyl)indole) (**44**), which occurs in sprouting barley, is typical of the simpler indole alkaloids. The indole skeleton is found in the more complex monoterpene-derived, calycanthus, evodia, carboline, and ergot alkaloids. The latter group of over six hundred compounds includes strychnine and reserpine (see Alkaloids).

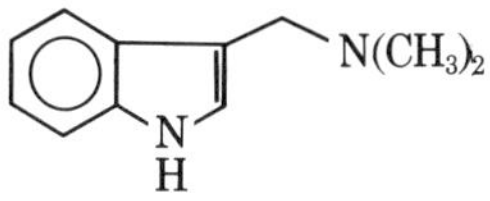

(**44**)

Indole compounds that carry substituents, especially a hydroxy group, on the benzene ring include serotonin (**45**), which is a vascoconstrictor hormone that plays a part in conducting impulses to the brain.

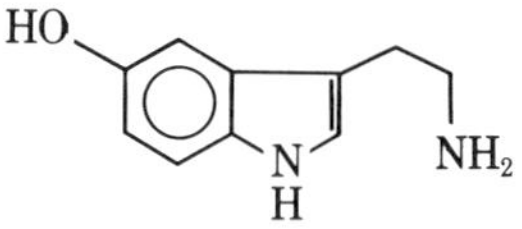

(**45**)

Bufotenine (**46**) is found in the skins of toads, toxic mushrooms, and West Indian snuff. Psilocybin (**47**) occurs in certain mushrooms. Both are known for their psychotropic effects.

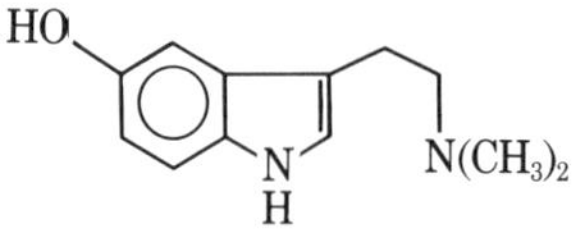

(**46**) 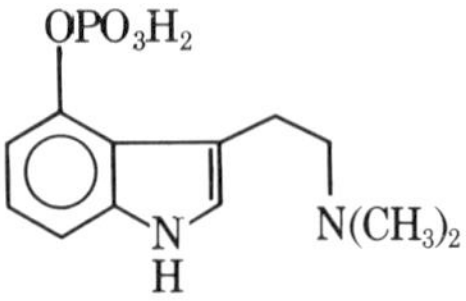 (**47**)

Derivatives of indole carrying a hydroxy or keto group on the pyrrole ring include indoxyl (**48**).

OH

N
H

(**48**)

Indican is the glucoside of indoxyl which occurs in *Indigofera* from which natural indigo was obtained by fermentation (qv) and oxidation. Synthetic methods for the manufacture of indigo also proceed through the intermediate, indoxyl (see Dyes, natural).

Isatin (**22**) (1980 price: $45–60/kg), an oxidation product of indigo, is hydrolyzed under alkaline conditions to *o*-aminophenylglyoxylic acid (**49**) which reacts with chloroacetone to form 3-hydroxy-2-methylquinoline-4-carboxylic acid (**50**). This is a convenient method for expanding the five-membered indole to the six-membered quinoline heterocyclic system.

O
CO_2Na
NH_2

(**49**)

COOH
OH
N
H
CH_3

(**50**)

BIBLIOGRAPHY

"Indole" in *ECT* 1st ed., Vol. 7, pp. 839–847, by W. C. Sumpter, Western Kentucky State College, and F. M. Miller, University of Maryland; "Indole" in *ECT* 2nd ed., Vol. 11, pp. 585–595, by W. C. Sumpter, Western Kentucky State College, and F. M. Miller, University of Maryland.

1. U.S. Pat. 2,442,952 (June 8, 1948), G. C. Kitchens (to Burton T. Bush, Inc.).
2. G. Yagil, *J. Chem.* **71,** 1034 (1967).
3. R. L. Hinman and J. Lang, *J. Am. Chem. Soc.* **86,** 3796 (1964).
4. R. L. Hinman and E. B. Whipple, *J. Am. Chem. Soc.* **84,** 2534 (1962).
5. G. F. Smitt in A. R. Katritzky, ed., *Advances in Heterocyclic Chemistry,* Vol. 2, Academic Press, Inc., New York, 1953, p. 300.
6. N. P. Buu-Hoi, *Justus Liebigs Ann. Chem.* **556,** 1 (1944).
7. K. Piers, C. Meimaroglou, R. V. Jardine, and R. K. Brown, *Can. J. Chem.* **41,** 2399 (1963).
8. G. Berti, A. Da Settimo, and E. Nannipieri, *J. Chem. Soc. C.,* 2145 (1968).
9. W. E. Noland, L. R. Smith, and K. R. Rush, *J. Org. Chem.* **30,** 3464 (1965).
10. A. P. Terent'ev and L. V. Tsymbal, *C. R. Acad. Sci. USSR* **55,** 833 (1947).
11. G. Plancher, *Chem. Ber.* **31,** 1496 (1898).
12. C. Zatti and A. Ferratini, *Chem. Ber.* **23,** 1359 (1890).
13. M. E. Speeter and W. C. Anthony, *J. Am. Chem. Soc.* **76,** 6208 (1954).
14. B. Oddo, *Gazz. Chim. Ital.* **43,** 190 (1913).
15. B. Reichert, *Die Mannich-Reaction,* Springer, West Berlin, FRG, 1959.
16. J. Szmuszkovicz, *J. Am. Chem. Soc.* **79,** 2819 (1957).
17. H. E. Johnson and D. G. Crosby, *J. Org. Chem.* **28,** 2030 (1963).
18. G. F. Smith, *J. Chem. Soc.,* 3842 (1954).

19. C. W. Rees and C. E. Smithen in A. R. Katritzky, ed., *Advances in Heterocyclic Chemistry,* Vol. 3, Academic Press, Inc., New York, 1964, p. 57.
20. P. Seidel, *Chem. Ber.* **83,** 20 (1950).
21. A. G. Perkin and F. Thomas, *J. Chem. Soc.* **95,** 793 (1909).
22. B. Witkop and H. Fiedler, *Justus Liebigs Ann. Chem.* **558,** 91, 1947; B. Hughes and H. Suschitzky, *J. Chem. Soc.,* 875 (1965).
23. C. B. Hudson and A. V. Robertson, *Aust. J. Chem.* **20,** 1935 (1967).
24. A. Smitt and J. H. P. Utley, *Chem. Commun.,* 427 (1965).
25. L. J. Dolby and G. W. Gribble, *J. Heterocyclic Chem.* **3,** 124 (1966).
26. S. O'Brien and D. C. C. Smith, *J. Chem. Soc.,* 4609 (1960).
27. W. J. Houlihan in A. Weissburger and E. C. Taylor, eds., *The Chemistry of Heterocyclic Compounds,* Vol. 25, Part 1, Wiley-Interscience, New York, 1972, p. 232.
28. Ref. 27, p. 317.
29. Ref. 27, p. 385.
30. Ref. 27, p. 396.
31. Y. Ito, K. Kobayashi, and T. Saegusa, *J. Am. Chem. Soc.* **99,** 3532 (1977).
32. Ger. Offen. 2,057,840 (May 21, 1971), A. D. Batcho and W. Liemgruber (to Hoffmann-La Roche).
33. E. E. Garcia and R. I. Fryer, *J. Heterocyclic Chem.* **11,** 219 (1974).
34. Ref. 27, p. 413.
35. Ref. 27, p. 436.
36. Ref. 27, p. 437.
37. Ref. 27, p. 461.

D. W. Bannister
Toms River Chemical Corp.

INDOPHENOL. See Sulfur dyes.

INDULINES. See Azine dyes.

INDUSTRIAL ANTIMICROBIAL AGENTS

Industrial antimicrobial agents are chemical compositions that are used to prevent microbiological contamination and deterioration of commercial products, materials, and systems. Their areas of application include cosmetics (qv), disinfectants and sanitizers, wood (qv) preservation, food and animal feeds, paint (qv), cooling water, metalworking fluids, hospital and medical uses, plastics and resins, petroleum, pulp (qv) and paper (qv), textiles (qv), latex, adhesives (qv), leather (qv) and hides, and paint slurries (1–4) (see Disinfectants and antiseptics; Food additives; Latex technology; Pet and other livestock feeds; Petroleum; Plastics, processing; Resins, natural; Water, treatment of swimming pools). The following general classes of antimicrobial agents are recognized in industry: phenolics; halogen compounds; quaternary ammonium compounds (qv); metal derivatives; amines (qv), alkanolamines (qv), and nitro derivatives; anilides; organosulfur and sulfur–nitrogen compounds; and miscellaneous compounds (see also Antibacterial agents; Antibiotics; Fungicides (agricultural)).

A given antimicrobial agent may either destroy all the microbe cells present or just prevent their further proliferation to numbers that would be significantly destructive to the substrate or system being protected. In sharp contrast to certain medicinal situations that require complete sterility, most industrial applications demand only that potentially destructive microorganisms are inhibited from proliferating to a detrimental level (see Sterile techniques). The terms, microbes and microorganisms, refer primarily to bacteria and fungi. Both of these groups are subdivided into two general subclasses: Gram-positive and Gram-negative bacteria (after Hans C. J. Gram, Danish researcher) and, among the fungi, molds and yeasts (qv). A number of use-orientated terms include antibiotic (which refers to a medicinal or therapeutic agent), preservative, disinfectant, antiseptic, antifoulant, slimicide, and mildewcide. A preservative is used for a variety of products, eg, food, cosmetics, paint, etc, and a disinfectant is employed for household needs, dairy machinery, etc.

Modes of Action

The mechanisms by which chemical agents exert antimicrobial activity depend upon effective contact between the chemical and the microorganism and involve disruptive interaction with a biochemical or physical component of the organism which is essential to its structure or metabolism (5–6). The targets may be a single enzyme, a cell membrane, intracellular systems, cytoplasm, or combinations of these, and the nature of the action is dependent on the organism and the environment in which the interaction occurs as well as on the antimicrobial agent. Quaternary ammonium agents exert a surface-active influence on the cytoplasmic membrane and thereby upset the osmotic balance and permeability that are essential to the life of the cell (7). Phenols apparently precipitate proteins (qv), and glutaraldehyde [*111-30-8*] and formaldehyde [*50-00-0*] (qv) act by alkylation of amino and sulfhydryl groups of proteins (8). Certain metals and chelating ligands interfere with enzyme systems (9).

Economic Aspects

Approximately 125 different industrial biocide products are supplied by ca 100 different companies (10). The growth in consumption of industrial antimicrobial agents in the United States is represented (according to application) in Table 1 (11–12).

The EPA regulates the use of any chemical substance that is manufactured, processed, or distributed for use as a pesticide (13–15). In compliance with a body of regulations promulgated by the EPA (16), a supplier of an industrial antimicrobial agent is required to provide data supporting claims that an agent is effective in a particular use and nonhazardous to users and the environment when used according to directions.

The most significant modification of the Federal Insecticide, Fungicide, and Rodenticide Act (FIFRA) (15) allows conditional registration which facilitates registration of a pesticide product that is "identical or substantially similar to" a currently registered pesticide and "new uses of existing registered pesticides," when "neither the product nor any of its uses would cause an unreasonable adverse effect on the environment" (16). The FDA exercises authority over food and food packaging, drugs, cosmetics, and devices for human use.

In general, one should expect the supplier to provide the common group of acute tests, ie, one-time dose responses: acute oral LD_{50}, acute dermal LD_{50}, skin irritation, eye irritation, and in some cases, acute inhalation, and TLV. The application, extent of exposure of humans, and other factors may dictate that a number of subacute or chronic, long-term tests be performed to establish safety. There is one particularly valuable reference source (17) and an on-line computerized data bank that can be consulted (18) (see Information retrieval).

Test Methods

There are two kinds of test methods: fundamental microbiological tests, which determine the inherent antimicrobial activity of the agent, and simulation tests, which establish the effectiveness in a particular application. By way of the former, by one modification or another, one determines the lowest concentration of the chemical that will inhibit or kill microorganisms of concern. Results usually are reported as MIC

Table 1. Estimated U.S. Sales of Industrial Antimicrobial Agents, 10^6 dollars

	1972[a]	1977[b]
industrial and institutional sanitizers and disinfectants	10	50
paint	10	45
wood preservatives	8	30
metalworking fluids	3	25
cooling water	8	20
paper	10	18
petroleum production	5	9
plastics and resins	2	8
all other	9	20
Total	*65*	*225*

[a] Ref. 11.
[b] Ref. 12.

or MLC (Minimum Inhibitory Concentration or Minimum Lethal Concentration). The simulation tests are designed to reflect the proposed use, to establish directions for use, and to support label claims. Contributions towards improving and standardizing test methods have been made by several organizations: Association of Official Analytical Chemists (AOAC) (19), the American Society for Testing and Materials (ASTM), the American Association of Textile Chemists and Colorists (AATCC), the Society for Industrial Microbiology (SIM), The American Society for Microbiology (ASM), and the government regulatory agencies (see Regulatory agencies). An overview is given in ref. 20, and discussions of laboratory procedures for assessing the potential of antimicrobial agents as industrial biocides (21), the Dipslide technique for microbiological testing of industrial fluids (22), and laboratory assessment of antibacterial activity (23) have been reported.

Choosing an Antimicrobial Agent

The agent is chosen which: (*1*) is claimed to combat the microorganism of concern to the degree desired, (*2*) appears to be physically and chemically compatible with the system, ie, will not upset desirable physical and chemical properties of the system and (conversely) will not be inactivated by the ingredients of the system, (*3*) maintains stability under use and storage conditions (pH, temperature, light, etc) for the required length of time, (*4*) is safe and nontoxic in handling, formulation, and use, (*5*) is environmentally acceptable, and (*6*) is economically acceptable and cost-effective.

Phenolics. The largest class of industrial antimicrobial agents used in the United States are the phenolic derivatives (see Table 2). In 1978, ca 28,200 metric tons of phenolic compounds were utilized in industrial applications, representing a $\$39.5 \times 10^6$ market at the manufacturer's level (30). The evolution of various phenolic derivatives, their mechanisms of action, influence of substituents on activity, test procedures, research activities, and other aspects have been reported (31). Although the specific antimicrobial activity of each phenolic derivative is partially dependent on the varying substituents, the phenolics generally exhibit broad-spectrum activity against Gram-positive and Gram-negative bacteria as well as against fungi (32). This property plus their relatively moderate cost, make them quite cost-effective (see Alkylphenols; Chlorophenols; Phenol).

Phenols are considered to be moderately to highly toxic. The chlorophenols can be absorbed through the skin in toxic levels and their dusts are very irritating to the skin, eyes, and respiratory tract. The formation of highly toxic chlorinated dibenzo-*p*-dioxins as impurities in the production of various chlorophenols has been reported (33). Environmental and human safety considerations, and the regulatory action associated with them, have prompted the withdrawal of six products from the market during recent years: 2,4,6-trichlorophenol, 2-chloro-4-phenylphenol, 2,3,4,6-tetrachlorophenol, pentachlorophenol (nonpurified), 4-chloro-2-cyclopentylphenol, and mixture of 4- and 6-chloro-2-phenylphenol. However, the acceptable products discussed below continue to keep the phenolic class of antimicrobial agents among the most widely used in industry; the chlorophenols have the largest volume of use.

Pentachlorophenol (PCP) represents by far the largest share of the chlorophenols market with an overall demand in 1977 of 20,700 t and 27,780 t in 1978 (Monsanto ceased production in 1978) (34). It is second only to creosote [*8021-39-4*] when used

Table 2. Phenolic Industrial Antimicrobial Agents

Common name	Structure	CAS Registry No.	Trade names	Producers	Applications	Ref.
pentachlorophenol (PCP)		*[87-86-5]*	Dowicide 7[a] 49-162	Dow Reichhold Vulcan	wood and paper construction materials leather paint textiles	24
sodium pentachloro-phenoxide		*[131-52-2]*	Dowicide G-ST	Dow	adhesives construction materials leather paint drilling muds pulp and paper textiles water treatment slimicide	25
2,4,5-trichlorophenol		*[95-95-4]*	Dowicide 2	Dow Vertac	adhesives rubber textiles paper-mill slimicide wood paint	26
sodium 2,4,5-trichloro-phenoxide		*[136-32-3]*	Dowicide B	Dow	adhesives cooling-water slimicide leather metalworking fluids	27
o-benzyl-*p*-chlorophenol		*[120-32-1]*	Santophen 1	Monsanto	disintectants hard-surface cleaners	

o-phenylphenol	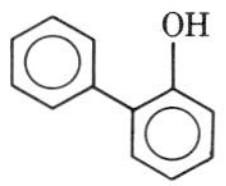 OH	[*90-43-7*]	Dowicide 1	Dow	hard-surface cleaners metalworking fluids automotive gaskets leather paint polish formulating	28
sodium *o*-phenylphenoxide	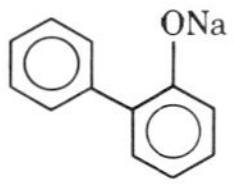 ONa	[*132-27-4*]	Dowicide A	Dow	adhesives ceramic glazes clay slips chemical toilets construction material feathers floor waxes colloidal graphite metalworking fluids paint textiles	29

[a] Current trade name, Pentachlorophenol DP-2.

as a wood preservative. Sales of pentachlorophenol were \$16 × 10^6 in the U.S. in 1978 for wood preservation (35). Wood is vulnerable to biodegradation by fungi, bacteria, and insects. The various constituents, eg, cellulose hemicellulose, lignin, and pentosans (mostly xylan), serve as nutrients for these organisms, and customary conditions of use and storage provide a supporting environment for proliferation of the organisms. Fungal degradation can occur on the surface in the form of soft rot or internally as white rot or brown rots. In 1973, ca 2.3 × 10^6 m^3 wood in the U.S. was treated with pentachlorophenol alone (36) and this application represented only one third of all wood treated with preservatives, including cross-ties, utility poles, lumber, fence posts, pilings, and wood pulp and chips. Wood preservatives generally are applied as organic solvent solutions or water emulsions by a pressure impregnation application or by nonpressure dipping, spraying, or brushing techniques.

Pentachlorophenol normally is applied as ca 7% organic solvent solution by various pressurized processes (37). Its minimum recommended use concentrations are 4.8–8.0 kg/m^3 (4–6.7 lb/100 gal) wood, depending upon the application (38). Pentachlorophenol is indicated for utility poles, posts, and other in-ground applications around industry and farm buildings. It gives cleaner surfaces than creosote treatments and thus can be painted. It is not used for marine pilings or railroad ties where creosote has certain advantages. Outside of wood preservation, pentachlorophenol has few additional uses. Although relatively inexpensive at ca \$1.00/kg, pentachlorophenol is considered a mature product with only a 2–3%/yr growth potential expected through 1981 (34).

Sodium pentachlorophenoxide, the water-soluble sodium salt of pentachlorophenol, is effective in a variety of applications. In 1978, \$2.5 × 10^6 of sodium pentachlorophenoxide was used for antibacterial and antifungal protection (35). Often employed in conjunction with pentachlorophenol or sodium 2,4,5-trichlorophenoxide [*136-32-3*], it is the most popular preservative for leather and hides; about a 0.5% concentration of the chlorophenols is effective. Other areas of utility for sodium pentachlorophenoxide include protection against the growth of bacteria in petroleum drilling muds and the control of algal, fungal, and bacterial slimes in industrial recirculating cooling water (see Water, industrial water treatment). Adhesives that are based on starch, and vegetable and animal proteins also are protected during their manufacture and shelf and service life with sodium pentachlorophenoxide (25).

2,4,5-Trichlorophenol (2,4,5-TCP) and its sodium salt accounted for a total of \$2.7 × 10^6 in sales in 1978 in industrial preservation uses (35). The areas of application, pattern of use, and use concentrations are similar to those of pentachlorophenol and its sodium salt (26–27). Concern has been shown over the potential toxicity hazard associated with 2,4,5-trichlorophenol. The EPA has issued a notice of rebuttable presumption against registration (RPAR) against 2,4,5-trichlorophenol (33). It has been found to contain small amounts of the highly toxic impurity, 2,3,7,8-tetrachlorodibenzo-*para*-dioxin (TCDD).

o-Benzyl-*p*-chlorophenol and *o*-phenylphenol are used in disinfectant and sanitizer applications where they compete with quaternary ammonium compounds and the most widely used active halogen products. These two phenolics are used primarily in detergent–disinfectant formulations for cleaning and disinfecting hard surface areas in hospitals, households, and various industries. About 1800 t \$6.1 × 10^6 worth, of *o*-benzyl-*p*-chlorophenol was used in 1978 in disinfectants (39). Toxicity and odor preclude the use of phenolics around food-related operations, and to some extent, their tendency to cause skin irritation has limited their use in general janitorial applications.

The use level and cost (100–750 ppm at \$0.60–2.60/m^3) (\$0.20–1.00/100 gal) for *o*-benzyl-*p*-chlorophenol (and *o*-phenylphenol) are somewhat higher than for quats or active halogens, but the slightly higher cost often is tolerated when it is important to achieve odor control, broad-spectrum disinfection of tuberculosis-causing bacteria, other Gram-positive and Gram-negative bacteria, or certain viruses and fungi. Both *o*-benzyl-*p*-chlorophenol and *o*-phenylphenol are stable toward a broad range of use and storage conditions but degrade to innocuous by-products in the environment.

o-Phenylphenol is the only nonchlorinated phenol of industrial significance as an antimicrobial agent (see Alkylphenols). *o*-Phenylphenol and its sodium salt were employed in a wide variety of applications totaling 2,200 t in the U.S. in 1978 valued at over \$7.7 $\times$ 10^6 (40). Two other subclasses of phenolic compounds possess significant antimicrobial activity: bis-phenols (eg, hexachlorophene [*70-30-4*] and dichlorophene [*97-23-4*]), and the parabens (eg, methyl *p*-hydroxybenzoate [*99-76-3*]) (see Disinfectants). The former are used almost exclusively in nonindustrial pharmaceutical applications, and the latter are useful in food, cosmetic, and pharmaceutical preservation.

Halogen Compounds. The halogens constitute a class of antimicrobial agents that are larger in commercial distribution than the phenolics. Many of their applications, eg, swimming-pool sanitizers, household and hospital disinfectants, and surgical scrubs are outside the scope of this article, but many industrial uses exist, especially for a number of organic halogen derivatives. In 1978, the consumption of halogen antimicrobial agents was estimated to be ca 35,000 t (\$108 $\times$ 10^6). Of the total, chloroisocyanurates contributed 62%; iodine compounds, 16%; and bromine compounds, 6% and various other products, 16% (41).

The halogen can be in the active or available form, in which it demonstrates antimicrobial activity because of its oxidizing capacity through a positive valence state. The halogen also contributes activity as a covalent substituent of a complex structure. The spectrum of commercial products ranges from the elements, to inorganic hypohalites, to complex halogen-substituted organic compositions. The elements and the hypohalites belong to the class of active or available halogen products, but the more complex halo-derivatives are of various types. A variety of *N*-halo or *C*-halo compositions, which also feature electronegative substitution on the same nitrogen or carbon atom, are considered active by virtue of the generation of hypohalite in water solution. Significant commercial organohalogen antimicrobial agents are shown in Table 3.

Chlorine Compounds. Liquid chlorine [*7782-50-5*] is used in potable water purification and swimming-pool sanitation. Hypochlorites are the oldest and most widely used disinfectants among active chlorine compounds (51). Principal products are calcium hypochlorite [*7778-54-3*], sodium hypochlorite [*7681-52-9*], lithium hypochlorite [*13840-33-0*], and chlorinated TSP [*56802-99-4*] (a 1:4 adduct of sodium hypochlorite and trisodium phosphate). All are marketed as solid products, except sodium hypochlorite which is sold as aqueous solutions at 5–40% concentrations (see also Chlorine oxygen acids and salts, chlorine monoxide, hypochlorous acid, and hypochlorites).

Chlorine dioxide [*10049-04-4*] is a unique agent used in potable-water purification, especially for taste and odor control (52) and is attracting considerable attention as a potential substitute for chlorine in order to avoid the conversion of organic pollutants to carcinogenic chlorinated hydrocarbons (eg, chloroform). Since this gaseous chemical cannot be stored, it is generated *in situ* as needed from sodium chlorite or sodium

Table 3. Industrial Halogen-Containing Antimicrobial Agents

Common name	Structure	CAS Registry No.	Trade names	Producers	Applications	Ref.
Chlorine compounds						
trichloroisocyanurate	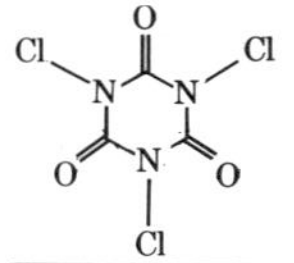	[*87-90-1*]	ACL-85 CDB-90 Pace	Monsanto FMC Olin	swimming-pool sanitizer disinfectant	42
sodium dichloroisocyanurate	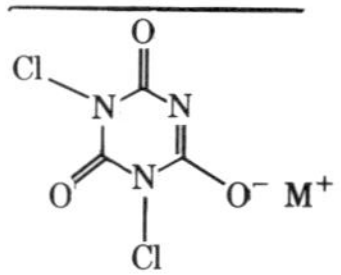 $O^{-}\ M^{+}$	[*2893-78-9*] [*51580-86-0*] (.2H$_2$O)	ACL-60 ACL-56 Clearon	Monsanto Monsanto FMC		
potassium dichloroisocyanurate		[*2244-21-5*]	ACL-59	Monsanto		
mono(trichloro)-tetra(potassium dichloroisocyanurate) 1:4		[*34651-95-1*]	ACL-66	Monsanto		
dichlorodimethylhydantoin	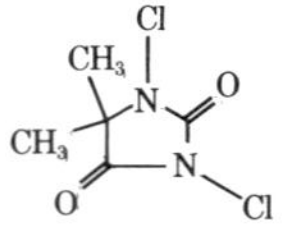	[*118-52-5*]	Halane	BASF Wyandotte	disinfectant	
sodium *p*-toluenesulfonchloramide	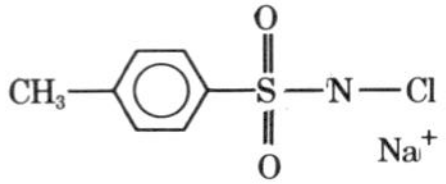 $CH_3-C_6H_4-SO_2-N(Cl)^{-}\ Na^{+}$	[*127-65-1*]	Chloroamine-T	EM Laboratories	germicide	
p-toluenesulfondichloramide	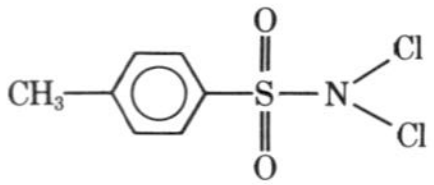 $CH_3-C_6H_4-SO_2-NCl_2$	[*473-34-7*]	Dichloramine-T	EM Laboratories	germicide	
Iodine compounds						
iodine–poly(vinylpyrrolidinone) complex	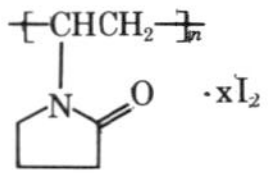 $\cdot xI_2$	[*25655-41-8*]	PVP–Iodine	GAF	germicide	
iodine–polyethylene glycol mono(nonylphenyl) ether complex	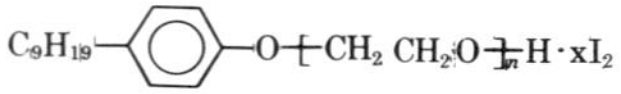 $C_9H_{19}-C_6H_4-O(CH_2CH_2O)_nH\cdot xI_2$	[*11096-42-7*]	Iosan Biopal VRO 20	West Chemical GAF	sanitizer disinfectant	43

p-tolyldiiodomethyl sulfone	$CH_3C_6H_4SO_2CHI_2$	[*20018-09-1*]	Amical 48	Abbott	paint preservative	44
					paint mildewcide	44
					water-based adhesives preservative	45
p-chlorophenyldiiodomethyl sulfone	$ClC_6H_4SO_2CHI_2$	[*20018-12-6*]	Amical 77	Abbott	water-based sealants preservative	45
Bromine compounds						
bromochlorodimethylhydantoin	1-bromo-3-chloro-5,5-dimethylhydantoin ring (N–Cl, N–Br, two CH_3, two C=O)	[*126-06-7*]	DiHalo Sticks	Great Lakes	swimming-pool disinfectant	46
2,2′-dibromo-3-nitrilopropionamide	$N{\equiv}C{-}CBr_2{-}C(=O){-}NH_2$	[*10222-01-2*]	X-7287L	Dow	cooling-water slimicide; secondary oil recovery; metalworking fluids	46
bis(1,4-bromoacetoxy)-2-butene	$BrCH_2C(=O){-}OCH_2CH{=}CHCH_2O{-}C(=O)CH_2Br$	[*20679-58-7*]	Slimacide V-10	Vineland	paper-mill slimicide	47
1,2-dibromo-2,4-dicyanobutane	$BrCH_2CBr(CN)CH_2CH_2CN$	[*35691-65-7*]	Tektamer 38	Merck	paint; adhesives; latex emulsions; joint cements	48
2-bromo-2-nitropropane-1,3-diol	$HOCH_2CBr(NO_2)CH_2OH$	[*52-51-7*]	Onyxide 500	Onyx	cosmetic preservative	49
benzyl bromoacetate	$C_6H_5CH_2OC(=O)CH_2Br$	[*5437-45-6*]	Merbac 35	Merck	paint; latexes	50

chlorate. It is also used in the paper industry to prevent slimes in paper-mill white-water systems (52) as well as in its primary use for wood-pulp bleaching (see Bleaching agents).

Among the organic chlorine derivatives used in various industrial applications, the chloroisocyanurates hold the largest share of the market. The five chloroisocyanurates shown in Table 3 have the greatest commercial significance. All are crystalline materials, are sold in powder or granular form, and have high available chlorine contents (from 59–90%). They exhibit a chlorinous odor and variable water solubilities depending upon whether or not the product is an alkali metal salt (see Cyanuric and isocyanuric acids). The chlorine is active or available, but is bound to the triazinetrione nucleus. In water solutions, the chlorine is liberated as free available chlorine.

Some 86% of the $66.7 × 10^6 sales of the chloroisocyanurates is consumed as swimming-pool disinfectants (42). Other uses include laundry bleaches and disinfectants in dishwasher detergents, scouring powders, and other household and industrial products. Prices of the various chloroisocyanurates are $2.50–3.50/kg and, on a cost-performance basis, the products are considered highly competitive. In addition to the chloroisocyanurates, a few other organic chloramine derivatives include 1,3-dichloro-5,5-dimethylhydantoin, chloramine-T, and dichloramine-T (see Chloramines and bromamines).

Iodine Compounds. The use of elemental iodine [*7553-56-2*] as an antiseptic dates back to 1839 (53). It is used today for various medicinal purposes. The combination of iodine with various solubilizing polymers led to a class of new compositions known as iodophors, which dominate the market once satisfied by simple alcoholic or aqueous iodine solutions. The iodine complexes with either nonionic surfactants, eg, polyethyleneglycol mono(nonylphenyl) ether, or poly(vinylpyrrolidinone) (PVP). The complexes function by rapidly liberating free iodine in water solutions. They exhibit good activity against bacteria, molds, yeasts, protozoa, and many viruses. These iodophors are generally nontoxic (54), nonirritating, nonsensitizing, and noncorrosive to most metals (except silver and iron alloys).

The uses continue to be largely medicinal, though some iodophors are used in industrial sanitation and disinfection in hospitals, building maintenance, and food-processing operations (43). There has been some interest in the use of iodine for purification of potable water and swimming pools (55). Two other iodine-containing compounds, *p*-tolyldiiodomethyl sulfone and *p*-chlorophenyldiiodomethyl sulfone have been recommended as preservatives (44–45).

Bromine Compounds. Although there is evidence that elemental bromine [*7726-95-6*] exhibits the same antimicrobial activity as chlorine and iodine (47), there is little evidence of extensive commercialization (partly owing to the handling problems of bromine liquid and vapor). However, the organic bromine derivatives shown in Table 3 are used in swimming pools, paper mills, and cooling towers, and to a lesser extent, as preservatives for latexes, paints, and cosmetics. The products have adequate solubility and stability in water but do exhibit some potential for skin irritation.

Bromochlorodimethylhydantoin holds the largest share of the bromoorganic antimicrobial market, with most of the sales ($2.6 × 10^6) applied to swimming-pool sanitation (46). This product owes part of its activity to active chlorine. 2,2-Dibromo-3-nitrilopropionamide exhibits broad spectrum control of microorganisms in aqueous systems. Its principal use is in cooling-water systems, and its annual sales are ca $1 × 10^6 (46). Annual sales of bis(1-4-bromoacetoxy)-2-butene are ca $1 × 10^6

(47) and 1,2-dibromo-2,4-dicyanobutane was registered with the EPA in 1978 as a bacteriostat (48); its annual consumption is valued at less than \$1 × 10^6. The activity of 2-bromo-2-nitropropane-1,3-diol may be partly attributable to the bromine substitution. Benzyl bromoacetate is a recommended raw material for paints, especially poly(vinyl acrylate) and acrylic–latex emulsions (50).

Quaternary Ammonium Compounds (Quats). This category represents one of the largest, most diverse classes of antimicrobial agents in use; ca 10,900 t were consumed in 1978 at a value of about \$38 × 10^6 (56). Chemically, the products may be represented by the general formula:

$$\left[\begin{array}{c} R^1 \\ | \\ R^4\!-\!N\!-\!R^2 \\ | \\ R^3 \end{array}\right]^+ X^-$$

The nitrogen atom carries four covalently bound substituents that give it a cationic charge. The R groups may be almost any organic substituent that allows carbon–nitrogen bonding with any permutation of like and unlike R groups. The nitrogen atom may be part of a heterocyclic, aromatic structure, thereby automatically fixing three substituents. The counterion is almost always chloride. A few commercial products contain, as their anion, bromide, sulfamate, or saccharinate (see Quaternary ammonium compounds).

The commercially significant quats, as illustrated in Table 4, fall into the following categories: monoalkyltrimethyl-, dialkyldimethyl-, monoalkyldimethylbenzyl- (referred to as benzalkonium), alkyl-heterocyclic-aromatic-, bis-quaternary, and polymeric quaternary ammonium salts. The alkyl substituent usually is a long-chain hydrocarbon (C_{12}–C_{18}) and most commercial products feature a mixture of C_{12}, C_{14}, C_{16}, and C_{18} or fractions thereof, in proportions approximately of the sources from which they are derived. One quat, Q-5700, features a 3-(trimethoxysilyl) group on one of the alkyls, which apparently imparts long-term substantivity to textiles (68). Another, Bardac LF, was designed to overcome the foaming characteristic of most quats which is detrimental in certain applications (69).

At low concentrations, commercial quats are bacteriostatic, fungistatic, algistatic, sporostatic, and tuberculostatic; at medium concentrations, quats are bactericidal, fungicidal, algicidal, and virucidal against lipophilic viruses; quats are not tuberculocidal, sporocidal, or virucidal against hydrophilic viruses, even at high concentrations (70–71).

Detergent-compatible quats, which can both clean and disinfect (57), have their most popular use as hard-surface, disinfectant–sanitizers for floors, walls, and equipment surfaces in health-care facilities, schools, public facilities, and households. Comprehensive data supporting the efficacy of quats in the disinfection–cleaning operation are given in refs. 72–73. The development of aerosol formulations (74) led to spray deodorants and disinfectants (see Cosmetics).

Quats also have been proposed for sanitizing food and beverage utensils in restaurants (75) and food-processing equipment in various industries, eg, dairy (58), egg processing (see Eggs) (59), fishing (see Aquaculture) (60), brewing (see Beer) (61), sugar refining (62), and laundry (63). Commercially, the most significant uses (other than

Table 4. **Industrial Quaternary Antimicrobial Agents**

Common name	Structure	R	CAS Registry No.	Trade names	Producers	Applications	Ref.
benzalkonium chloride	$C_6H_5CH_2N^+(CH_3)_2R\ Cl^-$	C_{12-18}	[*68391-01-5*]	BTC-100	Onyx	germicide	57
				Maquat MC-1416	Mason	disinfectant	57
				Bio-Quat 50-24	Bio-Lab		
		C_{12-16}	[*68424-85-1*]	Hyamine-3500	Rohm and Haas	sanitizer	58–63
				Roccal II	Hilton-Davis	swimming-pool algicide	64
		C_{12-16}	[*139-08-2*]	Barquat MB 50	Lonza		
				Dibactol	Hexcel	cooling water	65
						pulp and paper	66
dialkyldimethyl–ammonium chloride	$RN^+(CH_3)_2R\ Cl^-$	C_8	[*5538-94-3*]	Bardac LF	Lonza	cooling water	65
						secondary oil recovery	67
						laundry sanitizer	63
						swimming-pool algicide	64
cetyltrimethyl–ammonium bromide	$CH_3N^+(CH_3)_2R\ Br^-$	C_{10}	[*7173-51-5*]	BTC-1010	Onyx	disinfectant	57
		C_{16}	[*57-09-0*]	Bardac 22	Lonza	sanitizer	58–63
				Bromat	Hexcel	antiseptic	57
cetylpyridinium chloride (CPC)	$C_5H_5N^+{-}R\ Cl^-$	C_{16}	[*123-03-5*]		Hexcel	disinfectant	57
3-(trimethoxysilyl)-propyldimethyl-octadecylammonium chloride	$(CH_3O)_3Si(CH_2)_3N^+(CH_3)_2R\ Cl^-$	C_{18}	[*27668-52-6*]	Q-5700	Dow-Corning	textiles	68

for disinfection) are as cooling-water slimicides (65), swimming-pool algicides (64), paper-mill slimicides (66), and secondary-oil-recovery bactericides (67). A unique application of a silicone-substituted quat as a permanent textile treatment has been cited (68).

Toxicologically, the quaternary ammonium products are considered to be less hazardous than certain competitive agents, eg, phenolics. Although concentrations of 10% or more can cause severe irritation of skin, mucosa, and eyes, and poisoning upon oral consumption, the effective use concentrations are well below hazardous levels. Furthermore, there appears to be no concern for chronic or cumulative effects; the acute oral LD_{50}s for several quats range from about 230–730 mg/kg (76–77). It has been suggested that 0.1% is the highest concentration that will not cause primary skin irritation or sensitization (77). Comments by the FDA refer to safe handling practices related to treatment of food-contact surfaces (78). A number of quats are approved for food-related uses, eg, paper slimicides (2ICFR 176.300), adhesives (2ICFR 175.105), and sanitizing food plants (2ICFR 178.1010) (79).

Metal-Containing Compounds. The metals, whose organic and inorganic derivatives are recognized as industrially significant antimicrobial agents, are mercury, arsenic, tin, copper, zinc, silver, chromium, barium, and selenium. Compositions range from simple inorganic salts to complex organometallic (qv) structures (see Table 5), and in some cases, a contribution is made by the nonmetallic portion. Commercially, in 1978, ca 4500 t ($24 × 10^6) of organometallics (94) and ca 13,600 t ($27 × 10^6) of inorganics (95) were consumed. Among the organometallics, mercurials accounted for the largest dollar volume (ca $11 × 10^6), arsenicals and tin compounds each accounted for $5–6 × 10^6 of the market. Sales of copper products amounted to ca 10^6 dollars (94). In 1978, chromated copper arsenate, cuprous oxide, and barium metaborate accounted for ca $20-, $3.5-, and $1 × 10^6, respectively (95).

Mercurials. The history, activity, and use of antimicrobial mercury compounds has been reported (96). Two mercurial agents constitute most of the production in the U.S. About 700 t each of phenylmercuric acetate (PMA) in various forms and di(phenylmercury)dodecenyl succinate (PMDS) were consumed in 1978 (97). In-can paint preservation and paint mildewcide activity accounts for ca 99% of the consumption and a small amount is used as textile and leather mildewcides. Phenylmercuric acetate is available as a 100% powder (mp, 149°C) (containing 60% mercury) and as a 30% soluble formulation (containing 18% mercury). The former is insoluble in water but soluble in hydrocarbon solvents, whereas PMA (30%) is soluble in water but only slightly soluble in organic solvents.

The antibacterial and antifungal activity of mercurials at low concentrations, their lack of color and odor, and their low cost have made them the most popular in-can preservatives and paint-film mildewcides of the past 25 yr. As a paint preservative, PMA or PMDS can be used at 0.01–0.10%. As a paint-film mildewcide, the use concentration range is about 0.1–1.5% (80).

Mercurials are among the most toxic antimicrobial agents in use and cause toxic effects through skin absorption, inhalation, or ingestion. Special handling precautions are indicated by suppliers. Use of phenyl mercurials is limited to water-based paints. Under the Clean Water Act of 1977, national effluent limits will become law by 1981 and should include pretreatment standards (98). If the mercury limitations are severe, it may be impossible to use mercurials in paint-formulation operations or other similar industries (see Mercury compounds).

Table 5. Industrial Metal-Containing Antimicrobial Agents

Common name	Structure	CAS Registry No.	Trade names	Producers	Applications	Ref.
Mercurials						
phenylmercuric acetate	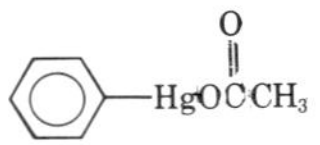	*[62-38-4]*	PMA-100	Cosan	paint preservative	80
			Troysan PMA	Troy	paint mildewcide leather	80
di(phenylmercury)-dodecenyl succinate (PMDS)	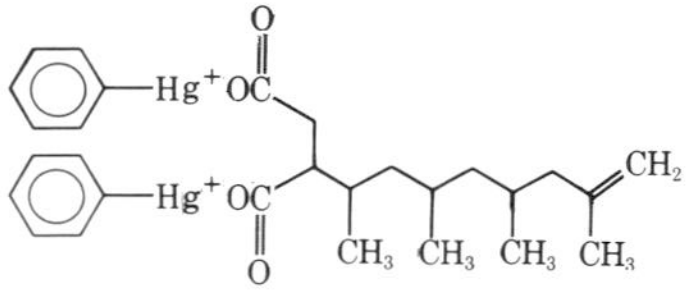	*[27236-65-3]*	Super-Ad-It	Tenneco	paint preservative paint mildewcide textiles	80 80
Arsenicals						
10,10′-oxybisphenoxy-arsine	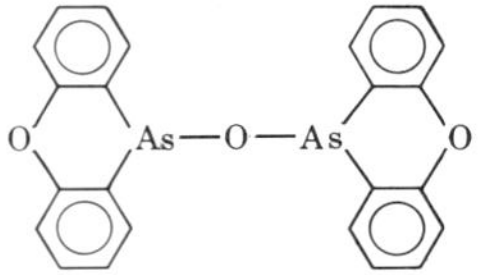	*[58-36-6]*	Vinyzene BP-5 Durotex	Ventron Ventron	plastics textiles	81 82
Organotins						
tributyltin oxide	$(n\text{-}C_4H_9)_3Sn-O-Sn(C_4H_9\text{-}n)_3$	*[56-35-9]*	Cotin 300 Keycide X-10 Intercide 340A	Cosan Ferro Interstab	antifoulant paint latex paint plastics	83 84 84
tributyltin fluoride	$(n\text{-}C_4H_9)_3SnF$	*[1983-10-4]*	Biomet 204	M&T	antifoulant paint	83
Copper compounds						
copper 8-quinolinate	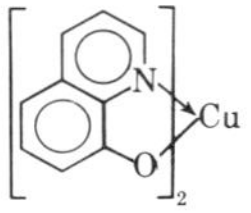	*[10380-28-6]*	Cumilate Quindex	Kocide Ventron Tenneco Troy	textiles wood plastics	85–87 88

copper naphthenate	$(C_6H_{11}COC(=O)-O)_2Cu$	[*1338-02-9*]	Nuodex	Tenneco Troy Witco	textiles wood	85–87
chromated copper arsenate	$Cu_3(AsO_4)_2 \cdot NaCr_2O_7$	[*37337-13-6*]		Koppers	wood	85–87
ammoniacal copper arsenate	$CuNH_4AsO_4$	[*32680-29-8*]	All Weather Wood	Osmose	wood	85–87
cuprous oxide	Cu_2O	[*1317-39-1*]		American-Chemet SCM-Glidden	antifoulant paint	
Miscellaneous						
barium metaborate	$BaO \cdot B_2O_3$	[*13701-59-2*]	Busan 11-M1	Buckman Labs	paint mildewcide	89–90
selenium sulfide	SeS_2	[*7488-56-4*]	Selsun	Abbott Labs	antidandruff shampoos	91
zinc oxide	ZnO	[*1314-13-2*]		Allied New Jersey Zinc St. Joe Minerals	paint mildewcide	92–93

Arsenicals. 10,10′-Oxybisphenoxyarsine is the only organoarsenic product of importance as an industrial antimicrobial agent (see Table 5). The active ingredient is a colorless microcrystalline powder, but it is sold only in dilute formulations to minimize toxicity. Vinyzene solutions are used primarily for the plastic industry; Vinyzene is sold at 1% or 2% concentration in various plasticizers, eg, epoxidized soybean oil and various phthalate esters (qv), and at 5% as a pelletized solid polymeric resin (81). A 1% solution in 1,4-butanediol is provided for use in polyurethanes (88). The Durotex solution that is used in the textile industry is 2% active and is formulated with nonionic, cationic, or anionic emulsifiers (82). Although many arsenic compounds are generally recognized as being toxic, the 10,10-oxybisphenoxyarsine products are claimed to be less toxic because the arsenic is organically bound. Nonetheless, when undiluted, the active ingredient is a skin and eye irritant; however, the products are diluted in order to minimize toxicity (see Arsenic compounds).

Organotins. Organotin derivatives of the R_3SnX type are very active antifungal agents (83) (where R represents alkyl or aryl and X represents a group linked to the tin through an atom other than carbon, eg, halogen, hydroxyl, or organic radical linked through oxygen). The most widely used organotin antimicrobial agents are tributyltin oxide (TBTO) and tributyltin fluoride (TBTF).

Tributyltin oxide is a clear, light-amber liquid and is insoluble in water but soluble in many organic solvents. It can be emulsified in aqueous systems, and formulations are available commercially with 12–100% active ingredient. Tributyltin fluoride is offered as a 97.5% active product: a white powder which also is essentially insoluble in water and slightly soluble in organic solvents. Both products are very active antifungal agents, effective at a few ppm in preventing slime in cooling-water systems. Since they are also active against barnacles, algae, tubeworms, and hydroids, they are used as antifoulant agents in marine paints (see Coatings, marine), but the high concentrations required (5–20%) make them less cost-competitive.

In 1978, 230 t of TBTO was consumed in cooling-water applications; ca 140 t of combined TBTF and TBTO was used in antifoulant paints; another 180 t of the various organotin agents were used in plastics and miscellaneous applications (84). Overall consumption totalled about 5.7×10^6 in sales. An approach to controlled-release antifoulant activity, involving the chemical bonding of tin antifoulant agents to the paint resin, was developed in 1976 (99) and has been evaluated (100). The organotins are considered to be moderately toxic orally and potentially irritating to the skin and eyes (101) (see Tin compounds).

Copper Compounds. Copper 8-quinolinate and copper naphthenate are used in mildewproofing textiles and almost exclusively for military applications (eg, tents, tarpaulins, and sandbags). The high concentration levels required, 0.1–1%, and the intense green color make them unsuitable for most consumer products. Their use in plastics has been limited to plasticized poly(vinyl chloride) (PVC) (88). Although there was a surge in consumption during the early 1970s (resulting from military activities in tropical areas), the consumption in 1978 was only ca 500 t for copper naphthenate and nearly the same for copper 8-quinolinate (102). A small amount of copper 8-quinolinate is used in wood preservation. Copper chelated complexes, eg, copper citrate [*866-82-0*] and copper ethylenediaminetetraacetic acid (EDTA) [*54453-03-1*], are used almost exclusively as swimming pool algicides; but altogether, their use amounts to only ca 50 t (102). Most of the varied copper soaps, eg, copper oleate [*1120-44-1*], and the cuprammonium derivatives have been used for many years in preserving textiles, wood, and cellulosics (85–86) (see Copper compounds).

Copper 8-quinolate, a discreet chelated complex, has very low solubility in water but very often is formulated with water repellants to minimize leaching. It has a characteristic yellow-green color but it has no odor. Copper naphthenate is a blue-green, waxy solid composed of a mixture of copper salts of naphthenic acids; it is soluble in common organic solvents but not in water.

Toxicity to humans of copper 8-quinolinate and copper naphthenate is low, and it does not irritate skin. Copper-containing antimicrobials require EPA registration but face no adverse regulatory action at this time.

Two inorganic copper salts are used in large volumes in singular applications, copper arsenate in wood preservation, and cuprous oxide as a marine antifoulant. Copper arsenate is available in two forms: chromated and ammoniacal (see Table 5). Both are applied as water-borne salts that are injected into wood where they react with cellulose and establish long-term protection against fungi, marine organisms, termites, and other insects (87). The chromated form is used predominantly (9100 t in 1978 (95)). Applications require from 4.0–40 kg/m^3 of treated material (103). The arsenic in these compositions attracted a rebuttable presumption against registration (RPAR) of these copper arsenates in 1978, and the EPA is evaluating additional data to determine whether the risks outweigh the benefits of the products.

Cuprous oxide is a popular marine antifoulant, is economical to use, and is specified for military antifouling paints. It is used at levels of 30–70% (based on the total weight of paint) and prevents the growth of a multitude of marine organisms on ships' hulls by slowly releasing its antimicrobial activity through the paint surface. Cuprous oxide is sometimes used in combination with other antifoulants, eg, tributyltin oxide.

Barium metaborate is used exclusively as a paint mildewcide in water-based paints (89–90) and generally at levels of 2.4–12 kg/m^3 (2–10 lb/100 gal.) paint (103). Selenium sulfide (Selsun) is used exclusively as an antidandruff agent at 1% concentrations in a shampoo base (91). The material is provided as a water-insoluble yellow-brown powder with a 100% assay. Zinc oxide often has been considered a paint mildewstat and has been used for many years in 30–50% concentrations as a pigment–mildewcide combination (92–93). However, the antimicrobial activity of zinc oxide has been debated, with one investigator suggesting that zinc oxide is effective because it hardens the paint film (104) and another demonstrating that zinc oxide inhibits respiration of fungal spores, thereby preventing growth, but does not kill the cells (105). A few other heavy metals, eg, silver and its compounds, have limited use as industrial antimicrobial agents (106–107).

Anilides. Trichlorocarbanilide (TCC) is the only anilide that enjoys sizable sales as an antimicrobial agent; it is used almost exclusively in deodorant bar soaps (1800 t/yr) (108). The FDA's Advisory Review Panel on over-the-counter antimicrobial drug products critically reviewed appropriate data regarding the safety of salicylanilides and carbanilides, and found considerable evidence of photosensitization by halogenated salicylanilides (109). The toxicology has been discussed in various articles (110–111). Subsequently the FDA placed halogenated salicylanilides in Category 2 as not safe and effective for use in drug products and cosmetics (109).

TCC is sold as a powder, is white to off-white in color, and has a minimum assay of 97%. It is essentially insoluble in water and is moderately soluble in various organic solvents. The antimicrobial activity of TCC is high against Gram-positive bacteria but not against Gram-negative bacteria and fungi (112). TCC and other industrial anilide antimicrobial agents are listed in Table 6.

Table 6. Industrial Anilide Antimicrobial Agents

Common name	Structure	CAS Registry No.	Trade names	Producers	Applications	Ref.
trichlorocarbanilide (TCC)	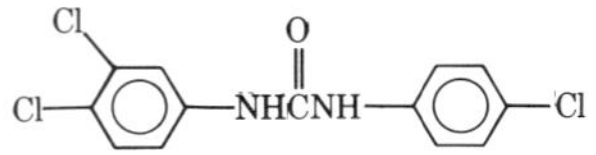 Cl, Cl, O, NHCNH, Cl	[*101-20-2*]		Monsanto	soap germicide	108
tribromosalicylanilide	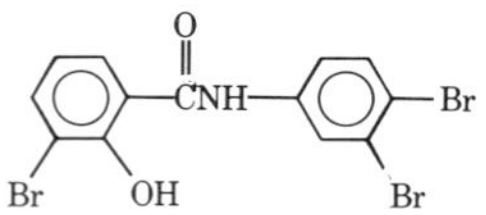 O, CNH, Br, Br, OH, Br	[*38848-67-8*]	Temasept IV	Hexcel	soap germicide	108
salicylanilide	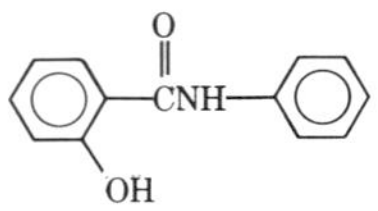 O, CNH, OH	[*87-17-2*]	Shirlan	DuPont	textiles leather	
trifluoromethyldichlorocarbanilide	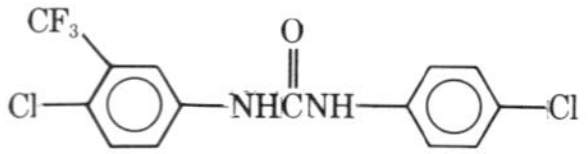 CF_3, Cl, O, NHCNH, Cl	[*369-77-7*]	Iragasan CF_3	Ciba-Geigy	soap germicide	108

Amines, Alkanolamines, and Nitro Compounds. The nitrogen atom is implicated in the bioactivity of a variety of organonitrogen compositions, eg, halogen, quaternary ammonium, anilide, and nitrogen–sulfur derivatives. Approximately 6800 t (\$18 × 10^6) of cyclic and linear amines and certain alkanol and nitro derivatives were consumed in 1978 (113). The commercially significant products are identified in Table 7; they range from simple fatty alkyl-substituted alkylenediamines, to alkanol-substituted linear amines, to complex cyclic structures (see Amines; Diamines; Alkanolamines).

The simple fatty alkyl ethylene- and propylenediamines, when undiluted, usually are waxy solids or viscous liquids, depending upon the length and mix of the alkyl substituents (usually coco-alkyl) (see Amines, fatty). They are used as antimicrobial agents almost exclusively in secondary oil recovery and circulating cooling-water systems, where they not only effectively control sulfate-reducing bacteria but function as corrosion inhibitors (114). The products are inexpensive, averaging ca \$2.20/kg, and are considered nontoxic and environmentally acceptable at their effective concentration levels (118).

N-[(α)-(1-nitroethyl)benzyl]ethylenediamine is a new product and is claimed to be effective at 2–2.5 ppm against slime-forming bacteria and fungi in pulp and paper-mill systems (115). It is approved by the FDA for use in the manufacture of paper food-packaging materials. Environmentally, the product is said to be nonpersistent in effluent waters (115).

2-(Hydroxymethyl)aminoethanol (119) and 2-(hydroxymethyl)amino-2-methylpropanol are companion products that are sold as water-soluble preservatives for latex paints, resin emulsions, adhesives, and wall-joint cements. Recommended use levels range from 0.05–0.2%, the products sell for under \$3.30/kg, and an estimated total of 3600 t was sold in 1978 (115).

Among the cyclic compositions, hexahydro-1,3,5-triethyl-*s*-triazine (120) and hexahydro-1,3,5-tris(2-hydroxyethyl)-*s*-triazine (121), enjoy ca 50% of amine sales; the latter compound accounted for ca 2600 t, the former accounted for ca 600 t in 1978 (113). The tris(2-hydroxyethyl) derivative, which is marketed as a 78.5% concentrate, is used almost exclusively to preserve metalworking fluids, particularly the soluble-oil emulsion type. The triethyl analogue, marketed as a 95% concentrate, is used to preserve latex paints and metalworking fluids, but its largest use is in preserving pigment slurries (116).

Bioban P-1487 is a mixture of two nitro-substituted morpholine derivatives (see Table 7). It is marketed as a 90% concentrate and is used exclusively as a metalworking-fluid preservative at concentration levels of 0.05–0.10% (117). About 500 t (\$2.9 × 10^6) was consumed in 1978 (122). Two other nitrogen-containing antimicrobial agents, 1,3-dimethylol-5,5-dimethylhydantoin [*6440-58-0*] and *N*,*N*″-methylene bis[*N*′-(1-hydroxymethyl)-2,5-dioxo-4-imidazolinyl] urea [*39236-46-9*], are used primarily as cosmetic preservatives.

Organosulfur and Sulfur–Nitrogen Compounds. The development of several sulfur-containing antimicrobial agents for crop protection are discussed in a recent review (123). These antimicrobial agents also bear a nitrogen-containing functionality. The significant commercial products are illustrated in Table 8.

Methylenebisthiocyanate (MBT) has the simplest composition of those compounds listed in Table 8 and is one of the most widely used paper-mill and cooling-water slimicides. By 1978, MBT was consumed at a level of ca 300 t (138). Despite its

Table 7. Industrial Amine, Alkanolamine, and Nitro-Containing Antimicrobial Agents

Common name	Structure	CAS Registry No.	Trade names	Producers	Applications	Ref.
Linear						
N-coco-trimethylenediamine	$RNHCH_2CH_2CH_2NH_2$ $R = C_8—C_{18}$	*[61791-63-7]*		Armak Ashland	secondary oil recovery cooling-water bactericide	114 114
			DIAM 21	Enenco Inc. General Mills	paper	
N[α-(1-nitroethyl)-benzyl]-ethylenediamine	$C_6H_5—CH(CH(CH_3)NO_2)NHCH_2CH_2NH_2$	*[14762-38-0]*	Metasol J-26	Merck	paper-mill slimicide	115
2-(hydroxymethyl)-aminoethanol	$HOCH_2NHCH_2CH_2OH$	*[34375-28-5]*	Troysan 174	Troy	latex paint resin emulsions	
2-(hydroxymethyl)-amino-2-methyl-propanol	$HOCH_2NHC(CH_3)_2CH_2OH$	*[52299-20-4]*	Troysan 192	Troy	adhesives wall-joint cements	
Cyclic						
hexahydro-1,3,5-tris-(2-hydroxyethyl)-*s*-triazine	hexahydrotriazine ring, N-substituents: $HOCH_2CH_2—$, $—CH_2CH_2OH$, $—CH_2CH_2OH$	*[4719-04-4]*	Onyxide 200 Grotan	Millmaster-Onyx Lehn & Fink	metalworking fluids	
hexahydro-1,3,5-triethyl-*s*-triazine	hexahydrotriazine ring, N-substituents: $CH_3CH_2—$, $—CH_2CH_3$, $—CH_2CH_3$	*[7779-27-3]*	Vancide TH	Vanderbilt	paint preservative rubber latex adhesives	116
4-(2-nitrobutyl)-morpholine and 4,4′-(2-ethyl-2-nitrotrimethylene)-dimorpholine	morpholino–$CH_2CH(NO_2)CH_2CH_3$ + morpholino–$CH_2C(NO_2)(CH_2CH_3)CH_2$–morpholino	*[37304-88-4]*	Bioban P-1487	IMC Corp.	metalworking fluids	117

Table 8. Industrial Organo-Sulfur and Sulfur–Nitrogen Antimicrobial Agents

Common name	Structure	CAS Registry No.	Trade names	Producers	Applications	Ref.
Bisthiocyanates						
methylenebisthiocyanate (MBT)	$NCSCH_2SCN$	*[6317-18-6]*	Cytox 3522 Biocide N-948	American Cyanamid Stauffer	paper-mill slimicide	124
vinylene-bisthiocyanate	$NCSCH{=}CHSCN$	*[14150-71-1]*	Cytox 3711	American Cyanamid	cooling-water slimicide	125
chloroethylenebis-thiocyanate	$NCSCH(Cl)CH_2SCN$	*[24689-89-2]*	Cytox 3810	American Cyanamid	secondary oil recovery	126
Dithiocarbamates						
sodium dimethyldithiocarbamate (Sodam)	$(CH_3)_2NC(=S)SNa$	*[128-04-1]*	Thiostat Thiostop N	Uniroyal	paint mildewcide slimicide	
disodium ethylenebisdithiocarbamate (Nabam)	$NaSC(=O)NHCH_2CH_2NHC(=O)SNa$	*[142-59-6]*	Dithane D 14	Rohm and Haas	cooling-water slimicide	
zinc dimethyldithiocarbamate (Ziram)	$(CH_3)_2NC(=S)SZnSC(=S)N(CH_3)_2$	*[137-30-4]*	Vancide 51Z	R. T. Vanderbilt	paper-mill slimicide	
Sulfones						
bis(trichloromethyl) sulfone	$Cl_3C{-}S(=O)_2{-}CCl_3$	*[3064-70-8]*	Biocide N-1386	Stauffer	paper-mill slimicide cooling-water slimicide secondary oil recovery	127 127

Table 8 (*continued*)

Common name	Structure	CAS Registry No.	Trade names	Producers	Applications	Ref.
trans-1,2-bis(*n*-propylsulfonyl)-ethylene	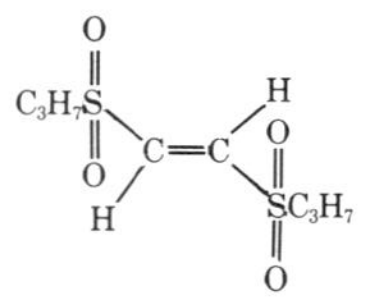	[*1113-14-0*]	Vancide PA	R. T. Vanderbilt	paint mildewcide	128
2,3,5,6-tetrachloro-4-(methylsulfonyl)-pyridine	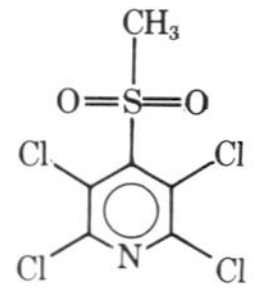	[*1308-52-6*]	Dowicil S-13	Dow	latex paint mildewcide	129
2,3,5-trichloro-4-(propylsulfonyl)pyridine	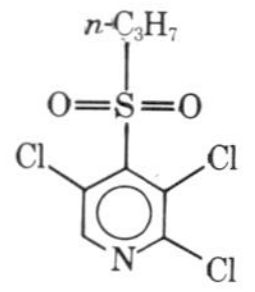	[*38827-35-9*]	Dowicil A-40	Dow	acrylic paint preservative	130
Nitrogen-sulfur heterocyclics						
tetrahydro-3,5-dimethyl-2*H*-1,3,5-thiadiazine-2-thione (DMTT)	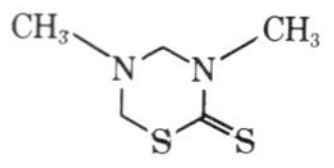	[*533-74-4*]	Metasol D3T Biocide N-521	Merck Stauffer	paper-mill slimicide paint cooling-water slimicide adhesives	131–132
sodium pyridinethione	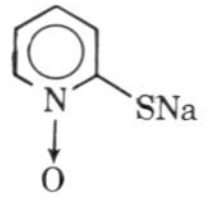	[*3811-73-2*]	Sodium Omadine	Olin	metalworking fluids cosmetics	

zinc pyridinethione	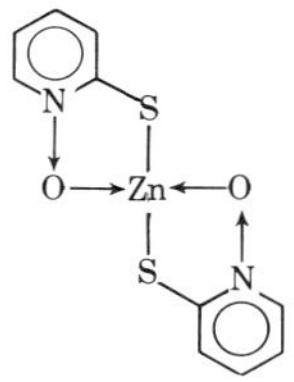	[*13463-41-7*]	Zinc Omadine	Olin	antidandruff agent cosmetics	
1,2-benzisothiazoline-3-one	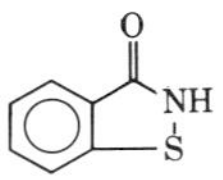	[*2634-33-5*]	Proxcel CRL	ICI America	latex paint preservative pigment slurries adhesives metalworking fluids	
2-*n*-octyl-4-isothiazolin-3-one	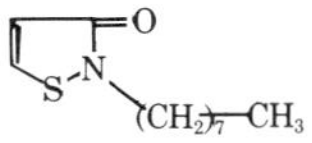 $(CH_2)_7-CH_3$	[*26530-20-1*]	Skane M-8	Rohm and Haas	latex paint mildewcide	133
			Kathon LP	Rohm and Haas	leather hides	134
2-(4-thiazolyl)benzimidazole	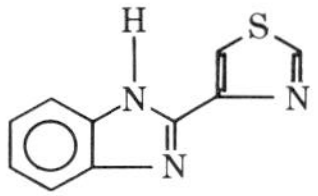	[*148-79-8*]	Metalsol TK-100	Merck	latex paint mildewcide soap wrappers	135
N-(trichloromethylthio)-4-cyclohexene-1,2-dicarboximide (Captan)	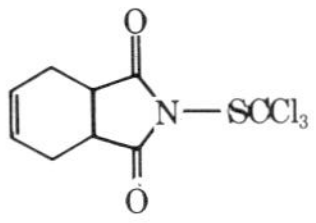 $N-SCCl_3$	[*133-06-2*]	Vancide 89RE Orthocide	R. T. Vanderbilt Chevron Chemical	cosmetics paint lacquer resins	136
N-(trichloromethylthio)phthalimide (Folpet)	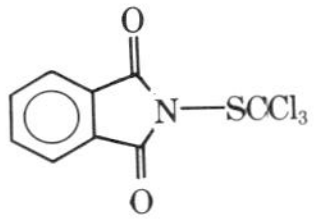 $N-SCCl_3$	[*133-07-3*]	Cosan P Phaltan Fungitrol 11	Cosan	plastics	88
				Chevron Chemical Tenneco	paint resins	137

lachrimator properties and high oral toxicity, methylenebisthiocyanate can be handled with moderate caution and shows high activity against slime-forming bacteria as well as action against spore-formers and fungi (124,139). Two other thiocyanate derivatives, vinylenebisthiocyanate (125) and chloroethylenebisthiocyanate (126), exhibit wide-spectrum performance against bacteria, fungi, yeasts, and algae in industrial water systems. The performance characteristics of organic thiocyanates have been described (140).

The dithiocarbamate derivatives, eg, ethylenebis(dithiocarbamate) (141), which originally were developed in the 1940s as rubber accelerators (142), constitute another large class of organosulfur–nitrogen antimicrobial agents (see Fungicides). Disodium ethylenebis(dithiocarbamate) (Nabam) and zinc ethylenebis(dithiocarbamate) (Zineb) were used (500 t in 1978) as cooling-water and paper-mill slimicides as well as fungicides for rubber, latex paints, and textiles (143). However, the simpler zinc dimethyldithiocarbamate, in a mixture with zinc 2-mercaptobenzothiazole, is used as a preservative in the textile area and elsewhere (144). The teratogenicity of the ethylene analogue zinc salt (Zineb) has been studied in rats and the compound is not considered a significant hazard (145). No other serious toxicity has been cited against the other dithiocarbamates.

Bistrichloromethyl sulfone is used with the MBT organo sulfur derivative as a paper-mill and cooling-water slimicide (127). Total consumption of the chlorosulfone for both applications in 1978 was 200 t ($900,000) (146). *Trans*-1,2-bis(*n*-propylsulfonyl)ethene has been used as a fungicide in oil-based and alkyd paints (128); it is a white, insoluble powder with low toxicity and it is recommended for use at 0.5–1.0% (147). 2,3,5,6-Tetrachloro-4-methylsulfonylpyridine is used almost exclusively, but in limited volume, as a latex-paint mildewcide (129). It is claimed to be more effective against fungi and Gram-positive bacteria than against Gram-negative bacteria (148). A companion product, 2,3,5 trichloro-4-propylsulfonylpyridine, also is offered for use as an antifungal preservative for exterior acrylic-base paint at levels of 15–30 kg/m^3 (12.5–25 lb/100 gal) (130).

Heterocyclic nitrogen–sulfur compositions constitute one of the most advanced classes of antimicrobial agents. Consumption of a variety of pyridine derivatives, isothiazolines, benzimidazoles, thiadiazines, and other structures amounted to ca 1800 t ($16 × 10^6) in 1978 (143). Tetrahydro-3,5-dimethyl-2*H*-1,3,5-thiadiazine-2-thione (DMTT) has the largest consumption (ca 500 t (143)) and is used most widely in paper-mill slimicides and as a preservative in pigment slurries; however, it has limited use in cooling water and adhesives (131–132). It is claimed to be active against fungi and Gram-negative bacteria, which possibly is a result of the hydrolysis product, methyl isothiocyanate rather than of the composition (149).

The sodium and zinc salts of 2-mercaptopyridine-1-oxide represent the largest dollar volume ($10 × 10^6 in 1978) in the nitrogen–sulfur segment of the industry (143). This is largely because of the widespread use of the zinc complex as an antidandruff agent. The zinc derivative is marketed as a 48% aqueous dispersion and a 100% powder and has limited use as a preservative for cosmetics and for other industrial products in addition to the antidandruff use (150). The sodium salt (as a 40% aqueous solution) is used mostly as a preservative for metalworking fluids; although as a solution and as a 100% powder it does have limited use as a preservative for cosmetics and several other industrial products (150). EPA registrations have been established for both products as well as several FDA approvals for products containing these pyridinethiones.

1,2-Benzisothiazoline-3-one is used primarily as a preservative for latex paints, and latexes; it is used less in pigment slurries, adhesives, and metalworking fluids. The product is available as an off-white, thermally stable powder that exhibits low toxicity (151). 2-*n*-Octyl-4-isothiazolin-3-one is used almost exclusively as a latex paint mildewcide (133), to the extent of ca 100 t in 1978 (143). A small percentage of the total, however, represents its application as a mildewcide for hides and leather processing (134). 2-(4-Thiazolyl)benzimidazole has achieved limited use as a latex paint (film) mildewcide and as a fungicide for soap wrappers and other coatings and for postharvest fruit treatment (135). A new product, Grotan K [*55965-84-9*], which is being promoted for use as a metalworking-fluid preservative, is a mixture of the hydrochlorides of 5-chloro-2-methyl-4-isothiazoline-3-one and 2-methyl-4-isothiazoline-3-one (152).

The two trichloromethylthiocarboximides, *N*-(trichloromethylthio)-4-cyclohexene-1,2-dicarboximide (Captan) and the *N*-(trichloromethylthio)phthalimide (Folpet) have been used as agricultural fungicides but have limited use in industrial areas. A highly purified form of Captan is recommended as a fungicide preservative for cosmetics and topical pharmaceuticals (136). Folpet was evaluated as a fungicide in poly(vinyl chloride) (PVC) film formulations and was effective in most cases (137). Both Captan and Folpet are used in plastics as the undiluted active compounds at concentrations of 0.25–0.5%; they provide microbial resistance to the formulations but do not prevent surface growth (88). Despite their extensive use over the past 30 yr, the potential teratogenetic hazards of these two organosulfur derivatives still is questioned (153).

Miscellaneous Compounds. A few antimicrobial agents are unique in their functional classification. In general, each has one significant main use and, in most cases, only limited market significance as an industrial antimicrobial agent. The significant products are shown in Table 9. Organic acids and their salts constitute a sizable category among antimicrobial agents. However, a large fraction of the 24,400 t ($36.7 $\times 10^6$) in 1978 was consumed by the food industry. Sorbic acid [*110-44-1*] (qv), its potassium salt [*590-00-1*], and the sodium [*137-40-6*] and calcium [*4075-81-4*] salts of propionic acid are in this category. Benzoic acid [*65-85-0*] (qv) and its sodium salt [*532-32-1*] do have some industrial use as preservatives for adhesives, although their main areas of utility are for pharmaceutical, cosmetic, and toiletry products. They are recognized as safe by the FDA and may be used in direct contact with food in concentrations of up to 0.1%. Salicylic acid [*69-72-7*](qv) and undecylenic acid [*112-38-9*] are used exclusively as preservatives for pharmaceutical products. The various alkyl esters of *p*-hydroxybenzoic acid [*99-96-7*] also are not employed as industrial preservatives but are used widely in cosmetics, food, and pharmaceutical applications.

Formaldehyde, glutaraldehyde, and acrolein (see Acrolein and derivatives) are moderately significant antimicrobial agents. Only acrolein is used in the industrial area where it has been employed as a slimicide in paper manufacturing and in oil recovery operations in the field. Acrolein is a highly irritating, highly toxic, volatile liquid that must be handled with great care. The TLV for safe, continued 8-h exposure is 0.1 ppm by volume in air.

A mixture of two glycol borates, 2,2′ - oxybis(4,4,6 - trimethyl - 1,3,2-dioxaborinane) [*14697-50-8*] and 2,2′-(1-methyltrimethylenedioxy) bis(4-methyl-1,3,2-dioxaborinane) [*2665-13-6*], has been successfully employed to preserve jet-aircraft and diesel fuels as well as heating oil (154). The dioxaborinane mixture is claimed to be

Table 9. Miscellaneous Industrial Antimicrobial Agents

Common name	Structure	CAS Registry No.	Trade names	Producers	Applications	Ref.
acrolein	$CH_2{=}CHCH{=}O$	*[107-02-8]*	Aqualin	Shell Chemical	paper-mill slimicide oil recovery	
mixed dioxaborinanes	two dioxaborinane rings (CH_3; CH_3, CH_3) joined by B—O—B		Biobor JF	U.S. Borax	hydrocarbon fuels	154
2,6-dimethyl, 1,3-dioxanol-4 acetate	dioxane ring with CH_3, CH_3 and $OCCH_3$ (C=O)	*[828-00-2]*	Givguard-DXM	Givaudan	textiles	155
tris-(hydroxymethyl)-nitromethane	$HOCH_2C(NO_2)(CH_2OH)CH_2OH$	*[126-11-4]*	Tris-Nitro	Commercial Solvents	cutting fluids	156
tetrachloroisophthalonitrile	benzene ring with Cl, Cl, Cl, Cl, CN, CN	*[1897-45-6]*	Nopcocide N-96	Diamond Shamrock	latex paints	
1-(3-chloroallyl)-3,5,7-triaza-1-azoniaadamantane chloride	triazaazoniaadamantane cage, N^+–$CH_2CH{=}CHCl$, Cl^-	*[4080-31-3]*	Dowicil 100	Dow	adhesives floor waxes latex emulsions metalworking fluids paint	157

specifically active against the mold *Cladosporium resinae*, the chief microbial offender in fuel storage situations. Current sales are ca 100 t/yr and apply primarily to aircraft and fuel tanks of naval vessels. 6-Acetoxy-2,4-dimethyl-*m*-dioxane is used to prevent microbial contamination in textile processing solutions. It is available as a 93% active water-soluble liquid and is used at a concentration of 0.1% in emulsions to inhibit discoloration and deterioration of both the textiles and the treating solutions which can result from bacterial and fungal attack (155). A nitroparaffin derivative, tris-(hydroxymethyl)nitromethane, is an effective antimicrobial agent for preserving cutting fluids (156). In 1978, consumption amounted to 1100 t ($\$1.3 \times 10^6$) (122). The product is available as both a 50% aqueous solution and as a white crystalline powder. Tetrachloroisophthalonitrile is used mainly as a mildewcide for aqueous emulsion/latex paints. Consumption in 1978 was ca 300 t ($\$3 \times 10^6$) (157). The compound is offered as a 96% active off-white powder (158). Although it exhibits a low order of acute toxicity, the product should be handled with caution as it may produce temporary allergic rash on exposed skin and redness of the eyes (158). 1-(3-Chloroallyl)-3,5,7-triaza-1-azoniaadamantane chloride is used in a variety of industrial applications including adhesives floor waxes, latex emulsions, metalworking fluids, and paint (159). The ionic adamantane derivative is offered in three different solid forms of varying purity. In latex paints, the derivative is employed at concentrations from 0.05–0.2% where it acts by gradually releasing formalin (160).

BIBLIOGRAPHY

"Industrial Fungicides" in *ECT* 1st ed., under "Fungicides," Vol. 6, pp. 991–995, by R. G. H. Siu, Quartermaster General Laboratories, U.S. Army; "Industrial Fungicides" in *ECT* 2nd ed., under "Fungicides," Vol. 10, pp. 228–236, by N. Joe Turner, Boyce Thompson Institute for Plant Research, Inc.

1. R. A. Payne and J. P. Stekla, *Biocides U.S.A. 1979,* C. H. Kline & Co., Inc., Fairfield, N.J., 1979.
2. S. S. Block, ed., *Disinfection, Sterilization and Preservation,* 2nd ed., Lea and Febiger, Philadelphia, Pa., 1977.
3. F. J. Buono and G. A. Trautenberg "Biocides" in N. M. Bikales, ed., *Encyclopedia of Polymer Science and Technology,* Suppl. 1, Wiley-Interscience, New York, 1976, pp. 95–115.
4. J. H. Conkey, *Tappi* **52,** 2312 (1969).
5. A. Albert, *Selective Toxicity: The Physicochemical Basis of Therapy,* 5th ed., Halsted Press, a division of John Wiley & Sons, Inc., New York, 1979.
6. H. B. Kostenbauder in Ref. 2, pp. 912–932.
7. J. F. Gardner in Ref. 2, p. 889.
8. Ref. 1, p. 888.
9. Ref. 1, pp. 890–891.
10. Ref. 3, pp. 111–114.
11. *Chem. Mark. Rep.* **203**(3), 7 (1973).
12. *Soap Cosmet. Chem. Spec.* **54**(6), 74B (1978).
13. *Federal Insecticide, Fungicide, and Rodenticide Act* (61 Stat. 163; 7 U.S.C. 121 note, 135, 135 note, 135a, 135a note, 135 b–f, 135 f note, 135 g, 135 g note, 135 h, 135 h note, 135 i, 135 j, 135 k), U.S. Government Printing Office, Washington, D.C., 1947.
14. *Federal Environmental Pesticide Control Act, Public Law 92-516* (86 Stat. 973, 7 U.S.C. 136-136y: 15 U.S.C. 1261, 1471; 21 U.S.C. 321, 346A), U.S. Government Printing Office, Washington, D.C., 1972.
15. *Federal Insecticide, Fungicide, and Rodenticide Act, Public Law 95-396,* interim printing, U.S. Environmental Protection Agency, Washington, D.C., Sept. 30, 1978.
16. *Fed. Regist.* **44,** 27932 (May 11, 1979).

17. R. J. Lewis, Sr. and R. L. Tatken, ed., *Registry of Toxic Effects of Chemical Substances,* 8th ed., NIOSH, Publication No. 79-100, GPO Stock No. 017-033-00346-7, Cincinnati, Ohio.
18. *National Library of Medicine (NLM) ELHILL "Toxline" Data Base,* MEDLARS Management Section, Bethesda, Md.
19. *Official Methods of Analysis of the Association of Official Analytical Chemists,* 12th ed., Association of Official Analytical Chemists, Washington, D.C., 1975, pp. 57–71.
20. W. G. Roessler in Ref. 2, pp. 3–10.
21. M. Singer, *Process Biochem.* **11**(6), 30 (1976).
22. C. Genner, *Process Biochem.* **11**(6), 40 (1976).
23. D. F. Spooner and G. Sykes in J. R. Norris and D. W. Ribbons, ed., *Methods in Microbiology,* Vol. 7B, Academic Press, Inc., New York, 1972, pp. 211–76.
24. *Dowicide 7 Antimicrobial,* Form No. 192-89-72, Section I-6, The Dow Chemical Company, Midland, Mich., 1976.
25. *Dowicide G-ST Antimicrobial,* Form No. 192-475-76, Section I-11, The Dow Chemical Company, Midland, Mich., 1976.
26. *Dowicide 2 Antimicrobial,* Form No. 192-477-76, Section I-2, The Dow Chemical Company, Midland, Mich., 1976.
27. *Dowicide B Antimicrobial,* Form No. 192-478-76, Section I-10, The Dow Chemical Company, Midland, Mich., 1976.
28. *Dowicide 1 Antimicrobial,* Form No. 192-84-69, Section I-1, The Dow Chemical Company, Midland, Mich., 1969.
29. *Dowicide A Antimicrobial,* Form No. 192-82-69, Section I-9, The Dow Chemical Company, Midland, Mich., 1969.
30. Ref. 1, p. 101.
31. R. F. Prindle and E. S. Wright in Ref. 2, pp. 219–251.
32. Ref. 31, pp. 221–223.
33. *Chem. Week* **123**(7), 19 (1978).
34. *Chem. Mark. Rep.* **215**(1), 9 (1978).
35. Ref. 1, p. 107.
36. *Chemical Economics Handbook,* Stanford Research Institute, Menlo Park, Calif., Oct. 1976, p. 573.5007 M.
37. S. S. Block in Ref. 2, pp. 790–791.
38. Ref. 1, pp. 254–255.
39. Ref. 1, p. 234.
40. Ref. 1, p. 120.
41. Ref. 1, p. 81.
42. Ref. 1, p. 84.
43. L. Gershenfeld in Ref. 2, pp. 196–218.
44. *Amical Preservatives for Protective Coatings,* Bulletin No. 73-7, Abbott Laboratories, North Chicago, Ill., July 2, 1973.
45. *Adhes. Age* **17**(9), 49 (1974).
46. Ref. 1, p. 93.
47. A. S. Behrman, *Water is Everybody's Business,* Doubleday & Co., Inc., New York, 1968, Chapt. 7.
48. Ref. 1, p. 94.
49. *Onyxide 500,* Data File Bulletin 7671, Onyx Chemical Company, Division of Millmaster Onyx Corporation, Jersey City, N.J., 1971.
50. *Merbac-35 to Inhibit the Growth of Microorganisms in Aqueous Coatings,* Product Data Bulletin PL-1004, Merck Chemical Division, Merck & Co., Rahway, N.J., 1972.
51. M. A. Lesser, *Soap Sanit. Chem.* **25**(8), 119, 139 (1949).
52. Ref. 2, p. 184.
53. J. Davies, *Selection in Pathology and Surgery,* Longmans, Orme, Brown, Green & Longmans, London, Part II, 1839.
54. H. A. Shelanski and M. V. Shelanski, *J. Int. Coll. Surg.* **25,** 727 (1956).
55. Ref. 2, pp. 209–210.
56. Ref. 1, p. 137.
57. O. Rahn and W. P. Van Eseltine, *Ann. Rev. Microbiol.* **1,** 173 (1947).
58. R. G. Puhle, *Soap Sanit. Chem.* **26**(12), 133, 135, 137, 139 (1950).
59. V. J. Penniston and L. R. Hedrick, *Science* **101,** 362 (1945).

60. D. K. Tressler, *Fishing Gaz.*, 66 (Jan. 1947).
61. G. J. Lehn and R. L. Vignolo, *Brew. Dig.* **21,** 41 (1946).
62. U.S. Pat. 3,694,262 (Sept. 26, 1972), J. A. Casey.
63. E. G. Shay and A. Petrocci, *41st Annual Meeting of Soap and Detergent Assoc.*, New York, Jan. 24–26, 1968.
64. H. J. Antonides and W. S. Tanner, *Appl. Microbiol.* **9,** 572 (1961).
65. Ref. 1, p. 333.
66. U.S. Pat. 3,231,509 (Jan. 25, 1966), B. F. Shema (to Betz Laboratories, Inc.).
67. U.S. Pat. 2,733,206 (Jan. 31, 1956), J. H. Prusick and V. P. Gregory (to Armour & Co.).
68. P. A. Walters, E. A. Abbott, and A. J. Isquith, *Appl. Microbiol.* **25,** 253 (1973).
69. *Barquat and Bardac Quaternary Ammonium Compounds,* Product Information Bulletin L-40, 11/72-2, Lonza Inc., Fair Lawn, N.J., 1972.
70. Ref. 2, pp. 329–331.
71. Ref. 2, pp. 331–335.
72. R. B. Kundsin and C. W. Walter, *Appl. Microbiol.* **9**(2), 167 (1961).
73. E. Lenhart, A. Petrocci, and R. Gardon, *Mod. Sanit. Build. Maint.* **5**(12), 27 (1961).
74. U.S. Pat. 3,287,214 (Nov. 22, 1966), F. G. Taylor and R. F. Prindle (to Sterling Drug Inc.).
75. A. J. Krog and C. G. Marshall, *Am. J. Public Health* **30,** 341 (1940).
76. H. A. Shelanski, *Soap Sanit. Chem.* **25**(2), 125 (1949).
77. J. K. Finnegan and J. B. Dienna, *Soap Sanit. Chem.* **30**(2), 147, 149, 151, 153, 157, 173, 175 (1954).
78. Ref. 2, p. 337.
79. Ref. 1, p. 141.
80. Ref. 1, p. 167.
81. *Vinyzene BP-5,* Bulletin No. 30-RP, Ventron Corporation, Chemicals Division, Beverly, Mass., 1973.
82. Ref. 1, p. 173.
83. G. J. M. van der Kerk and J. G. A. Luijten, *J. Appl. Chem.* (*London*) **4,** 314 (1954).
84. Ref. 1, p. 179.
85. U.S. Pat. 1,443,602 (Jan. 30, 1923), M. G. Weber.
86. U.S. Pat. 1,482,416 (Feb. 5, 1924), W. O. Snelling.
87. Ref. 2, p. 793.
88. E. L. Cadmus "Preservatives" in *Modern Plastics Encyclopedia 1978–1979,* Vol. 55, No. 10A, McGraw-Hill Inc., New York, 1978, pp. 218, 220.
89. U.S. Pat. 2,818,344 (Dec. 31, 1957), S. J. Buckman (to Buckman Laboratories, Inc.).
90. R. T. Ross, *Develop. Ind. Microbiol.* **6,** 149 (1964); *Am. Paint J.* **55,** 23 (1971).
91. Ref. 1, p. 157.
92. H. A. Gardner, L. P. Hart, and G. G. Sward, *Natl. Paint Varnish Lacquer Assoc. Circ. No. 448,* 17 (1934).
93. W. G. Vannoy, *Off. Dig. Fed. Paint Varnish Prod. Clubs* **277,** 163 (1948).
94. Ref. 1, p. 165.
95. Ref. 1, p. 153.
96. N. Grier in Ref. 2, pp. 361–394.
97. Ref. 1, p. 169.
98. *Chem. Eng.* **86**(2), 60 (1979).
99. M. D. Steele and R. W. Drisko, *J. Coat. Technol.* **48**(616), 59 (1976).
100. *Chem. Week* **125**(4), 42 (1979).
101. C. K. Banks "Tin Compounds" in A. Standen, ed., *Encyclopedia of Chemical Technology,* 2nd ed., Vol. 20, John Wiley & Sons, Inc., New York, 1969, p. 321.
102. Ref. 1, p. 184.
103. Ref. 1, p. 156.
104. R. C. Harnden, *Paint Oil Chem. Rev.* **108,** 32 (1945).
105. S. B. Salvin, *Ind. Eng. Chem.* **36,** 336 (1944).
106. N. Grier in Ref. 2, pp. 395–407.
107. A. J. Salle in Ref. 2, pp. 408–416.
108. Ref. 1, p. 203.
109. *Fed. Regist.* **39,** 33102 (Sept. 13, 1974).
110. D. S. Wilkinson, *Brit. J. Dermatol.* **73,** 213 (1961); *Brit. J. Dermatol.* **74,** 295 (1962).
111. L. J. Vinson and R. S. Flatt, *J. Invest. Dermatol.* **38,** 327 (1962).

112. *Trichlorocarbanilide (TCC),* Bulletin No. FC-4A, Monsanto Company, St. Louis, Mo., Jan. 1973.
113. Ref. 1, p. 189.
114. E. Lagarde, *Ann. Inst. Pasteur* **100,** 368 (1961).
115. Ref. 1, p. 198.
116. Ref. 1, p. 191.
117. *Bioban P-1487,* Technical Data Sheet No. 19, IMC Chemical Group, Inc., NP Division, Hillside, Ill.
118. Ref. 1, pp. 195–196.
119. *Troysan 174,* Technical Bulletin No. FP-124, 5M-10/71, Troy Chemical Corporation, Newark, N.J., Oct. 1971.
120. *Vancide TH Industrial Preservative,* Technical Data Sheet R-A-V-31, Form VTH-1g, R. T. Vanderbilt Company, Inc., New York, 1972.
121. *Onyxcide 200, Technical Data Sheet,* Onyx Chemical Company, Division of Millmaster Onyx Corp., Jersey City, N.J., Nov. 1972.
122. Ref. 1, p. 194.
123. R. J. Lukens in Ref. 2, pp. 863–865.
124. Ref. 2, p. 810.
125. *Cytox 3711 Industrial Biocide, Water Treatment Chemicals,* Data Bulletin 8-1416-200, American Cyanamid Company, Water Treating Chemicals Dept., Wayne, N.J., July 1969.
126. *Cytox 3810 Industrial Biocide, Water Treating Chemicals,* Data Bulletin 9-1326-200, American Cyanamid Company, Water Treating Chemicals Dept., Wayne, N.J., Jan. 1970.
127. *Stauffer N-1386, Product Data Sheet 1344-000-02C,* Stauffer Chemical Company, Specialty Chemical Division, Westport, Conn.
128. S. Hart, *Am. Paint J.* **55,** 21 (1971).
129. *Dowicil S13 Antimicrobial,* Form No. 192-108-69, Section I-15, The Dow Chemical Company, Midland, Mich., 1969.
130. *Dowicil A-40 Antimicrobial,* Form No. 192-300-74, Section I-16, The Dow Chemical Company, Midland, Mich., 1974.
131. *"Metasol D3T," Product No. 91535, Product Data Bulletin PL-2011 1272,* Merck & Co., Inc., Merck Chemical Division, Rahway, N.J., June 1972.
132. *Biocide N-521, Product Sheet 1344-000-02A,* Stauffer Chemical Company, Specialty Chemical Division, Westport, Conn., Aug. 1972.
133. *Skane M-8—New Paint Mildewcide,* Bulletin C-334, Rohm and Haas Company, Philadelphia, Pa., May 1972.
134. *Kathon LP—Mildewcide for Hide and Leather Processing,* Bulletin LE-47A, Form 5177B, Rohm and Haas Company, Philadelphia, Pa.
135. *Metasol TK-100,* Product Data Bulletin, PL-96-01-969, Merck & Co., Inc., Merck Chemical Division, Rahway, N.J., Sept. 1969.
136. *Vancide 89RE, Technical Bulletin No. 91,* R. T. Vanderbilt Company, Inc., New York.
137. A. M. Kaplan, M. Greenberger and T. M. Wendt, *Polym. Eng. Sci.* **10,** 241 (1970).
138. Ref. 1, p. 147.
139. *Cytox 3522 Industrial Biocide,* Water Treatment Section, Data Bulletin 7-1341-300, American Cyanamid Company, Industrial Chemicals Division, Wayne, N.J., Oct. 1967.
140. D. C. Wehner and C. F. Hinz, *Develop. Ind. Microbiol.* **12,** 404 (1971).
141. D. C. Torgeson "Agricultural Fungicides" in A. Standen, ed., *Encyclopedia of Chemical Technology,* 2nd ed., Vol. 10, John Wiley & Sons, Inc., New York, 1966, p. 224.
142. G. D. Thorn and R. A. Ludwig, *The Dithiocarbamates and Related Compounds,* Elsevier Publishing Co., Inc., New York, 1962.
143. Ref. 1, pp. 146–149.
144. *Vancide 51Z-Wettable Powder Mildew Inhibition Agent, Preservative and Slime Control Agent for Industry,* Technical Data Sheet No. V51Z-1F, R-A-V-56, R. T. Vanderbilt Company, Inc., New York, Apr. 1972.
145. T. Petrova-Vergieva and L. Ivanova-Tchemishanska, *Food Cosmet. Toxicol.* **11,** 239 (1973).
146. Ref. 1, pp. 335, 405.
147. *Vancide PA: Non-Metallic Mold Inhibitor for Paint,* Technical Data Sheet 12-69-1500, R. T. Vanderbilt Company, Inc., New York, July 1967.
148. P. A. Wolf and F. J. Bobalek, *Appl. Microbiol.* **15,** 1376 (1967).
149. Ref. 141, pp. 224–225.

150. *Zinc OMADINE and Sodium OMADINE,* Technical Bulletin AD-1372-973, Olin Corporation, Designed Products Division, Stamford, Conn., Sept. 1973.
151. *Proxcel CRL—Microbiostat for In-Can Preservation of Paints and Latex Emulsions,* Technical Bulletin LP-80 (Revision of Bulletin 325), ICI American Inc., Atlas Chemicals Division, Wilmington, Del., 1973.
152. *Grotan K, Product Data Bulletin,* Sterling Drug Inc., Lehn & Fink Industrial Products Division.
153. K. P. Shea, *Environment* **14**(1), 22, 29 (1972).
154. *Biobor JF (Mixed Dioxaborinanes),* Service Bulletin DP205, U.S. Borax & Chemical Corp., Los Angeles, Calif.
155. *Dioxin (Now Givguard-DXN)—Brand of Dimethoxane Antimicrobial for Textile Chemical Specialties,* Technical Bulletin, Givaudan Corporation, Sindar Division, New York.
156. E. O. Bennett and H. N. Futch, *Lubr. Eng.* **16**, 228 (1960).
157. Ref. 1, p. 97.
158. *Nopcocide N-96, Technical Microbiocide for Control of Bacteria and Fungi,* Nopco Technical Bulletin PAD-8 Rev., Diamond Shamrock Chemical Company, Nopco Chemical Division, Morristown, N.J., Oct. 1972.
159. *Dowicil 100 Antimicrobial,* Form No. 122-220-68, Section I-14, The Dow Chemical Company, Midland, Mich., 1968.
160. Ref. 2, p. 806.

SAMUEL I. TROTZ
JAMES J. PITTS
Olin Corporation

INDUSTRIAL HYGIENE AND TOXICOLOGY

Early humans used the toxic effects of animal venoms and poisonous plants in hunting and warfare. The Ebers papyrus, perhaps our earliest medical record (ca 1500 BC), lists more than 800 recipes, many containing poisons such as hemlock. Industrial hygiene begins in the fourth century, BC, with Hippocrates' account of lead toxicity in the mining industry. Some 500 years later, Pliny the Elder recognized the dangers encountered in handling zinc and sulfur, whereas Galen, a Greek physician residing in Rome in the second century, described the dangers of acid mists to copper miners.

Thereafter, little attention was paid to occupational diseases until the 18th century. Sir George Baker correctly attributed Devonshire colic to lead in the cider industry, whereas Percival Pott recognized soot as a cause of scrotal cancer and promoted the passage of the Chimney-Sweepers Act of 1788 in England. The English Factory Acts of 1833 are considered the first effective legislation concerning the health and safety of the working population. England's lead was followed by passage of Workmen's Compensation Acts by various European nations.

In the United States, an early 20th century champion of workers' welfare was Alice Hamilton, a physician (1). Her efforts mark the beginning of an increasing public awareness of occupational medical problems. In 1908, the U.S. government passed a compensation act for certain civil employees, and in 1911, the first state compensation laws were passed. These Workmen's Compensation laws were important in the development of industrial hygiene in the United States, as employers found it more economical to protect the workers by environmental control.

Recent Federal Laws

By 1948, all the states had compensation laws, and in recent years, four important pieces of legislation were passed (see also Regulatory agencies).

The Metal and Nonmetallic Mine Safety Act of 1966 formulates health and safety standards for metal and nonmetallic mining. It also requires annual reports of all accidents, injuries, and occupational diseases occurring in mines.

The Federal Coal Mine Health and Safety Act of 1969 attempts to provide the highest degree of health protection for the miner, and creates an advisory committee to study mine problems.

The Occupational Safety and Health Act of 1970 intends to "assure so far as possible every working man and woman in the nation safe and healthful working conditions and to preserve our human resources." Under its terms, the federal government is authorized to promulgate occupational safety and health standards applicable to any business affecting interstate commerce. Responsibility for enforcing such standards rests with the Occupational Safety and Health Administration (OSHA) in the Department of Labor. However, the National Institute for Occupational Safety and Health (NIOSH) in the Department of Health, Education, and Welfare, is responsible for research and education and training programs for effectuating the purposes of this act.

The U.S. Toxic Substances Control Act (TOSCA), which took effect Jan. 1, 1977, assigns to the EPA the responsibility of preparing an inventory of all chemicals manufactured in or imported into the United States (2–8). This inventory is expected to provide a profile of the U.S. chemical industry, and to make it possible to regulate new chemicals. A manufacturer intending to market a new chemical must notify the EPA, and supply data on possible hazards. Taking costs and benefits into account, the EPA is supposed to decide whether the new chemical represents an unreasonable risk to human health or the environment. Furthermore, the EPA is empowered to ban certain chemicals that have been found to be harmful, as for example, nonessential aerosol uses of certain chlorofluorocarbons (see also Aerosols). For the convenience of industry and the public, the Industry Assistance Office of the Office of Toxic Substances (OTS) has set up a toll-free number (800-424-9065) which gives information on the status of chemicals.

In addition, increasing public concern with the effect of toxins in air, food, and water has led to the following legislation during the period 1972–1976: Federal Environmental Pesticide Control Act, 1972; Federal Insecticide, Fungicide, and Rodenticide Act, 1972; Federal Water Pollution Control Act, 1972; Safe Drinking Water Act, 1974; Resource Conservation and Recovery Act, 1976; Clean Air Act, 1970.

Industrial Hygiene

The profession of industrial hygiene embraces the total realm of control, including recognition and evaluation of those factors of the environment originating in the place of work which may cause illness, discomfort, or lack of well-being, either among workers or among the community as a whole (9–10).

Industrial hygiene begins with the recognition of industrial health problems such as biological problems caused by insects and mites, molds, yeasts and fungi, bacteria, and viruses; chemical problems caused by liquids, dusts, fumes, mists, vapors, or gases;

energy problems caused by electromagnetic and ionizing radiations, noise, vibration, and temperature and pressure extremes; and ergonomic problems caused by body position in relation to task, monotony, repetitive motion, boredom, work pressure, anxiety, and failure.

After recognition of the hazard, the work atmosphere is evaluated for long- and short-range effects on health.

Finally, corrective measures are developed such as the replacement of harmful or toxic materials; changing work processes; new ventilation procedures; increasing distance and time between exposures to radiation; introduction of water to reduce dust emissions in certain fields, eg, mining; maintenance of "good housekeeping" and adequate methods of waste disposal; and use of proper protective working clothes, including respirators and ear protectors when necessary (see also Air pollution; Noise pollution).

Evaluation. The four general steps in the evaluation of the occupational environment are recognition of potential hazards, preparation of a field study, the conduct of the field study and interpretation of survey results.

Recognition of potential hazards includes an inventory of physical and chemical agents encountered in a specific process, a periodic review of different activities in a work area, and a study of existing control measures. In the preparation of a field study, the proper instruments are selected and checked, and the appropriate analytical methods are developed. When conducting a field study, the manner of sampling is of great importance. Finally, the results of the survey are interpreted. Comparison should be made with health standards and previously obtained data.

The contaminants in the atmosphere must be accurately and precisely identified by the latest analytical methods and instrumentation (11–12) (see also Analytical methods).

The following procedures are used (13) in surveying an occupational environment.

Survey Checklist. Determination of purpose and scope of study:

- comprehensive industrial hygiene survey
- evaluation of exposures of limited group of workers to specific agent
- determination of compliance with specific recognized standards
- evaluation of effectiveness of engineering controls
- response to specific complaint

Discussion with appropriate representatives of management and labor.

Analysis of plant operations and pertinent data:

- process flow sheets and plant layout
- raw materials, intermediates, by-products, and products
- toxicological information
- list of job classifications and the environmental stresses to which workers may be exposed
- activities associated with job classifications
- status of workers' health
- administrative and engineering control measures
- reports of previous studies
- determination of health hazards associated with plant operations

Preparation for field study:

- determination of chemical and physical agents to be evaluated
- estimation of range of contaminant concentrations

- review or development of sampling and analytical methods, including sensitivity and specificity
- collection and calibration of field equipment
- assembly of personal protective equipment (hard hat, safety glasses, goggles, hearing protection, respiratory protection, safety shoes, coveralls, gloves, etc)
- preparation of tentative sampling schedule

Field study:
- check of process operating schedule
- notification of management and labor of field study
- organization of monitoring or sampling units
- each sample should be identified by:
  - number
  - description
  - time sampling began
  - flow rate of sampled air
  - time sampling ended
  - any other significant information or observation (eg, process upsets, ventilation system not operating, etc)
- dismantling of sampling units
- all samples (filters, liquid solutions, charcoal or silica gel tubes, etc) requiring subsequent laboratory analyses should be sealed and labeled

Interpretation of results:
- analysis of samples
- determination of:
  - TWA exposures of job classification
  - peak exposures of workers
  - statistical reliability of data
- comparison of sampling results with applicable industrial hygiene standards

Discussion of survey results with representatives of management and labor.

Implementation of corrective action comprised of:
- engineering controls (isolation, ventilation, etc)
- administrative controls (job rotation, reduced work time)
- personal protection
- biological sampling program
- medical surveillance

Determination of need for evaluation of:
- air pollution
- water pollution
- solid waste disposal
- safety

Control Procedures. ***Administrative Control.*** By limiting the amount of time a worker is exposed to a particular substance or group of substances, the 8-h average exposure is kept below the permissible limit. For example, the airborne contaminant limit for oil mist is 5 mg oil/m^3 air based on an 8-h exposure. Therefore, a worker exposed to a level of 10 mg/m^3 is limited to 4 h/d. The remainder of the shift must be spent in an area free of oil mist. This principle can be extended to noise exposure. This method may be inapplicable because of work schedules, union agreements, or worker training requirements.

Engineering Control. Engineering and process controls fall into four categories.

Process Change. Many processes used today were developed when water was inexpensive, air pollution was disregarded, and employee protection a minor concern. Today, the emphasis has shifted and control of air and water contaminants has greatly increased costs. As a result of Federal legislation and union efforts, employee health and welfare are of primary concern. In process redesign, pollution and industrial hygiene problems must be minimized. An example of the reduction of air contamination by a process change is the replacement of spray painting by dip painting. Low-speed oscillating sanders have replaced high-speed disk sanders in order to control lead dust generated in grinding automobile solder seams.

Substitution. A product with low toxicity can often be substituted for a more toxic product to bring an excessive exposure into acceptable limits. For example, the replacement of carbon tetrachloride (exposure limit 10 ppm) by methylchloroform (exposure limit 350 ppm), or the use of harmless powders in foundries to replace parting compounds containing free silica.

Isolation. Control by isolation, or containment, is usually reserved for highly toxic materials. OSHA defines an isolated system as a fully enclosed structure, other than the vessel of containment, which is impervious to the contaminant and prevents its loss to the environment in case of leakage or spillage. A glove box is an example of an isolated system. In a less strict sense, isolation is the imposition of a barrier around a physical hazard. Storage of physical chemicals in a ventilated cabinet is an example. Pumps can often be isolated in an area so ventilated that small packing leaks are of no consequence, and significant ones can be repaired without a serious inhalation hazard.

Ventilation. General ventilation controls contaminants by dilution with large volumes of air which must be exhausted and replaced (see also Exhaust control, industrial; Air pollution control methods). This operation can be expensive in cold climates and should not be used as a primary control method. It is more economical to control worker exposure by local ventilation, with general ventilation as a secondary exhaust system. Local exhaust prevents dispersal of contaminants by capturing them at or near their point of generation, requiring lower and more economical air flow rates. With local exhaust, the complete exhaust systems consist of hoods, ductwork, air-cleaning device, and the fan (see Fans and blowers).

Hood design and location is crucial to the functioning of a local exhaust system. The enclosing hood provides the best control with the lowest flow of exhaust air. However, many processes require physical access, so that hood enclosure is not feasible. Open hoods require heavier air exhaust and are subject to interference by cross drafts. Escape of contaminant from an open hood can be reduced by tank covers, foams, beads, or other floating material.

Ductwork provides a channel for flow of contaminated air from a hood to the point of discharge. Duct diameters must permit adequate transport velocity for any particles entrained in the air stream.

Air-cleaning devices are required on most local exhaust systems in order to meet air pollution regulations. With increased concern for conservation of energy, recirculation of air is often considered. However, with regard to contaminant control, recirculation has traditionally been discouraged, because incorrect operation or poor maintenance causes contaminated air to be returned to the workplace. Clean make-up air decreases exposure if it is discharged at the workplace rather than from a plenum at ceiling height.

The fan, the last component of local exhaust, should be placed downstream from the air-cleaning device to handle clean air. In this manner, fan blades remain clean and a backward-curved blade can be used for high efficiency at low power. If a fan must be located in the dirty airstream, radial or paddle-wheel-type blades are best, because they are less likely to clog.

Industrial Toxicology

Industrial toxicology is largely concerned with the development and use of toxicity information as a means of predicting safe or harmful amounts of materials which may be encountered in the workplace, home, or the general environment. Knowledge of the toxicity of industrial chemicals is extensive (11,14–15).

Definitions. LD_{50}, *average lethal dose,* is the dosage that, when administered to animals, kills half (50%) of them. It is usually expressed in mg toxicant per kg body weight, and the test route (oral, intravenous, or subcutaneous) is usually given (16).

LC_{50}, *average lethal concentration,* is that concentration of a toxicant in air which, when administered to test animals for a definite time period, kills half (50%) of them. It is expressed in ppm for the specific time period (16).

TDL_0, *toxic dose low,* is the lowest known dose of a substance that has produced any toxic, carcinogenic, mutagenic, teratogenetic, or neoplastic (tumorous) effects (17).

TCL_0, *toxic concentration low,* is the lowest concentration of a substance that is reported to have any toxic effect (17).

TLV, threshold limit value, is an estimate of the average safe toxicant concentration that can be tolerated on a repetitive basis, usually an 8-h period on a day-to-day basis (16).

TWA, time-weighted average, is a mathematical expression summing the products of toxicant concentrations and durations of exposure to those toxicants, and dividing by the total exposure time. In simple terms, the concentrations of the various toxicants are multiplied by the duration of exposure to each individual toxicant, the results are added, and then divided by the time of exposure (16).

Inhalation. The adult human lung has an enormous gas–tissue interface (90 m^2 total surface, 70 m^2 alveolar surface) (18). This large surface, together with the blood capillary network surface of 140 m^2, with its continuous blood flow, makes possible an extremely rapid rate of absorption of many substances from the air in the alveolar portion of the lungs into the blood stream. Methods of evaluating the hazard of inhaled environmental substances have been studied more intensively than percutaneous hazards (19–21). The vast literature on inhalation toxicology includes a number of excellent monographs (22–25).

The blood output of the heart is about 5 L/min. The combination of the large surface area and rapid passage of a very thin layer of blood over the surface makes it possible for substances in the air sacs to pass rapidly by simple diffusion into, or out of, the circulating blood plasma. After passing through the lung, the blood is pumped by the left strium and ventricle to the general circulation; thus, an inhaled substance can be directly passed to areas such as the brain or kidney, without necessarily passing through the liver, as is the case with ingested materials.

The concentration of gases or vapors that dissolve in a liquid is directly proportional to its concentration in the free space above the liquid (Henry's law). This holds true for a gas in the alveolar air (air sacs) in contact with blood, and a characteristic equilibrium ratio exists for each substance called the blood–air distribution coefficient (DC):

$$DC = \frac{\text{mg/L blood}}{\text{mg/L alveolar air}}$$

This ratio remains constant over a wide range of concentrations. The partial pressure of the gas in the alveolar air determines the concentration in the blood up to the limits of solubility in blood. Materials that are water-soluble tend to have high distribution coefficients (eg, acetone = 330), whereas those that are less soluble in water have lower values (eg, ethyl ether = 15, benzene = 6.6).

Vapors that have high DC values generally build up slowly in the blood and die away slowly on cessation of exposure. Those with low DC values build rapidly to a maximum, and die away rapidly when exposure ceases. The lung-ventilation rate determines the saturation rate of the blood and tissues for materials having a high DC, whereas the circulation rate is of much greater importance for substances with a low DC. This type of information is not only invaluable in understanding the nature of a toxic hazard, but is used increasingly as a means of establishing the extent of an individual's recent environmental exposure.

In respect to aerosol hazards, the anatomy and aerodynamics of the respiratory tract and the physical properties of the particles become of the utmost significance in determining the deposition, retention, and removal from the lung.

Adsorption of a gas or vapor on a particle may increase the local reaction (26). The ability of the lung to clear itself of particles deposited in the ciliated lining of the bronchioles, bronchi, and trachea results in many types of particles being transported from the lung to the mouth and thence to the stomach. If the substance is capable of being absorbed from the intestinal tract, then this may be the route of entry into the systemic circulation.

The factors governing the sites of deposition, retention, distribution, and final health effects of solid and liquid particulates differ in most respects from those for gases, fumes, and vapors. However, liquids adsorbed on particulates and submicrometer particles 0.05 μm in diameter that may act as gases adsorbed on particulates are the exceptions.

The aerodynamic diameter of particulates governs exposure of the respiratory system (27). Fiber geometry is also important in relation to certain toxicologic properties, eg, in the induction of mesotheliomas by asbestos particles (see also Asbestos).

In order to simplify calculations of the deposition pattern of aerosols, the manifold compartments of the respiratory system are reduced to the nasopharyngeal, tracheobronchial, and pulmonary compartment. If the particulates are assumed to be present as log-normal distributions and three tidal volumes are used, a table can be developed showing the amount of particles deposited in each of the three compartments according to unit-density sizes, ranging from 0.01 to 10 μm. As expected, no particles less than 0.6 μm were deposited in the nasopharynx at any of the three breathing rates, whereas practically all of the particles greater than 6 μm were deposited at this site at all breathing rates. Thus, the main site of deposition of the smaller particles is the lung; deposition in the tracheobronchial compartment never exceeds 25 to 30% of the total particles inspired, even at the smallest sizes (ca 0.01 μm) and the slowest breathing rate. Although various degrees of mouth-breathing would upset the calculations of deposition in the upper respiratory tract, it is not believed to seriously affect pulmonary deposition.

The interpretation of inhalation toxicity data requires expert knowledge of the methods, and a wide experience with many types of substances. Approaches to conducting experimental toxicological studies vary. The Federal government program provides for continual review of test procedures and for their modification as new data become available. A report of four governmental agencies, CPSC (Consumer Product Safety Commission), EPA, FDA, and OSHA, issued in 1979 (28) seems to represent a consensus of approved methods. These agencies form the Interagency Regulatory Liaison Group (IRLG) which is an attempt to share their research, avoid duplicate regulations, and set consistent standards.

Skin Contact. The skin has ca 1.86 m^2 of surface area for possible contact with harmful substances (29). It retains a large portion of the total available body water and has, in addition to its protective and excretory functions, a principal role in controlling heat exchange with the environment. During the past forty years, methods for evaluating systemic hazards from skin contact have been applied on a large scale (30–32), and it has now become common practice to determine the acute toxicity of a new material by dermal application as well as by oral or inhalation routes. The National Research Council has prepared a useful summary (33) of methods for evaluation of skin irritancy, with particular reference to household substances. Current reports of IRLG should be consulted routinely for any changes in recommendations.

The most commonly employed methods for determining acute dermal toxicity (34) generally use rabbits or guinea pigs (35). Substances are applied to the clipped skin in varying quantities and held in place for 24 hours by a sleeve of impervious plastic sheeting or rubber dam. Observations are made for at least two weeks, much as in the case of the acute oral LD_{50}. Such tests allow an estimate of the hazard of serious systemic effects by dermal contact, and they may give an idea of the rapidity of absorption. Generally they do not provide a quantitative measure of the percentage of the applied dose that has penetrated the skin. Some substances may be absorbed into the skin but not penetrate it. Quantitative data can be obtained in a variety of ways by tracer substances or measuring their concentration in blood or excretion in urine, and by in vitro methods (36–37).

When a substance makes contact with the skin, the latter and its associated film of lipid can act as an effective barrier against penetration, injury, or other forms of disturbance; the substance can react with the skin surface and cause primary irritation (dermatitis); it can penetrate the skin and react with tissue protein, resulting in skin sensitization; and it can penetrate the skin, enter the blood stream, and act as a possible systemic poison.

Serious and even fatal poisonings have occurred from brief exposures of confined areas of the skin to highly toxic substances. If the skin has abrasions, cuts, and lacerations, the possibility of penetration is increased.

The importance of appreciable skin contact in industrial exposures is shown by substances such as benzidine, which have negligible vapor pressure, but are readily absorbed through the skin (see Amines, aromatic). The skin provides the major route of entry for such substances; for this reason, no air standard has been set. Exposure is controlled by biological monitoring, which involves analysis of a fluid such as urine, blood, tears, perspiration, etc, or body components, eg, hair or nails.

Prediction of the ability of a substance to cause human skin sensitization (skin allergy) is of great practical significance and the literature is extensive. Guinea pigs can be induced to respond to repeated contact with certain simple chemical compounds in a manner similar to humans (38); this phenomenon involves an immune process. After repeated contacts with the chemical, during which time no inflammatory reactions can be seen, the skin develops an altered capacity to react to the substance even in very minute amounts, and in any area. The process is thought to involve formation of a modified skin protein that in some way induces antibody formation, probably in the regional lymph nodes. This form of allergy is known as delayed hypersensitivity.

Human patch testing to predict the likelihood of skin sensitization is carried out by repeated topical application. Formerly it was thought essential that large numbers of subjects must be used for statistical reliability, but this has been questioned. Some dermatologists believe that the value of such tests can be improved by using mild skin irritation, more repeated exposures, better occlusion of the area, etc (39–40).

Ingestion. In the industrial environment, ingestion is the least important of the three modes of entry to the body. Few substances can be ingested, and the frequency and degree of contact are very limited. Furthermore, and most important, toxicity by mouth is generally of a lower order than that by inhalation.

Modes of Action. Toxic materials, after entering the body, may exert their effects by physical, chemical, or physiological (enzymatic) means or a combination of these (11,15).

Carcinogens. In 1975, NIOSH listed 1500 chemical substances suspected of having carcinogenic effects (41–42). In the past four years, this list has grown to more than 2000.

Substances designated as carcinogenic by FDA and EPA are suspect only, as determined by animal experiments, whereas substances considered occupational carcinogens have been identified by epidemiologic studies or based on actual experience (43). Leads may be gathered from short-term tests of mutagenicity (44), the most important of which is the Ames test (45–49), which shows a strong correlation between carcinogenicity and mutagenicity. This test, which currently costs between $300 and $1000, is used routinely by a number of companies for premarket testing of new products. Suggestive evidence can also be obtained from comparative chemical structures (50).

Criteria of risk estimates are presented by the IRLG (28) in their report on the identification of potential carcinogens. Conditions for estimations of risk and determination of threshold of response of occupational carcinogens are more favorable than those concerning carcinogens in the general environment.

Exposure-Limit Values. Although it is the responsibility of OSHA to establish legal standards for the workplace, many organizations in the United States determine TLV data or their equivalents. Probably the most prominent of these is the American Conference of Governmental Industrial Hygienists (ACGIH) which publishes a list every year (51). These TLVs represent concentrations of airborne substances and physical agents below which workers may be exposed eight hours per day, forty hours per week, without adverse effect. Medical surveillance is recommended to detect workers who are oversensitive to specific chemicals or physical agents. These workers should only be exposed to these substances under special protection. Ceiling values represent exposure limits related to substances that are fast acting and whose threshold limits are more appropriately based on a particular biologic response. In instances where the cutaneous route is an important source of absorption, a substance is so labeled since the TLV refers only to inhalation as the source of entry of the agents into the body.

The American National Standards Institute, Inc. (ANSI) publishes consensus standards of acceptable concentrations for chemical and physical agents. These standards are used to design engineering procedures for the prevention of objectionable levels of chemical and physical agents in the work environment. Acceptable concentration values are presented in terms of a time-weighted 8-h workday, acceptable ceiling concentrations within an 8-h workday, and acceptable maximum peak concentrations for short specified durations. ANSI-acceptable concentrations are values below which ill effects are unlikely. However, these values are not to be used as the basis for establishing the presence of occupational disease.

The Committee on Toxicology of the National Research Council (National Academy of Sciences and National Academy of Engineering) publishes a list of recommended emergency exposure limits (52). This list is not intended as a guide in the maintenance of healthful working environments, but for the planning of the management of emergencies.

The American Industrial Hygiene Association (AIHA) publishes a *Hygienic Guides Series* covering an extensive list of chemicals including significant physical properties, hygienic standards (limits for 8-h, time-weighted exposures, short-exposure tolerance, and atmospheric concentrations immediately hazardous to life), toxic properties (exposure via inhalation, ingestion, or skin contact), industrial uses, hazards and their recommended controls, and medical information (emergency treatment and special procedures). In 1976, AIHA established a new technical committee, the Workplace Environmental Exposure Limits Committee (WEEL) (53), to define workplace environmental limits for chemical substances and physical stresses for which no TLV, hygienic guide, ANSI standard, or other limit existed. Substances may be submitted for evaluation.

The responsibility of developing and publishing criteria dealing with toxic materials and harmful physical agents rests with NIOSH. These criteria describe safe levels of exposure for various periods of employment (54). In addition, NIOSH conducts and publishes research, including industry-wide studies, which lead to the development of standards. The recommended OSHA and NIOSH standards are presented in Table 1 (55).

The NIOSH Registry of Toxic Effects of Chemical Substances (RTECS), formerly the Toxic Substances List, is a compendium of toxicity data abstracted from the scientific literature. The 1978 edition contains 124,247 name entries: 33,929 are different substances with their associated toxicity data, and 90,318 are synonyms. This edition also includes primary skin and eye irritation data (58).

The RTECS is also produced in microfiche on a quarterly basis. This facilitates data storage and processsing, and allows NIOSH to update economically and produce the RTECS file four times per year; ca 1500 new substances are added to the file each quarter. Subscriptions to the quarterly microfiche edition of the Registry are available. Another data bank for the toxicologist and pathologist conducting bioassays is the TOXSYS marketed by Beckman Instruments (59).

Table 1. Summary of NIOSH Recommendations for Occupational Health Standards[a] October, 1978

Substance	CAS Registry Number	Current OSHA environmental standard	NIOSH recommendation for environmental exposure limit[b]	Health effect considered	Comments
acetylene	[*74-86-2*]	2500-ppm (10% of lower explosive limit)	no exposure in excess of 2500-ppm (2662 mg/m^3)	indirect asphyxia	employers to check for, and inform employees of contaminants such as arsine and phosphine
acrylamide	[*79-06-1*]	0.3 mg/m^3, 8-h TWA (skin)	0.3 mg/m^3 TWA	skin, eye, nervous system effects	skin and eye contact to be prevented
acrylonitrile	[*107-13-1*]	20-ppm, 8-h TWA (skin)	not greater than 4-ppm by recommended method (8.7 mg/m^3)	lung and bowel cancer	chest x-ray required, first-aid and medical kits to be available during use; hazardous liquid, skin
aldrin/dieldrin[c]	[*309-00-2*], [*60-57-1*]	0.25 mg/m^3, 8-h TWA (skin)	lowest reliably detectable level; 0.15 mg/m^3 TWA by NIOSH-validated method; skin contact to be prevented	cancer	
alkanes (C_5–C_8)		1000-ppm, 8-h TWA, pentane; 500-ppm, 8-h TWA, *n*-hexane, *n*-heptane, octane	350 mg/m^3 TWA (120-ppm pentane; 100-ppm hexane; 85-ppm heptane; 75-ppm octane) mixtures to be not greater than 350 mg/m^3 TWA; 1800 mg/m^3 ceiling singly or mixtures (15-min)	skin and nervous system effects	action level defined as 200 mg/m^3 for these substances
allyl chloride	[*107-05-1*]	1-ppm, 8-h TWA	1-ppm TWA (3.1 mg/m^3); 3-ppm ceiling (9.3 mg/m^3); (15-min)	liver, kidney, lung effects	urine, blood, and pulmonary function testing required
ammonia	[*7664-41-7*]	50-ppm, 8-h TWA	50-ppm ceiling (34.8 mg/m^3) (5-min)	airway irritation	hazardous liquid; eye damage
antimony	[*7440-36-0*]	0.5 mg/m^3, 8-h TWA	0.5 mg/m^3 TWA	irritation; heart and lung effects	chest x-ray, pulmonary function, and electrocardiogram testing required
arsenic inorganic	[*7440-38-2*]	0.5 mg/m^3 TWA	2 μm as/m^3 ceiling (15-min)	dermatitis, lung and lymphatic cancer	chest x-rays required
asbestos	[*1332-21-4*]	2,000,000 fibers/m^3 8-h TWA, 10,000,000 fibers/m^3 ceiling	100,000 fibers/m^3 over 5 μm TWA; 500,000 fibers/m^3 over 5 μm ceiling; (15-min)	asbestosis, lung cancer	Federal standard promulgated July 7, 1972; TWA lowered July 1, 1976
asphalt fumes		2.5 mg/m^3	5 mg/m^3 ceiling (15-min)	eye and respiratory irritation	hazardous substance, skin

Table 1 (*continued*)

Substance	CAS Registry Number	Current OSHA environmental standard	NIOSH recommendation for environmental exposure limit[b]	Health effect considered	Comments
benzene	[*71-43-2*]	10-ppm, 8-h TWA; 25-ppm acceptable ceiling; 50-ppm maximum ceiling (10-min)[d]	1-ppm ceiling (3.2 mg/m^3) (60-min)	blood changes including leukemia	blood testing required
benzoyl peroxide	[*94-36-0*]	5-mg/m^3, 8-h TWA	5 mg/m^3 TWA	airway and eye irritation, skin effects	
benzyl chloride	[*100-44-7*]	1-ppm (5 mg/m^3), 8-h TWA	5 mg/m^3 ceiling (15-min)	irritation; skin and eye effects	chest x-ray and pulmonary function testing required
beryllium	[*7440-41-7*]	2 $\mu m/m^3$, 8-h TWA; 5-$\mu m/m^3$, acceptable ceiling; 25 $\mu m/m^3$ maximum ceiling (30-min)	0.5 $\mu m/m^3$ (130-min)	lung cancer	pulmonary function chest x-ray, and sputum cytology required
boron trifluoride	[*7637-07-2*]	1-ppm ceiling	none recommended	respiratory system effects	adequate procedures for sampling and analysis not available; pulmonary function testing required
cadmium	[*7440-43-9*]	0.1 mg/m^3, 8-h TWA; 0.3 mg/m^3 ceiling (fume; erroneously published as 3 mg/m^3) 0.2 mg/m^3, 8-h TWA; 0.6 mg/m^3 ceiling (dust)	40 μm CD/m^3, TWA; 200 μm CD/m^3 ceiling (15-min)	lung and kidney effects	urine and pulmonary function testing required
carbaryl	[*63-25-2*]	5 mg/m^3, 8-h TWA	5 mg/m^3, TWA	nervous and reproductive system effects	medical warnings of possible effects on reproductive system and minimum exposure during pregnancy required; skin and eye contact to be prevented
carbon black	[*1333-86-4*]	3.5 mg/m^3, 8-h TWA	3.5 mg/m^3 TWA; 0.1 mg/m^3 TWA in presence of polycyclic aromatic hydrocarbons	lung, heart, and skin effects; cancer	chest x-rays, pulmonary function testing, ECG, and sputum cytology required
carbon dioxide	[*124-38-9*]	5000-ppm, 8-h TWA	10,000-ppm TWA (18,000 mg/m^3) 30,000-ppm ceiling (54,000 mg/m^3) (10-min)	respiratory effects	

carbon disulfide	*[75-15-0]*	20-ppm, 8-h TWA; 30-ppm acceptable ceiling; 100-ppm maximum ceiling	1-ppm TWA (3 mg/m^3); 10-ppm ceiling (30 mg/m^3) (15-min)	heart, nervous, and reproductive system effects	employees to be advised of potential effects on reproductive system
carbon monoxide	*[630-08-0]*	50-ppm, 8-h TWA	35-ppm TWA (40 mg/m^3); 200-ppm ceiling (229 mg/m^3)	heart effects	
carbon tetrachloride	*[56-23-5]*	10-ppm, 8-h TWA; 25-ppm acceptable ceiling; 200-ppm maximum ceiling (5-min in 4-h)	2-ppm ceiling (12.6 mg/m^3) (60-min)	liver cancer	
chlorine	*[7782-50-5]*	1-ppm, 8-h TWA	0.5-ppm ceiling (1.45 mg/m^3) (15-min)	eye/airway irritation	chest x-rays required
chloroform	*[67-66-3]*	50-ppm ceiling	2-ppm ceiling (9.78 mg/m^3) (60-min)	liver or kidney tumors and central nervous system effects	current Federal standard should be TWA: published as "C" in error
chloroprene	*[126-99-8]*	25-ppm, 8-h TWA	1-ppm ceiling (3.6 mg/m^3) (15-min)	reproductive effects; potential for cancer	chest x-ray and pulmonary function testing required; medical warnings to workers concerning reproductive effects in animals; pregnant workers to be counseled about continuing work with chloroprene
chromic acid	*[1308-14-1]*	1 mg/10 m^3 ceiling	0.05 mg CrO_3/m^3 TWA; 0.1 mg/m^3 ceiling (15-min)	nasal ulceration	
chromium(VI)		100 μm/10 m^3 ceiling	1 μm/m^3 for carcinogenic Cr(IV); 25 μm/m^3 Cr(VI); 50 μm/m^3 ceiling (15-min)	lung cancer, skin ulcers, lung irritation	employer must demonstrate absence of carcinogenic Cr(VI); x-ray required
chrysene[c]	*[218-01-9]*	none	to be controlled as an occupational carcinogen	cancer	control recommendations also included for polycyclic aromatic hydrocarbons
coal-gasification plants		existing Federal occupational exposure limits or NIOSH-recommended limits to be enforced		various effects depending on substances present; carcinogenic potential	extensive work practice and control procedures recommended; document limited to those technologies most likely to be operational in the United States
coal-tar products			0.1 mg/m^3 TWA, (cyclohexane-extractable fraction)	lung and skin cancer	includes coal tar, creosote, and coal tar pitch; pulmonary function testing, chest x-rays, and sputum cytology required

Table 1 (*continued*)

Substance	CAS Registry Number	Current OSHA environmental standard	NIOSH recommendation for environmental exposure limit[b]	Health effect considered	Comments
coke-oven emissions		150 $\mu m/m^3$ TWA	work practices to minimize exposure to emissions	lung cancer	sputum cytology and chest x-ray required; Federal standard promulgated Oct. 22, 1976
cotton dust		200 $\mu m/m^3$, 8-h TWA (yarn manufacturing); 750 $\mu m/m^3$, 8-h TWA (slashing and weaving operations); 500 $\mu m/m^3$, 8-h TWA (all other operations)	200 $\mu m/m^3$ lint-free cotton dust	pulmonary disease (byssinosis)	pulmonary function testing required; Federal standard promulgated June 23, 1978
cresol	*[93-51-6]*	22 mg/m^3, 8-h TWA (skin)	10 mg/m^3 TWA	skin, liver, kidney, and pancreas effects	applies to mixtures of cresols and cresylic acid; hazardous substance, skin and eyes possible delayed effects
cyanide, and HCN cyanide salts	*[74-90-8]*	10-ppm, 8-h TWA (alkali cyanides); cyanide 5 mg CN/m^3 (skin)	5 mg CN/m^3 ceiling (4.7-ppm) (10-min)	thyroid, blood, respiratory system effects	concurrent measurement required for HCN when measuring for cyanide salt; trained first-aid personnel and first-aid kits to be available during use.
DDT[c]	*[50-29-3]*	1 mg/m^3, 8-h TWA (skin)	lowest reliably detectable level; 0.5 mg/m^2 TWA by NIOSH validated method; skin contact to be avoided	cancer	hazardous liquid, skin and eye
decomposition products of fluorocarbon		none	none recommended	lung effects, polymer fume fever	workroom air to be monitored for inorganic fluorides and hydrogen fluoride
dibromochloro-propane	*[96-12-8]*	1-ppb, 8-h TWA; eye and skin contact to be avoided	10-ppb ceiling (0.1 mg/m^3 (30-min)	sterility; renal and liver effects	medical warnings to workers of reproductive system abnormalities including sterility, and notice of findings of cancer in animals following direct gastric application; Federal standard promulgated March 17, 1978

diisocyanates	*[2647-62-5]* *[101-68-8]* *[822-06-0]* *[3173-72-6]* *[4098-71-9]* *[5124-30-1]*	0.02-ppm (0.14 mg/m^3 ceiling (TDI); 0.02-ppm (0.2 mg/m^3) ceiling (MDI)	all values are in μm/m^3 toluene diisocyanate (TDI): 35 TWA, 140 ceiling; diphenylmethane diisocyanate (MDI): 50 TWA, 200 ceiling; hexamethylene diisocyanate (HDI): 35 TWA, 140 ceiling; naphthalene diisocyanate (NDI): 40 TWA, 170 ceiling; isophorone diisocyanate (IPDI): 45 TWA, 180 ceiling; dicyclohexylmethane 4,4′-diisocyanate (hydrogenated MDI): 55 TWA, 210 ceiling; other diisocyanates to be controlled to 20-ppb ceiling and 5-ppb TWA	respiratory effects and sensitization; irritation	chest x-ray and pulmonary function testing required
dinitro-*o*-cresol	*[534-52-1]*	0.2 mg/m^3 8-h TWA (skin)	0.2 mg/m^3 TWA	CNS and metabolic effects	blood and urine monitoring required; hazardous substance, skin and eyes; possible delayed effects
dioxane	*[123-91-1]*	100-ppm, 8-h TWA (skin)	1-ppm ceiling (3.6 mg/m^3) (30-min)	liver and kidney effects; cancer	blood and urine testing required; hazardous liquid, skin
epichlorohydrin	*[106-89-8]*	5-ppm, 8-h TWA (20 mg/m^3)	2 mg/m^3 TWA; 19 mg/m^3 ceiling (15-min)	skin, kidney, liver, and respiratory system effects	medical warning infertility effects required; hazardous liquid, skin
ethylene dibromide	*[106-93-4]*	20-ppm, 8-h TWA; 30-ppm acceptable ceiling; 50-ppm maximum peak (5-min)	1 mg/m^3 ceiling (0.13-ppm) (15-min)	damage to skin, eyes, heart, liver, spleen, respiratory and central nervous systems, potential for cancer and mutagenesis	medical warnings to workers of potential reproductive abnormalities and cancer following direct administration in animals; hazardous liquid, contact to be prevented
ethylene dichloride	*[107-06-2]*	50-ppm, 8-h TWA; 100-ppm acceptable ceiling; 200-ppm maximum ceiling (5-min in 3 h)	1-ppm TWA (4 mg/m^3); 2-ppm ceiling (2 mg/m^3) (15-min)	cancer; nervous system, respiratory, heart, and liver effects	nursing infants at risk

Table 1 (*continued*)

Substance	CAS Registry Number	Current OSHA environmental standard	NIOSH recommendation for environmental exposure limit[b]	Health effect considered	Comments
ethylene oxide[c]	[*75-21-8*]	90 mg/m^3 (50-ppm), 8-h TWA	90 mg/m^3 (50-ppm) TWA; 135 mg/m^3 (75-ppm) ceiling (15-min)	mutagenesis; cancer	recommendations for blood monitoring and medical counseling concerning mutations found in animal tests
ethylene-thiourea[a]	[*96-45-7*]	none	use of encapsulated form in industry; worker exposure to be minimized	carcinogenesis and teratogenesis	workers to be informed of carcinogenic and teratogenetic hazards; special attention to be given to thyroid function tests
fibrous glass		15 mg/m^3 total dust; 5 mg/m^3 respirable fraction (nuisance dust)	3,000,000 fibers/m^3 TWA (fibers <3.5 μm dia and >10 μm length); 5 mg/m^3 TWA (total fibrous glass)	eye and skin and airway effects	NIOSH recommends this limit also apply to other chemical fibers
fluorides, inorganic		2.5 mg/m^3, 8-h TWA	2.5 mg F/m^3 TWA	kidney and bone effects	urine monitoring required
formaldehyde	[*50-00-0*]	3-ppm, 8-h TWA; 5-ppm acceptable ceiling; 10-ppm maximum ceiling (30-min)	1.2 mg/m^3 ceiling (1-ppm) (30-min)	irritation, lung effects	hazardous liquid, skin
glycidyl ethers	[*106-92-3*] [*2426-08-6*] [*4016-14-2*] [*122-60-1*]	45 mg/m^3 ceiling (allyl glycidyl ether, AGE); 270 mg/m^3, 8-h TWA (*n*-butyl glycidyl ether); 2.8 mg/m^3, 8-h TWA (di(2,3-epoxypropyl) ether, DGE); 240 mg/m^3, 8-h TWA (isopropyl glycidyl ether, IGE); 60 mg/m^3, 8-h TWA (phenyl glycidyl ether, PGE)	45 mg/m^3 (AGE); 30 mg/m^3 (BGE); 1 mg/m^3 (DGE); 240 mg/m^3 (IGE); 5 mg/m^3 (PGE); all ceiling values (15-min)	skin, mucous membrane effects, sensitization potential, tumorigenesis and mutagenesis	possible additive effects with mixtures
hydrazines	[*302-01-2*] [*57-14-7*] [*100-63-0*] [*60-34-1*]	1.3 mg/m^3 (hydrazine), 1.0 mg/m^3 (1,1-dimethylhydrazine), 22 mg/mm (phenylhydrazine) as 8-h TWA values; 0.35 mg/m^3 ceiling (methylhydrazine)	0.04 mg/m^3 (hydrazine), 0.15 mg/m^3 (1,1-dimethylhydrazine), 0.6 mg/m^3 (phenylhydrazine), 0.08 mg/m^3 (methylhydrazine); all as ceiling values (120-min)	liver, blood, eye, skin effects; cancer	blood and urine monitoring, chest x-ray required; bowel examination required for some workers
hydrogen fluoride	[*7664-39-3*]	3-ppm, 8-h TWA	2.5 mg F/m^3 TWA (ca 6-ppm); 5.0 mg/m^3 ceiling (ca 12-ppm) (15 min, fluoride ion)	skin/eye/airway irritation; bone effects	pelvic x-ray (male) and urine testing required

hydrogen sulfide	*[7783-06-4]*	20-ppm acceptable ceiling; 50-ppm max ceiling (10-min)	15 mg/m³ ceiling (ca 10-ppm) (10-min)	irritation; severe acute effects, nervous and respiratory systems	continuous monitoring required if potential exists for exposure to 70 mg/m³ or greater; evacuation required at this level
hydroquinone	*[123-31-9]*	2 mg/m³, 8-h TWA	2 mg/m³ ceiling (15-min)	eye and skin effects	special provisions for darkroom use
isopropyl alcohol	*[67-63-0]*	400-ppm, 8-h TWA	400-ppm TWA, (984 mg/m³); 800-ppm ceiling (1968 mg/m³) (15-min)	mucous membrane irritation; possible cancer threat in manufacturing process	more stringent work practices and medical surveillance for manufacturing workers
kepone	*[143-50-0]*	none	1 μm/m³ ceiling (15-min)	nervous system effects; liver cancer	liver function testing required
ketones	*[67-64-1]*	acetone, 2400 mg/m³	590 mg/m³ TWA	irritation; liver, kidney, and nervous system effects	urinalysis required; warning of nervous system effects in workers exposed to methyl *n*-butyl ketone
	[78-93-3]	methyl ethyl ketone, 590 mg/m³	590 mg/m³ TWA		
	[107-87-9]	methyl *N*-propyl ketone, 700 mg/m³	530 mg/m³ TWA		
	[591-78-6]	methyl *N*-butyl ketone, 410 mg/m³	4 mg/m³ TWA		
	[110-43-0]	methyl *N*-amyl ketone, 465 mg/m³	465 mg/m³ TWA		
	[108-10-1]	methyl isobutyl ketone, 410 mg/m³	200 mg/m³ TWA		
	[110-12-3]	methyl isoamyl ketone, none	230 mg/m³ TWA		
	[108-83-8]	diisobutyl ketone, 290 mg/m³	140 mg/m³ TWA		
	[108-94-1]	cyclohexanone 200 mg/m³	100 mg/m³ TWA		
	[141-79-7]	mesityl oxide 100 mg/m³	40 mg/m³ TWA		
	[123-42-2]	diacetone alcohol 240 mg/m³	240 mg/m³ TWA		
	[78-59-1]	isophorone 140 mg/m³ all values 8-h TWA	23 mg/m³ TWA		
lead	*[7439-92-1]*	50 μg/m³, 8-h TWA[e]	less than 100 μm/m³	kidney, blood, and nervous system effects	air level to be maintained so that worker blood lead remains at or below 0.060 mg/100 g; blood monitoring required

Table 1 (*continued*)

Substance	CAS Registry Number	Current OSHA environmental standard	NIOSH recommendation for environmental exposure limit[b]	Health effect considered	Comments
malathion	[*121-75-5*]	15 mg/m^3, 8-h TWA	15 mg/m^3 TWA	nervous system effects	skin contact to be prevented; blood monitoring required
mercury, inorganic	[*7439-97-6*]	0.1 mg/m^3 ceiling	0.05 mg/m^3 TWA	central nervous system and mental effects	
methyl alcohol	[*67-56-1*]	200-ppm TWA	200-ppm TWA (262 mg/m^3); 800-ppm ceiling (1048 mg/m^3) (15-min)	blindness; metabolic acidosis	
4,4′-methylene-bis(2-chloro-aniline[c]	[*101-14-4*]	none—standard remanded by court	3 μm/m^3 TWA; skin contact to be avoided	cancer	chest x-ray, blood and urine testing required
methyl parathion	[*298-00-0*]	none	0.2 mg/m^3 TWA	nervous system effects	skin contact to be prevented; blood monitoring required
methylene chloride	[*75-09-2*]	500-ppm, 8-h TWA; 1000-ppm acceptable ceiling; 2000-ppm maximum (5-min in 2 h)	75-ppm TWA (261 mg/m^3); 500-ppm ceiling (1740 mg/m^3) to be lowered in presence of carbon monoxide	central nervous system effects; carbon monoxide toxicity	blood monitoring required
nickel carbonyl[c]	[*13463-39-3*]	7 μm/m^3 (1-ppb), 8-h TWA	7 μm/m^3 (1-ppb) TWA	cancer	recommendations for chest x-ray, pulmonary function, and urine monitoring
nickel, inorganic and compounds	[*7440-02-0*]	1 mg/m^3, 8-h TWA	15 μm Ni/m^3 TWA	skin effects; lung and nasal cancer	chest x-ray and pulmonary function testing required
nitric acid	[*7697-37-2*]	2-ppm, 8-h TWA	2-ppm TWA (5 mg/m^3)	dental erosion, nasal/lung irritation	hazardous liquid, eyes and skin, chest x-ray required
nitriles	[*75-05-8*] [*3333-52-6*] [*78-82-0*] [*107-12-0*] [*111-69-3*] [*110-61-2*]	70 mg/m^3 (40-ppm), 8-h TWA (acetonitrile); 3 mg/m^3, (0.5-ppm), 8-h TWA, (skin) (tetramethylsuccinonitrile)	to be TWA values: acetonitrile: 34 mg/m^3 (20-ppm); *N*-butyronitrile [*109-74-0*]: 22 mg/m^3 (8-ppm); isobutyronitrile: 22 mg/m^3 (8-ppm); propionitrile: 14 mg/m^3 (6-ppm); malononitrile: 8 mg/m^3 (3-ppm); adiponitrile: 18 mg/m^3 (4-ppm); succinonitrile: 20 mg/m^3 (6-ppm)	hepatic, renal, respiratory, cardiovascular, gastrointestinal and nervous system effects	chest x-ray and pulmonary function testing required; trained personnel and first-aid kits to be available during use; hazardous substances, skin and eyes

	[75-86-5] [107-16-4]		to be ceiling values (15-min): acetone cyanohydrin: 4 mg/m^3 (1-ppm); glycolonitrile: 5 mg/m^3 (2-ppm); tetramethyl-succinonitrile: 6 mg/m^3 (1-ppm); when present as mixtures or with other sources of cyanide, exposure to be considered additive and environmental limit to be calculated		
nitrogen oxides		NO_2: 5-ppm, 8-h TWA	NO_2, 1-ppm (1.8 mg/m^3) ceiling (15-min)	airway effects	pulmonary function testing required
		NO: 25-ppm, 8-h TWA	NO, 25-ppm TWA (30 mg/m^3)	blood effects	
nitroglycerin: ethylene glycol dinitrate (EDGN)	[55-63-0] [628-96-6]	2 mg/m^3 (skin) 8-h TWA (nitroglycerin); 1 mg/m^3 (skin) ceiling, (EDGN)	0.1 mg/m^3 ceiling (20-min)	circulatory system effects	skin contact to be prevented; recommended limit for either substance alone or mixtures
organotin compounds		0.1 mg tin/m^3 8-h TWA	0.1 mg tin/m^3 TWA	eye, skin, liver, nervous system, and heart effects	chest x-ray, blood and urine monitoring, eye tests, heart examination, and nervous system testing required liquid, skin and eyes
parathion	[56-38-2]	0.11 mg/m^3 TWA	0.05 mg/m^3 TWA	nervous system effects	skin contact to be prevented; blood monitoring required
pesticide manufacturing and formulation		current OSHA or previously recommended NIOSH levels to be followed; stringent work practice and medical surveillance requirements to be instituted; pesticides considered in 3 groups based on toxicity; skin contact to be prevented		wide range of toxicities considered; nervous and reproductive system effects; cancer	blood monitoring requiring for some groups; medical warning of reproductive system effects in some compounds
phenol	[108-95-7]	5-ppm, 8-h TWA (skin)	20 mg/m^3 TWA (5.2 ppm); 60 mg/m^3 ceiling (15.6 ppm) (15-min)	skin, eye, CNS, liver, and kidney effects	hazardous substance, skin, and eyes
phosgene	[75-44-5]	0.1-ppm, 8-h TWA	0.1-ppm TWA (0.4 mg/m^3); 0.2-ppm ceiling (0.8 mg/m^3) (15-min)	airway effects	pulmonary function testing and x-ray required

Table 1 (*continued*)

Substance	CAS Registry Number	Current OSHA environmental standard	NIOSH recommendation for environmental exposure limit[b]	Health effect considered	Comments
polychlorinated biphenyls		1 mg/m^3, 8-h TWA (42% chlorine); 0.5 mg/m^3, 8-h TWA (54% chlorine)	1 μm/m^3 TWA	cancer; skin, liver, and reproductive effects	blood testing required, medical warning of adverse effects to be given to women workers of childbearing age and nursing mothers
refined petroleum solvent		500-ppm, 8-h TWA (2950 mg/m^3) (Stoddard solvents)	350 mg/m^3 TWA; 1800 mg/m^3 ceiling (15-min)	skin, lung, and nerve irritation	blood and urine monitoring required; action level for petroleum ether, rubber solvent, naphtha be 200 mg/m^3 TWA; action level for mineral spirits and Stoddard solvent to be 350 mg/m^3 TWA; action level for kerosene to be 100 mg/m^3 TWA; hazardous substance, skin
silica, crystalline		250/%SiO + 5 in MPPCF, or 10 mg/%SiO + 2 (respirable quartz	50 μm/m TWA, respirable-free silica	chronic lung disease (silicosis)	x-ray, pulmonary function testing required
sodium hydroxide	[*1310-73-2*]	2 mg/m^3, 8-h TWA	2 mg/m^3 ceiling (15-min)	airway irritation	hazardous liquid, eyes and skin
sulfur dioxide	[*7446-09-5*]	5-ppm, 8-h TWA	0.5-ppm TWA (1.3 mg/m^3)	respiratory effects	pulmonary function testing required
sulfuric acid	[*7664-93-9*]	1 mg/m^3, 8-h TWA	1 mg/m^3 TWA	pulmonary irritation	hazardous liquid, eyes and skin
1,1,2,2-tetra-chloroethane	[*79-34-5*]	5-ppm; 8-h TWA (skin)	1-ppm TWA (6.87 mg/m^3)	liver, gastrointestinal, and nervous system effects	skin contact to be prevented; blood monitoring required
tetrachloro-ethylene	[*127-18-4*]	100-ppm, 8-h TWA; 200-ppm acceptable maximum ceiling; 300-ppm maximum ceiling (5-min in 3 h)	50-ppm TWA (339 mg/m^3); 100-ppm ceiling (678 mg/m^3) (15-min)	nervous system, heart, respiratory, liver effects	medical warning of possible congenital abnormalities required
thiols: N-alkane mono, cyclohexane, and benzene	[*109-79-5*] [*74-93-1*] [*75-08-1*]	10-ppm, 8-h TWA (butylmercaptan); 10-ppm ceiling (methyl and ethyl mercaptan)	ceilings 15-min: benzenethiol [*108-95-5*], 0.5 mg/m^3 (0.1-ppm); 1-methanethiol, 1.0 mg/m^3 (0.5-ppm); ethane-	irritation, eye, skin, blood, and nervous system effects	hazardous substance, skin, blood and urine monitoring required

	[107-03-9] [109-79-5] [110-66-7] [111-31-9] [1639-09-4] [111-88-6] [1455-21-6] [143-10-2] [112-55-0] [2917-26-2] [2885-00-9] [1569-69-3]		thiol, 1.3 mg/m^3 (0.5-ppm); 1-propanethiol, 1.6 mg/m^3 (0.5-ppm); 1-butanethiol, 1.8 mg/m^3 (0.5-ppm); 1-pentanethiol, 2.1 mg/m^3 (0.5-ppm); 1-hexanethiol, 2.4 mg/m^3 (0.5-ppm); 1-heptanethiol, 2.7 mg/m^3 (0.5-ppm); 1-octanethiol, 3.0 mg/m^3 (0.5-ppm); 1-nonanethiol, 3.3 mg/m^3 (0.5-ppm); 1-decanethiol, 3.6 mg/m^3 (0.5-ppm); 1-undecanethiol, 3.9 mg/m^3 (0.5-ppm); 1-dodecanethiol, 4.1 mg/m^3 (0.5-ppm); 1-hexadecanethiol, 5.3 mg/m^3 (0.5-ppm); 1-octadecanethiol, 5.9 mg/m^3 (0.5-ppm); cyclohexanethiol, 2.4 mg/m^3 (0.5-ppm); mixtures of thiols to be controlled by calculation of equivalent concentrations		
o-tolidine	[119-93-7]	none	20 μm/m^3 ceiling skin contact to be prevented	nasal irritation; cancer	urine testing required; quarterly urine monitoring recommended
toluene	[108-88-3]	200-ppm, 8-h TWA; 300-ppm acceptable ceiling; 500-ppm maximum ceiling (10-min)	100-ppm TWA (375 mg/m^3); 200-ppm ceiling (750 mg/m^3) (10-min)	central nervous system depressant	
toluene diisocyanate	[26471-62-5]	0.02-ppm ceiling	0.005-ppm TWA (0.035 mg/m^3); 0.02 ppm ceiling (0.14 mg/m^3) (20-min)	airway effects	chest x-ray blood tests, pulmonary function testing required; recommendations revised—see diisocyanates
1,1,1-trichloro-ethane	[71-55-6]	350-ppm, 8-h TWA	350-ppm ceiling (1910 mg/m^3) (15-min)	nervous system, liver, and heart effects	action level set at 200-ppm TWA; medical warning of possible congenital abnormalities required
trichloro-ethylene	[79-01-6]	100-ppm, 8-h TWA; 200-ppm acceptable ceiling; 300-ppm maximum ceiling; (5-min in any 2 h)	25-ppm (134 mg/m^3) TWA	central nervous system depressant; cancer	workers to be warned of hazards

Table 1 (*continued*)

Substance	CAS Registry Number	Current OSHA environmental standard	NIOSH recommendation for environmental exposure limit[b]	Health effect considered	Comments
tungsten and cemented tungsten carbide	*[7440-33-7]* *[12070-13-2]*	none	insoluble tungsten: 5 mg/m^3 TWA; soluble tungsten, 1 mg W/m^3 TWA; dust of cemented tungsten carbide (>2% cobalt), 0.1 mg Co/m^3 TWA: dust of cemented tungsten carbide (>0.3% nickel): 15 μm Ni/m^3 TWA	lung and skin effects	pulmonary function testing and chest x-ray required
vanadium	*[7440-62-2]*	vanadium pentoxide (dust): 0.5 mg/m^3 ceiling; (fume) 0.1 mg/m^3 ceiling ferrovanadium; 1 mg/m^3, 8-h TWA	vanadium compounds: 0.05 mg/m^3 ceiling (15-min); metallic vanadium and vanadium carbide: 1 mg/m^3 TWA	eye, skin, and lung effects	pulmonary function testing and chest x-ray required
vinyl acetate	*[108-05-4]*	none	15 mg/m^3 (4-ppm) ceiling (15-min)	irritation	
vinyl chloride	*[75-01-4]*	1-ppm, 8-hr TWA; 5-ppm ceiling, (15-min sample)	minimum detectable level; 1-ppm ceiling (2.55 mg/m^3) (15-min)	liver cancer	standard promulgated October 4, 1974; liver function testing required
vinyl halides		1-ppm, 8-h TWA; 5-ppm ceiling (15-min)	as promulgated for vinyl chloride in 29 CFR 1910.1017 with eventual goal of zero exposure	cancer	document includes vinyl chloride, vinylidene chloride, vinyl bromide, vinyl fluoride, and vinylidene fluoride monomers
waste anesthetic gases and vapors		none for substances when used as anesthetic agents	2-ppm ceiling (halogenated anesthetic agents) (1-h); 25-ppm TWA during periods of use (nitrous oxide)	reproductive effects and audiovisual performance decrements	employees to be advised of potential effects; abnormal outcome of pregnancies of employees and spouse to be documented
xylene	*[1330-20-2]*	100-ppm, 8-h TWA	100-ppm TWA (434 mg/m^3); 200-ppm ceiling (868 mg/m^3) (10-min)	central nervous system depressant; airway irritation	
zinc oxide	*[1314-13-2]*	5 mg/m^3; 8-h TWA	5 mg/m^3 TWA, 15 mg/m^3 ceiling (15-min)	metal fume fever	

[a] Ref. 55.
[b] NIOSH TWA recommendations based on up to a 10-h exposure unless otherwise noted.
[c] Special hazard review.
[d] Reduction of standard to 1 ppm rejected by U.S. Supreme Court (56).
[e] Ref. 57.

The RTECS is also available as a real-time, interactive computer data base from the National Library of Medicine (NLM) Medical Literature Analysis and Retrieval System (MEDLARS), and the National Institutes of Health/Environmental Protection Agency (NIH/EPA) Chemical Information System (CIS). These systems permit users to search the RTECS file and to compile subfiles tailored to their particular needs. These systems are updated on a quarterly basis.

Monitoring and Measurements. Current Federal regulations require the testing of industrial discharge with fish, daphnia, or other aquatic organism to determine whether toxic chemicals are being released. The Microtox system is a new system introduced by Beckman instruments (60). It is based on a strain of luminescent bacteria known as *Photobacterium fischeri* (see Chemiluminescence). These bacteria are highly sensitive to toxicants and the decrease in light emission linearly related to the concentration of the toxicant can be measured. Gas chromatographic/mass spectrometer systems are available for the analysis of organic compounds in wastewaters, eg, the OWA-20 and OWA-30 by Finnegan Instruments (61). An ir spectrophotometer, equipped with an adjustable, long path-length gas cell is used to analyze most of the 400 gases and vapors for which maximum tolerable limits have been established (see also Trace and residue analysis). National/Dräger multigas detectors and detector tubes (National Mine Service Co.) have been certified by NIOSH.

BIBLIOGRAPHY

"Industrial Hygiene and Toxicology" in *ECT* 1st ed., Vol. 7, pp. 847–870, by C. H. Hine, University of California, and L. Lewis, Industrial and Hygiene Associates; "Industrial Toxicology" in *ECT* 2nd ed., Vol. 11, pp. 595–610, by David W. Fassett, Eastman Kodak Company.

1. A. Hamilton, *Industrial Toxicology,* Harper and Bros., New York, 1934; *Exploring the Dangerous Trades,* Little, Brown and Co., Boston, Mass., 1943.
2. *The U.S. Toxic Substances Control Act,* PL 94-469, Oct. 11, 1976.
3. R. M. Druley and G. L. Ordway, *The Toxic Substances Control Act,* Bureau of National Affairs, Inc., Washington, D.C., 1977.
4. G. S. Dominguez, *Guidebook: The Toxic Substances Control Act,* CRC Press, Cleveland, Ohio, 1977.
5. *Report on the Toxic Substances Control Act,* U.S. House Interstate and Foreign Commerce Committee, Report No. 94-1341, 1976.
6. *Report on the Toxic Substances Control Act,* U.S. Senate Commerce Committee Report No. 94-698.
7. *Fed. Regist.* **43,** (Jan. 18, 1978).
8. *Fed. Regist.* **42,** (Dec. 23, 1977).
9. *AIHA J.* **20,** 5 (1959).
10. G. D. Clayton, *The Industrial Environment—Its Evaluation and Control,* U.S. Dept. of Health, Education, and Welfare, PHS, Superintendent of Documents, Washington, D.C., 1973, Chapt. 1.
11. G. D. Clayton and F. E. Clayton, eds., *Patty's Industrial Hygiene and Toxicology,* John Wiley & Sons, Inc., New York, 1978.
12. M. Katz, ed., *Methods of Air Sampling and Analysis,* 2nd ed., APHA InterSociety Committee, Washington, D.C., 1977.
13. R. D. Soule, *The Industrial Environment—Its Evaluation and Control,* U.S. Dept. of Health, Education, and Welfare, PHS, 1973, Chapt. 5.
14. L. J. Casarett and J. Doull, eds., *Toxicology, The Basic Science of Poisons,* Macmillan Publishing Co., Inc., New York, 1975.
15. G. D. Clayton and F. E. Clayton, eds., *Patty's Industrial Hygiene and Toxicology,* Vols. 2A, 2B, and 2C, John Wiley & Sons, Inc., New York, 1980.
16. *Chem. Anal.* **67**(5), 2 (1977).
17. M. J. Wallace, *Chem. Eng.,* 73 (Apr. 24, 1978).

18. H. E. Stokinger "Routes of Entry" in *Occupational Diseases,* rev. ed., U.S. Dept. Health, Education, and Welfare, PHS, NIOSH, Supt. of Documents, Washington, D.C., 1977.
19. Task Group on Lung Dynamics, *Health Phys.* **12,** 173 (1966).
20. R. E. Albert, M. Lippmann, and W. Briscoe, *Arch. Environ. Health* **18,** 738 (1969).
21. F. D. Malkinson in W. Montagna and W. C. Lobitz, eds., *The Epidermis,* Academic Press, Inc., New York, 1964, Chapt. 21.
22. T. F. Hatch and P. Gross, *Pulmonary Deposition and Retention of Inhaled Aerosols,* Academic Press, Inc., New York, 1964.
23. C. N. Davies, ed., *Inhaled Particles and Vapours,* Pergamon Press, Inc., New York, 1961.
24. E. M. Papper and R. J. Kitz, eds., *Uptake and Distribution of Anesthetic Agents,* McGraw-Hill Book Co., New York, 1963.
25. B. D.Dinman in G. D. Clayton and F. E. Clayton, eds., *Patty's Industrial Hygiene and Toxicology,* John Wiley & Sons, Inc., New York, 1978, Chapt. 6.
26. M. O. Amdur, "Aerosols on the Response to Irritant Gases" in C. N. Davies, ed., *Inhaled Particles and Vapours,* Pergamon Press, Inc., New York, 1961.
27. G. W. Wright, in G. D. Clayton and F. E. Clayton, eds., *Patty's Industrial Hygiene and Toxicology,* Vol. 1, John Wiley & Sons, Inc., New York, 1978, Chapt. 7, Sect. 7.
28. *Scientific Bases for Identifying Potential Carcinogens and Estimating Their Risks,* Report of Inter-agency Reg. Liaison Group, Feb. 6, 1979.
29. D. J. Birmingham in *The Industrial Environment—Its Evaluation and Control,* U.S. Dept. of Health, Education and Welfare, PHS, Supt. of Documents, Washington, D.C., 1973, Chapt. 34.
30. S. Rothman, "Percutaneous Absorption" in *Physiology and Biochemistry of the Skin,* Chicago University Press, Chicago, Ill., 1954.
31. F. D. Malkinson "Permeability of the Stratum Corneum" W. Montagna and W. C. Lobitz, eds., in *The Epidermis,* Academic Press, Inc., New York, 1964.
32. E. Cronin and R. B. Stoughton, *Arch. Dermatol.* **86,** 265, 1962.
33. Committee on Toxicology, *Principles and Procedures for Evaluating the Toxicity of Household Substances,* Pub. No. 1138, National Academy of Sciences, National Research Council, Washington, D.C., 1964.
34. J. H. Draize, G. Woodard, and H. O. Calvery, *J. Pharmacol. Exptl. Therap.* **82,** 377 (1944).
35. R. L. Roudabush and co-workers, *Toxicol. Appl. Pharmacol.* **7,** 559 (1965).
36. I. H. Blank, *J. Occup. Med.* **2,** 6 (1960).
37. R. B. Stoughton, *Toxicol. Appl. Pharmacol.* (Suppl. 2) 7, (1965).
38. K. Landsteiner, *The Specificity of Serological Reactions,* Harvard University Press, Cambridge, Mass., 1945.
39. D. W. Fassett, *Am. Dyest. Rep.* **52**(17), 39 (1963).
40. C. D. Calnan and co-workers "Methods of Evaluating Contact Sensitizers" in T. H. Sternberg and V. D. Newcomer, eds., *The Evaluation of Therapeutic Agents and Cosmetics,* McGraw-Hill Book Co., New York, 1964.
41. H. E. Christensen and T. T. Luginbyhl, eds., *Suspected Carcinogens,* U.S. DHEW, PHS, Superintendent of Documents, U.S. Government Printing Office, Washington, D.C., 1975.
42. *Chem. Eng. News* **56**(33), 19 (July 31, 1978).
43. H. E. Stokinger "Occupational Carcinogenesis" in G. D. Clayton and F. E. Clayton, eds., in *Patty's Industrial Hygiene and Toxicology,* Vol. 2B, John Wiley & Sons, Inc., New York, in production.
44. J. McCann and B. N. Ames, *Ann. N.Y. Acad. Sci.* **271,** (1976).
45. B. N. Ames, *Science* **204,** 587 (1979).
46. B. N. Ames, J. McCann, and E. Yamasaki, *Mutat. Res.* **31,** 347 (1975).
47. J. McCann, E. Choi, E. Yamasaki, and B. N. Ames, *Proc. Nat. Acad. Sci. USA* **72,** 5135 (1975).
48. I. E. Mattern and H. Greim, *Mutat. Res.* **53,** 369 (1978).
49. F. J. DeSerres and M. D. Shelby, *Science* **203,** 563 (1979).
50. K. Bridbord, J. K. Wagoner, and H. P. Blejer "Chemical Carcinogens" in *Occupational Diseases, A Guide to their Recognition,* rev. ed., U.S. Govt. Printing Office, Washington, D.C., June 1977.
51. *Threshold Limit Values for Chemical Substances in Workroom Air Adopted by ACGIH for 1978,* ACGIH, Cincinnati, Oh., 1978.
52. *Emergency Exposure Limits Recommended by National Academy of Science,* National Research Council Committee on Toxicity, Washington, D.C.
53. *AIHA J.* **41,** 2, A22 (Feb. 1980).

54. *NIOSH/OSHA Pocket Guide to Chemical Hazards,* DHEW (NIOSH) Publication No. 78-210, U.S. Govt. Printing Office, Washington, D.C., Sept. 1978.
55. *Summary of NIOSH Recommendations for Occupational Health Standards,* U.S. DHEW, PHS, U.S. Govt. Printing Office, Washington, D.C., Oct. 1978.
56. *Chem. Eng. News* **58**(27), 4 (1980).
57. *Chem. Week* 15 (Aug. 27, 1980).
58. *NIOSH Registry of Toxic Effects of Chemical Substances,* GPS No. 017-033-00346-7, Supt. of Docs., U.S. Govt. Printing Office, Washington, D.C., 1978.
59. *Science* **203,** 1328 (1979).
60. *Science* **203,** 1327 (1979).
61. *Chem. Eng. News,* 42 (Sept. 25, 1978).

General References

Ref. 11 is also a general reference.

D. Schuetzle, ed., *Monitoring Toxic Substances,* ACS Symposium Series No. 94, American Chemical Society, Washington, D.C., 1979.

Toxic Substances Control II, Government Institutes, Inc., Washington, D.C., 1978.

Toxic Substances Laws & Regulations, 1977, Government Institutes, Inc., Washington, D.C., 1977.

Toxic Substances Control Act, Chemical Substances Inventory, Industry Assistance Office TS-799, EPA, Washington, D.C., 1979.

Occupational Health & Safety Regulation, Government Institutes, Inc., Washington, D.C., 1978.

OSHA Candidate List, OSHA, Washington, D.C., 1980.

N. I. Sax, ed., *Dangerous Properties of Industrial Materials,* 4th ed., Van Nostrand Reinhold Company, New York, 1975.

C. E. Searle, ed., *Chemical Carcinogens,* ACS Monograph 173, American Chemical Society, Washington, D.C., 1977.

M. A. Mehlman, R. E. Shapiro, and H. Blumenthal, eds., *Advances in Modern Toxicology,* John Wiley & Sons, Inc., New York, 1978.

Hazardous & Toxic Wastes, Hazmat Publishing Co., Kutztown, Pa., 1980.

The Hazardous Substances Guide Series, Vols. 1–4, J. J. Keller & Associates, Inc., Neenah, Wis., 1979.

GEORGE D. CLAYTON
Clayton Environmental Consultants, Inc.

INFORMATION RETRIEVAL

The information explosion is more than a catch phrase to describe the proliferation of technical information in recent years. For many chemists, it marks a destruction, a detonation if you will, of the traditional ways of finding chemical facts with a resultant information fallout, the unexpected, often bewildering products of this sudden surge. Information is more abundant and more specialized than ever before; but chemists, amid this vast wealth of information, are often unable to obtain easily the particular facts they need. This article is for those individuals; it identifies and describes some of the main resources currently available and the methods of information retrieval as of January 1980. Certain sections also are of interest to information and business professionals. The first section gives an overview of information retrieval: the technology that has developed to handle the wealth of scientific material, its advantages and disadvantages, the consequences for chemists and the chemical industry, and the prospects for the future. The second section describes the particular characteristics of technical information and outlines the productive routes and useful resources for chemical fact-finding; it reviews the main abstracting and indexing services, handbooks, physical data compilations, encyclopedias and treatises, main computerized data bases, and on-line vendors of chemical information. The discussion of business information in the third section shows that technical–economic, business, and management literature differs from the highly structured technical literature and that it varies according to political, social, and legal environments.

DEVELOPMENTS IN INFORMATION-RETRIEVAL TECHNOLOGY

To discuss the advances in information retrieval requires familiarity with the terms used for the sources of searchable information. Primary sources represent either new information, new interpretations of old knowledge, or new compilations of known information. They include books, journals, and patents. Secondary sources serve as locators for primary information—they are the organizers and the condensers of primary literature. These include encyclopedias, dictionaries, and handbooks. Chemical Abstracts Service (CAS) is an example of an abstracting and indexing service that organizes the primary periodical literature. CAS has emerged as the single, most important secondary source of information for chemists and, through international agreements, has indeed become responsive to an international community. Tertiary sources direct the user to both the primary and secondary sources; among tertiary sources are guides and directories.

For many years, chemical indexing and abstracting services, in processing the large number of publications in the field of chemistry, provided a way to keep track of developments in particular fields. Thus they enabled chemists to reduce the time spent on literature searching. However, the chemical literature grew so rapidly that the chemist no longer could rely on indexes and abstracts. Also, the interdisciplinary nature of the work being done made it difficult to locate pertinent information that might have appeared in another subject category. Computerized retrieval of information pointed a way to satisfy these needs.

Searching for Information. The computer has become indispensable in information retrieval. There is, however, a wealth of material that predates computer-generated information. In the 1960s computers generated indexes to the literature and, in specialized systems, identified structures. These indexes were produced as paper products for manual searching. Extensive use of the computer to actually store the indexes for manipulation really started about 1970. The computer offers the ability to retrieve current information in ways never before possible, but until technology develops ways of providing access to the accumulated knowledge of the past, the traditional retrieval techniques are still needed.

The computer is an essential tool in the management of information as well. Chemical-information systems with their associated hardware and software deal with the collection, storage, and interpretation of chemical data. The hardware of these systems includes the computer itself and the physical equipment involved in the storage and processing of data. The software is the particular collection of computer programs that enable the computer to perform particular tasks. Together they provide the technology to handle large quantities of chemical data and to allow timely and inexpensive access to information. To understand the level to which this technology has developed and its affect on the working chemist, it is necessary to know the kinds of information available for retrieval and the physical forms that they take.

In the late 1960s the computer-readable tapes that were used in the production of printed publications of the main abstracting and indexing services began to be used for alerting purposes. This was the beginning of a significant change in information retrieval. These tapes were manipulated further to produce a variety of products. They became searchable in batch-mode; ie, they could be queried, and then scanned by the computer in one operation. Then, on-line systems were developed. In on-line systems, data is transmitted directly between a computer and remote terminals and is immediately processed by the computer. These systems provided a way to search the tapes interactively. The person at the terminal, in direct communication with the computer, could refine the query during the scanning process. The development of these systems led to a proliferation of machine-readable files called data bases.

Because purchasing and maintaining these data bases for in-house use was expensive, companies and search services emerged that collected data bases and sold access to them either for a fee based on use or by subscription. Several of these systems are accessible by dialing directly or by dialing one of the commercial data transmission networks such as TYMNET (TYMESHARE), TELENET, TELEX, or TWX. Once connected to the retrieval system, the user can look at the contents of that search service's data bases. This process of examining data bases at remote locations using a computer terminal is called on-line searching.

Several types of files are searchable on line. Bibliographic data bases provide references to source material. To identify relevant source material, the user chooses appropriate keywords and phrases, links them together using the words AND, OR, and NOT (Boolean logic), and interactively refines the query while scanning the results. Relevant references can be printed immediately on line, or they can be printed off line, ie, not printed during the direct communication with the computer, but afterward, at the remote computer center. Nonbibliographic or numerical data bases allow the user to retrieve specific data directly and offer the additional capabilities of adding and manipulating data for analysis. On-line dictionary and vocabulary files, useful for specific information, also guide the user to formulas, synonyms, particular spellings of words, and particular terms for optimum retrieval of information from other on-line data bases.

Automated systems offer two modes for information retrieval: retrospective searching, by which the user identifies material on a given subject; and current awareness, by which the user identifies the most recent material on a given subject to keep abreast of the latest developments. This activity, called selective dissemination of information (SDI), is a continual searching service.

Access to Documents. The availability of original articles is an increasing problem. Because retrieval systems readily provide references to international and interdisciplinary source material, libraries often receive requests for full documents from obscure journals or periodicals to which they do not subscribe. Furthermore, documents that are available in libraries are requested and photocopied more often.

Photocopying practices and the distribution of full copies of documents are now affected by a new United States copyright law. This law discusses in detail what individuals may copy for themselves and what an information center, library, or research center may copy for its clientele. Details of the new law are reviewed in ref. 1 (see Trademarks and copyrights). Other countries are considering similar laws.

Manipulating Chemical Data for Retrieval. Retrieving chemical information requires manipulation of textual material and chemical structures. Analysis of text involves extracting data and concepts from a document and expressing them in words or phrases. The bibliographic data elements by which a document is stored and retrieved include its title, author(s), subject, journal sources, date, and reference number. Keywords and descriptors identify the contents of the document and, unlike strictly bibliographic information, require intellectual effort to be useful. Comprehensive, in-depth indexing requires much intellectual effort but also yields the most useful tool for identifying relevant material. Technology is developing to have computers do more of the indexing work.

There have been many efforts to devise systems to represent the nature of chemical substances including systematic nomenclature and stereochemical drawings. Because structural diagrams are extremely difficult to convey in written text, systematic names have been developed. These names are unique textual representations of chemical structures, but they can be cumbersome and difficult to prepare correctly because of the number of rules that govern the naming process. Chemists revert to structural diagrams to define the chemical structures of compounds unequivocally. The most commonly used systems for the retrieval of structures are topological systems and linear-notation systems. The latter use codes to represent chemical fragments and the linking of atoms in the molecule; examples are the Wiswesser Line Notation and the IUPAC (International Union of Pure and Applied Chemistry)/Dyson Notation system, which parallels chemical nomenclature. Topological systems use connection tables or matrix representations of the topology of the chemical structures.

Registry Numbers. Chemical Abstracts Service's Registry Numbers serve as a useful bridge between textual and structural information. A Registry Number is a computer-checkable serial number assigned to a chemical compound by CAS in sequential order when the compound initially enters the registry system. This system, which originated in 1965, has registered over five million substances and additional compounds are being registered at the rate of over three hundred fifty thousand per year. A Registry Number has no chemical significance, but it is the unique identifier for a particular compound that links various textual identifiers of the substance (ie, nomenclature) with the molecular structure.

Registry Numbers for a particular substance can be found in the CAS *Registry Handbook,* or in the CAS *Chemical Substance Index, Formula, Index,* or *General Index.* They may be found also in chemical-dictionary files. Registry Numbers have appeared regularly in *Chemical Abstracts* volume indexes since 1972 and are an integral part of the CA SEARCH data base. The Registry Number appears as a data element in an ever increasing number of on-line chemical files and technical journals, and it is even required in reporting to the EPA. In the *Encyclopedia,* Registry Numbers have been given for title compounds in articles and for all compounds in the index as a concise means of substance identification. The use of Registry Numbers as a link between on-line bibliographic files is a refined retrieval technique not previously available.

Handling Proprietary Information. In today's high technology environment, knowledge of the surrounding technology and some means for communicating that technology are essential; information is an integral part of the utilization of technology. This applies also to proprietary information, that body of knowledge collected and held privately by particular organizations. Just as published information is publicly available, a company's information must be available to those within that company who need it. The emphasis, however, must be on the transfer of information rather than on the transfer of paper. The process of transferring proprietary information is cyclic and involves several steps: (*1*) information generation, the technical activity that gives rise to data that is or will be of value to others; (*2*) information preparation, the making of a permanent record of the data; (*3*) information processing, the collection, indexing, storage, and distribution of the data; (*4*) information retrieval, the search for specific, requested data; (*5*) information analysis, the organization of data to improve its applicability; and (*6*) information application, the technical activity that arises from the transfer of data and that, in turn, leads to the generation of new data. These six processes are interdependent. The objective in handling proprietary information is to make the step between the generation and application of information as transparent as possible to the user and to eliminate the barriers (eg, location, organizational structure) between the generator of information and the user who wants to apply what was generated.

Advanced information and office equipment technology has evolved to meet these objectives. Historically, proprietary information was stored in internal files of some type; these files employed a classified arrangement ranging from very simple author files to very complex and elaborate hierarchical breakdowns of subjects. Now computer-produced indexes eliminate the need for manually listing the places where documents are filed. And because information needed in the course of business is not based just in one document, the need to identify multiple documents, to extract particular data, and to use merged information is served by improved technology.

Text processors, systems for capturing information in digital form when it is initially typed, are replacing typewriters. They permit information to be revised, stored, and manipulated readily. Although they were initially sold as improved typewriters, they are now being developed to accept and store information in digital form for processing and retrieval.

Telecommunications between machines are providing new services and options. Letters and reports typed in one location can be printed at remote locations. Electronic mail provides immediate delivery of messages without the use of a messenger. Printing-on-demand allows information stored in digital form to be printed as needed in a variety of ways.

All of these technical developments affect the handling of proprietary information, and work going on right now will affect it further. For example, the video disk currently used in entertainment, advertising, and publishing may be applied in the field of information storage and retrieval (see Recording disks). Another technology that bears watching is the voice-activated computer. In the current technology, most terminals have typewriter keyboards for entering information, but the voice-activated computer eliminates keystroking, opening the door to the office of the future in which a computer terminal on a desk top is as common as the telephone is today. This is indeed a projection into the future, but in view of the pace of technological development, it is worthy of consideration.

Consequences of Advanced Information Technology. There exists today an identifiable information industry. The ability to identify and obtain information through the use of computers is so powerful and has grown so fast that its effect is being felt throughout the world. The growth of an information industry based on repackaging the products of abstracting and indexing organizations is a new phenomenon. The repackaging of proprietary information within industrial organizations is occurring in a similar fashion and, of necessity, is causing the growth of information centers in large industries. Smaller organizations that cannot afford to have in-house capabilities use an information broker to supply their needs and may also use the expertise of the information consultant to organize their private collections. That information is a marketable commodity, with all its attendant professionals, is a new concept.

Sophisticated information technology often requires the use of a trained specialist. Because research and development is expensive, information must be used to avoid duplication of effort and to keep scientists abreast of developments that can affect their work. Consequently, these functions have been transferred largely to a specialist who can not only gather information efficiently but also manage information to serve the needs of the users. This specialist must be technically competent and have the ability to evaluate information.

Although the benefits of on-line retrieval are clearly evident, there is concern that users should be in direct contact with the information they need. The phrases friendly systems, friendly interfaces, and user-oriented systems are beginning to be heard. At one time, users went directly to their information sources with little effort. In fact, many chemists had their own subscriptions to *Chemical Abstracts* and other secondary materials. They considered it part of their obligation to maintain knowledge of developments relative to their work. As the generation of published literature escalated and the journals proliferated, the sheer volume of material appeared to be endless. Chemists found their task impossible using the old method of manual scanning. The advent of computerized current-awareness tools gave them a way to review literature and indeed made it possible once again to keep abreast of new developments. However, information retrieval continues to be a difficult task for average users without the assistance of information intermediaries. Work is in progress to develop systems better oriented to users.

Before 1950, graduating chemists and engineers could look forward to applying their training throughout their careers. Now there is the potential of technical obsolescence of scientists and engineers. To thwart this possibility, the aggressive use of information is essential.

The following section addresses the matter of scientific and technical information retrieval using traditional as well as computerized methods and resources.

Bibliography

J. A. Luedke and co-workers, "Numerical Databases and Systems" in M. E. Williams ed., *Annu. Rev. Inf. Sci. Technol.* **12,** 119–181 (1977).

J. M. Morris and E. A. Elkins, *Library Searching: Resources and Strategies with Examples from the Environmental Sciences,* Jeffrey Norton Publishers, Inc., New York, 1978.

B. H. Weil, "Authorized Services for Supplying Photocopies and for Collection of Payments for In-House Photocopying under the New Copyright Law," in N. B. Glick and F. Simora, eds., *The Bowker Annual of Library and Book Trade Information,* 23rd ed., R. R. Bowker Company, New York, 1978, pp. 33–41.

M. E. Williams, "On-line Retrieval—Today and Tomorrow," *On-Line Rev.* **2**(4), 353–366 (1978).

SCIENTIFIC AND TECHNICAL INFORMATION FOR BASIC AND APPLIED RESEARCH

Studies of scientific and technical information users disclose interesting findings. Radwin found that only 10.7% of library users needed exhaustive information; 39.9% needed single facts or data; and 44.3% needed background information. Direct use of the literature produced maximum idea generation. However, problem solvers found three information-gathering techniques to be equivalent: personal interaction, literature use, and experimentation and analysis. This duality of information needs for idea generation and for problem solving is similar to Garfield's description of scientific and technical information needs as recovery and discovery, or Radwin's description of known and unknown needs.

Two-part division is used here, dealing first with the fact-finding or problem-solving needs of chemists at bench and production activities, and then with the comprehensive and conceptual needs for basic research or research planning.

Selected Bibliography of User Studies

T. J. Allen, *Managing the Flow of Technology: Technology Transfer and the Dissemination of Technological Information Within the R&D Organization,* MIT Press, Cambridge, Mass., 1977.

A. K. Chakrabarti, "Information Use and Training of Industrial Scientists and Engineers," *Libr. Sci. Slant Doc.* **15,** Paper N (June, 1978).

S. Crawford, "Information Needs and Uses" in M. E. Williams, ed., *Annu. Rev. Inf. Sci. Technol.* **13,** 61–81 (1978).

E. Garfield, C. E. Granito, and A. E. Petrarca, "Information Retrieval Services and Methods" in A. Standen, ed., *Kirk-Othmer Encyclopedia of Chemical Technology,* 2nd ed., Supplement Volume, Interscience Publishers, a division of John Wiley & Sons, Inc., New York, 1971, pp. 510–535.

K. Holt, "Information and Need Analysis and Idea Generation," *Res. Manage.* **18**(3), 24–27 (1975).

J. Martyn, "Information Needs and Uses" in C. A. Cuadra ed., *Annu. Rev. Inf. Sci. Technol.* **9,** 3–23 (1974).

P. S. Nagpaul and S. Pruthi, "Problem-Solving and Idea-Generation in R&D: the Role of Informal Communication," *R&D Manage.* **9**(3), 147–149 (1979).

E. Radwin, "Field Survey of Information Needs of Industry Sci/Tech Users" in C. W. Husbands and R. Tighe, eds., *Information Revolution: Proceedings of the 38th Annual Meeting of the American Society for Information Science,* American Society for Information Science, Washington, D.C., **12,** 41–42 (1975).

B. H. Weil, "Benefits from Researcher Use of the Published Literature at the Exxon Research Center" in E. Jackson, ed., *Special Librarianship; A New Reader,* Scarecrow Press, Metuchen, N.J., 1980.

F. W. Wolek, "Uses and Benefits of Technical Information Systems," *Res. Manage.* **20**(5), 37–41 (1977).

Bibliography of Chemical Literature Sources

Some readers may wish to learn more about chemical information resources. The following bibliography lists some current literature guides:

R. T. Bottle, *Use of the Chemical Literature,* 3rd ed., Butterworths Publishers, Inc., Woburn, Mass., 1979.

C. Chen, *Scientific and Technical Information Sources,* MIT Press, Cambridge, Mass., 1977; an extensive bibliography with citations of book reviews and reference librarians.

R. E. Maizell, *How to Find Chemical Information: A Guide for Practicing Chemists, Teachers, and Students,* Wiley-Interscience, New York, 1979; descriptive guide for chemists, mixed with sections for librarians.

D. B. Owen and M. M. Hanchey, *Indexes and Abstracts in Science and Technology: A Descriptive Guide,* Scarecrow Press, Metuchen, N.J., 1974.

T. P. Peck, ed., *Chemical Industries Information Sources (Management Information Guide Series 29),* Gale Research Company, Detroit, Mich., 1978; lists resources in chemistry, chemical engineering, and associated industries; this is a wide-ranging guide to organizations as well as literature, including classic literature.

S. H. Wilen, *Use of the Chemical Literature, An Introduction to Chemical Information Retrieval,* ACS Audio Course, American Chemical Society, Washington, D.C., 1978.

H. M. Woodburn, *Using the Chemical Literature: A Practical Guide (Library and Information Science Series,* Vol. 11), Marcel Dekker, Inc., New York, 1974; a self-teaching tool; this informative guide to some main resources for obtaining chemical information discusses manual retrieval and current-awareness techniques in a clear and practical manner; it deals only with those resources known to cause problems to users.

Information in the Laboratory

At the bench level, chemists are looking for facts, often urgently because an experiment is already in progress. Besides facts, experimental procedures or techniques are used at the bench. Until now the information resources for these needs were in traditional book form and were usually kept in the chemist's personal library or, in the case of larger data compilations, series on techniques, and encyclopedias, in the organization's library.

There is now a trend toward delivering factual, handbook information in electronic form via the laboratory's computer terminal, already in use for automated instrument control and data analysis. For the chemical industry, delivery of handbook information is workable if the data, equations, and the like are accompanied by design modules that allow retrieved information to be manipulated to produce an answer (2). In contrast, simple electronic reference systems are not as useful.

The discussion here concentrates on a selected list of universally available, printed resources. The literature guides listed in the bibliography and the reader's reference librarian or information specialist will be helpful in locating the most appropriate and up-to-date resources for the reader's particular needs.

Handbooks. Handbooks are the most convenient reference tools for the laboratory. They contain just the kind of information required in the middle of things—physical constants and mathematical tables. The data may or may not be evaluated but they are usually reliable—the hard core of chemistry.

Bibliography

American Chemical Society, Committee on Analytical Reagents, *Reagent Chemicals: American Chemical Society Specifications,* 5th ed., American Chemical Society, Washington, D.C., 1974.

J. A. Dean, ed., *Lange's Handbook of Chemistry,* 12th ed., McGraw-Hill Book Co., New York, 1979; includes physical constants, thermodynamic properties, and mathematical tables and equations; some self-instructional units; designed for desk use.
A. J. Gordon and R. A. Ford, *The Chemist's Companion; A Handbook of Practical Data, Techniques, and References,* Wiley-Interscience, New York, 1973; designed for easy use in the lab.
F. A. Lowenheim and M. K. Moran, *Faith, Keyes, and Clark's Industrial Chemicals,* 4th ed., John Wiley & Sons, Inc., New York, 1975.
R. H. Perry and C. H. Chilton, *Chemical Engineers' Handbook,* 5th ed., McGraw-Hill Book Co., New York, 1973; completely revised to assist engineers in producing efficient designs based on scientific advances.
J. Pinkava, *Handbook of Laboratory Unit Operations for Chemists and Chemical Engineers,* Gordon and Breach Science Publishers, New York, 1971.
G. J. Shugar, *Chemical Technicians' Ready Reference Handbook,* McGraw-Hill Book Co., New York, 1973.
R. C. Weast, ed., *CRC Handbook of Chemistry and Physics,* 59th ed., CRC Press, Inc., West Palm Beach, Florida, 1978–1979; original edition 1918; lists sources of critical data; revised annually; physical constants, spectral data, thermodynamic properties, conversion factors, references to numeric data projects, useful index.

There are many highly specialized handbooks. The *Subject Guide to Books in Print* (R. R. Bowker, New York, annual publication) and *Handbooks and Tables in Science and Technology* (R. H. Powell, ed., Oryx Press, 1980) are useful in identifying relevant ones in individual areas. A few examples are the following:

G. L. Baughman, *Synthetic Fuels Data Handbook,* 2nd ed., Cameron Engineers, Inc., Denver, Colorado, 1978; synthetic fuels from oil shale, coal, and oil sands.
C. A. Harper, *Handbook of Plastics and Elastomers,* McGraw-Hill Book Co., New York, 1975.
I. Mellan, *Industrial Solvents Handbook,* Noyes Data Corp., Park Ridge, N.J., 1977; useful in solvent selection or replacement.
Synthetic Organic Chemical Manufacturers Association, *SOCMA Handbook; Commercial Organic Chemical Names,* American Chemical Society, Washington, D.C., 1965; lists 6300 industrial organic compounds, giving structure, nomenclature, and Chemical Abstracts Service Registry Number.

Physical Data Compilations. Using data compilations can save chemists' time if they provide quality information (3). The institutions that produce these collections spend considerable effort in evaluating data. The complex organizational schemes designed to accommodate updates require the user to read the introduction carefully.

The following list represents only a small sample of the useful physical data collections; these have proved their value over many years.

International Compendium of Numerical Data Projects, a Survey and Analysis, CODATA-The Committee on Data for Science and Technology of the International Council of Scientific Unions, Springer-Verlag, New York, 1969.
E. W. Washburn, ed., *International Critical Tables of Numerical Data—Physics, Chemistry and Technology,* published for the National Research Council by McGraw-Hill Book Co., New York, 1926–1933; 7 data volumes and 1 index volume; an old but significant and trusted data source.
Landolt-Bornstein Zahlenwerte und Funktionen aus Physik, Chemie, Astronomie, Geophysik und Technik, 6 Aufl., Springer-Verlag, Berlin, 1950–.
K. H. Hellwege, ed., *Landolt-Bornstein Zahlenwerte und Funktionen aus Naturwissenschaften und Technik,* Neue Serie, Springer-Verlag, New York, 1961–; this series contains reliable data but no index: to locate information one selects a volume by reading the name on the spine and identifies the likely table from the contents; in the new series, side-by-side English and German are used; the old series is entirely in German.

The National Standard Reference Data System (NSRDS) of the National Bureau of Standards (NBS) was established to coordinate the data-compiling activities of several governmental agencies. The publications from this program are listed in the *Handbook of Chemistry and Physics,* and much of the information is now published in the *Journal of Physical and Chemical Reference Data,* as well as in NSRDS-NBS publications.

The Center for Information and Numerical Data Analysis and Synthesis (CINDAS), Purdue University, Y. S. Touloukian, director, operates the Thermophysical Properties Research Center.

Thermophysical Properties of Matter, 13 vols., IFI/Plenum, New York, 1970–1977; covers thermal conductivity, specific heat, radiative properties, thermal diffusivity, viscosity, and thermal expansion; supplements are being published.
Thermophysical Properties Research Literature Retrieval Guide, Y. S. Touloukian, IFI/Plenum, New York, 1967: *Basic Ed.,* 3 volumes, 1967; *Supplement I,* 6 volumes, 1973; *Supplement II,* 6 volumes, 1979; the *Retrieval Guide,* which is complementary to the TRC data tables, provides extensively indexed access to world literature on thermophysical properties; literature from 1920 to mid 1964 is covered in the *Basic Ed.;* mid 1964 to 1971 in *Supplement I,* and 1971–1977 in *Supplement II.*

Thermodynamics Research Center (TRC) at Texas A&M University, B. J. Zwolinski, director, participates in NSRDS and has two major projects. The arrangement of these publications is complex.

B. J. Zwolinski and co-eds., *Comprehensive Index of API 44—TRC Selected Data on Thermodynamics and Spectroscopy, TRC Publication #100,* 2nd ed., 1974; conventional index to all TRC publications.
Selected Values of Properties of Hydrocarbons and Related Compounds; Thermodynamics Research Center, American Petroleum Institute Research Project 44; this series provides evaluated physical, thermodynamic, and spectral data for thousands of compounds.
Selected Values of Properties of Chemical Compounds; Thermodynamics Research Center Project; covers organic compounds other than the hydrocarbons included in *API 44.*

Bibliography

JANAF (Joint Army–Navy–Air Force) Thermochemical Tables, 2nd ed., *NSRDS-NBS 37,* Office of Standard Reference Data, National Bureau of Standards, Superintendent of Documents, U.S. Government Printing Office, Washington, D.C., 1971; critical compilation of tables of thermodynamic properties of propellant–combustion products and inorganics; supplements published in *J. Phys. Chem. Ref. Data.*
Sadtler Standard Spectra (INFRARED), Sadtler Research Laboratory, Philadelphia, Pa.; regularly updated looseleaf service.
A. Seidell, *Solubilities of Inorganic and Metal-Organic Compounds,* 4th ed., Vol. 1, D. Van Nostrand Co., Inc., New York, 1958; Vol. 2, American Chemical Society, Washington, D.C., 1965; unevaluated; aqueous and nonaqueous solvent systems; references to data sources.
H. Stephen and T. Stephen, *Solubilities of Inorganic and Organic Compounds,* 2 vols., Pergamon Press, Ltd., Oxford, Eng.; distributed by the Macmillan Co., New York, 1963.
G. W. C. Kaye and T. H. Laby, *Tables of Physical and Chemical Constants and Some Mathematical Functions,* 14th ed., Longman Group Ltd., London, Eng., 1973; all tabulated values now in SI units; aim is to provide data in a convenient size and at a moderate price.

Encyclopedias and Treatises. Encyclopedias provide background information. The treatises, by dealing in-depth, can help the nonspecialist become more knowledgeable. Treatises generally are descriptive, contain evaluated data, discuss methodologies, and are likely to have better indexes than handbooks or monographs.

C. H. Bamford and C. F. H. Tipper, *Comprehensive Chemical Kinetics,* 18 vols., American Elsevier Publishing Company, New York, 1969–1977; multivolume sections; coverage is thorough; Section 1—Practice and Theory; Section 2—Homogeneous Decomposition and Isomerization Reactions; Section 3—Inorganic Reactions; Section 4—Organic Reactions; Section 5—Polymerization Reactions; Section 6—Oxidation and Combustion Reactions; Section 7—Selected Elementary Reactions.
D. H. R. Barton and W. D. Ollis, eds., *Comprehensive Organic Chemistry: The Synthesis and Reactions of Organic Compounds,* 6 vols., Pergamon Press, New York, 1979; the expanding literature created a need for a new, rapidly-published work in organic chemistry. One technique the editors used to accomplish this goal was not to treat theoretical organic chemistry separately, since this area changes first. This is a six-

volume work, with the sixth containing the indexes, with references to review articles. The indexes are organized by formula, subject, author, reaction, and reagents, and provides pointers to the contents of volumes 1–5, the references in those volumes, and significant literature through mid 1978.

Chemical Technology: An Encyclopedic Treatment; The Economic Application of Modern Technological Developments, 8 vols., Barnes & Noble, Inc., New York, 1968–1973; for lay persons and technologists.

S. Coffey, ed., *Rodd's Chemistry of Carbon Compounds, A Modern Comprehensive Treatise,* 2nd ed., Elsevier Publishing Company, New York, 1964–; five multipart volumes arranged using the Richter classification.

Dictionary of Organic Compounds (Heilbron): The constitution and physical, chemical, and other properties of the principal carbon compounds and their derivatives together with relevant literature references, 4th ed., 5 vols., Oxford University Press, New York, 1965; updated by supplements; constitution and physical and chemical properties as extracted from the literature.

J. E. Faraday, ed., *Encyclopedia of Hydrocarbon Compounds,* 13 vols., Chemical Publishing Co., New York, 1946–1954.

Kirk-Othmer Encyclopedia of Chemical Technology, 2nd ed., A. Standen, ed., 1963–1972; 3rd ed., 1978–, M. Grayson and D. Eckroth, eds., Wiley-Interscience, New York; 12 volumes published by end 1980; annual indexes; 25 vols to be published in 3rd ed., by 1984 including cumulative index.

I. M. Kolthoff and P. J. Elving, eds., *Treatise on Analytical Chemistry,* John Wiley & Sons, Inc., New York, 1959–; a comprehensive review series in three parts: I—Theory and Practice; II—Analytical Chemistry of Inorganic and Organic Compounds; III—Analytical Chemistry in Industry; started in the 1960s and continuing.

Encyclopedia of Chemical Processing and Design, Marcel Dekker, Inc., New York, 1976 (J. J. McKetta and W. A. Cunningham, eds.); intended to assist designers and developers of chemical processes and products in practical design problems; 11 volumes published by end of 1980.

J. W. Mellor, *A Comprehensive Treatise on Inorganic and Theoretical Chemistry,* Longman's, Green & Co., Ltd.; John Wiley & Sons, Inc., New York, 16 vols. (main series), 1922–1937; supplements, 1956–1972; arranged by periodic table, this work presents information in the order: history, preparation, properties, hydride, oxide, halides, sulfide, sulfate, carbonate, nitrate, phosphate, references; each primary and supplementary volume has an index; volume 16 is a general index to the main series.

E. Josephy and F. Radt, eds., *Elsevier's Encyclopedia of Organic Chemistry,* Springer-Verlag, Berlin, 1948–1965; based on structural skeleton, this encyclopedia offers compact presentation of data; updated by supplements.

F. D. Snell and L. S. Ettre, *Encyclopedia of Industrial Chemical Analysis,* 20 vols., Wiley-Interscience, New York, 1966–1974; Vols. 1–3 cover general techniques, Vols. 4–19 are specific, and Vol. 20 is the index to Vols. 4–19; coverage includes individual compounds of commercial significance (acetone, dioxane), elements and their compounds (aluminum), chemical classes (amines), compounds by end use (adhesives), and compounds by industrial use (food products).

J. C. Bailar, Jr., H. J. Emeleus, R. Nyholm, and A. F. Trotman-Dickenson, eds., *Comprehensive Inorganic Chemistry,* 5 vols., Pergamon Press, New York, 1973; the presentation in each compound or family is compact with much data in tabular form and references as footnotes; the index is quite complete.

Ullmann's Encyklopaedie der Technischen Chemie, 4th ed., Verlag Chemie International, Deerfield Beach, Fl., 1972–; 25 vols. to be published in the 4th ed.

Beilstein and Gmelin. The Beilstein and Gmelin chemical treatises hold a special place in the minds of chemists because they present evaluated information on a comprehensive scale. Both are compound-based but have different organizational schemes. The value of these handbooks is immense, and retrieval is rapid; instructional information is available for each.

Beilstein's Handbuch der Organischen Chemie, simply known as *Beilstein,* is a comprehensive reference source of carbon-containing compounds. First published in 1881–1882 in two volumes by F. K. Beilstein, the handbook now comprises more than 190 volumes. More than 100 chemists and physicists are employed at the Beilstein Institute to review critically all primary publications containing information on carbon compounds. The editor is R. Luckenbach. In addition, the chemical literature is evaluated, and data are checked to ensure that misleading or erroneous results are not documented. *Beilstein* is a key source to factual chemical information, free from the inaccuracies and redundancies of the open chemical literature.

The *Beilstein Handbook* began as 27 volumes of the *Basic Series,* using the Beilstein system to order carbon compounds logically. This system was developed by P. Jacobson and B. Prager in 1907 and continues to be used as the organic chemical classification system by which information is collected and presented. Details of the Beilstein system are found in the opening pages of the *Basic Series'* Volume 1, which should be consulted to ensure proper usage of the handbook. Four series of supplemental volumes have been added to the *Basic Series* since 1910, and a fifth series covering the literature from 1960 to 1979 is currently being prepared. Continuity is one of the most valued features of the Beilstein system. New information on a compound already reported is published in the same volume of later *Supplement Series.* Thus, once the volume and system number are identified, all other information in *Beilstein* on that compound can be located easily.

Beilstein gives specific data on and references to nearly four million (4×10^6) compounds of known constitution. The description of such compounds includes their composition, configuration, natural occurrence and isolation from natural products, preparation and purification, structural and energy parameters, physical properties, chemical properties, their characterization and analysis, and their salts and addition compounds. A booklet entitled *How to Use Beilstein* is available free of charge from the publisher, Springer-Verlag, Berlin, Heidelberg, and New York. Woodburn (4) includes a helpful chapter on using *Beilstein.*

The Gmelin Handbuch der Anorganischen Chemie, or simply *Gmelin,* is the most authoritative and comprehensive treatise on inorganic compounds. The first edition was published in 1817 by L. Gmelin. The current, eighth edition of 365 volumes is published in Frankfurt by the Gmelin-Institut für Anorganische Chemie und Grenzgebiete, an institute of the Max Planck Society for the Advancement of Science. The editor is E. Fluck. Like *Beilstein, Gmelin* is a compilation of data from primary literature that has been critically evaluated and presented in coherent groups of related subject matter.

Each volume and its supplement discuss an element and its compounds, their formation and occurrence, methods of preparation, and physical and chemical properties. The classification system used in *Gmelin* is based on a numerical sequence of the elements by system number. Information about a compound is found under that element in the compound having the highest system number. One key to all the elements and compounds of the eighth edition is the formula index, now in preparation. The Gmelin Institute is progressively translating its entire collection from German to English; the complete handbook eventually will be available in English.

The value of the *Gmelin Handbook* must not be underestimated. The current edition contains reevaluated material from the older editions and thus represents a comprehensive documentation of the entire body of inorganic chemical literature. Material is presented in a form that allows comparative assessments and includes references to allied concepts and critical reviews. The handbook is described in detail in the Institute's brochure *Was ist der Gmelin?,* available in either English or German from the Gmelin Institute, Frankfurt. H. M. Woodburn has described the original scope and use of the handbook (5).

Bibliography

F. Beilstein, *Handbuch der Organischen Chemie,* 4th ed., Springer-Verlag, Berlin.
How to use Beilstein, Beilstein Handbook of Organic Chemistry, Beilstein Institute, Springer-Verlag, New York.
R. Luckenbach, "Der Beilstein," *Chemtech* **9**(10), 612–621 (1979).
H. M. Woodburn, *Using the Chemical Literature: A Practical Guide* (*Library and Information Science Series,* Volume 11), Marcel Dekker, Inc., New York, 1974.
Gmelin's Handbuch der Anorganischen Chemie, 8th ed., Springer-Verlag, New York.

Experimental Methods. Chemists have a long tradition of publishing methodological compilations. These make it easy to locate synthetic preparations or discussions of laboratory techniques at a more practical level than found in encyclopedias.

Weissberger's *Techniques of Chemistry* series (6), a successor to his *Technique of Organic Chemistry,* is organized into several multipart volumes: Vol. 1—Physical Methods of Chemistry, A. Weissberger and B. W. Rossiter, eds., 1971–1977; Vol. 2—Organic Solvents: Physical Properties and Methods Purification, J. A. Riddick and W. B. Bunger, eds., 1971; Vol. 3—Photochromism, G. H. Brown, ed., 1971; Vol. 4—Elucidation of Organic Structures by Physical and Chemical Methods, K. W. Bentley and G. W. Kirby, eds., 1972–1973; Vol. 5—Technique of Electroorganic Synthesis, N. L. Weinberg, ed., 1974–1975; Vol. 6—Investigation of Rates and Mechanisms of Reactions, E. S. Lewis and G. G. Hammes, eds., 1974; Vol. 7—Membranes in Separation, S. Hwang and K. Kammermayer, eds., 1975; Vol. 8—Solutions and Solubilities, M. R. J. Dack, ed., 1975–1976; Vol. 9—Chemical Experimentation under Extreme Conditions, B. W. Rossiter, ed., 1980; Vol. 10—Applications of Biochemical Systems in Organic Chemistry, J. B. Jones, C. J. Sih, and D. Perlman, eds., 1976; Vol. 11—Contemporary Liquid Chromatography, R. P. W. Scott, ed., 1976; Vol. 12—Separation and Purification, E. S. Perry and A. Weissberger, eds., 1978; Vol. 13—Laboratory Engineering and Manipulation, A. Weissberger and E. S. Perry, eds., 1979; Vol. 14—Thin Layer Chromatography, J. G. Kirchner, ed., 1978; Vol. 15—Theory and Applications of Electron Spin Resonance, W. Gordy, ed., 1979.
E. Muller, ed., *Houben-Weyl Methoden der Organischen Chemie,* 4th ed., Thieme, Stuttgart, FRG, 1952–; originated early in the century as a critical survey of laboratory methods; gives physical and analytical methods as well as preparations, transformations, and theory.
Organic Reactions, John Wiley & Sons, Inc., New York, 1942–; annual series providing critical discussion of widely used organic reactions.
Organic Syntheses, John Wiley & Sons, Inc., New York, 1921–; annual volumes published since 1921; the ten-year collective volumes are revised as required; emphasis is now on model procedures for reaction types, rather than on preparations of specific organic compounds.
Theilheimer's *Synthetic Methods of Organic Chemistry,* (S. Karger), a series based on reaction schemes, and Buehler and Pearson's *Survey of Organic Synthesis* (Wiley) are series that help the organic chemist locate experimental methodology rapidly.

Dictionaries. There are many technical dictionaries used to define specific terms and, to a limited degree, chemical structures. In chemistry, the line between dictionaries and encyclopedias is thin but the term dictionary here refers to the one-volume works. Technical foreign language dictionaries are not covered here.

Bibliography

H. Bennett, ed., *Concise Chemical and Technical Dictionary,* 3rd ed., Chemical Publishing Company, New York, 1974.
T. C. Collocott and A. B. Dobson, eds., *Chambers Dictionary of Science and Technology Terms,* successor to *Chambers Technical Dictionary,* W&R Chambers, Ltd., Edinburgh, UK, 1974.
G. L. Clark and G. G. Hawley, eds., *The Encyclopedia of Chemistry,* 2nd ed., Reinhold Publishing Corp., New York, 1966.
D. M. Considine, ed., *Van Nostrand's Scientific Encyclopedia,* 5th ed., Van Nostrand Reinhold Company, New York, 1976; in addition to chemistry, this covers physics, materials sciences, energy technology, earth and space sciences, life science, mathematics, and information science; references included.
W. Gardner, E. I. Cooke, and R. W. I. Cooke, eds., *Handbook of Chemical Synonyms and Trade Names,* 8th ed., CRC Press, Cleveland, Ohio, 1978.

J. Grant, ed., *Hackh's Chemical Dictionary,* 4th ed., McGraw-Hill Book Co., New York, 1969.
D. N. Lapedes, ed., *McGraw-Hill Dictionary of Scientific and Technical Terms,* 2nd ed., McGraw-Hill Book Co., New York, 1978.
G. G. Hawley, ed., *Condensed Chemical Dictionary,* 9th ed., Van Nostrand-Reinhold Corp., New York, 1977.
M. Windholz, S. Budavari, L. Y. Stroumtsos, and M. Noether Fertig, eds., *The Merck Index,* 9th ed., Merck & Co., Inc., Rahway, N.J., 1976; descriptive information on 10,000 chemicals, drugs, and biologicals.

Reviews. One rapid way for chemists to keep current in several areas or to bring their knowledge to the state-of-art level is to read reviews. In fact, on-line data bases can be searched using REVIEW as a keyword or document type. Publishers even provide indexes to reviews. The Institute for Scientific Information publishes one of these semiannually and cumulates it annually in *Index to Scientific Reviews.*

There are several review journals: *Accounts of Chemical Research,* American Chemical Society; *Angewandte Chemie—International Edition,* in English, Verlag Chemie; *Chemical Society Reviews,* The Chemical Society; *Chemical Reviews,* American Chemical Society; and *CRC Critical Review Series,* CRC Press.

A number of review series are published, often entitled *Advances in . . . ,* or *Progress in . . . .* The Chemical Society publishes two such series, *Annual Reports on the Progress of Chemistry,* and *Specialist Periodical Reports.* The Society of the Chemical Industry publishes a similar series called *Report on the Progress of Applied Chemistry.*

Miscellaneous. There are more sources of information than we can discuss in one article. Some of these are governmental reports, dissertations, monographs, standards, technical brochures, catalogues, and conference papers. Because abstracting and indexing tools provide access to these also, it is now no more difficult to identify a conference paper than to identify an article in the *Journal of the American Chemical Society.*

Technical Information, Retrospective and Comprehensive

So far this article has discussed techniques of information retrieval, the importance of information in all aspects of the chemist's work, and the kinds of resources used for fact finding. Reviewed here are some main resources for retrospective searching—exhaustive and background, manual and computerized—and current awareness as printed bulletins or personalized, computer-produced, SDI profiles. The review begins with a discussion of search services.

Lockheed–DIALOG. Lockheed Information Service (LIS) in Palo Alto, California, was established in 1965 as a division of Lockheed Missiles and Space Corporation. LIS designed a retrieval system called RECON (REmote CONsole) for the document-citation collection of the National Aeronautics and Space Administration (NASA) and further developed it into the DIALOG on-line information system. Commercial operation began in 1972 providing access to three data bases—ERIC (Educational Resources Information Center), CAIN (Cataloging and Indexing data base of the National Agricultural Library), and PANDEX (the on-line version of the *PANDEX Current Index to Scientific and Technical Literature*). The current collection contains more than 100 data bases in a wide range of disciplines, including directories and statistical files as well as bibliographic data bases, and with access to over thirty million records in total. The cost of searching is based on connect time to the particular data base, that is, the actual time elapsed from sign-on to sign-off from a certain computer file. No charge is made until the system is used.

System Development Corporation–ORBIT. System Development Corporation (SDC), founded in 1956 in Santa Monica, California, designed a system called COLEX for the Foreign Technology Division of the United States Air Force System Command. This system developed into a series of ELHILL programs and finally into ORBIT (On-line Retrieval of Bibliographic Information Timesharing). Commercial operation of the SDC Search Service began in 1973 offering access to the three data bases, ERIC, CHEMCON (Chemical Condensates), and MEDLINE (Medlars on line). Currently there are 65 data bases providing access to over twenty-five million records. On-line searching is charged by connect hour.

Bibliographic Retrieval Service–STAIRS. Established in 1976 in Scotia, N.Y., the Bibliographic Retrieval Service (BRS) is the newest of the United States on-line search services. It was formed to offer on-line service affordable to small organizations and public organizations such as state libraries. BRS uses a modification of the STAIRS (Storage and Information Retrieval System) software system, developed by International Business Machines (IBM). It provides access to 30 data bases, 9 of which are unique to BRS, bringing the total number of searchable records to nearly sixteen million. Coverage is expanding to include on-line access to standard reference works. BRS files are used on a subscription basis, ie, the user is entitled to a specific number of connect hours that may be used at any time during the year. Reduced rates are offered for increased usage. These rates, however, are exclusive of communications charges, any applicable data-base royalty charges, and optional services such as off-line searching, printing, and SDI.

National Library of Medicine–MEDLARS. The National Library of Medicine in Bethesda, Maryland, began experiments with on-line bibliographic searching in 1967. Time-consuming, expensive batch searches of their *Index Medicus* data base were run using the Medical Literature Analysis and Retrieval System (MEDLARS), a system devised originally to produce *Index Medicus.* Because MEDLARS continued to grow with the expansion of the biomedical literature and the increased number of demand searches, a faster and inexpensive system was needed. SDC was hired to develop the on-line searching portion of MEDLARS; the ELHILL III system has emerged as the operating system since 1975. Currently there are 18 data bases accessible on line, 14 of which are available only through NLM. Five data bases available for off-line searching contain citations, and often abstracts, of the biomedical literature going back to the mid to late 1960s. A total of five million records are accessible. NLM is relatively inexpensive and offers reduced rates for usage outside of prime time.

Department of Energy–RECON. The DOE–RECON (Department of Energy–REmote CONsole) system was initially developed by Lockheed for NASA. The Atomic Energy Commission purchased the system to support the *Nuclear Science Abstracts* data base, and the system was installed at Oak Ridge National Laboratory (ORNL) in 1970–1971, where it operated experimentally for about four years while the on-line retrieval software was improved. Dial-up access to DOE–RECON has been available since 1975. The current version contains 21 unclassified data bases, including a few trial data bases that will be retained if use is sufficient. Most of the data bases are prepared by governmental offices and are unique to DOE–RECON. The DOE–RECON system differs from the commercial search services in that it is restricted to use by DOE personnel, holders of DOE-sponsored research contracts, and other federal agencies engaged in work related to DOE's programs.

Canada Institute for Scientific Information–CAN–OLE. The Canada Institute for Scientific and Technical Information (CISTI) was formed by the National Research Council of Canada in 1974 to provide Canadian researchers, technologists, and industries with scientific and technical information. The Institute operates a range of information services, including a computerized current-awareness service (CAN–SDI) and an on-line reference system (CAN–OLE). In cooperation with the U.S. National Library of Medicine, it provides on-line access in Canada to the medical literature (MEDLINE). CAN–OLE currently contains 12 data bases, which can be interactively searched in either English or French. Paper copies of the references retrieved may be ordered from CISTI. Billing for the service is on a connect-hour basis. Through the DATAPAC communications network of the Trans-Canada Telephone System, there is access to these information services from over 50 cities.

QL Systems Ltd.–Shared Information Service. QL Systems Ltd. was an outgrowth of the QUIC–LAW Project at Queen's University, Ontario, Canada. The project was established in 1968 to investigate possible use of computers by lawyers, particularly in legal research. Owing to its rapid growth in the early seventies, the project was incorporated as a separate entity outside the University. Since 1973, QLS has been dedicated to developing and maintaining the first Canadian commercial information-retrieval system, the Shared Information Service. Over 30 data bases are currently available, including some full-text files. The subject coverage ranges from the environment to business to law with most of the legal and governmental files in both English and French. On-line searching is billed per search, for each data-base selection, and by a communications charge per on-line hour in the United States; some data bases are also subject to a royalty charge per on-line hour.

EURONET–DIANE. EURONET is the data-transmission network set up by the postal and telecommunications authorities of European Economic Communities (EEC) to provide on-line access to host computers and their data bases. DIANE (Direct Information Access Network for Europe) is a group of host information services that offer their data bases through EURONET. A few of these are BLAISE (British Library Automated Information Service); Datacentralen; DIMDI (Deutches Institut für Medizinische Dokumentation und Information); EPO; Infoline; and IRS (Information Retrieval Service), the oldest European service.

Chemical Abstracts Service. One might think it unnecessary to discuss *Chemical Abstracts* (*CA*) in an article on information retrieval for chemists. It is the one tool chemists learn about as undergraduates. However, beginning in the early 1960s Chemical Abstracts Service (CAS) formulated new methods for carrying out its mission—the abstracting and indexing of the world's chemical literature. The volume of published literature was exploding, and the power of the computer was becoming an everyday reality.

The full effect of the growth in scientific and technical literature is clear when we see that in 1979 Chemical Abstracts Service published nearly 437,000 abstracts and covered 14,000 periodicals. This means that approximately 16,800 abstracts were issued every two weeks; no one chemist can cope with that. Even the indexes have grown in depth and size. The *8th Collective Index* comprises 35 books; the *9th Collective Index,* 57 books; and the *10th Collective Index* (1977–1981) is expected to have between 71 and 75 volumes of indexes alone.

International cooperation is the first of the three principles CAS adopted to manage information in a changing environment. The chemical societies of The Federal Republic of Germany, the United Kingdom, France, and Japan provide abstracts from their national chemical publications or pay a share of production costs, in exchange for certain use and distribution rights. The second principle is the data-base concept, through which CAS has established multiple, compatible files that allow new and timely services. CAS maintains three types of files: bibliographic information; abstract text; and data elements for subject indexes. The third principle is to use unique, noninformative, identifying numbers for chemicals for *CA* production (structure, nomenclature), and for information retrieving and linking—Registry Numbers.

CAS Indexes. *Chemical Abstracts* has indexes of varying depth and type. In 1963 *CA* added a permuted keyword–subject index and patent concordance to its author and numerical patent indexes for the weekly issues.

The volume semiannual indexes and the collective indexes, now issued every five years, are much more extensive. These indexes are as follows: Author Index, Formula Index, Index of Ring Systems, Numerical Patent Index, Patent Concordance, General Subject Index, and Chemical Substance Index.

It is important to consult the *CA Index Guide,* first issued in 1969, because cross references, scope notes, and synonyms are no longer printed in the subject or substance indexes. The *Registry Handbook* and its *Update,* arranged in Registry Number order, tell the *CA* systematic name, the molecular formula, and any changes in registration or cross references. Occasionally, multiple registrations need to be resolved, or changes made based on further structural elucidation. The *Parent Compound Handbook* replaces the old *Ring Index.* It has a broader scope and six routes of access: the *Ring Analysis Index,* the *Ring Substructure Index,* the *Parent Name Index,* the *Wiswesser Line Notation Index,* the *Parent Formula Index,* and the *Parent Registry Number Index.*

CAS Computer Searching. *CA* indexes can also be searched by on-line, interactive, computer techniques from several commercial and national services (SDC, Lockheed, BRS in the U.S.; EURONET–DIANE in Europe; and CISTI in Canada), all based on CA SEARCH. Although the time span accessible (late 1960s or early 1970s to the present) is limited compared to the manual search of indexes back to 1907, computers can search rapidly, allow for iterative and heuristic querying, search across information types (author and subject and patent number) to search information types not accessible in the manual indexes (organizational author, document type, word fragments), and provide substructure searching based on nomenclature and registry information. In addition, the use of Boolean logic enables one to structure complex queries, in contrast to searching printed indexes one heading at a time. Another limitation of computer-produced search results is the lack of abstracts.

CAS Current Awareness. CAS has a number of current-awareness services. When individual issues became unwieldy, *CA* established five section groups, which could be subscribed to separately. These divide the 80 sections into biochemistry, organic chemistry, macromolecular chemistry, applied chemistry and chemical engineering, and physical and analytical chemistry.

Chemical Titles (*CT*), the first computer-produced publication, is a general publication that has a permuted key-word-in-context (KWIC) Index to article titles from over 700 journals of pure and applied chemistry and chemical engineering. It also contains a bibliographic section organized by journal and an author index. *CT* is a timely publication since CAS prepares it from galley proofs.

Personalized SDI profiles based on *CA,* a computer technique discussed earlier, are available in many organizations, internal, government-sponsored, and commercial.

CA Selects is a new service that reflects CAS's responsiveness to consumer demands and the flexibility of its automated systems. *CA Selects* are topical current-awareness bulletins that draw across *CA* sections, include the abstracts, and are printed with *CA* quality. The bulletins average 136 abstracts on 12 pages. The computer-searching profiles use all the information in the CAS data base: bibliographic data, keyword phrases, abstract text, chemical substance and general subject index entries, CAS Registry Numbers, and molecular formulas. Some examples of topics are forensic chemistry, ion exchange, and radiation chemistry, but offerings change with demand.

The *CA Selects* concept has been extended to serve individual organizations. A company may work with *CA* to define a similar product for internal use. Registry Numbers may be used to monitor new information on company products.

Future. CAS is planning to test their substructure searching system during 1980–1981 and to make publicly available the first version of the system by 1982. If successful, it will offer chemists access to the entire registry, now over five million compounds.

CAS Bibliography

A. J. Beach, H. F. Dabek, Jr., and N. L. Hosansky, "Chemical Reaction Information Retrieval from Chemical Abstracts Service Publications and Services," *J. Chem. Inf. Comput. Sci.* **19**(3), 149–155 (1979).

J. E. Blake, V. J. Mathias, and J. Patton, "CA Selects—A Specialized Current Awareness Service," *J. Chem. Inf. Comput. Sci.* **18**(4), 187–190 (1978).

D. L. Dayton, M. J. Fletcher, C. W. Moulton, J. J. Pollock, and A. Zamora, "Comparison of Retrieval Effectiveness of CA Condensates (CA Con) and CA Subject Index Alert (CASIA)," *J. Chem. Inf. Comput. Sci.* **17**(1), 20–28 (1977).

P. G. Dittmar, R. E. Stobaugh, and C. E. Watson, "The Chemical Abstracts Service Chemical Registry System, I. General Design," *J. Chem. Inf. Comput. Sci.* **16,** 111–121 (1976).

R. G. Dunn, W. Fisanick, and A. Zamora, "A Chemical Substructure Search System Based on Chemical Abstracts Nomenclature," *J. Chem. Inf. Comput. Sci.* **17**(4), 212–218 (1977).

M. D. Huffenberger and R. L. Wigington, "Chemical Abstracts Service Approach to Management of Large Data Bases," *J. Chem. Inf. Comput. Sci.* **15**(1), 43–47 (1975).

J. W. Lundeen, "Experimental Use of a Search Profile to Derive a Specialized Information Base from the CAS Data Base," *175th ACS National Meeting, Anaheim, Calif., March 1978.*

I. R. McKinley and A. K. Kent, "ACS/CS Cooperation Agreement," *Chem. Br.* **12**(1), 4–6 (1976).

C. W. Moulton, "Fossil Fuels in Chemical Abstracts," *J. Chem. Inf. Comput. Sci.* **19**(2), 83–86 (1979).

D. C. Myers, J. A. Rathbun, F. A. Tate, and D. W. Weisgerber, "Bridging and Interlinking the Information Resources," *J. Chem. Inf. Comput. Sci.* **16**(1), 16–19 (1976).

R. E. O'Dette, "The CAS Data Base Concept," *J. Chem. Inf. Comput. Sci.* **15**(3), 165–169 (1975).

O. B. Ramsay, "Chemical Abstracts: An Introduction to its Effective Use," *Tape/Slide Audio Course,* American Chemical Society, Washington, D.C., 1979.

L. G. Wade and R. G. M. Cosgrave, "User's Reaction to a Corporate-Designed Current Awareness Bulletin," *Chem. Inf. Comput. Sci.* **22**(3), 179–181 (1980).

M. E. Williams, "Analysis of Terminology in Various CAS Data Files as Access Points for Retrieval," *J. Chem. Inf. Comput. Sci.* **17**(1), 16–20 (1977).

Institute for Scientific Information. The Institute for Scientific Information (ISI) offers a number of information tools for the chemist. The level of complication varies to serve a variety of needs. Most chemists are already familiar with the current-awareness series, *Current Contents.* These reproduce, by permission, the contents pages of core journals in particular disciplines, in weekly, pocket-sized publications for personal scanning. Each issue has a subject index, an index to the principal author of each paper, and an address directory. *Current Contents, Physical and Chemical Sciences* is the primary publication in the series for chemists, but others, such as *Life Sciences* or *Engineering and Technology,* may be of interest.

Current Abstracts of Chemistry (*CAC*), published weekly, provides chemists with a guide to chemical research and technology. It contains abstracts of articles from 107 core journals reporting the synthesis, isolation, and identification of new compounds. Graphic and narrative abstracts are prepared from the original article or selected from the source journal. Extensive use of flow diagrams facilitates rapid scanning, and use-profile and technique-data symbols alert the user to chemical activity of a compound and to analytical procedures used by the investigator (see Fig. 1). The technique-data symbol also highlights articles containing new synthetic methods. Details on these methods appear in the monthly publication, *Current Chemical Reactions* (*CCR*). *CAC* includes reaction schemes, experimental data, bibliographic information, authors' abstracts, and an index section containing journal, author, permuted subject terms, and corporate addresses.

Index Chemicus (*IC*), the companion index to *CAC*, is incorporated into the weekly issues. It contains the following sections: molecular formula, author, subject, biological activity, and an alert to labeled compounds. A list of the journal issues covered each week appears on the back cover. All of the above indexes are cumulated quarterly and annually, the cumulations include a rotaform index of molecular formulas.

All new compounds are encoded using the Wiswesser Line Notation (WLN). These are then cumulated and permuted monthly and annually and are available on microfilm as the *Chemical Substructure Index* (*CSI*).

A personalized current-awareness service that is a product of this family of ISI products is the *Automatic New Structure Alert* (*ANSA*).

Current Chemical Reactions, published monthly, is a guide to new and newly modified reactions and syntheses reported in current journal literature. Flow charts and complete bibliographic information accompany every entry (see Fig. 2). Authors' abstracts and product yields are provided when they are given in the source article. Also included are descriptions of the type of reaction, highlights of techniques used in analyzing compounds, and notices of explosive reactions. Bibliographic information on review articles containing reactions and syntheses is also included. Reports of reactions and syntheses are taken from 107 organic chemistry and pharmaceutical journals. Review articles are drawn from these same source journals plus additional journals and selected books. A list of titles of journals and books covered appears periodically throughout the year in *Current Chemical Reactions*. Each monthly issue of *Current Chemical Reactions* is indexed by author, journal, author affiliation, and permuted subject entry. These four indexes are cumulated annually.

A third family of ISI products is based on their *Science Citation Index* (*SCI*). The printed journal issues bimonthly, with annual and five-year cumulations. The *SCI*, by providing the unique ability to identify papers citing an earlier key paper, allows users to search forward in time. It identifies follow-up work, criticism, work in an interdisciplinary area, and work in a new area with undefined vocabulary. The printed product is in three parts—the *Citation Index*, the *Source Index* (author index), and the *Permuterm Subject Index*. All three of these search approaches are available to users of the on-line version of *SCI*, which is updated weekly and is thus one of the most current of all retrieval resources.

The SDI Service based on the *SCI* is called Automatic Subject Citation Alert (ASCA). The familiar subject or author terms can be included in the search profile, along with cited author or cited reference search keys. ISI also publishes a series of packaged SDIs in narrow areas called ASCATOPICS.

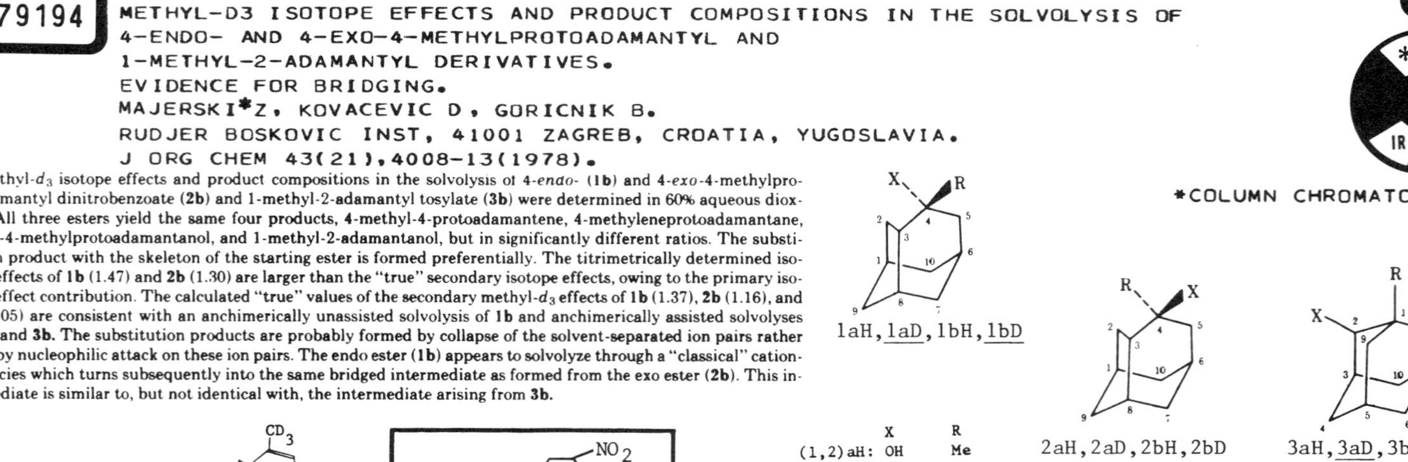

279194 PROTOADAMANTYL-ADAMANTYL REARRANGEMENT.
METHYL-D3 ISOTOPE EFFECTS AND PRODUCT COMPOSITIONS IN THE SOLVOLYSIS OF 4-ENDO- AND 4-EXO-4-METHYLPROTOADAMANTYL AND 1-METHYL-2-ADAMANTYL DERIVATIVES.
EVIDENCE FOR BRIDGING.
MAJERSKI*Z, KOVACEVIC D, GORICNIK B.
RUDJER BOSKOVIC INST, 41001 ZAGREB, CROATIA, YUGOSLAVIA.
J ORG CHEM 43(21),4008-13(1978).

*COLUMN CHROMATOGRAPHY

Methyl-d_3 isotope effects and product compositions in the solvolysis of 4-*endo*- (**1b**) and 4-*exo*-4-methylprotoadamantyl dinitrobenzoate (**2b**) and 1-methyl-2-adamantyl tosylate (**3b**) were determined in 60% aqueous dioxane. All three esters yield the same four products, 4-methyl-4-protoadamantene, 4-methyleneprotoadamantane, 4-*exo*-4-methylprotoadamantanol, and 1-methyl-2-adamantanol, but in significantly different ratios. The substitution product with the skeleton of the starting ester is formed preferentially. The titrimetrically determined isotope effects of **1b** (1.47) and **2b** (1.30) are larger than the "true" secondary isotope effects, owing to the primary isotope effect contribution. The calculated "true" values of the secondary methyl-d_3 effects of **1b** (1.37), **2b** (1.16), and **3b** (1.05) are consistent with an anchimerically unassisted solvolysis of **1b** and anchimerically assisted solvolyses of **2b** and **3b**. The substitution products are probably formed by collapse of the solvent-separated ion pairs rather than by nucleophilic attack on these ion pairs. The endo ester (**1b**) appears to solvolyze through a "classical" cationic species which turns subsequently into the same bridged intermediate as formed from the exo ester (**2b**). This intermediate is similar to, but not identical with, the intermediate arising from **3b**.

	X	R
(1,2) aH:	OH	Me
aD:	OH	CD_3
bH:	ODNB	Me
bD:	ODNB	CD_3

	X	R
3aH:	OH	Me
aD:	OH	CD_3
bH:	OTs	Me
bD:	OTs	CD_3

Figure 1. Sample abstract from *Current Abstracts of Chemistry* (7). Courtesy of Institute for Scientific Information.

TYPICAL ABSTRACT from
CURRENT CHEMICAL REACTIONS™

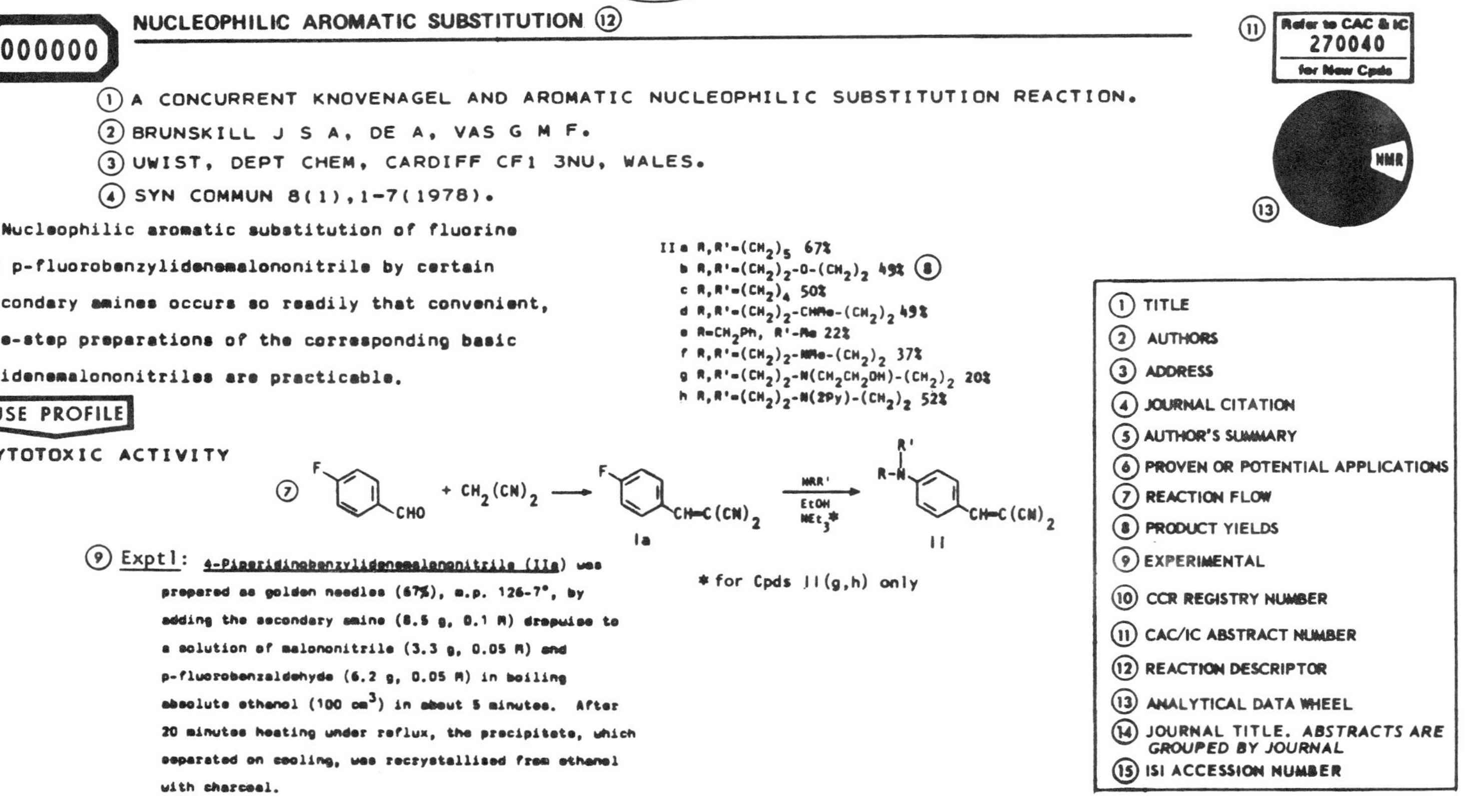

(14) Syn. Commun. 8(1), 1978 (15) EN 955

(10) 000000 NUCLEOPHILIC AROMATIC SUBSTITUTION (12)

(11) Refer to CAC & IC 270040 for New Cpds

(13) NMR

(1) A CONCURRENT KNOVENAGEL AND AROMATIC NUCLEOPHILIC SUBSTITUTION REACTION.
(2) BRUNSKILL J S A, DE A, VAS G M F.
(3) UWIST, DEPT CHEM, CARDIFF CF1 3NU, WALES.
(4) SYN COMMUN 8(1),1-7(1978).

(5) Nucleophilic aromatic substitution of fluorine in p-fluorobenzylidenemalononitrile by certain secondary amines occurs so readily that convenient, one-step preparations of the corresponding basic ylidenemalononitriles are practicable.

(6) USE PROFILE
CYTOTOXIC ACTIVITY

(7) p-F-C6H4-CHO + $CH_2(CN)_2$ → p-F-C6H4-CH=C(CN)$_2$ (Ia) —NRR', EtOH, NEt_3*→ p-(RR'N)-C6H4-CH=C(CN)$_2$ (II)

IIa R,R'=$(CH_2)_5$ 67%
b R,R'=$(CH_2)_2$-O-$(CH_2)_2$ 49% (8)
c R,R'=$(CH_2)_4$ 50%
d R,R'=$(CH_2)_2$-CHMe-$(CH_2)_2$ 49%
e R=CH_2Ph, R'=Me 22%
f R,R'=$(CH_2)_2$-NMe-$(CH_2)_2$ 37%
g R,R'=$(CH_2)_2$-N(CH_2CH_2OH)-$(CH_2)_2$ 20%
h R,R'=$(CH_2)_2$-N(2Py)-$(CH_2)_2$ 52%

* for Cpds II(g,h) only

(9) Exptl: 4-Piperidinobenzylidenemalononitrile (IIa) was prepared as golden needles (67%), m.p. 126-7°, by adding the secondary amine (8.5 g, 0.1 M) dropwise to a solution of malononitrile (3.3 g, 0.05 M) and p-fluorobenzaldehyde (6.2 g, 0.05 M) in boiling absolute ethanol (100 cm^3) in about 5 minutes. After 20 minutes heating under reflux, the precipitate, which separated on cooling, was recrystallised from ethanol with charcoal.

(1) TITLE
(2) AUTHORS
(3) ADDRESS
(4) JOURNAL CITATION
(5) AUTHOR'S SUMMARY
(6) PROVEN OR POTENTIAL APPLICATIONS
(7) REACTION FLOW
(8) PRODUCT YIELDS
(9) EXPERIMENTAL
(10) CCR REGISTRY NUMBER
(11) CAC/IC ABSTRACT NUMBER
(12) REACTION DESCRIPTOR
(13) ANALYTICAL DATA WHEEL
(14) JOURNAL TITLE. *ABSTRACTS ARE GROUPED BY JOURNAL*
(15) ISI ACCESSION NUMBER

Figure 2. Typical abstract from *Current Chemical Reactions*. Courtesy of Institute for Scientific Information.

There have been many studies using the *SCI* to identify key scientists, core journals, citation networks, and award winners. A sampling of references follows.

P. W. Brennen and W. P. Davey, "Citation Analysis in Literature of Tropical Medicine," *Bull. Med. Libr. Assoc.* **66**(1), 24–30 (1978).
A. E. Cawkell, "Evaluating Scientific Journals with Journal Citation Reports—Case Study in Acoustics," *J. Am. Soc. Inf. Sci.* **29**(1), 41–46 (1978).
R. A. V. Diener, "Transnational Information Flow as Assessed by Citation Analysis," *J. Am. Soc. Inf. Sci.* **30**(2), 114–115 (1979).
P. Ellis, G. Hepburn, and C. Oppenheim, "Studies on Patent Citation Networks," *J. Doc.* **34**(1), 12–20 (1978).
E. Garfield, "Citation Indexing for Studying Science," *Nature* **227**, 669–671 (1970).
E. Garfield and A. E. Cawkell, "Location of Milestone Papers Through Citation Networks," *J. Libr. Hist.* **5**(2), 184–188 (1970).
H. C. Liu, "Faculty Citation and Quality of Graduate Engineering Departments," *Eng. Ed.* **68**, 739–741 (1978).
N. Wade, "Citation Analysis—New Tool for Science Administrators," *Science* **188**, 429–432 (1975).

Patents. In general, patents are an underutilized source of technical information. The significance of this statement has been shown in a study that found that the information in 71% of U.S. patents is never disclosed in the nonpatent literature and that the information content of another 13% is only partially disclosed (8). The patent information that finally is published in journals generally issues several years after the patent (see also Patents, literature).

Patent literature affects research and development in three areas as: a fruitful source of information regarding the economic potential of existing or near future art; a teacher of the art; and a guide to application areas for marketing activities (9). An extension of this commentary is the ability to mine the richness of patent information data bases as intelligence tools. For instance, patent information can be used to predict trends or to analyze competitors' research and development in a given area (10–11).

The immense volume of patent literature may be intimidating. There are 100,000–150,000 basic (first disclosed) chemical patents issued annually throughout the world, and 150,000 equivalent chemical patents (12). But patents are technical publications: once the reader is comfortable with their format and jargon (13), patents have more experimental detail than most journal articles (14).

To use patent information effectively generally requires the assistance of experienced information professionals. Each service described below approaches a search from a different perspective. Searching caveats are based on the following: each service has different rules of coverage (subject and country); each uses a different depth of the U.S. or International Patent Classifications; each may index end-product chemicals only or include reactants and intermediates; and each may or may not enhance the titles. References 12 and 15–17 provide in-depth discussions of these differences.

A new journal, *World Patent Information,* began publication in 1979 (K. G. Saur Verlag). For those interested in patent policy, a collection of papers from a recent ACS symposium has been published (18), and the journal *Research Management* frequently discusses the issue.

National Bulletins. Individual national patent offices, and now the European Patent Office (EPO) as well, issue bulletins disclosing the title, assignee, and abstract of approved applications. In the United States, the *Official Gazette,* published weekly on Tuesdays, is organized by classification. About 30% of the patents reported are chemical (13).

Derwent. Derwent Publications, Ltd. (19) covers patents issued in 24 countries, plus EPO patents and international Patent Cooperation Treaty (PCT) patents. Some of these are from slow-issue countries, which examine applications, and some are from fast-publishing ones. Derwent's special value is the comprehensiveness and timeliness of its service. Consequently, many large industrial companies subscribe to its services. Derwent *World Patents Index* (*WPI*) contains information on all chemical areas since 1970. For pharmaceutical, agricultural, and polymer-related chemicals, the service extends back to different points in the 1960s.

For keeping abreast of new inventions, Derwent targets two bulletins for chemists. Twelve *Alerting Bulletins—Classified* provide very rapid notice of disclosures with brief abstracts of all basic patents and all examined equivalents. The *Profile Booklets* are a few weeks behind the *Alerting Bulletins* but include longer abstracts and cover more defined subject areas. Two other bulletin types used by patent professionals and chemists are the *Alerting Bulletins—Country Order* and the *Basic Abstract Journal.*

Derwent has other alerting services designed for patent attorneys and numerous searching products—manual, microform, and computerized. In addition to complex subject-coding techniques, access by title word, country, patentee, classification, and patent family is possible. Although bibliographic searching is excellent, subject searchability varies. There are now two sources for on-line access: SDC in the United States and INFOLINE, a UK host on the EURONET–DIANE System. For a fee, Derwent's Technical Services Division performs searches for subscribers.

Chemical Abstracts. *Chemical Abstracts* does not treat patents as legal documents, so their coverage is selective; to be covered a patent must disclose new chemical information. In some countries, *CA* covers all basics; in others, only patents issued to nationals. But CA's chemical indexing is not restricted to end products. It indexes all substances, uses, and methods of manufacture included in the claims and additional uses or substances if actually prepared (12,17,20–21).

IFI–Plenum. IFI–Plenum has several levels of patent-retrieval services. The most readily accessible is the CLAIMS family of data bases from the Lockheed-DIALOG on-line service. Together these cover an extended time period but the indexing is shallow.

CLAIMS–CHEM. 1950–1970 bibliographic information, including U.S. classification codes, for United States chemical patents. Revised annually to accommodate U.S. Patent and Trademark Office (PTO) reclassification.

CLAIMS–U.S. PATENTS. 1971–1977 bibliographic information as above but for all U.S. patents.

CLAIMS–U.S. PATENT ABSTRACTS. Claims from the *Official Gazette* are now included, with chemical structures in linear form and the presence of mechanical drawings noted; covers the time period 1978 to the present.

CLAIMS–U.S. PATENT ABSTRACTS WEEKLY. Unedited, rapid information, for use until the monthly edited file is added to CLAIMS–U.S. PATENT ABSTRACTS.

CLAIMS–CLASS. A keyword index to the U.S. PTO classification manual.

IFI–Plenum also has a search service for use of their COMPREHENSIVE DATA BASE, which adds deep uniterm indexing and chemical fragmentation coding (from DuPont) to the data base (starting 1950). In addition, they publish a *UNITERM Index,* now available on line, and the *IFI Assignee List.*

APIPAT. The American Petroleum Institute Patent data base (APIPAT), dating from 1964, is available on SDC (22). Although its scope is relatively narrow, the indexing is done from a controlled vocabulary. This vocabulary provides a simple fragment or chemical-aspects system that minimizes false retrieval by linkages of the aspects while allowing for some structure searching. The use of role indicators allows a chemical to be searched as a starting material or as an end product.

In conjunction with API's retrospective data base, a number of patent alerts are produced in cooperation with Derwent for use by the scientists in API's subscribing organizations.

INPADOC. INPADOC is a joint venture produced by the Austrian government and the World Intellectual Property Organization (WIPO). It obtains, standardizes, and compiles the bibliographic details of patents from 46 national patent offices and two international organizations. The data base contains 7.5 million patents dating from 1968. It can be used to locate individual inventors or corporate patentees. However, subject information can be obtained only through the broad and imprecise medium of international patent classes, with only the patent title as an indication of the patent's content. There are no abstracts. One of INPADOC's prime uses is for keeping track of equivalent members of a patent family in countries around the world. The INPADOC products are marketed in the United States by IFI–Plenum Data Corporation. In 1980, the most recent six-weeks of INPADOC became a searchable file on DIALOG.

Search Check. Search Check, an information service based in Arlington, Virginia, applies the concept of citation searching to patents. Every U.S. patent issued since 1947, and the patents cited therein, are in the Search Check data base. This service is very helpful for locating references that the patent examiners cite. At this time, Search Check runs the computer search for customers on a one-time or contract basis.

Bibliography

H. W. Grace, *A Handbook on Patents,* Charles Knight & Co., Ltd., London, Eng., 1971.

S. M. Kaback, "A User's Experience with the Derwent Patent Files," *J. Chem. Inf. Comput. Sci.* **17**(3), 143–148 (1977).

E. J. Saxl, "The Significance of a Patent," *Am. Lab.* **10**(9), 31–35 (1979).

H. Skolnick, "Historical Aspects of Patent Systems," *J. Chem. Inf. Comput. Sci.* **17**(3), 114–121 (1977).

J. T. Maynard, *Understanding Chemical Patents: A Guide for the Inventor,* American Chemical Society, Washington, D.C., 1978.

J. F. Mezieres, "The European Patent—Henceforth a Reality," *Chemtech* **8**(11), 658–661 (1978).

F. Newby, *How to Find Out About Patents,* Pergamon Press, New York, 1967.

C. Oppenheim, "Recent Changes in Patent Law and Their Implication for Information Services and Information Scientists," *J. Doc.* **34**(3), 217–219 (1978).

W. Pilch and W. Wratschko, "INPADOC: A Computerized Patent Documentation System," *J. Chem. Inf. Comput. Sci.* **18**(2), 69–75 (1978).

E. S. Turner, "Patent Literature," in A. Standen, ed., *Kirk-Othmer Encyclopedia of Chemical Technology,* 2nd ed., Vol. 14, Wiley-Interscience, New York, 1967, pp. 583–635 (see also Vol. 16, 3rd ed.).

The NIH-EPA Chemical Information System. The still developing Chemical Information System (CIS) of the National Institutes of Health and the Environmental Protection Agency (NIH–EPA) is an excellent example of computer utilization to help the chemist in the laboratory. Several United States agencies were motivated to use computers to assist both the chemist and the regulator in identifying substances, solving problems, relating chemical structure to biological or environmental behavior, and responding to emergencies. CIS is a key tool for the Interagency Regulatory Liaison Group (IRLG), consisting of the EPA, FDA, OSHA, and the Consumer Product Safety Commission. The system is fully interactive, has referral, identification, characterization, calculation, and graphics features, and contains chemical, biological, and bibliographic data bases.

The Structure and Nomenclature Search System (SANSS) is the heart of CIS (see Fig. 3) (23). SANSS works on a unified data base, built from many files that identify chemicals of commercial importance or known biological activity (ie, drugs, pesticides, commodity chemicals, food additives), and therefore relevant to these regulatory agencies (see Table 1). The CAS Registry Number is used as the unifier for the various files. Reported or regulated substances that are structurally undefined get pseudo-Registry Numbers. Ultimately, because of the overlap in files, and because of the limited number of chemicals actually used commercially, the file is expected to total 175,000–200,000 substances.

The search techniques of SANSS employ the following data elements to identify particular compounds: nomenclature, ring, fragment, molecular formula, substructure, and full structure.

The display options are by chemical structure: *CAS Collective Index Names;* synonyms, common or trade names; molecular formulas; and list of files containing the substance. One can also retrieve by CAS Registry Number. Once Registry Numbers are located, one can search the other CIS data bases. Conversely, the chemist can use the mass-spectral system to identify an unknown, and use the Registry Number in SANSS to retrieve the structure.

Among the chemical-data components currently available in CIS are the Mass Spectral Search System, X-Ray Crystallographic Search System, X-Ray Single Crystal Search System, and Carbon-13 Nuclear Magnetic Resonance Spectral Search System. Examples of the biologically oriented components are *Registry of Toxic Effects of Chemical Substances,* Ames Test, and AQUATOX. Files containing *Federal Register* notices and mass-spectrometry bulletins are bibliographic. One last component category is the Mathematical Modeling Laboratory, where the data from other components are used to test predictive theories of structure–activity relationships.

Bibliography

G. W. A. Milne and S. R. Heller, "The MSDC/EPA/NIH Mass Spectral Search System," *Am. Lab.* 8(9), 43–54 (1976).

S. R. Heller, G. W. A. Milne, and R. J. Feldman, "Quality Control of Chemical Data Bases," *J. Chem. Inf. Comput. Sci.* **16**(4), 232–233 (1976).

R. J. Feldmann, G. W. A. Milne, S. R. Heller, A. Fein, J. A. Miller, and B. Koch, "An Interactive Substructure Search System," *J. Chem. Inf. Comput. Sci.* **17**(3), 157–163 (1977).

G. W. A. Milne and S. R. Heller, "The NIH-EPA Chemical Information System," in D. H. Smith, ed., *Computer-Assisted Structure Elucidation, ACS Symposium Series 54,* American Chemical Society, Washington, D.C., 1977, pp. 26–45.

S. R. Heller and G. W. A. Milne, "The NIH–EPA Chemical Information System," in W. J. Howe, M. M. Milne, and A. F. Pennell, eds., *Retrieval of Medicinal Chemical Information, ACS Symposium Series 84,* American Chemical Society, Washington, D.C., 1978, pp. 144–167.

S. R. Heller, G. W. A. Milne, and R. J. Feldmann, "A Computer-Based Chemical Information System," *Science* **195,** 253–259 (1977).

G. W. A. Milne, S. R. Heller, A. E. Fein, E. F. Frees, R. G. Marquart, J. A. McGill, J. A. Miller, and D. S. Spiers, "The NIH–EPA Structure and Nomenclature Search System," *J. Chem. Inf. Comput. Sci.* **18**(4), 181–186 (1978).

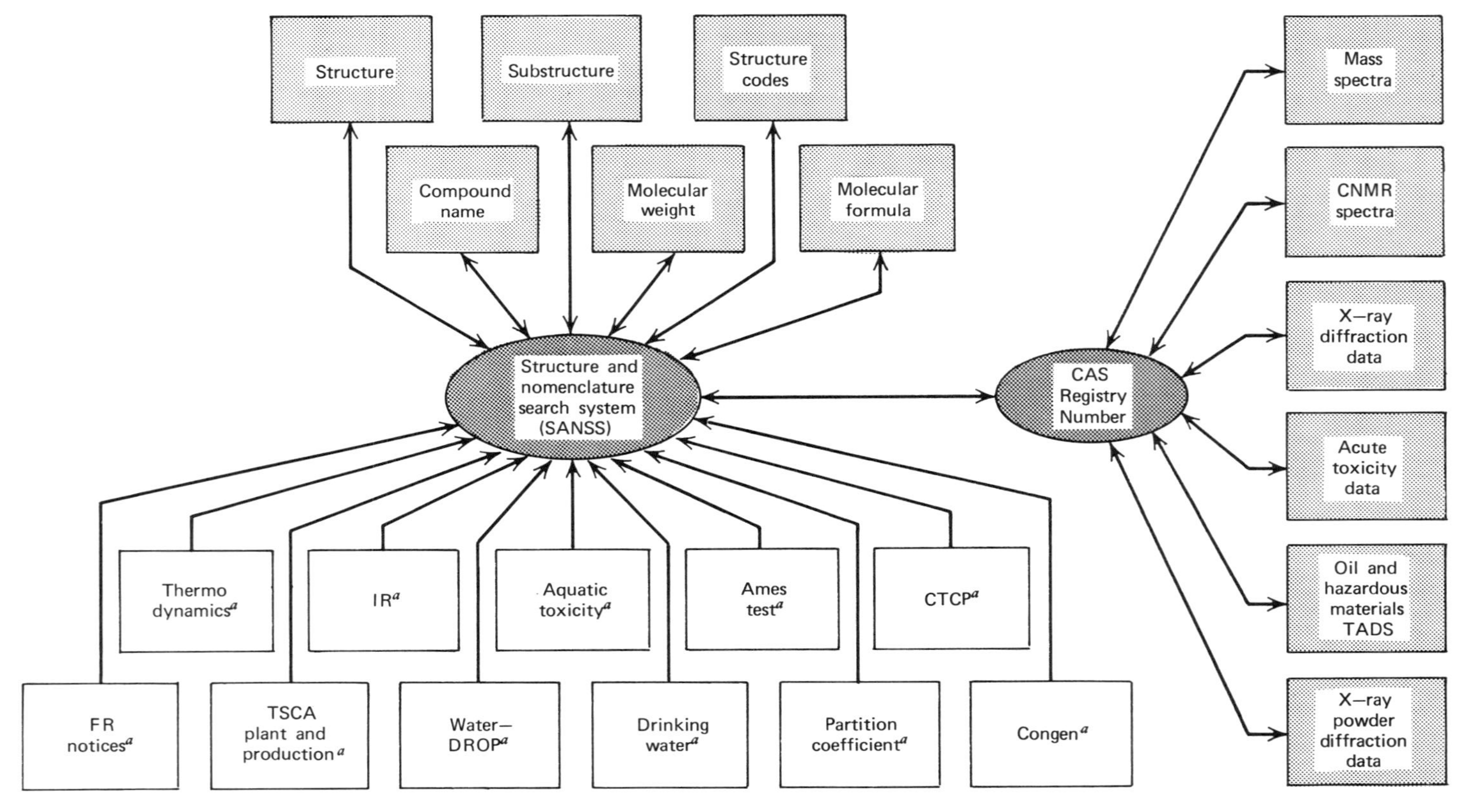

Figure 3. The NIH–EPA Chemical Information System (23). [a] Under development. Courtesy of the American Chemical Society.

Table 1. SANSS Files[a]

SANSS file	Substances	SANSS file	Substances
TSCA inventory	43,278	NBS x-ray crystal	18,338
CIS mass spectrometry	25,560	EPA effluent guidelines	125
CIS carbon-13 nmr spectrometry	3,805	EPA organic chemical producers	375
EPA pesticides active ingredients	1,453	IPC chemical product	104
EPA OHM/TADS	858	IPC chemical plant	103
Cambridge x-ray crystal	14,854	NSF chemicals list	225
Merck Index	8,959	EROICA thermodynamics	4,492
EPA pesticides analytical reference standards	473	PHS149 carcinogenic activity	4,447
EPA STORET	234	NIOSH RTECS	19,882
EPA chemical spills	577	NIOSH NOHS	4,560
EPA AEROS SOTDAT	572	ORNL EMIC	4,030
NIMH psychotropic drugs	2,039	ORNL ETIC	3,241
EPA AEROS SAROAD	65	EPA selected organic air pollutants	579
NBS proton affinities	440	Clean Air Act Section 112	4
CPCS CHEMRIC	890	EPA/NCTR study (1976)	91
EPA pesticides registered inert ingredients	735	EPA environmental carcinogen assessment program	21
NBS gaseous ions	3,167	EPA restricted use pesticides	23
NFPA hazardous chemicals	396	EPA compounds for mutagenicity evaluation	25
FDA/EPA pesticides reference standards	613	CIIT priority chemicals lists (toxicological)	27
U.S. International Trade Commission	9,193	NMFS survey of trace elements	15

[a] Ref. 23.

S. R. Heller and G. W. A. Milne, "The NIH–EPA Chemical Information System," *Environ. Sci. Technol.* **13**(7), 798–803 (1979).

S. R. Heller and G. W. A. Milne, "Chemcorner," *Database* **2**(3), 69–79 (1979).

H. J. Bernstein and L. C. Andrews, "The NIH–EPA Chemical Information System," *Database* **2**(1), 35–49 (1979).

Other Bibliographic Data Bases. There are many data bases of interest to chemists beyond those directed to chemistry. Data bases available in the United States are listed in ref. 24. The number of data bases listed is overwhelming, however, the quality and variety should encourage chemists to use these resources.

Careful selection of data bases, based on the needs of each particular query, can save time. For example, if a user is looking for an overview of an area or an issue or wants to sense the amount of activity in a field, searching the mission-oriented data bases, such as ENVIROLINE or ASFA (Aquatic Sciences and Fisheries Abstracts), may be more efficient than searching the core displinary data bases such as CA SEARCH or BIOSIS PREVIEWS (on-line versions of *Chemical Abstracts* and *Biological Abstracts,* respectively). On the other hand, to achieve any degree of comprehensiveness, multidata-base searching is necessary. Despite some overlap, the indexing perspective of each service varies and so retrieval varies. When many files are searched, overlap is annoying. Techniques to eliminate redundancy in the final result now are being developed.

Other systems to combat the confusion resulting from the richness of the available data bases are in use or are being evaluated. Selector systems allow identification of the data bases that contain terms or parameters relevant to the search. Vocabulary-switching systems translate the original search terms (search strategy) into the language of other data bases.

On-line Chemical Dictionary Files. On-line chemical dictionary data bases, companions to both bibliographic and numerical data bases, are not traditional dictionaries at all. They do not define chemicals but are designed to assist chemists in identifying CAS Registry Numbers and all the names and synonyms of chemicals. They are primarily identifier or locator tools with some substructure-searching capability. They thus enhance comprehensive searching in other data bases. Because the four dictionary files now available can be expected to change in scope, and because others surely will be developed soon, the following discussion of features is general. The file names, in the order they became available, are CHEMLINE of NLM, CHEMNAME of Lockheed, SANSS of NIH–EPA, and CHEMDEX of SDC. The Lockheed, NLM, and SDC chemical dictionaries are based on information from the *Chemical Abstracts Registry Nomenclature File,* which each search service has enhanced beyond the straightforward Registry Number, name (8th or 9th collective index), synonyms, trade names, molecular formulae, and cross-reference Registry Number. One can do a complex substructure search based on name fragments, such as dichloro or di or chloro, or by molecular formula and fragment, including element count or element presence. Searching is by position in the periodic table, ring counts, ring sizes, ring-element analysis, and even keyword or category (ie, polymer, alloy). CHEMLINE adds the Wiswesser Line Notation and pointers that specify the presence of the chemical in other NLM data bases.

Bibliography

R. E. Buntrock, "Chemcorner," *Database* **2**(1), 33–34 (1979).

S. R. Heller and G. W. A. Milne, "Chemcorner," *Database* **2**(3), 71 (1979).

J. Kasperko, "Online Chemical Dictionaries," *Database* **2**(3), 24–35 (1979).

P. E. Pothier, "Substructure Searching in CHEMLINE," *Online (Weston, Conn.)* **1**(2), 23–25 (1977).

R. J. Schultheisz, D. F. Walker, and K. L. Kannan, "Design and Implementation of an On-line Chemical Dictionary (CHEMLINE)," *J. Am. Soc. Inf. Sci.* **29**(4), 173–179 (1978).

Nonbibliographic Data Bases. The newest aspect of the information explosion is the increasing number of data bases that contain numerical information such as physical properties, product consumption, or social statistical data. These are not only fact-finding systems but also tools for manipulating facts for analyses or evaluation. Users can even add their own data and merge or compare it with that in the data base. Unfortunately, most nonbibliographic data bases are in different search-service systems from the bibliographic ones. Thus the software is different. But over the next few years there will be more numerical data bases, they will be larger, and more will contain evaluated data. Cuadra Associates, Inc., publishes a comprehensive and regularly updated list of numerical data bases (25).

These information resources are intended for use in everyday work. The opportunity to interact directly with nonbibliographic systems is available to the individual chemist. The chemical-information specialist will continue to use these systems in an intermediary role, and to act as informer, trainer, and coordinator for others' use of nonbibliographic services.

Thus, the tremendous wealth of scientific and technical information contained in the ever increasing number of information resources can be seen. This proliferation of literature increases the likehood of not finding the information needed. Don't be concerned. Consult a librarian or information specialist. These information professionals can help in a search or obtain pertinent information. Although information is becoming more obscure to end users it is, at the same time, becoming more focused or specialized. The evidence of this is the growth and development of business information sources useful to chemists and chemical technologists.

BUSINESS INFORMATION

Although the nature of technical information is fundamentally universal, economic information remains essentially national and is often private or sensitive as well. Indeed, each country conducts its own affairs according to its individual history, traditions, social organization, political environment, and particular attitudes. The result is that this information is confined and subject to local constraints and, therefore, when accessible, is available in a variety of forms.

National authorities, however, as well as regional or international organizations, have become increasingly aware of the importance of economic information. Although it is vital to private, regional, and national economic development, it remains scattered, sometimes insufficiently known, yet abundant. The many sources of economic information have been the object of various surveys but no satisfactory effort has been undertaken yet to organize it suitably.

To conduct and manage its business, the chemical industry requires vast amounts of primary information to be selected, analyzed, organized, and transformed for use by management. The principal areas of important nontechnical information are the political, social, legal, and economic environments; past, present, and future markets; competition; and finances, prices, taxes, and tariffs.

The Political, Social, Legal, and Economic Environments. Every country presents a unique environment for the conduct of business. Attitudes toward chemical business, whether national or foreign, differ among countries and change with time. Political and social aspects that can influence chemical operations include governmental stability, its current and anticipated attitudes toward both national and foreign business, the possibility that it will enter into a specific chemical business through part ownership or nationalization, and ways in which it might favor local manufacturing industries rather than foreign companies or enterprises. Duties, taxes, port charges, import and export restrictions, and price regulations are factors essential to any contemplated or operating business. Labor practices and laws, trade union attitudes, levels and trends of unemployment, as well as the outlook for inflation and foreseeable levels of wage increases require careful consideration. Recent, potential, and long-range growth of the economy in general and in specific businesses, corresponding consumption, balance of payments, and monetary policies are crucial.

Markets. The understanding of markets whether past, present, or future, requires a comprehensive body of information. This information includes the list of all significant products that are manufactured, imported, exported, sold, and potentially pertinent to the business. Product information includes forms, grades, types, and specifications as well as sales, prices, and regulations concerning storage, transportation, labeling, and environmental aspects. Materials that have been or could be replacements for products also need to be known. This in turn calls attention to factors that bear on the possible vulnerability of any one product to price, local restrictions or new regulations, public health, or public opinion.

An understanding of the market requires knowing the supply and demand for each individual product or group of products within each geographical area contemplated for manufacturing, importing, and trading. This information covers the current year and past periods of five to ten or more years depending on particular products and situations. It is expressed in terms of physical volume and prices. Yearly, quarterly, or monthly figures may be required depending on seasonal variations that may affect forecasts.

Likewise, the historical and forecasted demands for products by main users are important because they may reveal characteristics that indicate whether the demand is cyclic and price-sensitive. They may reveal that the end-use product is vulnerable to imports or other products manufactured at lower cost.

Forecasting demand requires various kinds of data, eg, product supply and demand, end-use breakdown, consuming industries, and overall economic growth. These data should go back as far as possible; it should, by a rule of thumb, cover a period at least twice as long as the period of the forecast.

Forecasting accounts for supplies from domestic sources, imports from foreign countries, and exports to other regions. This means that current and expected production capacities compared to imports in countries that provide export outlets for domestic manufacture must be thoroughly reviewed. A detailed demand-and-supply analysis in terms of major manufacturers, newcomers, importers, consuming industries, and customers requires further information, which is described below.

Companies, Manufacturers, and Customers. A company's success depends upon careful, objective analysis of the industry and end-use markets and trends, competition, significant manufacturing and technical trends, the projected availability of feedstocks, and energy requirements. Therefore, company information must include financial performance and manufacturing capabilities and capacities. In this respect, the technical-business analyst requires reliable data on current, historical, and projected production capacities by product. Data compilations regarding specific plants must include information such as owning companies, geographical location, licensed processes, engineering design, contractor who built the plant, start-up date, and dates of debottlenecking, expansions, and shutdowns. General industrial conditions that influence prices must be reviewed. These include new suppliers, technological changes, increases in average industrial-plant size, required capital expenditures for new capacities, and likely increases of capital with time. General industrial conditions are indicative not only of current and likely utilization rates but also of trends relating to manufacturers securing their own captive outlets or to customers providing their own source of supply. The position, strengths, and weaknesses of companies are measured in financial terms such as sales and profit trends, capital expenditure, cash flow, interest levels, return on employed capital, and value added per employee.

The Different Aspects of Information. The data requirements for technical-business information can be succinctly summarized as *macroeconomic:* world, region, country, and area—economic, monetary, tariffs, taxes, controls, commerce prices, cost, gross national product (GNP), gross domestic product (GDP), population, movement of capital, and employment; and *sectorial:* company, supplier, customer, country, government—production, industrial indexes, technology, markets, distribution, organization, and structure.

Sources

Political, Social, Legal, and Economic Environments. Because of national differences every country has a unique set of information sources on the economic, legal, social, and political environment. Access to these sources is easy in countries such as the United States or the countries of the European Economic Community; elsewhere it is much more difficult; in countries with centrally planned economies, such information is minimally accessible.

Therefore, a first resource is the international and regional organizations that have systematically collected and organized data and information over several years. Organizations like the United Nations (UN), the Organization for Economic Cooperation and Development (OECD), the European Economic Community (EEC), and the International Monetary Fund (IMF) issue valuable publications for the technical-business analyst. These publications include monographs, specialized periodicals, statistical yearbooks, as well as monthly and quarterly statistical bulletins and conference proceedings.

Information about international organizations appear in publications such as refs. 26–27. The latter provides not only detailed information on international organizations, institutions, or associations but also basic information on individual countries and purely national organizations whether governmental, political, industrial, judicial, or educational. The information for each country includes general and statistical surveys of its recent history together with an overview of the economy. Also included is a directory of the diplomatic corps, the press and publishers, finance, banks, trade unions, chambers of commerce, and universities.

All international organizations publish enormous amounts of material that may not be fully covered in any one bibliography. Although each organization issues various indexes and catalogues, an overview of all that is available in any specific field is difficult. Some useful references, source guides, and information-providing services are listed below.

United Nations. Addresses: United Nations Publications, Palais des Nations, 1211 Geneva 10, Switzerland; United Nations Publications, Sales Section, Room A-3315, New York, N.Y. 10017, U.S.

A Guide to the Use of United Nations Documents Including Reference to the Specialized Agencies and Special UN Bodies, B. Brimmer and co-workers, Oceana Publications, Dobbs Ferry, New York, 1962, is an excellent key to the literature published by the UN. Likewise, *UNDOC Current Index: United Nations Document Index,* Dag Hammerskjold Library, United Nations, New York, published bimonthly, the *United Nations Document Index (UNDI),* Cumulated Index to Vol. 1–13, KTO Press, Millwood, New York, 1974, and *United Nations Publications in Print 1979–1980,* United Nations Sales Section, New York, *The United Nations Document Index,* and the *Catalogue: United National Publications* are useful for locating and learning the contents of publications that can be purchased from the UN.

The publications of the United Nations Statistical Office pertinent to the chemical sector in particular are: *Demographic Yearbook; Labour Supply and Migration* (by regions); *World Trade Annual; Yearbook of National Accounts Statistics; Statistical Yearbook; Monthly Bulletin of Statistics;* and *The Growth of World Industry* (published every two to three years).

The Department of Economic and Social Affairs publishes *World Economic Survey* (annual).

Each regional economic commission issues it own publications. The Economic Commission for Europe publishes *Economic Survey of Europe* (annual); *Economic Bulletin of Europe* (annual); and *Structure and Change in European Industry, 1977.*

The Economic Commission for Asia and the Far East publishes the *Statistical Yearbook for Asia and the Far East.* The Economic Commission for Latin America publishes the *Analyses on Projections of Economic Development.* The Economic Commission for Africa publishes the *Foreign Trade Statistics of Africa* and the *African Economic Indicators.* The UN also publishes an annual series of *Studies on Selected Development Problems in Various Countries in the Middle East.*

The International Labour Organization (ILO). Address: 1211 Geneva 22, Switzerland. The International Labour Organization (ILO), a specialized United Nations agency, deals with the improvement of living and working conditions of the member countries. ILO issues its *Catalogue of ILO Publications* irregularly. Information on new publications can be found in the *International Labour Review.* The publications of ILO that are most pertinent for industry are: *Yearbook of Labour Statistics; Bulletin of Labour Statistics* (quarterly); and *Social and Labour Bulletin* (quarterly). *The Social and Labour Bulletin* provides information on such topics as social policy, labor legislation, labor relations, collective agreements, employment, working conditions, wages, social security, equal opportunity, multinational enterprises, and trade unions on a country-by-country basis. ILO also publishes a series of special reports on topics such as productivity and workers' involvement in management.

The Food and Agricultural Organization (FAO). Address: Viale della Tezme di Caracalla, 00100 Rome, Italy. This UN agency issues considerable amount of economic information on food, agriculture, fisheries, and forestry; this information is referenced in the *FAO Documentation: Current Index* and the *Catalogue of FAO Publications.* Important FAO statistical publications include the following: *FAO Trade Yearbook; FAO Production Yearbook;* and *Monthly Bulletin of Agricultural Economics and Statistics.* FAO also issues useful reviews and bulletins: *FAO Commodity Review and Outlook; Annual Fertilizer Review; Per Caput Fibre Consumption;* and *Commodity Bulletin Series: Impact of Synthetics on Jute and Allied Fibres.*

The International Monetary Fund (IMF). Address: 19th and H Streets, N.W., Washington, D.C. 20431, U.S. The IMF issues several publications for the corporate financial expert. Information on foreign-exchange-rate systems for individual countries appears in *Annual Report: Exchange Arrangements and Exchange Restrictions.* International statistics on all aspects of domestic and international finance appears in *International Financial Statistics Yearbook,* which provides annual data for the 1949–1978 period. IMF also publishes a *Balance of Payments Yearbook* and a quarterly entitled *Finance and Development.*

The General Agreement on Tariffs and Trade (GATT). Address: Villa le Bocage, Palais des Nations, 1211 Geneva 10, Switzerland. Trade information on Eastern European countries that are members of the COMECON (Council of Mutual Economic Assistance, also known as CMEA) is provided by GATT in the annual publication, *International Trade,* which generally covers the world.

The Organization for Economic Cooperation and Development (OECD). Address: 2, Rue André-Pascal, 75 775 Paris Cedex 16, France. Every two to three years, the Organization for Economic Cooperation and Development issues *Catalogue of Publications,* which is updated in the interim by supplements. Because the primary objective of the OECD is to encourage the trade, economic growth, and improvement of living standards of the member countries (ie, Western Europe, Yugoslavia, Canada, Japan, the United States, Australia, and New Zealand), this organization publishes several items that bear on economists' and business analysts' activities. OECD's Department of Economics and Statistics produces an impressive number of statistical tables and graphs each year. Its major publications are:

Industrial Production, Historical Statistics (1955–1971; 1960–1975).

Indicators of Industrial Activity (quarterly), which has been published since 1979 and replaces the former *Industrial Production* (quarterly supplement) and the *Short Term Economic Indicators for Manufacturing Industry,* is jointly issued by the Industry Division and the Economic Statistics and National Accounts Division and surveys the economic development of industries. It is divided into two sections: statistical data and qualitative business-trends. *Main Economic Indicators, Historical Statistics* (1955–1971; 1960–1975) provides monthly, quarterly, and annual data on gross national product, gross domestic product, industrial production, trade, unemployment, money supply, savings, consumer prices, consumers' expenditures, governmental expenditure, and the like.

Main Economic Indicators (monthly) contains data on recent economic developments of member countries, the long-term outlooks, and is divided into three sections: indicators by subject, indicators by country, and price indexes (data are presented both in graphical and tabular form).

OECD Economic Outlook (July, December) reviews and analyzes economic trends and short-term forecasts that bear on the world's economy (supplemented by occasional studies on topics such as capital movements, income distribution, international competition, and the like).

OECD Economic Surveys are annual economic surveys for each member country that analyze general trends in demand and activity, balance of payments, and economic policy in fiscal, monetary, employment, and industrial terms; the short-term outlook also provides international comparisons.

National Accounts of OECD Countries are published in two separate volumes: the first gives for each country the main aggregates in the form of graphs, the growth in real terms and implicit price deflators, the gross domestic product, and its main components from 1960 to 1975, the main aggregates in national currencies from 1950 to 1975 as well as main components of final expenditure, provides comparative tables of national accounts in U.S. dollars, volume and price indexes, population, and exchange rates; the second volume gives detailed statistics for the last twelve years on domestic product and consumption, gross domestic product, composition of gross capital formation, income, and outlay transactions of government and households, financing of gross capital formation, external transactions, and distribution of national disposable income.

OECD Financial Statistics (June, December) is a financial tool that assembles large amounts of data from individual national publications in different languages, provides information on capital operations, financial transactions, interest rates, markets for long-term securities, lending and borrowing operations of groups of financial institutions, governmental finances, transactions of each economy with the rest of the world, rates of the European currency market, international bond issues, and security issues.

The Chemical Industry (yearly) breaks down the chemical industry according to the Standard International Trade Classification (SITC) and provides production indexes and trends for the major branches of the chemical industry, turnover, value added and investment data, labor, price indexes, trade figures, and production and consumption data.

The OECD also issues, under the title *Energy Balances of OECD Countries* (1960–1974; 1974–1976), statistical data on energy that is conveniently expressed in common units allowing total-energy estimation, forecasting, and studies of conservation and substitution. The balances are calculated largely on the basis of data published in the following:

Statistics of Energy (yearly).

Oil Statistics (yearly), which provides detailed statistics on the supply and disposal of crude oil and natural gas, on feedstocks, and their major derivatives, on production and consumption by end-use sectors, on source of imports, and on destinations of exports.

Quarterly Oil Statistics, which provides timely and detailed statistics on oil supply and demand, ie, balances of production, trade, refinery throughput, final consumption, stock levels, and trade data.

Trade by Commodities—Market Summaries: Imports and *Trade by Commodities—Market Summaries: Exports,* which give foreign trade statistics covering all goods with indication of imports' origin and exports' destination. For any individual SITC code, values are in U.S. dollars, and quantities are in metric units.

The European Economic Community (EEC). Because the EEC has not selected a unique location for its headquarters (they operate from Brussels, Luxembourg, and Strasbourg), locating documents released by any one of the principal organs is difficult. The principal publishing groups are the Commission of the European Communities (CEC), the Council of Ministers, the European Parliament or Assembly, the Court of Justice, The European Coal and Steel Community (ECSC), The European Atomic Energy Community (Euratom), and the European Investment Bank. The addresses of EEC information offices in member countries and in some nonmember countries such as Japan and the United States, are listed in the *Europa Yearbook.* These information offices provide valuable assistance and are an excellent starting point.

EEC publications are issued in all of the official languages of the Community: Danish, Dutch, English, French, German, and Italian. The list of these publications appears in the *Publications of the European Communities: Catalogue* and in *Official Publications,* issued by the information offices in the member countries.

The main periodicals published by the Commission of the European Communities are the following: *Official Journal of the European Communities: Series L* (Legislation), which gives community secondary legislation coming into force (at least daily); *Series C* (Communications), which gives details on draft secondary legislation, information of general interest, summaries of the proceedings of the European Assembly and the Court of Justice (almost daily); and *Bulletin of the European Communities* (monthly), issued jointly by the ECSC, EEC, and Euratom. This bulletin contains three parts: (*1*) special features; (*2*) current activities such as economic and monetary policies, internal market and industrial affairs, competition, finance and taxation, employment, customs, environment, agriculture, energy, transport, research, and development management information; (*3*) documentation on European units of account, additional references in the *Official Journal,* infringement procedures, public opinion in the Community, publications of the European Communities. *European*

Economy (March, July, November) gives short-term economic trends and prospects with reports and studies on problems of current interest, reviews on the economies of each member state, and a statistical index containing the main economic indicators. *European Economy—Supplements* (monthly): *Series A* gives recent economic trends with tables and graphs; *Series B* gives business-survey results; *Series C* gives consumer-survey results. *Graphs and Notes of the Economic Situation in the Community* (monthly) gives data and graphs on industrial production, trade, employment, consumer and wholesale prices, governmental budgets, long-term interest rates, and the like. *Report of the Results of the Business Surveys* details results of surveys carried out among heads of enterprises in the community (three per year).

The Statistical Office of the European Communities publishes many series on behalf of the different organs of the Community. *Eurostatistics* (monthly) provides data for short-term economic analyses on national accounts, population, employment, output of products, trade, prices, balance of payments, industrial production, opinions in industry, and financial statistics. *Eurostat Industrial Short-Term Trends* (monthly) provides graphs and data on production, turnover, new orders, number of employees, wages and salaries, by country for key sectors and products of the economy. *Eurostat National Accounts ESA, Aggregates* (yearly) provides historical data covering fifteen or more years, in the form of comparative tables and country tables, and in terms of gross domestic product and market prices, compensation of employees, consumption of fixed capital, net operating surplus of the economy, net national disposable income, net national savings, lending and borrowing, final domestic uses, consumption, capital formation, trade, price indexes, population, and employment. *Eurostat Energy Statistics Yearbook* gives a detailed survey of the energy economics for the nine-country Community, the six-country Community, and for each member country over a period of ten to sixteen years. Its quarterly bulletin provides data on the overall energy balance-sheet for main items of energy supply for the whole community and for each member country; it provides monthly data for each energy source. *Iron and Steel Yearbook and Iron and Steel Bulletin* (bimonthly) provide data on production, employment and wages, orders and deliveries, and trade. *Eurostat Industrial Statistics Yearbook* provides data on production, employment and wages, orders and deliveries, and trade and provides historical series of industrial-production indexes back to 1968 and data on the production of basic materials and manufactured goods covering approximately 500 items back to 1958. *Eurostat Analytical Tables of Foreign Trade NIMEX* (yearly) represents a statistical breakdown of the Community's foreign trade based on the EEC's Common Customs Tariff (CCT), which was derived from the 1955 Brussels Tariff Nomenclature. NIMEX (Nomenclature of Imports and Exports) has about 6500 headings; a CCT-NIMEX correlation is provided. *Eurostat Monthly External Trade Bulletin* provides data for broader categories of products with an indication of trends in trade of the EEC countries. *ACP: Statistical Yearbook* is a selection of the main demographic, economic, and social indicators relating to the ACP countries, ie, those that have signed the Lome Convention. The data, which extend back five to ten years, cover population, national accounts, transport, trade, external aid, food supply, finance, balance of payments, and agricultural and industrial production.

Industrial Associations and Confederations. In a number of countries, companies in the same industrial sector are often members of trade associations or employers' organizations. These associations publish much valuable information in the form of periodical bulletins, occasional publications, reports, yearbooks, directories, and statistical data, all available to member associations and companies. These organizations and the information they release can provide good insight into the individual countries' developments, trends, and industrial sectors.

Most associations have their own information and documentation services that assist outside users as well as member companies. Full names, addresses, and useful details on industrial associations can be found in the *Europa Yearbook* and the *Directory of European Associations,* 2nd ed., Part 1, National Industrial, Trade and Professional Associations, 1976, Part 2, National Learned Scientific and Technical Societies, 1979, I. G. Anderson and G. P. Henderson, eds., CBD Research Ltd., Beckenham, Eng. Other useful sources are the *Encyclopedia of Associations,* 14th ed., N. Yakes and D. Akey, eds., 3 vols., Gale Research Co., Detroit, Mich., 1980; the *National Trade and Professional Associations of the United States and Canada, and Labor Unions,* 15th ed., C. Colgate, Jr. and P. Broida, eds., Columbia Books, Inc., Washington, D.C., 1980; the *World Guide to Trade Associations,* M. Zils, ed., K. G. Saur Publishing Co., New York, 1980.

National Sources. There is an impressive wealth of economic, statistical, and historical information in the governmental or official publications of each country. Unfortunately for the potential business user, the sheer volume of data makes it difficult to use. Presentation in several languages, in numerous reporting systems, nomenclatures, classifications and codes do not easily lend themselves to useful comparisons. Although international organizations undertake the formidable task of collecting and bringing data together, its usability is still a problem.

In the United States, as in other countries, finding and obtaining governmental publications require considerable expertise. A key to this vast body of information is the *Monthly Catalog of U.S. Government Publications* and the yearly *Directory of U.S. Government Periodicals and Subscriptions Publications.*

The various organs and committees of the United States Congress are sources of fundamental information on the nation and its rules, practices, laws, and policies. The Joint Economic Committee, together with its subcommittees, is responsible for reports and studies on price levels, economic growth, employment, international exchange, foreign economic and trade policies, federal regulations, and economic statistics. Likewise, the various standing committees of the Senate and the House of Representatives deal with all aspects of the nation's economic activity from a legal viewpoint. Select and investigating committees have been responsible for publishing some of the most authoritative documents on the economy.

In Europe, the picture is more complicated because information is published in the national languages and according to the particular rules and practices of the individual countries. Foreign users should seek advice and assistance from national trade associations, chambers of commerce, embassy services, national information centers, and professional information services. Examples of these associations and information centers are the following: ASLIB (Association of Special Libraries and Information Bureaux) in the United Kingdom; the Association Belge de Documentation in Belgium; the French Association des Documentalistes et Bibliothécaires Specialisés; the Deutsche Gesellschaft für Dokumentation in the FRG; the Dansk Teknisk Oplysningstjeneste, which provides assistance to industry in Denmark; the NOBIN (Stichting Nederlands Orgaan, voor de Bevordering vande Informatieverzorging) in the Netherlands. Full names and addresses can be obtained from directories, from the secretariat of the International Federation for Documentation in The Hague, or from the UNESCO Division of General Information Programme in Paris.

The United Kingdom has a long-standing tradition of record keeping. Accordingly, the abundance of source material in the UK is not surprising. British governmental publications are correspondingly numerous, and a useful starting point in locating them is *An Introduction to British Government Publications* by J. G. Olle. Information on official publications concerning the economy appears in the *Brief Guide to Official Publications* published by Her Majesty's Stationery Office (HMSO). These bibliographic tools allow access not only to official publications of direct interest to industry but also to all documents issued by Parliament since 1801.

Until ten years ago the UK Board of Trade issued census publications covering five-year periods. It has now established the Business Statistics Office (BSO) which publishes more current data in its *Business Monitors*. These publications are series of monthly, quarterly, and yearly issues based on information collected by the BSO from industrial firms. These series are described in detail in the *Guide to Short-Term Statistics of Manufacturers' Sales* (Business Monitor PQ 1001) and in the *CSO Guide to Official Statistics* (CSO = Central Statistics Office) published by HMSO. The quarterly publication provides comprehensive and up-to-date sales figures for over 4000 products. In particular, the *Business Monitor Manufacturers' Sales* provides detailed figures on the sales and production of UK establishments that employ 25 or more persons and whose main activity is manufacturing chemicals. It also provides data on exports, imports, employment, production indexes, and wholesale prices. The *Annual Census of Production Monitors* covers practically every sector of the industry and includes data such as total purchases and sales, stocks, work in progress, capital expenditure, employment and wages. The activities of BSO are described in detail in ref. 28. In all these publications, items are classified under the SITC classification system. *The Guide to the Classification for Overseas Trade Statistics,* also published by HMSO, will greatly assist the user. Additional statistical publications of the UK and their issuing agencies appear in Table 2.

In France, as in the United Kingdom, sources of business information are numerous. The Institut National de Statistique et d'Etudes Economiques or INSEE is the organization responsible for collecting, organizing, and publishing economic information. It issues statistical series and short-, middle-, and long-term forecasts. A main publication is the *INSEE Annuaire Statistique de la France,* a statistics yearbook containing tabular and graphical data for the last ten years and covering population, use of natural resources, production of goods including chemicals and allied products, services, prices and wages, and foreign trade and consumption. An outline of the scope and use of the various INSEE publications is given in ref. 29. The INSEE monthlies and other useful statistical publications of France appear in Table 2.

The Benelux countries (Belgium, the Netherlands, and Luxembourg) are unique in the concentration of their main domestic and foreign companies in a small geographical area. These companies, having built large-scale plants for the production of exports, need socio-economic information relating to the countries in which the industries operate and to the importing countries. In this respect, trade statistics are a key instrument. An exhaustive description of economic information in Belgium is given in ref. 30.

The Dutch Ministry of Economic Affairs maintains the Economic Information Service backed by a library of over 80,000 books, many directories and manuals, and

Table 2. National Statistical Series

Country	Issuing authority	Publication or report	Frequency
United States	Bureau of Census	*Statistical Abstract of the United States*	yearly
		Census of Business	every 5 yr
		Census of Manufacturers	every 5 yr
		Census of Production	every 5 yr
		Census of Transportation	every 5 yr
		Census of Government	every 5 yr
	Office of Business Economics	*Survey of Current Business*	monthly
	U.S. Department of Commerce, Industry and Trade Administration	*International Economic Indicators*	quarterly
	Bureau of International Commerce	*Overseas Business Reports*	irregular
		Market for U.S. Products	irregular
		International Commerce	irregular
	Department of Commerce	*Business Service Checklist*	weekly
		Monthly Catalog of U.S. Government Publications	monthly
	Department of Labor	*Monthly Labor Review*	monthly
	Bureau of Labor Statistics	*BLS Reports*	irregular
		BLS Bulletins	irregular
		Manpower Report	yearly
	Congressional Information Service	*American Statistics Index: A Comprehensive Guide and Index to the Statistical Publications of the U.S. Government*	yearly
	U.S. Government Publications Office	*Business Conditions Digest*	monthly
	U.S. Council of Economic Advisors	*Economic Indicators*	quarterly
	Board of Governors Federal Reserve System	*Industrial Production*	monthly
		Federal Reserve Bulletin	monthly
United Kingdom	Business Statistics Office (BSO)	*Business Monitors*	monthly, quarterly, yearly
	Central Statistical Office (CSO)	*Monthly Digest of Statistics*	monthly
		Annual Abstracts of Statistics	yearly
	Department of Trade and Industry	*Overseas Trade Statistics of the United Kingdom*	monthly, yearly, every 2 yr
		Trade and Industry	weekly
		British Business	weekly
		Statistics of Trade through United Kingdom Parts	quarterly
France	Institut National de Statistique et d'Etudes Economiques (INSEE)	*Economie & Statistiques*	monthly
		Tendances de la Conjoncture-Graphiques Mensuels	monthly
	Ministère de l'Industrie (Ministry of Industry)	*Annuaire de Statistiques Industrielles*	yearly
		Bulletin Mensuel de Statistiques Industrielles	monthly

Table 2 (*continued*)

Country	Issuing authority	Publication or report	Frequency
	Direction Générale des Douanes et Droits Indirects	*Statistiques du Commerce Extérieur*	yearly
Belgium	Institut National de Statistiques (INS)	*Annuaire Statistique de la Belgique*	yearly
		Statistiques Industrielles	monthly
		Communiqué Hebdomadaire–Weekbericht	weekly
		Statistique Fiscale des Revenus	yearly
	Banque Nationale de Belgique (National Bank of Belgium)	*Statistiques Economiques de Belgique*	monthly
	Ministère des Affaires Economiques (Ministry of Economy)	*Bulletin Mensuel du Commerce Extérieur*	monthly
The Netherlands	Central Bureau of Statistics (CBS)	*Statistical Yearbook (Jaarcijfers)*	yearly
		Statistisch Zakboek	yearly
		Maandstatistiek van de Industrie	monthly
		Maandschrift	monthly
		Statistisch Bulletin	irregular
		Chemische Industrie-Produktiestatistieken	yearly
		Rubberverwerkende Industrie-Produktiestatistieken	irregular
		Maandstatistiek van Buitenlandse Handel	monthly
Federal Republic of Germany	Statistisches Bundesamt	*Statistisches Jahrbuch für die Bundesrepublik Deutschland*	yearly
		Produzierendes Gewerbe	quarterly
		Aussenhandel nach Waren und Ländern	monthly
	IFO-Institut für Wirtschaftsforschung	*Wirtschafts-Konjunktur*	monthly
		IFO Schnelldienst	weekly
	Federal Minister of Economics	*The Economic Situation in the Federal Republic of Germany*	monthly
	Verband der Chemischen Industrie e.V.	*Jahresbericht*	yearly
		Chemiewirtschaft in Zahlen	yearly
Italy	Instituto Centrale di Statistica (ISTAT)	*Annuario Statistico Italiano*	yearly
		Bollettino Mensile di Statistica	monthly
		Annuario di Statistiche Industriali	yearly
		Indicatori Mensili	monthly
		Notiziario ISTAT	monthly
		Statistica Mensile del Commercio con l'Estero	monthly
	Instituto Nazionale per lo Studio della Congiuntura (National Institute for Economic Research)	*Congiuntura Italiana*	monthly
		Quaderni Analitici	biweekly
Norway	Nordisk Råd (Nordic Council)	*Yearbook of Nordic Statistics*	yearly

Table 2 (*continued*)

Country	Issuing authority	Publication or report	Frequency
		Statistik Årbok	yearly
		Utenrikshandel	yearly
Denmark		*Statistik Årbog*	yearly
		Varestatistik	yearly
		Danmarks Vareindførsel og-udførsel	yearly
Finland		*Suomen Tilastollinen Vuosikirja*	yearly
		Ulkomaankauppa	yearly
Sweden		*Statistik Årsbok*	yearly
		Utrikeshandel	yearly
Switzerland	Eidgenössisches Statistisches Amt (Federal Bureau of Statistics)	*Statistisches Jahrbuch der Schweiz-Annuaire Statistique de la Suisse*	yearly
	Eidgenössische Oberzoll-direktion (Customs)	*Jahresstatistik des Aussenhandels der Schweiz*	yearly
Austria	Österreichisches Statistisches Zentralamt	*Statistisches Handbuch*	yearly
		Industriestatistik	yearly
		Der Aussenhandel Österreichs	yearly
Spain	Instituto Nacional de Estadística	*Anuario Estadístico de España*	yearly
		Estadísticas de Producción Industrial	monthly, yearly
	Comisión Asesora y de Estudios Técnicos de la Industria Química Española	*La Industria Química en España*	yearly
		Situación y Perspectivas de la Industria Química Española	yearly
		Estadística del Comercio Exterior de España	yearly
Portugal	Instituto Nacional de Estatística	*Anuário Estatístico*	yearly
		Estatísticas Industriais	yearly
		Estatísticas do Comércio Externo	yearly
Greece	National Statistical Service of Greece	*Statistical Yearbook of Greece*	yearly
USSR	Council for Mutual Economic Assistance	*Statistical Yearbook*	yearly

over 3000 periodicals. The Service publishes *Economic Titles,* a semimonthly bulletin on the economy of the Netherlands and of other countries and regions. The Service carries out literature searches and provides rapid reference service on request. This main source of economic information in the Netherlands and also those in Belgium are described in ref. 31.

In the Federal Republic of Germany, the Statistisches Bundesamt in Wiesbaden is the central organization that by law issues data on a national level. Different ministries, in their regional and local offices, also collect and publish statistics. As a result there are hundreds of German-language publications available to the public. The main source publications are named in Table 2.

Foreign trade statistics for imports and exports are more readily obtainable from authorities that have computerized their operations. The user may either subscribe to the monthly update tapes and process these in-house or obtain monthly computer printouts that provide data on the commodities of interest. The names and addresses of some of the major foreign trade-data sources are the following: Institut National de Statistiques, Rue de Louvain 44, 1000-Bruxelles, Belgium; Direction Générale des Douanes, Centre de Renseignements Statistiques, 192 Rue St. Honoré, 75056 Paris, France; Statistisches Bundesamt, Gustav Stresemann—Ring, 11, Postfach 5528, 6200 Wiesbaden, FRG; Centraal Bureau voor de Statistiek, Statistieken van de Buitenlandse Handel, Prinses Beatrix laan 428, Voorburg, The Netherlands; Instituto Centrale di Statistica, Direzione Generale dei Servici Tecnici, Via Cesare Balbo 16, 00 100 Rome, Italy; and H. M. Customs and Excise Statistical Office, Portcullis House, 27 Victoria Avenue, Southand-on-Sea SS2 6AL, UK.

Information on finance, taxes, and monetary economics may cover broad as well as specific subject matter; public finance encompasses governmental and industrial activities, national issues such as nationalization or the financing of local projects, budgeting, fund raising, taxation of corporate or private income, the restriction on movement of funds abroad, currency stability and rates of exchange, reevaluation and devaluation, and inflation. Numerous specialized publications provide rapid and reliable information on these matters.

The main banks in many countries are known for their publications and some for their outstanding information services. They publish bulletins on the financial and economic situation, domestic and foreign, and reviews on the outlook and trends of various industrial sectors. The *Bankers' Almanac and Yearbook,* Thomas Skinner and Co., Ltd., London, covers banks on an international scale. Some noteworthy bank publications are:

Belgium: Banque Nationale—*Bulletin;* Banque Bruxelles-Lambert—*Bulletin Commercial, Bulletin Financiere, Reports from Brussels;* Societe Générale de Banque—*Bulletin de la Société Générale de Banque, Conjoncture Boursière;* Kredietbank—*Weekly Bulletin;* Banque de Paris et des Pays-Bas—*Parisbas Belgique Présente.*

France: Banque de France—*Enquête Mensuelle de la Conjoncture, Bulletin Trimestriel de la Banque de France, Cahiers Economiques et Monetaires.*

Germany: Deutsches Bundesbank—*Monthly Report of the Deutsches Bundesbank, Statistische Beihefte zu den Monatsberichten der Deutschen Bundesbank.*

Italy: Banca d'Italia—*Bolletino Statistico Trimestrale, Contributi alla Ricerca Economica;* Banca Nazionale del Lavoro—*Italian Trends;* Banco di Roma—*Review of the Economic Conditions in Italy.*

Spain: Banco de España—*Boletin Estadístico, Informe Anual;* Banco de Bilbao—*Servex.*

The Netherlands: Amsterdam-Rotterdam Bank—*Amro Economisch Bulletin, Dutch Economy in Figures, Netherlands Economic Report.*

United Kingdom: Bank of England—*Quarterly Review, Bank of England Quarterly Bulletin.*

United States: Citibank—*Monthly Economic Letter;* Morgan Guaranty Trust Co.—*World Financial Markets.*

Denmark: Danmarks Nationalbank—*Danmarks Nationalbank Report and Accounts.*

In addition to the publications from official bodies and specialized organizations, newspapers and periodicals are primary sources of current information. Press agencies and national information offices of each country may provide guidance in the selection of titles most appropriate to the user's requirements. Publications pertinent to chemical industries are listed in Table 3. A minimal reading list is: *The Financial Times; The Journal of Commerce* (U.S. Edition); *The Journal of Commerce* (European edition); *The Wall Street Journal; The Frankfurter Allgemeine Zeitung; Il Corriere della Serra; Les Echos; Le Monde; Nachrichten für Aussenhandel; L'Echo de la Bourse-AGEFI;* and *Neue Zürcher Zeitung.* The following references are most useful for further details and source material.

DAFSA—Centre Français du Commerce Extérieur, *L'Information Economique et Financière en Europe—Répertoire des Sources,* DAFSA (S.A. de Documentation et d'Analyses Financieres), Paris, Fr.

Eleonore de Dampierre et François Charpentier, *Les Sources d'Information Economique et Financière: Etats-Unis, Royaume-Uni, France,* Center for Business Information, Paris, Fr.

J. Fletcher, ed., *The Use of Economics Literature,* Butterworths Pub., Inc., Woburn, Mass., 1971; a most illuminating and instructive book.

Published Data on European Industrial Markets, Industrial Aids, Ltd., London, Eng. 1975; an outstanding directory of source material and its contents.

L. Daniells, *Business Information Sources,* Center for Business Information, University of California Press, Berkeley, Calif., 1976.

J. M. Harvey, *Statistics—Europe: Sources for Social, Economic and Market Research,* 3rd ed., CBD Research Ltd., Beckenham, Eng., 1976, Gale Research Company, Detroit, Mich.

J. M. Harvey, *Statistics—Europe: Guide for the Market Research to 34 Countries of Europe,* 3rd ed., International Publications Service, New York, 1976.

B. Lawrence, "Preliminary Project Evaluation—Any Technologist Can Do It. Guide to Chemical Business Information Sources," *Chemtech* **5,** 678–681 (1975).

K. D. Vernon, ed., *Use of Management and Business Literature,* Butterworths Pub., Inc., Woburn, Mass., 1976.

N. Roberts, *Use of Social Sciences Literature, (Information Sources in Sciences and Technology Series),* Butterworths Pub., Inc., Woburn, Mass., 1976.

F. C. Pieper, *SISCIS, Subject Index to Sources of Comparative International Statistics,* CBD Research, Ltd., Beckenham, Eng., 1978; very useful.

The British Overseas Trade Board publishes a series of pamphlets entitled *Hints to Businessmen (Country Name)* that contains general information on the country that business visitors may need to know, economic factors, import and exchange control regulations, methods of doing business, and a list of governmental and commercial organizations.

Sources of European Economic Information, 2nd ed., compiled by Cambridge Information and Research Services, Ltd., Gower Press, distributed by Unipub, New York, 1978, is an easy-to-use reference to source publications.

Market and Product Information. Many resources are available for obtaining information on markets and products. In-house market researchers should seek the assistance of professional marketing consultants when the information required cannot be readily prepared in-house or when these consultants have the needed expertise in fields outside the concerns of the user. Much information is also obtainable from official agencies described above, particularly industrial associations, chambers of commerce, and from periodicals, trade magazines, and the press. Specialized and specific trade associations are important sources of information. Among the most important of these are the Association of British Pharmaceutical Industry, the Paint Maker Association of Great Britain, the British Agrochemicals Association, the Fertilizer Manufacturers' Association, the Hydrocarbon Solvents Association, the British Compressed Gases Association, the British Colour Makers Association, and the National Sulfuric Acid Manufacturers' Association (see Market and marketing research).

The individual countries' foreign-trade offices are a source of market information abroad. Main offices include Office Belge du Commerce Extérieur (Belgium), Centre Français du Commerce Extérieur (France), Economische Voorlichtingsdienst (the Netherlands), Bundesstelle für Aussenhandels Information (FRG), and Office Suisse d'Expansion Commerciale (Switzerland). These offices collect international information on economies, markets, and products from periodicals, dailies, and original documents. This information is then indexed, stored, and disseminated as printed publications or microfiche. Their libraries store large collections of foreign trade statistics, catalogs, directories, and reference material and well organized market files, business opportunities files, and foreign importers files. These offices are also a source of business and marketing information on countries in the Middle East, Africa, and particularly countries with centrally planned economies. In this respect, the daily, *Nachrichten für Aussenhandel,* published by the German Bundesstelle für Aussenhandels Information, which provides statistical and national supplements, is outstanding. The foreign-trade offices also have documentation and information centers from which assistance may be obtained, the addresses and scope of services of which are provided in the DAFSA *L'Information Economique et Financière en Europe—Répertoire des Sources.*

Chambers of commerce, special agencies, and offices established in most countries are another source of market information. Names and activities of these organizations appear in publications such as the *Encyclopedia of Business Information Sources,* 3rd ed., P. Wasserman and co-eds., Gale Research Company, Detroit, Mich., 1977; *Marketing and Management: A World Register of Organizations,* I. G. Anderson, CBD Research, Ltd., London, Eng., 1969; International Publications Service, New York, 1969; *Published Data on European Industrial Markets,* 3rd ed., Industrial Aids, Ltd., London, Eng., 1974.

Marketing research firms and consultants are unique sources of market information. The *Directory of U.S. and Canadian Marketing Surveys and Services,* K. Di Cioccio, and V. Kollonitsch, eds., Charles H. Kline & Co., Inc., Fairfield, N.J., 1976, lists about 1500 multiclient marketing reports and describes the scope of services of 125 consulting firms. This publication is supplemented by the *International Directory of Published Market Research* compiled by the British Overseas Trade Board, Ltd., London, Eng., 1976. It has a detailed alphabetical product index; studies and publications are entered under the British Standard Industrial Classification scheme so that they are grouped by subject. Products and services covered by these studies are briefly described and include geographical coverage, date, source, and price.

The following directories, guides, handbooks, business research reports, product studies, and reports series are useful for markets and products information.

D. Degen and T. E. Miller, eds., *Findex: the Directory of Market Research Reports, Studies and Surveys, 1979,* Information Clearing House, Inc., New York; reference guide to more than 2500 commercially available market and business research reports produced internationally.

Management Information Guide Series, Gale Research Co., Detroit, Mich.; deals with specialized areas of business and industry.

L. J. Wheeler, ed., *International Business and Foreign Trade Information Sources,* Gale Research Co., Detroit, Mich., 1968.

J. V. Kopycinski, *Textile Industry Information Sources,* Gale Research Co., Detroit, Mich., 1964.

Food and Beverage Industries: A Bibliography and Guidebook, A. C. Vara, Gale Research, Detroit, Mich., 1970.

T. Landau, ed., *European Directory of Market Research Surveys,* Gower Press, distributed by Teakfield, Ltd., Hampshire, Eng.; Unipub, New York, 1976; references about 1500 market research reports published since 1972.

Table 3. Publications Pertinent to Chemical Industries

Country	Frequency	Publication
Belgium	dailies	*L'Echo de la Bourse-AGEFI*
		Europe
		Financieel Economische de Tijd
		Le Lloyd Anversois
		Official Journal of the EEC
	weeklies	*Banque Bruxelles Lambert*
		Bulletin Commercial
		Bulletin Financiere
		European Report
		Kredietbank Bulletin
		Le Marché
	fortnightlies	*Fédération des Entreprises de Belgique—Bulletin*
	monthlies	*Aperçu de l'Evolution Economique*
		Banque Bruxelles Lambert—Bulletin de Conjoncture
		Banque Nationale de Belgique—Bulletin
		Belgian Business
		Belgian Trade Review
		Commerce in Belgium
		Distribution d'Aujourd'hui
	quarterlies	*Annales des Sciences Economiques Appliquées*
		Cahiers Economiques de Bruxelles
		Recherches Economiques de Louvain
	annual	*Banque Bruxelles Lambert—Rapport Annuel*
	irregular	*Economic Situation in the Community*
France	dailies	*Les Echos*
		Le Monde
	weeklies	*M.O.C.I. Moniteur du Commerce International*
		L'Usine Nouvelles
	fortnightlies	*Arab Oil and Gas*
		Chimie Actualités
		L'Observateur de l'OECD
		OECD Observer
		Parfums, Cosmétiques et Arômes
		Revue Générale Africaine
	monthlies	*La Corderie Française*
		Emballages
		L'Expansion
		Industrie des Plastiques et Elastoméres
		L'Industrie Textile
		Information Chimie
		Information d'Outre-Mer
		Main Economic Indicators
		Le Nouvel Economiste
		L'Officiel des Plastiques et du Caoutchouc
		Plastiques Modernes et Elastoméres
	quarterlies	*European Business*
	irregular	*OECD Economic Outlook*
		Revue du Marché Commun
FRG	dailies	*Frankfurter Allgemeine Zeitung*
		Handelsblatt
		Nachrichten für Aussenhandel
	fortnightlies	*Bundesgesundheitsblatt*
		Europa Chemie
	monthlies	*Chemie Ingenieur Technik*

Table 3 (*continued*)

Country	Frequency	Publication
		Chemische Industrie
		Chemie Fasern—Textile Industrie
		Erdöl und Kohle
		Farbe und Lack
		German International
		Die Gummibereifung
		Kautschuk und Gummi-Kunstsoffe
		Kunststoffberater-Rundschau
		Kunststoffe—German Plastics
		Melliand Textilbericht
		O E L
		Plastverarbeiter
		Wirtschaft und Statistik
Israel	quarterlies	*Reviews on Environmental Health*
Italy	dailies	*Corriere della Sera*
		24 Ore
	fortnightlies	*Industria Chimica*
		L'Industria Italiana del Plastici
	monthlies	*Chimica e l'Industria*
		Espansione
		L'Industria della Gomma
		L'Industria della Vernice
		Materie Plastiche
Japan	weeklies	*Japan Chemical Week*
	fortnightlies	*Japan Plastics*
	monthlies	*Chemical Economy and Engineering Review*
		Japan Textile News
		Plastics Industry News (Japan)
The Netherlands	weeklies	*Chemisch Weekblad*
		Chempress
		Wereldmarkt
	monthlies	*Plastica*
	quarterlies	*Common Market Law Review*
Spain	weeklies	*Actualidad Económica*
		El Economista
	monthlies	*I.Q. (Industria Química)*
Sweden	weeklies	*Veckans Affarer*
Switzerland	dailies	*Neue Zürcher Zeitung*
	weeklies	*Chemische Rundschau*
United Kingdom	dailies	*Financial Times*
		Journal of Commerce
		Lloyd's List
		Lloyd's Shipping Index
	weeklies	*Chemical Age*
		Chemist and Druggist
		The Economist
		European Chemical News
		Fairplay
		IMS Pharmaceutical Market Letter
		Investors Chronicle
		Marine Week
		Marketing in Europe
		Metal Bulletin
		Middle East Economic Digest

Table 3 (*continued*)

Country	Frequency	Publication
		New Scientist
		Pharmaceutical Journal
		Plastics and Rubber Weekly
		Scrip
		Shoe and Leather News
	fortnightlies	*Africa Confidential*
		Agra Europe
		Chemical Insight
		Eurolaw Commercial Intelligence
		International Dyer
		International Flavours and Food Additives
		International Pest Control
		Nitrogen
		Petroleum Times
		Phosphorous and Potassium
		Polymers, Paint, and Colour Journal
		Sulphur
		Wool Record and Textile World
	monthlies	*British Plastics and Rubber*
		Confectionary Production
		Continental Paint and Resin News
		Cordage, Canvas and Jute
		Dairy Industries International
		European Industrial Relations Reviews
		European Plastics News
		European Rubber Journal
		Fertilizer International
		Food Manufacture
		Food Processing Industries
		Furniture Manufacturer
		Information—CEE
		Journal of Flour and Feed Milling
		Manufacturing Chemist and Aerosol News
		New African
		Packaging News
		Packaging Review
		Petroleum Economist
		Pigment and Resin Technology
		Reinforced Plastics
		Soap, Perfumery, and Cosmetics
		Soft Drinks Trade Journal
		Textile Month
		Textile Institute and Industry
	quarterlies	*British Ink Maker*
		Futures
		Rubber Trends
	irregular	*Plastics Today (ICI)*
United States	dailies	*Wall Street Journal*
		Journal of Commerce
	weeklies	*Business Europe*
		Business International
		Business Week
		Chemical and Engineering News
		Chemical Marketing Reporter
		Chemical Week
		Feedstuffs

Table 3 (*continued*)

Country	Frequency	Publication
		The Guardian
		Oil and Gas Journal
		Petrochemical News
	fortnightlies	*Business Eastern Europe*
		Chemical Engineering
		Fortune
		Naval Stores Review
	monthlies	*Adhesives Age*
		Chemical Engineering Progress
		Chemtech
		Cotton
		Cosmetics and Toiletries
		Drug and Cosmetic Industry
		Farm Chemicals
		Food Technology
		Household and Personal Products Industry
		Hydrocarbon Processing
		Insulation
		Modern Packaging
		Modern Plastics International
		Modern Textiles
		National Safety News
		Near East Business
		Pulp and Paper International
		Rubber World
		Soap, Cosmetics, and Chemical Specialties
		Textile Chemist and Colorist
		Textile Organon

Process Evaluation/Research Planning (PERP reports), Chem Systems, Inc.; comprehensive studies, each covering a type of chemical product and the markets by national regions; updated by the *PERP Quarterly Reports.*

Orthoxylene–Phthalic Anhydride Annual World Survey, the *para-Xylene–DMT/PTA Annual World Survey,* and the *World Naphtha Survey Capacity Tables–Petrochemical Plants* are an interesting series of product studies offered by the HYPLAN Consulting Group.

Eastern Europe—Its Impact on the West European Chemical Industry, 1978, is a unique HYPLAN report that contains information on the production of base petrochemicals, first and second order petrochemical derivatives, and the main product in Bulgaria, Czechoslovakia, the German Democratic Republic, Hungary, Poland, Romania, the USSR, and Yugoslavia.

Petrochemical Units in Western Europe and *Petrochemical Units in the OPEC and OAPEC Countries,* Institute Français du Petrole, Paris; updated occasionally.

Industria Petrolchimica Europea (IPE), Parpinelli-Tecnon, Italy.

The *Chemeurop–EEC* report series, SEMA (Metra International), provides statistical data for about 130 chemicals and derivatives in terms of production, producers, location, current and projected capacities, and total consumption breakdown by main uses.

SRI International offers diverse services to its subscribers from a wide range of industries. SRI's Business Information Program provides in-depth studies and forecasts covering business, economic, political, social, and technological topics on a regional or international basis. SRI developed a number of information resources specifically for the chemical, petroleum, and engineering sectors. These include the following:

Process Economics Program (PEP) report series, SRI International; evaluates the processes for the production of chemicals and petroleum products.

Chemical Economics Handbook (CEH), SRI International; provides data on production, sales, exports, imports, consumption, stocks and shipments of raw materials, primary and intermediate chemicals, or end products such as fertilizers or polymers.

The *World Hydrocarbons Program*, 1979, developed by SRI International; series of reports on approximately 60 main chemicals of interest to the international petrochemical industry.

Directory of Chemical Producers—USA and *Directory of Chemical Producers—Western Europe*, 1979, SRI International provide information on companies and products, list chemicals by manufacturer and plants by location.

The National Economic Development Office (NEDO) in the UK has issued several reviews on industrial sectors and monographs on manufacturing industries in the UK and in EEC countries. Some of the titles are listed in T. Landau's *European Directory of Market Research Surveys*.

Principal Sources of Marketing Information, a booklet by C. Hull of the *Times (London)* Information and Marketing Intelligence Unit, Oct. 1975, brings together the principal sources of information on markets for several industries.

Sources and Production Economics of Chemical Products 1979, compiled and edited by *Chemical Engineering*, McGraw-Hill Publications Company, New York, 1979.

European Chemical Industry Handbook, Hedderwick Stirling Grumbar, UK, 1979; a compilation of data on production, trade, and finance of the chemical industry.

Handbooks and annual installments of these and other series rapidly become dated. Therefore, specialized information services must continuously capture, index, disseminate, and store current business, economic, marketing, and industrial information from periodicals such as those listed in Table 2. Although this list is not exhaustive, it does show the most useful sources of business information. The daily publications, *Platt's Oilgram Price Report*, which indicates the prices of petroleum products internationally, and *Petrochemical Scan*, which covers the base petrochemicals and is distributed via telex networks, both include timely informational briefs on major events that influence petroleum products.

Company Information. Although available information on companies is diverse, the most common needs are for company names, addresses, and products. Information requested next is usually for a company's performance including data such as turnover, sales, profit before and after taxes, capital expenditures, number of employees, ownership, number of subsidiaries, and facilities. Intercompany comparisons may require that the financial results be analyzed in more detail thus requiring the examination of balance sheets and historical data to reveal trends.

Company names, addresses, ownership, and products appear in directories and handbooks that have international or national coverage. Some of these are listed below.

Who Owns Whom, Dun & Bradstreet, Ltd., London, Eng. lists the parent and associated companies in a series of volumes covering the world.

The Worldwide Chemical Directory, prepared by ECN Chemical Data Services, IPC Business Press Ltd., New York, 1977 provides the address, telex, and the telephone numbers of chemical companies and a summary of their activities.

Chem Sources—U.S.A., 21st ed., Directories Publishing Co., Inc., Flemington, N.J., 1980, and *Chem Sources—Europe*, Chemical Sources Europe, Mountain Lakes, N.J., 1980 provide information on manufacturers or suppliers of chemical products.

OPD Chemical Buyers Directory, 67th ed., Schnell Publishing Co., Inc., New York, 1979 (appearing once a year as part of an annual subscription to *Chemical Marketing Reporter*) provides information on chemicals, their suppliers, shipping, and storage.

1980 Chemical Week Buyers' Guide Issue, McGraw-Hill, Inc., New York (appearing once a year as part of an annual subscription to *Chemical Week*) gives information on chemicals, new materials, specialties, suppliers, trade names, company addresses, and directors' names.

Directory of Chemical Producers—United States and *Directory of Chemical Producers—Western Europe*, SRI International, Menlo Park, Calif., 1979; provide company names, addresses, ownerships, subsidiaries, plants, and products and list the chemicals by manufacturing company and plants by location.

KOMPASS, Kompass Register, Ltd., Croydon, Eng.; series of directories for individual countries, especially those in Europe; products and company indexes, supplier names, addresses, and business activities, employees, and a trade name index are provided.
Directory of West European Chemical Producers, 1977–1978 ed., Chemical Information Services, Ltd., Oceanside, N.Y. provides information for about 28,000 products and their suppliers.
European Chemical Buyers' Guide, IPC Business Press, Ltd., New York, 1979 provides product, company and supplier, and trade name indexes.

More detailed information on the ownership, shareholders, capital, industrial operations, and finances appears in directories such as the *Financial Times International Business and Companies Year Book: 1978/79,* B. E. Donovan, ed., The Financial Times, Ltd., London, Eng., 1978–1979. This book quotes the top companies worldwide, their addresses, subsidiaries, ownership, capital results, main products, and gives a brief overview of the individual countries' economies.

Comprehensive information on companies appears in the following sources:

R. P. Hanson, ed., *Moody's Industrial Manual,* Moody's Investors Service, Inc., New York, 1979 covers all United States industrial sectors.
J. Love, ed., *Jane's Major Companies of Europe 1979–80,* Macdonald and Jane's Publishers, Ltd., London, Eng.; Franklin Watts, Inc., New York, 1979.

A number of official agencies and specialized services provide information on corporations and companies for investment, financial analyses, and accounting and legal matters. Because the information they deal with is required by law, these agencies are reliable sources of these particular types of business data. In the United States this type of information is filed with the Securities and Exchange Commission (SEC). The SEC issues several comprehensive reports, all of which would not be possible to cover in detail here. Also worth mentioning are the 10-K annual business and financial reports. These identify the main products and services of a company and summarize its operations in detail for the last five fiscal years. The 10-Q quarterly financial reports filed by most companies provide current information on their financial position during the year. The 8-K reports inform shareholders of important events or changes such as acquisition of new assets, increase or decrease in the amount of securities outstanding, and the like.

In the United Kingdom, companies registered in England and Wales are required to file certain information with the Companies Registration Office and Companies House; those registered in Scotland must file with the Register of Companies. In France, similar information can be obtained from the official publication, *Bulletin des Annonces Légales Obligatoires* (BALO); in Germany, Verlag Hoppenstedt publishes economic and financial information on companies. Names and addresses of agencies and services that publish or file company information appear in DAFSA's *L'Information Economique et Financière en Europe, Les Sources d'Information Economique et Financière: Etats-Unis, Royaume-Uni, France,* E. de Dampierre and F. Charpentier, Center for Business Information, and *Published Data on European Industrial Markets,* Industrial Aids, Ltd.

Two specialized services worthy of note are the DAFSA–Hoppenstedt *Information Internationales* and the British Extel Statistical Service, Ltd. The former provides information on large companies in loose-leaf form printed in French, German, and English. The latter publishes information on cards covering mostly British companies but including also some major European companies.

Of course, the daily press and periodicals such as those listed in Table 2 are sources for recent news. Secondary services also provide information on company activity internationally. The best known for the chemical industrial sector are Chemical Abstracts Service's *Chemical Industry Notes* (*CIN*), the American Petroleum Institute's *Petroleum/Energy Business News Index,* and Predicasts' *Prompt* and *F&S Index.*

Computerized Information Sources. Computerized information files, accessible on line from remote terminals, constitute a distinct body of information that has grown rapidly in recent years. Searching requires expertise, particularly in the area of business information, because of the diversity of available files. Many are bibliographic, but some technical-economic files, resembling the numerical or nonbibliographic files discussed in the previous section of this article, provide business information directly. Some systems offer manipulative and computational capabilities that allow the searcher to store data, enter data, merge or modify series of data, manipulate them, edit results on line, and print these in report format. An advantage of these systems is that they use natural-language commands; the searcher needs no programming training or knowledge of programming language. But because a large number of data bases are available, searchers should develop a familiarity with those files that most often meet their needs, although other files, upon exploration, may unexpectedly provide additional information. Searchers must be on the alert continuously for new files and services and must be able to assess their value to the organizations they serve.

Table 4 outlines some business-information services and data bases currently available. The data bases listed are those important to chemical business, management, and corporate policy. The listing excludes those data bases that are primarily scientific and technical in nature.

Nonbibliographic Business Data Bases. Data Resources, Inc. (DRI), in Boston, Mass., makes available information on the economies of several countries from data bases that contain time series from United States, foreign governmental and nongovernmental sources and organizations. DRI's services allow the user to relate external economic developments to the corporate planning of manufacturing companies in terms of future sales, cash flow, capital budgeting and investment, inventories, production, pricing, business cycles, or the implications of governmental decisions. Furthermore, DRI has developed economic models and forecasts for individual regions, ie, the United States, Canada, the European Economic Communities, and Japan.

DRI's source material includes economic, financial, and demographic series on the United States; the complete details of consumer, wholesale, and industrial price indexes as compiled by the Bureau of Labor Statistics; a complete set of OECD series pertaining to economic indicators, national income accounts, trade series, industrial production inclusive of chemicals, petroleum and coal products; and international financial statistics from the International Monetary Fund. DRI's data bases can be searched in a variety of ways, data can be manipulated on line, and several report-writing capabilities are available. Figure 4 illustrates a feature that permits the user to draw a three-month moving-average line from dispersed monthly figures.

Chase Econometrics provides access to several data bases and to econometric models with full simulation capability. These models allow the user to link the economies of the following countries: the United States, FRG, Belgium, France, Italy, the Netherlands, the United Kingdom, Spain, Sweden, Japan, Canada, Mexico, and Brazil. The system also comprises historical and forecast data bases. The historical data bases include detailed financial series and wholesale and consumer price series. The forecast data bases contain Chase Manhattan's estimates as derived from their models.

Chase Manhattan also has a computerized company data-base service, called EXSTAT from the Extel Statistical Services, Ltd., of London, which provides information on the United Kingdom's, continental European, and Australian companies.

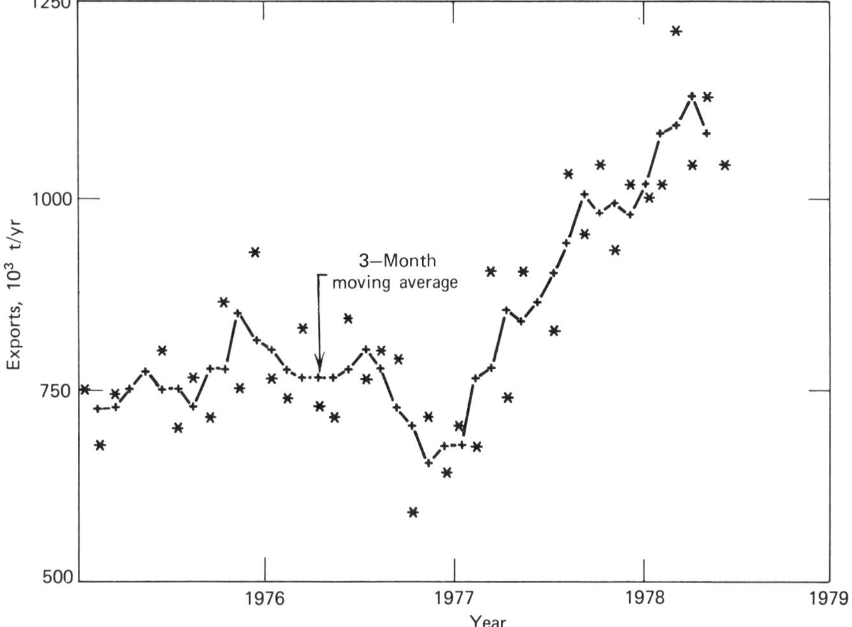

Figure 4. Exports of five main plastics. Courtesy of Data Resource, Inc.

The information comes from the same source as the Extel Cards service but can be analyzed interactively on line through Interactive Data Corporation's (IDC) time-sharing network.

Automatic Data Pioneering, Inc. (ADP), through ADP Network Services International and Information Services, offers files including the EXSTAT data base from Extel, COMPUSTAT, which provides financial information on the United States, COPPER DEVELOPMENT DATA BASE, McGraw-Hill's METALS WEEK, and the IPC CHEMICAL DATA BASE, which covers plant and international product information on chemicals.

Parpinelli-Tecnon, in Milan, Italy, computerized their petrochemicals file and made it accessible on line under the name EPICS (European Petrochemical Industry Computerized System). The statistics segment of the file provides the complete balance for each product for selected countries or regions from 1965 to the current year, for the next five years (forecasts), and an output eighth year. The plants section provides information on each selected plant or product in terms of producing company, location, feedstock(s), plant status, start-up date, license, contractor, and capacity for the same time period as the statistics section. Program features allow users to retrieve information on an entire company group (ie, the mother company with all its subsidiaries) or on all plants in the same location or region that manufacture a particular product. The program also allows the user to analyze supply-and-demand balances for each product by region, country, location, or individual company. In carrying out this analysis, users can manipulate the retrieved data by adding or replacing some of it with their own figures. An example of data provided by the EPICS data base is shown in Table 5.

Stanford Research Institute's (SRI) WORLD PETROCHEMICAL DATA BASE

Table 4. Business Data Bases

Data base name	Supplier	Coverage
ABI–INFORM	Data Courier, Inc.	business management and administration; 1971–present
ACCOUNTANTS' INDEX	American Institute of Certified Public Accountants	international accounting and finance; 1974–present
AFO (Analyse Financière sur Ordinateur)	CISI, France	financial data for quoted companies in France
APTIC	U.S. Environmental Protection Agency	air pollution in the broad sense, including social, political, legal, administrative, and technical aspects; 1966–1978
ASI (American Statistics Index)	Congressional Information Service, Inc.	abstracts and indexes of all U.S. federal statistical publications; 1973–present
BIPA (Banque d'Information Politique et d'Actualité)	La Documentation Française and Télesystèmes, France	several files of official French publications on a variety of economic and political issues
CBPI	Information Access of Toronto, Canada	Canadian business periodicals; 1975–present
CEE (Archivio della Giurisprudenza della Corte Di Giustizia della Comunita Europee)	Centro Elettronko di Documentazione della Corte Suprema di Cassazione, Italy	file on jurisprudence containing all the digests of the decisions of the EEC's Court of Justice
CELEX	Legal Service of the Commission of the European Economic Communities	legislative and legal information for member countries; anticipated in 1980; 1977–present
CIN (Chemical Industry Notes)	Chemical Abstracts Service	chemical processing industries; 1974–present;
CIS INDEX	Congressional Information Service, Inc.	current, comprehensive index to the entire U.S.; congressional working papers; 1970–present
COMPUSTAT (Files)	Standard and Poor's Compustat Services, Inc.	several nonbibliographic files containing financial information on the U.S. and Canada
COPPER DEVELOPMENT ASSOCIATION	Copper Development Association	nonbibliographic information on metals
CRECORD	Capitol Services, Inc.	current coverage of the *Congressional Record*—all proceedings of the U.S. Congress; 1976–present
CRONOS (Chronological Series)	Statistical Office of the Commission of the European Economics Communities	time series of the member countries of the European Economic Communities; may become a public file in 1980
DAFSA–LES LIAISONS FINANCIERES	Société Pour d'Informatique, France	information on corporate relationships between companies in France
DISCLOSURE	Disclosure Incorporated	securities and exchange Commission filings; 1977–present
DRI CAPSULE DATA BASE	Data Resources, Inc.	business and economics data
DRI CENTRAL DATA BANK	Data Resources, Inc.	macroeconomics, primarily U.S. data
ECONOMICS ABSTRACTS INTERNATIONAL	Learned Information, Ltd., UK	international coverage of markets, industries, country-specific economics and research in management and economics; 1974–present

Table 4 (*continued*)

Data base name	Supplier	Coverage
EIS INDUSTRIAL PLANTS	Economic Information Systems, Inc.	current nonbibliographic information on establishments representing over 90% of U.S. industrial activity; replaced three times per year
EIS NONMANUFACTURING ESTABLISHMENTS	Economic Information Systems, Inc.	current data on U.S. nonmanufacturing establishments employing 20 people or more; replaced three times per year
ENCYCLOPEDIA OF ASSOCIATIONS	Gale Research Company	detailed information on several thousand trade associations, professional societies, etc; current year
ENERGYLINE	Environment Information Center, Inc.	socio-economic, governmental policy and planning, current affairs, and scientific–technical aspects of energy; 1971–present
ENVIROLINE	Environmental Information Center, Inc.	international environmental information including management, planning, law, politics, as well as science and technology; 1971–present
EPB (Environmental Periodicals Bibliography)	Environmental Studies Institute	multidisciplinary index to environmental periodicals; 1973–present
EPICS (European Petrochemical Industry Computerized System)	Parpinelli-Tecnon, Italy	statistics and plant information
EUROLEX	European Law Center, Ltd., UK	full-text data base of the legal and business professions in Europe; anticipated in 1980
EXSTAT	Extel Statistical Services, Ltd., UK	international company information
FEDREG	Capitol Services, Inc.	citations from the U.S. *Federal Register;* multidisciplinary; 1977–present
FOREIGN TRADERS	U.S. Department of Commerce	directory of manufacturers, wholesalers, distributors, etc, in 130 countries that import or wish to import from the U.S.; a nonbibliographic file restricted to U.S. use; current 5 years
GPO MONTHLY CATALOG	U.S. Government Printing Office	reports, fact sheets, conference proceedings, etc, issued by all U.S. governmental agencies, including Congress; multidisciplinary; 1976–present
IBIS	CERVED, Italy	information on over 160,000 companies in about 130 countries
INTDB (International Data Base)	Chase Econometric Associates, Inc.	international nonbibliographic data base on commercial banks, government finance, international liquidity and transactions, and national income
IPC CHEMICAL DATA BASE	International Publishing Corporation, UK	international plant and product information for about 120 chemicals
ITIS	CERVED, Italy	economic outlook and commercial data on 90 countries

Table 4 (*continued*)

Data base name	Supplier	Coverage
KOMPASS-FRANCE	Kompass, and Société Pour d'Informatique, France	manufacturing companies, suppliers and products in France
LABORDOC	The International Labour Organization, Switzerland	international coverage of industry, economics, management, public finance, occupational safety and related fields from ILO publications; 1965–present
MAGAZINE INDEX	Information Access Corporation	multidisciplinary index to over 370 popular magazines; 1977–present
MANAGEMENT CONTENTS	Management Contents, Inc.	business and management related topics; 1974–present
METALS WEEK	McGraw-Hill	metal prices
NATIONAL NEWSPAPER INDEX	Information Access Corporation	front-to-back-page indexing of *The Christian Science Monitor, The New York Times*, and *The Wall Street Journal;* 1979–present
NEWSEARCH	Information Access Corporation	indexing of current newspapers, magazines, and periodicals; material is transferred to the NATIONAL NEWSPAPER INDEX and MAGAZINE INDEX at the end of each month; daily updates
NEW YORK TIMES INFORMATION BANK	New York Times Company	business, economics, politics, and law from major newspapers and magazines
PAIS INTERNATIONAL	Public Affairs Information Service, Inc.	international coverage of all fields of social science including political science, banking, public administration, law, policy, etc; 1976–present
P/E NEWS	American Petroleum Institute	extensive indexing of seven major periodicals in the petroleum and energy fields; 1975–present
POLLUTION ABSTRACTS	Data Courier, Inc.	citations to literature on pollution; 1970–present
PTS F&S INDEXES	Predicasts' Terminal System (PTS), Predicasts, Inc.	international company, product and industry information on acquisitions, mergers, new products or technology, socio-political factors, and summaries of forecasts; 1972–present
PTS FEDERAL INDEX	Predicasts Terminal Systems (PTS), Predicasts, Inc.	coverage of federal actions such as proposed regulations, bill introductions, contract awards, etc; 1976–present
PTS INTERNATIONAL ANNUAL TIME SERIES	Predicasts Terminal Systems (PTS), Predicasts, Inc.	time series including historical data from 1957 and projections for the future; all countries of the world are covered except the U.S.; economic, demographic, industrial, and product data; 1972–present
PTS INTERNATIONAL STATISTICAL ABSTRACTS	Predicasts Terminal System (PTS), Predicasts, Inc.	abstracts of published forecasts with historical data for all countries of the world except the U.S.
PTS PREDALERT	Predicasts Terminal System (PTS), Predicasts, Inc.	current month (records added on a weekly basis) corresponding to the cumulative files PTS PROMPT, PTS F&S INDEXES, and PTS FEDERAL INDEX

Table 4 (*continued*)

Data base name	Supplier	Coverage
PTS PROMPT	Predicasts Terminal System (PTS), Predicasts, Inc.	overview of markets and technology; marketing data for scientifically and technically oriented industries; current records in PTS PREDALERT; 1972–present
PTS US ANNUAL TIME SERIES	Predicasts Terminal System (PTS), Predicasts, Inc.	coverage as in PTS INTERNATIONAL ANNUAL TIME SERIES, but for the U.S. only; 1971–present
PTS US STATISTICAL ABSTRACTS	Predicasts Terminal System (PTS), Predicasts, Inc.	coverage as in PTS INTERNATIONAL STATISTICAL ABSTRACTS but for the U.S. only; 1971–present
QUEBEC-ACTUALITE	Microfor, Inc., Canada	French-language file of three Quebec newspapers, *Le Devoir, La Presse,* and *Le Soleil;* 1973–present
RAPRA ABSTRACTS	Rubber and Plastics Research Association of Great Britain	economic, commercial, technical, and research aspects of the rubber and plastics industries; 1972–present
SAFETY SCIENCE ABSTRACTS	Cambridge Scientific Abstracts	interdisciplinary coverage of safety literature; includes legislative regulations and their effect as well as technical aspects; 1975–present
SANI (Anagrafe Camerale ed Operativa delle Imprese Italiane)	CERVED, Italy	information on Italian companies
SGB (Société Générale de Banque de Belgique)	Société Pour d'Informatique, France	finance, administration, and management file developed by a major Belgian bank
SIBB (Archivio degli Indici dei Bollettini delle S.p.A. ed S.r.l.)	CERVED, Italy	financial information on registered Italian companies
TITLEX (Archivi dei Titoli della Legislazione Statale)	Centro Elettronico di Documentazione della Corte Suprema di Cassazione, Italy	titles of current Italian laws and decrees
TRIBUT (Archivio della Giurisprudenza della Commissioni Tributaria Centrale e delle Commissioni Tributarie di lo e 2o Grado)	Centro Elettronico di Documentazione della Corte Suprema di Cassazione, Italy	Italian tax law
USPSD (United States Political Science Documents)	University of Pittsburgh	scholarly articles in the broad area of political science; 1975–1977
WORLD PETROCHEMICAL DATA BASE	Stanford Research Institute	international production information

(WP Data Base) covers approximately 80 petrochemicals and derivatives. A computer program allows complex searching stragegies and allows search results to be edited on line. Information in the WP Data Base is keyed to individual products and subdivided according to country. Within each country, product information is expressed in terms of manufacturing company, plants in operation during the period from 1974

Table 5. Balance for Ethylene in The Netherlands, in 1979, Thousand Metric Tons

Company	Location	Production	Consumption	Balance	Consumption breakdown				
					LDPE[a]	HDPE[b]	EO[c]	EDC[d]	EB[e]
Limbourg									
DSM	Beek	524	336	+188	267	69			
Area total		*524*	*336*	*+188*	*267*	*69*			
West									
Dow	Terneuzen	758	422	+336	160		80		182
Gulf	Rotterdam	242	53	+189					53
Shell	Pernis	113	47	+66			47		
Shell	Moerdijk	403	137	+266			113		24
Group total		*516*	*184*	*+332*			*160*		*24*
ICI	Rozenburg		125	−125	125				
AKZO	Botlek		177	−177				177	
Area total		*1516*	*961*	*+555*	*285*		*240*	*177*	*259*
other derivatives			16						
Country total		*2040*	*1313*	*+727*	*552*	*69*	*240*	*177*	*259*

[a] LDPE = low density polyethylene.
[b] HDPE = high density polyethylene.
[c] EO = ethylene oxide.
[d] EDC = ethylene dichloride.
[e] EB = ethylene dibromide.

to the current year, and forecast figures to 1985, captive use of product, and raw materials. The supply-and-demand section of the WP Data Base provides the year-end capacities, production years, imports and exports, and consumption figures. Ownership relations and subsidiary names are searchable too. An example from SRI's WP Data Base of a supply-and-demand report for one product covering North America is shown in Table 6.

Parpinelli-Tecnon's EPICS and SRI's WP data base allow rapid searching, analysis, consolidation, and editing of data that otherwise can only be laboriously assembled from the large volume of published material. Furthermore, current information is available promptly from the on line files; the latest published information may be several months old.

Other Business Data Bases. Several bibliographic files for business information have been characterized by J. L. Hall in *On-Line Bibliographic Data Bases—1979 Directory* published by ASLIB, London, Eng., or by A. Tomberg, *EUSIDIC Data-Base Guide 1980,* Learned Information, Abingdon, Eng. These provide references to other publications on on-line data bases, both bibliographic and nonbibliographic.

A few files are of particular interest to chemical business and marketing. References to published articles, and often full abstracts, on the chemical industry's activities and economic outlook may be found in the American Petroleum Institute's APILIT and in P/E NEWS, which provides technical-economic information for the petrochemical sector. These files have been described in ref. 32. The chemical industries are also covered by files such as Chemical Abstracts Service's CIN (Chemical Industry Notes) and by Predicasts' PTS PROMPT and PTS F&S INDEXES which supplement each other. In addition, Predicasts offers the EIS INDUSTRIAL PLANTS file which is an on-line directory of manufacturing industries providing for each plant the name and address of the operating company and its parent company, the principal product, the market share, total sales, the employment-size class, and size of shipments. This file is described in ref. 33 where Predicasts' PTS PROMPT, PTS F&S INDEXES, and the PTS statistical data bases are also described.

Information on companies that file reports with SEC in the United States can be searched on the file DISCLOSURE, which is accessible through Lockheed's DIALOG system. Information on companies registered in Europe can be searched

Table 6. North American Supply and Demand for Low Density Polyethylene, Thousands of Metric Tons

Supply year[a]	Year-end capacity	Production	Imports	Exports	Apparent consumption	Consumption
1974	3242	2953	130	187	2896	2874
1975	3287	2415	60	151	2324	2312
1976	3465	2942	72	269	2745	2704
1977	4038	3270	119	286	3103	3077
1978	4264					3210
1979	4391					3390
1980	4947					3590
1981	5185					3801
1982	5480					4027
1983	5480					4270

[a] Data for 1980–1983 are estimated.

in the EXSTAT file. As of the end of 1979, the DAFSA LIAISONS FINANCIERES file, offered by the Société Pour d'Informatique (SPI), provides information on the financial relationships between companies. SPI also offers KOMPASS-FRANCE, which allows users to search for information on manufacturing companies, suppliers, and products in France, and a file developed by the Belgian bank, the Société Générale de Banque de Belgique (SGB), covering finance, administration, and management.

CISI in Paris offers the Analyse Financière sur Ordinateur (AFO) file which gives financial and stock-exchange data for French companies including five years of historical information.

The work of committees and subcommittees of the United States Congress is covered in the CIS INDEX data base, which includes industrial topics on both national and international governmental policies.

The Congressional Information Service developed the American Statistics Index (ASI), which is a key to all statistical publications of the United States government, eg, from the Bureau of Census, the Bureau of Labor Statistics, the National Center for Health Statistics, the National Center for Social Statistics, the Statistical Reporting Service, and the Department of Agriculture. The index also covers several statistical publications from other parts of the United States government.

CRECORD produced by Capitol Services, Inc., is an on-line index, with abstracts, of the *Congressional Record,* the official journal of the proceedings of the Congress. Capitol Services also developed FEDREG, the on-line file that provides references to the *Federal Register* on agriculture, business, economy, finance, environment, labor, resources, taxation, trade, and the like. The CIS INDEX, ASI, CRECORD, and FEDREG are all accessible on line through SDC's services.

Predicasts' PTS FEDERAL INDEX, accessible through Lockheed's DIALOG system, covers similar subjects from the *Washington Post,* the *Congressional Record,* the *Federal Register,* and *Commerce Business Daily.*

Information bases covering major dailies and periodicals are now available. A main feature of two of these, the NEW YORK TIMES INFORMATION BANK and NEWSEARCH, is their timeliness. The New York Times Company selects articles on business, economic, political, and legal topics from about 60 newspapers and magazines covering petroleum, petrochemicals, the chemical and allied industries. Some of the major publications which are covered include the *New York Times,* the *Wall Street Journal,* the *Washington Post, Business Week, Financial Times, Barron (Gadler and London), The Economist, Forbes, Fortune,* and *Journal of Commerce.* All front pages or main news articles are put on line within 24 h. This file is described in ref. 34.

NEWSEARCH, a daily reference file accessible via the DIALOG service, contains bibliographic records of news items and articles selected from publications such as the *Christian Science Monitor,* the *New York Times,* and the *Wall Street Journal* and over 370 popular American magazines. Information no longer current (more than 30 d old) is accessible from the MAGAZINE INDEX and NATIONAL NEWSPAPER INDEX files, which are updated monthly. These two files include such subject matter as company information, trademarks, trade names, legislation, business and finance, product evaluation, and current affairs whether national or international. The NATIONAL NEWSPAPER INDEX is accessible through the SDC and Lockheed services.

The number of business-information files is already considerable, and it grows as additional systems are computerized. Although the objective is to provide information promptly and effectively, the number of data bases and national languages will make it both easier in one sense and more difficult in another for information professionals to provide timely and comprehensive business data. Information professionals, while continuing to search several files for scattered bits and pieces of the puzzles they put together, will feel the need for a few comprehensive files concerned with the chemical sector that can bring together the information they need. Because such a utopian product is unlikely to appear in the next few years, business information will, without doubt, continue to be collected piecemeal from the wealth of sources worldwide.

BIBLIOGRAPHY

"Literature, Mechanized Searching" in *ECT* 1st ed., Vol. 8, pp. 449–467, by J. W. Perry, Bjorksten Research Laboratories, and R. S. Casey, W. A. Sheaffer Pen Co.; "Literature of Chemistry and Chemical Technology" in *ECT* 2nd ed., Vol. 12, pp. 500–529, by T. J. Devlin, Esso Production Research Co., and B. H. Weil, Esso Research and Engineering Co.; "Information Retrieval Services and Methods" in *ECT 2,* Suppl. Vol., pp. 510–535, by Eugene Garfield and Charles E. Granito, Institute for Scientific Information, and Anthony E. Petrarca, The Ohio State University.

1. D. F. Johnston, *Copyright Handbook,* R. R. Bowker Company, New York, 1978.
2. S. L. Burgoon, *The Viability and Impact of Electronic Storage and Delivery of Handbook-Type Information, PB 278072,* National Technical Information Service, Springfield, Va., 1978.
3. Y. S. Touloukian, *Bull. Am. Soc. Inf. Sci.* **5**(6), 44 (1979).
4. H. M. Woodburn, *Using the Chemical Literature: A Practical Guide,* Marcel Dekker, Inc., New York, 1974, pp. 111–123.
5. *Ibid.,* pp. 141–151.
6. A. Weissberger, ed., *Techniques of Chemistry,* Wiley-Interscience, New York, 1971–1979.
7. *Curr. Abstr. Chem.* **12**(3), (Jan. 17, 1979).
8. J. F. Terapane, *Chemtech* **8**(5), 272 (1978).
9. W. J. Bowman, *J. Chem. Inf. Comput. Sci.* **18**(2), 81 (1978).
10. H. M. Allcock and J. W. Lotz, *Chemtech* **8**(9), 532 (1978).
11. J. F. Terapane, *J. Chem. Inf. Comput. Sci.* **17**(3), 130 (1977).
12. S. M. Kaback, *Online (Weston, Conn.)* **2**(1), 16 (1978).
13. J. T. Maynard, *Chemtech* **8**(2), 91 (1978).
14. M. J. Marcus, *J. Chem. Inf. Comput. Sci.* **18**(2), 76 (1978).
15. K. M. Donovan and B. B. Wilhide, *J. Chem. Inf. Comput. Sci.* **17**(4), 139 (1977).
16. M. M. Duffey, *J. Chem. Inf. Comput. Sci.* **17**(3), 126 (1977).
17. R. G. Smith, L. P. Anderson, and S. K. Jackson, *J. Chem. Inf. Comput. Sci.* **17**(3), 148 (1977).
18. W. Marcy, ed., *Patent Policy: Government, Academic, and Industry Concepts, ACS Symposium Series 81,* American Chemical Society, Washington, D.C., 1978.
19. *Central Patents Index 1980,* Derwent Publications, Ltd., London, Eng., 1979.
20. R. J. Rowlett, Jr., *Chemtech* **9**(6), 348 (1979).
21. J. T. Maynard, *J. Chem. Inf. Comput. Sci.* **17**(3), 136 (1977).
22. A. Girard, S. M. Kaback, and K. Landsberg, *Database* **1**(2), 46 (1978).
23. S. R. Heller and G. W. A. Milne, *Environ. Sci. Technol.* **13**(7), 798 (1979).
24. M. Williams, *Computer-Readable Data Bases: A Directory and Data Sourcebook,* Knowledge Industry Publications, Inc., White Plains, N.Y., Oct. 1979; M. Williams, *Bull. Am. Soc. Inf. Sci.* **6**(2), 22 (1979).
25. R. N. Landau, J. Wanger, and M. C. Berger, eds., *Directory of On-Line Data Bases,* Cuadra Associates, Inc., 1980.
26. Union of International Associations (Brussels), ed., *Yearbook of International Organization 1978–1979,* 17th ed., International Publication Services, New York, 1978.
27. *Europa Yearbook, A World Survey,* 20th ed., 2 vols., Europa Publications, Ltd., London, Eng., 1979.

28. L. Huckfield, *Trade Ind.* **31**(9), 483 (June 2, 1978).
29. B. Niewenglowska, *ASLIB Proc.* **27**(11–12), 453 (1975).
30. M. Shollaert, *La Documentation Economique, Commerciale et Financière,* Colloque National de la Documentation, Association Belge de Documentation, Brussels, Belgium, May 9–10, 1974.
31. G. Bloch, *ASLIB Proc.* **27**(11–12), 459 (1975).
32. L. Rogalski, *J. Chem. Inf. Comput. Sci.* **18**(1), 9 (1978).
33. M. A. Burylo, *Online (Weston, Conn.)* **1**(3), 53 (1977); G. Sharp, *Online (Weston, Conn.)* **2**(1), 33 (1978).
34. D. R. Dolan, *Online (Weston, Conn.)* **2**(2), 26 (1978).

MARGARET H. GRAHAM
Exxon Research and Engineering Company

ALEXIS B. LAMY
Essochem Europe, Inc.

BARBARA LAWRENCE
Exxon Corporation

LORRAINE Y. STROUMTSOS
Exxon Research and Engineering Company

INFRARED DETECTORS. See Infrared technology; Photodetectors.

INFRARED TECHNOLOGY

The electromagnetic spectrum is divided arbitrarily into a number of wavelength regions, called bands, distinguished by the methods utilized to produce and detect the radiation. There is no fundamental difference between radiation in the different bands of the electromagnetic spectrum; they are all governed by the same laws and the only differences are those due to the differences in wavelength.

Thermography makes use of the infrared spectral band. At the short-wavelength end, the boundary lies at the limit of visual perception, in the deep red. At the long-wavelength end, it merges with the microwave radio wavelengths, in the millimeter range.

The infrared bands commonly are subdivided further into four lesser bands, the boundaries of which also are chosen arbitrarily. They include: the near infrared (0.75–3 μm), the short infrared (3–6 μm), the long infrared (6–15 μm), and the far infrared (15–1000 μm).

Blackbody Radiation

A blackbody is defined as an object that absorbs all radiation which impinges upon it at any wavelength. The apparent misnomer "black" relating to an object emitting radiation is explained by Kirchhoff's law, which states that a body capable of absorbing all radiation at any wavelength is equally capable in the emission of radiation.

The construction of a blackbody source is very simple, in principle. The radiative characteristics of an aperture in an isothermal cavity made of an opaque absorbing material represents almost exactly the property of a blackbody. A practical application of the principle to the construction of a perfect absorber of radiation consists of a box that is light-tight except for an aperture in one of the sides. Any radiation that then enters the hole is scattered and absorbed by repeated reflections so only an infinitesimal fraction can possibly escape. The blackness that is obtained at the aperture is nearly equal to a blackbody and almost perfect for all wavelengths.

An isothermal cavity with a suitable heater is termed a cavity radiator. Such a cavity heated to a uniform temperature generates blackbody radiation, the characteristics of which are Planckian, ie, determined solely by the temperature of the cavity. Such cavity radiators are commonly utilized as sources of radiation in temperature reference standards in the laboratory for calibrating thermal measurement instruments (see Temperature measurement).

If the temperature of a blackbody radiator increases over 525°C, the source begins to be visible so that it appears to be no longer black. This is the incipient red-heat temperature of the radiator, which then becomes orange or yellow as the temperature increases further. The definition of the so-called color temperature of an object is the temperature to which a blackbody would have to be heated to have the same appearance.

Each of the following expressions describes the radiation emitted from a blackbody.

Planck's Law. Planck described the spectral distribution of the radiation from a blackbody by the following equation (Fig. 1):

$$W_{\lambda b} = \frac{2\,\pi h c^2}{\lambda^5(e^{hc/\lambda kT} - 1)} \times 10^{-6} \text{ in W/(m}^2\cdot\mu\text{m)}$$

where $W_{\lambda b}$ = the blackbody spectral radiant emittance at wavelength λ, c = the velocity of light = 3×10^8 m/s, h = Planck's constant = 6.6×10^{-34} J·s, k = Boltzmann's constant = 1.4×10^{-23} J/K, T = the absolute temperature (K) of the blackbody, and λ = wavelength (μm).

Instruments used for measuring spectral emittance characteristics, eg, the spectroradiometer, must utilize a narrow band of radiation in order to register a reading. Thus, a value for spectral radiant emittance is meaningless unless the spectral interval is also specified.

Planck's formula, when plotted graphically for various temperatures, produces a family of curves. Following any particular Planck curve, the spectral emittance is zero at $\lambda = 0$, then increases rapidly to a maximum at a wavelength λ_{max}, and after passing, it approaches zero again at very long wavelengths. The higher the temperature, the shorter the wavelength at which the maximum occurs.

Wien's Displacement Law. Wien's formula differentiates Planck's formula with respect to λ, and finds the maximum:

$$\lambda_{max} = \frac{2898}{T}\,\mu\text{m}$$

It expresses mathematically the common observation that colors vary from red to

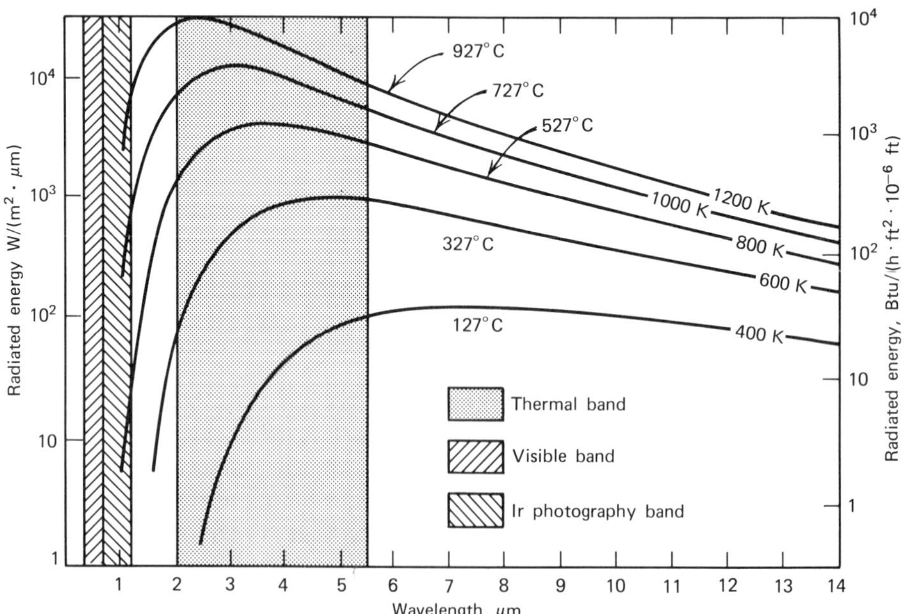

Figure 1. Energy distribution for a blackbody at various temperatures.

orange or yellow as the temperature of a thermal radiator increases. The wavelength of the color is the same as the wavelength calculated for λ_{max}. A good approximation of the value of λ_{max} for a given blackbody temperature is obtained by applying the rule-of-thumb (3000 K). Thus, a very hot star such as Sirius (11,000 K), emitting bluish-white light, radiates with the peak of spectral radiant emittance occurring within the invisible ultraviolet spectrum, at wavelength 0.27 μm. The sun (ca 6000 K) emits yellow light, peaking at about 0.5 μm in the middle of the visible-light spectrum. At room temperature (300 K), the peak of radiant emittance lies at 9.7 μm, in the far infrared. At the temperature of liquid nitrogen (77 K), the maximum of the almost insignificant amount of radiant emittance occurs at 38 μm, in the extreme infrared wavelengths.

The Stefan-Boltzmann Law. The Stefan-Boltzmann law is obtained by integrating Planck's formula from $\lambda = 0$ to $\lambda = \infty$ to yield the total radiant emittance W_b of a blackbody:

$$W_b = \sigma T^4 \text{ W/m}^2$$

where σ = the Stefan-Boltzmann constant = 5.7×10^{-8} W/(m^2·T^4), T = K.

The Stefan-Boltzmann formula states that the total emissive power of a blackbody is proportional to the fourth power of its absolute temperature. Graphically, W_b represents the area under the Planck curve for a particular temperature. It can be shown that the radiant emittance in the interval $\lambda = 0$ to λ_{max} is only 25% of the total, which represents approximately the amount of the sun's radiation which lies inside the visible-light spectrum.

From the Stefan-Boltzmann formula it can be calculated that one kilowatt of power is radiated by the human body at a temperature of 300 K and an external surface area of ca 2 m^2. This great power loss could not be sustained if it were not for the compensating absorption of radiation from surrounding surfaces, at room temperatures that do not vary too drastically from the temperature of the body or, of course, the addition of clothing.

Graybody Radiation

Real objects almost never comply with the laws of blackbody radiators over an extended wavelength region, although they may approach the blackbody behavior in certain spectral intervals. For example, white paint appears perfectly white in the visible-light spectrum, but becomes distinctly gray at about 2 μm, and beyond 3 μm it is almost black.

There are three processes that can prevent a real object from acting like a blackbody: a fraction of the incident radiation α may be absorbed, a fraction ρ may be reflected, and a fraction τ may be transmitted. Since all of these factors are wavelength dependent, the subscript λ is used to imply the spectral dependence of their definitions. Thus: the spectral absorptance α_λ = the ratio of the spectral radiant power absorbed by an object to that incident upon it; the spectral reflectance ρ_λ = the ratio of the spectral radiant power reflected by an object to that incident upon it; and the spectral transmittance τ_λ = the ratio of the spectral radiant power transmitted through an object to that incident upon it.

The sum of these three factors must always add up to the whole at any wavelength:

$$\alpha_\lambda + \rho_\lambda + \tau_\lambda = 1$$

For opaque materials $\tau_\lambda = 0$, and the relation simplifies to

$$\alpha_\lambda + \rho_\lambda = 1$$

Another factor, called the emissivity, is required to describe the fraction ϵ of the radiant emittance of a blackbody produced by an object at a specific temperature. The spectral emissivity ϵ_λ = the ratio of the spectral radiant power from an object to that from a blackbody at the same temperature and wavelength.

Expressed mathematically, this can be written as the ratio of the spectral emittance of the object to that of the blackbody as follows:

$$\epsilon_\lambda = \frac{W_{\lambda o}}{W_{\lambda b}}$$

where $W_{\lambda o}$ = energy emitted by an object at a certain spectral interval; $W_{\lambda b}$ = energy emitted by a blackbody.

Generally speaking, there are three types of radiation source, distinguished by the ways in which the spectral emittance of each varies with wavelength: (*1*) a blackbody, for which $\epsilon_\lambda = \epsilon = 1$; (*2*) a graybody, for which $\epsilon_\lambda = \epsilon$ = constant less than 1; and (*3*) a selective radiator, for which ϵ_λ varies with wavelength.

According to Kirchhoff's law, for any material the spectral emissivity and spectral absorptance of a body are equal to any specified temperature and wavelength, ie, $\epsilon_\lambda = \alpha_\lambda$. For an opaque material ($\alpha_\lambda + \rho_\lambda = 1$): $\epsilon_\lambda + \rho_\lambda = 1$.

For highly polished materials, ϵ_λ approaches zero, so that for a perfect reflecting material (a perfect mirror) $\rho_\lambda = 1$.

Taking into account ϵ for a graybody radiator, the Stefan-Boltzmann formula becomes

$$\mathrm{W} = \epsilon\sigma T^4 \ \mathrm{W/m^2}$$

This states that the total emissive power of a graybody is the same as a blackbody at the same temperature—reduced in proportion to the value of ϵ for the graybody.

Typical Values of Emissivity. The values for ϵ obtained by using a thermal measurement instrument are, in effect, the average of ϵ_λ occurring over the infrared-wavelength interval utilized by that instrument. If ϵ_λ varies with the wavelength, ϵ (the average value) depends on the object temperature.

Unoxidized metals represent an extreme case of almost perfect opacity and high spectral reflectivity, which does not vary greatly with wavelength. Consequently, the emissivity of metals is low—only increasing with temperature. For nonmetals, emissivity tends to be high, and decreases with temperature.

Typical emissivities for a variety of common materials are shown in Table 1. The values are meant to be used only as a guide, however, because they depend upon the spectral response of the instrument used to obtain them.

Ir Detectors

An ir detector is a converter that absorbs ir energy and converts it to a signal, usually an electrical voltage or current. There are two principal types: thermal detectors and photon detectors. Thermal detectors have been conceived on the notion of the temperature rise produced in an absorbing receiver, such as in the pneumatic (Golay) cell, the thermocouple, the bolometer, and the new pyroelectric (capacitor) detector. The most important thermal detector today is the thermistor bolometer, which utilizes the change in resistance of a semiconductor (qv) film when it is heated by the radiation.

Typical of thermal detectors is the "flat" spectral response. If they have been properly blackened, the output signal remains practically constant over a very wide range of wavelengths. The main drawback with most thermal detectors is the comparatively slow response to radiation variations, due to thermal processes involved. The pyroelectric thermal detector, however, has relatively fast response owing to its use of the ferroelectric effect of certain crystals (see Ferroelectrics).

Two types of photon detector are of particular interest today: the photoconductive and the photovoltaic detectors (see Photodetectors; Photovoltaic cells). In a photoconductive detector, the gap energy is determined by the nature of the material itself, and the effect of photon absorption is to free the electronics and thereby increase the detector's conductivity. In photovoltaic detectors, the gap energy is also determined by the material, but the radiation-generated charge carriers are swept away by the electric field in a *p-n* junction, thereby directly producing a voltage rather than a change in conductivity.

High speed scanning of necessity requires a detector with a very short response time. The advantage of photon detectors is that they are more sensitive, and have a much shorter response time than thermal detectors. The drawbacks are that they have also a limited spectral response, and they require cooling for optimum sensitivity—generally to the temperature of liquid nitrogen (77 K). Some of them even must be cooled to the temperature of liquid helium (4.2 K) (see Cryogenics).

A widely accepted figure of merit for expressing the sensitivity of ir detectors is a quantity called the detectivity. The symbol for detectivity is D^*. D^* is a normalized figure of merit that is particularly convenient for comparing the performance of detectors having different electric bandwidths. D^* assumes higher values as detector sensitivity improves.

In the case of an ideal thermal detector that has a perfectly flat spectral response curve, it is sufficient to state a single value of detectivity D^*. However, the situation is more complicated with photon detectors, whose spectral response is not flat, and typically drops off to zero at the long-wavelength cut-off point. The D^* is, then, wavelength dependent and consequently bears the usual subscripts, D^*_λ. $D^*_{\lambda max}$ is noted at the peak of the spectral response curve.

The ultimate limit on detectivity is set by the radiation noise signal which is generated in a detector, resulting from the statistical fluctuation of the radiation received and reemitted by the detector itself. The noise signal is characterized by its random fluctuations in amplitude, frequency, and phase. The result on a display due to noise is the familiar TV "snow" on the picture screen. A detector where this noise sets the limit of detectivity is said to have background-limited performance (BLIP).

The theoretical limits of detectivity are now being approached by the newest semiconductor materials used in photon detectors, with peak detectivities at wavelengths beyond 40 μm.

Measurement Instruments

Some confusion has existed in the past concerning the term infrared photography, as contrasted with thermography (see Color photography). The distinction is one of wavelength. Conventional infrared film emulsions are sensitive to wavelengths no longer than 1.2 μm. For this reason, the wavelength-span 0.75–1.2 μm is the photographic infrared spectrum.

Table 1. Emissivities (Total Normal) of Various Common Materials

	Temperature, °C	Emissivity, ϵ
Metals and their oxides		
aluminum		
polished sheet	100	0.05
anodized sheet, chromic acid process	100	0.55
brass		
highly polished	100	0.03
oxidized	100	0.61
copper		
polished	100	0.05
heavily oxidized	20	0.78
gold, highly polished	100	0.02
iron		
cast, polished	40	0.21
cast, oxidized	100	0.64
sheet, heavily rusted	20	0.69
magnesium, polished	20	0.07
nickel		
electroplated, polished	20	0.05
oxidized	200	0.37
silver, polished	100	0.03
stainless steel (type 18-8)		
buffed	20	0.16
oxidized	60	0.85
steel		
polished	100	0.07
oxidized at 800°C	200	0.79
tin, commercial tin-plated sheet iron	100	0.07
Other materials		
brick, common red	20	0.93
carbon		
candle soot	20	0.95
graphite, filed surface	20	0.98
concrete	20	0.92
glass, polished plate	20	0.94
lacquer		
white	100	0.92
matte black	100	0.97
oil, lubricating (thin film on nickel base)		
nickel base alone	20	0.05
film thickness 0.025 mm	20	0.27
0.051 mm	20	0.46
0.125 mm	20	0.72
thick coating	20	0.82
paint, oil, average of 16 colors	100	0.94
paper, white bond	20	0.93
plaster, rough coat	20	0.91
sand	20	0.90
skin, human	32	0.98
soil		
dry	20	0.92
saturated with water	20	0.95

Table 1 (*continued*)

	Temperature, °C	Emissivity, ϵ
water		
distilled	20	0.96
ice; smooth	−10	0.96
frost crystals	−10	0.98
snow	−10	0.85
wood, planed oak	20	0.90

Beyond the 2 μm wavelength lies the so-called thermal infrared. Thus, infrared photography exploits the differences in the absorptive and emissive properties of surfaces. It depends upon the reflection of very short infrared wavelengths generated by outside sources such as the sun, or lamps that are much hotter than the object.

One of the most common instruments used for measuring thermal energy is the point radiometer (Fig. 2). This relatively simple device most often uses a thermal detector, such as a thermopile, or a pyroelectric detector. Generally, there are two major types of point radiometers. One type is a hand-held device, which is battery-powered and normally in a pistol shape. The second type is intended as a fixed unit, which is often permanently mounted in some type of process or manufacturing facility.

The hand-held unit is, of course, self-contained and displays its measurements directly. Displays vary widely from meters to digital LEDs or LCDs (see Digital displays; Light-emitting diodes; Liquid crystals). Although older units require some type of calibration before each use, more modern devices are self-calibrating. Adjustments often include emissivity or ambient temperature.

The fixed mounted unit normally has only an adjustment for emissivity and no

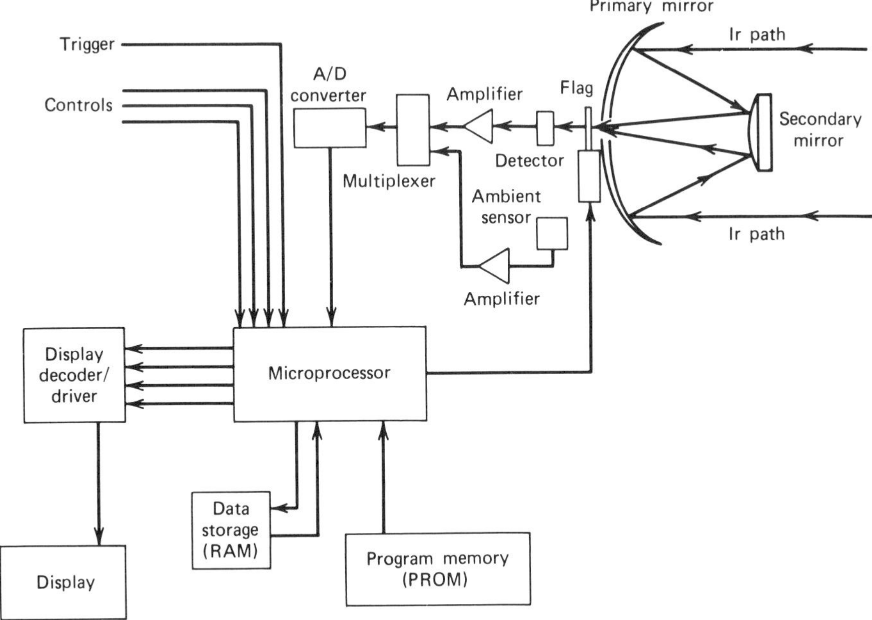

Figure 2. Schematic of a hand-held point radiometer.

display. Outputs from these devices are often fed to chart recorders, analogue-to-digital converters, or servo controllers.

Line-measurement systems are most often used in some type on-line process monitoring and control. Although these measurement systems vary widely, there are basically only two kinds of line-measurement systems. The first and most simple is one employing a radiometer which is mechanically traversed across the object. The resulting signal is displayed on a simple plotter processed in a computer, thus providing a means for a graphic presentation of the thermal distribution along the traversed line.

A somewhat more complex system (Fig. 3) involves scanning the object with a refractory or reflective electromechanical scanning system and displaying the resulting graph on an oscilloscope screen. This data may also be fed into a computer. Of the two types of line-scanning, traversing radiometer and electro-optical scanner, the first is by far most common. These systems are used in diverse manufacturing fields, but are very commonly found in monitoring paper machines, cement kilns, rubber processing and steel mills, and plastic-film extruders.

Although only recently introduced, nonmeasurement imaging systems have found many uses in the processing industry. These systems are most often very simple and less expensive than measuring systems and offer the user a clearly portable, rugged, and practical alternative for thermal imaging.

There are a number of types of simple imaging systems available, and they fall into two categories. The first, and still considered by many to be in a development stage, is based on the use of the pyroelectric vidicon tube (PEV). Similar to a television vidicon, the PEV is sensitive to infrared radiation in the 8–14 micrometer region. These systems offer high scanning rates with full TV compatibility. The disadvantage of these systems is that the detector must receive changing thermal information in order to

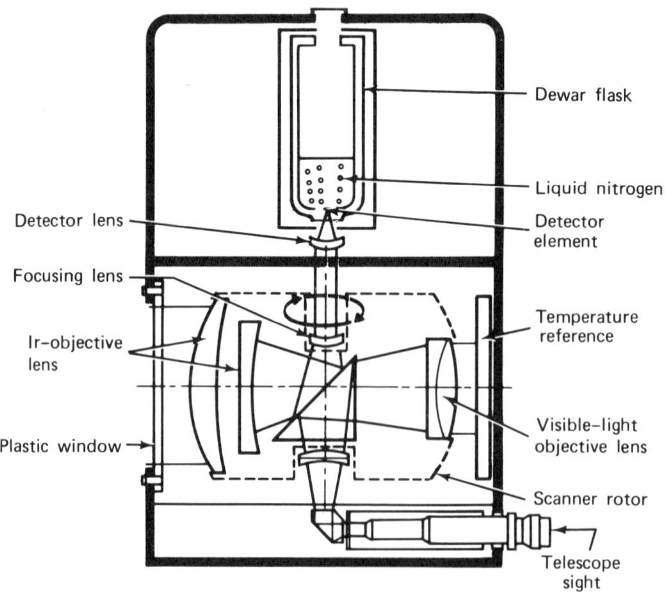

Figure 3. A line-scanning and measuring system.

generate an electrical signal. Thus, the unit must be panned continuously in order to maintain an image. An alternative means of operating such systems is to chop the field of view; however, this complicates the system and reduces its reliability, which already is normally limited to less than 2000 h.

The second type of system is based on an electro-optical-mechanical scanning system using multielement detectors. Such systems often offer a highly reliable package, often without the need for liquid nitrogen. The most modern of these devices utilize thermoelectric cooling which, when properly constructed, allows use of the system at any time and any place as long as the batteries are charged properly (Fig. 4).

Simple imaging systems are used most often in the trouble-shooting phase to identify hot spots or other overheating conditions where the major goal is to simply determine where the fault lies. Since they are simple, highly portable, and practical, they can be utilized by less qualified operators than required by the more complex measurement systems. When used in conjunction with a point radiometer, the combination of the two can result in a cost-effective means for also determining the temperature of the designated targets.

The combination of imaging and thermal measurement in the same unit has become increasingly important in petrochemical plants over the last few years. These systems not only offer the possibility of locating faults, but within the same system can identify the temperatures and the surrounding contours. Such systems employ electro-optical-mechanical scanners often using only a single-element detector. In order to gain the necessary fast response for imaging, the detectors are of the photon type, indium–antimonide or mercury–cadmium–telluride, and require liquid-nitrogen cooling (Fig. 5).

In spite of the intricate scanning technique, imaging systems with measurement

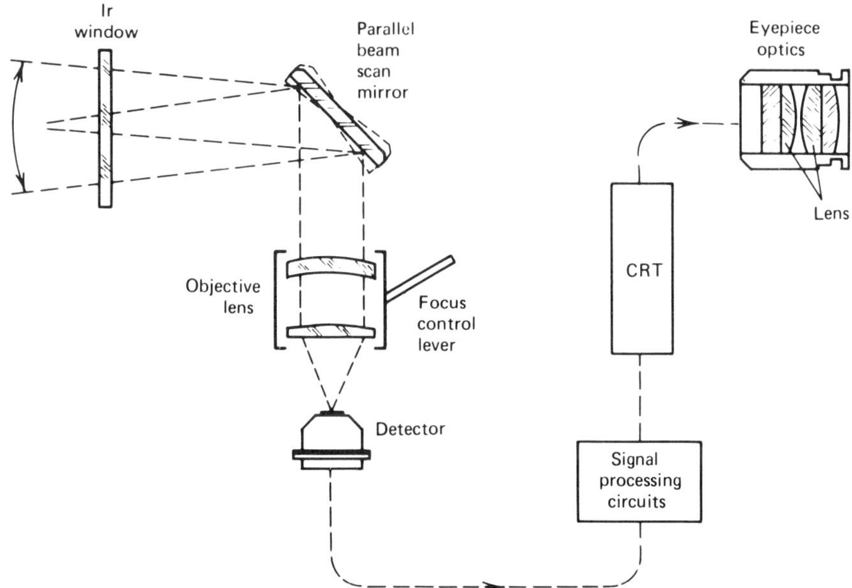

Figure 4. A nonmeasuring imaging system.

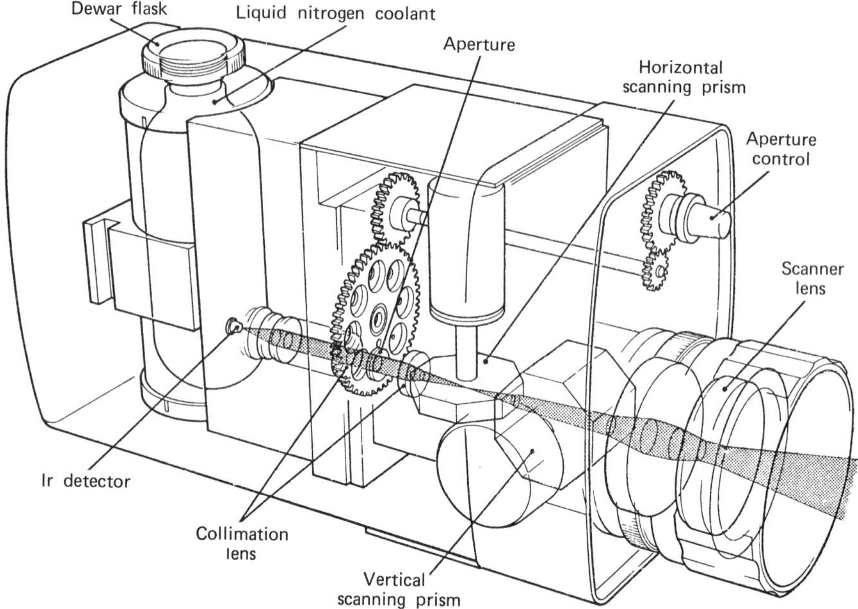

Figure 5. A thermal imaging and measurement system.

have proven to be very reliable and practical for field use. Measurements can be made on both a simple and a sophisticated basis, depending on the measurement requirement. These systems normally present the image in black and white, although color quantification is also available. Such quantification, however, reduces the portability of the system.

Specialized systems have been developed for the petrochemical industry. These systems often are modified so as to accommodate the high temperatures and radiative environments of furnaces for internal tube-temperature measurement. However, because of the varying environments based on different types of fuels used, the performance of infrared scanners for measurement purposes within a furnace vary greatly.

Shown in Table 2 are producers of various types of infrared measurement instruments.

Uses

Thermal Measurement, Principles of Application. The practical use of ir technology requires a broad spectrum of theoretical and applied aspects of infrared physics, optics, electronics, and signal processing. There are also a number of practical constraints affecting portability, power, environment, and cost effectiveness. Many of these factors have already been sorted out by manufacturers of thermal-sensing instruments, resulting in the many types of units described above.

For each thermal measurement application there is an optimum thermal measurement instrument. If the target is well known and the defect has already been identified, then the most useful instrument is a point radiometer. If the position of

Table 2. Various Types of Infrared Measurement Instruments

Point measurement instruments
AGA Infrared Systems AB, Danderyd, Sweden
Barnes Engineering Company, Stamford, Conn.
Raytek, Inc., Mountain View, Calif.
Simple imaging instruments
AGA Infrared Systems AB
Hughes Aircraft Company, Carlsbad, Calif.
M.E.L., Crawley, West Sussex, England
Line-scanning and measuring instruments
AGA Infrared Systems AB
Barnes Engineering Company, Stamford, Conn.
Daedalus Enterprises, Inc., Ann Arbor, Mich.
Thermal imaging and measuring instruments
AGA Infrared Systems AB
Jeol Ltd., Akishima, Tokyo, Japan
Inframetrics, Inc., Bedford, Mass.

the defect must be determined, then a thermal imaging system is required. If the defect must be clearly identified thermally and in absolute terms, an imaging system with measurement must be used.

Thermal measurement instrumentation measures only the surface radiation of an object. Thus, it is important to examine not only the surface condition, but also the ambient environment which surrounds the object. Surfaces that are heavily oxidized can radiate in an irregular fashion so as to obscure the absolute temperature-measurement accuracy. Other surfaces that have low emissivities reflect the ambient environment, again leading to false measurements. To determine the possibility of misleading information or the extent of error, the user should apply a piece of masking or electrical tape to the surface or spray the surface with black paint and test the difference between the sample area and the surrounding surface. Appreciable differences in temperature would indicate that an emissivity adjustment must be utilized when making the measurement. Emissivity values, based not only on the actual temperature being measured but the type of instrument being used, must be determined.

Although thermal measurements often are used to determine the surface temperature, in many applications it is the condition below the surface that is important. Thus, principles of thermal conductivity play an important role in the interpretation of remotely measured thermal information. For example, measuring the refractive condition of a furnace from the outside of the furnace is based on the notion that the surface has an essentially uniform core which radiates through a uniform refractory insulation, thus causing a uniform thermal pattern on the exterior surface (see Furnaces). Thus, irregular thermal patterns on the surface are caused by the nonuniform conductance of heat energy from the internal core to external surface. Such indications would indicate some defect in the refractory which has caused the irregular thermal pattern. Correct interpretation of thermal information requires structural and thermodynamic familiarity with the object being measured. Other situations are less complex, eg, electrical-component inspections in which the defect becomes evident as a function of the change in resistance as the component deteriorates. The change in resistance causes changes in the current, thus heating the component and allowing detection by thermal measurement.

In short, all uses of remote thermal measurement instrumentation are based on the notion that the defect is evident at the surface of the object. Any defect that would not result in a change in surface temperature could not be detected by remote thermal measurement.

Within the refinery and petrochemical industries there are five major application areas (see Petroleum refinery processes). In nearly all the application areas, thermal measurement instrumentation described above can be used. These application areas are electrical, refractory, product flow, furnace tube, and energy conservation. Of the five major areas, electrical and refractory are the most established. However, as energy costs increase and availability has become more unreliable, the uses of thermal measurement for energy conservation have expanded rapidly. Many applications which were previously justified on the basis of maintenance alone now have higher priority because of the need to identify major areas of energy loss. Applications such as product flow and furnace tube are somewhat less developed and often require more detailed component knowledge than the first two areas.

Electrical Power-Supply Systems. The use of thermal measurement instruments for electrical inspection is well established as a relatively simple, fast, and direct application.

As electrical components corrode or deteriorate (age), the electrical properties also change, mainly in electrical resistance. These changes usually are evident as increased surface temperatures. The component's temperatures, differentiated with respect to ambient temperature, can indicate a partial failure. Most components deteriorate gradually before complete failure. Consequently, electrical inspection can be performed at infrequent intervals, allowing for planned and regular schedules.

For most plants, electrical inspection is necessary only twice a year. An entire plant can be surveyed in one to two days, and sometimes in less than a day. Whenever possible, the inspection should be performed during the peak loading period. For example, if the plant uses extensive air conditioning (qv), it is advisable to perform electrical inspection in the summer. If the electrical consumption is relatively balanced throughout the year, the best time to perform inspections is when the outside climate is comfortably conducive to walking inspections.

With experience, it is advisable to develop a set of action priorities to correlate with various temperatures. Common decision rules used by electric-power companies in different countries are: (*1*) at 10°C and below (overheated): inspect and/or repair on routine basis; (*2*) at 10–35°C (overheated): inspect and/or repair as soon as practical; and (*3*) at 35°C and above (severely overheated): inspect and/or repair as soon as possible.

Besides preventing emergency shutdowns, electrical inspections allow for better inventory control. Ordinarily, the ordering of electrical replacement parts is based upon theoretical specifications obtained from the component manufacturers. With regular thermal inspection, electrical-parts inventory can be made to correspond more closely to the actual occurrence of defective components. Only those components known to be faulty are replaced, and costs are reduced in terms of time, labor, and materials.

Refractory and Insulation Conditions. Inspection of refractory and insulation conditions is based on the theory that a uniform internal temperature exists within a vessel and that the resulting exterior-surface temperature is a direct function of the heat conduction through the insulating medium and external wall (see Refractories; Insulation, thermal).

An ideal vessel would have a perfectly uniform temperature on its external surface. If a crack, or other defective conditions, existed in the insulating medium, the temperature of the exterior surface would increase proportionally to, and in the exact location of, the defect. The defect obviously would be based on the nonuniform conductance of heat from the interior to the exterior surface.

In normal practice, there are other structural variations that cause nonuniform heat patterns in addition to the defects, eg, grid work, refractory-type variations, port holes, cat walks, etc. However, even with these variations, refractory inspections are neither complex nor time-consuming.

Inspections for refractory and insulation defects should be performed before and after scheduled shutdowns and, of course, after all unscheduled shutdowns. Where ir-imaging inspection is substituted for temperature-sensitive paint (see Chromogenic materials, thermochromic and electrochromic), inspections should be carried out approximately every two months, depending on national regulations. The inspections are fast and efficient. Normally, a vessel can be scanned and measured in less than half an hour. Point measurements, taking only a few minutes, can be used for follow-up.

Product Flow. In most product-flow inspections, thermal energy is used as an indicator of thermally related defective conditions, eg, leakages, blockages, thinning conditions, corrosion and erosion. In each case, however, an analysis must be made to determine the correlation between the thermal information and the defect.

Since the diversity in product-flow applications is so extensive, a number of examples are given to illustrate the basic techniques.

The detection of leaking relief valves is based on the theory that when a product reaches the seat of a valve that is not properly sealed, the leaking product will heat or cool the outlet tube. To detect this phenomenon, the product should be at ambient temperature ±15°C. However, minute leakage is possible even at ambient temperature.

Outlet temperature can be correlated theoretically with actual leakage rate, but this is difficult because of the many variables involved, eg, wind conditions, mechanical heat conduction, and ambient temperature. Consequently, the application is performed most profitably when there is a bank of suspected relief valves and it is of interest to know which ones might be leaking.

Heat-exchanger inspections usually are performed to determine overall flow blockages or to indicate areas of extreme thinning conditions. The presence of overall blockages is normally due to corrosion in the tubes or deposits around them. Abnormal conditions are indicated by nonuniform heat flow across the heat exchanger (see Heat-exchange technology).

When internal-thinning conditions are suspected, infrared inspection may be used to simplify skin-thickness determinations. Once the thinnest area is identified, as a function of the highest temperature, the skin thickness is determined with the use of an ultrasonic device (see Ultrasonics).

In flue-gas ducting, or gas mains, deposits of particulates may accumulate in certain regions or in the complete main. The deposits act as a thermal insulation and are evident as cooler areas when inspecting from below, and as cooler levels from the side of the ducting. A sharp temperature gradient of the boundary indicates a horizontal layer of deposits, and a gradual temperature gradient indicates a concave form of the dust.

Determining tank liquid or sludge levels is a simple matter of noting the sharp difference in temperature at the liquid level (see Figs. 6**a**–**b**). The level is more distinct when the tank is heated by the sun for products at or cooler than ambient temperature than for products warmer than ambient temperature. If the tank is insulated and the product temperature is close to ambient (within 10°C), it may not be possible to determine the level.

Furnace Tubes. Internal furnace inspections may be performed to identify and measure tube hot spots, caused by local internal coking or flame licking, or to check the steam-decoking process. The inspection method should allow a rapid survey to identify thermal anomalies in the furnace. However, there are a number of variables that must be taken into account to achieve suitable measurement accuracy. The emissivity is the first important variable. Different types of tubes normally have somewhat different emissivities, although 0.91 is the most common value. Table 3 shows measured values of several common tubes.

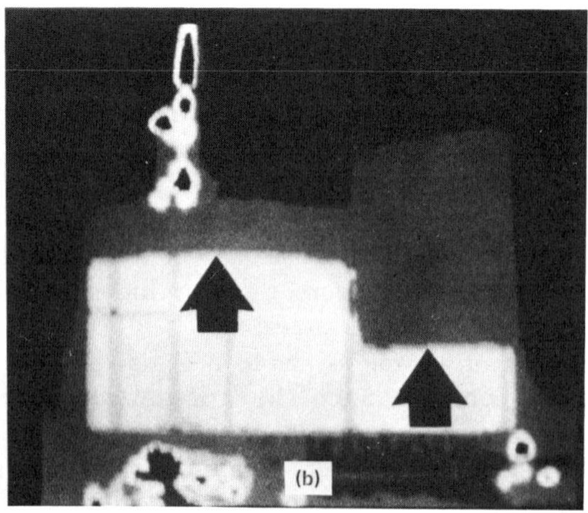

Figure 6. Infrared detection of tank liquid or sludge levels.

Table 3. Emissivities of Several Furnace Tubes

Composition	Emissivity	Temperature range, °C	ASTM Specification	
			Tube	Pipe
5% chromium			A 200	A 335
½% molybdenum	0.91	450–600	grade T.5	grade P.5
9% chromium			A 200	A 335
1% molybdenum	0.91	500–750	grade T.9	grade P.9
austenitic			A 312	A 312
Ti stabilized	0.91	500–620		
			grade TP.321	grade TP.321
25% chromium				
20% nickel	0.90	900–1030		grade
(centrifugally cast)				HK.40

Radiometric instruments measure the temperature of the surface of the tube. Scaling of the tubes, which may occur in some oil-fired furnaces, influences the measurement. Experiments are often needed to determine to what extent scaling may obscure a "true" hot spot due to internal coking. Theoretical estimation can be made of the effect of scale from the thermal conductivities of coke, tube, and scale. For example, if the particular scale has a thermal conductivity of the same order as coke, a 1-mm layer of scale probably will not obscure the effect of a 20-mm thick layer of coke.

Thermal measurements can be made in both gas- and oil-fired furnaces. In either case, special filters must be used to avoid interference from the atmosphere, unless the system is operating in the 8–14-micrometer band. However, special filtering is designed to look through gases and not particulates. In a burning fuel, heavy particulates act as radiators interfering with the temperature measurement, ie, they cause large fluctuations in the readings. Heavy particulates appear to be especially prevalent in furnaces burning coal or certain oils or when the burners are badly adjusted.

Energy Conservation. Thermal measurements have been used for routine plant maintenance to prevent power failures, locate defective refractory material, and otherwise optimize process efficiency. Prevention of unexpected shut-downs is a major goal. These same inspection tasks, justified on the basis of production benefits, also offer energy savings. Many other problems often considered maintenance problems are, in fact, energy problems. Whether these problems are attended to on the basis of energy or production efficiency is a matter of management priority. For example, most plants are highly motivated to keep production running, at whatever cost. With this attitude, use of thermography for energy conservation has a lower priority than its use for preventive maintenance. However, primary use of thermography for energy conservation will in most cases offer similar benefits. For example, an overheating electrical component located and replaced today could easily prevent a shutdown tomorrow, but it also can be said that the overheating component is wasting energy in addition to threatening a production shutdown.

The energy lost through radiation is proportional to the difference between the absolute object temperature to the fourth power and the absolute ambient temperature to the fourth power, whereas the energy lost through convection is proportional to the difference between the two temperatures. This means that as the object temperature rises, radiation dominates convection. The point at which the two terms are equal

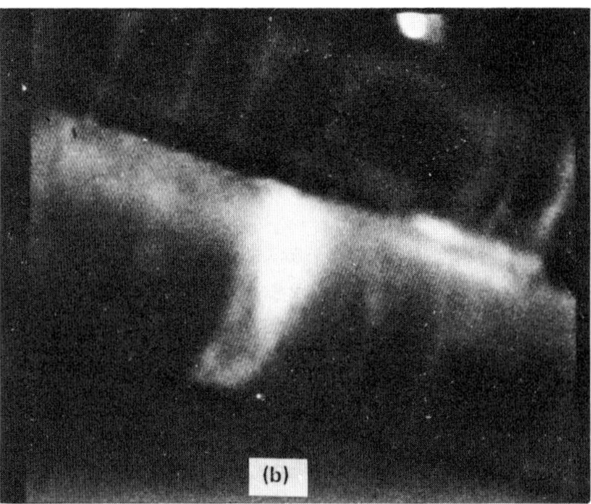

Figure 7. Infrared detection of defective insulation.

depends on the geometry of the object and the wind speed, but in general, radiation is much larger than convection for object temperatures over 100°C, if the ambient temperature is 20°C.

The following formula may be used to estimate the size of the two components, radiation r and convection cv:

$$Q_r = 5.67 \cdot \epsilon_o [(T_o/100)^4 - (T_o/100)^4] \text{ in W/m}^2$$

$$Q_{cv} = 0.533 \cdot (t_o - t_a)^{\cdot 5/4} \cdot \sqrt{\frac{V + 0.35}{0.35}} \text{ in W/m}^2$$

where T_o = object temperature, K; T_a = ambient temperature, K; ϵ_o = object emissivity; t_o = object temperature, °C; t_a = ambient temperature, °C; and V = wind velocity, m/s; or

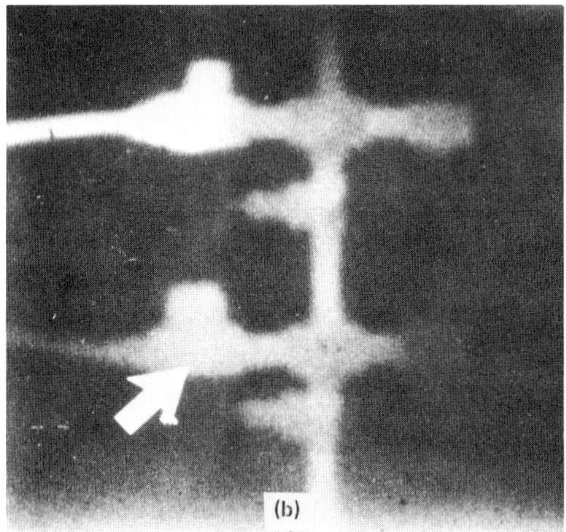

Figure 8. Infrared detection of leaking steam traps.

$$Q_r = 0.174 \cdot \epsilon_o [(T_o/100)^4 - (T_o/100)^4] \text{ in Btu/(ft}^2\cdot\text{h)}$$

$$Q_{cv} = 0.296 \cdot (t_o - t_a) \cdot {}^{5}\!/_{4} \cdot \sqrt{\frac{V + 68.9}{68.9}} \text{ in Btu/(ft}^2\cdot\text{h)}$$

where T_o = object temperature, °R; T_a = ambient temperature, °R; ϵ_o = object emissivity; t_o = object temperature, °F; t_a = ambient temperature, °F; and V = wind velocity, ft/min.

Thermal measurement is used as a means for both locating and quantifying energy losses. Once such information is obtained, management can use those data to set energy-saving priorities. The priorities determined are usually based on the cost-benefit parameters associated with each energy-conservation measure taken.

Many structures have been designed to optimize their primary function rather than to optimize their energy efficiency. Consequently, many structures exist today that are functioning properly (for their intended purpose), but are extremely inefficient with respect to energy consumption. Thermal inspection is an ideal technique for analyzing energy losses in existing structures. Not only can energy-wasting design features be identified, but once corrective action is taken, thermography can be used to determine how effective the corrective action has been (see Energy management).

Although many structures are designed properly, poor workmanship may still make them energy inefficient. For example, poor workmanship in installing insulation can completely negate the expected benefits. Unfortunately, workmanship has become a critical issue today and the use of thermography as a means of quality control for the installation of retrofit, insulation as well as in new construction, has become increasingly important. For example, it is increasingly common to use injected foam for retrofit insulation. With this technique, the installer cannot always be certain of the quality of work. Thermography provides not only the insulator, but also the owner, with concrete evidence of an insulation job properly and effectively completed (Fig. 7).

When structures have been designed and built in a correct manner, thermography is used, on a routine basis, to identify components that fail and waste energy because of aging and other environmental conditions, usually at a frequency related to their known failure potential (Fig. 8).

The list below summarizes the components and structures commonly inspected using thermal measurement and imaging instruments: *Design:* exhaust stacks, flue pipes; heating units (ovens, boilers, furnaces); buildings (offices, schools, administration); heating systems (ducts, vents, controls); process pipes, vessels, lines; steam and water lines; kilns; and cryogenic storage vessels; *Workmanship:* operational procedures; installation of refractory insulation; installation of foam insulation; installation of fiberglass insulation; replacement of parts, components; and maintenance procedures; and *Component failure:* steam traps; electrical substations; electrical power lines; electrical components; insulation, foam; insulation, refractory; seals, low and high temperature; doors, ports, windows; cooling towers, heat exchangers; plumbing lines; and motors, pumps, ventilators, bearings.

BIBLIOGRAPHY

General References

L. W. Bowden, *Manual of Remote Sensing,* American Society of Photogrammetry, Falls Church, Va., 1975.

R. D. Hudson, Jr. and J. W. Hudson, *Infrared Detectors,* Halsted Press, New York, 1975.

R. D. Hudson, Jr., *Infrared System Engineering,* Wiley-Interscience, New York, 1969.

F. A. Jenkins and E. H. White, *Fundamentals of Optics,* McGraw-Hill Book Co., Inc., New York, 1975.

P. W. Kruse, L. D. McGlaughlin, and R. B. McQuistan, *Elements of Infrared Technology Generation, Transmission and Detection,* John Wiley & Sons, Inc., New York, 1962.

M. J. Lloyd, *Thermal Imaging Systems,* Plenum Press, Inc., New York, 1975.

K. Seyrafi, *Electro-Optical Systems Analysis,* Electro-Optical Research Company, Los Angeles, Calif., 1973.

I. Simon, *Infrared Radiation,* The Commission on College Physics, D. Van Nostrand Co., Inc., N.J., 1966.

T. A. Smith, F. E. Jones, and R. P. Chasmar, *The Detection and Measurement of Infrared Radiation,* Oxford University Press, London, 1968.
R. Vanzetti, *Practical Applications of Infrared Techniques,* Wiley-Interscience, New York, 1972.
A. Vasko, *Infrared Radiation,* SNTL, Prague, 1968.
W. L. Wolfe and G. Zissis, *The Infrared Handbook,* Office of Naval Research, Washington, D.C., and Environmental Research Institute of Michigan, Ann Arbor, Mich.

CLIFTON WARREN
AGA Infrared Systems AB

INITIATORS

Initiators are chemical substances or energy sources that are used to initiate a chemical reaction and that are consumed during the initiation process. This is in contrast to catalysts, which are used in small quantities to either start or speed up a reaction but which are not consumed in the process, thereby allowing the catalyst to be re-used (see Catalysis). A catalyst is an acceptable and sometimes preferred description of certain ionic initiators, especially the cationic types, since initiation mechanisms are not always completely understood and the role of the counterion (which is not consumed in the initiation step) may be considered catalytic in nature.

The amount of chemical initiator required for reaction depends upon the type of chemical reaction and can vary from stoichiometric or greater quantities in certain nonchain-propagating reactions to very small quantities in chain-propagating reactions (eg, polymerizations) (see Polymerization mechanisms and processes). A chain-propagating reaction generally involves three steps: initiation, propagation, and termination. For example, in a vinyl monomer polymerization, these steps can be represented by the following equations:

$$\textit{Initiation: } I^* + M \rightarrow I - M^*$$

$$\textit{Propagation: } I - M^* + n\ M \rightarrow I - M_n M^*$$

$$\textit{Termination: } I - M_n M^* \rightarrow \text{dead polymer (eg, } I - M_{(n+1)}X)$$

where I* is the active-initiator species, M is the vinyl monomer, M* is the propagating active monomer species, and X is some terminating function which results from the reaction of M* with a terminating species.

The active-initiator species (I*) can be a free radical (a reactive intermediate carrying an unbonded or unpaired—lone or free—electron), cationic (positively charged species), anionic (negatively charged species), ion–radical (positively or negatively charged species carrying a free electron), a stable neutral molecule, or an energy source, eg, ultraviolet, gamma, and beta radiation.

Common vinyl monomers and the type of polymerizations they undergo are listed in Table 1. Some vinyl monomers (eg, styrene) can be initiated by free-radical, anionic, and cationic initiators. Others (eg, vinyl chloride) can be initiated only by free radicals

Table 1. Polymerization Methods of Some Common Vinyl Monomers

Monomer	Polymerization method		
	Free radical	Anionic	Cationic
ethylene	yes	yes	yes
butadiene	yes	yes	yes
styrene	yes	yes	yes
vinyl chloride	yes	no	no
vinyl acetate	yes	no	no
acrylic and methacrylic esters	yes	yes	no
acrylonitrile	yes	yes	no
isobutylene	no	no	yes
alkyl vinyl ethers	no	no	yes
N-vinylcarbazole	yes	no	yes
N-vinyl-2-pyrrolidinone	yes	no	yes

and still other monomers (eg, isobutylene) only undergo cationic polymerization (see Vinyl polymers).

Although initiators are used widely in the polymer industry, they also are employed for initiating a variety of other chemical reactions, eg, oxidation and autoxidation, chlorination, bromination, and nonpolymeric addition to double bonds. The same type of initiators generally are used for initiating polymerizations and nonpolymer chemical reactions.

Free-Radical Initiators

Since active-initiator free-radical species ($I^* = I\cdot$) are reactive intermediates having short lifetimes [half-lives ($t_{1/2}$) of less than 10^{-3} s] (1), they are generated during the initiation process from a precursor. (Free-radical initiator is used synonymously with the precursor or free-radical generator rather than with the free-radical species.) There are three general ways in which radicals can be produced: thermal bond homolysis, one-electron redox reactions, and irradiation processes. Although a variety of methods for generating free radicals (by one or more of the three methods above) are reported in the literature, commercial free-radical generators (or initiators) are, primarily, organic peroxides and aliphatic azo compounds, and some commercial processes are initiated directly by beta radiation or by uv radiation in the presence of sensitizers or uv-sensitive initiators. Some common free-radical-initiated processes include: vinyl monomer polymerizations and copolymerizations; cross-linking of thermoplastics (eg, polyethylene) and elastomers (see Elastomers, synthetic; Radiation curing) (eg, ethylene–propylene co- and terpolymers); curing resins (eg, unsaturated polyester–styrene blends); curing rubber (see Rubber, natural); grafting vinyl monomers onto polymer backbones; autoxidation of hydrocarbons (see Hydrocarbon oxidation); anti-Markovnikov additions to terminal olefins (qv) (eg, formation of primary mercaptans); halogenations; and telomerization. (Free-radical reactions are discussed in refs. 2–11.)

A typical chain-propagating, free-radical reaction is illustrated below for the anti-Markovnikov addition of hydrogen sulfide to a terminal olefin:

$$\textit{Initiation:}\ I\cdot + H_2S \rightarrow I{-}H + HS\cdot$$

$$\textit{Propagation:}\ HS\cdot + RCH{=}CH_2 \rightarrow R\dot{C}HCH_2SH$$

$$R\dot{C}HCH_2SH + H_2S \rightarrow RCH_2CH_2SH + HS\cdot$$

Peroxides. Organic peroxides have the general structure, ROOR′ or ROOH, and decompose thermally by the initial cleavage of the oxygen—oxygen bond to produce two free radicals:

$$ROOR' \xrightarrow{\Delta} RO\cdot + \cdot OR'$$

Certain organic peroxides also can be decomposed by specific promoters or activators. Such decompositions occur well below the peroxides' normal thermal decomposition temperatures and usually generate one free radical instead of two in a redox reaction:

$$ROOH + Fe^{2+} \rightarrow RO\cdot + OH^- + Fe^{3+}$$

There are more than 40 different organic peroxides (12–17) in more than 80 formulations that are produced as free-radical initiators primarily for the polymer and resin industries (see Peroxides and peroxy compounds, organic). Nine organic peroxides, produced commercially as free-radical initiators, are shown in Table 2 with their nonpromoted temperature–activity ranges which are expressed in 10-h half-life temperatures, ie, the temperatures at which the peroxides are 50% decomposed in 10 h. Peroxide half-lives usually are determined in solvents and the $t_{1/2}$ data are useful for comparing the activity of one peroxide against another provided that the half-lives were determined in the same solvent and at the same concentration and, preferably, that the peroxides are the same type. Certain organic peroxides are susceptible to radical-induced decomposition:

$$R'\cdot + ROOR \rightarrow ROR' + RO\cdot$$

Radical-induced decompositions of peroxides result in inefficiency in radical production since the peroxide decomposes without adding more radicals to the system. Such decompositions generally do not occur in vinyl monomer polymerizations because the monomer quickly scavenges the initially generated radicals. In nonscavenging environments (eg, nonolefinic solvents), induced decomposition occurs with those peroxides that are susceptible and becomes more pronounced as the concentration is increased. The homolysis of organic peroxides is a first-order reaction, whereas the radical-induced decomposition is a second-order reaction. Therefore, decomposition rates are significantly faster than the true first-order rates in those peroxide systems where induced decomposition also is occurring. Most peroxides decompose faster in more polar or polarizable solvents. This is true even if the peroxide is not generally susceptible to radical-induced decomposition. An example of this is illustrated by *tert*-butyl peroxypivalate [*927-07-1*] wherein the 10-h half-life temperature (0.2 *M*) varies from 62°C in decane (nonpolar), to 55°C in benzene (polarizable), to 53°C in methanol (polar). Thus, care must be exercised in applying half-life data to free radical-initiated systems, eg, vinyl monomer polymerizations. The major criterion used for selection of a peroxide initiator is the initiator's temperature–activity range (18).

From Table 2, it is seen that certain organic peroxides, eg, the diacyl peroxides and peroxyesters, show a wide variation in temperature activity, depending upon their structure; whereas other types, such as the peroxydicarbonates and monoperoxycar-

Table 2. Commercial Peroxide Classification

Peroxide type	Structure	10-h $t_{1/2}$, °C[a]
diacyl peroxides	$RC(=O)OOC(=O)R$	20–75
acetyl alkylsulfonyl peroxides	$RS(\rightarrow O)_2OOC(=O)CH_3$	32–42
dialkyl peroxydicarbonates	$ROC(=O)OOC(=O)OR$	49–51
tert-alkyl peroxyesters	$R'C(=O)OOR$	49–107
OO-tert-alkyl *O*-alkyl monoperoxycarbonates	$ROOC(=O)OR'$	90–100
di(*tert*-alkylperoxy)ketals	$(ROO)_2C(R')(R'')$	92–115
di-*tert*-alkyl peroxides	ROOR′	117–133
tert-alkyl hydroperoxides	ROOH	133–172
ketone peroxides	$(R)(R')C(OOH)_2$ + $HOO\text{–}C(R)(R')\text{–}OO\text{–}C(R)(R')\text{–}OOH$ + other structures	

[a] $t_{1/2}$ = 10 h.

bonates, show very little activity variation with structure changes. The diperoxyketals and dialkyl peroxides show a moderate variation in activity with differing structures. The acetyl alkylsulfonyl peroxides (eg, acetyl cyclohexanesulfonyl peroxide [*3179-56-4*]) are low temperature initiators that are used exclusively for initiating certain vinyl chloride polymerizations (19–20). With the hydroperoxides and ketone peroxides, the half-life generally is not important since these two classes of peroxides usually are employed with activators (or promoters) in redox systems in which first-order decomposition rates have no significance. Table 2 shows that commercial organic peroxides exist with nonpromoted temperature activity varying from well below RT to well above 100°C.

Structural differences that affect nonpromoted activity of peroxides are related to three factors: the relative stability of the radicals formed, ie, the more stable the radical that is formed, the less stable the peroxide; steric factors, ie, highly strained peroxides are less stable since decomposition relieves steric strain; and electronic effects where, eg, electron-donating functions generally destabilize and electron-withdrawing functions stabilize peroxides. These factors manifest themselves in the E and A factors in the Arrhenius first-order equation:

$$k = A \times \exp(-E/RT)$$

Some commercial diacyl peroxides and their 10-h half-lives (determined in benzene at 0.2 M) are listed in Table 3 (21). Although these peroxides cleave at the oxygen—oxygen bond, other bond cleavages (decarboxylations) can and do occur either simultaneously with or sequentially to oxygen—oxygen bond breaking:

$$\mathrm{R\overset{O}{\overset{\|}{C}}OO\overset{O}{\overset{\|}{C}}R} \longrightarrow 2\,\mathrm{R\overset{O}{\overset{\|}{C}}O\cdot} \longrightarrow 2\,\mathrm{R\cdot} + \mathrm{CO_2}$$

$$\mathrm{R\overset{O}{\overset{\|}{C}}OO\overset{O}{\overset{\|}{C}}R} \longrightarrow 2\,\mathrm{R\cdot} + \mathrm{CO_2}$$

The rate of decarboxylation primarily depends upon temperature, pressure, and the stability of the R· formed. The more stable the radical, the faster the decarboxylation and, with many diacyl peroxides, it occurs simultaneously with oxygen—oxygen bond breaking. Acyloxy radicals are known to form initially only from diacetyl peroxide and the benzoyl peroxides (because of the high reactivity of the corresponding decarboxylated methyl and phenyl radicals). Diacyl peroxides that are derived from nonalpha-branched carboxylic acid also may form acyloxy radicals initially, but they decarboxylate extremely fast and evidence for their formation is not conclusive.

Some commercial peroxyesters and their 10-h $t_{1/2}$s determined in 0.2 M benzene solutions are listed in Table 4 (22). Peroxyesters cleave at the oxygen—oxygen bond to supposedly generate acyloxy and alkoxy radicals:

$$\mathrm{R\overset{O}{\overset{\|}{C}}OOR'} \longrightarrow \mathrm{R\overset{O}{\overset{\|}{C}}O\cdot} + \cdot\mathrm{OR'}$$

Table 3. Commercial Diacyl Peroxides[a]

Name	CAS Registry No.	Structure	10-h $t_{1/2}$, °C[b]
dibenzoyl peroxide (BPO)	[*94-36-0*]	$\mathrm{C_6H_5\overset{O}{\overset{\|}{C}}OO\overset{O}{\overset{\|}{C}}C_6H_5}$	73
di(2,4-dichlorobenzoyl) peroxide	[*133-14-2*]	$\mathrm{(2,4\text{-}Cl_2C_6H_3)\overset{O}{\overset{\|}{C}}OO\overset{O}{\overset{\|}{C}}(C_6H_3Cl_2\text{-}2,4)}$	54
diacetyl peroxide	[*110-22-5*]	$\mathrm{CH_3\overset{O}{\overset{\|}{C}}OO\overset{O}{\overset{\|}{C}}CH_3}$	69
dilauroyl peroxide	[*105-74-8*]	$\mathrm{CH_3(CH_2)_{10}\overset{O}{\overset{\|}{C}}OO\overset{O}{\overset{\|}{C}}(CH_2)_{10}CH_3}$	62
diisobutyryl peroxide	[*3437-84-1*]	$\mathrm{(CH_3)_2CH\overset{O}{\overset{\|}{C}}OO\overset{O}{\overset{\|}{C}}CH(CH_3)_2}$	21

[a] Ref. 21.
[b] In benzene (0.2 M).

Table 4. Commercial Peroxyesters[a]

Name	CAS Registry No.	Structure	10-h $t_{1/2}$, °C[b]
tert-butyl perbenzoate	[*614-45-9*]	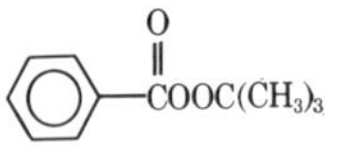 $C_6H_5COOC(CH_3)_3$ (C=O)	105
tert-butyl peracetate	[*107-71-1*]	 $CH_3COOC(CH_3)_3$ (C=O)	102
2,5-di(benzoylperoxy)-2,5-dimethylhexane	[*618-77-1*]	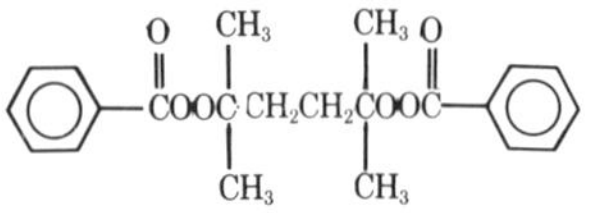 $C_6H_5COOC(CH_3)_2CH_2CH_2C(CH_3)_2OOCC_6H_5$ (C=O)	100
di-*tert*-butyl diperoxyazelate	[*16580-06-6*]	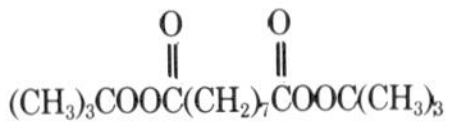 $(CH_3)_3COOC(CH_2)_7COOC(CH_3)_3$ (C=O)	99
tert-butyl peroxy-2-ethylhexanoate (*tert*-butyl peroctoate)	[*3006-82-4*]	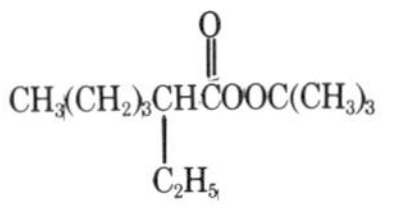 $CH_3(CH_2)_3CH(C_2H_5)COOC(CH_3)_3$ (C=O)	73
tert-amyl peroctoate	[*686-31-7*]	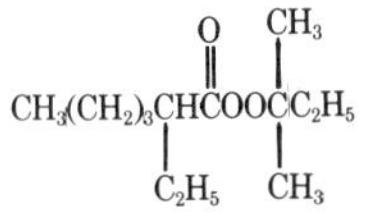 $CH_3(CH_2)_3CH(C_2H_5)COOC(CH_3)_2C_2H_5$ (C=O)	70
2,5-di(2-ethylhexanoylperoxy)-2,5-dimethylhexane	[*13052-09-0*]	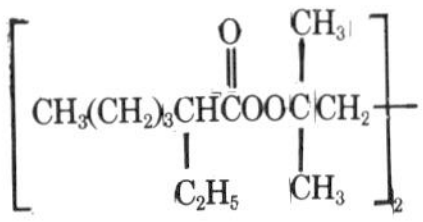 $[CH_3(CH_2)_3CH(C_2H_5)COOC(CH_3)_2CH_2-]_2$ (C=O)	67
tert-butyl peroxyneodecanoate	[*748-41-4*]	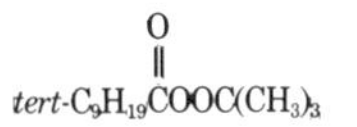 *tert*-$C_9H_{19}COOC(CH_3)_3$ (C=O)	47 49[c]

[a] Ref. 22.
[b] In benzene (0.2 *M*).
[c] In trichloroethylene (0.2 *M*).

The acyloxy radical can decarboxylate as noted above. Only tertiary alkyl peroxyesters are available commercially and, consequently, only tertiary alkoxy radicals are generated. Alkoxy radicals can undergo a beta-scission reaction:

$$R{-}\overset{\overset{R'}{|}}{\underset{\underset{R''}{|}}{C}}O\cdot \longrightarrow R'\overset{\overset{O}{\|}}{C}R'' + R\cdot$$

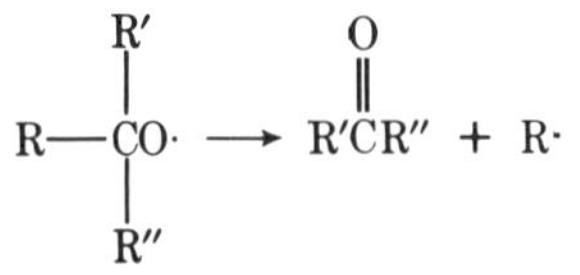

One of the R groups splits off to form a ketone and a new alkyl radical. The group that splits off is the one that forms the most stable radical.If the radical that splits off is more stable than the alkoxy radical, the beta-scission reaction will be fast and the predominant initiating species may be the alkyl rather than the alkoxy radical. The beta-scission reaction also is temperature-dependent, ie, more scission occurs if the alkoxy radical is generated at high temperatures.

The relative stability of radicals can be correlated with the hydrogen-bond dissociation energies of the parent compound (23):

Parent compound	*Bond dissociation energy, kJ/mol (kcal/mol)*
$(R)_3C—H$	ca 381 (ca 91)
$(R)_2CH—H$	ca 397 (ca 95)
$RCH_2—H$	ca 410 (ca 98)
$CH_3—H$	ca 435 (ca 104)
RO—H	ca 439 (ca 105)
$RC(=O)OO—H$	ca 469 (ca 112)
$C_6H_5—H$	ca 469 (ca 112)
HO—H	ca 498 (ca 119)

The higher the bond dissociation energy, the less stable (more reactive) the corresponding radical is that was formed by removing the hydrogen atom. Thus, tertiary alkyl radicals are more stable than secondary alkyl radicals which are more stable than primary alkyl radicals, and the hydroxyl radical is the most reactive. The methyl radical is more reactive than other primary alkyl radicals and is about as reactive as alkoxy radicals. This has important consequences in designing or choosing a peroxide initiator for hydrogen (H) abstraction reactions (eg, grafting, cross-linking) where more reactive radicals are required.

OO-tert-Butyl *O*-alkyl monoperoxycarbonates (eg, *OO-tert*-butyl *O*-isopropyl monoperoxycarbonate [*2372-21-6*]) are another class of peroxides that generate alkoxy radicals:

$$(CH_3)_3COOC(=O)OR \longrightarrow (CH_3)_3CO\cdot + \cdot OC(=O)OR$$

The other radical generated is an alkoxycarbonyloxy radical, and the nature of the R group has practically no effect on the activity. No matter what the R group is, the 10-h half-life (as 0.2 *M* benzene solutions) remains at 99–100°C for the *OO-tert*-butyl monoperoxycarbonates, which are the only monoperoxycarbonates of commercial significance (22). Lower temperature active monoperoxycarbonates can be obtained by employing other *OO-tert*-alkyl groups and by observing the decreasing order of stability: *tert*-butyl > *tert*-amyl > *tert*-octyl > *tert*-cumyl. This same order of activity is observed in all *tert*-alkylperoxy compounds, eg, *tert*-alkyl peroxyesters, di(*tert*-alkylperoxy)ketals, and di-*tert*-alkyl peroxides (ie, the *tert*-butylperoxy derivatives are the most stable).

Some commercially available di(*tert*-butylperoxy)ketals and their 10-h half-lives in benzene are listed in Table 5 (24). These diperoxy compounds decompose thermally to produce four radicals, two of which can be tertiary alkoxy:

$$[(CH_3)_3COO]_2C(R)(R') \xrightarrow{\Delta} 2\,(CH_3)_3CO\cdot \text{ plus other fragments, including } R\cdot + R'\cdot + CO_2$$

Some commercially available dialkyl peroxides and their 10-h $t_{1/2}$s in benzene are listed in Table 6 (25). Dialkyl peroxides initially cleave at the peroxy—oxygen bond and thereby supposedly generate two alkoxy radicals. The alkoxy radicals here are more prone to the beta-scission reaction because the dialkyl peroxides are among the most stable of the commercial organic peroxides and, therefore, the alkoxy radicals are generated at higher temperatures where beta scission is faster.

Table 5. Commercial Diperoxyketals[a]

Name	CAS Registry No.	Structure	10-h $t_{1/2}$, °C[b]
ethyl 3,3-di(*tert*-butylperoxy)butyrate	[*55794-20-2*]	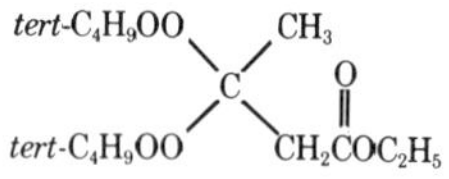 $(tert\text{-}C_4H_9OO)_2C(CH_3)CH_2COOC_2H_5$	111
2,2-di(*tert*-butylperoxy)butane	[*2167-23-9*]	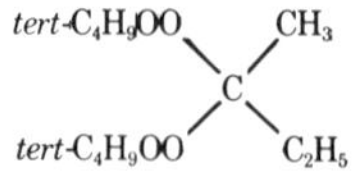 $(tert\text{-}C_4H_9OO)_2C(CH_3)C_2H_5$	104
1,1-di(*tert*-butylperoxy)cyclohexane	[*3006-86-8*]	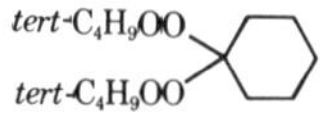 $(tert\text{-}C_4H_9OO)_2$-cyclohexane	95
1,1-di(*tert*-butylperoxy)-3,3,5-trimethylcyclohexane	[*6731-36-8*]	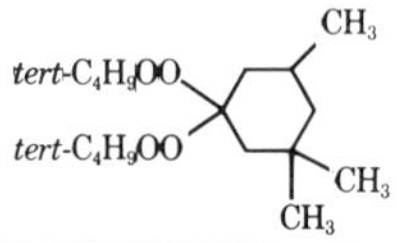 $(tert\text{-}C_4H_9OO)_2$-cyclohexane with CH_3, CH_3, CH_3	92

[a] Ref. 24.
[b] In benzene (0.2 *M*).

Table 6. Commercial Dialkyl Peroxides[a]

Name	CAS Registry No.	Structure	10-h $t_{1/2}$, °C[b]
2,5-di(*tert*-butylperoxy)-2,5-dimethylhex-3-yne	[*1068-27-5*]	$tert\text{-}C_4H_9OOC(CH_3)_2C{\equiv}CC(CH_3)_2OO\text{-}tert\text{-}C_4H_9$	128
di-*tert*-butyl peroxide	[*110-05-4*]	$tert\text{-}C_4H_9OOC_4H_9\text{-}t$	126
2,5-di(*tert*-butylperoxy)-2,5-dimethylhexane	[*78-63-7*]	$tert\text{-}C_4H_9OOC(CH_3)_2CH_2CH_2C(CH_3)_2OO\text{-}tert\text{-}C_4H_9$	119
dicumyl peroxide	[*80-43-3*]	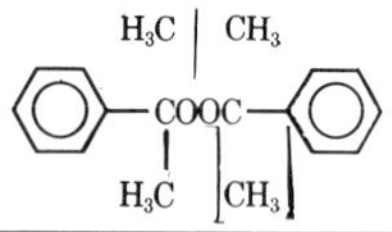 $C_6H_5C(CH_3)_2OOC(CH_3)_2C_6H_5$	115

[a] Ref. 25.
[b] In benzene (0.2 *M*).

Some commercially available dialkyl peroxydicarbonates and their 10-h half lives determined in 0.2 *M* trichloroethylene (TCE) solutions are listed in Table 7 (26). These peroxides are active at low temperatures and initially cleave to two alkoxycarbonyloxy radicals which subsequently may decarboxylate to alkoxy radicals:

$$\text{ROC(O)OOC(O)OR} \longrightarrow \text{ROC(O)O}\cdot + \cdot\text{OC(O)OR} \longrightarrow \text{RO}\cdot + CO_2$$

Dialkyl peroxydicarbonates are very susceptible to radical-induced decomposition:

$$R'O\cdot + RO\overset{O}{\overset{\|}{C}}O O\overset{O}{\overset{\|}{C}}{-}OR \longrightarrow R'OH + \;>C{=}O + 2\,CO_2 + RO\cdot$$

Decomposition rate studies on dialkyl peroxydicarbonates show dramatic solvent effects which primarily result from their susceptibility to induced decomposition and give a decreasing order of stability in solvents of: TCE > saturated hydrocarbons > benzene > alcohols. Decomposition rates are slowest in TCE most probably because TCE is a radical scavenger and, thereby, prevents radical-induced decomposition. Table 7 shows that the nature of the alkyl group has no effect on $t_{1/2}$ in dialkyl peroxydicarbonates. Despite the fact that the alkyl groups are primary, secondary, and cyclic, all of these peroxides have the same 10-h half-life in TCE.

Commercially available (27) *tert*-alkyl hydroperoxides, ROOH, include: *tert*-butyl hydroperoxide [*75-91-2*], *tert*-cumyl (or cumene) hydroperoxide [*80-15-9*], diisopropylbenzene monohydroperoxide [*98-49-7*], *p*-menthane hydroperoxide [*80-47-7*], and 2,5-dihydroperoxy-2,5-dimethylhexane [*3025-88-5*]. Hydroperoxides can decompose thermally to form (initially) an alkoxy and a hydroxy radical:

$$ROOH \rightarrow RO\cdot + \cdot OH$$

However, because of their high temperature, first-order decomposition rates (10-h $t_{1/2}$s 133–172°C) and their extreme sensitivity to radical-induced decompositions, they have only limited use as thermal initiators. The O—H bond in hydroperoxides is weak [368–377 kJ/mol (88.0–90.1 kcal/mol) bond dissociation energy] and is readily attacked by most radicals to initiate induced decomposition:

$$ROOH + R'\cdot \rightarrow R'{-}H + ROO\cdot$$

The fate of ROO· depends upon the environment but generally produces other radicals that can attack undecomposed ROOH and, thereby, carry on the induced decomposition chain. Radicals also can attack undecomposed peroxide by displacement on peroxy–oxygen:

$$ROOH + R'\cdot \rightarrow ROR' + \cdot OH \text{ or } R'OH + RO\cdot$$

Table 7. Commercial Peroxydicarbonates[a]

Name	CAS Registry No.	Structure	10-h $t_{1/2}$, °C[b]
di-*n*-propyl peroxydicarbonate	[*16066-38-9*]	$(n\text{-}C_3H_7OC(=O)O{-})_2$	50
diisopropyl peroxydicarbonate	[*105-64-6*]	$(i\text{-}C_3H_7OC(=O)O{-})_2$	50
dicyclohexyl peroxydicarbonate	[*1561-49-5*]	$(C_6H_{11}{-}OC(=O)O{-})_2$	50
dicetyl peroxydicarbonate	[*26322-14-5*]	$(CH_3(CH_2)_{15}OC(=O)O{-})_2$	50

[a] Ref. 26.

[b] In trichloroethylene (0.2 *M*).

This is the type of induced decomposition that occurs with other peroxide classes, eg, the dialkyl peroxydicarbonates and diacyl peroxides.

Hydroperoxides are more widely used as initiators in low temperature applications (≤RT) by using transition metal, M, salts as activators by electron-transfer (redox) mechanisms:

$$ROOH + M^{n+} \rightarrow RO\cdot + OH^- + M^{(n+1)+}$$

$$ROOH + M^{(n+1)+} \rightarrow ROO\cdot + H^+ + M^{n+}$$

Either oxidation state of a transition metal (Fe, Mn, V, Cu, Co, etc) can decompose the hydroperoxide. Thus, a small amount of transition metal ion can decompose a large amount of hydroperoxide; therefore, transition-metal promoters should never be premixed with the peroxide. Also, trace transition-metal impurities are to be avoided with such peroxides.

Transition metal ions also react with the generated radicals to convert them to ions:

$$RO\cdot + M^{n+} \rightarrow RO^- + M^{(n+1)+}$$

This equation is one example of several possible radical–transition metal ion interactions. The significance of these interactions is that the radical is destroyed and can no longer initiate radical reactions; consequently, the optimum use levels of transition metals are very low. When too much transition metal is used, the peroxide decomposes quickly but radical efficiency is impaired.

Ketone peroxides are mixtures of peroxides and are composed primarily of the two structures shown in Table 2 and are marketed as solutions in some inert solvent such as dimethyl phthalate. By far the most popular ketone peroxide [*1338-23-4*] formulation is the one derived from methyl ethyl ketone although smaller quantities of the peroxides of methyl isobutyl ketone [*28056-59-9*], cyclohexanone [*12262-58-7*], and 2,4-pentanedione [*37187-22-7*] are used commercially (28). Ketone peroxides are used primarily in room-temperature initiated curing of unsaturated polyester resins (containing styrene monomer) using transition-metal promoters (cobalt naphthenate being the most popular), since these peroxides contain the hydroperoxy (—OOH) grouping and, thus, behave much like the hydroperoxides (except that they are mixtures with somewhat lower thermal stability).

Other peroxides also are used with promoters to lower decomposition temperatures, usually at some sacrifice in radical efficiency. The most widely used system is the dibenzoyl peroxide (BPO)–dimethylaniline (DMA) combination for unsaturated polyester resin blends (eg, in putty for autobody repair kits). Here, the aromatic tertiary amine promoter attacks the BPO to form initially an ion pair of *N*-benzoyloxydimethylanilinium and benzoate ions which decomposes to a dimethylaniline radical cation and a benzoyloxy radical which, in turn, initiate the cure reaction (29):

$$BPO + DMA \longrightarrow \left[C_6H_5-\overset{\displaystyle CH_3}{\underset{\displaystyle CH_3}{\overset{|}{\underset{|}{N}}}}-O\overset{\displaystyle O}{\overset{\|}{C}}C_6H_5 \right]^+ \; {}^-O\overset{\displaystyle O}{\overset{\|}{C}}C_6H_5 \longrightarrow C_6H_5\overset{\displaystyle CH_3}{\underset{\displaystyle CH_3}{\overset{|}{\underset{|}{N}}}}{}^{\cdot +} + \cdot O\overset{\displaystyle O}{\overset{\|}{C}}C_6H_5 + C_6H_5\overset{\displaystyle O}{\overset{\|}{C}}O^-$$

Whereas the BPO–DMA system works well for curing unsaturated polyester blends, it is not a very efficient system for initiating vinyl monomer polymerizations and, therefore, generally is not used in such applications.

Other peroxide redox systems that have been used, primarily in initiating emulsion polymerizations of vinyl monomers, are BPO–ferrous ammonium sulfate; hydrogen peroxide–ferrous ammonium sulfate; hydrogen peroxide–dodecyl mercaptan; potassium peroxydisulfate–sodium bisulfite; and potassium peroxydisulfate–dodecyl mercaptan (30–31). Potassium peroxydisulfate, $K_2S_2O_8$ [*7727-21-1*] (or the corresponding sodium or ammonium salt), is an inorganic peroxide that is used widely in emulsion polymerizations (eg, of latexes, rubbers, etc) usually in combination with a reducing agent, as illustrated above. In the absence of reducing agents, it decomposes to give sulfate ion radicals:

$$S_2O_8^{2-} \rightarrow 2 \cdot SO_4^{-}$$

With transition metal activators, the initiation process is postulated as:

$$S_2O_8^{2-} + Fe^{2+} \rightarrow Fe^{3+} + SO_4^{2-} + \cdot SO_4^{-}$$

The reaction with mercaptans is believed to generate initiating sulfur radicals:

$$K_2S_2O_8 + 2\,RSH \rightarrow 2\,KHSO_4 + 2\,RS\cdot$$

Hydrogen peroxide, in combination with reducing agents (transition metals), also is used in those applications where its high water- and low oil-solubility is not a problem or has been overcome.

The selection and use of peroxide initiators has been reviewed in refs. 29 and 32 and a review of peroxide initiators for vinyl chloride polymerizations has been reviewed in ref. 19 (see Peroxides and peroxy compounds, inorganic).

When handling and using peroxide initiators, care should be exercised since they are thermally sensitive and will decompose (sometimes violently) when exposed to excessive temperatures when they are in their pure or highly concentrated states. However, they are useful as initiators because of their thermal instability. What may be a safe temperature for one peroxide can be an unsafe temperature for another, since peroxide initiators encompass a wide activity range. Some peroxides are shock or friction sensitive in the pure state, and such peroxides are desensitized by formulating them into solutions, pastes, or powders with inert diluents. Carefully scrutinize all manufacturer's literature prior to handling and using specific peroxide initiators and review peroxide safety literature generally (20–22,24–28,32–36).

Azo Compounds. The commercially available azo initiators are the symmetrical (**1**) and unsymmetrical (**2**) azonitriles:

$$\underset{\text{CN}}{\overset{\text{R}'}{\text{RC}}}\text{—N=N—}\underset{\text{CN}}{\overset{\text{R}'}{\text{CR}}} \qquad (\mathbf{1})$$

$$\underset{\text{CH}_3}{\overset{\text{CH}_3}{\text{CH}_3\text{C}}}\text{—N=N—}\underset{\text{CN}}{\overset{\text{R}}{\text{CR}'}} \qquad (\mathbf{2})$$

The symmetrical azonitriles (**1**) are solids with somewhat limited solubility (37–39) and the unsymmetrical *tert*-butylazonitriles (**2**) are liquids or low melting solids that are completely miscible in all common organic solvents, including petroleum hydrocarbons (40–44). Azonitriles are not susceptible to radical-induced decompositions (45) and their decomposition rates are little affected by the environment. In contrast to most peroxides, azonitrile decomposition rates show only minor solvent effects (37–39), and are not affected by transition metals, acids, bases, and other contaminants.

Thus, azonitrile decomposition rates are predictable. They are ideally suited as thermal initiators for curing resins that contain a variety of extraneous materials without affecting cure rates and displaying exceptional shelf (or pot) lives at ambient storage temperatures (45). Some commercial azonitriles and their 10-h half-lives are listed in Table 8. The symmetrical azonitriles are less stable than the corresponding unsymmetrical azonitriles, eg, 2,2′-azobis(2-methylpropionitrile) has a 10-h half-life at 64°C, whereas the analogous 2-(*tert*-butylazo)-2-methylpropionitrile has a 10-h half-life at 79°C. Azo initiators decompose thermally by cleavage of two carbon—nitrogen bonds to form two alkyl radicals and nitrogen:

$$RN{=}NR' \rightarrow R\cdot + N_2 + \cdot R'$$

The alkyl radicals generated from commercial azo initiators are all tertiary alkyl

Table 8. Commercial Azo Initiators[a]

Name	CAS Registry No.	Structure	10-h $t_{1/2}$, °C
2,2′-azobis[2,4-dimethyl]pentanenitrile	*[4419-11-8]*	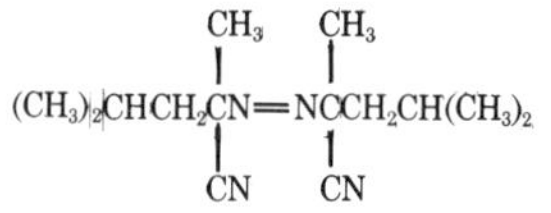 $(CH_3)_2CHCH_2C(CH_3)(CN)N{=}NC(CH_3)(CN)CH_2CH(CH_3)_2$	52
2-(*tert*-butylazo)-4-methoxy-2,4-dimethylpentanenitrile	*[55912-17-9]*	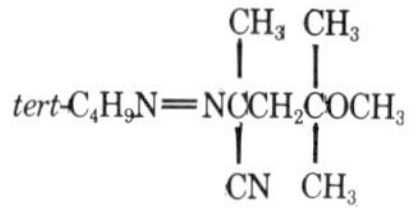 $tert\text{-}C_4H_9N{=}NC(CH_3)(CN)CH_2C(CH_3)_2OCH_3$	55
2,2′-azobis(isobutyronitrile)	*[78-67-1]*	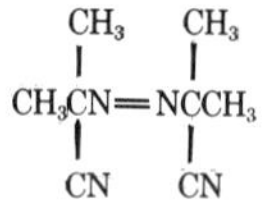 $CH_3C(CH_3)(CN)N{=}NC(CH_3)(CN)CH_3$	64
2-(*tert*-butylazo)-2,4-dimethylpentanenitrile	*[55912-18-0]*	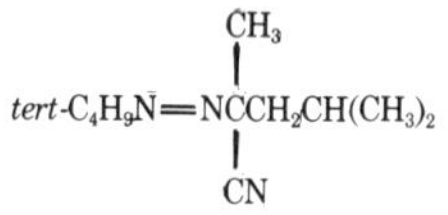 $tert\text{-}C_4H_9N{=}NC(CH_3)(CN)CH_2CH(CH_3)_2$	70
2-(*tert*-butylazo)isobutyronitrile	*[25149-46-6]*	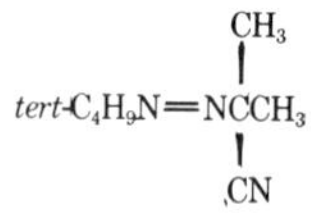 $tert\text{-}C_4H_9N{=}NC(CH_3)(CN)CH_3$	79
2-(*tert*-butylazo)-2-methylbutanenitrile	*[52235-20-8]*	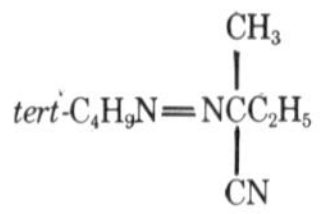 $tert\text{-}C_4H_9N{=}NC(CH_3)(CN)C_2H_5$	82
1,1-azobis-cyclohexanecarbonitrile	*[2094-98-6]*	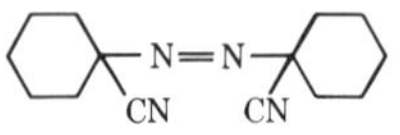 $C_6H_{10}(CN){-}N{=}N{-}C_6H_{10}(CN)$	88
1-(*tert*-amylazo)cyclohexanecarbonitrile	*[55912-19-1]*	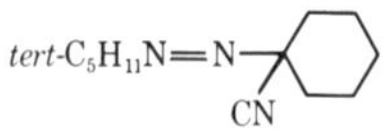 $tert\text{-}C_5H_{11}N{=}N{-}C_6H_{10}(CN)$	94
1-(*tert*-butylazo)cyclohexanecarbonitrile	*[25149-47-7]*	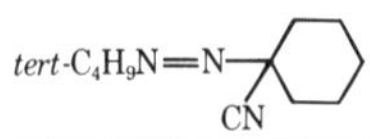 $tert\text{-}C_4H_9N{=}N{-}C_6H_{10}(CN)$	96

[a] Refs. 37–45.

radicals. As noted above, tertiary alkyl radicals are relatively more stable than most radicals generated from peroxide initiators. Thus, when these azonitriles are used as initiators for vinyl monomer polymerizations, the primary initiator radicals generally do not attack growing polymer backbones; such attack can occur with peroxide initiators. This results in more linear polymers with less long-chain branching.

Care should be exercised in handling and using azo initiators in their pure and highly concentrated states since they are thermally sensitive and can decompose extremely rapidly when overheated. The same cautions that apply to peroxides also should be applied to handling and using azo initiators and manufacturer's safety literature should be read carefully (38,40–44). Azonitrile decomposition products include organonitriles; some are toxic to varying degrees depending upon their specific molecular weights and cyano function content. Thus, potential toxicity hazards associated with the decomposition products of azonitriles must be taken into account. Such hazards are present primarily when pure or highly concentrated azonitrile solutions are decomposed in poorly ventilated areas. The chemistry of aliphatic azo compounds is reviewed in refs. 46–49.

Other. A large number of individual chemical methods for generating free radicals (not using peroxides or azo compounds) are reported in the literature. Specific examples include reactions of: alkylmercuric halides and sodium borohydride (50); molybdenum carbonyl and carbon tetrachloride (51); malonic acid and Mn^{3+} systems (52); triethylaluminum and cuprous chloride (Cu_2Cl_2) (53); manganic hydroxide and hydrazine (54); cupric sulfate and hydrazine (55); *N*-alkylhydroxylamines and Ti^{3+} systems (56); copper complexes and CCl_4 (57); transition metals and organic halides (58); nickel and anhydrides (59); diazonium salts and transition metals (60); aralkyl halides and silver (61); iodobenzene and magnesium (62); permanganate oxidation of carbanions (63); and thermolysis of transition-metal carboxylates (64); *N*-nitrosohydroxylamines (65); *N*-nitrosoacetanilide (66); and certain hexasubstituted ethanes (67–71). Most of these free-radical generating systems cannot compete with the peroxide (or azo) initiators, especially in the polymer industry, primarily because of cost and/or efficiency; although some may be well-suited for initiating specific free-radical reactions or polymerizations.

Initiation Through Radiation. Free-radical reactions and polymerizations (primarily curing systems) have been initiated by high energy radiation, ie, x-rays, gamma rays, and beta rays (72). These radiation processes fragment chemical bonds into ions as well as free radicals. Ionizing radiation is often used in referring to these processes (see Radiation curing).

Initiating free-radical reactions with uv radiation is widely used in industrial processes (see Photochemical technology). In contrast to high energy radiation processes where the energy of the radiation alone is sufficient to initiate reactions, initiation by uv irradiation almost always requires the presence of a photoinitiator, ie, a chemical compound or compounds that generate initiating free radicals when subjected to uv radiation. There are two types of photoinitiator systems: those that produce initiator radicals by intermolecular H abstraction and those that produce initiator radicals by intramolecular photocleavage (73–76).

In the case of intermolecular H abstraction, a second component (H donor) is required. Typical H donors have an active H atom positioned alpha to an oxygen or nitrogen (eg, alcohols, R_2CHOH; ethers, R_2CHOR; and tertiary amines, R_2CHNR_2, or the active H atom directly attached to sulfur, eg, thiols, RSH). Some photoinitiators

Table 9. Photoinitiators that Undergo Intermolecular H Abstraction

Name	CAS Registry No.	Structure
benzophenone	[*119-61-9*]	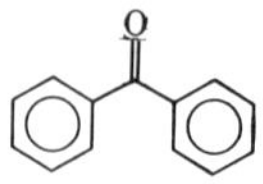
xanthone (X = O) thioxanthone (X = S) 2-chlorothioxanthone	[*90-47-1*] [*492-22-8*] [*86-39-5*]	X
4,4′-bis(*N*,*N*′-dimethylamino)-benzophenone (Michler's ketone)	[*90-94-8*]	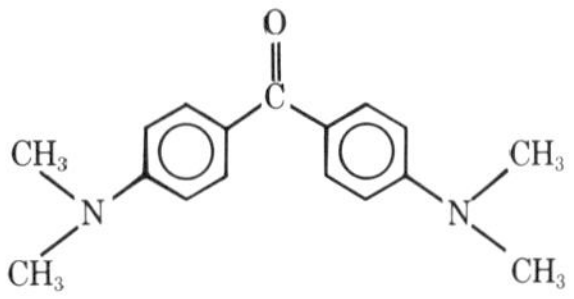 CH_3, N, CH_3, CH_3, N, CH_3
benzil	[*134-81-6*]	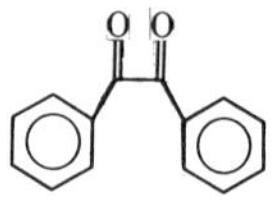
9,10-phenanthraquinone	[*84-11-7*]	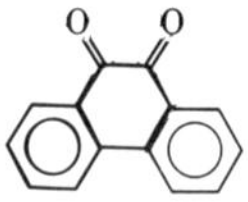
9,10-anthraquinone	[*84-65-1*]	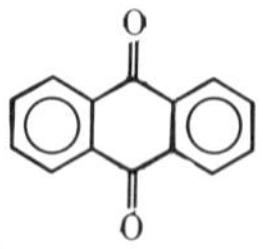

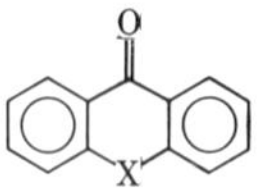

that undergo intermolecular H abstraction from the H donor upon excitation by uv radiation are listed in Table 9. An illustration of this process is depicted below using benzophenone as the photoinitiator and an alcohol as the H donor:

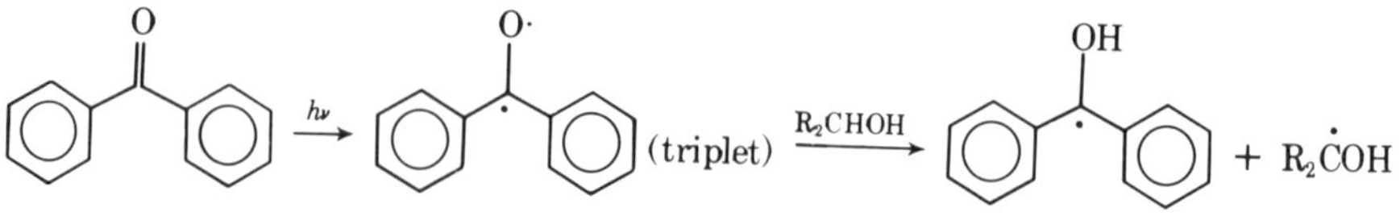

$$(C_6H_5)_2C{=}O \xrightarrow{h\nu} (C_6H_5)_2\dot{C}{-}O\cdot \text{ (triplet)} \xrightarrow{R_2CHOH} (C_6H_5)_2\dot{C}{-}OH + R_2\dot{C}OH$$

The benzophenone is excited to the triplet ketone (a diradical) which abstracts an alpha H atom from the alcohol and results in the formation of two initiating radicals. With amine H donors, an electron transfer may precede the H-transfer process from a triplet exciplex of the benzophenone and the amine:

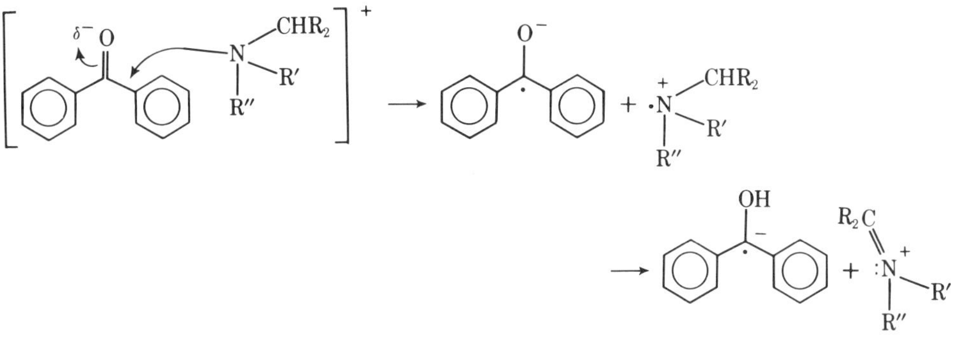

Some photoinitiators that undergo intramolecular photocleavage are listed in Table 10. These initiators are subject to a Norrish Type I cleavage directly to two initiating radicals upon photoexcitation, as illustrated below for a benzoin ether:

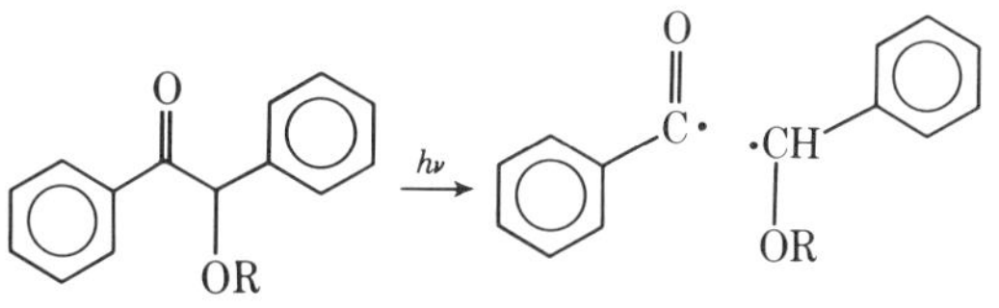

Table 10. Photoinitiators that Undergo Intramolecular Photocleavage

Name	CAS Registry No.	Structure
benzoin ethers		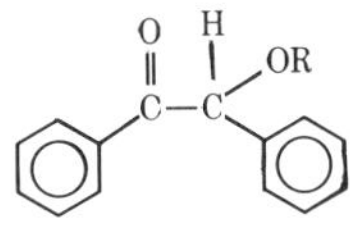
α,α-dimethoxy-α-phenylacetophenone	[24650-42-8]	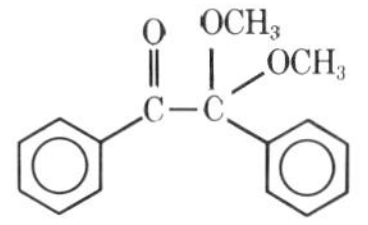
α,α-diethoxyacetophenone	[6175-45-7]	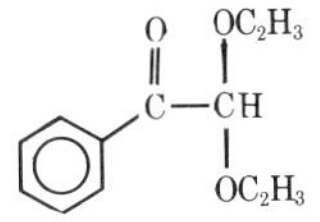
1-phenyl-1,2-propanedione-2-*O*-benzoyl oxime	[17292-57-8]	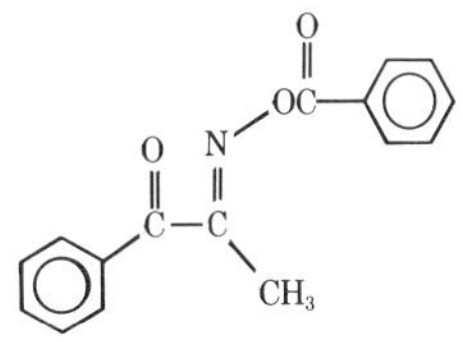

In many photoinitiated processes, a photosensitizer may be used. A photosensitizer absorbs light and subsequently transfers the absorbed energy to an energy acceptor which then may produce initiator radicals by H abstraction or by photocleavage. The energy-transfer agent (photosensitizer) usually undergoes no net change. A variety of photosensitizers have been used such as eosin, chlorophyll, methylene blue, and thioxanthone. In photosensitized processes, the energy acceptor often is referred to as a co-initiator. Photoco-initiators do not absorb light but accept energy from the excited photosensitizer which distinguishes them from the photoinitiators listed in Tables 8 and 9. Typical co-initiators that undergo H abstraction are the H donors mentioned above. An example of a co-initiator undergoing photocleavage is quinoline-8-sulfonyl chloride photosensitized by thioxanthone (73) (see also Dyes, sensitizing).

The peroxide and azo thermal-free-radical initiators discussed above also are photochemically unstable and have been used as radical sources at well below their normal thermal decomposition temperatures. However, their industrial use as photoinitiators has been limited because their light absorption characteristics frequently are unsuitable and their obvious potential complication of slow decomposition, which leads to poor shelf-life and unreproducible photoactivity in given formulations (74). Further information on photoinitiators can be found in ref. 77 (this reference uses the term sensitizer rather than photoinitiator).

Ionic Initiators

Ionic initiators are compounds that are capable of initiating ionic polymerization reactions. The words catalyst and initiator have been used frequently without further differentiation. The relationship of catalyst to catalysis is different than that of initiator to initiation. Ionic polymerization can be better defined in terms of catalysts and catalysis (31). This is especially true of cationic polymerization reactions.

Anionic. Alkali metals and aromatic complexes of alkali metals have been used as initiators in anionic polymerization (78) (see Elastomers, synthetic). The mechanism of initiation involves an electron transfer step (as depicted below), which results in the formation of a radical anion.

$$Li + CH_2{=}CHCH{=}CH_2 \rightarrow \dot{C}H_2CH{=}CH\bar{C}H_2Li^+$$

$$2\,\dot{C}H_2CH{=}CH\bar{C}H_2Li^+ \rightarrow {}^+Li\bar{C}H_2CH{=}CHCH_2CH_2CH{=}CH\bar{C}H_2Li^+$$

This can subsequently couple to form a dianion, which is capable of propagating the chain. The radical anion also can participate in a second electron transfer step (as shown below) and still form a dianion species:

$$CH_2CH{=}CH\bar{C}H_2Li^+ + Li \rightarrow {}^+Li\bar{C}H_2CH{=}CH\bar{C}H_2Li^+$$

The polymerization system, in general, and the alkali metal, in particular, influences the magnitude of these two individual reactions.

Various organolithium compounds are commonly used as initiators in anionic polymerization. Their solubility in a variety of solvents has contributed to their versatility. These initiators function by direct anionic attack, without the electron-transfer step, resulting in a monofunctional propagating species:

$$CH_3(CH_2)_3Li + CH_2{=}CHCH{=}CH_2 \rightarrow CH_3(CH_2)_3CH_2CH{=}CH\bar{C}H_2Li^+$$

Anionic polymerization systems are characterized by the absence of chain-transfer and termination reactions. However, this is true only if the system is free of reactive impurities.

The type of initiator and solvent system that are used often control the polymer microstructure and, therefore, contribute significantly to the importance of anionic polymerization even on an industrial scale (78–80).

Cationic. Cationic polymerization can be initiated by using a variety of compounds (81–84). It is preferred to refer to them as catalysts, since a co-catalyst generally is necessary to initiate the polymerization reaction. The initiation mechanism is quite complex and is influenced strongly by the individual polymerization system. Catalysts that are used in cationic polymerization can be classified as follows: protonic or Brønsted acids; aprotonic acids, eg, the various Lewis acids and Friedel–Crafts halides; stable carbonium ion salts, eg, antimony and aminium salts; and substances capable of generating cations—iodine is by far the most well-known compound in this category. Catalysts, such as BF_3, $AlCl_3$, AlR_3, $SnCl_4$, $TiCl_4$, and $SbCl_5$ (ie, aprotonic acids), generally can be used with most cationically active monomers. They often are used in combination with a co-catalyst (such as water), a protic acid, or an alkyl halide.

Considerable work has been done in the area of photoinitiated cationic polymerization. Various inorganic and organometallic salts have been found to be effective photoinitiators. Aryldiazonium and diaryliodonium salts are the most popular since they can be used with a wide variety of cationically active monomer systems (85).

BIBLIOGRAPHY

1. D. Griller and K. U. Ingold, *Acc. Chem. Res.* **9,** 13 (1976).
2. W. A. Pryor, ed., *Organic Free Radicals, ACS Symposium Series 69,* American Chemical Society, Washington, D.C., 1978; J. M. Hay, *Reactive Free Radicals,* Academic Press, Inc., New York, 1974.
3. J. K. Kochi, ed., *Free Radicals,* Vols. I–II, John Wiley & Sons, Inc., New York, 1973.
4. C. Walling, *Free Radicals in Solution,* John Wiley & Sons, Inc., New York, 1957.
5. W. G. Lloyd, *Chem. Tech.,* 371 (June 1971); 687 (Nov. 1971).
6. J. M. Tedder and J. C. Walton, *Acc. Chem. Res.* **9,** 183 (1976).
7. W. A. Pryor, *Chem. Eng. News,* 70 (Jan. 15, 1968).
8. I. P. Gragerov, *Russ. Chem. Rev.* **38,** 626 (1969).
9. K. U. Ingold, *Free Radical Substitution Reactions,* Wiley-Interscience, New York, 1971.
10. E. S. Huyser, *Free Radical Chain Reactions,* Wiley-Interscience, New York, 1970.
11. G. Sosnovsky, *Free Radical Reactions in Preparative Organic Chemistry,* MacMillan, New York, 1964.
12. A. G. Davies, *Organic Peroxides,* Butterworths, London, Eng., 1961.
13. E. G. E. Hawkins, *Organic Peroxides,* E. and F. F. Spon Ltd., London, Eng., 1961.
14. A. V. Tobolsky and R. B. Mesrobian, *Organic Peroxides,* Interscience Publishers, New York, 1954.
15. D. Swern, ed., *Organic Peroxides,* Vol. I, Wiley-Interscience, New York, 1970.
16. D. Swern, ed., *Organic Peroxides,* Vol. II, Wiley-Interscience, New York, 1971.
17. D. Swern, ed., *Organic Peroxides,* Vol. III, Wiley-Interscience, New York, 1972.
18. *Technical Publication 30.30, Evaluation of Organic Peroxides from Half-Life Data,* Lucidol Division, Pennwalt Corporation, Buffalo, N.Y., 1977.
19. *Technical Publication 30.90, Free Radical Initiators for the Suspension Polymerization of Vinyl Chloride,* Lucidol Division, Pennwalt Corporation, Buffalo, N.Y., 1976.
20. *Organosulfonyl Peroxides,* product bulletin, Lucidol Division, Pennwalt Corporation, Buffalo, N.Y., 1977.
21. *Diacyl Peroxides,* product bulletin, Lucidol Division, Pennwalt Corporation, Buffalo, N.Y., 1977.
22. *Peroxyesters,* product bulletin, Lucidol Division, Pennwalt Corporation, Buffalo, N.Y., 1977.
23. A. A. Zavitsas, *J. Am. Chem. Soc.* **94,** 2779.88 (1972); R. T. Sanderson, *J. Am. Chem. Soc.* **97,** 1367 (1975); D. M. Golden and S. W. Benson, *Chem. Rev.* **69,** 125 (1969); P. Gray, A. A. Herod, and A. Jones, *Chem. Rev.* **71,** 247 (1971).

24. *Peroxyketals,* product bulletin, Lucidol Division, Pennwalt Corporation, Buffalo, N.Y., 1977.
25. *Dialkyl Peroxides,* product bulletin, Lucidol Division, Pennwalt Corporation, Buffalo, N.Y., 1977.
26. *Peroxydicarbonates,* product bulletin, Lucidol Division, Pennwalt Corporation, Buffalo, N.Y., 1977.
27. *Tertiary Alkyl Hydroperoxides,* product bulletin, Lucidol Division, Pennwalt Corporation, Buffalo, N.Y., 1977.
28. *Ketone Peroxides,* product bulletin, Lucidol Division, Pennwalt Corporation, Buffalo, N.Y., 1977.
29. C. S. Sheppard and V. R. Kamath, *Polym. Eng. Sci.* **19,** 597 (1979).
30. E. W. Duck, "Emulsion Polymerization" in N. M. Bikales, ed., *Encyclopedia of Polymer Science and Technology,* Vol. 5, Interscience Publishers, a division of John Wiley & Sons, Inc., New York, 1966, pp. 801–859.
31. H. F. Mark, "Catalysis" in N. M. Bikales, ed., *Encyclopedia of Polymer Science and Technology,* Vol. 3, Interscience Publishers, a division of John Wiley & Sons, Inc., New York, 1965, pp. 26–34.
32. O. L. Mageli and J. R. Kolczynski, "Peroxy Compounds" in N. M. Bikales, ed., *Encyclopedia of Polymer Science and Technology,* Vol. 9, Interscience Publishers, a division of John Wiley & Sons, Inc., New York, 1968, pp. 814–841.
33. *Technical Publication 30.40, Organic Peroxides: Their Safe Handling and Use,* Lucidol Division, Pennwalt Corporation, Buffalo, N.Y., 1977.
34. *Technical Publication 30.43, Safe Handling, Storage and Transportation of Peroxides Requiring Refrigeration,* Lucidol Division, Pennwalt Corporation, Buffalo, N.Y., 1976.
35. *Suggested Relative Hazard Classification of Organic Peroxides,* technical publication, Organic Peroxide Producers Safety Division, The Society of the Plastics Industry, Inc., New York, 1975.
36. *Technical Bulletin No. 19, The Storage and Handling of Organic Peroxides in the Reinforced Polyester Fabricating Plant,* Organic Peroxide Producers Safety Division, The Society of the Plastics Industry, Inc., New York, 1975 revision.
37. Ref. 4, pp. 511–516.
38. *Dupont Vazo® Vinyl Polymerization Catalyst (Azobisisobutyronitrile),* product information bulletin, E. I. du Pont de Nemours & Co., Inc., Wilmington, Del.
39. R. Zand, "Azo Catalysts" in N. M. Bikales, ed., *Encyclopedia of Polymer Science and Technology,* Vol. 2, Interscience Publishers, a division of John Wiley & Sons, Inc., New York, 1965, pp. 278–295.
40. *LUAZO® 82, 2-t-Butylazo-2-cyanobutane,* product bulletin, Lucidol Division, Pennwalt Corporation, Buffalo, N.Y., 1979.
41. *LUAZO® 79, 2-t-Butylazo-2-cyanopropane,* product bulletin, Lucidol Division, Pennwalt Corporation, Buffalo, N.Y., 1979.
42. *LUAZO® 96, 1-t-Butylazo-1-cyanocyclohexane,* product bulletin, Lucidol Division, Pennwalt Corporation, Buffalo, N.Y., 1979.
43. *LUAZO® 70, 2-t-Butylazo-2-cyano-4-methylpentane,* product bulletin, Lucidol Division, Pennwalt Corporation, Buffalo, N.Y., 1979.
44. *LUAZO® 55, 2-t-Butylazo-2-cyano-4-methoxy-4-methylpentane,* product bulletin, Lucidol Division, Pennwalt Corporation, Buffalo, N.Y., 1975.
45. *New Azo Initiators for Effective Curing of Unsaturated Polyesters,* technical publication, Lucidol Division, Pennwalt Corporation, Buffalo, N.Y., 1974.
46. S. Patai, ed., *The Chemistry of the Hydrazo, Azo, and Azoxy Groups,* Parts 1 and 2, John Wiley & Sons, Inc., New York, 1975.
47. H. Zollinger, *Azo and Diazo Chemistry, Aliphatic and Aromatic Compounds,* Wiley-Interscience, New York, 1961, Chapters 9 and 12.
48. C. G. Overberger, J.-P. Anselme, and J. G. Lombardino, *Organic Compounds with Nitrogen—Nitrogen Bonds,* The Ronald Press Company, New York, 1966, Chapt. 4.
49. P. A. S. Smith, *The Chemistry of Open-Chain Organic Nitrogen Compounds,* Vol. II, W. A. Benjamin, Inc., New York, 1966, Chapter 11.
50. C. L. Hill and G. M. Whitesides, *J. Am. Chem. Soc.* **96,** 870 (1974).
51. C. H. Bamford, G. C. Eastmond, and F. J. T. Fildes, *Proc. R. Soc. Ser. A* **326**(1567), 431, 453 (1972).
52. N. G. Devi and V. Mahadevan, *Die Makromol. Chem.* **152,** 177 (1972).
53. W. Kawai, M. Ogawa, and T. Ichihashi, *J. Polym. Sci. A-1* **9,** 1599 (1971).
54. C. W. Brown and H. M. Longbottom, *J. Appl. Polym. Sci.* **14,** 2927 (1970).
55. J. Bond and P. I. Lee, *J. Polym. Sci. A-1* **6,** 2621 (1968).
56. N. H. Anderson and R. O. C. Norman, *J. Chem. Soc. B,* 993 (1971).

57. M. Imoto, K. Ueda, and K. Takemoto, *Die Makromol. Chem.* **138,** 11 (1970); J. Barton, V. Durdovic and V. Horanska, *Die Makromol. Chem.* **133,** 205 (1970).
58. C. H. Bamford, *Eur. Polym. J.* (Suppl.), 1 (1969); C. M. M. daSilva Correa and W. A. Waters, *J. Chem. Soc. C,* 1880 (1968).
59. T. Otsu and S. Kubota, *Polym. Rep.* (147), 18 (1970).
60. S. C. Dickerman and co-workers, *J. Org. Chem.* **34,** 714 (1969).
61. M. Kinoshita, N. Yoshizumi, and M. Imoto, *Die Makromol. Chem.* **127,** 185 (1969).
62. M. Ya. Turkina and I. P. Gragerov, *J. Org. Chem. USSR* **5,** 575 (1969).
63. S. V. Kulkarni, A. E. Fiebig, and R. Filler, *Chem. Ind.* **11,** 364 (1970).
64. E. I. Heiba, R. M. Dessau, and W. J. Koehl, Jr., *J. Am. Chem. Soc.* **91,** 138 (1969).
65. T. Koenig and J. M. Owens, *J. Am. Chem. Soc.* **96,** 4052 (1974); T. Koenig and W. R. Mabey, *J. Am. Chem. Soc.* **92,** 3804 (1970).
66. L. Benati, A. Tundo, and G. Zanardi, *J. Chem. Soc. Chem. Commun.,* 590 (1972); P. Spagnolo and co-workers, *J. Chem. Soc. Perkin I,* 93 (1972).
67. E. J. Walsh, Jr. and co-workers, *Tetrahedron Lett.* **1,** 77 (1968).
68. Can. Pat. 905,992 (July 25, 1972), H. A. P. deJongh and C. R. H. I. deJonge (to Akzo N.V.).
69. U.S. Pat. 4,036,898 (July 19, 1977), H. H. J. Oosterwijk, E. P. Magre, and W. M. Beyleveld (to Akzona Inc.).
70. U.S. Pat. 3,948,858 (Apr. 6, 1976), U. E. Wiersum (to Akzo N.V.).
71. U.S. Pat. 4,124,763 (Nov. 7, 1978), W. J. Mijs, C. H. V. Dusseau, and H. J. M. Sinnige (to Akzona Inc.).
72. A. Chapiro, R. B. Fox, and R. F. Cozzens, "Radiation-Induced Reactions" in N. M. Bikales, ed., *Encyclopedia of Polymer Science and Technology,* Vol. 11, Interscience Publishers, a division of John Wiley & Sons, Inc., New York, 1969, pp. 702–760.
73. S. P. Pappas, ed., *UV Curing: Science and Technology,* Technology Marketing Corporation, Stamford, Conn., 1978, Chapter 1.
74. A. Ledwith, *Pure Appl. Chem.* **49,** 431 (1977)
75. A. Ledwith, *J. Oil Colour Chem. Assoc.* **59**(5), 157 (1976).
76. A. Pryce, *J. Oil Colour Chem. Assoc.* **59**(5), 166 (1976).
77. G. Oster, "Photopolymerization and Photocrosslinking" in N. M. Bikales, ed., *Encyclopedia of Polymer Science and Technology,* Vol. 10, Interscience Publishers, a division of John Wiley & Sons, Inc., New York, pp. 145–156.
78. M. Morton and L. J. Fetters, *Rubber Chem. Technol.* **48,** 359 (1975).
79. N. G. Gaylord and S. S. Dixit, *Macromolecular Rev.* **8,** 51 (1974).
80. C. G. Overberger and H. Mark, eds., *International Symposium on Macromolecules, Polymer Symposia No. 62,* John Wiley & Sons, Inc., New York, 1978.
81. *Makromol. Chem.* **174**(Special Issue), 1017 (1974).
82. J. P. Kennedy, *Cationic Polymerization of Olefins: A Critical Inventory,* Wiley-Interscience, New York, 1975.
83. J. P. Kennedy, ed., *Fourth International Symposium on Cationic Polymerization,* Wiley-Interscience, New York, 1976.
84. N. G. Gaylord, *Macromolecular Rev.* **4,** 183 (1970).
85. J. V. Crivello in ref. 73, Chapt. 2.

CHESTER S. SHEPPARD
VASANTH KAMATH
Pennwalt Corporation

INKS

According to most historians, writing ink was first prepared and used by the Chinese and the Egyptians as early as 2600 BC. These early inks were probably composed of carbonaceous materials such as lamp black or soot mixed with animal glue or vegetable oil vehicles. Reference is also made to the Chinese invention of solid ink blocks and pellets similar to India ink as it is known today, which had its origin during the period 220–419 AD. Writing-ink formulation became a highly developed art under the Chinese, who were printing from hand-cut blocks in the 11th century AD, 400 years before Gutenberg introduced movable type in Europe. Writing inks differ from printing inks in that the latter are generally applied to a substrate by means of a printing press. Printing inks as supplied to the graphic arts industry are used in much greater volume by far as compared to writing inks. This article is divided into a discussion of printing inks, followed by miscellaneous categories of ink, including ink for ball-point pens, with which by far the greatest amount of ink writing is done in the United States. The number of printing-ink manufacturing establishments in the United States is approximately 700. This includes some 100 captive ink plants. The value of the total printing-ink production in the United States was approximately 1×10^6 in 1979 (see also Printing processes; Reprography).

Printing Inks—General Considerations

Printing ink is a mixture of coloring matter dispersed or dissolved in a vehicle or carrier, which forms a fluid or paste which can be printed on a substrate and dried. The colorants used are generally pigments, toners, and dyes, or combinations of these materials, which are selected to provide color contrast with the background on which the ink is printed. The vehicle used acts as a carrier for the colorant during the printing operation, and in most cases, serves to bind the colorant to the substrate.

Printing inks are applied in thin films on many substrates such as paper, paperboard, metal sheets and metallic foil, plastic films and molded plastic articles, textiles, and glass (see Film and sheeting materials; Paper). Printing inks can be designed to have decorative, protective, or communicative functions. In some cases, combinations of these functions are achieved.

It is generally conceded that there are four classes of printing ink, which vary considerably in physical appearance, composition, method of application, and drying mechanism. They are letterpress and lithographic (litho) inks, which are commonly called oil inks or paste inks, usually of a pasty consistency, and flexographic (flexo) and rotogravure (gravure) inks, which are referred to as solvent or liquid inks (1–3).

Four properties of inks are of cardinal importance—drying, printability (which is largely a function of the rheology of the ink), color, and use properties. Use properties are those considerations that determine how printed substrates function throughout all processing and usage from the time of printing throughout the useful life of the printed product.

Drying. The term drying may be defined as any process that results in the transformation of a fluid-printing ink film into a more or less rigid form. An ink is considered dry when a print does not stick or transfer to another substrate placed in contact with it.

Drying is accomplished by one or more of the following physical or chemical mechanisms: absorption, evaporation, precipitation, oxidation, polymerization, cold setting, gelation, and radiation curing.

Absorption. Some inks dry by penetration or absorption into the pores of the printed stock—a blotter or sponge effect. This is accomplished by the gross penetration of the ink into the large pores of the substrate; the partial separation of the vehicle from the pigment; and the diffusion of the vehicle throughout the paper.

The ability of an ink to penetrate into paper depends on the number and size of the air spaces present in the paper, the affinity or receptivity of the stock for the ink, and the mobility of the ink.

Solvent Evaporation. Drying of ink by evaporation of solvent is generally achieved by application of heat, depending upon the volatility of the solvent employed in the ink formulation. Web-offset and letterpress inks which dry by solvent evaporation are of the heat-set type. All flexographic and rotogravure inks contain more volatile solvents than the heat-set publication inks referred to above and do not require as much heat to effect drying. In either case, a balance must be maintained between satisfactory drying of the ink on the printed substrate and adequate stability, or resistance to drying, on the distributing rollers of the printing press.

Resin Precipitation. An ink may also be caused to dry by precipitation of its binder rather than by evaporation of solvent. This is accomplished by adding a diluent, such as water in the form of steam, to a hygroscopic water-miscible solvent ink system, which will cause the solubility of the resin in the ink film to decrease sharply and cause it to precipitate when its tolerance for the diluent is reached. Further drying is accomplished by absorption of the freed solvents into the stock and then by evaporation.

Oxidation. Inks that dry by oxidation behave much like paint films and dry by means of drying oils. They also contain metallic driers, which catalyze the absorption of oxygen by the drying oil (see Driers and metallic soaps; Drying oils; Paint).

Cold Setting. Hot-melt inks are applied to a substrate in a molten state and upon cooling form a dry ink film.

Gelation. Drying by gelation is accomplished by dispersing fine particles of a high molecular weight polymer, such as poly(vinyl chloride) resin, in a plasticizer for the polymer. At room temperature the dispersed polymer behaves like a pigment but at elevated temperatures it solvates, causing an immediate gel formation which is dry to the touch.

Polymerization. Thermal polymerization or curing of an ink film at elevated temperatures can follow many different chemical paths. Condensation and cross-linking reactions may be accomplished with or without the use of catalysts. However, this method of drying generally has not been used widely for printing inks, except for metal and glass decoration.

Radiation Curing. Ultraviolet and electron-beam curing of inks and coatings has progressed markedly in recent years owing largely to the fact that these drying mechanisms produce little or no effluent. They are similar processes in that both involve polymerization by a free-radical mechanism and very high rates of reaction can be achieved with relatively low energy inputs. Uv inks contain a photochemically reactive vehicle species, colorant, and a photosensitizer which initiates the reaction upon exposure to the uv radiation source. Electron-beam inks do not require a photoinitiator, the higher density-energy of the electrons is sufficient to create large numbers of free radicals in the vehicle system (see Radiation curing).

Infrared Drying. Infrared drying is used in some heat-set and evaporative solvent-ink applications where the wavelengths emitted by the radiator can, in principle, be tuned in on the optimum absorbency of the ink vehicle, thereby heating it selectively and not the substrate on which it is printed.

Radio Frequency and Microwave Drying. There are similar mechanisms differing in wavelengths generated and hardware required. Rf operates in the 30–100 MHz range and microwave drying in the 1000–3000 MHz range. Both generate electromagnetic waves which intersect and cause molecular excitation of polar molecules used in the ink film, causing rapid heating and drying to take place (4–5).

It may be noted that with the exception of oxidation- and polymerization-curing mechanisms, all other drying processes described depend upon physical phenomena. The complexity of organic reactions occurring in the nonphysical methods also appear to limit the speed of reaction which can be expected, so that the physical methods generally provide faster drying. Certain specialized methods of drying, such as ionic interaction and decomposition reactions, offer the fastest drying mechanisms but they have not been reduced to practice at this time.

Rheology. The rheology of inks is of the greatest importance. In a Newtonian liquid, any stress produces a flow, and the rate of flow is proportional to the stress. But inks are generally non-Newtonian and the most common terms used by the ink maker to describe rheology are: *viscosity* (resistance to flow); *yield value* (point at which a liquid starts to flow under stress); *thixotropy* (decreasing viscosity with increasing agitation); and *dilatancy* (opposite of thixotropy). Mayonnaise is an example of a material having a high yield value but a low viscosity (6–8).

The ink-distribution systems of flexo and gravure printing presses are very simple and do not provide the means to distribute and level highly viscous inks. Therefore, the viscosity must be low, on the order of 50–100 mPa·s (= cP). Yield value must be low also to permit pumping of ink from reservoir to fountain. Thixotropy should also be avoided for the same reason.

Letterpress and offset inks can vary in viscosity from under 500 mPa·s for a letterpress-type news ink to over 500 Pa·s (5000 P) for special litho ink formulations. The viscous nature of letterpress and litho inks has forced press designers to use a multitude of rollers in the ink distribution unit to ensure uniform and adequate transfer of ink to the printing plate.

In addition to the importance of proper ink rheology as related to roller-to-plate transfer, it is rheology, again, that controls, in large part, the fidelity of printing, drying speed, holdout, and trapping properties obtained on the substrate. In most cases, higher press speeds require lower viscosity inks and vice versa. Low viscosity ink is preferred for fine-line flexography and shallow-cell gravure printing. Printing smooth, dense solids can best be achieved using higher viscosity ink.

Rheology is also an important color-strength determinant. Over-pigmentation leads to a high yield value, thixotropic ink, thereby forcing a compromise between color intensity and rheology.

There are many varieties of instruments designed for measuring viscosity and other rheological properties of inks. These include capillary, orifice, rotational, and other miscellaneous types. Modern ink-rheology measurements are made on instruments such as the Shirley-Ferranti Cone-Plate viscometer, Brookfield Torsion viscometer, and the Haake, Laray, Interchemical, Stormer, and Band viscometers. The Inkometer and Tackoscope are also used to record tackiness as well as ink misting and roller stability (see Rheological measurements).

Color. Color is the third common denominator for all types of ink formulation and may very well be the most important one because it has such a tremendous psychological impact on the ultimate consumer. Color appears in three different dimensions referred to as hue, saturation or chroma, and value (see Color).

Coloring Materials. Pigments are used not only for color but also for other physical properties such as bulk, opacity, specific gravity, viscosity, yield value, and printing qualities. Different pigments of the same color have different permanency to light, heat, chemicals, and bleed or staining in water, oil, alcohol, fat, grease, acid, or alkali (9–10).

Judicious selection of pigment is also required, according to the use of the ink, in considering subsequent operations such as varnishing, waxing, lacquering, or laminating.

Inks make use of inorganic pigments and of dyes, most commonly in the form of pigments, that is to say, rendered insoluble in one of a number of ways (see Pigments, organic). Five dye families that are of interest to ink manufacturers are azo, triphenylmethane, anthraquinone, vat, and phthalocyanine (see Dyes and dye intermediates; Dyes—application) (11–12).

Flushed Colors. Many pigments, although dry when purchased, are prepared in a water phase and are not wetted easily by oil phases. In the process known as flushing, pigments are prepared that obviate this difficulty. Phase transfer is generally accomplished by mixing the pigment presscake with a water-immiscible vehicle in a heavy duty sigma blade (dough) mixer. Most of the water separates and can be removed, although complete dehydration requires further processing, usually heating under agitation for several hours in a vacuum.

Flushed colors result in higher gloss and color strength, and better rheology and economy than their dry-pigment counterparts. Their use is expanding constantly.

Color Concentrates, Easily Dispersible Pigments. In order to provide greater versatility and, at the same time, ease of use, pigment manufacturers offer pigments in the form of color concentrates. These are liquids or pastes with 35–65% of pigment content. Pigments in this form are very well dispersed, requiring little if any mechanical effort to achieve the desired color value or fineness of grind. The concentrates often contain monomeric and resinous dispersants plus a solvent. They are classified according to the solvent used.

Easily dispersible pigments are produced by coating freshly precipitated pigment in the slurry tank with a resinous and/or monomeric dispersant and, eventually, by drying and pulverizing the solids. Such pigments are also called stir-ins as they require minimum grinding effort in order to develop the full color value.

Toners are full strength undiluted pigments used to strengthen tinctorially weak pigments. Occasionally, dyes are utilized as toners.

The most common pigments used in ink manufacture are listed below.

Black Pigments. The only black pigment used to an appreciable extent in inks is carbon black (qv). It is used in newsprint and publication printing, and therefore, in very large quantities. Black pigments are offered in fluffy or bead form.

White Pigments. Opaque white pigments commonly used in inks (listed in order of decreasing opacity) are titanium dioxide, zinc sulfide, and zinc oxide. TiO_2 is by far the most popular white pigment. Mixtures of these are often made with the various colored pigments to add opacity or to lighten the hue.

Transparent pigments (extenders) commonly used in inks (listed in order of decreasing transparency) are alumina hydrate (a product of the approximate composition $5Al_2O_3.2SO_3.xH_2O$), magnesium carbonate (magnesia), calcium carbonate, blanc fixe (precipitated barium sulfate), barita (natural barium sulfate), and clay. Extenders are used to reduce the color strength and change the rheology of inks.

Inorganic Color Pigments. Chrome yellow is generally lead chromate, modified with other lead compounds to produce several shades. Chrome orange and molybdate orange are also modified lead compounds. All of the chrome colors are quite fast to light, opaque, and have high specific gravity. They, however, can be toxic and their uses are restricted. They can not be used on articles intended for use by children or in inks that could contact food.

Cadmium yellow, orange, and red are very lightfast pigments and have excellent soap and alkali resistance. Cadmium–mercury reds come in a range of shades from a bright to a deep red. Vermilion red is a brilliant mercury sulfide pigment of high specific gravity. Their use is also restricted as a result of mercury content which is considered toxic.

Iron blue (a complex cyanide) is made in a number of shades such as Milori blue, bronze blue, and Prussian blue. It has poor alkali resistance. Ultramarine blue is very permanent to light but has poor acid resistance.

Organic Color Pigments. Yellow pigments consist mainly of yellow lakes, Hansa yellows, and diarylide yellows. (Lakes are produced from dyes by depositing them on alumina hydrate.) Hansa yellows are very strong and permanent pigments. They are very resistant to most chemicals. They are produced from azo dyes in a number of hues. Diarylide yellows are also very strong, usually not as fast to light as Hansa yellows, but more transparent pigments. They are produced from 3,3′-dichlorobenzidine, a carcinogen that must be handled with care (according to EPA regulations).

Orange pigments include dianisidine orange, benzidine orange, and Persian orange lakes.

Red pigments range in shade from orange to deep maroon, from dirty reds to the very clean, brilliant reds. A few of the more common types are the following. (*1*) The para reds, toluidines, and fire reds are fairly fast to light and semitransparent. They are used in poster and label inks. (*2*) The lithol reds are available in a wide range of shades with moderate permanency to light. (*3*) Phloxine or eosine lakes are very clean, brilliant reds, not fast to light, used in process printing on books and magazines. They also represent toxicity problems owing to lead content. (*4*) Rhodamine reds are brilliant and have much better lightfastness. They are quite costly. (*5*) Red Lake C (CI Pigment Red 53; CI 15585) is used for making the very brilliant orange shades or warm reds. (*6*) Madder lakes have resistance to bleeding and permanency to light. (*7*) Watchung reds or Permanent Red 2B (CI Pigment Red 48; CI 15865) pigments are fairly clean and have fair bleed and product resistance. (*8*) Naphthol reds are relatively permanent and are soap-resistant. They are available in yellow and blue shades.

Blue Pigments. Peacock blue is a clean, brilliant, greenish shade of blue. It is fugitive and is now seldom used. The permanent types are quite expensive.

Phthalocyanine blues are the most useful of the greenish shades. They have excellent lightfastness, very good resistance to most chemicals, and are used in process work. Their precursor crude is, in part, imported.

Victoria blue is the clean, red shade royal blue. It is only moderately fast to light and not too popular an ink pigment.

Alkali blues of deep reddish shades find extensive use in all types of inks, often for toning blacks. Alkali blue toners are supplied in flushed form.

Purple Pigments. Methyl violet is the most commonly used, also for toning purposes. Dioxazine violet, an expensive color, is used where permanence and resistance properties are needed.

Green Pigments. Phthalocyanine green is a permanent and clean pigment. Green Lake is one of the triphenylmethane colors and has its disadvantages. It is difficult to grind, requires more drier to achieve scuff resistant ink film and occasionally bleeds in fat or plasticizer.

Daylight Fluorescent Pigments. These are used in silk screen, gravure, flexographic, letterpress, and offset inks. They are difficult to handle and require heavy films of ink to be effective. They are generally not very fast to light (see Luminescent materials).

Silver Powder. This is usually aluminum flake, and "gold" is generally a mixture of brass and copper flakes, and possibly other metals. The gold effect can also be achieved by toning aluminum flake with a transparent yellow pigment (diarylide yellow).

Other Ingredients

***Driers* (*qv*).** These are soaps of cobalt, manganese, and lead formed with organic acids such as linolenic, naphthenic, and octanoic acids. They catalyze oxidation of drying oils, and thus should be used in inks that dry by oxidation.

***Waxes* (*qv*).** These are mainly dispersions of polyethylenes, hydrocarbons, or vegetable and animal waxes in the vehicle system used. They impart slip and scuff-resistance to ink films. Lately, polyolefinic waxes are also used directly in micropulverized form.

***Antioxidants* (*qv*).** Antioxidants of universal acceptance are eugenol and ionol. They retard premature oxidation of inks on the press at 1% or lower concentrations.

Miscellaneous Additives. These include lubricants, surface-energy reducing agents, thickeners, gellants, defoamers, wetting agents, and shorteners.

Letterpress, Dry Offset, and Lithograph (Paste) Inks

Historically, paste inks were developed much earlier than flexographic and rotogravure (liquid) inks. Since they utilize distinctly different solvents and application techniques, it has been customary to treat them separately.

Dry offset inks are not treated separately since they are quite similar to letterpress inks, except for 20–30% higher color strength and somewhat higher viscosity (6–8).

Table 1 summarizes properties of inks of the three classes, divided among eleven use classifications.

Letterpress and Litho Newsprint Inks. Letterpress newsprint inks are based mainly on mineral oil, which sometimes is combined with a resin. They dry by penetration into the absorbent paper stock. Most litho newsprint inks are based on vehicles that contain rosin-, fossil-, or hydrocarbon-type resins dissolved in aliphatic hydrocarbon solvents (see Hydrocarbon resins; Terpenoids; Resins, natural). The solvents used are fractions of petroleum in the 220–280°C boiling range. Mineral oil is used in these inks to a lesser extent, mainly in the nondrying versions. Mineral oil renders long press stability as it is somewhat higher boiling than the solvents. The heatset litho newsprint inks are heat-dried in ovens of various makes. HVHA (high velocity hot air) dryers

Table 1. Letterpress, Lithographic, and Dry-Offset Inks[a]

Ink type and use	Ink class, printing method	Drying manner	Application, stock feed	Ink consistency	Approx printing speed	Typical vehicle composition
news (papers and comics)	LP, litho	pen, HS	web-fed	liquid medium paste	300–450 m/min	min oil, res–S–O
publications (magazines, periodicals)	LP, litho	HS, QS, uv	web-fed, sheet-fed	soft–medium pastes, medium pastes	300–370 m/min, 3500–7500 sph	res–S–O, acr
commercial (display, document, pamphlet, label)	LP, litho	QS, Ox, ir, uv	sheet-fed	medium pastes	3500–7500 sph	res–O–S, acr
business documents and forms (duplicator, narrow web)	litho	QS, HS, ir	sheet-fed, web-fed	medium pastes	2000–4000 sph, 10–180 m/min	res–S–O, res–S–O–min. oil
folding cartons (food, soap, drug, liquid, parts, detergent)	LP, litho, DO	QS, Ox, uv	sheet-fed	medium–heavy pastes	3500–8000 sph	res–O–S, acr
containers and boxes (corrugated, linear)	LP	Ox, MS, pen, FW	sheet-fed	medium–heavy pastes	2500–5000 sph	res–O, res–G
books (school, directories, religious, library)	LP, litho, DO	Ox, pen	sheet-fed, web-fed	soft–medium pastes	2000–7000 sph, 120–240 m/min	res–O–S, res–min. oil–S
bags (paper, cloth, plastic)	LP, DO	HS, Ox, MS	web-fed	medium pastes	80–180 m/min	res–S–O, res–O, res–G
wrappers (bread, box, produce)	LP	MS, HS	web-fed	medium pastes	120–240 m/min	res–G, res–S
metal containers (beverage and food cans, pails, drums, displays)	litho, DO	HC, uv	sheet-fed, pr. can	medium–heavy pastes	2000–5000 sph, 100–500 cpm	alkyd–S, res–O–S, alkyd–amines–S, acr
plastic (cups, bottles, displays, book covers)	DO, litho	HS, Ox, uv	pr. can, sheet-fed	medium–heavy pastes	100–200 cpm, 2000–4000 sph	res–S, res–O–S, acr

[a] Abbreviations: LP, letterpress; DO, dry offset; pen, penetration; HS, heat-set; ir, infrared; Ox, oxidation; QS, quick-set; FW, fiber wetting; MS, moisture set; uv, ultraviolet; HC, heat cure; pr. can, preformed can; acr, acrylate oligomers; sph, sheets per h; cpm, containers per min; min. oil, mineral oil; res, resin; S, solvent; O, oil; G, glycol.

are most commonly used but infrared and gas-flame types also are employed occasionally on older presses.

Rotary Heat-Set Publication and Commercial Inks. Most rotary heat-set publication inks for both letterpress and litho presses are based on vehicles containing natural or semisynthetic resins (often gilsonite–hydrocarbon resins in blacks or rosin esters with glycerol or pentaerythritol with hydrocarbon resin and occasionally phenolic modified resin in colors) dissolved in 4°C fractions of aliphatic hydrocarbons that boil between 200–280°C. 5–10% of alkyd is normally contained in the litho versions (see Alkyd resins). These inks dry in 1 s or less by solvent evaporation or by solubilization of dispersed resins at the web temperature of 90–180°C which is achieved in one of the three dryers mentioned above.

Sheet-Fed Commercial Inks. A large segment of commercial printing is done on sheet-fed presses mainly by the litho process. Inks for these presses are based on vehicles containing phenolic, maleic-modified, or rosin-ester resins dissolved in vegetable drying oils and diluted with hydrocarbon solvents. The better inks contain alkyds in place of the oils or, occasionally, these are cross-linked via diisocyanates forming urethane linkages. On coated stocks, most sheet-fed inks quick-set first to a tackfree state by coagulation and solvent separation and then fully dry by oxidation. Air (oxygen) upon reaction with unsaturated oils forms a highly cross-linked (polymerized) tough ink film. Most commonly used oils are linseed, soya, Chinawood, and tall oils. Recently, new acrylic resins have been developed for use in quickset inks. They offer nonskinning properties and prolonged press stability.

Document-Reproduction (Duplicator) and Business Form Inks. Duplicator sheet-fed machines require very press-stable, yet quick setting inks. They must also possess fine lithographic properties of high tolerance for water-fountain solution and of dense and sharp printing properties on a wide variety of papers. The inks contain drying-oil alkyds or plain bodied oils, along with lithographic-type (nonpolar) resins and high boiling (280–370°C) hydrocarbon solvents. The rubber-type inks are based on cyclized (depolymerized) synthetic rubber which exhibits exceptionally fast setting. This property results from rapid release of the solvent.

Business-form inks closely resemble the lithographic heatset news inks (first group); however, they are higher in viscosity and often contain some mineral oil. Recently, business forms have been printed by the jet-dot method (see Ink Jet Printing). The inks are based on glycols, water, and dyes.

Folding-Carton Inks. The majority of folding-carton inks are based on various quick-set vehicles (see above). However, when maximum gloss and high rub and product (ie, grease, soap) resistance are required, they contain glossy oleo–resinous vehicles. These vehicles are composed of phenolic, phenolic-modified, and maleic modified and/or esterified rosin resins which have been dissolved in drying oils or alkyds or both. These vehicles dry by oxidation to form tough, glossy films which, unfortunately, become difficult to print over at a later time. Ultraviolet-light-cured inks are also being used in litho printing of high quality folding cartons (see Uv and Electron-Beam Systems).

Corrugated and Kraft-Liner Container Inks. Ink vehicles for corrugated and kraft-liner containers often consist of oeloresinous materials. They dry by oxidation. In the instances, however, where very fast drying is required and gloss is not of major importance, moisture-set or fiber-wetting inks are employed. Moisture-set inks are based on several acid-bearing resins (fumarated rosin) or their part-esters dissolved in diethylene, dipropylene, or triethylene glycols. They dry by precipitation of the resin solids upon the absorption of moisture by the hygroscopic glycols. Moisture-insensitive fiber-wetting varieties of these inks are prepared by forming alkaline salts of the acid resins. They dry by preferential wetting of the cellulose fiber by the polar glycols (hexylene, di- and tripropylene glycols).

Book, Bag, and Wrapper Inks. Vehicles in book, bag, and wrapper inks essentially are very similar to those already discussed under the commercial and publication inks.

Metal Container Inks. Ink vehicles for metal containers that are printed on special sheet-fed litho presses are based mainly on blends of oleo-resinous and heat-set varnishes containing the previously mentioned resins, alkyds, and oils (see Folding-Carton Inks). Some of these (high bake versions) are composed of nondrying alkyds and melamine resin. Metal-decorating inks dry during a 10 min cure at 150–200°C in 6–30 m gas-fired ovens. Polymerization, oxidation, and cross-linking reactions account for their drying and hardening. Ultraviolet-cured inks also have become very popular in metal decorating (see below).

Preformed Aluminum or Steel Containers. Ink vehicles for dry-offset printing of preformed aluminum or steel containers are essentially based on polyester vehicles used in conjunction with melamine cross-linkers. (Recently developed short-cycle ovens are now used which dry inks in 2–10 s at a temperature as high as 340°C.) The rheology of these inks must be adjusted to the unique geometry of the press. Ferranti, Haake, IC Rotational, or Laray viscometers are being utilized to determine the viscosity and yield value of the inks. In addition to these instruments, the Inkometer is used to determine tack value and press stability. Desired rheological properties are achieved by the use of gellants as well as pigments.

Plastics. Vehicles in inks for plastics (polyethylene, polystyrene, vinyl) are based on hard-drying oleoresinous varnishes which sometimes are diluted with hydrocarbon solvents. Dry-offset inks for polystyrene employ vehicles of somewhat more polar nature. Polyester or other synthetic resins (acrylic) dissolved in glycol ethers and/or esters represent a frequent choice. Uv inks are widely used for decoration of preformed plastic containers.

Manufacture. Paste inks are produced in two ways: (*1*) by mixing predispersed (preground) or flushed pigment concentrates with vehicles, solvents, oils, and compounds, or (*2*) by mixing dry pigments or resin coated pigments with vehicles and compounds and then grinding them on ink mills. Some typical ink formulas are given in Table 2.

Mixing is done in sigma-blade (dough) mixers of various sizes, in plain change-can mixers, or in large agitated tanks which can hold up to 4500 kg. A variety of equipment is being employed, from simple agitator to high speed, saw-bladed or concentric, double-blade dispersor in an attempt to achieve the desired dispersion. Baker-Perkins dough mixers are steam-jacketed, thus they allow hot (90–120°C) mastication mixing (see Mixing and blending).

Milling is done on three- or five-roller horizontal or vertical mills and in shot mills. Softer, more fluid inks are ground in shot mills (vertical and horizontal), in ball mills, and occasionally, in colloid mills.

The three-roller mill is commonly used for grinding viscous paste inks. Its steel rollers are polished and accurately machined to form a crown in the center. A definite speed differential exists in rotation of each of the rollers with respect to each other.

Table 2. Typical Ink Formulas[a]

Parts	Component
LP newsprint black	
10 or 15	carbon black
2 or 3	induline toner
85 or 67	mineral oil (1 Pa·s or 10 P)
3 or 15	mineral oil (low viscosity)
100 100	*Total*
Quick-set litho black	
16 or 18	carbon black
4 or 6	alkali blue toner
63 or 58	phenolic resin–oil (HC solvent 280°C)
10 or 6	bodied linseed oil or alkyd
2 or 2	cobalt linoleate
5 or 10	aliphatic HC solvent (280°C)
100 100	*Total*
Lp moisture-set light blue	
25	titanium dioxide
3	alkali blue
9	iron blue
50	fumarated rosin–TEG
10	HC wax compound
3	TEG
100	*Total*

Parts	Component
HS publication blue	
14	phthalocyanine blue
10	alumina hydrate
35	pentaerythritol rosin ester–HC solvent
30	HC resin in HC solvent
6	polyethylene wax compound
5	aliphatic HC solvent (240°C)
100	*Total*
Uv cure ink	
25	permanent 2B red
5	clay or aluminum hydrate
50	epoxy diacrylate oligomer
10	benzophenone
5	wax compound
5	trimethylolpropane triacrylate
100	*Total*
Metal decorating litho white (sheet fed)	
50 or 55	titanium dioxide
45 or 25	drying alkyd
0 or 15	hard resin–linseed oil
1 or 2	manganese naphthenate
4 or 3	aliphatic HC solvent (280°C)
100 100	*Total*

[a] Abbreviations: LP, letterpress; HC, hydrocarbon, TEG, triethylene glycol; HS, heat set.

They also can be set to provide greater or lesser clearance between them. The dispersion of the pigment is accomplished by shearing forces generated by this differential speed as well as by the closeness of roller setting. The shot mills break pigment particles by hammer action of the speeding shots. The smaller the size of the shot and the higher loading and the more complex (tortuous) path in the chamber, the better the dispersion. In order to accommodate the more viscous inks, some mills have a cooled agitator shaft and are equipped with protruding fingers in the chamber.

Finished inks are packed in metal cans (0.5, 1, 2, or 5 kg), in metal or plastic pails (8, 11, and 19 L), or in 114- or 190-L metal or fiber drums and, occasionally, in tote bins. The more fluid inks (news, flexo, or gravure) usually are delivered in tank cars.

Ink vehicles are produced in separate resin-varnish plants. Here 1,900–19,000-L reactor kettles are equipped with powerful agitators, reflux condensers, liquid traps, gas and liquid pumps, and ducts for material charge. Heating is done by multistage Dowtherm or hot-oil jackets, or in some installations, electric (induction) heaters. Older units employ banks of gas-fired ceramic-cup burners. Temperatures and reaction rates are automatically monitored and controlled. Finished vehicles are filtered routinely via cone filters. In order to comply with EPA rules, reactor exhaust is scrubbed or incinerated. Operators should wear protective gear, such as respirators, safety glasses, etc, in order to avoid any health hazards or injury. The plant air is analyzed routinely for the presence of dust or volatiles from organic compounds.

Synthesized resins can be cast in solid form from the reactor kettles or more typically, converted directly into fluid varnishes in a single operation, often employing special blending kettles for this purpose. Recently, reactors used for vinyl-type polymers are of continuous (pipe) design which normally is more economical.

Control and Testing. Control of inks is done by examining their color, strength, hue, tack, rheology, drying rate, stability, and product proofness. Elaborate control equipment and laboratory testing procedures are employed to test the finished inks. Weather-Ometers, Fade-Ometers, glossmeters, colorimeters, spectrophotometers, viscometers, rub testers, and occasionally, gas chromatographs are employed to check production batches or to pretest new submissions or raw materials. Proofing presses and sometimes pilot presses are utilized by ink manufacturers to control the production and test new formulations. The trend toward higher printing speeds and quality, and toward greater economy and reduced air pollution necessitates vigorous ink research and development effort. New regulations, changing cost of raw materials, different substrates, and faster presses require constant updating of the formulations. Considerable research and development time and expenditures have been devoted to lower-temperature drying and nonpolluting inks. Ultraviolet cure for inks and waterborne inks for plastics represent the major programs and radical deviations from the traditional systems (see Uv System).

Lithographic Inks—Special Considerations. As with inks for all other printing processes, lithographic inks must be deposited on the substrate in selected areas only, eg, as required to produce an image. The method of achieving this selectivity, however, is considerably different in the lithographic process. The image on the plate is neither raised nor lowered but is on the same plane as the nonimage. Nevertheless, the ink must be kept from depositing on the nonimage area of the printing plate (see Printing processes) (6–8).

The ability to attain this image selectivity is governed by the physicochemical nature of the process-involving plate, fountain solution, and ink all acting together in a ternary dynamic system. It is this situation that imposes certain requirements upon lithographic inks not common to other printing inks.

Thus, the first requirement is that the vehicle be water-immiscible; indeed, mere water immiscibility is not enough but the interfacial relationship between ink and fountain solution must also prevent easy emulsification. Second, the vehicle must wet the pigment sufficiently well so as to preclude any possibility of the pigment's transferring to the aqueous fountain solution.

Obviously, these two primary characteristics are achieved by proper selection of vehicle and pigment. In general, heat-bodied drying oils, oil-extended alkyds, and solutions of various types of nonpolar resins in water-immiscible solvents serve as satisfactory vehicles. Pigments are selected for their dispersibility in the appropriate vehicle, their tinctorial strength, and their nonbleeding characteristics into the acidic fountain solution. Of course, consideration must be given to any particular use requirements involved, such as resistance to high temperature bake, product resistance and lightfastness.

The requirements of a particular lithographic ink are further governed by the nature of the press itself, such as whether it is single- or multicolor, sheet-fed or web-fed, or whether the ink is aimed at metal decorating. As sheet-fed presses increase in size, they also increase in linear printing speed. Web-offset presses also operate at higher linear printing speeds than the sheet machines. The linear printing speed, therefore, must be considered in adjusting the rheology of the ink; ie, the higher the speed, the lower the ink tack should be. This is necessary to prevent the substrate from sticking to the offset blanket, which can result in picking or tearing the sheet or pulling it from the grippers, or in web breaks in the case of web-fed presses.

Good performance of the ink on the plate is of paramount importance. Aside from the intrinsic characteristics of the ink by reason of the specific formulation, the most important extrinsic factor is the condition of the lithographic plate, particularly that of the nonimage area. The ink must transfer a sharp, clean image to the blanket and thence to the substrate. This means that no ink is permitted to appear in the nonimage area (if this occurs, it is known as scumming). Scumming may result from such causes as improper desensitization of the nonimage area, incorrect fountain solution, or poor dampening, as well as improper surface characteristics of the ink. Color can appear in the nonimage area for other reasons also, such as by running too little fountain solution (a condition known as catchup), by pigment transfer to the aqueous phase or by the ink itself having too low a viscosity.

The web-fed lithographic process (web offset) is one of the fastest growing printing methods. Heatset and paper-penetrating inks are being offered for this application. The presses range from full width (91 cm) to half web and narrow web (30 cm) types, operating from 120 to 490 m/min. In some cases, letterpresses are being partly converted to a hybrid lithographic process called direct litho (di-litho). Here, the printing is done directly from lithographic plate onto the paper, rather than indirectly via a blanket cylinder in classic lithography. Considerable research on the lithographic system led to the introduction of printing plates that are sufficiently selective to the ink in the image and nonimage areas to be able to eliminate the need for water-fountain solution. This variation, however, found only modest acceptance by the industry. Another recent development is the energy conserving low temperature drying (LTD) ink formulation based on the dispersed polymer principle or on co-solvent approaches.

Radiation-Cure Inks

In response to air-pollution regulations, the shortage or high cost of natural gas and because of a general desire to conserve energy, new ink formulations–drying systems have emerged which are based on radiant energy: ultraviolet, electron-beam, and infrared radiation. Lithographic and letterpress inks are now available for web-fed and sheet-fed presses that will cure–dry instantly when exposed to radiators of the required intensity. A brief description of each system follows.

Uv System. In the most simple form, uv inks consist of a reactive (monomeric or oligomeric acrylate) vehicle, photoinitiator (benzophenone), and colorants (conventional pigments). With the absorption of photons from a mercury lamp, the sensitizer becomes excited. It then undergoes electronic transitions and, thereby, becomes very reactive. In this state, it initiates free-radical reaction among the reactive species of the vehicle. This results in hardening (drying) of the ink film. General quality, stability, and safety difficulties must be considered when the manufacture and use of uv inks is attempted. They also tend to be more costly than the conventional types. In spite of these shortcomings, uv inks are now commercial for printing of paper, plastic, and metal because of the environmental and energy conservation benefits (13–17).

Electron-Beam System. Electron-beam-cured inks are similar to their uv counterparts. Electron irradiation, however, does not require photoinitiators in the formula as it readily cross-links the vinyl-rich vehicle, often requiring only 10–20 kGy ((1–2) $\times 10^6$ rad) of energy. The irradiators must be well shielded with a lead plate to be harmless. The production units are still very costly as the electron-beam system is only in the early development stage for printing applications. At present only a few installations exist. The great energy conservation potential of electron-beam cure (10% of the heat energy) makes the system quite attractive to printers and convertors (15,18).

Ir System. Ir radiation imparts instant intense heat to the substrate and to the ink. It causes evaporation of solvent in heatset inks or quicksetting of sheet-fed inks. The latter application is relatively new and especially effective for inks that gel and penetrate readily into the stock. For many years, gas-fired ir-cup irradiators have been used on web-litho and letterpress machines with good success. The most recent innovation, however, is based on new, electrically (rather than gas) energized elements with sufficiently broad wavelength output to render the beam essentially nonselective to colored inks (see Infrared technology).

Flexographic and Rotogravure Inks

Flexographic and rotogravure inks have many elements in common with inks used for other printing processes. They constitute a distinct class, however, because of the specific characteristics of their printing processes, their application and ingredients, and methods of manufacture. The colorants, pigments, and dyes are the same as those used for letterpress, offset, and other methods. (A possible exception is the use of special surface treatment to provide better wetting and dispersion in flexographic and rotogravure-ink vehicles.) However, the flexographic and rotogravure inks differ from other inks in that they are of very low viscosity, they almost always dry by evaporation of highly volatile solvents, and generally they are produced by different manufacturing processes from those used for other inks (3,6,8).

Solvents. Flexographic and rotogravure printing processes require the use of very volatile solvents such as low boiling point alcohols, esters, aliphatic and aromatic hydrocarbons, ketones, and water. However, the flexographic process requires solvents that do not attack rubber rollers and printing plates, whereas there is no such limitation to gravure. Table 3 shows relevant properties of a number of solvents; those marked are limited to gravure.

High volatility is needed to provide fast enough drying to allow subsequent impressions on multicolor work and to prevent sticking or blocking in rewinds.

The high boiling solvents used for heat-set letterpress and offset inks would cause distortion and, therefore, cannot be used for printing plastic substrates usually printed by flexography and rotogravure.

Gravure inks usually are made to dry faster than flexographic inks because each impression must be completely dry before a subsequent impression. Flexographic inks do not have to be completely dry to receive a second impression but trapping (overprinting) is best when there is a great difference in dryness between the first-impression ink and the second-impression ink at the point of the second impression. This is usually controlled through variation in drying speed of the inks, with the first color printed being the most volatile.

Solvents for inks used for odor-sensitive packaging must be of low residual odor

Table 3. Typical Physical Constants of Commercially Used Flexo and Gravure Solvents[a]

Solvent	Comparative drying time[b]	Boiling range, °C	Flash point, °C	Quantity at 20°C, g/L
Alcohols				
methyl alcohol	10	64–65	16	791
ethyl alcohol, 99% anhydrous	18	75–80	18	791
ethyl alcohol, 95%	20	75–80	18	815
isopropyl alcohol, 99%	27	81–83	21	785
n-propyl alcohol	55	95–98	29	803
sec-butyl alcohol	63	99–100	29	801
isobutyl alcohol	77	106–109	39	801
n-butyl alcohol	125	116–119	46	809
Aliphatic naphthas[a]				
hexane	7	66–70	−18[c]	683
fast diluent naphtha	7	60–82	−18[c]	689
heptane (tolu-sol)	15	93–104	−18[c]	725
lacquer diluent	17	93–116	−1[c]	743
octane	31	102–110	−1	743
VM & P (varnish makers' and painters') naphtha	51	121–149	10	755
mineral spirits	600	154–204	38	779
Aromatic hydrocarbons[d]				
toluene	26	109–112	7	868
xylene	88	135–143	27	870
Esters[e]				
methyl acetate, 80%	5	53–59	−1[c]	899
ethyl acetate, 88%	10	72–80	6	887
isopropyl acetate	12	84–90	16	869
n-propyl acetate	22	95–103	18	881
sec-butyl acetate	33	104–117	32	860
isobutyl acetate	36	110–119	31	863
n-butyl acetate	63	115–130	39	875
amyl acetate	100	120–150	47	863
Cellosolve acetate	330	145–165	49	971
Glycol ethers				
Dowanol PM	88	118–126	38	917
methyl Cellosolve	130	123–126	46	962
Cellosolve	190	132–137	54	928
butyl Cellosolve	1000	166–173	74	804
Ketones[f]				
acetone	5	56–57	−9[c]	791
methyl ethyl ketone	11	78–81	−1[c]	804
methyl isobutyl ketone	37	114–117	24	801
methyl isoamyl ketone	150	143–146	43	813
cyclohexanone	270	130–173	54	944
Nitroparaffin				
2-nitropropane	50	119–120	39	987

[a] Acceptable for flexo if used with buna rubber plates and rollers.
[b] The drying-time comparisons are based on the use of ethyl acetate as standard at 10. Acetone is about twice as fast as ethyl acetate and is listed as 5.
[c] At very low temperatures, flash-point determinations are inconsistent; flash occurs below the temperature shown.
[d] Cannot be used in flexo.
[e] Should not exceed 25% of flexo solvent.
[f] Flexo usage restricted to butyl rubber plates and rollers.

to prevent product contamination. Hydrocarbons, in particular, should be selected to have narrow boiling ranges and negligible mercaptan content.

Solvent volatility, like rheology, is varied to attain the best printability with different types of work and printing speed. Slow drying particularly improves printability of fine detail flexography and gravure highlights. It is thus apparent that the high volatility required for high speed is often compromised with printability. Most flexographic and gravure inks operate best when formulated with a combination of solvents. However, the use of solvent blends causes a problem because the solvent evaporates from ink fountains and must be replaced to maintain viscosity. The constituent solvents in a blend seldom evaporate in the same proportion as they were initially in the ink. Therefore, makeup solvent should be proportioned to duplicate the evaporating proportion.

The evaporation of solvent mixtures can be quite complex, because of the interaction of the solvents. In some cases, such as ethyl acetate and toluene, the more volatile solvent, ethyl acetate, is more volatile from mixtures of all compositions; but ethanol (bp 78.3°C) is more volatile from mixtures containing mostly ethanol, whereas from mixtures containing a low percentage of toluene (bp 110.6°C), the toluene shows enhanced volatility.

Solvent combinations are selected for each type of ink on the basis of printability, drying speed, economy, and odor. Examples of specific flexographic and gravure ink systems are given in the sections dealing with specific inks.

Most volatile solvents constitute a fire hazard, requiring both printing plants and ink-manufacturing facilities that are constructed with fire- and explosionproof equipment.

Resins or Polymers and Other Vehicle Solids. The selection of resins or polymers for flexographic and rotogravure inks depends on the printing process, the solvent, and most important, the substrate to be printed and the ultimate use of the printed matter. Detailed examples are covered under the sections dealing specifically with flexographic inks and rotogravure inks. However, certain general requirements are common to both processes.

Printing Process. The printing process is a resin selection determinant because flexographic rubber rollers and plates preclude the use of aromatic hydrocarbons and limit the use of aliphatics, esters, and ketones. Therefore, resins soluble in aromatics only cannot be used in flexographic inks. Esters and ketones should be used in low concentrations. Chlorinated rubber is a typical example of a resin limited to gravure inks.

Solubility. Resins or polymers must be soluble readily and should provide low viscosity solutions at high solids (25–50%). Generally, the better the solubility, the better the print quality. Certain polymers, such as poly(vinyl acetate), tend to string or cobweb, causing poor printing.

Substrate. Paper and other stocks which partially absorb and mechanically hold inks do not present an adhesion problem. Certain paper or paperboard coatings and many plastic films, however, require that the ink have a chemical affinity for the substrate. This is achieved by proper resin selection. For instance, soluble vinyls are usually used in ink formulations for vinyl substrates.

Ultimate Use. The ink vehicle determines the resistance properties of a printed substrate. This indicates that the ultimate use of printed matter is an important resin determinant. For instance, a packaging material used for greasy foodstuffs could not be printed with an ink based on aliphatic-soluble resins because grease acts as a solvent causing the print to smear. Similarly, cellophane which is to be heat-sealed should be printed with inks based on resins of high enough melting point to withstand the heat-seal operation. Inks for food packaging and other odor-sensitive applications should not be formulated with odoriferous resins such as phenolics.

Auxiliary Vehicle Components. Plasticizers are used in flexographic and rotogravure inks to provide ink film flexibility and aid adhesion. Excessive use causes a soft film and blocking.

Waxes and lubricants are often added in small quantities to provide scuff resistance.

Manufacture. Manufacturing processes can be related to two operations—vehicle preparation and color dispersion. Vehicle preparation can be as basic as polymerization of resins or as simple as cold dissolving of vehicle solids in appropriate solvents. Therefore, vehicle equipment includes autoclaves for polymerization reactions and high speed mixers for simple dissolving.

Pigment dispersion usually is done in ball mills which lend themselves to volatile fluid formulations. Pebble mills using porcelain balls and linings are used for white and light colors because steel ball mills, although more effective, cause discoloration. Darker colors such as reds, blues, and blacks are usually made in steel ball mills.

Color concentrates and pigment resin dispersions called chips are made in dough mixers and two-roll rubber mills. These high shear methods often result in better dispersion and consequently higher gloss than is achieved with ball mills. Rubber-mill chips are dissolved similarly to resins to provide color concentrates. Dough mixer and chip concentrates must be diluted with solvent and other vehicles to make finished inks. Sand milling is becoming more widely used in both flexo and gravure ink manufacture. Other high speed dispersing units, such as the Morehouse, Cowles, Kady, and others, are also used to advantage.

Flexographic Inks. Flexographic inks comprise an almost infinite variety of specific formulations for specific purposes. They can, however, be classified into five general composition categories: (*1*) alcohol-dilutable inks containing nitrocellulose as a main constituent of the vehicle; (*2*) polyamide inks; (*3*) dye inks; (*4*) acrylic inks; and (*5*) water-based inks. Inks can also be classified by intended use. Examples include paper or carton board, surface-printed foil and plastic films, and reverse-printed plastic films to be subsequently laminated.

Flexographic inks are supplied at a higher viscosity than they are used, and they must be diluted (reduced) just before use. A normal diluting solvent, or a fast-drying solvent may be used, according to conditions.

Alcohol-Dilutable Inks Containing Nitrocellulose as a Main Constituent of the Vehicle. Nitrocellulose (cellulose nitrate) frequently is combined with other resins and plasticizers to satisfy a variety of needs. These inks are used on paper, cartonboard, glassine, foil, cellophane, cellulose acetate, ethyl cellulose, polyester, polystyrene, and, occasionally, treated polyethylene. Dilutability with solvents primarily containing ethanol is a desirable characteristic because it permits the use of highly resilient, long wearing natural-rubber printing plates. The best solubility and consequent printability results from a mixture of 80–90% ethanol, the remainder comprising an ester such as ethyl, isopropyl, or *n*-propyl acetate, and a glycol ether. The proportion of ester and glycol ether determines drying rate and is, therefore, selected on the basis of press speed and color sequence in multicolor work.

Nitrocellulose-based flexographic inks usually have a little less gloss than the best grades of polyamide type inks. Most nitrocellulose inks have fair to good heat resistance, generally 120–180°C. Nitrocellulose can be combined with acrylic ketone and polyester resins to provide compatibility with laminated adhesives and extruded polyethylene. Adhesive compatibility is required when ink is sandwiched between films to be used as packaging for snacks and other products requiring moisture and gas-barrier properties.

A typical surface-printing nitrocellulose ink formula is: titanium dioxide, 35 wt %; anhydrous ethanol, 30 wt %; esterified fumarated resin, 10 wt %; *n*-propyl acetate, 10 wt %; 0.25-s SS (viscosity measured by Zahn cup) nitrocellulose, 7 wt %; propylene glycol monomethyl ether (retarder), 5 wt %; dibutyl phthalate, 2 wt %; microcrystalline wax, 1 wt %. A typical laminating nitrocellulose ink formula is: titanium dioxide, 35 wt %; anhydrous ethanol, 32 wt %; ketone resin, 18 wt %; *n*-propyl acetate, 8 wt %; 0.25-s SS cellulose nitrate, 7 wt %. The diluents for both are: normal diluent, 90% anhydrous ethanol, 10% *n*-propyl alcohol; fast drying diluent, ethyl acetate; retarder, propylene glycol monomethyl ether (1-methoxy-2-propanol; not for laminating inks); slow drying diluent, 90% *n*-propanol, 10% *n*-propyl acetate. Dry nitrocellulose is explosive; extreme caution should be employed in use and storage. It should be stored in a well-ventilated, cool, isolated place, not in thick sections but in small units with space for ventilation and dissipation of any accumulated heat, not near steam pipes or other source of heat. In case of fire, combat by flooding with water. If fire gets out of control, move personnel out and let it burn. Do not close up space where nitrocellulose is burning; the accumulation of very hot flammable gases explodes violently on contact with air. Avoid inhaling brown fumes from decomposing nitrocellulose.

Polyamide Inks. The most widely used family of flexographic printing inks for printing plastic packaging films are formulated with polyamide resins. Co-solvent polyamide resins are most soluble in a mixture of equal parts of alcohol and aliphatic hydrocarbons. The most common blend is normal propyl alcohol and a low residual odor narrow-cut VM & P (varnish makers' and painters') naphtha. Buna rubber printing plates should be used to accommodate the hydrocarbon solvent.

Co-solvent polyamide resins largely have been replaced with alcohol-soluble varieties. Alcohol-soluble polyamides have the advantage of improved nitrocellulose compatibility and printability with natural-rubber plates. Alcohol-soluble polyamide inks have replaced co-solvent types to prevent odor caused by absorbed hydrocarbon solvents.

A typical co-solvent-type polyamide ink is: *n*-propyl alcohol, 30 wt %; narrow-cut VM & P naphtha, 30 wt %; polyamide resin, 24 wt %; phthalocyanine blue, 15 wt %; polyethylene wax, 1 wt %; normal diluent, 50% *n*-propyl alcohol, 50% VM & P naphtha; fast-drying diluent, 50% isopropyl alcohol, 50% heptane range hydrocarbon (bp 93–113°C).

A typical alcohol-reducible-type polyamide ink is: ethyl alcohol, 33 wt %; alcohol soluble polyamide resin, 28 wt %; phthalocyanine blue, 15 wt %; *n*-propyl alcohol, 15 wt %; *n*-propyl acetate, 8 wt %; polyethylene wax 1 wt %; normal diluent, 90 wt % ethyl alcohol, 10 wt % *n*-propyl acetate; and slow diluent, 90 wt % *n*-propyl alcohol, 10 wt % propyl acetate.

Where heat resistance is not a primary consideration, polyamide-based inks are used almost universally for printing treated polyethylene and treated polypropylene films, poly(vinylidene chloride)-coated films of all types, polyester films, cellulose acetate films, oriented and expanded polystyrene films, and polyamide films. Moderate heat resistance is achieved by replacing part of alcohol-soluble polyamide with nitrocellulose. Polyamide inks containing nitrocellulose have somewhat less gloss than inks containing only polyamide resin.

Slow solvent–low odor hydrocarbons, moderately less volatile than VM & P naphtha, are not generally available. Therefore, the best way to retard drying speed is by increased proportion of n-propyl alcohol or by modification with a resin that has slow solvent release (esterified fumarated rosin) (see Polyamides).

Dye Inks. The original flexographic inks were based on dyes dissolved in alcohol and mixed with shellac. The dyes, usually of triphenylmethane type, provide brilliant, intense colors for paper, glassine, and foil. Permanence and bleeding of these inks have been improved through laking with phenolic resin but dye inks are still not considered for packaging films.

A typical dye-ink formula is: methyl violet, 15 wt %; phenolic resin, 30 wt %; ethyl alcohol, 55 wt %; normal diluent, ethanol; slow-drying diluent, glycol ester; fast-drying diluent, ethyl acetate.

Acrylic Inks. Acrylic-type flexographic inks are most widely used where substantial heat resistance, together with their unique properties, are required for printing a wide range of poly(vinyl chloride) and poly(vinylidene chloride) films and coatings, oriented polystyrene, rubber hydrochloride, some grades of polyester films, some grades of cellulose acetate films, polyamide films, and some oriented polypropylene films.

Acrylic inks usually have less gloss than nitrocellulose-based inks. Dried ink films based on acrylic resins are usually very hard and tough with excellent abrasion resistance.

Most of the flexographic inks in this category require a mixture of a minimum of 20% ester solvent with anhydrous proprietary ethanol. In some cases, higher percentages of ester solvent are recommended to obtain good printability. Because of this, either natural-rubber printing plates and rollers or butyl-rubber printing plates and rollers often are recommended. In addition to excellent adhesion on many plastic film surfaces, acrylic-type flexographic inks are distinguished for good block resistance, excellent grease resistance, excellent water resistance, and many other difficult product requirements. With suitably formulated inks, these qualities are frequently combined with good heat-sealing characteristics. Partial replacement of acrylic resin with nitrocellulose provides even better heat-sealing characteristics, heat resistance and compatibility with laminating adhesives.

A typical acrylic ink formula is: ethanol, 32 wt %; titanium dioxide, 30 wt %; alcohol-soluble acrylic resin, 20 wt %; n-propyl acetate, 17 wt %; microcrystalline wax, 1 wt %; normal diluent, 65 wt % ethanol, 35 wt % n-propyl acetate; slow-drying diluent, glycol ether; and fast-drying diluent, ethyl acetate.

Water-Based Inks. Approximately 30% of all flexographic inks use water as their primary solvent and diluent. Until recently, they contained vehicles based on alkali-soluble proteins such as casein and alpha protein and partially esterified fumurated rosin and shellac. Carboxylated acrylic resins, usually containing styrene, have largely replaced natural resins because they provide better abrasion and water resistance. Ammonia or other volatile amines are used to solubilize carboxylated resins and form resin salts. The volatile alkali evaporates from the ink film, rendering the printed matter water-resistant.

Water inks differ from acrylic water paints which contain latices. Latices historically have resulted in high yield value, thixotropic inks having poor printing properties. Newer latex concepts, however, have been useful. They benefit ink formulation by increasing vehicle solids. High molecular weight compositions can be varied to provide improved physical properties such as adhesion, gloss, and product resistance.

Water inks historically have been used for a variety of paper and paper board. The excellent heat resistance of proteins make that type of ink particularly useful for kraft liner board which must later withstand the heat of a corrugating process. Main advantages of water inks include excellent press stability printing quality, heat resistance, absence of fire hazard, and the convenience and economy of water for reduction and washup. The historical disadvantages include low gloss and slow drying which limited their use to absorbent stocks. The increasing cost of petroleum-derived organic solvents and air-pollution regulations have led to the development of water inks having excellent gloss, water resistance, and adhesion to foil and plastic film.

A typical low cost, water-based ink formula is: water, 62 wt %; barium lithol red, 15 wt %; isolated soya protein, 5 wt %; octyl alcohol (defoamer), 5 wt %; ammonium hydroxide (density, 0.90 g/cm^3) (29.1 wt % NH_3), 2 wt %; microcrystalline wax, 1 wt %. A typical better wet and dry rub resistance, water-based ink formula is: water, 63.5 wt %; carboxylated acrylic resin, 18 wt %; barium lithol red, 15 wt %; ammonium hydroxide (density, 0.90 g/cm^3) (29.1 wt % NH_3), 2 wt %; polyethylene wax, 1 wt %; fatty acid ester or silicone defoamer, 0.5 wt %. The diluents for both are: normal diluent, water; slow-drying diluent, glycol ether or glycol and fast-drying diluent, ethanol.

Rotogravure Inks. Since there is no rubber contact with solvents in gravure ink formulations, it is permissible to use solvents such as ketones and aromatic hydrocarbons which cannot be tolerated in flexo inks. This provides the gravure ink formulator with much greater latitude with regard to binder selection and constitutes the principal difference between gravure and flexo inks. In other respects the compositions used are on the whole similar (2,6,8).

Straight polyamide ink formulations have been reported to cause trouble on gravure equipment because of poor wiping characteristics. This objection has been compounded by the inability of straight polyamide inks to dry hard enough between color stations to work well at high press speeds. Additions of nitrocellulose to such an ink will improve both performance characteristics. Gravure printing of polyethylene film is limited to only a few converters which have mastered the stretchiness and poor registration problems with polyethylene by using carrier webs—sometimes referred to as slip sheeting. The use of polyamides in protective overlacquers applied by gravure appears to be more important than its use in gravure inks at this time.

Nitrocellulose-based inks are widely used for rotogravure printing of the same variety of plastic films mentioned in the section on flexographic inks. The main difference here again is that the rotogravure process does not involve any solvent restrictions so that solvents, such as isopropyl acetate, toluene, etc, can be used. Alcohol frequently is used as a part of the blend but in minor quantities.

Acrylic-type rotogravure inks are quite widely used for the same combination of properties, product resistance, and use requirements as those listed for flexographic inks. However, the thinner or solvent mixtures for the acrylic inks in rotogravure printing normally consist of ester solvents such as isopropyl acetate. Little or no alcohol is used in the rotogravure version of the acrylic-type ink formulations.

Another important type of rotogravure ink system for flexible packaging application is that based on chlorinated rubber binder dissolved in aromatic hydrocarbon solvent such as toluene. Inks of this type cannot be used on flexographic printing equipment because of the undesirable solvent swelling of rubber plates and rollers that would result.

As with flexo inks, gravure ink formulations must take into account the following end-use characteristics: adhesion; blocking; alkali resistance; odor; grease and oil resistance; moisture-vapor transmission; paraffin bleed; acid resistance; scuff resistance; heat resistance; and product resistance.

To achieve the above-mentioned end-use characteristics, one must know which substrate is to be printed, ie, paper, film, foil, or board, and then one can select an ink containing the proper binder and solvent combination to satisfy the requirements. All of the various gravure ink systems are not necessarily compatible with one another. Because of this, the system shown in Table 4 has been devised to avoid intermingling incompatible types.

Rotogravure publication inks represent another large segment of the total production of gravure inks made and used in the United States. Inks of this type are used in printing paper substrates used in books, magazines, catalogues, Sunday supplements, preprints, comics, etc. Hydrocarbon solvent systems are generally used for the sake of economy. The binders used are aliphatic or aromatic hydrocarbon-soluble modified rosin types having good solvent release to ensure fast drying and low residual odor. Good gloss and printability are also prime requisites. Publication printing press speeds in the range of 550 m/min are not considered unusual now and progress is being made toward the 760 m/min mark.

A typical rotogravure publication ink (type A blue) formula is: lactol spirits (aliphatic hydrocarbon solvent), 37 wt %; metallated rosin, 35 wt %; clay, 17 wt %; phthalocyanine blue pigment, 11 wt %.

Miscellaneous Inks

Screen-Process Inks. These inks, often known in the past as silk-screen inks, are printed on the substrate by being forced through a screen stencil by means of a squeegee. For many years this was a hand-printing operation, but it has now become largely mechanized. Screen-process inks are dispersions of pigments in vehicles which are, for the most part, solutions of resins in solvents of the boiling range of VM & P naphtha. Drying is usually by evaporation, but in some cases it is a combination of oxidation and evaporation. Various types of binders are used such as rosin esters, phenolics, cellulose derivatives, vinyls and oleoresinous varnishes, depending on the film properties desired. After premixing, the ink is ground on a three-roll mill. The resulting ink should be short and soft so as not to drag on the squeegee and to release the substrate cleanly after the print is made.

The screen-process method of printing can apply a thicker film of ink to the substrate than other printing processes, and for this reason, is particularly adapted to applications where maximum opacity is desired, or where fluorescent pigments are used, or for printing etch-resists for printed circuits. Screen process also is used for printing of both board and paper posters, metal signs, glass, ceramics, and plastics.

Table 4. Gravure Ink Selection Systems

Type	Resin or binder used	Solvents used	Applications
A	metallated rosins, gilsonite, and other hydrocarbon solvent resins	low-cost aliphatic hydrocarbons such as hexane, textile spirits, lactol spirits, VM & P naphtha, mineral spirits	newspaper supplement, catalogue preprint, and similar publication printing
B	ethyl cellulose plus other modifying resins	50% aromatic hydrocarbon solvent 50% aliphatic hydrocarbon solvent	same as above except designed for better performance on coated stock
C	nitrocellulose modified with resins and plasticizers	esters, ketones usually extended with aromatic hydrocarbon solvent diluents such as toluene, xylene	for printing on all papers, films, foils, and paperboard including nitrocellulose-coated cellophane, glassine, acetate, metallized paper, etc
D	polyamides	usually a 50/50 blend of alcohol and aliphatic or aromatic hydrocarbon solvent	printing on foil, paper, boards, polymer-coated cellophane, polyethylene, polyester, and other specialty films; also for hard, tough, glossy, overlacquer application
E	nitrocellulose or ethyl cellulose plus alcohol-soluble resin modifiers and plasticizers	ethanol or other alcohols plus ester solvents such as ethyl acetate	dye inks and pigmented inks for stocks described under type C
T	chlorinated rubber plus other modifying resins and plasticizers	usually aromatic hydrocarbon solvent such as toluene or xylene	for nitrocellulose-coated cellophane, foil, paper and board for labels, wrappers, and cartons; also glossy overlacquers
W	natural or synthetic resins such as shellac, casein, maleated rosins, etc	water plus alcohols when required; ammonia or other	for absorbent stocks such as liner board, wallpaper, gift wrap, laminating inks, board to be waxed, etc
X	any other nonrecognized type required	alkali used also for solubilization of binder	all other miscellaneous applications

The substrate need not be flat as the process will also accommodate round, oval, or other irregularly shaped objects.

Mimeograph Inks. Mimeograph printing is a stencil-printing process used primarily for office duplicating work for reproducing typewritten copy. The film stencil is cut by typewriter or by means of a hand stylus and wrapped around the ink-filled drum of a mimeograph printing machine. Mimeograph inks are therefore rather fluid and nonoxidizing but must dry quickly by absorption into the stock. They are generally made by dispersing carbon black into a glycol vehicle containing one or more of a cellulose binder, castor oil, and lanolin.

Stamp-Pad Inks. These inks are impregnated into a cloth or foam rubber pad and transferred by pressure to rubber type which is then stamped or impressed against the substrate. The inks must be completely nondrying in the pad and yet dry by rapid penetration into the paper. Since it is desirable that the total ink soak into the stock, dyes, such as induline black, are used rather than pigments. The vehicles used are usually glycols, although other solvents may be used when the ink is for stamping on metal. In the latter case, phenol or cresol is incorporated into the formula.

Collotype Inks. These inks are similar to lithographic inks simply because both printing methods are planographic processes. The collotype plate, also called a photogelatin plate, is coated with sensitized gelatin, which is then exposed through an unscreened negative. The gelatin is hardened in proportion to the light intensity through the various areas. When saturated with glycerol and water, the light-hardened portions take ink, whereas the unhardened portions do not. Dampening rollers are not used and the plate is kept damp by moisture from the atmosphere as well as by renewing the glycerol from time to time. The fact that no halftone screens are used results in continuous-tone printing. The inks are formulated similarly to air-drying lithographic inks using oleoresinous vehicles but have a higher concentration of pigment and are much heavier in body.

Ball-Point Inks. These inks are medium-viscosity semi-Newtonian fluids of high tinctorial strength which must be slow drying and free of particles so as to continue to feed to the paper without clogging. Drying on the paper is accomplished by rapid penetration and some evaporation. These properties are obtained by strong dye solutions and pigment dispersions in vehicles containing oleic acid and castor oil or a sulfonamide plasticizer. Rheology of these inks exhibits modest thixotropy which prevents their leakage through the openings around the ball.

Water-based writing inks consist of very fine pigment dispersions in aqueous media containing small amounts of glycol or glycerol and a dispersing aid. They dry mainly by evaporation and quick wetting of cellulosic fibers in paper substrates.

Steel-Die Inks. Steel-plate printing is an intaglio process inks where the image is etched in continuous nonscreened lines on a steel plate. The ink is applied in a heavy layer to fill the engraving and then wiped off the nonprinting surface with paper or cloth, leaving the ink only in the engraving which is pressed against the paper with rather strong pressure to deposit the ink on the paper. This process is used for high-quality stationery, stock certificates, and paper currency. Owing to the thick film that can be deposited, high strength formulations are not required, but the body of the ink is somewhat short so as to wipe cleanly. Drying is by oxidation or by evaporation of solvent. The pigment, including a large percentage of colorless extender pigment, is dispersed on a three-roll mill in a vehicle composed of heat-bodied drying oil or oleoresinous vehicle, sometimes in combination with a resin–solvent-type vehicle. Owing to the thick film of ink deposited, sufficient drying oil must be present to prevent brittleness on aging.

Electrostatic Inks. Electrostatic printing is accomplished by causing charged colored particles to move in an electrostatic field to a substrate in the form of an image. The image may be formed by a screen stencil or by a gravure cylinder. One of the primary advantages of this process is its ability to print through gaps and thus without pressure. The electrostatic ink, also called an electrostatic toner, is a powder composed of pigment dispersed in a resin. The particles must have the proper electrical properties, particle-size range, and be free-flowing. After the image is deposited on the substrate, it is heat- or solvent-fused to a continuous film. Pigments and resins must be chosen to meet the application requirements and at the same time to satisfy the physical and chemical resistance requirements of the process (see Electrophotography).

Decalcomania Inks. Decalcomania is a transfer method of printing. The design is first printed on a temporary base by lithography, letterpress, gravure, or screen process, depending upon the detail and thickness of image desired. Usually the temporary base is paper which has been coated with a water-soluble material. The inks must dry completely on the surface by oxidation or solvent evaporation because the treated paper has no ink absorbency. After the initial printing, the design is transferred to the permanent base by direct contact and soaking with water. Obviously, the formulation of decalcomania inks is governed by the particular printing process employed in printing the transfer paper. Decalcomanias for ceramics require pigments that may be heated to high temperatures. Further, most decalcomanias should use pigments that are fast to light because many are subsequently transferred to outdoor signs or to store windows. Vehicles consist of oleoresinous varnishes containing metallic driers or of resin–solvent types.

Hot-Transfer Inks. Hot-transfer printing is similar to the decalcomania process in that the printing is first done on a temporary substrate, but heat is used as the transfer mechanism rather than water solubility. One type of hot-transfer ink is made with heat-fusible resins and waxes to be transferred to cloth. These inks should penetrate into the cloth and not be affected by subsequent washing of the fabric. In a more recent development for the packaging industry, labels are printed on a web of special coated paper by conventional printing such as gravure. This web is then fed to automatic labeling equipment. The relationship between the paper web and the ink is such that the ink is immediately released by heat and transferred to the surface of the package, such as plastic bottles, to which it then adheres as a permanent label.

Ink-Jet Printing. This is a nonconventional form of printing that has the potential advantage of including all the necessary elements of a printed page in some digitized form in computer memory, thereby eliminating the step of platemaking. One principle of operation involves the issuance of liquid ink from an orifice at very high speed to form a jet which is then broken up by ultrasonic energy to produce uniform droplets that can be charged electrically. These droplets can be deflected electrostatically into a catcher, while the uncharged droplets continue in flight to form dots on the printing surface to construct images. Other principles of operation include pressurized, single-deflected jet; pressurized, multiple-deflected jet; nonpressurized, single-undeflected jet; and nonpressurized, multiple-deflected jet.

Ink-jet technology provides a means of fast, dependable, high quality, single-copy printing. Its nonimpact nature permits printing on uneven surfaces and delicate materials. Computer compatibility permits the encoding of repetitive and nonrepetitive information. Special coated papers are not necessary for ink-jet printing.

The inks formulated for jet printing must be quite fluid, stable, and free of any particles that could cause clogging of the jet nozzles, and be capable of depositing and adhering to a substrate with a minimum of character fogging. They are generally formulated with soluble dye colorant in a suitable resin–solvent vehicle (19–20).

Environmental Considerations

Government regulations are becoming increasingly stringent and have drastically altered the manufacture and use of printing inks and their ingredients (21). The effect of these controls has necessitated the investment of millions of dollars for research to change ink formulation techniques as well as methods of printing.

OSHA has enforced strict, and in some cases, prohibitive control of materials such as lead and chromium salts. The FDA, although not controlling inks that are not in direct contact with food, influences formulations used on meat and poultry packages. The USDA documents acceptability of all such products. Listing of all nonvehicle ingredients under *Title 21, Code of Federal Regulations* is a specific requirement. The Consumer Products Safety Commission limits the amount of lead in children's products (22).

The EPA most significantly affects ink formulation through TSCA, the *Clean Air Act of 1970* and amendments of 1977, the *Resources Conservation and Recovery Act* (RCRA) controlling disposal of waste material, and the *Clean Water Act.* TSCA established an inventory of chemical materials acceptable for manufacture. New chemical compounds require a complex and expensive procedure to gain a manufacturing permit. The *Clean Air Act* has imposed severe restrictions on fuming of common lithographic web-offset and letterpress publication inks. Exhaust of volatile organic compounds (VOC) used as solvent in rotogravure and flexographic ink printing is severely limited because they react in the atmosphere to form ozone and consequent smog. The most economical method to overcome air pollution is through the replacement of VOC with water. Solvent recovery via carbon adsorption and incineration equipment are currently used to satisfy state and Federal emission requirements.

Hazardous waste is defined as anything flammable, corrosive, toxic, or reactive. Many colorants in ink contain lead, chromium and organic compounds defined by RCRA as toxic. Volatile solvents in flexographic and gravure inks are flammable. The *Clean Water Act* controls sewage, severely limiting disposal of wash from corrugated board presses and chemical waste from manufacture of gravure cylinders.

BIBLIOGRAPHY

"Ink, Writing" in *ECT* 1st ed., Vol. 7, pp. 870–877, by Robert S. Casey, W. A. Sheaffer Pen Co.; "Printing Ink" in *ECT* 1st ed., Vol. 11, pp. 149–163, by Andries Voet, J. M. Huber Corporation; "Inks" in *ECT* 2nd ed., Vol. 11, pp. 611–632, by Hugh Dunn, Bohdan V. Burachinsky, James K. Ely, and Paul W. Greubel, Interchemical Corporation.

1. *Printing Ink Handbook,* National Association of Printing Ink Manufacturers, New York, 1975.
2. *Rotogravure and Rotogravure Ink,* Gravure Technical Association and Champlain Co., New York, 1957.
3. D. E. Bissett, *The Printing Ink Manual,* 3rd ed., Northwood Publications, Ltd., London, Eng., 1979.
4. J. F. Trembley and C. M. Loring, Jr., *TAPPI* **52**(10), 18 (1969).
5. H. J. Dunn, *Radio Frequency–Microwave Drying of Water-Based Gravure Inks, Gravure Bulletin No. 4,* Gravure Technical Association, New York, Winter 1976.
6. E. A. Apps, ed., *Ink Technology for Printers and Students,* 3 Vols., Chemical Publishing Co., Inc., New York, 1964.
7. H. J. Wolfe, *Printing and Litho Inks,* 6th ed., MacNair-Dorland Co., New York, 1967.
8. F. A. Askew, ed., *Printing Ink Manual,* 2nd ed., W. Heffer & Sons, Ltd., Cambridge, Eng., 1969.
9. L. M. Larsen, *Industrial Printing Inks,* Reinhold Publishing Corp., New York, 1962.
10. C. A. Smith, *Mod. Paint Coat.* **69**(7) 30; (8), 47; (9), 61 (July–Aug.–Sept. 1979).
11. T. C. Patton, ed., *Pigment Handbook,* 3 Vols., John Wiley & Sons, Inc., New York, 1973.
12. W. M. Morgans, *Pigments for Paints and Inks,* SITA, Manchester, Eng., 1977.
13. R. F. Bock, *Novel Methods for Drying Inks, GATF Progress Report No. 86,* Pittsburgh, Pa., Dec. 1970.
14. P. W. Gruebel, *Print. Plates Mag.* **60**(5), 3 (1974).
15. A. M. Wells, *Printing Inks—Recent Developments,* Noyes Data Corporation, Park Ridge, N.J., 1976.

16. H. Rubin, *UV Curing—Role of Radiant Absorption by Photoinitiators,* TAGA, Rochester, N.Y., 1976, p. 275.
17. S. P. Pappas, ed., *UV Curing: Science and Technology,* Technology Marketing Corporation, 1978.
18. S. V. Noblo and F. P. Tripp, "Electron Curing for Gravure Applications," *Gravure Technical Association Convention, New York, April 11–13, 1978.*
19. P. L. Duffield, *TAGA Proceedings,* Rochester, N.Y., 1974, p. 116.
20. E. W. Zaleski, *TAGA Proceedings,* Rochester N.Y., 1975, p. 345.
21. J. Quarles, *Environ. Rep. Monogr. 28* **10**(1), (May 4, 1979).
22. K. A. Bownes, *Food Drug Packag.* **37**(7), 32 (Oct. 1977).

BOHDAN V. BURACHINSKY
HUGH DUNN
JAMES K. ELY
Inmont Corporation

INORGANIC HIGH POLYMERS

The area of inorganic polymers is very extensive. Many inorganic compounds exist in the solid state as three-dimensional networks or two-dimensional layer structures. Inorganic polymers include solid metals, ionic crystals, ceramics, silica, silicates, etc. Such substances embrace a wide part of traditional inorganic chemistry and are adequately discussed elsewhere (see Ceramics; Silica; etc). In addition to these inorganic polymeric materials, there are inorganic polymers that exhibit elastomeric and plastic properties. The polymer backbones consist essentially of inorganic elements and occasionally have organic side groups to modify properties.

In general, inorganic high polymers offer unique property profiles not available with conventional organic polymers. Polysiloxanes are prime examples of inorganic high polymers. They possess a unique combination of high temperature stability and excellent low temperature elastomeric properties. Siloxanes are commercially important inorganic high polymers (see Silicon compounds, silicones).

Poly(phosphazenes)

Poly(phosphazenes) are polymers having alternating phosphorus and nitrogen atoms in the polymer backbone:

$$-\!\!\left(\begin{array}{c} \mathrm{Y} \\ | \\ \mathrm{P{=}N} \\ | \\ \mathrm{Y} \end{array}\right)_{\!n}\!\!-$$

where Y can be halogen, pseudohalogen, or alkyl, aryl, alkoxy, aryloxy, arylamino, or alkylamino groups (see Phosphorus compounds; Fluorine compounds, organic, fluoroalkoxyphosphazenes) (1–3).

Physical Properties. ***Transition Temperatures.*** The two main transitions in polymers are the glass–rubber transition (T_g) and the crystalline melting point (T_m). The T_g is the most important parameter of an amorphous polymer because it determines whether the material will be a hard solid or an elastomer at specific use temperature ranges and at what temperature its behavior pattern changes. In addition to a T_g and T_m, many poly(phosphazenes) exhibit another first-order transition. This transition, referred to as T_1, represents the transformation from a crystalline to mesomorphic state (4). The term mesomorphic is used to suggest an intermediate level of organization between crystalline and the glassy or liquid states. Data on the thermal transitions determined by dta and dsc (differential scanning calorimetry) of a large number of poly(phosphazenes) appear in Table 1.

The dichloro and difluorophosphazene polymers exhibit low glass-transition temperatures (1). Copolymers of trifluoroethoxy and heptafluorobutoxy (50:50) groups are elastomeric with a glass transition of −77°C (5). Poly(aryloxyphosphazenes) exhibit a T_g, T_1, and T_m. The useful properties of these polymers are dominated by the position of T_1. The softening of the polymer at T_1 sets the upper limit for useful properties as a structural material (4).

The T_gs for poly(arylaminophosphazenes) are rather high compared with those of the aryloxy polymers. The polymers studied have no simple melting transitions (6).

Thermal Stability. Thermogravimetric analysis (tga) of several poly(phosphazenes) exhibit the onset of weight loss at temperatures above 300°C. However, tga data overestimate the thermal stability of poly(phosphazenes) and the weight loss more accurately reflects the boiling points of oligomers formed by equilibration at lower temperatures (7–9).

Detailed studies of thermal stability were made by examining the influence of

Table 1. Transition Temperatures of Poly(phosphazenes)

Y	CAS Registry No.	T_g, °C	T_1, °C	T_m, °C
F	[*27290-47-7*]	−96		−68
Cl	[*26085-02-9*]	−63		−30
Br	[*56529-01-2*]	−15		270
OCH_3	[*60998-43-8*]	−76		
OCH_2CH_3	[*60495-46-7*]	−84		
OCH_2CF_3	[*28212-50-2*]	−66		242
$OCH_2CF_3/OCH_2(CF_2)_3F$	[*52902-20-2*]	−77		
$OCH_2(CF_2)_2H/OCH_2(CF_2)_6H$	[*52902-19-9*]	−60		
C_6H_5O	[*28212-48-8*]	6	160	390
p-$CH_3C_6H_4O$	[*52233-91-7*]	0	152	340
p-ClC_6H_4O	[*52233-66-6*]	4	169	365
m-ClC_6H_4O	[*52233-67-7*]	−24	90	370
C_6H_5NH	[*57650-68-7*]	105		
p-$CH_3C_6H_4NH$	[*57629-02-4*]	97		
m-$CH_3C_6H_4NH$	[*57629-03-5*]	76		
p-$CH_3OC_6H_4NH$	[*57629-07-9*]	92		
p-ClC_6H_4NH	[*57629-08-0*]	85		
m-ClC_6H_4NH	[*57629-09-1*]	80		

heat-aging time and temperature on the molecular weight as measured by intrinsic viscosity and gel permeation chromatography. For example, poly(diphenoxyphosphazene) exhibited thermally induced chain cleavage at 100°C. Above 150°C the depolymerization was very rapid (7–8).

The decomposition temperature range as determined by tga in air for poly(arylaminophosphazenes) is 220–266°C (6).

Manufacture. Poly(dichlorophosphazene) is prepared by the thermal polymerization of hexachlorocyclotriphosphazene:

$$(NPCl_2)_3 \xrightarrow[\text{in vacuo}]{250\ ^\circ\text{C}} \left[P(Cl)_2{=}N \right]_n$$

Generally a cross-linked rubbery material is obtained (1). However, under carefully controlled reaction conditions and purification of trimer, the linear polymer can be prepared. It is important to terminate the polymerization reaction at an intermediate stage (250°C for 24–48 h) in order to prevent extensive branching and cross-linking (1–3). Thus, linear polymers with degrees of polymerization up to 15,000 are obtained in yields of 15–70%. The purification procedure is straightforward (3). The polymer is soluble in dry benzene or toluene. Cross-linked material is insoluble and is removed by filtration. The solution is then poured into pentane or light petroleum to precipitate the polymer. Cyclic material and low molecular weight oligomers are not precipitated and can be removed easily. This polymer is hydrolytically unstable and degrades slowly upon exposure to a moist environment. Replacement of the chloro groups has resulted in improved stability.

Alkoxyphosphazene and fluoroalkoxyphosphazene polymers are prepared by treating poly(dichlorophosphazene) with a slight excess of sodium alkoxide in refluxing benzene–tetrahydrofuran (60–80°C). Yields of polymer are often 60–80% (5,10).

$$\left[P(Cl)_2{=}N \right]_n \xrightarrow{2\ \text{NaOR}} \left[P(OR)_2{=}N \right]_n + 2\ \text{NaCl}$$

More rigorous conditions are required for the preparation of poly(aryloxyphosphazenes). Temperatures greater than 100°C are necessary to obtain complete substitution with aryloxy groups (10–11). However, at higher temperatures polymer degradation can lead to lower yields. Generally, yields of ca 30–80% are obtained.

The preparation of poly(diaminophosphazenes) is carried out in the presence of an excess of an organoamine which functions both as a nucleophile and a hydrochloride acceptor (12–13). A tertiary amine also may be added as a hydrochloride acceptor. Lower molecular weights and low yields (12–30%) are usually obtained which suggest decomposition during the substitution reaction.

At ambient temperatures, the use of methylamine results in cross-linked polymer (12). However, at lower temperatures linear polymer is obtained (12). Straight-chain primary aliphatic amines and meta- and para-substituted anilines give completely substituted linear polymer.

$$-\!\!\left(\begin{array}{c}Cl\\|\\P\\|\\Cl\end{array}\!=\!N\right)_{\!n}\!- + 2\,RNH_2 \longrightarrow -\!\!\left(\begin{array}{c}NHR\\|\\P\\|\\NHR\end{array}\!=\!N\right)_{\!n}\!-$$

Secondary amines such as dimethylamine and piperidine give completely substituted polymer. Bulkier secondary amines and branched primary amines give incomplete substitution (12).

$$-\!\!\left(\begin{array}{c}Cl\\|\\P\\|\\Cl\end{array}\!=\!N\right)_{\!n}\!- + 2\,R_2NH \longrightarrow -\!\!\left(\begin{array}{c}NR_2\\|\\P\\|\\Cl\end{array}\!=\!N\right)_{\!n}\!-$$

This mono-substituted polyphosphazene can be subjected to further nucleophilic substitution with a less hindered amine to yield mixed substituted derivatives (12).

$$-\!\!\left(\begin{array}{c}NR_2\\|\\P\\|\\Cl\end{array}\!=\!N\right)_{\!n}\!- + R'NH_2 \longrightarrow -\!\!\left(\begin{array}{c}NR_2\\|\\P\\|\\NHR'\end{array}\!=\!N\right)_{\!n}\!-$$

Properties of Vulcanizates. The poly(phosphazene) copolymers and terpolymers containing equal amounts of trifluoroethoxy and octafluoropentoxy substituents are amorphous with a low T_g (−67°C) and have been studied in some detail (13–14). The terpolymer contains a small amount of a substituent which facilitates vulcanization (15).

Formulation Parameters. Typical elastomer formulations consist of three major components: reinforcing and extending fillers, a high activity magnesium oxide for efficient cure and enhanced thermal stability, and a peroxide curing agent. In addition, auxiliary agents such as silanes, stabilizers, process aids, and co-agents have been used in specialized applications (16).

Heat Aging. The thermal stability of vulcanizates was evaluated in heat-aging studies in air at 135–204°C (14,16). Excellent retention of properties was exhibited after heat aging at 135°C and 149°C—even after 1000 h, 78 and 76% of the tensile strength remained, respectively. At 177°C and 204°C the tensile strengths decreased more rapidly. After 240 h in air at 177°C and 204°C, only 56 and 21% of the tensile strength was retained, respectively.

Solvent Resistance. Fluoroalkoxyphosphazene vulcanizates have good to excellent resistance to lubricants, fuels, and hydraulic fluid at ambient and elevated temperatures. For example, a vulcanizate displayed excellent retention of mechanical properties, low volume swell, and hardness change after more than 1000 h of immersion in a lubricant at 125°C (16).

Hydrolysis Resistance. The hydrolytic stability of vulcanizates was demonstrated by exposure to 100% rh at 100°C for nearly 900 h. The vulcanizates retained almost 85% of their original tensile properties (16).

Flammability. Poly(phosphazenes) exhibit good to excellent flame-retardant properties. The oxygen index values are 24–65. There is only moderate smoke evolution (3) (see Flame retardants, phosphorus compounds).

Uses. The unusual resistance to oils, gasoline, jet fuel, and hydraulic fluids (qv), as well as the low temperature flexibility of fluorophosphazene elastomers (see Elastomers synthetic, fluorinated), have led to utility in the preparation of O-rings, gaskets, and hydrocarbon fuel hoses (14). Low smoke, flame-retardant foams (qv) and wire coatings have been prepared using poly(aryloxyphosphazene) copolymers (3).

Poly(carborane–siloxanes)

Carborane–siloxanes are a family of polymers that have the linear structure (1)

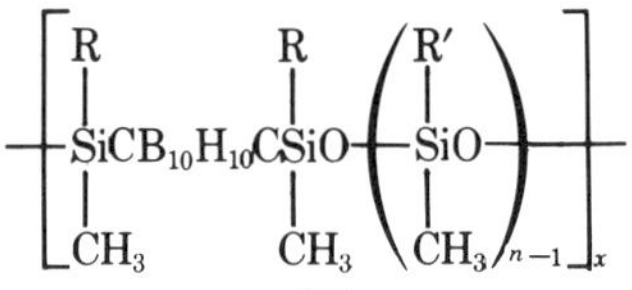

$$\left[\begin{array}{c} R \\ | \\ -SiCB_{10}H_{10}CSiO- \\ | \\ CH_3 \end{array} \!\!\! \begin{array}{c} R \\ | \\ \\ | \\ CH_3 \end{array} \left(\begin{array}{c} R' \\ | \\ SiO \\ | \\ CH_3 \end{array} \right)_{n-1} \right]_x$$

(1)

where R and R′ can be methyl or fluoroalkyl, and in addition R′ can be phenyl. The term carborane is commonly used in a generic sense to describe compounds composed of boron, hydrogen, and carbon whose molecular geometries are polyhedra or polyhedral fragments (see Boron compounds, boron hydrides, carboranes, and their metalloderivatives). Several families of carboranes exist with the general formulas $CB_nH_{(n+2)}$, $CB_nH_{(n+4)}$, $C_2B_nH_{(n+2)}$, etc (17). Of these families, the neutral, closed, polyhedral $C_2B_nH_{(n+2)}$ (n = 5 and 10) species have been used in the preparation of carborane–siloxane polymers (17–18). Polymers prepared from 1,7-dicarba-*closo*-dodecacarborane-12, $C_2B_{10}H_{12}$, have been studied extensively. The interest in carborane–siloxanes centers around the need for new polymers having enhanced flame resistance, and greater thermal and oxidative stability (17). Indeed, the incorporation of the m-carborane moiety into the siloxane backbone has resulted in significant enhancement of properties (see Silicon compounds, silicones).

The nomenclature used for these polymers is similar to that used in silicone chemistry. In this convenient shorthand, the letter D designates that the silicon atoms are difunctional (ie, two oxygen atoms attached to it). Numerical subscripts on the letter indicate the number of siloxane moieties

$$-\!\!\left(\begin{array}{c} | \\ SiO \\ | \end{array}\right)_{m}\!\!-$$

Physical Properties. ***Transition Temperatures.*** Physical transitions, T_g and T_m, for a series of linear carborane–siloxane polymers (1), have been examined and the data appear in Table 2.

For the homologous series of carborane–dimethylsiloxanes [(1) R = R′ = CH_3, n = 1 through 5], the glass-transition temperature decreases with increasing siloxane content (ie, greater n) (19). The D_1 and D_2 polymers have crystalline melting points. This crystalline phase adversely affects elastomeric properties. In the case of the D_2 polymer, this crystallinity can be eliminated by the incorporation of phenyl groups on the polymer backbone (R′ = C_6H_5). Thus, phenyl modification results in a completely amorphous system; however, the T_g has risen from −42 to −12°C. Partial phenyl modification (R′ = CH_3/C_6H_5 [67/33]) produces an amorphous polymer with a low T_g (−37°C) (20). The D_3, D_4, and D_5 polymers are amorphous with decreasing T_gs of −68, −75, and −88°C, respectively. The trifluoropropyl-modified polymers are amorphous and exhibit decreasing T_g with increasing n and increasing T_g with increasing $CH_2CH_2CF_3$ content (21–22). For the D_2-m-series, the glass transition temperature increased with the number of methyltrifluoropropyl groups. The —$CH_2CH_2CF_3$ group is bulkier than the —CH_3 group; therefore, as the methyl groups are replaced with —$CH_2CH_2CF_3$, the polymer chain mobility decreased and the T_g increased. This is consistent with literature data, which report the T_g of polymethyltrifluoropropylsiloxane as −75°C, compared to −123°C for polydimethylsiloxane (22).

Thermal Stability. Thermogravimetric analyses (tga) in an inert atmosphere reveal that rapid weight loss does not occur until heating above 400°C for carborane–siloxanes. For the homologous series of carborane–dimethylsiloxanes, the stability of the polymers increases with decreasing siloxane content per repeat unit (19,23). The total weight loss by 700°C appears in Table 2.

The phenyl-modified polymers show a decrease in weight loss compared to their all-methyl analogues (20). It has been reported that the presence of phenyl groups

Table 2. Transitions and Stability of Carborane–Siloxanes

n	R	R′	CAS Registry No.	T_g, °C	T_m, °C[a]	Loss at 700°C, wt %
1	CH_3		[*26635-30-3*]	25	260	20
2	CH_3	CH_3	[*31762-68-2*]	−42	90	32
2	CH_3	C_6H_5		−12	A	5
2	CH_3	CH_3 (33) C_6H_5 (67)		−22	A	5
2	CH_3	CH_3 (67) C_6H_5 (33)		−27	A	5
3	CH_3	CH_3	[*26793-80-6*]	−68	A	45
4	CH_3	CH_3	[*41578-83-0*]	−75	A	48
5	CH_3	CH_3	[*32473-92-0*]	−88	A	60
1	$CH_2CH_2CF_3$			28	A	
2	CH_3	$CH_2CH_2CF_3$	[*62725-81-9*]	−29	A	25
2	$CH_2CH_2CF_3$	CH_3	[*62725-80-8*]	−12	A	59
2	$CH_2CH_2CF_3$	$CH_2CH_2CF_3$	[*62725-82-0*]	−3	A	69
3	$CH_2CH_2CF_3$	$CH_2CH_2CF_3$	[*41655-25-8*]	−15	A	78

[a] A = Absent.

in carborane–siloxanes leads to cross-linking and less loss of weight. Trifluoropropyl modification leads to increased weight loss (21–22).

Manufacture. Different structures of carborane–siloxane polymers are obtainable via several synthetic routes which have been developed. The D_3 and D_5 polymers can be prepared by hydrolysis condensation of a carborane-based siloxane (eq. 1) (24).

$$Cl\left(\begin{array}{c}R'\\|\\SiO\\|\\CH_3\end{array}\right)_{\left(\frac{n-1}{2}\right)}\begin{array}{c}R\\|\\SiCB_{10}H_{10}CSi\\|\\CH_3\end{array}\left(\begin{array}{c}R'\\|\\OSi\\|\\CH_3\end{array}\right)_{\left(\frac{n-1}{2}\right)}Cl \xrightarrow{H_2O} (\mathbf{1}) \qquad (1)$$

The D_4, D_5, and D_6 polymers are prepared by the cohydrolysis condensation shown in equation 2 (24).

$$\begin{array}{c}R'\;\;R\qquad\qquad R\;\;R'\\ClSiOSiCB_{10}H_{10}CSiOSiCl\\H_3C\;\;CH_3\qquad H_3C\;\;CH_3\end{array} + \begin{array}{c}R'\\|\\ClSi\\|\\CH_3\end{array}\left(\begin{array}{c}R'\\|\\OSi\\|\\CH_3\end{array}\right)_{n-4}Cl \xrightarrow{H_2O} (\mathbf{1}) \qquad (2)$$

D_2 polymer can be prepared according to equation 3 where X is an *N*-phenyl-*N′*-tetramethyleneureido group (25).

$$\begin{array}{c}R\qquad\qquad R\\|\qquad\qquad |\\HOSiCB_{10}H_{10}CSiOH\\|\qquad\qquad |\\CH_3\qquad\quad CH_3\end{array} + \begin{array}{c}R'\\|\\XSiX\\|\\CH_3\end{array} \longrightarrow (\mathbf{1}) \qquad (3)$$

In addition, the D_2 polymer can be prepared directly from 1,7-carborane (**2**) in a facile, one-pot synthesis (26). Compound (**2**) is first converted to its dilithio salt (**3**) by reaction with *n*-butyllithium (eq. 4).

$$\underset{(\mathbf{2})}{HCB_{10}H_{10}CH} + 2\,CH_3(CH_2)_3Li \rightarrow \underset{(\mathbf{3})}{LiCB_{10}H_{10}CLi} + CH_3(CH_2)_2CH_3 \qquad (4)$$

The dilithio salt is then treated with a 1,5-dichlorotrisiloxane (**4**) to give polymer (eq. 5).

$$(\mathbf{3}) + \underset{(\mathbf{4})}{\begin{array}{c}H_3C\;\;R\;\;CH_3\\|\quad|\quad|\\ClSiOSiOSiCl\\|\quad|\quad|\\H_3C\;\;CH_3CH_3\end{array}} \longrightarrow \left(CB_{10}H_{10}\begin{array}{c}H_3C\;\;R\;\;CH_3\\|\quad|\quad|\\CSiOSiOSi\\|\quad|\quad|\\H_3C\;\;CH_3CH_3\end{array}\right)_n \qquad (5)$$

Equation 6 shows the ferric chloride-catalyzed synthesis of the D_1 polymer (27).

$$\begin{array}{c}R\qquad\qquad R\\|\qquad\qquad |\\ClSiCB_{10}H_{10}CSiCl\\|\qquad\qquad |\\CH_3\qquad\quad CH_3\end{array} + \begin{array}{c}R\qquad\qquad R\\|\qquad\qquad |\\CH_3OSiCB_{10}H_{10}CSiOCH_3\\|\qquad\qquad |\\CH_3\qquad\quad CH_3\end{array} \xrightarrow[\Delta]{FeCl_3} (\mathbf{1}) \qquad (6)$$

Properties of Vulcanizates. The amorphous polymers have useful elastomeric properties and can be formulated with fillers and other additives and vulcanized using standard silicone technology.

Formulation Parameters. The phenyl-modified polymers possess the optimum combination of high temperature and elastomeric properties and were used in the study of formulation parameters (17). These variables can have an important effect on the thermal stability and property profile of vulcanized systems. For example, the use of reinforcing silicas, peroxide content, and oxidative stabilizers are important (25,28–29). However, polymer–silica interactions have the most pronounced effect on retaining properties during high temperature aging studies. Heat aging at 315°C in air of a sample reinforced with hydrophilic silica (amorphous) leads to a rapid loss of elongation after 150 h. However, modifying the silica by trimethylsilylation (hydrophobic silica) results in substantial improvements; even after 1000 h at 315°C in air, elastomeric properties are retained (30).

The adverse effect of the hydrophilic silica is attributed to the condensation reaction of surface silanol groups on the silica and phenylsilane moieties on the polymer backbone. This results in increased cross-linking via formation of siloxane bonds between the polymer and silica.

Solvent Resistance. Resistance to solvents is important in several end-use situations. The degree of swelling in various solvents can be modified by changes in the polymer backbone. The incorporation of trifluoropropyl groups onto the D_2 polymer leads to substantial decreases in swelling (22). In vulcanized systems, the swelling decreased with increasing $CH_2CH_2CF_3$ content as shown in Table 3.

Hydrolysis Resistance. Exposure of a D_2-vulcanizate to 91% rh at 38°C for 14 d does not indicate any deterioration of properties (31). Immersion of a D_2-vulcanizate in water for seven days at ambient temperature and for 24 h at 100°C showed no significant change in tensile strength or elongation.

Flammability. The oxygen index (OI), the minimum amount of oxygen (expressed as percent by volume) necessary to sustain combustion of a material, is a measure of flammability characteristics of polymers. A D_2-carborane–siloxane vulcanizate compounded with 30 phr (parts per hundred of resin) silica exhibited an OI of 62 (25). For comparison, silicone rubber has an OI of 30–33.

Electrical Properties. Dielectric constants and loss factors (ASTM 150) for D_2- and D_4-vulcanizates are comparable with other insulating materials, such as conventional silicone, and butadiene-styrene polymers (31).

Table 3. Swelling as a Function of Polymer Modification[a]

n	R	R′	Swelling, %	
			Toluene	Reference fuel B
2	CH_3	CH_3 (67) C_6H_5 (33)	160	113
2	CH_3	$CH_2CH_2CF_3$	98	80
2	$CH_2CH_2CF_3$	CH_3	63	43
2	$CH_2CH_2CF_3$	$CH_2CH_2CF_3$	23	20

[a] ASTM D 471; vulcanizates with 30 phr (parts per hundred of resin) silica.

Uses. The high temperature capabilities of these polymers have utility as liquid phases in gas chromatography. Carborane–siloxanes have been fabricated into O-rings, gaskets, and wire coatings which are capable of performing at temperatures exceeding 300°C (18,25,31) (see Heat-resistant polymers).

Poly(sulfur Nitride)

Poly(sulfur nitride) [*56422-03-8*] or polythiazyl, $(SN)_x$, is composed of nonmetallic elements. However, it exhibits metallic electronic conducting properties. At low temperatures, it becomes superconducting (see Polymers, conductive; Superconducting materials) (32–34).

Poly(sulfur nitride) is not a new material; however, its unusual physical properties were not recognized until 1973 (33). Since then, investigations have focused on its electrical properties.

Properties. Polymeric sulfur nitride is a crystalline, fibrous material that is soft and malleable and can be flattened readily by mild pressure. Its density is 2.30 g/cm^3 (35).

Structure. Electron-micrograph studies of $(SN)_x$ reveal that the crystals are composed of layers of fibers stacked parallel to each other along the long axis of the crystal (36). Where the crystals are separated mechanically, long fibrous strands were apparent. These crystals are highly anisotropic and are easily cleaved along a plane parallel to the fibers.

Single crystals of analytically pure $(SN)_x$ were examined in x-ray studies (37). Poly(sulfur nitride) consists of an almost planar chain of alternating sulfur and nitrogen atoms.

```
S(c)—N(c)       S(a)—N(a)
        \       /       \       /
         S(b)—N(b)       S(d)—N(d)
```

There are four SN units per unit cell which has a = 0.4153(6), b = 0.4439(5), and c = 0.7637(12) nm, β = 109.7°(1), and d_c = 2.30 g/cm^3. The refined structure (R = 0.11) has as its major feature $(SN)_x$ chains with intrachain distances of S(a)–N(a) = 0.1593(5), S(a)–N(b) = 0.1628(7), S(a)–S(b) = 0.2789(2), N(a)–N(b) = 0.2576(7), and S(a)–N(c) = 0.2864(5) nm and bond angles of S–N–S = 119.9°(4) and N–S–N 106.2°(2). The S–N bond lengths are all very similar, and they correspond to a sulfur–nitrogen bond order that is intermediate between those expected for a single and a double bond.

Brominated poly(sulfur nitride) has the approximate formula of $(SNBr_{0.4})_x$ (38). This material is crystalline and fibrous and has a density of 2.67 g/cm^3. X-ray studies show considerable defect structure.

Thermal Stability. When $(SN)_x$ was heated in a sealed tube *in vacuo* at approximately 140°C, it decomposed to sulfur, nitrogen, and other species (35).

Electrical Properties. The room temperature conductivity of $(SN)_x$ crystals parallel to the fibers can be as high as approximately 3700/(Ω·cm) (35). The conductivity increases over 200 times on lowering the temperature to 4.2 K. At 0.26 K, the crystals become superconducting (39).

Brominated poly(sulfur nitride), $(SNBr_{0.4})_x$, exhibits an average room temperature conductivity of 3.8×10^4/(Ω·cm) in the direction parallel to the $(SN)_x$ fibers and

8/(Ω·cm) perpendicular to the fibers (38). The conductivity increased approximately 90 times on lowering the temperature to 4.2 K.

Manufacture. Poly(sulfur nitride) has been prepared by thermal decomposition of S_4N_4 into S_2N_2 which is transformed to $(SN)_x$ by solid-state polymerization. This material was first prepared in 1910 by passing S_4N_4 vapor over silver gauze or quartz at 100–300°C (40–41). Several workers have noticed the explosive hazards and sensitivity to friction of S_2N_2 (34). However, this danger was reported to be minimized by conditions under which crystal growth of $(SN)_x$ can occur (41).

Pure $(SN)_x$ is prepared by pumping S_4N_4 vapor from solid S_4N_4 at 85°C through silver wool at 220°C and the S_2N_2 formed is condensed on a liquid-nitrogen-cooled cold finger (33,42). The S_2N_2 is then slowly sublimed from the cold finger into a trap with rectangular walls which is immersed in a 0°C bath. After allowing the S_2N_2 to form crystals of a certain shape, the trap temperature is raised to room temperature. The initially colorless, monoclinic crystals of S_2N_2 rapidly turn intense blue-black and become paramagnetic. After several hours the crystals change spontaneously to a bright, lustrous golden color. The crystals are left at room temperature for two days after which all traces of unpolymerized S_2N_2 are removed under reduced pressure at 75°C.

$(SN)_x$ can be depolymerized at ca 145°C to a gaseous, very reactive isomer of S_4N_4. When this vapor comes in contact with a solid surface at ambient temperatures, it repolymerizes spontaneously to give golden $(SN)_x$ films.

Crystals of $(SN)_x$ in 8 kPa (60 mm Hg) of bromine vapor react at ambient temperatures to give shiny black crystals which have a lustrous blue-purple tinge (38). After the crystals are subjected to reduce pressure for 30 min, they have the structure $(SNBr_y)_x$ where y is 0.40 ± 0.02 (38).

Uses. Polymeric sulfur nitride has been used as an electrode material in aqueous solution (43–44). The $(SN)_x$ was found to be remarkably stable in aqueous solutions under a wide variety of conditions and a suitable electrode material for chemical modification.

Polysilanes

Polysilanes or polysilylenes are polymers composed of disubstituted silicon atoms:

$$-\!\!\left(\begin{array}{c} R \\ | \\ \mathrm{Si} \\ | \\ R \end{array}\right)\!\!-_n$$

Physical Properties. Compared to polysiloxanes, polysilanes are less stable and more rigid. The relatively low bond strength of the silicon–silicon bond results in lower thermal stability. Hydrogen substitution on silicon results in increased oxidative sensitivity. For example, $(SiH_2)_n$ is spontaneously flammable at room temperature in air (45).

Crystallinity. Poly(dimethylsilane) high polymer with an average degree of polymerization of about 600 is a crystalline powder. X-ray diffraction patterns indicate around 80% crystalline fractions (46). Thermal analysis reveals a melting point around 375°C (46). Poly(diphenylsilane) does not melt but sinters (47).

Density. The density of poly(dimethylsilane) is 0.97 g/cm^3 (46).

Solubility. Lower molecular weight poly(dimethylsilanes) are soluble; however, higher polymers are insoluble in common solvents but dissolve at temperatures above 200°C (46). Copolymers of dimethylsilane and ethylmethylsilane are insoluble in octane. Copolymers of dimethylsilane and methylpropylsilane ranged from partially soluble to completely soluble in octane (47). Poly(diphenylsilane) is insoluble in common solvents (48).

Manufacture. The parent polysilane, $Si_nH_{(2n+2)}$, has been prepared from the reaction of acidic reagents with metallic silicides in an inert atmosphere (45,49–50). The yields are usually poor. The reaction of glacial acetic acid with calcium monosilicide gives a higher molecular weight polysilane (45):

$$2n\ CH_3\overset{O}{\overset{\|}{C}}OH + n\ Ca_2Si \xrightarrow[H_2O]{O_2} \left(\overset{H}{\underset{H}{Si}} \right)_n + n\ Ca(O\overset{O}{\overset{\|}{C}}CH_3)_2 \cdot Ca(OH)_2$$

Poly(dichlorosilanes) have been prepared by the passage of silicon tetrachloride in an inert gas diluent through a hot tube at 1000–1100°C (51):

$$n\ SiCl_4 \xrightarrow{\Delta} \left(\overset{Cl}{\underset{Cl}{Si}} \right)_n$$

Poly(dimethylsilanes) have been prepared by coupling dichlorodimethylsilane with alkali metals in benzene or octane (46,52):
Higher molecular-weight polymers were prepared using very pure dichlorodiphenylsilane (46).

$$n\ Cl\overset{CH_3}{\underset{CH_3}{Si}}Cl + 2n\ Na \longrightarrow \left(\overset{CH_3}{\underset{CH_3}{Si}} \right)_n + 2n\ NaCl$$

Mixtures of dimethyl- and ethylmethyldichlorosilane and dimethyl- and methylpropyldichlorosilane with sodium metal dispersed in hot octane gave copolymers [*74578-04-4*] and [*71092-17-6*], respectively (47). In general $\overline{M}_n$ varied from 25,000 to 50,000.

Poly(diphenylsilane) was prepared from the reaction of diphenyldichlorosilane with lithium in tetrahydrofuran to give a dilithium compound. Subsequent reaction with an equivalent of diphenyldichlorosilane gave polymer (52):

$$Cl\overset{C_6H_5}{\underset{C_6H_5}{Si}}Cl + Li \longrightarrow \text{Li intermediate} \xrightarrow{Cl\overset{C_6H_5}{\underset{C_6H_5}{Si}}Cl} \left(\overset{C_6H_5}{\underset{C_6H_5}{Si}} \right)_n$$

Properties of Organosilane Chains. Catenated organosilane chains possess unusual electronic qualities which resemble those of conjugated carbon chains (53–54). The electronic properties of high polymers are still under investigation (47).

Uses. Polysilanes can be pyrolyzed to form microcrystalline beta-silicon carbide fibers (see Carbides) (55). The use of polysilanes as binders for inorganic, heat-resistant materials has been reported (56).

MISCELLANEOUS

Poly(metallo-siloxanes)

Metallo-siloxanes are siloxane polymers with another metallic atom in the polymer backbone (see Metal-containing polymers). General structures are:

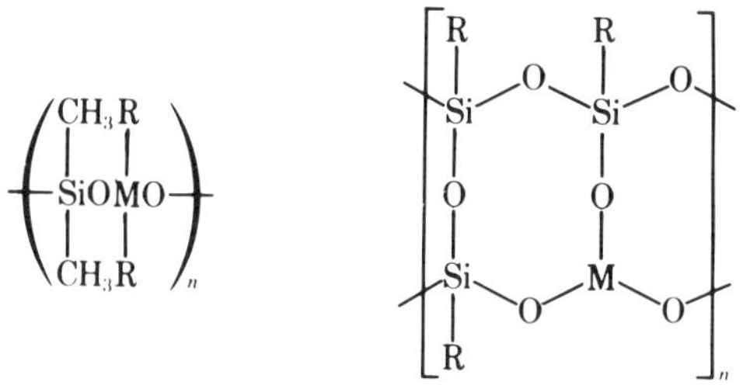

$$\left(\begin{array}{cc} CH_3 & R \\ | & | \\ Si\!O & M\!O \\ | & | \\ CH_3 & R \end{array} \right)_n$$

$$\left[\begin{array}{ccccc} R & & R & & \\ | & & | & & \\ -Si & -O- & Si & -O- & \\ | & & | & & \\ O & & O & & \\ | & & | & & \\ -Si & -O- & M & -O- & \\ | & & & & \\ R & & & & \end{array} \right]_n$$

where M can be magnesium, aluminum, titanium, tin, boron, nickel, arsenic, lead, or antimony, etc (57–58).

Numerous reactions are available for the preparation of poly(metallo-siloxanes). For example, the reaction of a dihydroxy siloxane with a metal alkoxide gives a linear polymer:

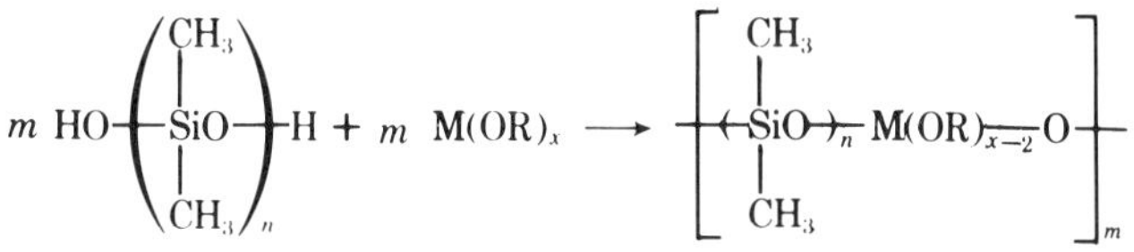

$$m\ HO\!\left(\begin{array}{c} CH_3 \\ | \\ SiO \\ | \\ CH_3 \end{array} \right)_{\!n}\!H + m\ M(OR)_x \longrightarrow \left[\left(\begin{array}{c} CH_3 \\ | \\ SiO \\ | \\ CH_3 \end{array} \right)_{\!n} M(OR)_{x-2}O \right]_m$$

The reaction of a metal chloride with disodium disilanolate yields a metallo-siloxane polymer.

$$n\ NaO\begin{array}{c} R \\ | \\ Si \\ | \\ R \end{array}ONa + n\ MCl_2 \longrightarrow \left(\begin{array}{c} R \\ | \\ Si \\ | \\ R \end{array} OMO \right)_n$$

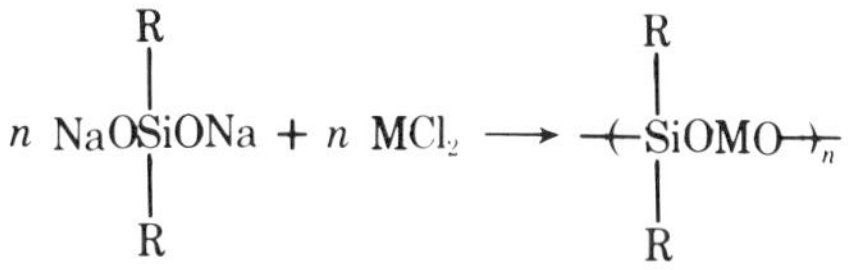

The metallo-siloxane polymers reported in the literature are either linear or cross-linked. Some of the polymers are well characterized and some are not. Some polymers are mostly siloxanes with small amounts of metal atoms in the backbone. In general, the thermal and hydrolytical stability are not as good as conventional silicones.

Boron–Nitrogen Polymers

Work on boron–nitrogen polymers has centered around the polyborazynes $(XBNY)_n$ and the polyborazenes $(X_2BNY_2)_n$ and polymers containing borazine rings (59). The polymers containing borazine rings can be linked directly together; by oxygen or sulfur atoms;

$$\text{or by } -\overset{\overset{\large R}{|}}{N}-, \quad -\overset{|}{N}-R-\overset{|}{N}-, \quad -O-R-O-, \text{ etc}$$

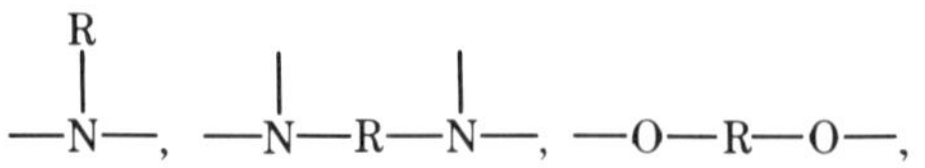

In general, boron–nitrogen polymers exhibit outstanding thermal stability, however, their hydrolytic stability is very poor (see Boron compounds).

Poly(metal Phosphinates)

Polymeric metal phosphinates are prepared by the reaction of an appropriate metal compound with a phosphinic acid, RR′P(O)OH (60). Typical synthesis are:

$$M(OR)_2 + 2\ R_2'\overset{\overset{O}{\|}}{P}OH \longrightarrow [M(OPR_2'O)_2]_n$$

$$MCl_2 + 2\ NaO\overset{\overset{O}{\|}}{P}R_2 \longrightarrow [M(OPR_2O)_2]_n$$

Polyphosphinates containing aluminum, cobalt, chromium, nickel, titanium, zinc, etc, have been prepared.

Single crystal x-ray studies have suggested that some polymers consist of alternating single and triple bridges (60):

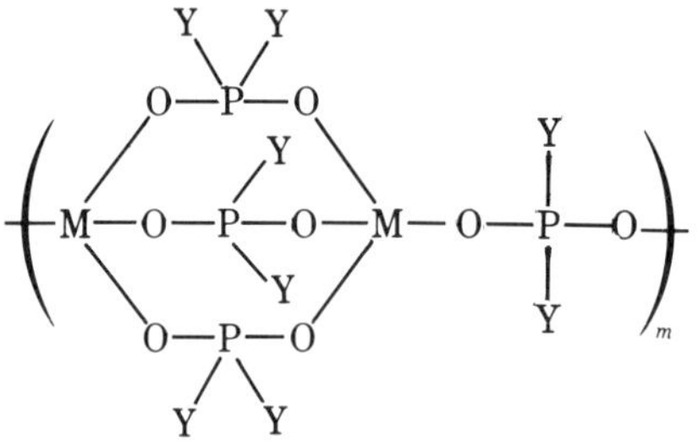

$$\left(-M\begin{matrix} O-P(Y_2)-O \\ O-P(Y_2)-O \\ O-P(Y_2)-O \end{matrix}M-O-\underset{\underset{Y}{|}}{\overset{\overset{Y}{|}}{P}}-O- \right)_m$$

Poly(metal phosphinates) have shown utility in coatings and as additives.

Poly(arylenesiloxanes)

This class of organometallic polymers has the general structure:

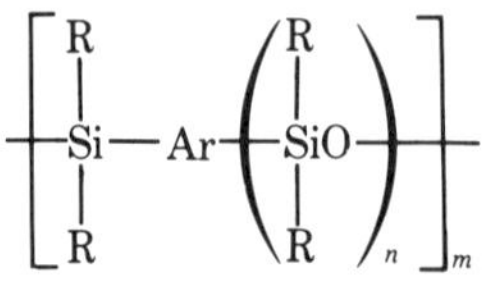

$$\left[-\underset{\underset{R}{|}}{\overset{\overset{R}{|}}{Si}}-Ar-\left(\underset{\underset{R}{|}}{\overset{\overset{R}{|}}{Si}}O- \right)_n \right]_m$$

The incorporation of the silylarylene moiety into the siloxane backbone results in increased T_g, improved thermal stability, and more resistance to cleavage under hydrolysis (61–63).

Poly(xylylenesiloxanes) have xylene moieties in the siloxane chain:

$$\left[\begin{array}{c}\mathrm{R}\\ |\\ \mathrm{SiCH_2}\\ |\\ \mathrm{R}\end{array}\!\!-\!\mathrm{C_6H_4}\!-\!\begin{array}{c}\mathrm{R}\\ |\\ \mathrm{CH_2SiO}\\ |\\ \mathrm{R}\end{array}\left(\begin{array}{c}\mathrm{R}\\ |\\ \mathrm{SiO}\\ |\\ \mathrm{R}\end{array}\right)_n\right]_m$$

Compared to the phenylene analogues, these polymers generally have lower T_gs, increased thermal stability, and decreased oxidative stability (64–65).

BIBLIOGRAPHY

"Inorganic High Polymers" in *ECT* 2nd ed., Vol. 11, pp. 632–650, by B. P. Block, Pennsalt Chemicals Corporation.

1. H. R. Allcock, *Angew. Chem. Int. Ed. Engl.* **16,** 147 (1977).
2. H. R. Allcock, *Science* **193,** 1214 (1976).
3. R. E. Singler, N. S. Schneider, and G. L. Hagnauer, *Polym. Eng. Sci.* **15,** 321 (1975).
4. N. S. Schneider, C. R. Desper, R. E. Singler, M. N. Alexander, and P. L. Sagalyn, C. E. Carraher, in J. E. Sheats and C. U. Pittman, eds., *Organometallic Polymers,* Academic Press, New York, 1978, p. 271.
5. S. H. Rose, *J. Polym. Sci.* **B6,** 837 (1968).
6. J. E. White and R. E. Singler, *J. Polym. Sci.* **15,** 1169 (1977).
7. H. R. Allcock, G. Y. Moore, and W. J. Cook, *Macromolecules* **7,** 571 (1974).
8. H. R. Allcock and W. J. Cook, *Macromolecules* **7,** 284 (1974).
9. G. Allen, C. J. Lewis, and S. M. Todd, *Polymer* **11,** 44 (1970).
10. H. R. Allcock and R. L. Kugel, *J. Am. Chem. Soc.* **87,** 4216 (1965).
11. R. E. Singler, G. L. Hagnauer, N. S. Schneider, B. R. LaLiberte, R. E. Sacher, and R. W. Matton, *J. Polym. Sci. Polym. Chem. Ed.* **12,** 433 (1974).
12. H. R. Allcock, W. J. Cook, and D. P. Mack, *Inorg. Chem.* **11,** 2584 (1972).
13. J. E. White, R. E. Singler, and S. A. Leone, *J. Polym. Sci. Polym. Chem. Ed.* **13,** 2531 (1975).
14. G. S. Kyker and T. A. Antkowiak, *Rubber Chem. Technol.* **47,** 32 (1974).
15. P. Touchet and P. E. Gatza, *J. Elastomers Plast.* **9,** 3 (1977).
16. J. C. Vicic and K. A. Reynard, *J. Appl. Polym. Sci.* **21,** 3185 (1977).
17. E. N. Peters, *J. Macromol. Sci. Rev. Macromol. Chem.* **C17,** 173 (1979).
18. R. W. Williams, *Pure Appl. Chem.* **29,** 569 (1972).
19. M. B. Roller and J. K. Gillham, *Polym. Eng. Sci.* **8,** 567 (1974).
20. E. N. Peters, J. H. Kawakami, G. T. Kwiatkowski, D. W. McNeil, and E. Hedaya, *J. Polym. Sci. Polym. Phys. Ed.* **15,** 723 (1977).
21. R. N. Scott, K. O. Knollmueller, H. Hooks, Jr., and J. F. Sieckhaus, *J. Polym. Sci. A-1* **10,** 2303 (1972).
22. E. N. Peters, D. D. Stewart, J. J. Bohan, R. Moffitt, and C. D. Beard, *J. Polym. Sci. Polym. Chem. Ed.* **15,** 973 (1977).
23. K. A. Andrianov, S.-S. A. Pavlova, I. V. Zhuravleva, Yu, I. Tolchinskii, and B. A. Astapov, *Polym. Sci. USSR* **19,** 1037 (1977).
24. K. O. Knollmueller, R. N. Scott, H. Kwasnik, and J. R. Sieckhaus, *J. Polym. Sci. A-1* **9,** 1071 (1971).
25. E. N. Peters, E. Hedaya, J. H. Kawakami, G. T. Kwiatkowski, D. W. McNeil, and R. W. Tulis, *Rubber Chem. Technol.* **48,** 14 (1975).
26. E. N. Peters and D. D. Stewart, *J. Polym. Sci. Polym. Let. Ed.* **17,** 405 (1979).
27. S. Papetti, B. B. Schaeffer, A. P. Gray, and T. L. Heying, *J. Polym. Sci. A-1* **4,** 1623 (1966).
28. E. N. Peters, D. D. Stewart, J. J. Bohan, G. T. Kwiatkowski, C. D. Beard, R. Moffitt, and E. Hedaya, *J. Elastomers Plast.* **9,** 177 (1977).

29. E. N. Peters, D. D. Stewart, J. J. Bohan, and D. W. McNeil, *J. Elastomers Plast.* **10,** 29 (1978).
30. E. N. Peters, *Am. Chem. Soc. Div. Org. Coat. Plast. Chem. Pap.* **40,** 440 (1979).
31. H. A. Schroeder, O. G. Schaffling, T. B. Larchan, F. F. Trulla, and T. L. Heying, *Rubber Chem. Technol.* **39,** 1184 (1966).
32. H.-P. Geserich and L. Pintschovius, *Adv. Solid State Phys.* **16,** 65 (1976).
33. V. V. Walatka, M. M. Labes, and J. H. Perlstein, *Phys. Rev. Letters* **31,** 1139 (1973).
34. M. M. Labes, P. Love, and L. F. Nichols, *Chem. Revs.* **79,** 1 (1979).
35. A. G. MacDiarmid, C. M. Mikulski, M. S. Saran, P. J. Russo, M. J. Cohen, A. A. Bright, A. F. Garito, and A. J. Heeger in R. B. King ed., *Inorganic Compounds with Unusual Properties,* American Chemical Society, 1976, Chapt. 6.
36. R. H. Baughman, P. A. Apgar, R. R. Chance, A. G. MacDiarmid, and A. F. Garito, *J. Chem. Phys.* **66,** 401 (1977).
37. M. J. Cohen, A. F. Garito, A. J. Heeger, A. G. MacDiarmid, C. M. Mikulski, M. S. Saran, and J. Kleppinger, *J. Am. Chem. Soc.* **98,** 3844 (1976).
38. M. Akhtar, C. K. Chiang, M. J. Cohen, A. J. Heeger, J. Kleppinger, A. G. MacDiarmid, J. Milliken, M. J. Moran, and D. L. Peebles in C. E. Carraher, J. E. Sheats, and C. U. Pittman, eds., *Organometallic Polymers,* Academic Press, New York, 1978, p. 301.
39. R. L. Greene, W. D. Gill, L. J. Azevedo, W. G. Clark, and G. Deutscher, *Lect. Notes Phys.* **65,** 603 (1977).
40. F. P. Burt, *J. Chem. Soc.,* 1171 (1910).
41. M. Boudeulle, A. Douillard, P. Michel, and G. Vallet, *C. R. Acad. Sci. Ser. C* **272,** 2137 (1971).
42. A. G. MacDiarmid, C. M. Mikulski, P. J. Russo, M. S. Saran, A. F. Garito, and A. J. Heeger, *J. Am. Chem. Soc.* **97,** 6358 (1975).
43. R. J. Nowak, H. B. Mark, Jr., A. C. MacDiarmid, *J. Chem. Soc. Chem. Commun.,* 9 (1977).
44. R. J. Nowak, W. Kutner, and H. B. Mark, Jr., *J. Electrochem. Soc.* **135,** 232 (1978).
45. R. Schwarz and P. Heinrich, *Z. Anorg. Chem.* **221,** 277 (1935).
46. J. P. Wesson and T. C. Williams, *J. Polym. Sci. Polym. Chem. Ed.* **17,** 2833 (1979).
47. J. P. Wesson and T. C. Williams, *U.S. N.T.I.S. AD/A Rep., No. 063075,* National Technical Information Service, Springfield, Va., 1978.
48. H. J. Winkler and H. Gilman, *J. Org. Chem.* **27,** 254 (1962).
49. F. Feher, G. Kuhlborsch, and H. Luhleigh, *Z. Anorg. U. Allg. Chem.* **303,** 283 (1960).
50. W. C. Johnson and S. Isenberg, *J. Am. Chem. Soc.* **57,** 1349 (1935).
51. A. Pflugmacher and I. Rohrmann, *Z. Anorg. Chem.* **290,** 101 (1957).
52. C. A. Burkhard, *J. Am. Chem. Soc.* **71,** 963 (1949).
53. R. West, *J. Polym. Sci. C* **29,** 65 (1970).
54. W. G. Boberski and A. L. Allred, *J. Am. Chem. Soc.* **96,** 1244 (1974).
55. S. Yajima, J. Hayashi, M. Omori, and K. Okamura, *Nature* **261,** 683 (1976).
56. Jpn. Pat. 78-80,500 (July 15, 1978) S. Takami, O. Okamoto, S. Matayoshi, and M. Aoyama (to Shin-Etsu Chemical Industry Co., Ltd.).
57. K. A. Andrianov, *Metalorganic Polymers,* Interscience, New York, 1965.
58. K. A. Andrianov, *Inorg. Macromol. Rev.* **1,** 33 (1970).
59. I. B. Atkinson and B. R. Currell, *Inorg. Macromol. Rev.* **1,** 203 (1971).
60. B. P. Block, *Inorg. Macromol. Rev.* **1,** 115 (1970).
61. W. R. Dunnavant, *Inorg. Macromol. Rev.* **1,** 165 (1971).
62. R. E. Burks, Jr., E. R. Covington, M. V. Jackson, and J. E. Curry, *J. Polym. Sci. Polym. Chem. Ed.* **11,** 319 (1973).
63. L. W. Breed, R. L. Elliot, and M. E. Whitehead, *J. Polym. Sci. A-1* **5,** 2745 (1967).
64. H. Rosenberg and E. Choe in C. E. Carraher, Jr., J. E. Sheats, and C. U. Pittman, Jr., eds., *Organometallic Polymers,* Academic Press, New York, 1978, p. 239.
65. I. J. Goldfarb, E. Choe, and H. Rosenberg, ref. 64, p. 249.

General References

F. G. A. Stone and W. A. G. Graham, eds., *Inorganic Polymers,* Academic Press, New York, 1962.
M. F. Lappert and G. J. Leigh, eds., *Developments in Inorganic Polymer Chemistry,* Elsevier, New York, 1962.
D. N. Hunter, *Inorganic Polymers,* John Wiley & Sons, Inc., New York, 1963.
N. H. Ray, *Inorganic Polymers,* Academic Press, New York, 1978.

Inorganic Polymers, Special Publication No. 15, The Chemical Society (London), Burlington House, London, 1961.
A. H. Gerber and E. F. McInerney, *Survey of Inorganic Polymers, NTIS NASA CR-159563,* 1979.

EDWARD N. PETERS
Union Carbide Corporation

INORGANIC REFRACTORY FIBERS. See Refractory fibers.

INOSITOL. See Vitamins.

INSECT CONTROL TECHNOLOGY

Insects are the most numerous of living organisms and the nearly one million (10^6) described species constitute approximately 70% of all animal species. Of these, about 1% are considered significant pests; they attack humans and/or their domestic animals; transmit human, animal, and plant diseases; destroy structures; and compete for available supplies of food and fibers. In the United States, at least 600 species of insects are important pests. Estimates suggest that the total annual loss to agriculture in the U.S. is ca 10% of production, and worldwide agricultural losses are about 14% of production (1–2).

The losses resulting from the depredations by insect vectors of human and animal diseases include malaria, which is transmitted by the bites of ca 80 species of *Anopheles* mosquitoes, and is responsible for about 100 million clinical cases and perhaps one million deaths annually. At least 250 million persons suffer from lymphatic filariasis which is caused by the nematode parasites, *Wuchereria bancrofti* and *Brugia malayi*, and is transmitted largely by the common house mosquito, *Culex fatigans*. The number of persons at risk from this predominately urban disease in the world has doubled in 20 years. Trachoma, a viral disease transmitted by the housefly, is the most common causes of blindness, infecting an estimated 40 million people. American trypanosomiasis (or Chagas' disease) infects ca 7 million people; the causative agent, *Trypanosoma cruzi* is transmitted, through the bites of ca 95 species of blood-sucking Triatominae bugs. The tsetse fly (*Glossina* spp.) vectors of African trypanosomal diseases of humans and cattle infest ca 11.7×10^6 km^2 of Central Africa (one and one-half times the area of the contiguous U.S.). The biting blackfly, *Simulium damnosum,* transmits the filarian, *Onchocerca volvulus,* to humans and infects ca one million persons in seven countries of the Volta River Basin in Africa. About 70,000 of these victims are considered blind and this disease transmission has greatly retarded the agricultural and economic development of this area (3). Other important human diseases transmitted by insects include plague by the oriental rat flea, *Xenopsylla cheopis*, and other fleas, and endemic typhus by the human louse, *Pediculus corporis*.

Role of Chemicals. Plant-derived insecticides, eg, nicotine, rotenone, veratrines, and pyrethrum, have been used since antiquity, but the major development of insecticides for plant protection began in ca 1865 with the use of paris green as an arsenical stomach poison for chewing insects. The improved arsenical, lead arsenate, was introduced in 1892 and calcium arsenate in 1907; their combined production in the U.S. was applied predominantly for cotton (qv) insect control. Cryolite was introduced as a stomach poison insecticide in 1928, to avoid objectionable arsenical residues on fruits and vegetables. These stomach poison insecticides were effective only against chewing insects and had little or no contact action. The arsenicals had the grave disadvantages of high toxicity to humans and domestic animals, considerable phytotoxicity to plants, and extreme environmental persistence (4–5).

The development of DDT insecticide in 1939 began an era of chemical pest control that saw the synthesis and evaluation of hundreds of thousands of synthetic organic chemicals as insecticides. DDT, with its efficient contact insecticidal action, largely replaced the arsenicals and hundreds of new uses were developed (2), including the global malaria eradication program of the World Health Organization. DDT production in the United States reached a maximum of 77,800 metric tons in 1961 (6). Large-scale usage of the other organochlorine insecticides, benzene hexachloride, the cyclodienes, and chlorinated camphenes, rapidly followed, and as insecticide resistance ensued, the organophosphorus insecticides were introduced. By 1975, more than 300 chemical compounds were marketed as insecticides in the U.S. and annual production was 302,000 t (7), valued at $765,324,000. About 59% of the total insecticide use in the United States in 1976 was by farmers (7), and U.S. consumption of insecticides was ca 60% of world insecticide production. Three major classes of synthetic organic insecticides (organochlorine, organophosphorus, and carbamate) dominated usage with the rapidly developing newer synthetic pyrethroids. As recently as 1966, organochlorines accounted for 60% of all farm insecticides but, because of objectionable environmental pollution and insect pest resistance, usage dropped to 46% in 1971, and to 29% by 1976. The use of the more acutely toxic organophosphorus insecticides rapidly increased to 49% of the total by 1976, and employment of these more biodegradable compounds has substantially decreased residue problems but has caused increased acute health hazards to pesticide users. Estimated U.S. production (10^3 t of active ingredients) of those insecticides having the largest volume use in 1972 was: toxaphene, 23; carbaryl, DDT, and methyl parathion, 20; malathion, 16, chlordane, 11, parathion, 6.8; aldrin, diazinon, and methoxychlor, 4.5; carbofuran, disulfoton, and phorate, 3.6; heptachlor and metalkamate, 2.7; dicofol, azinphosmethyl, and fensulfothion, 2; fonofos, ethion, and ronnel, 1.4; and carbophenothion, naled, dimethoate, aldicarb, endosulfan, chlorobenzilate, and cruformate, about 1 (8). Since 1972, use of DDT, aldrin, chlordane, and heptachlor have been greatly curtailed by federal environmental restrictions, and production and use of methyl parathion, carbaryl, carbofuran, and methomyl have increased.

The average benefit/cost ratio of insecticide use in agriculture ranges from three to five dollars return for every dollar invested by the farmer for insect control. This estimate is overoptimistic as it neglects important externalities, eg, insect pest resistance, environmental pollution, and the cost of federal and state regulation of insecticide use. Nevertheless, the use of insecticides often represents the difference between profitable crop production and no marketable crop at all. A familiar example is the corn earworm, *Heliothis zea*, which in many areas routinely damages 90–100% of the ears of sweet corn. A simple dust application of insecticide to the silks of individual ears gives complete protection. Marketable apples scarcely can be produced anywhere in the world without insecticidal treatment to prevent attack by the codling moth, *Laspeyresia pomonella*.

Insecticides have reduced the hidden toll exerted by endemic pest populations on crop plants. Insecticides have provided the only practical means for curbing the ravages of more than 150 destructive plant pathogens transmitted by the feeding of aphids and leafhoppers. Applications of the systemic insecticides, phorate [*298-08-2*] and dimethoate [*60-51-5*], at 0.42 kg/ha to wheat in New Zealand to control the cereal aphid *Rhopalosiphum padi* resulted in 23–54% increases in yields by decreasing the spread of yellow-dwarf virus disease (see Wheat and other cereal grains). Similar treatments of sugar beets in California increased sugar yields by 1344 kg/ha or 10%, by reducing the attacks of mites, aphids, and leafhoppers, and by decreasing the transmission of virus yellows (see Sugar, beet sugar).

The value of insecticides in controlling human and animal diseases has been dramatic. Before the onset of widespread insecticide resistance, DDT successfully controlled the insect vectors of the pathogens causing malaria, typhus, yellow fever, and plague. Between 1942 and 1952, the public health uses of DDT saved no less than 5 million lives and prevented no less than 100 million illnesses. Vector control schemes against filariasis using fenthion and against onchocerciasis using temephos larvicides are giving promising results.

Integrated Pest Management (IPM). The widespread use of chemical insecticides since 1946 has resulted in increasing difficulties in practical insect control. These difficulties include hereditary selection of races of more than 400 insect pests that are resistant to one or more classes of insecticides and some to every available material; resurgences of pests and outbreaks of secondary pests that result from elimination of natural enemies by the use of broad spectrum biocides; adverse human health effects from injudicious use of highly toxic insecticides; pollution of virtually every segment of the environment by persistent, lipophilic organochlorines; and exponentially increasing costs of new insecticides. Therefore, an ecologically based insect control strategy relying on multiple insect control interventions with minimal disturbances to the ecosystem provides the only effective way in which to deal with major insect control problems in agricultural and public health. Such a general strategy is under worldwide development as integrated pest management (IPM). IPM relies heavily on protection and conservation of the natural enemies, parasites, predators, and diseases that regulate or balance populations of insect pests. Thus, IPM rejects the regular or preventive use of broad spectrum insecticides and the general philosophy of eradication of insect pest species which has proved unworkable. IPM programs are based upon the concept of the economic threshold or that level of increasing insect pest population at which control measures should be applied to prevent economic injury, ie, where loss caused by the pest equals the cost of control. Under this concept, many conventional insecticide applications become unnecessary and often damaging because of their needless destruction of natural enemies (9–10).

IPM represents a highly skilled technology that almost always reduces the need for chemical insecticides by 50–90% or more over conventional spray programs. By encouraging natural enemies—parasites, predators, and diseases—IPM practices greatly decrease the rigor of natural selection by pesticides that is responsible for resistance. These natural enemies also prevent the great fluctuations and surges in insect pest populations observed after injudicious use of broad spectrum insecticides. Under the IPM concept, insecticides generally are to be used when other practices are inadequate and the pest population reaches the economic threshold. In order to make the concept effective, insecticides must be used in as selective a manner as possible, with minimal disturbances to all other elements of the ecosystem. In IPM systems, insecticides with resistant host plants, biological control, cultural control, attractants, and repellents are the hardware for the systems approach to regulation of pest densities.

Pesticide Management. Essential concerns in pesticide management include pesticide residues in food and in the environment and the impact of these on humans. Pesticide management in IPM programs rejects the routine spray schedule approach to insect control and most of the uses of insecticides as forms of crop production insurance where they are applied independently of any certain knowledge the pest damage will occur. Pesticide management governs applications of insecticides carefully chosen for selectivity and minimal effects on nontarget species and the environment. The dominant regulating factor governing the timing of insecticide applications is the economic injury level, ie, the lowest pest population density that will cause economic damage or that point where the loss caused by the pest equals in value the cost of available control measures (9–10).

INSECTICIDE FORMULATION

The successful employment of any insecticide depends upon its proper formulation into a preparation that can be applied for insect control with safety to the applicator, animals, and plants. Insecticides are commonly formulated as dusts, water dispersions, emulsions, and solutions. The preparation and use of such formulations involves accessory agents such as dust carriers, solvents, emulsifiers, wetting and dispersing agents, stickers, and deodorants or masking agents (2,11).

Dusts are the simplest formulations and generally contain low concentrations, 0.1–20%, of the toxicant, although ground botanical preparations may be used undiluted. Therefore, the properties of the carrier largely determine the quality of the finished dust. Carriers commonly include organic flours, sulfur, silicon oxides, lime, gypsum, talc, pyrophyllite, bentonites, kaolins, attapulgite, and volcanic ash. Selection of the carrier is made on the basis of compatibility with the desired insecticide (including pH, moisture content, and stability), particle size, abrasiveness, absorbability, density, wettability, and cost. The mixture of the toxicant and diluent is made by a variety of simple operations, eg, milling, solvent impregnations, fusing, and grinding. The particle size usually ranges from 0.5–4.0 μm in diameter.

Wettable powders are prepared by blending the toxicant in high concentration, usually from 15–95%, with a dust carrier such as attapulgite which wets and suspends properly in water. One to two percent of a surface-active agent usually is added to improve the wetting and suspensibility of the powder. Sprays of wettable powders are used widely in agriculture because of their relative safety to plants.

Granulars are pelleted mixtures of toxicant, usually at 2.5–10%, and a dust carrier, eg, absorptive clay, bentonite, or diatomaceous earth, and commonly are 250–590 μm in particle size. They are prepared by impregnation of the carrier with a solution or slurry of the toxicant and are used principally for mosquito larviciding and soil applications.

Emulsives are solutions of toxicant in water-immiscible organic solvents, commonly at 15–50%, with a few percent of surface-active agent to promote emulsification, wetting, and spreading. The choice of solvent is predicated upon solvency, safety to plants and animals, volatility, flammability, compatibility, odor, and cost. The most commonly used solvents are kerosene, xylenes, and related petroleum fractions, methyl isobutyl ketone, and amyl acetate. Water emulsion sprays from such emulsive concentrates are widely used in plant protection and for household insect control.

Baits include mixtures of toxicant, usually at 1–5%, with a carrier especially attractive to the insect pest. Carriers include sugar for the houseflies, protein hydrolysates for fruit flies, bran for grasshoppers, and honey, chocolate, or peanut butter for ants.

Slow release formulations incorporate nonpersistent compounds, eg, methyl parathion, insect growth regulators, and sex pheromones, in a variety of granular, microencapsulated, and hollow-fiber preparations.

Application. The usefulness of any insecticide is substantially dependent upon its proper application and this is determined by the properties of the insecticide, the habits of the pest to be controlled, and the site of the application to be made. The three general methods of applying insecticides are spraying (with water or oil as the principal carrier), dusting (with a fine dry powder as the carrier), and fumigation (where the insecticide is applied as a gas) (2,11).

The proper choice and application of an insecticide for pest control are predicated upon factors, eg, the life history and ecology of the pest, the relation of pest population to economic damage, the effect of the insecticide on the pest or its plant or animal host, related organisms in the ecosystem, and proper timing of the application to prevent illegal residues at harvest and to avoid damaging of bees and other pollinating insects.

Sprays are the most common means of insecticide application and generally involve the use of water as the principal carrier, although volatile oils sometimes are used. With the older inorganic insecticides, suspensions in water were used at dilutions of 0.1–0.2%. The development of the more effective organic insecticides has allowed the widespread use of concentrate sprays in which the toxicant is contained at 10–98% and the amount of carrier to be applied is enormously reduced. The use of concentrate or ultralow-volume sprays has brought about a revolution in spray equipment away from hydraulic nozzles with coarse atomization producing droplets of 200–500 μm in diameter to air blast and other atomizing nozzles producing droplets of 30–80 μm in diameter. The use of concentrate sprays also has assisted the development of spraying by airplane and helicopter.

Aerosols (qv) are very finely divided sprays having droplet diameters of 1–30 μm. They are used almost entirely as space sprays for application to enclosures, particularly against flying insects. Aerosols are most conveniently applied by the familiar liquefied gas dispersion or bomb but can be generated on a larger scale by rotary atomizers or twin fluid atomizers.

Dusts are the simplest means of insecticide dispersal and are applied by introducing the finely divided carrier, with particles of 0.5–3.0 μm in diameter, into a moving air stream. In comparison with sprays, dusts adhere poorly to surfaces and cause serious drift problems away from the treatment area.

Registration and Regulation. The registration and sale of pesticides in the United States is controlled by the Federal Insecticide, Fungicide, and Rodenticide Act (FIFRA), Public Law 92-516 (Oct. 21, 1972) and its amendments, Public Law 94-140 (Nov. 28, 1975), and Public Law 95-396 (Sept. 30, 1978). This law requires registration of all pesticides with the EPA.

The application of any insecticide to food plants involves the possibility of a persisting residue which, in addition to determining the degree of residual protection from insect attack, may produce deleterious effects upon humans and animals that consume the treated produce. The range of persistence of the spray residue varies from tetraethyl, pyrophosphate, which is destroyed by moisture within a few hours through nicotine, pyrethrins, and rotenone, which are decomposed by light and air within a few days, to the more stable carbamate, pyrethroids, organophosphorus, and organochlorine insecticides, which may persist for weeks. Most persistent are the inorganic substances, eg, lead arsenate and cryolite, which are removed only by weathering or washing.

Residues of organic insecticides may penetrate and accumulate rapidly in the oily tissues of plants or animals. The residues on the surface are degraded by weathering, light, and volatilization, and those that penetrate into the living tissues are attacked by enzymes and acids. The net effect of all these attritions is expressed as the residue-persistence curve which follows first-order kinetics and is a linear function when plotted as the log of the amount of residue versus time. Such a plot gives the half-life of the insecticide on a particular substrate, which is relatively constant, and also indicates the number of days after application when the residue may be expected to reach any practical level. Under the Food, Drug, and Cosmetic Act (FDCA) of 1938 (as amended by the Miller Act of 1954), the responsibility for the safety of foods containing pesticide residues is assigned to the U.S. FDA.

Insecticides

Insecticides are chemicals that are used to control damage or annoyance from insects. Generally, control is achieved by poisoning the insects by oral ingestion of stomach poisons, by contact poisons that penetrate through the cuticle, or by fumigants that penetrate through the respiratory system. Ancillary chemicals also are employed in insect control and include attractants and repellents, which influence insect behavior, and chemosterilants which influence reproduction.

Stomach poisons generally are applied against insects with chewing mouthparts but, under certain conditions, they are effective against insects with sponging, siphoning, lapping, or sucking mouthparts. Principal application methods of stomach poisons are: (*1*) The food of the insect is covered thoroughly with the poison so that the insect cannot feed without ingesting it; (*2*) The poison is mixed with an attractant to form a poison bait that the insect will seek out and feed upon; (*3*) The poison (in finely divided form) is sprinkled over the runways of the insect and becomes lodged upon feet or antennae and is subsequently swallowed by the insect while it is cleaning these appendages with the mouthparts; and (*4*) The insecticide may be applied as a systemic poison that is absorbed and distributed through the tissues of the plant or animal host so that the insects feeding thereon are killed; by this means, sucking insects may be controlled with stomach poisons.

Contact poisons are the principal weapons against insects with sucking mouthparts which feed beneath the surface and are not affected by the stomach poisons as the latter commonly are applied. The contact poisons may penetrate the blood directly through the insect cuticle or by entrance through the spiracles of the respiratory system into a trachea. They owe much of their effectiveness against insects to the extraordinarily efficient absorptive properties of the insect cuticle for organic molecules so that the lethal dosage applied externally is almost equivalent to that when the poison is injected into the insect body cavity, eg, the topical and injected LD_{50} values for DDT to the cockroach, *Periplaneta americana*, are 10 and 8 mg/kg, respectively; whereas for the rat, the LD_{50} values are 3000 and 150 mg/kg, respectively, when applied intraperitoneally (ip). Contact insecticides may be applied directly to the insect or as residues to plant surfaces, animals, habitations, or other places frequented by insects. Such residues may kill the insect directly by action upon the insect tarsi.

Fumigants are gaseous poisons used to kill insects. Their application generally is limited to plants or products in tight enclosures, or to those that can be enclosed in gastight tents or wrappings, or to soil. Fumigants are effective against nearly all insects, regardless of the type of mouthparts, since the gas enters the insect body through the spiracles during respiration.

Attractants are substances that lure insects through olfactory stimulation. They may be food lures, sex lures, or oviposition lures. The incorporation of these attractants with insecticides, as in poison bran baits for grasshoppers, in ant syrups, and in sugar baits for flies, as well as specific sex pheromones, is an important means of insect control (see Hormones, sex hormones).

Repellents may be mildly poisonous or only offensive, thereby making food or living conditions unattractive for insects. They are used in a variety of ways, eg, poison barriers for chinch bugs and termites, repellents on plants or animals to prevent insect feeding (eg, mosquito repellents), and mothproofing agents to render woolens impervious to the feeding of clothes moths and carpet beetles (see Repellents).

Insecticides may be classified by their chemical nature and source of supply as inorganic compounds, organic compounds of plant origin, and synthetic organic compounds. The inorganic insecticides generally act only as stomach poisons and the plant derivatives act largely as contact poisons. The synthetic organic insecticides may have contact and stomach poison action and sometimes are used as fumigants.

Inorganic Stomach Poisons. ***Arsenicals.*** Various arsenicals have been widely used as stomach poisons for insects. Arsenous oxide [*1327-53-3*], As_2O_3, secured from flue dust after the roasting of various metallic ores, is the principal source of the arsenical insecticides. This oxide, water soluble up to 2.04 g/100 mL at 25°C, forms arsenous acid [*13768-07-5*], H_3AsO_3, a weak monobasic acid having a dissociation constant of $K_1 = 6.0 \times 10^{-10}$ at 25°C. Thus, it reacts with strong bases to give a solution of arsenite ion, $H_2AsO_3^-$. Three series of arsenite salts exist, and are derived from acids that apparently cannot be isolated in the free state: orthoarsenites as Na_3AsO_3 [*7631-89-2*], pyroarsenites as Na_4AsO_5 [*15195-05-8*], and metarsenites as $NaAsO_2$. Similarly, arsenic pentoxide [*1303-28-2*], As_2O_5, is water soluble to 150 g/100 mL at 16°C and forms tribasic orthoarsenic acid [*7778-39-4*] with successive dissociation constants of $K_1 = 2.5 \times 10^{-4}$, $K_2 = 5.6 \times 10^{-8}$, and $K_3 = 3 \times 10^{-13}$. Pyroarsenic acid [*13453-15-1*], $H_4As_2O_7$, and metarsenic acid [*13768-07-5*], $HAsO_3$, also are known and all three readily form metallic salts.

The insecticidal activity of an arsenical usually is directly related to the percentage of metallic arsenic it contains; although other metals in combination, eg, lead or copper, also may add to the toxicity. The phytotoxicity of an arsenical is directly related to the percent of water-soluble arsenic that can enter the living leaf tissue and poison it. Thus, the ideal arsenical is one having a high arsenic content, none of which is soluble in water, but all of which is readily soluble in the digestive fluids of the insect gut.

The arsenites (trivalent) are the least stable of the arsenicals and are the most toxic to insects and plants and have been used chiefly as the toxicants in baits. The arsenates (pentavalent), although less insecticidal, are more stable and safer to plants and are favored as general-purpose stomach poisons. The individual arsenicals that have been used as insecticides and some of their properties are shown in Table 1.

Lead arsenate, $PbHAsO_4$, was first developed in 1892 for the control of the gypsy moth, *Porthetria dispar*, in forests of the eastern United States. It is the safest and most stable of all arsenicals. Lead arsenate is a fluffy powder and wets and suspends well. In water, especially water containing strong alkali, it undergoes the following reaction, which produces plant burn:

$$5\,PbHAsO_4 + H_2O \rightarrow Pb_4(PbOH)(AsO_4)_3 + 2\,H_3AsO_4$$

The addition of lime or zinc sulfate to the water as a precipitant for the arsenic acid serves as a safener. Basic lead arsenate (a product of the approximate composition $Pb_4(PbOH)(AsO_4)_3.H_2O$) is preferred for use on delicate foliage in foggy or humid regions where the above reaction occurs.

The undiluted dust of calcium arsenate was effective against the cotton boll weevil, *Anthonomus grandis*, in 1920, and 38,163,000 kg was produced in the United States in 1942. The compound has had a resurgence of use for the control of races of boll weevil resistant to organic insecticides. Commercial calcium arsenates are mixtures of $Ca_3(AsO_4)_2$ and $CaHAsO_4$ with an excess of lime and calcium carbonate. They tend to decompose, as shown with lead arsenate above. A safe form of basic calcium arsenate is $(Ca_3(AsO_4)_2)_3.Ca(OH)_2$.

Table 1. Arsenical Insecticides

Compound	CAS Registry No.	Formula	As, %	H_2O soly, %	Oral LD_{50} rat, mg/kg	Uses
arsenic trioxide	[*1327-53-3*]	As_2O_3	75.7	1.2	138	poison bait, stock dip
calcium arsenate	[*7778-44-1*]	$Ca_3(AsO_4)_2$.	37.6		40–100	general stomach poison
calcium arsenate, basic	[*74966-85-1*]	$(Ca_3(AsO_4)_2)_3.Ca(OH)_2$	25	0.4–0.5		general stomach poison
copper acetoarsenite (paris green)	[*12540-34-0*]	$Cu(C_2H_3O_2)_2.3Cu(AsO_2)_2$	44.3	2–3	22	mosquito larvicide
copper arsenate, basic	[*12774-48-0*]	$Cu(CuOH)AsO_4$	26	0.1		
lead arsenate, acid	[*7784-40-9*]	$PbHAsO_4$	21.6	0.25	800	general stomach poison
lead arsenate, basic	[*62648-18-4*]	$Pb_4(PbOH)(AsO_4)_3$	14			on very tender foliage
sodium arsenite	[*7784-46-5*]	$NaAsO_2$	57.7	soluble	10–50	poison bait, stock dip, herbicide

Mode of Action. The arsenates produce regurgitation, torpor, and quiescence in insects. The epithelial cells of the midgut of poisoned insects generally are disintegrated with vacuolized cytoplasm and the chromatin of the nuclei is clumped. These are obvious manifestations of a more fundamental biochemical lesion as arsenical poisoning slowly decreases the oxygen consumption of the insect. This is associated with a reaction between the trivalent arsenic and the sulfhydryl (—SH) groups of glutathione and various enzymes, eg, pyruvate and α-ketoglutarate dehydrogenases. BAL (British antilewisite), or 2,3-dimercapto-1-propanol, which readily complexes trivalent arsenic, serves as antidote in both insects and mammals poisoned by arsenic.

Fluorides. The properties of fluoride compounds that have had insecticidal usage are listed in Table 2. They are salts of hydrofluoric acid, HF, fluorosilicic acid, H_2SiF_6, and fluoroaluminic acid, H_3AlF_6. The insecticidal properties of these compounds are approximately related to the fluorine content, and the solubility in the digestive juices of the insect. The compounds with high water solubility, eg, sodium fluoride and sodium fluorosilicate produce severe plant damage.

Cryolite, or sodium fluoroaluminate, Na_3AlF_6, occurs naturally as a mineral from Greenland and also is produced synthetically. The synthetic form is lighter and fluffier and is preferred. Because of its low water solubility and low mammalian toxicity, it has been used extensively as a stomach poison, dust, or spray for the control of the codling moth, Mexican bean beetle, tomato worm, flea beetles, and others. Cryolite, although relatively insoluble in water, is soluble in dilute acids and alkalies, and spray residues have been removed from apples by washing in dilute hydrochloric acid. With lime, cryolite undergoes the following reaction,

$$Na_3AlF_6 + 3\ Ca(OH)_2 \rightarrow Na_3AlO_3 + 3\ CaF_2 + 3\ H_2O$$

which results in plant injury.

Mode of Action. In insects, the fluoride compounds produce spasms, regurgitation, flaccid paralysis, and death. Fluoride inhibits the enzymes, eg, enolase, which require magnesium as a prosthetic group by precipitating a complex magnesium fluorophosphate; thus, it prevents phosphate transfer in oxidative metabolism.

Table 2. Fluorine Insecticides

Compound	CAS Registry No.	Formula	F content, %	H_2O soly, %	Oral LD_{50}, mg/kg: *Bombyx mori* L.	Oral LD_{50}, mg/kg: Rat	Uses
sodium fluoride	*[7681-49-4]*	NaF	45.2	4.3^{25}	110–150	200	cockroaches, chewing lice, toxicant in poison baits
sodium fluorosilicate	*[16893-85-9]*	Na_2SiF_6	60.6	$0.65^{17.5}$	100–130	125	mothproofing
barium fluorosilicate	*[17125-80-3]*	$BaSiF_6$	40.8	0.026^{17}	90–120	175	chewing insects on plants
sodium fluoroaluminate (cryolite)	*[15096-52-3]*	Na_3AlF_6	54.2	0.035[a] 0.061[b]	50–70	13,500	general stomach poison, spray and dust

[a] Natural.
[b] Synthetic.

Miscellaneous Inorganic Insecticides. Many inorganic substances have had limited usefulness in insect control but have been superseded by much safer organic insecticides. Borax [*1303-96-4*], $Na_2B_4O_7.10H_2O$, and sodium tetraborate [*12258-53-6*], $Na_2B_4O_7$, have been used to kill housefly maggots in manure or refuse; to prevent the breeding of mosquito larvae in water to be used only for laundering, or as glyceroboric acid [*36314-25-7*] to treat maggot-infested wounds in animals. Boric acid [*1303-96-4*], H_3BO_3, has been used as a stomach poison for cockroaches (see also Boron compounds, boric acid). Mercuric chloride [*7487-94-7*] or corrosive sublimate, $HgCl_2$, and mercurous chloride [*10112-91-1*] or calomel, Hg_2Cl_2, have been used as drenches, for such root-infesting insects as the cabbage maggot, *Hylemya brassicae*, the onion maggot, *H. antiqua,* and the larvae of flea beetle and fungus gnats.

Cuprous cyanide [*544-92-3*], CuCN, and zinc phosphide [*22569-71-7*], Zn_3P_2, have been used as stomach poisons for mosquito larvae and agricultural pests, and thallium sulfate [*7446-18-6*], Tl_2SO_4, combined at ca 0.5% with sugar, honey, chocolate, or peanut butter, is a very effective poison bait for ants. Sodium selenate [*13410-01-0*], Na_2SeO_4, has been used as a systemic insecticide applied to greenhouse soil for the control of red spider mites and aphids.

White phosphorus [*7723-14-0*], incorporated in sweet syrup, forms a useful bait for cockroaches. Silicic acid, SiO_2 or H_2SiO_3, very finely divided, is a rapidly acting desiccant that kills cockroaches, fleas, termites, and stored-grain pests by dehydration.

Sulfur and its compounds are among the oldest and most widely used pesticides. Elemental sulfur is especially effective as a dust for the control of mites attacking citrus, cotton, and field crops and as a protectant against chiggers, *Trombicula* spp, attacking humans. Sulfur also is a valuable fungicidal diluent for other dust insecticides and is used in wettable form as a spray mixture. Lime sulfur [*1344-81-6*], has been a standard dormant spray for the control of the San Jose scale, *Aspidiotus perniciosus*, and for other scales and various plant diseases. Lime sulfur is a water-soluble mixture of calcium pentasulfide, CaS_5, calcium tetrasulfide, CaS_4, and calcium thiosulfate, CaS_2O_3, that form with calcium sulfite, $CaSO_3$, which is relatively water insoluble, when lump or stone lime (1 part by wt), ground sulfur (2 parts), and water (ca 8 parts) are boiled together. The filtered stock material, which is a deep orange, malodorous liquid (sp gr 1.28) is diluted by ca 1 part to 7 or 8 parts of water (to sp gr 1.035) for dormant winter spraying and by ca 1 part to 49 parts of water (to sp gr 1.005) for summer spraying. Dehydrated dry lime sulfur is available commercially and is used after appropriate dilution with water. A typical preparation contains 65–70% CaS_4 and CaS_5, ca 5% CaS_2O_3, 5–10% free sulfur, and ca 20% inert ingredients.

Contact Poisons of Plant Origin. ***Nicotinoids.*** Nicotine from tobacco was one of the earliest insecticides and was recommended for use in 1763 as a tea for the destruction of aphids (12–13). Nicotine, L-1-methyl-2-(3′-pyridyl)pyrrolidine (bp 247°C, d 1.009 g/cm^3), is found in the leaves of *Nicotiana tobacum* and *N. rustica* in amounts ranging from 2–14%, and also is found in *Duboisia hopwoodii* and in *Aesclepias syriaca*. It occurs as the principal alkaloid along with small amounts of twelve other alkaloids of which nornicotine, 2-(3′-pyridyl)pyrrolidine (bp 270°C, d 1.07 g/cm^3), and anabasine, L-2-(3′-pyridyl)piperidine (bp 281°C, d 1.048), are of insecticidal importance (see Alkaloids). Nornicotine occurs as both the D and L forms, the former in *D. hopwoodii* and the latter commonly predominating in *Nicotiana*. Anabasine is the chief alkaloid of *Anabasis aphylla*, where it occurs from 1–2% in the shoots and

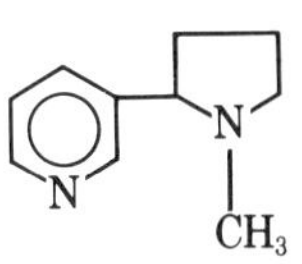

nicotine
[*54-11-3*]

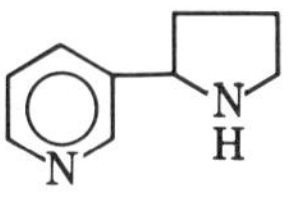

nornicotine
[*494-97-3*]

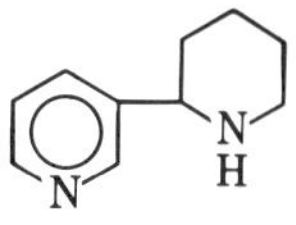

anabasine
[*494-52-0*]

is found to ca 1% in *Nicotiana glauca*. These nicotinoids are appreciably volatile (nicotine vapor pressure, 5.7 Pa or 0.0425 mm Hg at 25°C) and, although colorless liquids when pure, rapidly darken upon exposure to air. They are highly basic ($K_{b1} = 1 \times 10^{-6}$, $K_{b2} = 1 \times 10^{-11}$) and readily form salts with acids and many metals. Nicotine sulfate [*65-30-5*], $(C_{10}H_{14}N_2)_2.H_2SO_4$, is widely used as an insecticide because it is more stable and less volatile. Nicotine has an oral LD_{50} to the rat of 30 mg/kg and a dermal LD_{50} to the rabbit of 50 mg/kg.

Mode of Action. Nicotine and anabasine affect the ganglia of the insect central nervous system, facilitating transsynaptic conduction at low concentrations and blocking conduction at higher levels. The extent of ionization of the nicotinoids plays an important role in both their penetration through the ionic barrier of the nerve sheath to the site of action and in their interaction with the site of action, which is believed to be the acetylcholine receptor protein. There is a marked similarity in dimensions between acetylcholine and the nicotinium ion.

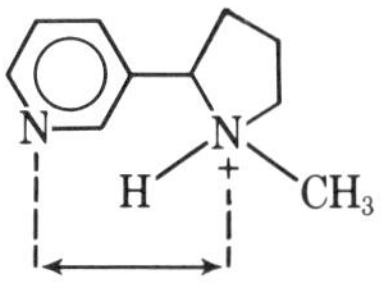

0.42 nm

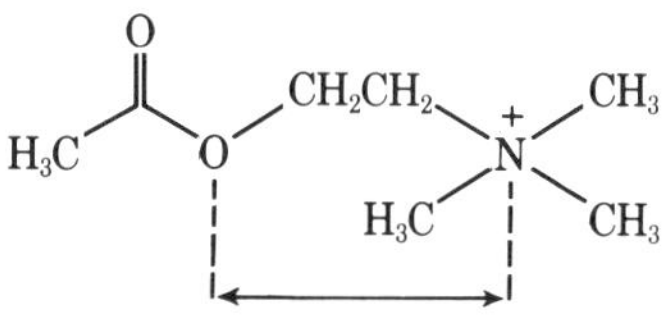

0.30–0.45 nm

Thus, nicotinoids that have the highest insecticidal action have the highest pK_a and, consequently, exist largely in the ionized form at physiological pH. This produces the anomaly that the compounds that are most highly ionized react most rapidly with the receptor protein, yet they are less able to penetrate through the ionic barrier surrounding the insect nerve synapse. This has been demonstrated by the lack of toxicity of the completely ionized nicotine methiodide and nicotine dimethiodide upon injection into *Musca*. The necessity for the nicotinoids to penetrate the insect nerve in the unionized state explains their relative lack of insecticidal action against many insects whose nervous tissues are protected by a more efficient ion barrier than that of the highly susceptible aphids.

Nicotine is used as a contact insecticide for aphids attacking fruits, vegetables, and ornamentals, and as a fumigant for greenhouse plants and poultry mites. Nicotine sulfate is safer and more convenient to handle and the free alkaloid is rapidly liberated by the addition of soap, hydrated lime, or ammonium hydroxide to the spray solution. Nicotine sprays commonly contain 0.05–0.06% nicotine, and nicotine dusts, 1–2% nicotine.

Pyrethroids. The insecticidal properties of pyrethrum from the ground flowers of *Chrysanthemum cinerariaefolium* and *C coccineum* result from six esters, the pyrethrins I and II, the cinerins I and II, and the jasmolins I and II, which are present in the flowers, mostly in the achenes, from 0.7–3% in selected strains (12–13). The structure of the pyrethrum esters is as follows:

		R	R′
pyrethrin I	[*121-21-1*]	CH_3	$—CH_2CH=CHCH=CH_2$
pyrethrin II	[*121-29-9*]	$—COOCH_3$	$—CH_2CH=CHCH=CH_2$
cinerin I	[*25402-06-6*]	CH_3	$—CH_2CH=CHCH_3$
cinerin II	[*121-20-0*]	$—COOCH_3$	$—CH_2CH=CHCH_3$
jasmolin I	[*4466-14-2*]	CH_3	$—CH_2CH=CHCH_2CH_3$
jasmolin II	[*1172-63-0*]	$—COOCH_3$	$—CH_2CH=CHCH_2CH_3$

The pyrethroid esters differ considerably in potency, as shown by the topical LD_{50} values (μg/g) to the house fly: pyrethrin I: 0.59, pyrethrin II: 0.38; cinerin I: 1.42, and cinerin II: 1.14. These esters have asymmetric carbons and double bonds in both the alcohol and acidic moieties. The naturally occurring forms are the D-trans acid esters of the D-cis alcohols. The active principles are extracted from the ground flowers by petroleum ether, ethylene dichloride, or other organic solvent, and freed from wax by adsorption on carbon to produce 90–100% pyrethrins. These insecticides are highly unstable to the action of light, air, moisture, and alkali, and residues deteriorate rapidly after application. Knowledge of the structure of the pyrethrins and their high cost stimulated the structural optimization of synthetic derivatives.

Mode of Action. The pyrethrins readily penetrate the insect cuticle as shown by the LD_{50} values to the cockroach, *Periplaneta:* topical, 6.5 mg/kg; injected, 6.0 mg/kg. Despite a great deal of investigation, the specific biochemical lesion of pyrethrin intoxication in insects is unknown. The pronounced effects of optical and cis-trans isomerism upon insecticidal activity and the rigorous structural requirement of the isobutenyl and cyclopropane moieties of the chrysanthemic acid suggest that biological activity is associated with a specific stereochemical interaction with a biological receptor and with the chemical and physical properties of the groups thus positioned. The pyrethroids may interact with a receptor in a three-point contact at the isobutylene group of the acid, the dimethylcyclopropane ring, and the unsaturated group of the keto alcohol. The pyrethrins, when applied to the insect nerve axon, produce a rhythmic spontaneous discharge at 0.01–0.1 ppm and a reversible blocking of conduction at 1.0 ppm. Paralyzed insects exhibit characteristic vacuolization of the nerve tissues.

A characteristic of pyrethrin action on insects is rapid knockdown followed by substantial recovery. Thus, an approximately threefold increase in dosage is required to produce a 24-h mortality equivalent to the 25-min knockdown of houseflies. This recovery from paralysis is the result of rapid enzymatic detoxication in the insect, apparently by microsomal oxidases. The detoxication enzymes are inhibited by a number of compounds, especially those of the methylenedioxyphenyl structure, eg, piperonyl butoxide or 3,4-methylenedioxy-6-propylbenzyl butyldiethylene glycol ether (bp 180°C at 1 mm, d 1.06 g/cm^3); sulfoxide or 1,2-methylenedioxy-4-[2-(octylsulfinyl)propyl]benzene (d 1.07 g/cm^3); sesamex or 2-(3,4-methylenedioxyphenoxy)-3,6,9-trioxaundecane (bp 137°C at 0.08 mm); and propyl isome or di-*n*-propyl-2-methyl-6,7-methylenedioxy-1,2,3,4-tetrahydronaphthalene-3,4-dicarboxylate (d 1.14 g/cm^3). These synergists, when used at 10:1, may activate pyrethrins I about 30 times and allethrin [*584-79-2*] 2–3 times.

Another type of compound with this same synergistic action but an entirely different structure is *N*-(2-ethylhexyl)-[2,2,1]-5-heptene-2,3-dicarboximide, or MGK 264 (bp 158°C at 2 mm; d 1.05 g/cm^3).

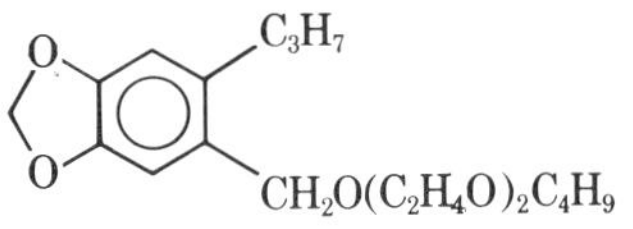

piperonyl butoxide [*51-03-6*]

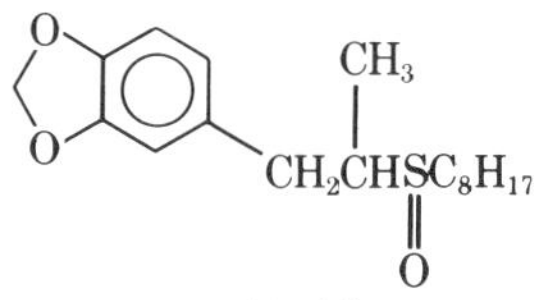

sulfoxide [*120-62-7*]

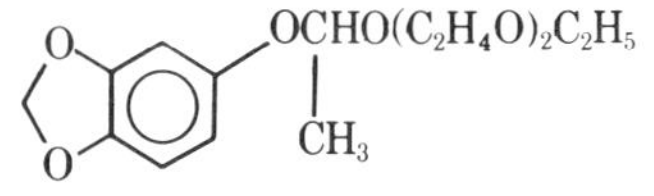

sesamex [*51-14-9*]

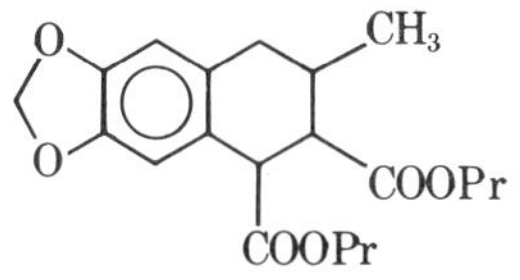

propyl isomer [*83-59-0*]

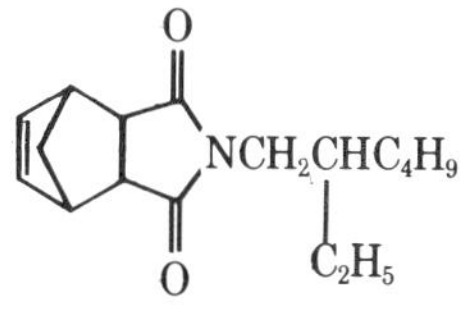

MGK 264 [*113-48-4*]

These pyrethrin synergists have considerable commercial importance in aerosol, and household and cattle sprays (see Aerosols); they also are effective with carbamate and other insecticides.

Because of their very rapid knockdown, the pyrethroids have been favored for use in household, fly, and cattle sprays at ca 0.03–0.1%, usually with 5–10 times this amount of synergist. Aerosol sprays commonly contain 0.04–0.25% pyrethroids plus synergist. These materials are of low mammalian toxicity (rat oral LD_{50} 580 mg/kg) when ingested because of rapid detoxication. The synergists and the combinations with pyrethroids also are of low toxicity; oral LD_{50} values to the rat are (mg/kg): piperonyl butoxide, 11,500; propyl isomer, 15,000; sulfoxide, 2000; and MGK 264, 2800.

Rotenoids. The use of rotenone-bearing roots as insecticides in the United States was developed as a result of Federal laws against residues of lead, arsenic, and fluorine upon edible produce. Rotenone is harmless to plants, highly toxic to many insects, and relatively innocuous to mammals. Sixty-eight species of plants, including twenty-one species of *Tephrosia, twelve of Derris, twelve of Lonchocarpus*, ten of *Milletia*, and several of *Mundulea*, have been reported to contain rotenone or rotenoids. The principal economic species are the various kinds of derris, *Derris elliptica* and *D. malaccensis*, from Malaya and the Indonesian islands, and cubé, or timbo, *Lonchocarpus utilis* and *L. urucu*, from South America. *Tephrosia* (*Cracca*) *virginiana* has been studied as a possible native source of rotenone.

A number of toxic constituents have been isolated from the roots and seeds of these plants, the most important of which is rotenone (mp 163°C (a dimorphic form exists, mp 181°C)). Rotenone has the following chemical structure:

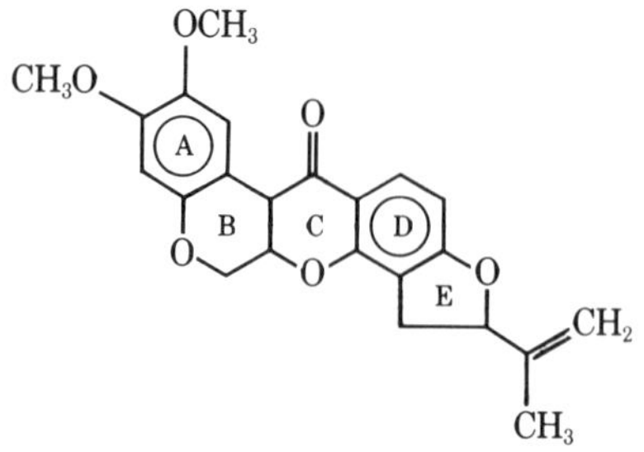

OCH3 CH3O A O B C O D O E CH2 CH3

rotenone
[*83-79-4*]

The other naturally occurring rotenoids are elliptone [*478-10-4*] (mp 159°C), which has a furan ring,

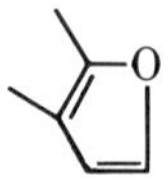

O

sumatrol [*82-10-0*] (mp 188°C), which is 5-hydroxyrotenone; malaccol [*478-07-9*] (mp 244°C), which is 5-hydroxyelliptone; L-α-toxicarol [*82-09-7*] (mp 101°C), which has a hydroxy group at carbon 5 and the ring,

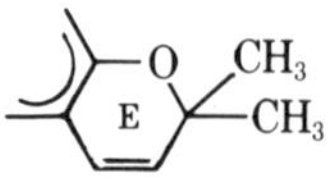

O CH3 E CH3

in place of ring *E* of rotenone; and deguelin [*522-17-8*] (mp 165–171°C), which has a hydrogen atom on carbon 5 in place of the hydroxyl group of toxicarol. A related material, tephrosin [*76-80-2*] (mp 197–198°C), which has a hydroxyl group on one of the carbon atoms between rings *B* and *C*, does not appear to occur naturally in derris resins but is an oxidation product of deguelin. All the naturally occurring rotenoids exist as levo forms. Rotenone is from five to ten times as effective insecticidally as the other rotenoids. The content of these materials in the various commercial plant species and varieties is variable: roots of *D. elliptica* averaging from 5–9% rotenone (max 13%) with up to 31% ether extractives, whereas *D. malaccensis* contains up to 4% rotenone and up to 27% extractives. *L. utilis* averages 8–11% rotenone as a maximum and 25% total extractives. Toxicarol and deguelin make up the bulk of the total extractives. When exposed to light and air, rotenone decomposes, changing from colorless through yellow to deep red with the formation of a hydroxyl derivative at *C*-7 and then dehydration to form dehydrorotenone with a C=C bond between rings *B* and *C*. These reactions render rotenone residues inactive after 5–10 d exposure to normal sunlight. Decomposition to dehydrorotenone is accelerated by the presence of alkali and rotenone should be considered incompatible with alkaline dusts, eg, lime, or with soaps and other alkaline wetting and spreading agents. Pure crystalline rotenone is prepared by extracting powdered rotenone-containing roots with a solvent, eg, ether or carbon tetrachloride, and concentrating the solution to produce crystallization. The pure rotenone has the following solubilities (g/100 mL of solution at 20°C): water, ca 0.00002; ethyl alcohol, 0.2; carbon tetrachloride, 0.6; amyl acetate, 1.6; xylene, 3.4; acetone, 6.6; benzene, 8.0; chlorobenzene, 13.5; ethylene dichloride, 33; and chloroform, 47.

Insects poisoned with rotenone exhibit a steady decline in oxygen consumption and the insecticide has been shown to have a specific action in interfering with the electron transport involved in the oxidation of reduced nicotinamide adenine dinucleotide (NADH) to nicotinamide adenine dinucleotide (NAD) by cytochrome *b*. Poisoning, therefore, inhibits the mitochondrial oxidation of Krebs-cycle intermediates which is catalyzed by NAD.

Rotenone-containing insecticides have been used as dusts of ground roots, dispersible powders, and emulsive extracts. Their principal uses have been for application to edible produce just prior to harvest and for the control of animal ectoparasites and cattle grubs.

Sabadilla. The seeds of sabadilla, or *Schoenocaulon officinale*, family Liliaceae, of South and Central America have been used in native louse powders for centuries and have attained some commercial importance as an insecticide. The related species, *S. drummondii* and *S. texanum*, which are indigenous to the United States, also contain the toxic principles which are a complex group of alkaloids called veratrine. These are present in sabadilla seed to 2–4%, and include about 13% cevadine [*62-59-9*], 10% veratridine [*71-62-5*], and lesser amounts of cevadilline [*1415-76-5*], and sabadine [*28903-30-2*]. The crude alkaloids have the following solubilities (g/100 mL of solvent): cold water, 0.055; boiling water, 0.1; alcohol, 35; chloroform, 144; ether, 24; and kerosene, 0.1–0.2. Sabadilla preparations are markedly activated by heating to 75–80°C for 4 h, moistening with 10% sodium carbonate solution, or by milling with hydrated lime—which is most effective. The toxic alkaloids are rapidly destroyed by the action of light and, thus, sabadilla, which is both a contact and a stomach poison, is safe to use on food crops within a few days of harvest. Cevadine, $C_{32}H_{49}O_9N$ (mp 205°C), the angelic acid ester of cevine, and veratridine, $C_{36}H_{51}O_{11}N$ (mp 160–180°C), the veratric acid ester of cevine,

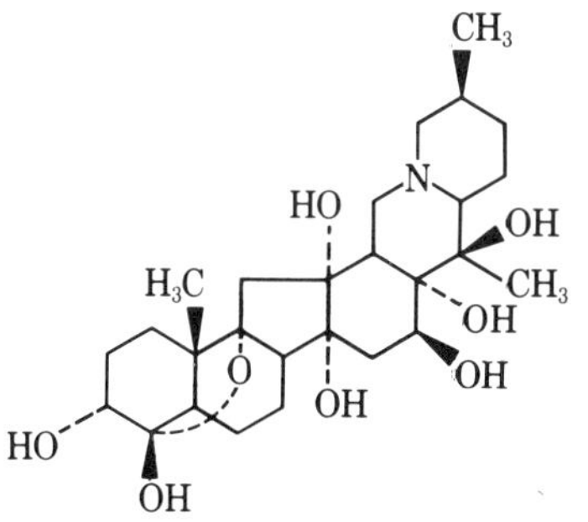

cevine
[*124-98-1*]

are the alkaloids that are responsible for the insecticidal action. These alkaloids are highly poisonous to mammals and may cause irritation of the eyes and respiratory tract and violent sneezing. The ground seeds have an oral LD_{50} to the rat of 5000 mg/kg. Sabadilla is used as a dust or wettable powder of the ground seeds for the control of plant-feeding Hemiptera, and with sugar as a toxic bait for thrips.

Ryania. The root and stem of the plant *Ryania speciosa,* family Flacourtiaceae, native to South America, contain from 0.16–0.2% of insecticidal components, the most important of which is the alkaloid, ryanodine, $C_{25}H_{35}O_9N$ (mp 219–220°C). This compound is effective as both a contact and a stomach poison. Ryanodine is soluble in water, methyl alcohol, and most organic solvents but not in petroleum oils. It is more stable to the action of air and light than pyrethrum or rotenone and has considerable residual action. Ryania has an oral LD_{50} to the rat of 750 mg/kg. The material has shown considerable promise in the control of the European corn borer and codling moth and is used as a wettable powder of ground stems or as a methanolic extract.

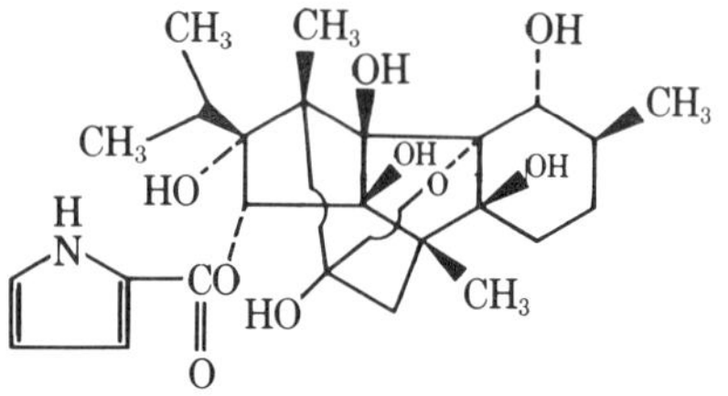

ryanodine
[*15662-33-9*]

Synthetic Organic Insecticides. ***Dinitrophenols.*** Dinitrophenols have a wide variety of biocidal actions and have been used as insecticides, acaricides, herbicides, and fungicides (see Fungicides (agricultural)). Dinitrocresol (DNOC) [*534-52-1*] was the first synthetic organic insecticide. The compounds in use today are all derivatives of 4,6-dinitro-2-alkylphenols and of their salts or esters. DNOC, 4,6-dinitro-*o*-cresol, mp 85°C, is a yellow, odorless compound (vapor pressure 6.9 mPa (0.052 μm Hg) at 25°C) with the following solubilities (g/100 g at 15°C): acetone, 100; benzene, 37; carbon tetrachloride, 2.4; ethanol, 4.3; spray oil, 5.9; and water, 0.013. Like the other dinitrophenols, it is a pseudoacid (K_a 2.5×10^{-6}) and readily forms water-soluble ammonium, potassium, and sodium salts. DNOC has rat oral and dermal LD_{50}s of ca 30 and 200 mg/kg, respectively. The free phenol has been used as an oil spray for grasshopper control and as a dust barrier for chinch bugs. The sodium salt is used as a dormant spray acaracide for mites, scales, and aphids which attack deciduous fruits.

Dinoseb [*88-85-7*], 4,6-dinitro-2-*sec*-butylphenol (mp 42°C), is a reddish liquid and has solubilities (g/100 g solvent at 25°C): ethanol, 23.5; petroleum oil, 8.8; and water, 0.073. The rat oral LD_{50} (mg/kg) is 37–50 mg/kg and the dermal LD_{50} is 80–200 mg/kg. The triethanolamine salt is used as a dormant spray for deciduous fruit pests. Binapacryl [*485-31-4*] (mp 65–70°C) is the 3-methyl-2-butenoate ester and is used as an acaricide in foliage applications. The rat oral and dermal LD_{50}s are 136–225 and >1000 mg/kg, respectively.

Dinocap [*39300-45-3*], 2-(6-methylheptyl)-4,6-dinitrophenyl crotonate (bp 138–140°C at 6.7 Pa (0.05 mm Hg); rat oral LD_{50}: 980–1190 mg/kg, dermal LD_{50}: >4700 mg/kg), is used as an acaricide and fungicide.

Mode of Action. The toxicity of the dinitrophenols results from their action as uncouplers of oxidative phosphorylation. They are, therefore, highly toxic to plants, insects, humans, and higher animals and should be used with great care. Poisoned insects undergo pronounced increases in rates of respiration (reaching 3–10 times normal) and die from metabolic exhaustion because of their inability to utilize the energy provided by respiration and glycolysis for conversion of inorganic phosphate to high energy phosphates, eg, adenosine triphosphate (ATP).

Organothiocyanates. The compound, 2-(2-butoxyethoxy)ethyl thiocyanate [*112-56-1*], $C_4H_9OCH_2CH_2OCH_2CH_2SCN$ (bp 124°C at 33 Pa (0.25 mm Hg); d^{25}, 0.915 g/cm^3) was the first widely used synthetic organic insecticide. It has a rat oral LD_{50} of 90–300 mg/kg and a dermal LD_{50} of 250–500 mg/kg. A variety of other compounds incorporating the thiocyano, —SCN, or isothiocyano, —CNS, groups have useful insecticidal properties. Isobornyl thiocyanoacetate [*115-31-1*] is an amber liquid (d^{25} 1.1465 g/cm^3; rat oral LD_{50}, 1600 mg/kg). It has been used at 2.5–5% in deodorized kerosene as a household and livestock spray. These compounds exemplify the concept of a conductaphoric group that produces an effective concentration of the molecule at the site of action, and a toxophoric group (—SCN) that produces the biochemical lesion.

DDT. DDT, 1,1,1-trichloro-2,2-bis(*p*-chlorophenyl)ethane, was first synthesized in 1874 and its insecticidal properties were discovered in 1939 (3,12,14).

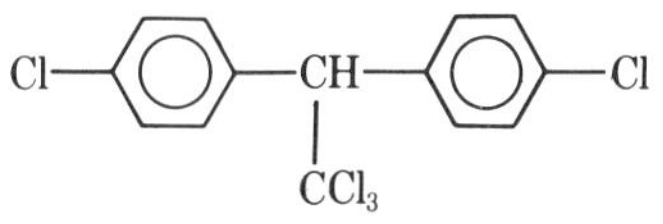

$$Cl-C_6H_4-CH(CCl_3)-C_6H_4-Cl$$

DDT
[*789-02-6*]

Technical DDT is a white amorphous powder, composed of up to 14 chemical compounds. These include, in addition to from 65–80% of the active *p,p′* DDT, 15–21% of nearly inactive 1,1,1-trichloro-2-(*o*-chlorophenyl)-2-(*p*-chlorophenyl)ethane [*789-02-6*], *o,p′* DDT (mp 73–74°C); up to 4% of 1,1-dichloro-2,2-bis(*p*-chlorophenyl)ethane [*72-54-8*], DDD or TDE; up to 1.5% of 1-(*o*-chlorophenyl)-2,2,2-trichloroethanol; and traces of 1,1,1-trichloro-2,2-bis(*o*-chlorophenyl)ethane, *o,o′* DDT, and bis(*p*-chlorophenyl) sulfone. Technical DDT should have a setting point of 89°C or above, and should contain 9.5–11% hydrolyzable chlorine. It melts over a range of 80–94°C. Chemically pure *p,p′* DDT consists of white needles (mp 109°C; density 1.6; vapor pressure ca 0.2 μPa (0.15 nm Hg) at 20°C). In alkaline solution, DDT is readily dehydrochlorinated to form a noninsecticidal product, 1,1-dichloro-2,2-bis(*p*-chlorophenyl)ethylene, DDE (mp 85°C). This compound may be oxidized to *p,p′*-dichlorobenzophenone by ultraviolet radiation catalysis. These two reactions apparently account for the decomposition of DDT residues. The dehydrochlorination reaction of DDT is catalyzed readily by traces of iron, aluminum, and chromium salts, but DDT is stable in the presence of aqueous alkali. Highly purified DDT is stable to heat and does not decompose below 195°C, but the technical material decomposes at ca 100°C because of the catalytic action of trace impurities. Most petroleum solvents inhibit the decomposition of DDT but chlorinated and nitrated solvents accelerate it. Pure DDT has the following approximate solubilities (g/100 mL of solvent at 27°C): water, 0.0000001; xylene, 60; carbon tetrachloride, 60; dioxane, 100; *o*-dichlorbenzene, 68; tetralin, 68; crude kerosene, 8; refined kerosene, 4; methylated naphthalenes, 40–60; mineral oil, 5; dichlorodifluoromethane, 2; dibutyl phthalate, 33; and acetone, 50.

Mode of Action. DDT has a highly specific insecticidal action affecting peripheral sensory organs of insects to produce violent trains of afferent impulses that cause hyperactivity and convulsions. The paralysis and death that ensue are thought to occur from metabolic exhaustion or from elaboration of a naturally occurring neurotoxin. The insecticidal action of DDT is facilitated by the ease of absorption through the insect cuticle, eg, the topical LD_{50} to the cockroach *Periplaneta americana*, is 10 mg/kg compared to the injected LD_{50} of 8 mg/kg. The primary biochemical lesion in the nerve axon seems to result from the peculiar stereochemistry of the molecule which is absorbed into a critical interface of the lipoprotein, eg, the sodium gate; or from specific inhibition of ATP'ase. The stereochemical configuration of the molecule is the determinative factor, as shown in Table 3.

Environmental. DDT is the most permanent and durable of the commonly used contact insecticides because of its insolubility in water, its very low vapor pressure, and its resistance to destruction by light and oxidation. Residues of DDT applied to indoor locations may remain effective for as long as a year, losing activity only when covered by accumulations of grease and dirt. Applications on foliage are less durable, owing to weathering and slow decomposition. The unusual stability of DDT and its high lipid/H_2O partitioning (>100,000) has resulted in many environmental problems, eg, soil persistence with a half-life of 2.5–10 yr, bioaccumulation from water to fish

Table 3. Effect of Structure on the Toxicity of DDT Analogues to Insects[a]

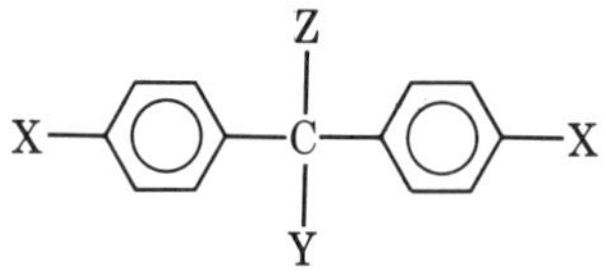

X	Y	Z	CAS Registry No.	Topical LD_{50} fly, *Musca domestica*, μg/g	LD_{50} mosquito larva, *Culex quinquefasciatus*, ppm
H	CCl_3	H	*[2971-22-4]*	>500	1.1
F	CCl_3	H (Gix)	*[475-26-3]*	5.0	0.074
Cl	CCl_3	H (DDT)	*[50-29-3]*	1.6	0.070
Br	CCl_3	H	*[2990-17-2]*	1.9	0.018
CH_3	CCl_3	H	*[4413-31-4]*	9.0	0.080
CH_3O	CCl_3	H (methoxychlor)	*[72-43-5]*	3.4	0.067
NO_2	CCl_3	H	*[2971-23-5]*	>500	>10
Cl	$HCCl_2$	H (TDE)	*[72-54-8]*	6.5	0.038
Cl	$=CCl_2$	—(DDE)	*[72-55-9]*	>500	>10
Cl	CF_3	H	*[361-07-9]*	>500	2.2
Cl	CBr_3	H	*[4399-08-0]*	>500	0.55
Cl	$C(CH_3)_3$	H	*[19685-37-1]*	30	
Cl	$CH(NO_2)CH_3$	H	*[117-27-1]*	8.5	0.064
Cl	CCl_3	Cl	*[3563-45-9]*	>500	>10
Cl	CCl_3	F	*[1545-65-9]*	125	0.092
Cl	CCl_3	D	*[13259-51-3]*	2.3	0.016

[a] Ref. 15.

at levels >1,000,000, transport through food chains, and ubiquitous tissue storage in humans and animals. These properties are shared by DDE, which is even more environmentally recalcitrant.

DDT is slowly converted *in vivo* by reductive dechlorination to DDD and by further dechlorinations to 4,4′-dichlorodiphenylacetic acid (DDA), the predominant excretory metabolite. Anaerobically, it may form 4,4′-dichlorodiphenylacetonitrile (DDCN). However, most DDT that enters the environment is sequestered as DDE, which is ubiquitously present in the body lipids of invertebrate and vertebrate animals—more than 2×10^6 t have been used. In humans, DDT is stored in body fats and is secreted in milk as DDT, DDD, and DDE with traces of their *o,p′*-isomers. Percent levels of these compounds in United States inhabitants average about 7 ppm but have declined from ca 12 ppm in 1970 as a result of sharply curtailed usage.

DDT is highly toxic to fish (LC_{50} for trout and blue gill, 0.002–0.008 ppm), and it is only moderately toxic to birds (oral LD_{50} mallard 1300 and pheasant >2240 mg/kg). However, widespread bird kills have resulted from bioconcentration of DDT through food chains, ie, from fish or earthworms. A major environmental problem has resulted from the specific effects of DDE on eggshell formation in raptorial birds where accumulation has caused decreases in shell thickness of 10–15%, resulting in widespread breakage. The rat oral and dermal LD_{50}s of DDT are 113 and 2000–3000 mg/kg, respectively.

Uses. DDT has been employed for control of hundreds of species of insect pests of orchard, garden, field, and forest. It was generally applied as a dust or as water sprays of suspensions or emulsions of DDT at 0.1–5%. At the peak of its usage in the United States, DDT was registered for use on 334 agricultural commodities but these registrations were cancelled in 1973 because of widespread environmental contamination, and the use of DDT is restricted to essential public health usage. Similar restrictions have been invoked in most countries of Europe and in Japan. DDT is still widely employed for control of insects of public health importance as a residual spray in dwellings for malaria control and as a dust applied to humans in mass delousing programs for typhus control. Thus, DDT has saved millions of lives and prevented hundreds of millions of illnesses. However, its effectiveness for these uses is substantially reduced because of insect resistance.

Analogues. Several closely related compounds have attained commercial importance as insecticides. DDD (TDE), 1,1-chloro-2,2-*bis*-(*p*-chlorophenyl)ethane, is almost indistinguishable from DDT—in which it occurs as an impurity—in properties and behavior. The technical product consists largely of the *p,p′* isomer (mp 110°C) and 7–8% of the *o,p′* isomer (mp 76°C). The rat oral and dermal LD_{50}s are 3400 and >10,000 mg/kg, respectively. DDD suffers from the same environmental and resistance problems as DDT. Perthane [*72-56-0*] is the diethyl analogue, 1,1-dichloro-2,2-bis-(*p*-ethylphenyl)ethane (mp 56°C; rat oral LD_{50} 8170 mg/kg). It is substantially biodegradable and has been developed for applications where very low acute and chronic toxicities are desired.

Technical-grade methoxychlor, 1,1,1-trichloro-2,2-bis-(*p*-methoxyphenyl)ethane, is a white powder consisting of ca 88% of the *p,p′*-isomer (mp 89°C (a dimorphic form melts at 78°C)). This compound gives a more rapid knockdown of many insects than DDT does and is of much lower acute and chronic toxicity (rat oral LD_{50} 6000 mg/kg). The methoxy groups are readily dealkylated *in vivo* by microsomal oxidases producing phenols that are readily eliminated. Thus, methoxychlor does not bioaccumulate as

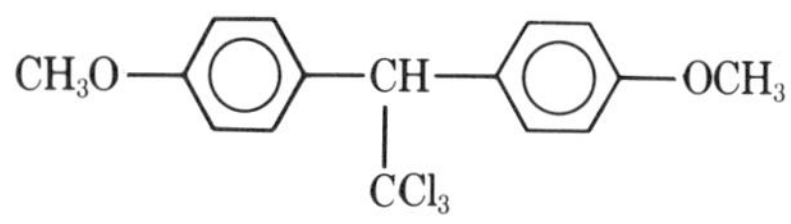

$CH_3O-C_6H_4-CH(CCl_3)-C_6H_4-OCH_3$

methoxychlor
[*72-43-5*]

DDT does and is favored for general environmental use. Unfortunately, however, insects that are resistant to DDT generally are cross-resistant to methoxychlor.

Benzene Hexachloride and Lindane. The active constituent of benzene hexachloride [*58-89-9*], is γ-1,2,3,4,5,6-hexachlorocyclohexane, or lindane, which has the following structural formula:

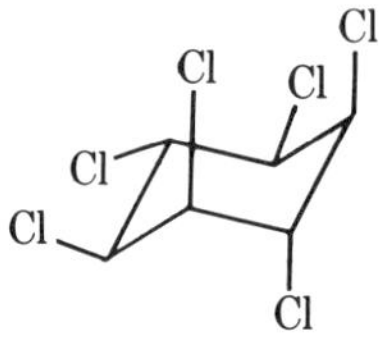

lindane
[*58-89-9*]

Benzene hexachloride is prepared by the chlorination of benzene in the presence of sunlight. The crude product is a grayish or brownish amorphous solid with a characteristic odor; it begins to melt at 65°C. It consists of 10–18% of the active γ isomer (configuration *aaaeee,* where *a* denotes axial and *e* equatorial) (mp 112°C) with at least four other nearly inactive stereoisomers. α-isomer (*aaeeee* or *aeeeea*) (mp 157°C), 55–70%; β isomer (*eeeeee*) (mp 309°C), 5–14%; δ isomer (*aeeeee*) (mp 138°C, 6–8%); ϵ isomer (*aeeaee*) (mp 219°C), 3–4%; and a trace of η isomer (*aeaaee*) (mp 90°C) (see Chlorocarbons, chlorinated benzenes). A heptachlorocyclohexane is present up to 4% as is a trace of an octachlorocyclohexane; both are insecticidally inactive. The γ isomer is by far the most toxic of the isomers, being from 500–1000 times as active as the α isomer and ca 5000–10,000 times as active as the δ isomer; the β isomer and ϵ isomer are nontoxic. The various isomers differ greatly in their solubilities, and pure γ isomer can be prepared by treating the crude product with methyl alcohol or acetic acid, in which the α and β isomers are nearly insoluble (leaving a marketable product containing 30–40% γ isomer), and then fractionally crystallizing the alcohol-soluble fraction from chloroform or by chromatographic adsorption. The pure γ isomer has a slight aromatic odor (d 1.85 g/cm^3; vapor pressure 1.3 mPa (9.4 nm Hg) at 20°C). It is very stable to the action of heat, light, and oxidation and can be burned without appreciable decomposition but readily decomposed by alkaline materials to form, principally, 1,2,4-trichlorobenzene and three moles of hydrogen chloride. The solubilities of the γ isomer (g/100 g of solvent at 20°C) are: acetone, 43; benzene, 29; carbon tetrachloride, 7; cyclohexanone, 37; ethyl acetate, 36; ethyl alcohol, 6; ethylene dichloride, 29; diesel oil, 4; deodorized kerosene, 2; dioxane, 31; mineral oil, 2; xylene, 25; and water, 0.001.

The crude benzene hexachloride has been used extensively as a soil poison, as a toxicant for grasshopper control and against cotton insects. Great care should be exercised in applying benzene hexachloride to edible crops or to soil in which such crops are to be grown, as severe off-flavors of the produce may result. This difficulty has been largely overcome by marketing the 99% pure γ isomer, lindane, for use on food crops, and as a household and public health insecticide. Lindane is of moderate toxicity to mammals with a rat oral and a dermal LD_{50} of ca 100 and 1000 mg/kg, respectively. Its mode of action is unknown but the specific toxicity of the γ isomer suggests that, like DDT, it may interact with the pores of the lipoprotein structure of the insect nerve, thereby causing distortion and consequent excitation of ionic transmission.

Chlorinated Terpenes. A group of incompletely characterized insecticidal compounds has been produced by the chlorination of the naturally occurring terpenes. Toxaphene [*8001-35-2*] is prepared by the chlorination of the bicyclic terpene, camphene, to contain 67–69% chlorine and has the empirical formula $C_{10}H_{10}Cl_8$. The technical product is a yellowish, semicrystalline gum (mp 65–90°C, d^{25} 1.64 g/m^3) and is a mixture of 175 polychloro derivatives. Toxaphene is unstable in the presence of alkali, upon prolonged exposure to sunlight, and at temperatures above 155°C, liberating hydrogen chloride and losing some of its insecticidal potency. It is very soluble in organic solvents; the solubilities (g/100 mL of solvent) for benzene, carbon tetrachloride, ethylene dichloride, and xylene are >450; for kerosene, 280; and for mineral oil, 55–60; but toxaphene is insoluble in water. The oral LD_{50} to the rat is 69 mg/kg.

The most active ingredients in technical toxaphene are 2,2,5-*endo*-6-*exo*-8,9,10-heptachlorobornane [*51775-36-1*] (mouse ip LD_{50} 6.6 mg/kg); and 2,2,5-*endo*-6-*exo*-8,9,9,10-octachlorobornane (mouse ip LD_{50} 3.1 mg/kg). Each constitutes ca 2–6% of the technical mixture.

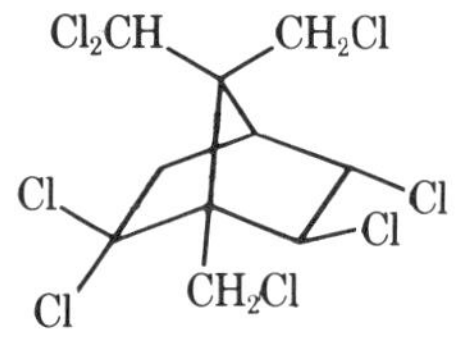

2,2,5-*endo*-6-*exo*-8,9,9,10-octachlorobornane
[*58002-18-9*]

Environmental. Toxaphene is extremely toxic to fish with LC_{50} values to trout and blue gill of 0.003–0.006 ppm. At water concentrations as low as 0.00005 ppm, toxaphene-treated fish suffer broken-back syndrome, a crippling collagen deformity. Bioaccumulation occurs from water to fish at levels up to 100,000-fold. Toxaphene also is highly toxic to birds (oral LD_{50} to pheasant 40 and to mallard 71 mg/kg). The soil persistence of toxaphene is difficult to assess because of the complex mixture but published estimates for half-life range from 2 mo–10 yr. Toxaphene has been found to be a carcinogen in rats and mice. Toxaphene is a broad spectrum, persistent pesticide that is widely used on cotton and other field crops.

Cyclodienes. The cyclodienes are polychlorinated cyclic hydrocarbons with endomethylene-bridged structures, prepared by the Diels-Alder diene reaction. The development of these materials resulted from the discovery in 1945 of chlordane, the chlorinated adduct of hexachlorocyclopentadiene and cyclopentadiene (qv) (3,12).

The chlorination of chlordene (hexachlorodicyclopentadiene) in CCl_4 adds two chlorine atoms across the double bond of the five-membered ring to form the two isomers of chlordane, or 2,3,4,5,6,7,8,8-octachloro-2,3,3a,4,7,7a-hexahydro-4,7-methanoindene, *α-trans* (mp 106.5°C) and *β-cis* (mp 104.5°C). The β-isomer has significantly greater insecticidal activity. Technical chlordane is an amber liquid (bp 175°C at 267 Pa (2 mm Hg), d 1.61 g/cm^3) which contains about 60% of these isomers plus variable amounts of heptachlor, or 1,4,5,6,7,8,8-heptachloro-3a,4,7,7a-tetrahydro-4,7-methanonindene (mp 95°C) and of chlordene. Technical chlordane is dehydrohalogenated in alkaline medium and is completely soluble in kerosene, xylene, esters, ketones, and ethers, but is insoluble in water. It has an oral LD_{50} to the rat of 457 mg/kg and a dermal LD_{50} to the rabbit of ca 750 mg/kg. Heptachlor, which is about 3–5 times as active insecticidally as chlordane, is more efficiently produced by chlorinating chlordene with SO_2Cl_2. The pure crystalline material (vapor pressure 0.04 Pa (0.3 μm Hg) at 25°C) has the following percent solubilities at 27°C: acetone, 75; benzene, 106; carbon tetrachloride, 112; cyclohexanone, 119; orthodichlorobenzene, 100; ethyl alcohol, 4.5; deodorized kerosene, 18.9; xylene, 102; and water, 0.0000056. Heptachlor has an oral LD_{50} to the rat of 90 mg/kg and a dermal LD_{50} to the rabbit of 2000 mg/kg. Heptachlor is readily oxidized to heptachlor epoxide [*1024-57-3*] (mp 159–160°C), which is somewhat more active: oral LD_{50} to the rat is 47 mg/kg. Hydrogenation of heptachlor produces β-dihydroheptachlor [*14168-01-5*] (mp 135°C), which retains high insecticidal activity with very low mammalian toxicity; oral LD_{50} for the rat is >5000 mg/kg.

Telodrin, 1,3,4,5,6,7,8,8-octachloro-3a,4,7,7a-tetrahydro-4,7-methanophthalan (mp 122–123°C, vapor pressure 0.4 mPa (3 nm Hg) at 20°C) is a closely related compound of similar activity. It has the following solubilities (g/100 mL at 25°C): acetone, 25; benzene, 38; carbon tetrachloride, 34; fuel oil, 2; toluene, 34; and xylene, 29. Telodrin has an acute oral LD_{50} to the rat of 8.7 mg/kg.

Many other variations of the Diels-Alder reaction with hexachlorocyclopentadiene produce useful insecticides. The adduct with bicycloheptadiene is aldrin, or 1,2,3,4,10,10-hexachloro-1,4,4a,5,8,8a-hexahydro-1,4-*endo*-*exo*-5,8-dimethanonaphthalene (mp 104°C). This compound has the following solubilities (g/100 mL at 25°C): acetone, 66; amyl acetate, 30; benzene, 83; carbon tetrachloride, 105; ethyl alcohol, 5; deodorized kerosene, 24; methyl ethyl ketone, 24; xylene, 92; and water, 0.000020. Aldrin has an oral LD_{50} to the rat of 67 mg/kg and a dermal LD_{50} to the rabbit of 150 mg/kg. Epoxidation of aldrin with peracetic or perbenzoic acid forms the 6,7-epoxy derivative, dieldrin (mp 176°C). This compound has the following solubilities

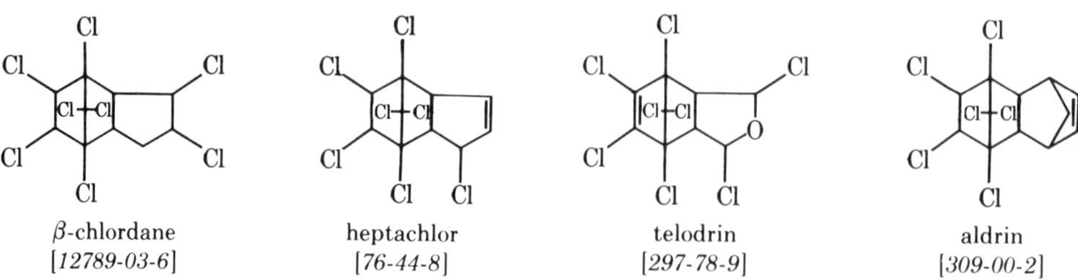

β-chlordane [*12789-03-6*]; heptachlor [*76-44-8*]; telodrin [*297-78-9*]; aldrin [*309-00-2*]

(g/100 mL at 25°C): acetone, 26; amyl acetate, 32; benzene, 56; carbon tetrachloride, 48; ethyl alcohol, 4; methyl ethyl ketone, 39; xylene, 52; and water, 0.000025. Dieldrin has an oral LD_{50} to the rat of 87 mg/kg and a dermal LD_{50} to the rabbit of 150 mg/kg.

Endrin, 1,2,3,4,10,10-hexachloro-6,7-epoxy-1,4,4a,5,6,7,8,8a-octahydro-1,4-endo,endo-5,8-dimethanonaphthalene (mp 245°C dec), is the endo, endo isomer of dieldrin and differs in the spatial arrangement of the two rings.

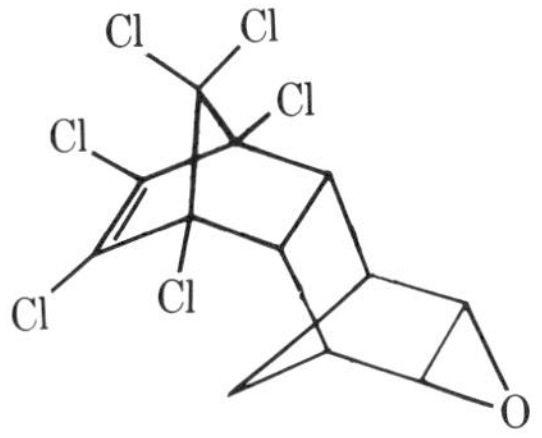

dieldrin
[60-57-1]

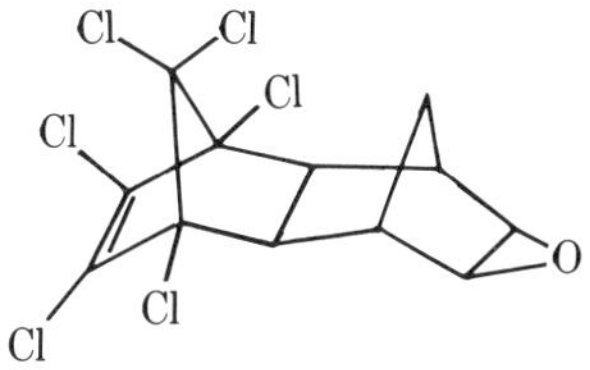

endrin
[72-20-8]

Endrin is less stable than dieldrin and has the following solubilities (g/100 mL at 25°C): acetone, 17; benzene, 13.8; carbon tetrachloride, 3.5; xylene, 18.3; and water, 0.000023. It has an oral LD_{50} to the rat of 10 mg/kg and a dermal LD_{50} of 15 mg/kg.

Endosulfan is the adduct of hexachlorocyclopentadiene and 1,4-dihydroxy-2-butene which reacts subsequently with $SOCl_2$ to produce 6,7,8,9,10,10-hexachloro-1,5,5a,6,9,9a-hexahydro-6,9-methano-2,4,3-benzodioxathiepin-3-oxide. The technical material is a brownish solid (mp 70–100°C), is insoluble in water, and is soluble to 50 g in 100 ml of xylene at 20°C. It has an oral LD_{50} to the rat of 110 mg/kg.

Technical endosulfan consists of about four parts of α isomer, (mp 108–109°C

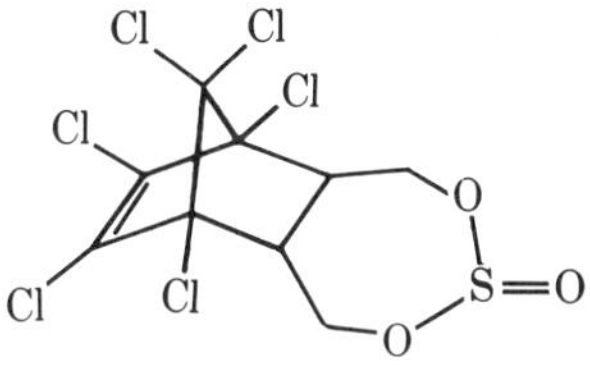

endosulfan
[115-29-7]

(cis with regard to the sulfite group)) and one part of β-trans isomer (mp 206–208°C). The α isomer, which is somewhat more insecticidal, is slowly converted to the more stable β form at high temperatures and both isomers are oxidized slowly in air and in biological systems and rapidly by peroxides or permanganates to endosulfan sulfate (mp 181–182°C). In acidic media, the isomers hydrolyze to form endosulfan alcohol (mp 203–205°C).

Mirex is 1,2,3,4,5,5,6,7,8,9,10,10-dodecachlorooctahydro-1,3,4-metheno-2*H*-cyclobuta-[*c*,*d*]-pentalene (mp 485°C; rat oral LD_{50} 306–600 mg/kg). Chlordecone, the corresponding 2-one (mp 349°C dec), is soluble in water to ca 0.4% at 100°C (rat oral LD_{50} 95–140 mg/kg). These two stomach poisons are highly lipophilic and are environmentally stable; mirex slowly forms chlordecone and photomirex [*39801-14-4*] (8-monohydromirex).

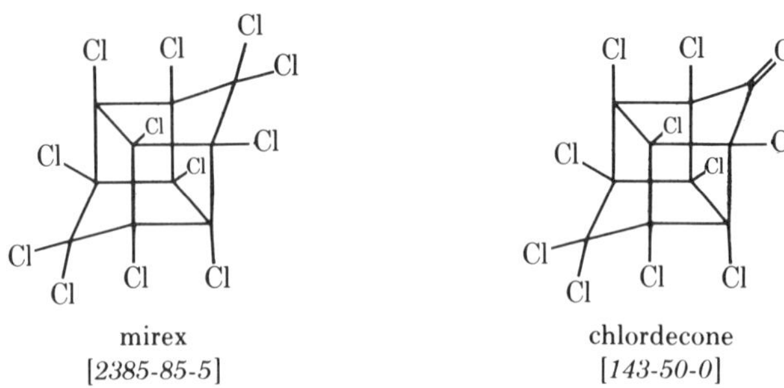

mirex
[*2385-85-5*]

chlordecone
[*143-50-0*]

Mode of Action. Compounds, eg, aldrin and heptachlor, which have unsubstituted double bonds, readily add oxygen to form epoxy derivatives. This is not only of importance in the synthesis of dieldrin but also markedly affects biological behavior. These epoxides form in the tissues of animals that ingest or absorb the compounds and are preferentially concentrated and stored in fats. They also form in and on plant tissues and in soils. As aldrin epoxide (dieldrin) and heptachlor epoxide are more stable (half-lives on alfalfa leaves of seven and eight days, respectively) than their parent materials (half lives of less than one day), *in situ* oxidation markedly affects the residual behavior of the insecticides.

The cyclodienes, like most other good contact insecticides, are efficiently absorbed by insect cuticle and, for aldrin, the LD_{50} values with the cockroach, *Periplaneta,* are 1.9 mg/kg topically and 1.5 mg/kg injected. Symptoms of poisoning include hypersensitivity, hyperactivity, convulsions, prostration, and death. The site of these disturbances is the ganglia of the central nervous system rather than the peripheral nerves as with DDT. The highest insecticidal action is shown by compounds containing both a polychlor center and an adjacent and properly oriented electronegative site, eg, a double bond or a Cl, O, S, or N atom.

Environmental. The high lipophilicity of the cyclodienes and the prolonged persistence of dieldrin and heptachlor epoxide (soil half-lives 2–10 yr) have resulted in severe environmental contamination. These compounds are bioaccumulated from water to fish up to 100,000–300,000-fold and are present everywhere in human fat and milk. Oxychlordane [*26880-48-8*], mirex, and chlordecone also are highly persistent and bioaccumulative. The cyclodienes are extremely toxic to fish, with LC_{50} values (ppm) to trout and bluegill of endrin, 0.001–0.002; endosulfan, 0.001–0.003; aldrin, 0.006–0.01; dieldrin, 0.003–0.015; heptachlor, 0.03–0.026; and chlordane, 0.022–0.095. The LC_{50} values to pheasant and mallard, respectively, are: aldrin, 16.8 and 520 mg/kg; dieldrin, 79 and 381 mg/kg; and endrin, 1.6 and 5.6 mg/kg. As shown by their rat oral LD_{50} values, they also are extremely toxic to small mammals; endrin has been used as a rodenticide (see Poisons, economic). These compounds are central nervous system convulsants; poisoned animals and humans exhibit repeated epileptiform convulsions and behavioristic changes. Aldrin, dieldrin, heptachlor, chlordane, mirex, and chlordecone have been shown to be animal carcinogens. Thus, the distribution of these insecticides has resulted in many problems of environmental quality.

Uses. Aldrin and heptachlor have been used widely as soil insecticides, for cotton insect control, for tree-fruit pests, and against grasshoppers. Chlordane is used for control of cockroaches, ants, termites, and certain pests of vegetable and field crops. Dieldrin is more residual and has been used in the treatment of houses to control the mosquito vectors of malaria and the Triatominae vectors of American trypanosomiasis, and for mothproofing. However, the carcinogenicity of these compounds to rats and mice, the extreme environmental stability of the oxidized products (dieldrin, heptachlor epoxide, and oxychlordane), and their ubiquitous presence in the human diet, body fat, and milk has resulted in their legal restriction in many countries. Uses of aldrin and dieldrin in the United States were cancelled in 1974 and most of those of heptachlor and chlordane were cancelled in 1978.

The uses of endrin for soil pests and cotton insects have been curtailed greatly. Endosulfan, which is substantially more biodegradable, is a broad spectrum insecticide for control of vegetable, fruit, field crop, and ornamental pests. Chlordecone registrations in baits for cockroach and ant control were cancelled in the United States in 1976 following evidence of severe human poisoning and environmental pollution. Mirex has been phased out as a stomach poison in baits for fire ant control because of its persistence and bioaccumulation.

Organophosphorus Insecticides. The discovery of the biological activity of the organophosphorus esters began with the chance observation in 1932 that casual exposure to the vapor of diethyl phosphorofluoridate produced strong cholinergic effects in humans. The practical development of these compounds as insecticides results from work in 1937 (16–18) when Bladan was devised as substitute for nicotine. Bladan contains, as its active principle, tetraethyl pyrophosphate. There followed the lime-stable compound parathion to control the Colorado potato beetle, *Leptinotarsa decemlineata,* schradan and demeton, the first practical systemic insecticides; azinphosmethyl, a broad-spectrum insecticide; trichlorfon, a stomach poison; coumaphos, for animal ectoparasites; and numerous others. These successes attracted many other investigators and as of 1975, ca 100 commercially successful organophosphorus insecticides were developed. A number of these are marketed in quantities of thousands of metric tons. It is estimated that more than 100,000 different organophosphorus compounds have been synthesized and evaluated as pesticides.

O—P toxicants are available with very short residual action eg, TEPP, mevinphos or with prolonged activity, eg, diazinon and azinphosmethyl. There are broad-spectrum insecticides, eg, parathion and malathion, and materials with highly specific action, eg, schradan. The unique properties of demeton and dimethoate have resulted in successful plant systemic insecticides and this activity has been further refined into seed and soil treatments with phorate, and disulfoton which will protect newly developed seedlings from insect attack. Compounds, eg, cruformate, can be fed to cattle to kill grubs living in the animals' bodies, whereas others, eg, trichlorfon, have pronounced activity as stomach poisons but almost no contact action. By taking advantage of differences in the processes of detoxication in the Insecta and Mammalia, compounds, eg, malathion and fenitrothion incorporate a high degree of insecticidal action and of safety to the human user and domestic animals.

The number of toxic compounds that can be made in conformity with the classic hypothetical phosphorylating agent (18) is limitless.

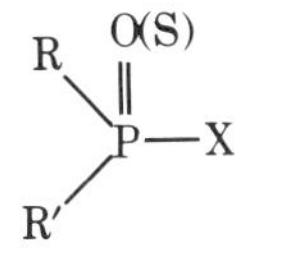

$$\begin{array}{l} R \quad O(S) \\ \;\;\diagdown \;\; \| \\ \quad\;\; P{-}X \\ \;\;\diagup \\ R' \end{array}$$

R and R′ are short-chain alkyl, alkoxy, alkylthio, or amide groups, and X is a labile leaving group or a group which can be metabolized *in vivo* to a labile entity.

A calculation based on ca 50 different groups for R and R′ and 10,000 different groups for X (a conservative estimate) suggests that there are at least 25,000,000 different toxic phosphorus compounds.

Tetraethyl pyrophosphate, TEPP, is produced in 20–40% yield from the reaction of phosphorus oxychloride or phosphorus pentoxide with triethyl phosphate, together with inactive ethyl polyphosphates. Pure tetraethyl pyrophosphate can be separated from this mixture by vacuum distillation and is a water-white liquid (bp 104–110°C at 11 Pa (0.08 mm Hg), d 1.1845 at 25°C). The pure compound decomposes above 200°C to produce ethylene and metaphosphoric acid, and various technical products undergo this decomposition above 140°C. TEPP is miscible in water, ethyl alcohol, ethyl acetate, carbon tetrachloride, benzene, xylene, and methylated naphthalenes but not in kerosene or mineral oils. Tetraethyl pyrophosphate hydrolyzes in water to produce two equivalents of nontoxic diethyl orthophosphoric acid; the times for 50 and 99% hydrolysis at 25°C are 6.8 and 45 h, respectively. This rapid destruction permits use on edible produce immediately before harvesting. All the insecticidal products containing TEPP are hygroscopic and should be protected from moisture in order to preserve the insecticidal activity. The principal uses of TEPP have been for the control of aphids and red spider mites on agricultural and ornamental crops and in greenhouses where it is effective at 0.005–0.02%. It is extremely poisonous to mammals, with oral and dermal LD_{50}s to the rat of 2 and 40 mg/kg, respectively. The material should be handled with extreme care and removed instantly by washing if it contacts the skin. A respirator should be worn at all times for spraying or dusting with TEPP.

Sulfotepp, tetraethyl dithionopyrophosphate, is a yellowish liquid with a garlic odor (bp 110–113°C at 27 Pa (0.2 mm Hg), d 1.196 g/cm^3). It is only ca 0.002% soluble in water, is soluble in organic solvents, and hydrolyzes only in alkaline solution. The compound has an oral LD_{50} to the rat of 5 mg/kg and is useful where a more stable product than TEPP is wanted, especially in aerosols or smokes for the control of greenhouse pests. The corresponding tetrapropyl analogue also has useful insecticidal properties.

Parathion, or *O,O*-diethyl *O*-*p*-nitrophenyl phosphorothionate, has, with its dimethyl analogue, become the most widely used of all the organophosphorus insecticides. The technical material is a dark-brown liquid (d 1.265 g/cm^3), which has an unpleasant garlic odor and is about 98% pure. The pure compound (mp 6°C and calculated bp 375°C), has a vapor pressure of 5.3 mPa (40 nm Hg) at 27°C. Parathion is very resistant to aqueous hydrolysis but is hydrolyzed by alkali to form inactive diethylphosphorothioic acid and *p*-nitrophenol. The times for 50% hydrolysis at 25°C are 120 days for a saturated aqueous solution and 8 h for a solution in limewater. At temperatures above 130°C, parathion slowly isomerizes to form *O,S*-diethyl *O*-*p*-nitrophenyl phosphorothioate [*597-88-6*], which is much less stable and less effective

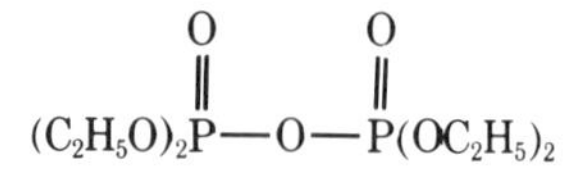

$$(C_2H_5O)_2\overset{\overset{\displaystyle O}{\|}}{P}-O-\overset{\overset{\displaystyle O}{\|}}{P}(OC_2H_5)_2$$

tetraethyl pyrophosphate
[*107-49-3*]

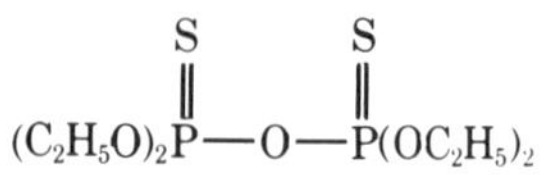

$$(C_2H_5O)_2\overset{\overset{\displaystyle S}{\|}}{P}-O-\overset{\overset{\displaystyle S}{\|}}{P}(OC_2H_5)_2$$

sulfotepp
[*3689-24-5*]

as an insecticide. Parathion is readily reduced to the nontoxic *O,O*-diethyl *O-p*-aminophenyl phosphorothioate and is oxidized with difficulty to the highly toxic diethyl *p*-nitrophenyl phosphate [*3735-01-1*]. Parathion is completely miscible in benzene, xylene, phthalates, and glycols, is almost insoluble in kerosene and mineral oils, and is soluble in water to ca 0.00002%.

Parathion has proved effective against a wider variety of insects than any other insecticide and generally is applied at concentrations of 0.01–0.1%. It has a short residual action and may be safely used on edible produce if applied 15–30 days before marketing.

The oral LD_{50} of parathion to the rat is ca 6–15 mg/kg and the dermal LD_{50} to the rabbit is ca 40–50 mg/kg. The compound is extremely toxic to humans, and many fatalities have resulted from its careless use. It may enter the body in toxic amounts through the mouth, nose, or skin, and great care should be exercised in handling the concentrated materials. The use of rubber gloves, goggles, a respirator, and other protective clothing is advisable. Any material spilled on the skin should be immediately removed with soap and water. In spraying and dusting with parathion, contaminated clothing should be changed frequently. Periodic estimation of blood cholinesterase levels is of value in the early detection of overexposure. In the case of acute poisoning by parathion or other organophosphorus insecticides, the victim should be given prompt medical attention, with the administration of therapeutic doses of atropine, which is a specific antidote. Parathion is of moderate chronic toxicity and is not stored in the mammalian body.

A limiting factor in the use of parathion has been its high toxicity to mammals, and much effort has been devoted to finding safer substitutes. Several closely related compounds have proved especially useful.

Methyl parathion, *O,O*-dimethyl *O-p*-nitrophenyl phosphorothionate, is a white solid (mp 36°C, d 1.358 g/cm^3). This compound is more widely used than parathion and is very similar in properties, but hydrolyzes and isomerizes more easily and, therefore, is less stable both in storage and as an insecticide residue. Methyl parathion is more effective than parathion against aphids and beetles and is somewhat less toxic to mammals; the oral LD_{50} to the rat is 14–42 mg/kg.

Two chlorinated analogues of methyl parathion have greatly reduced acute mammalian toxicities and are useful as household insecticides and for certain agri-

$(C_2H_5O)_2P(=S)O-C_6H_4-NO_2$

parathion
[*56-38-2*]

$(CH_3O)_2P(=S)O-C_6H_4-NO_2$

methyl parathion
[*298-00-0*]

$(CH_3O)_2P(=S)O-C_6H_3(Cl)-NO_2$

Chlorthion
[*500-20-8*]

$(CH_3O)_2P(=S)O-C_6H_3(Cl)-NO_2$

dicapthon
[*2463-84-5*]

cultural pests. Dicapthon, *O*-2-chloro-4-nitrophenyl *O,O*-dimethyl phosphorothionate, is a solid (mp 53°C, oral LD_{50} to the rat 500–1000 mg/kg). Chlorthion, *O*-3-chloro-4-nitrophenyl *O,O*-dimethyl phosphorothionate, is an oil (d^{20} 1.4330 g/cm^3, oral LD_{50} to the rat 1500 mg/kg).

Fenthion, *O,O*-dimethyl *O*-3-methyl-4-methylthiophenyl phosphorothionate, is a brownish liquid (bp 105°C at 1.3 Pa (0.01 mm Hg), d^{25} 1.245 g/cm^3, vapor pressure 0.29 mPa (2.2 nm Hg) at 20°C). It is soluble in organic solvents and insoluble in water and has an oral LD_{50} to the rat of 215 mg/kg. Fenthion is a persistent, general-purpose insecticide that is especially useful as a residual spray for fly and mosquito control and as a mosquito larvicide.

Ronnel, *O,O*-dimethyl *O*-2,4,5-trichlorophenyl phosphorothionate, is a white crystalline solid (mp 35–37°C), which is soluble in organic solvents and to ca 0.008% in water. It is a persistent, general-purpose insecticide (rat oral LD_{50} 906–3025 mg/kg) and is an animal systemic insecticide.

The closely related bromophos [*2104-96-3*], *O,O*-dimethyl-*O*-2,5-dichloro-4-bromophenyl phosphorothionate (mp 54°C, water solubility 0.004%, rat oral LD_{50} 4000 mg/kg), and iodofenfos [*18181-70-9*], *O,O*-dimethyl *O*-2,5-dichloro-4-iodophenyl phosphorothionate (mp 76°C, water solubility 0.0002%, rat oral LD_{50} 2100 mg/kg), are relatively safe insecticides for control of household pests.

Fenitrothion, *O,O*-dimethyl *O*-(3-methyl-4-nitrophenyl) phosphorothionate (bp 140–145°C at 13 Pa (0.1 mm Hg) d^{25} 1.3227 g/cm^3), is soluble in alcohols, ethers, ketones, and aromatic hydrocarbons and to 0.002%, in water. It has rat oral and dermal LD_{50}s of 500 and 3000 mg/kg, respectively. Fenitrothion is a general-purpose replacement for methyl parathion and is safe for use in the household, eg, for malaria control.

Chlorpyrifos, *O,O*-diethyl *O*-(3,5,6-trichloro-2-pyridyl) phosphorothionate (mp 42–43°C, vapor pressure 2.5 mPa (0.19 nm Hg) at 25°C), is soluble in methanol to 45%, in isooctane to 79%, and in water to 0.0002%. The rat oral and dermal LD_{50}s are 97–276 and 2000 mg/kg, respectively. Chlorpyrifos is used as a soil insecticide and for the control of household pests. The *O,O*-dimethyl analogue, chlorpyrifos methyl [*5598-13-0*] is much safer (rat oral LD_{50} 941–2140 mg/kg) and is used for aquatic insect control. Its P=O analogue is fospirate [*5598-52-7*], dimethyl 3,5,6-trichloro-2-pyridyl phosphate (rat oral LD_{50} 869 mg/kg).

Temephos, *O,O,O′,O′*-tetramethyl *O,O′*-thio-di-*p*-phenylene phosphorothionate (mp 30–30.5°C), is soluble in organic solvents and in water to 0.0000025%. Temephos is used as a larvicide for mosquito and blackfly control and has rat oral and dermal LD_{50}s of 100–200 and 4000 mg/kg, respectively.

Diazinon, *O,O*-diethyl *O*-2-isopropyl-4-methylpyrimidyl-6-phosphorothionate, is a brownish liquid (bp 83–84°C, d 1.11 g/cm^3), which is soluble in most organic sol-

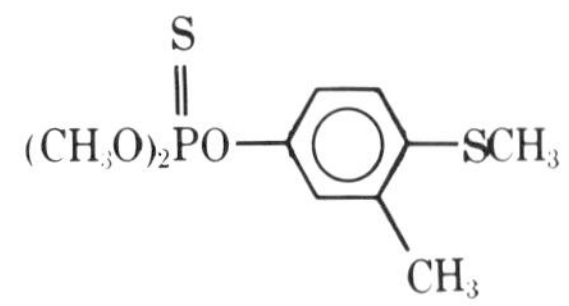

fenthion
[*55-38-9*]

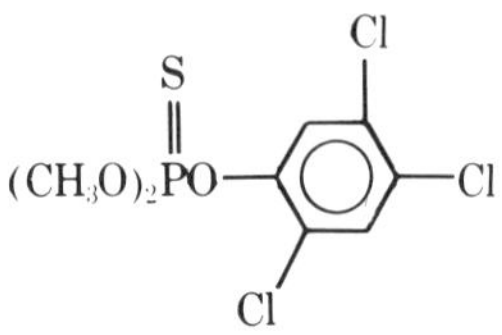

ronnel
[*299-84-3*]

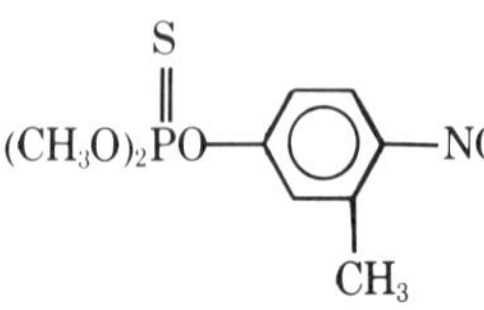

fenitrothion
[*122-14-5*]

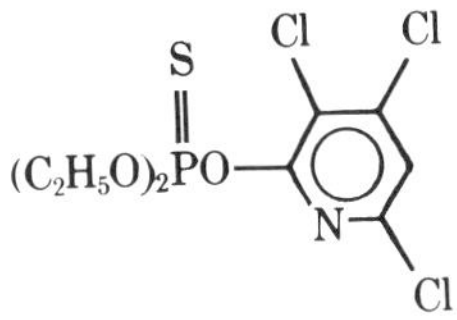

chlorpyrifos
[2921-88-2]

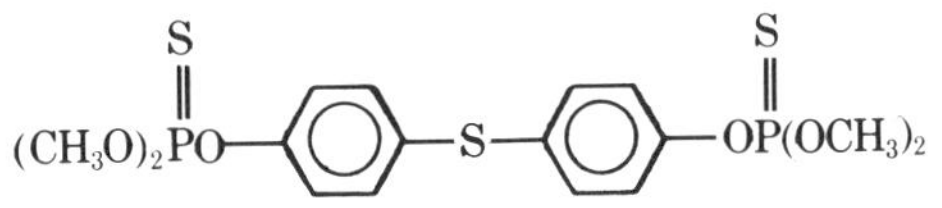

temephos
[3383-96-8]

vents but only to ca 0.004% in water. It has an oral LD_{50} to the rat of ca 100 mg/kg and a dermal LD_{50} to the rabbit of ca 1000 mg/kg. Diazinon is a persistent general-purpose insecticide used for the control of soil insects, of pests of vegetables, fruits, and field crops, and as a residual spray for dairy and livestock pests.

Azinphos methyl [*86-50-0*], *O,O*-dimethyl *S*-4-oxo,1,2,3-benzotriazin-3-(4*H*)-ylmethyl phosphorodithioate, is a white solid (mp 73–74°C, d^{20} 1.44). It is soluble in organic solvents but only to ca 0.003% in water and has an oral LD_{50} of 12 mg/kg and a dermal LD_{50} of 220 mg/kg to the rat. The *O,O*-diethyl analogue (mp 53°C) also is a commercial insecticide. Both azinphos methyl and azinphos ethyl [*2642-71-9*] are persistent broad-spectrum insecticides effective against pests of fruits, vegetables, cotton, and ornamentals.

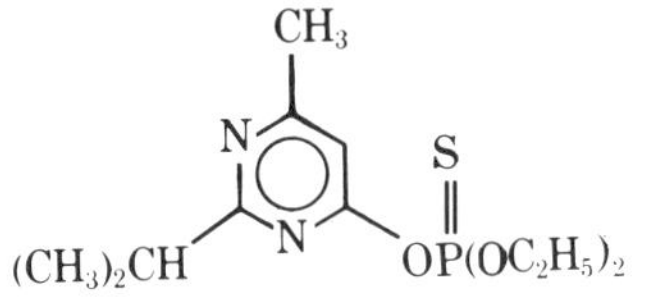

diazinon
[*333-41-5*]

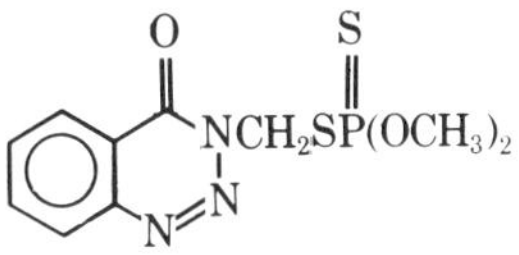

azinphosmethyl
[*86-50-0*]

Malathion, *O,O*-dimethyl *S*-(1,2-dicarbethoxyethyl)phosphorodithioate, is a brown liquid (bp 156–157°C at 93 Pa (0.7 mm Hg)). The technical material is 95–98% pure and has an unpleasant odor (d^{25} 1.23 g/cm^3). It is soluble in most organic solvents, slightly soluble in mineral oils, and soluble in water to 0.0145%. Malathion is easily hydrolyzed above pH 7.0 and below pH 5.0 and is incompatible with alkaline materials. The oral LD_{50} of malathion to the rat is 1375 mg/kg and the dermal LD_{50} is >5000 mg/kg. It is a persistent, general-purpose insecticide, especially suited for household, home garden, vegetable, and fruit insect control, and of public health importance for the control of mosquitoes, flies, and lice.

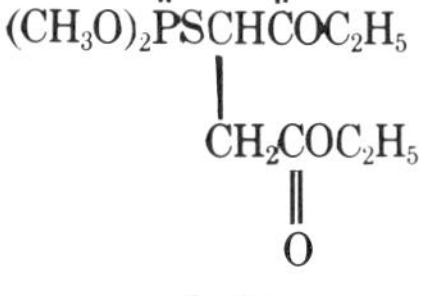

$$(CH_3O)_2P(=S)SCH(C(=O)OC_2H_5)CH_2C(=O)OC_2H_5$$

malathion
[*121-75-5*]

Trichlorfon, dimethyl 1-hydroxy-2-trichloroethylphosphonate,

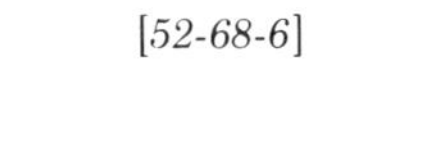

$$(CH_3O)_2P(=O)CH(OH)CCl_3$$

[*52-68-6*]

is a white solid (mp 83–84°C, d 1.73). It is soluble in water to 15% and in aromatic hydrocarbons, alcohols, and ketones. It has an oral LD_{50} to the rat of 450 mg/kg. Its butyl ester, butonate [*126-22-7*] or dimethyl 2,2,2-trichloro-1-(butyryloxy)ethylphosphonate, is an oily liquid with an oral LD_{50} of 1100 mg/kg to the rat and is a household insecticide. Trichlorfon is a stomach poison used in dry sugar bait for the control of flies and on foliage for chewing insects. Above pH 6, trichlorfon is rapidly converted to DDVP or dichlorvos, dimethyl dichlorvinyl phosphate,

$$(CH_3O)_2\overset{\overset{\displaystyle O}{\|}}{P}OCH{=}CCl_2$$

dichlorvos
[*62-73-7*]

a reaction which accounts for its toxic action. Trichlorfon thus has low persistence but also has neurotoxic properties. Dichlorvos is a colorless liquid (bp 120°C at 1.9 kPa (14 mm Hg), d 1.415 g/cm^3), and has an oral LD_{50} to the rat of 80 mg/kg. It has a half-life in water at pH 7.0 of ca 8 h. The compound is very effective in baits and aerosols for the rapid knockdown of flies, mosquitoes, moths, etc, and has fumigant action against household pests, where it is incorporated in a slow-release plastic strip and in pet collars against ectoparasites. Naled is brominated dichlorvos, or dimethyl 1,2-dibromo-2,2-dichloroethyl phosphate,

$$(CH_3O)_2\overset{\overset{\displaystyle O}{\|}}{P}OCHBrCBrCl_2$$

naled
[*300-76-5*]

(mp 26°C). It is insoluble in water and soluble in aromatic solvents and has an oral LD_{50} to the rat of 430 mg/kg. Naled is destructively hydrolyzed in water in about two days and provides a short-lived insecticide for use on plants, in the household, and for fly and mosquito control.

Dioxathion, 2,3-*p*-dioxanedithiol *S,S*-bis(*O,O*-diethylphosphorodithioate) (d 1.257 g/cm^3) is a brown liquid, is insoluble in water and soluble in organic solvents and has an oral LD_{50} to the rat of 110 mg/kg. It is used as an insecticide and acaricide for deciduous fruits.

Technical-grade EPN, *O*-ethyl *O*-*p*-nitrophenyl phenylphosphonothionate (mp 36°C), is a brown liquid (d 1.27, vapor pressure 4 Pa (0.03 mm Hg) at 100°C), is insoluble in water and soluble in organic solvents, and has an oral LD_{50} to the rat of 35 mg/kg. EPN is used for cotton insect control but has neurotoxic properties.

Phosmet, *O,O*-dimethyl *S*-phthalimidomethyl phosphorodithioate (mp 71–72°C), has an oral LD_{50} to the rat of 147 mg/kg. It is used on forage crops and deciduous fruits.

Dialifor or *S*-(2-chloro-1-phthalimidoethyl) *O,O*-diethyl phosphorothionate (mp 167–169°C, rat oral LD_{50} 5–71 mg/kg) is used to control codling moth, red spider mites, and other pests of deciduous fruits.

Carbophenothion, *O,O*-diethyl *S*-*p*-chlorophenylthiomethyl phosphorodithioate, is a light amber liquid (d 1.265 g/cm^3), and is insoluble in water and soluble in kerosene,

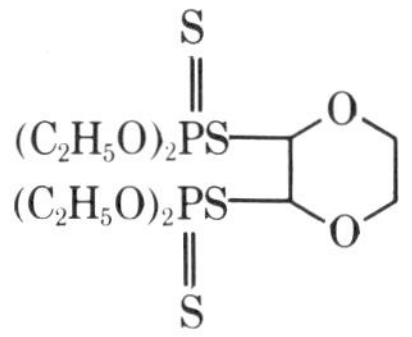

dioxathion
[*78-34-2*]

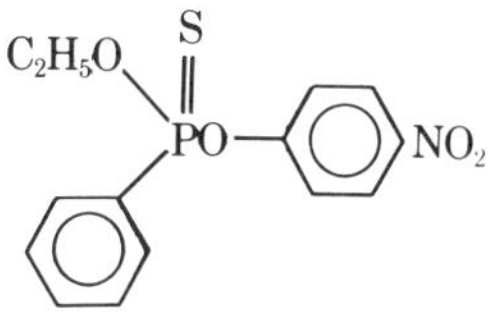

EPN
[*2104-64-5*]

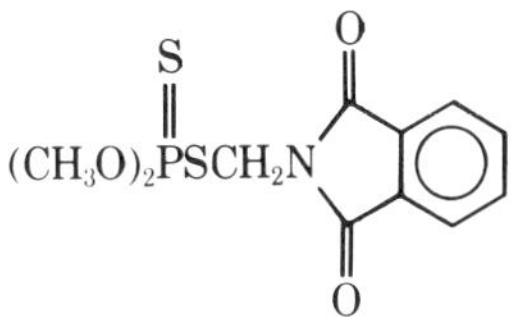

phosmet
[*732-11-6*]

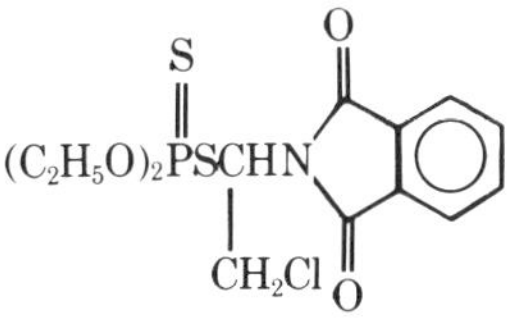

dialifor
[*10311-84-9*]

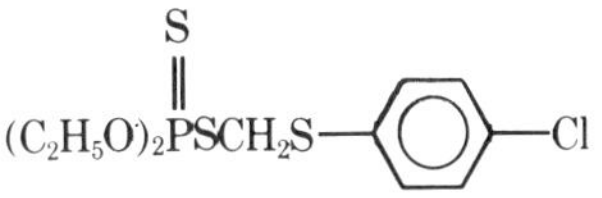

carbophenothion
[*786-19-6*]

xylene, alcohols, ketones, and esters. It has an oral LD_{50} to the rat of 28 mg/kg. The *O,O*-dimethyl ester, carbophenothion methyl [*953-17-3*] has a rat oral LD_{50} of 200 mg/kg and is a commercial insecticide. These are general-purpose insecticides used on fruits, vegetables, and forage.

Chlorfenvinphos, 2-chloro-1-(2,4-dichlorophenyl)vinyl diethyl phosphate (bp 168–170°C at 67 Pa (0.5 mm Hg), vapor pressure 23 μPa (0.17 nm Hg) at 25°C) is a general-purpose agricultural insecticide and has a rat oral LD_{50} of 12–56 mg/kg. The closely related 2-chloro-1-(2,4,5-trichlorophenyl)vinyl dimethyl phosphate or stirofos (mp 97–98°C, vapor pressure 5.6 μPa (0.042 nm Hg)) is water soluble to 0.011% and is one of the safest of all organophosphorus insecticides (rat oral LD_{50} 4000–5000 mg/kg). It is used for protection against stored product pests and vegetable and fruit insects.

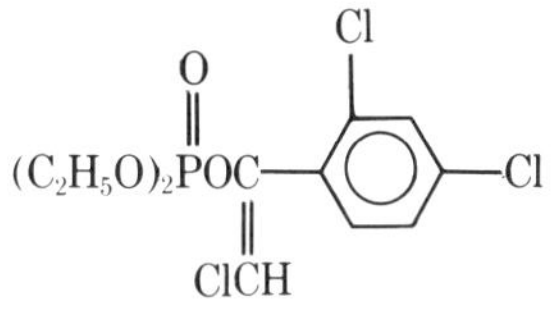

chlorfenvinphos
[*470-90-6*]

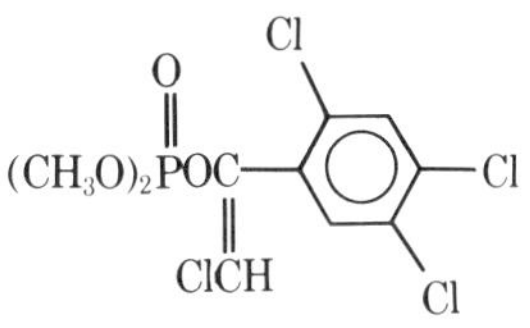

stirofos
[*22248-79-9*]

Methamidophos, *O,S*-dimethyl phosphoramidothioate (mp 39–41°C, vapor pressure 0.01 Pa (0.1 μm Hg)) is water miscible and has a rat oral LD_{50} of 19–21 mg/kg. It is used against insect pests of cotton, cole crops, lettuce, and potatoes. Its *N*-acyl derivative, acephate (mp 72–80°C), is water soluble to 65%, is of much lower acute toxicity (rat oral LD_{50} 866–945 mg/kg), and has similar uses.

Prophos, *O*-ethyl *S,S*-dipropyl phosphorodithioate, has rat oral and dermal LD_{50}s of 61 and 26 mg/kg, respectively. It is used to control pests of vegetables and ornamentals.

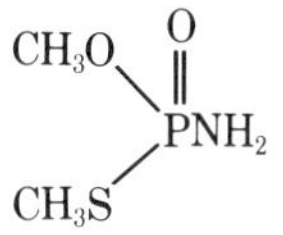

methamidophos
[*10265-92-6*]

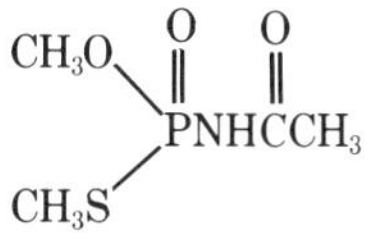

acephate
[*30560-19-1*]

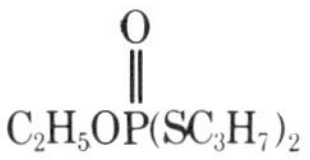

$C_2H_5OP(SC_3H_7)_2$

prophos
[*13194-48-4*]

Ethion, *O,O,O′,O′*-tetraethyl *S,S′*-methylenebis(phosphorodithioate),

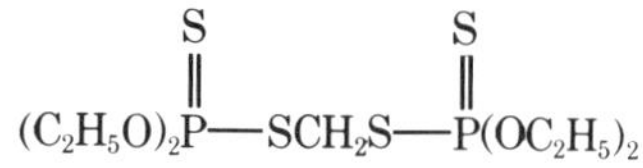

$$(C_2H_5O)_2\overset{\overset{\displaystyle S}{\|}}{P}—SCH_2S—\overset{\overset{\displaystyle S}{\|}}{P}(OC_2H_5)_2$$

ethion
[*563-12-2*]

is a liquid (mp −15 to −12°C, d 1.22 g/cm^3), is insoluble in water and soluble in most organic solvents. It has an oral LD_{50} to the rat of 96 mg/kg and a dermal LD_{50} to the rabbit of 915 mg/kg. Ethion is an acaricide and insecticide which is used on fruits and vegetables.

Systemic Insecticides for Plants. Systemic insecticides for plants (16,18–19), when applied to seeds, roots, or leaves of plants, are absorbed and translocated to the various plant parts in amounts lethal to feeding insects. This method of plant protection has the advantages of minimizing to some extent the inequalities of spray coverage, increasing the length of residual control by protection of the spray residue from attrition by weathering, protecting new plant growth formed subsequent to application, and having less damaging effects on beneficial predatory and pollinating insects. Systemic compounds may be applied as direct sprays to the foliage by drenching the soil (as in the irrigation water); by direct application to, or injection of, concentrates into the trunk or stem; by side-dressing about the root with granular or encapsulated material; or by treatment of seeds before planting. Because of the possible contamination of edible portions of the plant, the use of systemics on food crops must be predicated on a thorough knowledge of the nature and magnitude of toxic residues.

Schradan, or octamethyl pyrophosphoramide,

$$((CH_3)_2N)_2\overset{\overset{\displaystyle O}{\|}}{P}—O—\overset{\overset{\displaystyle O}{\|}}{P}(N(CH_3)_2)_2$$

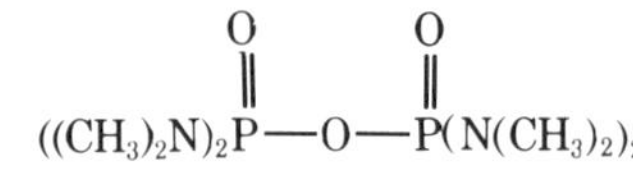

schradan
[*152-16-9*]

was among the first practical systemics. The technical material is a brown liquid (bp 154°C at 267 Pa (2 mm Hg), d^{25} 1.13 g/cm^3), which contains ca 40% of the octamethyl pyrophosphoramide and an equivalent amount of decamethyl triphosphoramide,

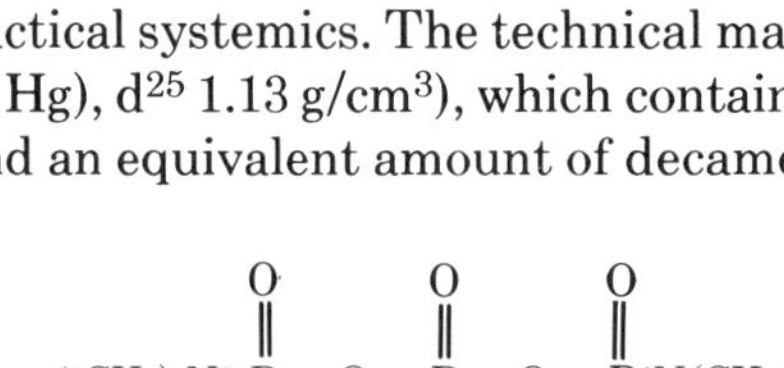

$$((CH_3)_2N)_2\overset{\overset{\displaystyle O}{\|}}{P}—O—\underset{\underset{\displaystyle N(CH_3)_2}{|}}{\overset{\overset{\displaystyle O}{\|}}{P}}—O—\overset{\overset{\displaystyle O}{\|}}{P}(N(CH_3)_2)_2$$

decamethyl triphosphoramide
[*3952-76-9*]

which is of about the same activity. The product is water miscible and also soluble in most organic solvents but only slightly soluble in petroleum hydrocarbons. Schradan is remarkably stable to alkaline hydrolysis but hydrolyzes much more rapidly in acids to form dimethylamine and phosphoric acid. It possesses only a weak contact action to most insects, although it is fairly toxic to Hemiptera and Homoptera and has been shown to be slowly converted in plants and rapidly in animals to the monoamide oxide, which is the toxicant. When absorbed into plants, schradan provides long-term protection from aphids and mites. The oral LD_{50} to the rat is ca 10 mg/kg, and the materials should be handled with the same precautions as described for parathion.

Dimefox, bis(dimethylamino)phosphoryl fluoride,

$$((CH_3)_2N)_2\overset{\overset{\displaystyle O}{\|}}{P}F$$

dimefox
[*115-26-4*]

is a colorless liquid (bp 67°C at 133 Pa (1 mm Hg), d 1.12 g/cm^3, vapor pressure 53 Pa (0.4 mm Hg) at 30°C). This compound is water soluble and has been used to control the mealybug vector of the swollen shoot disease of cacao by implanting soluble capsules about the roots. Because of its volatility, dimefox has a very short residual activity and is not suited for foliage application. The oral LD_{50} to the rat is 3.5 mg/kg.

Demeton [*8065-48-3*], is a mixture of two parts of *O,O*-diethyl *O*-2-(ethylthio)-ethyl phosphorothionate,

$$(C_2H_5O)_2\overset{\overset{\displaystyle S}{\|}}{P}OCH_2CH_2SC_2H_5$$

(bp 94°C at 533 Pa (4 mm Hg), d 1.119 g/cm^3, water solubility 0.0066%) and one part of *O,O*-diethyl *S*-2-(ethylthio)ethyl phosphorothiolate [*126-75-0*],

$$(C_2H_5O)_2\overset{\overset{\displaystyle O}{\|}}{P}SCH_2CH_2SC_2H_5$$

(bp 110°C at 53 Pa (0.4 mm Hg), d^{20} 1.132 g/cm^3, water solubility 0.2%). Technical demeton is a yellowish liquid (d 1.183 g/cm^3), is soluble in organic solvents, and is unstable in alkaline solution. It has an oral LD_{50} to the rat of 9 mg/kg and a dermal LD_{50} to the rabbit of 27 mg/kg. Demeton provides a long-lasting systemic insecticide that is absorbed rapidly by roots, stems, or foliage. In plant tissues, both isomers are rapidly oxidized to the sulfoxide and sulfone derivatives, as shown with phorate (see below).

Demeton methyl, *O,O*-dimethyl *S*-2-(ethylthio)ethyl phosphorothiolate,

$$(CH_3O)_2\overset{\overset{\displaystyle O}{\|}}{P}SCH_2CH_2SC_2H_5$$

demeton methyl
[*8022-00-2*]

has very similar properties and behavior to demeton and an oral LD_{50} to the rat of 80 mg/kg. Oxydemetonmethyl,

$$(CH_3O)_2\overset{\overset{\displaystyle O}{\|}}{P}SCH_2CH_2\overset{\overset{\displaystyle O}{\uparrow}}{S}C_2H_5$$

oxydemetonmethyl
[*301-12-2*]

is more water soluble and is a very active systemic that is especially suited for granular applications to the soil.

Dimethoate, *O,O*-dimethyl *S*-(*N*-methylcarbamoyl)methyl phosphorodithioate,

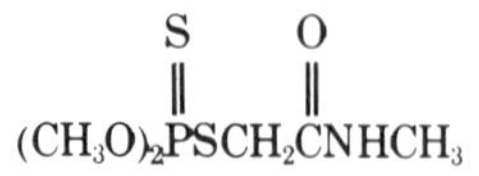

$$(CH_3O)_2\overset{S}{\overset{\|}{P}}SCH_2\overset{O}{\overset{\|}{C}}NHCH_3$$

dimethoate
[*60-51-5*]

is a white crystalline solid (mp 51°C). It is soluble in organic solvents and soluble in water to ca 7%. Dimethoate has an oral LD_{50} to the rat of 245 mg/kg. Because of its water solubility, it has proved especially useful as a persistent systemic for fruit fly larvae and for side-dressing of soil about plants.

Technical-grade mevinphos, dimethyl 2-carbomethoxy-1-methylvinyl phosphate,

$$(CH_3O)_2\overset{O}{\overset{\|}{P}}OC(CH_3){=}CH\overset{O}{\overset{\|}{C}}OCH_3$$

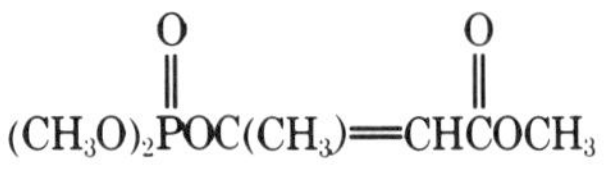

mevinphos
[*7786-34-7*]

is a colorless liquid, (bp 106–107.5°C at 133 Pa (1 mm Hg), d^{20} 1.25, vapor pressure 0.39 Pa (0.0029 mm Hg) at 21°C). It is miscible in water, xylene, and acetone, but is less than 5% soluble in kerosene. The technical product is composed of about two parts of the trans and one part of the cis isomer, the former of which is about ten times as active as latter. The oral LD_{50} to the rat is about 6.8 mg/kg. Mevinphos is especially useful for treatment of edible products close to harvest since it is rapidly dissipated by volatilization and enzymatic decomposition in the plant.

Phosphamidon, dimethyl-2-chloro-2-diethylcarbamoyl-1-methylvinyl phosphate

$$(CH_3O)_2\overset{O}{\overset{\|}{P}}O\underset{CH_3}{\underset{|}{C}}{=}\overset{Cl}{\overset{|}{C}}\underset{O}{\underset{\|}{C}}N(C_2H_5)_2$$

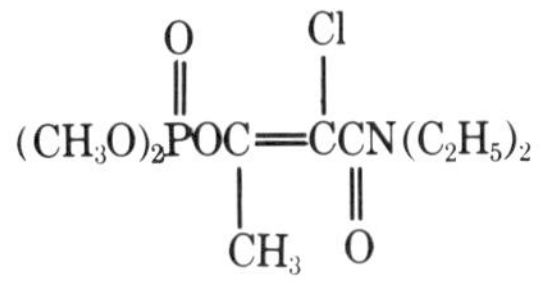

phosphamidon
[*13171-4-6*]

(bp 160°C at 200 Pa (1.5 mm Hg)) is miscible in water and organic solvents. It has an oral LD_{50} to the rat of 17 mg/kg. Phosphamidon is absorbed rapidly by plant surfaces and is decomposed quickly in the plant to provide a short-lived systemic.

Dicrotophos, 3-(dimethoxyphosphinyloxy)-*N,N*-dimethylcrotonamide,

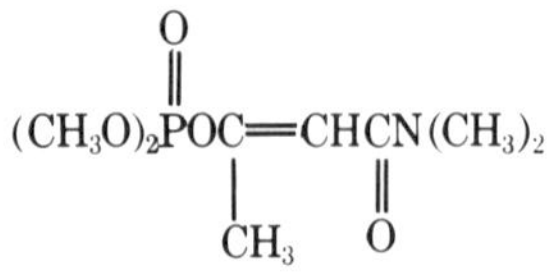

$$(CH_3O)_2\overset{O}{\overset{\|}{P}}O\underset{CH_3}{\underset{|}{C}}{=}CH\underset{O}{\underset{\|}{C}}N(CH_3)_2$$

dicrotophos
[*141-66-2*]

is a brown liquid, (bp 115–120°C at 0.133 Pa (0.001 mm Hg), d^{23} 1.19 g/cm^3) and is soluble in water, ketones, and alcohols. It has an acute oral LD_{50} to the rat of 27 mg/kg and a dermal LD_{50} to the rabbit of 112 mg/kg. Dicrotophos is especially useful for the systemic treatment by trunk injection of trees for the control of pests attacking bark and foliage. The corresponding *N*-methyl compound (rat oral LD_{50} 21 mg/kg) is monocrotophos [*6923-22-4*].

Seed and Soil Treatments. Several systemic compounds of low water solubility are especially suited for the protection of young plants from attack by mites, thrips, aphids, and leafhoppers by application as a 50% charcoal powder to the seeds of cotton, alfalfa, or sugar beets before planting; by granular application at time of transplanting; or as a granular side-dressing applied at planting (20).

Phorate, *O,O*-diethyl *S*-(ethylthio)methyl phosphorodithioate, is a yellowish liquid (bp 118–120°C at 107 Pa (0.8 mm Hg) d^{25} 1.167 g/cm^3). It is soluble in organic solvents and to 0.0085% in water. The oral LD_{50} to the rat is 3.7 mg/kg.

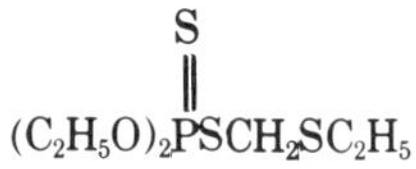

$$(C_2H_5O)_2P(=S)SCH_2SC_2H_5$$

phorate
[*298-02-2*]

In plant tissue, phorate is rapidly oxidized to the sulfoxide and sulfone metabolites that are responsible for the systemic toxicity of the compound. They are more water-soluble and less stable than the parent material and are decomposed in the plant to nontoxic hydrolysis products. The corresponding *S*-(*tert*-butyl) analogue is terbufos [*13071-79-9*] (rat oral LD_{50} 8 mg/kg), which is used widely for corn rootworm control.

Disulfoton, *O,O*-diethyl *S*-2-(ethylthio)ethyl phosphorodithioate, is the dithio analogue of demeton. This compound is a pale yellow liquid (bp 62°C at 1.3 Pa (0.01 mm Hg), d^{20} 1.144 g/cm^3). It is soluble in organic solvents and to 0.0066% in water. The oral LD_{50} to the rat is from 2–12.5 mg/kg. Disulfoton is oxidized in plant tissue in the same manner as phorate is; the final active metabolites are the same as the sulfoxide and sulfone metabolites of the demeton thiol isomer.

Morphothion, *O,O*-dimethyl *S*-2-(ethylthio)ethyl phosphorodithioate (rat oral LD_{50} 100 mg/kg), is the dithio analogue of demeton methyl. Fonofos, *O*-ethyl *S*-phenyl

$$(C_2H_5O)_2P(=S)SCH_2CH_2SC_2H_5$$

disulfoton
[*298-04-4*]

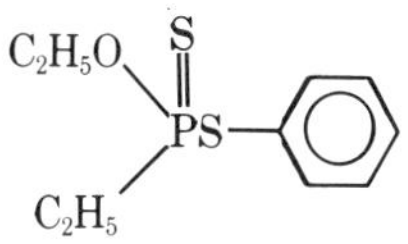

$$(C_2H_5O)(C_2H_5)P(=S)S-C_6H_5$$

fonofos
[*944-22-9*]

$$(CH_3O)_2P(=S)SCH_2CH_2SC_2H_5$$

morphothion
[*144-41-2*]

ethylphosphonodithioate (bp 130°C at 133 Pa (1 mm Hg), d 1.16, H_2O solubility 13 ppm, rat oral LD_{50} 8–16 mg/kg) is used as a soil insecticide for corn rootworm control.

Systemic Insecticides for Animals. When fed or topically applied to domestic animals, systemic insecticides will move through the body tissues in quantities lethal to such internal parasites as cattle grubs, screwworm larvae, and helminths and, over a shorter period, to external parasites, eg, horn and stable flies, mites, lice, and ticks, without causing injury to the host animal. The insecticides are slowly destroyed by enzymic action in the animal body, so that after a safe period of 60 days, the animal can be used for milk production or slaughtered for meat (see Chemotherapeutics, antiprotozoal).

Ronnel is effective when fed at 100 mg/kg of animal weight. Coumaphos, *O,O*-diethyl *O*-(3-chloro-4-methyl-2-oxo-2*H*-1-benzopyran-7-yl) phosphorothionate, is a tan crystalline solid (mp 90–92°C, vapor pressure, 0.13 μPa (0.1 nm Hg) at 20°C). It is insoluble in water and soluble in organic solvents and has an oral LD_{50} to the rat of 56–230 mg/kg. Coumaphos is applied as a 0.25–0.5% spray.

Crufomate, *O*-methyl *O*-(4-*tert*-butyl-2-chlorophenyl) methylphosphoramidate (mp 61°C) is soluble in most organic solvents and is insoluble in water. It is effective either as a 0.75% spray or when fed at 20–25 mg/kg of animal weight. Crufomate has an oral LD_{50} to the rat of 1000 mg/kg.

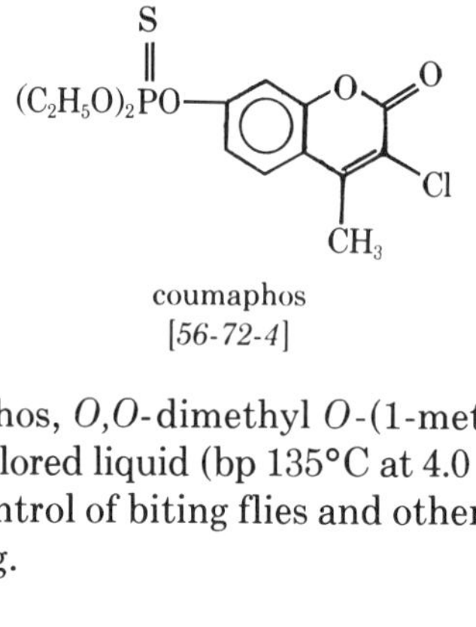

S ‖ $(C_2H_5O)_2PO$– [coumarin ring] O, O, Cl, CH_3

coumaphos
[*56-72-4*]

CH_3NH, O ‖ PO– [benzene ring] –$C(CH_3)_3$, CH_3O, Cl

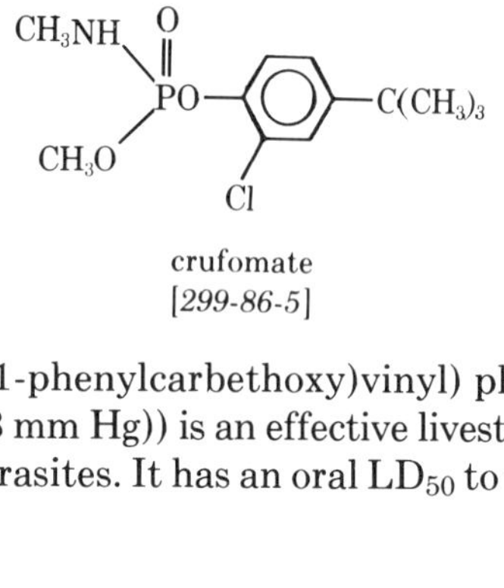

crufomate
[*299-86-5*]

Crotoxyphos, *O,O*-dimethyl *O*-(1-methyl-2-(1-phenylcarbethoxy)vinyl) phosphate, a straw-colored liquid (bp 135°C at 4.0 Pa (0.03 mm Hg)) is an effective livestock spray for the control of biting flies and other ectoparasites. It has an oral LD_{50} to the rat of 128 mg/kg.

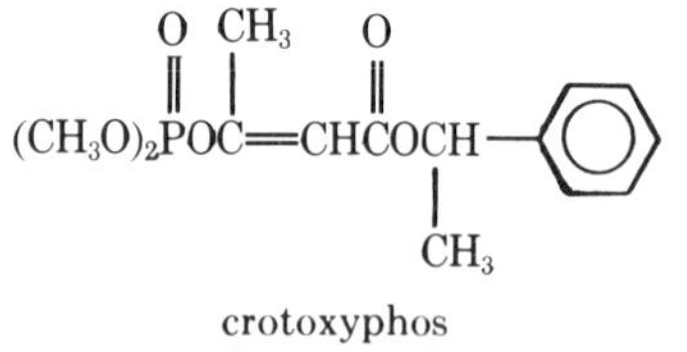

$(CH_3O)_2P(O)OC(CH_3)=CHC(O)OCH(CH_3)$–[benzene ring]

crotoxyphos
[*7700-17-6*]

Mode of Action. The organophosphorus compounds owe their biological activities to the capacity of the central P atom to phosphorylate the esteratic site of the enzyme, cholinesterase (ChE), which is an essential constituent of the nervous system not only of Insecta but also of all higher animals (16,21–22). The phosphorylated enzyme is irreversibly inhibited and, therefore, is no longer able to carry out its normal function of the rapid removal and destruction of the neurohormone, acetylcholine (ACh), from the nerve synapse. As a result, ACh accumulates and disrupts the normal

$$(A)\quad CH_3\overset{O}{\overset{\|}{C}}OCH_2CH_2\overset{+}{N}(CH_3)_3 + EH \underset{(1)}{\overset{K_E}{\rightleftharpoons}} CH_3\overset{O}{\overset{\|}{C}}OCH_2CH_2\overset{+}{N}(CH_3)_3\cdots EH \underset{(2)}{\overset{k_1}{\longrightarrow}}$$

$$CH_3\overset{O}{\overset{\|}{C}}E + HOCH_2CH_2\overset{+}{N}(CH_3)_3 \underset{(3)\ H_2O}{\overset{k_2}{\longrightarrow}} CH_3\overset{O}{\overset{\|}{C}}OH + EH$$

$$(B)\quad (RO)_2\overset{O}{\overset{\|}{P}}X + EH \underset{(1)}{\overset{K_E}{\rightleftharpoons}} (RO)_2\overset{O}{\overset{\|}{P}}X\cdots EH \underset{(2)}{\overset{k_1}{\longrightarrow}} (RO)_2\overset{O}{\overset{\|}{P}}E + HX \underset{(3)\ H_2O}{\overset{k_2}{\longrightarrow}} (RO)_2\overset{O}{\overset{\|}{P}}OH + EH$$

functioning of the nervous system, giving rise to the typical cholinergic symptoms associated in insects with O—P poisoning, ie, hyperactivity, tremors, convulsions, paralysis, and death. In higher animals, these cholinergic effects are translated into muscarinic effects, eg, nausea, salivation, lacrimation, and myosis; nicotinic effects, eg, muscular fasciculations; and central effects, eg, giddiness, tremulousness, coma, and convulsions (see Cholinesterase inhibitors).

The reaction between esterase and phosphorus inhibitor (*B*) is bimolecular, of the well known SN2 type, and represents the attack of a nucleophilic serine hydroxyl with a neighboring imidazole ring of a histidine residue at the active site, upon the electrophilic phosphorus atom, and mimics the normal three-step reaction that takes place between enzyme and substrate (*A*).

In the normal process (*A*), step (*3*) occurs very rapidly and step (*1*) is the rate-determining step, whereas in the inhibition process (*B*), step (*3*) occurs very slowly, generally over a matter of days, so that it is rate determining. Thus, it has been demonstrated with ChE that insecticides, eg, tetraethyl pyrophosphate and mevinphos, engage in first-order reactions with the enzyme; the inhibited enzyme is a relatively stable phosphorylated compound containing one mole of phosphorus per mole of enzyme; and as a result of the reaction, an equimolar quantity of alcoholic or acidic product HX is liberated.

The reactivity of the individual O—P insecticides is determined by the magnitude of the electrophilic character of the phosphorus atom, the strength of the bond P—X, and the steric effects of the substituents. The electrophilic nature of the central P atom is determined by the relative positions of the shared electron pairs between atoms bonded to phosphorus and is a function of the relative electronegativities of the two atoms in each bond (eg, P, 2.1; O, 3.5; S, 2.5; N, 3.0; and C, 2.5). Therefore, it is clear that in phosphate esters (P═O), the phosphorus is much more electrophilic and these are more reactive than phosphorothionate esters (P═S). The latter generally are so stable as to be relatively unreactive with ChE. They owe their biological activity to *in vivo* oxidation by a microsomal oxidase, a reaction that takes place in insect gut and fat body tissues and in the mammalian liver. A typical example is the oxidation of parathion to paraoxon:

$$(C_2H_5O)_2\overset{S}{\overset{\|}{P}}O\text{—}C_6H_4\text{—}NO_2 \xrightarrow{\text{oxidase, NADPH}} (C_2H_5O)_2\overset{O}{\overset{\|}{P}}O\text{—}C_6H_4\text{—}NO_2$$

parathion

paraoxon
[*311-45-5*]

Another type of *in vivo* microsomal oxidation is of importance in the lethal synthesis through which electron-donating groups, eg, CH_3S— are converted to the strongly electron-withdrawing groups, CH_3SO— and CH_3SO_2— (note the equivalence in toxicity between these three groups in Table 4). This type of reaction (with its built-in delay principle) is involved in the mode of action of certain systemic insecticides, eg, phorate and disulfoton, where both P=S and C_2H_5S— oxidation occur.

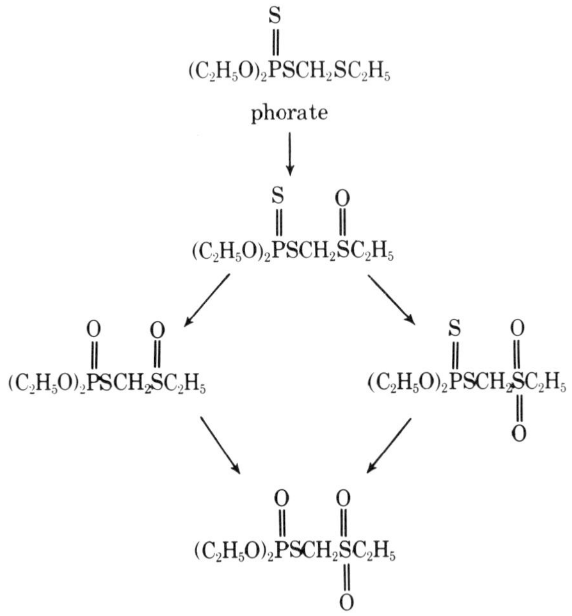

$$(C_2H_5O)_2P(=S)SCH_2SC_2H_5$$

phorate

↓

$$(C_2H_5O)_2P(=S)SCH_2S(=O)C_2H_5$$

↙ ↘

$$(C_2H_5O)_2P(=O)SCH_2S(=O)C_2H_5 \qquad (C_2H_5O)_2P(=S)SCH_2S(=O)_2C_2H_5$$

↘ ↙

$$(C_2H_5O)_2P(=O)SCH_2S(=O)_2C_2H_5$$

In this type of activation, which occurs in both animal and plant tissues, the original insecticide is relatively stable and can be translocated through plant tissues without destructive hydrolysis until the oxidation has occurred, which then makes the insecticide both highly toxic and relatively unstable so that it rapidly is hydrolyzed to nontoxic products.

The substituent groups on the phosphorus atom also strongly affect its electrophilic nature, mainly by inductive and mesomeric effects, which may be transmitted along chains of atoms and are positive (electrons repelled) or negative (electrons at-

Table 4. Relative Toxicity of Substituted Phenyl Diethyl Phosphates $(C_2H_5O)_2P(=O)O{-}C_6H_4{-}X$

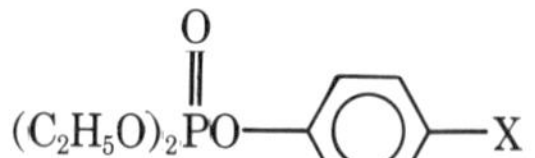

X	CAS Registry No.	K_E (fly ChE)	Topical LD_{50}, *Musca domestica*, μg/g	σ
NO_2	*[311-45-5]*	2.7×10^7	1.0	1.267
CH_3SO_2	*[6132-17-8]*	1.9×10^6	2.5	1.049
CN	*[6132-16-7]*	2.6×10^5	3.5	0.891
CH_3SO	*[6552-21-2]*	2.4×10^5	1.5	0.730
Cl	*[5076-63-1]*	2.3×10^3	150	0.227
H	*[2510-86-3]*	2.6×10	1500	0.0
CH_3S	*[3070-13-1]*	2.3×10^3	2.0	−0.047
CH_3	*[4877-08-1]*	2.0×10	>5000	−0.170
CH_3O	*[5076-68-6]*	2.9×10	>5000	−0.135

tracted). Studies with diethyl substituted-phenyl phosphates have shown a high degree of correlation between the quantitative measure of electron-withdrawing power of the substituent group, Hammett's σ, and anticholinesterase activity and toxicity (21) (see Table 4).

The alkyl and alkoxy substituents of phosphate or phosphonate esters also affect the phosphorylating ability of the compound through steric and inductive effects. A satisfactory correlation has been developed between the quantitative measure of these effects. Taft's σ^*, and anticholinesterase activity as well as toxicity (22) (see Table 5). Thus, long-chain and highly branched alkyl and alkoxy groups attached to phosphorus promote high stability and low biological activity.

Selective Toxicity of Organophosphorus Insecticides. Selectivity is implicit in the definition of insecticide as an agent for destroying insects. Nevertheless, many widely used products, eg, hydrogen cyanide and parathion, are highly toxic to nearly all animals and selectivity is apparent only through careful application and confinement to the area of treatment. The organophosphorus compounds, with their tremendous diversity of structures, offer the greatest promise for the development of true physiological selectivity. Such selectivity is important not only in providing safety for the user of pesticides and for plant and animal possessions but in providing ecological selectivity for wildlife and for the valuable beneficial insects—honey bees, pollinators, parasites, and predators (13,17,21,23).

The development of malathion in 1950 was an important milestone in the emer-

Table 5. Relative Toxicity of Substituted Ethyl *p*-Nitrophenyl Phosphonates

$$X\text{—}\overset{\overset{\displaystyle O}{\|}}{P}(OC_2H_5)(O\text{—}C_6H_4\text{—}NO_2)$$

R	CAS Registry No.	K_E (fly ChE)	Topical LD_{50}, *Musca domestica*, μg/g	σ^*
CH_3	[*3735-98-6*]	7.6×10^6	1.0	0.000
C_2H_5	[*546-71-4*]	1.5×10^7	1.2	−0.100
n-C_3H_7	[*3015-73-4*]	1.0×10^7	2.0	−0.115
n-C_4H_9	[*3015-74-5*]	4.0×10^6	2.4	−0.130
n-C_5H_{11}	[*3015-75-6*]	8.9×10^5	12.0	−0.145
$(CH_3)_2CH$	[*13538-10-8*]	1.1×10^5	28.0	−0.190
$(CH_3)_3C$	[*13538-14-2*]	6.7	>5000	−0.300

Table 6. Selective Toxicity of Parathion, Malathion, and Analogues

Insecticide	Topical LD_{50} housefly *Musca domestica*, μg/g	LC_{50} mosquito larvae *Culex quinquefasciatus*, ppm	Oral LD_{50} rat mg/kg
parathion	0.9	0.0032	6
methyl parathion	1.3	0.018	14
fenitrothion	2.6	0.0058	500
dicapthon	1.6	0.027	500
chlorthion	11.5	0.926	1500
fenthion	2.3	0.0045	215
ronnel	2.7	0.030	1740
malathion	27.0	0.081	1500

gence of selective insecticides. As illustrated in Table 6, this material is from one-half to one-twentieth as toxic to insects as parathion but is only about one two-hundredths as toxic to mammals. Its worldwide usage in quantities of thousands of metric tons in the home, garden, field, orchard, woodland, on animals, and in public-health programs has demonstrated substantial safety coupled with pest control effectiveness. The biochemical basis for the selectivity of malathion is its rapid detoxication in the mammalian liver, but not in the insect, through the attack of carboxyesterase enzymes upon the aliphatic ester moieties of the molecule.

Extraordinary selectivity has been accomplished with the parathion-type of insecticide by incorporating Cl or CH_3 groups in the meta position of the aryl ring. These groups interact sterically with acetylcholinesterase, increasing the affinity for the insect enzyme and decreasing it with the mammalian enzyme. Selectivity also is enhanced by net differences in the rates of microsomal oxidation of P=S to P=O and in hydrolytic detoxication between insects and mammals. The compounds, fenitrothion, fenthion, ronnel, dicapthon, and chlorthion (Table 6) are used widely in public health, household, and agricultural pest control.

Environmental. Organophosphorus insecticides are intrinsically reactive and readily degrade by oxidation and hydrolysis and in living organisms. Therefore, their use does not present serious problems of biomagnification and food-chain transfer. Soil persistence is low; half-lives range from malathion, 1–2 wk, to parathion, diazinon, and azinphos methyl, 3–6 mo. However, the organophosphorus insecticides are general biocides that are toxic to nearly all animal organisms. Oral LD_{50} values (mg/kg) to birds are: phorate: mallard 0.62, pheasant 7.1; parathion: mallard 2.0, pheasant 12; monocrotophos: mallard 4.8, pheasant 2.8; diazinon: mallard 3.5, pheasant 4.3; fenthion: mallard 5.9, pheasant 18; chlorpyrifos: mallard 75, pheasant 12; and azinphos methyl:

Table 7. Relative Toxicity of Substituted-Phenyl *N*-Methylcarbamates[a]

R	CAS Registry No.	I_{50} M, fly ChE	Topical LD_{50} fly, *Musca domestica*, μg/g		LC_{50} mosquito larva, *Culex quinquefasciatus*, ppm
			Alone	1:5 P.B.[b]	
H	*[1943-79-9]*	2×10^{-4}	500	38	>10
o-Cl	*[3942-54-9]*	5.0×10^{-6}	75	24	>10
m-Cl	*[4090-00-0]*	5.0×10^{-5}	500	36	>10
o-CH_3	*[1129-41-5]*	1.4×10^{-4}	500	85	10
m-CH_3	*[1128-78-5]*	1.4×10^{-5}	50	27	3.3
o-CH_3O	*[3938-24-7]*	3.7×10^{-5}	92.5	18	>10
m-CH_3O	*[3938-28-1]*	2.2×10^{-5}	90.0	14.5	10
o-CH_3S	*[3938-32-7]*	9.0×10^{-7}	48.5	14.0	3.9
m-CH_3S	*[3938-33-8]*	7.0×10^{-6}	8.5	6.5	1.5
o-$(CH_3)_2CH$	*[114-26-1]*	6.0×10^{-6}	95	24	0.56
m-$(CH_3)_2CH$	*[4089-99-0]*	3.4×10^{-7}	90	9	0.03
p-$(CH_3)_2CH$	*[64-00-6]*	7.0×10^{-3}	>500	500	>10
o-$(CH_3)_2CHO$	*[2631-40-5]*	6.9×10^{-7}	25.5	7	0.3
o-D-2-$C_2H_5CH(CH_3)$	*[75863-93-3]*	6.0×10^{-6}	515	41.0	2.12
o-L-2-$C_2H_5CH(CH_3)$	*[75863-94-4]*	1.0×10^{-6}	171	23.5	0.26

[a] Ref. 24.

[b] With five parts piperonyl butoxide synergist.

mallard 136, pheasant 75. These compounds are highly toxic to fish and LC_{50} values (ppm) are: phorate: bluegill 0.0055; parathion, bluegill 0.10, trout 0.047; diazinon: bluegill 0.052; trout 0.38; malathion: bluegill 0.11, trout 0.13; fenthion: bluegill 1.4, trout 0.93; and azinphos methyl: bluegill 0.005, trout 0.004. The organophosphorus insecticides are highly toxic to bees and to beneficial parasites and predators. Great care should be used in handling, applying, and storing these insecticides as the preponderance of insecticide poisonings throughout the world result from the use of parathion and related compounds.

Carbamates. Carbamates are synthetic relatives of the alkaloid physostigmine from *Physostigma venenosum* (see Alkaloids). Activity is associated with considerable variation in structure, and the individual compounds show a remarkable degree of insecticidal selectivity (24).

Carbaryl, 1-naphthyl *N*-methylcarbamate (mp 142°C, d^{20} 1.232 g/cm^3, vapor pressure 0.67 Pa (0.005 mm Hg) at 20°C) is ca 95% pure as the technical grade and has the following percent solubilities: carbon tetrachloride, 10; methyl isobutyl ketone, 20; petroleum ether, 20; xylene, 10; and water, 0.004. Carbaryl has a rat oral LD_{50} of 540 mg/kg and a dermal LD_{50} of >2000 mg/kg. It is a broad spectrum insecticide registered on more than 100 crops. Carbaryl is rapidly detoxified and eliminated in animal urine and is neither concentrated in fat nor secreted in the butterfat of milk. Thus, it is favored for application to food crops.

Carbofuran, 2,3-dihydro-2,2-dimethyl-7-benzofuranyl *N*-methylcarbamate (mp 150–152°C), is soluble in water to 0.07%. The rat oral and dermal LD_{50}s are 4.8 and >10,000 mg/kg, respectively. Carbofuran is a broad spectrum insecticide which is used as a soil insecticide.

Propoxur, 2-isopropoxyphenyl *N*-methylcarbamate (mp 91°C), is soluble in nonpolar solvents and in water to ca 0.1%. The rat oral and dermal LD_{50}s are 100 and >1000 mg/kg, respectively. Propoxur is used for the control of household insects, especially cockroaches, bedbugs, and mosquitoes, and is a replacement for DDT in malaria control.

Dioxacarb, *O*-(1,3-dioxolan-2-yl)-phenyl *N*-methylcarbamate (mp 114–115°C, vapor pressure 40 μ Pa (0.3 nmHg) at 20°C), is water soluble to ca 0.6%. The rat oral

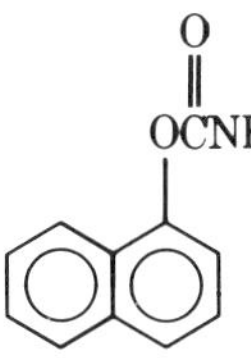

OCNHCH$_3$ (C=O)

carbaryl
[63-25-2]

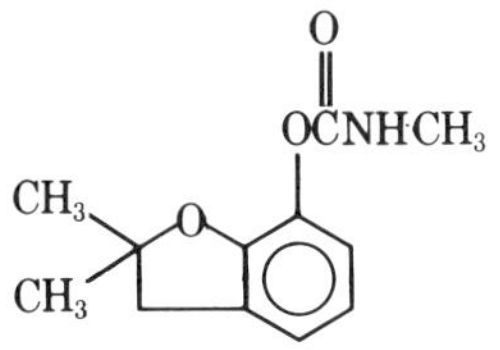

carbofuran
[1563-66-2]

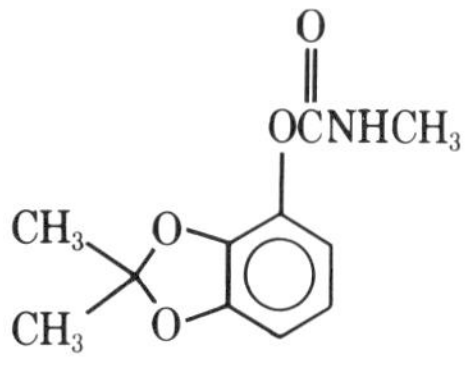

dioxacarb
[6988-21-2]

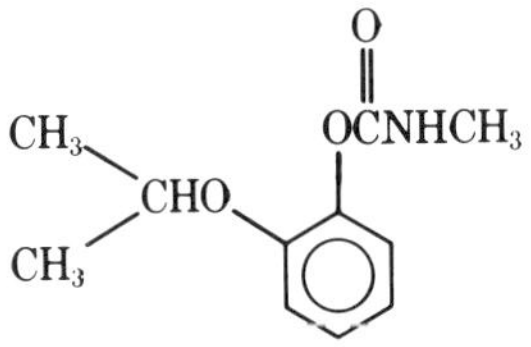

propoxur
[114-26-1]

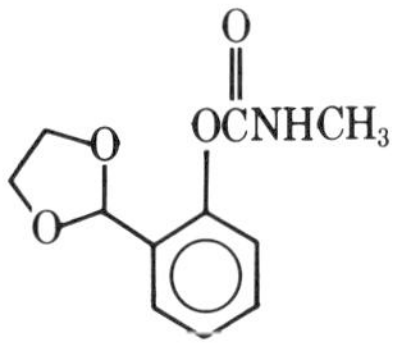

bendiocarb
[22781-23-3]

and dermal LD_{50}s are 125 and 3000 mg/kg, respectively. Dioxacarb is used to control cockroaches and a wide variety of household and stored-product pests.

Bendiocarb, 2,2-dimethyl-1,3-benzodioxol-4-yl *N*-methylcarbamate (mp 129–130°C, vapor pressure 0.7 mPa (5 nm Hg) at 25°C), is water soluble to ca 0.004%. The rat oral and dermal LD_{50}s are 179 and 1000 mg/kg, respectively. Bendiocarb is used to control cockroaches, other household pests, and soil insects.

Aldicarb, 2-methyl-2-(methylthio)-propionaldehyde *O*-methylcarbamoyl oxime (mp 99–101°C), has the following percent solubilities: water, 0.6; acetone, 30; chloroform, 35; ethanol, 25; and toluene, 10. The rat oral and dermal LD_{50}s are 0.93 and 5 mg/kg, respectively. Aldicarb is a broad spectrum systemic insecticide which is used for soil and seed treatments.

Methomyl, *S*-methyl *N*-(methylcarbamoyloxy)-thioacetimidate (mp 78–79°C, vapor pressure 7 mPa (50 nm Hg) at 20°C), is soluble in water and organic solvents. The rat oral LD_{50} is 17–26 mg/kg and the dermal LD_{50} >1500 mg/kg. It is widely used to control insects that attack vegetable crops.

Oxamyl, *S*-methyl 1-(dimethylcarbamoyl)-*N*-[(methylcarbamoyl)oxy]-thioformimidate, has rat oral and dermal LD_{50}s of 5.4 and 2960 mg/kg, respectively. Oxamyl is a systemic insecticide.

Pyrimicarb, 2-(dimethylamino)-5,6-dimethyl-4-pyrimidinyl *N,N*-dimethylcarbamate (mp 90.5°C, vapor pressure 4 mPa (0.3 nm Hg) at 30°C), has a water solubility of 0.27%. The rat oral and dermal LD_{50}s are 147 and >500 mg/kg, respectively. Pyrimicarb is a systemic aphicide.

Dimetilan, 2-(*N,N*-dimethylcarbamoyl)-3-methylpyrazol-5-yl *N,N*-dimethylcarbamate (mp 68–71°C), is used in sugar baits for housefly control. The rat oral and dermal LD_{50}s are 25–64 and 600 mg/kg, respectively.

Mexacarbate, 4-dimethylamino-3,5-xylyl *N*-methylcarbamate (mp 85°C), has rat oral and dermal LD_{50}s of 15–63 and >500 mg/kg, respectively. It has the following solubilities (g/100 g solvent) at 25°C: acetone, 162; benzene, 102; ethanol, 116; hexane,

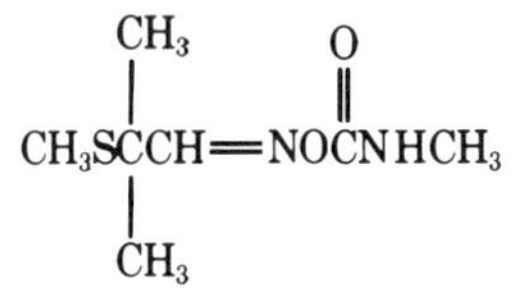

$CH_3SC(CH_3)_2CH{=}NOC(O)NHCH_3$

aldicarb
[*116-06-3*]

$CH_3SC(CH_3){=}NOC(O)NHCH_3$

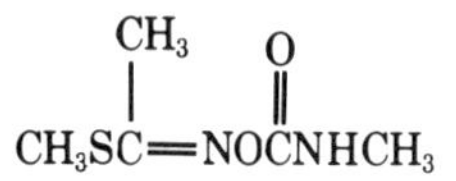

methomyl
[*16752-77-5*]

$CH_3SC(C(O)N(CH_3)_2){=}NOC(O)NHCH_3$

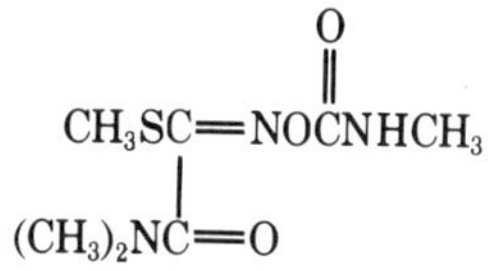

oxamyl
[*23135-22-0*]

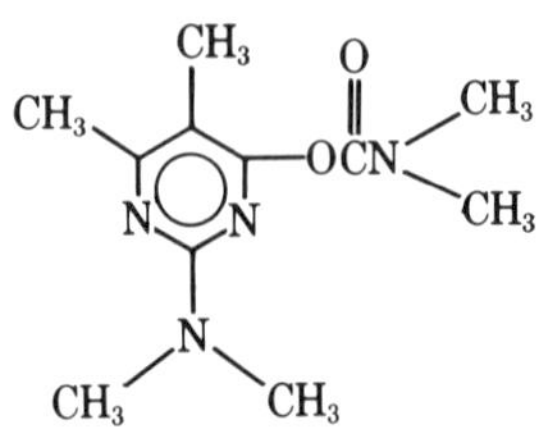

pyrimicarb
[*23103-98-2*]

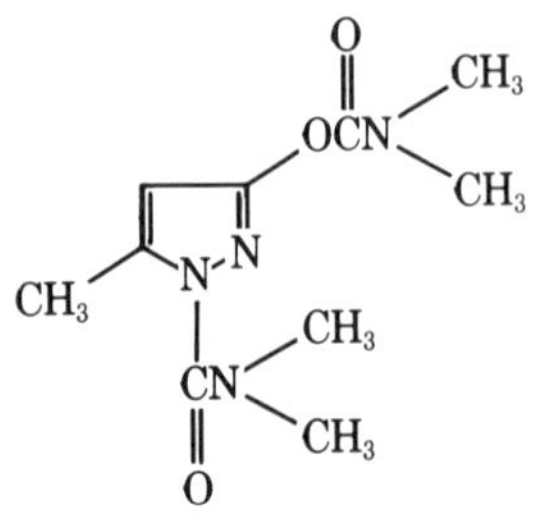

dimetilan
[*644-64-4*]

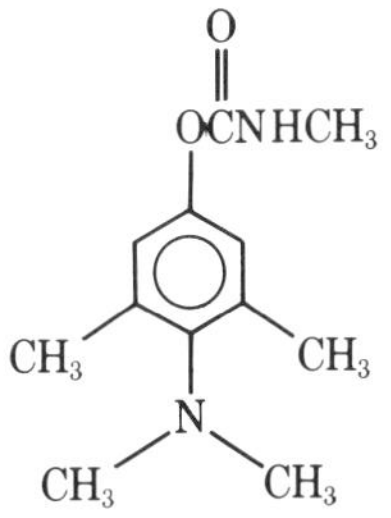

mexacarbate [315-18-4] aminocarb [2032-59-9] methiocarb [2032-65-7] formetanate [22259-30-9]

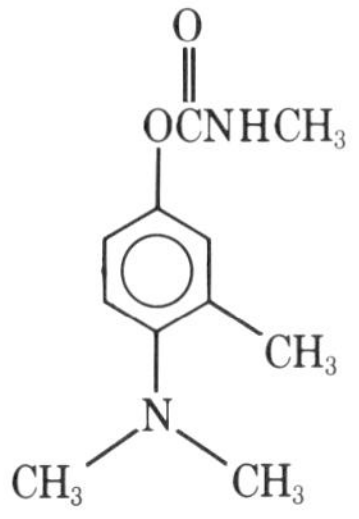

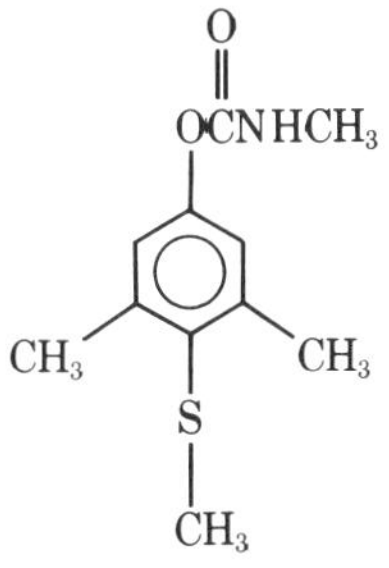

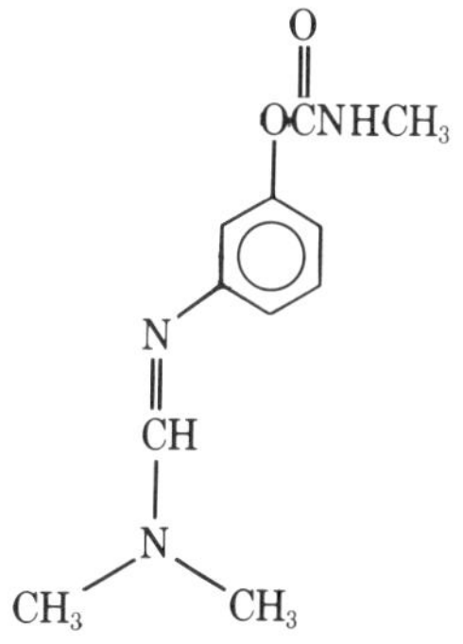

1.0; methylene chloride, 121; and water, 0.01. Like the other carbamates, it is unstable in alkaline solution and hydrolyzes rapidly above pH 11. Mexacarbate is particularly effective against snails, slugs, and lepidopterous larvae. Aminocarb is the closely related 4-dimethylamino-3-tolyl *N*-methylcarbamate (mp 93°C) and has a similar spectrum of action. The rat oral and dermal LD_{50}s are 30 and 275 mg/kg, respectively.

Methiocarb, 4-methylthio-3,5-xylyl *N*-methylcarbamate (mp 121°C), has rat oral and dermal LD_{50}s of 130 and >2000 mg/kg, respectively. Methiocarb is used to control insect pests of fruits and vegetables.

Formetanate, 3-dimethylaminoethyleneiminophenyl *N*-methylcarbamate hydrochloride, is a water-soluble acaricide that is used on deciduous fruits. The rat oral and dermal LD_{50}s are 15–26 and >5600 mg/kg, respectively.

A number of other substituted phenyl *N*-methylcarbamates have had limited insecticidal use. These include metalkamate [*8065-36-9*], a 1:4 mixture of *m*-(1-ethylpropyl)phenyl *N*-methylcarbamate [*672-04-8*] and *m*-(1-methylbutyl)phenyl *N*-methylcarbamate [*2282-34-0*], which is used as a soil insecticide. It has a rat oral LD_{50} of 87–170 mg/kg. Also included are promecarb [*2631-37-0*] (3-methyl-5-isopropylphenyl *N*-methylcarbamate), *o*-chlorophenyl *N*-methylcarbamate, *m*-tolyl *N*-methylcarbamate, and the 2-isopropylphenyl and 3-isopropylphenyl *N*-methylcarbamates (see Table 7).

Mode of Action. All of the insecticidal carbamates are cholinergic, and poisoned insects and animals exhibit violent convulsions and other neuromuscular disturbances. The compounds are strong carbamylating inhibitors of cholinesterase and may have

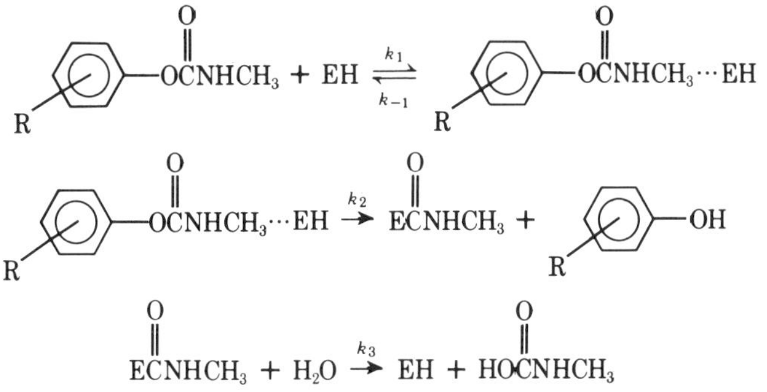

$$R\text{-}C_6H_4\text{-}OC(O)NHCH_3 + EH \underset{k_{-1}}{\overset{k_1}{\rightleftharpoons}} R\text{-}C_6H_4\text{-}OC(O)NHCH_3\cdots EH$$

$$R\text{-}C_6H_4\text{-}OC(O)NHCH_3\cdots EH \xrightarrow{k_2} EC(O)NHCH_3 + R\text{-}C_6H_4\text{-}OH$$

$$EC(O)NHCH_3 + H_2O \xrightarrow{k_3} EH + HOC(O)NHCH_3$$

a direct action on acetylcholine receptors because of their pronounced structural resemblance to acetylcholine. The importance of structural complementarity is shown by the differences in the activities of D- and L-*sec*-butylphenyl *N*-methylcarbamate and that of the *o*, *m*-, and *p*-isopropylphenyl *N*-methylcarbamates (Table 7).

The overall mechanism for carbamate interaction with cholinesterase is analogous to the normal three-step hydrolysis of acetylcholine. However, k_3 is much slower than with the acetylated enzyme.

Within some species of insects, eg, the housefly, detoxication of the carbamates takes place very rapidly through hydroxylation of ring and *N*-methyl groups and through hydrolysis. The methylenedioxy synergists, eg, piperonyl butoxide (see Synthetic Pyrethroids) largely prevent this detoxication by inhibiting the hydroxylation enzymes, as shown in Table 7.

Environmental. The *N*-methylcarbamates generally are biodegradable and of low soil persistence with half-lives for carbaryl and aldicarb of 1–2 wk and carbofuran of 1–4 mo. Certain carbamates are highly toxic to birds with the following oral LD_{50} values (mg/kg) carbofuran: mallard 0.40, pheasant 4.2; mexacarbate: mallard 3.0, pheasant 4.5; methomyl: mallard 16, pheasant 15 (compared to carbaryl: >2000 mg/kg). Fish toxicity of carbamates generally is low but these compounds are extremely toxic to bees. Carbamates with low LD_{50} values, including aldicarb, carbofuran, methomyl, oxamyl, and mexacarbate, can cause poisoning in higher animals and in humans.

Synthetic Pyrethroids. Knowledge of the chemical structures of the natural pyrethrins, their rapid and selective insecticidal action, and their high cost stimulated the search for effective synthetic derivatives (3,25). Allethrin, or DL-2-allyl-3-methylcyclopent-2-en-4-ol-1-onyl DL-*cis,trans*-chrysanthemate [*584-79-2*], was synthesized in 1949 and is produced commercially by a complex thirteen-step process. The product is a clear brownish liquid (d^{20} 1.005–1.015 g/cm^3) containing 75–95% of eight individual optical and geometric isomers whose identity, abundance, and selective toxicity are given in Table 8.

Substantial structural optimization has been applied and a number of highly effective synthetic pyrethroids have been marketed that possess unprecedented insecticidal action. The structures of these involve major departures from the natural pyrethrins in both acid and alcohol moieties.

Table 8. Allethrin Isomers with Their Abundance and Relative Toxicities to *Musca domestica*

Allylrethronyl portion	Chrysanthemate portion	Percent	Relative toxicity[a]
L-	D-*trans*-	12.4	0.58
D-	L-*trans*-	12.4	0.14
D-	D-*trans*-	22.8	3.37
L-	L-*trans*-	22.8	0.02
L-	D-*cis*-	8.0	0.33
D-	L-*cis*-	8.0	0.14
D-	D-*cis*-	6.8	1.77
L-	L-*cis*-	6.8	0.06

[a] The toxicity of DL-allylrethronyl DL-*cis,trans*-chrysanthemate is assumed to be equal to 1.00.

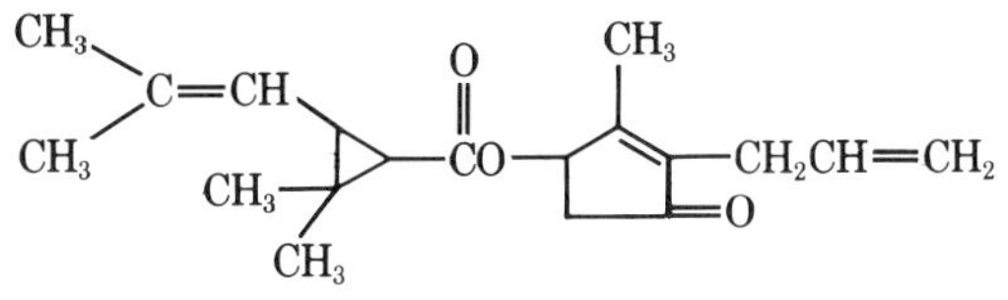

allethrin

Pyrethroids from Chrysanthemic Acid. Substantial insecticidal activity is found in the substituted benzyl chrysanthemates of which 2-chloro-3,4-methylenedioxybenzyl DL-*cis,trans*-chrysanthemate, barthrin [*70-43-9*], and 2,4-dimethylbenzyl DL-*cis,trans*-chrysanthemate, dimethrin, are best known. Dimethrin has outstanding mammalian safety and has been used as a mosquito larvicide that is safe for use in potable waters, eg, rain barrels and cisterns.

5-Benzyl-3-furylmethyl DL-*cis,trans*-chrysanthemate, resmethrin, is a pyrethroid of very high insecticidal activity and mammalian safety but of short persistence. Activity in this series is enhanced further by esterification with D-*trans*-chrysanthemic acid produced by microbial fermentation to bioresmethrin which is about 2.2 times as toxic to the housefly and 0.18 times as toxic to the rat as resmethrin. Greatly increased stability to light and air is found in the 3-phenoxybenzyl DL-*cis,trans*-chrysanthemate (phenothrin).

Pyrethroids with Modified Acid Components. The most recent developments in the pyrethroids have involved substantial changes in the structure of chrysanthemic acid and esterification with the optimized alcoholic moieties. The 3-phenoxybenzyl DL-*cis,trans*-3-(2,2-dichlorovinyl)-2,2-dimethylcyclopropanecarboxylate (permethrin) (mp 35°C, vapor pressure 45 μPa (0.34 nm Hg)) has outstanding insecticidal activity and markedly enhanced persistence. The α-cyano-3-phenoxybenzyl DL-*cis*-3-(2,2-dibromovinyl)-2,2-dimethylcyclopropanecarboxylate (decamethrin) seems to be the most active synthetic pyrethroid yet devised.

Further structural optimization involves the replacement of the cyclopropanecarboxylic acid moiety with acids, eg, α-isopropyl-p-chlorophenylacetic acid. Upon esterification with α-cyano-3-phenoxybenzyl alcohol, the resulting pyrethroid (fenvalerate) is a highly effective and relatively persistent insecticide.

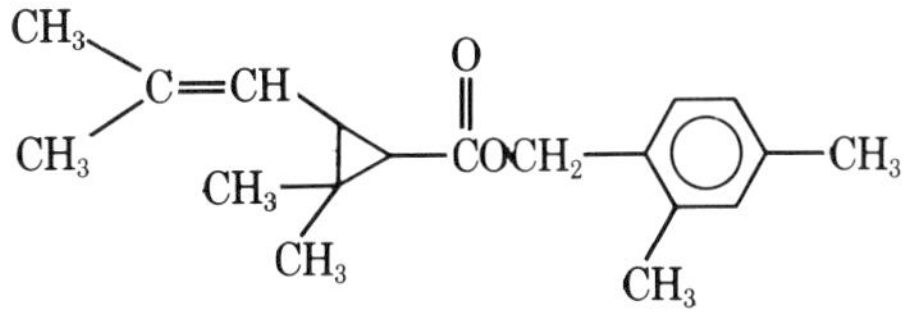

dimethrin

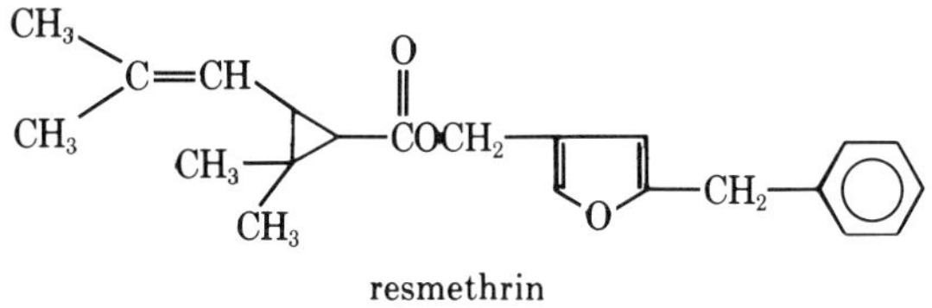

resmethrin

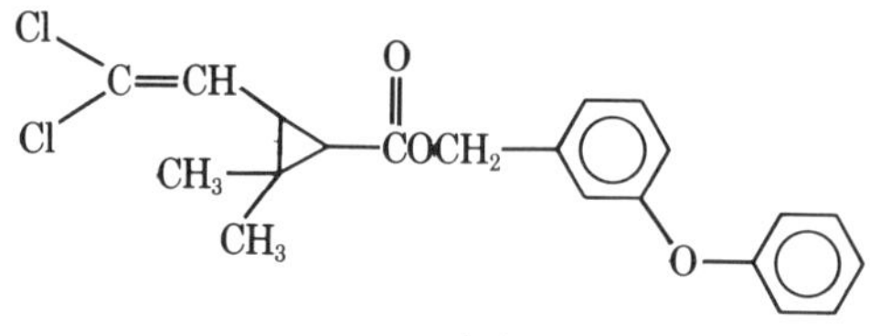

permethrin

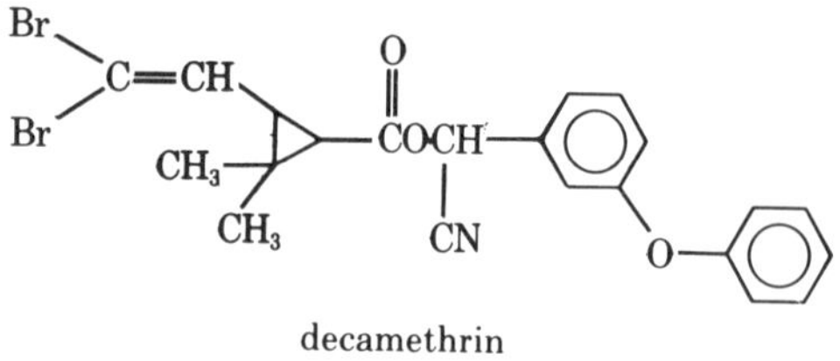

decamethrin

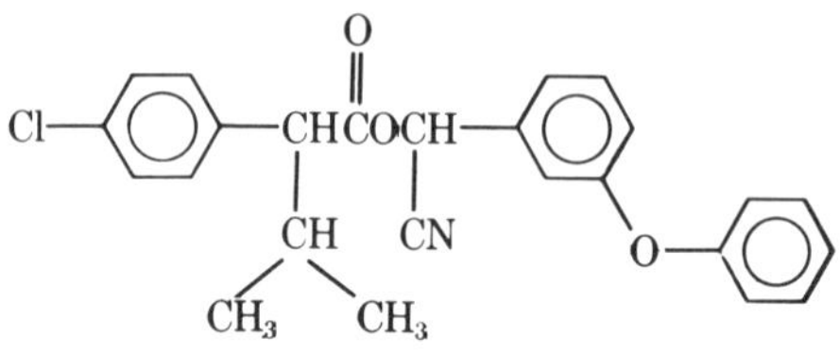

fenvalerate

Some of the biological properties of these new pyrethroids and their CAS Registry Numbers are shown in Table 9.

Environmental. The new synthetic pyrethroids offer substantial improvements in insecticide selectivity over the older organochlorine, organophosphorus, and carbamate insecticides. The quantities applied for plant protection are lower, 0.1–0.5 kg/ha and the insecticides generally have lower animal toxicity (Table 9). A notable exception is the extremely high fish toxicity, LC_{50} 1–10 ppb. Insecticides, eg, permethrin, decamethrin, and fenvalerate, are considerably more persistent than the natural pyrethrins and resmethrin because the degradophores in the acid and alcohol moieties have been removed in the structural optimization process, eg, replacement of dimethylvinyl by dihalovinyl and 2-pentadienylcyclopentenolone by 3-phenoxybenzyl. Thus, permethrin and fenvalerate may persist as foliage residues for 2–4 wk and soil residues for 1–2 mo. These insecticides are highly toxic to beneficial insects and should be used with great care.

Table 9. Biological Activity of Pyrethroids

Insecticide	CAS Registry No.	Topical LD_{50} housefly *Musca domestica*, $\mu g/g$	LC_{50} mosquito larvae *Culex quinquefasciatus*, ppm	Oral LD_{50} rat, mg/kg
pyrethrins (natural)		15		580
allethrin	[*584-79-2*]	8.5	0.14	770
dimethrin	[*70-38-2*]	50	0.088	>10,000
resmethrin	[*10453-86-8*]	0.8	0.0081	1400
bioresmethrin	[*28434-01-7*]	0.25		8000
phenothrin	[*26002-80-2*]	2.6	0.0083	5000
permethrin	[*52645-53-1*]	1.35	0.0039	600
biopermethrin	[*28434-01-7*]	0.85	0.0033	
decamethrin	[*52820-00-5*]	0.17	0.00010	125
fenvalerate	[*51630-58-1*]	4.0	0.0048	200

Insect Growth Regulators. The great majority of insecticides are neurotoxins and produce biochemical lesions at target sites, eg, acetylcholinesterase, that are common to both insects and vertebrates. For this reason, these insecticides often are lacking in selectivity, and thus man, domestic animals, and wildlife frequently are the unintended victims of insecticide use. The insect growth regulators that interfere with biochemical and physiological processes that are unique to the arthropods, eg, molting, ecdysis, and formation of the chitinous exoskeleton, are much more truly selective insecticides and there is an urgent need for their exploitation.

Juvenoids. The discovery of the insect juvenile hormone (JH or neotenin) of the corpus cardiacum paved the way for the synthesis of thousands of hormone mimics that act at the receptor sites of neotenin and, thus, affect ovarian development, adult morphogenesis, diapause, and other critical processes (3,26) (see Hormones, sex hormones). As shown in Table 10, neotenin is substantially insecticidal and prevents the emergence of physiologically competent adult insects when applied exogenously to immature insects. A considerable number of neotenin analogues are even more effective (see Table 10) and have substantial persistence to air, light, and bacterial action making them much more insecticidally active than the relatively ineffective neotenin.

Chitin Synthesis Inhibitors. Another type of insect growth regulator inhibits the formation of the insect chitinous exoskeleton and thus produces a critical biochemical lesion at the time of molting, pupation, or ecdysis. The best known of these compounds is diflubenzuron, 1-(4-chlorophenyl)-3-(2,6-difluorobenzoyl)urea (mp 239°C), which has a water solubility of 0.00002%.

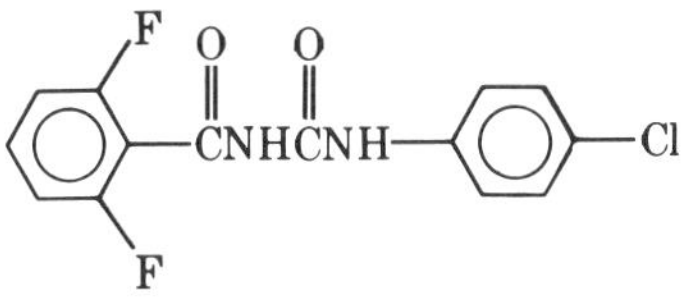

(2,6-F₂C₆H₃)–C(=O)NHC(=O)NH–C₆H₄–Cl

diflubenzuron
[*25367-38-9*]

This compound apparently inhibits the action of the enzyme, chitin synthetase, which polymerizes uridine–diphosphospho–acetylglucoseamine to chitin. It also has a specific ovicidal effect on the development of insect eggs. The rat oral LD_{50} of diflubenzuron is >4640 mg/kg and it is relatively specific in its action, affecting only arthropods.

Methoprene has some effective commercial usage as a mosquito larvicide and for horn-fly control (when fed to cattle) in manure. Compound R-20458 and similar compounds are used widely in China and Japan to prolong the last larval instar of the silkworm and thus increase silk production by 10–15%. As suggested in Table 10, these compounds are quite specific in their insecticidal action, although neotenin and its corresponding dimethyl analogue are widely distributed as insect juvenile hormones. The great virtue of these juvenoids for insect control is their very great safety to mammals; rat oral LD_{50} values of methoprene and hydroprene are >34,600 mg/kg and R-20458 >4000 mg/kg, respectively. Their major drawback for commercial pest control has been their rapid biodegradability, which has been corrected to some extent by slow-release formulations, and their high degree of pest specificity.

Table 10. Insecticidal Effectiveness of Some Juvenoids

Compound	CAS Registry No.	Structure	LD_{50} Values *Tenebrio molitor*, μg/pupa	*Galleria melonella*, μg/pupa	*Aedes aegypti*, ppm	*Musca domestica*, μg/pupa	*Myzus persicae*, μg/cm^2
neotenin	*[32766-80-6]*	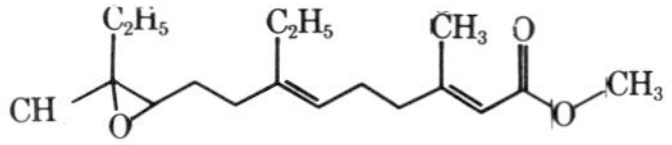	0.13	0.050	0.15	>100	>5
R-20458	*[37571-83-8]*	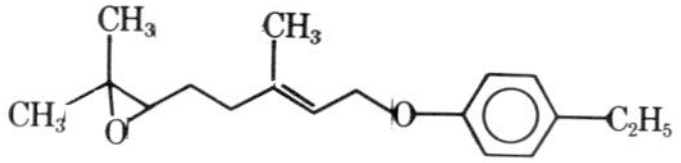	0.003	0.073	0.00047	2.9	>1
hydroprene	*[41096-46-2]*	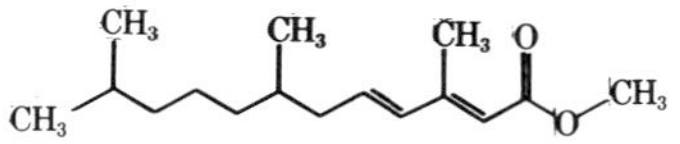	0.26	0.045	0.035	19	0.001
methoprene	*[40596-69-8]*	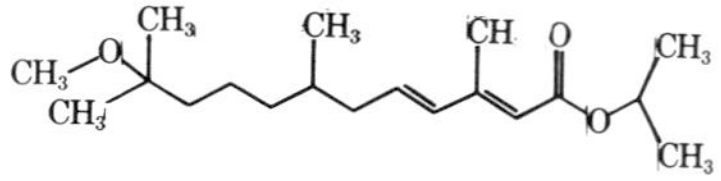	0.011	1.1	0.00011	0.005	0.010

Acaricides. Chemicals that are especially effective in controlling the mites and ticks of the order Acarina are acaricides (12). Most of the insecticidal chemicals discussed, with the exception of the dinitrophenol and organophosphorus insecticides, are not of practical value as acaricides. Applications of many of the chlorinated hydrocarbon insecticides have no effect on the phytophagous spider mites but often kill their predators and thus result in abnormally large populations of the spider mites. A number of acaricidal chemicals have come into widespread use that have almost specific toxicity to the mites but are inactive against insects. In general, these acaricides are highly stable compounds with comparatively prolonged residual action and low mammalian toxicity. Certain of the compounds described below are effective only as ovicides, killing the eggs and sometimes the newly emerged nymphs, and others are active against all the stages of the mites. These acaricides exhibit a considerable degree of specificity for various species of acarina and are most useful for the phytophagous Tetranychidae and Eriophyidae. Other acaricides that repel or kill the mites and ticks that attack man and animals are described later (see Repellents).

Chlorfenethol, di-(*p*-chlorophenyl)methylcarbinol, or 1,1-bis(*p*-chlorophenyl)-ethanol, is a white solid (mp 69.5–70°C). It is insoluble in water and has the following solubilities (g/100 mL of solvent at 25–30°C): petroleum ether, 4.3; ethanol, 125; and toluene, 110. The compound is readily dehydrated upon heating or in the presence of strong acids and forms the inactive 1,1-bis(*p*-chlorophenyl)ethylene. Chlorfenethol is active against all stages of mites. It has an oral LD_{50} to the rat of ca 200 mg/kg.

Chlorobenzilate, ethyl *p,p'*-dichlorobenzilate, is a yellowish viscous oil (bp 141–142°C at 8 Pa (0.06 mm Hg)). The technical material (d 1.2816 g/cm^3) contains ca 90% of the active compound, is insoluble in water, and is soluble to more than 40% in deodorized kerosene, benzene, and methyl alcohol. Chlorobenzilate is hydrolyzed in alkali and in strong acids to the inactive *p,p'*-dichlorobenzilic acid and ethanol. The compound is active against all stages of mites and has an oral LD_{50} to the rat of ca 1000 mg/kg.

Dicofol, 1,1-bis(*p*-chlorophenyl)-2,2,2-trichloroethanol, is a white crystalline solid (mp 78.5–79.5°C). This compound is insoluble in water and soluble in organic solvents, and in the presence of alkali, forms the inactive *p,p'*-dichlorobenzophenone and chloroform. Dicofol is a long-lasting acaricide and is active against all stages of mites. It has an oral LD_{50} to the rat of ca 1000 mg/kg.

Tetradifon, 2,4,5,4'-tetrachlorodiphenyl sulfone, is a crystalline solid (mp 148°C). It is insoluble in water and has the following solubilities (g/100 g of solvent at 18°C): petroleum ether, 0.4; ethyl acetate, 7.1; carbon tetrachloride, 1.6; methyl ethyl ketone, 10.5; and xylene, 11.5. Tetradifon is stable to the action of acids and alkalies, light and temperature, and has a very prolonged residual action. It is active against all stages of mites and has an oral LD_{50} to the rat of >14,700 mg/kg.

Sulphenone, *p*-chlorophenyl phenyl sulfone, is a white solid (mp 98°C). The

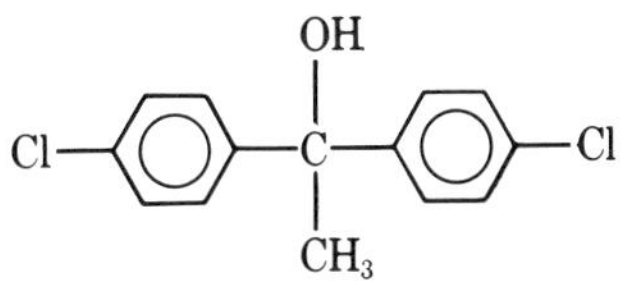

chlorfenethol
[*80-08-6*]

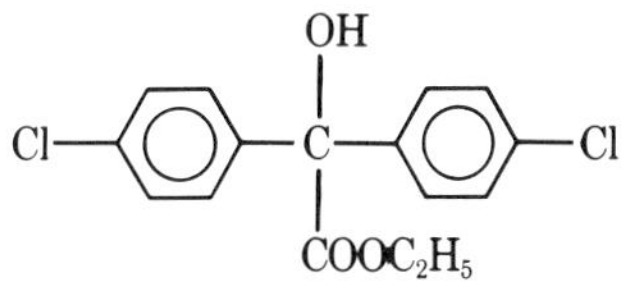

chlorobenzilate
[*510-15-6*]

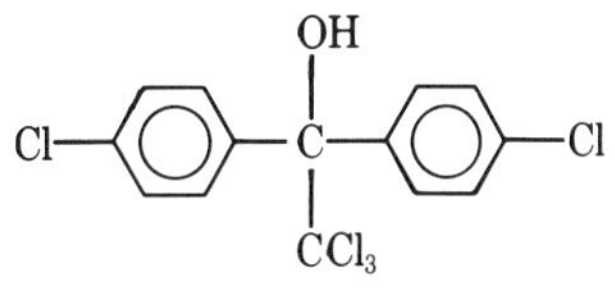

dicofol
[*54532-36-4*]

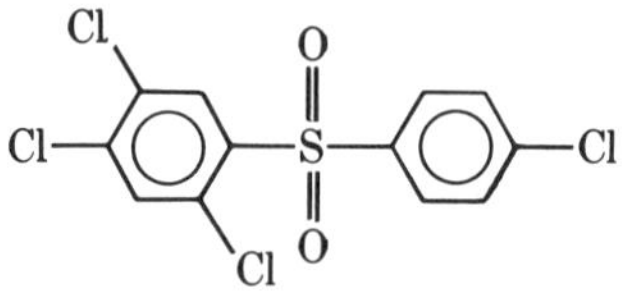

tetradifon
[*116-29-0*]

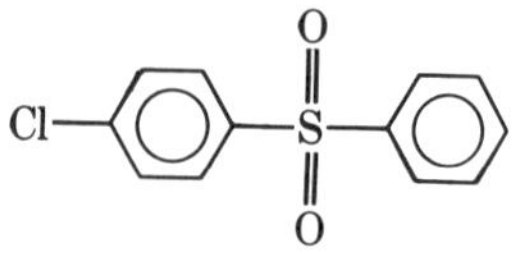

sulphenone
[*80-30-2*]

technical material consists of ca 80% of this compound, with small amounts of *o*-and *m*-isomers, bis(*p*-chlorophenyl) sulfone, and diphenyl sulfone. Sulphenone is insoluble in water and has the following solubilities (g/100 g of solvent at 20°C): hexane, 0.4; xylene, 18.2; carbon tetrachloride, 4.9; and acetone, 74.4. Sulphenone is effective against all stages of mites. Its oral LD_{50} to the rat is >2000 mg/kg.

Ovex, *p*-chlorophenyl *p*-chlorobenzenesulfonate, is a white solid (mp 86.5°C). It is insoluble in water and has the following solubilities (g/100 g of solvent at 25°C): kerosene, 2; carbon tetrachloride, 41; cyclohexanone, 110; ethyl alcohol, 1; ethylene dichloride, 110; and xylene, 78. Ovex is effective only as an ovicide. It has an oral LD_{50} to the rat of 2000 mg/kg.

Two analogues of ovex have similar solubilities, and ovicidal properties and, like ovex, are hydrolyzed in alkali to form the phenol and benzenesulfonate salt. Genite, 2,4-dichlorophenyl benzenesulfonate (mp 45–47°C, vapor pressure 36 mPa (0.27 μm Hg) at 30°C). The technical product is about 97% pure and has an oral LD_{50} to the rat of 1400 mg/kg. Fenson is *p*-chlorophenyl benzenesulfonate (mp 61–62°C, d 1.33 g/cm^3, rat oral LD_{50} 1600 mg/kg).

Chlorbenside is *p*-chlorobenzyl *p*-chlorophenyl sulfide (mp 74°C, vapor pressure 35 mPa (2.6 nm Hg) at 20°C). It is insoluble in water and has the following solubilities (g/100 g of solvent at 20°C): kerosene, 5–7.5; methyl ethyl ketone, 137; and xylene, 93. The technical product contains ca 90% *p,p′* isomer, 5% *o,p′* isomer, and 2.5% *m,p′* isomer. Chlorbenside is unaffected by reduction and by acid and alkaline hydrolysis, but it is readily oxidized to *p*-chlorobenzyl *p*-chlorophenyl sulfoxide [*7047-28-1*] (mp 125°C) and more slowly to *p*-chlorobenzyl *p*-chlorophenyl sulfone [*74512-22-4*] (mp 150°C). These reactions occur on the leaf surface, and the oxidation products are acaricidal but do not penetrate locally into the leaf tissue as does chlorbenside. Chlorbenside is active only against eggs and immature mites. The oral LD_{50} to the rat is >10,000 mg/kg.

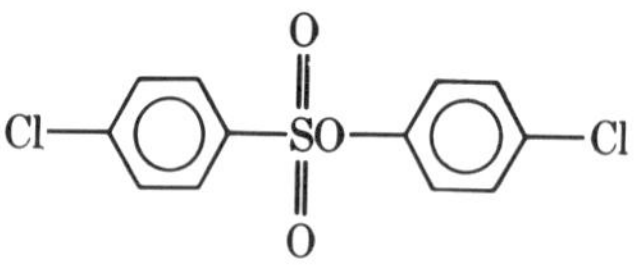

ovex
[*80-33-1*]

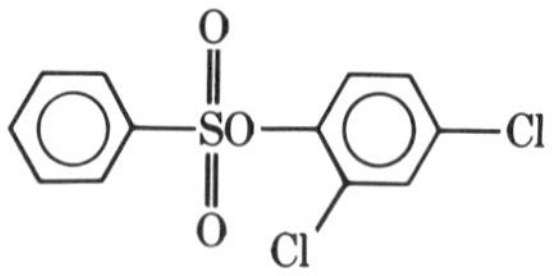

genite
[*97-16-5*]

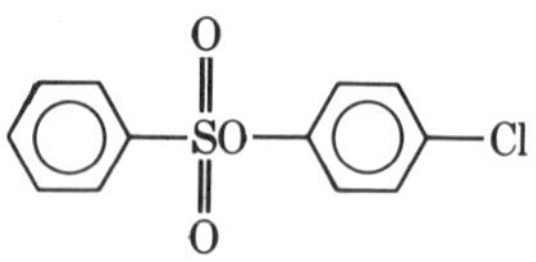

fenson
[*80-38-6*]

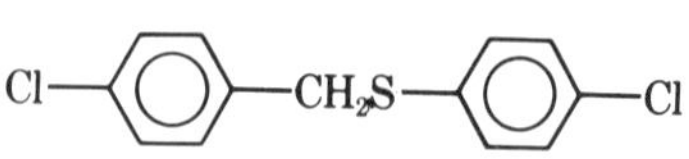

chlorbenside
[*103-17-3*]

Propargite, 2-(*p-tert*-butylphenoxy)cyclohexyl 2-propynyl sulfite, is a dark oil (d^{25} 1.1 g/cm^3), is insoluble in water, and hydrolyzes in alkaline solutions. The rat oral LD_{50} is 2200 mg/kg. Propargite is a widely used acaricide on fruits, vegetables, and row crops.

Oxythioquinox, 6-methyl-2,3-quinoxalinedithiol cyclic carbonate, (mp 172°C), is insoluble in water and soluble in organic solvents with a rat oral LD_{50} of 2500 mg/kg. It is used as an acaricide for tree fruits.

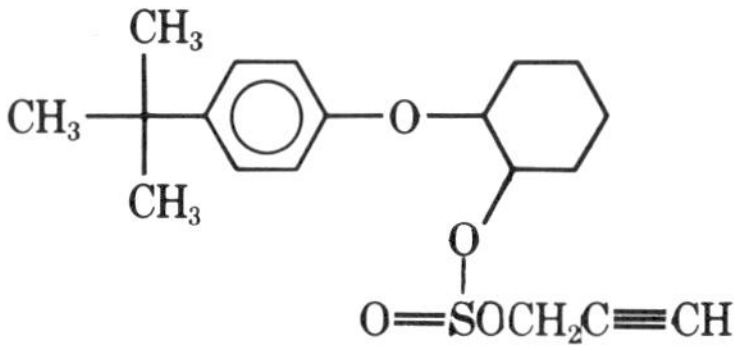

propargite
[*2312-35-8*]

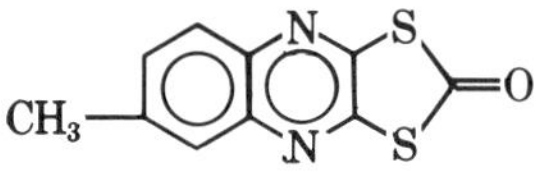

oxythioquinox
[*2439-01-2*]

Tricyclohexyltin hydroxide is an acaricide that is especially useful on deciduous fruits; it has a rat oral LD_{50} of 540 mg/kg. Hexakis-(β,β-dimethyl phenethyl) distannoxane [*13356-08-6*] (rat oral LD_{50} >2000 mg/kg) also is a commercial acaricide.

Chlordimeform, *N*′-(4-chloro-*o*-tolyl)-*N,N*-dimethylformamidine (mp 32°C) and its hydrochloride have been used as acaricides and ovicides. The rat acute oral LD_{50} range is 170–330 mg/kg. However, chlordimeform is an animal carcinogen and causes severe bladder injury in exposed humans. It is replaced by amitraz, *N*-methyl,

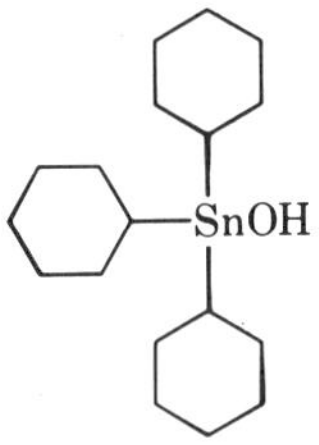

tricyclohexyltin hydroxide
[*13121-70-5*]

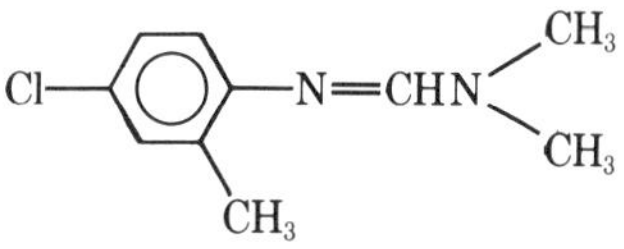

chlordimeform
[*6164-98-3*]

N′-2,4-xylyl-*N*-(*N*′-2,4-xylylformimidoyl)formamidine (rat oral and dermal LD_{50}s: 600–800 and 1600 mg/kg, respectively).

Pentac, bis(pentachloro-2,4-cyclopentadien-1-yl) (mp 122°C), is used as an acaricide on greenhouse and ornamental crops. The rat oral LD_{50} is 3160 mg/kg.

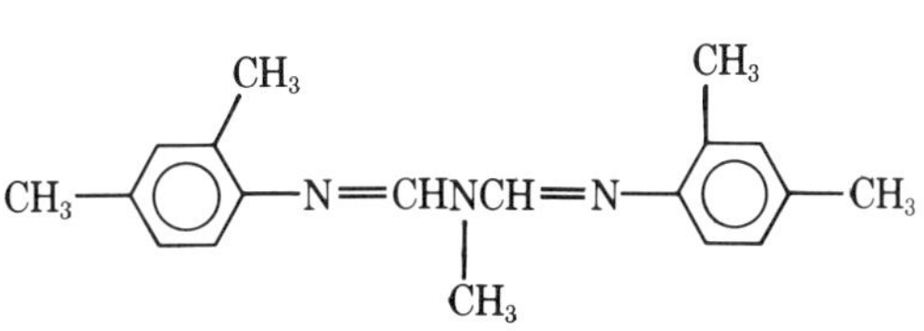

amitraz
[*33089-61-1*]

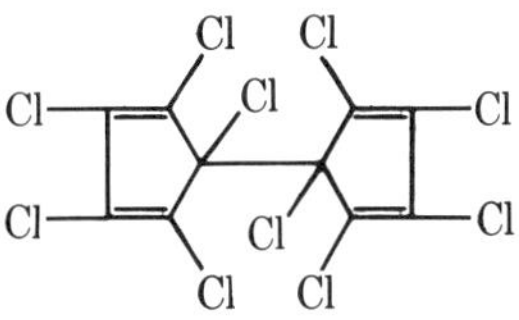

pentac
[*2227-17-0*]

Petroleum Oils. When satisfactorily stable kerosene–soap–water emulsions were produced in 1874, dormant (winter) oil sprays became widely used to control scale insects and mites. The first commercial emulsion or miscible oil was marketed in 1904 and by 1930 highly refined neutral or white oils, free from unsaturated hydrocarbons, acids, and highly volatile elements, were found to be safe when applied to plant foliage, thus greatly enlarging the area of usefulness of oil sprays (see Petroleum, petroleum products).

Petroleum oil sprays are used as insecticides for dormant sprays for the control of scale insects, mites, and insect eggs; summer foliage sprays for aphids, mealybugs, mites, thrips, psyllids, whiteflies, and scale insects; livestock sprays for the control of lice, fleas, and mites; and mosquito larvicides. They also are used as carriers for contact insecticides to increase their effectiveness. In order to understand the effects of spray oils upon insects and upon plants, the following properties must be considered:

Volatility or Distillation Range. The lower the volatility and the higher the boiling point, other properties being equal, the more effective the oil is in killing insects. For dormant spraying, the oil should distill ca 90% of its volume from 310–371°C and not over 2% at 110°C for 4 h. Unfortunately, the heavier oils are more toxic to plants so that it is important to find the lightest oil that will kill the insect pest or the heaviest that can be safely used on the plant to be sprayed.

The viscosity of the spray oil, as measured by the Saybolt test, also determines its safety on plants. Other properties being equal, oils of low viscosity are safer to use on foliage than those of high viscosity. For dormant sprays upon deciduous trees, oils with viscosities between 100 and 200 SUs at 37.8°C are considered satisfactory. A lower range is often used in colder and a higher range in warmer areas.

The degree of refinement of the spray oil is important as the presence of impurities, eg, unsaturated and aromatic hydrocarbons, which are chemically reactive and readily oxidized, causes the oil to become turbid and acid in reaction. For dormant spraying, 65–75% of these impurities must be removed and, for foliage spraying, 85–100% removal is necessary. The refining is accomplished chiefly by treatment with sulfuric acid and by washing to remove the resulting sludge. For any oil of unknown purity, the degree of refinement may be determined by treating a sample with sulfuric acid. If the unsaturated hydrocarbons have already been removed, the acid treatment does not remove any portion of the oil and it is said to be 100% unsulfonatable. Typical specifications for spray oils are shown in Table 11.

Table 11. Specifications for Spray Oils

Grade	Distillation at 336°C, %	Unsulfonated residue (UR), %	Viscosity at 37.8°C, SUs
dormant oil		90–92	90–120
foliage oil			
light	64–79	90	55–65
light–medium	52–61	92	60–75
medium	40–49	92	70–85
heavy–medium	28–37	92	80–95
heavy	10–25	94	90–105

Fumigants. Fumigants are chemicals that are distributed through space as gases and therefore, at a given temperature and pressure, must exist in the gaseous state in sufficient concentration to be lethal to the insect pest (22). This physical requirement greatly limits the number of insecticides that may be usefully employed as fumigants. Compounds boiling at about room temperature, eg, hydrogen cyanide, methyl bromide, and ethylene oxide, are the most useful general fumigants. For soil fumigation, however, the slower release of vapors from substances, eg, ethylene dibromide and β,β'-dichlorodiethyl ether, that boil as high as 180°C has proved effective. Other organic toxicants of relatively high vapor pressure, eg, naphthalene and *p*-dichlorobenzene, sublime readily enough to have special uses as fumigants; and contact insecticides, eg, azobenzene, lindane, dichlorvos, and mevinphos, may kill insects by vapor action under certain circumstances. The fumigant action of the more difficultly vaporizable insecticides is enhanced by the use of atomization, volatilization by heat, or burning in pyrotechnic mixtures.

The true insecticidal fumigants are described in Table 12. The major areas are mills, warehouses, grain elevators, groceries, museums, etc; stored food products either in bulk or in small packages; habitations for control of structural and household pests and for human or animal ectoparasites; soil fumigation for grubs, wireworms, ants, nematodes, etc; and living plants in greenhouses, nursery stock, or citrus and other tree crops.

The various fumigants often exhibit considerable specificity toward insect pests, as shown in Table 13. The proper choice for any control operation is determined not only by the effectiveness of the gas but by cost; safety to humans, animals, and plants; flammability; penetratability; effect on seed germination; and reactivity with furnishings. The fumigants may be used individually or in combination. Carbon tetrachloride often is incorporated with carbon disulfide, ethylene dichloride, or ethylene dibromide to decrease flammability, and carbon dioxide is used with ethylene oxide for the same purpose.

Dosage. The dosage of fumigant is commonly expressed as lb/1000 ft^3 or in mg/L. Successful fumigation results from the attainment of a critical Ct value, ie, that which will kill at least 99% of the insect population (product of the concentration of gas in mg/L × the exposure duration in h). Within moderate limits, therefore, the longer the exposure, the lower the concentration of gas necessary. In practice, the exposure time is limited by the escape of gas from the enclosure and convenience in treatment; the minimum concentration is limited by the ability of the pest to detoxify the gas at low dosages as rapidly as it is sorbed. The dosage and the attainment of a critical Ct value are dependent upon the rate of volatilization or generation of the fumigant and the nature of the commodity being fumigated. Some fumigants, eg, methyl bromide, are highly reactive with proteins and are rapidly sorbed by grain, flour, or seeds. From 10–32°C, the Ct value decreases by approximately one half for each 10°C increase in temperature.

Application. The fumigant is applied to an enclosure that is as gastight as possible. Low-boiling fumigants, eg, hydrogen cyanide, methyl bromide, ethylene oxide, methyl formate, and sulfuryl fluoride are obtained in cylinders of compressed or liquefied gas which is readily piped into the enclosure. Fumigants that are liquids at room temperature are volatilized by pouring onto cloths or spraying into the area to be treated. Elaborate metering mechanisms have been developed for the spot applications of fumigants to moving streams of grain in elevators. Forced recirculation of

Table 12. Properties of Fumigants

Name	CAS Registry No.	Formula	Boiling point, °C	Specific gravity: Liq $d_4^{20 a}$, g/mL	Specific gravity: Gas (air = 1)	Vapor pressure[a,b], kPa	Solubility[a], g/100 mL H_2O (at 20°C)	Flammability in air[c], vol %	Safe limit, ppm	Uses
acrylonitrile	[*107-13-1*]	$CH_2{=}CHCN$	78	0.797	1.8	11		3	20	mills, commodities
carbon disulfide	[*75-15-0*]	CS_2	46.3	1.263	2.6	41.9	0.22	1	20	household
carbon tetrachloride	[*56-23-5*]	CCl_4	76	1.595	5.3	12	0.08	nf	25	grain, carrier for other fumigants
chloropicrin	[*76-06-2*]	Cl_3CNO_2	112	1.651	5.7	2.7	0.19	nf		soil, grain
1,2-dibromo-3-chloropropane	[*96-12-8*]	$CH_2BrCHBrCH_2Cl$	199	2.08		0.08	0.12			soil
β,β′-dichlorodiethyl ether	[*111-44-4*]	$CH_2ClCH_2OCH_2CH_2Cl$	178	1.222	4.9	0.10	1.1	f	15	soil
1,1-dichloro-1-nitroethane	[*594-72-9*]	$CH_3CCl_2NO_2$	124	1.415		2.25	0.25		10	grain, stored products
1,2-dichloropropane	[*78-87-5*]	$CH_2ClCHClCH_3$	95.4	1.159		28	0.27		75	soil
trans-1,3-dichloropropene	[*542-75-6*]	$ClCH{=}CHCH_2Cl$	111	1.224		2.47	0.28	f		soil
ethylene chlorobromide	[*107-04-0*]	$ClCH_2CH_2Br$	107	1.689		5.3	0.7^{30}	nf		soil, commodities

ethylene dibromide	[*106-93-4*]	$BrCH_2CH_2Br$	131.6	2.172	6.5	1.0	0.34	nf	25	grain, fruits, vegetables
ethylene dichloride	[*107-06-2*]	$ClCH_2CH_2Cl$	83.5	1.257	3.4	10	0.87	6	100	grain, soil, household
ethyl formate	[*109-94-4*]	$HCOOC_2H_5$	54	0.917			10			dried fruits
ethylene oxide	[*75-21-8*]	$(CH_2)_2O$	10.7	0.887^{7}	1.5	146	∞	3	50	packaged foods
hydrogen cyanide	[*74-90-8*]	HCN	26	0.688	0.9	84.0	∞	6	10	general
β-methallyl chloride	[*563-47-3*]	$CH_2{=}C(CH_3)CH_2Cl$	72	0.925						
methyl bromide	[*74-83-9*]	CH_3Br	4.5	1.732^{0}	3.3	189.3	1.34^{25}	13.5	20	general
methyl formate	[*107-31-3*]	$HCOOCH_3$	32	0.974		83.2	30			dried fruits
naphthalene	[*91-20-3*]	$C_{10}H_8$	218 (mp 80)		4.4	0.01	0.003	f		fabric pests, greenhouse
p-dichlorobenzene	[*106-46-7*]	$C_6H_4Cl_2$	173.4 (mp 53)		5.1	0.1	0.008^{25}	nf	75	fabric pests
phosphine	[*7803-51-2*]	PH_3	−87.4	0.746^{-90}	1.2			2	0.05	grain
sulfuryl fluoride	[*2699-79-8*]	SO_2F_2	−55.2	1.342^{25}	3.5	1601.3	0.075^{25}	nf	100	structural pests
trichloroacetonitrile	[*545-06-2*]	Cl_3CCN	85	1.44^{25}						grain
trichloroethylene	[*79-01-6*]	$ClCH{=}CCl_2$	86.7	1.470^{15}	4.5	9.7	insol	nf		grain

[a] At 20°C unless otherwise indicated.
[b] To convert kPa to mm Hg, multiply by 7.5.
[c] nf = nonflammable; f = flammable.

Table 13. Comparative Toxicity of Fumigants, LC_{50} for 6 h, mg/L[a]

Fumigant	Granary weevil, *Sitophilus granarius*	Drugstore beetle, *Stegobium paniceum*	Confused flour beetle, *Tribolium confusum*	Bean weevil, *Acanthoscelides obtectus*	Saw-toothed grain weevil, *Oryzaephilus surinamensis*
acrylonitrile	2.0	1.7	3.0	1.1	0.8
carbon disulfide	43.0	42.0	75.0	29.0	40.0
chloropicrin	3.4	1.9	6.4	<1.5	<1.5
ethylene dibromide	3.0	2.8	3.4	10.2	0.9
ethylene dichloride	127.0	77.0	53.0	49.0	39.0
ethylene oxide	13.5	9.0	27.5	10.5	4.0
hydrogen cyanide	4.6	<0.4	0.8	0.9	<0.4
methallyl chloride	25.0	10.0	27.0	18.0	19.0
methyl bromide	4.8	4.4	9.2	4.2	4.4

[a] Ref. 22.

the fumigant through the commodity being treated improves the distribution of the gas and often is used for the fumigation of stored grains.

Certain fumigants may be controlled more readily if generated from relatively inert precursors. Hydrogen cyanide is rapidly released by treatment of sodium, potassium, or calcium cyanides with acid,

$$NaCN + H_2SO_4 \rightarrow NaHSO_4 + HCN\uparrow$$

or slowly upon exposure to moisture,

$$Ca(CN)_2 + 2\,H_2O \rightarrow Ca(OH)_2 + 2\,HCN\uparrow$$

Small packages of sodium cyanide may be dropped into containers of dilute acid to provide precisely controlled dosages of hydrogen cyanide for household or warehouse fumigation. Calcium cyanide absorbed on inert earth or in pellets may be spread in a thin layer on greenhouse walks or on paper on household floors; the hydrogen cyanide gas is generated by the moisture in the air.

The extremely toxic and flammable gas, phosphine, is safely and conveniently generated for the fumigation of grain in sacks or bins from 3-g tablets containing aluminum phosphide and ammonium carbamate which produce 1 g of phosphine in the presence of moisture.

$$AlP + 2\,NH_4O\overset{\overset{O}{\|}}{C}NH_2 + 3\,H_2O \longrightarrow PH_3\uparrow + Al(OH)_3 + 4\,NH_3 + 2\,CO_2$$

The organophosphorus ester, dichlorvos, is sufficiently volatile to be incorporated either in permeable plastic bottles or plastic strips which permit its controlled release for fumigation of cupboards and closets to control cockroaches, or as flea collars on pets.

Vacuum Fumigation. The molecular diffusion of a fumigant gas is inversely proportional to its molecular weight and is affected also by the number of collisions of the fumigant molecules with other gases present. The penetration and effectiveness of the fumigant may be improved greatly by using it in a partial vacuum. Vacuum fumigation of packaged foods, tobacco, spices, pharmaceuticals, and other commodities is carried out in steel chambers or vaults at a pressure of 2–23 kPa (15–175 mm Hg). The reduction in oxygen content makes the insects more susceptible to the effects of the fumigant and the addition of carbon dioxide frequently is used to stimulate the insects to respire at a higher rate and thus increases the sorption of the fumigant.

Tent Fumigation. Citrus and deciduous fruit trees have been fumigated for the control of scale insects for many years by hydrogen cyanide introduced under relatively gastight tents of 240–270 g/m^2 (7–8 oz/yd^2) army duck or of nylon impregnated with vinyl chloride–vinyl acetate copolymer. Methyl bromide fumigation of buildings for the control of termites or powder post beetles, or of mills or warehouses, or of bagged commodities for the control of stored-product insect pests is effected by wrapping the structure in polyethylene or polyvinyl chloride plastic sheeting that is 0.102–0.152 mm thick, rolled and clamped at the edges, and sealed at the bottom by soil or sand bags.

Soil Fumigation. This procedure has become a standard agricultural practice for the destruction of soil-inhabiting insects, nematodes, and fungi prior to planting, in orchard and ornamental plantings, and in turf. The fumigant may be injected directly ahead of the plow or applied by more elaborate soil injectors. Small plots are treated by pouring the fumigant into holes punched in the soil at regular intervals and covering immediately. More volatile soil fumigants, eg, methyl bromide are applied under plastic sheeting that is sealed at the edges with earth. The distribution of the soil fumigant involves vaporization outward from the point of application through air spaces in the soil and solution of the fumigant in soil water. The most efficient pattern for application of specific fumigants to various types of soils can be computed from the mathematical laws governing the diffusion of gases.

Environmental. The fumigants generally are highly reactive compounds that interact with vital biochemical processes within the target pest, usually by a bimolecular process, eg, alkylation. Their reactivity and high vapor pressure makes it difficult to prevent widespread contamination of the surrounding air, water, and soil. Soil fumigation that is practiced with highly persistent, lipophilic compounds (Table 12) at high dosages of $(8.97–44.8) \times 10^3$ kg/km^2 (80–400 lb/acre) has contaminated the treated food commodity and water in shallow wells that is used for domestic supply. Fumigants, eg, carbon tetrachloride, ethylene oxide, β,β'-dichlorodiethyl ether, ethylene dibromide, and 1,2-dibromo-3-chloropropane (DBCP) [*96-12-8*] are chemical carcinogens; the latter two compounds have produced azoospermia in workers exposed in both factory production and in agricultural operations. For these reasons, all U.S. usage of DBCP was cancelled in 1979.

Microbial Insecticides. Insects are attacked by a multitude of pathogens and ca 450 viruses, 80 bacteria, 460 fungi, 250 protozoa, and 20 rickettsial diseases are effective natural enemies (27–28). A number of these are adaptable for mechanical dissemination as microbial insecticides for the innoculation of insect populations, soils, fields, orchards, or forests with spores, microbial toxins, or virus suspensions. Microbial insecticides are highly specific in action against a few closely related pest species and generally are harmless to other animals. Possible disadvantages are their instability to light and air and the frequent dependence upon suitable climatic conditions and precise timing of applications for maximum effectiveness.

The classic utilization of a microbial insecticide is that of *Bacillus popillae* spores which are produced from infected larvae of the Japanese beetle and are diluted with talc to a standardized powder containing 10^8 spores/g. The powder is applied to soil (225–2250 kg/km^2) (2–20 lb/acre) and results in the very slow spread of the disease and in satisfactory control of the beetle larvae in turf within three years. Spores of *Bacillus thuringiensis* (*B.t.*) contain a crystalline endotoxin that is especially effective against the cabbage looper and other leaf-feeding caterpillars. Commercial preparations (*B.t.* insecticide) are standardized in reference to a standard strain of *B.t.* (E-61) that has a potency of 100 IU/mg. *B.t.* insecticide is used effectively on vegetables, rice, ornamentals, and shade trees. *Bacillus sphaericus* is highly specific for control of mosquito larvae.

A variety of insect viruses cause epizootics in insect populations. The first to be commercially developed in the U.S. is the nuclear polyhedrosis virus (NPV) of the cotton bollworms, *Heliothis* sp. Other commercial viral insecticides include the NPVs of the Lepidoptera *Prodenia*, *Spodoptera*, and *Trichoplusia*, and of the sawfly (Hymenoptera), *Neodiprion*. The fungal spores of *Beauveria bassiana* are standardized and used in China against the European corn borer and in the USSR against the Colorado potato beetle.

Genetic Control. Manipulation of the mechanisms of inheritance of the insect pest populations has occurred most successfully through the mass release of sterilized males, but a variety of other techniques have been studied, including the environmental use of chemosterilants and the mass introduction of deleterious mutations, eg, conditional lethals and chromosomal translocations (29–30) (see Genetic engineering).

Sterile-Male Release. The release of large numbers of sterile male insects into a population of virgin females has been shown (29) to reduce the number of fertile females in subsequent generations, as shown in Table 14. The most practical utilization of the sterile-male technique has been in the eradication of the screwworm fly, *Callitroga hominovorax*, from the Southeastern United States in 1959, which involved the release of ca 2×10^9 male flies, sterilized with gamma radiation from ^{60}Co, over an area of 181,000 km^2 for 18 mo.

The success of such a male-sterilization procedure is dependent upon a method of mass rearing of the insects, adequate dispersion of released sterile insects, a sterilization procedure which does not adversely affect mating behavior, an insect species in which mating occurs only once per lifetime or one in which the sperm from sterile males competes with those of fertile males, and a low insect pest population density or a means of reduction of the population to levels that make it possible to release a dominant population of sterile males. However, many species of pest insects do not conform to these criteria and the rearing and release of sterile males is an expensive procedure.

Table 14. Theoretical Effects of Release of Sterile Male Insects on a Natural Insect Population[a]

Generation	Wild female population	Population of sterile males released	Ratio, sterile males to wild females	Wild females mated to sterile males, %	Theoretical population of fertile females
0th	1,000,000	2,000,000	2:1	66.7	333,333
1st	333,333	2,000,000	6:1	85.7	47,619
2nd	47,619	2,000,000	42:1	97.7	1107
3rd	1107	2,000,000	1807:1	99.95	1

[a] Ref. 29.

Chemosterilants. The use of chemicals that would sterilize large segments of natural insect pest populations has been suggested. Several types of chemosterilants are known to produce adequate sterility in insects by preventing the production of ova or sperm, by causing death of sperm or ova, or by producing severe injury to the genetic material of sperm or ova so that the zygotes that are produced do not develop into mature progeny (31).

Antimetabolites include 5-fluorouracil and amethopterin, a folic acid antagonist, which produce sterility in female flies when fed at 0.01–0.05% in the diet.

Radiomimetic compounds include cancer chemotherapeutic compounds that incorporate the extremely reactive ethyleneimine group. Tepa or 1-tris(1-aziridinyl)-phosphine oxide (mp 41°C, rat oral LD_{50} 37 mg/kg) and its thionoanalogue, thiotepa [*52-24-4*] (mp 51.5°C) were among the first to be investigated. These compounds are alkylating agents for DNA and cause sterility in both sexes of the housefly, eg, when incorporated into food or applied topically at 0.1–1%. Mosquitoes and flies are sterilized by exposure to surface residues of 110–2690 mg/m^2. Other similar radiomimetic compounds include apholate or 2,2,4,4,6,6-hexa-(1-aziridinyl)-2,4,6-triphospho-1,3,5-triazine (mp 155°C, rat oral LD_{50} 90 mg/kg) and hempa or hexamethyltriphosphoramide (rat oral LD_{50} 2600 mg/kg).

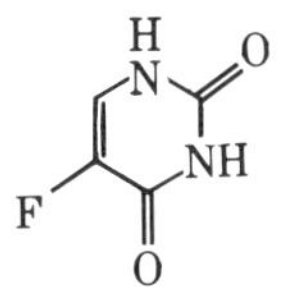

5-fluorouracil
[*51-21-8*]

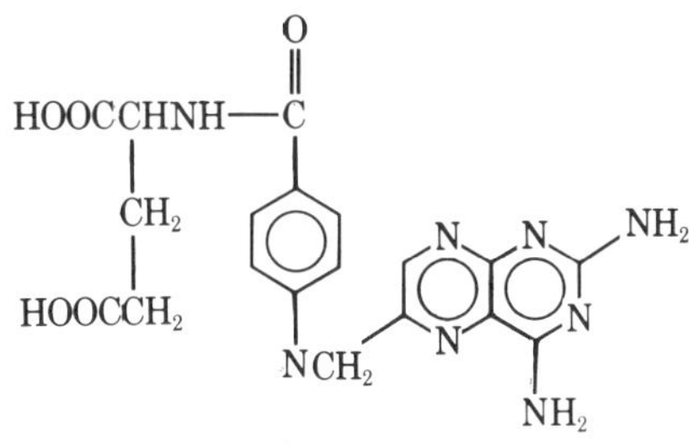

amethopterin
[*59-05-2*]

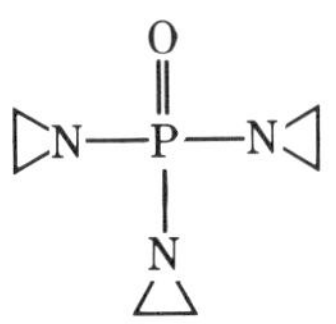

tepa
[*57-39-6*]

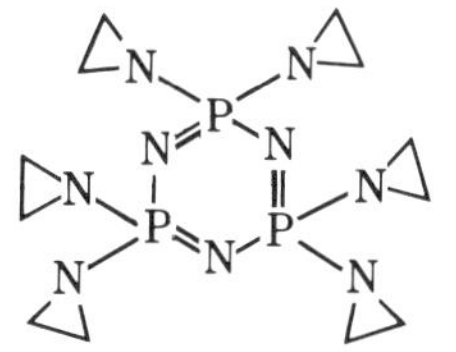

apholate
[*52-46-0*]

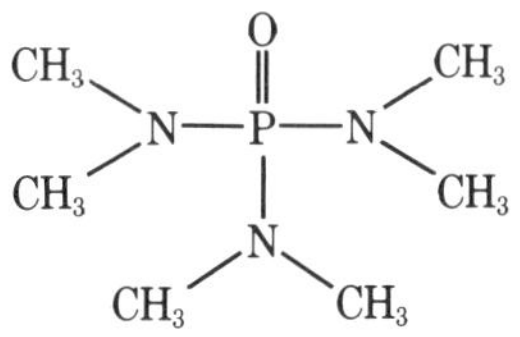

hempa
[*680-31-9*]

Unfortunately, all of these compounds, including hempa, are mutagens and strong carcinogens and the radiomimetic compounds produce degenerative changes in human germ plasm at dosages of 0.01 or less than the LD_{50}. Their casual and indiscriminate use in insect control, therefore, is precluded. Successful application of the chemosterilant principle must await the availability of more selective insect sterilants that do not specifically affect the universal DNA–RNA genetic mechanism.

Insect Resistance. A major problem in insect control results from the capability of many insect species to develop races that are sufficiently resistant to the action of insecticides so as to necessitate complete changes in control practices (32). This resistance results from the selection, through continuous and intensive use of an insecticide, of naturally occurring mutants possessing biochemical, physiological, or even behavioristic factors that confer some degree of immunity. This immunity is produced by genetic components and is inheritable. Dieldrin resistance in the mosquito *Anopheles gambiae*, for example, is attributed to the mutation of a single gene that is present in 0.4–6% of unselected wild populations. Several years of continuous selection by dieldrin residual spraying of dwellings in a malaria eradication program in Africa resulted in a 90% resistance factor in the anopheline population.

With the advent of DDT as a commercial insecticide in 1946, there were only 11 documented cases of resistance, including the codling moth, *Laspeyresia pomonella*, and the peach twig borer, *Anarsia lineatella*, to lead arsenate; the San Jose scale, *Aspidiotus perniciosus*, to lime sulfur; the cattle ticks, *Boophilus decoloratus* and *B microplus*, to sodium arsenite dip; the citrus thrips, *Scirtothrips citri*, and the gladiolus thrips, *Taenothrips simplex*, to potassium antimonyl tartrate; and the walnut husk fly, *Rhagoletis completa*, to cryolite. The widespread use of DDT introduced selection on a greatly increased scale and the number of well documented examples of insecticide resistance has increased: 12 by 1948, 16 by 1951, 25 by 1954, 76 by 1957, 137 by 1960, 159 by 1963, 224 by 1968, 364 by 1976, and 414 by 1979.

Cross resistance enables resistant species to survive exposure to related chemicals (eg, DDT resistance nonsusceptible to methoxychlor, lindane resistance nonsusceptible to dieldrin, or parathion resistance nonsusceptible to malathion) brought about by the action of a common detoxication system or target site insensitivity, eg, acetylcholinesterase. Multiple resistance is far more serious and most important pests now display resistance to a variety of classes of insecticides with differing modes of action and different detoxication pathways. Once selected, resistance genes in pest insect populations possess virtually limitless persistence in wild insect populations. Thus, DDT and cyclodiene resistance in Danish houseflies has persisted for more than 20 yr; when these compounds were tried again they became useless within 2 months. Multiple resistance reflects the past history of insecticide selection and precludes a return to those insecticides used previously, whereas cross resistance limits the choice of available insecticides. At least six insect pests have developed multiple resistance to the following groups of insecticides: DDT and methoxychlor, lindane and cyclodienes, organophosphates, carbamates, and pyrethroids. At least another 16 important pests have multiple resistance to organochlorines, organophosphates, and carbamates. Multiple resistance is known in at least 9 orders and 43 families of insects.

Insecticide resistance has had an enormous impact upon insect control practices. The corn rootworm beetles, *Diabrotica* spp. were controlled 1954–1964 by soil applications of aldrin and heptachlor. Resistance to these insecticides became almost total in the mid 1960s and the organochlorines, costing ca $1.80/mg, have been almost totally replaced by organophosphates costing ca $8.80/kg. Species of *Anopheles* mosquitoes that are vectors of malaria rapidly became resistant to DDT and dieldrin, which were the original insecticides used for residual house spraying. By 1975, 42 species of *Anopheles* were resistant to DDT and/or dieldrin (with 21 species resistant to both) (3,32). Substitute insecticides that are effective include malathion, propoxur and fenitrothion; yet these cost from 5–20 times as much per capita as did DDT and dieldrin. As a result, malaria control has become too costly for some developing countries; major recrudescences of the disease have occurred in India, from 125,000 cases in 1965 to 4×10^6 in 1975, and in Pakistan from 9500 in 1961 to ca 10^7 in 1975.

An insight into the biochemical mechanisms for resistance was gained by the demonstration that DDT-resistant houseflies contain greatly increased amounts of a detoxication enzyme, DDT dehydrochlorinase, which rapidly converts DDT to its noninsecticidal derivative, DDE.

$$\underset{\text{DDT}}{Cl{-}C_6H_4{-}\underset{\underset{CCl_3}{|}}{CH}{-}C_6H_4{-}Cl} \xrightarrow{\text{DDT dehydrochlorinase}} \underset{\text{DDE}}{Cl{-}C_6H_4{-}\underset{\underset{CCl_2}{\|}}{C}{-}C_6H_4{-}Cl}$$

Synergists, eg, the acaricides, chlorfenethol and dicofol, and *p*-chlorobenzene-*N*,*N*-dibutylsulfonamide [*127-59-3*], $ClC_6H_4SO_2N(C_4H_9)_2$, act as competitive inhibitors of DDT'ase and have been used temporarily as synergists for DDT against resistant insects. Other resistance mechanisms to DDT are known and include enhanced microsomal oxidation to the α-hydroxy derivative, dicofol; and knockdown resistance produced by the *kdr* (knockdown resistant) gene. Similarly, insect resistance to organophosphorus insecticides depends on enhanced enzymic detoxication. Malathion-resistant *Culex tarsalis* mosquitoes contain higher activity of carboxyesterase that detoxifies the insecticide

$$(CH_3O)_2\overset{\overset{S}{\|}}{P}S\underset{\underset{CH_2COOC_2H_5}{|}}{C}HCOOC_2H_5 \xrightarrow{\text{carboxyesterase}} (CH_3O)_2\overset{\overset{S}{\|}}{P}S\underset{\underset{CH_2COOC_2H_5}{|}}{C}HCOOH$$

and which can be inhibited by the insecticide, EPN, which is an effective synergist for malathion against the resistant mosquitoes. Pyrethroids and carbamates are detoxified in resistant insects by enhanced microsomal oxidase enzymes and these can be inhibited by the methylene dioxyphenyl synergists which may restore insecticidal activity. However, the use of synergists provides only temporary surcease to resistance problems and genetic selection to the combination soon occurs (as with DDT plus chlorfenethol where increased levels of DDT'ase soon evolved in the housefly). Moreover, the use of a mixture of chemicals markedly increases costs, complicates formulation of suitable products, and greatly increases the problems of persistent residues and human toxicology. Selection with organophosphorus and carbamate insecticides has resulted in the appearance of biochemically altered target-site acetylcholinesterase in houseflies, mites, cattle ticks, and rice leafhoppers. Such mutations are insensitive to whole families of insecticides.

The most practical means of dealing with the insecticide resistance problem is maintaining a careful check on the susceptibility of pest populations to insecticides using standard test methods. When it appears that an appreciable amount of resistance has resulted from the use of a given chemical, it should be replaced promptly with another that has been chosen carefully from genetic considerations affecting cross resistance and multiple resistance. However, the inexorable pressures of exponentially increasing numbers of resistant species and the dramatically increased developmental costs for new insecticides are making this strategy steadily less productive. The world is running out of effective insecticides for many important insect pests. Therefore, IPM programs are needed to retain the use of the many effective and relatively inexpensive chemicals now used. IPM with its multifactored intervention strategies involving cultural, biological, and chemical controls, provides the only satisfactory way to break up the selection process so that the lifetime of useful insecticides can be substantially prolonged.

Repellents and Attractants. ***Repellents.*** Repellents (qv) are substances that protect animals, plants, or products from insect attack by making food or living conditions unattractive or offensive. These substances, which may not be poisonous or only mildly toxic, are rarely, if ever, effective against all kinds of insects. Such chemicals can sometimes be employed to advantage where it is impossible to use an insecticide and may afford a greater or lesser degree of protection to manufactured products, growing plants, or the bodies of animals and humans. Among the many examples are the following:

(*1*) *Repellents against crawling insects.* Examples are the creosote lines used as barriers to the migration of chinch bugs; trichlorobenzene and other chemicals used to protect buildings from termites; heavy oils at the base of poultry roosts as a barrier to poultry mites; and certain chemical bands about tree trunks.

(*2*) *Repellents against the feeding of insects.* These include the application of bordeaux, lime, and similar washes to plants to ward off leafhoppers and some chewing insects; mosquito repellents and fly sprays to lessen the attacks of bloodsucking flies and mosquitoes; the application of sulfur to the body to keep chiggers from attacking; the use of smoke and smudges to repel biting flies; the chemical treatment of logs to keep beetle borers from destroying log cabins and other rustic work; and moth balls, oil of cedar, and mothproofing treatments to protect materials from attack by clothes moths and carpet beetles.

(*3*) *Repellents against the egg laying of insects.* Examples are the use of pine-tar oil and diphenylamine to keep screwworm flies from laying eggs about wounds of animals.

Bordeaux Mixture. Bordeaux mixture (see Fungicides) originated in France as a spray to control the downy mildew disease (caused by the fungus, *Peronospora viticola*) of grapes in ca 1882, and was first used in the U.S. in 1887. Although primarily a fungicide, bordeaux mixture is repellent to many insects, eg, flea beetles, leafhoppers, and potato psyllid, when sprayed over the leaves of plants. It is, to some extent, an ovicide and has some residual toxic effect upon the sap and, thereby, kills leafhoppers and psyllids for some days. Bordeaux mixture is produced by mixing hydrated lime, 3.6–4.5 kg (8–10 lb), and copper sulfate, 1.9–2.7 kg (4–6 lb), in 380 L (100 gal) of cold water to produce a precipitate of tetracupric sulfate, $4CuO.SO_3$. The colloid remains in suspension for several hours and when sprayed on foliage covers the leaves with a thin, tenacious film which gradually produces soluble copper.

Other compounds that have shown interesting repellent properties to plant-feeding insects include the fungicides, thiram, or tetramethylthiuram disulfide

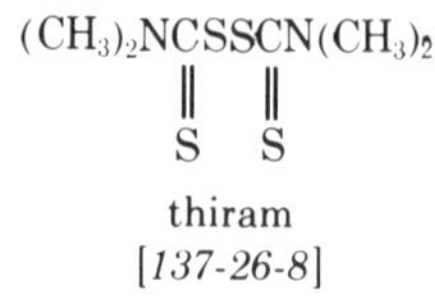

thiram
[*137-26-8*]

(mp 155°C), and nabam, or disodium ethylenebisdithiocarbamate,

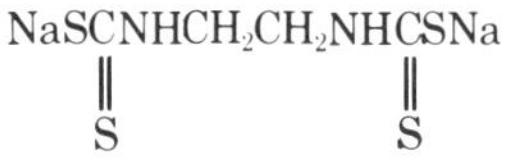

$$\underset{\substack{\| \\ S}}{NaSC}NHCH_2CH_2NH\underset{\substack{\| \\ S}}{CSNa}$$

nabam
[*142-59-6*]

which are especially effective against the Japanese beetle, *Popillia japonica*. The compound 4′-(dimethyltriazeno)acetanilide [*1933-50-0*] (mp 155°C) is an antifeeding compound for lepidopterous larvae.

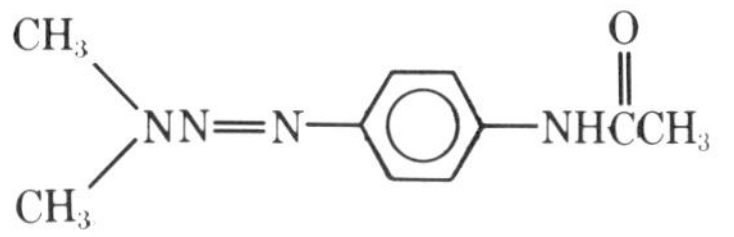

$$(CH_3)_2N{-}N{=}N{-}C_6H_4{-}NH\overset{\overset{O}{\|}}{C}CH_3$$

Repellents to Bloodsucking Insects. The limiting criteria for a good repellent against bloodsucking insects are, in order of importance, effective protection of the treated area for several hours, on all types of subjects, and under all climatic conditions; complete freedom from toxicity and irritation when regularly applied to human or animal skin; cosmetic acceptability, including freedom from unpleasant odor, taste, and touch, and harmlessness to clothing; protection against a wide variety of biting insects; and low cost and availability (33). Tests of many thousands of chemical compounds for repellent action to flies, mosquitoes, chiggers, fleas, and ticks have shown that, although many possess a significant degree of repellency to the various pests, few meet all the other requirements satisfactorily. The materials that have found practical application as repellents are listed in Table 15 with their properties and uses.

Repellent compounds exhibit wide differences in their activity against various species of mosquitoes and flies, as well as other biting arthropods, and it has been demonstrated that the overall protection against various pests is greatly extended by the use of mixtures, eg,

		Parts (by wt)
(*1*)	dimethyl phthalate	3
	Indalone [*532-34-3*]	1
	2-ethyl-1,3-hexanediol	1
(*2*)	dimethyl phthalate	4
	2-ethyl-1,3-hexanediol	3
	dimethyl carbate	3

These repellents also can be incorporated into various creams and lotions, eg,

		Parts (by wt)
(*3*)	dimethyl phthalate	7
	magnesium stearate	3
(*4*)	dimethyl phthalate	25
	white wax	19
	peanut oil	56

Table 15. Repellents for Mosquitoes, Flies, Mites, and Ticks[a]

Name	CAS Registry No.	Formula	Properties	Toxicity, oral LD_{50} rat, mg/kg	Uses
benzil	*[134-81-6]*	$C_6H_5C(=O)C(=O)C_6H_5$	solid, mp 95°C		clothing treatment for chiggers
benzyl benzoate	*[120-51-4]*	$C_6H_5C(=O)OCH_2C_6H_5$	oily liquid, bp 323°C, mp 21°C, d 1.12 g/cm^3	1900	clothing treatment for chiggers
2,3,4,5-bis(butyl-2-ene)-tetrahydrofurfural	*[126-15-8]*	(bicyclic tetrahydrofuran structure with CHO)	liquid, d 1.121 g/cm^3	2500	biting flies on cattle
butoxypolypropylene glycol	*[9003-13-8]*	$Bu^nO(CH_2CH(CH_3)O)_nCH_2CH(CH_3)OH$	liquid, 400 and 800 mol wt fractions, d 0.973–0.990 g/cm^3	11,200	fly repellent for cattle
N-butylacetanilide	*[91-49-6]*	$C_6H_5N(Bu^n)C(=O)CH_3$	liquid, bp 277–281°C, mp 22°C, d 0.99 g/cm^3	2830	clothing treatment for fleas, ticks
n-butyl 6,6-dimethyl-5,6-dihydro-1,4-pyrone-2-carboxylate	*[532-34-3]*	(dihydropyranone ring with 6,6-$(CH_3)_2$ and 2-$C(=O)OBu^n$)	brown liquid, bp 110–115°C/0.133 kPa[b], d 1.06 g/cm^3	7800	mosquito and fly repellent
dibutyl adipate	*[105-99-7]*	$Bu^nOC(=O)CH_2CH_2CH_2CH_2C(=O)OBu^n$	liquid, bp 193°C/1.9 kPa[b], d 0.965 g/cm^3	12,900	tick repellent
dibutyl phthalate	*[84-74-2]*	$C_6H_4(COOC_4H_9)_2$	liquid, bp 340°C, d 1.045 g/cm^3	21,000	clothing treatment for chiggers
di-*n*-butyl succinate	*[141-03-7]*	$Bu^nOC(=O)CH_2CH_2C(=O)OBu^n$	liquid, bp 108°C/0.53 kPa[b], mp −29°C	8000	fly repellent for cattle

N,N-diethyl-*m*-toluamide	[*134-62-3*]	$CH_3C_6H_4CON(C_2H_5)_2$	liquid, bp 111°C/0.133 kPa[a], d 0.996 g/cm^3	2000	general-purpose repellent
dimethyl carbate (*cis*-dimethyl bicyclo[2.2.1]-5-heptene-2,3-dicarboxylate)	[*39589-98-5*]	$C_7H_8(COOCH_3)_2$	solid, mp 40°C	1000	mosquito repellent
dimethyl phthalate	[*131-11-3*]	$C_6H_4(COOCH_3)_2$	liquid, bp 284°C, d^{25} 1.189 g/cm^3	8200	general-purpose repellent
2-ethyl-2-butyl-1,3-propanediol	[*115-84-4*]	$HOCH_2C(C_2H_5)(C_4H_9)CH_2OH$	solid, mp 40°C, d 0.931 g/cm^3	5040	mosquito repellent
2-ethyl-1,3-hexanediol	[*94-96-2*]	$HOCH_2CH(C_2H_5)CH(OH)CH_2CH_2CH_3$	liquid, bp 244°C, d 0.942 g/cm^3	2400	general-purpose repellent
di-*n*-propyl isocinchomeronate	[*136-45-8*]	$Pr^nOOC-C_5H_3N-COOPr^n$	liquid, bp 186°C/0.133 kPa[b], d 1.08 g/cm^3	6230	fly repellent for cattle
2-phenylcyclohexanol	[*1444-64-0*]	$C_6H_5C_6H_{10}OH$	solid, mp 41°C		general-purpose repellent
n-propyl *N,N*-diethylsuccinamate	[*5834-44-4*]	$C_3H_7OCOCH_2CH_2CON(C_2H_5)_2$	d 1.01 g/cm^3	6400	mosquito repellent

[a] Ref. 32.

[b] To convert kPa to mm Hg, multiply by 7.5.

Applications of the best materials will give from one to six hours protection against mosquitoes and biting flies.

Under many circumstances, the ideal applications of repellents are those made on clothing, gloves, and head nets where the protection time is extended to a week or more. Clothing applications may be made by rubbing or spraying the repellent on the cloth or by dipping the clothing in an emulsion of the repellent. The clothing-impregnation method is especially suited to military operations, and the formulation given in mixture (5) has become the standard clothing impregnant, M-1960, of the U.S. Armed Services for protection from mosquitoes, fleas, ticks, and chiggers.

		Parts
(5)	benzyl benzoate	3
	N-butylacetanilide	3
	2-ethyl-2-butyl-1,3-propanediol	3
	Tween 80	1

Clothing impregnation may be made from acetone solution or, preferably, from an emulsion of a mixture of 90% repellent and 10% of an emulsifier of the nonionic type. If this is unavailable, soft-soap, 6.3×10^{-2} L/L (½ pt/gal) H_2O, is effective. Such treatments are applied at the rate of 22 g/m^2 or ca 5% of the weight of the cloth, and will remain effective after several washings. Most of the repellents are solvents for lacquers and should not be applied to watch crystals, spectacle frames, synthetic fabrics, paints, and varnishes.

Mothproofing. Fabric pests cause from 200–$500 million damage annually in the United States to clothing, rugs, upholestry, and bookbindings (34). Ideally, such mothproofing materials should be fixed in the cloth during dyeing and should be completely effective against the insect pests, give good protection after repeated washings and dry cleanings, and be nonvolatile, stable to light, odorless, colorless, and nontoxic to humans. Several colorless dyestuffs possessing these properties can be applied to woolen goods during the dyeing operation and can be fixed in the fibers by chemical reactions with the protein. The two most widely used compounds are Eulan CN and Mitin FF. Both give protection effective for the lifetime of the article when applied at 1–3% of the weight of the cloth.

There are several simple processes for mothproofing which may be applied during

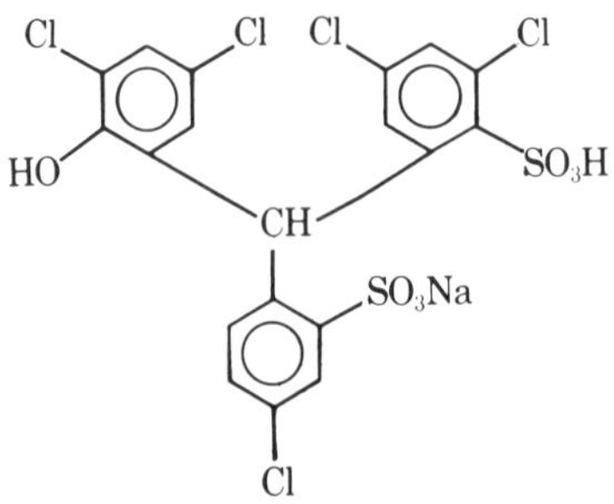

Eulan CN
[*4430-22-2*]

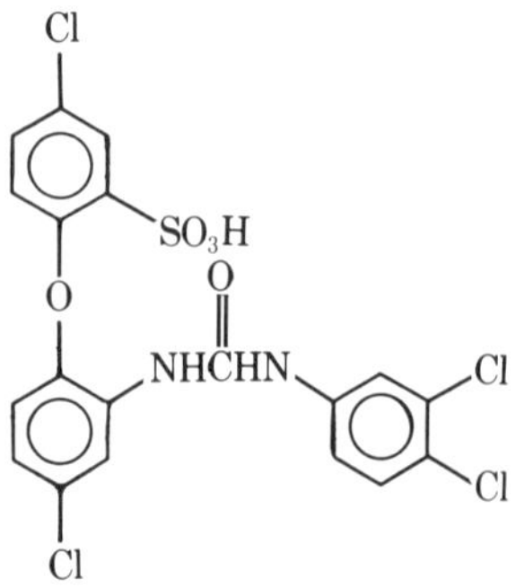

Mitin FF
[*3567-25-7*]

dry cleaning or by the housekeeper. Sodium fluorosilicate [*16893-85-9*], alone or with aluminum, magnesium, or ammonium salts, is an effective mothproofing agent when employed at 0.5–0.7% in water solution, preferably with 0.25–0.5% wetting agent. A popular formulation is stated to contain 0.6% sodium fluorosilicate, 0.3% potassium alum, and 0.03% oxalic acid. Such formulations may be applied by spraying and dipping the fabric and cannot be removed by dry cleaning. Fabrics impregnated with 0.25–0.75% DDT or 0.05% dieldrin, based on the dry weight of the cloth, are protected from moths and carpet beetles for periods of months to years if they are not washed or dry cleaned repeatedly. These applications can be made readily by spraying the cloth with 0.5% DDT or 0.05% dieldrin in a volatile solvent such as petroleum naphtha or by spraying or dipping the fabric in an aqueous emulsion of equivalent strength. The mothproofing treatment can be given during dry cleaning by adding the insecticide to the cleaning bath; it is invisible and has no effect on the properties of the cloth.

Wood Preservatives and Soil Poisons. Several chlorinated benzenes and phenols are used as repellents and soil poisons for termites and as wood preservatives against termites, powder-post beetles, and carpenter ants. Pentachlorophenol [*87-86-5*], C_6Cl_5OH, is a brownish crystalline material (mp 175–180°C) and has the following solubilities (g/100 g of solution at 20°C): carbon tetrachloride, 2; *o*-dichlorobenzene, 8.5; diesel oil, 3.1; pine oil, 32; Stoddard solvent, 1.5; and water, 0.0014. The sodium salt is water soluble to 26 g/100 ml and is converted to the free phenol below pH 6.8. *o*-Dichlorobenzene [*95-50-1*], $C_6H_4Cl_2$, is a colorless liquid (bp 179°C, d^{15} 1.305 g/cm^3). The technical product is a mixture of 75–85% *o*-isomer and 15–25% dissolved *p*-dichlorobenzene. Trichlorobenzene, $C_6H_3Cl_3$, is a colorless liquid (bp 205–250°C, d 1.460 g/cm^3). The technical product is a mixture of the 1,2,3 [*87-61-6*] and 1,2,4 [*120-82-1*] isomers.

Attractants. A number of chemicals have been identified that influence insect behavior as they search for food, oviposition sites, or mates (9,35–36). Many of these chemicals are used to attract insect pests into traps or to poison baits both for population control and for measurement of population densities, eg, for timing spray applications.

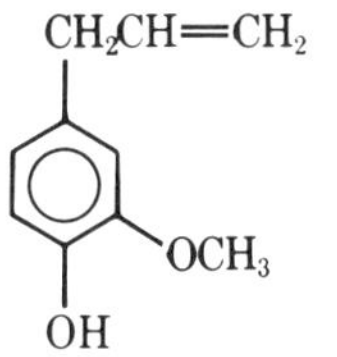

$CH_2CH{=}CH_2$ / OCH_3 / OH

eugenol [*97-53-0*]

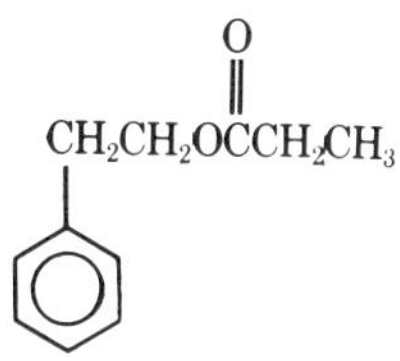

$CH_2CH_2OCCH_2CH_3$ (C=O)

phenethyl propionate [*122-70-3*]

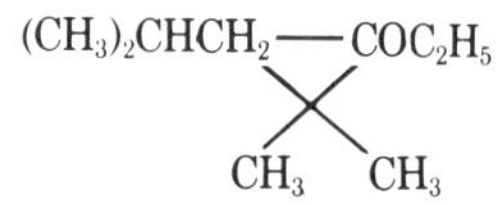

$(CH_3)_2CHCH_2$—COC_2H_5 / CH_3 CH_3

ethyl dimethylisobutylcyclopropane carboxylate [*33419-38-4*]

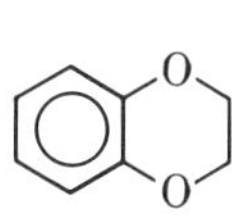

O / O

propyl benzodioxancarboxylate [*24902-02-1*]

Table 16. Sex Pheromones Used in Insect Control

Compound	CAS Registry No.	Structure	Species
cis-7,8-epoxy-2-methyloctadecane	[*29804-22-6*]	$CH_3(CH_2)_9CH\text{—}CH(CH_2)_4CH(CH_3)_2$ (epoxide O bridging the two CH)	gypsy moth, *Porthetria dispar*
trans-8,*trans*-10-dodecadienol	[*33956-49-9*]	$CH_3CH{=}CHCH{=}CH(CH_2)_6CH_2OH$	codling moth, *Laspeyresia pomonella*
cis-9-tetradecenal (with *cis*-11-hexadecenal)	[*53939-27-8*]	$CH_3CH_2CH_2CH_2CH{=}CH(CH_2)_7CHO$	tobacco budworm, *Heliothis virescens*
trans-11-tetradecenal	[*35746-21-5*]	$CH_3CH_2CH{=}CH(CH_2)_9CHO$	eastern spruce budworm, *Choristoneura fumiferana*
cis-11-hexadecenal	[*53939-28-9*]	$CH_3(CH_2)_3CH{=}CH(CH_2)_9CHO$	corn earworm, *Heliothis zea*
(*Z*,*Z*)-11,13-hexadecadienal	[*71317-73-2*]	$CH_3CH_2CH{=}CHCH{=}CH(CH_2)_9CHO$	navel orangeworm, *Amyelois transitella*
cis-7-dodecenyl acetate	[*14959-86-5*]	$CH_3(CH_2)_3CH{=}CH(CH_2)_6OC(=O)CH_3$	cabbage looper, *Trichoplusia ni*, alfalfa looper *Autographa californica*, soybean looper, *Pseudoplusia includens*
cis-8-dodecenyl acetate	[*28079-04-1*]	$CH_3(CH_2)_2CH{=}CH(CH_2)_7OC(=O)CH_3$	oriental fruitmoth, *Grapholitha molesta*
cis-9-dodecenyl acetate	[*16974-11-1*]	$CH_3CH_2CH{=}CH(CH_2)_8OC(=O)CH_3$	grape berry moth, *Paralobesia viteana*
cis-9-tetradecenyl acetate	[*16725-53-4*]	$CH_3(CH_2)_3CH{=}CH(CH_2)_8OC(=O)CH_3$	southern armyworm, *Spodoptera eridania* (with *cis*-9,*trans*-12-tetradecadienyl acetate beet armyworm, *Spodoptera exigua* fall armyworm, *Spodoptera frugiperda* (with *cis*-9,*trans*-12-tetradecadienyl acetate

cis-11-tetradecenyl acetate	[*20711-10-8*]	$CH_3CH_2CH{=}CH(CH_2)_{10}OC(=O)CH_3$	red-banded leafroller, *Argyrotaenia velutinana*
trans-11-tetradecenyl acetate (with *cis*-11)	[*39298-60-7*]	$CH_3CH_2CH{=}CH(CH_2)_{10}OC(=O)CH_3$	European corn borer, *Ostrinia nubilalis*
cis-9,*trans*-11-tetradecadienyl acetate (with *cis*-9,*trans*-12)	[*30562-09-5*]	$CH_3CH_2CH{=}CHCH{=}CH(CH_2)_8OC(=O)CH_3$	cotton leafworm, *Spodoptera littoralis*, *Spodoptera litura*
cis-9,*trans*-12-tetradecadienyl acetate	[*31654-77-0*]	$CH_3CH{=}CHCH_2CH{=}CH(CH_2)_8OC(=O)CH_3$	almond moth, *Cadra cautella* fig moth, *Cadra figulilella* Mediterranean flour moth, *Anagasta kuehniella* tobacco moth, *Ephestia elutella*
cis-7,*cis*-11-hexadecadienyl acetate (with *cis*-7-*trans*-11)	[*52287-99-5*]	$CH_3(CH_2)_3CH{=}CH(CH_2)_2CH{=}CH(CH_2)_6OC(=O)CH_3$	pink bollworm, *Pectinophora gossypiella* Angoumois grain moth, *Sitotroga cerealella*
cis-3,*cis*-13-octadecadienyl acetate	[*53120-27-7*]	$CH_3(CH_2)_3CH{=}CH(CH_2)_8CH{=}CH(CH_2)_2OC(=O)CH_3$	peach-tree borer, *Sanninoidea exitiosa*
trans-3,*cis*-13-octadecadienyl acetate	[*53120-26-6*]	$CH_3(CH_2)_3CH{=}CH(CH_2)_8CH{=}CH(CH_2)_2OC(=O)CH_3$	lesser peach-tree borer, *Synanthedon pictipes*

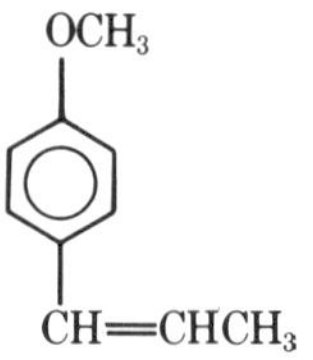

OCH_3 / $CH{=}CHCH_3$

anethole
[*104-46-1*]

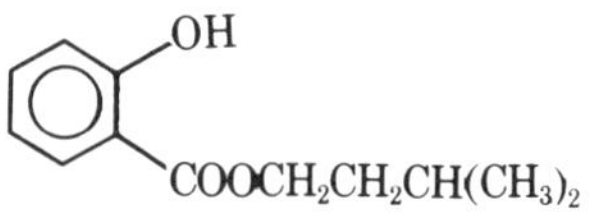

OH / $COOCH_2CH_2CH(CH_3)_2$

isoamyl salicylate
[*87-20-7*]

Food Lures. The Japanese beetle, *Popillia japonica,* is trapped by a mixture of eugenol [*97-53-0*] and phenethyl propionate [*122-70-3*]. Caproic acid [*142-62-1*] is used to attract the green June beetle, *Cotinus nitida*, ethyl 3-isobutyl-2,2-dimethyl cyclopropane carboxylate [*33419-38-4*], the coconut rhinoceros beetle, *Oryctes rhinoceros*; and propyl 1,4-benzodioxan-2-carboxylate [*24902-02-1*], the European chafer, *Amphimallon majalis*.

The use of fermenting sugars and syrups as attractants for moths and butterflies has been supplemented by a variety of essential oils, eg, anethole for the codling moth and isoamyl salicylate for the tomato and tobacco hornworm moths.

Attractants have proved especially effective for the fruit flies of the family Tephritidae. Methyl eugenol is strongly attractive to males of the oriental fruit fly, *Dacus dorsalis* and anisyl acetone for males of the melon fly, *D. cucurbitae*. Both of these attractants have been used in male annihilation programs by applying them to fiber board or string impregnated with a contact insecticide, eg, malathion, naled, or dichlorvos, using doses as low as 15 g of attractant and 1 g of toxicant per ha. Similarly the Mediterranean fruit fly *Ceratitis capitata* is strongly attracted to *t*-butyl 2-methyl-4-chlorocyclohexanecarboxylate (tri-med lure). Protein hydrolysates are used in bait sprays for a variety of these fruit flies and ammonia, which is liberated from a mixture of glycine and sodium hydroxide, is used to trap the walnut husk fly, *Rhagoletis completa*.

Yellow jackets, *Vespula* sp. are trapped with heptyl butyrate [*5870-93-9*], $CH_3(CH_2)_5CH_2OC(O)CH_2CH_2CH_3$, and metaldehyde [*108-62-3*], $[OCH(CH_3)]_4$, is used at 2–3% as an attractant in poison baits for snails and slugs.

Sex Pheromones. Sex pheromones are distributed widely throughout the Insecta and are specific chemicals or mixtures of chemicals generally secreted by glands in the terminal segments of the female abdomen, which attract males upwind from long distances (see Hormones, sex hormones). The pheromones typically are long-chain unsaturated alcohols, aldehydes, and esters and are among the most active biochemicals known; they are perceptible by the male in amounts as low as 10^{-7} μg. Most of the pheromone chemicals that have been used in IPM programs are identified in Table 16 and are available commercially.

The naturally occurring pheromones often are complex mixtures; that of the tobacco budworm, *Heliothis virescens,* contains—in addition to the major component, *cis*-11-hexadecenal—traces of *cis*-9-hexadecenal [*56219-04-6*], *cis*-7-hexadecenal [*56797-40-1*], *cis*-9-tetradecenal [*53939-27-8*], *cis*-11-hexadecenal [*56683-54-6*], hexadecanal [*629-80-1*], and tetradecanal [*124-25-4*]. The full spectrum of compounds is essential for the maximum male response.

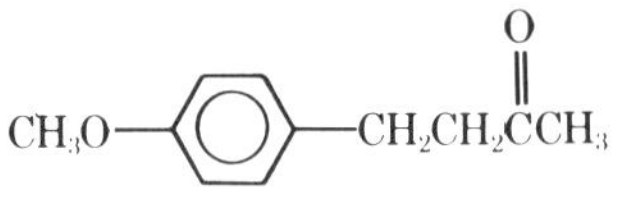

anisyl acetone
[*104-20-1*]

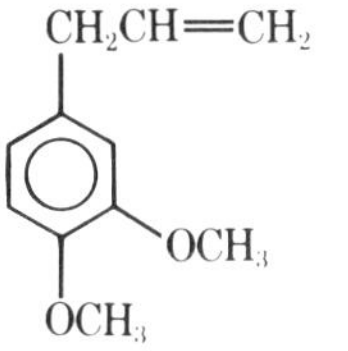

methyl eugenol
[*93-15-2*]

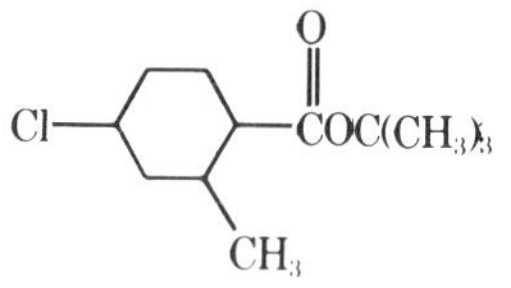

tri-med lure
[*12002-53-6*]

Aggregation Pheromones. The male cotton boll weevil, *Anthonomus grandis*, produces a four-component mixture, each compound of which is essential for effective aggregation; it is available commercially as grandlure [*11104-05-5*]:

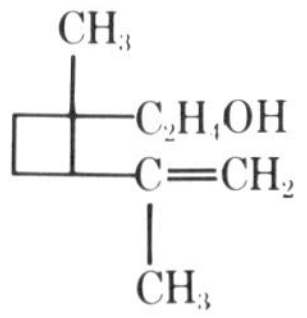

D-*cis*-2-isopropenyl-1-methylcyclobutaneethanol
[*30820-22-5*]

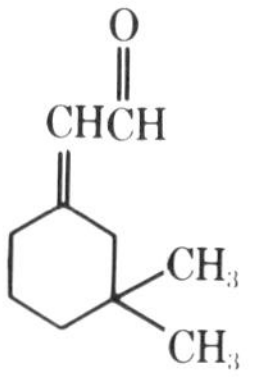

cis-, and *trans*-3,3-dimethyl-Δ′,α-cyclohexaneacetaldehyde
[*24393-78-0*] (*cis*-)
[*24398-89-8*] (*trans*-)

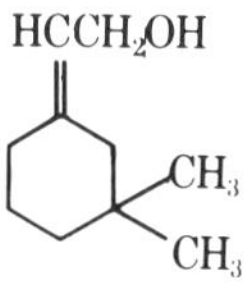

cis-3,3-dimethyl-Δ′,β-cyclohexaneethanol
[*26532-23-0*]

The aggregation pheromone of the smaller European bark beetle, *Scolytus multistriatus* (the principal vector of the Dutch elm disease pathogen, *Ceratocystis ulmi*), is a three-component mixture; two of the compounds are produced by the female and one by the tree:

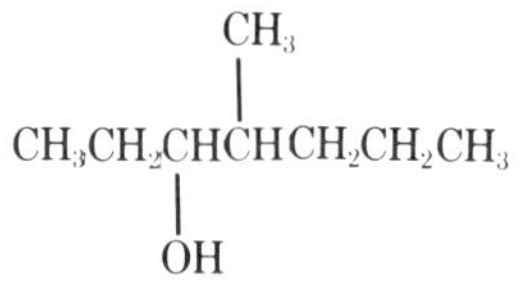

$CH_3CH_2CH(OH)CH(CH_3)CH_2CH_2CH_3$

4-methyl-3-heptanol
[*14979-39-6*]

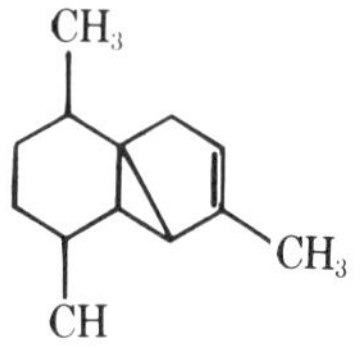

α-cubebene (tree)
[*17699-14-8*]

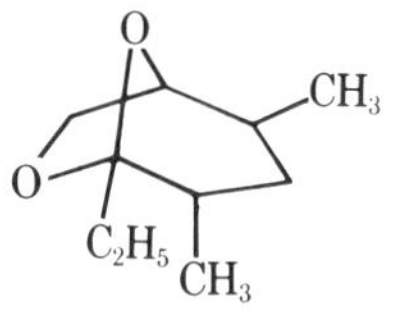

2,4-dimethyl-5-ethyl-6,8-dioxabicyclo[3.2.1]octane
[*54832-20-1*]

The aggregation pheromones of the boll weevil and various bark beetles have been used in IPM for both population monitoring and mass trapping.

An account of recent reviews is given in reference 37.

BIBLIOGRAPHY

"Insecticides" in *ECT* 1st ed., Vol. 7, pp. 881–908, by D. E. H. Frear, The Pennsylvania State College, and Myron A. Coler and Marion H. Peskin, Markite Company; "Insecticides" in *ECT* 2nd ed., Vol. 11, 677–738, by Robert L. Metcalf, University of California.

1. H. H. Cramer, *Plant Protection and World Food Production,* Pflanzenschutz-Nachrichten 20, Leverkusen, FRG, 1967.

2. C. L. Metcalf, W. P. Flint, and R. L. Metcalf, *Destructive and Useful Insects,* 4th ed., McGraw-Hill Book Co., Inc., New York, 1962.
3. R. L. Metcalf and J. J. McKelvey, Jr., eds., *The Future for Insecticides,* John Wiley & Sons, Inc., New York, 1976.
4. D. E. H. Frear, *Chemistry of the Pesticides,* 3rd ed., D. Van Nostrand Co. Inc., Princeton, N.J., 1955.
5. H. H. Shepard, *The Chemistry and Action of Insecticides,* McGraw-Hill Book Co., Inc., New York, 1951.
6. *The Pesticide Review,* Annual Report Agricultural Stabilization and Conservation Service, U.S. Department of Agriculture, Washington, D.C.
7. T. R. Eichers, P. A. Andrilenas, and T. W. Anderson, *Farmers' Use of Pesticides in 1976,* U.S. Department Agriculture, Agr. Econ. Rept. No. 418, Washington, D.C., 1978.
8. *Pest Control: An Assessment of Present and Alternative Technologies,* Vol. I, National Academy of Sciences, Washington, D.C., 1975.
9. R. L. Metcalf and W. H. Luckmann, eds., *Introduction to Insect Pest Management,* John Wiley & Sons, Inc., New York, 1975.
10. E. H. Smith and D. Pimentel, *Pest Control Strategies,* Academic Press, Inc., New York, 1978.
11. A. W. A. Brown, *Insect Control by Chemicals,* John Wiley & Sons, Inc., New York, 1951.
12. R. L. Metcalf, *Organic Insecticides,* Interscience Publishers, New York, 1955.
13. C. F. Wilkinson, ed., *Insecticide Biochemistry and Physiology,* Plenum Press, Inc., New York, 1976.
14. P. Müller ed., *DDT Insektizid Dichlorodiphenyltrichlorathan und seine Bedeutung,* Birkhauser Verlag, Basel, Switz., 1955.
15. A. W. A. Brown, *Ecology of Pesticides,* John Wiley & Sons, Inc., New York, 1977.
16. C. Fest and K. J. Schmidt, *The Chemistry of Organophosphorus Pesticides,* Springer-Verlag, New York, 1973.
17. R. D. O'Brien, *Toxic Phosphorus Esters,* Academic Press, Inc., New York, 1960.
18. G. Schrader, *Die Entwicklung neuer Insektizider Phosphorsaure-Ester,* Verlag Chemie, G.m.B.H. Weinheim, FRG, 1963.
19. W. E. Ripper, *Advances in Pest Control Research,* Vol. I, John Wiley & Sons, Inc., New York, 1957, p. 305.
20. H. T. Reynolds, *Advances in Pest Control Research,* Vol. II, John Wiley & Sons, Inc., New York, 1958, p. 135.
21. T. R. Fukuto, *Bull. WHO* **44,** 31 (1971).
22. D. L. Lindgren and L. C. Vincent, *Advances Pest Control Research,* Vol. V, John Wiley & Sons, Inc., New York, 1962, p. 85.
23. M. S. Quraishi, *Biochemical Insect Control,* John Wiley & Sons, Inc., New York, 1977.
24. R. L. Metcalf, *Bull. WHO* **44,** 43 (1971).
25. M. Elliott, *Bull. WHO* **44,** 315 (1971).
26. W. S. Bowers, *Bull. WHO* **44,** 381 (1971).
27. I. M. Hall, *Advances in Pest Control Research,* Vol. IV, John Wiley & Sons, Inc., New York, 1961, p. 1.
28. A. M. Heimpel, *Ann. Rev. Entomol.* **12,** 287 (1967).
29. R. C. Bushland, *Advances Pest Control Research,* Vol. III, John Wiley & Sons, Inc., New York, 1959, p. 1.
30. R. Pal and M. J. Whitten, eds., *The Use of Genetics in Insect Control,* Elsevier/North Holland Publishers, Amsterdam, 1974.
31. A. B. Borkovec, *Insect Chemiosterilants,* John Wiley & Sons, Inc., New York, 1966.
32. A. W. A. Brown and R. Pal, *Insecticide Resistance in Arthropods,* World Health Organization, Geneva, Switz., 1971.
33. G. F. Shambaugh, R. F. Brown, and J. J. Pratt, Jr., *Advances Pest Control Research,* Vol. I, John Wiley & Sons, Inc., New York, 1958, p. 277.
34. D. T. Waterhouse, ref. 20, p. 207.
35. N. Green, M. Beroza, and S. A. Hall, ref. 29, p. 129.

36. H. H. Shorey and J. J. McKelvey, Jr., eds., *Chemical Control of Insect Behavior*, John Wiley & Sons, Inc., New York, 1977.
37. R. Baker and D. A. Evans, *Chem. Br.* **16**, 412 (1980).

ROBERT L. METCALF
University of Illinois at Urbana-Champaign

INSTRUMENTATION AND CONTROL

Automatic control is self-correcting or feedback control, whereby some control instrument continuously monitors some output variable of the controlled process (eg, a plate temperature, or the heads composition of a distillation column), and compares this output with some established desired value. The instrument then uses any resulting error obtained from the comparison of the actual and desired value of the output variable to compute the required correction to the setting of the reflux valve, or other basic element of the piece of equipment being controlled. The value of the output variable then is returned to its desired level and is maintained there. Such servomechanism control should be distinguished from the common on–off type of control. This latter uses the presence or absence of an error to govern the application of a predetermined constant correction to the basic element of the equipment. The thermostatic control of a home refrigerator or furnace is an example of the latter operation. The servomechanism control is discussed in this article.

The design and use of a servomechanism control system requires a knowledge of each element of the control loop. For example, in the method illustrated in Figure 1, the distillate composition is regulated by controlling the distillate take-off rate and, thus, the reflux ratio of the column. The engineer must know the dynamic response or complete operating characteristics of each pictured device: the indicator or temperature bulb; the controller, including both the error detector and the correction computer; the control valve; the transmission characteristics of the connecting lines; and the operating characteristics of the column. Dynamic response or operating characteristics imply that one must be able to write the differential equations for the transient behavior of the process or its actions during periods of change of operating conditions, develop the transfer function of the process (another often simpler representation of the differential equations), or prepare an experimental or empirical representation of the same effects. The control engineer usually combines the devices shown in Figure 1 into a block diagram for convenience (see Fig. 2).

Because of time lags in long pneumatic lines from sensor to controller and the other delays in the process, time will elapse before knowledge of changes (ie, a drop) in output composition or temperature reaches the controller. When the controller notes the composition drop, it must compare it with the desired composition, compute how much and in what direction the distillate valve must be repositioned, and make the correction in the valve opening. Time also elapses before the effect of the valve correction on the output product composition can reach the output and thus be sensed. It is only then that the controller knows whether its first correction was accurate. At

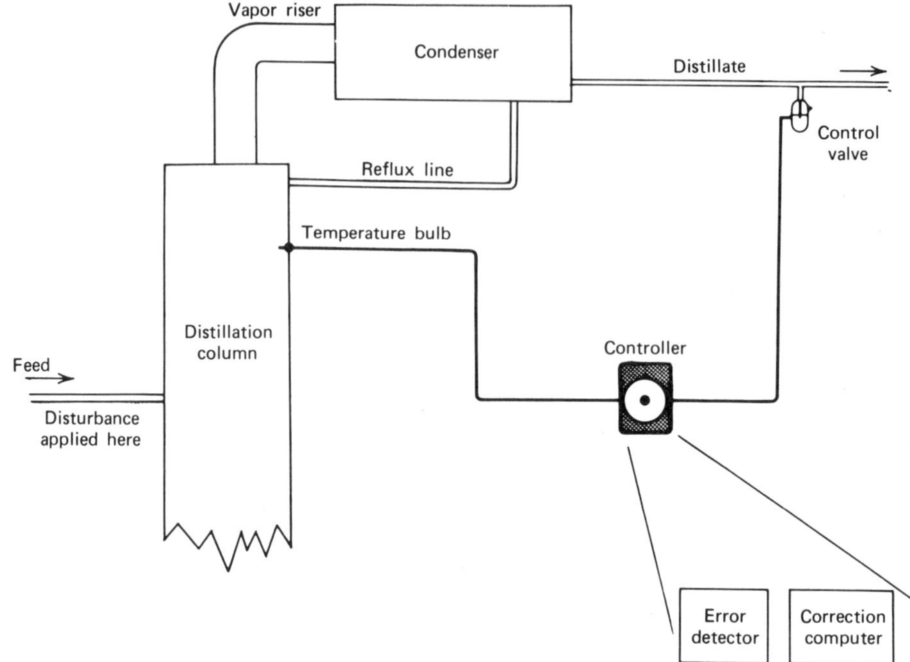

Figure 1. Servomechanism system in which a control loop is used to regulate the reflux ratio of the distillation column.

that time, it makes a further correction that will, after a time, cause another output composition change. The results of the second correction will be observed, and a third correction will be made, and so on.

The series of measuring, comparing, computing, and correcting actions will repeat through the controller and through the process in a closed chain of actions until the overhead composition is balanced at the desired value. This type of control is termed feedback control. Figure 2 shows the direction and path of this closed series of control actions.

Although the above example illustrates the basic principles involved, the actual attainment of automatic control of almost any industrial process, or other complicated device, usually is much more complex because of the speed of response, multivariable interaction, nonlinearities, response limitations, or other difficulties which may be present and the much higher accuracy or degree of control which usually is desired beyond that required for the simple process just mentioned.

In the automatic process, the control instrument continuously monitors some output variable of the controlled process, eg, a temperature, pressure, composition, etc, and it compares this output with some established desired value or set point of the controlled variable. The error resulting from this comparison is used by the instrument to compute a correction in the setting of the process control valve or other final control element in order to return the value of the output variable to its desired level and to maintain it there. If the set point is altered, the response of the control system in bringing the process to the new operating level is a servomechanism or self-correcting device. The action of holding the process at a previously established level when the process is opposed by external disturbances is that of a regulator.

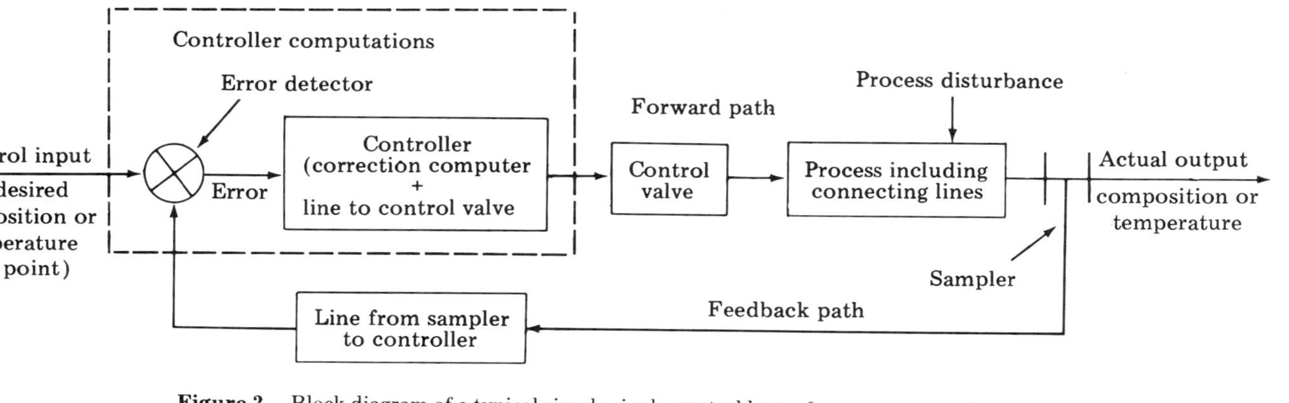

Figure 2. Block diagram of a typical simple single-control loop of a process control system.

In considering the possible responses of a system to various inputs or upsets, the control engineer usually confines the study to two types of response. The transient response, which is the action of the system when it is subjected to a step function, or other sudden change in operating point, which is applied as a forcing function or input to the system. The frequency response is the action of the system when subjected to a sinusoidally varying forcing function. A transient response is obtained when solving the differential equations of the system analytically or by means of electronic computers (qv). The frequency response is especially valuable for those systems that can be described by linear ordinary differential equations. There are several graphical shortcut methods that help greatly in the study of system responses which can be defined by such equations. Frequency response has been used in the experimental testing of systems, especially electrical ones.

Instrumentation of an Automatic Control System

The large number of variables of a typical industrial plant comprise a wide variety of flows, levels, temperatures, compositions, positions, and other parameters that are to be measured by the sensor elements of the control system. Such devices sense some physical, electrical, or chemical property of the variable under consideration and use the property value to develop an electrical, mechanical, or pneumatic signal that is representative of the magnitude of the variable in question. This signal then is acted upon by a transducer to convert it to one of the standard signal levels used in industrial plants: 21–103 kPa (3–15 psi) for pneumatic systems, and 1–4, 4–20, or 10–50 mA (0–5 V) for electrical systems. They also may be digitized at this point if the control system is to be digital. The sensing and conversion techniques have been described (1–5).

The signals, as continuous representations of the variables involved, are analogue signals in contrast to single on–off signals or digital signals. When analogue signals have been operated upon by an analogue-to-digital converter, they become a series of bits in feedback control or on–off signals and are also called digital signals even though several must be considered together to properly represent the converted analogue signal (usually 10–12 bits).

The size of most industrial plants necessitates connection of the transducer by long electrical cables to the controller unit. Quite often, the connecting cable must run close to a source of high voltage, resulting in noise pickup on the leads, which manifests itself as a voltage added to the transducer output or series-mode voltage. The series-mode voltage often is of a much higher frequency than the information frequency of the signal, and therefore can be attenuated sufficiently by a low-pass R-C (resistor-capacitor) filter between the connecting cable and the input selection switch. This and other techniques, when properly applied, usually can nullify the effects of such noise pickup. In other circumstances, there may be considerable external noise pickup at the sensor, eg, common-mode voltage noise or voltage. Again, several grounding and shielding techniques can counteract the effects of such pickups. They must, however, always be taken into proper consideration to obtain a satisfactory data collection and control system (6–7).

Any error between resulting sensed and set-point values is used by the controller to compute the correction to the controller output which goes to the valve or other system actuator. A typical algorithm by which the controller (either analogue or digital) computes its correction is as follows (8):

$$\text{output} = K_p e_n + \Sigma K_R e_n + K_D(e_n - e_{n-1}) + K_1$$

where K_P, K_D, and K_R are the proportional, derivative, and integral gains, respectively, of the usual analogue controller. Thus, the equation used most often for direct digital control is a direct model of that used by the analogue controller. The error, e_n, should be calculated as:

$$e_n = \pm \text{ (set point } - \text{ controlled variable)}$$

The option of using either a plus or minus sign in the error calculation should be provided for each control loop. An adjustable midrange constant K_1 should be included to provide for the possibility that proportional action only (ie, $K_D = 0$, $K_R = 0$) might be desired. The value of K_1 should be such that if the error and K_R were zero, the output would be midrange. The summation shown in the controller calculation should be considered over all previous iterations; therefore, the current value of $\Sigma K_R e$ must be stored between iterations. Proper computation factors should be built into the controller calculations to prevent improper operation of the system during initial connection of the control output to the process being controlled.

Once the correction has been computed, it must be carried by cables to another set of transducer units that convert the correction signals (from the controller) to actuating signals for the valve or other unit which changes the plant operating parameter to the values that are desired by the control system. Pneumatic valves are the most common final actuators in industrial plants because of their ruggedness and their small size relative to the power they exert (9).

If a digital computer is involved in the control system, the sensor signals usually are multiplexed to permit sequential sampling of their values by the computer; this is in contrast to the continuous action of the multiple analogue controllers that it replaces. Likewise, the output signals must be multiplexed to the operating valves.

In many cases, the variable that the operator desires to control, eg, a reactor yield, cannot be measured directly. However, it often is possible to develop a mathematical expression that relates reactor yield to several other reactor variables, eg, reactant flow rate, temperature, output composition, etc, thus permitting the desired value and reactor yield to be calculated. Such a calculation is called an inferred variable.

Analogue Control. The first theoretical analysis of regulator action was developed by James Clerk Maxwell in 1868 who presented a differential equation model of the Watt governor (10), and this work was generalized in 1877 (11). The next important theoretical study was made in 1922 of the steering system of the U.S. naval battleship, New Mexico (12).

Electrical feedback of long-distance telephone amplifiers was studied and described (13–15) and the principle of regeneration was announced in 1932. In 1934, Hazen derived his theory of servomechanisms (16). In 1935, Bailey developed the pneumatic controller, and the development of the first analogue computer by George A. Philbrick occurred in 1939 (17). These developments provided the necessary background material for the rapid development of control system theory and applications during World War II as exemplified by the Mark IX Skysweeper anti-aircraft gun, and the defensive gun-firing control system of the B-29 bomber.

Both theoretical and practical studies of automatic control were confined principally to single-loop studies similar to Figures 1 and 2 until the mid-1950s. The work involved continued development of the methods that first were proposed in the 1930s and were used with such important results in the succeeding two decades. Particularly important were studies of stability.

All of the above work involved the use of analogue control systems, ie, mechanical or electrical devices that simulated a desired action, ie, they were analogues of the action that a human operator would take in the same circumstance. Because of the limitations in complexity of the mechanical and electrical devices that were built in large quantities during this period and because of the limitations in the control systems theory existing at that time, nearly all of these systems were built to control single loops.

Computer Control. Digital computers removed the limitations on complexity imposed by the earlier analogue devices (18) and have become popular elements of industrial plant control systems (19).

The three classes of computer application to industrial plant control problems are supervisory or optimizing control, as exemplified in Figure 3; direct digital control, as defined in Figure 4; and hierarchy control, as illustrated in Figure 5 and which is a combination of the other two. It affects all levels of decision making in the plant.

As can be seen from Figure 3, supervisory or optimizing control places the computer in an external or secondary control loop to the primary plant-control system, which remains as the conventional plant instruments and individual electronic or pneumatic analogue controllers. The computer merely changes the set point, or level of control governed by the controllers, either directly or through manual intervention. Its task is to trim the plant operation to improve the economic return from its operation but not to affect its dynamic control. Thus, its overall gain potential is sharply limited. However, a malfunction of the computer cannot adversely affect the plant control. When the computer makes the set-point computation but does not change the plant controller and instead depends upon the plant operator for this final action, this is an open-loop control. However, where the computer merely samples the process variables and determines their correct operating levels but does not make the optimization calculations, process monitoring takes place.

Direct digital control uses the computer to replace a group of single-loop analogue controllers of the type in Figure 2 with a digital computer. This is done with the hope that the single computer will be less expensive than the many controllers it can supplant (20). The digital computer's computational ability also makes possible the application of more complex, advanced control techniques.

Hierarchy control, the most recent and most ambitious of the concepts, as shown in Figure 5, attempts to apply computers to all plant control situations at the same time. As such, it requires the best of computer capabilities and automatic control potentialities to carry out its function of integrating the plant operation completely from management decision to final valve movement. Most work in this area is in the development stages, although several installations that carry out many of the possible functions listed in the figures have been installed or are in the advanced planning stage. The hierarchy organization, which comprises the largest version of a recent further refinement of the hierarchy control concept, is outlined in Figure 5 (21). Level 1 and Level 2 machines are similar in function, differing only in the extent of the control functions assigned to each. The dedicated digital controller of Level 1 handles complex devices, including chemical analyzers, eg, chromatographs, and specialized feedforward and noninteracting, multivariable control setups. The direct digital controller of Level 2 handles a much larger number of three-mode and related control loops, as in Figure 2. It also communicates with the plant operators through a console, probably composed of cathode-ray tubes and a keyboard. Communication between all digital computers

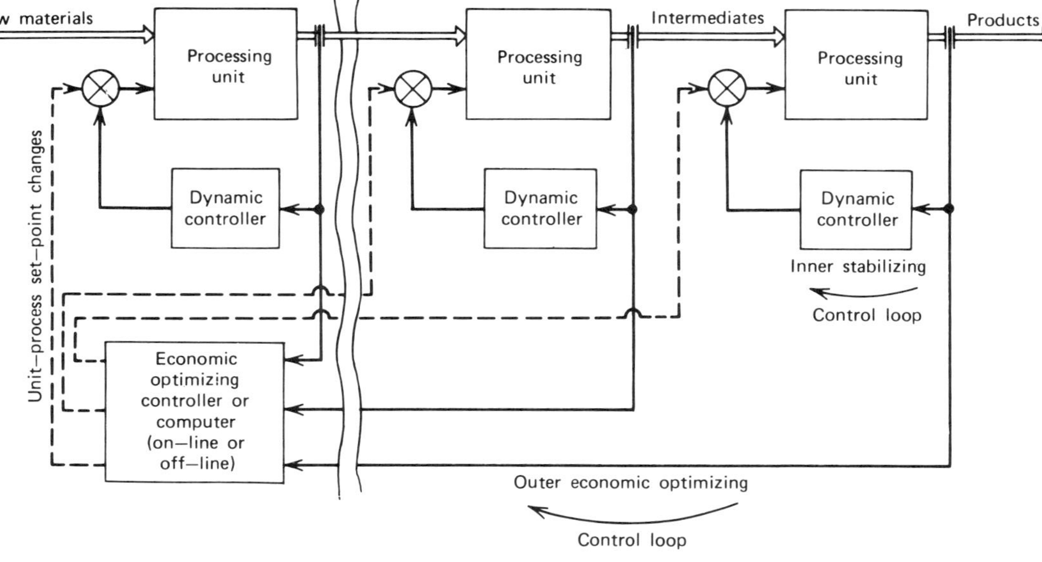

Figure 3. Schematic of the use of computer control as a plant optimizer and supervisory controller.

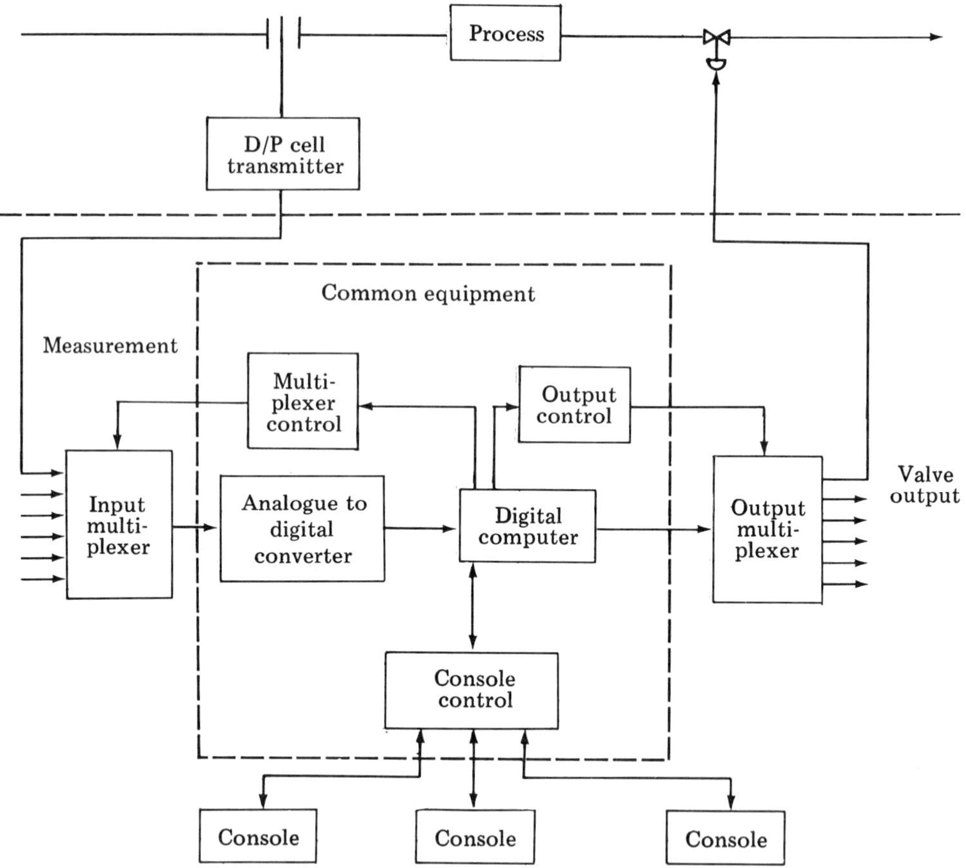

Figure 4. Block design of a direct digital control system. D/P = differential pressure.

and with all consoles is by digital signals. Connections to analogue signals only are required of the first- and second-level machines. Level 3 machines serve as supervisory computer control systems, as outlined above. Level 4 machines carry out production scheduling and management information functions (see Overall Plant and Company Control Systems).

Advanced Control Methods

Figures 1 and 2 define the standard feedback control techniques where the correction computation part of the control loop readjusts the position of the actuator of the process which generally is a control valve. There are severe limitations on the ability of the classical control methods, as exemplified by the single-loop controllers of Figures 1 and 2, whether they are implemented by analogue or digital methods. The digital computer provided the incentive for the beginning of a new era of development in automatic control theory. This class of developments has been labelled modern control in contrast to the classical control methods.

A short listing of some of the advanced control topics that have extended vastly

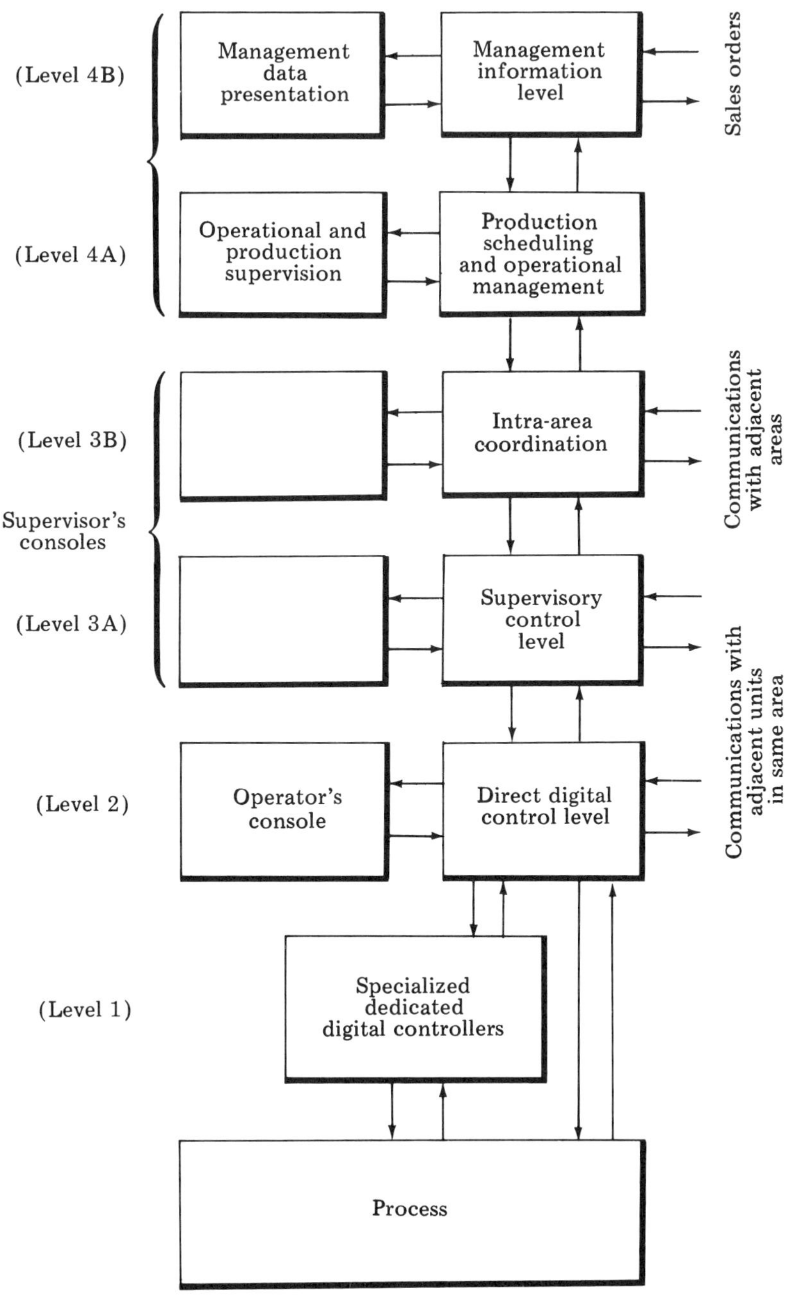

Figure 5. The hierarchical structure of the computer control system for the fully automated industrial plant.

the capabilities of industrial control systems follows. As indicated, a computer is almost always necessary for implementation.

Adaptive control is the capability of the control system to modify its own operation so as to achieve the best possible mode of operation. An adaptive system must be capable of performing the following functions: provide continuous information about the present state of the system or identify the process, compare present system performance to the desired or optimum performance and make a decision to change the system so as to achieve some previously defined optimum performance, and initiate a proper modification so as to drive the control system to the optimum. These three principles, identification, decision, and modification, are inherent in any adaptive system (22–26). Figure 6 illustrates these concepts by picturing the adaptive controller as in a secondary loop and modifying the operation of the regular systems controller which is located in the primary process-control loop.

Steady-state optimization allows the computer to determine for the process a new best operating level if exterior conditions require such changes in order to maintain the process operation at some optimum (usually economic) criterion. Such optimizations are computed under the assumption that the process is at a steady state and can be instantaneously transferred from one steady state to another. This assumption is necessary to transform all process operating equations to algebraic form for ready solution on the computer. Most supervisory control systems (Fig. 3) are of this type.

Dynamic optimizing control is the requirement that the control system operate so that a specific performance criterion is satisfied. This criterion usually is formulated so that the controlled system must move from the original position to a new position in the minimum possible time or at the minimum total cost (27–30). Dynamic optimization and control adds another level of sophistication to that of the steady-state optimization just mentioned. The process not only is maintained at its optimum performance level while in the steady-state regime, but the change from one operating level to another also is made so as best to satisfy the established control criteria, which

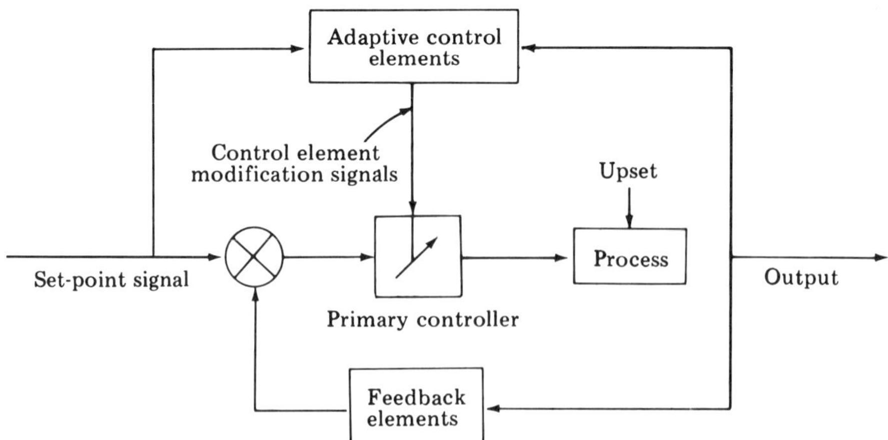

Figure 6. Adaptive control elements form a secondary-control loop around the conventional process-control loop.

usually is economic. The control is mainly of academic interest because of the extremely large and powerful computing capacity that is required to achieve it. However, its practical attainment is highly possible. It has particular potential application in cyclical catalytic processes and in the batch processing of plastics. A large computing capacity, both in speed and memory, is necessary since the equations to be solved in the mathematical model are sets of simultaneous differential equations in contrast to the algebraic equations of the steady-state optimization.

Learning control implies that the control system contains sufficient computational ability to develop representations of the mathematical model of the system being controlled and to modify its operation in order to compensate for the newly developed knowledge. Thus, the learning control system is a further development of the adaptive controller (26).

Multivariable, noninteracting control concerns large systems, the size of whose internal variables are dependent upon the values of other related variables of the process. The single-loop techniques of classical control theory do not suffice and more advanced techniques must be used to develop appropriate control systems for such processes (31–32). Figure 7 diagrams such a system. Feedforward control compensates for the dynamics and process delays of the process and helps the unit process controllers compensate for input upsets. The multivariable, noninteracting controller compensates for the effect of variations in flow of product *A* by action of unit controller 1 on product *B* and prevents interaction of the output variables of the process.

Modeling of Systems

The methods of classical control are capable of handling only very simple systems. For more sophisticated methods of analysis, mathematical models of the total system are necessary. They permit the extended calculations required to use the methods of advanced control. The basic need in plant automatic control engineering is for sufficient knowledge of industrial processing plants, their manufacturing tools and machinery, and their related distribution and economic systems, so that a satisfactory representation of their dynamic behavior can be characterized and computed along with any postulated external or internal influence.

Production of mathematical models of plants and of their related devices and processes for time-varying requirements and for time-dependent upsets depends upon process dynamics. Future progress in process dynamics studies and its application to process control and process design problems depends on developments in three areas: better methods of approximating complex responses by simple models, more complete and more exact characterizations of the nonlinear parameters of process systems, and the development and use of faster and more capable computer systems for process simulation use.

Design Techniques

Principles. In designing control systems, the engineer makes use of the concepts of the system's transient response, and/or its frequency response. Both have led to a series of design techniques applicable to relatively simple systems. For more complex systems, simulation techniques are necessary.

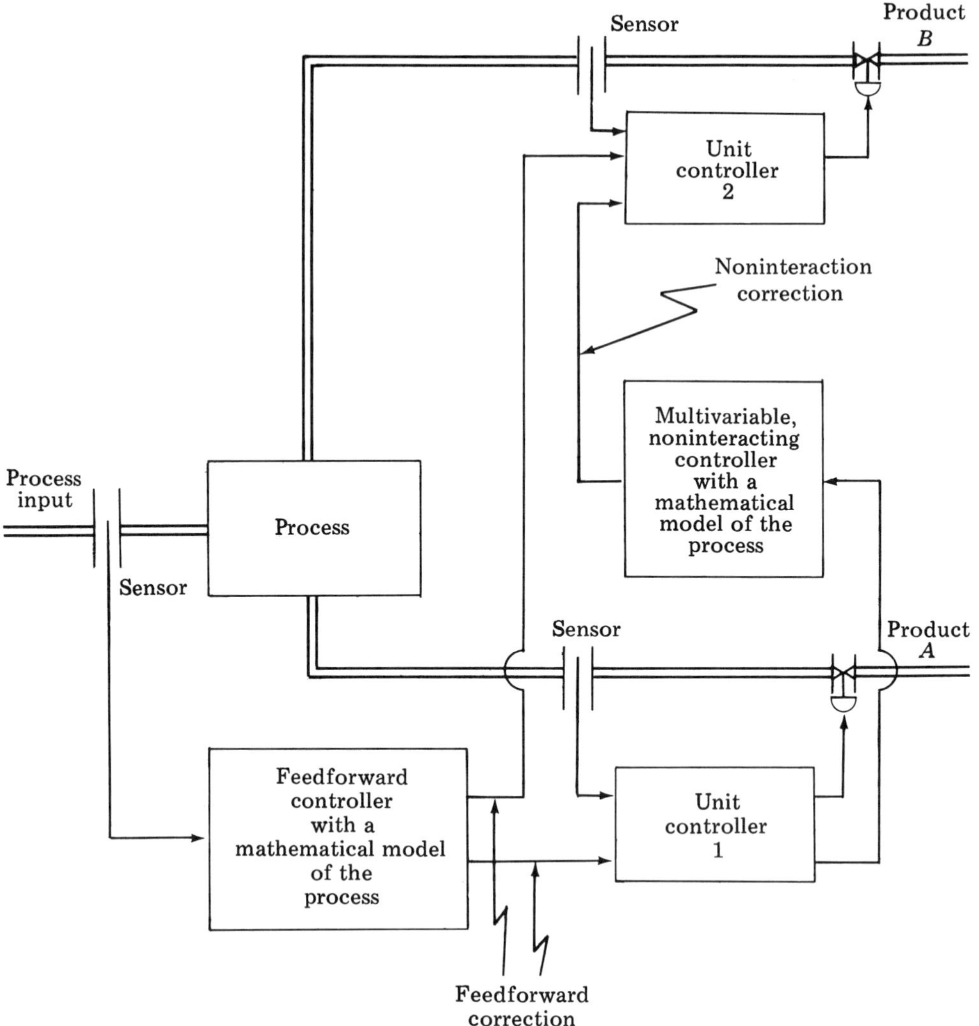

Figure 7. Diagram of a feedforward and multivariable, noninteracting controller system.

Transient Response. Transient response is the response of the system to a step function or other sudden change in operating point which is applied as a forcing function to the system. Depending on the characteristics of the system being forced, five types of curves may be obtained. These are shown in Figure 8 and are called, respectively, stable and overdamped, stable and critically damped, stable and underdamped, oscillatory or on the threshold of stability, and unstable (33).

Most fluid-flow and heat-transfer processes, eg, in the petroleum and chemical industries, give overdamped or critically damped curves when no controller is attached to the process (see Fluid mechanics; Heat-exchange technology, heat transfer; Petroleum, petroleum refinery processes—survey). An example of the underdamped curve is the action of a spring that is subjected to a displacement and then is released.

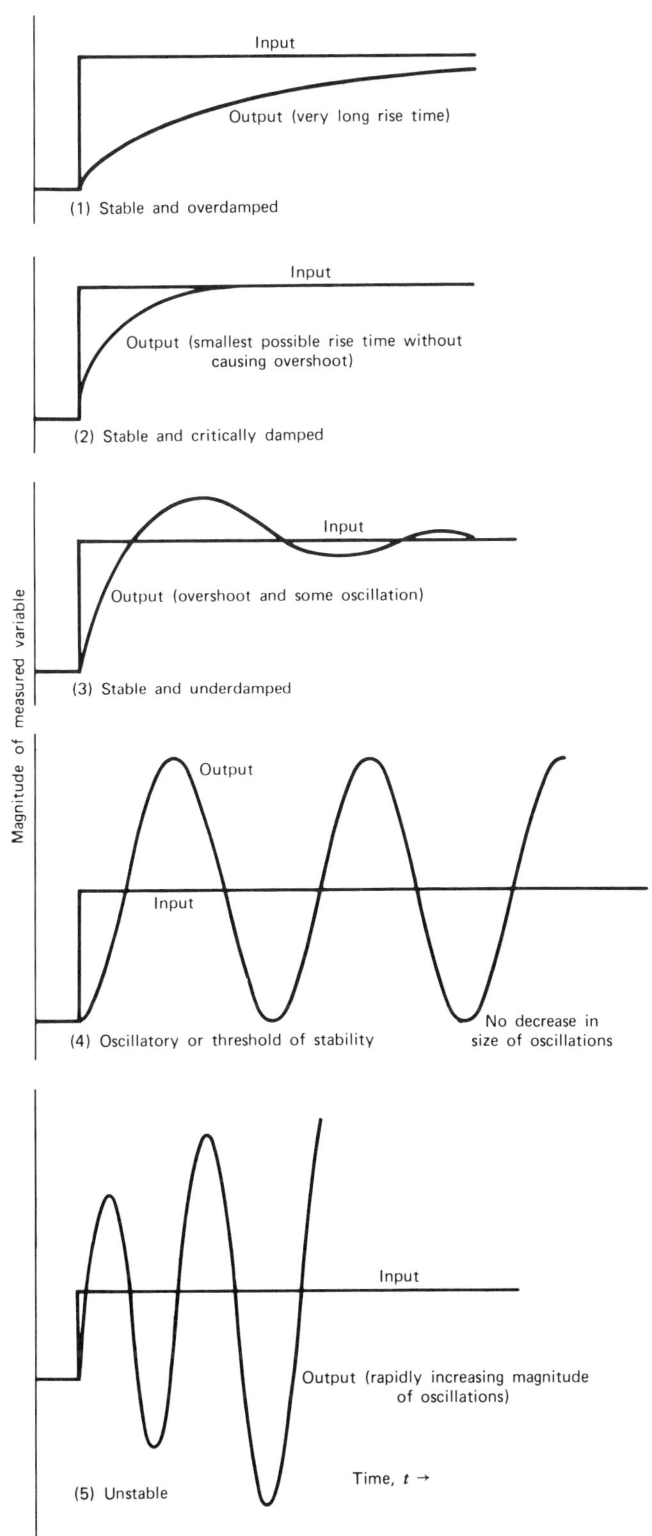

Figure 8. Various degrees of stability in transient response as achieved by automatic control systems.

When the spring friction is assumed to have a smaller and smaller value, the degree of oscillation increases until—for the case of the perfect spring, ie, no internal friction—the threshold stability condition or a permanently oscillatory system is obtained. The unstable condition normally is not present in purely physical systems that can be considered as passive systems but easily can be present in those involving exothermic chemical reactions. Instability requires activity of some sort and an internal source of energy to support the growing transient response.

Whenever a feedback loop and an automatic controller are added to a physical system, eg, whenever a closed-loop system is formed, the possibility of an oscillatory or unstable system is a danger if excessive values of the control constants are utilized. Too large a controller constant results in a correction whose effect is larger than that of the original disturbance. Usually, the effect of this larger correction is amplified by subsequent corrections resulting in ever increasing variations.

For a linear system, the size of the input step function has no bearing on the stability of the system. For nonlinear systems, however, the size of the step may be an important factor in stability.

The graphs of Figure 8 can be obtained experimentally. However, they usually are obtained from the general differential equations of the system as the particular solution with a step forcing function. Because of the nature of the mathematical representation of the step functions, the general equations of the system usually must be solved by the Laplace transform or by a similar operational calculus method when hand computation is to be used. However, the equations for processes fitted with a feedback loop and, particularly, those fitted with control equipment only can be solved by this method up to a relatively low degree of complexity. Beyond this, the methods of the Laplace transform break down from the shear size of the algebraic manipulations involved, and an electronic computer is necessary.

When the response of a physical system to the step input must be obtained experimentally, the step input can be simulated in the physical system by a rapidly opened valve, an on–off switch, or a sudden change in the set point of a controller. However, care must be used since this is the most severe upset to which a system can be exposed and, therefore, damage to the system is possible if the experiment is performed improperly. Such a response almost always is obtained from the closed-loop case, as in Figure 2.

The critically damped system is the most desirable for the control engineer to obtain. However, rather than take the chance of obtaining an overdamped system with resulting system sluggishness, he or she usually attempts to achieve a slightly underdamped response. There are many criteria that have been established to pick the best response. One common criterion is graphed in Figure 9. The first overshoot must be equal to or less than 20% of the input disturbance, and each successive oscillation must fit within an envelope defined by $\pm a/e^{t}$ or an exponential decay.

Frequency Response. The other method commonly used in determining process-controller system response and stability involves the application of a sine function disturbance (rather than a step function) to the input variable. The graph of magnitude versus time of the output variable is compared to a similar graph of the disturbance applied to the input variable to determine the magnitude of the peaks of the output as compared to the input, and the time differential between corresponding points on the two curves as compared to the time required for one complete cycle of the sine wave applied at the input. Thus, magnitude ratio, the ratio of the amplitude of the output

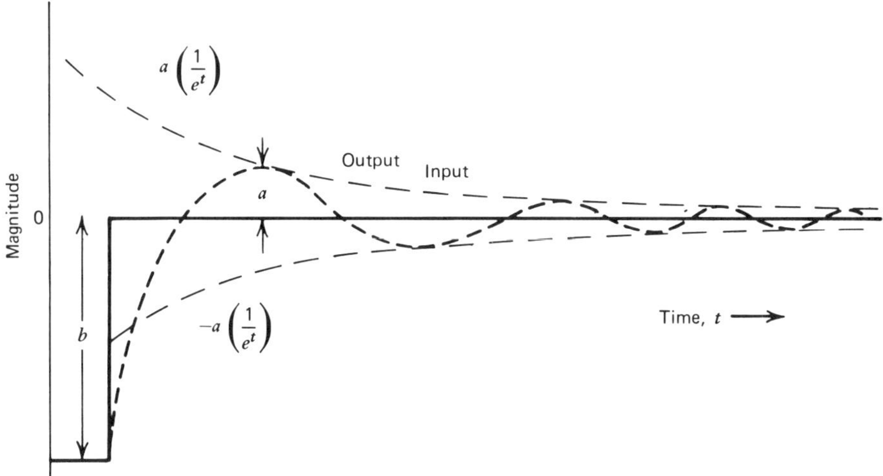

Figure 9. General stability criteria for transient response curves.

wave to that of the input wave, is determined. The value of this magnitude ratio may be greater or less than 1.0, depending upon the nature of the process and the control applied. The time variation between common points on the two waves is the phase shift and is expressed in degrees of arc where the time of the complete sine wave corresponds to 360° of arc. These concepts are diagrammed in Figure 10 and the open-loop case is described in Figure 11.

The magnitude ratio and the phase shift of the output variable are functions of the frequency of the applied sine wave and thus they must be determined for a wide range of frequencies in order to have value. However, they only need to be determined for the open-loop (no feedback) case, ie, without considering the effect of a feedback on the process and on the controller. This is possible since there are graphical methods for determining the effect of the feedback and controller upon the frequency response and is in contrast to a transient response where the equations must be resolved for each case to be investigated. For a transient response, the complete closed-loop

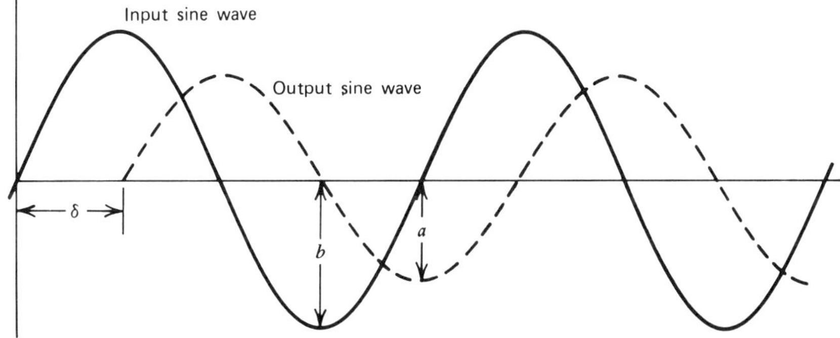

Figure 10. Diagram illustrating phase shift and magnitude ratio. Phase shift, δ = 90°; magnitude ratio, $M = a/b$.

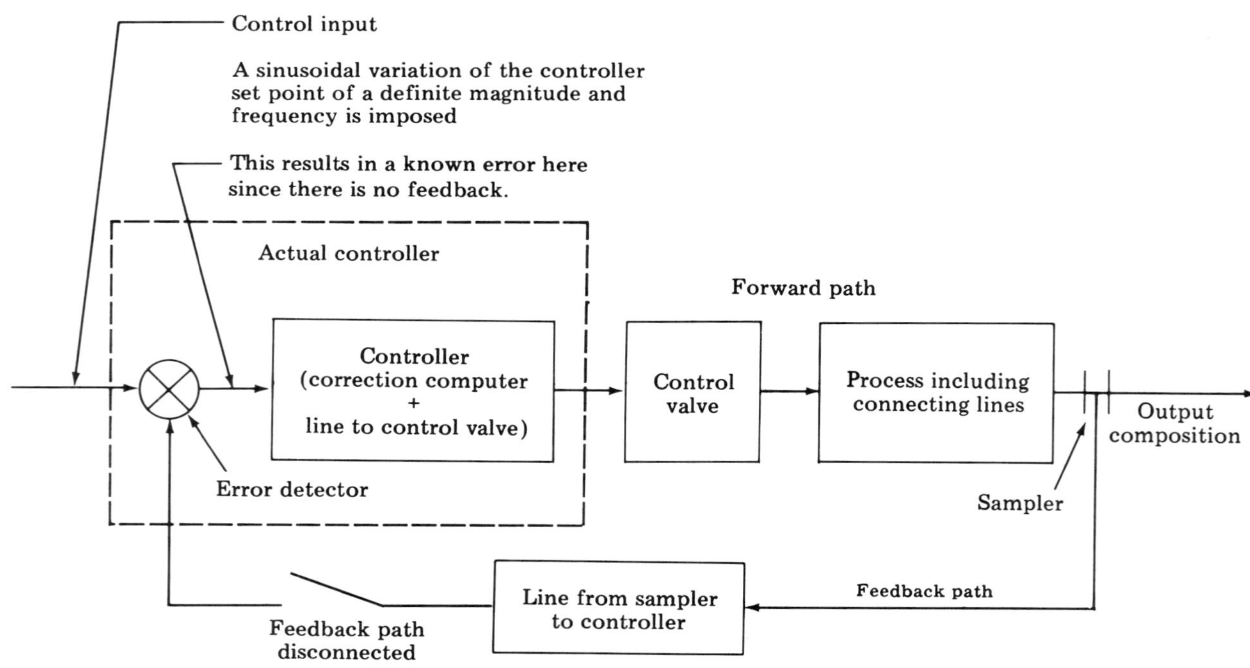

Figure 11. Scheme for obtaining the open-loop frequency response of the example of Figure 2.

equations must be used as the general equations of the process to obtain the desired mathematical solution.

Methods. The information obtained from the transient and frequency response experiments or from mathematical modelling is used to design the best control system for the process. The Nyquist method (13) and the Bode method (14) are the two methods most commonly used for plotting open-loop frequency response data in order to determine which controller to use on a process and to check the response and stability of the resulting process-controller complex. The Nyquist, or polar, diagram is the locus resulting from plotting the magnitude ratios obtained for each of a complete spectrum of applied frequencies versus the corresponding phase shifts obtained for each of the same frequencies. Two frequency-response curves are shown in Figure 12. Each is plotted for three values of the applied frequency ω where $\omega_2 > \omega_1 > \omega$. The phase shift δ is plotted clockwise and, at each frequency, the magnitude ratio M is measured along the radius. The criterion for stability is that, as one proceeds along the curve in the direction of increasing frequency, the point, $M = 1$, $\delta = 180°$, must lie to the left of the curve. Thus, curve A represents an unstable system, curve B a stable system.

The Bode or log-modulus plot consists of two graphs. The logarithm of the magnitude ratio is plotted on the ordinate versus the logarithm of the applied frequency on the abscissa. The phase shift is plotted in degrees against the logarithm of the applied frequency. To facilitate comparison, these graphs usually are plotted on the same sheet of paper with a common abscissa, as shown in Figure 13. The stability criterion is that the magnitude ratio curve must cross the zero line at a lower frequency than the phase-shift curve crosses the −180° line. Thus, Figure 13 represents a stable system.

In addition to the Nyquist and Bode plots, which are used almost exclusively for frequency-response studies, there is a third plotting method which has applications

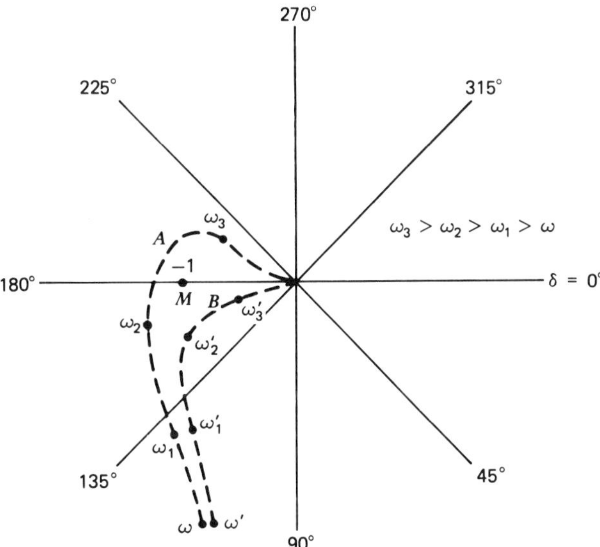

Figure 12. Nyquist plot of frequency-response curves. A, unstable. B, stable.

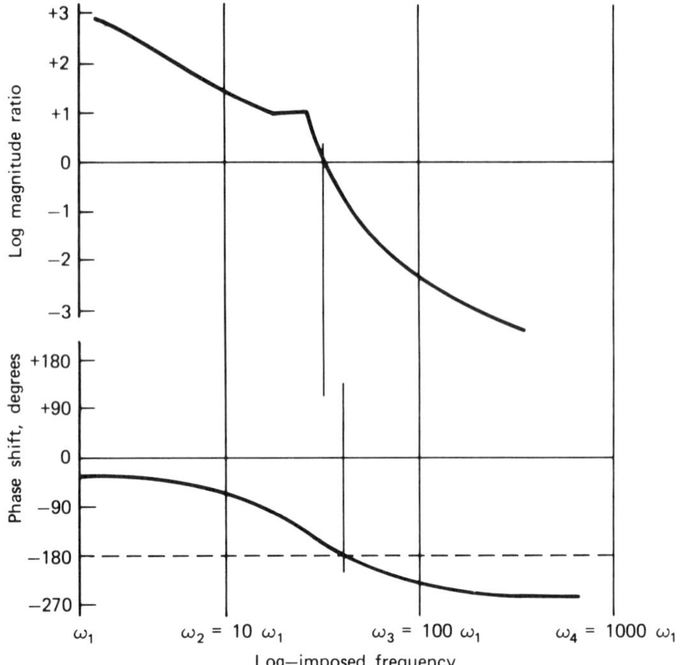

Figure 13. Stability on the Bode plot.

in both transient and frequency-response studies, eg, the Evans root-locus method (34). The roots of the characteristic equation of the open-loop system are plotted on a complex plane. For any set of roots, there is a curve or locus on the diagram which defines all the possible roots of the corresponding closed-loop equation. Each of the points plotted earlier is a point on the new locus and helps to establish the locus on the diagram. Stability determinations can be made as shown in Figure 14. There are graphic techniques that allow the closed-loop transient response and the frequency response to be determined directly from the diagram (34).

Each of the above graphs has the property that frequency-response results can be plotted directly from the differential equation without achieving a complete solution as is required for transient response. The same form of a linear differential equation always gives a graph of the same appearance regardless of the system whose equation is being graphed.

An added factor in favor of the plotting methods is the principle of superposition, which states that the response of a complex system is a product of the responses of those simpler elements from which the complex system can be considered to be made. As long as superposition can be considered as holding, ie, as long as the assumption of a linear system is valid, the responses of very complex systems are obtained easily. This factor is responsible for the tremendous popularity of the frequency-response method in the aircraft and electronic industries where the assumption of linear differential equations can be used extensively. Most of the controller functions, both theoretical and actual, lend themselves to representation on these graphs. Thus, the complete process system and the control elements can be evaluated on paper by these methods if the elements can be considered linear.

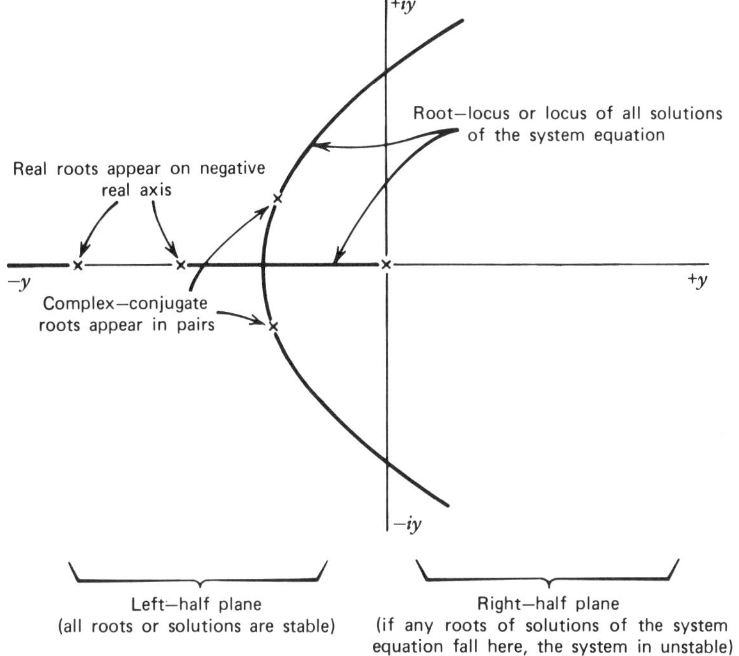

Figure 14. The Evans root-locus plot method.

If experimental data must be obtained, the frequency-response method does not lend itself well to evaluating industrial systems, since it requires a great deal of data at a wide range of frequencies to define a system's response. For an experimental study, transient response is more satisfactory, if care is taken in the application of severe inputs. Several methods have been developed to permit the calculation of frequency-response data from transient-response solutions.

General Procedure. The task of the control engineer is first to formulate the problem by drawing the block diagram of the system, including postulated control elements. If he or she can specify the differential equations of each element and if the equations are linear, he or she can proceed by one of the graphic methods of frequency response discussed above. Several trial control arrangements can be included in the system graphs. From this procedure, the best control arrangement and optimum controller settings are readily chosen (35–38).

If the equations are nonlinear, then a transient response, which probably must be obtained by using an analogue or digital computer, is needed. If the constants of some of the elements are not known, experimental tests may be the only method available to evaluate the system responses (39). In the extreme case, if the differential equations of the system cannot be written, a completely experimental approach, either through transient or frequency response, must be taken. An experimentally determined frequency-response graph for part of a system can be combined with the theoretically determined graphs for the other elements of the system and used in the same way as if the complete graph were obtained theoretically (40).

The use of analogue or digital computers through simulation is a powerful tool for design work on large-scale systems as is evidenced in its success in the aircraft and electronic fields. It is particularly important where nonlinearities occur in the equations of the system. Simulation is the exercise of the mathematical model of a physical process or business operation to determine how the process or operation will respond under the particular set of operating parameters used in the mathematical model. In the case of design, the effectiveness of a set of design parameters in the overall control of the plant or process under study is determined. By using a large number of simulation cases carried out with a wide variation of the parameters being studied, the optimum operating conditions and, thus, the optimum design parameters of the plant or process study can be found.

Overall Plant and Company Control Systems

Automatic control of a large, modern, industrial plant, whether achieved by a computer-based system or by conventional means, involves an extensive system for the automatic monitoring of numerous different variables operating under a very wide range of process dynamics. It requires the development of a large number of complex, usually nonlinear, relationships for the translation of the plant variable values into the required control correction commands. These control corrections must be transmitted to another large set of widely scattered actuation mechanisms of various types which, because of the nature of the manufacturing processes, may involve the direction of great energy expenditures. Plant personnel, both operating and management, must be kept aware of the current status of the plant and of each of its processes.

In addition, the industrial plant has the continual problem of adjusting its production to match its customers' needs—as expressed by the new-order stream being received—while maintaining a high plant productivity and the lowest practical production costs. This is handled in most cases through a manual, although computer-aided, production control system along with an in-process and finished-goods inventory.

It has been shown repeatedly that one of the major benefits of digital computer control systems in industrial plants has been in the role of a control-systems enforcer. In this mode, the computer's task is to assure continually that the control system equipment is carrying out the job that it was designed to do, so as to keep the units of the plant production system operating at some optimal level. Often the tasks carried out by these control systems are ones that a skilled and attentive operator could do. The difference is the degree of attentiveness that can be achieved over the long run.

As the overall requirements, both energy and productivity based, become more complex, more complex and capable control systems are necessary. To achieve these, the field will gravitate more and more toward digital computer-based systems to carry out the needed work. In order to obtain highly advanced control responses, an overall system with the following capabilities is needed:

(*1*) A tight control of each operating unit of the plant to assure that it is operating at its maximum efficiency of energy utilization and/or production capability based upon the production level set by the scheduling and coordination functions listed below. This control reacts directly to any emergencies that may occur in its own unit.

(*2*) A coordination system that determines and sets the production level of all units working together between inventory locations. This system assures that no unit exceeds the general area level and thus does not use excess energy or raw materials. The coordination system responds to the existence of emergencies or upsets in any of the units under its control by shutting down or systematically reducing the output in these and related units.

(3) A system capable of carrying out the scheduling for the plant from customer orders or management decision so as to produce the required products for these orders at the optimum combination of time, energy, and raw materials, ie, cost functions.

Because of the widening scope of authority of each of these requirements, they effectively become the distinct and separate levels of a superimposed control structure. In view of the amount of information that must be passed among the three levels of control, it appears that a distributed computational capability that is organized in a hierarchical fashion is necessary, eg, as in Figure 5.

Productivity Gains from Computer Control. The gains that computer control give fall into two categories: those for which an economic factor can be applied and those for which an economic factor can be determined only with difficulty or not at all. The former are called tangible benefits and include increased plant throughput or production, increased product quality, decreased energy requirements, decreased raw material requirements, decreased effluents and pollutants, manpower savings, reduced variability of product, less product giveaway, increased yields or improved product mix, and reduced rework. Examples of the second category, ie, the intangible benefits are: increased plant and personnel safety; better and more timely production, engineering, and management information; reduced maintenance requirements; flexibility of system in respect to changes or additions; increased customer service through faster response and better order monitoring; better response to emergencies; warning of impending catastrophic conditions in processing equipment; and better maintenance scheduling. Many of the latter are the most attractive to the process control engineer.

Types of computer systems are often divided into two classifications: control enforcement and control computation on the one hand and production scheduling and coordination on the other (41). The vast majority of applications have been in the former category. The latter includes the function of management information and communication with the company's staff functions, eg, accounting, purchasing, maintenance, etc. The first category generally covers the first three levels of the hierarchy illustrated in Figure 5 and the second involves the upper three levels with some overlap in the control levels.

Control Enforcement. Control enforcement is the determination of the best control strategy for a particular processing plant or piece of processing equipment under immediate conditions and the assurance that that control strategy is carried out. Both the determination of the strategy and its enforcement are vital.

Continuous-Process Plants. Continuous-production plants (generally flow-process plants) spend a major part of their operation in a level or steady-state condition. The gains that are made with computer control in these plants result from determining the bottleneck unit and controlling it to run at its production constraints at all times; and set the operating levels of all other units or elements to match the bottlenecked unit or process element, then control them to use minimum raw material and/or minimum energy, etc, while maintaining the required production level.

Intermittent or Batch-Process Plants. Control systems for intermittent or batch-process plants achieve their greatest gains in productivity by minimizing the processing time of a particular operation and by minimizing excursions from an optimal or best process path during the operating time. This does not necessarily mean the application of optimal path or minimum time in the modern control sense (although these can and will have great application). It may be a simple requirement such as attentive monitoring of process temperature or composition and halting the reaction at the correct point rather than permitting the reaction to continue with corresponding production losses from both reduced yield and increased processing time.

Mixed Cases. Many plants are mixtures of continuous and intermittent or batch units. In these cases, the continuous units run at their steady-state optimization goals, whereas the batch or intermittent units run at their dynamic control goals. The coordination of a mixed system must fall upon the production scheduling unit through its manipulation of plant in-process inventories and its setting of production levels for each unit.

Likewise, the operating period of a continuous plant includes periods when dynamic control is necessary, eg, during start-up and shut-down. In either case, those units operating continuously are treated as would the units of continuous plants and *vice versa.* The dynamic portions of the operation of continuous units should be handled as though they were dynamic units (ie, intermittent or batch).

Production Scheduling. On-line production scheduling is a critical factor only for the intermittently operated or batch plant in terms of increase in production. The continuous plant can be scheduled off-line for a considerable period in advance except for production scheduling for customer service.

Some basic assumptions involved in the productivity gains of computer control and, particularly, the control capability of analogue and manual systems follow.

(*1*) Results must be compared with the best of current practice that does not involve on-line computer systems, ie, with the best of conventional analogue control practice plus off-line computer computation as is appropriate for operator guides.

(*2*) Analogue control is mainly single-variable (one input–one output) control with some two-variable loops (cascade and ratio control) with a minimum of larger (3–4 variable) systems, eg, noninteracting, multivariable control.

(*3*) The human operator can coordinate 6–8 variables as a group if they are moving slowly. Because the operator's attention span is limited, he or she must concentrate on a smaller number (4–5) variables if they are moving rapidly.

(*4*) Control must not be considered as a substitute for less expensive engineering changes that may debottleneck a process plant (42).

Proposed rules for the degree of productivity and related gains that are readily possible in most industries from the several levels of computer control application incorporated in the hierarchy are included in Table 1. Increased production is presented as percentages of increased production (assuming no raw material or energy limitations) for the particular unit, plant, etc, under computer control. Energy savings as efficiency gains and as related to production levels are presented below.

Control Enforcement. (*1*) Control enforcement practices (ie, increasing plant energy efficiency) have the potential for saving ca 5% of the energy used with the conventionally controlled plant at the existing production level irrespective of plant sales. This percentage is independent of production level.

(*2*) Savings that are related to reduction of raw materials through better chemical

Table 1. Some Probable Gains in Production Level From Computer Control for the Sold-Out Plant

Category	Gains (%)
Control system selection	
continuous (overall steady-state optimization for increased production)	5
intermittent or batch (determination of optimum-process paths and process end points)	10
monitoring function as applied to both (alarms, warning of dangerous situations, maintenance prediction, etc)	0.5
Production scheduling	
continuous (production level)	0
intermittent or batch (coordination of cycles and better utilization of inventory, backup schedules, etc)	3

yields, reduced rework, etc, carry a correspondingly sized additional energy savings, ie, the productivity gains in Table 1 can be translated to energy savings in addition to item (*1*) above.

(*3*) Power-demand control is not an energy-saving system but a method of capitalizing on the working of utilities contracts.

Production Scheduling. Production scheduling can have an enormous effect upon the use of available waste gases, intermittent supplies, etc, as alternatives to purchased utilities. Likewise, the coordination of plant utility schedules with plant production schedules avoids the overproduction of steam, electricity, plant gases, etc, with corresponding wastage. The savings should amount to at least an additional 5% to that given in Table 1.

The productivity gain numbers are translatable to dollars by considering that plant cost in many industries is approximately the same in value as the selling price of a year's production of goods. Manpower savings can be considered as independent of any productivity gains, eg, increased production, reduced energy, etc, and can be taken as additions to the values related to the latter gains.

There are several excellent articles on techniques and procedures for determining potential gains (19,39,43–49) and for carrying them out (42,45,50–51). The gains from the control enforcement of fluid-process production units are covered (52–63), and the gains in various processes operating cyclically or intermittantly and those in a dynamic change from one operating level to another are treated (26,39,57,64–73). Optimization in production scheduling is not treated extensively because of its novelty in the on-line computer control field (74–78). The potential for computer control in energy conservation is discussed in refs. 79–86.

Operations Research. Operations research (qv) is a set mathematical technique (see Energy management) used to evaluate the best of several possible procedures in a given situation. It is useful for problems, eg, optimum production scheduling and shipping routing, where numerous possibilities may occur. If the possible choices involved in a particular problem can be expressed as a set of linear simultaneous algebraic equations that show the costs to be borne for each choice, the resulting set of simultaneous equations can be solved directly by the methods of matrix algebra to give the solution that will show the highest profit or that will give the least cost per item, or any other such criterion that may be desired as the basis of choice. This method is called linear programming (86) and is the basis for much of the steady-state optimization practiced in industry. Because of its ease of use and its resulting popularity, every effort is made by industry problem solvers to linearize the mathematical expression of their problems to permit the use of linear programming. Where linear programming is not possible, the methods of nonlinear programming or of simulation must be used. Nonlinear programming is the collection of a set of special techniques for solving special cases of nonlinearities (87–88).

For the large fraction of cases where neither linear nor nonlinear programming methods apply, no direct solution method is possible and the use of simulation techniques is necessary. Thus, by using a large number of simulation cases that are carried out under a wide range of variations of these operating parameters, simulation can be used to obtain the optimum operating conditions for the process under study. However, to avoid having to do almost an unlimited number of cases, some search method for choosing trial values of the parameters is necessary or a statistical technique must be used (89–90). Another method is to use the mathematical technique, the Monte Carlo trial and error procedure, to try various possible solutions to the family of equations. The results then are evaluated statistically to determine the trends in the resulting solutions in order to reach the optimum solution. Where the number of combinations of variables to be tried is moderate, search methods are recommended. Where the resulting combination is very large, statistical techniques must be used because the dimensionality involved becomes too complex for the search techniques.

These mathematical methods can be used for any situation where multiple choices of procedure are possible and where the resulting operations can be expressed mathematically, in order to pick that set of choices of procedures that will give the most profit, least cost, maximum production, or other optimum solution of the oper-

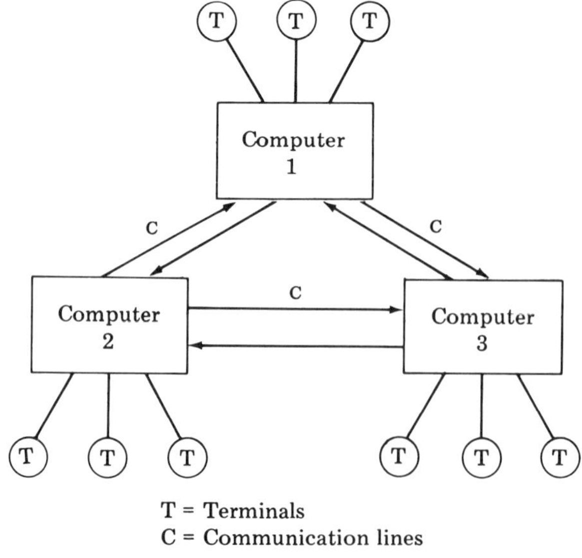

Figure 15. A fully connected distributed information network (3-node example).

ating criterion chosen by management for that particular case. Although business and manufacturing examples are used in this description, the techniques apply equally well to similar problems that exist in agriculture, health services, office functions, transportation—essentially any area of human endeavor.

Data-Base Management. In any hierarchy control system, there are a multitude of separate files that either can be stored in a single computer or distributed in the other computers in the system (91). The former is a central data-base system and the latter is a distributed data-base system. In a distributed data-base system, as shown in Figure 15, the files stored in one computer must be accessible by the other computers. An objective of a distributed system might be reduction in access delay by allowing multiple-access paths to the files. The total set of files stored in the system forms a distributed data base, and the computers and communication lines in the system are referred to as nodes and links, respectively. The study of the use of such file systems is called data-base management (92–94) (see Information retrieval).

Expectations

There are many possible paths that the development of control systems can take in the future, and in many cases, the paths appear clear:

(*1*) A major concern of all installations of computer control systems has been the reliability of the overall system. The computer involvement in overall plant control systems in contrast to the multiple single loops of the analogue system make the former more vulnerable. However, this will be countered by the use of multiple redundant-computer systems in place of single computers and of fault-tolerant and fault-remedying technique whenever possible.

(*2*) Computer programming complexity is the major bottleneck in computer control system development. Company proprietary considerations, both of the user and the vendor, will delay the development of remedies to this problem but they should occur because of the extreme economic incentives involved.

(*3*) Electronic techniques have the potential—however, unexploited—of drastically reducing the cost of industrial control systems. However, the needs of vendor companies for developmental capital and a lack of incentive on the part of user companies because of the relatively small percentage that instrumentation and control comprises of the total plant cost, probably will prevent such cost reductions. The vendor may choose to compensate for lowered electronic-component cost with a greater system complexity to increase reliability or to decrease programming costs.

(*4*) Hierarchical and distributed computer systems are the wave of the future. Their acceptance is driven by the low cost and relatively low capability of the microprocessor. At the same time, they are promoting a modularity and simplification of systems programming, which should be retained. Likewise, they tend to promote reliability by subdividing system tasks among several computers and thus lessening the impact of the failure of any one.

BIBLIOGRAPHY

"Instrumentation" in *ECT* 1st ed., Vol. 7, pp. 908–926, by John Procopi, Minneapolis-Honeywell Regulator Company; "Instrumentation" in *ECT* 2nd ed., Vol. 11, pp. 739–774, by Nicholas A. Fiorino, Honeywell, Inc.

1. D. M. Considine, ed., *Process Instruments and Controls Handbook,* 2nd ed., McGraw-Hill Book Company, New York, 1974.
2. G. C. Carroll, *Industrial Process Measuring Instruments,* McGraw-Hill Book Company, New York, 1962.
3. G. W. Ewing, *Instrumental Methods of Chemical Analysis,* McGraw-Hill Book Company, New York, 1954.
4. H. Von Koch and G. Ljungberg, *Instruments and Measurements,* Vol. 1, Academic Press, Inc., New York, 1961.
5. F. H. Raven, *Automatic Control Engineering,* 2nd ed., McGraw-Hill Book Company, New York, 1968.
6. J. Jusik, *Control Eng.* **10,** 61 (Aug. 1965).
7. W. N. Patterson "Intercabling Practices" in *Process Computer Data Book,* General Electric Company, Phoenix, Ariz. (Nov. 8, 1963).
8. J. B. Cox, L. J. Hellums, T. J. Williams, R. S. Banks, and G. J. Kirk, Jr., *ISA J.* **13,** 65 (Oct. 1966).
9. C. S. Beard, *Final Control Elements, Valves and Actuators,* Rembach Publications Division, Chilton and Co., Philadelphia, Pa., 1969.
10. J. C. Maxwell, *Proc. Roy. Soc. London* **16,** 270 (1868).
11. E. J. Routh, *Dynamics of a System of Rigid Bodies,* 3rd ed., Macmillan, London, 1877.
12. N. Minorsky, *J. Am. Soc. Nav. Eng.* **42,** 280 (1922).
13. H. Nyquist, *Bell Syst. Tech. J.* **11,** 126 (Jan. 1932).
14. H. W. Bode, *Bell Syst. Tech. J.* **19,** 421 (July 1940).
15. H. S. Black, *Bell Syst. Tech. J.* **13,** 1 (Jan. 1934).
16. H. L. Hazen, *J. Franklin. Inst.* **208,** 279 (1934).
17. H. M. Paynter, ed., *A Palimpsest of the Electronic Analog Art,* Geo. A. Phillrick Researchers, Inc., Boston, Mass., 1955.
18. M. V. Long and E. G. Holzmann, *Trans. ASME* **75,** 1373 (Oct. 1953).
19. E. S. Savas, *Computer Control of Industrial Processes,* McGraw-Hill Book Company, New York, 1965.
20. T. J. Williams and F. M. Ryan, *Progress in Direct Digital Control,* Instrument Society of America, Pittsburgh, Pa., 1969.
21. T. J. Williams, *Proceedings of the Seventh Triennial Congress of the International Federation of Automatic Control,* Pergamon Press, London, 1978, pp. 1393–1406.
22. V. W. Eveleigh, *Adaptive Control and Optimization Techniques,* McGraw-Hill Book Company, New York, 1967.
23. G. N. Saridis, *Self Organizing Control of Stoichiastic Systems,* Marcel Dekker, Inc., New York, 1977.
24. R. H. Marcus and J. O. Hougen, *Chem. Eng. Prog. Symp. Ser.* **59,** 70 (1963).
25. R. M. Bakke, *paper 3.2-1-64 presented at ISA Annual Conference and Exhibit,* Oct. 1964.
26. T. J. Williams, *DECHEMA Monogr.* **67,** 1222 (1955).
27. L. S. Pontryagin, V. G. Boltyanskii, R. V. Gamrelidze, and E. F. Mischenko, *The Mathematical Theory of Optimal Processes,* Wiley-Interscience, New York, 1962.
28. M. Athans and P. L. Falb, *Optimal Control,* McGraw-Hill Book Company, New York, 1966.
29. H. C. Lim, *Ind. Eng. Chem. Process. Des. Dev.* **8,** 334 (July 1969).
30. R. S. Novack and T. J. Williams, *Seminaries IRIA, Temps Reel, 1971,* IRIA, LeChesnay, Fr., 1974, pp. 177–269.
31. P. S. Buckley, *Techniques of Process Control,* John Wiley & Sons, Inc., New York, 1964.
32. F. A. Shinskey, *Process Control Systems,* McGraw-Hill Book Company, New York, 1967.
33. T. J. Williams and V. A. Lauher, *Automatic Control of Chemical and Petroleum Processes,* Gulf Publishing Company, Houston, Tex., 1961.
34. W. R. Evans, *Control System Dynamics,* McGraw-Hill Book Company, New York, 1954.
35. G. S. Brown and D. P. Campbell, *Principles of Servomechanisms,* John Wiley & Sons, Inc., New York, 1948.
36. D. P. Campbell, *Process Dynamics,* John Wiley & Sons, Inc., New York, 1958.
37. H. Chestnut and R. W. Mayer, *Servomechanism and Regulator System Design,* Vol. 1, John Wiley & Sons, Inc., New York, 1951.
38. J. G. Truxal, *Automatic Feedback Control Synthesis,* McGraw-Hill Book Company, Inc., New York, 1955.

39. T. J. Harrison, *Handbook of Industrial Control Computers,* John Wiley & Sons, Inc., New York, 1972.
40. R. C. Oldenbourg and H. Sartorius, *The Dynamics of Automatic Controls,* The American Society of Mechanical Engineers, New York, 1948.
41. T. J. Williams, *Computer Control and Its Effect on Industrial Productivity,* A. I. Johnson Memorial Lecture, Canadian Conference on Automatic Control, Montreal, Can., May 24, 1979.
42. B. C. Horn, *Chem. Eng. Prog.* **74,** 77 (June 1978).
43. C. E. Geisel, *TAPPI* **61,** 33 (Aug. 1978).
44. J. Hannigan, *TAPPI* **61,** 779 (Aug. 1978).
45. E. J. Kompass and T. J. Williams, eds., *On-Line Optimization Techniques in Industrial Control, Proceedings of the Fifth Annual Advanced Control Conference,* Technical Publishing Company, Chicago, Ill., 1979.
46. T. M. Stout, *Control Eng.* **13,** 87 (Sept. 1966).
47. T. M. Stout, *Instrum. Technol.* **16,** 56 (June 1969).
48. T. M. Stout and R. P. Cline, *Instrum. Technol.* **23,** 51 (Sept. 1976).
49. T. M. Stout in T. J. Harrison, ed., *MiniComputers in Industrial Control,* Instrument Society of America, Pittsburgh, Pa., 1978.
50. R. E. Griffith and R. A. Stewart, *Manage. Sci.* **7,** 379 (1961).
51. P. R. Latour, *Instrum. Technol.* **25,** 67 (July 1978).
52. R. A. Baxley, *Instrum. Technol.* **16,** 75 (Oct. 1969).
53. R. A. Baxley, *5th IFAC/IFIP International Conference on Digital Computer Applications to Process Control,* The Hague, Netherlands, June 1977.
54. R. D. Eisenhardt and T. J. Williams, *Control Eng.* **7,** 103 (Nov. 1970).
55. T. Q. Eliot and D. R. Longmire, *Chem. Eng.* **69,** 99 (Jan. 8, 1962).
56. W. J. Fiero and P. E. Kelly, *Hydrocarbon. Process.* **56,** 117 (Sept. 1977).
57. A. G. Goossens, M. Dente, and E. Ranzi, *Hydrocarbon Process.* **57,** 227 (Sept. 1978).
58. J. Gutzon and J. Sagures, *Brit. Chem. Eng.* **7,** 432 (June 1962).
59. F. P. Larmon, J. Van Reusel, and L. Viville, *5th IFAC/IFIP International Conference on Digital Computer Applications to Process Control,* The Hague, Netherlands, June 1977.
60. C. G. Laspe, *Oil Gas. J.* **66,** (Dec. 23, 1968).
61. C. G. Laspe, *Instrum. Technol.* **25,** 51 (May 1978).
62. R. J. LaSpisa, G. Stacy, and H. M. Sochan, *Oil Gas J.* **76,** 117 (June 19, 1978).
63. P. R. Latour, "Use of Steady State Optimization for Computer Control in the Process Industries," in Ref. 45.
64. *Digital Computer Control of an ICI Plant at Fleetwood,* Argus Project Report No. 1, Ferranti, Ltd., Wytheshawe, Manchester, Eng.
65. J. G. Balchen, "The Applicability of Modern Control Theory Related to Dynamic Optimization in Industry Today," in Ref. 45.
66. U. Borisson and R. Syding, *Automatica* **12,** 1 (1976).
67. L. Eriksson, *Pulp Pap. Mag. Can.* **71,** 55 (1970).
68. L. Eriksson, *Sven. Papperstidn.* **80**(18), (1977).
69. M. Fjeld, *Automatica* **14,** 107 (1978).
70. J. W. Gee and R. E. Chamberlain, *5th IFAC/IFIP International Conference on Digital Computer Applications to Process Control,* The Hague, Netherlands, June 1977.
71. A. L. Guisti, R. E. Otto, and T. J. Williams, *Control Eng.* **9,** 104 (June 1962).
72. P. R. Latour, L. B. Koppel, and D. R. Coughanowr, *IEC Proc. Des. Dev.* **6,** 452 (1967).
73. R. L. Mahra and C. H. Wells, *IFAC/IFIP Conf. on Dig. Comp. Appl. to Process Control,* ISA Publ., Helsinki, 1971.
74. R. L. Curtis in T. J. Harrison, ed., *Minicomputers in Industrial Control,* Instrument Society of America, Pittsburgh, Pa., 1978, pp. 225–278.
75. E. J. Kompass and T. J. Williams, eds., *Hierarchical and Distributed Computer Control, Proceedings of the Third Advanced Control Conference,* Dun-Donnelly Publishing Company, Chicago, Ill., 1976.
76. K. H. Lee and A. J. Koivo, *Dynamic Scheduling of Ingot Processing in Steel Production, Report Number 104,* Purdue Laboratory for Applied Industrial Control, Purdue University, West Lafayette, Ind., May 1979.

77. G. T. Mackulak and C. L. Moodie, *A Production Control Strategy for Hierarchical Multi-Objective Scheduling with Specific Application to Steel Manufacture, Report Number 112,* Purdue Laboratory for Applied Industrial Control, Purdue University, West Lafayette, Ind., May 1979.
78. L. L. Thompson, *The Business Case for Optimized Operation of a Pulp and Paper Mill With a Hierarchical Computer System,* International Business Machine Company, White Plains, N.Y.
79. T. E. Dy Liacco, *5th IFAC/IFIP International Conference on Digital Computer Applications to Process Control,* The Hague, Netherlands, June 1977.
80. C. G. Laspe, F. B. Smith, and R. A. Krall, *ISA paper 76-524, presented at ISA 76 International Conference and Exhibit,* Houston, Tex., Oct. 11–14, 1976.
81. N. Leffler and M. S. Shigemura, *1978 IFAC Conference,* Helsinki, June 1978.
82. N. Leffler, "Optimization of Cogeneration," *TAPPI Engineering Conference,* San Francisco, Calif., Sept. 1978.
83. F. G. Shinsky, *Energy Conservation Through Control,* Academic Press, Inc., New York, 1978, p. 79.
84. A. J. Tysso and C. Brembo, *Automatica* **14,** 213 (1978).
85. T. J. Williams, *paper presented at the National Computer Conference, Dallas, Tex., June 1977; Control Eng.* **24,** 71 (Sept. 1977); *Oil Gas J.* **75,** 83 (Dec. 1977).
86. G. H. Symonds, *Linear Programming: The Solution of Refinery Problems,* Esso Standard Oil Company, New York, 1955.
87. D. J. Wilde, *Optimum Seeking Methods,* Prentice Hall, Inc., Englewood Cliffs, N.J., 1964.
88. D. L. Magasarian, *Nonlinear Programming,* McGraw-Hill Book Company, New York, 1969.
89. R. Hooke and T. A. Jeeves, *J. Assoc. Comput. Mach.* **8,** 212 (1961).
90. M. J. D. Powell, *Comput. J.* **7,** 155 (1964).
91. D. S. Surh and M. P. Deisenroth, *Allocation of Files in a Hierarchical Information System for Industrial Control and Monitoring, Report Number 103,* Purdue Laboratory for Applied Industrial Control, Purdue University, West Lafayette, Ind., Mar. 1979.
92. G. Booth, *International Congress on Computers and Communication,* 1972, pp. 371–376.
93. W. Chu, *IEEE Trans. Comput.* **C-18,** 885 (Oct. 1969).
94. W. Chu and G. Ohlmacher, *Proceedings of ACM Conference,* 1974.

THEODORE J. WILLIAMS
Purdue University

INSULATION, ACOUSTIC

Acoustic insulation materials control sound and include four categories of ma terial: sound absorbent, sound reflective, damping, and vibration isolation materials.

Sound-Absorbent Materials

Porous materials are the most common materials used to provide sound absorption. As sound passes into the intercommunicating pores, air moves back and forth rapidly within the material, and sound energy is converted into heat by frictional and viscous forces at the air–material interface. Most porous sound-absorbing materials are applied to walls or ceilings in living or office spaces (architectural treatment) and in unoccupied structures, eg, machinery housings (acoustic lining) to control reflected sounds. In both cases, the amount of acoustic energy dissipated by a sound-absorbent material is a function of the physical properties of the material, the kind of sound field that the material is immersed in, and the physical properties and positioning of nearby reflecting surfaces. Special standard tests are used to measure the performance of sound-absorbent materials or products intended as architectural treatments or machine linings.

Porous materials also are used to reduce sound transmission. Generally, porous materials serve in a supplementary way in this application, because lightweight porous materials do not, by themselves, reduce sound transmission effectively. A different set of standard tests is used for evaluating products of this type.

Absorption Coefficients. The fundamental unit for describing sound-absorbing performance in both architectural and acoustic lining treatments is the absorption coefficient α which indicates the percentage of incident sound energy dissipated upon reflection during specific test conditions. An absorption coefficient of one implies that no sound energy is reflected under the test conditions; whereas a coefficient of zero implies that incident sound energy is reflected completely.

Acoustic performance is always dependent on the frequency of the sounds involved. Therefore, a series of absorption coefficients is needed to portray performance adequately. Normally, for a given material or product, six coefficients are measured—one for each of the six frequency bands (octave bands) from 125–4000 Hz. Table 1 provides the acoustic performance of some common building materials as determined under reverberation-room testing. Different absorption coefficients result from different kinds of tests; thus, the test results are meaningful only if the test method is specified also. The American National Standards Institute (ANSI) recognizes two standard tests (2–3): the impedance-tube method and the reverberation-room method. The former is the simplest and requires an apparatus consisting of a tube with a test specimen at one end and a loudspeaker at the other. A probe microphone is placed inside the tube and between the loudspeaker and the specimen. Sound that is emitted from the loudspeaker propagates toward the specimen and is reflected; the intensity of the reflection depends on how well the test material absorbs sound. A standing-wave pattern develops inside the tube, and the probe microphone establishes the nature of the pattern. The absorption coefficient is related directly to the standing-wave pattern through:

$$\alpha_N = 1 - \left[\frac{\log_{10}^{-1}(L/20) - 1}{\log_{10}^{1}(L/20) + 1}\right]^2$$

Table 1. Acoustic Performance of Selected Common Building Materials[a]

Material	Random-incidence sound absorption coefficients, α[b]					
	Octave band center frequency, Hz					
	125	250	500	1000	2000	4000
brick, unglazed	0.03	0.03	0.03	0.04	0.05	0.07
concrete block, coarse	0.36	0.44	0.31	0.29	0.39	0.25
concrete block, painted	0.10	0.05	0.06	0.07	0.09	0.08
concrete floor	0.01	0.01	0.02	0.02	0.02	0.02
gypsum board, 1.3 cm thick nailed to 5 × 10-cm studs, 41 cm on center	0.29	0.10	0.05	0.04	0.07	0.09

[a] Adapted from ref. 1.
[b] Determined by reverberation-room testing.

where the quantity L is the standing-wave ratio or difference, in decibels (dB) between the maximum sound pressure and minimum sound pressure of the standing wave. The subscript of α_N indicates that the absorption coefficient is determined only for sound incident in one direction: perpendicular (normal) to the absorbing material.

In most acoustic problems, the angle of incidence of sound is not restricted to any angle. Indoors, eg, sound travels in many directions and angles and can approach a condition of random incidence, ie, where an equal likelihood exists for incidence in any direction. The absorption coefficients for randomly incident sound are likely to be significantly greater than corresponding normal-incidence sound-absorption coefficients. Thus, impedance-tube performance measurements are used mainly for basic research and for special applications, such as those where sound is incident only in a perpendicular direction. The second method for measuring absorption coefficients is the reverberation-room method which generally is preferred over impedance-tube testing for design work. Measurements of the rate that sound decays in a room are made in order to establish a value for the room absorption A through the following relationship:

$$A = 0.9210\ Vd/c$$

where A is the amount of room absorption, m^2; V is the room volume, m^3; d is the rate of decay, dB/s; and c is the speed of sound, m/s. By subtracting the amount of absorption occurring in the empty room from the amount of absorption with the test specimen present, a value for the absorption provided by the test specimen is established. The coefficient for the test material then is determined:

$$\alpha_R = A_{test\ material}/S$$

where S is the surface area of test specimen, m^2. The subscript in α_R indicates that the procedure provides a random incidence value for absorption.

In the reverberation-room method, test patches are small, which ensures that the specimen does not significantly influence the diffuseness of the sound field. Some influence on the sound field is inevitable, however, and the net resultant test coefficients often exceed one. Some laboratories report values as measured, others round off coefficients that exceed 1 to 0.99, and still others apply a scaling factor to reduce all measured coefficients to 1 or below 1, with proportionately less reduction to lowest absorption coefficients. An especially important consideration in the test procedure is the mounting method used to support the acoustic material. Of the many standard mounts specified by ANSI, the three most commonly used in field installations are illustrated in Figure 1. Performance deviates from reported results if different mounting systems are used in an actual installation.

Other Units. The sabin is used to relate absorption to surface area. One metric sabin equals 1 m^2 of perfectly absorptive material ($\alpha = 1$). A 500-m^2 acoustic tile ceiling with an absorption coefficient of 0.70 at 500 Hz equals 350 m^2 (500 × 0.70) of perfectly absorbent ceiling and, thus, provides 350 sabins of absorption. The behavior of sound in a space can be predicted by how many sabins are present in that space.

The contribution to sound absorption of irregularly surfaced objects, eg, upholstered furniture, is assessed as A_{eq}—the number of equivalent sabins of absorption for each object. A_{eq} is determined by testing a grouping of the objects using the reverberation-room procedure (thereby determining the amount of absorption provided by the grouping), and then by dividing that amount by the number of objects evaluated. The amount of absorption provided is affected by how the objects are positioned relative to one another in the test.

Often, suppliers of acoustic products specify a noise reduction coefficient (NRC) for their material. The NRC is a shorthand performance indicator obtained by aver-

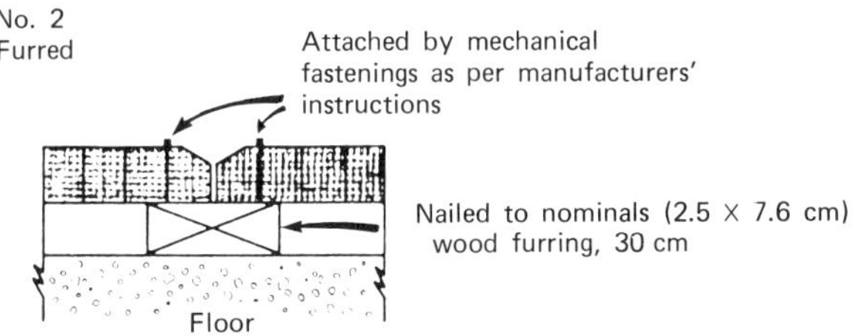

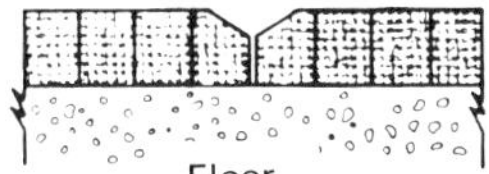

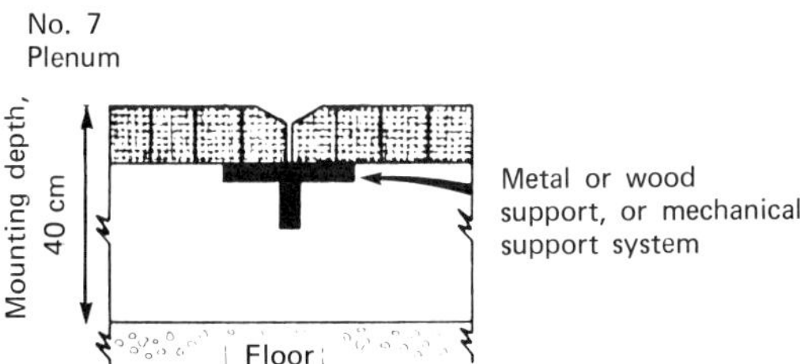

Figure 1. Three examples of standard mountings for sound-absorption tests (3).

aging absorption coefficients at 250, 500, 1000, and 2000 Hz and rounding the average to the nearest multiple of 0.05. Although the NRC is a simpler measure than a set of six absorption coefficients, its use is limited because materials with the same NRC can perform substantially differently in a particular octave band.

Another measure used for design purposes is the average absorption coefficient $\bar{\alpha}$; this measure is useful for describing the amount of absorption in a space that contains several kinds of absorbing materials:

$$\bar{\alpha} = \frac{S_1\alpha_1 + S_2\alpha_2 + \cdots S_n\alpha_n + A_1 + A_2 + \cdots A_n}{S}$$

where S is $S_1 + S_2 + \cdots S_n$; $S_1, S_2 \cdots S_n$ are the areas of the individual absorptive surfaces, m^2; $\alpha_1, \alpha_2 \cdots \alpha_n$ are the respective individual absorption coefficients of the absorptive surfaces, dimensionless; and $A_1, A_2 \cdots A_n$ are the number of metric sabins of absorption provided by the individual kinds of sound-absorptive objects, if any, in the space.

Noise Transmission Reduction. The acoustic performance of products that use porous sound-absorbing materials to help prevent the transmission of sound from one point to another is usually described in decibel ratings. Ratings of a product's acoustic performance include: *insertion loss*—the difference in the sound power level before and after installation of a product, at a reference position away from the product; *transmission loss*—the difference in the sound power level on two sides of a noise reduction element; and *attenuation*—the reduction in sound intensity between a sound source and a receiver location. Of these ratings, insertion loss is the most useful. The insertion loss of both acoustic ducts and prefabricated silencers is determined by a standard test procedure in which the change in the sound power delivered to a reverberation room is evaluated before and after a test specimen is inserted in a duct system that discharges into the room (4). Because both room volume and room absorption remain constant, the change in sound power is equivalent to the change in the sound pressure level. Tests can be conducted with or without air flow. The former is more meaningful, since performance varies according to both the direction and speed of air flow. With the exception of wall assemblies (see Materials), standard procedures do not exist for measuring the transmission loss or attenuation of other kinds of products that employ porous absorbing materials.

Materials. The characteristics that determine sound absorption are possessed by thousands of materials, including heavy woven fabrics, felts (qv), and fibrous thermal insulation materials (see Textiles; Insulation, thermal). Thus, many materials can serve as acoustic insulation; however, only a small portion of these materials are marketed intentionally for acoustical purposes.

Fibrous. The fibrous material that most often is associated with acoustic insulation is glass fiber (see Glass). Fibrous glass acoustic products have excellent thermal properties and are inexpensive. They have wide application in mufflers, wall insulation, and enclosure linings. The primary disadvantage of glass fiber is that the material tends to shred, settle, or erode in the presence of vibration or high velocity air flow. Therefore, acoustic materials are made with a wide variety of binders to hold the fibers together, and acoustic components often are faced with a low-flow resistance element to protect the absorbent substance from erosion. Raw glass fiber materials are processed into boards or blankets or bulk material. The bulk material is collected and cast into tiles or panels with tailored surface features or is used directly as fill material.

Rock wool (mineral fiber that is formed in a manner similar to that of glass fiber) is made from nonvirgin siliceous raw materials. It is used more often than glass fiber in acoustic applications in Europe. Refractory fibers (also formed in a manner similar to that of glass fibers) are available for high temperature acoustic applications (see Refractory fibers).

Felted products also are used as fibrous absorbent material. Wool (qv)—especially in combination with kapok or vegetable fibers—is mechanically bonded to form sheets that have sound-absorbing qualities (see Fibers, vegetable). Maximum dimensions are limited to a thickness of about 1 cm and a maximum density of about 0.07 g/cm^3. Felt often is used in surface treatments for machine enclosures, typewriters, and air conditioners. The material may be made with a flame retardant and combined with perforated vinyl plastic facings. Most felts are sold in thin, 0.5-cm sheets which perform well only at high frequencies.

Foamed. Plastic acoustic foams are made by two processes; both involve combining reactants that (simultaneously) produce a polymer (typically polyurethane) and generate a gas (see Urethane polymers). The gas expands the reacting mass and, eventually, the bubbles form contiguous polyhedrons. The foamed plastic can be used for acoustic material if the contact planes between the polyhedrons rupture and establish openings between the cells. In one process, a reacting mass is batch formulated, ie, allowed to form a large bun about 1-m thick that subsequently is cut into sheets. Pressure-sensitive adhesives, mass-loaded backings, or thin impervious plastic-sheet facings sometimes are applied after manufacture. In another process, the foam product is continuously cast and formed into a final thickness, and various substrates are applied as the foam is formed.

Other. Other materials that are used in porous acoustic products include wood and metal fibers and sintered metals. Wood (qv) fibers often are combined with binders and flame-retardant chemicals and sprayed on ceilings and walls (see Flame retardants). Ultimate performance of this treatment usually is more variable than other kinds of surface treatments, since surface density and thickness are less easily controlled. Metal fibers and sintered metals are made particularly for applications involving severe chemical or physical environments. Such products can be made with finely controlled physical properties.

Sound Intensity and Reverberation Control. ***Commercial Products.*** *Tiles and Boards.* Tiles and boards are prefabricated rigid products that are made of cellulose or mineral fiber (glass fiber or rock wool) and can be installed as a suspended ceiling or as a surface treatment. Many varieties are commercially available, and they differ in composition, texture, and in acoustic performance. Typical dimensions are 1.3–3.8 cm by 30.6–61 cm by 61–122 cm. Typical performance data for these products are given in Table 2. These products generally are soft and incapable of withstanding physical abuse. Precaution should be observed in repainting porous products: paint tends to clog the openings and, thus, reduce performance.

Many suppliers offer ceiling tile and board products that are integrated into a system that provides an array of mechanical services, eg, lighting, air conditioning, and heating elements, in combination with acoustic performance. One such system is illustrated in Figure 2.

Metal Pan Assemblies. Metal pan assemblies are perforated aluminum or steel pans that are suspended from the ceiling by T-bars. A pad of absorptive material nested above the pan provides the sound-absorptive element, and various surface finishes

Table 2. Typical Acoustic Performance of Ceiling Tiles and Boards[a]

Type of product	Thickness, cm	Random-incidence absorption coefficients, α[b] Octave band center frequency, Hz 125	250	500	1000	2000	4000	Mounting[c]
smooth-surfaced mineral fiber tile	1.6	0.27	0.29	0.70	0.82	0.67	0.55	plenum
perforated mineral fiber tile	1.6	0.42	0.38	0.57	0.83	0.78	0.62	plenum
fissured mineral fiber tile	1.6	0.26	0.33	0.57	0.83	0.84	0.79	plenum
fissured mineral cellulose fiber	1.9	0.44	0.41	0.65	0.83	0.85	0.91	plenum
lay-in panel	2.5	0.37	0.41	0.26	0.37	0.48	0.67	plenum
mineral fiber lay-in panel	2.5	0.06	0.25	0.68	0.97	0.99	0.92	direct contact
lay-in panel	2.5	0.72	0.70	0.82	0.98	0.94	0.84	plenum
perforated metal panel with mineral fiber pad	5.2	0.73	0.93	0.81	0.94	0.82	0.60	plenum

[a] Adapted from ref. 1.
[b] Determined by reverberation-room testing.
[c] See Figure 1.

can be used on the exposed pan surface. The pad element often is encased in a thin plastic bag to prevent insulation particles from falling through the perforations. The acoustic performance of one such product is included in Table 2.

Blanket, Sheet, and Board. Several suppliers provide porous absorbent materials in the form of flexible blankets or sheets or rigid boards that are suitable for direct use or for modification by fabricators of more complex noise-control products and that are available in a variety of densities and dimensions. Typical performance data for these products are given in Table 3 and they illustrate the general tendencies of better performance in the higher octave bands and increased low frequency performance for thicker versions of a given material (or when an air space is left between the material and the surface to which it is attached).

Unit Absorbers. There are two kinds of unit absorbers; one is applied to wall surfaces as an architectural treatment and the other is suspended from and across a ceiling for industrial applications. Architectural unit absorbers are made in several forms and most are covered with a decorative fabric facing. Many are attached to wall surfaces with special hardware, and a few are glued directly onto wall surfaces. Performance depends upon the thickness of the material, the depth of the air space behind it, and the spacing between the units. Performance most often is described in sabins of absorption per unit, for a specific arrangement of the units, or of equivalent sound absorption of the area covered with the product. Industrial unit absorbers also are made in several forms but most often as a sheet-metal-framed panel or plastic wire-mesh-encased tube. Triangular- or diamond-shaped cross sections also are available. Often, these products are refabricated versions of ceiling boards or other surface-type products. Typical performance data for unit absorbers are provided in Table 4.

Acoustic Roof Decks. The three kinds of roof-deck systems that employ acoustic insulation generally are used in one-story industrial buildings. Acoustic form boards for poured decks are supported by steel subpurlins and are used in place of the usual

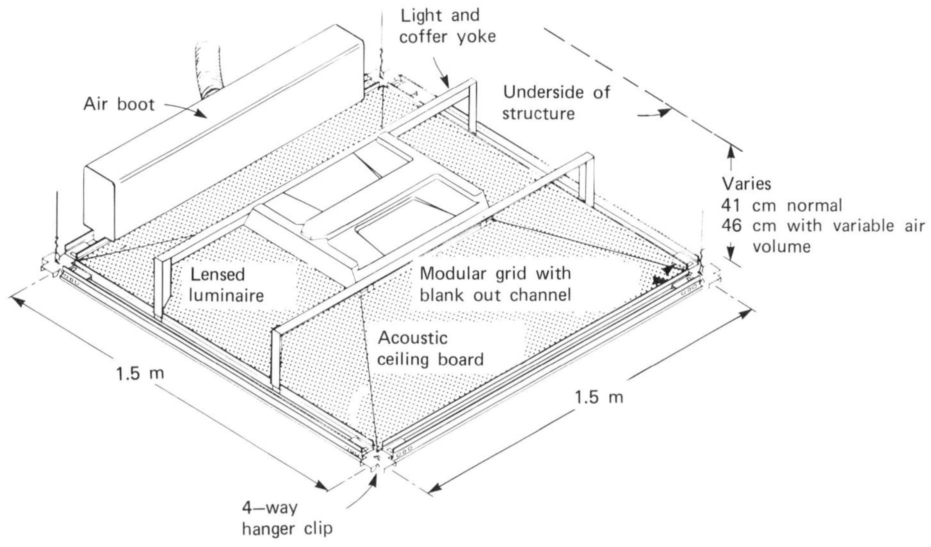

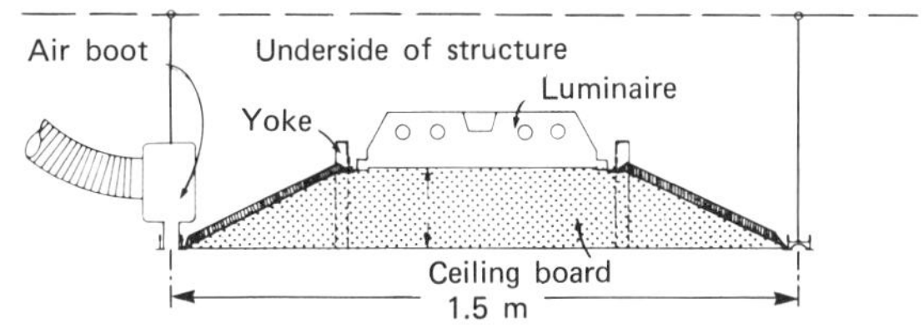

Figure 2. Integrated ceiling system. Courtesy of Owens-Corning Fiberglas.

nonacoustical form boards. Perforated roof-deck panels consist of hollow steel roof-deck panels, the underside of which are perforated, and an absorptive element (typically mineral wool) is placed inside the hollow part. These panels serve as the base for subsequent layers of roof insulation and built-up roofing. One such roof-deck panel is illustrated in Figure 3. Structural roof insulation consists of thick, rigid insulation that serves both as a structural roof decking material and an acoustic insulation element. Conventional roofing is applied over the decking.

Carpeting and Drapery. No carpet or drapery textile products are sold specifically as acoustic insulation; however, some laboratory-determined measurements are reported in refs. 5 and 6. In general, carpeting is most effective when the backing is permeable (rather than coated with starch or latex) or when the carpet has a thick pile and heavy underlay. Fiber content, pile structure, and yarn weight appear to have little effect on performance. Increased pile density appears to increase low frequency and decrease high frequency performance. Heavy drapery performs well acoustically, and the performance depends on percent fullness (the percent by which the width of the material making up the drape exceeds the width of the installed drape) and the distance of the drapes from the reflecting surface behind them.

Table 3. Typical Acoustic Performance of Porous Sound-Absorbing Products in Sheet Form[a]

			Random-incidence absorption coefficients[b]						
			Octave band center frequency, Hz						
Material	Thickness, cm	Mounting[c]	125	250	500	1000	2000	4000	NRC
plastic foam	1.3	direct contact	0.15	0.21	0.26	0.17	0.63	0.65	0.31
	2.5	direct contact	0.16	0.25	0.45	0.84	0.97	0.87	0.63
	5.1	direct contact	0.24	0.49	0.81	0.91	0.98	0.97	0.80
glass fiber	1.3	direct contact	0.04	0.13	0.32	0.46	0.61	0.73	0.40
	2.5	direct contact	0.13	0.30	0.64	0.76	0.78	0.82	0.60
	5.1	direct contact	0.27	0.81	0.99	0.99	0.99	0.99	0.95
	2.5	plenum	0.67	0.72	0.64	0.75	0.63	0.45	0.70
sprayed-on cellulose fiber	2.5	direct contact	0.08	0.29	0.75	0.98	0.93	0.76	0.75

[a] Adapted from ref. 1.
[b] Determined by reverberation-room testing.
[c] See Figure 1.

Table 4. Typical Acoustic Performance of Unit Absorbers

	Metric sabins per unit[a]					
	⅓ Octave band center frequency, Hz					
Type of unit and spacing	125	250	500	1000	2000	4000
31-cm dia cylinder 61 cm long, 122 cm on center	0.34	0.54	0.66	0.76	0.78	0.93
61 × 122 × 3.8 cm glass-fiber board wrapped in perforated vinyl film, 122 cm on center	0.39	0.61	0.91	1.24	1.26	1.24
29 × 41 × 5 cm porous glass blocks, 61 cm on center	0.00	0.10	0.20	0.20	0.20	0.20

[a] Determined by reverberation-room testing.

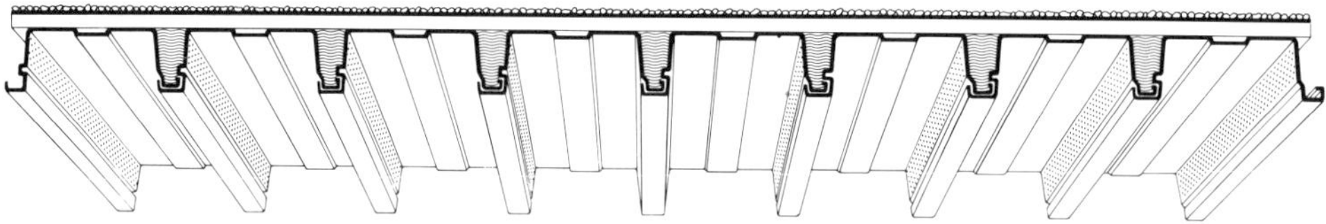

Figure 3. Example of perforated roof-deck panel. Courtesy of Inryco.

Sound Transmission Reduction. ***Commercial Products.*** *Mufflers.* The two principal categories of mufflers are reactive and dissipative mufflers. Reactive mufflers reflect sound to the point of origin. Their configuration determines their performance. Dissipative mufflers use sound-absorbing materials and are exemplified by parallel baffle silencers, prefabricated mufflers, and integral ducts.

Parallel baffle mufflers comprise rectangular air passages with one or more interior surfaces covered with porous absorbing material which is held in place by perforated sheet metal. A cutaway view of a typical parallel baffle muffler is shown in Figure 4. Parallel baffle mufflers are side-by-side groupings of lined ducts and are used primarily in air-handling-system noise problems (see Noise pollution). Performance is frequency variable and dependent on the depth of the absorbent lining, the width and length of the duct, and the flow resistance of the lining material. In each case, performance usually is described in terms of insertion loss as a function of frequency. The performance data of the muffler shown in Figure 4 is provided in Table 5.

Prefabricated mufflers consist of tubular or rectangular cross-sectional units that have industrial applications. There is an absorbent lining on the inner side of each

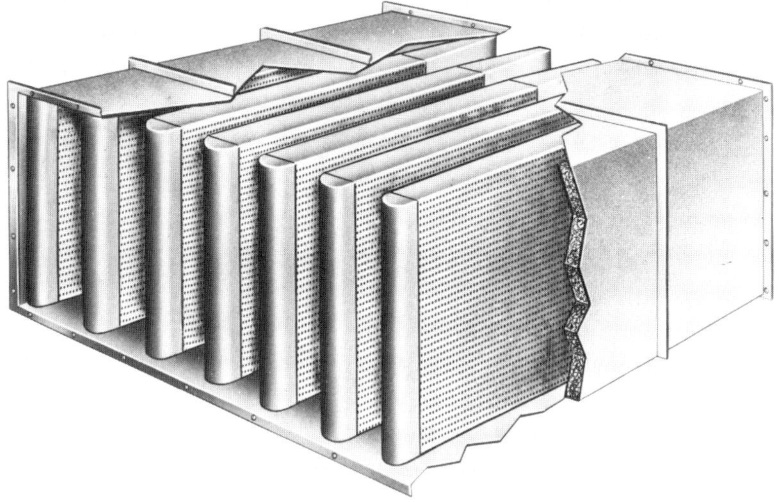

Figure 4. Parallel baffle muffler. Courtesy of Industrial Acoustics Company, Inc.

Table 5. Dynamic Insertion-Loss Ratings: Forward(+)/Reverse(−) Flow[a]

Muffler length, m[b]	Silencer face velocity, m/min	Dynamic insertion loss, dB: Octave band center frequency, Hz: 63	125	250	500	1000	2000	4000	8000
0.9	+1219	6	8	14	23	27	20	12	8
	−610	6	8	14	23	27	20	14	8
	+600	2	7	12	19	23	23	18	11
	+1219	2	6	11	15	22	23	18	11
1.5	−1219	9	12	21	34	43	33	22	9
	−610	8	11	18	32	42	33	22	11
	+610	6	10	18	30	42	34	23	14
	+1719	4	9	17	29	38	34	23	14
2.1	−1219	11	15	27	38	50	40	27	13
	−610	10	15	26	36	49	40	28	14
	+610	10	14	24	36	48	44	31	18
	+1219	9	12	22	36	47	44	31	19
3.0	−1219	14	24	36	44	55	50	34	19
	−610	13	21	35	38	55	50	37	20
	+610	12	20	34	37	55	52	43	22
	−1219	11	16	32	36	55	53	43	23

[a] Dynamic insertion losses are for 16°C. Up to ca 149°C, they are nominally correct.
[b] See Figure 4.

unit's shell and, often, additional absorbing material is contained within an axial element that extends the length of the unit. A cutaway view of one such muffler is shown in Figure 5.

Integral ducts and duct boards are glass-fiber boards (typically 2.5 cm thick) used for low velocity air-distribution systems. They are faced on the outside with a lightweight impervious material, eg, aluminum foil. The inner facing of the board may be coated with neoprene to help prevent surface erosion in the presence of air flow (see Elastomers, synthetic). These products provide thermal and acoustic insulation. The lightweight construction makes ductboard more susceptible to damage and less capable of preventing sound transmission through the duct walls. Integral ducts are preformed glass-fiber-lined ducts.

Modular Acoustic Panels. Modular acoustic panels typically are 10 cm thick, constructed with solid light-gauge steel backs and light-gauge steel framing, and faced with a perforated steel front. The interior is filled with acoustic insulation, typically mineral wool. Modular dimensions vary from manufacturer to manufacturer, as do methods of joining and erecting the panels. These products provide built-in absorption and transmission loss characteristics for industrial noise shelters or equipment enclosures. Most suppliers furnish elements for complete panel systems, including ventilation panels, lighting and electrical services, doors, windows, and special floating (vibration-isolated) floors.

Lagging–Wrapping Treatments. Application of lagging or wrapping treatments (ie, applying a layer of porous, absorbent material that is covered by an outer impervious material) reduces noise from pipes, valves, and vibrating surfaces, eg, gearboxes, and are used in aircraft fuselage interiors to attenuate noise that is generated outside the passenger cabin. These applications almost always employ a layer of impervious material, eg, asphalted paper, sheet metal, mass-loaded vinyl fabric, or other impervious materials on the quiet (outer) side of the application. The outer layer significantly improves performance for frequencies where the wavelength of sound in the material is small compared with the depth of the lining.

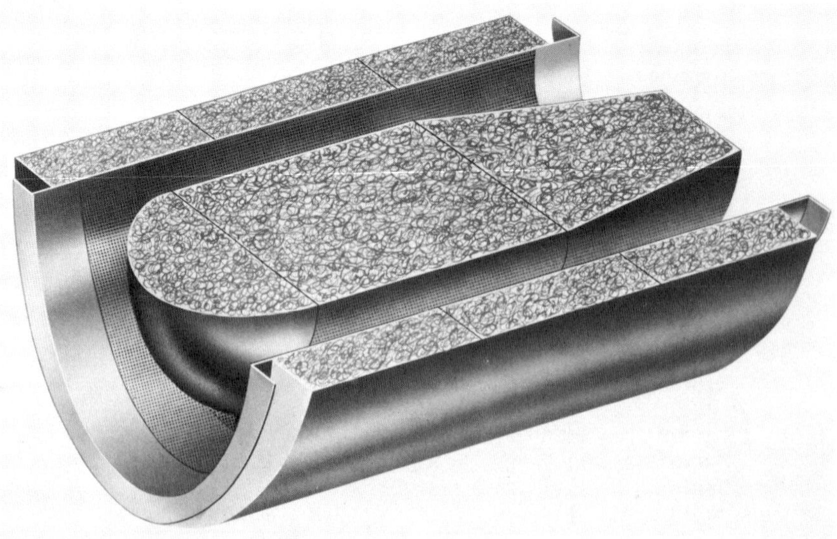

Figure 5. Cutaway of prefabricated muffler. Courtesy of Industrial Acoustics Company, Inc.

Nonacoustical Factors. A designer or noise-control engineer usually must select a particular material or system from a wide variety of effective products. Nonacoustic factors that are taken in account when selecting a particular material or system include esthetics, flame spread and fire endurance, toxicity, mechanical strength and durability, ease of handling and maintenance, light reflectance, weight, and dimensional stability for temperature and humidity.

Uses. Absorbent materials are used extensively in architectural applications to moderate the intensity of sound in a reverberant sound field and to control reverberation time. Products employing porous absorbing materials are also used to control sound transmission.

Sound Intensity. The relationship among the physical characteristics of a space, the acoustic energy emitted by a noise source, and the sound intensity caused by the noise of the source in that space is:

$$L_p(r) = L_W + 10 \log_{10} \left(\frac{Q_\theta}{4\, \pi r^2} + \frac{4}{R} \right)$$

where $L_p(r)$ is the sound pressure level at any position in the space at distance r from the sound source and away from the immediate vicinity of a reflecting surface. $L_p(r)$ is directly related to the sound intensity, and both are measured in dB relative to 20 μPa; L_W is the acoustic energy emitted by the source into the space (dB re 10^{-12} W); Q_θ is a directivity factor of the source in direction θ. A directivity factor of one applies to an omnidirectional source that is positioned away from any reflective surface (eg, suspended in the middle of a room). A directivity factor of two applies to an omnidirectional source placed on a single reflecting plane, eg, a floor; r is the distance from the source, m; and R is the room constant, m^2:

$$R = \left(\frac{S\overline{\alpha}_s}{1 - \overline{\alpha}_s} \cong S\overline{\alpha}_R \right)$$

where S is the surface area in space, m^2; and $\overline{\alpha}_s$ is the average statistical absorption coefficient of the surface materials—appropriate only for performance in a perfectly diffuse sound field. (The statistical absorption coefficient α_s—the theoretical random incidence absorption coefficient—is less than the laboratory-measured random incidence absorption coefficient α_R. Because α_s is a difficult quantity to measure and because the quantity R generally is close to $S\overline{\alpha}_R$, the general practice in architectural acoustics is to use $S\overline{\alpha}_R$ instead of R.)

Figure 6 is a plot of sound pressure level vs distance for spaces of varying R that has been normalized to the sound pressure level 0.3 m from the acoustic center of a noise source. At a distance of 0.3 m, the acoustic energy L_W of many noise emitters is approximately equal to the sound pressure level L_P. Figure 6 shows that there are two spatial regions surrounding any noise emitter. The first—the direct sound field—occurs where the term $Q_\theta/4\pi r^2$ is large when compared with $4/R$. In this region, L_P depends on L_W and the distance to the noise source; R has little impact on the sound intensity. The second region, which is farther from the noise source, is the reverberant sound field. Here, $4/R$ is the larger term in the parentheses and L_P depends on L_W and R; L_P remains constant at any position. In the reverberant sound field, the amount of absorption governs sound intensity and the following relationship applies:

$$L_P \sim 10 \log_{10} \frac{1}{R}$$

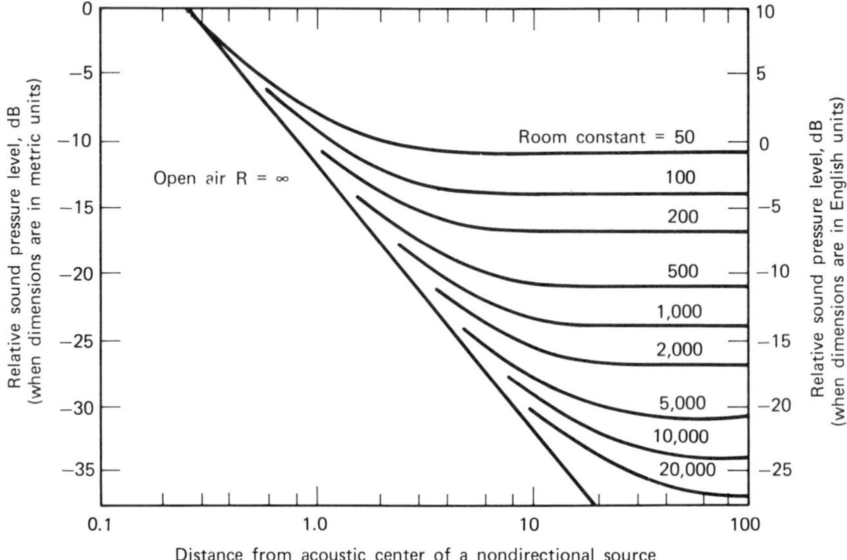

Figure 6. Approximate relationship between sound pressure level and distance, as a function of room constant.

Each doubling of the amount of absorption lowers L_P by 3 dB. Eventually, $Q_\theta/4\pi r^2$ becomes large when compared with $4/R$, thus an upper limit can be placed on possible sound reduction by adding absorption.

Reverberation Time. The reverberation time, the time it takes for an instantaneously stopped sound to decay by 60 dB from its initial intensity, is a measure of the acoustic character of a space. In general, longer reverberation times are desirable where the primary function of the space is as a music listening room, eg, symphonic halls, whereas shorter reverberation times are desirable for facilities where listening to speech is most important, eg, classrooms.

A fundamental relationship between reverberation time and the amount of absorption in a space is given by the sabin equation:

$$T_{60} = \frac{0.161\ V}{S\overline{\alpha}_R}$$

where T_{60} is the reverberation time, s; V is the room volume, m^3; S is the surface area in the room, m^2; and $\overline{\alpha}_R$ is the average random-incidence absorption coefficient of the surface materials. Thus, for a room with fixed volume, the reverberation time is inversely proportional to the amount of absorption in the space. By adding or subtracting absorption, the reverberation time can be raised or lowered. In modern architectural acoustics, more sophisticated equations take into account air absorption as well as acoustic effects not predicted by the sabin equation.

Sound Transmission. Ways in which porous absorbent materials are used to help reduce sound transmission from one point to another include the use of: muffler and lined-duct components, wrapping or lagging treatments, wall fill, and simple transmission-loss elements. The equations describing the behavior of sound propagating through structures are detailed in refs. 5 and 7.

Sound-Reflective Materials

The second category of acoustic insulation consists of materials that contain or redirect sound. When used in sound-containment systems, the physical properties of the intervening structure that isolates the noise source from the receiver usually are the most critical for preventing sound from being transmitted across the construction. Redirection of sound is provided by partial height, impervious, structural elements (barriers and shields). The performance of such structures usually depends on how much sound diffracts (bends) around their edges or on the effect of sound reflections from nearby surfaces. The procedures for estimating barrier performance can be found in refs. 5 and 7.

The performance measure of interest for structural elements that contain sound is the sound transmission loss *TL*. By definition, $TL = 10 \log_{10} W_i/W_t$, in dB, where W_i is the incident sound power (which induces structural deformations of the intervening structure), and W_t is the transmitted sound power (which results from the radiation of sound on the opposite side of a wall because of induced structural deformations). Structural stiffness, internal damping, and mass are attributes affecting the performance of a sound-reflective material. The *TL* values must be known in considerable detail to ensure that the transmission loss of a particular construction or material is characterized adequately. A standard set of *TL* data consists of 16 data points, one for each contiguous one-third octave band from 125 to 4000 Hz.

The preferred method for measuring the *TL* of space-dividing elements (walls, doors, floor–ceiling assemblies) is described in ref. 8. The test is applicable for estimating *TL* of constructions to be used in diffuse or near-diffuse sound fields. It is not applicable in qualifying the insulating performance of constructions, eg, tight-fitting machinery housings, constructions exposed to unidirectional sound, partial height barriers, and noncontinuous walls. It also is not applicable for the measurement of *TL* of *in situ* structures.

In the *TL* test, the specimen is mounted as a partition between two reverberation rooms. One room is a source room and the other is a receiver room. Transmission loss is found from the following:

$$TL = NR + 10 \log_{10} (S/A_2)$$

where S is the area of sound-transmitting surface of the specimen, m^2; A_2 is the sound absorption in the receiving room, metric sabins; and NR is the difference in space–time average sound pressure levels in the two rooms.

Random noise is introduced in a source room and is measured in a receiving room. Precautions are taken to ensure that flanking transmission (noise penetrating into the receiver room through paths other than through the test wall) is insignificant. The test specimen must include all construction elements in their normal size and in proportions that are typical of actual use. For *in-situ* conditions, another standard procedure (9) is used, which is similar to the above procedure but involves the determination of field performance. Measurements of NR, S, and A_2 are made in the field. The resultant field transmission-loss (FTL) values are invariably lower than laboratory *TL* data because of flanking paths (7,10).

Transmission-loss data are needed to fully describe the performance of a structural element exposed to airborne sound. Simpler performance measures can be used if the performance of an element exposed to speech sounds is being studied specifically.

Speech sound contains most of its acoustic energy in the 500–2000-Hz octave bands. The sound-transmission class (STC) is used to study performance in the speech octaves more closely. The STC rating is determined by comparing *TL* data with points on a reference STC contour, following rules stipulated in ref. 11. Transmission loss and STC ratings of several wall constructions are given in Table 6.

Transmission loss and STC ratings apply only to airborne sound attenuation. Structureborne sound can occur as a result of steady-state vibration that is induced by air conditioners, washing machines, etc, or as a result of impacts such as those caused while walking across a floor or by slamming a door.

The Impact Isolation Class (IIC) and Impact Noise Rating (INR) are used to rate the insulation value of a material or structure regarding impact-type sounds. The IIC is determined by placing a standard tapping machine to one side of a structure undergoing evaluation. The noise output of the tapping machine is measured on the opposite side of the structure. Details of the measurement procedure are given in refs. 12 and 13. The INR rank orders impact-noise isolation relative to an arbitrarily selected construction. To within about 2 dB, IIC approximates INR + 51. An INR rating of 0 provides just satisfactory isolation. Thus, an IIC of 51 is indicative of a just satisfactory performance. The IIC and INR ratings of several floor constructions are provided in Table 7. Impact-isolation ratings are highest for assemblies with soft finishes. For example, padding and carpeting a concrete floor can improve an IIC by as much as 20 dB. This reduction occurs because a soft finish provides an elastic layer that protracts the force–time history of the impact and, consequently, eliminates a large portion of the higher frequency components by the impact.

Damping Materials

Damping is any process that reduces vibration by converting the ordered mechanical energy of vibration into thermal energy. Interface friction, fluid viscosity, turbulence, and magnetic hysteresis are examples of the many ways of damping vibration. Only one damping method depends on material properties—mechanical hysteresis, otherwise known as material damping. The stresses in the material caused by vibratory motions are relieved partly as material strains develop, and a portion of the mechanical energy available for vibration is converted into heat in the process. Because no material is perfectly elastic, a molecular-level energy transformation occurs in all materials, including common structural materials. Damping materials, however, dissipate more strain energy than ordinary materials because, when under stress, they tend to flow like a very viscous liquid. This behavior is characteristic of viscoelastic materials used extensively to damp vibration. When properly joined to the surface of a vibrating element, the damping material participates in the vibration of the structure, is stressed, and dissipates vibrational energy.

Loss Factor. Material damping is characterized by the loss factor η, a dimensionless quantity:

$$\eta = D/2\pi W_0$$

where D is the energy dissipated per cycle of vibration, and W_0 is the average total energy of the vibrating system. (Many other measures of damping are in common use; all are related to one another and to the loss factor. For a fuller discussion of these other measures, see ref. 7.) The loss factor of most structural materials, eg, aluminum, steel,

Table 6. Acoustic Performance of Selected Materials and Constructions Used to Prevent Sound Transmission

		Transmission loss, dB															
		⅓ Octave band center frequency, Hz															
Material or construction	STC	125	160	200	250	315	400	500	630	800	1000	1250	1600	2000	2500	3150	4000
3-mm thick Vitran strips, 300-mm wide, double overlap	12	8	7	7	7	7	8	10	11	11	12	12	13	13	13	13	14
0.14-cm thick leaded vinyl fabric, 3.7 kg/m^2	23	11	11	12	14	15	16	18	20	23	24	26	28	30	32	34	35
1.3-cm thick gypsum board nailed to 5 × 10-cm studs on 41 or 61 cm center	32	9	13	23	27	26	31	34	35	39	42	44	46	47	41	36	40
1.3-cm thick gypsum board nailed to 5.2 × 10.2-cm studs on 41 or 61 cm center; 1.3-cm thick resilient glass fiberboard glued between studs and gypsum	37	14	21	31	32	36	41	44	49	52	54	55	54	50	51	56	63
15-cm hollow core, block wall, unpainted	43	33	34	34	34	37	38	40	42	45	47	48	50	52	52	49	48

Table 7. Impact Noise Isolation of Several Floor-Ceiling Constructions

Construction	IIC[a]	INR[b]
10-cm thick concrete slab, reinforced with 15 × 15-cm, 6 AWG[c], reinforcing mesh placed at the centerline horizontal plane of the slab; all surface cavities sealed with thin mortar mix	25	−26
15.2-cm thick reinforced concrete slab; on the floor side, 1.9-cm thick tongue-and-groove wood flooring nailed to 3.8 × 5-cm wooden battens, 31 cm on center, floating on 2.5-cm thick glass wool quilt; on the ceiling side, 1.3-cm layer of plaster	57	+6
5 × 20-cm wooden joists 41 cm on center; on the floor side, 2.2-cm tongue-and-groove flooring nailed to joists; on the ceiling side, 1-cm gypsum wallboard nailed to joists with joints sealed; total thickness, 24 cm	32	−19
5 × 25-cm wooden joists 41 cm on center with 7.6-cm thick mineral fiber batts stapled between joists; on the floor side, 1.3-cm thick plywood subfloor nailed 15 cm on center along edges and 25 cm on center in the field, building paper underlayment, 1.9-cm oak flooring nailed at each joist intersection and midway between joists; carpet (1.04 kg/m^2) with hair felt pad (0.95 kg/m^2) placed on flooring; on the ceiling side, 1.6-cm thick gypsum wallboard nailed 15 cm on center of joists; all joists taped and finished; total thickness: 32 cm	58	+7

[a] IIC = Impact Isolation Class.
[b] INR = Impact Noise Rating.
[c] AWG = American Wire Gauge.

concrete, and wood, is within 10^{-4}–10^{-2}. These values apply over a wide temperature range and throughout the audible frequency range. The loss factors of viscoelastic materials are orders of magnitude higher but also are both temperature and frequency sensitive. References 5 and 7 provide details about the loss factors of viscoelastic materials.

National standards do not exist for the measurement of damping. The most often used procedures involve the determination of decay rates of vibratory motion of test panels. A test panel is evaluated with and without the damping material applied to the panel to determine the damping properties of a material.

Extensional Damping. Two forms of material damping applications are used: extensional and constrained layer. In extensional damping, the material is applied directly as a surface treatment. As the damping material is flexed and extended, energy is dissipated. Generally, a damping material is applied in a continuous surface to ensure that the application participates in the surface motions. The composite system-loss factor depends on both the material properties and the geometry of the assembly. The composite loss factor approaches and maintains a maximum when the damping layer is about twice as thick as the substrate to which it is applied, and when the substrate is considerably stiffer than the damping material (as is typically the case).

Extensional damping materials are either compounds or sheet and sheetlike products. Compound products are sprayed, painted, or trowelled; the material provides the necessary adhesion to ensure that the damping properties are fully realized. Compounds are used when thick layers of damping are required or when the surface to be covered is irregularly shaped. Some damping compounds must be mixed before use. Others require heat treatments to cure the compound after it is applied. All require drying and curing time before the full damping capacity is realized, the actual time varies from product to product.

Sheet and sheetlike damping materials usually are available in a variety of thicknesses. Most can be cut to a desired size on-site, but a few are relatively stiff products cast into final form by the suppliers. All these products must be joined carefully to the quieted surface to obtain the fullest benefit. Delamination of the damping layer under severe physical or chemical–environmental stress poses a serious problem.

Constrained-Layer Damping. In constrained-layer damping, the damping material is attached to a base material and is sandwiched by an outer constraining layer. When such a system is subjected to flexure, the damping material is put into shear as well as bending. Energy losses associated with this form of damping are related primarily to the damping material which relieves the shear strains. The composite system-loss factor depends on the geometry of the structure, the moduli of elasticity of the system components, and the shear modulus of the damping material.

Thin films of damping material provide large amounts of energy dissipation. The inner layer of damping material in the sandwich construction often is only a micrometer thick and, generally, doubles as an adhesive layer to hold the substrate and constraining layers together. A few constrained-layer products are sold only with an outer layer and a middle adhesive layer. These are applied in the same manner as extensional damping sheets.

Uses. Damping materials control sound in two ways: by reducing the motion of a structure vibrating at a natural frequency of vibration, with consequent reduction in stress, structural fatigue, and noise radiation; and by reducing the propagation of flexural structureborne waves with consequent localization of vibration.

Resonant Vibration. Surfaces are induced to vibrate at their natural frequencies by impacts, eg, the ringing of a bell. However, ringing also results when water drips in a sink, gear teeth mesh, and hard objects fall onto hard surfaces (eg, when parts drop into a hopper). Damping materials reduce the amplitude of the vibration and the time it takes for the vibration to decay. The amplitude reduction at each natural frequency is equal to:

$$20 \log_{10} \frac{\text{the composite loss factor of the damped system}}{\text{the composite loss factor of the original undamped system}}$$

Structural vibrations also can be caused by means other than impacts; a surface can be driven into vibration at any frequency by an oscillatory force. These forced vibrations occur, eg, in structures exposed to sound or structures excited by motions of machine components. Generally, the forced vibrational response of the structure being excited is more complex than the free response associated with impacts. In particular, the vibration contains components at the driving frequency as well as at natural frequencies near the driving frequency. The overall response to forced vibration depends not only on damping properties but on the structural stiffness and mass.

Flexural-Wave Propagation. The ability of damping materials to retard the propagation of flexural waves is important in wall transmission loss. At and near the critical frequency of the wall (where the flexural structureborne wave excited by the sound travels at the speed of sound in air), the wavefronts of the airborne sound match well with the propagating structureborne waves and reinforce them, with subsequent reradiation of most of the incident sound. By retarding the propagation of the structureborne waves, damping results in better transmission-loss performance at and near the critical frequency. By retarding flexural-wave propagation, damping also localizes vibration. In industrial settings, vibrations often are used to assist in material flow processes, eg, to ease the flow of materials at hopper discharges or to transport lightweight parts along conveyors. In such cases, vibrations may propagate to other surfaces, creating unnecessary noise. Strategically applied damping material can reduce this noise.

Vibration Isolation Materials

Structural vibrations can be inhibited by introducing a vibration isolator in the path of the disturbance. A vibration isolator is an elastic material or structure that, through its dynamic properties, tends to prevent the transmission of vibration. The performance of an isolator is characterized by its transmissibility—the ratio of the force transmitted to the quiet side of the isolator compared with the driving force introduced by the motion of the object to be isolated:

$$\text{transmissibility} = \text{output force/input force}$$

Transmissibility varies according to frequency and is a function of the natural frequency of an isolator and its internal damping, assuming the isolator rests on a foundation that is considerably stiffer than the isolator. Figure 7 shows the transmissibility function for a family of simple isolators of different internal damping η. The figure illustrates the generally applicable aspect that transmissibility decreases only at driving frequencies f greater than $f_0\sqrt{2}$, where f_0 is the natural frequency of the isolator:

$$f_0 = \frac{1}{2\pi}(\sqrt{k/m})$$

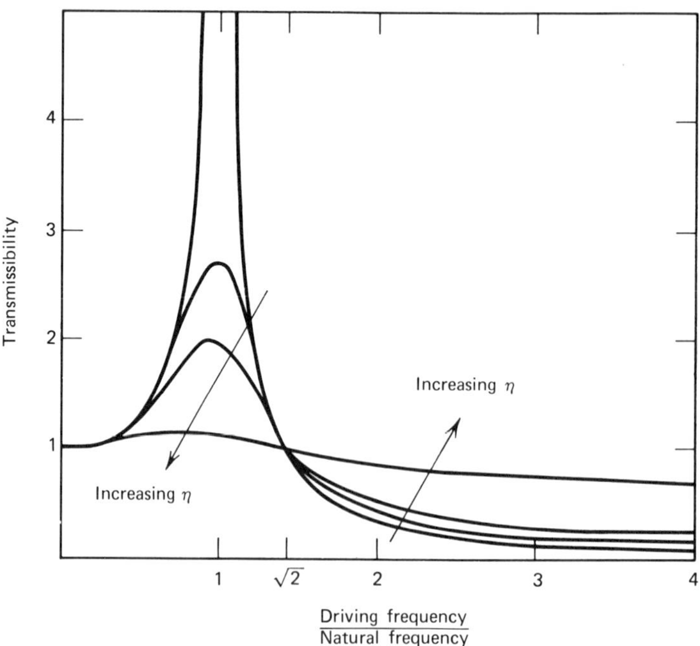

Figure 7. Transmissibility as a function of frequency ratio for single-degree-of-freedom isolators with different degrees of internal damping.

where k is the spring stiffness of the isolator, N/m (mN/m = dyn/cm); and m is the mass supported by the spring, kg. For most materials, the natural frequency of the isolator is related directly to the static deflection of the isolator through the relationship:

$$f_0 = 5/\sqrt{x}$$

where x is the static deflection, cm.

Isolators are chosen that have large static deflections to yield minimum transmissibility at the most critical driving frequency. The physical stability of the isolated system, especially during equipment starting and stopping, often governs the limits of acceptable static deflection. Vibration isolators are selected from the numerous varieties available according to their ability to carry an imposed load (including, if necessary, any lateral loading necessary to prevent the isolated equipment from

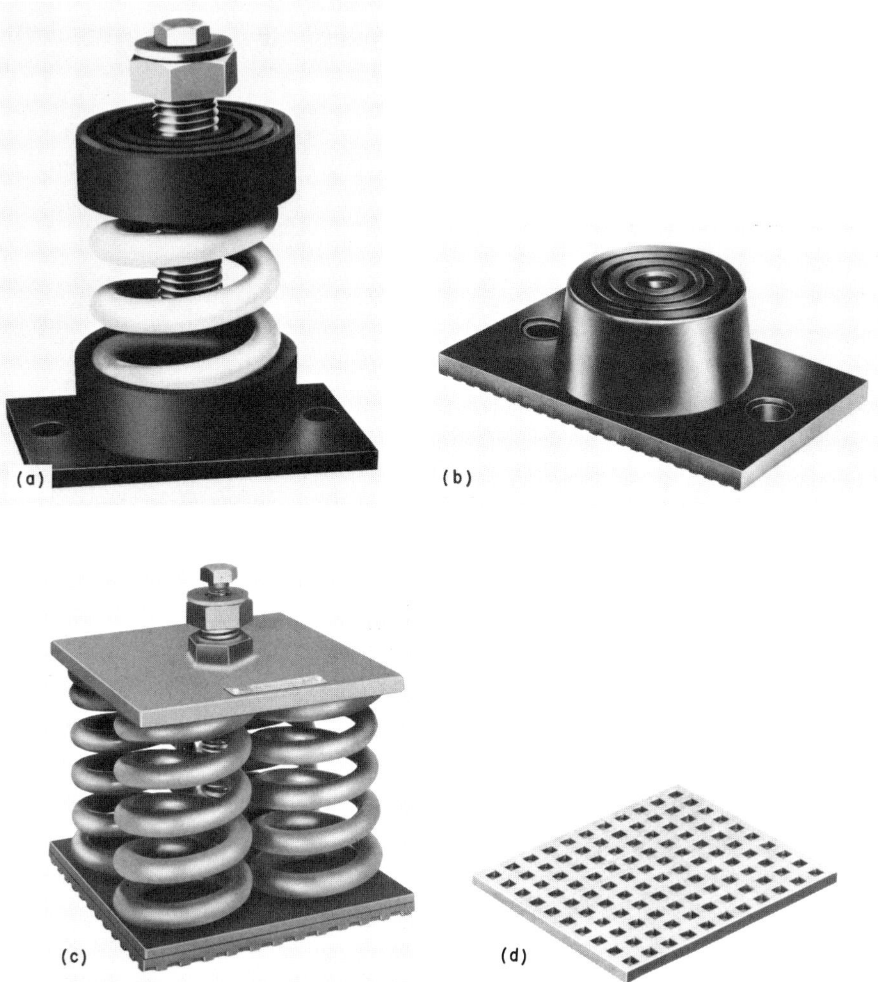

Figure 8. Vibration isolators: (**a**) single-spring mount with base plate; (**b**) neoprene mount; (**c**) multiple-spring mount; (**d**) neoprene waffle pad. Courtesy of Mason Industries, Inc.

walking or being sheared from its mounts); their ability to withstand the environment in which they are used (eg, extreme temperatures, chemical exposure, etc); and their ability to function satisfactorily acoustically. Several forms of vibration isolators are illustrated in Figure 8.

Uses. Vibration isolators are used extensively at support points of vibrating equipment and components, such as mechanical equipment, noisy pipes, and industrial machinery (see Noise pollution). When so used, they prevent vibrational energy from being transmitted as noise or to vibration-sensitive areas. Isolators also are used to insulate a noise- or vibration-sensitive machine or area from external vibrations. Examples of the latter include isolators used to support floating floors in recording studios and special mounts to protect delicate instrumentation.

Five resilient materials are used in isolators; metals (in the form of various kinds of springs), elastomers, cork (qv); glass fiber, and felt (in the form of pads and blocks). Spring isolators are available in a range of static deflections and are used in applications varying from protecting delicate scientific instruments to isolating vibrations of heavy industrial equipment weighing thousands of kilograms. An advantage in using metal springs is that they can be readily and repeatedly manufactured with predictable characteristics. They also can be used in severe chemical and physical environments. Two practical disadvantages are that metal springs have little inherent damping, and they tend to transmit high frequency vibrational energy. For these reasons, dashpots or other suitable damping mechanisms and high frequency pad-type vibration isolators often are used in conjunction with spring-type isolators.

Elastomeric isolators also are available in numerous forms, eg, molded products used both in compression or shear mounts. The ability of an elastomer to withstand enormous numbers of repeated cycles of vibration makes it an attractive material. When used in compression, the shape of the resilient element must be taken into account since the volume compressibility of solid rubber is very low. Space must be provided for lateral expansion. Ribbed pad construction often is used to satisfy this requirement. Natural rubber can be used in applications that do not involve exposure to hydrocarbons, ozone, or high temperature (see Rubber, natural). Neoprene and other synthetic rubbers are used in severe environments involving elevated temperatures or exposure to oils and other chemicals. Rubber also is used in the manufacture of a new type of vibration isolator that essentially is an air bladder which effectively isolates low frequency vibration. Elastomers also may be used in combination with ground cork to form isolator blocks or pads.

Composition board (ground cork and binders) has been used extensively for vibration isolation for over a hundred years, and it resists oils well. However, cork tends to compress with age under load, and its effectiveness is limited at high temperatures (>100°C). Furthermore, in compression, the stress–strain curve for cork is not linear, restricting its optimum performance range to loading at ca 49–147 kPa (7–21 psi).

Felt also is used as a vibration isolator material. Unlike the other materials, the natural frequency of a felt-pad isolator is more a function of the pad thickness than the static load. Felt is most useful in isolating frequencies >40 Hz. Glass-fiber isolation board is available in various densities. These boards tend to absorb moisture and other contaminants and, thus, they usually are supplied with sealed surfaces or are enclosed in plastic bags.

BIBLIOGRAPHY

1. *Compendium of Materials for Noise Control, HEW (NIOSH) Publication 75-165,* HEW, Washington, D.C., June 1975.
2. *Standard Test Method for Impedance and Absorption of Acoustical Materials by Impedance Tube Method, ANSI/ASTM C 384-77,* ASTM, Philadelphia, Pa., May 1978.
3. *Standard Method of Test for Sound Absorption of Acoustical Materials in Reverberation Rooms, ANSI/ASTM C 423-77,* ASTM, Philadelphia, Pa., May 1978.
4. *Standard Method of Testing Duct Liner Materials and Prefabricated Silencers for Acoustical and Airflow Performance, ANSI/ASTM E 477-73,* ASTM, Philadelphia, Pa., 1973.
5. C. M. Harris and C. E. Crede, eds., *Handbook of Noise Control,* 2nd ed., McGraw-Hill Book Co., New York, 1979.
6. M. Rettinger, *Acoustic Room Design and Noise Control,* Chemical Publishing Co., New York, 1968.
7. L. L. Beranek, ed., *Noise and Vibration Control,* McGraw-Hill Book Co., New York, 1971.
8. *Standard Method for Laboratory Measurement of Airborne Sound Transmission Loss of Building Partitions, ANSI/ASTM E 90-75,* ASTM, Philadelphia, Pa., Jan. 1978.
9. *Standard Test Method for Measurement of Airborne Sound Insulation in Buildings, ANSI/ASTM E 336-77,* ASTM, 1977.
10. R. D. Berendt and E. L. R. Corliss, *Quieting: A Practical Guide to Noise Control, NBS Handbook 119,* NBS, Washington, D.C., July 1976.
11. *Classification for Determination of Sound Transmission Class, ANSI/ASTM E 413-77,* ASTM, Philadelphia, Pa., 1973.
12. *Recommendation R717, Rating and Sound Insulation for Dwelling,* ISO, Washington, D.C., May 1968.
13. *Recommendation R140, Field and Laboratory Measurements of Airborne and Impact Sound Transmission,* ISO, Washington, D.C., Jan. 1977.

General References

Porous acoustical materials

G. E. Johnson, "How to Select Effective Lagging Configurations," *Sound Vib.* **11**(4), 28 (Apr. 1977).

Acoustical Ceilings: Use and Practice, Ceilings and Interior Systems Contractors Association, New York, 1978.

G. J. Sanders, "Silencers: Their Design and Application," *Sound Vib.* **2**(2), 6 (Feb. 1968).

W. A. Utley, A. Cumming, and H. D. Parbrook, "The Use of Absorbent Material in Double-Leaf Wall Constructions," *Sound Vib.* **9**(1), 90 (Jan. 1969).

H. N. LoFaro, "Engineers Guide to Felt," *Mat. Eng.* 34 (Nov. 1971).

Guide and Data Book, Systems, American Society of Heating, Refrigeration, and Air Conditioning Engineers, New York, 1976, Chapters 35, 40.

Guide and Data Book, Fundamentals, American Society of Heating, Refrigeration, and Air Conditioning Engineers, New York, 1977, Chapters 7, 13.

Sound reflective material

T. B. Heebink and J. B. Granthan, "Field/Laboratory STC Ratings of Wood-Framed Partitions," *Sound Vib.* **5**(10), 12 (Nov. 1971).

I. L. Ver, "Impact Noise Isolation of Composite Floors," *J. Acoust. Soc. Am.* **50**(4), 1043 (Nov. 1971).

N. K. D. Chondhury and P. S. Bhandari, "Impact Noise Ratings of Resilient Floors," *Acoustica* **26**(3), 135 (Mar. 1972).

R. D. Berendt and G. E. Winzer, *Sound Insulation of Wall, Floor, and Door Constructions, National Bureau of Standards Monograph 77,* NBS, Washington, D.C., Nov. 1964.

Standard Recommended Practice for Installation of Fixed Partitions of Light Frame Type for the Purpose

of Conserving Their Sound Insulation Efficiency, ANSI/ASTM E 497-76, ASTM, Philadelphia, Pa., 1976.
Standard Method for Laboratory Measurement of the Noise Reduction of Sound-Isolating Enclosures, ANSI/ASTM E 596-78, ASTM, Philadelphia, Pa., 1978.

Sound damping materials

S. M. Brown, "The Development and Testing of Vibration Damping Materials," *Sound Vib.* **11**(4), 28 (Apr. 1977).
J. J. Burns, "The Use of Asphalt-Mastics for Acoustical Damping," *Sound Vib.* **6**(10), 4, 6, 8 (Nov. 1972).

Vibration isolation

E. Rivin, "Vibration Isolation of Industrial Machinery—Basic Considerations," *Sound Vib.* **12**(11), 14 (Nov. 1978).

CHARLES R. JOKEL
Bolt, Beranek & Newman

INSULATION, ELECTRIC

PROPERTIES AND MATERIALS

Electrical devices generally consist of four separate components: conductors, a magnetic circuit, mechanical supports, and electrical insulation. Insulation confines, restrains, and directs the flow of electric currents. Gases, liquids, and solids are used to insulate electrical devices, depending on the specific needs of the system. Although the function of an insulator is to restrain current flow from a conductor to ground or to a lower potential, insulations often must provide mechanical support, protect the conductor from environmental degradation, and transfer heat to the surrounding environment. The insulating system influences the reliability of the device, and the type and quality of the insulation affects its cost, weight, size performance, and life.

Conductors and Insulators

All materials conduct electricity to some extent. When a voltage is applied to a material, the current flow is defined by Ohm's Law: $I = E/R$, where I is current (A), E is voltage (V), and R is resistance (Ω). Resistivities of materials vary over a broad range, from 1.7 $\mu\Omega$/cm for a good conductor, eg, copper, to 10^{18} Ω/cm for polystyrene (see Fig. 1). The conduction properties of metals differ from those of insulators primarily as a function of how the elements are bonded. The three types of bonds are metallic, covalent, and ionic.

Metals are molecules formed from electropositive elements on the left side of the

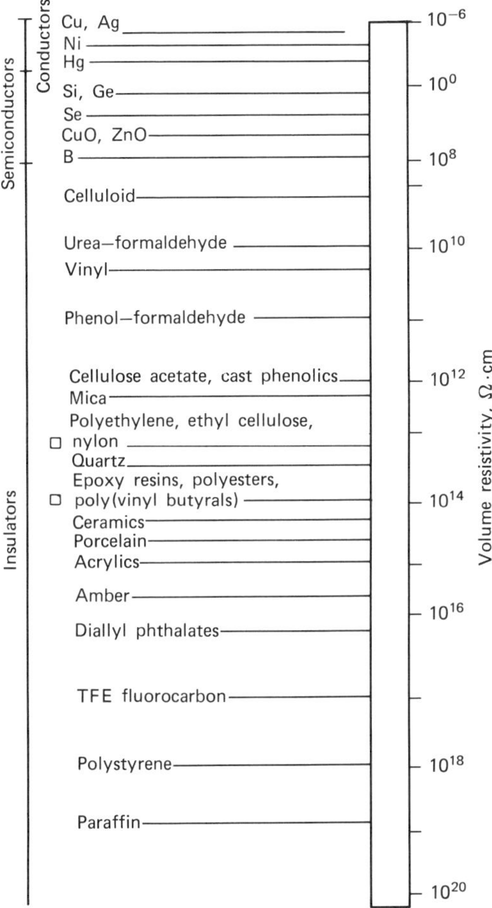

Figure 1. Resistivity of materials (1).

Periodic Table. In metals, the outer electrons (valence electrons) of an atom can be partly shared with an adjacent atom. However, these electrons are insufficient to satisfy either atom or to achieve stability. Electron-sharing bonds are formed and hold a collection of atoms together to form a molecule (metallic bonding). The atoms are highly oriented, resulting in a crystal lattice with the valence electrons moving continuously and randomly in the lattice.

When an electric field is applied to a metal, a drift motion of the valence electrons occurs, thereby establishing a current. The current density J is proportional to voltage as $J = \sigma E$, where σ is conductivity. J also is expressed by $J = nev$, where n is the number of valence electrons per unit volume of the metal, e is electronic charge, and v is the drift velocity. Therefore, $\sigma = nev/E$; conductivity is a function of the availability of electrons and of their drift velocity, both of which are determined by the lattice structure. Electrons moving in a metal collide with imperfections (eg, impurities, grain boundaries, lattice holes), and these collisions result in frictional energy absorption which increases the temperature of the metal. The increase in the ambient temperature of the metal causes greater vibration of the lattice, which increases the rate of electron collision. These cycles of vibration and collision lead to a decrease in the metal's conductivity.

The most common insulators are organic polymers, which are covalently bonded (see Polymers; Polymers, conductive). Covalent molecules are formed by the reaction of electronegative elements from the right side of the Periodic Table. In covalent molecules, the outer electrons are shared. Covalent molecules do not form lattice structures, and most polymers are amorphous rather than crystalline. Electrons, being tightly shared between atoms, generally are not available to move over the molecule and, hence, applied electric fields do not cause current flow. Therefore, the covalently bonded organic polymers are excellent insulators. The very low conductivities of polymers vary with the environment. When conditions, eg, moisture or increasing temperature, cause a softening of the polymer, the conductivity increases. Although the organic polymers are not crystalline, they often are highly polar or unsymmetrical. This occurs because the covalent bond occurs between unlike atoms and the electron sharing is not completely equal, the electrons spend more time closer to one nucleus than to the other. This polarity affects conductivity as described below.

In ionic bonding, one element donates an electron to another and both elements attain stable shells. The strength of the bond is related to the fact that the electron sharing has created a positive and a negative ion which are associated electrostatically. Examples of ionic molecules are NaCl, CaO, and $MgCl_2$. Ionically bonded molecules are excellent insulators while dry, but moisture dissociates the molecule and permits ionic conduction. The ionic molecules usually are highly polar and crystallize readily.

When an electrical insulating system is being designed, the most common insulating materials chosen are the covalently bonded organic polymers. The conductivity of polymers is a function of two factors: ionic conduction and dielectric absorption. The conductivity is greatly influenced by the type of material, viscosity, temperature, humidity, and field frequency.

Ionic Conduction. All polymers contain certain ionic species. These species result from impurities, residual catalysts, and additives intended to enhance performance, thermal breakdown, and oxidation. The ionic impurities conduct under voltage stress depending on their concentration and mobility. The mobility is influenced greatly by the environment. Water from high humidity environments diffuses into the polymers and increases the ionic conduction. At temperatures above the glass-transition temperature of the polymers, the flexibility of the polymer increases and causes the polymer structure to relax, which results in greater ion mobility. Flexible polymers have been shown to be more conducting than rigid polymers of the same chemical species. In liquids, increased temperatures decrease the viscosity and increase conduction. Therefore, thick liquids are better insulators than liquids of low viscosity.

Dielectric Absorption. When dielectric materials are electrically stressed in a d-c field, a current immediately flows. The current decreases with time until a steady-state condition of current flow is reached. If the dielectric is shorted, the current does not decrease immediately to zero but decays with time. These current flows differ with each polymer and the environment. Under a-c stress, the same phenomenon occurs two times during each cycle. The current flow is different from the capacitive effect of charging a capacitor and is known as dielectric absorption; dielectric absorption has three components; dipole orientation, interfacial polarization, and atomic and electric polarization.

Dipole Orientation. Organic polymers may be either nonpolar or polar depending on their molecular structure. Polymers, eg, polyethylene and polytetrafluoroethylene, are nonpolar because they are completely symmetrical. Polymers, eg, epoxies and polyesters (qv), display varying polarity depending on their asymmetry (see Epoxy resins). Various functional groups on the polymer chain, eg,

$$—NH_2,\ —OH,\ —\overset{\overset{\displaystyle O}{\|}}{C}NH_2,$$

affect the polarity as do sidebranching, random cross-linking, and imperfections in the polymer chain. When a voltage stress is applied to a nonpolar molecule, the molecule does not respond and does not conduct; however, when it is applied to a polar molecule, the dipoles rotate in the field to align with the charge, thereby conducting a current. Conduction stops when the molecule rotates as much as structure permits. In d-c fields, the conduction stops with the final alignment, but in a-c fields the conduction is influenced strongly by frequency; higher frequencies cause greater conductivity up to some frequency where the molecules do not have enough time to rotate before the polarity reverses. Factors affecting the dipole orientation are viscosity, temperature, degree of cross-linking, molecular structure, frequency, and moisture content.

Interfacial Polarization. Insulation systems rarely consist of a single pure material but usually are combinations of two or more insulators in series, eg, paper (qv), film, and a fluid (as in capacitors) and rubber and air (as in cables). In such combinations of insulators, many interfaces of the different materials exist with each material possessing a unique conductivity and dielectric constant. When the system is stressed in an electric field, charges accumulate at the interfaces (as in a capacitor) with consequent energy storage. This process is known as Maxwell-Wagner interfacial polarization and is strongest at very low frequencies.

Atomic and Electronic Polarization. Atomic and electronic polarizations are most prominent in gases, somewhat so in liquids, and of little effect in solids. Atomic polarization is the displacement of the nuclei of a molecule in an electric field, and electronic polarization results in the displacement of electrons toward the positive electrode. Most of the effects of these polarizations occur in the infrared and visible frequencies and relate to the optical index of refraction n by $\epsilon' \cong n^2$ (ϵ' = dielectric constant).

Insulator Properties

Dielectric Constant. The dielectric constant (ϵ') is a fundamental property of every insulating material. The dielectric constant of a polymer is the ratio of the capacitance of a capacitor with the polymer as the dielectric to a similar capacitor with a vacuum as the dielectric. The dielectric constant is determined by the contribution of the energy stored in the material. Typical values of ϵ' for polymers vary from 2–12, as compared with clean dry air or vacuum as 1. Inorganic compounds, eg, titanates, have dielectric constants of over 1000; when blended with polymers, these compounds permit insulations with high values of ϵ'.

The dielectric constant for a material is dependent on molecular polarity, field frequency, and temperature. Nonpolar materials do not vary greatly in ϵ' with frequency, but polar materials have a great dependency on frequency, as illustrated in Figure 2. At low frequencies, all dipoles that can rotate do so, and the dielectric constant is maximum at ϵ_e' (equilibrium value). As frequency increases, the motion of the dipoles increases until such rapid reversals are occurring that the molecules cannot respond, resulting in the value of ϵ_∞'. Note that this value is ca 2 and that the contribution to ϵ' below this is caused by atomic and electron effects. The mean values of $\epsilon_e' - \epsilon_\infty'/2$ defines the relaxation time γ which is the mean time for the molecule to rotate 180°.

In both nonpolar and polar materials, ϵ' decreases slightly with increasing temperatures as the polymer chains relax. At the glass transition temperature, a more abrupt change occurs and ϵ' continues to decline until the polymer melts or the degradation temperature is reached.

The voltage distribution across insulators in series is proportional to the insulation thickness and inversely proportional to the dielectric constant. Thus,

$$\frac{v_1}{v_t} \times 100 = \frac{t_1}{\epsilon_1'} \frac{(\epsilon_1' \epsilon_2' \cdots\cdots \epsilon_n')}{t_1\epsilon_2'\epsilon_n' + t_2\epsilon_1'\epsilon_n' + \cdots\cdot t_n\epsilon_1'\epsilon_2'},$$

where v_t = total average voltage across the entire system; v_1 = voltage across the first component of insulation in series; t_1 = thickness of that insulation; ϵ_1' = dielectric constant of that same insulation; and n = the nth component. Standard test methods for dielectric constants are described in ASTM D 150.

Dissipation Factor. When an a-c voltage is imposed on a dielectric material, a current flows and leads the voltage in time by an amount δ as shown in Figure 3. This current results in a power loss within the dielectric of $W = 2\pi f C_p \tan \delta E^2$ where f = frequency, C_p is the equivalent parallel capacitor, E is the voltage, and δ is the dissipation factor. Defined vectorially, the dissipation factor is the ratio of parallel reactance to parallel resistance in a capacitor having the dielectric material. These relationships

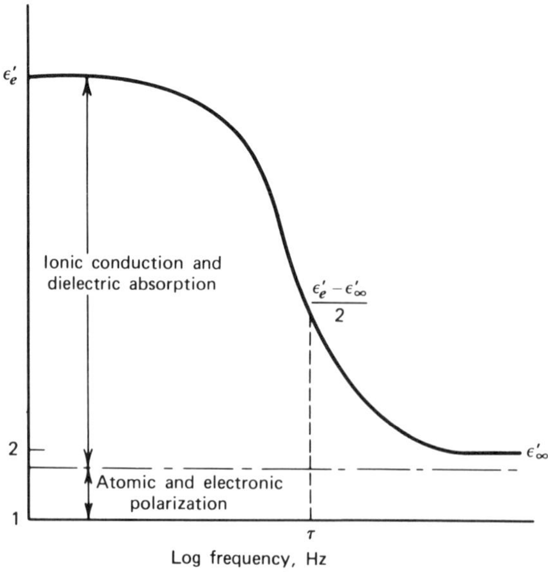

Figure 2. Variation of dielectric constant of polar molecules with frequency.

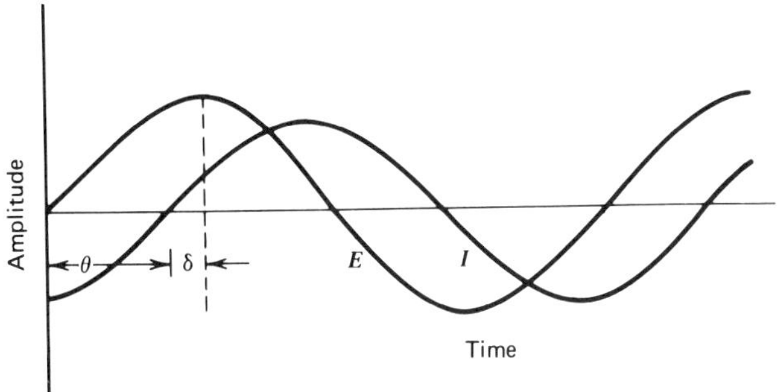

Figure 3. Relationship between the voltage (E) and the current (I) for a dielectric material in an a-c field, θ = phase angle and δ = loss angle.

are shown in Figure 4. The angle δ is the loss angle, θ is the phase angle, $\tan \delta$ is the dissipation factor, and $\cos \theta$ is the power factor.

Note that, for small values, the dissipation factor closely approximates the power factor. It is desirable to use dielectric materials with low values in an insulation system (if consistent with other system requirements) in order to limit the thermal losses in the dielectric. Thermal runaway can occur with poorly selected insulation systems wherein conduction losses in the dielectric material increase its temperature. The increasing temperature increases the dissipation factor which increases the losses, resulting in ever-increasing dissipation factor and thermal buildup, until dielectric failure occurs. Dissipation factor is affected by temperature, frequency, humidity, voltage, and partial discharges that occur in or on the insulation. Standard test methods are defined by ASTM D 150.

Loss Factor, Loss Index, Relative Loss Index. The loss factor ϵ'' is the product of the dissipation factor and the dielectric constant or the ratio of the unrecoverable energy ω' to the total energy ω that is introduced by an electrical stress E applied to an insulator. Therefore: $\epsilon'' = \omega'/\omega$, and $\omega = CV^2/2 = \epsilon' E^2/8\pi$, where C = capacitance of capacitor, V = voltage of capacitor, and E = electric field $\therefore \omega' = \epsilon' \tan \delta\ (\omega C_o V^2)$, and $\epsilon'' = \epsilon' \tan \delta$.

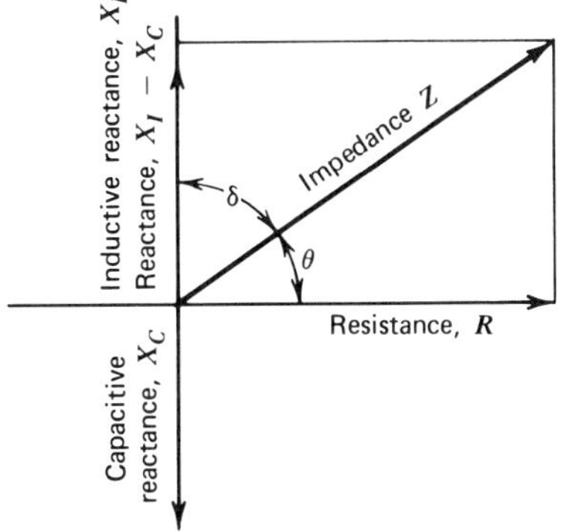

Figure 4. Vector diagram for a dielectric showing the relationship between the reactance and resistance (δ is the loss angle and θ is the phase angle).

Resistivity. Resistivity is the inverse value of conductivity. It commonly is expressed by: volume resistivity, surface resistivity, and insulation resistance as defined in ASTM D 257.

Volume Resistivity. Volume resistivity is the resistance between the faces of a unit cube of an insulator and is equal to $\rho = AR/\alpha$ where volume resistivity is $\rho\Omega$·cm, A is the cross-sectional area of the electrical path (cm^2), R is the measured resistance (Ω), and α is the length of the electrical path (cm). Values for insulators range from 10^8–10^{19} Ω·cm and are affected greatly by temperature and humidity.

Surface Resistivity. Surface resistivity is the ability of an insulator to resist the flow of a current on its surface. It is defined as PR/d where R is the measured resistance (Ω), d is the distance between the electrodes (cm), and P is a parameter of the guarded electrode (cm) given in ASTM D 257. The units of surface resistivity are expressed in Ω/square. The surface resistivity is dependent on the condition of the surface, its contamination with dirt or moisture, and its topography.

Insulation Resistance. Insulation resistance is measured on a device or configuration and is the integrated effect of volume and surface resistivity, usually expressed in MΩ. The quality of an insulation system can be judged by this measurement, and it is particularly useful for field evaluations.

Arc Resistance–Track Resistance. At times, insulators are subjected to electric arcs, scintillations, flashovers, or other conditions where the surface of the insulator suddenly is subjected to an electrical discharge. This discharge is not a surface corona but an air breakdown between two electrodes across the surface of an insulator. Because this breakdown can occur under a variety of voltage and current conditions and with many types of insulator contamination, a variety of measurements has been developed. A low current, high voltage test of low intensity on clean insulator surfaces is described in ASTM D 495. The test measures the time of exposure of the sample to an intermittent arc of increasing duration. The end point occurs when a carbonaceous conducting path forms. Typical values are 20 s for a poor insulator, eg, a phenolic, to more than 180 s for an arc-resistant grade of polyester. Three other tests (ASTM D 2132, D 2302, and D 2303) have been developed wherein the surface of the insulator is intentionally contaminated to heighten the tendency to form conducting paths. Other work has been done to evaluate devices and materials under field conditions, particularly out-of-doors, and one of these permits predictions of life expectancy under voltage stress (2).

Dielectric Breakdown. Every insulator possesses a few free electrons. Under the stress of an electric field, these electrons tend to move between the electrodes. A voltage level is reached wherein the electrons impact with atoms and molecules, therein dislodging other electrons and causing an electron avalanche or breakdown. Three theories regarding the basis of dielectric breakdown have been postulated: the thermal theory is based on a transport of current in certain portions of the insulation that have high conductivity resulting in ever-increasing current flow, temperature, and conductivity until failure occurs; a theory of ionic species generation within the insulator postulates ionic movement which uses energy and generates increasingly more ions until failure; and physical failure of the insulation occurs, opening direct paths for current flow and causing massive electrical failure.

In practice (at least in solids), probably all three conditions exist and often are initiated by the physical failure of the insulation. All of these conditions describe dielectric strength which is the voltage at failure divided by the thickness of the insulator

(MV/m). Values of dielectric strength are highly dependent on the density of the insulator, the electron-absorbing quality, the applied voltage and the rate of its application, thickness, the moisture content of the insulator, frequency, time, and temperature. Tests are defined by ASTM D 149. Dielectric strength values are useful in comparison studies, but values tend to be highly influenced by thicknesses—the thinnest insulators being best.

In gases, the values of breakdown are defined by Paschen's law (3) as shown in Figure 5 which relates the breakdown voltage of a gas to the product of pressure and electrode spacing for uniform fields at constant temperature. Note that a minimum breakdown of 335 V for air exists at very small spacings. In general, the electric strength of solids is greater than that of liquids which is greater than that of gases. The intrinsic dielectric strengths of some materials are shown in Table 1.

The breakdown of solid insulation decreases with increasing time. This characteristic is known as voltage endurance and is described in ASTM D 2275. Figure 6 gives a typical voltage endurance curve for a turbine generator coil insulated with a polyester resin and mica. Voltage endurance is controlled by the electrical stress concentration and partial discharges; both are internal and on the surface.

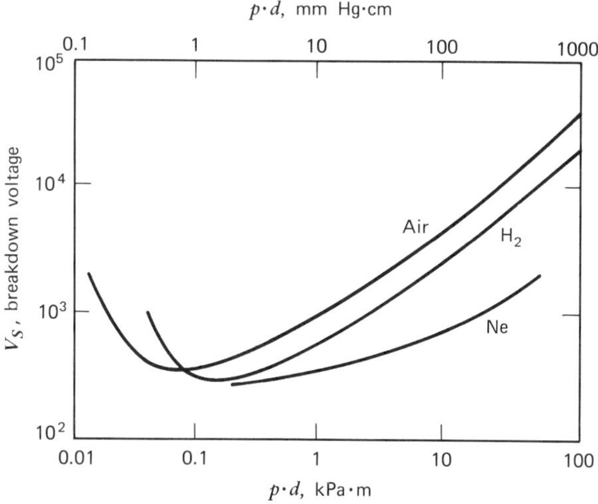

Figure 5. Paschen's curve for air, hydrogen, and neon; $p \cdot d$ = pressure × distance.

Table 1. Intrinsic Dielectric Strengths, MV/m [a]

Material	MV/m
air, 101.3 kPa[b] (1-cm gap)	3
sulfur hexafluoride, 101.3 kPa[b] (1-cm gap)[c]	8
air, 608 kPa[b] (1-cm gap)	15
sodium chloride crystal	150
Wemco C mineral oil	197
polyethylene	650
poly(methyl methacrylate)	984
mica, perpendicular to laminations	965

[a] Ref. 4.

[b] To convert kPa to atm, divide by 101.3.

[c] See Fluorine compounds, inorganic–sulfur.

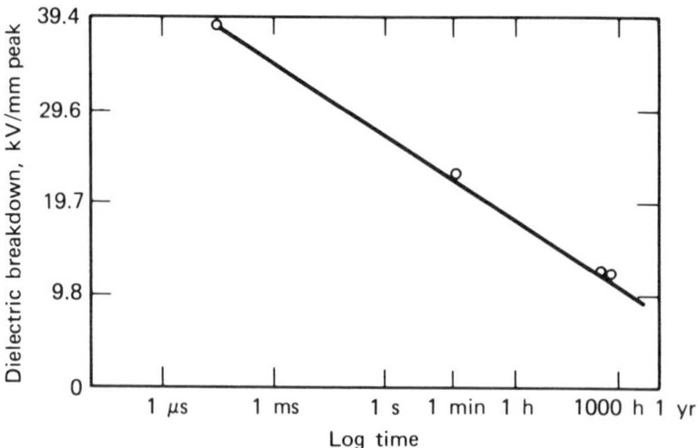

Figure 6. Voltage endurance curve for a modern machine insulation, illustrating reduction in breakdown level at long periods of dielectric stress (5).

A breakdown value frequently used in rating electrical equipment subject to lightning or switching surges is the basic impulse level (BIL). This is an arbitrary value associated with a device that describes the ability of the insulation system to withstand impulses. Associated with the value is a statement of the polarity and the wave form used (usually 1.2 μs rise and 50 μs fall). BIL values usually are given for transformers, circuit breakers, and associated transmission devices.

Partial Discharges—Corona. Almost all solid insulation systems have small holes or voids somewhere in the structure that have been formed as a result of entrapped gases, delaminations, shrinkage, discontinuities from differences in thermal expansion, thermal aging, or lack of adhesion. At voltages below breakdown, the gases in the voids ionize and electrical discharges occur. This value is the corona inception voltage and is defined (6) as: $V_I = 93{,}564(t/\epsilon_s^{'})^{0.46}$ ($V_I = 720(t/\epsilon_s^{'})^{0.46}$), where: V_I = corona inception voltage, t_1 = thickness in meters (mils); and $\epsilon_s^{'}$ = dielectric constant of solid insulators.

V_I is not an intrinsic property of the insulation materials but is related to the insulation system and its method of manufacture. The electric strength of a gas or liquid is lower than that of solids, and a gas pocket often is at a higher electric stress than is the solid. The stress level is related to the dielectric constants of the material (see Dielectric Constant) by the equation:

$$E_g = \frac{\epsilon_s^{'}}{\epsilon_g^{'}} E_s$$

where E = the voltage stress, ϵ' = the dielectric constant, and subscripts s and g refer to solid and gas, respectively.

In many cases, ϵ' for solids is greater than 2 and ϵ' for air is 1, resulting in an electrical stress in the gas that is at least twice that in the solid. If the inclusion is a liquid, partial discharges occur, even if the stress in the liquid is lower than the solid because of the inferior electric strength of the liquid. When the solid is void-free, partial discharges can occur on the surface, particularly at the sharp edges of an electrode where voltage stresses are very high. Partial discharges in air create ozone (qv), oxides of nitrogen, and decomposition products of the insulation resulting in corrosive acidic by-products which further degrade the insulation. If the corrosive by-products are present in sufficient quantities, they ultimately result in insulation failure. Most electrical equipment is designed to operate at voltages well below those where partial discharges occur but, in some high voltage devices, this is not economically feasible. Generators, transmission insulators, and bushings are devices in the insulation system that permit operation in the presence of partial discharges. Mica (see Micas) and glass (qv) (being inorganic and resistant to acids and oxidation) are particularly corona resistant and, hence, are used where electrical stress levels are high.

Treeing: Electric, Water, Electrochemical. High molecular weight polyethylene is used to fabricate cables for distribution circuits—much of which are buried in the ground—at stresses above 6 MV/m (see Olefin polymers). Voids in the cable easily experience corona and result in electrical treeing. Electric treeing is associated with high electric stresses and produces a branched, treelike series of interconnected channels in the insulation.

At electrical stresses of 2 MV/m, corona does not occur, and it was thought that cables from 5–35 kV at this lower stress level would have long lives. However, cable failures at these stress levels have occurred and have resulted in breakdown channels similar to those associated with electric treeing but with a structure that is more diffuse. This breakdown behavior is water or electrochemical treeing (7–8). It has been found that water treeing initiates at protrusions, imperfections, voids, or inclusions in the insulation, and that water must be present. Water treeing also is influenced by voltage stress, frequency, and to a small extent, temperature. The presence of the trees reduces the electric strength to as little as 15% of the original and can cause the dissipation factor to increase by a factor of 100. Cross-linked polyethylene and ethylene–propylene rubber (see Elastomers, synthetic) are substantially more resistant to water treeing than is high molecular weight polyethylene.

Thermal Conductivity. Thermal conductivity controls heat transfer from the conductor to the ambient environment. Heat is generated in the conductor, the magnetic circuit, and the insulator. The thermal conductivity of insulation systems is much lower than that of metals. Examples of the thermal conductivity of various materials are shown in Table 2.

Thin layers of air often occur in an insulation. This is particularly true in tapes that are applied in multiple layers. These tapes have much higher thermal conductivity in the longitudinal direction than they do in the transverse direction across the insulation and, frequently, these tapes are impregnated with a resin that removes the air and greatly improves the thermal conductivity.

The addition of fillers (qv) to resins improves their thermal conductivity. Metallic fillers are best but detract from the electrical properties. Inorganic fillers whose particles are rodlike or leaflike improve the thermal conductivity more than spherical particles because of the improved contact area from particle to particle. High filler contents must be reached for significant effect since the relationship between filler content and thermal conductivity is not linear. Consideration must be given to the viscosity of the resin as fillers are added since this property is also filler dependent. Fillers having optimum particle-size distribution can substantially improve thermal properties with minimal influence on viscosity (9).

Thermal Expansion. Many insulation applications depend on the adhesion of the insulation to conductors, the magnetic core, or the structural components (see Adhesives). Electrical components are enameled magnet wire, resin encapsulations, and cast structures. Large differences in the coefficient of thermal expansion can cause

Table 2. Thermal Conductivity of Materials

Thermal conductivity	(W/m·K)
copper (99.99% pure)	401[a]
	398[b]
aluminum (99.99% pure)	236[a]
	237[b]
stainless steel (various grades)	36.7[a]
std (300 series)	16.3[a]
porcelain (ceramic/glass)	3.38[b]
(quartz)	1.36[b]
cast epoxy, alumina filled (77%)	1.72[b]
cast epoxy, silica filled (74%)	1.22[b]
epoxy/glass-cloth laminate	0.30[b]
alkyd wire enamel	0.21[b]
mica tape (resin impregnated) Doryl/Kapton	0.17[b]
polyester film (Mylar)	0.038[a]
phenolic/cellulose-cloth laminate (phenol formaldehyde/cellulose paper)	0.0076[b]

[a] At 373 K.
[b] At 298 K.

debonding which introduces problems in heat transfer, corona, and partial-discharge formation. Coefficients can vary over several orders of magnitude from 2.13×10^{-4}/°C for polyethylene to 13×10^{-6}/°C for mica.

Physical Properties. Electrical insulations must have good physical properties to function satisfactorily since their role in the system often is mechanical as well as electrical. In addition to withstanding the mechanical forces of the manufacturing process, the insulators are designed for specific mechanical performance. For example, high tensile strengths are required of filament-wound rings that are used to support and restrain the end turns of turbine generators during short-circuit conditions. Slot wedges, which are used in motors to hold the coils in place against centrifugal forces, require laminates with high flexural strength. Many insulations must have great flexibility to permit their application to conductors; and tear strength and burst strength are required of film insulations.

Thermal Life. ***Oxidation.*** Where insulation systems are used in air environments, oxidation reactions become important and must be considered in determining the life of the insulation system. Most polymers oxidize, and the aliphatic resin systems are particularly oxidation sensitive. For example, polyolefins (polyethylene, polypropylene, etc) oxidize readily. These polymers are stabilized for oxidation by incorporating antioxidant compounds (usually phenolic or aromatic compounds) (see Antioxidants and antiozonants). When the antioxidants are exhausted, the polymer fails dramatically. For example, a polypropylene heated at 154°C for 90 days showed no change in appearance or in properties. On the 91st day, the polymer began to chalk and discolor as the antioxidant reached depletion. On the 95th day, the polymer completely discolored, began flaking apart, and failed by brittle fracture. A rough rule of thumb is that each 8°C increase in temperature doubles the rate of oxidation.

Oxidation causes several responses: production of volatile by-products which lead to partial discharges and to electrical breakdown; generation of corrosive by-products, mostly acidic chemicals, which attack the insulation system; and chemical changes to the molecular structure, which make the polymer more rigid and brittle, increasing its propensity to shrink, crack, or break prematurely.

Pyrolysis. Pyrolysis is the decomposition of organic polymers caused by the effects of heat exclusive of oxidation. Pyrolysis usually results in decreased molecular weight or even a reversion to the monomer and, even in air, occurs within the bulk of the material. Pyrolysis effects can be observed by conducting elevated-temperature tests in vacuum or in an inert gas, eg, nitrogen. Generally, two competing reactions occur, ie, cross-linking and depolymerization. Cross-linking is the tendency of a polymer (especially a thermoset polymer, but also thermoplastics) to form additional cross-linking bonds with thermal activation. Cross-linking occurs most extensively at temperatures below 250°C. Depolymerization is molecular fragmentation caused by high temperatures in which pendant molecular side chains and the main molecular structure fracture or become separated into smaller and smaller molecular fragments, resulting in increasingly lower molecular weight polymer structures. These effects become predominant for most polymers at ca 250°C but also occur at lower temperatures at very low rates.

Glass Transition Temperature (T_g). Most polymers used for electrical insulation are hard, glasslike materials that are rigid and inflexible. However, as temperatures are increased, polymer chains slip more and relax, permitting increased bending and flexibility. Each polymer has a critical temperature at which the polymer ceases to be rigid and strong and becomes comparatively flexible and rubberlike. This temperature represents a distinct energy input defined by sudden and abrupt changes in tan δ, index of refraction, hardness (qv), dielectric constant, physical strength, and conductivity. In thermoset polymers, the glass transition temperature is closely related to the polymerization (curing) conditions and usually exceeds 70°C. In thermoplastic polymers, the T_g often is below room temperature (ca −20°C).

Predicting Thermal Life. Many investigators have studied the integrated thermal effects of oxidation, pyrolysis, depolymerization, and chemical attack, and have concluded that aging is a chemical-rate phenomenon (10–11) and can be represented by the Arrhenius equation:

$$\text{rate} = Ae^{-E/RT}$$

where E = the activation energy for the reaction, R = gas constant per mole, T = absolute temperature, K, and A = constant, dependent on concentration. The equation can be rewritten as:

$$\ln L = \ln A + BT$$

where L = insulation life (d) and B = constant.

These relationships are used to predict the life of insulation materials from a limited number of tests at elevated temperatures. In practice, some limiting property value of an insulation is selected as an end point. This can be a physical or an electrical property, but often is 50% of the original electric strength. Samples are exposed to three or four temperatures in excess of use temperature, and the time to reach the end point is plotted so that the reciprocal of absolute temperature is the abscissa and the log of life is the ordinate.

A thermal life curve for some wire enamels is given in Figure 7. The Arrhenius method is of great use in predicting life for wire enamels, varnishes and tape-wrapped coils.

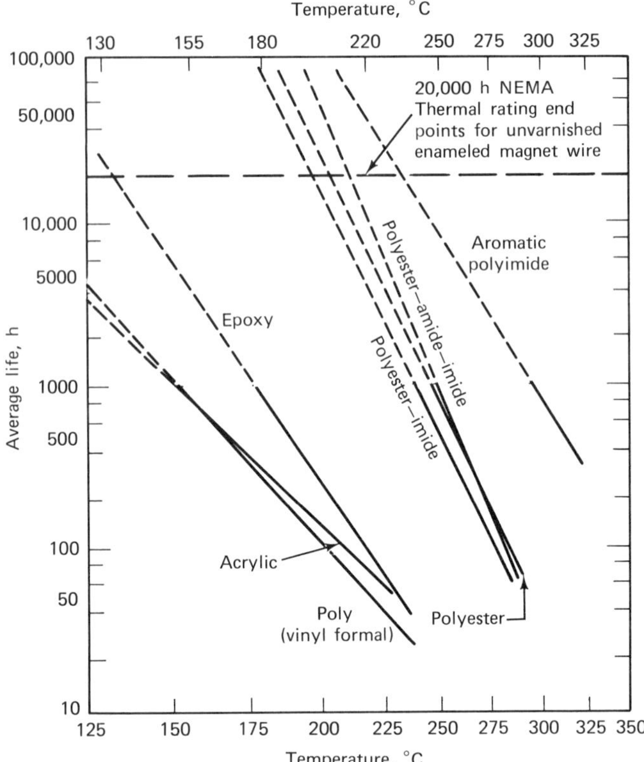

Figure 7. Thermal life curves of several different enameled magnet wires, tested per IEEE 57 (ASTM D 2307-68) on AWG 18 copper wire—unvarnished twisted pairs.

Another method that is useful for rough-screening estimates is the use of thermogravimetric analysis (tga) to establish a thermal index (TI) rating for insulations (12). The tga test establishes a plot of weight loss versus temperature. The thermal index is:

$$\mathrm{TI} = \frac{A + B}{2K}$$

where A = the temperature where a straight line drawn through the 50% and 20% weight-loss values intercepts the 0% weight-loss temperature, B = the temperature where the curve intercepts 50% weight loss, and K = a factor that is derived by using Kapton film as a standard in the tga test, eg, $240 = A(\text{Kapton}) + B(\text{Kapton})/2K$.

Because of the many possible interactions, insulations are not usually evaluated for thermal endurance by any single test. Instead, a variety of systems tests (functional tests) have been developed that test all the components at one time at elevated temperatures. These include motorettes, formettes, wrapped-bar tests, and motor-reversal tests. Almost all of these test methods use elevated temperature or high frequencies to accelerate the aging effects and Arrhenius plots to predict life. Performance must be compared to systems with known lives.

Chemical Resistance. Any insulation system design must provide for resistance to chemical environments to which the system may be exposed during manufacture or use. For example, many are exposed to cleaning solutions, fluxes, electroplating solutions, and paint and varnish solvents during manufacturing (see Paint and varnish removers). Most systems experience, in their lifetimes, accidental or intentional contact with a variety of chemicals, solvents, and detergents. The special cases of oxidation and pyrolysis also are chemical effects, but the chemical effects of concern are those that exhibit deleterious plasticizing, solvent, and stress-relief effects on polymers. One of the most significant of these is water which, in reaction with polymers, causes hydrolysis or a chemical reaction with the polymer chain and results in chain breakage and decreased properties. Several polymer families are particularly sensitive to hydrolysis including polyesters, some polyurethanes, and ureas (see Urethane polymers; Urea and urea derivatives). In general, those polymers that are derived from condensation reactions during their synthesis are most susceptible to hydrolysis. Other polymers absorb water and, although no reaction occurs, the water plasticizes the polymer thereby causing undue flexibility, a decrease in physical and electrical strengths and resistivity, and increased dissipation factors. In particular, the cellulose-based polymers are poor in water resistance. Certain polymer systems resist water well, eg, epoxies, fluorocarbons, poly(vinyl chloride), polyethylene, and polypropylene.

Organic solvents can be particularly harsh on many polymers. The thermoplastic resins often are dissolved by many organic solvents and, even when they do not dissolve, they can be locally crazed and cracked, particularly at areas of high mechanical stress. Many polymers used in power equipment are in contact with insulating oils and must be resistant to the attack and absorption of these fluids.

Corona or partial discharges in or on the insulation provide localized areas of chemical attack from the chemical species formed in the arc. For example, if air is present, quantities of ozone and oxides of nitrogen $(NO)_x$ are formed by the arc. Ozone is a powerful oxidant, and the nitrous oxides, in combination with water, form corrosive acidic compounds that readily attack many polymers. Another case of chemical attack is in hermetic refrigeration (qv) systems where the insulations used in the hermetic motors are immersed in oils and the fluorochlorocarbons are used as refrigerants. Insulators must be carefully selected and compounded to avoid deterioration. Special tests are performed to establish the resistance to hermetic environments.

Outdoor Weathering. Insulation designers sometimes are required to use materials in electrical applications that are exposed to the weather. Destructive influences are ultraviolet radiation and moisture. Few polymers can succeed in such outdoor environments:

Poly(vinyl chloride) materials (PVC) are weather resistant depending on the plasticizer system used. However, in electrical applications, few PVC materials are used, most being used in structural, home-building applications.

Elastomers, ie, butyl [*9010-85-9*], neoprene [*126-99-8*], and silicone [*68083-20-5*], can resist weather conditions and are used as cable and wire insulation.

Fluorocarbons are highly weather resistant but, because of their high cost, are seldom used solely for weather resistance.

Acrylics, eg, poly(methyl methacrylate) [*9011-14-7*] and other acrylic copolymers, weather well and are formulated into varnishes and other electrical components.

Acetals, eg, the polymers of Delrin (DuPont) and Celcon (Celanese) which are based on formaldehyde, are used in arc-resistant applications, eg, cut-outs.

Epoxies are used extensively in bushings and insulators in outdoor applications. The polymers, mostly cycloaliphatic epoxy resins, are combined with fillers which make them arc, track, and weather resistant. The track resistance usually is derived from aluminum oxide trihydrate [*21645-51-2*] (13) which, in the presence of the heat of an electric arc, breaks down and yields the water of hydration, thus cooling the arc and absorbing energy. In addition, the water reacts with carbon forming hydrogen and carbon monoxide from pyrolysis of the resin (the water-gas reaction):

$$H_2O + C \rightarrow H_2 + CO$$

This reaction uses more energy and interrupts the carbon path.

Flammability. Most organic polymers burn readily in air. However, the presence of the halogens (chlorine, fluorine, and bromine) and of nitrogen and phosphorus add to the flame resistance. Polymers, eg, PVC, can be ignited but extinguish quickly when the ignition source is removed. Polytetrafluoroethylene (TFE) polymers and copolymers cannot be ignited in air. Epoxies usually are made self-extinguishing by adding bromine (>39 wt %) to the molecule. Fillers add to the flame resistance, especially antimony trioxide [*1309-64-4*], which has a synergistic effect in the presence of chlorine (see Flame retardants).

Design of Insulation Systems

Economic Aspects. A total system cost includes the cost of the materials, the amount needed, the application process, and the life expectancy. Material costs are related to the polymers used, with polyolefins and phenolics (see Phenolic resins) being low in cost and the fluorocarbons and polyimides (qv) being extremely high. Insulation with high electric strength can be used in thinner sections than other materials; films of polyester, mica tape, and polyimide are very high in electric strength and fabric-based laminates rather low. Process costs can vary widely since labor-intensive processes or long process times result in high manufacturing costs. Other cost factors include the availability of the material, the safety factors to be used, and maximum temperatures being considered.

Health and Safety. When fully polymerized, organic polymers generally are inert and harmless. However, certain polymers or their monomers exhibit positive responses in the Ames test (see Industrial hygiene and toxicology). Notable among these are certain of the bisphenol-A epoxies. Many of the amines that are used to cure epoxies are harsh chemicals that, when in contact with skin, can cause dermatitis (see Amines). Polyester materials, although inert by themselves, often are copolymerized with monomeric styrene which is highly volatile and must be kept to low concentrations in the air. A common polyurethane co-reactant is MOCA (4,4′-methylene-bis-2-chloroaniline [*101-14-4*]) which is thought to be carcinogenic.

Polymers often are used with solvents to impregnate fabrics or papers. Some of these solvents are being investigated for their health effects and they include cresols, toluene (qv), and xylenes (see Xylenes and ethylbenzene).

A common insulating fluid was a polychlorinated biphenyl (Aroclor) [*1336-36-3*] which was used in transformers and capacitors for its excellent electrical strength and low flammability (see Chlorocarbons and chlorohydrocarbons). It resists biodegradation and remains in many fish and bird food chains. It has been removed from the market in the United States except for certain essential applications.

One solid insulation commonly used in electrical insulation systems was asbestos (qv). In the form of tapes and fabrics, it was used for its thermal and chemical stability. However, asbestos fibers were found to cause asbestosis, and the use of this material is strictly limited.

Vinyl chloride monomer is a gas and has been found to be carcinogenic. Its concentration in air is limited to 5 ppm by OSHA.

Electrical insulations sometimes are subjected to short-circuit conditions resulting in electrical arcs, discharges, and fires. Combustion of the insulation produces smoke, carbon dioxide, carbon monoxide, and other chemical species. To improve flammability properties, halogens are added to some insulations; combustion by-products also contain HCl or HBr which are toxic. Certain acrylics and polymers are available that generate very little smoke on combustion.

Gas Insulation

Gases are used as electrical insulation. Highly compressible gases have low conductivity and dielectric constants close to unity. The density is defined by the gas laws:

$$\text{density} = \frac{M}{22.4} \cdot \frac{P}{101.3} \cdot \frac{273}{T} \text{ g/L}$$

where M = molecular weight, g; P = gas pressure, kPa; and T = absolute temperature, K.

The dielectric strength of a gas is a basic property of the gas and is greatly affected by pressure (ASTM D 2477). Figure 8 shows d-c dielectric breakdown for several gases as a function of pressure (and, hence, density). This data is for uniform fields and the behavior changes radically in nonuniform fields; the latter condition is far more common. In nonuniform fields, the breakdown increases with pressure to a critical value beyond which the electric strength decreases. The d-c strength of gases is always higher than the a-c strength. For a given gas, the a-c strength increases slightly with increasing frequency (ca 10%) until the time of ½ cycle equals the transit time of the electrons, at which point the electric strength decreases sharply.

Air is a common dielectric gas (ASTM D 3283). Air insulation is used on most overhead transmission lines at voltages up to 700 kV, and it is often used in switchgear where blasts of compressed air extinguish arcs that form when the breaker contacts open. In many other low voltage applications, eg, switches and relays, air and a solid insulator are used in series to perform the insulating functions.

Nitrogen [*7727-37-9*] (qv) was widely used in cables in the past: however, its use there is decreasing. In transformers, dry nitrogen is often used to blanket the oil, thereby decreasing oxidation of the oil and other insulating components. Carbon dioxide [*124-38-9*] (qv) also has been used in transformers but for different reasons. Carbon dioxide is highly soluble in transformer oil; therefore, transformers often are shipped dry, with CO_2 impregnating the windings. At the use site, oil is admitted and whatever CO_2 remains is dissolved in the oil without forming sites for partial discharges to occur.

Hydrogen [*1333-74-0*] (qv) is used in turbine generators although (in those applications) it contributes only minor insulation functions. Its high thermal conductivity, high specific heat, and light weight make it desirable as a coolant. It is pumped

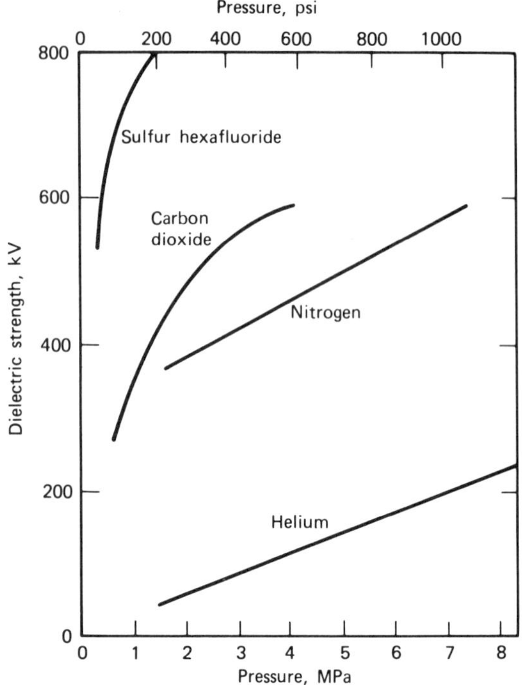

Figure 8. The effect of pressure on the d-c voltage strength of gases in a uniform field. Electrodes are stainless steel plates at a gap of 12.7 mm (14).

at 304 kPa (3 atm) through conduits embedded in the insulated coils to remove heat and maintain the insulation temperature below 135°C.

Electronegative gases are those that attract and hold electrons within their molecular structure and include: SF_6 [*2551-62-4*], CH_2Cl_2 [*75-09-2*], C_2Cl_4 [*56-23-5*], CF_4 (Freon 14 [*75-73-0*]), $C_2Cl_2F_4$ (Freon 114 [*76-14-2*]), $CClF_3$ (Freon 13 [*75-72-9*]), $C_2Cl_3F_3$ (Freon 113 [*76-13-1*]), CCl_2F_2 (Freon 12 [*75-71-8*]), $CHClF_2$ (Freon 22 [*75-45-6*]), and CCl_3F (Freon 11 [*75-69-4*]). Hence, they have high electric strengths (Fig. 8), inhibit partial discharges, and tend to extinguish power arcs. The most useful of these gases is sulfur hexafluoride (SF_6), which is described in ASTM D 2472.

Sulfur hexafluoride is used widely in power circuit breakers and metal-clad switchgear as an arc-interrupting medium. Because it is stable up to 150°C, the intense energy of the arc tends to decompose the SF_6 to lower fluorides of sulfur (SF_2 [*13814-25-0*], S_2F_2 [*13709-35-8*], SF_4 [*7783-60-0*], S_2F_{10} [*5714-22-7*]), and hydrogen fluoride (HF). These compounds are highly corrosive to many insulating components (especially silica-based compounds) and, although special absorber systems remove arced products, insulation systems that are to be exposed to arced SF_6 must be carefully evaluated. SF_6 is used in oil-impregnated paper cable, replacing nitrogen to obtain better corona starting voltages. It increases the corona initiation level by almost 40%, thereby giving a much higher safety factor at a given electrical stress.

SF_6 is used at 446 kPa (4.4 atm) in compressed gas-insulated transmission (CGIT) systems. It is selected for its electric strength in systems at 230 kV and (experimentally) in systems at 800 kV. Entire SF_6-insulated substations have been developed up to 500 kV and result in a great saving of space.

Conducting particles as small as 0.1 mm, which find their way into gas-insulated systems, tend to move in the electric field (particularly in a-c systems) and gravitate between the electrodes. The particles cause small discharges and ultimately result in power arcs and breakdown of the gas or solid insulating members. Elaborate methods have been developed to capture these particles in particle traps and to remove them from the electrically stressed area.

Liquid Insulation

The most common insulating liquids are largely aliphatic mineral oils consisting of $C_nH_{(2n+2)}$ and C_nH_{2n}. The oils usually are complex mixtures of straight and branched-chain compounds and cyclic structures. They may have sizable aromatic contents consisting mostly of benzene and naphthalene derivatives with a wide molecular weight range. Branched molecules inhibit crystallization and depress the freezing point. The oils are high boiling and have dielectric constants from 2–7. Specifications for mineral oil are given by ASTM D 3487. The conductivity of oils is related to their quality and the presence of impurities. Oils are used both for insulation and as a heat-transfer media. For a given application, the selection of the oil depends on the viscosity, pour point, electric strength, specific heat, thermal conductivity, flammability, and vapor pressure. The electric strength (ASTM D 877 and D 1816) varies with purity, suspended particles, electrode shape and spacing, time, and the moisture content (15). A water content of less than 20 ppm is desired since larger amounts rapidly decrease the strength as shown in Figure 9. Impulse strengths (ASTM D 3300) usually are ca three times the peak 60 Hz, one minute electric strength. Discharges in oil produce hydrogen, methane, and carbon particles, and often cause release of dissolved air. Oils oxidize at elevated temperatures (ASTM D 1313 and D 1698), producing sludges that may decrease heat transfer. The presence of copper and other alloys accelerates this effect. Antioxidants and stabilizers, eg, di-*tert*-butyl-*p*-cresol, are added to control the oxidation. The largest use of mineral oil is in transformers where their role is both electrical and as heat-transfer fluids. Paper insulation also is used with the oil in transformers to obtain high electric strengths. Typical properties of a transformer oil are shown in Table 3. Circuit breakers use oil both for electric strength between electrodes, and for quenching the power arcs. Oil that is used in cables is selected mostly for electric strength. Until 1960, the insulation in most power capacitors consisted of polychlorinated biphenyl (PCB) liquids and paper. This combination had excellent electrical and flammability properties which were necessary because of the high voltage stress used. During the 1960s, it was found that polypropylene film and paper are exceptionally stable with trichlorobiphenyl, and these composite structures became the norm until, in 1970, the lack of biodegradability of PCB was discovered. Efforts to develop fluids with good electric strength and corona resistance resulted in a variety of materials (see Table 3) including isopropylbiphenyl, isobutylmonochlorobiphenyl oxide, and mixtures of di-2-ethylhexyl phthalate and trichlorobenzene. In Japan, capacitors now use isopropylnaphthalene [*6158-45-8*] and dimethyl-1,1-diphenyl ethane [*719-79-9*].

Polychlorinated biphenyls were used in transformers for their low flammability properties, particularly for indoor applications. Silicones are being considered as a replacement for PCB, particularly dimethyl silicone which has a viscosity of 50 mm^2/s (= cSt) and a flash point (open cup) of 345°C. Another candidate is a specially refined

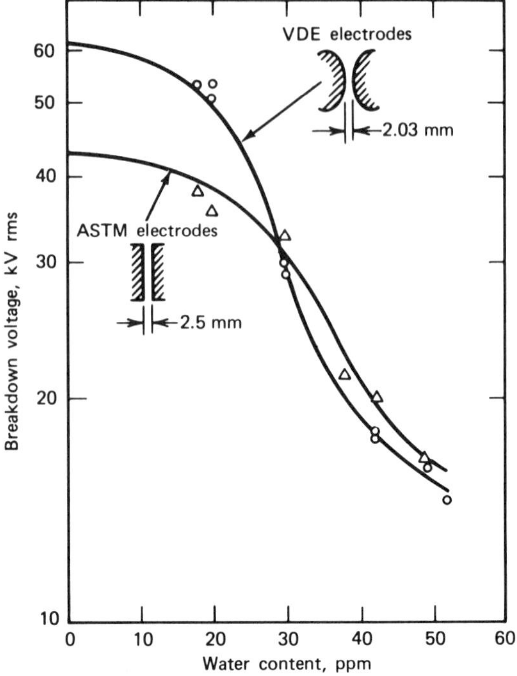

Figure 9. Breakdown voltage of transformer oil vs water content with ASTM and VDE electrodes. Rate of rise = 2 kV rms/s (rms = root mean square) (16).

high molecular weight aliphatic mineral oil with a viscosity of 310 mm^2/s (= cSt) and a flash point of 300°C. Neither of these fluids are completely satisfactory because arcs may cause explosions and because burning tends to be of long duration before the flame extinguishes. Much better performance in transformers has been experienced by the use of fluorocarbons of the general formula C_nF_{2n}, $C_nF_{(2n+2)}$, or $C_nF_{2n}O$ but, although performance is excellent, their cost is extremely high.

Flammability. Transformers have been developed that depend on tetrachloroethylene [*127-18-4*] (Perchlor; C_2Cl_4) or on trichlorotrifluoroethane [*354-58-5*] (CR 113). These are used as insulating fluids and as cooling gases. Solid insulations that are in contact with Perchlor must resist its solvency, since it was developed as a powerful solvent for dry-cleaning applications.

Solid Insulation

Solid insulations are the most common form of insulating materials and, even in those devices utilizing gases or liquids, solid materials are invariably used in series. Although many insulators are supplied by the manufacturer in their solid form, a large number of insulations are supplied as solutions, eg, B-stage (partly polymerized) or monomeric systems, which are applied to the device and polymerized to the final form. The latter case applies to impregnants and varnishes. Most solid insulations play a mechanical role as well as an electrical one. Insulations that separate high voltage members from earth are ground insulations. Insulating materials applied to wires,

Table 3. Properties of Dielectric Fluids[a]

Property	Insulating liquid							
	Mineral oil [8012-95-1]	Trichloro-biphenyl (a PCB) [25323-29-2]	Isopropyl-diphenyl [25460-78-2]	4-Isobutoxy-4′-chlorobiphenyl [41198-92-9]	Di-2-ethylhexyl phthalate −25[b]	Di-2-ethylhexyl phthalate [117-81-7]	Polydimethyl-siloxane [1646-73-7]	Hydro-carbon distillates[c]
relative dielectric constant (at 60 Hz, 100°C)	2.1	4.9	2.6	4.0	4.5	4.3	2.6	2.1
dissipation factor (at 60 Hz, 100°C), %	0.1	0.5	0.2	1.0	1.0	0.5	0.01	0.4
dielectric strength (ASTM D 877), kV	35	35	60	35	35	42	35	38
gas absorption coefficient (ASTM D 2300), μL/min	21	55	180	80	38	25	−13	−5
viscosity (at 37.8°C) mm^2/s (= cSt)	10	10	5.8	10.5	13	29	41	310
pour point, °C	−55	−19	−55	−45	−52	−45	−55	−20 to −30
flash point, °C	150	315	155	200	255	235	345	300
use	transformers, cables, circuit breakers	fire-resistant transformers	high voltage capacitors	high voltage capacitors	high voltage capacitors	low voltage capacitors	fire-resistant transformers	fire-resistant transformers

[a] Reprinted by permission of The Electrochemical Society, Inc., Ref. 17.
[b] 25% by weight of trichlorobenzene.
[c] Mol wt = 500–700.

cables, and conductors comprise turn insulation or conductor insulation. Sheet materials used to separate discrete layers of windings are layer insulation.

Enameled Magnet Wire. Enameled magnet wire is the basic building block of most insulation systems (17). The conductors are either copper or aluminum, both round and rectangular. The wire enamel is applied from solution and metered or wiped onto the surface of the conductor. It is then transported through long ovens (wire towers) where the solvent is evaporated and the polymers are polymerized to flexible films that encase the wires. The polymer films are thin (0.013–0.1 mm), very flexible, abrasion and cut-through resistant, and can withstand many chemicals. Polymers that are used include the traditional oleoresinous low cost resins, poly(vinyl formal), epoxies (used for water resistance), polyurethanes (for use in soldering applications), polyesters, aromatic polyimides (18), and copolymers of polyester amide–imides (19) which are used up to 180°C. Enameled wire often is overcoated with a second polymer to enhance its handling and winding properties. Many insulation applications require that the enameled conductors be coated with an insulating varnish to rigidify them and to restrain vibrations.

Powder-Coated Wire. Powder coatings (qv) are not as thin nor as high in electric strength as solvent-based enamels but they are easily applied, are low in cost, and have attractive insulating properties. Sprayed or coated onto hot wires, the powders stick to the surface, coalesce, flow, and polymerize into a coherent coating. They are based primarily on epoxy, polyester, and polyurethane polymers and are used in transformers, bus bars, Watt/hour meter coils, and as motor-slot insulation.

Cable Insulation. Polyolefins are the base resins in many distribution cables. The most common are polyethylene, cross-linked polyethylene, ethylene–propylene copolymer (EPDM) [*9010-79-1*], and, in Europe, poly(vinyl chloride). Underground systems range to 35 kV. Overhead cables generally are insulated with oil-impregnated paper.

Hook-Up Wire. Most hook-up wire is based on poly(vinyl chloride) (ASTM D 922 and D 2301), silicone, polyethylene, polytetrafluoroethylene [*9002-84-0*], and chlorofluoroethylene copolymers (see Fluorine compounds, organic). Heat-shrinkable tubing often is used to insulate joints and splices, and these are made from poly(vinyl chloride), poly(vinylidene fluoride) [*24937-79-9*], and cross-linked polyethylene.

Varnishes. Varnishes are used in electrical equipment to bond windings to achieve better vibration resistance. In addition, they resist chemical attack, moisture, and mechanical damage, and they greatly enhance heat transfer. Synthetic resins comprise most varnishes (20). The first of these were oil-modified alkyds in which the drying oils were triglycerides of long-chain and unsaturated fatty acids that react with the alkyd in a complicated series of oxidation and polymerization reactions. Drying oils include linseed, tung, oiticica, soybean, and castor (see Driers; Alkyd resins). These varnishes are being replaced by modified phenolics, phenolic alkyds, polyesters, epoxies, polyurethanes, silicones, and polyimides. Epoxies and urethanes are often chosen where chemical resistance is paramount. Silicones are used in applications where hot spots are 180°C or higher. However, silicones cannot be used in enclosed rotating d-c machines where they decompose slightly and deposit films of silica on the brushes and commutators, thereby insulating the surfaces and increasing the brush wear excessively.

For high temperature applications where silicones cannot be used, the polyimide varnishes are used because they do not cause brush wear and they are superior in

Table 4. Insulating Varnishes

Varnish	Solvent	Solids content, %	Viscosity Demmler #1, s	Specific gravity (21°C)	Dielectric strength, MV/m	Continuous use temperature, °C
oleoresinous	xylene	50	40	0.89	55.1	105
oil-modified asphalt	mineral spirits	50	50	0.90	53.2	105
oil-modified phenolic	toluene	46	80	0.93	65.0	130
oil-modified melamine–phenolic	toluene–alcohol	50	80	0.95	63.1	130
phenolic	alcohol	42	50	1.00	78.7	130
phenolic–alkyd	xylene	47	70	0.96	78.7	155
epoxy	xylene	50	110	0.96	90.6	155
epoxy-ester	xylene	50	100	0.96	90.6	155
polyester	xylene	60	30	0.98	118.1	155
polyurethane	xylene	50	20	0.95	118.1	120
polyimide	NMP[a]	50	80	0.96	118.1	220
silicone	xylene	49	50	1.02	66.9	200

[a] NMP = *N*-methylpyrolidinone.

Table 5. Varnish Test Methods

Title	ASTM Number
sampling and analysis of shellac	D 29
methods of testing varnishes used for electrical insulation	D 115
testing electrical insulating varnishes for 180°C and higher	D 1346
thermal endurance of flexible insulating varnishes	D 1932
bond strength of varnishes by helical coil method	D 2519
weight-loss test of varnishes	D 2756
solventless varnishes—gel time	D 3056
thermal degradation of electrical insulating of varnishes by helical-coil method	D 3145
thermal aging characteristics of electrical insulating varnishes applied over film-coated magnet wire	D 3251
test for reactive monomers of varnishes	D 3312
weight loss of varnishes	D 3377

thermal capability. The solvents used in varnishes control the viscosity and the coating ability, and include VM & P (Varnish Makers' and Painters') naphtha, mineral spirits, xylene, toluene, and alcohols. Properties of typical insulating varnishes are given in Table 4 and methods of testing insulating varnishes are provided in Table 5.

Paper. In 1977, more than 69,000 metric tons of paper was used in the United States and Canada (21). The transformer industry uses more of the cellulose in the form of conductor tapes, layer insulation, pressboard, and special paper (ie, creped kraft, tubes). The paper usually is saturated with mineral oil where the combination produces a system with very high electric strength. A variety of additives and chemical modifiers are added to the paper to increase its thermal stability (22–23).

Special grades of cellulose paper are used in capacitors. Carefully prepared papers made with specially refined pulps result in thin (0.01 mm) papers having few pinholes and free of ionic impurities. These papers are combined with special grades of polypropylene film and are saturated with various dielectric fluids. The insulation in capacitors is stressed higher than any other insulation system, ie, to as much as 55.1 MV/m.

Nomex is a polyamide synthetic paper used in aircraft generators, high performance motors, and special transformers (see Aramid fibers). Available as the virgin fiber and with mica flakes, the papers are useful to 180°C. The paper is nonmelting and self-extinguishing. Mica papers are used in tape forms to insulate generators and large motors. The tapes consist of a carrier fabric such as Dacron with mica flakes distributed on the carrier and bonded with a small amount of a semipolymerized resin (mica bond). These tapes are used in high voltage applications where partial discharges occur, since mica is highly resistant to corona discharges.

Other synthetic fibers are used in paper form including nylon, polyester, glass, polyethylene, and polypropylene (see Olefin fibers). Vulcanized fiber has poor electrical properties with high dissipation factor and low electrical strength. However, when it is exposed to electric arcs, large quantities of water are generated, cooling the arc and suppressing it in a reaction similar to that discussed for aluminum oxide trihydrate. The lack of carbonized surfaces makes the vulcanized fiber useful in fuses, circuit breakers, arc chutes, switchgear, and lightning arresters. It also is used as slot cells and in transformers.

Mica Tapes. Mica [*12001-26-2*] (qv) is a mineral with a laminated structure which permits the lamina to be split and separated, resulting in thin, flexible, and tough flakes. Two different materials are used: phlogophite [*12003-38-7*], ($K_2Mg_6Al_2Si_6O_{20}[OH]_4$) from Canada and muscovite [*1318-94-1*] ($K_2Al_4Al_2Si_6O_{20}[OH]_4$) from India, North America, and Africa. Resistivity of mica can be as high as 10^{16} Ω·cm and dissipation factors range from 0.0001 for muscovite to 0.003 for phlogophite. Breakdown strength is 352 MV/m in air and 696 MV/m in mineral oil. Mica tapes are used in turbine and hydro generators and in large motors in combination with low viscosity resins to provide coil insulations with high electric strength and excellent resistance to partial discharges.

Asbestos papers and sheets had been used widely in electrical machinery, but are being replaced by other materials because of the difficulty of complying with OSHA standards.

Films. Films are thin plates of polymer up to ca 254 μm thick. At greater thicknesses, they are referred to as sheets. Films are used in electrical devices mostly as tapes and either by themselves or combined with a reinforcing web or fabric which enhances their strength. The usefulness of films is derived from their flexibility which permits them to be wrapped on conductors. Films also have high electric strengths. Although any thermoplastic polymer can be made into a film, only several are in use in the electrical industry, and these are listed in Table 6.

Polypropylene films are used in power capacitors in combination with paper and an insulating dielectric fluid. Some capacitors are being made where the paper is eliminated: consequently, the surface of the film must be altered to increase the wetting by the insulating fluid.

The most widely used film is poly(ethylene terephthalate) (polyester); it has use in motors as slot cells, in special transformers, in electronic capacitors, and as tapes in motor windings. Polyester films are tough and have low moisture absorption and very high electric strength. These films are rated about 15–20°C better than cellulosic insulations and can be used in 135°C-insulation systems. Polyesters are attacked by highly alkaline systems and by exposure to hot water.

Table 6. Insulating Films

Properties	Cellulose acetate [9004-35-7]	Poly-ethylene [9002-88-4]	Polyimide [25038-81-7]	Poly-propylene [9003-07-0]	Poly-(ethylene tereph-thalate) [25038-59-9]	Poly-tetrafluoro-ethylene [9002-84-0]	Poly-(parabanic acid) [60092-26-4]
area factor $cm^2/(g \cdot mm)$	12,340	16,540	10,880	17,550	12,670	7,180	11,660
specific gravity	1.31	0.94	1.42	0.9	1.4	2.2	1.33
tensile strength, MPa[a]	113.1	24.1	172.4	68.9	206.8	27.6	110.3
elongation, %	70	650	70	1,000	120	350	10
tear strength, kN/m[b]	4.9	147	3.9	22	7.4	49	2.45
folding endurance, cycles	2000	>100,000	10,000	>100,000	14,000		25,000
water absorption (for 24 h), %	9	<0.01	2.9	0.005	0.8	0.00	2.8
dielectric constant							
at 1 kHz	3.6	2.2	3.5	2.1	3.1	2.1	3.4
at 1 MHz	3.2	2.2	3.4	2.1	3.0	2.1	
at 1 GHz	3.2	2.2		2.1	2.8	2.1	
dissipation factor							
at 1 kHz	0.013	0.0003	0.003	0.0003	0.0047	0.0002	0.004
at 1 MHz	0.038	0.0003	0.010	0.0003	0.016	0.0002	
at 1 GHz		0.0003		0.0003	0.003	0.0002	
dielectric strength, MV/m	197	19.7	275.6	118.1	275.6	16.9	224.4
volume resistivity, $\Omega \cdot cm$	10^{13}	10^{16}	10^{18}	10^{16}	10^{18}	10^{18}	10^{16}

[a] To convert MPa to psi, multiply by 145.
[b] To convert N/m to dyn/cm, multiply by 1000.

A new film based on poly(parabanic acid) (Tradlon) is being developed as an electrical film. Midway in properties and cost between polyesters and polyimides, it should have considerable use in applications where thermal stability up to 180°C is needed. It has low losses, high dielectric strength (>196.9 MV/m), and excellent mechanical properties. A foam grade of film exists with a dielectric constant of 1.6 and a dielectric strength of ca 24 MV/m.

Polytetrafluoroethylene films are used to insulate wires and as sheet insulation. They have great flexibility and are inert to all chemicals and have a use temperature of ca 200°C. Their surfaces are so inert that other components of the insulation system cannot bond to them, which often limits their use. Polyimide films (Kapton) have the highest thermal capabilities of all organic films. The films can be used continuously at 220°C, and have high dielectric strength and excellent tear and tensile strength. Polyimides are very resistant to chemicals except to highly polar solvents, eg, dimethylpyrrolidinone and alkaline solutions (see Film and sheeting materials).

Vacuum/Pressure-impregnating Resins (VPI). Some polymers, before polymerization, exist in the liquid form or can be dissolved in liquid monomers that co-react. These systems are solventless resins or liquid-reactive resins, and include epoxies, silicones, polyurethanes, acrylics, and unsaturated polyesters dissolved in monomers, eg, styrene or diallyl phthalate. These resins are used in the vacuum/pressure impregnation process (VPI) because of their low viscosities, useful polymerization characteristics, and excellent electrical and physical properties. VPI processes refer to methods used to insulate wound coils used in generators and large motors by alternatively using vacuum and pressure to cause total impregnation of the taped coils. The advantages of these systems include complete impregnation which enhances corona resistance, thermal conductivity, thermal stability, high dielectric strength, and low dissipation factor.

Casting and Potting Resins. Many of the same solventless resins discussed in the previous section also are applicable to the insulating techniques of casting, potting, and impregnating. In the casting process, a resin is compounded with mineral filler, catalyst, hardeners, and other additives. The resins that usually are selected are epoxies because they have low polymerization shrinkage, excellent electrical properties, and good adhesion to inserts and other components (see Epoxy resins). Fillers (qv) that are used include silica [*7631-86-9*], calcium carbonate [*471-34-1*], and aluminum oxide trihydrate [*21645-51-2*]. The latter is selected for its arc and track resistance, but all fillers contribute to viscosity control, thermal conductivity, and control of thermal expansion. Carefully machined molds are used to produce the desired shapes, and these molds are made of steel-, aluminum-, or fiberglass-reinforced polyesters. Cast components include bushings, stand-off insulators, weathersheds, and a variety of terminal boards and mechanical supports.

Potting resins are similar to cast resins in formulation. However, instead of using molds to produce a free-standing shape, the mold usually becomes a part of the insulating system and stays with the device. Potting resins encapsulate electronic components and insulate cable and bus-bar joints (see Embedding).

Laminates. Laminates are made by bonding layers of a reinforcing web. The reinforcements consist of fiberglass, paper, fabrics, or synthetic fibers. The bonding resins usually are thermosetting and consist of phenolic, melamine, polyester, epoxy, and silicone. Laminates have been standardized by NEMA (National Electrical Manufacturing Association) (24), and have a wide range of properties, defined by the selection of resin and reinforcements as shown in Table 7. Laminates are made by impregnating the reinforcing webs in treating towers, advancing them to the B stage, and pressing and polymerizing them to shape under heat and pressure. Similar techniques are used to produce tubes and rods. Laminates are used in transformer terminal boards, switchgear arc chutes, motor and generator slot wedges, motor bearings, structural supports, and spacers (see Laminated and reinforced plastics).

Copper-Clad Laminates. Copper-clad laminates used to fabricate printed circuit boards include a layer of copper foil bonded to one or both faces of the laminate during the pressing application. Commercial constructions are limited to paper-phenolic, glass-fabric epoxy, glass-mat polyester, and glass-fabric polyimide, but a few hybrid

Table 7. NEMA-Grade Laminates

	NEMA Grade No.						
	XXXP	C	G-5	G-7	G-11	N-1	GPO-3
	Resin						
	Phenolic	Phenolic	Melamine	Silicone	Epoxy	Phenolic	Polyester
	Reinforcement						
Property	Paper	Cotton fabric	Glass cloth	Glass cloth	Glass cloth	Nylon cloth	Glass mat
tensile strength (lengthwise), MPa[a]	85	69	255	159	275	59	65
tensile modulus (lengthwise), MPa[a]	6900	6900	15,900	12,400	17,300	2750	6900
hardness (Rockwell), M[b]	105	103	120	100	112	105	100
density, g/m^3	1.30	1.36	1.90	1.68	1.80	1.15	1.8
thermal expansion, 10^{-5}/°C	2.0	2.0	1.0	1.0	0.9		
thermal conductivity, W/(cm·K)	29.3	29.3	50.2	29.3	29.3		
NEMA temperature index (electrical −°C)	125	85		170	140		
electrical strength (step by step), MV/m	25.6		8.7	15.7	19.7	25.6	11.8
insulation resistance (ASTM D 257), GΩ	20		0.1	2.5	200	50	

[a] To convert MPa to psi, multiply by 145.
[b] M is a standard Rockwell hardness scale, see Hardness.

compositions also are made. These materials also are standardized by NEMA (24). In these applications, the circuit is applied to the laminate by etching away the unwanted foil which produces a circuit bonded to the laminate. Multilayer circuits are made by laminating several layers of prepared printed circuits with a bonding layer of B-staged resin on a reinforcing web. As many as 30 layers can be made but most contain no more than 10. Flexible printed circuits also are used by adhesively bonding foil to a film insulation, eg, polyester or polyimide, and processing them in a manner similar to that for laminates.

Molded Products. Molded insulating components are melted and injected into a closed mold under pressure where they solidify. Thermoset and thermoplastic polymers are used (see Plastic processing).

Thermoplastic polymers are those that are fully polymerized by the resin manufacturer and, because of the lack of cross-linking, can be repeatedly melted and frozen. Thermoplastic materials are shaped by injection molding, vacuum molding, and extrusion. They often are used in electronic applications as connectors and terminal boards. Other uses include switch bases, gears, cams, lenses, connectors, plugs, stand-off insulators, knobs, handles, and wire ties. Because they melt, they are limited to application temperatures of less than 90°C, but several polymers can be used to 135°C. Polymers used for insulating moldings include fluorocarbons, nylon, polycarbonates, polyethylene, polyimides, polyphenylene oxide-modified polymers, polypropylene, and poly(vinyl chloride). Typical properties of selected thermoplastic molding materials are listed in Table 8.

Table 8. Properties of Thermoplastic Molding Materials

Properties	Polyacetal	Acrylic	Nylon	Polycarbonate	Polyethylene (high density)	Polyester	Polyimide	Polypropylene	Polysulfone	Phenylene oxide-based resin	Polytetrafluoroethylene	Poly(vinyl chloride)
volume resistivity, Ω·cm	10^{14}	10^{14}	10^{15}	10^{16}	10^{16}	10^{15}	10^{16}	10^{16}	10^{17}	10^{13}	10^{18}	10^{14}
dielectric strength (step-by-step), MV/m	15.7	13.8	12.6	14.3	21.7	18.1	17.7	17.7	15.7	15.7	16.9	13.8
dielectric constant												
at 60 Hz	3.8	4.0	5.5	3.2	2.4	3.3	3.43	2.6	3.1	2.6	2.1	9.0
at 1 MHz	3.8	3.5	4.9	3.0	2.4	3.4	3.42	2.6	3.1	2.6	2.1	8.0
at 1 GHz	3.8	3.2	4.7	3.0	2.4	3.2	3.42	2.6	3.1	2.6	2.1	5.0
dissipation factor												
at 60 Hz	0.004	0.04	0.01	0.0009	0.0005	0.005	0.005	0.0005	0.0008	0.0004	0.0002	0.15
at 10^6 Hz	0.004	0.03	0.01	0.004	0.0005	0.03	0.005	0.0005	0.001		0.0002	0.15
at 10^9 Hz	0.004	0.02	0.03	0.01	0.0005		0.0018	0.0005	0.005	0.0009	0.0002	0.16
arc resistance, s	129	>200	140	120	200	192	230	185	122	75	>200	60
specific gravity	1.42	1.20	1.14	1.2	0.965	1.38	1.43	0.910	1.24	1.10	4.5	2.3
tensile strength, MPa[a]	62.1	75.8	95.5	65.5	37.9	56.5	117.2	38.0	70.3	75.8	31.0	24.1
elongation, %	15	10	320	100	100	250	10	700	100	80	400	450
tensile modulus, 10^2 MPa[a]	28.3	31.0	26.2	24.1	10.3	19.3	13.1	15.9	24.8	26.2	4.0	
compressive strength, MPa[a]	124.1	124.1	89.6	86.2	221	100.0	206.8	55.2	96.5	89.6	11.7	11.7
flexural strength, MPa[a]	96.5	117.2		93.1	6.9	115.1	193.1	55.2	106.2	103.4		
Izod impact strength, J/m[b]	74.7	26.7	213.5	854.1	1067.6	53.4	37.4	80.1	69.4	101.4	160.1	
hardness (Rockwell)[c]	R120	M105	R118	R118		M85	E99	R110	R120	R123		
thermal conductivity, W/(cm·K)	23.0	16.7	24.7	19.2	51.9	28.9	10.9	11.7	25.9	18.8	25.1	20.9
thermal expansion (per °C)	8.1	9	8	7	12	9.5	5	10	6	3	10	10
heat distortion temp (at 1.8 MPa[a]), °C	124	100	75	138	49	85	138	63		190	120	
maximum use temp, °C	85	88	120	120	115	120	180	160	170	130	220	100

[a] To convert MPa to psi, multiply by 145.
[b] To convert J/m to (ft·lbf)/in, divide by 53.38 (see ASTM D 256).
[c] E, M, and R are standard Rockwell hardness scales, see Hardness.

Table 9. Properties of Thermoset Molding Materials

Properties	Diallylphthalate Glass fiber	Diallylphthalate Mineral	Epoxy Glass fiber	Epoxy Mineral	Melamine α-Cellulose	Phenolic Wood flour	Polyester Glass fiber	Polyester Mineral	Silicone Glass fiber	Silicone Mineral	Urea Formaldehyde α-Cellulose
volume resistivity, Ω·cm	10^{14}	10^{13}	10^{14}	10^{14}	10^{14}	10^{10}	10^{15}	10^{14}	10^{14}	10^{14}	10^{12}
dielectric strength, MV/m											
short time	17.7	16.5	15.7	15.7	15.7	15.7	16.5	17.7	15.7	15.7	15.7
step-by-step	15.7	15.7	15.7	15.7	11.8	14.8	15.4	13.8	11.8	15.0	11.8
dielectric constant											
at 60 Hz	4.3	5.2	5.0	5.0	9.5	13	7.3	7.5	5.2	3.6	9.5
at 1 kHz	4.4	5.3	5.0	5.0	9.2	9.0	4.7	6.2	5.0	3.2	7.5
at 1 MHz	4.5	4.0	5.0	5.0	8.4	6.0	6.4	5.5	4.7	3.2	6.8
dissipation factor											
at 60 Hz	0.01	0.03	0.01	0.01	0.030	0.05	0.011	0.009	0.004	0.004	0.035
at 1 kHz	0.004	0.03	0.01	0.01	0.015	0.04	0.090	0.010	0.004	0.002	0.025
at 1 MHz	0.009	0.02	0.01	0.01	0.027	0.03	0.008	0.015	0.002	0.002	0.250
arc resistance, s	180	190	180	190	180	tracks	180	150	250	420	150
specific gravity	1.78	1.68	2.0	2.0	1.52	1.45	2.3	2.3	2.0	2.82	1.52
tensile strength, MPa[a]	75.8	60.0	206.8	103.4	89.6	68.9	68.9	55.2	34.5	24.1	89.7
elongation, %			4	1	0.9	0.8	3	0.9	2	1	1.0
tensile modulus, 10^2 MPa[a]	151.7	151.7	209.6		96.5	117.2	172.4	179.3			103.4
compressive strength, MPa[a]	241.3	220.6	276.0	276.0	310.3	248.2	206.8	172.4	103.4	124.1	310.3
flexural strength, MPa[a]	131.0	75.8	413.7	103.4	110.3	82.7	137.9	68.9	96.5	55.2	124.1
Izod impact strength, J/m[b]	533.8	24.0	533.8	21.4	18.7	32.0	854.1	26.7	800.7	18.7	21.4
hardness (Rockwell)[c]	M110	M103	M112	M112	M125	M128	B20	B50	M84	M90	M120
thermal conductivity, W/(cm·K)	20.9	29.2	16.7	41.8	29.3	19.7	41.8	62.8	31.4	46.0	29.3
thermal expansion (per °C)	3.6	4.2	3.5	5.0	4.0	4.5	3.3	5.0	0.8	3.1	8.2
heat distortion temp. (at 1.8 MPa[a]), °C	175	160	120	120	180	130	210	175	>300	>300	130

[a] To convert MPa to psi, multiply by 145.
[b] To convert J/m to (ft·lbf)/in, divide by 53.38 (see ASTM D 256).
[c] M and B are standard Rockwell hardness scales, see Hardness.

Thermoset polymers are those that only partially react during the resin manufacture. Final polymerization occurs in the molding operation, and the products are nonmelting, infusible, and highly cross-linked. Molding processes include compression molding, transfer molding, injection molding, and reaction-injection molding (RIM). Typical electric and electronic applications include circuit-breaker bases, arc chutes, terminals, structural supports, connectors, switchgear components, bus-bar spacers, ventilating fans, end bells, and tap changers. Molding compounds include diallyl phthalates, epoxies, phenolics, polyesters, polyurethanes, silicones, and urea–formaldehyde. Properties of selected thermoset molding materials with various fillers are listed in Table 9.

Elastomers. Elastomers (rubbers) are polymers having a limited cross-linking density and great flexibility and softness (see Elastomers, synthetic). The polymers often are used in hook-up wire and cable and to insulate electric and electronic devices. Polymers that are used are butyl copolymers of ethylene and propylene, fluoropolymers, neoprene, polysulfides, polyurethanes, and silicones. Applications include molded instrument transformers, meter coils, motor end windings, lead wires, and connectors.

Glasses and Ceramics. Glass has been used for many years in low voltage, electric and telephone distribution systems as stand-off insulators. Glass sheets drawn to a thickness of ca 25 μm are used as the dielectric in some capacitors. Drawn into fibers of 5 μm in diameter, glass is used as wire serving and in tapes and fabrics for a wide variety of insulation applications. Glass also is mixed with finely ground mica (1:2) and molded under heat and pressure to form molded insulators. Normally amorphous, glass can be caused to crystallize by including certain nucleating agents and using special heat-treating schedules. These products have exceptional flexural and impact strength and excellent thermal shock resistance.

Porcelain ceramics are made from natural refined clays mixed with silica and feldspars, molded to shape, and fired at high temperatures to vitrify them (see Ceramics). Porcelains have extensive use in power bushings, transmission-pin insulators, weathersheds, cut-out switches, line switches, circuit breaker bases, fuses, and lightning arresters. Although it has low porosity (<0.1%), porcelain usually is glazed with a glass coating to make the surface smooth, water repellent, and weatherproof (see Enamels). Thick films of 96% alumina [*1344-28-1*] or 99% beryllia [*1304-56-9*] are used for circuit substrates. Alumina is selected for cost and stability, but beryllia often is selected in applications where improved thermal conductivity is needed. At microwave frequencies, insulators of forsterite [*1343-88-1*] or steatite [*14807-96-6*] are used.

BIBLIOGRAPHY

"Dielectrics" in *ECT* 1st ed., Vol. 5, pp. 51–75, by E. B. Baker and W. C. Goggin, The Dow Chemical Company; "Insulation, Electric" in *ECT* 2nd ed., Vol. 11, pp. 774–801, by Jack Swiss, Westinghouse Electric Corporation.

1. C. A. Harper, *Fundamentals of Electrical Insulating Materials,* Vol. 20, No. 12, Industrial Research/Development, 1978.
2. T. W. Dakin and G. A. Mullen, *Accelerated Salt Fog Testing of Plastic Insulators for Outdoor High Voltage Application,* IEEE, A78 075-4, New York, Jan. 1978.
3. F. Paschen, *Ann. Phys.* **37,** 69 (1889).
4. J. Swiss and T. W. Dakin, *Westinghouse Eng.* **14,** 114 (1954).

5. G. L. Moses, *Insulation Engineering Fundamentals,* Lake Publishing Co., Lake Forest, Ill., 1958, p. 8.
6. T. W. Dakin, H. M. Philofsky, and Q. X. Owens, *Trans. AIEE* **73**(Pt. I), 155 (1954).
7. E. J. McMahon, *IEEE Trans. Elect. Insul.* **El-13,** 4 (Aug. 1978).
8. R. M. Eichorn, *IEEE Trans. Elect. Insul.* **El-12,** 2 (1977).
9. U.S. Pat. 3,434,087 (Mar. 18, 1969), C. F. Hoffmann (to Westinghouse Electric Co.).
10. T. W. Dakin, *Trans. Am. Inst. Elect. Eng.* **67**(Pt. I), 43 (1948).
11. L. C. Whitman and P. Dorgan, *Trans. Am. Inst. Elect. Eng.* **5-61,** 45 (June 1954).
12. P. M. DiCerbo, *Insul. Circuits* **21,** 21 (Feb. 1975).
13. R. S. Norman and A. A. Kessel, *AIEE* **77**(Pt. 111), 632 (1958).
14. F. M. Clark, *Insulating Materials for Design and Engineering Practice,* John Wiley & Sons, Inc., New York, 1962, pp. 58–130.
15. A. F. Rohlfs and F. J. Turner, *Trans. AIEE* **75-III,** 1439 (1956).
16. D. K. Fink, *Standard Handbook for Electrical Engineers,* McGraw-Hill Book Co., Inc., New York, 1978, Sect. 260, pp. 4–140.
17. T. W. Dakin, L. Mandelcorn, and R. N. Sampson, *J. Electrochem. Soc.* **126,** 55C (Feb. 1979).
18. U.S. Pat. 3,179,614 (Apr. 20, 1965), M. W. Edwards (to E. I. du Pont de Nemours & Co., Inc.).
19. F. F. Trunzo, *Insulation* **15,** 42 (Feb. 1970).
20. D. K. Fink, *Standard Handbook for Electrical Engineers,* McGraw-Hill Book Co., Inc., New York, 1978, Sect. 342, pp. 4–207.
21. H. R. Sheppard and A. C. Baker, *Cellulose Insulation for Transformers—An Overview,* Electric Power Research Institute Workshop, Monterey, Calif., July 1978.
22. J. G. Ford, M. C. Leonard, J. Swiss, and G. C. Gainer, *AIEE Trans.* **77**(Pt. III), 804 (1958).
23. M. F. Beavers, E. L. Raab, and J. C. Leslie, *AIEE Trans.* **79**(Pt. III), 64 (1960).
24. *Industrial Laminated Thermosetting Products,* National Electric Manufacturing Association, LI1, 1971 (R 1979).

R. N. SAMPSON
Westinghouse Electric Corporation

WIRE AND CABLE COVERINGS

Increasing awareness of environment and safety has emphasized the importance of electric wires and cables. The prevention of power-system blackouts and the mitigation of subway and building fires and nuclear-plant incidents is dependent on proper design, manufacture, testing, and installation of electric cables. The ultimate installation and performance of cable is not controlled completely by the cable manufacturer. During installation, cable may be inadvertently exposed to excessive bending, tensile, and compressive forces; cutting and abrasive surfaces; and improper jointing and terminating workmanship. Cable routes often are exposed to extreme temperature and moisture. In addition, cables must be capable of withstanding unexpected steam, oil, or chemical leaks; flash fires; or other short- or long-term changes in the ambient environment. Because cable routes usually are inaccessible to facilitate servicing, cables must withstand their installation environment for their lifetime with no inspection or preventive maintenance.

Specifications for wires and cables used in the United States are provided by ASTM; IEEE (The Institute of Electrical & Electronic Engineers); AEIC (Association of Edison Illuminating Companies); ICEA–NEMA (Insulated Cable Engineers Association–National Electrical Manufacturers Association); UL; REA (U.S. Department of Agriculture Rural Electrification Administration); ANSI; MIL (Department of Defense Military Cable Specifications); and IMSA (International Municipal Signal Association).

Cross-Linked (Thermoset) Insulations

Prior to World War II, the only insulating elastomer available to the wire and cable industry was natural rubber (NR) which, being an unsaturated polymer, limited the useful temperature and voltage rating of solid dielectric cables (see Rubber, Natural). During and subsequent to World War II, styrene–butadiene rubber (SBR) [*9003-55-8*], isobutylene–isoprene rubber (IIR) [*9010-85-9*], polyethylene (PE) [*9002-88-4*], and ethylene–propylene copolymer (EPM) [*9010-79-1*] and terpolymer (EPDM) [*9010-79-1*] elastomers were developed and were suitable for wire and cable insulation compounds (see Elastomers, synthetic). The final evaluation of insulation compounds must be made by each cable manufacturer in mixing and extrusion production trials to verify life characteristics and reproducibility in repetitive manufacture regarding the range of cable sizes required and the compatibility of the insulation in contact with other coverings.

Insulation compounds based on the polyolefin elastomers, PE and EPM/EPDM, because of their moisture, heat, and corona-discharge resistance, predominate in cable installations requiring the highest degree of reliability and long life (1) (see Olefin polymers). Compounds based on IIR and SBR achieved their peak usage when they were either the only elastomer available (SBR) or offered the best properties (IIR). These compounds are used in limited cases either because they offer economic advantage for noncritical cables or because the manufacturer, having developed specific recipes and processing, does not wish to change to new materials until forced to by economics or noncompetitive performance characteristics. Natural-rubber insulation compounds are no longer used. However, because natural rubber is a nonpetroleum-based, renewable natural resource, a combination of economic and availability conditions may occur that could lead to a resumption of its use for wire and cable.

The approximate composition of cross-linked insulation compounds is shown in Table 1, and the properties of these insulations are listed in Table 2. The insulations and properties are based on the use of compounds formulated for vulcanization or chemical cross-linking in steam (steam or hot gas for silicone rubber). Ionizing radiation with the use of triallyl cyanurate or ethylene dimethacrylate co-agents permit radiation curing of many polymers. This method of cross-linking in some cases provides economic or performance advantages in the finished product. For example, chemically (steam) cross-linked polyethylene (XLPE) compounds that have been formulated using a peroxide curing system are limited to low density polyethylene elastomer (mp 125°C) to prevent compound precuring during mixing. Compounds for radiation cross-linking may contain high density polyethylene (mp 140°C) because the co-agents can withstand high mixing temperatures and, as a result, radiation-induced cross-linked in-

Table 1. Composition of Typical Cross-linked Insulation Compounds[a]

Function in compound	Approximate content, %	Typical ingredients
elastomer	25–40	styrene–butadiene (SBR), Buna S
		isobutylene–isoprene (IIR), butyl
		ethylene–propylene (EPM)
		ethylene–propylene–diene (EPDM)
		poly(dimethyl siloxane) (silicone)
		polyethylene
filler and reinforcing agent	40–70	clay
		silica
		calcium carbonate
		barium sulfate
		talc
		vulcanized natural asphalts
		solid bitumen
		carbon black
vulcanization system	<5	sulfur
		selenium diethyl dithiocarbamate (Selenac)
		dibenzoylquinone dioxime (Dibenzo GMF)
		dicumyl peroxide (Dicup)
		mercaptobenzothiazole (MBT)
		zinc dimethyldithiocarbamate
		tetramethylthiuram monosulfide (Monex)
		zinc oxide
		litharge
		stearic acid
aging inhibitors	<10	zinc salt of mercaptobenzimidazole
		alkylated diphenylamine
		polymerized trimethyldihydroquinoline
processing aids	<2	resins, waxes, petroleum oils, plasticizers

[a] Ref. 2.

Table 2. Properties of Various Cross-linked Insulations

Property	Elastomer					
	SBR or NR	Butyl	Ethylene–propylene	Polyethylene	Silicone	Polyethylene
	Filler					
	Oil base	Mineral	Mineral	Unfilled	Mineral	Carbon
electrical at 20°C						
dielectric constant (max)	5.0	4.5	3.5	3.0	3.5	6.0
100 tan δ (max)	5.0	3.5	1.0	2.0	1.0	2.0
resistivity (min), ohm–cm	10^{14}	10^{15}	10^{15}	10^{15}	10^{14}	10^{14}
dielectric withstand reel test (min avg stress), kV/mm (V/mil)	3.9 (100)	3.9 (100)	5.9 (150)	7.9 (200)	3.9 (100)	2.9 (75)
sample dielectric strength, rapid rise, kV/mm (V/mil) avg stress	23.6 (600)	23.6 (600)	23.6 (600)	27.6 (700)	19.7 (500)	19.7 (500)
impulse, 1.5 × 40 μs wave (max stress), kV/mm (V/mil)	47.2 (1200)	47.2 (1200)	47.2 (1200)	59.1 (1500)	39.4 (1000)	31.5 (800)
mechanical						
tensile strength (min), MPa (psi)	3.1 (450)	4.1 (600)	4.8 (700)	12.4 (1800)	5.5 (800)	12.4 (1800)
elongation at rupture (min), %	250	350	300	300	250	300
cont. high temp limit, °C	75	85	90	90	125	90
resistance to						
oxidation	good	excellent	excellent	excellent	excellent	excellent
moisture	good	good	excellent	excellent	fair	excellent
corona discharge	good	good	excellent	fair	excellent	fair
flame	poor	poor	poor	poor	fair	fair
solvents	poor	poor	poor	excellent	fair	excellent
brittleness at low temp	poor	excellent	excellent	excellent	excellent	excellent
radiation	fair	fair	excellent	good	good	good

sulations containing high density resin have a higher deformation temperature. Radiation curing is limited to low voltage cables with thin insulation because of the impractical economic and safety problems associated with radiation equipment for thick insulation.

Only thermoset polyethylene (XLPE) has mechanical strength and weather and chemical resistance properties that are adequate to permit its application as an integral covering for low voltage use on single conductor cables in the smallest thickness recognized by UL. All other cables must be supplemented by an outer braid, jacket, or sheath for mechanical protection and to provide weather, flame, and chemical resistance. Even thermoset polyethylene sometimes requires a surface coating to provide the specified degree of flame retardance.

It is necessary to test insulation compounds by laboratory-accelerated methods in order to project their life expectancy. Accelerated methods also are necessary because the chemical industry has produced improved elastomers and compounding ingredients at such a rate that the wire and cable industry discontinues their use before their life is determined in actual service. The principal determinants of insulation life for both low and high voltage cables are oxidation and water absorption; both are temperature dependent and can be accelerated in laboratory tests.

Figure 1 shows the typical Arrhenius curve for insulation compounds using PE or EPM/EDPM polyolefin elastomers (see Insulation, electric—properties and materials). For comparison and to illustrate the progress that has been made in oxidation

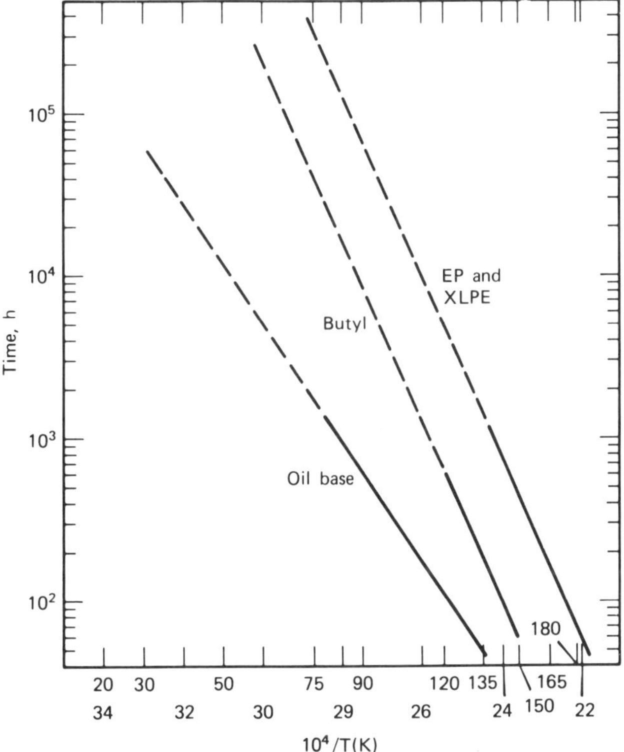

Figure 1. Typical Arrhenius curves. Time to 40% retention of elongation.

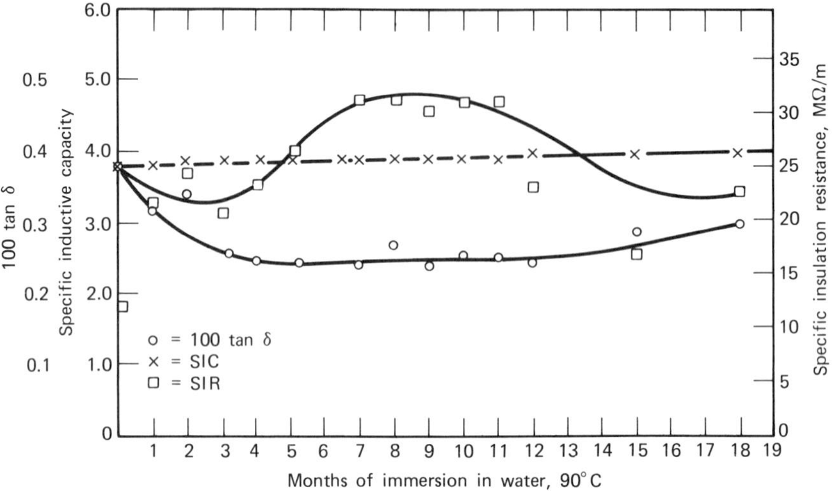

Figure 2. Moisture resistance for typical XLPE insulation (#14AWG 0.75 mm) XHW-grade compound, 60 Hz measurements at 3.3 kV/mm (80 V/mil) 600 V = 60 Hz continuous stress.

life, curves also are shown for oil-base rubber and butyl compounds that were in common use until they were replaced by the polyolefin elastomer insulations. The Arrhenius relationship is valid over the temperature range in which one chemical reaction is predominant in the aging process. It is common practice to extrapolate high temperature, accelerated data to rated or operating temperature to estimate service life. Problems arise when this technique does not demonstrate a desired life, eg, 40 yr (3.5×10^5 h), which frequently is the period used in the design of power plants. Various methods are used to project service life, eg, a parallel curve through the 40-yr life point, the use of a test-data end point which is not precisely measurable, or the use of an inferred service life by analogy to older materials of actual demonstrated life under the same service conditions. The Arrhenius technique is not consistently applied by all manufacturers and more verification by actual service life is needed before it can be applied according to a standard method for predicting long-term life for wire and cable. It is also common practice to measure the weight gain of insulation after submergence in hot water. This measurement is useful to characterize material and verify uniformity of production. Since service life depends on electrical performance, the electrical test is the significant qualification test.

Water-absorption tests are accelerated by high temperature and also by using specimens with the smallest practical insulation thickness, usually 0.51–1.0 mm. The relationship of insulation thickness to service life in water is exponential but its precise value has not been determined since it is variable with compound dispersion particularly for the more amorphous compounds. Test specimens are placed in water maintained at 50–90°C (depending on the grade of insulation), and the specific inductive capacity, tan δ (power factor), and insulation resistance are measured at intervals so that a plot will clearly indicate the rate of change (3). The samples usually are subjected to continuous voltage (600 V ac or dc) except when measurements are taken. Plots of such data are shown in Figure 2.

In addition to oxidation and moisture resistance, the ability to withstand corona discharge is a vital characteristic for cable operation above 2000 V. When air is subjected to a stress of ca 2.0 kV/mm (50 V/mil), it breaks down or ionizes. This ionization or corona causes an ionic bombardment of adjacent surfaces and breaks down atmospheric oxygen so that ozone is formed. In cables, air gaps can occur between conductor and insulation, the insulation and the outer ground covering, and in any voids or imperfections in the insulation. These gaps are capacitances that are electrically in series with the insulation capacitance. The voltage stress is distributed from the grounded covering across the capacitances in inverse proportion to their individual proportion of the total capacitance between the ground plane and the high voltage conductor. Corona discharge causes deterioration of insulation by erosion with consequent formation of tree patterns or tubular tracks resulting from ionic bombardment in some cases, or rapid oxidation by ozone causing fragmentation of insulation in others. The possibility of corona discharge exists for cables in circuits operating at voltages of 2300 and higher. 2300 V is the approximate dividing line between high and low voltage and is the lowest primary distribution voltage in the United States. Except for large individual loads operating at this voltage, it is standard practice to transform this or higher voltage primary power to utilization voltages of less than 600 V at which level corona discharge does not occur (4). In order to simulate the effect of both ozone and ionic bombardment on cable insulation, U-bend plate corona-discharge tests are made on high voltage insulated conductors before application of shielding. The insulated conductor is bent 180° into a U shape around a mandrel having a diameter that is about four to eight times greater than the cable diameter and subjected to 4 to 6 times rated voltage stress (Fig. 3).

With the use of tandem- and dual-extrusion equipment, which permit the application of stress-relieving and voltage-grading layers free of interface contamination and air gaps, the service voltage horizon for polyolefin elastomer compounds extends to extra high voltage (EHV) service (EPRI is funding research to develop equipment for the manufacture of cables and to provide for the testing of prototypes of 138 and 230 kV cables). The lower dielectric loss index (SIC $\times$ tan δ) of cross-linked polyethylene (XLPE) compared to EPM/EPDM compounds makes it a better candidate for EHV use. ICEA specifies 2.0% maximum power factor or tan δ for both compounds with maximum SIC of 3.5 for XLPE and 4.0 for EPM/EPDM compounds. However, values being obtained from compounds in use and the loss indexes are $2.5 \times 0.1\% = 0.25$ for XLPE and $3.0 \times 0.5\% = 1.5$ for EPM/EPDM. The greater loss index and lower dielectric and impulse strength of EPM/EPDM compounds will probably limit their use to cable rated up to 138 kV.

The tendency of electrochemical trees (a tubular channel with branching fissures of diminishing diameter in the pattern of a tree trunk and its branches) to form in XLPE after a period of time under voltage stress in the presence of moisture is the cause of frequent failures (5). Cables that have been sheathed in XLPE and installed in dry locations or those with an outer jacket so that the insulation is essentially dry have performed satisfactorily (6). The determination that moisture has a major effect on the service life of XLPE insulation has led to the development of equipment using pressurized heated inert gas instead of steam as the heat-transfer and pressure medium. The gas system eliminates the halo (an opaque concentric layer about midway in the thickness of steam-cured insulation) which consists of a large number of microvoids believed to be filled with water and peroxide decomposition products.

For typical insulations, ambient temperatures of 20°C for underground and 40°C

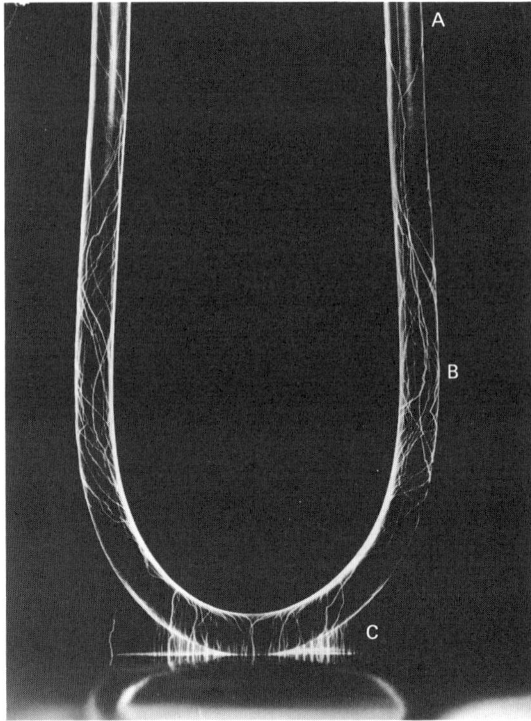

Figure 3. U-bend plate corona-discharge test. A, insulation without shield and conductor energized at 9.87 kV/mm (250 V/mil). B, corona-discharge streamers. C, grounded metal plate.

outdoors in sunlight are used. For power-generating and many industrial plants, 40°C ambient temperature is used for all installations. Current-carrying capacities are assigned by the National Electric Code and ICEA (7–11). Most modern insulations have a normal life expectancy of 40 yr with 90°C conductor temperature. The conductor temperature is the result of the temperature rise above ambient produced by the heat generated by the current (proportional to I^2R where I is the current and R is the resistance) plus heat from the insulation dielectric loss for high voltage cables. In many low voltage circuits, the maximum current permitted by I^2R loss would cause too large a voltage drop—particularly for longer length circuits—so that voltage drop IZ (Z being circuit impedance, the vectorial sum of resistance R and reactance X) becomes the limiting factor. Terminal equipment ratings limit the amperage rating for some circuits. In some applications, the ambient temperature is such that special heat-resistant insulations must be used and, even then, they are replaced frequently. Cables supplying power to steel-billet soaking pits and downhole leads for oil-well bottom-hole pumps are applications where cable must be replaced in a matter of months or a few years because cable manufacturers have not been able to develop cable insulation that can withstand the environment within the space or other design limitations imposed by the application. For some installations, costly cable designs are used to withstand possible emergency or overload conditions. U.S. Navy shipboard cables operate at 90°C or at lower conductor temperatures but are insulated with costly 125°C-rated silicone rubber to permit the cable to remain operative in fires. The silicone rubber burns to a nonconducting ash of silica so that electrical integrity is maintained and control systems remain functional.

Thermoplastic Insulations

Poly(vinyl Chloride). Poly(vinyl chloride) [*9002-86-2*] (PVC) is a dry hard material which in powder form is mixed with a plasticizer, stabilizer, and colorant (if desired) and is furnished in granulated form (see Vinyl polymers, vinyl chloride and poly(vinyl chloride)). Since the resin constitutes the major part of the mix, its inherent properties predominate in the finished compound. The compounding ingredients are varied to modify physical properties, eg, hardness, heat resistance, flame retardance, and low temperature flexibility. As low voltage insulation, PVC provides low cost resulting from its low material cost, its fast extrusion speed (>7.6 m/s on small wire compared with <2.5 m/s for thermoset insulation), and the fact that it is an integral covering requiring no further protection. The properties that permit its use as an integral covering are excellent mechanical and abrasion resistance at normal operating temperature; excellent moisture, solvent, chemical, flame, and weathering resistance; and fade resistance in colored compounds. Most appliances, homes, stores, and industrial plants have PVC-insulated wire and cable in their low voltage circuits. It is recognized by UL for use in wet locations with a maximum conductor operating temperature of 75°C (the same limit as for thermoset rubber) and in appliances for temperatures of up to 105°C. PVC is not well suited for high voltage cables because of its high dielectric losses.

In low voltage applications, the use of PVC is limited by two factors. First, the thermoplastic nature of PVC prevents its use in circuits of high short-circuit current capacity. Under fault conditions of high current, large footages of thermoplastic wire can be damaged permanently through insulation melting (12). Second, under continuous d-c voltage in wet locations, as in battery-operated control circuits, single-conductor PVC-insulated wire may fail as a result of electroendosmosis.

Polyethylene. The properties of low density and high density-high molecular weight, unfilled polyethylene, as well as polypropylene [*9003-07-0*] copolymer, are given in Table 3.

The dielectric constant of 2.3 for low density polyethylene is close to the theoretical minimum that can be obtained for a solid substance that is mechanically suitable for insulation use, and the dissipation factor is about as low as that of any available material. These properties, in combination with high dielectric strength, are ideal for high voltage insulation. Based on these properties and its low material and manufacturing costs, large amounts of concentric neutral underground-distribution primary (15,000–25,000 V) cable have been buried directly in the earth throughout the United States. Intolerable failure rates are being experienced on a large proportion of this cable after 5–7 yr of service. However, despite poor experience to date, the low cost of these cables is such that efforts are continuing to provide satisfactory service by means of additives to inhibit failures resulting from tracking and discharge in the polyethylene. The inability of polyethylene to tolerate minor imperfections, undetectable with production test methods, in the resin or its application to cables, have made it impractical for high voltage power, transmission cable use.

The original major application of polyethylene, low loss insulation in high frequency, coaxial cables for electronic use continues, and it is the insulation used for

Table 3. Typical Properties of Polyolefin Resins

Property	Polyethylene ASTM Type I	Polyethylene ASTM Type III	Polypropylene copolymer	ASTM test method
density, g/cm^3	0.910–0.925	0.941–0.959	0.905	D 1505
mp, °C	125	140	172	D 2117
tensile strength, MPa (psi)	11.7 (1700)	20.7 (3000)	24.1 (3500)	D 638
elongation, %	800	600	300	D 638
modulus in flexure, MPa (psi)	483 (70,000)	1551 (225,000)	1551 (225,000)	D 790
vicat softening temp, °C	102	127	135	D 1525
heat distortion temp (at 0.5 MPa (66 psi)), °C	47	77	105	D 648
coefficient of linear expansion, °C	1.0×10^{-4}	1.3×10^{-4}	0.8×10^{-4}	D 696
hardness, Shore A	60	68	77	D 2240
brittleness temp, °C	<−118	<−118	−25	D 746
water absorption (24 h, 3-mm thick), %	<0.01	<0.01	<0.02	D 570
limiting oxygen index, %	17	17	17	D 2863
dielectric constant (at 1.0 kHz)	2.30	2.32	2.25	D 1531
dissipation factor (at 1.0 kHz)	<0.0001	<0.0001	0.0002	D 1531
dielectric strength (for short time, 3-mm thick), kV/mm (V/mil)	31.5 (800)	23.6 (600)	31.5 (800)	D 149
volume resistivity, (50% rh and 23°C), ohm–cm	$>10^{16}$	$>10^{16}$	$>10^{16}$	D 257

ocean telephone and telegraph cables. Because of its low cost, mechanical strength, and ability to withstand continuous d-c voltage under wet conditions, it is being successfully used for direct buried cathodic-protection circuits. Large amounts of direct buried telephone cable are made with high density polyethylene insulation and low density polyethylene jackets (Fig. 4).

Telephone cable users experienced difficulty that results from moisture entering the cable core through mechanically caused breaches in the cable jacket or by vapor diffusing through the jacket and condensing in the core. The problem was solved by displacing the air in the cable with a polyethylene–petroleum jelly blend applied when the cable pairs are assembled. Since the jelly has a dielectric constant that is twice that of the displaced air, it was necessary to increase the conductor insulation thickness for equivalent capacitance with resultant increase in overall cable diameter. The diameter increase is eliminated by including in the insulation resin an azodicarbonamide which liberates nitrogen gas at the approximate 204°C extrusion temperature, thereby causing the insulation to be about 50% cellular and reducing its dielectric constant so that its thickness can be reduced and overall cable diameter is the same as for nonfilled cable. The insulation is applied in a dual extrusion head with the first head extruding the cellular insulation and the second head extruding a thin skin of noncellular insulation to minimize penetration of the jelly filler into the cellular insulation and to provide crush resistance.

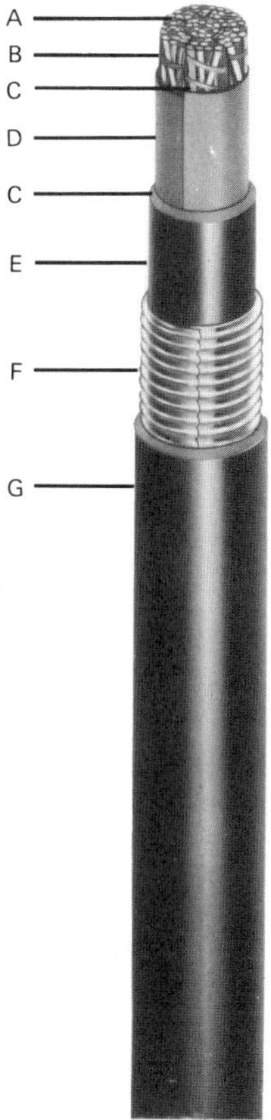

Figure 4. Direct-burial communication cable. A, solid copper conductors. B, pigment-colored polypropylene insulation. C, jelly-filling compound. D, nonhygroscopic core tape. E, inner low density polyethylene jacket. F, rodent and termite-barrier tape. G, outer low density polyethylene jacket.

Polypropylene. Some manufacturers prefer to use polypropylene copolymer instead of high density polyethylene for telephone cable insulation. The choice is based on mass volume cost (cost per kg × specific gravity). Polypropylene has physical and electric characteristics that are practically identical to high density polyethylene, is more resistant to cracking in ultraviolet light than noncarbon-filled polyethylene, and has a higher melting and deformation temperature (165°C) than polyethylene (the latter having 125°C for low density and 140°C for high density). The latter characteristics have led to its use as insulation for pump cables in oil wells where the bottom-hole temperature does not exceed 90°C.

Fluorinated Plastics. Poly(tetrafluoroethylene) (PTFE) and fluorinated ethylene–propylene copolymer (FEP) are thermoplastics with outstanding heat and chemical resistance and are suitable for use over a temperature range of −90 to 250°C; their properties are listed in Table 4. Their high cost, unique processing, and corona sensitivity have limited their use to situations where the service requirements demand their unique properties. Principal applications have been in military electronics, air-

Table 4. Typical Properties of Fluorinated Plastics[a]

Property	Polytetra-fluoroethylene (PTFE) [*9002-84-0*]	Fluorinated ethylene–propylene (FEP) [*9010-79-1*]	Poly(vinylidene fluoride) (PVF) [*24937-79-1*]	Poly(ethylene–*co*-chlorotri-fluoroethylene) (E-CTFE) [*25101-45-5*]	Poly(ethylene–*co*-tetrafluoro-ethylene) (ETFE) [*25038-71-5*]	ASTM test method
specific gravity	2.13–2.20	2.12–2.17	1.78	1.68	1.70	D 792
mp, °C	327	265	160	245	270	
tensile strength, MPa (psi)	24 (3500)	20 (2900)	45 (6500)	48 (7000)	45 (6500)	D 638
elongation, %	200–400	250–330	100–300	200	200	D 1708
tensile modulus, MPa (psi)	400 (58,000)	345 (50,000)	827 (120,000)	1655 (240,000)	827 (120,000)	D 638
hardness, Rockwell	D 50 (Shore)	R 25	D 80 (Shore)	R 93	R 50	D 785
heat distortion tapes (at 0.5 MPa (66 psi)), °C	121	72	150	115	106	D 648
coefficient of linear expansion (cm/cm)/°C × 10^{-5}	10.0	8.3–10.5	8.5	8.0	5.2	D 696
water absorption (24 h, 3-mm thick), %	<0.01	<0.01	0.04	<0.01	0.03	D 570
dielectric constant, 1.0 kHz	2.0	2.1	8.5	2.6	2.7	D 150
dissipation factor, 1.0 kHz	<0.002	<0.0003	0.018	0.002	0.0008	D 150
dielectric strength (for a short time, 3 mm) kV/mm (V/mil)	19 (480)	22 (560)	10 (260)	19 (480)	21 (530)	D 149
volume resistivity (50% RH and 23°C), ohm–cm	$>10^{17}$	$>10^{17}$	10^{14}	10^{14}	$>10^{15}$	D 257

[a] Ref. 13.

Table 5. UL Dynamic Cut-Through Using 1.59-mm Blade and #12 AWG Wire[a]

Covering	Total thickness, mm	Cut-through, kg	
		at 23°C	at 75°C
ETFE[b]	0.51	163	61
rubber-PCP (polychloroprene)	1.98	36	20
EPDM–CSPE (chlorosulfonated polyethylene)	1.98	51	17
XLPE	1.19	168	23
XLPVC	0.76	50	8.2

[a] Ref. 14.

[b] In addition, ETFE has excellent moisture, heat, radiation, chemical, and flame resistance. It is finding increasing use for control and instrumentation cables. The comparative properties of various fluorinated plastic insulations are listed in Table 4.

craft, and missiles where weight and space are valuable. In addition to outstanding heat and chemical resistance, these plastics have excellent mechanical and dielectric characteristics which permit miniaturization of the maze of internal wiring used in electronic computers and industrial controls with superior reliability. Their very low dielectric loss characteristics permit their use in high frequency cables.

A new fluorinated plastic, a copolymer of ethylene and tetrafluorethylene (ETFE) that has a useful temperature range of −90°C to 150°C, has become available. This plastic can be extruded on conventional extrusion equipment with Hastelloy barrel, screw, and tools to prevent corrosion. The mechanical performance of this insulation, which justifies the use of reduced thickness, is shown in Table 5 (see Fluorine compounds, organic).

Taped Insulations

Impregnated Paper. Many cables are covered with insulation applied in the form of tape. Until ca 1960, most cable that was rated 15,000 V and above was insulated with oil-impregnated paper. Aside from a few prototype installations of extruded solid-dielectric-insulated cables that were rated 115–138 kV, it is the standard insulation used at present for 115–550 kV circuits (15). For intermediate voltage (15–69 kV) cables, the use of extruded solid-dielectric insulation exceeds that of paper insulation. Several producers have discontinued manufacture because of the diminished market for intermediate-voltage paper-insulated cable. Paper cable has an unequalled service record extending to 40 yr and more and permits transmission of larger power loads through given size ducts than other insulations. Intermediate-voltage, paper-insulated cable will probably continue to be manufactured because of these desirable characteristics and the need to maintain existing installations. Future petroleum cost and availability conditions may make the use of intermediate-voltage paper-insulated lead-sheathed cables advantageous again since their petroleum content is substantially less than the solid dielectrics in use (see Paper).

Unimpregnated paper is wound spirally over the conductor in the form of tapes that are 19–25 mm in width and having a thickness of 0.10–0.20 mm. The thickness of insulation ranges from 4.2 mm for 15 kV cable to 31.8–34.0 mm for 500 kV cables. The paper is manufactured from chemical wood pulp (qv) by the sulfate process with care being taken to eliminate salts by extensive washing. The paper sheet is formed on either a cylinder or a Fourdrinier machine. The paper fibers are more uniformly distributed in cylinder-machine paper so that this paper is preferable for higher voltage cables. High density paper, ca 1.15 g/cm^3, sometimes is applied at the conductor for part of the total insulation thickness instead of paper having a density of ca 0.70 g/cm^3. The high density paper increases the dielectric and impulse strength of the cable.

The paper-taped conductor is placed in an autoclave where heat and vacuum are applied to draw off moisture. When the paper is dry, preheated impregnating fluid is admitted under pressure to force the impregnant into the paper. Mineral oil or polyisobutylene [*9003-27-4*] impregnants and modifications thereof are commonly used. The essential characteristics of the impregnant are low dielectric loss and oxidation resistance. For some lower-voltage cables, the paper tapes are impregnated before being wrapped on the conductor.

Because the dielectric properties of the oil-impregnated paper insulation are greatly affected by exposure to moisture, it is necessary that the insulated conductor be hermetically sealed. A lead sheath normally is employed as the enclosure for lower-voltage cables (see Metallic Protective Coverings). The continuous application of 0.1–1.4 MPa (15–200 psi) oil pressure to the insulation within the hermetic enclosure improves the dielectric properties of the insulation, permitting operation at high voltages. Since lead creeps at hoope stress greater than 0.9 MPa (131 psi) (resulting from 0.1 MPa (15 psi) oil pressure, either applied or from ca 10.7 m elevation change for cable with typical lead-sheath diameter), it is necessary to reinforce lead cable sheaths for internal oil pressure greater than 0.1 MPa (15 psi) by means of external tapes or by pulling the unsheathed insulated conductors into a steel pipe enclosure for high pressure systems (Fig. 5). Oil commonly is used as the pressure medium although gas sometimes is used because of its low cost compared to oil and the simplicity

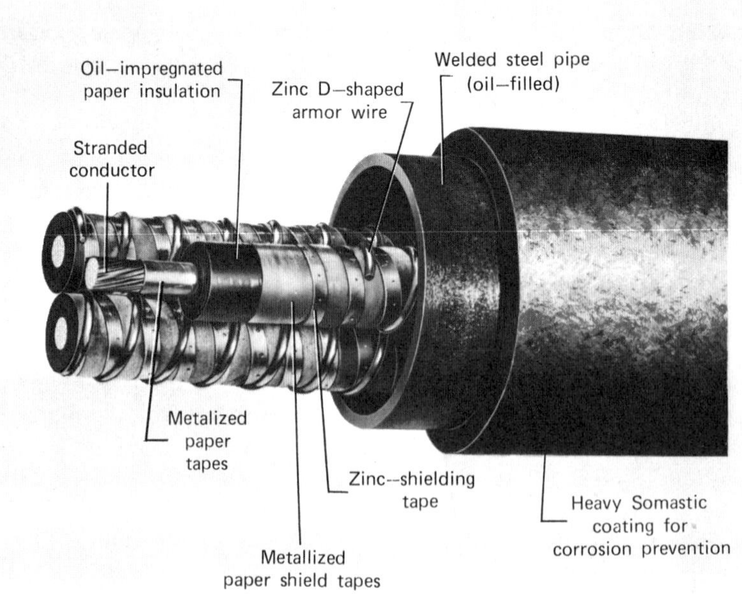

Figure 5. Underground pipe-type transmission cable.

of the equipment used to maintain pressure. However, it is not as effective in suppressing ionization within the insulation and the consequences of pressure loss are more severe than with oil. The details of construction and acceptance tests for oil-impregnated, paper-insulated cables are given in AEIC Specifications.

Dielectric loss (per unit length) W_{DL}, and charging current I_c, for cable are calculated in accordance with the following equations (16):

$$W_{DL} = e^2 \omega C_o k \tan \delta$$

$$I_c = e\omega C_s \text{ (assuming no dielectric loss)}$$

where e = voltage to neutral, $\omega = 2\pi \times$ frequency, C_o = capacitance per unit length with vacuum dielectric, k = ratio of dielectric constants of insulation and vacuum, and $C_s = C_o k \times$ cable length.

The dielectric loss per unit length (usually calculated on a per meter basis) occurs in the insulation in direct proportion to the dielectric constant k and dissipation factor ($\tan \delta$) of the insulation; the other factors in the formula are fixed by the system frequency and voltage and the cable size. The total charging current is the vector sum of the current needed to energize the cable capacitance (proportional to k and length) and the current needed to supply the dielectric loss (proportional to $\tan \delta$ and length). Since the charging current must flow from the power source at one end of the circuit along the length of the cable in order to reach each unit length of the insulation, an additional heat loss occurs in the conductor. This heat loss is the I^2R_{ac} loss (where I is the current and R_{ac} is the a-c resistance of the conductor) resulting from charging current flowing in the conductor and is analogous to the I^2R_{ac} loss that is produced by load current. As the circuit increases in length, an increasing portion of the cable's heat-dissipating capacity must be allocated for absorbing charging current I^2R_{ac} losses, leaving a reduced capacity available for dissipating load-current losses. Both the dielectric charging current and I^2R_{ac} loss increase with system voltage.

With the operating voltages (up to 345 kV) and typical circuit lengths, a significant proportion of the thermal capacity of cable systems with oil-impregnated paper insulation is needed to absorb charging current and dielectric heat losses, thus reducing the thermal capacity available for absorbing heat losses that result from load current (15). Therefore, load capability is reduced. Attempts are being made to reduce the dielectric constant k, which is ca 3.5, and the dissipation factor, which is on the order of $\tan \delta = 0.3\%$ for oil-impregnated paper insulation. The use of low loss polymeric tapes alone or laminated with paper may be possible if solutions are found for the problems of obtaining oil impregnation through the tapes and slippage and corona-discharge sensitivity of otherwise desirable polymers.

Dry Paper. Spirally wrapped, nonimpregnated kraft paper insulation was the standard insulation for telephone cables for many years and still is used to a considerable extent. Dry paper provides a thin, unusually tough, low cost insulation and has the low attenuation and capacitance characteristics required for telephone transmission. Telephone cable cores are vacuum dried and hermetically sealed with a lead sheath, or with a composite covering consisting of longitudinally applied aluminum or aluminum and steel tapes with a black polyethylene jacket overall. The metal tapes are corrugated across their width to permit bending the cable. The seam in the steel tape is soldered to make it moisture tight. Dry paper-insulated telephone cables are operated with 34–69 kPa (5–10 psi) of dry gas pressure, which is applied to the core to detect leaks in the sheath and to prevent the entrance of moisture in the core, until openings that occur in the sheath can be repaired.

In the paper-pulp insulating process, fifty or more wires are insulated continuously with wet kraft pulp and are dried to form a highly porous uniform coating of paper having excellent dielectric properties (17). The use of paper insulation enables the manufacture of telephone cables having more than 2000 pairs. A large amount of polyethylene- or polypropylene-insulated telephone cable is used for cables with pair counts that are less than 2000. The moisture-resistant nature of the polyethylene individual-conductor insulation is advantageous because these conductors can be fanned out and terminated in terminal boxes without the need of splicing in moisture-resistant pigtails which is required for cables with dry-paper insulated conductors. Hermetic splice enclosures also are not essential with polyethylene-insulated cables. However, the size of polyethylene-insulated cables becomes excessive as the number of pairs increases because the diameter of each individual conductor is greater than is the case with paper insulation.

Coated Fabric. Cotton (qv) sheeting in thicknesses of 0.13–0.30 mm and in widths of 91–122 cm that are impregnated by multiple dips into a bath of asphalt base, and tung- and linseed-oil varnish, and subsequently baked with heat is varnished cambric insulation. The varnish impregnation provides the electrical insulation; the cotton sheeting supports the varnish film. A thin film of mineral oil is applied between lapped layers to provide slippage of the tapes on one another when the cable is bent. Varnished cambric has been almost completely replaced by the extruded solid dielectrics. However, since it continues to be used for coils, motors, and other electrical apparatus insulation, the tape continues to be available.

Silicone rubber-coated glass (qv) tapes, made by calendering silicone rubber into square woven-glass sheeting and then vulcanizing, are suitable for continuous service at 175°C in cables that are rated from 600–15,000 V. Such tapes are the primary insulation in large-size U.S. Navy shipboard power cables (Fig. 6). Friction tape made with cotton sheeting is used as protective covering for cable splices and wrapping and bindings in electrical and mechanical applications. Friction tape is fully described

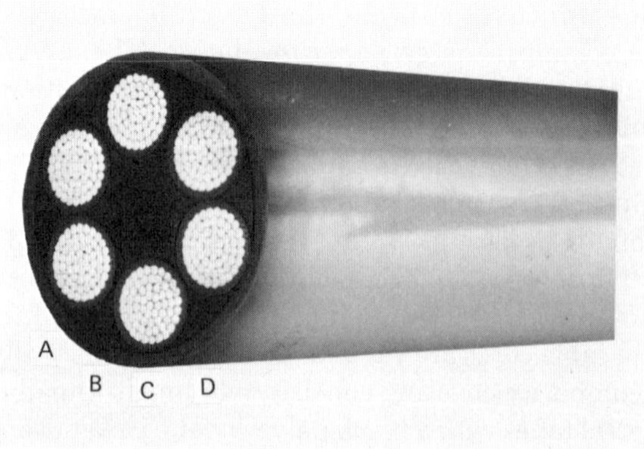

Figure 6. Shipboard cable (600 V rating). A, poly(vinyl chloride) jacket. B, polymeric water-blocking compound. C, silicone rubber-coated glass-tape insulation. D, stranded copper conductor with interstices water-blocked.

in ASTM Specification D 69. Wire nuts, snap-on plastic enclosures, and adhesive-coated plastic-film tape are being used for pigtail splicing on small-size low voltage wiring in dry locations instead of rubber and friction tape.

Splicing Tape. Poly(vinyl chloride) is calendered into a film 0.18–0.25-cm thick, and a thin coating of rubber-based adhesive is applied on one side. The film is slit into 1.9-cm wide strips and is wound into rolls for splicing applications. This pressure-sensitive PVC tape is in general use for low voltage wire and cable splicing. ASTM Specification D 1000 contains detailed requirements.

Unvulcanized-rubber insulation compounds are sheeted on a calender to a thickness of 0.076 mm and are used for field splices and repairs in rubber- and plastic-insulated and jacketed cables. The compounds used are similar to those in cable manufacture with modifications to make them tacky and self-fusing. For most applications, it is not practical or necessary to vulcanize splicing tapes after they are applied. Tapes made with vulcanizable compounds are used for splices and repairs in heavy-duty portable cables which must be vulcanized to have sufficient mechanical strength. The increasing cost of hand labor required for making taped cable splices and terminations in shielded, high voltage cables and the possibility of workmanship errors has led to the development of prefabricated terminators and splices wherein the stress-relief layers and shielding tape connectors are factory made and tested so that field installation involves only the stripping back of cable components prior to slipping on the device. These devices are designed for specific or limited cable diameters and sizes. The use of hand-wrapped tapes continues because they can be used regardless of cable size and rating.

Component Tapes. Polyester and polypropylene film (see Film and sheeting materials) that is slit into tape is used commonly as a component of thermoplastic-insulated and -jacketed cables. Because of their high strength, a thickness of 0.02–0.05 mm is adequate, resulting in reduced diameter. Such tapes are not particularly suited for thermoset-insulated and -jacketed cables because thermoset materials do not bond to them during vulcanization as they do to fabric tapes that are filled with rubberlike compounds.

Polyester and polyimide films (see Polyimides) are manufactured with a thin coating of thermoplastic that has a lower melting point, eg, polyethylene with polyester and fluorinated ethylene–propylene with polyimide film. The coated film is wrapped on conductors and is passed through an oven in which the lower melting point layer is melted and fused. The use of such tapes permits the manufacture of wire with thin coverings composed of 0.15 mm or less of high performance films that cannot be extruded directly.

Paper tape that has been supplemented with addition of conducting carbon black during the papermaking process (so as to make it semiconducting) or laminated with aluminum foil is used for electrostatic shielding in oil-impregnated, paper-insulated cable. These tapes are compatible with, and permit the free flow of, the impregnant during the impregnating process.

Coated Insulations

Enamels. Resin-insulated wires as distinguished from heavier conductors are insulated by passing clean, smooth wire through a bath of liquid enamel and a baking chamber and repeating the process as often as necessary to build the desired thickness of enamel, usually less than 0.05 mm on wire sizes AWG 40–8. Insulated wires of this type possess an excellent space factor (ratio of conductor volume to total volume) and are used wherever the space requirements are severe.

Magnet-wire enamels are selected on the basis of their mechanical properties, their ability to adhere firmly to the conductor to which they are applied, and their aging characteristics. In most applications, the enamel acts as insulation between the turns of the coil structure. In the manufacture of coils, the enamel must possess toughness, a high value of abrasion resistance, flexibility, and the ability to sustain a high mechanical load without plastic flow or cut-through. No one type of wire enamel can be accepted for general use in all environments, at all temperatures, and for all voltages and frequencies. Most of the more important enamels have peculiarities which promote their use over competing materials in certain limited fields of application. For example, some enamels can be soldered, whereas others require scraping or abrading of the coating before soldered connections can be made. Types of enamels are described in Table 6.

In the manufacture of a coil structure, a bonding varnish is applied by dipping and baking the complete coil. An alternative method is to use an overcoated wire; a standard-grade enamel wire is coated with a thin film of a heat-softening resin. After being wound, the coil is heated to a predetermined temperature which causes the overcoat resin to melt and fuse and to form a coil-binding adhesive.

For some uses, magnet wire is wrapped with fibrous threads or tapes of paper, cotton, glass, nylon, or asbestos which then are saturated with enamel and baked. Such wrappings provide a greater thickness of covering than is obtained with the enamel alone; they are used for greater resistance to winding hazards and for overload protection.

Metallic Protective Coverings

Aluminum. Aluminum [*7429-90-5*] or aluminum alloy strips that are 0.5–1.0-mm thick and 1.3–2.5-cm wide are used frequently as the mechanical protective covering

Table 6. Wire Enamels[a]

Chemical type	Base resin	Modifier	Max temp of general usability, °C
oleoresin	natural	tung oil	105
formal resins	poly(vinyl formal)	phenol-formaldehyde	105
nylon	polyamide		105
polyurethane (solderable)	polyurethane	polyester resin	105
polyurethane (nonsolderable)	polyurethane		125–130
epoxy	polyepoxide	polyester resin	130
acrylic	acrylonitrile and acrylate resins		105–130
polyester	terephthalate		130–155
silicone	silicone	terephthalate resin	155–180
fluorocarbon	polytetrafluoroethylene		180
polyimide	polyimide		220

[a] Ref. 18.

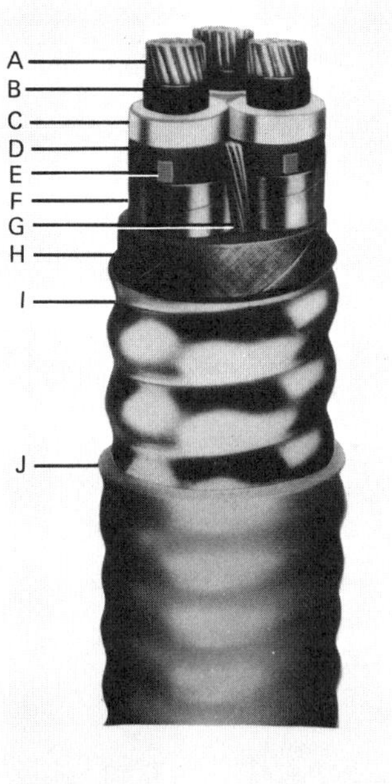

Figure 7. Industrial power-distribution cable. A, copper or aluminum conductor. B, Semicon EP strand shield. C, Okoguard EP insulation. D, Semicon EP insulation shield. E, phase-identification strip. F, copper-shielding tape. G, grounding conductor. H, fillers and binder tape. I, welded corrugated aluminum sheath. J, poly(vinyl chloride) jacket.

on large-size industrial-plant power distribution cables. Copper-free aluminum alloys have the best corrosion resistance for this application.

Cables with a sheath formed from a folded aluminum strip with the butting edges having a continuous longitudinal weld to make it impervious are in common use. For diameters <2.5 cm, the sheath can be swaged after welding to a snug fit on the cable core. The smooth sheath is desirable for installations, eg, food-processing lines where it is necessary to wash down and avoid the entrapment of dirt. For other installations where greater flexibility is advantageous, the welded sheath is passed through rollers that form corrugations ca 2.5–8.1-mm deep to fit the cable core firmly and provide flexibility resulting from the accordion effect of the corrugations when flexed. Such cables are preferred for many chemical plant, paper mill, and hazardous locations services because of the lightweight sealed sheath (Fig. 7).

The direct extrusion of aluminum sheaths has not been commercially practical because of equipment cost and complications and problems with billet welds and press stops. Cold-rolled aluminum strip 0.2 mm laminated on both sides with 0.05 mm ethylene–acrylic acid copolymer is applied over cable cores as a flat helical tape. When a polyethylene jacket is extruded over the aluminum strip, the extrusion heat causes the polyethylene to fuse to the copolymer on the strip and the copolymer layers at the strip overlap to fuse together. The combination of polyethylene jacket and fused strip provides moisture impermeability approaching that of seamless aluminum and the flexibility of polyethylene.

Copper and Copper Alloys. Copper [*7440-50-8*] is used for various types of protective coverings. An alloy of 90% copper and 10% zinc (commercial bronze), as well as other bronze alloys, can be used (see Copper alloys). They are applied as a helically wrapped tape or strip 0.13-mm thick, over rubber, varnish cloth, and oil-impregnated paper insulations on higher voltage power cables where they provide a nonmagnetic, electrostatic shield. These tapes usually are tin coated for corrosion protection when used with insulations and jackets containing sulfur. Commercial bronze and copper tapes are used in a similar manner to protect cables against damage by termites, teredos, and other insects and rodents in underground and submarine installations, particularly in warm climates (see Insect control technology). It is used in thicknesses of 0.51 mm and greater, applied either as a flat tape or corrugated and interlocked, for mechanical protection as an overall covering where it is desired to provide better corrosion resistance than is obtained with galvanized steel or aluminum. Tape that consists of a 0.05-mm layer of stainless steel which is metallurgically bonded to 0.05-mm layers of copper on both sides sometimes is used instead of 0.25-mm bronze tape on direct buried communication cables. The stainless steel layer provides protection against rodents and the copper layers provide a low resistance path to ground for lightning surges.

Commercial bronze is used for the basket-weave, wire-armor braid on cables for commercial ships where it provides a flexible covering with good salt-water corrosion resistance.

Fine copper wire, having a 0.13–0.25-mm diameter, is used in basket-weave braids to provide a flexible electrostatic shield over the insulation on flexible cables, eg, high voltage portable power cables, high frequency radar cables, and microphone cables (Fig. 8). Tin-coated copper wires generally are used for shielding braids in rubber-insulated cables in order to prevent corrosion of the copper wire if sulfur is present in the insulation jacket. Uncoated copper is used whenever possible to avoid the high cost of tin.

Because its 1083°C melting point would damage most insulated cable cores, direct extrusion of copper sheath is not practical. Copper-sheathed cables are made using

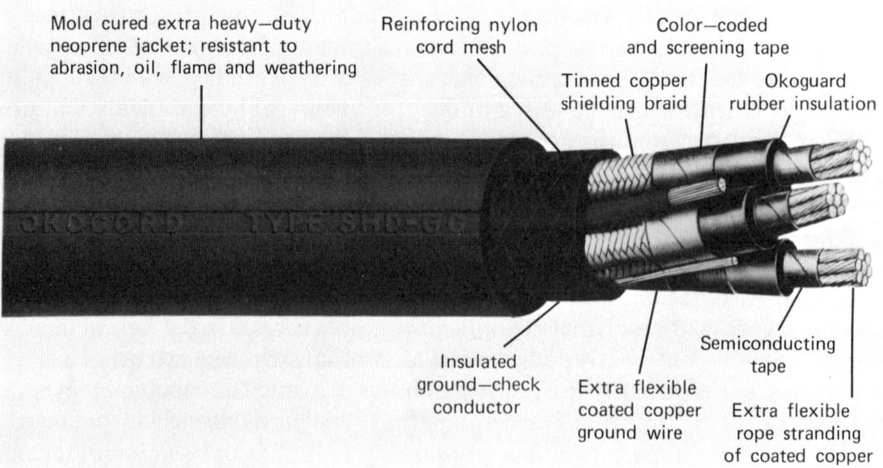

Figure 8. High voltage portable power cable.

a European process and starting with short billets consisting of the desired number of conductors embedded in dry magnesia and encased in a copper tube sheath (19). The assembly is drawn through dies which simultaneously reduce the diameter and elongate the conductors, magnesia, and copper tube to the final desired size. The cable is used for high temperature applications and where the utmost in fire resistance is necessary. The cable is difficult to terminate because of the hygroscopic nature of magnesia insulation (see Electrical connectors). Oxidation and embrittlement of the copper sheath becomes a problem with continuous temperature above 150°C. Smooth and corrugated copper sheaths can be applied to cable core by the folded strip and longitudinally welded method.

A nonmagnetic type of the copper–nickel alloy, Monel, sometimes is used where the best corrosion resistance and long life are desired for electrostatic shielding tape on high voltage rubber-insulated submarine cables. Interlocked armor formed of Monel metal also is used on cables exposed to severe corrosion conditions, eg, electrically operated bottom-hole oil- and brine-well pumps.

Lead. Lead [*7439-92-1*] sheaths are used with rubber-, varnished-cambric-, and paper-insulated cables as mechanical protection and to prevent entrance of moisture or loss of insulation impregnant. Lead is particularly suited because it may be applied by hydraulic extrusion presses in long, continuous lengths. It forms a flexible covering, allowing the cable to be bent easily during handling and installation. The use of lead as the sheathing metal allows the cable to be installed aerially or in most underground concrete-duct systems without serious corrosion difficulties. In underground installations, where severe conditions exist or electrolysis from stray electric currents is liable to cause sheath damage, the lead sheath may be protected by a covering of a fabric-reinforced, neoprene compound, which is applied in the form of tapes, or by an extruded jacket of polyethylene. Details of design and acceptance tests covering lead sheaths for power cables may be found in specifications issued by ICEA and AEIC.

The chemical classifications of pig lead, which may be used for cable sheathing, are outlined in ASTM B 29. In order to have the stability and mechanical properties required of a sheathing metal under service conditions, the lead must contain small additions of alloying metals. The most frequently used metals are 0.06% copper and 0.005% silver in combination. For lead-sheathed cables that are subjected to vibrational stresses in service, 0.75–1.0% antimonial or 2.0% tin alloys are used (see Insulation, acoustic). The addition of 50–100 ppm of sodium to molten pure lead purges it of oxides and impurities, which are removed as dross, and results in a more uniform sheath structure.

Steel. Large installations of high voltage cables in welded steel (qv) pipe have been made for more than 40 yr. The interior of the steel pipe is thoroughly cleaned and coated with epoxy enamel to prevent rusting and to reduce cable-pulling tension. The exterior is sand blasted, coated with an adhesive and jacketed with a 1.5-mm high density polyethylene jacket or Somastic coating of sand, bitumen, and asphalt for corrosion protection; and is subjected to a spark-detector test. Twelve-m lengths of pipe are buried directly in the earth and are welded for the circuit length. Welds are checked by x ray or fluoroscope, and heat shrinkable polyolefin or bitumastic coatings are applied for corrosion protection on the weld area. Nonleaded, oil-impregnated, paper-insulated cables are pulled into the pipe which then is filled with insulating oil or dry nitrogen gas. The oil or gas in the pipe is maintained at 1.40 MPa (200 psi) pressure during operation of the cables. The steel pipe provides a moisture-proof covering and takes the place of a lead sheath, which would not be strong enough at such a high pressure.

A large amount of wire and cable used in buildings is made with a spirally wrapped, interlocked, galvanized steel tape outer covering, BX. A similar protective covering is used on wires and cables for electric-power generating stations, for aerial power cables, and for electrical circuits on machinery. Flat steel tapes or strips—both plain and galvanized—are used extensively for protective coverings, particularly on cables buried directly in the ground. Two such steel tapes are helically wrapped on the cable, and applications of a layer of asphalt-saturated jute underneath and another over the steel tapes help to prevent corrosion in underground service. The steel tape is a mild steel having a tensile strength of ca 276–483 MPa (40,000–70,000 psi).

Galvanized mild-steel wires, wrapped helically around the cable core, are used for protection and to provide longitudinal strength on submarine, borehole, shaft, vertical-riser, and dredge cables. The tensile strength of the wires is from 345–483 MPa (50,000–70,000 psi). However, though galvanized, the mild-steel wire armor becomes corroded in salt water so that the service life of the cable is shortened considerably. In order to extend the life of the armor wires, it is a frequent practice to extrude a layer of polyethylene, 1.0-mm thick, over the individual wires before applying them to the cable.

Zinc. Notwithstanding its excellent corrosion protection as a galvanizing coating on steel wires and tapes for buried and submarine cables, zinc [*7440-66-6*]—as a pure metal or in zinc-rich alloys—is chemically active with weak acids and alkalies and is soluble in distilled water. Because of this chemical activity, the use of zinc and zinc-rich alloys is limited to dry locations, eg, for shielding tape in lead-sheathed cables, shielding tape, and D-shape skid wires on conductors for pipe-type, paper-insulated cables. Zinc tapes are satisfactory when used in lead-sheathed cables where there is no risk of contact with corrosive liquid or where the lead sheath will carry fault currents.

Nonmetallic Protective Coverings

Fibrous. ***Braids.*** Fibrous braids are used for optimum mechanical protection, color-coding and where presently available jacketing materials do not have adequate heat resistance, in which case, glass or asbestos braids are used. The use of asbestos braids has practically disappeared because of its carcinogenic effects (20); however, they are mechanically superior to the usual glass-braid substitute.

Servings. Servings consisting of roves of plied fibrous materials wrapped helically around wires and cables are used for separation and bedding purposes. Cotton, jute, nylon, and polypropylene are the most common. Dry cotton servings are used over the conductor under the insulation of flexible cords to impart increased flexibility and permit easy removal of the insulation and connections. Cotton or nylon servings are embedded in the rubberlike jacket or applied over the core under the jacket of large portable cables to reinforce the jacket against tearing and cutting. Jute servings thoroughly saturated with asphalt are used for bedding and corrosion protection of metallic coverings on direct buried, underground and submarine cables. Jute has low, bulk density so that it can absorb considerable quantities of saturant (eg, asphalt), can provide cushioning to absorb the forces involved in laying the cables, and protection against cutting by sharp projections, eg, rocks, shells, and gravel. Asphalt provides low cost corrosion protection for lead sheaths and metallic tapes and wire servings. Nylon instead of jute serving often is used over flat tapes or armor wires because it has better long-term resistance to rotting. In addition to corrosion protection, asphalt-saturated jute or nylon servings over wire-armored submarine cables protect against birdcaging of the wire armor and provide a friction surface for tensioning and braking the cable while it is being laid. Polypropylene roving should not be used for the outer servings because of its elastic memory which causes it to ravel at places where it is cut.

Fillers (qv) must satisfy the manufacturing requirements of making cable round, preventing deformation of individual conductor insulation, and being nonhygroscopic, flame-retardant, and inexpensive. For cables with thermoset overall jackets, soft unvulcanized elastomeric fillers that flow and conform to the interstitial spaces are used, even though they are costly. For other cables having thermoplastic or metallic outer coverings so that the cable core is not subjected to a second vulcanization, jute or wadded paper that is either dry or saturated with flame- and water-repellant compounds frequently are used. Polypropylene fillers that have been plyed from foamed fiber to increase their bulk also are used.

Thermoset Jackets. ***Neoprene.*** Neoprene [*31727-55-6*] (polychloroprene, PCP) was the first synthetic elastomer used in the wire and cable industry. It is not suitable as electrical insulation but its outstanding sunlight, water, flame, chemical, and oil resistance and its mechanical toughness led to its adoption as a cable jacket replacing cotton braids, natural-rubber jackets, and lead sheaths on rubber cables.

Chlorosulfonated Polyethylene. In cables for fixed installations, neoprene is being replaced by chlorosulfonated polyethylene [*9008-08-6*] (Hypalon), (CSPE), which has greater heat and moisture resistance than neoprene and sufficiently good electrical characteristics so that a jacket of CSPE replaces part of the conductor insulation (Fig. 9). Pigmented jackets with CSPE are superior in fade resistance to colored neoprene jackets and are used for open-pit, portable mining cables when color coding is desired.

Polyacrylonitrile (PAN). Jacket compounds containing polyacrylonitrile have superior oil and heat resistance and are used for oil-well cable jackets. PAN and poly(vinyl chloride) are blended and used as the base material for jacket compounds as a substitute for neoprene, particularly where fade-resistant colored jackets are needed. (see Acrylonitrile polymers).

Chlorinated Polyethylene. Compounds are formulated with chlorinated poly-only limited use in electric cables because its economic and performance characteristics in comparison with previously developed jackets still are under evaluation. The various elastomers used for insulation and jacket compounds are compounded with high-structure conductive carbon black for use as conductive jackets for stress relief and grading layers (21).

Thermoplastic Jackets. ***Poly(vinyl Chloride).*** Poly(vinyl chloride) (PVC) is one of the major materials for overall jacketing of fixed installation cables used at normal temperatures. Low material and processing costs combined with good performance properties are the reasons for this acceptance. Important uses are mechanical and corrosion protection for interlocked armor and metallic tape-shielded power-distribution cables (see Vinyl polymers).

Polyethylene. Polyethylene (PE) is used for cable jackets where moisture resistance is the primary requirement, eg, for nonleaded dry paper or plastic-insulated telephone cables. Such PE-jacketed cable types are used as a corrosion-protective

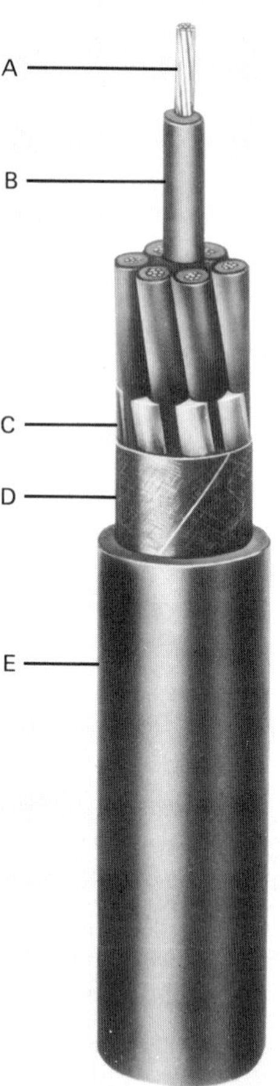

Figure 9. Generating-plant control cable. A, coated copper conductor. B, heat-, flame-, and moisture-resistant polyolefin insulation. C, nonhygroscopic, flame-resistant fillers. D, core binder tape. E, chlorosulfonated polyethylene jacket.

covering for lead-sheathed underground power cables where their low coefficient of friction facilitates pulling the cable into ducts. If cathodic protection is applied to such cables, the polyethylene jacket is able to resist the electro-osmotic effect of the d-c voltage and its high insulation resistance minimizes the rectifier capacity needed to energize the system. It is important that polyethylene resin with high molecular weight (ca 30,000) be used to prevent stress cracking in the presence of pulling lubricants and that about 2% of carbon black be added to the polyethylene to prevent ultraviolet-radiation cracking where the covering is exposed to weather.

High density polyethylene duct is extruded in long lengths over either single concentric neutral primary cables or two or more secondary cables. The cables occupy about 50% or less of the duct cross-sectional area and can be withdrawn and replaced if necessary. In diameters up to 7.6 cm, the duct and cable assembly is sufficiently flexible so that it can be coiled on standard cable reels and shipped for site installation with substantial savings in labor cost (Fig. 10) (see Olefin polymers).

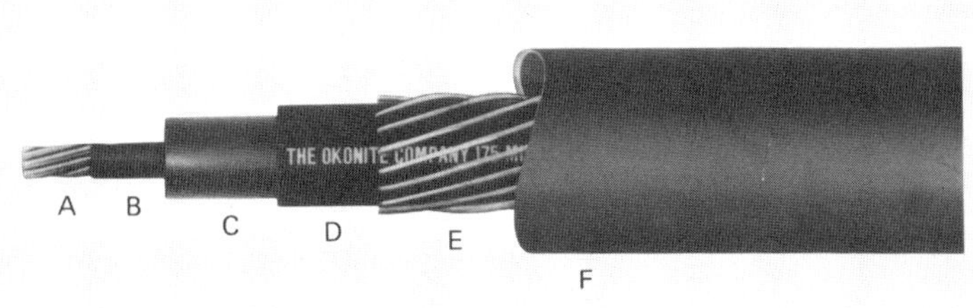

Figure 10. Primary distribution cable with preassembled duct. A, high voltage primary conductor. B, Semicon EP strand shield. C, Okoguard EP insulation. D, Semicon EP insulation shield. E, concentric return conductor. F, high density polyethylene duct.

Table 7. U.S. Sales of Wire and Cable Coverings[a]

Year	10^6 $
1973	4253
1974	5356
1975	3910
1976	4541
1977	5341
1978	
electronic	569.2
telephone and telegraph	1701.8
magnet wire	825.7
power wire	754.5
control and signal	127.2
building	982.4
appliance wire and cordage	578.6
other	464.8
Total (1978)	*6003.2*

[a] Ref. 22.

Nylon. Nylon (polyamide) jackets, because of their mechanical toughness, are used to permit a reduction in primary insulation thickness. Their application usually is limited to small sizes because of difficulties in extruding the thin jackets desired on larger cables (see Polyamides).

Thermoplastic Elastomers. Thermoplastic elastomers (TPE) are polyolefin thermoplastic rubbers that are a form of crystalline polyethylene and are supplied by the producers with a combination of the following properties, not all of which can be optimized in any one grade: flame resistance, low temperature performance, good abrasion resistance, excellent low temperature properties, good hydrocarbon resistance, and color stability. TPE is undergoing evaluation of its processing and cost factors as a wire and cable covering (see Elastomers, synthetic).

Economic Aspects

U.S. sales of wire and cable insulation by ca 150 companies (22) from 1973–1978 are listed in Table 7. Sales of other insulated materials include automotive, elevator, weatherproof and service drop cable, airframe, shipboard, and missile ground-support cables.

Analytical Test Methods

The tests of the quality and suitability of cable coverings can be broadly categorized as destructive or qualification tests and nondestructive or quality assurance tests. Destructive or qualification tests are only performed for initial design or to qualify design or component changes. The use of heat from nuclear fission to produce steam for electric power generation involves special qualification tests for wires and cables. Demonstration is required by the Nuclear Regulatory Commission (NRC) that wires and cables essential for the safe shutdown of the reactor in the event of a loss of coolant accident (LOCA) be capable of performing their function at any time during the expected reactor life (see Nuclear reactors). The requirements as defined by IEEE involve simulation of Arrhenius aging at rated temperature for design life, usually forty years; exposure to accumulated radiation dosage for normal life plus accident dosage (total of 2 MGy (200 Mrads) for plants in use in the United States); and exposure to the expected high temperature steam and chemical environment within the reactor containment building during LOCA, with rated current and voltage applied. The testing of materials to determine their flammability, corrosive off-gasing and smoke generation characteristics is of major interest (23). ASTM D 2863 specification describes the method for determining the oxygen index of materials for flammability comparison (see Flame retardants). In order to evaluate the flame propagating and self-extinguishing characteristics of complete cable designs, tray flame tests described in IEEE Standard 383 are made. Accelerated water immersion and aging tests involving months and years of exposure are destructive qualification tests that are performed as part of insulation or cable design and are repeated occasionally to verify that there has been no change over time or to qualify changes from the original design. It is necessary to monitor regular production to assure that the original quality is being reproduced uniformly.

Nondestructive quality assurance tests are necessary to maintain uniform quality by repetitive testing of in-process and finished cable. Spark tests that are carried out by applying voltage to the insulation surface with the conductor grounded during the insulating process provide the first assurance of freedom from holes and imperfections in the insulation (24). Visual and manual tests and inspections are made during subsequent manufacturing processes. These tests and inspections are not mandated by industry specifications but are performed by the manufacturer to assure that the completed cable will be satisfactory in final acceptance tests. Acceptance tests for power and control cables include electrical voltage resistance and partial discharge tests, minimum insulation resistance and maximum conductor resistance tests. Cables with nonmetallic coverings are submerged in water to provide a ground plane for electrical tests and metallic covered cables are tested in air. Conductor continuity, mutual capacity, and attenuation tests are performed by automatic scanning equipment on multiple-pair communication cables. Short samples are cut from production lengths on a sampling-plan basis for examination of dimensional construction, centering of conductors and cable core, and performance of physical tests of insulation and jacket (see Nondestructive testing).

Information on the testing of insulated electric cables is contained in the literature.

Nomenclature

AEIC	= Association of Edison Illuminating Companies
ANSI	= American National Standards Institute
ASTM	= American Society for Testing & Materials
AWG	= American wire gauge
CSPE	= chlorosulfonated polyethylene
EHV	= extra high voltage
E–CTFE	= ethylene–chlorotrifluoroethylene copolymer
EPDM	= ethylene–propylene terpolymer
EPM	= ethylene–propylene copolymer
EPRI	= Electric Power Research Institute
ETFE	= ethylene–tetrafluoroethylene copolymer
FEP	= fluorinated ethylene–propylene copolymer
I	= current
Ic	= charging current
ICEA	= Insulated Cable Engineer's Association
ICEA–NEMA	= Insulated Cable Engineer's Association–National Electrical Manufacturer's Association
IEEE	= Institute of Electrical & Electronic Engineers
IIR	= isobutylene–isoprene rubber
IMSA	= Internal Municipal Signal Association
LOCA	= loss of coolant accident
MIL	= U.S. Dept. of Defense, Military Cable Specifications
NR	= natural rubber
NRC	= Nuclear Regulatory Commission
PCP	= polychloroprene
PE	= polyethylene
PTFE	= poly(tetrafluoroethylene)
PVC	= poly(vinyl chloride)
PVF	= poly(vinylidene fluoride)
R	= resistance
REA	= U.S. Dept. of Agriculture, Rural Electrification Administration
SBR	= styrene–butadiene rubber
SIC	= specific inductive capacity
SIR	= specific insulation resistance
$\tan \delta$	= dissipation factor of insulation
TPE	= thermoplastic elastomers
UL	= Underwriters' Laboratories, Inc.
W_{DL}	= dielectric loss (per unit length)
X	= reactance
XLPE	= cross-linked polyethylene
Z	= impedance

BIBLIOGRAPHY

"Cable Coverings" in *ECT* 1st ed., Vol. 2, pp. 697–716, by E. D. Youmans, The Okonite Co.; "Electric Wire and Cable Coverings" under "Insulation, Electric" in *ECT* 2nd ed., Vol. 11, pp. 802–822, by John Hogan, The Okonite Co.

1. R. B. Blodgatt and R. G. Fisher, *IEEE Trans. PAS Paper 68TP651-PWR,* 1968.
2. R. O. Babbit, *The Vanderbilt Rubber Handbook,* R. T. Vanderbilt Co., Norwalk, Conn., 1978.
3. E. C. DeBaene and C. A. Anderson, *IEEE Trans. PAS Paper 54-399,* 1954.
4. E. B. Paine and H. A. Brown, *IEEE Trans.* **59,** 39 (1940).
5. R. M. Eichhorn, *IEEE Trans. Electr. Insul.* **EI-12** (Feb. 1977).
6. G. Bahder, C. Katz, J. Lawson, and W. Vahlstrom, *IEEE Trans. PAS, Paper T73-496-7,* 1973.
7. *National Electrical Code NFPA No. 70-1978.*
8. *Power Cable Ampacities,* Vols. I and II, ICEA Publication No. P-46-426.
9. *Ampacities—Cables in Open Top Cable Trays,* ICEA Publication No. P-54-440.
10. *Ampacities for Single Conductor Solid Dielectric Power Cable 15 through 69 kV,* ICEA Publication No. P-53-426.
11. J. H. Neher and M. H. McGrath, *The Calculation of the Temperature Rise and Load Capability of Cable Systems, IEEE PAS TP 57-660,* 1957.
12. R. C. Graham, *IEEE Trans. PAS Paper 55-444,* 1955.
13. *Insul. Circuits,* 18 (June 1980).
14. J. R. Perkins, *Electr. World* (Oct. 1, 1975).
15. A. S. Brookes, *ERC-Manufacturers 500/550 kV Cable Research Project,* IEEE 72 CHO 609-0-PWR, May 25, 1972.
16. A. R. Von Hippel, *Dielectric Materials and Applications,* Technology Press of Massachusetts Institute of Technology, Cambridge, Mass., and John Wiley & Sons, Inc., New York, 1954.
17. A. S. Windeler, *IEEE Trans. C&E, Paper 54-104,* 1954.
18. F. M. Clark, *Insulating Materials for Design and Engineering Practice,* John Wiley & Sons, Inc., New York, 1962.
19. C. A. Jordan, G. S. Eager, Jr., *IEEE Trans. PAS Paper 55-54,* 1955.
20. *Fed. Regist.* **37**(202), Para. 1910.19.
21. R. C. Mildner, *IEEE Trans. Power Appar. Syst.* **PAS 89,** 313 (Feb. 1970).
22. *Current Industrial Reports, Insulated Wire and Cable,* U.S. Dept. of Commerce, Bureau of the Census, MA-33L(78)-1, 1978.
23. J. O. Punderson, *Toxicity and Fire Safety of Wire Insulations—A State of The Art Review,* IEEE ICC TG 12-36, Nov. 8, 1976.
24. H. H. Clinton and T. W. Stewart, *Wire Wire Prod.* (Mar. 1968).

JOHN E. HOGAN
The Okonite Co.

INSULATION, THERMAL

Thermal insulation is defined by ASTM C 168 as a material or assembly of materials that is used primarily to provide resistance to heat flow. Thus, its distinguishing feature is high thermal resistance, which is measured by the ratio of the difference between the average temperatures of two surfaces to the steady-state heat flux that is in common through them (time rate of heat flow per unit area of one surface). Thermal resistance is related to thickness and is expressed as $K \cdot m^2/W$. Thermal properties vary with temperature and must be quoted at a specific mean temperature (1–2).

A low thermal conductivity (high thermal resistivity) is a relative concept depending upon the application, since other physical properties are also important in a practical insulation. Materials with a thermal conductivity of 0.014–0.029 W/(m·K) at 24°C are available but may not have the strength, temperature range, flame retardancy, or price that are appropriate for use in, for example, furnaces where direct-flame exposure to temperatures up to 1650°C is required; instead, insulating firebrick with a conductivity of 0.144–0.433 W/(m·K) would be the insulating material of choice. Thermal insulation may be specified for a number of purposes: to conserve energy and therefore, costs; to increase the comfort of living spaces; to facilitate control of the temperature of a process; to reduce the temperature of the shell of a pressure vessel; to control the external temperature of the insulated space in order to avoid danger to personnel; to protect structural members from damage by high temperature; to reduce the temperature of working spaces; and, in the case of equipment that is operated below ambient temperature, to prevent condensation or icing within the structure. Published values of thermal conductivity (1,3–4) are, in general, only typical and may or may not apply to the particular material and application under consideration. Final decisions should be made on the basis of information and recommendations of reputable manufacturers or as the result of unbiased tests. Properties of insulating materials (other than thermal conductivity) include strength, hardness (qv), compressibility, specific heat, resistance to high or low temperature, thermal or moisture coefficients of expansion, water-vapor permeance, etc. The choice of an insulating material for a given purpose usually involves a compromise among desirable properties. For example, insulating firebrick must have considerable strength and be resistant to the high temperature to which it is to be exposed. The desirable properties of high thermal resistance and light weight must be sacrificed to some extent to achieve the necessary properties of strength and temperature resistance.

Reflective insulations, eg, polished or bright metallic surfaces, depend upon the low thermal emittance of their surfaces and subdivision of the air space for their resistance to heat flow. To be effective in retarding heat flow by radiation, the reflective surface must face an air space (1,3,5).

A wide variety of materials have been used as insulations; some of the more common ones include cork (qv) and cork products; mineral fiber in various forms; expanded perlite or vermiculite; wood fibers (qv) in loose or felted forms or fabricated into boards; foamed plastics (qv); various types of paper (qv) (macerated loose fill, plain, creped, or corrugated); aluminum foil; calcium silicates; cellular glass (qv); diatomite (qv) products; ground newsprint; and porous clay (qv) products (for high temperatures).

Principles of Heat Transfer

The mechanisms by which heat is transferred through a thermal insulation or under pressure of a thermal gradient across the insulation, are radiation, gas conduction, gas convection, and solid conduction (see Heat-exchange technology, heat transfer). Each mechanism obeys its own laws and each, in a gas-filled insulation, is operative at all temperatures. The relative importance of the mechanisms depends on the physical properties of the insulation as well as on environmental conditions (6–7). The most favorable design is adjusted so as to give minimum total heat transfer.

Gas Conduction. Heat is carried by gas conduction in the form of kinetic energy of the gas molecules (7–8). Molecules near the hot surface transfer part of their energy by colliding with neighboring molecules of lower energy. The heat thus conducted through a gas is proportional to the product of the mean free path of the gas molecules and the gas density. Because the mean free path varies inversely with pressure and the density varies directly with pressure, the heat transferred by conduction in a free gas is independent of pressure. In many air-filled insulations at atmospheric pressure, this is the major form of heat transfer. As the gas pressure is reduced, a point is reached at which the mean free path of the gas molecules is larger than the interstices in the insulation. Below this pressure, gas conduction decreases as the pressure decreases. Insulations that are composed of very fine silica [*7631-86-9*] (qv) particles have interstices that are so small that gas conduction is reduced, even at atmospheric pressure. However, this reduction is achieved at the expense of higher density than is found in ordinary fibrous insulations.

Gas Convection. Heat is transferred by convective gas currents flowing in open spaces or connecting pores measuring 1–3 mm in diameter (9–10). The transfer is small in smaller pores because the driving force that sets the convective current in motion depends on the temperature difference across the pore. In low-density, fibrous insulations and in building or appliance insulations of density $<8\ kg/m^3$, convective transfer may represent up to 20% of the total transfer. In most insulations of higher density, convection is a minor heat-transfer process.

Solid Conduction. Heat that is transported by solid conduction passes through the unit solids (eg, spheres, cylinders, or plates) that comprise the insulation and across the unit contacts. Heat flow that arises from solid conduction is expressed as the product of the conductivity of the solid in its bulk state and of a contact factor describing the nature of the flow between the particles. This factor is influenced by porosity, particle shape and orientation, the pressure applied to the insulation, and to a limited extent, the particle size and elastic constants (11–14).

Many simple models have been proposed to find the contact factor for different systems. Best agreement is obtained with foams, in which solid conduction is along a three-dimensional network that is uninterrupted by high-resistance thermal contacts (10,12,14).

Radiation. Radiant-heat transfer through insulation takes place by multiple scattering and by absorption and reemission. Radiation is multiply scattered by the units comprising the insulation and multiply reflected by the radiation shields, if any (15–16). Unlike the other mechanisms of heat transfer, the scattering process returns some of the incident heat energy to its source, which greatly complicates the calculation of radiation conductivity. Characteristic of this type of radiant-heat transfer is that the wavelength distribution of energy entering the heat sink is the same as that of the heat source and is modified only by the amount of absorption in the insulation.

In absorption and reemission radiant-heat transfer (8,17), the energy is absorbed by a layer of the insulation and is reemitted in part. The part that is not emitted is lost to gas conduction and convection and solid conduction. The emitted heat, unlike that from the scattering process, has the spectral distribution of the temperature of the insulation layer. The decrease in intensity of the emitted energy is governed by the emissivity of the layer, which is a function of the infrared emission spectrum of the layer at the particular temperature. One half of the reemitted radiation travels back toward the source; the other half travels toward the heat sink. The reemitted energy from each layer is subject to multiple scattering and reflection from particles and radiation shields and to reabsorption by other particles.

In steady-state applications, the distinction between radiant heat that is unabsorbed despite multiple scattering, and heat that is transferred by absorption and reemission, is one of degree rather than kind. In transient heating, this is not the case. The scattered radiation enters the heat sink immediately, but the absorbed and reemitted heat experiences a time lag in transit—as do all modes of heat conduction (8). Thus, the thermal diffusivity of an insulation contains a contribution from the absorbed radiation but not from the purely scattered radiation. Two separate temperature waves often are not observed in an actual insulation because of the mixing between radiation types that normally takes place.

Effective Conductivity. When solid conduction plus any or all of the other modes of conduction are present, the effective conductivity of an insulation arises from a mixture of parallel and series heat transfers (8,17–18). The two types of transfer are interlinked since an increment of heat arriving at any site in the insulation has more than one way to be conducted. The separate solid conductivity of the insulation is altered by the presence of gas conduction and radiation. Many models have been proposed to demonstrate how to add the contributions from the separate conduction modes in terms of the porosity, particle shape and orientation, and the ratios of the conductivities (14,17–20).

The fundamental equation for heat transfer by radiation between two parallel surfaces that are separated by an empty space is

$$q = \delta EA(T_r^4 - T_i^4)$$

where q is the time rate of heat flow, δ is the Stefan constant of radiation, A is the area of the radiating or irradiated surface, T_r and T_i are the absolute temperatures of the radiating and irradiated surfaces, and E is a factor involving the emissive characteristics of the surfaces. When there is insulation between the surfaces, the amount of radiation is reduced and the expression becomes more complicated. Generally, at room temperature and below, the radiation contribution to the effective conductivity is small. Nevertheless, in evacuated insulations of low density (low solid conduction), the radiation component, although small, may be a significant fraction of the whole. Because cryogenic insulations are of increasing commercial importance, efforts have been made to reduce the radiation component by using reflective metal shields or insulating powders mixed with metal flakes (see Cryogenics).

Above 200°C, radiation generally is the dominant mechanism of heat transfer. Reduction in radiation can be achieved by proper-sized units that effect optimum multiple scattering and absorption (21) and by radiation shields made from highly reflective metals (14,22). Metal coatings on refractory particles, platelets, and fibers have been reported (22), and represent an attempt to increase reflection. Radiation also is important in low-density insulation, eg, fiberglass and plastic foams. At ambient temperatures, it has been found that for the low-density products, the concept of thermal conductivity is not completely valid at small thicknesses. It is necessary to measure thermal resistance at incremental thicknesses to be certain that an added increment of thickness will add equal thermal resistance (23–25).

Two or more components often are used in an insulation. Since absorptive capacity depends on the ir spectrum, a second component may be needed to block a window in the ir spectrum of the first component which causes the component to be transparent to a range of wavelengths. Particularly in high temperature insulations where a large temperature drop across the insulation is encountered, there is exposure to a wide range of wavelengths. Such an insulation may be divided into two or more layers that are effective for the temperature range to which they are exposed.

Materials

Thermal insulations are used at temperatures from −200°C or lower to 2200°C. For convenience, this range of temperatures is divided into five parts: below 0°C; from 0–150°C; from 150–300°C; from 300–650°C; and over 650°C. Table 1 presents a listing of blocks, board, and pipe insulation products for use from −268 to 1040°C. Further information may be obtained in refs. 3 and 26.

Below 0°C. With the growth of cryogenics, the need for improved low temperature insulations has increased greatly. Quick-freezing and low temperature chemical processes have pushed to lower limits the temperature at which insulations must function. Modern processes such as the production of liquid oxygen require operating temperatures below −150°C.

Table 1. Thermal Conductivities of Block, Board, and Pipe Insulation

Material	Temp range or max, °C	Thermal conductivity, W/(m·K) −20°C	0°C	24°C	100°C	200°C	400°C	540°C	ASTM specification
polyurethane	−73 to 110	0.026	0.024	0.025					C 591
polystyrene			0.031	0.035					C 578
cellular elastomeric		0.040	0.041	0.043					C 534
cork pipe insulation				0.048					C 640
cellulose fiber board				0.055					C 208
mineral fiber									
blanket	204		0.035	0.037	0.051				C 553
block	204			0.041	0.052	0.071			C 612
board	982					0.083	0.116		C 612
pipe insulation	650			0.045	0.058	0.075			C 547
cellular glass	−268 to 427			0.060	0.074	0.100			C 552
calcium silicate									
block	649				0.066	0.079	0.106		C 533
board (577 kg/m³ or 36 pcf)	650				0.123	0.126			C 656
diatomaceous earth	870					0.100	0.109	0.117	C 517
diatomaceous earth	1040					0.108	0.121	0.130	C 517
expanded perlite	816					0.079	0.105		C 610

Mineral wool is made from molten slag, rock, or glass (or from selected combinations of these ingredients), by blowing, drawing, or other means of fabricating it into fine fibers. For low-temperature insulation, it usually is supplied in the form of blocks or boards bonded with asphalt or synthetic resins or as flexible or semirigid blankets (see Resins). For cold-storage rooms and other applications, for which a degree of strength is advantageous, blocks or boards are used. Blankets are more frequently employed to insulate refrigerated cars and trucks, household refrigerators, low temperature food-storage boxes, and other small or portable equipment (see Refrigeration). Pipe insulation is available either as preformed sections or as blankets that are suitable for wrapping. Loose mineral wool or granulated mineral wool is sometimes used when a fill insulation is desired.

Cellular glass is made by foaming softened glass and permitting it to solidify to a product having many small, individually sealed voids. It has the advantage of considerable structural strength and it is practically impervious to the passage of water vapor.

Foamed plastics (qv) are used extensively at temperatures of up to 120°C because of their low density, conductivity, and resistance to moisture. The most common materials are polystyrene [*9003-53-6*], isocyanurates, and polyurethane [*27416-86-0*] (see Styrene plastics; Urethane polymers). Others, eg, poly(vinyl chloride) [*9002-86-2*] and phenolic foams, are coming into use (see Vinyl polymers; Phenolic resins). Polyurethane, with an aged thermal conductivity as low as 0.016 W/(m·K) at 23°C and the capability of foaming it at the installation site, has attracted considerable attention. Polyurethane is used widely in refrigerated railroad units. Urea-based foams (urea–formaldehyde [*9011-05-6*]) have proliferated for dwelling sidewall insulation (see Amino resins). However, problems of gas emission resulting from improper formulation and of increased heat flow resulting from shrinkage cracking exist (27). A survey of building construction insulations is described in ref. 7.

Perlite is a ground volcanic ore to which water is added which causes the ore to expand at high temperatures into cellular particles of controlled density and size. For low-temperature insulation, perlite is manufactured in uniformly sized spheres of low density. In this form, the powder flows freely and may be used as pouring insulation for cold boxes, cryogenic-liquid storage tanks, and refrigerated cars. Owing to its flowability, expanded perlite is particularly suitable in applications where the insulation must be removed periodically for equipment inspection and repair.

Silica aerogel has been offered and used principally for low-temperature insulation although it is said to be stable at up to 700°C. It is a finely divided powder with a remarkably low thermal conductivity. Its greatest uses have been in high-efficiency insulations (see Silica).

Reflective insulations, eg, aluminum foil, aluminum sheets, and coated steel are used as low temperature insulations—alone or in multiple layers—or as facings for bulk insulations. Multiple sheets of reflective insulation have the unique advantage of low heat capacity, which is important in cryogenic applications and equipment where rapid changes in temperature are desirable (28). Aluminum sheets and sheets of coated steel are impervious to the passage of water vapor if the joints are perfectly sealed. Very thin aluminum foils cannot always be depended upon as vapor barriers because of the possibility of pinholes and of small tears that may occur during application. The most effective cryogenic insulation is a series of spaced, reflecting shields made from low emissivity metal coatings or sheets having low density spacers to minimize the solid conduction. Such a structure can produce a thermal conductivity below 0.0014 W/(m·K). Although useful in simple geometries, difficulties in fabrication may arise when this insulation is applied on the outside of complex shapes.

Multiple-layer composites of mineral wool and reflective metals greatly reduce heat transfer at mean temperatures at or above 250°C where heat transfer by radiation is a significant part of the total heat transfer. When the air in the layers is replaced by denser gases, or evacuated, improvement in low temperature heat transfer is significant. Dense gas, eg, Freon, is used in urethane foam to improve heat-flow properties (see Fluorine compounds, organic, fluorinated hydrocarbons).

Temperatures From 0–150°C. ***Residential Applications.*** Mineral wool, which is in the form of batts or blankets or that is granulated for pneumatic application, is the most common building insulation in the United States. For walls, blankets and batts are supplied in various thicknesses up to the full stud thickness, and in various widths to fit between studs. Floor or ceiling batts and blankets have thicknesses of up to 305 mm. The batts and blankets usually are supplied with an integral vapor barrier which must be carefully applied to the warm side in order to be effective.

Some modern woodframe houses use 50 × 150-mm studs with 140-mm wall cavity spaces that can accommodate R 3.34 K·m^2/W (R-19) insulation. The use of foamed plastic sheathing in walls is increasing. A 25-mm thick plastic foam sheathing has a thermal resistance of 0.88 K·m^2/W (R-5) as compared to 0.07 for 8-mm plywood sheathing. Standards based on thermal resistance have been established for mineral wool, and they vary from a minimum R 1.94 (R-11) for the wall of a home to R 7.05 (R-40) for a ceiling of a high-quality home. These correspond to thicknesses of 89 mm and 305 mm in typical batts (29).

Macerated paper has come into increasing use as a blown insulation for filling attic joist spaces in homes. The cost is low and chemical treatments for reduction of combustibility and surface-flame spread have been improved greatly. Since the materials used for these treatments often are water-soluble, and condensation may occur in the insulated area, care should be exercised in selection to be sure the desirable low flammability is permanent. The thermal resistivity is better than that of mineral fiber and less thickness is needed.

Expanded vermiculite is poured into the stud spaces or other wall cavities of buildings and between top-floor ceiling joints.

Insulating board, made principally of wood or bagasse fibers, is used as sheathing and roof insulation. Along with its considerable structural strength, ease of handling, and economical application, its insulating value is in excess of the materials it replaces (eg, plywood) (see Laminated wood-based composites).

Insulated glass has a polyester film coating further coated with a thin layer of metal that passes visible light rays but reflects infrared heat waves (see Laminated materials, glass). The film functions like the middle pane in triple-pane insulating glass (30).

Aluminum foil is applied between the studs, rafters, or joints of buildings, either alone or supported by paper, gypsum board, or insulating board or by the paper attached to batts or blankets of mass-insulating material. To be thermally effective, the foil must face an air space. Three layers are commonly rated as R 1.6 (R-9).

Industrial Applications. Insulation of tanks, stills, ovens, process piping, hot-water heaters and piping, low-pressure steam equipment, etc. Mineral wool in the form of blankets and usually reinforced by wire mesh, expanded metal, or other reinforcing material, is used on heated tanks, towers, ovens, dryers, and other equipment. Heated pipe is insulated with mineral wool in the form of preformed sections or blankets. Asbestos millboard, composed of asbestos fiber with binders and fillers, is used frequently when a board product is indicated. Its value as a thermal insulation usually is subordinate to its use as a fire-resistant surfacing material. Boards of other mineral fibers are penetrating this market because of the health problems associated with asbestos (qv). Insulating cements composed of mineral wool, vermiculite, or calcium silicate combined with appropriate binders, are used where a trowelable insulation is required. They are furnished dry and are mixed with water at the application site. The thermal conductivity of an insulating cement usually is higher than that of a preformed product of similar composition. Physical properties of cements vary with mixing and application techniques (see Cement).

From 150–300°C. The most widely used insulation for temperatures up to 300°C has been mineral fiber or calcium silicate [*1344-96-3*]. They are made in blocks of various sizes, in sectional pipe form, in segmental form for pipes of large diameter, and in custom shapes. They are used for the insulation of ovens, boilers, steam pipes, stills, breechings, etc. Mineral-wool blankets are used up to ca 540°C, and insulating cements also are used in this temperature range where a trowelable insulation is required.

From 300–650°C. Calcium silicate insulation is composed principally of the reaction product of lime and silica which is combined with reinforcing fiber. The silica used in manufacture is in the form of ground sand or diatomite (qv). When in the form of blocks or pipe insulation, this material is used at temperatures up to 650°C. Typical applications are for high pressure boilers, furnaces, high pressure steam lines, and similar equipment operating below the temperature limit of the insulation. A higher temperature formulation has been offered for temperatures up to 1000°C.

High temperature mineral wool is available for use in temperatures of up to 1700°C. The combination of light weight and low thermal conductivity is very attractive for some aerospace applications. The compositions of the slags vary with increasing temperature. These fibers are produced in lower fiber diameters and higher bulk densities than the usual mineral-wool product which may be suited only to 900°C.

Above 650°C. Above 650°C, the requirements for insulation become more exacting. In the semirefractory range from 650–1100°C, chemical stability considerations are important. The thermal performance of an insulation and the physical integrity are affected by sintering that occurs within the insulation, by shrinkage, and by structural breakdown.

Diatomite, when bonded with suitable binders and molded to form blocks and pipe insulation, is used for temperatures from 650–1100°C. Typical applications are for furnaces, kilns, high pressure boilers, high pressure steam lines, metallurgical furnaces, regenerators, roasters, flues, and stacks. This insulation frequently is used with calcium silicate as a double-layer combination.

Insulating brick, which is made principally of diatomite, is used to back up refractories in furnace applications when rapid changes in temperature are not encountered. They are not suitable for direct exposure to flame. These bricks of high-silica content have unusually high compressive strength and are used under firebrick in the floor of furnaces where load-carrying ability is required.

Insulating firebrick is fire-clay brick and is made porous by burning away organic material. Firebrick is suitable for direct-flame exposure where conditions favoring abrasion or slagging are not severe. They are classified by ASTM into eight groups according to bulk density and probable behavior in service, as shown by ASTM C 155-76.

At temperatures above 1100°C, refractory metals and oxides (or graphites), and high temperature mineral fiber are used. The low pressure to which these insulations often are exposed helps them to retain chemical stability and eliminates gas conduction and convection as a heat-transfer means.

Although the same types of insulations may be used at high temperatures as at low temperature, there are restrictions. Powders are limited by their sintering temperature, and they must be contained in small compartments to prevent settling and redistribution owing to mechanical forces. Fibers are more resistant to settling and are more heat stable, but many refractory fibers (qv) begin to show dimensional instability in the felted form above 1100°C. As the temperature is increased, the fibers devitrify and lose strength and eventually sinter to produce a dense, rigid, ceramic body. This significantly reduces thermal performance. Felts can be preconditioned at service temperature, but this is not always satisfactory.

Heat-Retardation Systems. Heat-retardation systems include absorption and radiation systems (31–32). Absorption methods include ablation, transpiration cooling, and heat sinks. Ablation cooling allows each layer of the insulation to be heated by the incoming flux through the temperature range in which it disintegrates (ablates), thereby exposing a cooler surface underneath. The ablating layer absorbs heat owing to its specific heat, by physical changes (eg, melting, vaporization, or sublimation), or chemical changes (eg, depolymerization), by the transpiration cooling effect (see below) of the gases produced, and by radiation from its surface. Absorption by heat capacity is present at all heat-flux levels; the other heat-transfer processes increase in importance as the heat flux increases. Some of the best ablation materials are plastics that are reinforced with synthetic or mineral fiber and that have a high resin content. Cork compositions also may be used as a short-term ablation material. Synthetic-fiber (nylon, etc) reinforcement is used at the highest heat fluxes because decomposition of the synthetic fiber and resin produces hydrogen in large volumes. Hydrogen has a high heat content per kg and provides superior heat absorption as well as transpiration cooling (33). The minimum heat flux at which equilibrium ablation conditions can be maintained is dependent on both the resin and the reinforcing fiber (see Ablative materials).

Transpiration cooling is the absorption of heat by cooler gases that flow from the cold surface through a hotter, porous layer. The coolant may be a liquid that is pumped under pressure. The porous layer may be a metal or ceramic foam with interconnecting pores or it may be the result of prior disintegration. Transpiration cooling is the most effective method for dealing with high heat fluxes, but because of the piping required, it incurs a severe weight penalty and rarely is used.

Heat sinks depend on the stability of the system in storing large amounts of heat. For high heat fluxes, heat sinks are made of ceramic refractory materials (eg, zirconia, alumina, quartz) of low conductivity and high heat capacity (low thermal diffusivity) so as to block the flow of heat into the rest of the system. Heat sinks can be solids or liquids. Because the total heat capacity of the system up to its melting point is a criterion, heat-sink absorption is limited to short-term exposures. Reinforced plastics are good heat-sink materials; although, because they begin to volatilize above 250°C, their use is restricted to the low heat-flux range. Heat sinks are suited for fluxes below 3 W/m^2 (1 Btu/h·ft^2); ablation methods for fluxes of 30–300 W/m^2 (9.5–95 Btu/h·ft^2); and transpiration cooling for very high fluxes (300–3000 W/m^2). Combinations of these methods invariably occur in any thermal-protection system.

Radiation systems guard against long-term exposure to heat fluxes below ca 16 W/m^2 (5.1 Btu/h·ft^2). The radiation system requires that the surface of the insulation facing the source attain a sufficiently high equilibrium temperature to radiate considerable heat back toward the source. The maximum surface temperature is governed by the material. Surface-temperature limits for radiation materials are from 1100–1400°C. The material should have a low thermal conductivity but not necessarily a low thermal diffusivity since, after the first few minutes of operation, the system attains a steady-state condition.

Alumina or zirconia bonded to a metal shell provides a means of radiation protection as well as of oxidation resistance. The high melting temperature and low thermal diffusivity permit a surface buildup of heat followed by re-radiation. Some radiation systems make use of metal-clad insulations that are comprised of refractory felts or other forms which are encased in thin metal envelopes (31). These formations are used around jet engines and fuel tanks. Refractory felts made from aluminosilicate fibers range in density from 48–480 kg/m^3 and can withstand continuous service at 1100°C. Improved aluminosilicate fibers are available for service to 1500°C. Felts made from silica microfibers are stable to 1100°C and are available in a range of densities. Felts can be made in thicknesses, rigidities, and shapes to fit almost any contour. The temperature limit of a felted layer depends partly on the mechanical abuse to which the material is subjected—the limit decreasing as the severity of vibration increases. Low density, semirigid boards of ultrapure silica fibers can be preconditioned to have a high dimensional stability at temperatures approaching the softening temperature. Reflecting shields of tantalum and other refractory metals can be interposed between refractory-fiber blankets in order to improve the thermal conductance (6) (see Refractory fibers).

Unusually low thermal conductivities and thermal diffusivities are attained by a family of insulations made from fibers that are mixed with ultrafine silicate powders and radiation-blocking agents. Because the insulation pore size is less than the mean free path of air molecules, the thermal conductivity is lower than that of still air. Available as flexible felts or rigid blocks and with compressive strengths from 1.4–3.5 MPa (200–500 psi) (31), these insulations provide superior protection in metal-clad systems. They can operate continuously at 1000°C and at higher temperature under transient conditions.

Vapor Barriers. Whenever the cold side of an insulated structure is at a temperature below that of the condensation temperature of the vapor on the warm side, protection against vapor infiltration must be provided. In most cases this problem arises with respect to water vapor, but in very low temperature applications, carbon dioxide or even oxygen may condense within the insulation or on the cold surface. Unless the material is compatible with liquid oxygen, such condensation could cause an explosion.

If a vapor-pressure difference exists across the surface surrounding the insulation, infiltration of vapor may occur even though the surface is airtight. Many materials that have very high resistances to the flow of air have comparatively low resistances to the flow of vapors. A difference in the total pressure across the surface is not necessary for vapor infiltration. If there is a difference in the partial pressure of the vapor, infiltration occurs unless the surface is impermeable to the vapor to which it is exposed.

The vapor barrier must be impermeable to carbon dioxide, oxygen, or other condensable gases to which it may be exposed, depending upon the minimum temperature. Solid metal shells are effective but in many instances are prohibitively expensive. If metal shells are used, the welds or other joints must be as impermeable as the metal. For less severe conditions, membranes having low permeability to vapors are used to surround the insulation on the warm side and to prevent or minimize the gas infiltration. Insulation surrounding structures that prevent the vapor from leaving the cold side of the insulation, require especially good vapor barriers. The walls of refrigerated rooms may be adequately protected by a somewhat more permeable barrier if the inside (cold) surface of the wall is highly permeable. In such a case, the vapor passes through the inner surface and is condensed or frozen on the cooling coils of the refrigerating equipment. However, because this condensation or freezing adds to the refrigeration load, these walls should be as effectively vapor-proofed as is economically feasible (34–35).

Vapor barriers should always be applied to the warm surfaces of the walls or ceilings of buildings that experience severe winter weather. The need for vapor barriers appears to be greater in insulated than in uninsulated buildings, but laboratory experiments and field surveys have shown that condensation can occur in uninsulated walls as well. For home construction, a vapor barrier having a permeance of ≤ 0.66 metric perms (1 U.S. perm) is considered satisfactory (36); vapor barrier requirements for low temperature applications have not been well established, but a requirement of 0.03 metric perm has been suggested. A metric perm is 1 $g/24\ h{\cdot}m^2{\cdot}mm$ Hg or 8.70×10^{-11} $kg/(Pa{\cdot}s{\cdot}m^2)$. For homogeneous materials, permeance at a single temperature and relative humidity is inversely proportional to thickness (3,36–37).

Production

Estimated U.S. production capacity of materials used for insulation from below 0°C to 150°C are listed in Table 2.

The total market for insulation materials included for all temperatures has been estimated at $\$(2–3) \times 10^9$ as of 1978 and was divided approximately equally between residential and industrial applications.

Economic Aspects

It is estimated that the U.S. chemical process industry alone could save 266×10^{15} J/yr (2.52×10^{14} Btu/yr) by increasing its use of insulation and by improving system maintenance (39–41). With rapidly increasing fuel costs and decreasing fuel availability, the economic justification for increased thermal insulation for energy conservation can be demonstrated readily. Although as insulation design must still be checked for such other factors as adequate worker protection from excessive surface temperature, prevention of process-line freeze-up, elimination of surface condensation and control of process, the economics of energy conservation are becoming the predominant criteria for industrial insulation designs.

Life-cycle costing, as it applies to thermal insulation design, considers the total

Table 2. 1980 Estimated Insulation Capacity[a]

Material	Production × 10^6 units
slag wool, m^2 [b]	
batt and blanket	110.6
loose fill	79.1
fiberglass, m^2 [b]	
batt and blanket	1047
loose fill	76.2
cellulose loose fill kg (m^2)[c]	2221 (496)
polyurethane and isocyanurate	
board, m^2	124.1
on-site foaming, m^3	13.0
polystyrene, m^2	124.4
perlite, expanded, kg	181
vermiculite, expanded, kg	233
reflective, m^2	80.5

[a] Refs. 7 and 37.
[b] R 1.9 (R-11) equivalent.
[c] R 3.3 (R-19) equivalent.

costs of a particular application over its expected life (42–43). For a given thickness of insulation, the two costs that are involved are those of the insulation as installed and of the energy that is lost through the applied thickness. A simplified procedure for calculating the economic thickness of industrial insulation (ETI) has been developed (44).

Thermal Insulation Testing

The thermal conductivity of insulation materials that are in the form of blocks, sheets, slabs, blankets, loose-fill insulations, and cements, when designed for mean temperatures up to ca 1000°C, usually is measured by a guarded hot-plate apparatus (3–4) as standardized under ASTM C 177. Methods for measuring thermal performance of pipe insulation have been described (45–46) and have been standardized under ASTM C 335. Considerable effort over the past thirty years has been devoted to more rapid methods of measuring thermal conductivity (47–51). ASTM C 518 has been promulgated for the heat-flow meter apparatus. This apparatus usually is built for measurements at one mean temperature which can be from 10–600°C. The thermal conductivity probe also has been used for rapid thermal performance determinations (52). Data can also be obtained after a 15-min exposure and, as the temperature rise is limited, determinations may be made for materials that are not dry (53).

Heat flow through a wall, floor, roof, or other built-up structure (54) can be calculated from the known thermal conductivities of the component materials or by employing a standardized guarded hot box as described in ASTM C 236. The guarded hot box is probably the best means of determining the value of reflective insulations for building applications as they are applied. The calibrated hot box (55) used for measuring heat flow through assemblies, is generally larger than the guarded hot box, and is not as subject to module size limitations. The calibrated hot box may be used to evaluate larger or more representative sections of wall or roof structures than a guarded hot box. Methods of measuring the thermal performance of high temperature insulations, up to a mean temperature of 1300°C have been described (4), and one has been standardized under ASTM C 182 and C 201. The design of thermal conductivity testing apparatus for temperatures that are above 1100°C involves refractory materials and refined methods of temperature measurement and heat-flux control. Several radial heat-flow calorimeter designs that incorporate tantalum or other refractory metals, are available for operation in controlled atmospheres to temperatures above 2200°C (22,25).

Transient methods for measurement of thermal conductivity (25) generally involve a measurement of time needed by an initial temperature wave to pass through part of the specimen. The time of the passage is controlled by the thermal diffusivity α of the specimen

$$\alpha = \lambda/\rho C$$

where λ is the thermal conductivity, ρ is the density, and C is the specific heat. The units of diffusivity are area per unit time. The temperature pulse is accelerated when the thermal conductivity is high and is slowed when the heat capacity C is high. The quantity measured in this method is the thermal diffusivity. The thermal conductivity is obtained by substituting for the specific heat, which is measured separately, in the formula.

Transient, or nonsteady, heat flow occurs in insulations that are exposed to temperatures that change rapidly compared with the diffusion times. Nonsteady-state methods of measurement may therefore give more insight into the ultimate performance of some insulations than steady-state methods (25,56).

Field Performance Measurements. Checking field performance of insulation is also important. Insulation designs are based on data generated under carefully controlled laboratory conditions, following the prescribed detailed procedures referenced (57). The questions may be raised as to how well actual insulation installations perform under field conditions as compared with the theoretical design based on laboratory data. Whether it arises from possible faulty original workmanship, or deteriorations resulting from vapor barrier failure, physical abuse or plain use, the question is legitimate. Unfortunately there are no nondestructive devices that can be used to measure the field thermal performance of insulation systems.

Much work is in progress and several techniques show promise. The problems are accentuated by the need to cope with the dynamic real world where temperature equilibrium is seldom reached.

Most devices for field surveying thermal performance rely on inferences based on surface temperature. A failure of the underlying insulation will result in increased local heat flow through the failed area, and consequently an increased surface temperature difference compared with that over normal insulation. Although the difference in temperature between the surface and the ambient is a direct function of heat flow, it is also related to surface air flow, and the surface radiation emitted and heat flux. Thus surface-temperature measurement can be a misleading indication of insulation effectiveness. However, a number of instruments have been developed to determine relative surface temperature based on emitted ir radiation. Ir thermography has been used to survey and locate insulation faults in buildings, process piping and tanks, kilns and ovens, and refrigerated equipment (39–40) (see Infrared technology).

In some chemical processes, it is necessary to place the insulation on the inside of pressure vessels in order to keep the temperature of the vessel walls below the critical temperatures. In such applications, of which catalytic hydrogenation of a gas is an example, the insulation should not contaminate the product and it should not have a catalytic action to produce unwanted reactions. The possible corrosive effect of an insulation on structural materials must be guarded against, especially in cases where moisture may be present (see Corrosion and corrosion inhibitors).

Failures of stainless steel vessels and pipe in chemical plants have been attributed to stress corrosion that has been accelerated by the presence of even minute quantities of chlorides in insulation materials (58). Protective coatings, eg, sodium silicate, have been used on the insulation. Test methods and recommended practices for studying stress corrosion are discussed in ASTM C 692 (59).

Health and Safety

Not until 1968, when polychlorinated biphenyl (PCB)-contaminated cooking oil affected more than 1000 persons, was attention directed to the hazards of PCB. Since then, contamination of fish in the Great Lakes has been found excessive and human fat content of 6 ppm of PCB has been recorded in the Federal Republic of Germany. Sale of PCB has been greatly restricted since 1972 (see Chlorocarbons).

The combination of cigarette smoking and respirable asbestos fiber has been found to compound the health dangers. Asbestos fiber has been removed from the formulation for the majority of soft and friable insulations, eg, insulation cements or low density insulation. Asbestos-containing insulation should be removed with great care.

The toxic gases that may be released during formation of urea-based organic foams or during combustion of organic material are of concern. When urea-based foams are generated, formaldehyde gas may be released. In 1979, Massachusetts banned urea–formaldehyde foam because of the possible gas release.

Fire Hazards. Combustion of all organic substances results in production of carbon monoxide and carbon dioxide. Carbon monoxide has been found to cause more than 60% of the fire deaths in the United States. Organic materials may also, when burned, release HCl, HCN, SO_2, and other gases known to be deleterious to health. Thus, organic materials, eg, solid or foamed plastics, wood fiberboard, or loose-fill cellulosic insulation, must be carefully controlled during production to achieve the desired flame-retardant properties (see Flame retardants).

Many insulations that are noncombustible can absorb flammable chemicals which could cause the insulation material to burn. Where such exposures are possible, care should be taken to protect against absorption or nonabsorbtive insulations, eg, cellular glass, should be used.

BIBLIOGRAPHY

"Insulation, Thermal" in *ECT* 1st ed., Vol. 7, pp. 927–935, by Charles B. Bradley, Johns-Manville Research Center; "Insulation, Thermal" in *ECT* 2nd ed., Vol. 11, pp. 823–838, by R. H. Neisel and H. F. Remde, Johns-Manville Research & Engineering Center.

1. R. P. Tye, *Thermal Conductivity,* Vol. I, Academic Press, Inc., New York, 1969.
2. J. R. Welty, *Engineering Heat Transfer,* John Wiley & Sons, Inc., New York, 1974.
3. *ASHRAE Handbook and Product Directory, 1977 Fundamentals,* American Society of Heating, Refrigerating and Air Conditioning Engineers, Inc., New York, 1977.

4. J. B. Loser, C. E. Moeller, and M. B. Thompson, *Thermophysical Properties of Thermal Insulating Materials,* MLTDR 64-5 (AD 601535), Clearinghouse for Federal Scientific and Technical Information, Alexandria, Va., 1964.
5. H. E. Robinson, L. A. Cosgrove, and F. J. Powell, "Thermal Resistance of Airspaces and Fibrous Insulations Bounded by Reflective Surfaces," *Natl. Bureau of Standards Building Materials and Structures Rept. BMS 151,* U.S. Govt. Printing Office, Washington, D.C., 1957.
6. A. E. Wechsler and P. E. Glaser, *Investigation of the Thermal Properties of High Temperature Insulation Materials,* ASD TDR 63-574 (AD 420193), Clearinghouse for Federal Scientific and Technical Information, Alexandria, Va., 1963.
7. R. P. Tye and co-workers, *Current Thermal Insulation Materials and Systems for Building Application,* Department of Energy Report BNL-50862 UC95d, 1978.
8. J. D. Verschoor and P. Greebler, *Trans. ASME* **74,** 961 (1952).
9. E. R. G. Eckert, *Introduction to Heat and Mass Transfer,* McGraw-Hill Book Co., Inc., New York, 1950.
10. R. E. Skochdopole, *Chem. Eng. Prog.* **57,** 5 (1961).
11. H. M. Strong, F. P. Bundy, and H. P. Bovenkerk, *J. Appl. Phys.* **31,** 39 (1960).
12. R. H. Hardy, *Ind. Eng. Chem. Proc. Des. Dev.* **3,** 117 (1964).
13. G. N. Dulnev and Z. Sigalova, *Int. Chem. Eng.* **5,** 218 (1965).
14. P. E. Glaser, A. E. Wechsler, I. Simon, and J. Berhowitz, *Investigation of Materials for Vacuum Insulation up to 4000°F,* ADS TR62-88 (AD 274742), Clearinghouse for Federal Scientific and Technical Information, Alexandria, Va., 1962.
15. M. A. Heasley and B. Baldwin, *Am. Inst. Aeronaut. Astronaut. J.* **2,** 2180 (1964).
16. R. Viskanta, *J. Heat Trans.* **87,** 143 (1965).
17. D. Dunii and J. M. Smith, *AIChE J.* **6,** 71 (1960).
18. J. C. Harper and A. F. El Sahrigi, *Ind. Eng. Chem. Fundam.* **3,** 318 (1964).
19. C. L. Johnson and D. J. Hollweger, *Non-Evacuated Cryogenic Thermal Insulation Studies,* ML TDR 64-260 (AD 607891), Clearinghouse for Federal Scientific and Technical Information, Alexandria, Va., 1964.
20. M. J. Laubitz, *Can. J. Phys.* **37,** 798 (1959).
21. B. K. Larkin and S. W. Churchill, *AIChE J.* **5,** 467 (1959).
22. A. E. Wechsler and M. A. Kritz, *AFML TR-65-138, Contract AF 33(615)1335,* Clearinghouse for Federal Scientific and Technical Information, Alexandria, Va., 1962.
23. *Heat Transmission Measurements in Thermal Insulations,* American Society for Testing and Materials, STP 544, Philadelphia Pa., 1974.
24. *Standard Test Method for Steady State Thermal Transmission Properties by Means of the Guarded Hot Plate,* Appendix XI to ASTM C177-76, ASTM, Philadelphia, Pa., 1976.
25. *Thermal Transmission Measurements of Insulation,* ASTM STP660, Philadelphia, Pa., 1978.
26. J. F. Malloy, *Thermal Insulation,* Van Nostrand-Reinhold Co., New York, 1969.
27. *Use of Materials Bulletin No. 74,* Department of Housing and Urban Development, Oct. 13, 1977.
28. R. Kropschott, *Cryogenics,* 171 (1961).
29. *Minimum Property Standards for One and Two Living Units,* Federal Housing Administration, U.S. Govt. Printing Office, Washington, D.C., 1965.
30. *Chem. Week* 21 (Aug. 6, 1980).
31. J. O. Collins and S. Speil, *Mater. Des. Eng.* **53,** 114 (1961).
32. A. D. Little, *Thermal Insulation Systems,* NASA SP5027, 1967.
33. I. J. Gruntfest and L. H. Shenker, "Behavior of Materials at Very High Temperatures," *Proceedings Ann. Meet. Am. Chem. Soc.,* San Francisco, Calif., Apr. 1958.
34. H. M. Whippo and B. T. Armberg, *Survey and Analysis of the Vapor Transmission Properties of Building Materials,* U.S. Dept. of Commerce, Office of Technical Services, Washington, D.C., 1955.
35. *Standard Recommended Practice for Selection of Vapor Barriers for Thermal Insulations,* ASTM C755-73, Philadelphia, Pa., 1973.
36. *Condensation Control in Dwelling Construction,* Housing and Home Finance Agency, U.S. Govt. Printing Office, Washington, D.C., 1949.
37. G. Trainor, *Missiles Rockets* **10,** 41 (1962).
38. *U.S. Residential Insulation Industry,* Office of Business Research and Analysis of the U.S. Department of Commerce, Aug. 1977.
39. J. K. Eklund and D. Baeu, *Ind. Res.* (Apr. 1975).
40. *Ind. Heat.* **XLV,** 34 (June 1978).

41. A. B. Hill and D. V. Bevers, *Chemtech* **9,** 4 (Apr. 1979).
42. J. A. White, M. H. Agee, and K. E. Case, *Principles of Engineering Economic Analysis,* John Wiley & Sons, Inc., New York, 1977.
43. L. Busch, *Plant Eng.* **33,** 19 (Sept. 20, 1979).
44. *Economic Thickness for Industrial Insulation,* FEA Conservation Paper No. 46, USGPO St. No. 041-018-0-00115-8, 1976.
45. L. B. McMillan, *Trans. ASME* **37,** 921 (1915).
46. L. R. Kimball, *Thermal Conductance of Pipe Insulation,* ASTM STP 544, Philadelphia, Pa., 1974.
47. J. D. Verschoor and A. Wilbur, *Trans. Am. Soc. Heat. Vent. Eng.* **60,** 329 (1954).
48. D. L. Lang, *Am. Soc. Test. Mater. Bull.* **216,** 58 (1956).
49. R. H. Norris and N. D. Fitzroy, *Mater. Res. Stand.* **1,** 727 (1961).
50. R. C. Huebscher, L. F. Schutrum, and G. V. Parmelee, *Trans. Am. Soc. Heat. Vent. Eng.* **58,** 275 (1952).
51. C. M. Pelanne and C. B. Bradley, *Mater. Res. Stand.* **2,** 549 (1962).
52. D. D'Eustachio and R. E. Schreiner, *Trans. Am. Soc. Heat. Vent. Eng.* **58,** 331 (1952).
53. F. A. Joy, ASTM STP217, Philadelphia, Pa., 1957.
54. C. E. Lund and R. M. Lander, *ASHRAE J.* **3**(3), 47 (1961).
55. J. R. Mumaw, *Calibrated Hot Box: An Effective Means for Measuring Thermal Conductance in Large Wall Sections,* ASTM STP544, Philadelphia, Pa., 1974.
56. J. J. Catani and S. E. Goodwin, *ACI J.* **73,** 83 (Feb. 1976).
57. *ASTM Book of Standards,* Pt. 18, American Society for Testing Materials, Philadelphia, Pa., 1979.
58. H. R. Copson and C. F. Cheng, *Corrosion* **13,** 3971 (1957).
59. A. W. Dana, Jr., *ASTM Bull.* **225,** 46 (Oct. 5, 1957).

R. H. Neisel
J. D. Verschoor
Johns-Manville Sales Corporation

INSULIN AND OTHER ANTIDIABETIC AGENTS

The word diabetes stems from the Greek word meaning "to pass through a siphon," and the word mellitus, which means sweet or honey. The term diabetes was first applied by Aretaeus, an ancient physician, to people who seemed to melt away through the passage of a large volume of water. Much later, Thomas Willis tasted their urine and found it was sweet, hence diabetes mellitus.

Since the causes of diabetes are not understood, no totally satisfactory definition of the disease is possible. It is certainly a generalized chronic metabolic condition which, fully developed, may be characterized by hyperglycemia (elevated blood glucose), glycosuria (elevated urine glucose), increased protein breakdown, and ketosis and acidosis (elevated blood ketones: β-hydroxybutyric acid and acetoacetic acid). Although definitions sometimes vary, with normal fasting glucose concentrations of blood are assumed to be 65–95 mg/100 mL with the blood glucose rarely rising above 130 mg/100 mL following a meal. At these blood glucose concentrations, normal subjects do not exhibit glucose in the urine. If the disease is prolonged, usually it is complicated by degenerative disease of the blood vessels, the retina, the kidney, and the nervous system. In general, this metabolic state resembles that produced in experimental animals deprived of insulin, and abnormalities of insulin secretion or function are usually present in the condition in humans.

Although the disease is very heterogeneous with respect to age onset, severity, treatment, associated endocrine, autoimmune, and metabolic factors, etc, diabetic patients may be classified as either ketoacidosis-prone (juvenile, insulin-requiring) or ketoacidosis-resistant (adult, maturity-onset, noninsulin-requiring). The ketoacidosis-prone patient generally experiences the onset of the disease prior to age 20 and is dependent upon exogenous insulin to prevent ketoacidosis. Ketoacidosis-resistant patients generally experience the onset of diabetes after age 40, tend to be overweight, and do not require insulin.

Treatment at present offers only diet alone, diet with oral hypoglycemic (blood glucose-lowering) agents, or diet with insulin. Exercise is a valuable adjunct therapy. The present treatment capability, however, is far from satisfactory. Each type of treatment is predicated on an adequate amount of effective insulin, endogenous or exogenous. The diabetic successfully treated with diet alone succeeds because the amount of nutrient intake more closely corresponds with the ability to provide sufficient insulin of one's own. When the current oral agents are effective, they also utilize endogenous insulin. When necessary, exogenous insulin is injected, but in this instance, the patient must titrate the 24-h insulin requirement, considering in advance diet, activity, psychological problems, incipient infections, and other factors. Each of the above forms of diabetic therapy are discussed in greater detail below.

The U.S. market for antidiabetic drugs is very large and is comprised of only prescription products. Consolidated sales of pharmaceutical products prescribed for diabetic therapy is currently in excess of $222,000,000/yr with approximately ½ of the market being insulin products. The predicted deficiency of insulin by 1980 has not materialized. Genetic engineering (qv) will increase the supply. The market for dietary products is more difficult to estimate and is currently undergoing a process of rapid expansion.

Dietary Therapy

The important role of diet in the management of diabetes mellitus has been emphasized in several reports. In 1971, the University Group Diabetes Program (UGDP) reported results of a multicenter study showing that diet alone may be more effective in prolonging the life of the diabetic than other therapeutic agents. Although this study is still controversial, few dispute the conclusions concerned with diet (1). In 1975, The National Commission on Diabetes noted that the chance for diabetes mellitus doubles for every 20% of excessive weight. Hence, there is strong implication that diet may play a role in the pathogenesis of diabetes (2).

The dietary recommendations for diabetic persons are in most respects the same as for nondiabetic persons and are based on sound principles of nutrition. The most important objective of dietary treatment is control of total caloric intake to achieve desired body weight (see Table 1). Generally speaking, for the adult this means a daily intake of 150–220 g of carbohydrate, 60–100 g of protein, 60–100 g of fat—for a total caloric intake of 1400 to 2100 calories (kcal) or 5.86–8.79 MJ. Patients with insulin-resistant diabetes (plasma-insulin concentrations greater than normal and/or respond subnormally to exogenous insulin) frequently are obese and require hypocaloric diets.

Table 1. Estimating Desirable Body Weight

Build	Women	Men
medium frame	45 kg for first 1.5 m of height, plus 0.9 kg for each additional centimeter	48 kg for first 1.5 m of height plus 1.07 kg for each additional centimeter
small frame	subtract 10%	subtract 10%
large frame	add 10%	add 10%

Patients with failure of the beta cell (insulin-secreting cells of pancreas) require insulin therapy. Insulin therapy should always be integrated with the patient's meal and activity schedule; the primary dietary rule for the insulin-dependent diabetic is to follow a regular schedule of food intake.

Diabetes is not caused by a high intake of refined sugar. High carbohydrate diets may be beneficial to many diabetics. The current view of the American Diabetes Association is that some liberalization in carbohydrate intake is recommended, preferably as complex carbohydrate (starch associated with fiber) (see Dietary fiber) (3). This does not imply that unlimited ingestions of carbohydrate, particularly as sugars, is advocated. The carbohydrate calories (kcal or 4.184 kJ) must be limited and its intake so timed as not to overwhelm the endogenous insulin reserves or swamp the injected insulin. Fats should comprise no more than 35% of the caloric intake of the obese diabetic. Hypertriglyceridemia in the obese diabetic is best treated with hypocaloric diets that contain reduced quantities of carbohydrates. Obviously, no universal dietary rules exist because of the variation of individual patients. Thus dietary therapy for the diabetic must be worked out in conjunction with a physician and integrated with other therapy needed to achieve normal blood glucose (normoglycemia). A more detailed description of the nutritional management of diabetes mellitus, including the use of artificial sweeteners, meal planning, etc, is in ref. 4 (see Sweeteners).

Insulin

In 1889, von Mering and Minkowski demonstrated conclusively that the removal of the pancreas from healthy dogs results in a condition that resembles human diabetes mellitus (5). From that time, many workers throughout the world attempted to discover the factor in the pancreas that regulated carbohydrate metabolism. In the classical work of Banting and Best in 1921 (6), the hormone named insulin [*9004-10-8*] was isolated from an aqueous ethanolic extract of pancreas, and the effectiveness in the control of human diabetes was demonstrated (see Hormones). The larger portion of the pancreas consists of glandular tissue which secretes digestive enzymes, but there are also isolated groups of cells, called islets of Langerhans, the beta cells of which produce the insulin.

Insulin is the hormone that promotes the processes by which the various tissues in all parts of the body may use glucose, either as a fuel for liberation of energy, or as storage forms of energy such as glycogen, or fat. In the diabetic subject, exogenous insulin restores, temporarily, the ability to utilize carbohydrates and fats in a comparatively satisfactory manner; the concentration of sugar in the blood may be confined within normal limits; the urine becomes free of sugar and ketone bodies; and diabetic acidosis and coma are prevented. Conversely, with too much insulin, serious or dangerous symptoms from hypoglycemia may result, causing sweating, hunger, incoherence, convulsions, coma, and possibly death. Glucose administration relieves the symptoms of overdosage, and the diabetic patient often carries some source of glucose to alleviate hypoglycemia. Glucagon or epinephrine may be administered when hypoglycemia is severe (see Epinephrine and norepinephrine).

Chemical Properties. Insulin (monomer) with a molecular weight of about 6000 (bovine insulin, 5734) was the first protein whose structure was elucidated in a brilliant series of investigations by Sanger and co-workers in the period 1945–1953 (7). The structure of bovine insulin [*11070-73-8*] (**1**) is shown below:

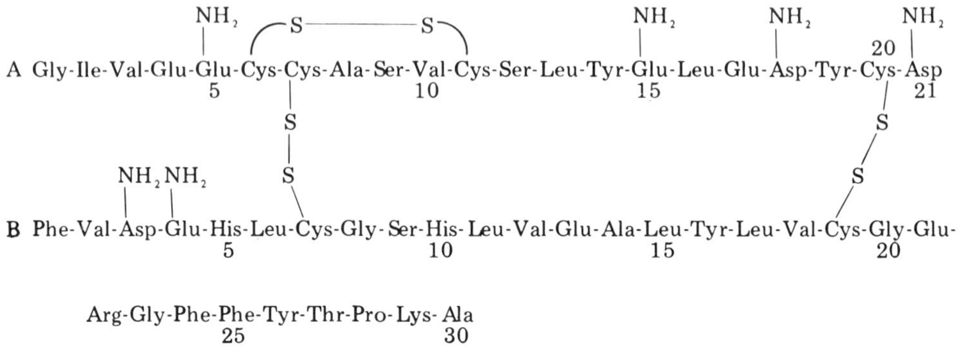

```
                    NH2   S-----------S                 NH2         NH2         NH2
                     |   /             \                 |           |       20  |
A Gly-Ile-Val-Glu-Glu-Cys-Cys-Ala-Ser-Val-Cys-Ser-Leu-Tyr-Glu-Leu-Glu-Asp-Tyr-Cys-Asp
                   5       |        10                  15                    /   21
                           S                                                 S
                           |                                                /
         NH2 NH2           S                                               S
          |   |             \                                             /
B Phe-Val-Asp-Glu-His-Leu-Cys-Gly-Ser-His-Leu-Val-Glu-Ala-Leu-Tyr-Leu-Val-Cys-Gly-Glu-
                    5              10                  15                    20

      Arg-Gly-Phe-Phe-Tyr-Thr-Pro-Lys-Ala
              25                  30
```

(1) bovine insulin

It consists of 2 polypeptide chains with specific amino acid sequences. The A chain has 21 amino acids, and the basic B chain has 30 amino acids. These chains are linked by 2 disulfide (—S—S—) bonds of cystine, with an additional cystine bridge within the A chain. In aqueous solution, the insulin monomer polymerizes to form macromolecules of 12,000 or 36,000 mol wt, depending on pH, temperature, and concentration. The isoelectric point of insulin is 5.3. It has been shown that about 20% of intact insulin is a right-handed α-helix and like other proteins is levorotatory. Preparations of crystalline insulin often contain about 0.5% zinc, the physiological function of which is unknown. The binding of the zinc to the *N*-terminal amino groups and the imidazoyl groups of insulin has been suggested to facilitate the formation of high molecular weight insulin complexes.

The arrangement of the amino acids in each of the two chains of insulin has been determined for insulins from several species of animals, including humans. Because the species differences are rather small, involving only amino acids 8–10 of the A chain, antibodies to insulin preparations can be prepared in some animals. Insulin is produced in the β-cells of the islets of Langerhans from a single helical chain, which has been named proinsulin [*11062-00-3*]. Insulin is produced from proinsulin by removal of the midportion of the helix (the connecting peptide, or C-peptide). Older, less pure, preparations of insulin were contaminated with as much as 3–5% proinsulin and C-peptide.

Preparation. Commercial insulin is obtained by a combination of acid–alcohol extractions, isoelectric precipitations, and separation of the hormone as an insoluble salt. In one preparation, frozen bovine or porcine pancreas glands in 45–68 kg lots are ground and placed in large extraction tanks filled with cold acidified 50–85% ethanol (pH 2.5–4.0) to extract the insulin (8). The mixtures are centrifuged, extracted, and pH adjusted to 5.5–8.5. The solution is then filtered and its pH is readjusted to 3 with sulfuric acid. By a subsequent evaporation and heat treatment, the fats are separated, the crude insulin solution is filtered through diatomaceous earth, and the clean filtrate is purified by the addition of zinc to produce crystals in an amount of ca 3100–4000 units of insulin per kilogram of glands processed. Bovine insulin has also been completely synthesized chemically but is not commercially available (8). Advances have also been made in the bacterial synthesis of human insulin [*11061-68-0*] using recombinant-DNA techniques (see Genetic engineering) (9). It can be expected that the use of this technique may be of commercial significance.

Distribution, Metabolism, and Excretion. The distribution of insulin in the body approximates the volume of extracellular fluid. Insulin has been detected in lymph, bile, and urine, although less than 10% is excreted by the kidney, where it is probably filtered and reabsorbed. The organs of major importance in the degradation of the hormone are the liver and kidney. The intact insulin molecule is more resistant to proteolysis than are the A and B chains. Consequently, the sequence of events in insulin degradation is probably first the rupture of the cystine bridges and then the splitting of amino acids (proteolysis) from the A and B chains. A number of proteolytic enzymes present in various tissues, including liver and kidney, have the capacity to hydrolyze the A and B chains of insulin.

Bioassay and Units. Insulin preparations must be bioassayed in order to ascertain their physiological activity. The application of modern physico-chemical techniques in recent years has been used to determine the purity of insulin but has not yet replaced the biological methods for the determination of potency. The bioassay used is based on two different doses of insulin standard and two doses of the test preparation in a cross-over design using rabbits as test animals. On each occasion, individual blood-glucose concentrations are determined at 1 and 2.5 h after injection. In other parts of the world, a mouse convulsant test also is used. The potency is expressed in USP units and the standard of comparison is USP zinc insulin [*8049-62-5*] crystals reference standard. The potency of the reference standard crystals is indicated on the label of each preparation. It is of the order of 24 units/mg.

Adverse Reactions. Insulin overdosage causes weakness, headache and fatigue, nervousness, tremor, pallor or flushing, and profuse sweating—a characteristic sign. If not treated immediately, the condition may lead to insulin reaction or shock. Allergic reactions to insulin are fairly common, especially local reactions at the site of injection which occur 10 times more frequently than systemic generalized reactions. The presence of insulin antibodies is in some cases important in the development of insulin resistance, a condition where patients require more insulin (>200 units daily) to control their disease. In immunologic insulin resistance, changing the species source from bovine, or mixed bovine–porcine, to porcine often reduces the amount of insulin required.

Individual Evaluations. The seven forms of insulin currently marketed in the United States differ with respect to the time of onset and duration of action. They may be divided into rapid-, intermediate-, and long-acting forms (see Table 2). Globin zinc, isophane (NPH), and protamine zinc insulins are conjugated with large protein molecules. As a result, their absorption from subcutaneous sites is delayed and their duration of action is prolonged. Absorption is also delayed by the large particle size and crystalline form of extended insulin zinc suspension (Ultralente insulin). Amorphous Semilente insulin, with a smaller particle size, is more rapidly absorbed and shorter

Table 2. Properties of Various Preparations of Insulin

Type	Preparation	Units/mL	Appearance	Animal source	Buffer	Protein modifier		Onset of action, h[a]	Interval to maximal action, h[a]	Approximate duration of action, h[a]
						Type	mg/100 Units			
rapid	insulin injection, USP (regular insulin, crystalline zinc)	40, 80, 100, 500	clear	porcine, mixed bovine–porcine	none	none		<1	2–3	5–7
	prompt insulin zinc suspension, USP (Semilente insulin)	40, 80, 100	turbid	bovine, mixed bovine–porcine	acetate	none		<1	4–6	12–16
	globin zinc insulin injection, USP (globin)	40, 80, 100	clear	mixed bovine–porcine	none	globin	3.8	1–2	6–10	12–18
intermediate	isophane insulin suspension, USP (NPH insulin, isophane, NPH Iletin	40, 80, 100	turbid	mixed bovine–porcine	phosphate	protamine	0.5	2	8–12	18–24
	insulin zinc suspension, USP (Lente)	40, 80, 100	turbid	bovine, mixed bovine–porcine	acetate	none		2–4	8–12	18–24
	protamine zinc insulin suspension	40, 80, 100	turbid	mixed bovine–porcine	phosphate	protamine	1.2	4–6	16–18	36
long	extended insulin zinc suspension (Ultralente)	40, 80, 100	turbid	bovine, mixed bovine–porcine	acetate	none		4–6	16–18	36

[a] These figures are representative of subcutaneous administration. The values are expected to vary over a relatively wider range, depending on the dose and the individual patient. The onset and duration are influenced by such factors as concentration and volume of insulin injected, depth and site of injection, possible binding by antibodies, and state of capillary permeability (10).

acting. Lente insulin, which is intermediate acting, is a combination of 70% Ultralente and 30% Semilente insulin.

Because the early data indicated that acid pH improved the stability of regular insulin, from the time of its introduction it has been given an acid pH as opposed to the modified insulins which were neutral. However, changes in the purification process removed substances which in the past degraded at neutral pH and allowed the preparation of unbuffered regular insulin with neutral pH (7.4). This preparation, termed neutral regular insulin (NRI), exhibits full potency for up to two years when stored at room temperature. The main clinical advantage of NRI is that it may be mixed with the modified insulins in any proportion desired to given optimal control of blood glucose over the entire day.

Eli Lilly and Company has produced a commercial insulin which, on gel chromatography, yields a profile consisting chiefly of a single peak. Therefore, it has been called single-peak insulin. The improvement afforded by this preparation, which is 99% insulin plus desamido and arginine insulins, is a reduction of higher molecular weight substances as evidenced by a proinsulin content reduced to $1/10$ the concentration of insulins not purified by similar chromatography. All insulins currently produced are single-peak. Single-peak insulin can also be further purified by DEAE cellulose chromatography to yield single-component or monocomponent insulin, which is >99% pure insulin. Single-component forms of insulin have recently become commercially available. Both of the above purified insulins are much less immunogenic than USP insulin. Patients with an allergy to insulin or those who develop lipodystrophy (atrophy of fat) at the site of insulin injection are benefited greatly by use of these forms.

Insulin U 100 is commercially available. This more concentrated insulin can replace U 40 and U 80 forms following adequate patient-education programs. Its use obviates serious dosage errors produced by using U 40 syringes with U 80 insulin or *vice versa*.

Insulin Injection USP (Iletin). Insulin injection [*9004-10-8*] (insulin, insulin hydrochloride, Iletin) is a sterile, acidified or neutral solution of the active principle of the pancreas that effects the metabolism of glucose. When not more than 100 USP units is present per mL, it is a colorless or almost colorless liquid (with 500 units/mL it may be straw-colored). It is substantially free from turbidity and from insoluble matter and it contains 0.1–0.25% wt/vol (mg/100 mL) of either phenol or cresol and 1.4–1.8% wt/vol of glycerol. Its pH, determined potentiometrically, is 2.5–3.5 for acidified injection, and 7.0–7.8 for neutral injection.

The specific therapeutic use of insulin is in the treatment of diabetes mellitus. Dosage varies with the individual case. Insulin must be given by hypodermic injection (the hormone is destroyed in the gastrointestinal tract). Diabetic individuals are trained to inject themselves. For this purpose a special syringe measuring the dosage of insulin directly in units is employed. Several novel systems for insulin delivery (implantable artificial pancreas and capillary-infusion pump) are under clinical evaluation and may have a strong impact on diabetic therapy.

This rapid-acting agent has a short duration of action and is the only insulin preparation that may be given intravenously as well as subcutaneously (see Table 1). Insulin injection may be mixed in the same syringe with isophane (NPH) insulin without alteration of the response to either form. It also may be mixed with insulin zinc suspension (lente insulin) if the ratio of regular to lente insulin does not exceed

1:1. These preparations have no advantage over the intermediate forms of insulin. It is the insulin of choice in the presence of unstable diabetes when complications such as infection, shock, or surgical trauma occur.

The interval from a hypodermic injection of insulin until its action is about one hour, but it is longer with the protein–insulin complexes. The duration of action is relatively short and the plasma elimination half-life is ca 40 min. The duration of action is not linearly proportional to the size of the dose, but is a simple function of the $\log_{10}$ of the dose, ie, insulin is inactivated in the body at a rate proportional to the amount in the body at the time (if 1 unit will last 4 h, 10 units will last 8 h). Since the usual duration is 2–4 h, 2–4 daily doses should be given for proper control of severe diabetes. This is timed ordinarily a few minutes before the ingestion of food, in order to avoid an unpleasant reduction of the blood glucose level.

Insulin is occasionally employed in nondiabetic cases. For example, it has been used to produce convulsive shock seizures for the treatment of certain psychiatric cases. It is also used in the treatment of underweight individuals, to stimulate the appetite by lowering the level of blood sugar. However, such uses of insulin are rare.

Prompt Insulin Zinc Suspension USP (Semilente Iletin). Semilente Iletin [*8049-62-5*] is a sterile suspension of insulin in buffered water for injection, modified by the addition of zinc chloride in a manner such that the solid phase of the suspension is amorphous. It is an almost colorless suspension of particles that have no uniform shape with a maximum size not to exceed 2 μm. The suspension contains for each 100 units of insulin, 0.20–0.25 mg of zinc (of which 40–65% is in the supernatant liquid), 0.15–0.17% (wt/vol, ie, mg/100 mL) of sodium acetate, 0.65–0.75% sodium chloride, and 0.09–0.11% of methylparaben.

Globin Zinc Insulin Injection USP. Globin zinc insulin [*9004-21-1*] is a sterile, almost colorless, solution of insulin modified by the addition of zinc chloride and globin. The globin used is obtained from bovine blood.

In the preparation of globin zinc insulin injection, the amount of insulin used is sufficient to provide either 40, 80, or 100 USP insulin units for each mL of the injection. It contains from 1.3–1.7% (wt/vol) of glycerol and either from 0.15–0.20% (wt/vol) of cresol or from 0.20–0.26% (wt/vol) of phenol, 0.25–0.35 mg of zinc and 3.6–4 mg of globin (calculated as 6 times the nitrogen content of the globin) for each 100 USP insulin units. Its pH is 3.4–3.8, determined potentiometrically. This preparation is indicated for patients who require more than one daily injection of regular insulin, those whose condition cannot be controlled by other forms of insulin, or those who are sensitive to protamine. This insulin is not recommended for the treatment of diabetic ketoacidosis.

Isophane Insulin Suspension USP. Isophane insulin [*9004-17-5*] (NPH insulin or NPH Iletin) is a sterile suspension of zinc insulin crystals and protamine sulfate in buffered water for injection, combined in a manner such that the solid phase of the suspension consists of crystals composed of insulin, protamine, and zinc. The protamine sulfate is prepared from the sperm or from the mature testes of fish (*Onchorhynchus suckley*, or *Salmo linne* (Fam. Salmonidae)).

Isophane insulin is a white suspension of rod-shaped crystals approximately 30-μm long, and free from large aggregates of crystals after being subjected to moderate agitation. It contains either 1.4–1.8% (wt/vol) of glycerol, 0.15–0.17% (wt/vol) of metacresol, and 0.06–0.07% (wt/vol) of phenol, or 1.4–1.8% (wt/vol) of glycerol, and 0.20–0.25% (wt/vol) of phenol. It also contains 0.15–0.25% (wt/vol) of diabasic sodium phosphate, 0.01–0.04 mg of zinc, and 0.3–0.6 mg of protamine for each USP insulin unit. The insoluble matter in the suspension is crystalline and contains not more than traces of amorphous material. Its pH is 7.1–7.4, determined potentiometrically.

Isophane insulin is never given intravenously. Hypoglycemic reactions in mid-to-late afternoon may be less obvious in onset, more prolonged, and more frequent than with rapid-acting preparations because of the prolonged effect of the dose.

Insulin Zinc Suspension USP (Lente Insulin). Lente insulin [*8049-62-5*] (Lente Iletin) is a sterile suspension of insulin in buffered water for injection, modified by the addition of zinc chloride in a manner such that the solid phase of the suspension consists of a mixture of crystalline and amorphous insulin in a ratio of ca 7 parts of crystals to 3 parts of amorphous material. Each mL is prepared from sufficient insulin to provide either 40, 80, or 100 USP insulin units of insulin activity.

Lente insulin is an almost colorless suspension of a mixture of characteristic crystals predominantly 10–40 μm in maximum dimension, and many particles that have no uniform shape and do not exceed 2 μm in maximum dimension. It contains 0.15–0.17% (wt/vol) of sodium acetate, 0.65 to 0.75% (wt/vol) of sodium chloride, 0.09–0.11% (wt/vol) of methylparaben, and 0.20–0.25 mg of zinc of which 40–65% is in the supernatant liquid. Its pH is 7.1–7.5.

The advantage of zinc insulin suspension is its freedom from foreign proteins, eg, globin or protamine, to which certain patients are sensitive.

Protamine Zinc Insulin Suspension USP. Protamine zinc insulin [*9004-17-5*] is a sterile suspension of insulin in buffered water for injection, modified by the addition of zinc chloride and protamine sulfate. The protamine sulfate is prepared as described for isophane insulin suspension.

Protamine zinc insulin is white, or almost white, free from large particles upon moderate agitation, and it must contain 1.4–1.8% (wt/vol) of glycerol, and either 0.18–0.22% (wt/vol) of cresol or 0.22–0.28% (wt/vol) of phenol. It also contains 0.15–0.25% (wt/vol) of Na_2HPO_4. It must contain 0.15–0.25 mg of zinc and 1–1.5 mg of protamine for each 100 USP insulin units. Its pH, determined potentiometrically, is 7.1 and 7.4.

Protamine zinc insulin and the more rapidly acting insulin have been mixed in various proportions and in different ways. Such variations produce mixtures that give intermediate patterns of speed and duration of physiological activity. This method, however, makes it possible through the long-acting insulin to take care of the sugar requirements during the night, and the short-acting insulin causes full utilization of food taken during the day. It thus permits avoiding large doses of the long-acting insulin such as might cause hypoglycemia in the early morning hours. However, the need to individualize the insulin schedule makes commercially fixed mixtures impractical.

Extended Insulin Zinc Suspension USP (Ultralente Insulin). Ultralente Iletin [*8049-62-5*] is a sterile suspension of insulin in buffered water for injection, modified by the addition of zinc chloride in a manner such that the solid phase of the suspension is crystalline. In its preparation, sufficient insulin is used to provide either 40, 80, or 100 USP insulin units for each mL of the suspension.

Ultralente Insulin is an almost colorless suspension of a mixture of characteristic crystals, the maximum dimension of which is predominantly 10–40 μm. It contains, for each 100 USP units of insulin, 0.20–0.25 mg of zinc (of which 40–65% is in the supernatant liquid), and not more than 0.70 mg of nitrogen. It also contains 0.15–0.17%

(wt/vol) of sodium acetate, 0.65–0.75% (wt/vol) of sodium chloride, and 0.09–0.11% (wt/vol) of methylparaben.

Oral Antidiabetic Agents

Two groups of drugs form the basis of the current oral therapy of diabetes mellitus: the arylsulfonylureas, known simply as the sulfonylureas, and the biguanides. Since October, 1977, the only oral agents available in the United States have been the sulfonylureas, although biguanides still are used in other parts of the world.

Up until 1977, the only U.S.-marketed biguanide was phenformin [*114-86-3*], which is available in the U.S. only under special circumstances.

Sulfonylureas. Shown in Table 3 are those agents currently used for therapy in the United States and their properties. The sulfonylurea radical,

$$—SO_2NH\overset{\overset{O}{\|}}{C}NH—,$$

should be noted since it confers the hypoglycemic activity on the compounds in the group. This was first noted in the early 1940s when several patients died in hypoglycemic coma after testing a synthetic sulfonamide used to treat typhoid. In 1955 the first sulfonylurea, carbutamide [*339-43-5*] (**2**), came onto the European market as a blood-sugar lowering agent, soon followed by tolbutamide (**3**). Substituents on the benzene and the urea groups effect differences in potency, duration of action, and

$$H_2N—C_6H_4—SO_2NHCONH(CH_2)_3CH_3$$

(**2**)

toxicity. They are rapidly absorbed from the upper gastrointestinal tract and are transported in the blood as protein-bound complexes. As they are released from protein-binding sites, the free (unbound) form becomes available for diffusion into tissues and to its sites of action.

Metabolism, when it occurs, takes place in the liver; in only some instances, the metabolites are biologically active. Differing rates and completeness of absorption, degrees of protein binding, rates of metabolism and renal excretion, and the relative hypoglycemic potencies of the drugs and their metabolites determine the propensity

Table 3. Sulfonylureas Available in the United States

CAS Registry Number	Acetohexamide (**4**) [*968-81-0*]	Chlorpropamide (**5**) [*94-20-2*]	Tolazamide (**6**) [*1156-19-0*]	Tolbutamide (**3**) [*64-77-7*]
daily effective dose[a]	0.25–1.5 g	125–750 mg	0.1–1 g	0.5–3 g
plasma half-life, h[b]	6–8	30–36	7	4–6
tablet sizes, mg	250, 500	100, 250	100, 250, 500	500, 1000
duration, h	12–24	60	≤24	6–12

[a] With all compounds except chlorpropamide, divided doses are used when the larger dosage is required.

[b] Biological half-life regarding hypoglycemic potential may be longer.

(see Table 3) of the various compounds to produce hypoglycemia. The mode of action of the sulfonylureas results from a direct action on pancreatic beta cells to release insulin, at least when acutely administered. As yet, the exact mechanism for their hypoglycemic action during chronic administration is undefined. Since these compounds require a pancreas capable of responding to stimuli, their most effective use is in patients with recent-onset diabetes (of less than 10 yr duration) which started in middle to old age and is mild (requiring 40 units of insulin daily or less) rather than severe. Almost without exception, the sulfonylureas have been useless in the juvenile-onset (insulin-requiring) diabetic. In general, the sulfonylureas are effective in inverse proportion to the insulin requirement of the patient, and conversely, the more nearly normal the patient, the better the hypoglycemic response. In the presence of acute complications such as fever, severe trauma, or infections, diabetics may also require appropriate doses of insulin to ensure continued control. During the early years of their use, the sulfonylureas were generally assumed to be the treatment of choice for mild maturity-onset diabetes. However, it has become increasingly apparent that substantial numbers of patients who initially respond to the sulfonylureas later become refractory to them (secondary failures).

A second generation of sulfonylureas with potency as high as one thousand times that of tolbutamide (**3**), but without an established quantitative difference in mechanism of action, is marketed in other parts of the world, but is not yet available for use in the United States. The following contains a more complete description of only the sulfonylureas currently marketed in the U.S.

Acetohexamide USP (Dymelor). Acetohexamide, 1-[(*p*-acetylphenyl)sulfonyl]-3-cyclohexylurea (**4**), mol wt 324.42, is a white, practically odorless crystalline powder from 90% ethanol, mp 188–190°C (from dilute ethanol, mp 175–177°C). It is practically insoluble in water and in ether, soluble in pyridine and in dilute solutions of alkali hydroxides, and slightly soluble in alcohol and in chloroform.

Preparation. Acetohexamide is prepared by adding *p*-acetylbenzenesulfonamide to anhydrous potassium carbonate, and the resulting potassium salt of the sulfonamide is treated with cyclohexyl isocyanate. After removal of the acetone, the residue (potassium salt of acetohexamide) is dissolved in water and acidified with hydrochloric acid to precipitate the acetohexamide. Purification is by recrystallization from aqueous ethanol (11–12).

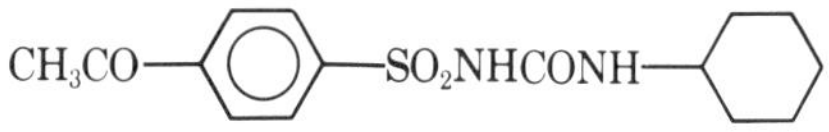

CH_3CO—⟨benzene ring⟩—$SO_2NHCONH$—⟨cyclohexyl⟩

(**4**)

Toxicity. The incidence of untoward effects is low and the reactions are reversible when acetohexamide is discontinued. Relatively severe hypoglycemic reactions have been observed occasionally in patients given large doses for a prolonged period without close observation. It is contraindicated in patients with hyperglycemia and glycosuria due to primary renal disease, and in those who are hypersensitive to sulfonylurea compounds. Table 4 lists LD_{50}s for chlorpropamide and tolbutamide. See Table 3 for duration of action and daily maintenance dose.

Use. Acetohexamide is similar to other hypoglycemic agents in the sulfonylurea class in being used for the treatment of mild, adult-onset (noninsulin-requiring) diabetes. Like other sulfonylureas, it is not effective in juvenile or insulin-requiring ketotic patients. However, since it is the only one with uricosuric properties, some clinicians

Table 4. LD_{50}s of Chlorpropamide and Tolbutamide [a]

Compound	Mode of administration	LD_{50}, g/kg Mouse	Rat
chlorpropamide	oral	1.7	2.4
	oral	0.7	0.9
tolbutamide	oral	2.5	4.0
	intravenous	0.6	
	subcutaneous	0.75	
tolbutamide sodium	oral	1.8	
	intravenous	0.6	
	subcutaneous	0.75	

[a] Ref. 13. Ref. 13 also has chronic toxicity information.

prefer this agent for the diabetic with gout. Much of the activity is ascribable to a metabolite, hydroxyhexamide, which has a plasma half-life of about 5 h.

***Chlorpropamide USP* (*Diabinese*).** Chlorpropamide, 1-[(*p*-chlorophenyl)sulfonyl]-3-propylurea (**5**), mol wt 276.75, is a white, crystalline powder, having a slight odor, mp 127–129°, uv max (0.01 *N* HCl): 232.5 nm (ϵ = 16,500). It is soluble in water at pH 6–2.2 mg/mL and practically insoluble at pH 7.3, soluble in alcohol, and sparingly soluble in chloroform, ether, and benzene.

$$Cl-C_6H_4-SO_2NHCONHCH_2CH_2CH_3$$

(**5**)

Preparation. p-Chlorobenzenesulfonamide undergoes addition to propyl isocyanate by warming a solution consisting of equimolar quantities of the two reactants in a suitable inert solvent (14).

Use. Its use is limited to patients with stable, mild to moderately severe diabetes mellitus who still have some residual pancreatic beta-cell function. If more than 40 units of insulin are required per day, the patient usually will not respond to chloropropamide. Refractoriness has been reported less frequently with this sulfonylurea. Because of the drug's long half-life (see Table 3), maximal accumulation and effect may take one to two weeks, and several weeks may be required for complete elimination of the drug from the body. In contrast to tolbutamide (**3**), chloropropamide (**5**) is not metabolically altered to a significant degree. The side effects are mainly the same type as those from tolbutamide but have a somewhat higher incidence. They include cholestatic jaundice, diarrhea, allergic reactions and dermatoses, leukopenia, agranulocytosis, and thrombocytopenia. Chlorpropamide increases the endogenous release of vasopressin and thus causes water retention with resultant hyponatremia and hypoosmolality. This could be life-threatening in patients with a tendency to retain water (eg, in those with congestive heart failure, cirrhosis of the liver, etc).

***Tolazamide USP* (*Tolinase*).** Tolazamide, 1-(hexahydro-1*H*-azepin-1-yl)-3-(*p*-tolysulfonyl)urea, (**6**), mol wt 311.40, is a white to off-white, crystalline powder, odorless, or having a slight odor, mp 170–173°C, with a p*Ka* −3.6 at 25°C and 5.68 at 37.5°C. It is very slightly soluble in water, freely soluble in chloroform, soluble in acetone, and slightly soluble in alcohol.

Preparation. Tolazamide may be synthesized by the method described in ref. 15. Methyl *p*-tolylsulfonylcarbamate undergoes an ammonolysis-type reaction with 1-aminohexamethyleneimine (16).

CH_3—C_6H_4—$SO_2NHCONH$—N(hexamethyleneimine ring)

(6)

Use. This potent hypoglycemic agent lacks antidiuretic action and may be especially useful in the treatment of patients who have a tendency to retain water. Approximately 35% of diabetic patients assumed to be primary or secondary failures with other sulfonylurea agents or with phenformin have responded satisfactorily to this agent. Tolazamide is metabolized to a number of hypoglycemic substances that are largely excreted by the kidney. The total incidence of side effects is about 5%; about 2% of the patients find it necessary to discontinue the drug. The side effects are similar to those of tolbutamide.

Tolbutamide USP (Orinase). Tolbutamide (**3**), 1-butyl-3-(*p*-tolylsulfonyl)urea, mol wt 270.35, is a white to off-white, practically odorless, crystalline powder, having a slightly bitter taste, mp 126–132°C. It is practically insoluble in water, and soluble in alcohol and chloroform.

CH_3—C_6H_4—$SO_2NHCONH(CH_2)_3CH_3$

(3)

Preparation. Condensation of the *p*-toluenesulfonamide with ethyl chloroformate in the presence of pyridine or another suitable basic catalyst yields ethyl *N*-*p*-toluenesulfonylcarbamate (**7**). Aminolysis of (**7**) with butylamine in ethylene glycol monomethyl ether solution yields tolbutamide (**3**). The solvent is removed by distillation and water is added to the residue. The crude solid tolbutamide thus obtained is dissolved in dilute ammonia solution, decolorized with charcoal, precipitated with HCl and crystallized from 50% ethanol (17).

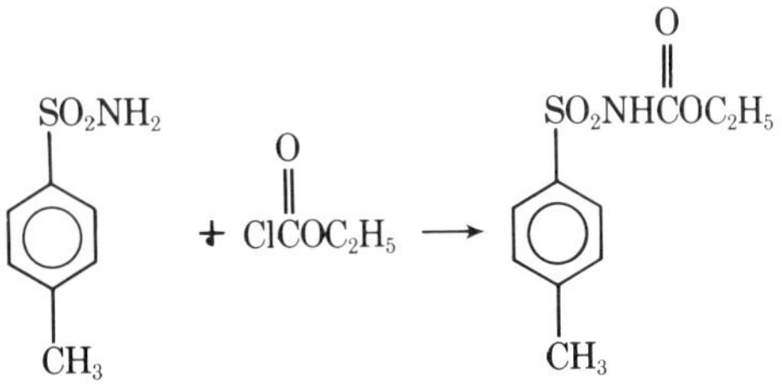

$$CH_3C_6H_4SO_2NH_2 + ClC(=O)OC_2H_5 \longrightarrow CH_3C_6H_4SO_2NHC(=O)OC_2H_5$$

(7)

Toxicity. Toxic reactions to tolbutamide have been infrequent and mild (see Table 4). The incidence of side effects was analyzed in 9168 cases (18). The total incidence of side effects was 3.2%, and the drug was withdrawn in 1.5% of the patients. Toxic effects include gastrointestinal upset, weakness, headache, leukopenia tinnitus, agranulocytosis, thrombocytopenia, paresthesia, allergic reactions, and alcohol intolerance. Impaired liver function may occur, and the drug is contraindicated in the presence of liver damage. As with chlorpropamide, water retention and hyponatruria may result from use of (**3**).

Uses. Tolbutamide has the same actions, uses, and limitations as other sulfonylurea compounds. This drug is of greatest value in patients who, because of poor general physical status, should receive a rapidly acting compound (see Table 3). Tolbutamide is metabolized by oxidation of the *p*-methyl group. The resulting totally inactive metabolite can be recovered in the urine within 24 h in amounts accountable for as much as 75% of the administered dose.

Tolbutamide Sodium USP (Orinase Diagnostic). Orinase Diagnostic [*473-41-6*], *N*-[(butylamino)carbonyl]-4-methylbenzenesulfonamide, monosodium salt, mol wt 292.33, is a white to off-white, practically odorless, crystalline powder, having a slightly bitter taste. It is freely soluble in water, soluble in alcohol and chloroform, and very slightly soluble in ether. It can be prepared by dissolving tolbutamide in aqueous NaOH.

Uses. Because of its water solubility, tolbutamide sodium may be given intravenously (1 g of tolbutamide equivalent over a period of 2 to 3 min). By this route, its rapid onset of action lends itself to the diagnosis of diabetes mellitus in persons in whom the usual studies are equivocal. The normal person responds with a more rapid and intense drop in blood glucose than does the diabetic, especially during the first hour after injection. Persons with pancreatic insulinoma respond with a prolonged hypoglycemia, so that the drug may also be used diagnostically when that condition is suspected. Tolbutamide sodium is an irritant, and thrombophlebitis and thrombosis of the vein occur in ca 1–2% of recipients.

Biguanides. ***Phenformin.*** The only commercially available U.S. preparation in the biguanide series of hypoglycemic agents was phenformin (**8**). Its action in reducing hyperglycemia occurs without promoting insulin release, and may include delayed gastrointestinal absorption of nutrients as well as direct promotion of glucose uptake by peripheral tissues. The drug did not lower the fasting blood glucose of nondiabetics. Phenformin found a most secure position in combination with sulfonylureas. The sulfonylureas stimulating the release of insulin which presumably was augmented by the action of the biguanide. Phenformin had a place in the treatment of diabetes when the sulfonylureas proved ineffective because the addition of biguanide permitted continued oral treatment for a definite segment of the diabetes population. In 1977 it was estimated that 336,000 patients were using phenformin alone or in combination with other drugs. Other than a high incidence of gastrointestinal side effects, there was little or no reported toxicity until almost universal agreement developed that the drug produced lactic acidosis. This life-threatening disorder is characterized by a significant reduction in arterial pH in the presence of lactate accumulation in the extracellular fluids. Because of this, in July, 1977, the FDA ordered general marketing of phenformin to cease within 90 d, branding it an imminent hazard and citing its possible implication in the deaths of 50–700 cases yearly. It was admitted that there was "a small group of diabetics in whom the use of insulin poses special hardships" and for whom this consideration outweighed the risks of this "imminent hazard." Although it is not generally available, phenformin can be used by physicians who apply for an IND (investigational new drug) status. IND is the new classification for phenformin, in which case the manufacturer provides the drug free of charge after shipment has been authorized by the FDA.

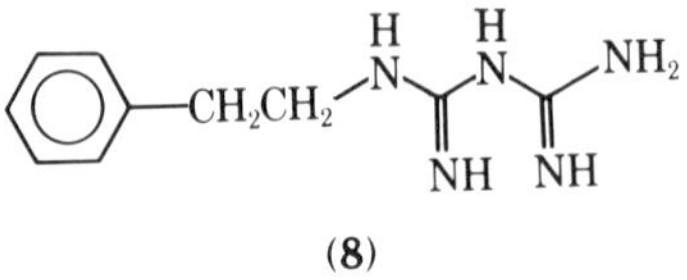

$C_6H_5CH_2CH_2NHC(=NH)NHC(=NH)NH_2$

(8)

Phenformin Hydrochloride USP (DBI, Metrol). Phenformin hydrochloride [*834-28-6*], 1-phenethylbiguanide, *N*-(2-phenylethyl)imidodicarbonimidic diamide (**9**), is a white to off-white, odorless, crystalline powder with a bitter taste, 175–178°C. It is freely soluble in water and alcohol, and practically insoluble in chloroform, ether, and hexane. Its pH in solution (1 in 40) is 6.0–7.0.

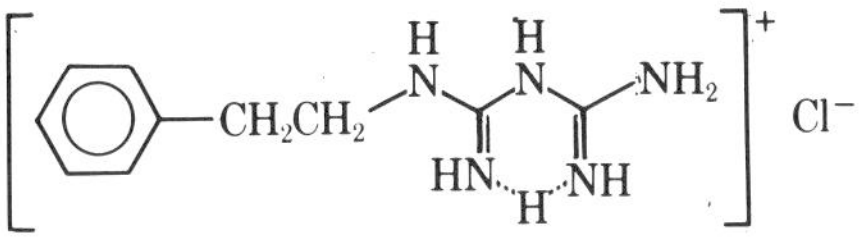

$$\left[C_6H_5{-}CH_2CH_2{-}NH{-}C({=}NH){-}NH{-}C({=}NH){-}NH_2 \right]^+ \; Cl^-$$

(**9**)

Preparation. Equimolar quantities of phenethylamine and dicyandiamide are mixed and neutralized with aqueous HCl. After refluxing for 3 h, water is removed by distillation until the mixture boils (at about 129°C). The crude product may be crystallized from isopropanol (19).

Uses. Indications for use were essentially those of the sulfonylureas. In addition, phenformin was often chosen for use in the obese diabetic because of the allegation that it might facilitate weight loss. However, double-blind studies have failed to confirm this finding. Both primary and secondary failures with the sulfonylureas were sometimes reversed when phenformin was added to the regimen. The drug is metabolized in the liver to the 1-(*p*-hydroxyphenethyl)biguanide and its *O*-glucuronide conjugate. The major route of excretion is the kidney. Duration of action of the drug is 6–14 h.

Miscellaneous

Two other agents used to regulate blood glucose are diazoxide USP [*364-98-7*] (**10**) and glucagon USP [*16941-32-5*] (20).

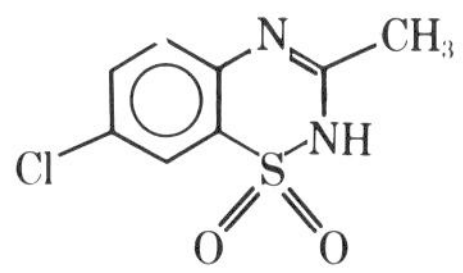

(**10**)

BIBLIOGRAPHY

"Insulin" in *ECT* 1st ed., Vol. 1, pp. 935–941, by H. F. Jensen, Army Medical Research Laboratory, Fort Knox; "Insulin" in *ECT* 2nd ed., Vol. 11, pp. 838–846, by Paul Turi, Sandoz, Inc.

1. C. R. Klimt and co-workers, *Diabetes* **19**(Supplement), 747 (1970); C. L. Meinert and co-workers, *Diabetes* **19**(Supplement), 789 (1970).
2. *Diabetic Forecast* **28,** 6 (1975).
3. F. Q. Nuttall and J. D. Brunzell, *Diabetes* **28,** 1027 (1979).
4. J. S. Skyler in H. M. Katzen and R. J. Mahler, eds., *Advances in Modern Nutrition,* Vol. 2, Hemisphere Publishing Co., Washington, D.C., 1978.
5. J. von Mering and O. Minkowski, *Arch. Exp. Pathol. u. Pharmacol.* **26,** 371 (1889).
6. F. G. Banting and C. H. Best, *J. Lab. Clin. Med.* **7,** 464 (1921).

7. F. Sanger and E. O. P. Thompson, *Biochem. J.* **53,** 353, 366 (1953).
8. Y. T. Kung and co-workers, *Sci. Sin.* **14,** 1710 (1965).
9. *Chem. Eng. News* **58**(30), 12 (1980).
10. *Diabetes Mellitus: Diagnosis and Treatment,* Vol. 3, American Diabetes Association, New York, 1971.
11. Brit. Pat. 912,789 (Dec. 12, 1962), (to Eli Lilly Co.).
12. U.S. Pat. 3,320,312 (May 16, 1967), M. V. Sigal, Jr. and A. M. Van Arendonk (to Eli Lilly Co.).
13. S. Podolsky, ed., *Symposium on Diabetes Mellitus, Medical Clinics of North America,* Vol. 62, W. B. Saunders and Company, Philadelphia, Pa., 1978.
14. Brit. Pat. 853,555 (Nov. 9, 1960), (to Chas. Pfizer Co.).
15. J. B. Wright and R. E. Willette, *Med. Pharm. Chem.* **5,** 815 (1962).
16. U.S. Pat. 3,063,903 (Nov. 13, 1962), J. B. Wright (to Upjohn Co.).
17. U.S. Pat. 2,968,158 (Jan. 17, 1961), H. Walter and co-workers (to Upjohn Co.).
18. C. J. O'Donovan, *Curr. Ther. Res.* **1,** 69 (1959).
19. U.S. Pats. 2,961,377 (Nov. 22, 1960), and 3,057,780 (Oct. 9, 1962), L. Seymour, L. Shapiro, and L. Freedman (to U.S. Vitamin and Pharmaceutical Corp.).
20. *AMA Drug Evaluations,* 4th ed., American Medical Association, John Wiley & Sons, Inc., New York, 1980, pp. 739–758.

General References

E. H. Frieden, *Remington's Pharmaceutical Sciences,* 15th ed., Mack Publishing Co., Easton, Pa., 1975.

J. A. Galloway, *New Horizons in the Care of Diabetes Mellitus,* Geigy Symposia Series CXVI, Birmingham, Alabama, Mar., 1973.

L. S. Goodman and A. Gilman, eds., *The Pharmacological Basis of Therapeutics,* 5th ed., MacMillian Publishing Co., New York, 1975.

A. J. Lewis, ed., *Modern Drug Encyclopedia and Therapeutic Index,* York Medical Books, New York, 1977.

The United States Pharmacopeia XX (*USP XX–NFXV*), The United States Pharmacopeial Convention, Inc., Rockville, Md., 1980.

G. F. TUTWILER
McNeil Laboratories

INTEGRATED CIRCUITS

Silicon integrated circuits are pervasive in all modern electronic systems, most of which would not exist without the availability of inexpensive high-performance integrated circuits (ICs).

The value of total world consumption of ICs has grown from less than 1×10^6 in 1960 to greater than 6×10^9 in 1979. The production of a modern, complex IC requires a coordinated effort covering three technical areas: the system design determines the overall system architecture, partitioning the system into sub-systems that can be put together with one or more ICs; the circuit design specifies the components with available IC elements; and the technology provides the capability of fabricating the IC (1–3). This review is concerned with the latter area and only with silicon IC technology.

Other materials, however, are used for fabricating ICs, most notably gallium arsenide (4), but the role played by these materials is small compared to silicon and usually is reserved for special applications.

An IC is comprised of passive and active electronic devices that are interconnected to provide a desired electronic function (5). The devices are formed by selectively introducing impurities into the silicon, so as to effect local change in its electrical properties. Two active devices are the metal-oxide semiconductor (MOS) transistor and the bipolar junction transistor; cross sections of these are shown in Figure 1. These devices divide silicon IC technology into two branches, MOS and bipolar; usually both devices are not used on the same IC because of the fabrication complications that would result. The different impurity (electrical) regions, n (excess electrons) and p (excess holes), are introduced into the silicon from the same side of the wafer; silicon IC technology is concerned with the processes used to form the required impurity regions and to interconnect them. The interconnection processes involve the formation and patterning of appropriate conducting and insulating layers (thin films) on the silicon wafer. The operation of IC devices is covered in the literature (5–7).

Bipolar devices (see Figs. 1**b** and 1**c**) require a separate isolation region to prevent interactions between devices, whereas MOS devices are self-isolating. Until recently, all bipolar ICs were junction-isolated with the p^+ isolation region being reverse-biased with respect to the n-type collector. In this case, the isolation region must be separated from the active device region and considerable silicon area is wasted. With oxide isolation, the isolation region can be merged directly into the device structure and, thereby, in the silicon area used and significantly improving device performance because parasitic capacitance is reduced (8–9).

The MOS device of Figure 1**a** is typical of those in use. It is an n-channel transistor and, by applying an appropriate voltage to the gate electrode, an n-type conducting region can be induced in the silicon which connects the n^+ source and drain regions. Self-isolation is achieved in normal operation by not allowing the n^+ source and drain to become forward-biased with respect to the p-type substrate. This self-isolation has meant that, in general, MOS ICs can be made with higher packing densities (less silicon area for a given electronic function) than bipolar ICs. The recent thrust in MOS technology has been to further increase packing density by reducing the minimum feature size used and by providing greater interconnection flexibility with increased amounts of interconnection.

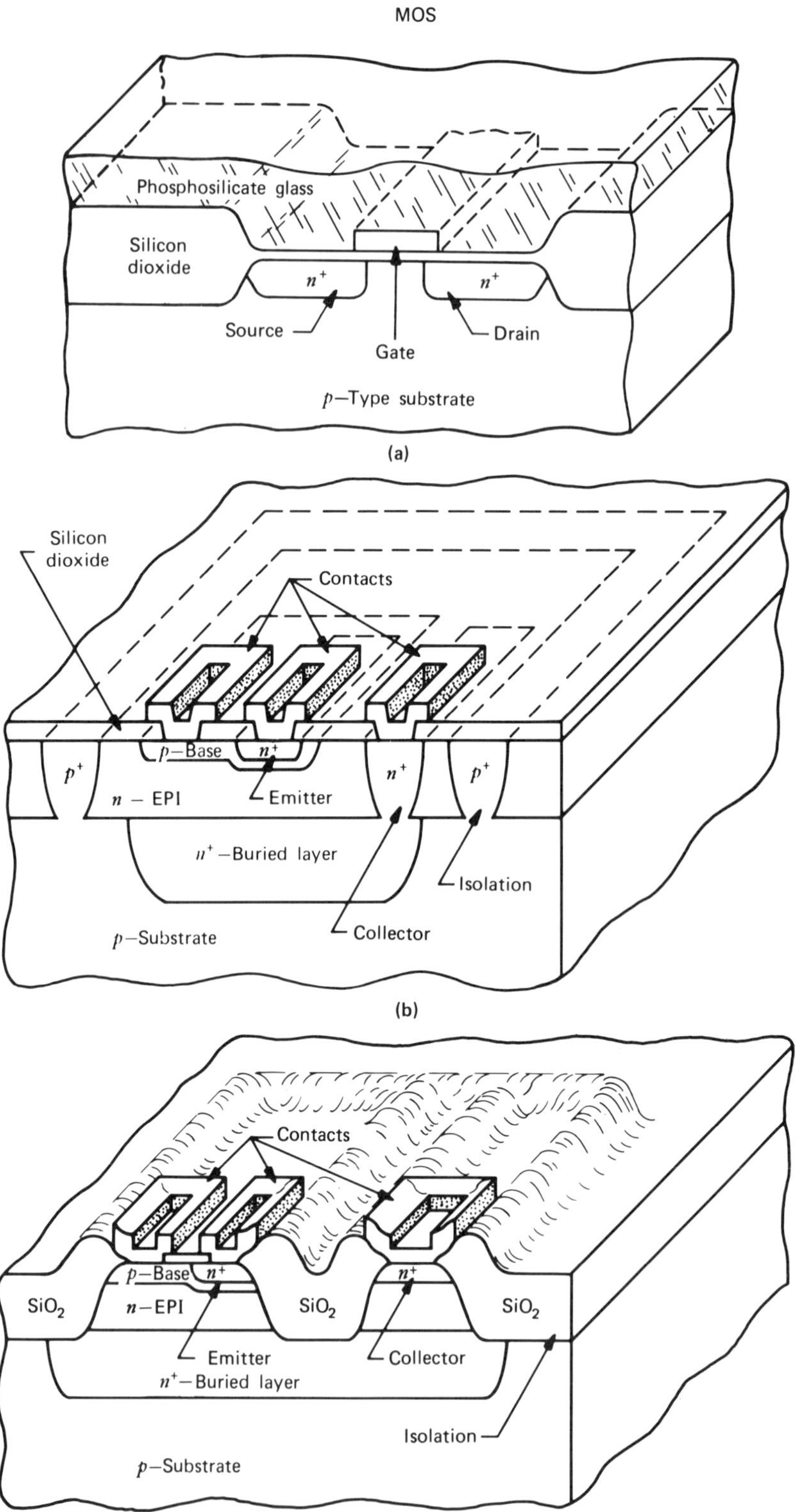

Figure 1. Cross-sections of electronic devices used in ICs. (**a**) MOS transistor, (**b**) bipolar transistor, junction isolated, (**c**) bipolar transistor, oxide isolated.

Manufacture and Processing

Planar Process. The majority of ICs are manufactured according to the planar process, as is illustrated in Figure 2 (4,6,10). This process is repeated as many times as necessary to form the various impurity regions required. The key to the planar process is the layer of silicon dioxide (which can be thermally grown with excellent uniformity and reproducibility for thicknesses from 0.01–2.0 μm). The thermally grown silicon dioxide is an insulator, and the diffusion coefficient of impurities of interest for selectively doping silicon are significantly lower for silicon dioxide than for silicon. Thus, silicon dioxide can be used as a diffusion mask for forming localized impurity regions in silicon. To serve as a diffusion mask, the silicon dioxide must be patterned by photolithography.

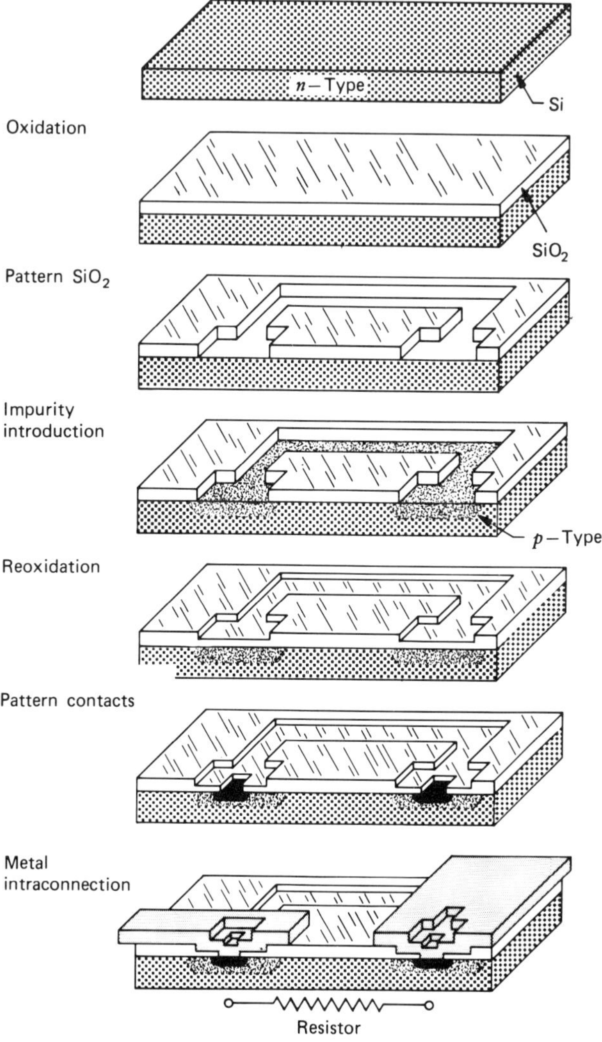

Figure 2. Pictorial view of the planar process used almost exclusively for silicon IC fabrication.

Contamination control (3) must be extended to the chemicals, gases, and water that are used; to the ambient air in the processing facility; and to the employees involved in the fabrication procedures. The latter two can be controlled if the processing is done in a limited-access, clean room in which the air is continuously filtered to remove particulates and in which the temperature and humidity are controlled (11). A successful alternative to a clean room is the use of clean hoods that are placed over critical work areas and which provide continuous filtering of the air (see Sterile techniques). Modern facilities use reverse osmosis (qv) and ion-exchange systems for providing high quality water (12) (see Ion exchange). The chemicals required have impurity levels lower than a few parts per million and in many cases are filtered at point of use.

Silicon Materials. The substrates that are used for silicon ICs are single-crystal, silicon wafers which are precisely aligned to crystallographic axes. An extremely high purity and degree of crystalline perfection are required for optimum performance. As the degree of integration and circuit area have increased, the preparation of silicon wafers has become more demanding because economic fabrication of larger circuits requires larger wafers with increased purity and crystalline perfection; 76- and 100-mm diameter wafers, 0.5-mm thick, are predominant.

The flow chart of Figure 3 is an overview of the wafer-fabrication process (13). The starting point is the production of raw silicon by reduction from its oxide. An electric-arc furnace is used at 2000°C and the metallurgical-grade silicon that is produced has a purity of 98–99% (see Furnaces, electric). The silicon then is pulverized

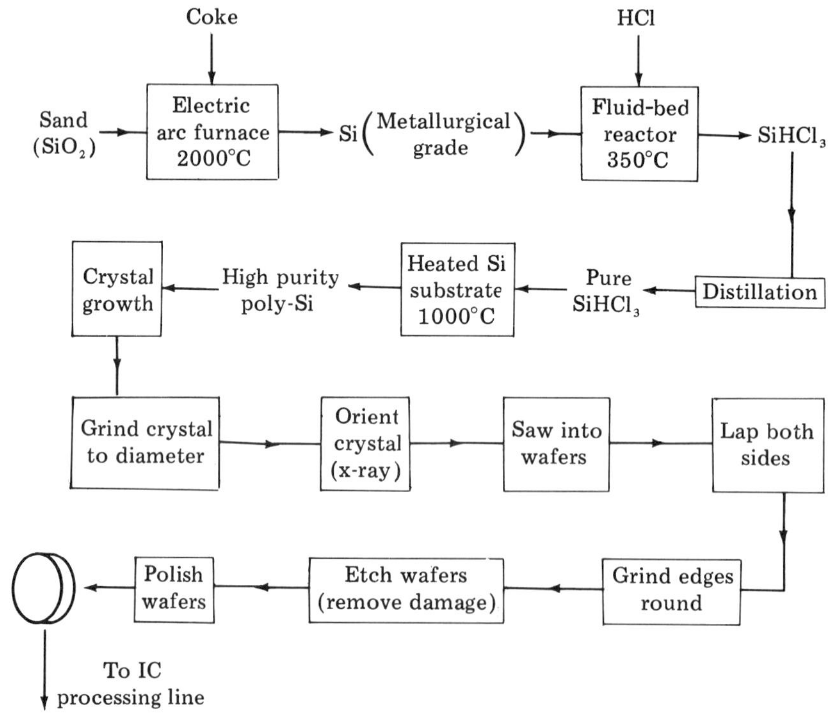

Figure 3. Flow chart of the process used to convert sand into single-crystal silicon wafers.

for use in a fluid-bed reactor where it reacts with hydrogen chloride at 350°C to form trichlorosilane ($SiHCl_3$), which is refined by distillation. High-purity, bulk, polycrystalline silicon, which has a level of electrically active impurities of ≤ 1 ppba [parts per billion (10^9) atomic], is produced by reducing the trichlorosilane with hydrogen on a resistance-heated (1000°C), thin rod of high-purity silicon. (The ppba purity is required because deliberately added impurities in amounts as low as 10^{14} molecules/cm^3 are used to control the electrical properties of the silicon.)

Two crystal-growth techniques, Czochralski and float zone, are used. In the former, polycrystalline silicon is melted in a quartz crucible and crystal growth is started using a seed crystal of desired orientation which slowly is pulled from the melt. Appropriate dopants can be added to the melt to control the resistivity of the single-crystal silicon. A small percentage of the quartz crucible dissolves in the silicon melt, resulting in $\cong 10^{17}$ atoms/cm^3 of oxygen in the finished crystal. Most IC wafers originate from Czochralski crystals and there is some evidence that the incorporated oxygen may be beneficial because it forms localized silicon dioxide precipitates which can act as gettering centers for impurities and defects during IC processing (14). The float-zone method is a containerless growth technique in which a localized region of the polycrystalline rod is heated by r-f induction to temperatures at which crystal growth can occur; the growth is initiated by placement of a seed crystal against one end of the poly rod. Much higher resistivity crystals can be produced with the float-zone technique, because no detectable oxygen is incorporated into the crystal during growth. Gas-phase impurity doping can be performed either during crystal growth or during the formation of the poly rod. For high resistivity, n-type crystals (>100 Ω-cm), nuclear-transmutation doping is used. Slices are irradiated with thermal neutrons which convert ^{30}Si to ^{31}P, an n-type dopant. ^{30}Si occurs at a natural concentration of $\cong 3\%$.

Centerless grinding is carried out to produce rods of the desired diameter and precision. The ground rods are aligned to crystal axes by x-ray diffraction and wafers of the desired orientation are cut from the rods using an inner-diameter saw. The wafers are mechanically lapped and polished to the desired thickness and surface finish. An additional edge-grinding step has been introduced to provide a radius of curvature to the edge of the wafer. Edge grinding eliminates sharp edges that tend to chip during processing. All of the above mechanical operations, however, introduce surface damage to the silicon crystal. Effective chemical etching techniques have been developed to remove this damage and to leave a highly polished surface of essentially perfect crystallinity.

All advanced MOS ICs are fabricated with (100)-oriented silicon wafers because the electrical properties of the (100) silicon–silicon dioxide interface are optimum for MOS devices (lowest surface charge and surface state density). For bipolar ICs, both (111) and (100) orientations are used, depending upon the specific application.

Epitaxial growth, the vapor-phase growth of single-crystal, silicon films upon single-crystal, silicon substrates, is used widely in bipolar ICs (13,15). The importance of epitaxy arises from the ease with which the resistivity (doping level) and conductivity type (n or p) of the epitaxial film can be controlled independent of the properties of the substrate. Sharp transition regions (<1 μm) between different doping levels and conductivity types are possible. In the bipolar structures of Figure 1, the moderately doped n-type collector regions are epitaxial films that are grown on lightly doped, p-type substrates that contain heavily doped, buried collector regions. This transistor structure minimizes collector resistance, and epitaxial growth is absolutely essential to its fabrication.

Epitaxial growth is carried out in a reactor that usually has quartz walls. The wafers are placed on a graphite susceptor and are heated to 1000–1200°C by r-f induction. The various source gases used for Si epitaxy are $SiCl_4$, $SiHCl_3$, SiH_2Cl_2, and SiH_4. The reaction which results in the growth of Si films when using $SiCl_4$ is:

$$SiCl_4\ (g) + 2\ H_2\ (g) \xrightleftharpoons{1250°C} Si\ (s) + 4\ HCl\ (g)$$

This reaction is reversible and usually, as a cleaning procedure, the first step is gas-phase etching of the silicon surface with HCl. Doping of the growing epitaxial film is accomplished by introducing gases, eg, PH_3, AsH_3, and B_2H_6, that contain the appropriate dopants into the reactor. For (111) orientation, the crystalline perfection and surface morphology of the epitaxial layer are significantly improved if the wafers are misoriented a few degrees off the asymmetric (111) plane (16). Control of film thickness to within ±5% and resistivity to within ±15% are within the capability of modern epitaxial technology. Future trends are the use of infrared heating and deposition at subatmospheric pressures (see Film deposition techniques).

Silicon Oxidation. Silicon ICs and the planar processing technology rely upon the complementary properties of silicon and silicon dioxide (see Table 1). Various techniques exist for the formation of thin films of silicon dioxide upon silicon substrates and include anodization, chemical vapor deposition (CVD), and sputter deposition; but the majority of thin silicon dioxide layers are formed by thermal oxidation which yields thin silicon dioxide films of excellent uniformity, high passivation quality, and low pinhole density. Furthermore, the films are amorphous and are characterized by an open random lattice of the SiO_4 tetrahedra.

Prior to any silicon substrate oxidation process, the surface of the wafers must be carefully cleaned, which involves organic and particulate removal, strong oxidizing treatment, chemical oxide removal, and drying (17).

Table 1. Physical Properties of SiO_2 and Si

Property	SiO_2	Si
density, g/cm^3	2.27	2.33
dielectric constant	3.9	11.7
mp, °C	≅1700	1415
breakdown field, V/μm	≅600	≅30
specific heat, J/(g·K)[a]	1.0	0.7
thermal conductivity, W/(m·K)	1.4	150
thermal diffusivity, cm^2/s	0.006	0.9
linear coefficient of expansion, 10^{-6} K	0.5	2.5
etch rate, μm/min (in buffered HF at 27°C): 34.6% NH_4F, 6.8% HF, 58.6% H_2O	≅0.10	≅0
band gap, eV	9	1.11
resistivity, Ω/cm	$>10^{16}$ [b]	10^{-3} to 10^5 [c]

[a] To convert J to cal, divide by 4.184.
[b] Insulator.
[c] Semiconductor.

Growth Kinetics. Species used in the oxidation of silicon substrates (18) are: oxygen, steam, and atmospheres of mixtures of these two gases with nitrogen or argon. A simple kinematic model which explains the observed oxidation behavior of silicon has been proposed (19). Assuming that the oxidizing species diffuses through the growing silicon dioxide film and reacts with the silicon substrate at a rate that is dependent upon the concentration of the oxidant, then

$$x^2 + Ax = Bt$$

where x = the silicon dioxide film thickness, t = the time of oxidation, and A and B = the kinetic growth constants that are dependent upon gas pressure, temperature, and crystal orientation. When $x \ll A$, a linear growth condition exists:

$$Ax \cong Bt$$

and for the complementary case, $x \gg A$, a parabolic growth condition exists:

$$x^2 \cong Bt$$

Figure 4 contains the growth characteristics of silicon dioxide films in dry oxygen and steam atmospheres, respectively, at atmospheric pressure for (100)-oriented silicon substrates.

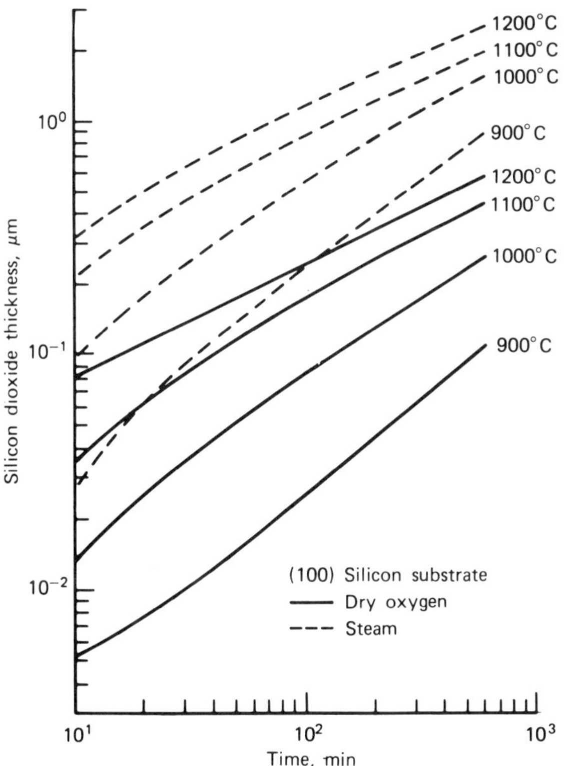

Figure 4. Silicon dioxide thickness versus time with temperature and atmosphere as parameters.

Conditions. Typical oxidations are conducted from 900–1200°C at 101.3 kPa (1 atm) in baffled, open-ended, horizontal resistance furnaces (see Furnaces, electric). Silicon substrates usually are handled in a batch mode by use of cassette handlers. For oxidations conducted above 900°C, care must be taken to prevent wafer warpage and crystal dislocations (slip) caused by nonuniform heating and cooling. Either the substrates can be slowly inserted and withdrawn from furnaces or the furnace temperature can be damped from 850°C; direct insertion at 850°C produces no crystal damage.

With the continuing reduction in minimum device feature-size and larger scales of integration, silicon oxidation at a lower temperature (700–900°C) is desirable. To offset the reduced oxidation rate, high pressure oxidation [1.0–2.0 MPa (10–25 atm)] has been incorporated into fabrication processes (20).

Film Characterization. A variety of capacitance–voltage techniques (21–26) permit determination of insulator breakdown strength, alkali ionic-charge density, silicon–silicon dioxide interface state, and fixed charge densities, and appropriate remedial processing procedures may be instituted to yield acceptable values. In particular, small additions (1.0%) of HCl gas to the oxidizing atmospheres improves breakdown strength ($\cong 8 \times 10^6$ V/cm) and reduces the concentration of the principal alkali-ion contaminant, sodium, to acceptable levels ($\leq 10^{10}$ Na^+/cm^2) (27). Further, post-oxidation annealing in hydrogen and nitrogen is used to reduce the interface state and fixed charge densities (28).

Photolithography. Photolithography is used to transfer patterns from a mask containing circuit-design information to thin films on the surface of a silicon wafer. The pattern transfer is accomplished with a photoresist, an ultraviolet light-sensitive organic polymer (29) (see Photoreactive polymers). A wafer that is coated with photoresist is illuminated through a mask and the mask pattern is transferred to the photoresist by chemical developers. Further pattern transfer is accomplished by appropriate etchants. The photoresist must be resistant to the etches used for patterning thin films of thermally grown silicon dioxide; the vapor-deposited insulators, eg, silicon dioxide and silicon nitride; and the deposited conductors, eg, aluminum, gold, and polycrystalline silicon. The masking process usually is repeated 5–10 times in the fabrication of an IC.

The photolithographic patterning process is depicted in Figure 5 for the case where a pattern is being defined in silicon dioxide. The pattern in the silicon dioxide is etched chemically by high purity (electronic-grade) hydrofluoric acid that is buffered with ammonium bifluoride using the exposed and developed photoresist as a mask. The chemical etching is isotropic and, as a result, the photoresist mask is undercut a distance at least equal to the thickness of the silicon dioxide. The patterning step typified in Figure 5 must occur uniformly across a wafer having a diameter up to 100 mm and reproducibly from wafer-to-wafer for feature sizes down to 2 μm.

The parameters used for assessing the quality of an IC photolithographic process are:

Minimum feature size and size control. The minimum feature-size capability determines the optimum integration level and the performance of ICs.

Level-to-level registration. The registration is accomplished by leaving fiducial marks (alignment features) in the thin film being patterned and using them to align one or more subsequent mask levels; registration tolerances from 1–2 μm are practical. However, for these tolerances to be usable, an image size control of ±0.5 μm or better

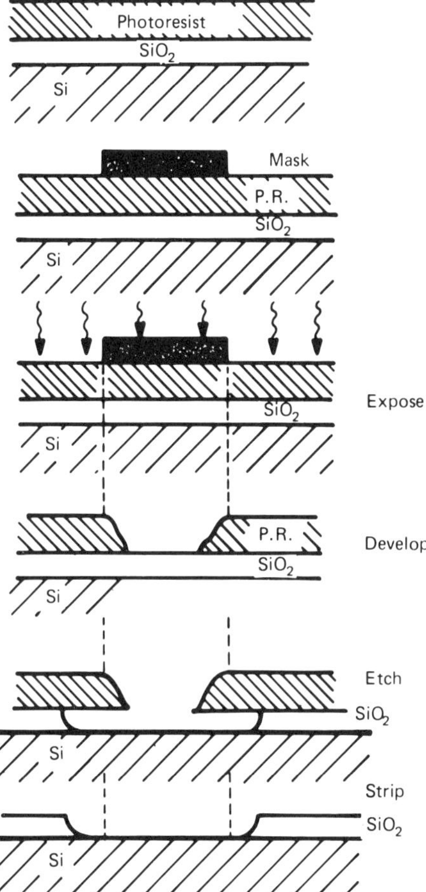

Figure 5. Schematic illustration of the photolithographic patterning process used for defining features in silicon dioxide.

is required. Electro-optical systems for automatic alignment are being developed in an effort to reduce the tolerance range and to provide reproducible, operator-independent alignment (30).

Pattern fidelity or defect density. The photolithographic defect density determines the largest chip size (31) that is economical.

Photoresist. There are two types of photoresist, positive and negative. Positive resists become more soluble in the developer in exposed areas relative to unexposed areas, whereas negative resists become less soluble in exposed areas. The detailed formulations of commercial resists are highly proprietary (32).

A negative photoresist is a linear polymer containing a few percent of a light-sensitive molecule (an activator or sensitizer). The activator absorbs incident radiation and promotes cross-linking of the polymer, thereby increasing its molecular weight and leading to insolubility in exposed areas. The developer is a solvent for the polymer and washes away unexposed areas. It also penetrates the cross-linked, exposed areas and causes swelling during development. The advantages of the negative resist are:

good adhesion to silicon dioxide and metal surfaces, good etch-resistance in both alkaline and acid solutions, low cost [\$16–18/L (\$60 to 70/gal)], and high sensitivity. The disadvantage is swelling during development which limits resolution (minimum feature size) to 4 or 5 times the resist thickness or ca 4–5 μm (since resist thicknesses of 1 μm are required to ensure a low pinhole density and adequate coverage over steps on the surface of the silicon wafer).

A positive resist is a polymer containing up to 25% of an inhibitor that prevents wetting and attack by the developer. Upon optical exposure, the inhibitor is destroyed, and exposed areas become soluble in the developer, an alkaline, aqueous-based solvent for the polymer. The developer does not penetrate unexposed resist but does dissolve it at a nonzero and, usually, nonnegligible rate. The advantage of the positive resist is its good resolution, which is approximately equal to the resist thickness. The disadvantages are: poor adhesion, poor alkaline etch resistance, high cost [\$66/L (\$250/gal)], significantly less sensitivity, and critical development requirements.

For feature sizes $\geq$4 μm, the negative resist has several advantages over the positive and, therefore, it is used extensively. For feature sizes below 4 μm, the poor resolution capability of the negative resist becomes a severe disadvantage and the positive resist is favored.

Process. The masking operation of Figure 5 is accomplished using the following process:

Wafer-surface preparation. The wafer surface must be cleaned properly to ensure adequate resist wetting and adhesion.

Photoresist application. The resist is applied individually to each wafer. The thickness used is a compromise between resolution, which is best with thin resists, and step coverage and pinhole protection, which is best with thick resists (32). Typical thicknesses are from 0.8–1.5 μm. Adhesion promoters or primers, eg, hexamethyldisilazane, sometimes are used, especially with a positive resist, and they are applied immediately prior to resist application.

Pre-exposure bake. This bake eliminates residual solvents from the resist, promotes adhesion, and hardens the resist. Temperature and time depend upon the specific resist being used but generally are from 70–90°C and 10–30 min, respectively.

Exposure. The resist is exposed using a light source which provides optical energy over a wavelength band that is contained within the resist adsorption spectrum, usually from 0.3–0.45 μm. Exposure energy (intensity × time) is critical in accurately replicating the mask pattern in the resist. Other parameters of the exposing system are uniform illumination, collimation, and degree of coherence. Several techniques have been developed for carrying out the resist exposure and they are illustrated in Figure 6. The oldest and probably most widely used technique is contact printing in which the wafer is pushed into intimate contact with the mask and then the mask is flooded with light. In proximity printing, the mask and wafer are separated by a gap, nominally from 10–25 μm. In projection printing, a high quality lens or mirror system is used to project the mask image onto the wafer surface. For the fabrication of large, dense ICs, projection printing is rapidly becoming the preferred manufacturing technique. With non-contact printing, defect generation from the abrasive effect of particles is minimized and mask damage is eliminated.

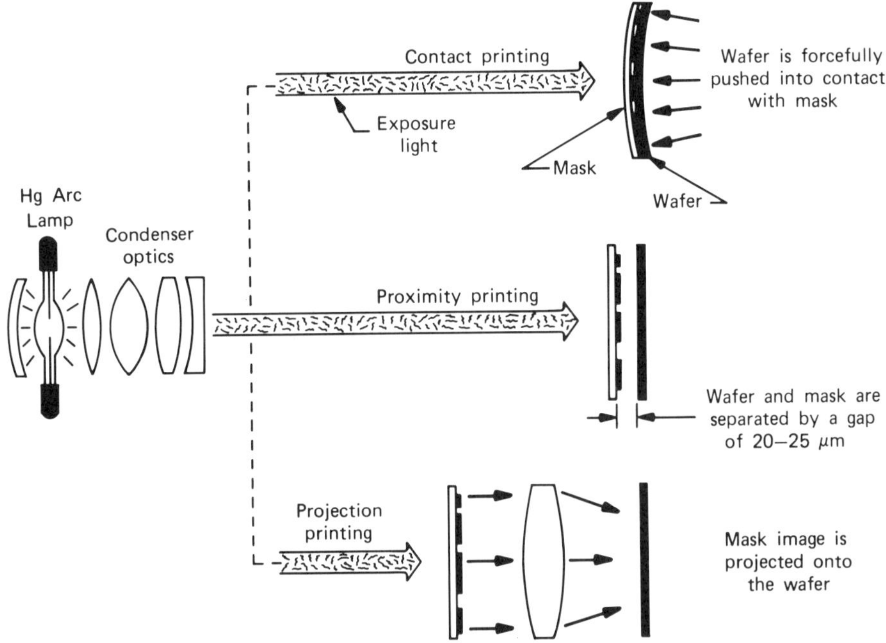

Figure 6. Schematic illustration of different photolithographic exposure techniques.

Development. Spray development is used for negative resists because it gives cleaner and more accurate images than an immersion development process. The same equipment that is used for coating is used for spray developing. The development process is critical because the final image size in the resist depends upon the development time (32) and because for positive resist the unexposed resist is dissolved at a nonzero rate.

Post-exposure bake. This bake improves surface adhesion, increases chemical and pinhole resistance, and drives off volatiles that have been retained from development. Temperatures and times from 80–150°C and 10–30 min, respectively, are typical.

Resist stripping. Chemical strippers generally are used, eg, a hot (80–90°C) mixture of sulfuric acid and hydrogen peroxide, and proprietary strippers for a given resist usually are available from the resist manufacturer. Commercial plasma equipment also is available for dry stripping in which an oxygen plasma is used to oxidize the resist to volatile compounds.

Photolithographic development is aimed at further reductions in the minimum feature size and most IC manufacturers have an extensive research and development effort on fineline lithography (<2 μm feature size) (33). Reduction of the minimum feature size to below 1.5 μm is being explored and may include a projection step and repeat on a wafer (34); the use of soft x-rays in place of near uv light—this technique appears to have a minimum feature size capability of 0.5–1.0 μm (35); and direct electron-beam exposure of the photoresist, which gives the best resolution, probably from 0.1–0.2 μm (36).

Masks. The masks are fabricated on transparent glass substrates which are usually square, large enough to cover the wafer diameter that is being used, and having a thickness from 1.5–2.5 mm. The opaque material in which the mask pattern is defined is deposited on the glass substrate, and two materials are in widespread use, photographic emulsion (4.0 μm thick) and chromium (0.08 μm thick) (see Photography). The emulsion can be patterned directly by exposure to light whereas a photoresist is used to pattern the chrome. Since the emulsion is a soft material, mask damage during use in IC processing is a constant concern, especially if contact printing is used. In this case, emulsion masks usually are printed a limited number of times and then are discarded. The advantages of emulsion masks are easy fabrication and low cost. Chrome masks are much less susceptible to damage, can be effectively cleaned and reused, and provide much better edge acuity than emulsion masks; however, they are more expensive and more difficult to fabricate.

Mask patterns are generated using either a computer-controlled light spot or electron beam. A flow chart is given in Figure 7 for the fabrication of a chrome mask using electron-beam exposure (37). Electron-beam-generated masks are widely used primarily because of their better resolution and absence of registration errors which are introduced by the step-and-repeat process.

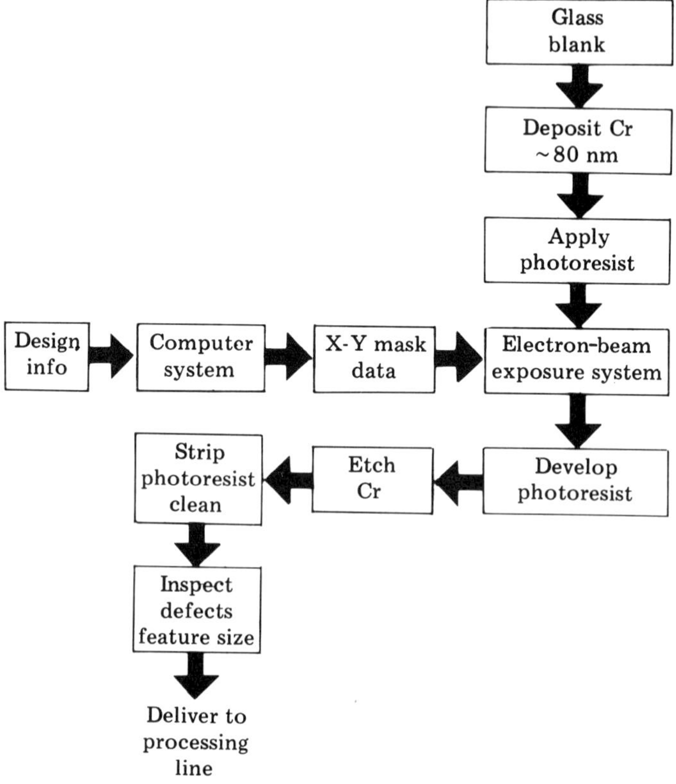

Figure 7. Flow chart illustrating a process for fabricating chrome masks.

Impurity Doping of Silicon. Planar ICs consist of an assembly of semiconductor devices, ie, transistors, diodes, and resistors, that are fabricated on the surface of a silicon substrate and are variously interconnected. Both the characteristics and the electrical isolation between the various devices are accomplished by means of the rectifying property of the p–n junction. IC technology is based upon the ability to selectively alter the impurity concentration near the surface of the silicon substrate and create the desired p–n junction structure.

Diffusion Theory. Diffusion theory is based upon the conservation of matter as expressed by the equation:

$$\nabla \cdot \bar{j}(\bar{\sigma},t) = -\frac{\partial}{\partial t} N(\bar{\sigma},t)$$

where $\bar{j}$ = the impurity particle current density; $N(\bar{\sigma},t)$ = the impurity concentration; and $\bar{\sigma},t$ = the position and time coordinate, respectively. In general, for a semiconducting material:

$$\bar{j}(\bar{\sigma},t) = -D(N(\bar{\sigma},t))\nabla N(\bar{\sigma},t) \pm \mu N(\bar{\sigma},t)\bar{E}(\bar{\sigma},t)$$

where $D(N(\bar{\sigma},t))$ = the concentration-dependent diffusion coefficient; μ = the impurity mobility; and $\bar{E}(\bar{\sigma},t)$ = the local electric field. The field-dependent particle current $[\mu N(\bar{\sigma},t)\bar{E}(\bar{\sigma},t)]$ is positive for a positively charged particle and negative for a negatively charged particle. For diffusion at high concentrations $\{N(\bar{\sigma},t) \geq 10^{20}$ molecules/cm$^3\}$, the local electric field and concentration-dependent, diffusion coefficient affects must be considered. However, it is advantageous to consider the diffusion equation in the one-dimensional case, given the approximations that the diffusion coefficient is independent of concentration and the electric field is negligible:

$$\frac{D\partial^2 N(x,t)}{\partial x^2} = \frac{\partial N(x,t)}{\partial t}$$

With respect to technological applications, there are two solutions associated with this equation.

Case I. Constant surface concentration $N(0,t) = N_0$:

$$N(x,t) = N_0 ERFC(x/2\sqrt{Dt})$$

where = $ERFC(Y) \triangleq 1 - (2/\sqrt{\pi})\int_0^Y e^{-z^2}dz$ and is defined as the complementary error function.

Case II. Constant impurity dose $\int N(x,t)dx$ = constant:

$$N(x,t) = \frac{Q}{2\sqrt{\pi Dt}} \exp\left[\frac{-(x-R)^2}{4\,Dt}\right]$$

where R = the position of the peak of the profile and Q = a constant that is proportional to the impurity dose.

Impurities. The elements of columns 3 and 5 of the periodic table form shallow electron-acceptor and donor states, respectively, when incorporated in a silicon crystal. At temperatures above −193°C, these states are ionized, giving rise to either mobile holes or electrons and, thereby, modifying the electrical conductivity of the silicon. Figures 8**a** and 8**b** contain solubility limit and diffusion coefficient versus temperature data for the technologically important elements (38–39).

The proper choice of an impurity is determined by the specific requirements as-

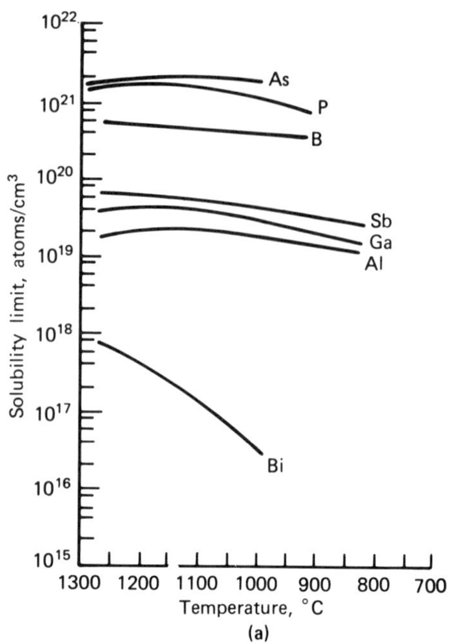

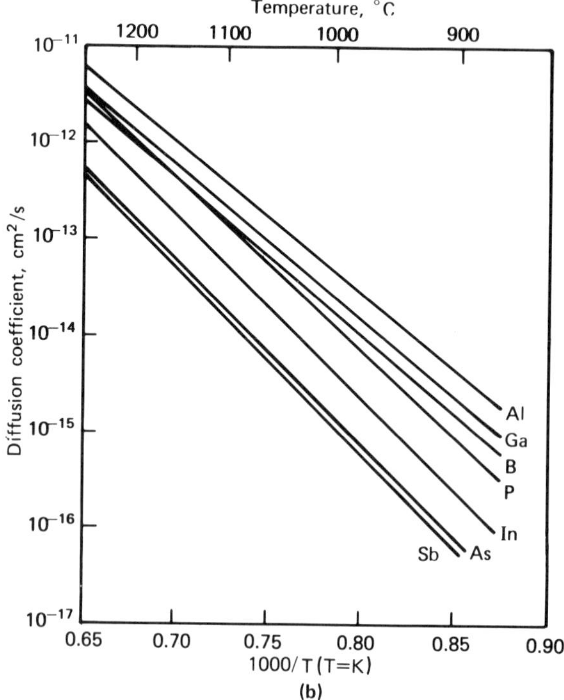

Figure 8. Solubility (**a**) and diffusion coefficient (**b**) for common silicon dopants versus temperature.

sociated with the IC fabrication sequence and the desired impurity profile. In general, low diffusion coefficients and high solubility limits are desired, making boron (acceptor) and phosphorus, arsenic, and antimony (donors) the most common impurities that are used commercially.

The diffusion coefficients for the common impurities are listed for both silicon and silicon dioxide in Table 2. The low values associated with silicon dioxide, coupled with its ease of formation by thermal oxidation and of pattern definition, have led to an almost universal adoption of this material as a diffusion mask for selective impurity diffusion into the surface of a silicon substrate (40).

Table 2. Diffusion Rates of Common Impurities in Silicon and Silicon Dioxide (At 1100°C)

Impurity	Silicon, cm^2/s	Silicon dioxide, cm^2/s
boron	2.0×10^{-13}	5×10^{-17}
phosphorus	2.0×10^{-13}	$<3.5 \times 10^{-16}$
arsenic	2.0×10^{-14}	$<1.0 \times 10^{-15}$
antimony	2.0×10^{-14}	1×10^{-16}

Techniques. The predeposition step consists of introducing a quantity of impurity atoms into the selectively masked silicon substrate. The drive-in step consists of thermally treating the substrate in an inert or oxidizing environment in order to establish the desired diffusion profile. In general, the final profile is of the form:

$$N(x,t) = \frac{Q}{2\sqrt{\pi Dt}} \exp\left[\frac{-(x-R)^2}{4\,Dt}\right]$$

Gas-Phase Doping. The silicon substrate is exposed to an inert atmosphere containing a determined partial pressure of impurity atoms. This technique allows one to vary surface concentration readily and has been used extensively for arsenic diffusions (41).

Glass-Source Doping. A silicon dioxide-based glass containing the desired impurity is deposited onto the surface of the silicon substrate, which is then heat treated to obtain the required predeposition profile (3). The doped glass may be formed either by a low temperature (450°C) chemical vapor deposition technique or a high temperature (900°C) growth technique conducted in the oxidizing atmosphere that contains the impurity.

Ion Implantation. Ion implantation (qv) (42–44) consists of exposing the silicon substrate to an accelerated, mass-analyzed beam of impurity ions. The beam penetrates the substrate where it is stopped and forms a characteristic profile. The spatial profile is given by the equation:

$$N(x) = \frac{Q}{\sqrt{2\pi}}\,\frac{1}{\sigma(I,E)} \exp\left[\frac{(x-R(I,E))^2}{2\sigma^2(I,E)}\right]$$

where E = the acceleration energy of the ion; and $R(I,E)$, $\sigma(I,E)$ = the energy dependent range and straggle for the implanted species I (45).

Examples of typical impurity profiles that are found in MOS and bipolar transistors are given in Figures 9 and 10, respectively. Ion implantation is the commonly used predeposition technique for all diffusion profiles in MOS transistors. Both

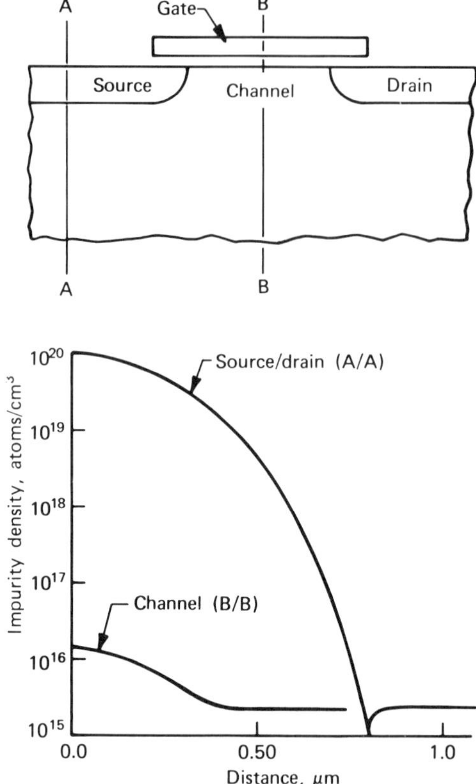

Figure 9. Typical doping profiles versus depth into the silicon for an MOS transistor.

thermal diffusion and ion implantation are used in bipolar transistors. Ion implantation (qv) is the preferred technique for high-performance structures having shallow bases and emitters.

Chemical Vapor Deposition. A variety of thin film materials, eg, refractory metals and insulators, are optimumly prepared by chemical vapor deposition (CVD) (46).

The kinetics of the CVD reaction are controlled by the five steps of the overall reaction, ie: transport of the reactants to the substrate surface; adsorption of the reactants on the surface; chemical reaction and diffusion on the surface; desorption of the by-product gases from the surface; and transport of the by-product gases away from the substrate (46).

Many CVD reactions may be performed over a wide temperature range. The optimum temperature of deposition should result in uniform step coverage. If the deposition temperature is too low and surface diffusion effects are negligible, the deposited film will be noncontinuous over steps, whereas at higher deposition temperatures, increased surface diffusion will result in uniform step coverage.

Film nonuniformity arises from nonuniformities in the mass transfer of gaseous reactants to the wafer surface and from the depletion of the reactant gases and accumulation of the by-product gases. These causes can be minimized by deposition system design; consequently, two types of reactors have evolved. In the hot-wall system the entire reaction chamber is heated to the reaction temperature T_R, thereby providing

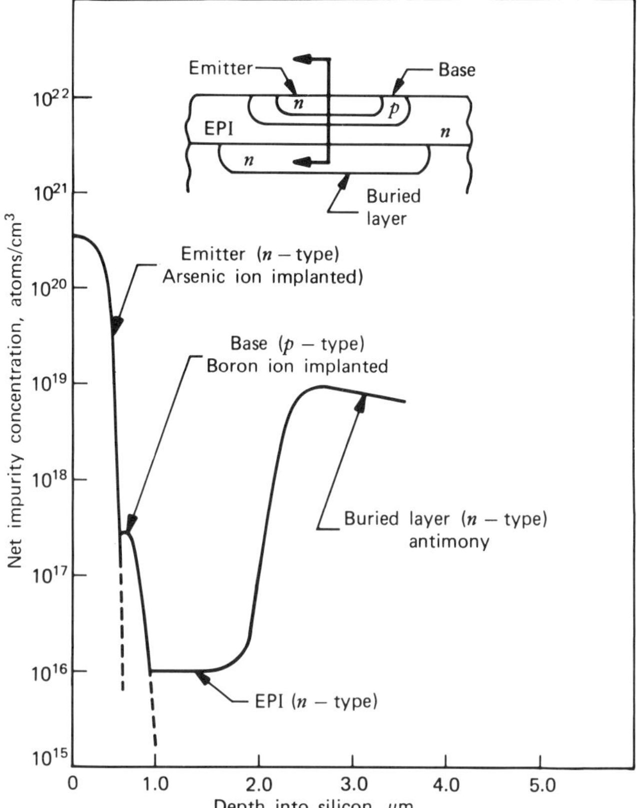

Figure 10. Typical doping profiles versus depth into the silicon for a bipolar transistor.

good temperature control. The disadvantages of this system are that the reaction occurs on the chamber walls, thus depleting the reactant gases and affecting uniformity; and the reaction can occur in the gas phase, thus creating particulates and affecting cosmetic quality.

In cold-wall systems, only the substrates are heated, either by r-f induction or by contact to a heated substrate holder, thereby minimizing reactant depletion effects and particulate problems. The disadvantage of this system is the lack of good temperature control.

Advances in CVD technology also have been made by the development of low pressure, hot-wall reactor systems (47).

These systems effect excellent uniformity (better than ±5%) in a large batch (100 to 200) of silicon wafers and in deposition rates (0.010 μm/min) that are comparable to atmospheric systems, resulting in greater processing throughout (see Film deposition techniques).

Applications and Materials. The deposition of silicon (15,65) has two principal applications in IC fabrication: the formation of epitaxial layers on single-crystal silicon substrates, and the formation of polycrystalline films on insulators. The polycrystalline films can be selectively diffused and patterned to form resistors, diodes, capacitors, and gate electrodes for MOS devices. Epitaxial reactors are of the cold-wall- and low pressure-types; hot-wall reactors are used for polycrystalline silicon deposition.

Silicon nitride has widespread use in IC fabrication because the diffusivity of sodium in it is extremely low at device-operating temperatures (≤125°C), thereby making it an ideal material for passivating ICs, and it has an extremely slow rate of oxidation at elevated temperatures (eg, ca 0.10 the rate of silicon), enabling it to be used as a selective oxidation mask during IC fabrication (46).

CVD layers of silicon dioxide are formed when it is impossible to form them by thermal oxidation, eg, in the case of overcoating metals (46).

Silicon dioxide-based glasses also are widely used in IC fabrication and serve as impurity sources for diffusion and as an insulation and passivation layer. The three common glasses that are used in the first application are the arsenosilicate, borosilicate, and phosphosilicate glasses. These glasses are prepared by incorporation of the respective volatile gases, ie, $AsBr_3$, $AsCl_3$, AsH_3; BBr_3, B_2H_6; and PH_3, $P_2O_3Cl_4$, of the desired impurity levels into the CVD reaction (46).

At device operating temperatures, phosphosilicate glass, containing greater than two atom percent phosphorus, possesses an extremely low value for the diffusivity of sodium, thereby making it an excellent passivation layer. Also, the glass becomes plastic at 1000°C and can flow or be fire polished on the surface of the silicon substrate (48).

Plasma Processing. Plasma etching (49–50) has been developed and can be controlled to be either isotropic or anistropic; the latter produces no undercutting of the mask. Plasma etching involves lower material costs (since the gases that are used generally are less expensive than the chemicals that are used in wet processing) and lower disposal costs (since the gaseous by-products of plasma processing usually are cheaper to dispose of than the by-products of wet processing).

A plasma is an ionized gas in which the positive and negative space charges are nearly balanced. For semiconductor applications, reactors that can be evacuated with a suitable vacuum pump and backfilled with the desired gas—usually to 10^{-1}–10^{3} Pa (10^{-3}–10^{1} mm Hg)—are used. The plasma is formed and sustained by r-f electrical energy. Plasma processing has two limiting regimes: sputtering and plasma deposition or etching.

Sputtering and Reactive-Ion Deposition and Etching. Sputtering is a physical process in which ions extracted from an inert gas plasma are accelerated to 0.5–5 keV and are bombarded upon the solid target material that is to be deposited (51). During impact, the ions transfer their energy to target atoms and cause them to be ejected. The most widely used reactor configuration is a parallel-plate-diode system (51).

Although sputtering can be used to deposit thin films of most materials, it has its widest application in the deposition of metals, especially refractory metals, eg, titanium, platinum, and tungsten, which are difficult to deposit by other means. For sputter etching, slices that contain material to be etched are placed on the cathode and the anode serves as a catcher plate (52). A masking material is used that has a much lower sputter etch rate than the materials being etched. Sputter etching is anisotropic, ie, the mask pattern can be transferred to the etched material with no undercutting; this is an advantage as compared to wet etching, which is isotropic, especially for fine-line (<4 μm) circuits. Sputter etching also is an effective cleaning step prior to sputter deposition. A problematic aspect of sputtering processes in IC fabrication is the high particle energies involved, which can cause irreparable damage to the semiconductor devices (53).

Reactive-ion processes are an extension of sputtering processes. The inert gas that is used for generating the plasma in sputtering is replaced by a gas that reacts with the material that is to be deposited or etched (52). If titanium or silicon are sputtered in a nitrogen atmosphere, then under the proper plasma conditions, titanium nitride, a metal, or silicon nitride (as an insulator) can be deposited. Similarly, for etching, if a gas is used that reacts with the material to be etched and forms volatile compounds, then higher etch rates and/or higher selectivity with respect to other materials may be obtained.

Plasma Deposition and Etching. Plasma-assisted deposition and etching are processes in which the function of the plasma is to generate reactive chemical species and which generally are carried out in a parallel-plate radial-flow reactor, as shown in Figure 11 (54–55). This reactor design provides the uniformity and the temperature and plasma control that are necessary for IC fabrication, and it has become the standard for critical deposition and etching processes.

Plasma etching processes have been developed for most thin films that are used in IC fabrication; a list of materials, etch gases, and typical etch rates are given in Table 3.

Establishing an adequate etch selectivity between the material being etched and underlying layers can be a problem with plasma etching (58). For thermally grown, silicon dioxide layers, the etch selectivity with respect to silicon using wet chemical etching, eg, hydrofluoric acid, is infinite, whereas with plasma etching, the best selectivities that have been reported are ≤ 15.

Plasma-assisted deposition has been used to deposit silicon dioxide, silicon nitride, and polycrystalline silicon (59). The most widely used material is plasma-deposited silicon nitride which is deposited from a mixture of silane, ammonia, and argon and/or nitrogen at temperatures from 300–350°C and which is used as a metallization overcoat and an insulating layer in multilevel metal structures (60). The salient feature of

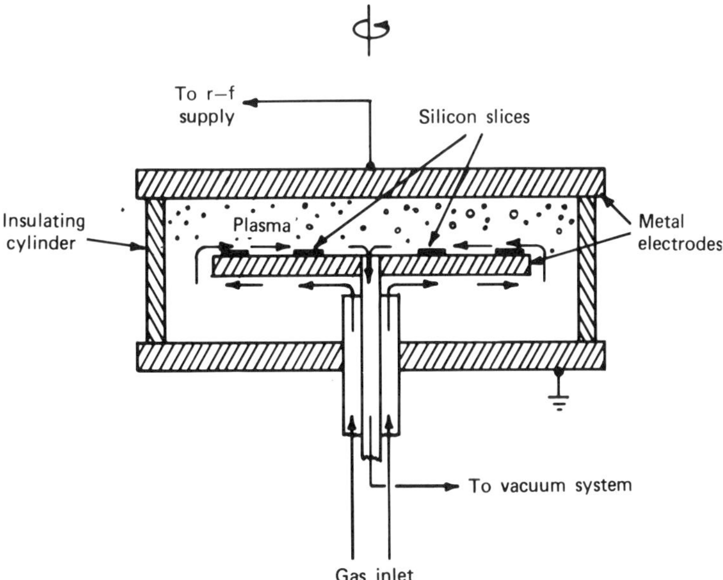

Figure 11. Horizontal parallel-plate radial-flow plasma reactor.

Table 3. Plasma Etch Gases for IC Materials[a]

Material	Etch gas	Typical etch rates, nm/min
thermal SiO_2	CF_4–O_2	10
	C_2F_6–CF_3Cl	30
	C_2F_6–CHF_3	70
phosphosilicate glass	C_2F_6–CHF_3	120
	C_2F_6–CF_3Cl	30
Si_3N_4	CF_4–O_2	70
	C_2F_6–CF_3Cl	30
	C_2F_6–CHF_3	70
plasma-deposited SiN	CF_4–O_2	150–300
	C_2F_6–CF_3Cl	50
	C_2F_6–CHF_3	70
single-crystal Si	CF_4–O_2	100
	C_2F_6–CF_3Cl	90
	C_2F_6–CHF_3	10
vapor-deposited poly-Si	C_2F_6–CHF_3	15
	C_2F_6–CF_3Cl	100
aluminum	CCl_4–Ar–H_2	100
	CCl_4–Cl_2	100

[a] Refs. 56 and 57.

plasma deposition is that, at low deposition temperatures (<400°C), it provides films with properties that usually are obtained only at much higher temperatures, eg, plasma silicon nitride provides excellent coverage over harsh topology.

Metallization. Metallization provides contact to the silicon and interconnects devices. Silicon contact is made at windows that are etched in a final glass layer which is formed over the entire wafer, either by thermal oxidation or vapor deposition, after all impurity regions have been defined. The metal is deposited uniformly and then patterned, the pattern being determined by the circuit function.

Aluminum is the most widely used metallization (61) and, for MOS ICs, it is predominant. For bipolar, two systems are used: one is based on aluminum but is modified to meet unique bipolar requirements (62) and the other is based on gold (63). The most commonly used IC metal systems are described in Table 4 and sheet resistances and typical thicknesses of the materials are given in Table 5 (see Electrical connectors).

Table 4. IC Metal Systems

MOS ICs	Bipolar ICs			Material function
Al, poly-Si	PtSi	PtSi	*1st Level*	Si contact
		Ti		glue layer
	Ti, W	TiN		barrier layer
poly-Si	Al	Pt		interconnection I
SiO_2 (P-doped)	SiO_2 (P-doped)	SiN		interlevel dielectric
SiO_2				
		Ti	*2nd Level*	glue layer
	Ti, W	TiN, Pt		barrier layer
Al	Al	Au		interconnection II

Table 5. Conductor Materials

Material	Sheet resistance, Ω/square	Thickness, μm	Use
Al, Au	0.02–0.04	1.0	interconnection, ohmic contact (Al)
Pt, W, Mo	0.2–0.5	0.3–0.6	barrier, interconnection
Ti, TiN	100–200	<0.2	glue layer, barrier
PtSi	5–10	<0.2	ohmic contacts, Schottky diodes
poly-Si	>15	0.4–0.8	gate electrode, interconnection

Silicon Contact. The metallization must provide ohmic, low resistance contact to the silicon and the capability for forming Schottky diodes (rectifying contacts with bipolar). The electrical characteristics of the contacts must be stable under any further thermal processing and over the expected life of the IC.

When a metal is deposited on lightly doped ($<10^{17}$ molecules/cm^3 silicon), a Schottky barrier diode or rectifying contact is formed (64). A potential barrier ϕ_{Bn} exists between the metal and the silicon, and charge carriers must surmount the barrier by thermonic emission so that current can flow. The current I voltage V characteristics are given by

$$I = AA^{**}T^2 \exp(q\phi_{Bn}/kT)\,[\exp(qV/kT - 1)]$$

where A is the contact area, A^{**} is the effective Richardson constant (= 112 or 32 A/(cm·K)^2 for electrons or holes), T is the absolute temperature, q is the electronic charge, and k is Boltzmann's constant. For positive voltages, the potential barrier is decreased and the current increases exponentially with voltage. For negative voltages, the current is independent of voltage until a critical field is reached in the silicon, at which point breakdown occurs as a result of avalanche multiplication, and large currents can flow. For voltages greater than the breakdown voltage, the above equation is not valid. The barrier height for contact to p-type silicon, ϕ_{Bp}, is given by

$$\phi_{Bp} = E_G - \phi_{Bn}$$

where E_G is the energy gap between the valence and conduction bands and is equal to 1.11 eV for silicon.

Ohmic contact is obtained by increasing the doping density under the contact region which decreases the width of the potential barrier between the metal and the silicon. At high enough doping densities ($>10^{18}$ molecules/cm^3), the barrier becomes thin enough to allow quantum mechanical tunnelling of carriers through the barrier, resulting in an ohmic contact. The current voltage characteristics are approximated by

$$I = V(A/R_c)$$

where R_c is the specific contact resistance, which is determined largely by the doping density in the contact region, and A is the area of the contact. Typical values of R_c for doping densities that are $>10^{19}$ molecules/cm^3 are 100 Ω·μm^2.

Interconnections. For interconnections, the metallization must be easily patterned, provide a low sheet resistance to minimize voltage drops, have good adherence to the underlying insulator, and provide adequate coverage over steps in the insulator surface. High levels of integration also require a multilevel capability, that is, two or more levels of metal interconnection with each level separated by an insulating, interlevel dielectric. Connections between levels are made through selective vias in the interlevel dielectric. Reliability considerations indicate that the metal or the metal in conjunction with an appropriate overcoated protection layer be resistant to corrosion and that the metal have good electromigration resistance (61). The final requirement is bondability to allow connection of an IC to the outside, non-IC environment.

MOS Metallization. Polycrystalline silicon is used both as an active gate material and as a level of interconnection (65). It is deposited uniformly by CVD and is defined by chemical or plasma etching. A typical chemical etch is a dilute aqueous solution of nitric and hydrofluoric acids. A photoresist is used as a mask for polycrystalline silicon definition and for all patterning steps required in the MOS metallization process. CVD is used for the interlevel-dielectric, phosphorus-doped (6–8 wt % P) silicon dioxide. The phosphorus doping allows the interlevel dielectric to flow at temperatures from 1000–1100°C (48). The flowing provides a smooth outer surface and eliminates top-level, aluminum step-coverage problems caused by first level topology. Chemical or plasma etching can be used to form vias through the interlevel dielectric and to pattern the aluminum which is deposited uniformly after via definition. Dilute aqueous solutions of phosphoric, nitric, and acetic acids are used to etch aluminum chemically. The aluminum adheres well to silicon dioxide and provides a sheet resistance of $\cong 0.03\ \Omega$/square for a nominal thickness of 1 μm.

To ensure good contact between the silicon and the aluminum, a short (<30 min) heat treatment (sintering) from 400–550°C usually is used after aluminum patterning (66).

Ohmic contact to n regions is achieved by ensuring that the surface doping in the contact areas is $>10^{18}$ molecules/cm^3. For p regions, ohmic contact is made easily because aluminum is a p-type dopant in silicon with a solubility that is $>10^{18}$ molecules/cm^3 over the temperature range used for sintering. The aluminum is usually overcoated with a physically protective dielectric layer—which is either low-temperature (<500°C), vapor-deposited, phosphorus-doped silicon dioxide or plasma-deposited silicon nitride—and vias to bonding pads are opened through this overcoat.

Bipolar Metallization. The metallization for bipolar ICs is determined largely by the necessity of forming Schottky diodes with stable and reproducible characteristics to lightly doped ($<10^{17}$ molecules/cm^3 n-type silicon). The diodes are connected in parallel with the base-collector junctions of npn transistors in saturating logic circuits to increase switching speed by preventing excess charge storage (67). With aluminum, problems have been encountered with variations in ϕ_{Bn} upon sintering and further thermal treatments (68–69). These problems, ascribed to silicon–aluminum interactions, have been eliminated by forming a layer of a metal silicide—usually platinum silicide—in the contact window (70). Platinum silicide Schottky diodes are stable and reproducible: $\phi_{Bn} = 0.85$ eV. An additional complication arises because aluminum and platinum silicide form intermetallics that lead to unstable Schottky diode characteristics (71). To prevent intermetallic formation, a barrier layer of titanium–tungsten is sputter deposited prior to aluminum deposition (62,72). The titanium–tungsten is chemically or plasma etched after aluminum patterning with the aluminum being used as a mask.

Multilevel wiring capability is provided by two levels of aluminum separated by

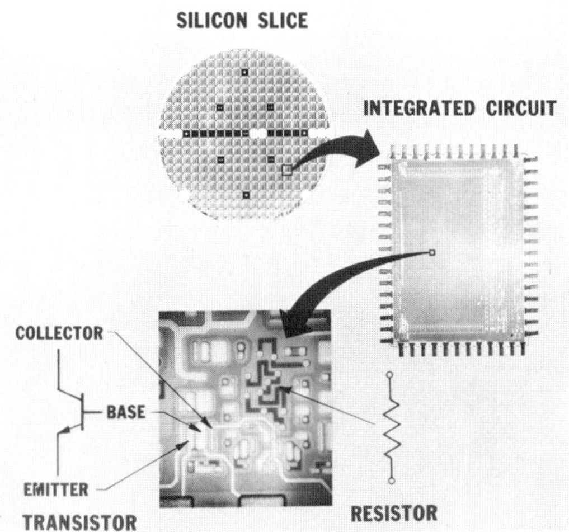

Figure 12. A photograph of a silicon slice (76 mm in diameter) containing many ICs, a photograph of an individual IC (4.5 mm × 3.3 mm), and a photomicrograph of typical IC circuit elements (magnification 400×).

an insulating, interlevel dielectric. The most widely used interlevel dielectric is a low temperature (400°C), vapor-deposited, phosphorus-doped silicon dioxide. Sputter-deposited silicon dioxide also has been used. The techniques described for aluminum patterning are used on both levels. The insulation integrity of the interlevel dielectric over the edges of first level metal must be ensured to prevent unwanted shorts in crossover areas. To mitigate these problems, the etching used for patterning the first-level aluminum can be modified to provide a slope to the edge of the aluminum line (61).

Another metal system that is used for bipolar ICs is the gold-based, beam-lead metallization, developed as an integral part of an interconnection technology in which many chips can be economically bonded to a metallized ceramic (63). The chip bonding is accomplished with gold beam leads (see Fig. 12) that extend over the edges of the separated chips. Gold is very ductile, can be thermocompression bonded very easily, has a high electrical conductivity, and is resistant to oxidation and corrosion. Gold cannot be used by itself for the chip interconnections, because it has poor adherence to insulators, eg, silicon dioxide and silicon nitride. A layer of titanium is used to provide adherence; however, titanium and gold form problematic intermetallics and, therefore, a barrier layer of platinum is used to prevent intermetallic formation (72). For multilevel capability, titanium–platinum is used as a first-level conductor, plasma-deposited silicon nitride is used as an interlevel dielectric, and titanium–platinum–gold is the second-level metal. Sputter etching is used for pattern definition on both levels and the vias are plasma etched (60). Scanning electron micrographs of this structure are shown in Figure 13.

Fabrication Sequence. The fabrication process for an IC involves, in a specified sequence, the process technologies reviewed in the previous sections with the specific sequence and details of the processing steps being determined by the technology of the device being fabricated. The n-channel, silicon-gate fabrication technology (see Fig. 1**a**) is illustrated in Figure 14 and describes the complete fabrication sequence (65,73). This is the technology used for the majority of n-channel MOS ICs.

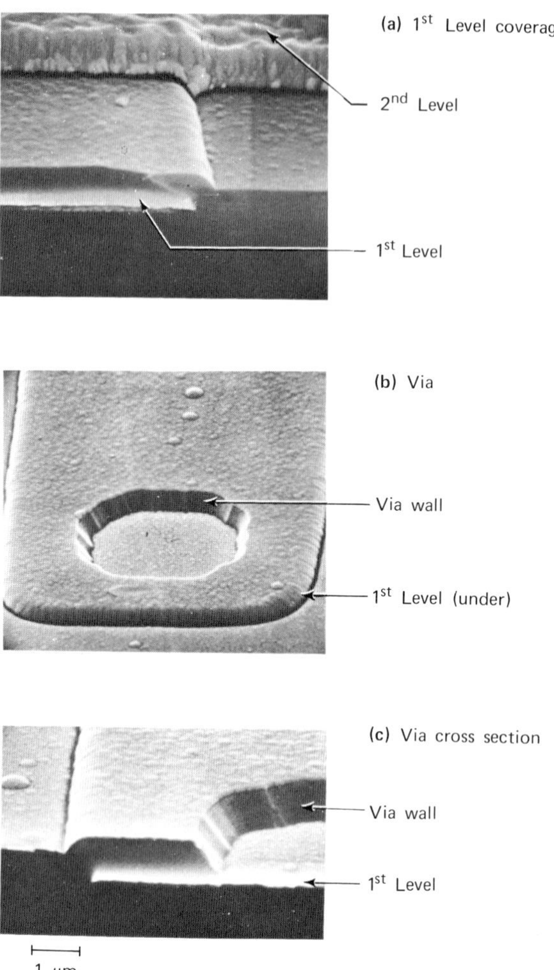

Figure 13. Scanning electron micrograph cross-section of a two-level metal structure. First level is Ti–Pt, interlevel dielectric is plasma-deposited SiN, and second level is Ti–Pt–Au.

Post-Wafer Processing. Testing and packaging are the final processes required to complete the manufacture of an IC after slice fabrication (74).

Computer-controlled equipment (test sets) is used to determine which ICs on a silicon wafer perform their intended functions. The test set automatically tests every potentially useful circuit on a wafer, and acceptable or nonacceptable circuits are identified, usually by an ink dot.

There are two general techniques used for packaging or assembling ICs. In one, several ICs are assembled with other electrical components on a common substrate to form a system. However, most ICs are packaged individually and the packages are interconnected on one or more printed wiring boards to form an electronic system.

A final electrical test is performed after packaging to weed out those ICs that were damaged during the separation and packaging operation. Burn-in, which involves

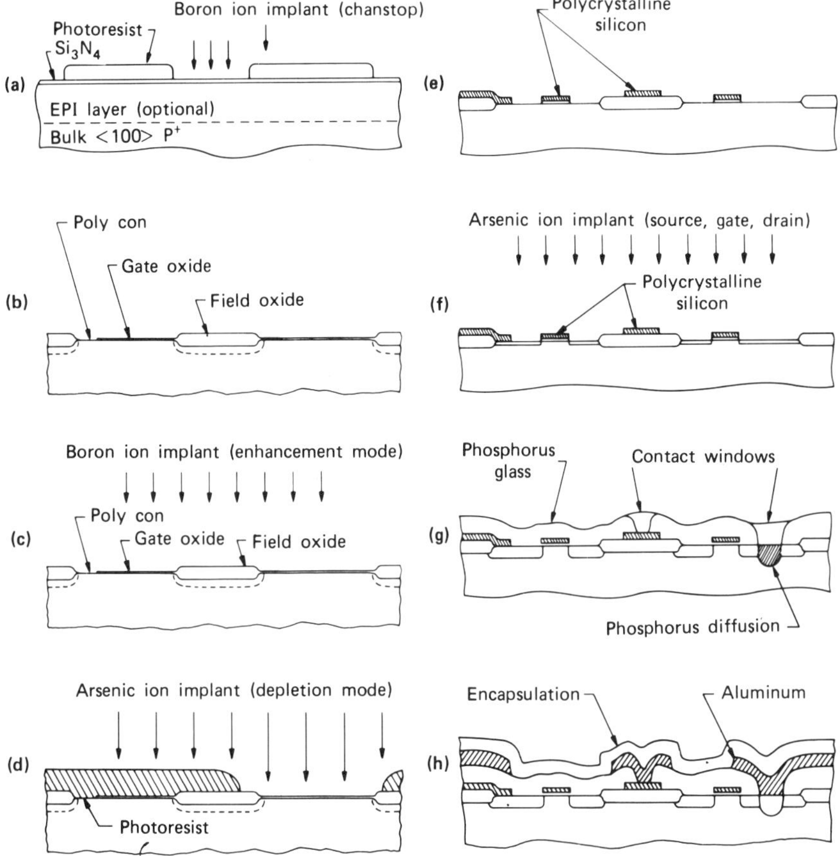

Figure 14. Fabrication sequence for a *n*-channel, MOS IC.

operating the ICs at elevated temperatures for a given time, may be carried out to guarantee high quality ICs. Burn-in is equivalent to accelerated aging and can be used to eliminate weak or unstable ICs that would otherwise fail in operational use after a short period of time.

Economic Aspects

By 1982, the world market for ICs is expected to pass $\$10 \times 10^9$ (75–76). In 1979, U.S.-based manufacturers held a two-thirds share of the estimated $\$6 \times 10^9$ market. It was estimated that, in 1979, ICs accounted for ~4.5% of the value of all electronic equipment, up from less than 1% in 1970. This percentage is expected to double by 1990. A principal market for ICs in the 1970s was computers and computer-based equipment. Computers will continue to be an important market in the 1980s and two other markets which are expected to burgeon are telecommunications and automotive electronics.

The top five U.S.-based producers of ICs for the merchant market in terms of dollar sales in 1979 were (in order of decreasing dollar sales): Texas Instruments,

Motorola, Intel, National Semiconductors, and Fairchild Semiconductors. In addition to the merchant markets, many systems manufacturers have established their own in-house capability for producing ICs for internal consumption, eg, International Business Machines, Western Electric, Hewlett-Packard, and Honeywell. The greatest threat to U.S. producers may come from the Japanese who have made, as national goals, leadership in IC technology and a concomitantly large share of the IC market.

There has been a continual decrease in cost per electronic function made possible by the increasing use of ICs, eg, the cost of hand-held calculators. Many identical ICs are fabricated simultaneously on a single slice or wafer of silicon (as indicated in Figure 12) and, through many of the fabrication steps, many slices ($\cong$100) are processed simultaneously, which results in low cost per IC.

To a large measure, the goal of minimum cost per electronic function determines the optimum size or complexity of an IC. The cost of an electronic system primarily is determined by the cost of the electronic components and the cost of interconnecting these components (31,78). The first increases as the complexity of an IC increases because increased complexity implies larger silicon area per IC and, thus, a higher probability that a manufacturing-induced defect will cause an individual IC to be defective. The interconnection costs decrease as IC complexity is increased because there are fewer ICs to interconnect. There is an optimum complexity which gives a minimum cost per electronic function, and complexity has doubled every year since 1960 (77–79).

Three factors have been important in moving the minimum cost per electronic function towards ICs of greater complexity. First, manufacturing processes have been perfected which has permitted ICs of larger area to be fabricated with acceptable yields. Second, the silicon area required per electronic function has been decreasing. Third, the minimum feature size used for defining the individual elements on an IC has decreased ca tenfold since 1960.

Health and Safety

Silicon is not classified as a hazardous or toxic material in its bulk crystal form. However, many of the chemicals used in IC fabrication are extremely hazardous or toxic and include liquid acids, eg, hydrochloric acid, hydrofluoric acid, nitric acid, and phosphoric acid; hydrocarbons, acetone, trichloroethylene, and photoresist solvents; and gases, eg, chlorine, phosphine, and silane (see Silicon compounds).

BIBLIOGRAPHY

1. D. J. Hamilton and W. G. Howard, *Basic Integrated Circuit Engineering,* McGraw-Hill, Inc., New York, 1975.
2. A. B. Glaser and G. E. Subak-Sharpe, *Integrated Circuit Engineering,* Addison-Wesley Publishing Co., Reading, Mass., 1977.
3. P. E. Gise and R. E. Glanehard, *Semiconductor and Integrated Circuit Fabrication Techniques,* Reston Publishing Co., Inc., Reston, Va., 1979.
4. R. C. Eden, B. M. Welch, R. Zucca, and S. I. Long, *IEEE Trans. Electron Devices* **ED-26,** 299 (1979).
5. J. D. Meindl, *Sci. Am.* **237,** 70 (1977).
6. A. S. Grove, *Physics and Technology of Semiconductor Devices,* John Wiley & Sons, Inc., New York, 1967.

7. R. S. Muller and T. I. Kamins, *Device Electronics for Integrated Circuits,* John Wiley & Sons, Inc., New York, 1977.
8. J. A. Appels, E. Kool, M. M. Patten, J. J. H. Schalorje, and W. H. C. G. VerKuylen, *Philips Res. Rep.* **25,** 118 (1970).
9. D. Peltzer and B. Herndon, *Electronics* **44,** 53 (1971).
10. B. E. Deal and J. M. Early, *J. Electrochem. Soc.* **126,** 20C (1979).
11. P. W. Morrison, ed., *Environmental Control in Electronic Manufacturing,* Van Nostrand Reinhold Co., New York, 1973.
12. C. E. Reid in V. Merten, ed., *Desalination by Reverse Osmosis,* M.I.T. Press, Cambridge, Mass., 1965; J. H. Myers, *Report #GA-9771,* Gulf General Atomic, Inc., San Diego, Calif., 1970.
13. H. Herrmann, H. Herzer, and E. Sirtl, *Festkoerperprobleme* **XV,** 279 (1975).
14. S. M. Hu, *Appl. Phys. Lett.* **31,** 53 (1977).
15. J. Bloem and L. J. Giling in E. Kaldis, ed., *Current Topics in Materials Science,* Vol. 1, North Holland Publishing Co., 147, 1978.
16. S. K. Tung, *J. Electrochem. Soc.* **112,** 436 (1965).
17. W. Keen, *RCA Rev.* **31,** 187 (1970).
18. B. E. Deal, *J. Electrochem. Soc.* **110,** 527 (1963).
19. B. E. Deal and A. S. Grove, *J. Appl. Phys.* **36,** 3770 (1965).
20. R. J. Zeto, N. O. Korolkoff, and S. Marshall, *Solid State Technol.* **22,** July, 1979, p. 62.
21. A. S. Grove, B. E. Deal, E. H. Snow, and C. T. Sah, *Solid State Electron.* **8,** 145 (1965).
22. C. E. Young, *J. Appl. Phys.* **32,** 329 (1961).
23. A. S. Grove, E. H. Snow, B. E. Deal, and C. T. Sah, *J. Appl. Phys.* **35,** 2458 (1964).
24. E. H. Nicolian and A. Goetzberger, *Solid State Electron.* **8,** 321 (1965).
25. B. E. Deal, M. Sklar, A. S. Grove, and E. H. Snow, *J. Electrochem. Soc.* **114,** 266 (1967).
26. B. E. Deal, *J. Electrochem. Soc.* **121,** 198 (1974).
27. R. J. Kriegler, Y. C. Cheng, and D. R. Colton, *J. Electrochem. Soc.* **119,** 388 (1972).
28. Y. T. Yeow, D. R. Lamb, and S. D. Brotherton, *J. Phys. D. Appl. Phys.* **8,** 1495 (1975).
29. W. S. DeForest, *Photoresist Materials and Processes,* McGraw Hill Inc., New York, N.Y., 1975.
30. G. Bouwhuiv and S. Wittekock, *IEEE Trans. Electron Devices* **ED-26,** 723 (1979).
31. B. T. Murphy, *Proc. Inst. Electr. Eng.* **52,** 1537 (1964).
32. L. F. Thompson and R. E. Kerwin, *Annu. Rev. Mater. Sci.* **6,** 267 (1976).
33. J. Lyman, *Electronics* **52,** 105 (April 12, 1979).
34. H. Binder and M. Lacombat, *IEEE Trans. Electronic Devices* **ED-26,** 698 (1979).
35. D. Maydan, G. A. Coquin, J. R. Maldorado, J. Somekh, D. Y. You, and G. N. Taylor, *IEEE Trans. Electronic Devices* **ED-22,** 492 (1975).
36. H. C. Pfeiffer, *IEEE Trans. Electronic Devices* **ED-26,** 663 (1979).
37. D. R. Herriott, R. J. Collier, D. S. Alles, and J. H. Stafford, *IEEE Trans. Electronic Devices* **ED-22,** 385 (1975).
38. F. A. Trumbore, *Bell Syst. Tech. J.* **39,** 205 (1960).
39. H. F. Wolf, *Silicon Semiconductor Data,* Pergamon Press, Oxford, Eng., 1969.
40. C. J. Frosch and L. Derrick, *J. Electrochem. Soc.* **104,** 547 (1957).
41. S. Ohkawa, Y. Nakayima, and Y. Fukukawa, *Jpn. J. Appl. Phys.* **14,** 458 (1975).
42. J. Stephen, *Radio Electron Eng.* **42,** 265 (1972).
43. J. C. Plunkett and J. L. Stone, *Solid State Technol.* **18,** 49 (Dec., 1975).
44. R. J. Duchynski, *Solid State Technol.* **20,** 53 (Nov., 1977).
45. W. S. Johnson and J. F. Gibbons, *Projected Range Statistics in Semiconductors,* University Press, Stanford, Calif., 1970.
46. W. A. Bryant, *J. Mater. Sci.* **12,** 1285 (1977).
47. R. S. Rosler, *Solid State Technol.* **20,** 63 (April, 1977).
48. U.S. Pat. 3,825,442 (July 23, 1974), G. E. Moore.
49. *Solid State Technol.* **21** (April, 1978); issue devoted to plasma technology.
50. T. C. Penn, *IEEE Trans. Electronic Devices* **ED-26,** 640 (1979).
51. L. I. Maissel and R. Glang, eds., *Handbook of Thin Film Technology,* McGraw Hill, Inc., New York, Chapts. 3 and 4, 1970.
52. C. M. Melliar-Smith, *J. Vac. Sci. Technol.* **13,** 1008 (1976).
53. W. D. Ryden, E. F. Labuda, and J. T. Clemens in H. Hughs and M. J. Rand, eds., *Etching for Pattern Definition,* Electrochemical Society, Inc., Princeton, N.J., 1976, p. 144.
54. A. R. Reinberg, *Electrochem. Soc. Extended Abstracts* **74-1,** 19 (1974).

55. A. R. Reinberg, ref. 53, p. 91.
56. C. J. Mogab and W. R. Harsbarger, *Electronics* **51,** 117 (1978).
57. C. M. Meilliar-Smith and C. J. Mogab in J. L. Vossen and W. Kern, eds., *Thin Film Processes,* Academic Press, New York, 1979.
58. J. W. Coburn, H. F. Winters, and T. J. Chuang, *J. Appl. Phys.* **48,** 3532 (1977).
59. M. J. Rand, *J. Vac. Sci. Technol.* **16,** 420 (1979).
60. W. D. Ryden and E. F. Labuda, *IEEE J. Solid State Ckts.* **SC-12,** 376 (1977).
61. A. J. Learn, *J. Electrochem. Soc.* **123,** 894 (1976).
62. P. B. Ghate, J. C. Blair, C. R. Fuller, and G. E. McGuire, *Thin Solid Films* **53,** 117 (1978).
63. M. P. Lepselter, *Bell Syst. Tech. J.* **45,** 233 (1966).
64. S. M. Sze, *Physics of Semiconductor Devices,* John Wiley & Sons, Inc., New York, Chapt. 8, 1969.
65. F. Faggin and T. Klein, *Solid State Elect.* **13,** 1125 (1970).
66. L. A. Berthoud, *Thin Solid Films* **43,** 319 (1977).
67. Ref. 64, p. 410.
68. T. M. Reith and J. D. Schick, *Appl. Phys. Lett.* **25,** 524 (1974).
69. H. C. Cord, *Solid State Commun.* **16,** 87 (1975).
70. M. P. Lepselter and J. M. Andrews in B. Schwartz, ed., *Ohmic Contacts to Semiconductors,* Electrochemical Society, Inc., Princeton, N.J., 159 (1969).
71. H. H. Hosack, *Appl. Phys. Lett.* **21,** 256 (1972).
72. M. A. Nicolet, *Thin Solid Films* **52,** 415 (1978).
73. J. T. Clemens, R. H. Doklan, and J. J. Nolen, *Technical Digest 1975 IEEE International Electron Device Meeting,* p. 299.
74. Ref. 2, Chapt. 10.
75. *Bus. Week,* 66 (Dec. 3, 1979).
76. *Status 79, A Report on the Integrated Circuit Industry,* Integrated Circuit Engineering Corp., Scottsdale, Ariz.
77. R. N. Noyce, *Science* **195,** 1102 (1977).
78. G. E. Moore, *Technical Digest 1975,* IEEE International Electron Device Meeting, pp. 11–13.
79. G. E. Moore, *IEEE Spectrum* **16,** 30 (1979).

E. F. LABUDA
J. T. CLEMENS
Bell Telephone Laboratories, Incorporated

IODINE AND IODINE COMPOUNDS

Iodine

Iodine [*7553-56-2*] is a nonmetallic element belonging to the halogen family in group VIIA of the periodic table. It is the heaviest common member of this family and the only one that is solid at ordinary temperatures. The principal valence numbers are −1, +1, +3, +5, and +7. An oxide of composition IO_2 with an average valence number of +4 is also known. Compounds are known in all these oxidation states (with the single exception of +4) that are thermodynamically stable with respect to their constituent elements and can be formed from them. Examples are KI, ICl, ICl_3, IF_5, and Na_5IO_6.

Iodine is a simple element, and only one stable atomic species with a mass number of 127 is known, but 10 or more radioisotopes have been prepared artificially (see Isotopes).

Iodine was discovered in the early 19th century by Courtois, a Frenchman engaged in the production of potassium nitrate from niter beds for Napoleon's armies. He found that his copper vessels were corroded by the liquors, an effect that he traced to an impurity in the soda ash derived from seaweed, which he used in his process. By adding sulfuric acid to the liquors, this impurity could be isolated as a black powder that yielded a violet vapor when heated which condensed in the form of brilliant crystalline flakes. Gay-Lussac recognized it as a new element, gave it its name, and prepared a number of its compounds.

The first practical application of iodine was in medicine, when a Swiss physician discovered that tincture of iodine is a remedy for goiter and effected some remarkable cures (1). The treatment of wounds with this tincture was introduced in 1828, and its first application to battle wounds probably occurred during the American Civil War.

Since that time the value of iodine as a laboratory reagent for both research and analytical chemistry was widely recognized. Such well known synthetic methods as the Hofmann alkylation of amines (1850), and those of Williamson (1851), Wurtz (1855), and Grignard (1900), were originally based on the use of iodine compounds, and applications of iodine in analysis became so numerous that this use is called iodometry (see also Analytical methods).

An early commercial application was in photography. Daguerre's process, introduced in 1839, utilized a silver plate that had been exposed to iodine vapor as the sensitive material. It was succeeded by the wet-plate process in which the silver iodide was formed in a collodion layer coated on glass plates.

The first commercial source of iodine was the ashes of seaweed. Today, the chief sources of iodine are nitrate deposits in Chile and brines in Japan and the United States.

The presence of iodine in the nitrate deposits of Chile was first reported in 1840. Chile's production of crude iodine is associated with the production of potassium and sodium nitrates. Chile was the world's largest iodine producer until the late 1960s when it was overtaken by Japan. Iodine-enriched brines associated with natural gas wells were discovered in Japan after World War II.

The first production of iodine in the United States was from Pacific coast seaweeds during World War I. In the late 1920s, it was discovered in brines from oil wells in Louisiana and California. Since the early 1960s iodine has been produced from brines in Midland, Michigan, which consist of a complex salt type, and are not associated with oil and gas. In 1977, iodine production was started in Oklahoma from brines originating in subsurface formations. This area has since become the main source of U.S. iodine production.

According to the U.S. Geological Survey, iodine is the forty-seventh most abundant element in the earth's crust, counting the rare earths as a single element. It is widely distributed in nature (2–3), occurring in rocks, soils, and underground brines in small quantities; seawater contains about 0.05 ppm. Iodine has been identified in less than a dozen minerals; lautarite [*7789-80-2*] (anhydrous calcium iodate), the form in which iodine occurs in the Chilean nitrate deposits, is probably the most important of these.

Despite the low concentration of iodine in seawater, some seaweeds, notably the brown varieties, can extract and accumulate it. Those of the *Laminaria* family are the richest, containing up to 0.45% on a dry basis, followed by another brown variety, *Fucus*.

The most important commercial source of iodine occurs in subsurface brines. Iodine is present in many subsurface brines; its concentrations rarely exceed 100 mg/L, and are often less than 10 mg/L. The highest recorded concentration of iodine in natural brines or waters occurs in certain subsurface brines of Oklahoma where it exceeds 1.4 g/L (4).

Physical Properties. In massive form, iodine is a soft bluish-black solid, whereas the familiar resublimed material consists of almost opaque, doubly refracting orthorhombic crystals with a nearly metallic luster, a high index of refraction, and pronounced pleochroism. Both forms yield the characteristic violet-colored vapor when heated, hence the name iodine from the Greek, *ioeides,* violet-colored. The physical properties of solid, liquid, and gaseous iodine are given in Table 1.

Iodine is slightly soluble in water, the solubility increasing with the temperature (see Table 2); no hydrate is formed. Water is somewhat soluble in liquid iodine and may be completely miscible at 300°C (5). Iodine is also soluble in aqueous iodide solutions owing to the formation of polyiodide ions. Some of these solutions are highly concentrated; for example, that in equilibrium with solid iodine and $KI_7.H_2O$ [*74685-93-1*] at 25°C contains 67.8% iodine, 25.6% potassium iodide, and 6.6% water. However, if large cations such as cesium, substituted ammonium, and iodonium are present, the increased solubility may be limited owing to precipitation of sparingly soluble polyiodides. Iodine is also more soluble in solutions of chlorides, bromides, and other salts than in pure water, but these salts have much less effect than the iodides.

Iodine dissolves in many organic solvents (see Table 3). The color of the solutions varies with the nature of the solvent. Thus, it is violet in aliphatic hydrocarbons and carbon tetrachloride, and brown or reddish brown in alcohols, ethers, and benzene (6–8). The formation of molecular complexes was first demonstrated by studies (9) of the visible and ultraviolet spectra of solutions of iodine in benzene and other aromatic solvents. Most significant was the discovery of an intense ultraviolet absorption band near 300 μm attributable neither to the solvent nor to iodine, but to a 1:1 complex. The absorption band characteristic of various iodine–solvent complexes, the so-called

Table 1. Physical Properties of Iodine

Properties	Solid[a]	Liquid[b]	Gaseous
melting point, °C	113.6		
boiling point, °C		184	
critical temperature, °C		512	
critical pressure (estd), MPa[c]		11.75	
density at 101.3 kPa[d], g/cm³, d_4^{20}	4.940		
d_4^{60}	4.886		
d^{120}		3.960	
d^{180}		3.736	
vapor density at 101.3 kPa[d], g/L			6.75
av cubic coeff of expansion, per °C (0–113.6°C)	2.81×10^{-4}		
crystal structure, 4 mol I_2 per unit cell	orthorhombic		
unit cell dimensions at 18°C, nm			
a	0.47761		
b	0.72501		
c	0.97711		
entropy at 25°C or 298.2 K, J/(mol·K)[e]	116.73		62.25
specific heat, J/(g·K)			
25–113.6°C	$0.21163 + 19.615 \times 10^{-5}\, t$		
113.6–184°C		0.3163	
25–1200°C			0.1464
heat of fusion at 113.6°C, J/g[e]	62.13		
heat of sublimation at 113.6°C, J/g[e]	238.24		
heat of vaporization at bp, J/g[e]		164.35	
viscosity, mm²/s (= cSt), at 116°C		0.5727	
at 184°C		0.3785	
vapor pressure[a,b], kPa[d], at 25°C	0.041		
at 113.6°C	12.07		
thermal conductivity at 24.4°C, W/(m·K)	0.421		
electrical resistivity, Ω–cm,			
at 25°C	5.85×10^6		
at 140°C		1.1×10^5	
dielec constant at 23°C	10.3		
at 118°C		11.08	
refractive index, n_D	3.34		

[a] For the solid, between 0 and 113.6°C, $\log p_{kPa} = -(3410.71/T) - 0.3523 \log T - 1.301 \times 10^{-3}\, T + 11.3140$[d].
[b] For the liquid, between 113.6 and 186°C, $\log p_{kPa} = -(2300.24/T) + 7.0249$[d].
[c] To convert MPa to atm, divide by 0.101.
[d] To convert kPa to mm Hg, multiply by 7.50.
[e] To convert J to cal, divide by 4.184.

charge-transfer band has a molecular extinction coefficient of the order of 10^4 and a half-width typically of 6000/cm. The interaction of solvent with iodine implies a certain amount of charge transfer in the ground state of the complex. Accordingly, the vibrational properties of the complex reflect new vibrational motions and changes in bond strengths of the two molecular units (8). This usually occurs near the visible region and is the cause of typically intense colors (10–11).

Iodine can be sublimed readily because of its high vapor pressure below its melting

Table 2. Solubility of Iodine in Water

Temp, °C	Soly, g/kg H_2O	Temp, °C	Soly, g/kg H_2O
0	0.162	60	1.06
20	0.293	70	1.51
25	0.340	80	2.17
30	0.399	90	3.12
40	0.549	100	4.48
50	0.769	110	6.65

Table 3. Solubilities of Iodine in Organic Solvents at 25°C

Solvent	Soly, g/kg solvent	Solvent	Soly, g/kg solvent
benzene	164.0	ethylene chloride	57.6
bromoform	65.9	ethylidene chloride	39.8
carbon disulfide	197.0	glycerol	9.7
carbon tetrachloride	19.2	*n*-heptane	17.3
chloroform	49.7	*n*-hexane	13.2
cyclohexane	27.9	isobutyl alcohol	97
cis-dichloroethylene	38.3	isooctane(2,2,4-trimethylpentane)	13.2
trans-dichloroethylene	37.6	mesitylene(1,3,5-trimethylbenzene)	253.1
2,2-dimethylbutane	13.9	pentachloroethane	69[a]
ethyl acetate	157	perfluoroheptane	0.12
ethyl alcohol	271.7	*sym*-tetrachloroethane	62[a]
ethyl bromide	146	tetrachloroethylene	61[a]
ethyl cyanide	141	toluene	182.5
ethyl ether	337.3	trichloroethylene	79[a]
ethylene bromide	115.1	*p*-xylene	198.3

[a] Grams per liter of solution.

point, but it also exists as a black mobile liquid at atmospheric pressure over a range of more than 60°C. It is a good solvent for the iodides of the alkali metals and ammonium bases. It also dissolves sulfur, selenium, the covalent iodides of such metals as aluminum, tin, and titanium, and many organic substances.

Solutions of the alkali-metal and ammonium iodides in liquid iodine are good conductors of electricity, comparable with fused salts and aqueous solutions of strong acids. The liquid is therefore a polar solvent of considerable ionizing power, whereas its own electrical conductivity indicates that it is appreciably ionized, probably into I^+ and I_3^- (triiodide). Iodine thus resembles water in this respect. The metal iodides and polyiodides are bases, whereas the iodine halides are acids (12).

Iodine vapor is characterized by its familiar violet color and by its unusually high specific gravity, approximately nine times that of air. It is made up of diatomic molecules at low temperatures; at moderately elevated temperatures, dissociation becomes appreciable. The concentration of monatomic molecules, for example, is 1.4% at 600°C and 101.3 kPa (1 atm) total pressure. It is fluorescent at low pressures and rotates the plane of polarized light when placed in a magnetic field. It is also thermoluminescent, emitting visible light when heated to 500°C or higher.

Chemical Properties. Iodine, like the other halogens, is very active chemically, but less violent in its action. As the heaviest of the common halogens, it has a lower electronegativity than the others. As a result, the iodides are less stable than the other halides and the iodine can be readily replaced by the other halogens. On the other hand, its oxides and other compounds in which iodine is in a positive valence state are much more stable than those of the other halogens, and iodine can replace chlorine and bromine in such compounds. Indeed, iodine has slight metallic properties. Its crystals have an almost metallic luster, it is a much better electrical conductor than either bromine or chlorine, and its +1 and +3 valence states have distinct basic characteristics.

Iodine forms binary compounds with all other elements except sulfur, selenium, and the noble gases. It does not react directly with carbon, nitrogen, or oxygen, and only at high temperature with platinum, but it does react with most other elements. It is likely to react slowly with metals that form nonvolatile iodides, especially if they are in a massive state. Magnesium strip, for example, has been heated up to 600°C in iodine vapor without appreciable signs of reaction. On the other hand, metals like aluminum, titanium, and zirconium that form volatile iodides, react rapidly, especially when finely divided. Copper and silver are exceptions as they are readily attacked even though their iodides are nonvolatile. In general, the products formed by these reactions correspond to those formed by the other halogens. However, a number of elements with multiple valences, such as copper and iron, do not form iodides corresponding to the highest chlorides.

Iodine does not form compounds with carbon monoxide, nitric oxide, or sulfur dioxide corresponding to carbonyl, nitrosyl, and sulfuryl chlorides.

Although iodine vapor does not react directly with oxygen, a mixture of the two, passed over heated alkaline earth oxides or carbonates, forms the corresponding normal paraperiodates:

$$10\,CaO + 2\,I_2 + 7\,O_2 \rightarrow \underset{[13763\text{-}58\text{-}1]}{2\,Ca_5(IO_6)_2}$$

Iodine, like the other halogens, reacts with hydrocarbons to form iodo compounds, but in general, the equilibria are unfavorable, since the displacement step with the iodine atom is endothermic by as much as 127.7 J (30.5 cal) for methane and 25.1 J (6.0 cal) for toluene. This is so to such an extent that hydrogen iodide can be used to reduce an alkyl iodide to the hydrocarbon plus molecular iodine.

However, complete iodination can, in some cases, be achieved by preventing the formation of free hydrogen iodide by the addition of an oxidizing agent, reaction in the presence of a base, or combination with mercuric salts (13–14). An oxidizing agent has the advantage that all of the iodine is available for substitution.

If the substance to be iodinated is a base, it acts as its own neutralizing agent; thus, aniline can be iodinated directly (with a 45% yield), but the yield is almost doubled by the addition of sodium bicarbonate or mercuric oxide (13).

Iodine adds to unsaturated compounds to form polyiodo derivatives.

$$CH_2{=}CH_2 + I_2 \rightarrow \underset{[624\text{-}73\text{-}7]}{CH_2ICH_2I}$$

Complete saturation is not always attained, and the reactions are often readily reversible. The equilibrium in the ethylene reaction has been studied over a range of temperatures. The presence of free iodine sometimes favors the decomposition of polyiodo compounds. Ethylene iodide, for example, melts at 81.5°C in the absence of iodine, but in its presence decomposes into ethylene and iodine at 45°C.

Small quantities of iodine have a pronounced affect on the rates of many organic reactions including halogenations, dehydrohydroxylations, isomerizations, and pyrolytic decompositions, and have therefore been used to catalyze such reactions, sometimes with positive results (15).

Reactions in Aqueous Media. Aqueous solutions of iodine are hydrolyzed to a much smaller extent than those of the other halogens. The equilibrium constant for the reaction

$$I_2 + H_2O \rightarrow HIO + H^+ + I^-$$

is only about 5×10^{-15} at 25°C. An aqueous solution containing iodide, iodate, and free iodine or triiodide ion, has a pH of ca 7; thus, a mixture of iodide and iodate can be used to determine the free-acid content of a solution by titrating the liberated iodine.

Iodine is a mild oxidizing agent in acid solution, with an equilibrium potential of the iodine–iodide ion couple of −0.5345 V at 25°C (16). It readily oxidizes sulfite to sulfate, thiosulfate to tetrathionate, and stannous and titanous salts to stannic and titanic salts. On the other hand, ferric and cupric salts and compounds of vanadium, chromium, and manganese in their highest valence states are reduced in acid solution by iodide ion with the liberation of free iodine. Solutions of many oxidizing agents, such as chlorine, bromine, nitrous acid, and moderate concentrations of nitric acid, especially when hot, also liberate iodine from iodides, but cold neutral and slightly acid nitrate solutions do not react.

Some of these reactions are reversible under the proper conditions. In dilute solution, iodine completely oxidizes sulfur dioxide to sulfuric acid, but iodides reduce concentrated sulfuric acid to sulfur dioxide, sulfur, and even hydrogen sulfide, with the liberation of free iodine. Because of this reaction, hydrogen iodide cannot be made by treatment of a metallic iodide with concentrated sulfuric acid.

Iodine itself can be oxidized in acid solution. Concentrated nitric acid and, in more dilute solutions, chlorine, bromine, chlorates, bromates, and permanganates convert it to iodate.

$$3\,I_2 + 10\,HNO_3 \rightarrow 6\,HIO_3 + 10\,NO + 2\,H_2O$$

$$I_2 + 2\,ClO_3^- \rightarrow 2\,IO_3^- + Cl_2$$

In strong hydrochloric acid, the foregoing reagents, as well as iodic acid, oxidize iodine to iodine monochloride or to the ICl_2^- ion.

$$2\,I_2 + HIO_3 + 5\,HCl \rightarrow 5\,ICl + 3\,H_2O$$

In neutral or slightly alkaline solutions, iodine becomes somewhat more strongly oxidizing and converts trivalent arsenic to the pentavalent state. In strongly alkaline solutions, it is a powerful oxidizing agent owing to the formation of hypoiodite ion in accordance with the reaction

$$I_2 + 2\,OH^- \rightarrow I^- + IO^- + H_2O$$

Such solutions are strong iodinating agents for organic compounds, particularly aromatic compounds. However, they must be prepared as required by adding iodine directly to a mixture of alkali and the compound to be oxidized, or used shortly after preparation, as the hypoiodite rapidly decomposes to iodate and iodide, a mixture which is much less strongly oxidizing, especially in alkaline solutions.

Iodine can also be oxidized in alkaline solutions. Solutions of hypochlorite or hypobromite convert it to iodate, whereas chlorine passed into a solution of iodine and alkali oxidizes the iodine to periodate.

Some iodine reactions which take place in aqueous media do not occur in nonaqueous media. Iodine and sulfur dioxide do not react when dissolved in a mixture of anhydrous methanol and pyridine, but in the presence of water the following reaction takes place:

$$SO_2 + I_2 + 2\,H_2O \rightarrow H_2SO_4 + 2\,HI$$

This is the basis of the Karl Fischer reagent for the determination of water.

Production and Processing. In the production of iodine from brine, the first step is the clarification of the brine to remove oil and other suspended material. In one process, a silver nitrate solution is added to precipitate silver iodide, which is filtered and treated with scrap iron to form metallic silver and a solution of ferrous iodide. The silver is redissolved in nitric acid for recycle, and the solution is treated with chlorine to liberate the iodine.

In another process, chlorine is added after clarification to liberate the iodine as the free element in solution. This solution is then passed over bales of copper wire, and insoluble cuprous iodide precipitates. At intervals, the bales are agitated with water to separate the adhering iodide; the bales are then recycled. The cuprous iodide suspended in the water is filtered, dried, and shipped as such.

Much of the U.S. iodine production employs a chlorine-oxidation, air-blowout method for the recovery of iodine as the free element from natural, subsurface brines in Oklahoma and Michigan. This process is similar to that for the recovery of bromine from seawater (see Bromine). The brines, containing in excess of 100 ppm iodine in Oklahoma and 30–40 ppm iodine in Michigan, are acidified with sulfuric acid and treated with a slight excess of chlorine to liberate the iodine in a denuding tower, where the brine gives up its iodine to a countercurrent stream of air. The iodine-enriched air then passes to a second tower where the iodine is absorbed by a solution of hydriodic and sulfuric acids. This solution is treated with sulfur dioxide to reduce the iodine to hydriodic acid; part is drawn off to a reactor for the recovery of iodine, and the remainder is recirculated to the absorption tower. The liquor in the reactor is treated with chlorine and the liberated iodine is settled, filtered, and melted in a kettle under concentrated sulfuric acid. It is then either poured onto a drum flaker or cast into pigs. Brine stripped of iodine is returned to its source.

This process is also used in Japan as one of the principal recovery techniques for subsurface brines containing 60–100 ppm iodine. Here the brine is coproduced with natural gas and separated at the surface.

The fourth and newest process uses ion-exchange resins on brines which have been oxidized to liberate iodine. The liberated iodine in the form of polyiodide is adsorbed on an anion-exchange resin (Amberlite IRA-400). When the ion-exchange resin is saturated, it is discharged from the bottom of the column and then transferred to the elutriation column. Iodine is elutriated (or desorbed) with a caustic solution followed by sodium chloride. The regenerated resin is returned to the adsorption column. The elutriant, rich in iodide and iodate ions, is acidified and oxidized to precipitate iodine. The crude iodine is then separated in a centrifuge and purified with hot sulfuric acid or refined by sublimation. In Japan, two plants using this process went on stream during the period 1963–1966.

The process used by the Chilean nitrate industry differs from the others since the iodine is present as iodate. The iodine is extracted from caliche as sodium iodate, along with sodium nitrate. The iodate accumulates in mother liquors during crystallization of the nitrate, attaining a concentration of about 6 g/L. Part is then drawn off and treated with the stoichiometric quantity of sodium bisulfite solution.

$$2\ NaIO_3 + 6\ NaHSO_3 \rightarrow 2\ NaI + 3\ Na_2SO_4 + 3\ H_2SO_4$$

To this solution, now acidic because of the bisulfite reaction, is added an equivalent quantity of fresh mother liquor to liberate the iodine.

$$5\ NaI + NaIO_3 + 3\ H_2SO_4 \rightarrow 3\ I_2 + 3\ Na_2SO_4 + 3\ H_2O$$

The precipitated iodine is filtered in bag filters and the iodine-free mother liquor returned to the nitrate leaching cycle after neutralization of any excess acid with crude soda ash. The iodine cake is washed with water, pressed to reduce the moisture content, broken up, and sublimed in concrete-lined iron retorts connected to condensers made of glazed sewer tile. The product is crushed and packed in polyethylene-lined fiber drums.

Materials of Construction. High silicon iron, Stellite 6, Hastelloy C, and stainless steels types 304, 309, 316, and 317 have low corrosion rates when completely immersed in an aqueous solution containing 5% I_2 and 7.5% KI at 25°C.

Other metals and alloys are less satisfactory. A 5% solution of iodine in 95% alcohol is more severe in its attack on the stainless steels, but the rate is reduced by the addition of potassium iodide. Iron and steel are badly corroded under most conditions, but medium-carbon steel (0.21% carbon) has an extremely low corrosion rate in a solution of 26.8 g/L of iodine in benzene when air and moisture are rigorously excluded (17).

Hastelloy B gives promising results when exposed to liquid and gaseous iodine up to 400°C; molybdenum seems capable of withstanding even higher temperatures. High silicon iron, lead-coated steel, and hard lead have been recommended as materials of construction for the sublimation of iodine (18).

Niobium, tantalum, molybdenum, and their alloys are resistant to corrosion in a range from 20 to 270°C in moist iodine, liquid iodine, and gaseous iodine (19). Most metals are strongly corroded in the presence of $H_2 + I_2 + HI$ at 370–700°C, ie, under conditions of HI synthesis (see below).

Among nonmetallic materials, glass, chemical stoneware, enameled steel, acid-proof brick, carbon, graphite, and wood are resistant to iodine and its solutions under suitable conditions, but carbon and graphite may be subject to attack. Polytetrafluoroethylene withstands liquid iodine and its vapor up to 200°C although it discolors; cloths made of Saran, a vinylidene chloride polymer, have lasted for several years when used in the filtration of iodine recovered from oil-well brines (20).

Economic Aspects. Iodine plant locations are dictated primarily by the availability of natural brines or bitterns containing adequate amounts of iodine. In 1979, the United States had two iodine-producing plants, in Midland, Michigan (Dow Chemical Company), and Woodward, Oklahoma (PPG Industries and Amoco Production Company). In the former, iodine is recovered as a coproduct with salt, potash, bromine, and calcium and magnesium salts, whereas in the latter, it is the sole product recovered from subsurface brines (see also Chemicals from brine).

Investment and maintenance costs are relatively high for a new iodine plant because of the deep wells required for brine production and disposal and the corrosive nature of the plant streams. The principal material cost is for chlorine: 1 kg is required for 1.4–1.8 kg of iodine produced. Chlorine requirements in the air blow-out process for iodine are higher than those for bromine, since iodine is converted to HI, which requires a second oxidation step to crystalline iodine for separation. The main operational expenses are for pumping the brine to the surface, air blowing the iodine, and returning the iodine to the subsurface.

World production of iodine was ca 10,900 metric tons in 1977, a decline of ca 300 t from 1976, basically accounted for by a decline in Japanese output. Still, Japan provided 56% of the world total, compared to 15% from Chile, 16% from the USSR, and less than 13% from the United States (21). Imports of crude iodine into the United States are given in Table 4. By 1977, the United States increased its production capacity by more than 900 t/yr.

Japanese production of crude iodine declined to 6000 t in 1978, 2% less than 1977 output. Economic difficulties prevented full-capacity output and expansion (23). The 1980 price of crude iodine was quoted at $13.00/kg in drum lots (24).

Annual U.S. consumption amounted to 2500–3500 t in recent years, or ca 30% of the world's output. Table 5 gives the U.S. consumption for the manufacture of iodine chemicals. Table 6 gives the distribution of consumption.

Specifications and Standards; Shipping. Commercial crude iodine normally has a minimum purity of 99.5%. Impurities in the Chilean product are chiefly water, sulfuric acid, and insoluble materials. The *U.S. Pharmacopoeia XX* (25) specifies an iodine content not less than 99.8% with a maximum of 0.028% total bromine and chlorine and 0.05% nonvolatile matter. The specifications of the Committee on Analytical Reagents of the American Chemical Society allow a maximum of 0.005% total bromine and chlorine and 0.01% nonvolatile matter.

Crude iodine is packed in double polyethylene-lined fiber drums containing 45 to 90 kg. There is no specific freight classification; it is shipped as Chemicals, NOIBN, and requires no special label. Resublimed iodine is also shipped in 11.3-kg lined fiber drums, and in 0.11-, 0.45-, and 2.26-kg bottles.

Table 4. U.S. Imports of Crude Iodine[a]

Country	1975[b]	1976	1977[c]
Chile	166	300	700
Japan	2243	2640	2445
Total	*2409*	*2940*	*3145*

[a] Refs. 21–22.
[b] At $4.85/kg (av).
[c] At $4.40/kg (av).

Analytical Methods. Most analytical methods use the oxidizing power of active iodine for its determination. The results are generally expressed as an equivalent concentration of elemental iodine. The choice of method for the determination of iodide in water depends primarily on the concentration range to be determined. The catalytic reduction photometric method (26–27) is applicable to waters containing iodide concentrations of 80 μg/L or less. The leuco crystal violet method (26) is better suited for determining iodide concentrations of 50 μg/L or greater. This method can also be used to measure iodide in the presence of iodine.

In the absence of chlorides and bromides, iodides can be determined by the Volhard method, ie, by adding an excess of standard silver nitrate solution and back titrating with standard thiocyanate solution, using ferric alum as an indicator.

The amperometric titration method measures iodine residuals over 7 mg/L. It is a special adaptation of the polarographic principle wherein iodine is determined on the titrator using an acetate buffer (pH 4.0), potassium iodide solution (to improve the sharpness of the endpoint), and phenylarsine oxide titrant (26).

In dilute acid solution, iodine is usually determined quantitatively as the free element by titration with a standard sodium thiosulfate solution or sodium arsenite with a starch indicator. The iodine is oxidized to iodate with bromine or permanganate, excess alkali iodide is added, and the liberated iodine titrated. By this method, six equivalents of iodine are titrated for each equivalent originally present. The equivalence also may be determined electrometrically by the so-called dead-stop technique (28).

Table 5. U.S. Consumption of Crude Iodine, 1977[a]

Chemical	Number of plants	Crude iodine consumed	
		t	% of total
resublimed iodine	6	230	9
potassium iodide	7	562	21
other inorganic compounds	14	810	30
organic compounds	20	1080	40
Total	*31*[b]	*2682*	*100*

[a] Ref. 22.
[b] Plants producing more than one product are counted only once in the total.

Table 6. Distribution of U.S. Crude Iodine Consumption, 1978

Use	Percent
animal feed additives	18
catalysts	22
inks and colorants	14
pharmaceuticals	15
photographic film	3
sanitary and industrial disinfectants	8
stabilizers	15
miscellaneous	5

Free iodine is detected by the violet colors of its vapor and solutions in carbon tetrachloride, chloroform, or carbon disulfide (27). Colorimetric methods for the determination of small quantities involve the starch–iodine complex, the α-naphthoflavone reagent (29), the I_3^- ion formed in aqueous solution, or solutions of iodine in organic solvents (28).

The reaction of iodide ion with chlorine or bromine is the basis of the well-known iodometric method for the determination of available residual chlorine or bromine. For example, chlorine liberates free iodine from potassium iodide solutions at pH 8 or less. The liberated iodine is titrated with a standard solution of sodium thiosulfate, with a starch indicator, preferably at pH 3 to 4. The iodometric method can also be used for the determination of ozone, chlorine dioxide, dissolved oxygen, and sulfide (30–31).

Potassium iodate is a powerful oxidizing agent and is used in the determination of arsenic, antimony, mercury, thallium, tin, hydrazine, and vanadium (31). The determination of iodate itself is achieved by reduction to iodide using sulfurous acid followed by precipitation with silver nitrate solution as silver iodide. Since periodates are also reduced by sulfurous acid they are determined similarly (32). In both cases, the liberated iodine is titrated with a thiosulfate or arsenite solution (33). Thermogravimetric techniques for iodates, periodates, and the iodides of palladium, silver, lead, copper, and thallium are also available (34–35).

Organic iodine can be determined after converting the iodine into an inorganic form by heating the compound with fuming nitric acid in a sealed strong glass tube (Carius method); sodium peroxide in a combustion bomb; or a mixture of chromic acid sulfuric acid, followed by addition of phosphorous acid and distillation into a solution of sodium arsenite (36–37).

Health and Safety Factors; Regulations. Iodine is much safer to handle at ordinary temperatures than the other halogens, because it is a solid and its vapor pressure is only 1 kPa (7.5 mm Hg) at 25°C, compared to ca 28.7 kPa (215 mm Hg) for bromine and 700 kPa (6.91 atm) for chlorine. However, the vapor is irritating to the lungs and eyes and prolonged exposure to a high concentration should be avoided. Short contact of solid iodine with the skin produces a coloration similar to that obtained when tincture of iodine is applied to a wound, and prolonged contact may be harmful.

Dilute solutions of iodine have been prescribed by physicians for many years and may be taken internally in small quantities without ill effects. However, iodine itself and its more concentrated solutions, including the USP tincture, are classed as poisons because of their irritating and toxic effects on the gastrointestinal tract. Recommended antidotes are a sodium thiosulfate solution (photographers' hypo), a 1–5% aqueous salt solution, or starch, flour, or egg white mixed with water, followed by an emetic of mustard. If these are unavailable, mushes made of bread, or other starch foods may be used. If necessary, demulcent drinks of eggs or milk can be administered. Medical attention should be provided.

When a reactive organic substance is present, as in iodination reactions, iodine and its solutions can be mixed safely with ammonia solutions. However, under certain conditions, contact of gaseous ammonia or its solutions with free iodine and its halogen derivatives in any form should be avoided to prevent formation of the explosive nitrogen iodide [*13444-85-4*], NI_3. If this dark brown to black solid forms, it should be discarded while wet or destroyed by adding a reducing agent, such as ethanol, or more of the compound to be iodinated. Mixtures containing iodine and ammonium salts under neutral or acid conditions appear to be entirely safe.

Toxicity test data on hydrogen iodide and hydriodic acid have not been reported. The same precautions should be observed as when handling hydrogen chloride or bromide and their solutions.

In general, iodides are acutely toxic only when ingested in large amounts. Chronic ingestion or absorption through the skin may produce iodism, manifested by a rash, nasal drip, or headache. In severe cases, weakness, anemia, and general depression may occur (38).

Care should be taken in handling the organic iodine compounds. Again, they have the advantage that their vapor pressures are usually lower than those of the corresponding chloro and bromo compounds, but a study (39) indicates that the toxicity of methyl iodide vapor is similar to that of methyl bromide, and it should be handled with the same precautions (see also Bromine compounds).

Contamination of oxygenated compounds such as the iodates and periodates with organic matter or other combustible material should be avoided as such mixtures are likely to be explosive.

The ACGIH has established 0.1 ppm (1 mg/m^3) as the TLV for iodine. Air monitoring is necessary to detect iodine at this low concentration. Iodine vapor, even in low concentrations, is irritating to the respiratory tract, eyes, and to a lesser extent, the skin. Concentrations as low as 0.1 ppm in the air may cause some eye irritation upon prolonged exposure. Concentrations higher than 0.1 ppm cause increasingly severe irritation to the eyes and the respiratory tract, and ultimately may lead to pulmonary edema (40).

Based on its irritant nature, as well as its toxicity, OSHA has set the maximum concentration (ceiling value) of iodine vapor permitted in the working atmosphere as 0.1 ppm (41).

Regulations. At present, the EPA is reviewing its policy with regard to iodine in drinking water (42) (see also Water, municipal water treatment). The basis for current enforcement is a 1972 guideline which states that iodine imposes little, if any, hazard to consumers if the iodine-treated water is consumed for less than three weeks. In spite of these guidelines, a number of small towns in New Mexico have used iodine to disinfect their water supplies for extended periods at concentrations of 0.5–1.0 ppm with no apparent ill effects (42).

Inorganic Iodine Compounds

Iodides. Iodides of sulfur, selenium, Co(III), and those corresponding to the highest chlorides of antimony, arsenic, copper, gold, iron, lead, molybdenum, phosphorus, rhenium, thallium, tungsten, and vanadium, are unknown. However, such iodides can sometimes be stabilized by complexing the cations with solvate groups, such as NH_3 and H_2O. Thus $Cu(NH_3)_4I_2$ [*74685-94-2*] and $Co(NH_3)_6I_3$ [*13841-85-5*] are known.

The melting and boiling points of the covalent iodides of such elements as carbon, silicon, titanium, and beryllium are usually higher than those of the corresponding chlorides and bromides, but those of the ionic iodides of elements like sodium and potassium are lower. There is a gradual change from one condition to the other as the ionic character of the bond increases. Because of the difference in size between the chloride and iodide ions (Cl^- = 0.181 nm, I^- = 0.220 nm), iodides and chlorides only rarely have the same crystal structure. Among the halides of bivalent elements, for example, it is known to be the case only for barium iodide [*13718-50-8*] and nickel iodide [*13462-90-3*].

In general, the iodides are very soluble in water and many are hygroscopic. However, a few are insoluble, eg, cuprous [*7681-65-4*], lead [*10101-63-0*], silver [*7783-96-2*], and mercurous [*15385-57-6*] iodides. With the exception of cuprous iodide, they dissolve in concentrated alkali iodide solutions. Some iodides are more stable in contact with water than the chlorides and bromides. A number of iodides are also soluble in nonaqueous solvents, eg, mercuric iodide [*1344-45-2*] is more soluble in ethanol and acetone than in water. Potassium iodide [*7681-11-0*] is much more soluble in acetone than the chloride. It also dissolves in liquid sulfur dioxide and ammonia.

The iodides have less tendency to form complexes than the other halides, but complexes with mercury and a number of other elements, for example, cadmium and platinum, are known. Among the iodomercurates are $KHgI_3$ [*7783-33-7*], the basis of Nessler's reagent for the detection of ammonia and Mayer's reagent for alkaloids, and Cu_2HgI_4 [*13876-85-2*] and Ag_2HgI_4 [*7784-03-4*], which undergo allotropic transformations at 70 and 50°C, respectively, with marked changes of color, and which have been used in thermosensitive paints (see Chromogenic materials). Silver iodomercurate(II) has the highest ionic electric conductivity at ordinary temperatures of any known solid.

The iodides vary widely in their stability to heat. Nitrogen triiodide, NI_3, and its ammoniate, $NI_3.NH_3$ [*14014-86-9*], formed by treating solid iodine or KI_3 [*7790-42-3*] solutions with excess ammonia or by passing ammonia over some of the less stable iododibromides (43), when dry, decompose explosively at ordinary temperatures under the slightest shock. The phosphorus iodides (P_2I_4 [*13455-00-0*] and PI_3 [*13455-01-1*]) are probably largely dissociated at their boiling points, whereas those of the noble metals, except silver and mercury, decompose at relatively low temperatures. The partial pressure of iodine vapor over nickel iodide is greater than 270 kPa (2.66 atm) at 710°C and ca 3 kPa (0.03 atm) over cobalt(II) iodide [*15238-00-3*] at 685°C. Between 900 and 1700°C, the vapors of the iodides of titanium(II), vanadium(II), chromium(II), iron(II), copper(I), zirconium(IV), hafnium(IV), thorium(IV), and uranium(IV) dissociate to produce solid or liquid metal and free iodine. On the other hand, the iodides of the alkali and alkaline earth metals are stable even at their boiling points.

The iodides of the alkali metals and those of the heavier alkaline earths are resistant to oxygen on heating, but most others can be roasted to oxide in air or oxygen. The vapors of the more volatile iodides, such as those of aluminum and titanium(II) actually burn in air. The iodides resemble the sulfides in this respect, with the important difference that the iodine is volatilized, not as an oxide, but as the free element, which can be recovered as such.

Chlorine and bromine readily displace iodine from the iodides, converting them to the corresponding chlorides and bromides. The iodides of the iron-group metals, iron, cobalt, and nickel, are quantitatively converted to the corresponding carbonyls when heated with carbon monoxide at 200–250°C, and 270 kPa (2.66 atm) in the presence of metallic copper or silver.

Alkali iodides, dissolved in acetone and other organic solvents, react with aliphatic chloro and bromo compounds to give the corresponding iodo compounds, a convenient method of preparation for many of these substances. The alkali chlorides and bromides precipitate and can be filtered. Since the iodo compounds formed are frequently more reactive than the original halo compounds, this reaction is the basis of an important use of the alkali iodides *in situ* as catalysts for reactions between chloro and bromo compounds and metals, salts, amines, and other substances in acetone and other solvents. Addition of a small quantity of alkali iodide forms an equivalent amount of iodo compound which then reacts further and regenerates the iodide for another cycle (15).

Arylsulfonyl radicals, such as *p*-tolylsulfonyl, can be replaced by iodine in the same manner as chloro and bromo groups. An iodo derivative of cotton has been prepared by this reaction.

Hydrogen Iodide. Hydrogen iodide [*10034-85-2*], HI, is a colorless gas which fumes strongly in moist air. Its physical properties are given in Table 7.

The classical preparation of hydrogen iodide, in other than small quantities, is by direct union of hydrogen and iodine vapor in the presence of a platinum catalyst (44). Commercially, it is prepared by the reaction of H_2S or hydrazine and iodine.

$$2\,I_2 + N_2H_4 \xrightarrow{H_2O} 4\,HI + N_2$$

The latter is nearly quantitative. More recently, an electrolytic method of producing aqueous hydrogen iodide solutions has been patented (45). The electrolytic route involves feeding iodine to an aqueous catholyte liquor in an electrolytic cell, passing current from an anode through an electrically conductive anolyte and catholyte. Hydrogen iodide of the desired concentration is produced in the catholyte from which it is collected. As more HI is formed, more of the solid iodine being fed can be solubilized. Aqueous hydrogen iodide solutions for commercial use contain generally 40–55 wt % HI. In the electrolytic route, the concentration of HI may exceed the azeotropic concentration (56.9% HI) and pure HI in gaseous form can be recovered from the cell. Smaller quantities can be made by carefully adding water to a mixture of red phosphorus and iodine. A concentrated solution is obtained by passing hydrogen sulfide into a suspension of iodine in water, boiling the resulting solution to expel excess sulfide, and filtering the precipitated sulfur through glass wool or a fritted glass plate.

Table 7. Physical Properties of Hydrogen Iodide

Property	Value
mp, °C	−50.8
bp, °C	−35.3
critical temp, °C	150
critical pressure, MPa[a]	8.3
density	
liq at −4.7°C, g/cm^3	2.85
gas at 25°C, g/L	5.23
sp heat, kJ/(kg·K)[b]	
liq at −37°C	0.463
gas at 25°C, constant pressure	0.228
soly at 10°C, g/100 g H_2O	234

[a] To convert MPa to atm, divide by 0.101.
[b] To convert J to cal, divide by 4.184.

Hydrogen iodide and water form an azeotropic mixture, sp gr 1.70, which, at 101.3 kPa (1 atm) pressure, contains 56.9% HI and boils at 127°C; the composition of this mixture varies with the pressure.

Hydrogen iodide, when dissolved in water, forms only H^+ (or H_3O^+) and I^- ions, and, unlike ammonia, exhibits no tendency to take up a hydrogen ion to form an iodonium cation, H_2I^+, or salts like H_2ICl. However, many stable, crystalline iodonium salts containing organic radicals in place of the hydrogens have been prepared (46).

When heated, hydrogen iodide gas dissociates to some extent into hydrogen and iodine. Despite this, there are no reports that it is combustible. It reacts with many metals, including even copper, mercury, and silver, to form iodides with the liberation of hydrogen. It adds to double bonds in organic compounds more readily than the other hydrogen halides and reacts with organic iodo compounds to replace the iodine with hydrogen.

$$CH_3I + HI \rightarrow CH_4 + I_2$$

The concentrated aqueous solution, hydriodic acid, when pure, is a colorless, highly corrosive liquid which fumes in moist air and has a strongly acid odor. The technical grade, 47% HI, sp gr 1.5, may have a brown color owing to the presence of free iodine liberated by the action of air, but this can be prevented by the addition of a small amount of hypophosphorous acid.

Solutions of hydriodic acid, like other solutions containing iodide ion, can dissolve large quantities of iodine. A solution saturated with iodine at 25°C contains 72.85% I_2, 18.1% HI, and 9.05% H_2O, and has a specific gravity of 3.28 (44).

Hydriodic acid is thermodynamically more stable than hydrogen iodide gas because of the large negative free energy of solution (−414 kJ/kg or −98.06 kcal/kg). It is one of the strongest acids and dissolves metals, oxides, carbonates, and salts of other weak nonoxidizing acids with the formation of iodides. The concentrated acid dissolves even metallic silver and silver iodide.

Hydriodic acid reacts with the lower aliphatic alcohols to yield the corresponding iodo compounds. With higher alcohols and other hydroxy compounds, other products may be formed.

Iodine Halides. Iodine forms six well defined compounds with the other halides: ICl, ICl_3, IBr [*7789-33-5*], IBr_3 [*7789-58-4*], IF_5 [*7783-66-6*], and IF_7 [*16921-96-3*] (see also Fluorine compounds, inorganic; Bromine compounds).

Iodine Monochloride. Solid iodine monochloride [*7790-99-0*] exists in two modifications: stable ruby-red needles, d_4^0 3.86 g/cm^3; mp, 27.3°C; sp heat, 0.347 kJ/(kg·K) (0.083 kcal/(kg·K)), heat of fusion, 47.46 kJ/kg (11.39 kcal/kg); vapor pressure at 25°C, 3.8 kPa (28 mm Hg); and metastable brownish red tablets, d_4^0 3.66 g/cm^3; mp, 13.9°C. The liquid is brownish red and closely resembles bromine in appearance; d^{29} 3.10 g/cm^3; viscosity, 1.21 mm^2/s (= cSt) at 35°C; electrical conductivity, 4.60×10^{-3} at 35°C; sp heat, 0.661 kJ/(kg·K) (0.158 kcal/(kg·K)), bp, 100°C (dec); heat of vaporization, 256.4 kJ/kg (61.28 kcal/kg). It is readily supercooled below its melting point. Liquid iodine monochloride is a polar solvent and dissolves potassium and ammonium chlorides to form conducting solutions. It also dissolves large quantities of iodine and is miscible in all proportions with bromine, carbon tetrachloride, and acetic acid. It dissolves undecomposed in moderately concentrated hydrochloric acid to form a yellow solution containing $HICl_2$ [*43413-52-1*] but dilute solutions are hydrolyzed to free iodine and hydrochloric and iodic acids. When vaporized, it is largely dissociated into iodine and chlorine.

Iodine monochloride reacts with inorganic materials very much like a mixture of its elements. With organic compounds it yields iodo derivatives and hydrogen chloride. Aniline and its derivatives are readily iodinated (14). It adds rapidly to double bonds, the chlorine tending to attach itself to the carbon linked to the smallest number of hydrogens. Reactions of the iodine halides with organic compounds are less exothermic than those of the more active halogens, but they tend to start more readily because of lower activation energies (47). This is the basis for the use of iodine as a halogenation catalyst (15).

Iodine monochloride is best prepared by the direct reaction of iodine with liquid chlorine (48). Aqueous solutions are obtained by treating a suspension of iodine in moderately strong hydrochloric acid with chlorine gas or iodic acid. The liquid vigorously attacks rubber and cork, and burns skin.

Iodine Trichloride. Iodine trichloride [*865-44-1*], ICl_3, forms orange-yellow needles, d_4^{15}, 3.111 g/cm^3; it decomposes at 65°C at 101.3 kPa (1 atm) and melts at 101°C (1.62 MPa or 16 atm); electrical conductivity, 8.60×10^{-3} at 102°C. It is prepared by adding iodine to liquid chlorine (48). It is soluble in benzene, carbon tetrachloride, and in concentrated hydrochloric acid, forming $HICl_4.4H_2O$ [*12694-70-1*], but is hydrolyzed in dilute solutions like the monochloride. In general, it reacts similarly to the monochloride; with acetylene it forms chlorovinyl iododichloride [*18964-25-5*], $C_2H_2Cl_3I$.

$$HC{\equiv}CH + ICl_3 \rightarrow ClCH{=}CHICl_2$$

This compound contains a double bond and two active chlorine atoms (those attached to the iodine), and can be converted to the corresponding iodonium salt, one of the few known examples of this type among aliphatic compounds.

Polyhalides. Iodine and its halides combine with the halides of the alkali metals and other strong bases to form a series of complex salts. These substances, which have formulas such as RbI_3 [*12298-69-0*], $KI_3.H_2O$ [*7790-42-3*], $N(CH_3)_4I_9$ [*3345-37-7*], $N(C_2H_5)_4IBr_2$ [*20445-98-1*], KIBrCl [*15859-97-9*], and $KICl_4$ [*14323-44-5*], form crystals, usually orthorhombic, ranging in color from deep black through red and orange to yellow and even white (KIF_6) [*20916-97-6*]. The more stable ones crystallize in the anhydrous condition. The partial pressures of iodine and its halides over some of these compounds are very much lower than their vapor pressures.

Iodine Oxides and Derivatives. Iodine does not react directly with oxygen, but three oxides, IO_2 [*13494-92-3*], I_4O_9 [*73560-00-6*], and I_2O_5 [*12029-98-0*], have been prepared by indirect methods; derivatives of three others, I_2O [*39319-71-6*], I_2O_3 [*11085-17-9*], and I_2O_7 [*12055-74-2*], are known. The dioxide, lemon-yellow crystals, sp gr 4.2, decomposing at 85°C, and I_4O_9, a yellow hygroscopic powder, decomposing at 75°C, are of only theoretical interest.

The unknown iodine monoxide [*39319-71-6*], I_2O, is shown by a study of its derivatives to be essentially basic in character. Hypoiodous acid [*14332-21-9*], HOI, dissociation constant 4.5×10^{-13}, is so weak that it has been called iodine hydroxide. Solutions of its alkali metal salts, prepared by dissolving iodine in hydroxide solutions, are powerful oxidizing agents, but decompose rapidly to iodate and iodide (49). No record has been found of the preparation of a pure hypoiodite. A solution of iodine nitrate [*14696-81-2*], INO_3, in alcohol, prepared by mixing solutions of iodine and silver nitrate in accordance with the reaction

$$I_2 + AgNO_3 \rightarrow INO_3 + AgI$$

deposits iodine cathode when electrolyzed. The I^+ cation can be stabilized by coordination with one or two molecules of pyridine or its homologues and a number of crystalline salts of such cations, including $I(C_5H_5N)NO_3$, $(I(C_5H_5N)_2)NO_3$ [*39001-42-8*], and those of many organic acids, have been prepared (50–52).

The unknown iodine sesquioxide, I_2O_3, is also basic in its properties (50). Its most interesting derivative is the Masson and Race reagent, a suspension of iodine(III) oxysulfate [*11085-17-9*], $I_2O_3.SO_3$, prepared by adding iodine and iodine pentoxide (in the proportions to form this oxide) to concentrated sulfuric acid. It converts aromatic hydrocarbons and their more stable derivatives to the corresponding iodoso and iodonium compounds in high yields (46).

Iodine pentoxide, I_2O_5, a white crystalline solid, sp gr 4.980, decomposes into its elements at 275°C. It is by far the most stable of the halogen oxides (28). It is prepared by dehydration of iodic acid and is available commercially. It is very soluble in water, reforming the acid. It is one of the few substances that quantitatively oxidize carbon monoxide to carbon dioxide at ordinary temperatures and is used for that purpose.

$$I_2O_5 + 5\,CO \rightarrow I_2 + 5\,CO_2$$

Iodic Acid. Iodic acid [*7782-68-5*], HIO_3, colorless rhombohedral crystals (see Table 8) is the acid corresponding to iodine pentoxide and is commercially available. It differs considerably from chloric and bromic acids. It is a stable crystalline solid; a relatively weak acid, with a dissociation constant of 0.164 at 25°C; and, despite the indications of its formula, it forms acid and poly salts of the types $KH(IO_3)_2$ [*13455-24-8*], $KH_2I_3O_9$ [*20717-58-2*] and KI_3O_8 [*74869-66-2*]. Experimental evidence indicates that concentrated solutions of the acid contain species other than HIO_3, H^+, and IO_3^- possibly $(HIO_3)_2$ and $(HIO_3)_3$, but their exact nature has not yet been determined.

Table 8. Physical Properties of Iodic Acid, and Potassium and Sodium Iodate

Property	Iodic acid	Sodium iodate	Potassium iodate
mp, °C	110 (dec)		
density, g/cm^3	4.650	4.206	3.979
soly, g/100 g H_2O	310	9.0 at 20°C	9.16 at 25°C
		34.0 at 100°C	32.2 at 100°C

Aqueous solutions of iodic acid are energetic oxidizing agents (53), although not as powerful as chloric and bromic acids. They release iodine from iodides, convert ferrous to ferric salts and, in the presence of strong mineral acids such as H_2SO_4, oxidize organic compounds, eg, oxalic acids. Alkaline solutions are much less active. The following data for iodate couples in acid and alkaline solutions, respectively, at 25°C are reported (16):

$$I_2 + 6\,H_2O \rightarrow 2\,IO_3^- + 12\,H^+ + 10\,e \qquad E° = -\,1.195\text{ V}$$

$$6\,OH^- + I^- \rightarrow IO_3^- + 3\,H_2O + 6\,e \qquad E_{B°} = -0.26\ V$$

where $E_{B°}$ = half-cell potential.

Iodic acid is prepared by oxidation of iodine with fuming nitric acid or by electrolytic oxidation of a suspension of iodine in water in a diaphragm cell (54). When heated, the crystals decompose with partial melting at 110°C to form HI_3O_8 [*12134-99-5*], which yields I_2O_5 at 170°C.

Iodic acid forms salts with most metals. Those of the alkali metals and magnesium are soluble in water. The rest are sparingly soluble or insoluble, but with the exception of those of the tetravalent metals, they are soluble in dilute nitric acid. The alkali iodates are moderately resistant to heat, the potassium salt melting undecomposed, but they eventually decompose to iodides and oxygen. Barium iodate [*10567-69-8*], when heated, forms normal paraperiodate, iodine, and oxygen, whereas the other iodates yield oxide, iodine, and oxygen.

Sodium iodate [*7681-55-2*], $NaIO_3$, colorless rhombohedral crystals (see Table 8), and potassium iodate [*7758-05-6*], KIO_3, colorless monoclinic crystals (see Table 8) can be prepared readily from the iodides by electrolytic oxidation, or by passing chlorine through their alkaline solutions. They are available commercially in small quantities but are potentially available in much larger quantities as they are byproducts of one method of manufacture of the alkali iodides (20). The potassium salt, when treated with strong hydrofluoric acid, forms potassium difluoroiodate [*16087-90-4*] KIO_2F_2, colorless rhombic tablets, sp gr 3.71. Potassium acid iodate [*13455-24-8*], $KH(IO_3)_2$, colorless monoclinic or rhombohedral cyrstals, soly 9.15 pts/100 pts H_2O at 50°C, is also available commercially as it is used as an alkalimetric standard.

Periodic Acid and Periodates. Although the oxide I_2O_7 is unknown, three acids derived from it, namely metaperiodic acid [*13444-71-8*], HIO_4, dimesoperiodic acid [*14922-00-0*], $H_4I_2O_9$, and paraperiodic acid [*10450-60-9*] (considered by some to be the ortho acid), H_5IO_6, are known. The first two are formed by dehydration of the last, which is the only one stable in aqueous solutions.

Periodic acid [*10450-60-9*], H_5IO_6, colorless monoclinic crystals, mp 130°C (dec) is commercially available. It is very soluble in water, ca 371 g/100 g H_2O at 25°C (55).

Aqueous solutions of paraperiodic acid are complex. Studies (56) indicate that the important substances present at 25°C are H_5IO_6, H^+, $H_4IO_6^-$, IO_4^-, and $H_3IO_6^{2-}$. The solutions are moderately stable, and decompose only slowly into iodic acid and ozonized oxygen, a reaction which can be accelerated by the presence of colloidal platinum.

Periodic acid is prepared by electrolytic oxidation of iodic acid in a diaphragm cell (54,57). Alternatively, an alkaline solution of sodium iodate is oxidized with chlorine and the resulting sodium periodate converted to the acid via the barium salt (48).

Periodic acid differs in practically all respects from perchloric acid, the only other known member of its class. It is a relatively weak polybasic acid, a powerful oxidizing agent even in dilute aqueous solutions, and forms heteropoly acids with molybdic and similar acids.

The oxidizing power of periodic acid in solution varies with the acidity. In acid solutions, it is one of the most powerful oxidizing agents known and readily converts manganous salts quantitatively to permanganate, but in strongly alkaline solutions it is less powerfully oxidizing than hypochlorite. Estimates for the periodate–iodate couple in acid and alkaline solutions, respectively, at 25°C are as follows:

$$3\,H_2O + IO_3^- \rightarrow H_5IO_6 + H^+ + 2\,e \qquad E° = \text{ca } -1.7\text{ V}$$

$$3\,OH^- + IO_3^- \rightarrow H_3IO_6^{2-} + 2\,e \qquad E_{B°} = \text{ca } -0.7\text{ V}$$

The oxidizing potential of periodic acid solutions can therefore be varied over a relatively wide range by changing the pH. Solutions of periodic acid react with many organic compounds containing hydroxy, carbonyl, and amino groups to form iodic acid and oxidation products of the organic compounds.

Periodates of most metals have been prepared. In general, they are insoluble in water, but soluble in dilute nitric acid. When heated, they decompose into iodate and oxygen or into oxide, iodine, and oxygen, but their stability increases with the ratio of basic oxide to I_2O_7. Thus, $NaIO_4$ [*7790-28-5*], decomposes at about 300°C into sodium iodate and oxygen, but sodium paraperiodate [*18122-77-0*], Na_5IO_6, requires a temperature of over 800°C and can actually be prepared at lower temperatures by passing oxygen over a mixture of sodium iodide and oxide or hydroxide (48). The normal alkaline earth paraperiodates, eg, $Ca_5(IO_6)_2$ [*13763-58-1*], are formed similarly by passing a mixture of iodine vapor and oxygen over the heated oxides or carbonates of these metals.

Sodium metaperiodate, $NaIO_4$, colorless tetragonal crystals; d_4^{26}, 3.865 g/cm^3; dec, 300–400°C; is stable in contact with its aqueous solution above 34.5°C, at which temperature 27.1 g dissolves in 100 g water. Below this temperature, the stable phase is $NaIO_4.3H_2O$ [*13472-31-6*] or $NaH_4IO_6.H_2O$ [*34410-79-2*], colorless rhombohedral crystals; d_4^{18} 3.219 g/cm^3; solubility 18.78 pts in 100 pts of water at 25°C. Potassium metaperiodate [*7790-21-8*], KIO_4, colorless tetragonal crystals, d_4^{13}, 3.618 g/cm^3; dec, 300°C; is sparingly soluble in water, 0.51 g dissolving in 100 g at 25°C. Aqueous solutions of both salts have an acid reaction owing to hydrolysis of the metaperiodate ion to H^+ and $H_3IO_6^{2-}$ [*16211-75-9*], trihydrogen periodate. The solubilities of the sodium and potassium salts are exactly the reverse of each other. $NaIO_4$ is soluble easily in water and acid solutions, but insoluble in strong sodium hydroxide solutions, whereas KIO_4 is very soluble in strong potassium hydroxide (58).

Organic Iodine Compounds

Organic iodine compounds (59) differ markedly from their chlorine and bromine analogues. They have a higher density, lower vapor pressure, greater reactivity, and a correspondingly lower stability. Because of a tendency to lose hydriodic acid and the relatively high cost of iodine compared with that of bromine or chlorine, their use is limited to specialty applications such as pharmaceutical and organic intermediates for special syntheses. Aliphatic iodine compounds are formed by the reaction of an alcohol with phosphorus triiodide or hydriodic acid; the addition of iodine monochloride or monobromide and, in a few cases, iodine itself to an olefin; replacement reactions in which the organic compound containing chlorine or bromine is heated with an alkali iodide in a suitable solvent; and the reaction of triphenyl phosphite with methyl iodide and an alcohol. For the preparation of aromatic iodine compounds, oxidizing agents such as nitric acid, fuming sulfuric acid, or mercuric oxide react with elemental iodine and the aromatic system. Organic iodine compounds are used in relatively small quantities in industry.

Methyl Iodide. Methyl iodide, CH_3I (iodomethane), is a colorless, pungent liquid; physical properties are given in Table 9. Like other alkyl iodides, it gradually liberates free iodine which may be removed by shaking with a reducing agent such as sodium sulfite, or by storing over copper turnings or silver powder. It is slightly soluble in water and soluble in alcohol, ethyl ether, and carbon tetrachloride. It forms a constant-boiling mixture (bp 39°C) with 93 wt % methyl alcohol.

Methyl iodide is prepared by the reaction of methanol with phosphorus and iodine (60–65) or by the action of dimethyl sulfate with an aqueous iodine slurry containing a reducing agent such as powdered iron or sodium bisulfite (66–68). Other methods involve reaction of methyl alcohol with hydriodic acid, potassium iodide with methyl *p*-toluenesulfonate (69–70), and aluminum and iodine with methyl phenyl ether (71). Methyl iodide forms in high yield in the reaction of methyl alcohol with iodine and diborane (63).

Methyl iodide is an effective methylating agent, but for industrial applications, methyl chloride or bromide are preferred because of their lower cost. A number of organometallic compounds made from methyl iodide are used as intermediates in organic syntheses. Methyl mercuric iodide [*143-36-2*] (iodomethyl mercury iodide), is produced from mercury and methyl iodide in sunlight. Methylmagnesium iodide [*917-64-6*] (iodomethyl magnesium iodide), is easily formed by the reaction of methyl iodide with magnesium in ether; methyl iodide and lithium produce methyllithium; the reaction with mercury–sodium amalgam gives dimethylmercury. With cacodyl, $(CH_3)_2As_2(CH_3)_2$, iododimethylarsine, $(CH_3)_2AsI$ (dimethylarsinous iodide) [*676-75-5*], and tetramethylarsonium iodide [*5814-20-0*], $(CH_3)_4AsI$, are obtained; the latter reacts with dimethylzinc forming $(CH_3)_5As$. With amines, methyl iodide gives methylamine derivatives; with an excess of methyl iodide the quaternary ammonium iodide is formed.

$$RNH_2 \xrightarrow{CH_3I} RNHCH_3 \xrightarrow{CH_3I} RN(CH_3)_2 \xrightarrow{CH_3I} R\overset{+}{N}(CH_3)_3\ I^-$$

The action of a base on quaternary ammonium salts is the basis of the Hofmann elimination reaction for the preparation of olefins.

Table 9. Physical Properties of Organic Iodine Compounds

Property	Methyl iodide [*74-88-4*]	Methylene iodide [*75-11-6*]	Iodoform [*75-47-8*]	Ethyl iodide [*75-03-6*]	Iodo-benzene [*591-50-4*]
mp, °C	−66.1	5	119	−108.5	−31.4
bp, °C	42.5	180		72.2	188.6
		106–107[a]			
density, d_4^{20}, g/cm^3	2.279	3.325	4.008	1.933	1.832
n_D	1.5290^{20}	1.7538^{24}		1.5137^{20}	1.621
soly at 20°C, g/100 mL H_2O	1.4	1.42	0.01[b]	0.4	insol

[a] At 9.1 kPa (68 mm Hg).
[b] At 25°C; 7.8 g in 100 mL ethanol; 13.6 g in 100 mL ether.

Quaternary phosphonium salts result from the addition of methyl iodide to trisubstituted phosphines. These compounds are intermediates in the preparation of alkenes by the Wittig reaction.

$$R_3P + CH_3I \rightarrow R_3\overset{+}{P}CH_3I^-$$

quaternary phosphonium salt

$$R_3\overset{+}{P}CH_3I^- \xrightarrow[\text{ether}]{n\text{-}C_4H_9Li} R_3P{=}CH_2 \xrightarrow{(CH_3)_2C{=}O} (CH_3)_2C{=}CH_2 + R_3PO$$

Complexes are formed with dimethyl sulfide and dimethyl sulfoxide

$$(CH_3)_2S + CH_3I \rightarrow (CH_3)_3\overset{+}{S}I^-$$

trimethylsulfonium iodide [*2181-42-2*]

$$(CH_3)_2SO + CH_3I \rightarrow (CH_3)_3\overset{+}{S}(O)I^-$$

trimethylsulfoxonium iodide

Alkylation of compounds with active methylene groups occurs readily with methyl iodide. Such reactions are important in the preparation of pharmaceutical intermediates and in organic syntheses. Methyl alkyl ethers are obtained with sodium alkoxides and mixed sulfides with thiols. Aromatic hydrocarbons are synthesized by the Wurtz-Fittig reaction of an aryl halide, sodium, and methyl iodide.

Methyl iodide may be identified as methylisothiourea picrate ($C_6H_2(NO_2)_3OH$).$HNC(SCH_3)NH_2$, mp, 224°C (72). Thiourea (1 g), the halide (1 g), and 100 mL ethyl alcohol are refluxed for 5 min, and picric acid is added.

Exposure to liquid methyl iodide or to the vapor should be avoided. Contact of the iodide with the skin produces burning, itching, and blisters (73). Signs of poisoning are nausea, vomiting, diarrhea, oliguria, vertigo, slurred speech, and visual disturbance. In several cases, ataxia, drowsiness, and coma ensue. Several fatalities have been reported (74). Adequate ventilation and avoidance of the vapors is essential.

Methyl iodide has been used to study sulfur linkages in vulcanized rubber (75). The methyl iodide reacts with simple sulfur compounds at rates that depend upon the type of sulfur bond and the residues attached to the sulfur.

In 1979, three companies reported a total production of less than 23 metric tons at $4.08/kg in 145-kg drums.

Methylene Iodide. Methylene iodide, CH_2I_2 (diiodomethane), is a liquid with a yellow to brown color due to traces of iodine. Its physical properties are given in Table 9. Methylene iodide is slightly soluble in water and soluble in alcohol and ether.

It is prepared conveniently by the reaction of sodium arsenite and iodoform with sodium hydroxide (76).

$$CHI_3 + Na_3AsO_3 + NaOH \rightarrow CH_2I_2 + NaI + Na_3AsO_4$$

Other methods of preparation involve the action of iodine, sodium ethoxide, and hydriodic acid on iodoform, and the oxidation of iodoacetic acid [*64-69-7*] with potassium persulfate, $K_2S_2O_8$ (77). It can also be prepared from methylene chloride and potassium iodide (78).

Methylene iodide in ether reacts with magnesium to give methylenebis(magnesium iodide) [*27329-48-2*], $CH_2(MgI)_2$ (diiodo-μ-methylenemagnesium iodide). This decomposes with water to give methane. With mercury, iodomethyl mercuric iodide [*141-51-5*], iodo(iodomethyl)mercury, ICH_2HgI, and di(iodomercuric)methane [*27329-48-2*] (methylenebis(iodomercury)), $CH_2(HgI)_2$, are formed. Iodomethylation occurs with compounds containing an active methylene group (79). With malonic ester and excess sodium methylate, methylene malonic ester, $CH_2{=}C(COOC_2H_5)_2$ is obtained. Methylene iodide is an effective reagent for ring-closure reactions achieved by alkylation of active methylene compounds. Cyclopropane derivatives may be prepared from olefins using methylene iodide and a zinc–copper couple (80).

$$R_2C{=}CR_2 + CH_2I_2 + Zn \longrightarrow \text{(cyclopropane with } R,R \text{ on each of two carbons)} + ZnI_2$$

Methylene iodide is used in refractive index work and as a heavy medium for gravity separations in the laboratory (see Gravity concentration).

Iodoform. Iodoform, CHI_3 (triiodomethane), is a yellow crystalline solid with a characteristic pungent odor. It has long been of medicinal interest and was formerly used extensively because of its germicidal action, but has been displaced by more effective techniques. Table 9 gives its physical properties. It is soluble in chloroform, benzene, and glycerol.

Iodoform may be prepared by treating any compound containing an acetyl group linked to hydrogen or carbon with alkali and iodine. The reaction is so characteristic that it serves as a test to determine microgram quantities of organic compounds (81). Iodoform is manufactured by electrolysis of acetone or alcohol solutions of a mineral iodide in the presence of sodium carbonate. It is easily prepared in excellent yield by the redistribution reaction of chloroform and methyl iodide (82).

Iodoform oxidizes arsenites to arsenates, antimonites to antimonates, and stannites to stannates.

$$CHI_3 + Na_3AsO_3 + NaOH \rightarrow CH_2I_2 + Na_3AsO_4 + NaI$$

Iodoform is determined by treating the substance to be analyzed with silver nitrate, ether, and nitric acid. Silver iodide is precipitated quantitatively, filtered, washed, dried, and weighed.

Ethyl Iodide. Ethyl iodide, C_2H_5I (iodoethane), is a colorless liquid which, like methyl iodide and other organic iodine compounds, darkens due to free iodine. Its physical properties are given in Table 9. Ethyl iodide is sparingly soluble in water, but completely miscible with ethyl alcohol and ether, and partially miscible with benzene and carbon tetrachloride. Its chemical properties and reactions are similar in nature and scope to those of methyl iodide.

Ethyl iodide may be prepared by reactions similar to those used for the preparation of methyl iodide. It is used in the production of pharmaceuticals and as an intermediate for the preparation of tin(IV) by extraction of stannic iodide (83).

Iodobenzene. Iodobenzene, C_6H_5I (phenyl iodide), a pale-yellow liquid, is almost insoluble in water but soluble in ethyl alcohol, ether, and chloroform. Its physical constants are given in Table 9.

Iodobenzene is readily prepared by the reaction of iodine and benzene in the presence of an oxidizing agent. It may also be prepared from benzenediazonium sulfate and potassium iodide (84).

Perhaps the most important use for iodobenzene is in the synthesis of iodine compounds in which iodide exists in a positive oxidation state (85), such as the iodoso compounds, RIO, the iodoxy compounds, RIO_2, and the iodonium salts, R_2IX. All these types are characteristic of iodine. Thus, iodobenzene reacts with chlorine to form iodobenzene dichloride [*932-72-9*], $C_6H_5ICl_2$ (dichlorophenyl iodide), whereas the other halobenzenes do not react with chlorine in this manner. Iodosobenzene [*536-80-1*], C_6H_5IO (iodosylbenzene), is formed by the reaction of iodobenzene dichloride with an alkali. It is also formed by iodobenzene diacetate [*3240-34-4*] and a strong base (86). Iodosobenzene diacetate is made from iodobenzene and 30% peroxyacetic acid. Iodoxy compounds can be made from iodoso compounds by heating in the presence of water followed by steam distillation to remove the volatile iodine compound formed in the same reaction.

$$2\ o\text{-}CH_3C_6H_4IO \rightarrow o\text{-}CH_3C_6H_4IO_2 + o\text{-}CH_3C_6H_4I$$

[*64297-59-2*] [*16825-70-0*] [*615-37-2*]

The positive iodine compounds have been extensively studied but have no industrial applications.

Uses

At present, well over half of the iodine produced is utilized in industrial applications, but significant amounts also find use in nutrition, sanitation, and medicine.

Industrial. In photography (qv), one of the oldest uses, the sensitive silver salt in rapid negative emulsions contains up to 7% or more of the iodide. Some dyes contain iodine, including 4′,5′-diiodofluorescein [*38577-97-8*] rose bengal [*11121-48-5*] and erythrosin [*15905-32-5*] (see Xanthene dyes), and members of the cyanine group (see Cyanine dyes). Erythrosin (tetraiodofluorescein) is an orthochromatic sensitizer for photographic emulsions and also a certified food colorant (see Colorants for foods, drugs, and cosmetics). Iodine is also utilized in the production of high purity metals such as titanium, silicon, hafnium, and zirconium (87).

Iodine and its compounds are very active catalysts for many reactions (88). When heated with small amounts of iodine, rosins, tall oil, and other wood products are converted to more stable forms (89–90). Iodine has been used with a tin salt as a catalyst in the hydrogenation of coal and its distillation products (91–92), and has been recommended as a catalyst for the production of drying oils from unsaturated animal fats (93–94). Nickel iodide has been used in Germany as a catalyst for the addition of carbon monoxide to organic compounds, presumably because the iodine catalyzes the conversion of the nickel salt to nickel carbonyl even in aqueous solutions. Alkalimetal and other iodides are effective catalysts in reactions involving aliphatic chloro and bromo compounds, such as the preparation of cyclopropane from 1,3-dichloropropane and metallic zinc (95). Iodine is used in the dehydrogenation of butane and butylene to 1,3-butadiene (qv) (96).

Iodine and certain iodides, for example, titanium tetraiodide [*7720-83-4*], are employed for producing stereospecific polymers, such as polybutadiene rubber (97). Iodine catalyzes the conversion of amorphous selenium to the black, semiconducting "metallic" modification, and is used for this purpose in the manufacture of photoelectric cells and electric rectifiers.

A solution of iodine in toluene has been recommended for the separation of olefins from paraffin hydrocarbons and from each other (98). The olefins dissolve when the gases are washed with this solution and are recovered separately by controlled absorption and stripping.

An iodine compound has been added to the alkali electrode of a rechargeable dry cell to prevent formation of excessive gas pressure upon recharge (99).

A spectacular use of an iodine compound is that of silver iodide as a smoke for the seeding of clouds to induce rainfall. The addition of cadmium or lead iodide to electric-generator and motor brushes greatly improves their efficiency and life under extreme conditions of low humidity and high altitudes and has also given effective service on high speed mainline electric locomotives (100).

Iodine can be utilized in the treatment of naphtha to yield a motor fuel with an improved octane number (101).

Ozone formation is inhibited or reduced by the addition of ten parts of iodine added to 100 million (10^8) parts of smog; ozone is the chief irritating substance in smog (102). The toxicity of the reaction products was not determined (see Air pollution).

The addition of iodine to an aromatic hydrocarbon such as *n*-butylbenzene results in the formation of charge-transfer complexes which display outstanding effectiveness as lubricants for hard-to-lubricate metals (103), such as titanium or steels (see also Lubrication and lubricants). Iodine also shows excellent utility for controlling bacterial growth in cutting-oil emulsions (104).

Sodium iodate improves the quality of bread made from certain flours (see Bakery processes and leavening agents).

Nutrition. Because of the requirements of the thyroid gland, a normal person needs about 75 mg iodine per year. To ensure this supply, it is added to salt for human consumption and to animal feedstuffs. The use of iodized salt has markedly reduced the incidence of goiter in certain iodine-deficient areas of the United States and elsewhere.

In the United States, average daily iodine intake now ranges between 200 and 700 μg and varies considerably from region to region (105). It is very difficult to define the normal range of iodine intake in humans and, despite efforts to provide iodine supplementation in many geographic areas of the world, endemic iodine deficiency and its attendant goiter remain a world health problem (106). Exposure to excess iodine may sometimes lead to the development of thyroid disease. This unusual type of iodide-induced goiter has been found, for example, in approximately 10% of the population of a Japanese island where fishermen and their families consume large quantities of an iodine-rich seaweed and have an iodine intake as high as 200 mg/d (105). On the other hand, relatively low iodine intakes of approximately 50 μg/d without goiter endemia, have been found in parts of western Europe. Whenever the iodine intake is constantly less than 50 μg/d, thyroxine (T_4) synthesis is reduced, with a compensatory increase in thyrotropin (TSH) secretion. This leads to increased trapping of iodide by the thyroid, a compensatory increase in the serum ratio of triiodothyronine to thyroxine (T_3/T_4), and goiter (107) (see also Hormones).

Iodized protein in stock and fowl feeds prevents various diseases and increases yields of milk and eggs. The four iodine derivatives accounting for nearly all iodine consumption in this area are potassium iodide, calcium iodate [*7789-80-2*], $Ca(IO_3)_2$, ethylenediamine dihydriodide [*5700-49-2*], (EDDI), $HI.NH_2CH_2CH_2NH_2.HI$, 1,2-ethanediamine dihydriodide, and iodophores (iodine complexes) (see Food additives; Pet and other livestock feeds).

Thyroid Hormones. Iodine enters the thyroid follicular cells as inorganic iodide and is transformed through a series of metabolic steps into the thyroid hormones, thyroxine [*51-48-9*], *O*-(4-hydroxy-3,5-diiodophenyl) 3,5-diiodo-L-thyrosine (T_4) and 3,5,3′-triiodothyronine [*6893-02-3*], *O*-(4-hydroxy-3-iodophenyl)-3,5-diiodo-L-thyrosine (T_3). The steps in this metabolic sequence may be characterized as follows: (*1*) active transport of iodide, (*2*) iodination of tyrosyl residues of thyroglobulin [*9010-34-8*], (*3*) coupling of iodotyrosine molecules within thyroglobulin to form T_4 and T_3, (*4*) proteolysis of thyroglobulin, with release of free iodotyrosines and iodothyronines, and secretion of iodothyronines into the blood, and (*5*) deiodination of iodotyrosines within the thyroid and reutilization of the liberated iodide (3) (see also Thyroid and antithyroid preparations).

Medicinals. Iodine and iodine-containing compounds and preparations are employed extensively in medicine, eg, as antiseptics, as drugs administered in different combinations in the prophylaxis and treatment of certain diseases, and as therapeutic agents in various thyroid dyscrasias and other abnormalities.

Radioactive iodine has been utilized successfully for the treatment of cancer of the thyroid (108) (see Radioactive drugs). Patients have remained apparently cancer-free for periods up to thirteen years following surgery. Treatment with radioactive iodine has also been used to treat patients with certain types of heart disease, such as angina pectoris, congestive heart failure, tachycardia, and auricular fibrillation (109). The iodine is taken up by the thyroid gland, the body's metabolism is slowed, and the work load on the heart decreased. Organic compounds containing iodine in the molecule are employed as x-ray contrast media (see Radiopaques).

Iodine-131 is the most widely available and used isotope for the diagnosis and therapy of thyroid disorders (see Radioactive tracers). The half-life ($t_{1/2}$) is 8.05 days with a maximum beta energy of 0.606 MeV and gamma energy of 364 keV. Its chief disadvantage lies in the fact that 90% of its radioactive emissions are beta particles which contribute no diagnostically useful information for scintillation and imaging of the thyroid. Iodine-125 has a physical half-life of 60 days and decays by electron capture with emission of 27-keV and 35-keV photons which are useful in delineation of superficial lesions. Iodine-123 is considered the agent of choice for imaging and uptake measurements despite the high cost. It's half-life is only 13 h and it emits a 159-keV photon (110).

Sanitation. Small amounts of iodine are used in detergent sanitizers complexed with nonionic surfactants (111). These so-called iodophors display the germicidal properties of iodine, are free of odor, do not stain, and have low irritating properties. They are used as disinfectants in dairies, laboratories, and food-processing plants, and for the sanitation of dishes in restaurants. The reaction product of lanolin and iodine shows utility as a germicide (112).

A polyvinyl pyrrolidinone–iodine complex [*25655-41-8*], PVP–iodine, has been used extensively in hospitals and elsewhere because of its germicidal, bactericidal, fungicidal, and generally disinfecting properties (113). It is sold as a solution which contains about 10% available, or active, iodine and about 5% inactive iodine, in the form of iodide ion (see Disinfectants and antiseptics; Industrial antimicrobial agents).

Water Purification. Iodine effectively disinfects water against bacteria, viruses, and cysts. Globaline tablets were developed for the disinfection of small or individual water supplies in the U.S. Army during World War II. Studies of the disinfection of public water supplies with iodine were initiated in 1963 at three prisons in Lowell, Florida (114). These studies showed that concentrations up to 5 ppm of iodine were not deleterious to health and that 1 ppm was sufficient to safely disinfect a water supply at the high pH values encountered in many treated supplies, with virtually no loss of effectiveness because of iodate build-up. Little is known whether adverse effects could occur through long-term intake of water treated with disinfection levels of iodine. Administration of approximately 2 mg of iodide in a single dosage induces the Wolff-Chaikoff effect (115), ie, the inhibitory effect of iodine on hormone formation by the thyroid. Commercially available iodinators control potentially dangerous organisms by passing a side stream of water through a bed of crystalline iodine at pH 7.5 and 20–26°C to provide 0.5 ppm iodine in a water supply; chlorine (9–12 ppm) is added to the water entering the treatment plant (see Water, municipal water treatment).

Iodine may also be used as an effective microbicide for swimming pools (114,116–117). It acts like free chlorine and is superior to chloramines in bacterial, virucidal, and cysticidal efficiency (see also Water, treatment of swimming pools). In addition, iodine shows decreased reactivity with organic matter, less variation in microbicidal efficiency over the typical pH range observed in pools, and potential for regeneration of free iodine through application of a suitable oxidant (eg, chlorine, ozone, or potassium monopersulfate, $KHSO_5$). Other advantages over chlorine are said to be longer life, lack of odor, no bleaching action, and minimal eye irritation. Unfortunately, dependable control of algae proliferation in well illuminated swimming pools has not been accomplished through the use of iodine alone. Controlled field experiments have demonstrated that compounds such as prometryne and terbutryne are effective algicides compatible (nonreactive) with free iodine as a microbicide (118). Chlorine can be used in combination with iodine as a means of controlling algae and to reoxidize residual iodide to free iodine.

BIBLIOGRAPHY

"Iodine and Iodine Compounds" in *ECT* 1st ed., Vol. 7, pp. 942–972, by A. C. Loonam, Deutsch and Loonam, for Chilean Iodine Educational Bureau, Inc., and M. G. Gergel and M. Revelise, Columbia Organic Chemicals Co., Inc.; "Iodine Preparations" in *ECT* 1st ed., Vol. 7, pp. 972–981, by L. Gershenfeld, Philadelphia College of Pharmacy and Science; "Iodine and Iodine Compounds" in *ECT* 2nd ed., Vol. 11, pp. 847–870, by A. W. Hart, The Dow Chemical Co. and M. G. Gergel and J. Clark, Columbia Organic Chemicals Co., University of South Carolina.

1. S. C. Werner and S. H. Ingbar, eds., *The Thyroid, A Fundamental and Clinical Text,* 4th ed., Harper and Row, Hagerstown, Md., 1978, p. 532.

2. J. F. McClendon, *Iodine and the Incidence of Goiter,* University of Minnesota Press, Minneapolis, Minn., 1939, pp. 8–40.
3. Ref. 1, pp. 3–4, 31.
4. A. G. Collins, *Chem. Geol.* **4,** 169 (1969).
5. F. C. Kracek, *J. Phys. Chem.* **35,** 417 (1931).
6. J. Kleinberg and A. W. Davidson, *Chem. Rev.* **42,** 601 (1948).
7. J. H. Hildebrand, H. A. Benesi, and L. M. Mower, *J. Am. Chem. Soc.* **72,** 1017 (1950).
8. J. C. Bailar, Jr., H. J. Emeleus, R. Nyholm, and A. F. Trotman-Dickenson, eds., *Comprehensive Inorganic Chemistry,* 1st ed., Pergamon Press, Ltd., Elmsford, N.Y., 1973, pp. 1197–1229.
9. H. A. Benesi and J. H. Hildebrand, *J. Am. Chem. Soc.* **71,** 2703 (1949).
10. M. J. Blander and M. F. Fox, *Chem. Rev.* **70,** 59 (1970).
11. F. A. Cotton and G. Wilkinson, *Advanced Inorganic Chemistry, A Comprehensive Text,* 3rd ed., Interscience Publishers, a division of John Wiley & Sons, Inc., New York, 1972, pp. 463–464.
12. G. Jander and K. H. Bandlow, *Z. Phys. Chem. (Leipzig)* **A191,** 321 (1943).
13. L. Jurd, *Aust. J. Sci. Res. Ser. A* **2,** 111 (1949).
14. R. G. Shepherd and C. E. Fellows, *J. Am. Chem. Soc.* **70,** 157 (1948).
15. *Iodine, Its Properties and Technical Applications,* Chilean Iodine Educational Bureau, Inc., New York, 1951.
16. W. M. Latimer, *The Oxidation States of the Elements and Their Potentials in Aqueous Solutions,* Prentice-Hall, Inc., Englewood Cliffs, N.J., 1938, pp. 56–60.
17. L. G. Gindin and M. V. Pavlova, *Dokl. Akad. Nauk SSSR* **69,** 377 (1949).
18. I. Ya. Klinov, *Khim. Prom.* (10–11), 17 (1944); *Chem. Abstr.* **40,** 2097 (1946).
19. V. I. Ginzburg and O. I. Kabakova, *Zashch Met.* **5,** 627 (1969).
20. F. G. Sawyer, F. G. Ohman, and F. E. Lush, *Ind. Eng. Chem.* **41,** 1547 (1949); Industrial and Engineering Chemistry Staff, eds., *Modern Chemical Processes,* Vol. 1, Reinhold Publishing Corp., New York, 1950, p. 26.
21. K. P. Wang, *Mineral Facts and Problems,* Bulletin 667, U.S. Dept. of Interior, Bureau of Mines, Washington, D.C., 1975, pp. 515–523.
22. *Minerals Yearbook,* U.S. Dept. of Interior, Bureau of Mines, U.S. Govt. Printing Office, Washington, D.C., 1977.
23. *Mineral Industry Surveys, Annual Advanced Summary, Iodine in 1978,* U.S. Dept. of Interior, Bureau of Mines, Washington, D.C., Aug. 31, 1979, pp. 1–5.
24. *Chem. Mark. Rep.,* 19, 36 (June 30, 1980).
25. *The United States Pharmacopeia XX, (USP XX-NF XV),* The United States Pharmacopeial Convention, Inc., Rockville, Md., 1980.
26. M. C. Rand, A. E. Greenberg, and M. J. Taras, eds., *Standard Methods for the Examination of Water and Waste Water,* 14th ed., APHA, AWWA, and WPCF, Washington, D.C., 1975, pp. 396–406.
27. V. A. Stenger and L. S. Ettre in F. D. Snell and L. S. Ettre, ed., *Encyclopedia of Industrial Chemical Analysis,* Vol. 14, Interscience Publishers, a division of John Wiley and Sons, Inc., New York, 1971, pp. 584–606.
28. A. J. Downs and C. J. Adams in Ref. 8, pp. 1230–1231.
29. L. Gershenfeld and B. Within, *Science* **110,** 2854 (1954).
30. Ref. 26, pp. 316–321, 350, 456, 505.
31. J. Bassett, R. C. Denney, G. H. Jeffery, and J. Mendham, eds., *Vogel's Textbook of Quantitative Inorganic Analysis,* 4th ed., Longham Inc., New York, 1978, pp. 370–390.
32. Ref. 31, p. 496.
33. F. D. Snell and L. S. Ettre, eds., *Encyclopedia of Industrial Chemical Analysis,* Vol. 14, Interscience Publishers, a division of John Wiley & Sons, Inc., New York, 1971, pp. 607–613.
34. R. J. Meyer and E. Pietsch in *Gmelin's Handbuch der Anorganischen Chemie, Iodine,* System No. 8, Verlag Chemie, Berlin, 1933, pp. 218–219.
35. C. Duval, *Inorganic Thermogravimetric Analysis,* 2nd ed., Elsevier Publishing Co., New York, 1963, pp. 521–523.
36. A. L. Chaney, *Anal. Chem.* **22,** 939 (1950).
37. J. W. Thomas and co-workers, *Anal. Chem.* **22,** 726 (1950).
38. N. I. Sax, *Dangerous Properties of Industrial Materials,* 4th ed., Van Nostrand Reinhold Co., New York, 1975, p. 32.
39. M. Buckell, *Brit. J. Ind. Med.* **7,** 122 (1950).
40. *Occupational Health and Safety,* Vol. 1, International Labor Office, Geneva, 1971.

41. *29 Code of Federal Regulations 1910.1000,* Table Z-1, July 1, 1978.
42. *Am. City County* **93,** 32 (1978).
43. H. W. Cremer and D. R. Duncan, *J. Chem. Soc.,* 2750 (1930).
44. C. F. Powell and I. E. Campbell, *J. Am. Chem. Soc.* **69,** 1227 (1947).
45. U.S. Pat. 4,053,376 (Oct. 11, 1977), W. W. Carlin (to PPG Industries, Inc.).
46. R. B. Sandin, *Chem. Rev.* **32,** 249 (1943).
47. P. H. Groggins, ed., *Unit Processes in Organic Synthesis,* 2nd ed., McGraw-Hill Book Co., Inc., New York, 1938, pp. 156–157, 202–204, 207–208.
48. H. S. Booth, ed., *Inorganic Syntheses,* Vol. 1, McGraw-Hill Book Co., Inc., New York, 1939, pp. 165–175.
49. M. L. Josien and C. Courtial, *Bull. Soc. Chim. Fr.,* 374 (1949).
50. J. Kleinberg, *J. Chem. Educ.* **23,** 559 (1946).
51. R. A. Zingaro and co-workers, *J. Am. Chem. Soc.* **71,** 575 (1949).
52. R. A. Zingaro, C. A. Vander Werf, and J. Kleinberg, *J. Am. Chem. Soc.* **72,** 5341 (1950).
53. W. Hurka, *Mikrochem. ver. Mikrochim. Acta* **31,** 83 (1943).
54. H. H. Willard and R. R. Ralston, *Trans. Electrochem. Soc.* **62,** 239 (1932).
55. P. P. Gyani and B. P. Gyani, *J. Indian Chem. Soc.* **26,** 239 (1949).
56. C. E. Crouthamel, A. M. Hayes, and D. S. Martin, *J. Am. Chem. Soc.* **73,** 82 (1951).
57. G. F. Smith, *Analytical Applications of Periodic Acid and Iodic Acid and Their Salts,* 5th rev. ed., G. Frederick Smith Chemical Co., Columbus, Oh., 1950.
58. A. E. Hill, *J. Am. Chem. Soc.* **50,** 2678 (1928).
59. D. F. Banks, *Chem. Rev.* **66,** 243 (1966).
60. R. Adams and V. Voorhees, *J. Am. Chem. Soc.* **41,** 706 (1919).
61. H. S. King, *Proc. Trans. Nova Scotia Inst. Sci.* **16**(2), 87 (1924).
62. H. S. King in A. H. Blatt, ed., *Organic Syntheses,* Coll. Vol. II, John Wiley & Sons, Inc., New York, 1943, p. 399.
63. N. Nagai, *J. Pharm. Soc. Jpn.* **407,** 1 (1916).
64. R. B. Reynolds and H. Atkins, *J. Am. Chem. Soc.* **51,** 280 (1929).
65. Brit. Pat. 565,452 (Nov. 10, 1944), A. I. Vogel.
66. U.S. Pat. 2,899,471 (Aug. 11, 1959), F. R. Huber and L. M. Schenck (to General Aniline and Film Corp.).
67. U.S. Pat. 3,053,910 (Sept. 11, 1962), F. R. Huber and L. M. Schenck (to General Aniline and Film Corp.).
68. Brit. Pat. 879,350 (Oct. 11, 1961), (to General Aniline and Film Corp.).
69. P. H. Peacock and B. K. Nenon, *Quart. J. Indian Chem. Soc.* **2,** 404 (1925).
70. V. Rodionov, *Bull. Soc. Chim.* **39,** 323 (1926).
71. M. T. Dangyan, *J. Gen. Chem. USSR* **11,** 1215 (1941).
72. W. J. Levy and N. Campbell, *J. Chem. Soc.,* 1442 (1939).
73. M. Buckell, *Brit. J. Ind. Med.* **7,** 122 (1950).
74. A. Garland and F. E. Camps, *Brit. J. Ind. Med.* **2,** 209 (1945).
75. M. L. Selker and A. B. Kemp, *Ind. Eng. Chem.* **36,** 16 (1944).
76. R. Adams and C. S. Maruez in H. Gilman and A. H. Blatt, eds., *Organic Synthesis,* 2nd ed., Coll. Vol. I, John Wiley & Sons, Inc., New York, 1943, p. 359.
77. W. H. Perkin, Jr. and H. A. Scarborough, *J. Chem. Soc.* **119,** 1400 (1921).
78. G. Panopovlos and A. Petzetakis, *Chem. Ztg.* **54,** 310 (1930).
79. R. Valters and G. Vanags, *Dokl. Akad. Nauk SSSR* **146,** 359 (1962).
80. E. P. Blanchard and H. E. Simmons, *J. Am. Chem. Soc.* **86,** 1337 (1964).
81. A. G. Kallianos and J. D. Molo, *Anal. Chem.* **34,** 1174 (1962).
82. H. Soroos and J. B. Hinkamp, *J. Am. Chem. Soc.* **67,** 1642 (1945).
83. A. D. Paul and J. A. Gibson, *Anal. Chem.* **36,** 2321 (1964).
84. H. J. Lucas and E. R. Kennedy in A. H. Blatt, ed., *Organic Syntheses,* Coll. Vol. II, John Wiley & Sons, Inc., New York, 1943, p. 351.
85. R. B. Sandin, *Chem. Rev.* **32,** 249 (1943).
86. J. G. Sharefkin and H. Saltzman in B. C. McKusick, ed., *Organic Syntheses,* Vol. 43, John Wiley & Sons, Inc., New York, 1963, pp. 60, 62, 65.
87. R. F. Rolsten, *Iodine Metals and Metal Iodides,* John Wiley & Sons, Inc., New York, 1961.
88. *Iodine Abstracts and Reviews,* Chilean Iodine Educational Bureau, Inc., New York.

89. U.S. Pat. 2,299,577 (Oct. 20, 1942), T. Hasselstrom and E. A. Brenhan (to G and A Laboratories, Inc.).
90. U.S. Pat. 2,311,386 (Feb. 16, 1943), T. Hasselstrom (to G and A Laboratories, Inc.).
91. K. Gordon, *Chem. Age (London)* **55,** 761, 795 (1949).
92. U.S. Pat. 2,452,271 (Mar. 15, 1949), H. R. Storch and L. I. Hirst (to the U.S.A.).
93. U.S. Pats. 2,411,111–2,411,113 (Nov. 12, 1946), A. W. Ralston and O. Turinsky (to Armour and Co.).
94. U.S. Pat. 2,498,133 (Feb. 1, 1950), A. W. Ralston, O. Turinsky, and L. Van Akkeren (to Armour and Co.).
95. H. B. Haas and co-workers, *Ind. Eng. Chem.* **28,** 1178 (1936).
96. U.S. Pat. 3,106,590 (Oct. 8, 1963), C. W. Bittner (to Shell Oil Co.); U.S. Pat. 3,119,881 (Jan. 28, 1964), R. L. Hodgson (to Shell Oil Co.).
97. Brit. Pat. 920,244 (Mar. 6, 1963), (to Phillips Petroleum Co.); Brit. Pat. 931,440 (July 17, 1963), (to Phillips Petroleum Co.).
98. U.S. Pats. 2,392,739–2,392,740 (Jan. 8, 1946), J. T. Horeczy and E. F. Wadley (to Standard Oil Development Corp.).
99. U.S. Pat. 2,991,325 (July 4, 1961), K. Kordesch (to Union Carbide Corp.).
100. H. M. Elsey, *Am. Inst. Chem. Engrs. Paper 45-108,* (1945).
101. U.S. Pat. 2,921,013 (Jan. 12, 1960), R. D. Mullineaux and J. H. Raley (to Shell Development Co.).
102. F. W. Hamilton, M. Levine, and E. Simon, *Science* **140,** 190 (1963); *Chem. Eng. News* **40**(39), 68 (1962).
103. U.S. Pat. 3,228,880 (Jan. 11, 1966), R. S. Owens and R. W. Roberts (to General Electric Co.).
104. U.S. Pat. 3,244,630 (Apr. 5, 1966), M. I. Shiekh (to Ford Motor Co.).
105. Ref. 1, pp. 528–530.
106. Ref. 1, p. 537.
107. *Brit. Med. J.*, 1566 (1977).
108. T. P. Haynie, M. M. Nofat, and W. H. Beierwalter, *J. Am. Med. Assoc.* **183,** 303 (1963); *Sci. News Lett.* **83**(7), 112 (1963).
109. E. Corday, H. L. Jaffe, and D. W. Irving, *Am. J. Cardiol.* **6,** 952 (1960); *Sci. News Lett.* **79**(2), 24 (1961).
110. Ref. 1, p. 297.
111. C. A. Lawrence, C. M. Carpenter, and A. W. C. Naylor-Foote, *J. Am. Pharm. Assoc. Sci. Ed.* **46,** 500 (1957); J. L. Wilson, W. G. Mizuno, and C. S. Bloomberg, *Soap Chem. Spec.* **36**(12), 100, 105, 141 (1960).
112. U.S. Pat. 3,152,951 (Oct. 31, 1964), S. D. Perlman.
113. U.S. Pat. 4,128,633 (Dec. 5, 1978), D. H. Lorenz and E. P. Williams (to GAF Corp.).
114. A. P. Black and co-workers, *J. AWWA* **60**(1), 69 (1968).
115. A. G. Vagenakis and L. E. Braverman, *Symposium on Current Concepts of Thyroid Disease, Medical Clinics of North America,* Vol. 59, 1975, p. 1075.
116. S. L. Chang and J. C. Morris, *Ind. Eng. Chem.* **45,** 1109 (1953).
117. A. P. Black, J. B. Lackey, and E. S. Lackey, *Am. J. Public Health* **49,** 1060 (1959); O. E. Byrd and co-workers, *Public Health Rep.* **78,** 393 (1963); M. S. Favero and C. H. Drake, *Public Health Rep.* **79,** 251 (1964).
118. E. L. Nilson and R. F. Unz, *Appl. Environ. Microbiol.* **34,** 815 (1977).

CHARLES J. MAZAC*
PPG Industries

* Current employment: Celanese Chemical Company, Inc.

IODINE VALUE. See Fats and fatty oils; Carboxylic acids.

IODOACETIC ACID. See Acetic acid.

IODOFLUOROHYDROCARBONS. See Fluorine compounds, organic.

ION EXCHANGE

Ion exchange is the reversible interchange of ions between a solid and a liquid in which there is no permanent change in the structure of the solid, which is the ion-exchange material. Ion exchange is used in water softening and deionization. It also provides a method of separation that is useful in many chemical processes and in analyses. It has special utility in chemical synthesis, medical research, food processing (qv), mining, agriculture, and a variety of other areas.

The utility of ion exchange rests with the ability to use and reuse the ion-exchange materials. For example, in water softening:

$$2\,\overline{R}Na^{+} + Ca^{2+} \rightleftharpoons \overline{R}_2Ca^{2+} + 2\,Na^{+}$$

The exchanger $\overline{R}$ in the sodium-ion form is able to exchange for calcium and, thus, to remove calcium from hard water and to replace it with an equivalent quantity of sodium. Subsequently, the calcium-loaded resin may be treated with a sodium chloride solution, regenerating the sodium form so that it is ready for another cycle of operation. The regeneration reaction is reversible; the ion exchanger is not permanently changed. Millions of liters of water may be softened per cubic meter of resin during an operating period of many years.

Ion exchange occurs in a variety of substances, eg, silicates, phosphates, fluorides, humus, cellulose (qv), wool (qv), proteins (qv), alumina, resins, lignin (qv), living cells, glass (qv), barium sulfate, and silver chloride. However, there are uses for ion exchange materials that depend on properties other than the interchange of ions between liquid and solid phases. Properties and uses of ion exchange are discussed in refs. 1–10. Ion exchange has been used on an industrial basis since ca 1910 with the introduction of water softening using natural and, later, synthetic zeolites (see Molecular sieves). Sulfonated coal, developed for industrial water treatment, was the first ion-exchange material that was stable at low pH. The introduction of synthetic organic ion-exchange resins in 1935 resulted from the synthesis (11) of phenolic condensation products containing either sulfonic or amine groups which could be used for the reversible exchange of cations or anions. Recent synthesis descriptions are given in refs. 12–14 (see Phenolic resins).

A variety of functional groups have been added to the condensation or addition polymers used as the backbone structures. Porosity and particle size have been controlled by conditions of polymerization. Physical and chemical stability have been modified and improved. As a result of these advances, the inorganic exchangers (mineral, greensand, and zeolites) have been almost completely displaced by the resinous types except for some analytical and specialized applications. However, the synthetic zeolites are used as molecular sieves.

Physical Properties of Resins

Conventional ion-exchange materials contain ion-active sites throughout their structure with a uniform distribution of activity, as a first approximation. A cation-exchange resin with a negatively charged matrix and exchangeable positive ions (cations) is shown in Figure 1. Ion-exchange materials are sold as granules or spheres of the proper size and uniformity to meet the needs of a particular application. The majority is prepared and sold in spherical (bead) form, from about 40 μm to 1.2 mm (400–16 mesh) in diameter. Crush strength of individual beads can exceed 200 g/bead. In the water-swollen state, ion-exchange resins typically show a specific gravity of 1.1–1.5. The bulk density as installed in a column includes a normal 35–40% voids volume for a spherical product. Bulk densities in the range of 560–960 g/L (35–60 lb/ft^3) are typical for wet resinous products. Ion-exchange material structures should be able to resist breakage that may result from handling or from osmotic shock (a rapid change in solution environment).

Hydraulic. ***Bed Expansion.*** Backwashing of an ion-exchange column to remove fine particles of exchanger and dirt impurities and to reduce the pressure drop across the column is a standard operating practice. Backwashing above a certain minimum rate causes partial fluidization of the resin bed, an increase in the void volume between the resin particles, and an increase in the bulk volume occupied by the resin. This increase in bed volume (bed expansion) is measured at a specified cross-sectional flow rate and temperature. Typical data are shown in Figure 2.

Pressure Drop. The pressure drop for the required flow through an exchanger bed is related to particle size, shape, uniformity, and compressibility as well as to solution viscosity and flow rate. The pressure drop characteristics of a spherical cation exchanger of two size ranges are shown in Figure 3.

Chemical Properties of Resins

Capacity. Ion-exchange capacity may be expressed in a number of ways. Total capacity, ie, the total number of sites available for exchange, normally is determined after converting the resin by chemical regeneration techniques to a given ionic form. The ion then may be chemically removed from a measured quantity of the resin and quantitatively determined in solution by any one of several conventional analytical methods. Total capacity may be expressed on a dry weight, wet weight, or wet volume basis. The dry weight capacity (meq/g—milliequivalents per gram, dry weight) of the resin is the most precise measure of the extent of chemical substitution during synthesis, as may be mathematically determined by comparison with a theoretical 1:1 reaction. For example, one sulfonic acid group per benzene ring on a styrene–8% divinylbenzene (DVB) resin would show a dry weight capacity of about 5.35 meq/g. The wet weight capacity (meq/g wet resin) includes consideration of the solvent (water) uptake. The water uptake of a resin and, thus, its wet weight and wet volume capacities are dependent on the nature of the polymer backbone as well as on the environment in which the sample is placed. Variations of dry weight and wet volume capacities with cross-linkage are shown in Figure 4 for a sulfonic resin.

Operating capacity is a measure of the useful performance obtained with the ion-exchange material when it is operating in a column under a prescribed set of conditions. It is determined by the inherent (total) capacity of the resin, the level of regeneration (the extent to which the resin has been converted to the proper ionic form), the composition of solution treated, resin specificity for one ion with respect to another, the flow rates through the column, temperature, particle size, and numerous other factors. An example is shown in Figure 5 for the case of water softening with a standard sulfonic resin at several regenerant levels.

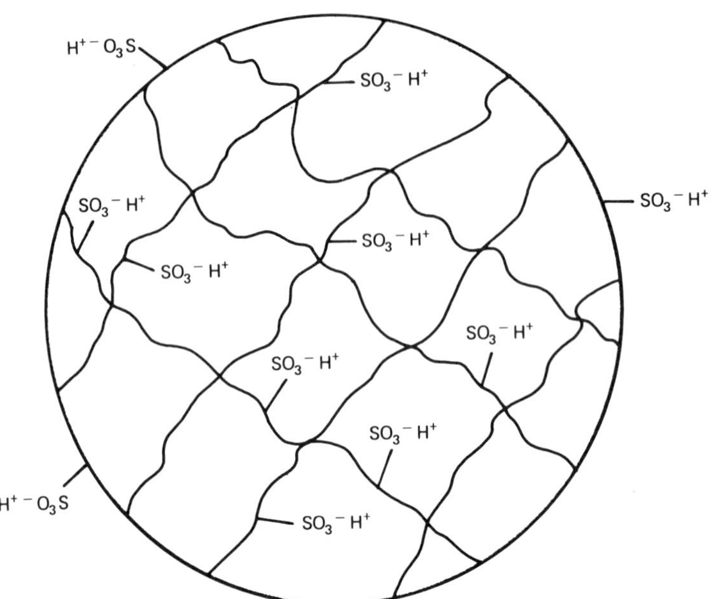

Figure 1. Cation-exchange resin schematic showing negatively charged matrix and exchangeable positive ions.

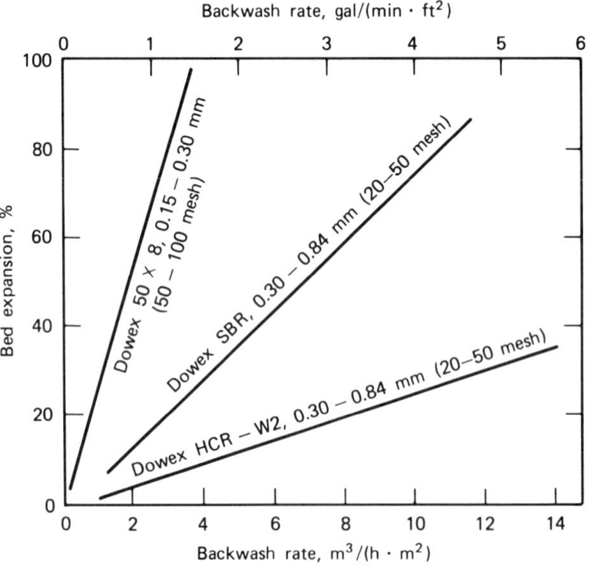

Figure 2. Backwash expansion characteristics at 25°C.

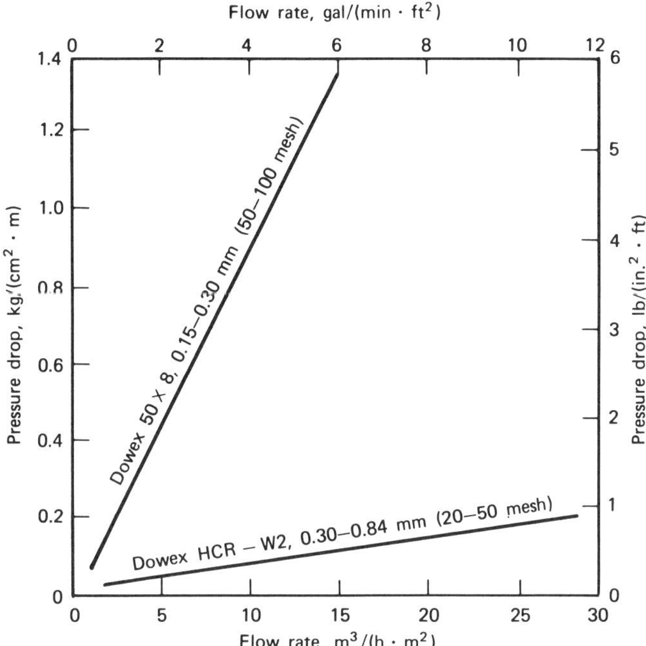

Figure 3. Head loss characteristics at 25°C.

Swelling Equilibria. The solvent retention capacity of an ion exchanger is a reproducible, equilibrium quantity that is dependent upon ion-exchange capacity, ionic form, the kind of solvent used, the composition of the solution, the extent of cross-linkage, relative humidity, and temperature (1). Water swelling of an ion exchanger is primarily a hydration of the fixed ionic groups and increases with an increase in capacity to the limits imposed by the polymer network. Resin volumes change with conversion to ionic forms of differing degrees of hydration; thus, for the sulfonic acid-type cation exchanger, there is a volume change with the monovalent ion species, $Li^+ > Na^+ > K^+ > Cs^+ > Ag^+$. With polyvalent ions, hydration is reduced by the cross-linking action; therefore, $Na^+ > Ca^{2+} > Al^{3+}$. In more concentrated solutions, less solvent is taken up, owing to greater osmotic pressure. This may be illustrated with Figure 6 in which water swelling of the hydrogen form of a sulfonic acid resin is related to cross-linkage and to the concentration of the hydrochloric acid solution (15). The increase in resin swelling with increase in temperature, though not great, is significant in certain operations. The water uptake of resins that are not immersed in water is dependent on relative humidity (see Fig. 7) especially as 100% relative humidity is approached (16).

Ionic Equilibria. Ion-exchange reactions are reversible. By washing a resin with an excess of electrolyte, the resin can be converted entirely to the desired salt form:

$$\overline{\mathrm{R}}A^+ + B^+ \rightarrow \overline{\mathrm{R}}B^+ + A^+$$

However, with a limited quantity of solution B^+ in batch contact, a reproducible equilibrium is established which is dependent on the proportions of A^+ and B^+ and

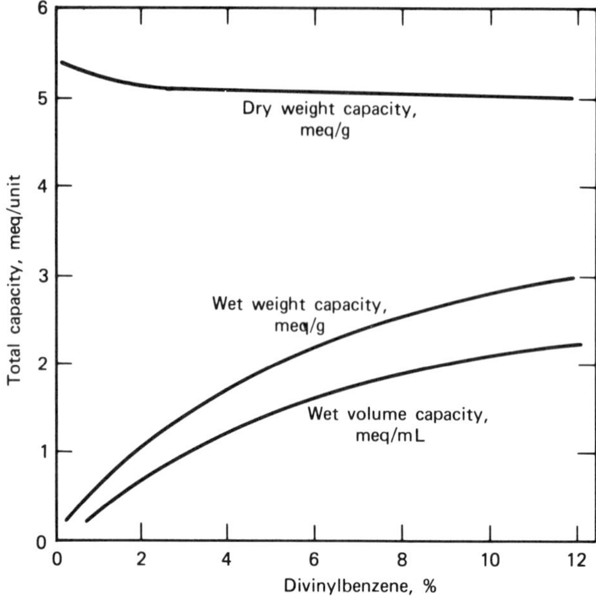

Figure 4. Total capacity vs cross-linkage (% divinylbenzene) polystyrene sulfonic acid resin, H^+ form.

on the selectivity of the resin. The selectivity coefficient, K_A^B, for this reaction is given by:

$$K_A^B = \frac{\overline{m}_B}{\overline{m}_A} \cdot \frac{m_A}{m_B}$$

where m and $\overline{m}$ refer to ionic concentrations in solution and resin phase, respectively. (This selectivity coefficient is not the same as the equilibrium constant which is related to activities rather than concentrations. A corrected selectivity coefficient is obtained by multiplying K_A^B by $\overline{\gamma}B\gamma A/\overline{\gamma}A\gamma B$ where γ and $\overline{\gamma}$ refer to the activity coefficients at equilibrium in solution and in the resin, respectively.)

The ion-exchange resin is regarded as a soluble electrolyte solution that is restrained only by the network cross-linkages (17). For example, in the case of cation resins, anions are almost completely excluded from the resin phase. The converse is true for anion resins. The ionic activity product of any permeant electrolyte is the same within the resin as it is without. Another theory of equilibrium bases resin selectivities on the swelling of the polymer network that accompanies the incorporation of different ions (18). The resin is selective for that ion (hydrated ion) occupying the least volume. The equation is that of Hooke's law.

A refined treatment of the above-mentioned theories includes both an activity coefficient term for chemical interaction and a swelling term:

$$\ln K_A^B = \ln (\overline{\gamma}A/\overline{\gamma}B) - \ln (\gamma A/\gamma B) + \P(\overline{V}_A - \overline{V}_B)/\mathrm{RT}$$

where ¶ is the swelling or osmotic pressure and $\overline{V}_A$ and $\overline{V}_B$ are the partial molal volumes of ion A and ion B within the resin. Many other factors, eg, ion pairing, solubilization, and salting-out effects influence the equilibrium. In addition, the resin phase

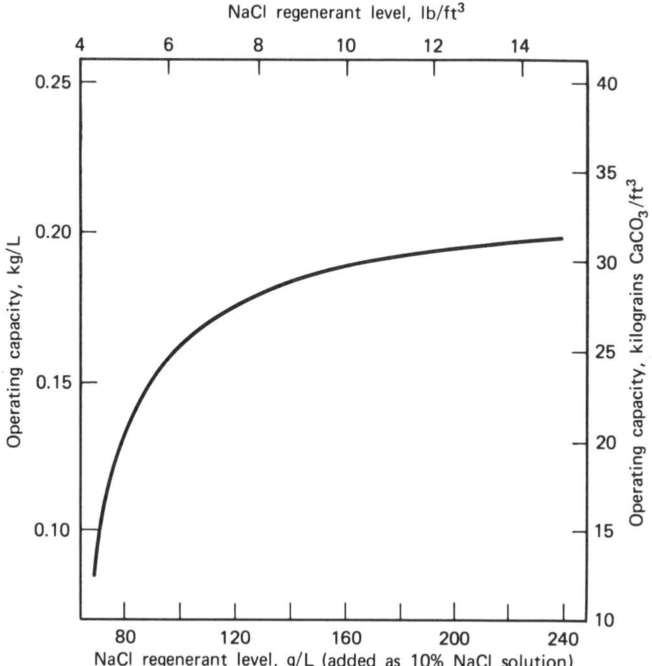

Figure 5. Operating capacity vs regenerant level for sodium-cycle operation, sulfonic acid resin.

is not truly homogeneous at the molecular level. The nonexchanging ion (co-ion) and neutral molecules may influence the exchanging ion. In addition, neutral molecules are sorbed to varying degrees (see Ion Exclusion Purification).

Kinetics. The speed with which ion exchange takes place determines the portion of the equilibrium capacity that may be utilized in a practical dynamic situation. In column operations, rates determine the extent of deviation from the theoretical elution curve. It is the very slow kinetics as well as solubility problems of some of the inorganic exchangers that limit their use in competition with the resinous types.

Overall exchange rates may be influenced by a change in solvent nature and content, particle size, temperature, and the functional group (kind and concentration). Exchange rates are most rapid in water systems, and become increasingly slow with less polar solvents. This is true because the solutes are more highly ionized in the more polar solvents as are the ion-exchange resins. Similarly, exchange rates are more rapid in lower cross-linked resins of the same inherent dry-basis capacity because of their higher moisture content. The ion-exchange process involves diffusion through the film of solution that is in close contact with the resins and diffusion within the resin particle. Film diffusion is rate-controlling at low concentrations and particle diffusion is rate-controlling at high concentrations (see Diffusion separation methods). Whether film diffusion or particle diffusion is the rate-controlling mechanism, the particle size of the resin also is a determining factor. Elevated temperatures, of course, increase exchange rates (see Chemical Processing; Sugar Separations and Purifications). The kind of functional group and its degree of dissociation under a given set of conditions greatly affects exchange rates where particle diffusion is controlling, but has no affect on film diffusion. Furthermore, a high degree of substitution of the functional groups on the inert polymer matrix directly enhances overall reaction rates.

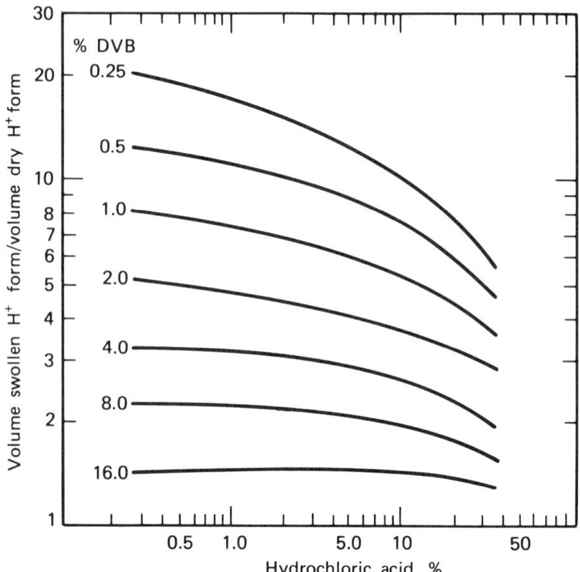

Figure 6. Swelling of polystyrene sulfonic acid resin (H^+ form) in hydrochloric acid. Parameter: % divinylbenzene (cross-linkage).

Stability. Extremely strong oxidizing agents, eg, boiling nitric acid or chromic acid/nitric acid mixtures, rapidly degrade the polymer matrix. Slower degradation with oxygen may be induced catalytically. For this reason, certain metal ions, eg, iron, manganese, and copper, should be minimized in an oxidizing solution. With cation exchangers, attack is principally on the polymer backbone; with anion exchangers, attack first occurs on the more susceptible functional groups. Highly cross-linked structures have an extended useful life because of the great number of sites that must be attacked before swelling reduces the useful volume-based capacity and produces unacceptable physical properties, eg, crush-strength reduction and pressure-drop increase.

The limits of thermal stability are imposed by the strength of the carbon-nitrogen bond in the case of anion resins. This strength is sensitive to pH, ie, a low pH favors enhanced stability. The quaternary ammonium salts are the least stable; a temperature limitation of 50°C often is recommended for hydroxide cycle operations (see Quaternary ammonium compounds). The tertiary amines are the most stable; good performance may be maintained up to 100°C (see Amines). Cation resin stability also is dependent on pH; the stability to hydrolysis of the carbon–sulfur bond diminishes with a lowering of pH. Sulfonic acid resins cannot be recommended for use above 150°C in the presence of water. Salt forms of anhydrous cation resins are stable to temperatures up to at least 250°C where depolymerization commences.

Attack by gamma radiation can cause a variety of simultaneous reactions, eg, polymerization, depolymerization, oxidation, and rupture of carbon–sulfur or carbon–nitrogen bonds. The observed results include gas evolution, bead swelling, weight loss, and appearance of weak-acid capacity.

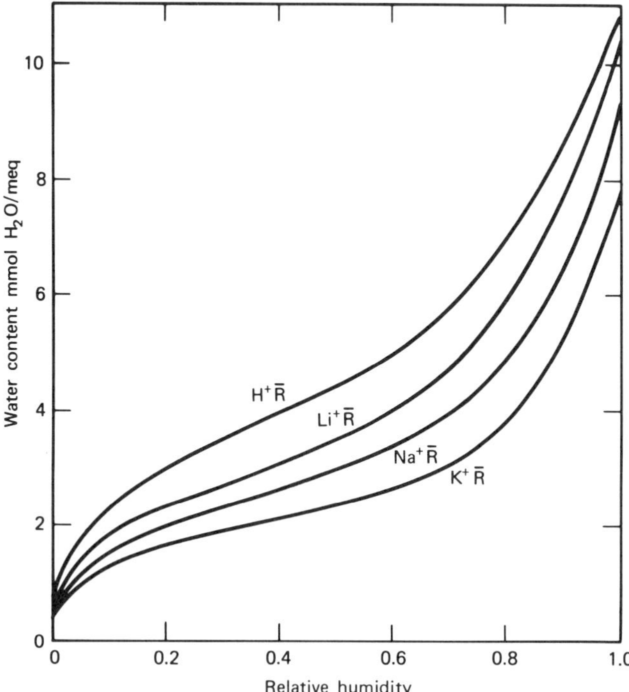

Figure 7. Water-sorption isotherms for alkali-metal salt forms and nominal 8% divinylbenzene, polystyrene sulfonic acid cation-exchange resin.

Materials

Inorganic. Inorganic ion-exchange materials (19) include both the naturally occurring materials such as the mineral zeolites (eg, sodalite and clinoptilolite), the greensands, and clays (eg, the montmorillonite group), and synthetic products such as the gel zeolites, the hydrous oxides of polyvalent metals (eg, hydrated zirconium oxide), and the insoluble salts of polybasic acids with polyvalent metals (eg, zirconium phosphate).

Synthetic Organic Products. ***Cation-Exchange Resins.*** The weak-acid, cation-exchange resins are based primarily on acrylic or methacrylic acid that has been cross-linked with a difunctional monomer, eg, DVB. The manufacturing process may start with the ester of the acid in suspension polymerization, followed by hydrolysis of the resulting product to produce the functional acid group. Other weak-acid resins have been made with phenolic or phosphonic functional groups but are not commercially significant (see Acrylic acid; Methacrylic acid).

The weak-acid resin has a high affinity for the hydrogen ion and, thus, is easily regenerated with strong acids. This property, however, limits the region in which salt splitting can occur to above pH 4. The acid-regenerated resin exhibits a high inherent capacity for the alkaline earth metals only when those metal ions are associated with alkalinity. Only limited capacities for the alkali metals are obtained with alkalinity other than hydroxide being present. No significant salt splitting occurs with neutral salts. However, when the resin is not protonated (eg, if it has been neutralized with

CH_3
|
$CH{=}CH_2$ + $CH{=}CH_2$ $\xrightarrow[\text{catalyst}]{\text{polymerization}}$
|
COOR
methacrylate

$\cdots{-}C(CH_3)(COOR)CH_2CHCH_2{-}\cdots$ (with $\cdots{-}CHCH_2{-}\cdots$) $\xrightarrow[OH^-]{\text{hydrolysis}}$ $\cdots{-}C(CH_3)(COO^- Na^+)CH_2CHCH_2{-}\cdots$ (with $\cdots{-}CHCH_2{-}\cdots$)

sodium hydroxide), softening is effective, even in the presence of high salt background.

The strong-acid resins of commercial significance are sulfonated copolymers of styrene and DVB (see Styrene plastics; Vinyl polymers). Sulfuric acid, sulfur trioxide, and chlorosulfonic acid have each been utilized for sulfonation.

$CH{=}CH_2$ (styrene) + $CH{=}CH_2$, $CH{=}CH_2$ (divinylbenzene) $\xrightarrow[\text{catalyst}]{\text{polymerization}}$

$\cdots{-}CHCH_2CHCH_2{-}\cdots$, $\cdots{-}CHCH_2{-}\cdots$ $\xrightarrow[\text{swelling agent}]{\text{sulfonating acid}}$ $\cdots{-}CHCH_2CHCH_2{-}\cdots$, $SO_3^-H^+$, $SO_3^-H^+$, $\cdots{-}CHCH_2{-}\cdots$

These materials are characterized by their ability to exchange cations or split neutral salts and are useful across the entire pH range.

Anion-Exchange Resins and Acid Adsorbers. The ability of the weak-base resins to sorb acids depends on their own basicity and the pK of the acid involved. A variety of base strengths are obtained depending on the nature of the amine functionality. Primary, secondary, and tertiary amine functionality, or mixtures of them, can be put into various structures ranging from epichlorohydrin–amine condensates and acrylic polymers, to styrene–DVB copolymers. These resins are capable of sorbing strong acids in good capacity but are limited by kinetics. In some instances, kinetics have been improved by the incorporation of about 10% strong-base capacity. The resulting resins are intermediate-base resins; they combine the good capacities of the weak-base resins with the superior kinetics required for acceptable rinse, eg, following caustic regeneration when used for water demineralization (see Water).

Improvement in the kinetic properties of weak-base resins may be obtained by modification of the copolymer matrix with the appropriate diluent during copolymer manufacture to produce a macroporous structure with true porosity and high surface area. This structure has almost completely displaced the gel structure used in the original styrene–DVB-based, weak-base resins.

Strong-base, anion-exchange resins—especially those based on styrene–DVB copolymer—are classed as type I and type II. Type I is a quaternized amine product made by the reaction of trimethylamine with the copolymer after chloromethylation with chloromethyl methyl ether (CMME).

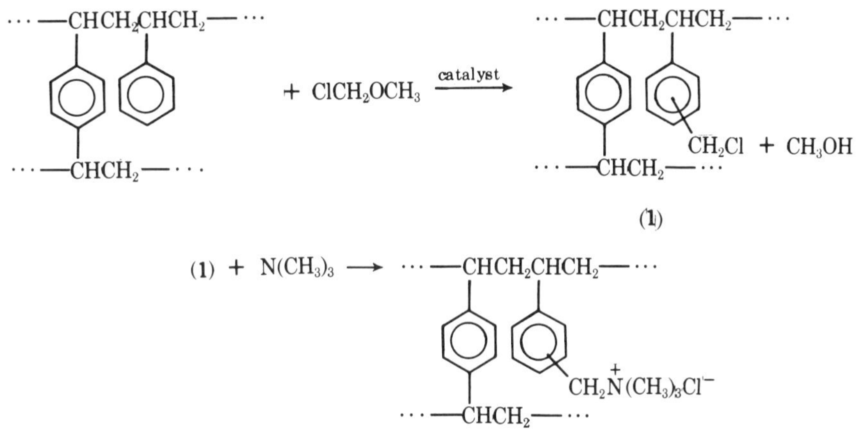

$$\cdots\!-\!CHCH_2CHCH_2\!-\!\cdots \text{ (styrene–DVB copolymer, } C_6H_4 \text{ and } C_6H_5 \text{ rings; } \cdots\!-\!CHCH_2\!-\!\cdots) + ClCH_2OCH_3 \xrightarrow{\text{catalyst}} \cdots\!-\!CHCH_2CHCH_2\!-\!\cdots \text{ (ring bearing } CH_2Cl) + CH_3OH$$

(1)

$$(1) + N(CH_3)_3 \longrightarrow \cdots\!-\!CHCH_2CHCH_2\!-\!\cdots \text{ (ring bearing } CH_2\overset{+}{N}(CH_3)_3Cl^-) ;\ \cdots\!-\!CHCH_2\!-\!\cdots$$

The type I functional group is the most strongly basic functional group available and has the greatest affinity for the weak acids that commonly are removed during a water demineralization process (eg, silicic acid and carbonic acid). However, the efficiency of regeneration of the resin to the hydroxide form is somewhat lower, particularly when the resin is exhausted with monovalent anions, eg, chloride and nitrate. The caustic regeneration efficiency of a type II resin is considerably greater than that of type I. Type II functionality is obtained by the reaction of the styrene–DVB copolymer with dimethylethanolamine. This quaternary amine has lower basicity than that of the type I resin, yet it is high enough to remove the weak-acid anions for most applications. The chemical stability of the type II resins is not as good as that of the type I resins, the type I resins being favored for the high temperature applications where chemical stability differences are most apparent. Quaternary amine functionality has been introduced into pyridine and acrylate polymers with limited commercial application (see Pyridine and pyridine derivatives).

Other Functional Groups. Ion-exchange resins with special functional groups have been made for specific applications. Of interest to the hydrometallurgical industry are a variety of resins having chelating ability and which are particularly applicable for the selective exchange of various heavy metals from alkaline earth and alkali metal solutions (20) (see Chelating agents). The iminodiacetic acid functional group (21) has been added to styrene–DVB gel or macroporous matrices. The resulting resins form metal chelate complexes having stability constants similar to the soluble chelate counterpart, iminodiacetic acid. Resins that are highly selective for copper in the presence of iron in solution incorporate pyridine derivatives and are operable at lower pH values than are normally realized with the iminodiacetic acid-type of chelate resin

(22). Mercury-ion-selective resins containing sulfhydryl groups have been produced to solve a specific pollution problem associated with the presence of mercury in wastewaters from various plants utilizing mercury compounds. The mercury-selective resins must have very high stability constants since mercury can exist in wastewater in a variety of forms ranging from metallic mercury, oxide or sulfide colloids, mercuric or mercurous cation, or complex mercuric chloride anions, etc (23).

Ion-retardation resins have been synthesized by polymerizing a functional monomer within the structure of an ion-exchange resin of the opposite charge. For example, the commercially available ion-retardation resin used for the removal of salt from caustic is made by polymerizing acrylic acid within a type I strong-base, anion-exchange resin (24–25). The resulting internally neutralized functional groups no longer exhibit normal salt-splitting, ion-exchange capacity, but have the ability to hold ion pairs by association. This capacity is a function of the concentration of the solution and, thus, is readily reversed by water washing (26) (see Ion Retardation).

In a concerted effort to desalt brackish water by ion-exchange processes without the cost of chemical regenerants, polymer structures have been synthesized with differing degrees of ionization and, thus, of capacity at different temperatures, yielding thermally regenerable ion-exchange resins (27–28). Commercially available resins capable of about 80% salt reduction from softened brackish waters consist of composite resins containing weak acid and weak base functionality within the same bead. The components exist as tiny particles or fragments in the composite structure but lack the intertwining on the molecular scale that is characteristic of the ion-retardation resins (see Water, supply and desalination).

A resin that is capable of selective removal of boron from water and brines is commercially synthesized by attaching an *N*-methyl glucamine functionality to a styrene–DVB matrix in much the same manner as other anion-exchange resins are produced (29). This resin also exhibits an affinity for iron as ferrate and is capable of use for the removal of iron from caustic.

Polymer Matrix. The structure and porosity of an ion-exchange resin are determined principally by the conditions of polymerization of the backbone polymer. Porosity determines the size of the species, molecule or ion, that may enter a specific structure and its rate of diffusion and exchange. There also is a strong interrelationship between the equilibrium properties of swelling and ionic selectivity.

In the most usual mass polymerization (by suspension) systems, the ratio of monofunctional and difunctional monomers and nonpolymerizable diluents may be varied over a wide range. For example, a conventional gel-type, styrene-based, ion exchanger is built on a matrix prepared by copolymerizing styrene and commercial DVB at various ratios. Commercial DVB contains approximately 55% DVB, 35% ethylvinylbenzene, and 10% saturates (principally diethylbenzene and naphthalene). The effect of the diethylbenzene normally is small because of the low level at which it is added.

In these systems, porosity is inversely related to the DVB cross-linking. However, the polymer structure has no appreciable porosity until it is swollen in a suitable solvating medium. Cross-linking polymers, as well as the ion-exchange resins derived from them, swell to a well-defined and reproducible degree in the appropriate solvent system, such as toluene for the copolymer or water for the ion exchanger. Porosity may be measured in terms of the volume increase on exposure to the solvent. This is plotted against cross-linkage in Figure 8.

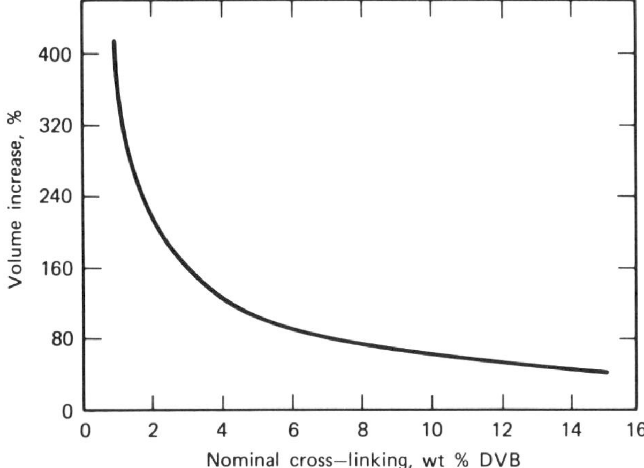

Figure 8. Solvent swelling of styrene–divinylbenzene copolymer beads in toluene at 70°C.

Mass polymers that are prepared in the presence of large amounts of nonpolymerizable diluents were first utilized in the preparation of ion-exchange membranes. Macroporous (macroreticular) ion-exchange resins have become prominent since ca 1960 because of some of their interesting and unusual properties (30–31). These resins have pores of a considerably larger size than those of the more conventional gel-type resin. Pore diameters are up to several hundred nanometers and their surface area may reach 500 m^2/g or higher. The resins also are made in bead form with special modifications of the kind and ratios of diluents. The macroporous polymers exhibit little volume change in solvent if prepared with a sufficient degree of cross-linkage. The pores may be varied tremendously in both size and uniformity. In general, good solvents give more uniform and smaller pores than do the poorer solvents. Pore diameters up to 0.1 μm or larger are possible (32).

Ion-exchange resins that are produced from these macroporous polymers exhibit small volume change on transfer from a highly polar to an essentially nonpolar medium. Much larger molecules may enter the bulk resin structure by virtue of its porosity. Because of the composition, ie, the high cross-linkage, high selectivity for one ion with respect to another is realized. Apparent oxidation stability is improved because these products initially are so highly cross-linked that the effects of a given amount of chain scission are less apparent. However, at similar cross-linkages, macroporous resins have greater exposure to potential oxidants resulting from the greater porosity and surface area.

Poorer regeneration efficiencies, lower capacities, and higher regeneration costs are the penalties paid for the use of the macroporous resins. Macroporous, ion-exchange resins may be used as catalysts, particularly in nonpolar media where standard resins do not perform satisfactorily because of their inaccessibility to the reactants (see Catalysis).

Porosity in ion exchangers may be varied other than by changes in content of difunctional monomer or by macroporous polymer techniques (33). Various interpolymer systems have been devised where entanglement reduces swelling. Cross-linking may occur during subsequent chemical reactions, eg, in the case of chloromethylation (34). Inert fillers (qv) may be introduced and subsequently removed by dissolution. Synthesis of ion-exchange resins has been reported in refs. 12–14.

Contactors. ***Batch.*** Batch contactors provide a single equilibrium stage per reactor. Thus, batch contact is limited to reactions that go to completion in one stage (eg, a neutralization reaction) or in relatively few stages. In the latter case, the allowable number of contact stages is based on economics. The batch contactor can be a simple stirred reactor and usually has a strainer or filter which allows separation of the resin from the reaction mass at the completion of the equilibrium reaction. Difficulties may arise if regeneration requires a greater number of equilibrium stages than the service or exhaustion portion of the cycle. One advantage of the batch contactor is its ability to handle slurries.

Packed Column. Packed column contactors provide multiple equilibrium stages in a single contactor. Conditions of flow and bed depth usually can be chosen to provide a sufficient number of stages in a packed column to force the equilibrium reaction to completion in the desired direction. The main components of a packed column are shown in Figure 9. The reaction zone proceeds down the column as the upper layer of resin reaches equilibrium with the influent solution, that is, as it becomes exhausted. At the end of the exhaustion or work cycle, the resin is backwashed to reestablish a clean, uniformly packed bed, which is regenerated with the appropriate chemical solution to put the resin into the original ionic form and is rinsed free of excess regenerant

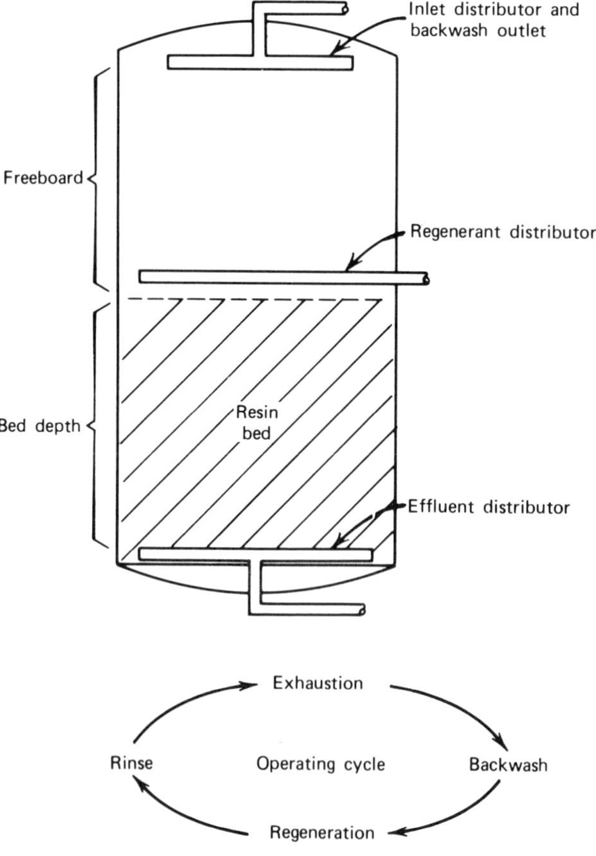

Figure 9. Fixed-bed column contactor.

chemical. The work cycle then may be repeated. Typically, this nonproductive regeneration-cycle time is a small fraction of the total operating cycle. When the ionic load to the resin is such that the regeneration-cycle time is as great, or nearly so, as the work-cycle time, other techniques (eg, semicontinuous contactor use) become economically attractive. Column contactors, when operated as packed beds, are poorly suited to the handling of slurries because of the excellent filtration characteristics of such packed beds.

The phenomenon of ion leakage is encountered with cocurrent regeneration of packed columns and may be markedly reduced by regenerating the column in a counterflow manner. Ion leakage is the amount of the ion that is being removed from solution which appears in the column effluent during the course of the subsequent exhaustion run. Figure 10 illustrates leakage caused by reexchange of nonregenerated ions during the work cycle in a cocurrently regenerated resin bed and shows how counterflow regeneration substantially reduces this leakage. Reexchange of the contaminating Na^+ ion from the base of the coflow-regenerated bed occurs when the incoming salt solution (feed) is converted to the corresponding dilute acid. When this acid solution contacts the sodium ion, reexchange of hydrogen ions for sodium ions occurs, and the sodium ions exit the column as leakage.

In order to obtain low leakage levels from a counterflow, regenerated resin system, the contaminating ions must be kept from the effluent end of the column during regeneration and rinse. This requires avoidance of conditions that would disrupt the packed-bed configuration. Backwash frequency also must be minimized which, in turn,

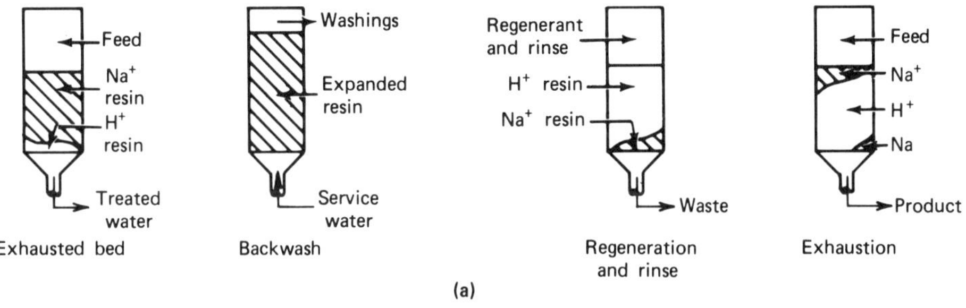

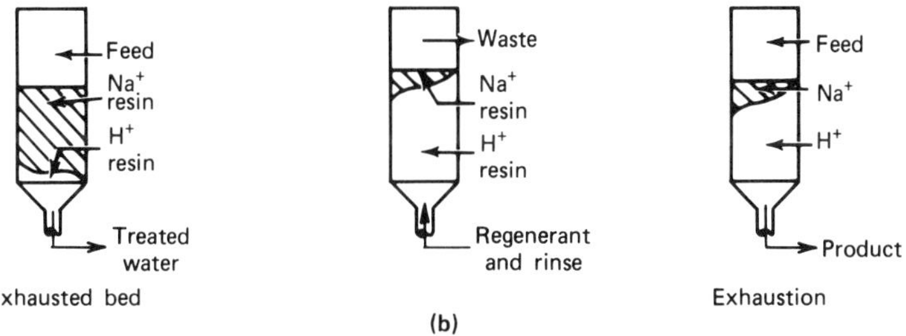

Figure 10. Ion leakage—(**a**) coflow and (**b**) counterflow regenerated fixed-bed column contactors (H^+ form cation resin; Na^+ removal).

requires that feed, regenerant, and rinse solutions be sufficiently clear so as to minimize particulate filtration and buildup within the bed. Entrained gases that disrupt the bed cause channeling of flow through the bed which, in turn, requires backwashing to reestablish the uniform packed bed configuration. Rinse water must be free of the contaminating ions. The leakage of ions also can be eliminated by carrying out the demineralization operation in a mixed bed of hydrogen and hydroxide forms of the cation- and anion-exchange resins, respectively. Contaminating sodium ion has been dispersed throughout the bed during mixing of the two resins and, when dilute acid is formed from the influent salt (feed) solution, it is immediately neutralized by the OH^- of the anion-exchange resin and forms water. Water is not ionized sufficiently to reexchange H^+ ions for the sodium ions in the resin; thus, significantly reduced leakage is experienced.

Figure 11 shows the steps involved in the regeneration of a mixed bed: the initial loading of the appropriate amount of matched resin pairs to ensure proper position of the interface between the resins following backwash separation; backwash separation; maintenance of proper flow balance during application of regenerants to minimize regenerant crossover; and remixing of the resin with air in such a manner to prevent air entrapment and to minimize restratification. A unique type of fixed bed that sometimes is used for condensate demineralization (polishing) is a powdered resin that is precoated on a septum filter (35). High-flow-rate filtration of condensate is obtained in the same time as mixed-bed demineralization in a resin bed that is as shallow as 0.63 cm. Since very little resin is involved, the process is limited to removing traces of ions from solutions of high purity. Regeneration of the resin is not practical in this configuration.

Semicontinuous and Continuous. Semicontinuous and continuous contactors operate as intermittently moving packed beds as typified by the Higgins contactor (36) (see Fig. 12), or as fluidized staged (compartmented) columns as typified by the Himsley contactor (37–39) (see Fig. 13). Flow is countercurrent in each type of continuous or semicontinuous contactor, and their use is in increased resin utilization and high chemical efficiency.

In the Higgins unit, the resin is hydraulically moved as a consolidated resin bed up through the contacting zone. The movement of resin is intermittent and out of phase

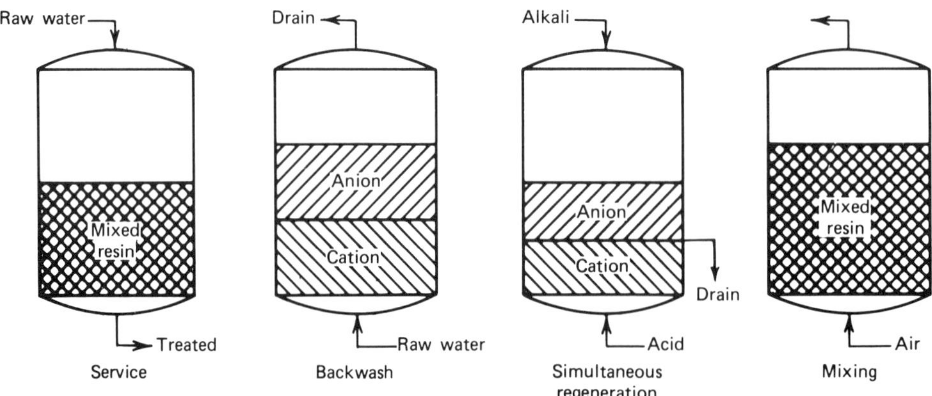

Figure 11. Regeneration of a mixed bed.

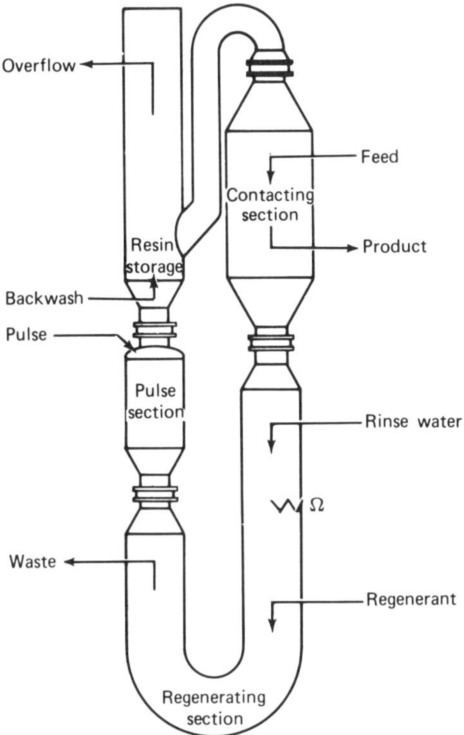

Figure 12. Higgins contactor.

with the flow of solutions that are to be treated. Solution flow is in the direction opposite to resin movement except for the brief period of resin advancement when it is cocurrent. The result of the operation is a near approach to steady-state operations within the contactor. All process steps, such as loading, rinsing, and regeneration, take place in one unit. Commercial installations include those for phosphoric pickle acid recovery, water softening, and ammonium nitrate recovery.

Fluidized staged column contactors are designed for specific feed compositions and effluent requirements. Each reaction stage must be supplied in the form of a tray or compartment. Resin is moved from one compartment or stage to the next countercurrent to the feed solution flow on a time basis that is consistent with the rate of loading to equilibrium in each stage. Commercially available equipment differs in the manner of resin transfer. For example, the Himsley contactor is designed to operate continuously by transferring the resin from stage to stage down the column, one stage at a time while feed flow continues. This is accomplished by recirculating the flow of feed solution around an individual stage by pumping the feed solution from below the stage to above the stage at such a rate that the net flow through the stage is downward. When complete, the process is repeated on the next higher stage. Other contactors of this type transfer the resin intermittently by interrupting the feed flow for a 5–10-s interval and simultaneously drawing liquid downward through all the stages. Regeneration usually is carried out countercurrently in a packed-column contactor.

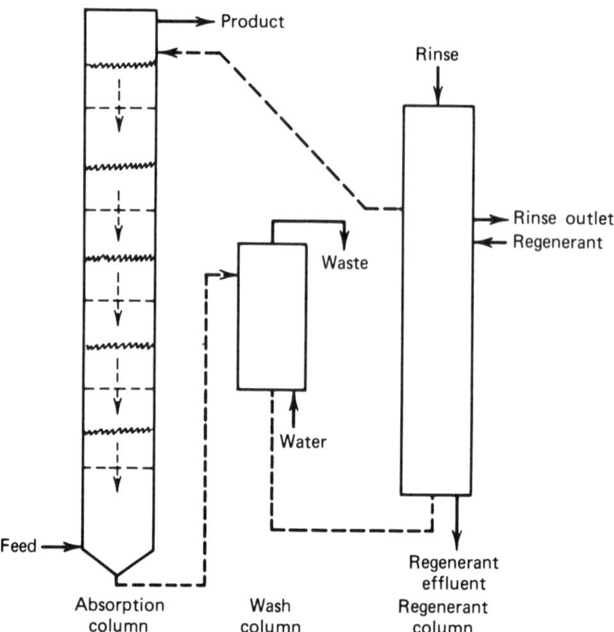

Figure 13. Himsley fluidized, staged column contactor: - - -, resin flow; —, solution flow.

Manufacture and Processing

The manufacture of ion-exchange resins is a combination of processes including the preparation of a cross-linked bead copolymer by suspension polymerization, sulfonation, or chloromethylation and the amination of the copolymer and an ionic conversion for preparation of special forms. A schematic layout of an ion-exchange production facility is given in Figure 14.

The copolymerization of styrene with varying amounts of DVB is carried out in an agitated vessel, in suspension in an aqueous phase. Suspending agents and reactor agitation control the particle size and particle size distribution of the resulting spherical copolymer beads. It is at this point that the properties of the resin can be modified by changes in DVB content or by the addition of a diluent. Following the copolymerization under controlled temperature, catalyst, and agitation conditions, the copolymer is transported to a vessel where it is washed with water of controlled quality to remove the suspending agents. It is dried by centrifuging and/or thermal drying and then is sent to a screening operation to obtain the desired particle size distribution.

Cation-Exchange Resin Plant. A cation-exchange resin plant, based on the sulfonation of a styrene–DVB copolymer, typically involves a vessel, with agitator, into which the copolymer is charged along with a sulfonation acid. The temperature and acid ratios are adjusted to cause the sulfonation reaction to occur to the degree desired by the resin manufacturer. At the end of the sulfonation reaction, the excess sulfuric acid is removed by washing. The resin is rinsed free of the last traces of acid and the resulting hydrogen-form resin is sent to packaging or to a neutralization tank where sodium hydroxide is added under controlled conditions to produce the sodium-form, cation-exchange resin. Following neutralization, the resin is washed with controlled-quality water prior to transfer to the packaging operation.

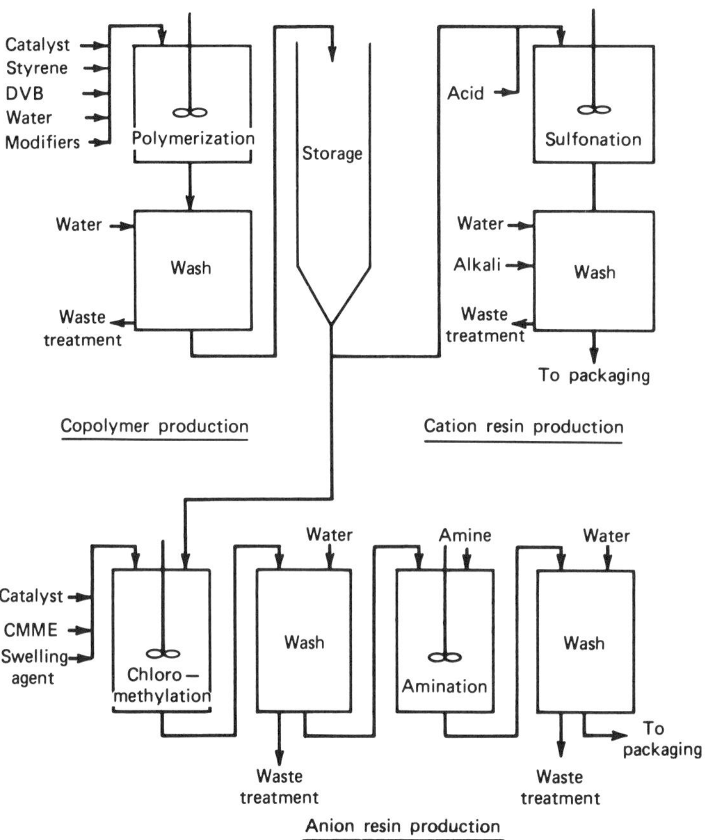

Figure 14. Ion-exchange production processes.

Anion-Exchange Resin Plant. An anion-exchange resin plant is considerably more complex than either a copolymer- or a cation-exchange resin facility. In addition to a chloromethylation facility, a unit for the manufacture of the CMME that is used in the chloromethylation reaction usually is employed. CMME is formed from the reaction of hydrochloric acid, methanol, and formaldehyde under closely controlled conditions. The resulting ether is used in the reaction kettle with the appropriate catalyst and solvent. The chloromethylated copolymer is not isolated because of the unusual toxicity and carcinogenic hazard of the CMME (including the possible contaminant bischloromethyl ether) and the high reactivity of the chloromethylated copolymer beads. These beads are washed to remove the excess organics and CMME prior to being contacted with the appropriate amine in an amination vessel. The excess amine is removed from the reaction vessel following the amination reaction. Hydrochloric acid is added to the aminated resin to convert it completely to the chloride form and the resin is washed with appropriate-quality water to remove residual nonreacted components. The weak-base resin again may be neutralized prior to packaging.

Waste Disposal. During each of the operations, substantial amounts of organic and inorganic suspending agents, solvents, and reactants must be processed for recovery and for disposal as wastes. Because these wastes are not always environmentally acceptable as they come directly from the process, significant distillation, neutralization, scrubbing, and filtration operations are incorporated in the ion-exchange process equipment. In some cases, wastes can be purified and concentrated for recycle in the system. These recovery processes not only control waste to the environment, they ensure the quality of the reactants being recycled. Packaging of ion-exchange materials typically includes the dewatering of the ion-exchange materials from a wash tank or blending station and packaging of the resin into various sized containers.

Production

A list of the commercial producers of granular ion-exchange materials is given in Table 1.

Table 1. Producers of Synthetic Ion-Exchange Resins[a]

Company	Country	Tradename(s)
Akzo	Holland	Imac
Bayer	FRG	Lewatit
Chemolimfex	Hungary	Varion
Diamond Shamrock	United States	Duolite
Diaprosim	France	Duolite
Dow	United States	Dowex
Ionac	United States	Ionac
Mitsubishi	Japan	Diaion
Montecatini	Italy	Kastel
Ostion	Czechoslovakia	Ostion
Permutit	UK	Zeocarb, Deacidite, Zerolit
Permutit A.G.	FRG	Orzelith, Permutit
Resindion	Italy	Relite
Rohm & Haas	United States	Amberlite
Wolfen	GDR	Wofatit
	USSR	AW^-, AV^-, KB^-, KU^-

[a] Does not include ion-exchange (resin) membranes.

Economic Aspects

Ion-exchange resin sales have experienced an increased growth rate in dollar value in the past five years, which has resulted partly from inflationary effects as shown in Figure 15. The 1973 oil embargo had a dramatic effect on resin prices since styrene pricing is directly affected by the price of benzene from which it is made. Availability also was severely curtailed during this period and prices rose accordingly. The latter half of the decade was characterized by a dollar value increase which resulted from a combination of inflationary pressures and real volume growth. The volume growth (Fig. 16) partly results from the increased need to process poorer quality water, the resurgence of the uranium processing industry, and the additional use of ion-exchange systems for sugar processing. The noncommunist market outside the United States for ion exchange is estimated to be about equal to that in the United States (in dollar value). However, the United States uses a significantly greater proportion of strong acid, cation-exchange resins than the rest of the world. Approximately 80% of the resins sold in the United States go to home and industrial water treatment with the balance

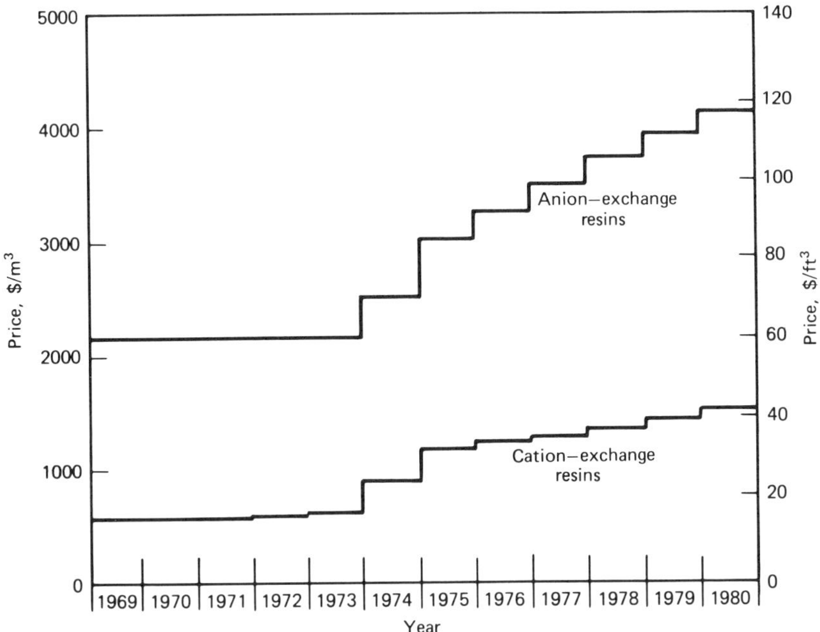

Figure 15. United States ion-exchange resin pricing trends—bulk prices by type.

divided between chemical processing, extractive metallurgy (qv), and food and pharmaceutical processing. A requirement for a good ion-exchange material is that it be available at such a price that its use in appropriate equipment is economically favorable over competing processes, eg, distillation, crystallization, dialysis, reverse osmosis, electrodialysis, solvent extraction, etc.

Specifications, Standards, and Quality Control

The critical properties of resins that are covered by specifications include particle size limits, water-retention capacity, total ion-exchange capacity, and percentage whole beads. ASTM is developing standard methods of analysis and specifications. FDA and TSCA regulations apply to ion-exchange resins that are to be used in processing of food or potable water.

Storage

Ion-exchange resins normally are shipped in the water-swollen form and should be stored so that moisture within the beads is retained. Resin beads that have dried out become free flowing and some shrinkage of the beads is evident. Dry beads reswell very rapidly when placed in aqueous or other polar solutions, with consequent strain which may lead to bead breakage.

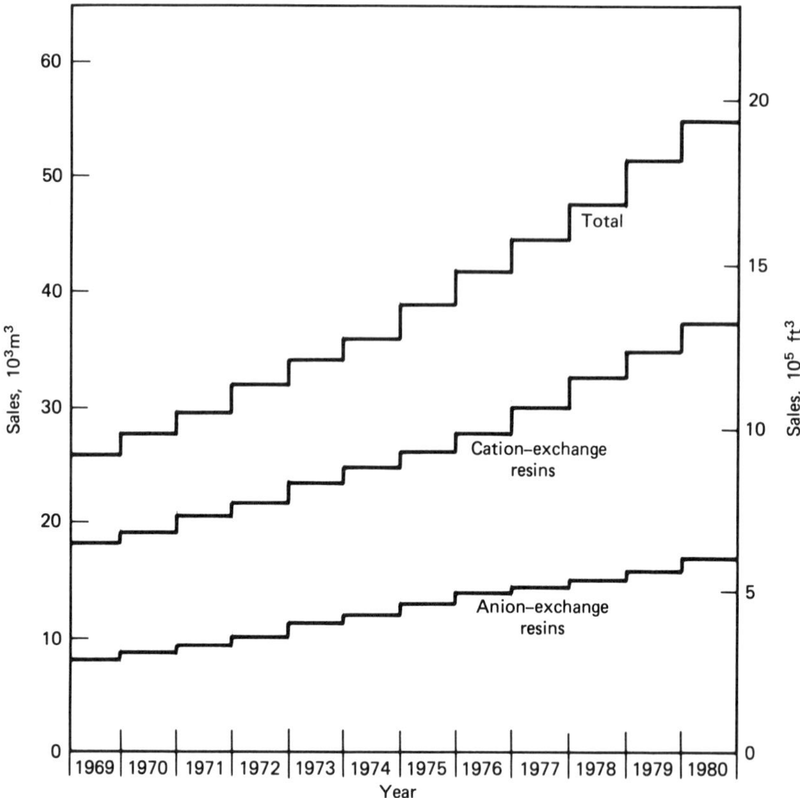

Figure 16. United States ion-exchange resin sales trend—bulk basis by type.

Health and Safety

As manufactured, the only resin ionic forms that present any significant chemical safety hazards are the hydrogen form of the cation-exchange resins and the hydroxide form of the anion-exchange resins; these resins produce acid or base, respectively, when contacted with salt solution. Such toxic ions as chromate, cyanide, or heavy metals may be concentrated by exchange from operating systems. An additional potential hazard which may be encountered during the use of ion-exchange resins is the susceptibility that organic material has to decomposition reactions with strong oxidizing agents, eg, nitric acid. Care should be taken to keep floors and table surfaces free of ion-exchange resin beads since they behave as plastic ball bearings which may cause accidents, eg, falls.

Uses

Water Treatment. Water quality control processes have been discussed (40). Water softening accounts for the major tonnage of resin sales. Hard waters, which contain principally calcium and magnesium ions, cause scale in power plant boilers, water pipes, and domestic cooking utensils. Hard waters also cause soap (qv) precipitation which forms an undesirable gray curd and a waste of soap. Water softening involves the interchange of one hard ion (calcium) for another soft ion of like charge (sodium) on the resin. Typically, hard water is passed through a bed of a sodium cation-exchange resin and is softened.

$$2\,\overline{R}Na^{+} + Ca^{2+} \rightarrow \overline{R}_2Ca^{2+} + 2\,Na^{+}$$

Regeneration of the exchanger involves the passage of a fairly concentrated (6–20%) solution of sodium chloride through the resin.

$$\overline{R}_2Ca^{2+} + 2\,Na^{+} \rightarrow 2\,\overline{R}Na^{+} + Ca^{2+}$$

Softening Brackish Waters. Strong-acid, cation-exchange resins can be used in softening brackish water that contains up to ca 8 g/L (0.5 lb/ft^3) dissolved solids by using beds in series with counterflow regeneration and with large excesses of salt regenerant applied to the second unit series. Leakage increases with TDS (total dissolved solids) concentration. In the 5–8 mg/L range of TDS, weak-acid, cation resin replaces the second strong-acid, cation-resin bed in the system. The latter resin effectively removes all calcium and magnesium that is present in the effluent of the first bed. Weak acid cation-exchange resins may be used alone for softening waters up to seawater salinities. Regeneration of the weak-acid resin is accomplished in two steps: hydrochloric acid is applied to rid the bed of hardness and sodium hydroxide is applied for neutralization.

Dealkalization. Many industrial processes require that hardness and alkalinity be removed from a raw water before the water is used in the process. Several processes involving ion exchange are used for softening and dealkalizing.

The split-stream process involves passing a portion of the raw water through a hydrogen-form, cation exchanger and the remainder through a sodium-form exchanger. The streams from the two units are blended to produce a neutral effluent. This process provides removal of dissolved solids that are equivalent to the amount of alkalinity in the raw water. The lime-zeolite process involves adding lime to the raw water followed by sodium softening. Dissolved solids are reduced to the extent of the calcium and magnesium bicarbonates present in the feedwater. The alkalinity in the raw water, which is associated with sodium, is not removed unless calcium chloride is added. This increases the dissolved solids in the effluent. The hot lime-zeolite process reduces oxygen and silica in addition to hardness and TDS (to the extent of bicarbonate hardness) in the raw water. A third process involves softening the entire raw water stream and neutralizing the alkalinity in the softener effluent with acid. This process does not decrease the total dissolved solids. Sodium softening followed by chloride-anion dealkalizing involves passing the raw water through a zeolite softener to remove hardness, then through a type II anion-exchange resin that is in the chloride form to remove alkalinity. Dissolved solids are removed to the extent of the alkalinity in the raw water by passing the raw water through a bed of weak-acid, hydrogen-form resin. The 100% utilization of regenerant acid that is characteristic of this process decreases operating costs and greatly minimizes the waste disposal problem. A weak-acid resin creates no free mineral acidity in the effluent when regenerated at a level of not more than 100–105% of the theoretically required amounts for the cations picked up.

Demineralization. Ion-exchange demineralization (41) is a two-step process involving treatment with both cation- and anion-exchange resins. For the series operation, water is passed first through a column of strong-acid, cation-exchange resin that is in the hydrogen form ($\overline{R}H^+$) to exchange the cation in solution for hydrogen ions:

$$\overline{R}H^+ + C^+ \rightarrow \overline{R}C^+ + H^+$$

where C^+ represents common cations, eg Ca^{2+}, Mg^{2+}, and Na^+. This effluent is passed to a column of anion-exchange resin in the hydroxide form ($\overline{R}OH^-$) to replace anions in solutions with hydroxide:

$$\overline{R}OH^- + A^- \rightarrow \overline{R}A^- + OH^-$$

where A^- represents common anions, eg, Cl^-, SO_4^{2-}, and NO_3^-. The hydrogen ions from the cation resin neutralize the hydroxide ions from the anion resin:

$$H^+ + OH^- \rightarrow H_2O$$

The net effect is the removal of electrolytes and a yield of purified water.

Alternatively, the impure water may be passed through an intimately mixed bed of cation- and anion-exchange resins where both types of exchange occur simultaneously:

$$\overline{R}H^+ + \overline{R}OH^- + C^+A^- \rightarrow \overline{R}C^+ + \overline{R}A^- + H_2O$$

The choice of the ion-exchange system for demineralization depends on the water quality desired, operating and capital economics, and composition of the raw water. Figure 17 shows the demineralization systems that commonly are employed.

Condensate Polishing. Mixed-bed, ion-exchange resins have been used in deep-bed filter demineralizers for reduction of particulate matter and dissolved contaminants in utility power-plant condensates. The working part of a deep-bed demineralizer polisher (and of the precoat filter–demineralizer) is the mixture of ion-exchange resins. Styrene–DVB-based resins have been used for this service.

Nitrate Removal. Ion exchange is used for the selective removal of nitrates from nitrate-polluted waters to below 10 mg/L (42). Strong-base, anion-exchange resins operating in the chloride ion form (salt solution regenerated) have been successfully used for this service.

Waste Treatment. ***Radioactive.*** Radiation waste systems in nuclear power plants include ion-exchange systems for the removal of trace quantities of radioactive nuclides from water that will be released to the environment. The primary resin system used is the mixed bed of synthetic organic cation- and anion-exchange resins which usually are based on the styrene polymer matrix. Powdered resin mixtures of the same type of materials also are used in some boiling-water reactor (BWR) systems (see Nuclear reactors).

Corrosion Inhibitor Chromate Recovery. Chromate-based corrosion inhibitors are used for protection of metals that are used in water-cooled, heat exchanging equipment and, in many instances, are considered the most effective inhibiting treatment for such systems (see Corrosion and corrosion inhibitors). A practical

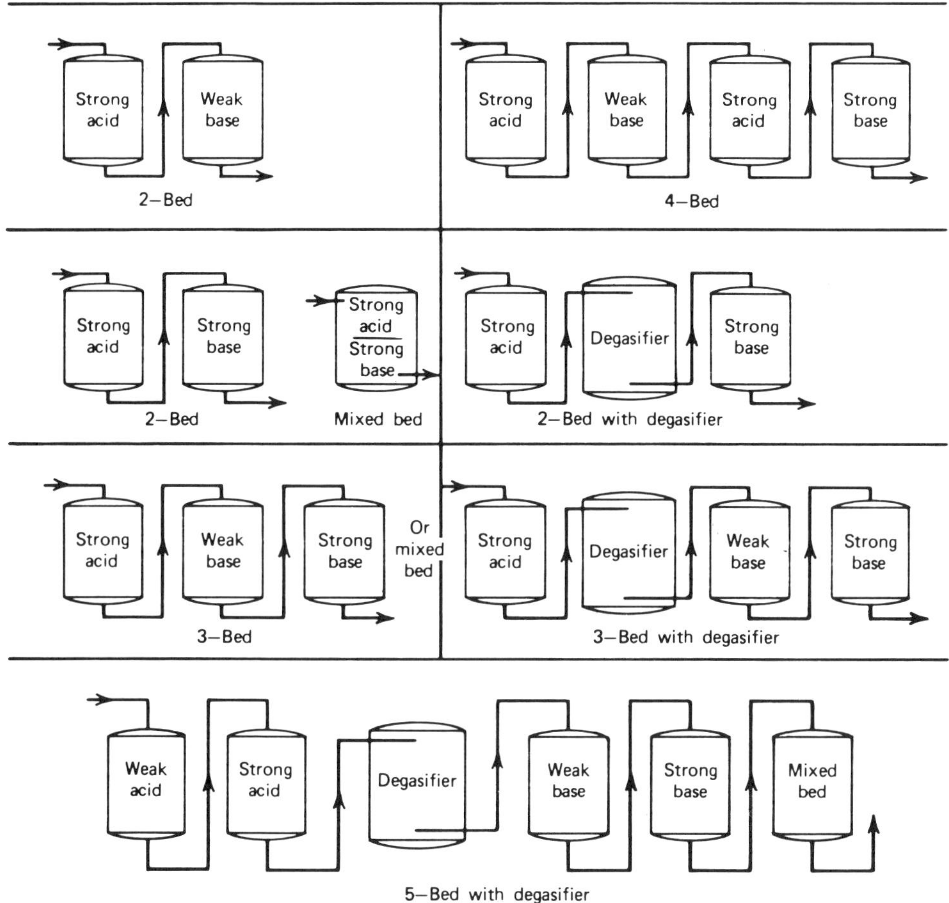

Figure 17. Various demineralization systems.

method of recovering sodium chromate from cooling-tower blowdown wastes by simple anion exchange has been developed: chromate ion is selectively removed from these wastes by the salt form of a strong-base resin which is housed in equipment that is similar to conventional softening or dealkalizing apparatus.

Chromate Recovery. Chromic acid plating solutions and anodizing solutions become fouled with metals (eg, aluminum, iron, magnesium, copper) during normal plating operations. Trivalent chromium also can be formed by the reduction of the chromic acid during this operation and is a contaminant. Macroporous, strong-acid, cation exchange is employed to remove these metal cations from the plating bath solution, thereby allowing recycling of the purified acid to the process (see Recycling).

Rinse-Water Recycling. Rinse waters from plating systems may contain in excess of 20 mg/L of chromic acid and, therefore, constitute a waste problem. Demineralization eliminates these impurities and allows recycling of rinse water of suitable quality. This is carried out with a demineralization system involving a strong-acid, cation-exchange resin followed by use of a weak-base, anion-exchange resin.

Chemical Processing. ***Catalysis.*** Since ion-exchange resins are solid, insoluble (but reactive) acids, bases, or salts, they may replace alkalis, acids, and metal ion catalysts in hydrolysis, inversion, esterification (qv), hydration or dehydration, polymerization, hydroxylation, and epoxidation (qv) reactions. The use of ion-exchange resins as acid–base catalysts has the following advantages: (*1*) Catalyst removal is simplified. The ion-exchange resin is separated easily from the products of reaction by decantation or filtration. (*2*) Repeated reuse. Since the catalyst which has been filtered from the reaction mass may be reused in additional reaction batches, the cost contribution of an ion-exchange resin may be below that for soluble catalysts. It also is possible to construct a continuous reactor without continuous addition of catalyst by employing a fixed bed of ion-exchange resin. (*3*) More selective. Frequently, side reactions are reduced or eliminated and product color often is superior. Reactants that are sensitive to homogeneous acidic and basic catalyst can be used safely with catalyst resins. In some cases, it is possible to obtain reaction intermediates that cannot be isolated when soluble catalysts are used. (*4*) Ion-exchange resins are confined to a reaction zone and it is not necessary to have piping and pumps that are constructed of special alloys or with special linings. The resins are transferred easily from one column to another as a slurry.

Purification. Purification by ion exchange is used to remove contaminating acids, alkalis, salts, or mixtures thereof, from nonionized or slightly ionized organic or inorganic substances. This use is feasible in aqueous, semiaqueous, or nonaqueous solvent systems. The processes may be equally applicable to organic and inorganic compounds (43–44). Examples of purification operations by ion exchange include formic acid removal from 50% formaldehyde solutions; removal of amines from methanol that has been synthesized from carbon monoxide and hydrogen; removal of iron from steel pickling operations using 15–20% phosphoric acid; purification of aluminum bright-dip baths that contain from 75–80% phosphoric acid; and iron removal in the purification of hydrochloric acid by the use of an anion-exchange resin that extracts the negatively charged iron chloride complex (45).

Metal Extraction, Separation, and Concentration. In aqueous or solvent mixtures containing large amounts of contaminants and small amounts of a desired solute, ion-exchange resins may be used to selectively isolate and concentrate the desired solute, eg, in the recovery of uranium from sulfuric acid leach solution utilizing strong-base, anion-exchange resins.

In copper recovery, the pickling of copper with dilute sulfuric acid normally is accompanied by copper and acid leaching into the rinse streams. Strong-acid, cation-exchange resin in the hydrogen form is used to retain the copper and the dilute sulfuric acid can be removed using a macroporous, weak-base resin. The recovery of nickel from nickel (chloride or sulfate) plating-bath rinse waters is accomplished in a similar operation. In the recovery of precious metals, a host of operations involve the use of solutions (eg, cyanide or chloride) that complex with the metals involved and allow for recovery of the metal complex on an anion-exchange resin.

Ion-Exclusion Purification. Ion exclusion permits separation of ionized materials from nonionized or slightly ionized organic or inorganic substances when both are present in aqueous or semiaqueous solution (46). It is used to purify grossly contaminated substances at low cost using water as the regenerating medium.

Ion Retardation. Ion retardation is a separation process based on a resin that contains both fixed cationic- and anionic-exchange groups within the resin structure (24). These exchange sites exist in such close proximity to each other that they partially neutralize the electrical charges, which results in sorbed ion pairs that are held only weakly since they must compete for the exchange sites with fixed counterions. As a result, they may be displaced by water. Like ion exclusion, ion retardation may be used for separation of salts from organic solutes. By virtue of differing anion affinities, ion retardation also may separate mixtures of inorganic, highly ionized electrolytes.

Desiccation. Ion-exchange resins, particularly strong-acid, cation-exchange resins in the dry state, are useful as desiccants (47). The resin may be viewed as an insoluble matrix to which are attached charged ions that have a high affinity for water of hydration and that are responsible for the drying activity. The sodium and potassium forms of strong-acid, cation-exchange resins are satisfactory for reusable applications. Ion-exchange resins show their greatest capability as desiccants in the drying of hydrophobic solvents, eg, hydrocarbons and chlorinated hydrocarbons (see Drying agents).

Sugar Separations and Purifications. ***Sucrose Decolorization and Deashing.*** Hydrogen-form cation exchangers have been applied to remove trace metal impurities, to catalytically invert sucrose to glucose and levulose components, and to convert salts to free acids as the first step in complete demineralization cation–anion treatments (see Sugar). Two-bed and mixed-bed demineralization treatment is utilized in liquid-sugar purification where mineral ash, color, and organic acid and proteinaceous matter removal is required. Ion-exclusion processes also are applied to raw beet and corn syrups for gross color and ash removal. Certain strongly basic, porous, anion-exchange resins may be used to remove organic color bodies from aqueous organic streams, eg, glycerol, sugar, and other polyols. The color that has been sorbed is readily eluted from the resin by a simple treatment with dilute salt brines.

Glucose–Fructose Separation. Fructose and high fructose-containing syrups are finding use as acceptable substitutes for sucrose. These syrups have qualities and sweetness similar to that of sucrose solutions and are sweeter on a weight basis; thus, they can be used to reduce the quantity of sweetener needed in a given application. The nonionic separation process, which uses ion-exchange resins in a cyclic process that is similar to the ion-exclusion process, provides a means of producing a product that is enriched in fructose and, therefore, sweeter (see Sweeteners; Syrups).

Sorbitol and Mannitol Purification. Other corn-sugar products, sorbitol and mannitol, are made from dextrose by hydrogenation. These materials usually are contaminated with color, salts, and other color precursors which must be removed from the finished product. Ion-exchange demineralization is used for removal of these impurities. Two-bed roughing demineralization generally is followed by mixed-bed polishing units (see Alcohols, polyhydric).

Analysis. Detailed presentations on the use of ion-exchange resins in analytical applications have been published (48–53). The recent development of ion chromatography provides a rapid, automated chromatographic method for the separation of various ions with determination by conductimetric detection (54–56).

Medicine. Ion-exchange resins are useful as carriers for medicinal materials and are useful in slow-release applications (see Pharmaceuticals, controlled release). In some select cases, the ion-exchange resin has the medicinal affect desired, eg, weak-base resins have been used as antacids. Cholestyramine, a strong-base, anion-exchange resin of closely controlled porosity, when dried and ground, is used to bind bile acids in the treatment of patients with high levels of blood cholesterol.

Pharmaceuticals. Ion-exchange resins also are used as diagnostic tools. Analyses are carried out for determination of amino acids in plasma, thyroxine in serum, ammonia in plasma, and methylmalonic acid in urine. Weak-acid, cation-exchange resins are used in the isolation and purification of streptomycin and neomycin and other similar antibiotics (qv).

BIBLIOGRAPHY

"Ion Exchange" in *ECT* 1st ed., Vol. 8, pp. 1–17, by Robert Kunin, Rohm & Haas Company; "Ion Exchange" in *ECT* 2nd ed., Vol. 11, pp. 871–899, by R. M. Wheaton and A. H. Seamster, The Dow Chemical Company.

1. F. Helfferich, *Ion Exchange,* McGraw Hill, New York, 1962.
2. F. C. Nachod and J. Schubert, eds., *Ion Exchange,* Academic Press, Inc., New York, 1949.
3. F. C. Nachod and J. Schubert, eds., *Ion Exchange Technology,* Academic Press, Inc., New York, 1956.
4. C. Calmon and T. R. E. Kressman, eds., *Ion Exchange in Organic and Biochemistry,* Interscience Publishers, New York, 1957.
5. R. Kunin, *Ion Exchange Resins,* Robert E. Krieger Publishing Co., Huntington, N.J., 1972.
6. J. A. Marinsky and Y. Marcus, eds., *Ion Exchange and Solvent Extraction,* Vols. I–III, V–VII, Marcel Dekker, Inc., New York, 1966–1977.
7. J. A. Kitchener, *Ion Exchange Resins,* John Wiley & Sons, Inc., New York, 1957.
8. J. E. Salmon and D. K. Hale, *Ion Exchange: A Laboratory Manual,* Academic Press, Inc., New York, 1959.
9. K. Dorfner, *Ion Exchangers,* Ann Arbor Science Publishers, Ann Arbor, Mich., 1973.
10. R. W. Grimsham and C. E. Harland, *Ion Exchange: Introduction to Theory and Practice, Monographs for Teachers No. 29,* The Chemical Society, London, Eng., 1975; C. Calmon and H. Goed, eds., *Ion-Exchange for Pollution Controls,* Vols. I and II, CRC Press, Inc., Boca Raton, Fla., 1979; R. E. Anderson, "Ion Exchange Separations," in P. A. Schweitzer, ed., *Separation Techniques for Chemical Engineers,* McGraw Hill Book Co., New York, 1979, pp. 359–414.
11. B. A. Adams and E. L. Holmes, *J. Soc. Chem. Ind.* **54,** 1-6T (1935).
12. R. M. Wheaton and M. J. Hatch in J. A. Marinsky, ed., *Ion Exchange,* Vol. 2, Marcel Dekker, Inc., New York, 1969, p. 91.
13. G. Schmuckler and S. Goldstein in ref. 6, Vol. VII, 1977, p. 1.
14. V. A. Davankov, S. V. Rogozhin, and M. P. Tsyurupa in ref. 6, Vol. VII, 1977, p. 29.
15. W. C. Bauman, "Swelling and Exchange Equilibria in Cation Exchange Resins of the Sulfonated Polystyrene Type," *Papers 23rd Natl. Colloid Symp., Minneapolis, Minn., June 6, 1949* (unpublished).
16. G. E. Boyd and B. A. Soldano, *Z. Elektrochem.* **57,** 162 (1953).
17. W. C. Bauman and J. Eichhorn, *J. Am. Chem. Soc.* **69,** 2830 (1947).
18. H. P. Gregor, *J. Am. Chem. Soc.* **70,** 1293 (1948).
19. A. Clearfield, G. H. Nancollas, and R. H. Blessing in ref. 6, Vol. V, 1973, p. 1.
20. D. C. Kennedy, *Chem. Eng.,* (June 16, 1980).
21. R. Rosset, *Bull. Soc. Chim. Fr.,* (1), 59 (1966).
22. K. C. Jones and R. R. Grinstead, *Chem. Ind.,* 637 (Aug. 6, 1977).
23. T. W. Clarkson, H. Small, and T. Norseth, *Arch. Environ. Health* **26,** 173 (Apr. 1973).
24. M. J. Hatch and J. A. Dillon, *Ind. Eng. Chem. Process Des. Dev.* **2,** 253 (1963).
25. U.S. Pat. 4,150,205 (Apr. 17, 1979), R. M. Wheaton (to The Dow Chemical Co.).
26. U.S. Pat. 4,154,801 (May 15, 1979), R. M. Wheaton (to The Dow Chemical Co.).
27. U.S. Pat. 3,425,939 (Feb. 4, 1969), D. E. Weiss and B. A. Bolto (to Commonwealth Scientific & Industrial Research Organization).
28. B. A. Bolto, D. E. Weiss, A. F. G. Cope, and G. K. Stephens, *Adv. Thermally Regenerated Ion Exchange, Soc. Chem. Ind. Conf., Cambridge, Eng.* (1976).
29. U.S. Pat. 2,813,838 (Nov. 19, 1957), W. R. Lyman and A. F. Preuss, Jr. (to Rohm & Haas).

30. J. R. Millar, D. G. Smith, W. E. Marr, and T. R. E. Kressman, *J. Chem. Soc.,* 183 (1963).
31. U.S. Pat. 3,549,562 (Dec. 22, 1970), M. Mindick and J. Svarz (to The Dow Chemical Co.).
32. W. G. Lloyd and T. Alfrey, Jr., *J. Polym. Sci.* **62,** 301 (1962).
33. J. R. Millar, *J. Chem. Soc.,* 1311 (1960).
34. R. E. Anderson, *Ind. Eng. Chem. Prod. Res. Dev.* **3,** 85 (1964).
35. J. H. Duff and J. A. Levendusky, *Proc. Am. Power Conf.* **24,** 739 (1962).
36. U.S. Pat. 3,580,842 (May 25, 1971), I. R. Higgins (to Chemical Separations Corporation).
37. Can. Pat. 980,467 (Dec. 23, 1975), A. Himsley (to Himsley Engineering).
38. U.S. Pat. 3,549,526 (Dec. 22, 1970), H. Brown.
39. U.S. Pat. 3,551,118 (Dec. 29, 1970), F. L. D. Cloete and M. Streat (to National Research Development Corporation, London).
40. W. J. Weber, Jr., "Physicochemical Processes for Water Quality Control" in *Ion Exchange,* Wiley-Interscience, New York, 1972, Chapt. 6.
41. S. B. Applebaum, *Demineralization by Ion Exchange,* Academic Press, Inc., New York, 1968.
42. M. Sheinker and J. Cudoluto, *Public Works,* 71 (June 1977).
43. A. Weissberger and E. S. Perry, *Techniques of Chemistry,* Vol. 13, John Wiley & Sons, Inc., New York, 1978, Chapt. 4.
44. F. J. Wolf, *Separation Methods in Organic Chemistry and Biochemistry,* Academic Press, Inc., New York, 1969.
45. K. A. Kraus and G. E. Moore, *J. Am. Chem. Soc.* **71,** 3263 (1949).
46. R. M. Wheaton and W. C. Bauman, *Ind. Eng. Chem.* **45,** 228 (1953).
47. C. E. Wymore, *Ind. Eng. Chem. Prod. Res. Dev.* **1,** 173 (1962).
48. O. Samuelson, *Ion Exchangers in Analytical Chemistry,* John Wiley & Sons, Inc., New York, 1963.
49. J. Inczedy, *Analytical Applications of Ion Exchangers,* Pergamon Press, New York, 1966.
50. W. Rieman, III, and R. N. Sargent, *Physical Methods in Chemical Analysis,* Vol. IV, Academic Press, Inc., New York, 1961.
51. W. Rieman, III, and H. F. Walton, *Ion Exchange in Analytical Chemistry,* Pergamon Press, New York, 1970.
52. H. F. Walton, *Ion Exchange Chromatography,* Dowden, Hutchinson and Ross, Stroudsburg, Penn., 1976.
53. J. X. Khym, *Analytical Ion Exchange Procedures in Chemistry and Biology,* Prentice Hall, Englewood Cliffs, N.J., 1974.
54. H. Small, T. S. Stevens, and W. C. Bauman, *Anal. Chem.* **47,** 1803 (1975).
55. J. D. Mulik and E. Sawicki, *Environ. Sci. Technol.* **13**(7), (July 1979).
56. A. L. Robinson, *Science* **208,** 164 (1980).

R. M. WHEATON
L. J. LEFEVRE
Dow Chemical U.S.A.

ION IMPLANTATION

Ion implantation is a process for injecting atoms of any element into any solid material to selected depths and concentrations in order to form an alloy or other solid mixture that has a different composition from the original solid and that, therefore, exhibits different and sometimes highly desirable chemical and physical properties. An ion-implantation accelerator, which is illustrated schematically in Figure 1, is similar to an isotope separator but has an added final acceleration stage (100–1000 keV) for the selected ions. This added energy causes substantial penetration of the ions into the lattice of the target material. Atoms of the selected chemical element are ionized by collisions with electrons in an electrical discharge in a gas and at low pressure. These ions pass through an orifice into a high-vacuum region where they are accelerated by an electric field (from the extraction electrode) to a moderate energy (10–30 keV) and where they are analyzed by a magnetic field according to isotope mass. In Figure 1 the magnet current is set so that the most abundant isotope of chromium (^{52}Cr) passes through the analyzer output slit. The selected ions are then accelerated to the desired energy, refocused by a quadrupole lens (not shown in Figure 1), deflected by a scanner system, collimated by a defining aperture, and allowed to strike the target. When the ions penetrate the target lattice, they lose energy through collisions with lattice atoms and come to rest.

Table 1 lists typical ranges of technical and economic parameters that apply to ion implantation that is performed to modify the chemical, optical, or mechanical properties of materials.

Ion implantation sometimes is confused with ion plating (see Film deposition techniques; Metallic coatings). Ion plating is a coating process that occurs in a glow

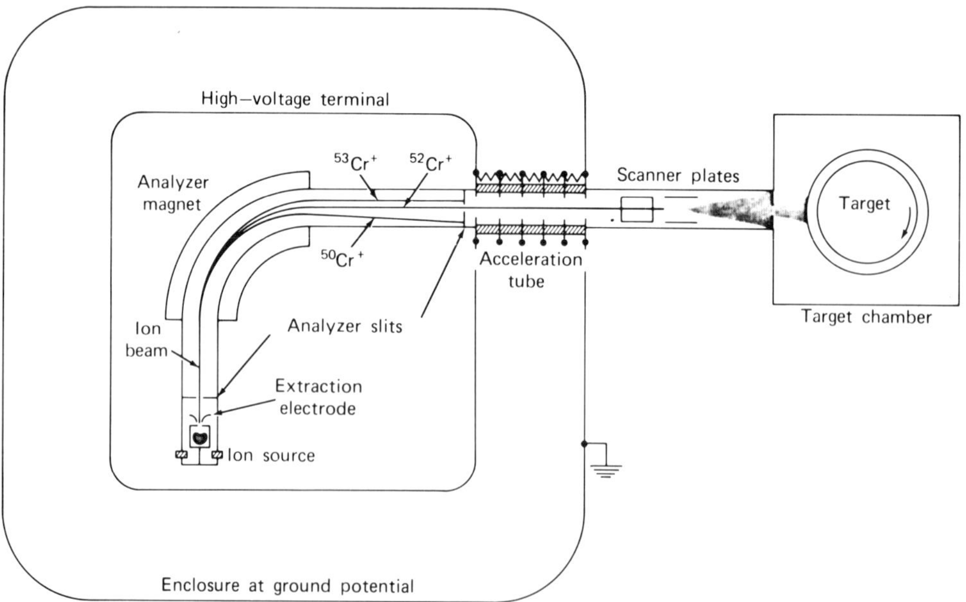

Figure 1. A schematic drawing of an ion-implantation machine.

Table 1. Technical and Economic Parameters and the Ranges of Their Values

Parameters	Ranges
ion species	any element or combination of elements
target material	any solid material
target preparation	clean and smooth surface
target dimensions, m	<2
target temperature, °C	−196–300
pressure in beam tubes, Pa[a]	10^{-5}–10^{-3}
potential on terminal, kV	10–1000
depth of implanted ions, μm	0.01–1
ion-beam current	1 nA–20 mA
ion-beam cross section, cm^2	0.1–1
ion fluence or dose, atoms/cm^2	10^{16}–10^{18}
relative concentration, atom %	1–50
surface erosion, nm	1–100
capital investment, $	200,000–400,000
power requirement, kVA (kW)	5–20
floor space requirement, m^2	10–20
implantation costs, $/$cm^2$	0.01–1

[a] To convert Pa to mm Hg, divide by 133.3.

discharge in a gas at a pressure of a few Pascals (1 Pa = 0.75 centitorr), and the energies of the ions (and neutral atoms) as they strike the surface are from 0.2–1 keV. Ion implantation does not produce a coating, and there is no sharp interface between the implanted region and the substrate.

Three U.S. patents (1–3) were issued in the late 1950s on the ion-implantation technique. Since the early 1970s, ion implantation has been used widely by the semiconductor device industry to introduce dopants into many types of devices, including large-scale integrated circuits (qv), such as are used in pocket calculators. The reasons for this widespread use are, principally, the reproducibility and controllability of the process; it is used mainly as a predeposition technique and is followed by thermal diffusion. Also, during the early 1970s, a few laboratories began to experiment with the modification of the chemical, optical, and mechanical properties of materials by means of ion implantation.

Implanters

An ion implanter has four principal parts, ie, the ion source, mass analyzer, acceleration tube, and target chamber.

Ion Sources. For implantations that are performed to modify the chemical, optical, or mechanical properties of a material, substantial concentrations of implanted ions are required (typically ca 20 atom %). Such concentrations require a large number of implanted ions per unit area (or fluence or dose). Therefore, a crucial characteristic of such an ion source is that it produce large ion-beam currents.

There are two principal types of ion source: gas discharge and sputter sources. The gas discharge is used for elements that exist naturally as gases or that form gaseous compounds. There are three methods of producing and sustaining the gas discharge. One method involves a hot filament as a source of electrons and a d-c power supply to accelerate the electrons; another method involves a cold cathode; and the third method (uncommon in implanters) involves the direct coupling of radio-frequency energy to the gas. A constant magnetic field sometimes is applied to concentrate the discharge, and a furnace sometimes is used to vaporize solids that have moderate melting points.

The sputter ion source produces ions from elements that are not gases and that do not form gas compounds. The sputter source produces a discharge in an inert gas (eg, argon) with a hot filament and a potential difference and then uses the argon ions that are produced to sputter atoms from a solid sample of the source material. The sputter atoms then are ionized by collisions in the same manner as the argon atoms are ionized.

Mass Analyzers. The ion beam from a source usually contains impurities, eg, vacuum system contaminants from residual air, from vacuum pumps, or from solid components of the source. When an ion that does not exist as an elemental gas is produced in an ion source, the ion beam contains even more undesired components. For example, a sputter source produces more argon ions (or ions of another gas used to sustain the discharge) than sputtered ions; and an ion source that uses a gas compound, eg, BF_3, produces several kinds of molecular ions from various combinations of the atoms that are present.

There are several types of mass analyzers: magnetic, crossed electric and magnetic fields, and quadrupole; the magnetic mass analyzer is the most common. For a magnetic analyzer, shown schematically in Figure 1, the product of magnetic field H and the path radius of curvature r is proportional to the square root of the ratio of ion mass m to charge q

$$Hr \propto V^{1/2}(m/q)^{1/2} \tag{1}$$

One cannot fix merely the magnetic field for a given set of values of r and V and expect to find the pure desired ions passing through the analyzer output slit for three reasons:

(*1*) Molecular ions may have the same mass-to-charge ratio m/q as the desired atomic ions. For example, the molecule CH_4^+ has the same nominal mass as O^+, and similarly for CO^+, N_2^+, and Si^+.

(*2*) Atomic ions may possess the desired value of m/q by having twice the mass and twice the charge of the desired ions, eg, Si^{2+} corresponds to the same nominal field setting as N^+.

(*3*) Charge exchange may occur as the result of an inelastic collision of a beam ion with a residual gas molecule in the vacuum system. For example, if an N_2^+ ion is extracted and then collides with a gas molecule, breaking up into two N^+ ions between the preacceleration stage and the analyzer, it simulates $^7Li^+$. A simple formula can be derived to predict the apparent mass-to-charge ratio of a beam component resulting from such charge exchange between acceleration (subscript a) and analysis (subscript b):

$$\left(\frac{m}{q}\right)_{\text{apparent}} = \left(\frac{m_b q_a}{m_a q_b}\right)^2 \left(\frac{m_a}{q_a}\right)_{\text{actual}} \tag{2}$$

For the case of presumed unit charge for all ions, the apparent mass of a given beam is

$$m_{\text{apparent}} = (m_b/m_a)^2\, m_a = (m_b/m_a)m_b \tag{3}$$

Charge exchange also may produce a neutral beam. If such charge exchange occurs downstream from the analyzer magnet, then it can result in dose errors, because the dose is measured by an integration of the current that is conveyed to the target by the beam. The usual technique for removing such a neutral beam is to install a small magnet downstream from the main magnet to deflect only the desired beam into the target chamber.

Because of these uncertainties and because identification of the implanted ions is essential, consideration must be given to the entire mass spectrum, eg, the known relative abundances of isotopes. Furthermore, the existence of a particular atomic-beam current sometimes implies the existence of an associated diatomic beam current, the magnitude of which is predictable from experience.

Acceleration Tubes. The function of an acceleration tube is to provide an evacuated path for the ions and an electric field to accelerate them. The electric field is shaped so as to focus the beam of selected ions from the mass analyzer and to impede the backstreaming, secondary electrons. In most implanters, the acceleration tube is placed after the analyzer magnet (as illustrated in Figure 1). In some machines (usually low current), the acceleration tube is placed before the analyzer magnet. Preacceleration analysis results in a smaller analyzing magnet and power supply; no high-voltage power wasted in accelerating unwanted ions; less self-defocusing of beam in the acceleration tube; lower level of x rays from backstreaming electrons; relaxed high-voltage stability requirements; and obviation of analyzer magnet readjustment with each change of high voltage. Advantages of postacceleration analysis are obviation of the isolation transformer for the analyzer-magnet power supply, simpler beam alignment because the analyzer magnet is readily accessible, and easier installation of multiple-beam tubes and target chambers.

Target Chambers. Target chambers provide line-of-sight, high-vacuum access to each area to be implanted. It is customary, but not necessary, for the beam to strike the target at normal incidence. At nonnormal incidence, the penetration is smaller, and target sputtering is greater.

A target chamber with auxiliary equipment provides some or all of the following functions: ion-beam-current integration (by which the dose is measured); secondary-electron suppression (to avoid ion-beam-current errors); vapor condensation (by means of a cold tube containing liquid nitrogen); differential pumping (high vacuum to inhibit target contamination); beam sweeping (to scan the target, as an electron beam in a kinescope); target manipulation (to expose every part of the target to the beam); target cooling (using a heat-transfer system to dissipate ion-beam power); auxiliary magnet (to separate the desired beam from any neutral beam); and automatic target changing (by means of vacuum locks and a conveyor).

Ion-Target Interactions

Atomic Collisions. The collisions between the incident ion and lattice atoms involve two types of energy transfer: elastic transfer in which the two partially screened nuclei experience each other's Coulomb field, thereby transferring momentum in the process as in a billiard-ball collision, and inelastic transfer in which energy is transferred to the electrons of the struck atoms. A binary collision model (4–5) is used to calculate the magnitudes of the energy and momentum transfers; the elastic and inelastic contributions are assumed to be independent, with no interference between the two. (This assumption is not strictly true, but it provides results that are tractable.) The nature of the interaction is strongly dependent on the instantaneous energy of the incident ion E_1; the atomic number of the incident ion Z_1; and the atomic number of the target atom Z_2. Generally, for large values of E_1 and for small values of Z_1 (eg, a 100-keV helium ion), the predominant mode of energy transfer is inelastic; for relatively low values of E_1 and for large values of Z_1 (eg, a 10-keV molybdenum ion), the elastic mode of energy transfer predominates. Figure 2 illustrates the variations of the energy loss per unit path length as a function of E_1 for two values of Z_1 and for each kind of collision in a target of iron (6–8). The crossover point for incident iron ions is ca 500 keV, but for nitrogen ions, it is ca 30 keV.

For low ion energies and for a hard elastic interaction potential, ϕ, between ion and atom (ie, if ϕ falls off more sharply than r^{-2}, as is assumed for the ion-atom interaction), the elastic-energy-transfer rate increases with ion energy for the same reason as with billiard ball collisions. However, unless the potential is infinitely hard, at sufficiently high speeds for the incident ion the interaction time is so short that the atom cannot be fully accelerated during the interaction (except for rare, head-on collisions, for which the interaction potential is hard). (The momentum transferred is $\int F dt$.) Thus, the energy loss rate decreases for higher ion energies as illustrated in Figure 2. The inelastic energy transfer rates have a constant slope of 0.5, implying that the inelastic transfer rate is proportional to $E_1^{1/2}$ or to the ion velocity, v. This relationship comes from the model of the inelastic interaction. When the electron shells of incident ion and host atom overlap, some electrons change their atomic identification and transfer momentum mv to the electronic structure of the recipient atom. The effective inelastic force between the two atoms is equal to the product of the electron

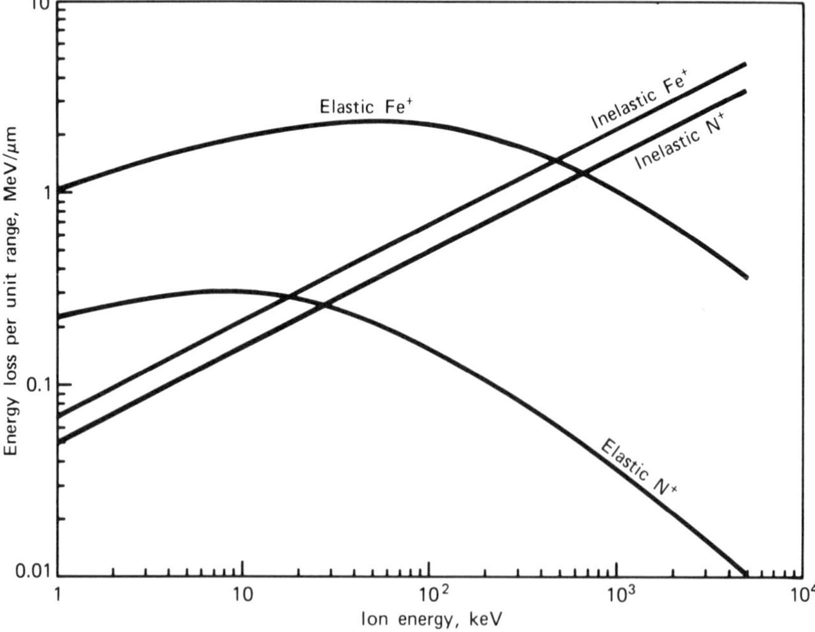

Figure 2. Rate of energy loss of iron ions and nitrogen ions in an iron sample as a function of instantaneous ion energy (4–8).

transfer rate and mv. Therefore, the inelastic energy transfer is proportional to $\int mv \cdot dR$, where R is the relative position vector.

Ion Range. Because the incident ion is deflected by elastic collisions, its path is not straight. Variations in the range (the projected length of the path onto the beam's initial velocity vector) from ion to ion (with all conditions the same) arise because of statistical fluctuations in the energy losses in elastic collisions and in the direction of the ion after a collision. Those collisions that involve greater energy transfers also involve greater deflections. The inelastic collisions are assumed to constitute a smooth braking force and, hence, do not contribute to the straggling in range. If the projection of the initial velocity vector is divided into equal increments or bins and the number of ion-path projections that terminate in each such bin are counted, a range distribution curve may be obtained.

The shape of the distribution depends on the ion species, incident ion energy, and target material. For a given ion species and target material, the skewness may be either toward the right or the left, depending on the incident ion energy. For implantation into semiconductor devices, the skewness may be important because tails on the distribution may lead to free carriers existing in regions where free carriers are not desired, but for implantations performed to modify the chemical or mechanical properties of materials, the precise shape of the distribution is not so important. Hence, the distribution in range, ie, in concentration of implanted ions, usually is assumed to have a gaussian shape.

Figure 3 shows the calculated distribution of ranges for several incident energies of iron ions implanted into an iron sample (calculation based on references 7 and 8). Figure 3 suggests the possibility of tailoring the depth concentration profile of implanted ions so as to achieve almost any desired smoothly varying or constant profile.

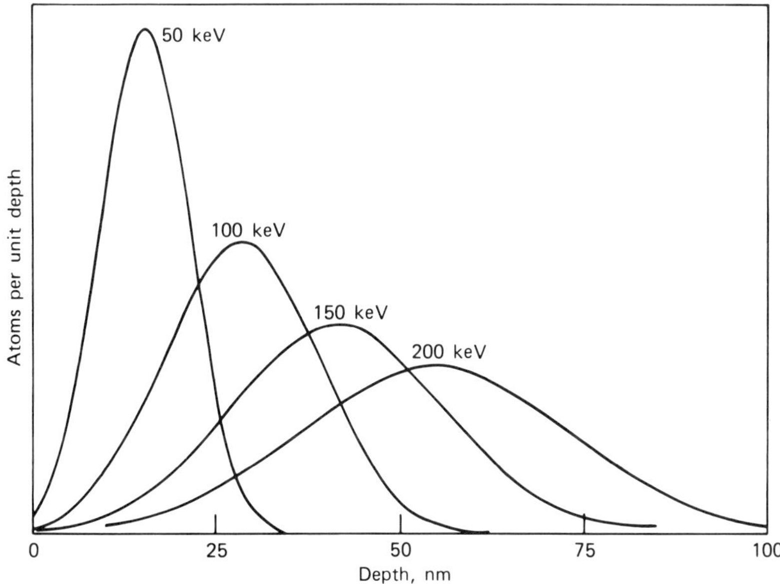

Figure 3. The calculated range distributions for iron ions of various incident energies E_i in an iron sample at normal incidences (based on references 7 and 8).

Of course, the range is dependent on Z_1 and Z_2 as well as incident ion energy E_i. Sputtering of the surface influences the final depth–concentration profile of the implanted atoms. Several computer programs are available, and a few tabulations of results have been published, providing values of mean range and standard deviation for any combination of ion and solid material (see General references).

Channeling. In metals, grains (which are miniature crystals) typically have random orientations; hence, for any region much larger than a grain, the implicit assumption of random arrangement of atoms is reasonable. However, if the incident ions are within a small angle (typically, 1°) of a low index direction in a single-crystal target, then most of the ions are steered (9) through the channel. No displacements occur; therefore, the only energy transfers are inelastic (electronic) and, in a perfect crystal, the ion's range may be several times the normal value. If a defect is present (eg, an interstitial vacancy, dislocation, impurity atom, or thermal perturbation), the defect is likely to dechannel the ion.

Because typical commercial materials are not single crystals and because large implantation doses destroy the regularity if the target is a single crystal, channeling is important in ion implantation only in special cases, such as where deep implants of small doses are desired (eg, radioactive ions or dopants for luminescence or for certain semiconductor devices). Sometimes, however, channeling is hard to avoid (eg, with low-energy, heavy ions in a single-crystal target) because the critical angle may be several degrees and there are many directions that produce channeling.

Radiation Damage. To displace an atom from its lattice site requires the expenditure of a minimum amount of energy (20–40 eV) for most materials; the average displacement energy E_d is ca 25 eV. When a collision between an incident ion and a target atom causes an energy transfer that is much greater than E_d, the struck atom collides with other atoms, knocking some of them out of their lattice sites. This process creates a cascade of collisions producing secondary and tertiary knock-on atoms. Thus, an ion leaves a path of havoc in its wake, and the residual effect of this havoc is radiation damage. However, this term is somewhat misleading because the overall result may be beneficial instead of detrimental. For example, the ion beam tends to clean the surface (by means of sputtering); and the interstitial atoms in the implanted layer produce compressive stresses which appear to play a role that is somewhat analogous to those in prestressed concrete.

The total number of displaced atoms depends on the average displacement energy E_d; the ion's initial energy E_i; the relative masses of the colliding atoms; and the dose or fluence of implanted ions. The flux of incident ions is the number of ions per unit area per unit time striking the target, and the fluence (dose) is the integration of flux over time. Therefore, fluence has dimensions of ions per unit area. For the situation in which the bombarding species and bombarded species are the same, the number of displacements per incident ion N_d is given by the approximation

$$N_d = 0.8\,\nu(E)E_i/2E_d, \tag{4}$$

which is a variation of the Kinchin-Pease formula (10). The energy partition function, $\nu(E)$, represents the fraction of the total energy transfer that goes into elastic collisions (11). The 2 in equation 4 is a statistical factor that arises from the model; the final displacement collision in each branch of the cascade results in two atoms, each of which possesses a kinetic energy of E_d or less.

Radiation damage usually is not specifically annealed out of metals after implantation, but it must be annealed from semiconductor devices because the disorder causes a drastic reduction in minority-carrier lifetime. The annealing is performed from 300 to 900°C, depending on the ion species, semiconductor material, attached materials (eg, as aluminum leads), implantation energy, and dose; but some devices are annealed by laser or electron beams (see Lasers). During the annealing process, the amorphous layer may recrystallize epitaxially onto the undisturbed substrate. Whereas a high-dose implantation into a semiconductor material usually leaves the material in an amorphous state, similar implantations into metals do not ordinarily leave the material amorphous as discussed above. The reason for this different response is the nature of the atomic bonds. Covalent bonds are especially vulnerable.

For implantations in alloys in order to enhance their resistance to corrosion or wear, the residual radiation damage may be beneficial and, hence, is not annealed out. The implantation process leaves many atoms (both implanted and host) in interstitial sites in the implanted layer (typically, a few hundred atoms thick), and these interstitials constitute wedges. As a result, the layer tends to expand, but the expansion is restricted by the much thicker substrate and by the surface energy. The result is a compressive stress near the surface, and this compressive stress may inhibit the initiation or propagation of cracks, thereby accounting, in part, for the observed improvements in mechanical properties of the material. Another reason that radiation damage may be beneficial is that ion implantation in metals may produce an amorphous surface layer (12), and some amorphous metals produced by means other than ion implantation possess greater strength (13) and greater corrosion resistance (14) than the corresponding crystalline alloys (see Glassy metals).

Sputtering. Whenever the cascade of knock-on atoms intersects the surface, an atom may sputter or leave the surface if its energy exceeds the surface binding energy (usually from 2–5 eV). The number of sputtered atoms per incident ion is the sputtering yield (S). Sputtering is a significant factor for heavy implantation doses, which are common for metal targets, because it may affect the depth concentration profile of implanted ions and may limit the maximum achievable concentration of implanted ions. The sputtering yield depends on ion energy, species, flux, angle of incidence, target material, dose, crystal state, and surface binding energy.

Ion Energy. The dependence of S on incident ion energy E_i may be understood in terms of elastic collisions and the binding energy of a surface atom. A given target atom may receive a maximum amount of energy (determined by classical mechanics) from the incident ion and, if this energy is less than the surface binding energy, no sputtering can occur. Thus, a sputtering threshold energy E_t is expected. At higher energies, S rises with E_i because collisions involving less than the maximum energy transfer begin to contribute to the process. As E_i continues to increase, an energy eventually is reached at which the rate of elastic energy transfer decreases (as in Figure 2). The collision cascade near the surface subsides, leading to a decrease in sputtering.

Ion Species. Figure 2 shows that a high-Z incident ion (eg, iron) transfers a larger fraction of its energy by way of elastic collisions than a low-Z ion (eg, nitrogen). Hence, S increases as a function of Z_1.

Angle of Incidence. As the angle of incidence, θ (the angle between the incident-velocity vector and the normal to the surface), varies from zero, S is expected to be proportional to sec θ since, for a given instantaneous depth of the incident ion in the target, the path length of the ion is proportional to sec θ. However, because of the multiple scattering implicit in the sputtering process, the sputtering yield (15) depends on $(\sec \theta)^x$, where x is slightly greater than one; and the precise dependence is a function of the incident and target atoms. This sec θ dependence breaks down for large values of θ because of the escape of incident ions from the surface. For single-crystal materials, the relationship between θ and channeling directions influences S.

Target Material. The sputtering yield depends on the mass of the target atoms because the energy transferred in a collision depends on this mass. A more critical dependence is the approximately inverse proportionality between S and the surface binding energy, which depends on surface topography, grain characteristics, crystal plane and crystal axis orientation within the grain, and other surface conditions, as well as atomic number and basic lattice constants.

Dose. If the target atoms and incident ions are not the same species, then the composition of the surface changes as a function of time (or dose or fluence); in order to predict the sputtering at any given time, the near-surface composition at that time must be considered.

The conditions that lead to large sputtering yields are a low value of target surface binding energy, a high value of Z_1, an intermediate value of E_i (ca 100 keV), and a high value of θ (but not more than ca 70°). Figure 4 illustrates some of these dependences and gives measured values of S for different ions that are incident on polycrystalline copper (16). An example of the significance of sputtering yield follows. From Figure 4 it can be seen that 50-keV argon ions striking polycrystalline copper, at normal in-

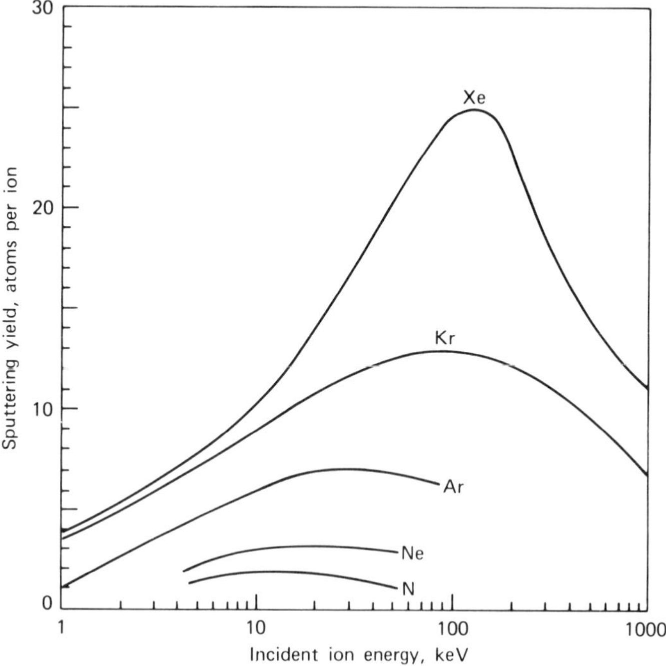

Figure 4. The sputtering yield S for a polycrystalline copper target as a function of incident ion energy E_i for various ion species at normal incidence (16). These curves have the same general shape as the elastic energy-loss curves of Figure 2.

cidence, exhibit a sputtering yield of ca 7. Therefore, for a fluence of 10^{17} argon ions/cm^2 on copper, ca 7×10^{17} copper atoms/cm^2 (equivalent to 360 atom layers) are sputtered. Although this dimensional change is not measurable by ordinary instruments, it is several times the range of 50-keV argon ions in copper (95 atom layers); it removes all of the atoms initially implanted if the implantation energy is held constant.

Sputtering tends to clean the surface being bombarded and (under certain conditions) tends to polish the surface. As the surface erodes during bombardment, lower layers become exposed; so if the bombardment lasts long enough, the layer in which the initial ions were implanted is exposed. This effect can be an advantage where a substantial concentration of implanted ions is wanted on the surface for anticorrosion purposes. Surface analysis techniques can be used with sputtering to measure concentrations of elements as a function of depth (see Analytical methods). Sputtering leads to a saturation concentration of implanted ions. The larger the fraction of implanted atoms in the target, the more implanted atoms are sputtered until, eventually (after the erosion depth becomes a few times the original implantation depth), an equilibrium is reached at which one previously implanted atom is sputtered for each new atom that is implanted. If S (sputtering yield) $\gg 1$, then the equilibrium relative concentration of implanted atoms is $1/S$, or $100/S$ atom %. This effect can be beneficial, for example, in smoothing out concentration fluctuations in the implantation of an object having a complex shape for which it is difficult to assure uniform irradiation. (A complicating consideration is that, if an object has a complex shape, resulting in different angles of incidence at different locations, then S is a function of location, and the equilibrium concentration also is a function of location.) Although the equilibrium relative concentration of implanted atoms is $100/S$ atom %, the equilibrium value can be exceeded substantially by decreasing the incident ion energy as a function of time such that the newly implanted ions always come to rest in the same physical layer and by stopping the implantation process when the eroding surface is near or at this physical layer.

Power Dissipation. The power density of the beam at the target can be substantial. For example, a 0.5 mA beam of 200-kV ions has a power of 100 W. If the beam has a cross section of 0.5 cm^2, then the power transport density is 200 W/cm^2. Therefore, precautions must be taken to prevent unwanted temperature excursions of the target; eg, scanning the beam over the front of the target and rotating the target about its own axis. Such techniques can keep the average target temperature within tolerable limits (eg, 100°C).

The power dissipation problem is exacerbated by the vacuum environment of the target: the principal mode of heat transfer may be radiation, and polished metals are poor radiators. Conduction transfer is significant only if the target makes good contact with a material that is soft enough to conform to the microscopic contours of the target, that itself is a good conductor of heat, and that is thermally coupled to a heat sink.

Economic Aspects

Estimates of the production-line cost (depreciation, overhead, materials, and labor) of implanting metals depend on several assumptions (including the ion current available and the fluence needed). For a fluence of 10^{17} ions/cm^2 of an element such as chromium into a steel subcomponent (which is a typical dose for many nonsemiconductor applications), the cost is ca \$0.10/cm^2. This cost rate, which may decrease significantly as the technology advances, implies that the ion implantation of small or medium-sized parts is cost-effective where reliability is important or where the implantation substantially increases the time between replacements and where a substantial amount of labor is involved in the replacement.

Uses

Corrosion Resistance. Alloys produced by ion implantation generally have the same corrosion properties as conventional bulk alloys of about the same composition (17–19). For example, the implantation of chromium and nickel into iron to about the same concentrations as found in stainless steels provides about the same corrosion behavior. Therefore, a stainless steel on and near the surface of an easily corrodible steel that possesses the required bulk properties can be produced. Since ion implantation is essentially a brute force (athermal) process involving individual atoms, the process is not restricted by the laws of thermodynamics governing equilibrium processes (see also Metallic coatings, explosively clad metals). For example, the implanted atoms can be placed at any desired location and to any desired concentration in a solid material (within the limitations of available ion energy) without being restricted by diffusivity constants or solubility constants (20–21). Also ion implantation can, under some circumstances, produce an amorphous alloy, which generally possesses greater corrosion resistance than the corresponding crystalline alloy (14). Experiments have shown that the implantation of both chromium and molybdenum into M50 steel produces even greater resistance to both pitting and crevice corrosion than implantation of either one alone (22).

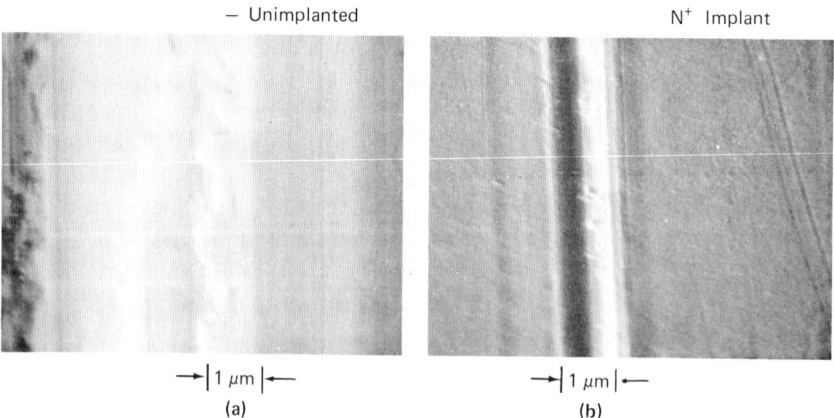

Figure 5. Scanning electron micrographs of wear tracks on a cylinder of type 304 stainless steel after wear test with a crossed stationary cylinder of type 416 stainless steel, lubricated by a jet engine oil. Half the rotating cylinder was implanted with nitrogen ions and half was left unimplanted. (**a**) Unimplanted track (a wide, deep apparently soft furrow). (**b**) Implanted track (a narrow shallow groove that appears to have been cut instead of worn).

Wear Resistance and Friction Reduction. Several experiments have shown that the implantation of selected species of ions (eg, nitrogen or titanium) into a variety of materials enhances their resistance to sliding wear under lubrication conditions (see Bearing materials). Pin-on-disk wear tests (23) of En40B nitriding steel, 440C stainless steel, and mild steel disks implanted with nitrogen, boron, or molybdenum ions and lubricated with white spirit show lower wear rates than unimplanted specimens by roughly a factor of 30. Crossed-cylinder wear tests with a polyester jet-engine oil following the implantation of nitrogen ions into AISI-416 and AISI-304 stainless steels give a similar result (24). Figure 5 shows a portion of a wear track for an unimplanted sample of AISI-416 steel and a portion of a wear track with all conditions the same except that the rotating cylinder had been implanted with nitrogen ions. It appears that the implantation of elements that individually or collectively constitute solid lubricants (eg, tin or molybdenum and sulfur) reduces sliding friction (25–26). Implantation of the wear surfaces of the following tools increases their lifetimes by factors from 2–12: paper slitters, acetate punches, taps for drilling plastic, slitters for synthetic rubber, tool inserts, forming tools, dies for copper rod, drawing dies, and dies for steel wire (27).

Fatigue Resistance. Metal fatigue, leading to high-cycle failure of a part, appears to be initiated by a surface or subsurface crack that eventually undergoes catastrophic propagation. The implantation of 18/8/1 stainless steel with nitrogen ions greatly increases its fatigue lifetime (28). Flat samples of maraging steel (martensitic steel tempered to provide optimum workability) gives similar results. Experiments also were performed on rods of AISI-1018 low-carbon steel, which were implanted and aged at room temperature for 4 mo or at 100°C for 6 h. The implanted samples had a fatigue lifetime of roughly 100 times that of the unimplanted samples (29). The implantation of carbon into titanium alloy (6% aluminum and 4% vanadium) increases the endurance limit by 20% and the lifetime by a factor of 4 or 5 (30).

Other. The index of refraction of optical-grade silicon can be modified as a function of depth by the implantation of nitrogen ions (31). The implantation of boron into beryllium (which is used in gas-support bearings in gyroscopes because of its low density) to a relative concentration of 40 atom % followed by annealing produces an apparent increase in hardness by a factor of 6 (32). As the implanted concentration of nitrogen ions in molybdenum increases from 1 to 23 atom %, the critical temperature for superconductivity increases from 1.5 to 9.2 K (33) (see Superconducting materials).

BIBLIOGRAPHY

1. U.S. Pat. 2,750,541 (June 12, 1956), R. S. Ohl (to Bell Telephone Laboratories).
2. U.S. Pat. 2,787,564 (April 2, 1957), W. Shockley (to Bell Telephone Laboratories).
3. U.S. Pat. 2,842,466 (July 8, 1958), J. W. Moyer (to General Electric Co.).
4. J. Lindhard, V. Nielsen, and M. Scharff, *Kgl. Danske Videnskab. Selskab Mat.-Fys. Medd.* **36**(10), (1968). This is a fundamental treatment of elastic energy transfer.
5. M. Robinson and I. M. Torrens, *Phys. Rev.* **B9,** 5008 (1974).
6. J. Lindhard, *Kgl. Danske Videnskab. Selskab Mat.-Fys. Medd.* **28**(8), (1954); O. B. Firsov, *Sov. Phys. JETP* **9,** 1076 (1959). These are fundamental treatments of inelastic energy transfer.
7. J. Lindhard, M. Scharff, and H. E. Schiøtt, *Kgl. Danske Videnskab. Selskab Mat.-Fys. Medd.* **33**(14), (1963). This is a fundamental treatment of ion range.
8. I. Manning and G. P. Mueller, *Comp. Phys. Comm.* **7,** 85 (1974).
9. M. T. Robinson and O. S. Oen, *Phys. Rev.* **132,** 2385 (1963).

10. G. H. Kinchin and R. S. Pease, *Rep. Prog. Phys.* **18,** 1 (1955).
11. M. J. Norgett, M. T. Robinson, and I. M. Torrens, *Nucl. Eng. Des.* **33,** 50 (1975).
12. A. Ali, W. A. Grant, and P. J. Grundy, *Phil. Mag.* **37B,** 353 (1978); *Radiat. Eff.* **34,** 251 (1977).
13. T. Masumoto and R. Maddin, *Mater. Sci. Eng.* **19,** 1 (1975).
14. T. M. Devine and L. Wells, *Scr. Met.* **10,** 309 (1976); M. Naka, K. Hashimoto, and T. Masumoto, *Corrosion* **32,** 146 (1976); K. Hashimoto, K. Osada, T. Masumoto, and S. Shimodaira, *Corros. Sci.* **16,** 71 (1976).
15. Adapted from K. B. Cheney and E. T. Pitkin, *J. Appl. Phys.* **36,** 3542 (1965).
16. Adapted from H. H. Andersen and H. L. Bay, *Radiat. Eff.* **19,** 139 (1973); **13,** 67 (1972); O. Almen and G. Bruce, *Nucl. Inst. Methods* **11,** 257 (1961); G. Dupp and A. Scharmann, *Z. Physik* **192,** 284 (1966); **194,** 448 (1966); and F. Keywell, *Phys. Rev.* **97,** 1611 (1955).
17. V. Ashworth, W. A. Grant, R. P. M. Procter, and T. C. Wellington, *Corros. Sci.* **16,** 393 (1976).
18. B. D. Sartwell, A. B. Campbell, and P. B. Needham, in F. Chernow, J. A. Borders, and D. K. Brice, eds., *Ion Implantation in Semiconductors, 1976,* Plenum Press, New York and London, 1977, pp. 201–211.
19. B. S. Covino, Jr., B. D. Sartwell, and P. B. Needham, Jr., *J. Electrochem. Soc.* **125,** 366 (1978).
20. A. G. Cullis, J. A. Borders, J. K. Hirvonen, and J. M. Poate, *Phil. Mag.* **B37,** 615 (1978).
21. V. Ashworth, W. A. Grant, R. P. M. Procter, and E. J. Wright, *Corros. Sci.* **18,** 681 (1978).
22. Y. F. Wang, C. R. Clayton, G. K. Hubler, W. H. Lucke, and J. K. Hirvonen, *Thin Solid Films* **63,** 1 (1979).
23. N. E. W. Hartley, G. Dearnaley, J. F. Turner, and J. Saunders, in S. T. Picraux, E. P. EerNisse, and F. L. Vook, eds., *Applications of Ion Beams to Metals,* Plenum Press, New York and London, 1974, pp. 123–138.
24. J. K. Hirvonen, *J. Vac. Sci. Technol.* **15,** 1662 (1978).
25. N. E. W. Hartley, G. Dearnaley, and J. F. Turner, in B. L. Crowder, ed., *Ion Implantation in Semiconductors and Other Materials,* Plenum Press, New York and London, 1973, pp. 423–436.
26. H. M. Pollock, *Brit. J. Appl. Phys.* **D11,** 39 (1978).
27. N. E. W. Hartley, *Inst. Metall. Ref.* **1101-78-Y,** 197 (1978); *AERE Report* (Harwell, England) **R-9065 U/C** (1978).
28. N. E. W. Hartley, *Inst. Phys. Conf. Ser.* **28,** 210 (1976).
29. W. W. Hu, C. R. Clayton, H. Herman, and J. K. Hirvonen, *Scr. Met.* **12,** 697 (1978).
30. R. G. Vardiman, private communication, 1979.
31. G. K. Hubler, P. R. Malmberg, T. P. Smith, III, *J. Appl. Phys.* **50,** 7147 (1979).
32. R. A. Kant, *Thin Solid Films* **63,** 27 (1979).
33. O. Meyer, *Inst. Phys. Conf. Ser.* **28,** 168 (1976).

General References

Books

G. Dearnaley, J. H. Freeman, R. S. Nelson, and J. Stephen, *Ion Implantation,* North-Holland Pub. Co., Amsterdam and London, 1973. An excellent, thorough, and detailed account of the field of ion implantation; extensive bibliography.

P. D. Townsend, J. C. Kelly, and N. E. W. Hartley, *Ion Implantation, Sputtering and their Applications,* Academic Press, London, New York, and San Francisco, 1976. A very readable, relatively short account with good selection of topics.

J. K. Hirvonen, ed., *Ion Implantation,* Academic Press, New York and London, 1980. The first comprehensive account of the applications of the ion implantation of metals.

Ion Implantation as a New Surface Treatment Technology, report issued by the NRC Committee on Ion Implantation and Competing New Surface-Treatment Technologies to the U.S. Dept. of Defense, Washington, D.C., 1979.

Proceedings of International Conferences

F. H. Eisen and L. T. Chadderton, eds., *Ion Implantation,* Gordon and Breach, London, New York, and Paris, 1971.

I. Ruge and J. Graul, eds., *Ion Implantation in Semiconductors,* Springer-Verlag, Berlin, Heidelberg, and New York, 1971.
B. L. Crowder, ed., *Ion Implantation in Semiconductors and Other Materials,* Plenum Press, New York and London, 1973.
S. T. Picraux, E. P. EerNisse, and F. L. Vook, eds., *Applications of Ion Beams to Metals,* Plenum Press, New York and London, 1974.
S. Namba, ed., *Ion Implantation in Semiconductors,* Plenum Press, New York and London, 1975.
G. Carter, J. S. Colligon, and W. A. Grant, eds., *Applications of Ion Beams to Materials 1975,* The Institute of Physics, London and Bristol, 1976.
F. Chernow, J. A. Borders, and D. K. Brice, eds., *Ion Implantation in Semiconductors 1976,* Plenum Press, New York and London, 1977.
W. Brown, ed., *AVS Symposium on Ion Implantation, New Prospects for Materials Modification, J. Vac. Sci. Technol.* **15,** 1629 (1978).
J. Gyulai, T. Lohner, and E. Pásztor, eds., *Ion Beam Modification of Materials,* Vols. I–III, Central Research Institute for Physics, Budapest, 1979.

Tabulations of Ion Ranges and Related Information, Listed Chronologically

J. F. Gibbons and W. S. Johnson, *Projected Range Statistics in Semiconductors,* Stanford University Book Store, Palo Alto, Calif., 1969.
L. C. Northcliffe and R. F. Schilling, *Nuclear Data Tables* **A7,** 233–463 (1970).
B. J. Smith, *AERE Report R-6660,* Harwell, U.K., 1971.
D. K. Brice, *Ion Implantation Range and Energy Deposition Distributions,* Vol. 1, *High Incident Ion Energies,* Plenum Press, New York, Washington, and London, 1975.
K. B. Winterbon, *Ion Implantation Range and Energy Deposition Distributions,* Vol. 2, *Low Incident Ion Energies,* Plenum Press, New York and London, 1975.
J. F. Gibbons, W. S. Johnson, and S. W. Mylroie, *Projected Range Statistics,* 2nd ed., Halsted Press, a div. of John Wiley & Sons, Inc., New York, 1975.

JAMES W. BUTLER
Naval Research Laboratory

IONOMERS. See Ionomers, Supplement volume.

IONOPHORES. See Antibiotics, polyethers; Antibiotics, polypeptides; Chelating agents.

ION-SELECTIVE ELECTRODES

Ion-selective electrodes (ISEs) are membrane electrodes as compared to the older and better known metal/metal-salt redox electrodes. ISEs are electrochemical devices whose voltage output at virtually zero current is directly related to the concentration of some species which, generally, is in solution.

The first true ISE was the glass pH electrode which responds to hydrogen-ion activity (1,2). First developed in 1906–9, it did not come into general usage until the early 1930s, when the availability of vacuum-tube voltmeters (instead of quadrant electrometers) made the measurement of hydrogen-ion activity convenient (3). The use of glass electrodes for sodium instead of pH was first reported in 1923 (4), but it was not until 1957 that a systematic search of glass compositions was carried out and the most widely used sodium glass (NAS 11-18) was reported (5). No further highly selective glass (qv) electrodes have been found, but the search by Ross for a calcium-selective glass led to the conclusion that a more mobile phase was required. This resulted in the invention of the calcium-selective electrode in 1964 (6), the first of a series of liquid-membrane electrodes, which later included nitrate and water-hardness electrodes. Most recently, the liquid-membrane approach has been extended by the development of neutral-carrier electrodes, beginning with the potassium electrode in 1966 (7) and continuing with an improved calcium electrode (8).

Modern solid-state ISEs, which are based on crystalline solids rather than glass, date back to the work of Kolthoff in 1937 (9) when it was shown that silver halide membranes could function as the sensing element in electrodes. However, such membranes were light sensitive, and it was not until the discovery in 1966 that the addition of Ag_2S to the halide eliminated the light sensitivity and made such electrodes become feasible commercially (10). There followed a series of solid-state membrane electrodes for sulfide (11); for copper, cadmium, and lead (12); for cyanide (13); and for fluoride (14). Fluoride was particularly important because it was the first nonsilver-based, solid-state membrane electrode, and because fluoride analyses were difficult by other techniques. Also included as ISEs in this article are several gas-sensing electrodes, including the Severinghause CO_2 electrode (15) and the ammonia electrode (16). Some of the commercially available ISEs and their characteristics are listed in Table 1 (see also Hydrogen-ion activity).

Theory of Operation

A hypothetical arrangement, in which two solutions of different concentrations of the same salt are separated by a solvent-impermeable membrane, is illustrated in Figure 1. If the membrane were permeable to the salt, both solutions, at equilibrium, would contain the same concentration of the salt. On the other hand, if the membrane were permeable to only one species, say A^+, then the situation becomes somewhat more complex. There is a flux of A^+ in both directions, with the rate in each direction being proportional to the thermodynamic activity. This results in an uneven charge distribution across the membrane, with a consequential building up of a potential across the membrane in a direction which opposes the greater flux. Positive charges would begin to accumulate on the more dilute side of the membrane, and a negative charge would begin to accumulate on the more concentrated side where there are more un-

Table 1. Commercially Available Electrodes

Electrode	Type	Concentration, M	Typical slope, mV	Major interferences	Typical uses
ammonia	gas-sensing	$1–10^{-6}$	−57	volatile amines	waste streams, seawater, Kjeldahl digestions
bromide	solid-state	$1–10^{-6}$	−56	S^{2-}, I^-, CN^-, high levels of Cl^- and NH_3	grains, plant tissues
cadmium	solid-state	$1–10^{-7}$	+28	Ag^+, Hg^{2+} and Cu^{2+}, high levels of Cd^{2+} and Fe^{3+}	plating baths
calcium	liquid-membrane	$1–10^{-6}$	+28	0.2 M Na^+ gives 10% error at 10^{-3} M Ca^{2+}	serum, soils, milk
carbon dioxide	gas-sensing	$10^{-2}–10^{-4}$	+56	volatile weak acids	blood, groundwaters
chloride	solid-state	$1–10^{-6}$	−56	S^{2-}, Br^-, I^-, Cu^{2+}	boiler feedwater, foods
cupric	solid-state	satd–10^{-8}	+26	S^{2-}, Ag^+, Hg^{2+}, high levels of Cl^-, Br^-, Fe^{3+}, and Cd^{2+}	natural waters, plating baths
cyanide	solid-state	$10^{-2}–10^{-6}$	−58	S^{2-}, Ag^+, Hg^{2+}, I^-	plating baths, waste streams
fluoride	solid-state	satd–10^{-6}	−58	OH^-	drinking water, stack gases, bone and materials, urine, soils
fluoroborate	liquid-membrane	satd–10^{-6}	−57	NO_3^-, I^-	plating baths
iodide	solid-state	$1–5 \times 10^{-8}$	−56	S^{2-}	milk, feeds, pharmaceuticals
lead	solid-state	$1–10^{-6}$	+25	Hg^{2+}, Ag^+, Cu^{2+}, high levels of Cd^{2+}, and Fe^{3+}	sulfate titrations
nitrate	liquid-membrane	$1–7 \times 10^{-6}$	−56	I^-, CN^-, Br^-, high levels of Cl^-	soils, fertilizers, plant tissues, drinking water
pH	glass	pH 0–14	+59	high levels of Na^+ in alkaline range	industrial process control, blood, wastewater, plating baths, etc
potassium	liquid-membrane	$1–10^{-6}$	+56	Cs^+, NH_4^+, H^+	serum, soils, wine
silver/sulfide	solid-state	Ag^+: $1–10^{-7}$ S^{2-}: $1–10^{-7}$	+56 −28	Hg^{2+}	photographic solutions, pulping liquors, wastewater, soils
sodium	glass	satd–10^{-7}	+56	Ag^+, H^+	serum, boiler feedwater, foods
water hardness	liquid-membrane	$10^{-2}–6 \times 10^{-6}$	+24	responds to most M^{2+}	water treatment

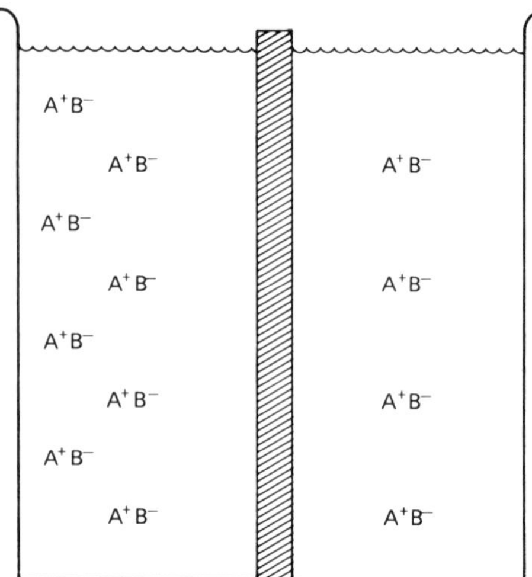

Figure 1. Hypothetical system in which two salts are separated by a solvent-impermeable membrane.

matched B^- ions. In time, depending on the mobility and number of charge carriers, the resulting potential would exactly balance the net flow in each direction, and no further net flow of charge would occur. In a system in which no current can flow, the total number of ions required to develop this opposing potential is extremely small compared to usual solution concentrations. This potential difference is of the same form as that described for redox electrodes by Nernst (15). The Nernst potential can be observed by inserting a pair of voltage probes (one into each solution) and by observing the output on a suitable voltmeter. In practice, one of the probes is placed inside the electrode—of which the sensing element is the membrane described above, and the second probe is placed in a solution of fixed composition, which is connected electrolytically to the sample solution. Such an arrangement for the glass pH electrode is shown in Figure 2.

For an ideal membrane, which is permeable to only one species, and for ideal voltage-measuring probes, the Nernst equation may be written as:

$$E = E^0 + (RT/nF) \ln A$$

where E = the observed potential; E^0 is a constant potential which varies with the choice of reference electrodes; R is the gas constant in joules; T is the temperature in K; n is the integral value of the charge on the measured species, A; F is the Faraday constant; and A is the thermodynamic activity of the measured species. For a tenfold change in A, E changes by a factor of 60 mV/n; this is referred to as the electrode slope. (Because probes often are nonideal, an additional term, E_j, which results from the junction potential at the electrolytic junction, should be added. Most ISEs are used in situations where differences in junction potential between standards and samples are not significant; junction potentials are not discussed in this article).

The properties and performance of the electrode depend on the composition and choice of membrane material. This provides a convenient way to discuss the various types.

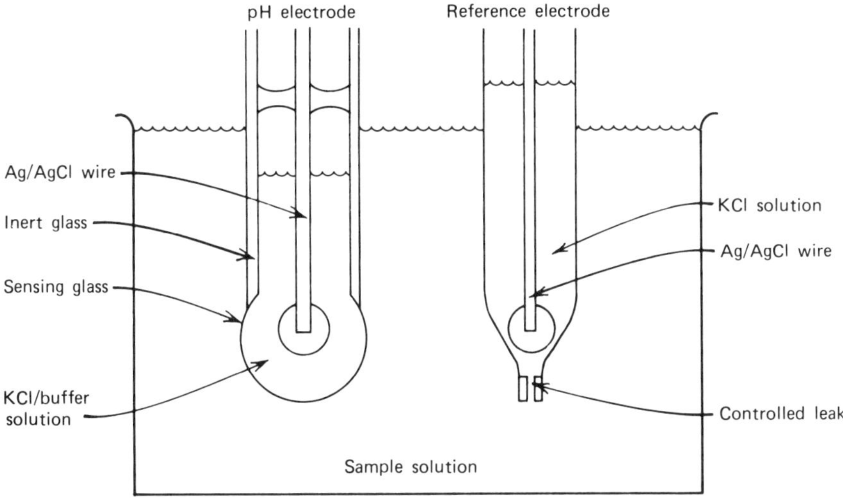

Figure 2. Glass pH electrodes are typical of ISEs.

Types of Electrodes

Glass. Glass electrodes consist of a three-dimensional silicate network containing occasional SiO^- sites. Because of the relatively open network structure, it is possible for monovalent cations to move from site to site. Although this provides a convenient pathway for monovalent cations, anions cannot enter and glass electrodes usually have a response that is very close to theoretical. Because polyvalent cations are held at more than one site simultaneously, their mobility in glass is very low, and no commercially-useful divalent glass electrode has been found.

Since monovalent cations are mobile in the silicate network, the selectivity of glass electrodes depends upon ion-exchange sites at the surface. In the case of pH electrodes, there is general agreement that a hydrated silica layer occurs at the surface and is responsible for the selectivity. For sodium electrodes, the substitution of some aluminum for silica results in more acidic sites and less tendency to bind hydrogen ion (5). At the optimum level of aluminum, the glass still shows a preference for H^+ ion, but less than that of pH glasses by a factor of at least 10^6.

The fact that the selectivity is determined by an ion-exchange process at the surface results in a gradual interference effect (response to a species other than the one for which the electrode is intended) which depends on the ratio of the two competing species. This can be seen in Figure 3, which shows the response of the sodium electrode as a function of pH. As the hydrogen-ion activity increases, the level at which the electrode ceases responding to sodium also increases. In this instance, the electrode shows a preference for hydrogen ion of ca 100:1 over sodium. In general, for glass electrodes, a modified form of the Nernst equation may be written which takes into account the effect of interferences:

$$E = E^0 + (RT/nF) \ln (A + K_s Bn/m)$$

where K_s is the selectivity constant and m is the charge on ion B. (The IUPAC practice for this potentiometric selectivity constant between ions A and B is to write it as $K^{\text{pot}}_{A,B}$).

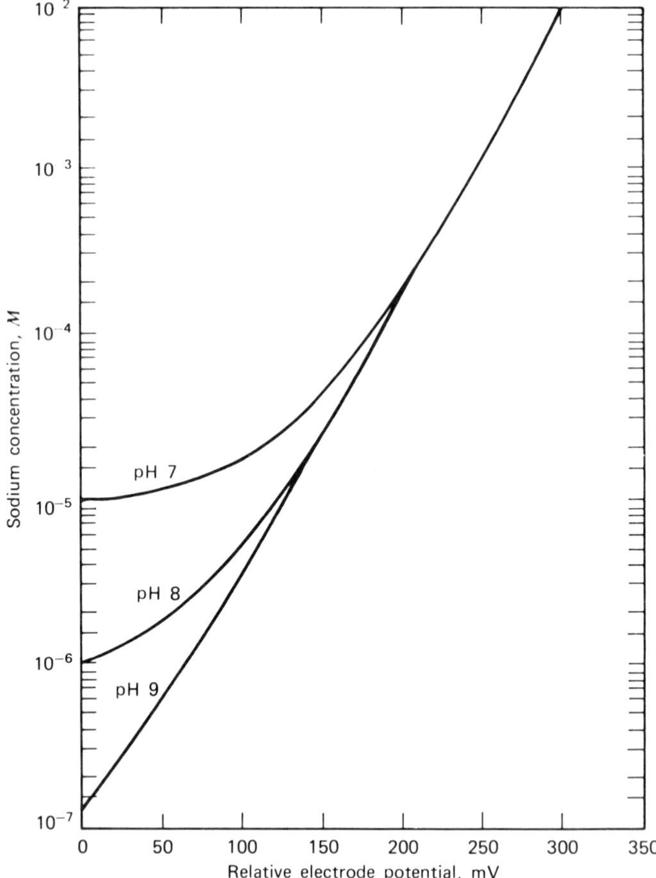

Figure 3. Response of sodium glass electrode to sodium ion as a function of solution pH.

Most values of K_s are only approximate and the constant may vary depending on the ratio of ions present, the ionic strength of the solution, the temperature, and other factors. Nonetheless, values of K_s are helpful in estimating whether or not an electrode method may be used. (Although values of K_s that are well below 1.0 are highly desirable, larger values may be tolerated if there are ways of chemically removing the interferences. In the above example of the sodium electrode, raising the pH of the solution with ammonia or organic amines permits sodium measurements down to the part per billion (109) level, and this is done in several commercial boiler-water monitors that use sodium electrodes).

Liquid-Membrane. Liquid-membrane electrodes are similar, both in mechanism and in response to interferences, to the glass electrodes. The difference is that in liquid-membrane electrodes, the ion-exchange sites are dissolved in a high-molecular-weight, water-insoluble organic solvent and are themselves mobile. For example, in the calcium electrode, a high molecular weight, partially esterified, organic phosphoric acid is used as the ion exchanger, which is dissolved in a phosphate ester, eg, di-*n*-octyl phenylphosphonate (6) or tri-*n*-pentyl phosphate. The substitution of decanol for the phosphonate as the solvent in the calcium electrode results in a marked change in selectivity and in an electrode that is nearly equally selective to calcium and magnesium (16). Neutral-carrier exchangers (see below) often are used with 2-nitrophenyl *n*-octyl ether or dibutyl sebacate.

In early versions of these electrodes, the solvent and ion exchanger were impregnated in microporous cellulose acetate, which separated the sample from the internal reference solution. The solvent was first used as a plasticizer in a membrane made from poly(vinyl chloride) in 1970 (17) and most commercial membranes are of this type.

In the nitrate electrode, the "site" consists of a cationic metal complex of bathophenanthroline and nickel (18). This compound has a marked preference for soft anions (eg, perchlorate or iodide as opposed to chloride and fluoride). The most effective nitrate ion exchangers have either been quaternary ammonium compounds, eg, Aliquat 336 (methyltricaprylammonium chloride) or compounds based on metal salts of bathophenanthroline (18).

The liquid-membrane potassium electrode (7) is unusual in that, instead of an ion exchanger, the potassium ion appears to be transported across the membrane by an electrically neutral compound, the antibiotic, valinomycin. This carrier, or ionophor, is a cyclic polypeptide with an extremely high selectivity for potassium over sodium. It is generally accepted that this selectivity is based on the ability of valinomycin to form a lipophilic cage around the potassium ion, into which other monovalent cations do not fit. This high selectivity makes it possible to measure potassium in biological fluids, eg, in serum (19) or whole blood. Because it is possible, in principle, to tailor the chemical nature and shape of the sites for any particular ion, neutral-carrier, liquid-membrane electrodes are the most active research area for highly selective electrodes (20).

For those applications where high selectivity is not required, many other liquid membrane electrodes have been used. The concentration of relatively pure acetylsalicylic acid in aqueous solutions could be measured by dissolving, eg, a high molecular weight quaternary ammonium salt of acetylsalicylic acid in a high-boiling, water-insoluble, organic liquid (eg, di-*n*-octyl phthalate) and use this as the basis of an electrode. There is extensive literature on such devices (21) (see Salicylic acid).

Liquid-membrane electrodes are made by dissolving poly(vinyl chloride), a plasticizer, and an ion exchanger in excess tetrahydrofuran. This solution is cast onto glass plates and is dried slowly; the result is a membrane film which can be handled and which is either cemented or mechanically sealed to the end of a poly(vinyl chloride) tube.

Solid-State. With the exception of the fluoride electrode, all of the commercial solid-state electrodes are silver-based. In a number of silver compounds (eg, the silver halides and silver sulfide), there are more lattice sites available than silver ions that are necessary for charge neutralization. Since the silver ion is a small monovalent ion, it is able to move between sites with only a modest activation energy. Since a number of silver salts are quite water-insoluble, they are almost ideal membrane materials. Any of them can be used as specific electrodes for the silver ion. In addition, for silver sulfide and the silver halides, in the absence of added silver, the Ag^+ activity at the membrane surface is a function of the activity of the anion in solution. For example, for silver sulfide, which has a solubility product of ca 10^{-50}, the following equilibrium can be written:

$$Ag_2S \rightleftharpoons 2\,Ag^+ + S^{2-}$$

In a solution that contains no added silver but does contain sulfide, the silver-ion activity is related to the sulfide-ion activity:

$$Ag^+ = \left[\frac{K_{sp}}{S^{2-}}\right]^{1/2}$$

From the user's point of view, it is not possible to distinguish this mechanism from a membrane that responds directly to sulfide ions. When used in silver solutions, the electrode has a slope of 60 millivolts ($n = 1$ in the Nernst equation) and, when used in sulfide solutions, it has a slope of 29 millivolts ($n = 2$).

The use of secondary equilibria has been extended by preparing a membrane consisting of a mixture of silver sulfide and several other metal sulfides. For example, a mixture of silver sulfide and cupric sulfide yields an electrode that responds with a theoretical slope to cupric ions (12). The hypothesis which led to their development was that, in the absence of added silver or sulfide ions, the activity of the cupric ion would fix the sulfide level at the membrane interface (since silver sulfide is less soluble than cupric sulfide) and this, in turn, would fix the silver level at the interface. Similar reasoning led to the development of cadmium and lead electrodes. However, the same reasoning does not lead to usable zinc or nickel electrodes, presumably because of kinetic factors.

The cyanide ISE also has a silver-based membrane, but it is not a truly Nernstian device. The membrane, which consists of a mixture of silver iodide and silver sulfide, reacts slowly with dilute cyanide solutions at the interface to provide a level of silver ions that is related to the cyanide activity in the solution:

$$AgI + 2\,CN^- \rightarrow Ag(CN)_2^- + I^-$$

$$AgI \rightleftharpoons Ag^+ + I^-$$

A cyanide electrode is sensitive to stirring and the membrane material is slowly attacked by the solution in which it is placed. An alternative method of measuring cyanide (22) involves using the highly insoluble silver sulfide electrode as a silver sensor. A small amount of silver cyanide complex is added to the solution. The extent of dissociation of the complex, and the resulting silver activity in the solution, are fixed by the level of free cyanide:

$$Ag(CN)_2^- \rightleftharpoons Ag^+ + 2\,CN^-$$

The silver activity varies as the square of the cyanide activity, and this method has a 120 mV slope for cyanide ion.

The silver-based electrodes behave differently in regard to selectivity than the glass and liquid-membrane electrodes. As the membranes are not permeable to other ionic species, there is no gradual onset of interference by other ions. However, because of solubility-product equilibria, it is possible to convert the membrane surface to another compound if a sufficient level of another species is present in the sample. Thus, very small amounts of sulfide would convert the surface of a silver iodide membrane to silver sulfide, resulting in the inability to measure iodide. In the same way, a 500-fold excess of chloride would convert a silver bromide surface to a silver chloride surface, since the ratio of the AgCl/AgBr solubility products is ca 200 at 25°C. The onset of such interferences is fairly sharp and can be calculated from the solubility-product equilibria.

Most of the silver-based ISEs are made by pressing pellets of the appropriate material at high pressure and by cementing them into a plastic tube (which usually is of epoxy resin). Other approaches have been described in the literature, eg, mixing the silver materials with silicone rubber (23) or rubbing the mixture of sulfides on a graphite rod or tip (24). The mechanism appears to be identical in all of these cases, and commercial electrodes are available using the latter approach.

The fluoride electrode, which has a sensing membrane made of a rare earth fluoride (eg, LaF_3), with or without doping to increase conductivity, is the only nonsilver-based membrane material to achieve commercial success. Hexagonally-crystallizing rare earth fluorides (LaF_3, CeF_3, NdF_3, and SmF_3) have a lattice structure in which there are alternate layers of LaF^{2+} and F^- ions; the latter require only a small activation energy for mobility.

Because of the small size of the fluoride ion, the membrane does not appear to be permeable to any other species and, using an electrical potential to force other anions (eg, OH^-) into the lattice, results in crystal breakage. Hydroxide ion, at levels of about tenfold greater than the amount of fluoride present, does cause an interference, which is believed to result from a surface reaction involving the formation of mixed lanthanum hydroxide/fluoride complexes. This hydroxide interference is normally handled by adjustment of the solution pH. In the silver-based systems, solubility-product equilibria permit preparation of electrodes for halides and for some divalent metals. The corresponding equilibria are too slow in the case of fluoride-based compounds; thus, there are no fluoride-based electrodes for the rare-earth metal ions, or for lead (based on double equilibria) eg, LaF_3/PbF_2.

The fluoride electrode is made from disks that are sliced from a rod of single-crystal LaF_3, which then is sealed to an epoxy tube. The sealing techniques are highly proprietary, and require a good deal of skill, since there is a large mismatch in expansion coefficient between the lathanum fluoride crystal and the epoxy tube. The requirement that the seal be highly resistant to solvents and strong chemicals prevents the use of most cements that are ductile or elastic enough to contain the thermally-induced stresses. At present, the cement is usually the weakest link in determining the resistance of the electrode to various solvent acids. Both epoxy and silicone rubber seals have been used, but the epoxy types usually are superior.

Manufacture of the silver-based, solid-state electrodes is more complex, since it is difficult to assure the proper stoichometry in the crystal. For example, in the manufacture of the cadmium sulfide/silver sulfide membrane, the procedure that was patented for bulk co-precipitation (12) directed that a mixture of the corresponding metal solutions be added to an excess of sodium sulfide with vigorous stirring. The purpose of this procedure was to ensure that an excess of sulfide was present throughout the precipitation. The excess sodium sulfide subsequently was washed out. Steps (eg, proper washing to remove sodium sulfide without oxidizing any of the extremely fine precipitate) must be done with great care. Typically, after washing with water and acetone, the precipitates are dried and powdered. They are then pressed in a die at high pressure. In some instances, the pressing is done at elevated temperatures, but this may lead to some decomposition or reaction with the die materials, and generally is not recommended.

Gas-Sensing. In gas-sensing electrodes, the sensing electrode is placed in a small amount of solution which is separated from the sample by a water-impermeable, gas-permeable membrane. In the case of the CO_2 electrode, a pH electrode and a reference electrode are in an extremely small amount of solution on one side of a silicone rubber membrane, and the sample solution is on the other side. CO_2 diffuses through the silicone rubber and changes the pH of the solution on the inside, according to the equilibrium:

$$CO_2 + H_2O \rightarrow HCO_3^- + H^+$$

If its level is fixed by placing a relatively large amount of bicarbonate ion in the inside solution, the pH of that solution will vary directly with the CO_2 vapor pressure in the external sample.

The principle of diffusion of a neutral species through a membrane followed by a reaction that can be sensed by an electrode led to the development of a number of other gas-sensing electrodes, of which only the ammonia electrode has assumed significant commercial importance. This electrode operates in identical fashion to the CO_2 electrode, with gaseous NH_3 diffusing through the membrane from an alkaline solution to establish the following equilibrium on the inside of the membrane:

$$NH_3 + H_2O \rightleftharpoons NH_4^+ + OH^-$$

A high level of NH_4^+ is maintained inside the membrane. Both the CO_2 and NH_3 electrodes have a 60-mV slope, since each molecule of gas produces one OH^- or H^+ ion. As might be expected, however, the slopes are in the opposite direction.

Manufacture of gas-sensing electrodes usually begins with modified pH electrodes that have carefully controlled radii of curvature. Because most measurements are made in the middle pH range, low resistance glasses that have poor sodium rejection are satisfactory. The outer membrane usually is a microporous Teflon or polyethylene membrane. A small amount of filling solution is introduced between the glass and Teflon membranes. In the ammonia electrode, the filling solution contains 0.1 *M* ammonium chloride and an anion that limits the upper end of the ammonium concentration. For example, picric acid has been added to the filling solution so that if the ammonium level exceeds 0.1 *M* as a result of evaporation, ammonium picrate precipitates and holds the ammonium level constant. If fresh water is diffused into the filling solution, the ammonium picrate redissolves.

Other. A number of other types of electrodes have been proposed but have not been applied commercially. Among these are tertiary, silver-based systems for sulfate (25), silicone rubber membranes for sulfate and phosphate (26–27), and enzyme electrodes. The latter are similar to the gas-sensing electrodes, but the reaction with a neutral species (eg, glucose or urea) takes place enzymatically within a water-permeable membrane, and the reaction product is detected by a suitable electrode (28). Although the approach appears quite promising, the long-term stability of the immobilized enzymes has not been adequate for widespread commercial usage (see Enzymes, immobilized). There also has been interest in miniaturized electrodes, in which the membrane is applied directly to the gate of a field-effect transistor (Chemfets) (29). The performance is similar to other ISEs except for lifetimes, which often are shorter. Although the approach offers a possibility for savings by mass manufacture, there are no commercial applications as of 1980.

Manufacturers

Of the worldwide manufacturers, the largest is Orion Research Inc. of Cambridge, Mass. They make the most complete line of electrodes, hold most of the patents in the field and have from 70–80% of the world market. The second largest manufacturer is Radelkis of Budapest, Hungary. Radelkis supplies virtually all of the electrodes sold in Eastern Europe and the USSR. They have been selling in Western Europe and may be marketing in the United States under local trade names. Manufacturers in Western Europe are the Phillips Company in Eindhoven, Holland, and Radiometer of Denmark. In the United States, Corning Glass Works offers a complete line of solid-state and gas electrodes, but is manufacturing only part of the line. HNU Systems, Inc. in Newton, Mass. offers a complete line of electrodes, all of which appear to be of their own manufacture.

In the United States, most of the suppliers of pH electrodes (Leeds & Northrup, Beckman, Foxboro, etc) offer ISEs but, in most cases, these companies appear to be making their own glass electrodes and to be obtaining the others from one or more of the suppliers listed above. In many cases, the private-label electrodes may be tailored for specific applications and manufactured to more stringent specifications than the general laboratory line offered by the manufacturers.

Economic Aspects

It is estimated that, exclusive of pH electrodes, ca $\$10^7$ worth of ISEs were sold worldwide in 1978. The world market for pH electrodes is believed to be an additional $\$(50–60) \times 10^6$ (see Hydrogen-ion activity). Laboratory versions of ISEs generally sell in the \$175–300 price range; glass pH electrodes are \$30–75. Laboratory meters for both range from \$200–1600.

The fluoride electrode has the highest worldwide unit sales, followed closely by the glass sodium electrode, and then, probably, by the ammonia gas-sensing electrode. The two most popular liquid-membrane types are the nitrate and potassium electrodes, and these outsell the remaining solid-state electrodes. The two most widely used solid state electrodes are the silver/sulfide and the chloride electrodes. The metal-detecting, solid-state electrodes are next, and the least widely used electrode probably is the liquid-membrane, perchlorate electrode.

Analytical Methodology

Activity Measurements. ISEs can be used for making activity measurements (pIon), by techniques that are analogous to the measurement of pH. The electrode is standardized in solutions of known activity, and the activity of an unknown solution is measured. Since the electrodes always respond to activity and not to concentration, activity measurements are straightforward. Although the exact analogues of NBS pH buffers have not been developed for most ions, several papers have been published on the subject (30). Such measurements should be useful in work involving ionic equilibria (eg, metal-ion chelation) and are of potential importance in such applications as preventing boiler-scale formation or in predicting precipitation of potassium tartrate in wine.

Concentration Measurements. In general, almost all ISEs are used for concentration measurements rather than for activity. There are a number of related procedures that can be used to obtain concentration results, and the choice of the appropriate method generally depends on the application.

Direct Calibration. Direct calibration can be used where the background remains relatively constant; standards are prepared using exactly the same background as is used in the unknown. For example, to measure chloride in tomato soup, one would start with salt-free soup and add varying amounts of salt to establish the initial calibration curve. It should be recalled that, since electrodes follow the Nernst equation, calibration curves are prepared by plotting the millivolt readings against the log of the concentration (see Fig. 3). Direct-reading instrumentation is widely available.

In many instances, the background varies, which results either in a change in the ionic strength (which affects the activity, see Fig. 4), or in the species of interest being complexed to a varying extent with other ions in the solution. For those ions whose measurement can be performed without interference in the presence of a large excess of some other ion, the noninterfering ion can be added in a large amount as a way of swamping out variations in ionic strength. In the chloride case above, more than one type of soup could be measured without recalibration by mixing a small amount of soup with a large volume of a soluble nitrate salt.

If the ion of interest is partly or completely complexed, an error will result, since electrodes respond only to the "free" fraction. To overcome this, it sometimes is possible to add a decomplexing or masking agent. The most notable example of this approach occurs in the measurement of fluoride in drinking water (see Water, municipal water treatment). Aluminum often is present in varying amounts as a result of alum treatment of the water, and a variable fraction of the fluoride is complexed with the

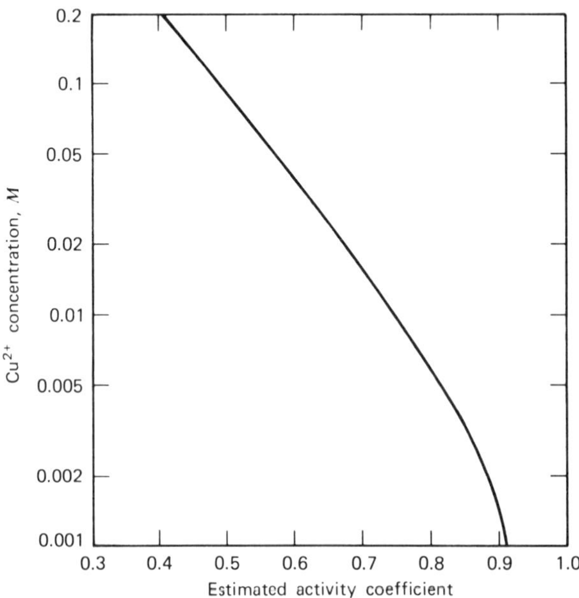

Figure 4. Estimated variation of the activity coefficient for cupric ion as a function of solution total ionic strength.

aluminum. This is overcome by the use of TISAB (total ionic strength adjustment buffer) (31) which contains a high level of chloride ion to provide an unvarying ionic strength background, and citrate or CDTA [(1,2-cyclohexylenedinitrilo)tetraacetic acid] to preferentially complex the aluminum and to free the fluoride. TISAB also contains an acetate pH buffer, since the fluoride electrode is affected by high levels of hydroxide. In these direct calibration procedures, the same level of background must be added to all samples and standards.

Increment Methods. Known increment methods are based on the concept that the activity value A is different from the concentration value C by some factor γ which, in the following case, includes both ionic strength and complexing effects, ie,

$$A = \gamma C$$

If the value of γ were known, then the observed activity readings could be multiplied by an appropriate factor to give the desired concentration value. In known increment methods, γ is estimated either by adding a known amount of the species being measured and observing the new reading or by removing a known portion of the measured species. These procedures are called *known addition* and *known subtraction*. Both are based on the assumption that the value of γ is not changed by the increment procedure.

The method may be understood by reference to the equations below: where C_x is the initial unknown concentration, γ has the significance discussed above, and C_a is the known concentration increment.

$$E_1 = E^0 + S \log (\gamma C_x)$$

$$E_2 = E^0 + S \log (\gamma C_x + \gamma C_a)$$

Instrumentation is available to perform the necessary calculations for this method, including corrections for the change in volume that results from the addition of the increment.

It follows, from the requirement that γ not change, that complexing agents should be present in sufficiently large excess so that the addition does not shift the equilibrium to a measurable extent. This condition does not hold in the case of drinking water, for example, where roughly comparable levels of fluoride and aluminum often are present. To use a known addition procedure in this case (rather than the TISAB procedure), it would be necessary, in principle, to add a large amount of aluminum, eg, ca a 100-fold excess over the fluoride. (Actually, the Al/F equilibria often are slow, and the TISAB procedure is better).

Titration. ISEs lend themselves particularly well to titration procedures. They may serve as detectors either for the ion being measured or for the titrant being added. Titrations take longer than direct measurement but generally yield improved precision because of the large change in millivolt reading which occurs near the end point. Examples of such titrations that may be done with an ISE are the determination of molybdate or sulfate by titrating with lead perchlorate, of thioglycolate with silver ion, or of copper with Na-EDTA (ethylenediaminetetraacetic acid). Extensive lists of possible titrations have been published (32).

As in the case of pH titrations, it is sometimes possible to measure more than one species in a single titration, eg, a mixture of copper and zinc ions can be titrated with EDTA. The copper reacts first, decreasing the copper activity in the solution until virtually all of the copper has been complexed. Additional EDTA does not significantly change the copper activity below this point until all of the zinc has been complexed, because the zinc–EDTA complex forms and keeps the free EDTA level low. A new end point is observed when all of the zinc is consumed and, consequently, the copper activity is further decreased.

A small amount of Cu–EDTA can be added to a solution of calcium before titrating. The Cu–EDTA serves as an indicator in a complexometric titration (33). Thus, it is possible to measure species like nickel, zinc, or manganese, for which there is no ISE. With this method, it is possible to obtain a better titration of calcium by EDTA with the copper electrode than with the calcium electrode. When the calcium activity begins to diminish during the titration, the calcium electrode begins to respond to the added sodium from the disodium EDTA that is used as the titrant. If the copper electrode is used, it responds indirectly to the free EDTA level but there is no sodium ion interference.

Uses

Agriculture and Food. The nitrate electrode is used for soil nitrate measurements to indicate soil fertilizer requirements (see Fertilizers). In general, dried soil is extracted with an aqueous solution containing a noninterfering salt (eg, an acetate buffer) and the measurement is made directly. In some regions, fertilizer requirements are based on nitrate measurements of vegetation, using the electrode and a similar extraction procedure on dried, ground plant material. The salt content of foods (eg, margarine, peanut butter, potato chips, sugar syrup, and tomato paste are measured using the chloride electrode. Interference from some unknown proteinlike materials was reported to be a problem but this can be overcome if the measurements are made in an acidic background. With the rapidly rising cost of silver, direct measurements of chloride by electrode can represent a considerable cost savings over the usual silver nitrate titrations.

Biomedicine. The principal applications are the measurements of Na^+ and K^+ in whole blood and the measurement of ionized calcium in blood serum (see Blood). Automated laboratory instruments are available for both measurements (see Biomedical instrumentation). There has also been a large amount of work on the measurement of fluoride in body fluids, following the treatment of teeth or the use of fluorinated organic anesthetics (qv). The chloride electrode has been used in screening for cystic fibrosis, a disease which has, as one symptom, increased sodium chloride concentration in the sweat. A flush-mounted chloride electrode and a battery-operated meter (for safety) were especially developed for this purpose (34).

Electroplating. The cyanide electrode is used to measure cyanide in plating baths (after dilution to 10^{-3} M or less) and in waste streams. In many cases, it is possible to measure (by titration) both free and total cyanide, often including cyanide complexed to metals. A commercial cyanide waste-stream monitor uses the silver/sulfide electrode (22). Fluoride is measured in chrome-plating baths (after dilution with excess sodium acetate solution). Many of the metals that are used in plating (eg, zinc, cadmium, nickel, copper, and silver) can be measured either directly or indirectly by titration techniques (see Electroplating; Recycling).

Steam and Power. The purity of feed water and steam (qv) condensate in the parts per billion (10^9) range is monitored by sodium electrodes. A 1978 survey indicated that nearly half of the plants having high pressure steam turbines owned at least one such monitor (35).

Petroleum Refining. Chloride is measured at low levels in water that is used to desalt crude oil, and NH_3, H_2S, HCN, and HCl that are trapped by sour-water scrubbers are measured by the appropriate electrode. Fluoride electrodes have been used in catalytic cracking installations to detect leakage of HF into the cooling water side of heat exchangers (see Heat-exchange technology).

Industrial. Direct measurement of water hardness (see Water), both in the laboratory and on-line, is accomplished by using the divalent cation electrode. Breakthrough of ion-exchange columns usually is detected with the calcium electrode, because it has much greater selectivity against sodium than the divalent electrode. Sulfide is an important measurement in paper and pulp mills, and both sodium and sulfide are monitored in black, green, and white liquors. Cyanide and ammonia measurements are common in the wastewater from steel mills. Fluoride also has been measured in stack gases or as emissions from aluminum manufacturing. Electrodes also are used to determine the fluoride level of the urine of workers in those plants.

The main area of use remains overwhelmingly in the laboratory, although a small portion has industrial on-line applications. The recent trend has been for increased electrode sales for measurements in quality and production control laboratories; each electrode is used to make many measurements on closely related samples, rather than for general purpose or research use. (The research applications account for a disproportionately large fraction of new publications in the field). The growth areas are in instruments that use electrodes as the sensors, eg, in industrial monitoring and biochemical measurement (see Instrumentation and control).

BIBLIOGRAPHY

1. M. Cremer, *Z. Biol.* **47,** 562 (1906).
2. F. Haber and Z. Klemensiewitz, *Z. Physik. Chem.* **67,** 385 (1909).
3. C. Morton, *Trans. Faraday Soc.* **24,** 14 (1928); L. W. Elder, Jr., *J. Am. Chem. Soc.* **51,** 3266 (1929); H. M. Partridge, *J. Am. Chem. Soc.* **51,** 1 (1929).
4. K. Horovitz, *Z. Physik* **15,** 369 (1923).
5. G. Eisenman, D. O. Rudin, and J. U. Casby, *Science* **126,** 831 (1947).
6. U.S. Pat. 3,429,785 (Feb. 25, 1969), J. W. Ross (to Corning Glass Works).
7. Z. Stefanac and W. Simon, *Chimia* **20,** 436 (1966).
8. M. Oehme, M. Kessler and W. Simon, *Chimia* **30,** 204 (1976); D. Ammann, M. Gueggi, E. Pretsch, and W. Simon, *Anal. Lett.* **8,** 790 (1975).
9. I. M. Kolthoff and H. L. Sanders, *J. Am. Chem. Soc.* **59,** 416 (1937).
10. U.S. Pat. 3,563,874 (Feb. 16, 1971), J. W. Ross, M. S. Frant, and J. H. Riseman (to Orion Research, Inc.).
11. U.S. Pat. 3,672,962 (June 27, 1972), M. S. Frant and J. W. Ross (to Orion Research, Inc.).
12. U.S. Pat. 3,591,464 (July 6, 1971), M. S. Frant and J. W. Ross (to Orion Research, Inc.).
13. *Orion Research Newsletter* **6,** 1 (1974).
14. U.S. Pat. 3,431,182 (March 4, 1969), M. S. Frant (to Orion Research, Inc.).
15. W. Nernst, *Z. Physik. Chem.* **2,** 613 (1888); 4, 129 (1889).
16. J. W. Ross, *Science* **156,** 1378 (1967).
17. G. J. Moody, T. B. Oke, and J. D. R. Thomas, *Analyst* **95,** 910 (1970).
18. U.S. Pat. 3,483,112 (Dec. 9, 1969), J. W. Ross (to Orion Research, Inc.).
19. M. S. Frant and J. W. Ross, *Science* **167,** 987 (1970).

20. W. Simon, E. Pretsch, D. Ammann, W. E. Morf, M. Gueggi, R. Bissig, and M. Kessler, *Pure Appl. Chem.* **44,** 613 (1975).
21. R. P. Buck, *Anal. Chem.* **50,** 17R (1978); **48,** 23R (1976); R. P. Buck, *Crit. Rev. Anal. Chem.* **5,** 23 (1975).
22. U.S. Pat. 3,923,608 (Dec. 2, 1975), M. S. Frant and J. W. Ross (to Orion Research, Inc.).
23. E. Pungor, *Anal. Chem.* **39,** 28A (1967).
24. J. Ruzicka and C. G. Lamm, *Anal. Chim. Acta* **53,** 206 (1971).
25. G. A. Rechnitz, G. H. Fricke, and M. S. Mohan, *Anal. Chem.* **44,** 1098 (1972).
26. E. Pungor, J. Havas, and K. Toth, *Instrum. Control Syst.* **38,** 105 (1965); G. A. Rechnitz, Z. F. Lin, and S. B. Zamochnik, *Anal. Lett.* **1,** 29 (1969).
27. E. Pungor, J. Havas, and K. Toth, *Z. Chem.* **5,** 9 (1965).
28. G. G. Guilbault, *Pure Appl. Chem.* **25,** 727 (1971).
29. P. Cheung, *Proceedings of Conference at Case-WRU, March 1977,* CRC Press, Cleveland, Ohio, 1977.
30. R. A. Durst and R. G. Bates, *Natl. Bur. Standards (U.S.) Spec. Publ.* **450,** 247 (1977).
31. M. S. Frant and J. W. Ross, *Anal. Chem.* **40,** 1169 (1968).
32. *Analytical Methods Guide,* Orion Research, Inc., Cambridge, Mass., 9th ed., 1978.
33. J. W. Ross and M. S. Frant, *Anal. Chem.* **41,** 1900 (1969).
34. L. Kapito and H. Schwachman, *Pediatrics* **43**(5), 794 (1969).
35. B. W. Bussert, R. M. Curran, and G. C. Gould, *ASME/IEEE Power Gen. Conf., Dallas, Texas, Sept. 10–14, 1978.*

General References

Ref. 21.

J. Koryta, *Ion-Selective Electrodes,* Cambridge Univ. Press, Cambridge, Eng., 1975; G. J. Moody and J. D. R. Thomas, *Sensitive Ion-Selective Electrodes,* Merrow Publ. Co., Watford, U.K., 1971; P. L. Bailey, *Analysis with Ion-Selective Electrodes,* Heyden, London, 1976.

J. Vesely, D. Weiss, and K. Stulik, *Analysis with Ion-Selective Electrodes,* Ellis Horwood Ltd., Chichester, Eng., 1978.

G. Baiulescu and V. V. Cosofret, *Applications of Ion-Selective Membranes in Organic Analysis,* Ellis Horwood Ltd., Chichester, Eng., 1977.

Bibliography, Orion Research, Inc., Cambridge, Mass., 1979.

MARTIN S. FRANT
Foxboro Analytical
A Division of The Foxboro Company

IRIDIUM. See Platinum-group metals.

IRON

Iron [*7439-89-6*], Fe, atomic number 26, is in group VIII of the periodic table and is between manganese and cobalt in period 4. It has a valence of +3 or +2 and combines readily with other elements. Iron has four stable isotopes, of mass 54 (6.04%), 56 (91.57%), 57 (2.11%), and 58 (0.28%); it is the fourth most abundant element in the earth's crust, outranked only by aluminum, silicon, and oxygen. Although gold, silver, copper, brass, and bronze were in common use before iron, it was not until man discovered how to extract iron from its ores that civilization developed rapidly.

Pure iron is a silvery white, relatively soft metal that has a body-centered cubic (bcc) crystal lattice at normal temperatures. Native metallic iron rarely is found in nature because iron combines readily with other elements, eg, oxygen and sulfur. Iron oxides are the most prevalent natural form of iron. Generally, these iron oxides (iron ores) are reduced to iron in a blast furnace (see Iron by direct reduction). The iron from the blast furnace is refined in steelmaking furnaces to make steel (qv).

The earliest record of human's use of iron dates to ca 2000 BC (1) in Egypt, Asia Minor, Assyria, China, and India. It is almost certain, however, that the first iron to be used was not processed but was obtained from meteorites (2).

One of the few places where native iron is found is in Greenland, where it occurs as very small grains or nodules in basalt (an iron-bearing igneous rock) that erupted through beds of coal.

Processed iron first was produced around 1300 BC. It is presumed that the first iron was made accidentally as a result of very hot fires built on top of some iron-bearing rocks or soil. The iron-bearing rocks could have been reduced to iron by being heated in the presence of hot charcoal and in the absence of air. Upon raking out the ashes, the first ironmaker probably found a spongelike chunk of hard but malleable metal with considerable slag in its pores. Reduced iron sponge had to be hammered and squeezed while still hot to expel most of the slag in order to make effective use of the metal. This hammering and working process produced wrought iron.

The first furnaces specifically made for smelting iron ore were low shaft furnaces: low boxlike hearths made of stone, open at the top, and with an opening near the bottom for air intake. The Catalan hearth furnace for making wrought iron is described in writings from Central Europe in the twelfth and thirteenth centuries AD. These Catalan furnaces were low stone shaft furnaces with hearth dimensions of ca $60 \times 60 \times 75$ cm. As better blowing devices were invented and the height of the furnaces was increased, it is probable that, when the fires became hot, liquid, high-carbon iron, which was not malleable, was produced and could be used to make cast-iron articles. The Stuckofen or old high bloomery appeared in Germany in ca 1300 AD. This type of furnace was 3–5 m high and enclosed a tapered vertical shaft that was 1–1.2 m in diameter. Small openings near the bottom were provided for nozzles (tuyeres) that permitted air, which was supplied by bellows, to be blown into the furnace. This type of furnace had essentially the same fundamental design as modern blast furnaces.

Eventually, processes were developed for converting the high-carbon product of the blast furnace to wrought iron. In the puddling process, the blast-furnace pig iron first was melted and then the silicon, manganese, and carbon were oxidized by the hot gases and the iron oxides that were added. During this operation the puddler continually stirred the bath. As the decarbonization approached completion, the metal began to solidify. The series of hammering and squeezing operations separated the slag from the iron to produce wrought iron.

In the United States, the first ironworks was built at Jamestown, Virginia, in 1619. The Hammersmith furnace near Saugus, Massachusetts, was built in 1645 and operated until 1675. This early American ironworks has been restored and is called the Saugus Iron Works. Iron blast furnaces appeared in many localities where there were deposits of iron ore. Small bodies of iron ore in New Jersey, Connecticut, Massachusetts, Pennsylvania, and New York formed the basis of many small colonial furnaces. Ironmaking in the United States did not expand rapidly until after the Revolutionary War. Then, as the colonists moved westward, their need for iron prompted the establishment of ironworks near the new settlements. A blast furnace built by Jacob Anschutz in 1796 was the beginning of the great iron and steel center in Pittsburgh, Pennsylvania.

The early U.S. blast furnaces of the nineteenth century usually were in the form of a truncated cone or pyramid, 6–9 m in height, enclosing a shaft 1.2–2.4 m in maximum diameter, and constructed of stone. The output of these furnaces was from 1–6 metric tons per day. In the next fifty years, many changes were made in furnace design and operation. The shaft was enclosed in a metal shell, the top was closed to prevent the escape of top gases, the blast air was preheated, and furnaces were enlarged. By 1900, a single blast furnace made as much as 200 t/d. Improvements continued to be made but, generally, modern large furnaces are only larger than their nineteenth century ancestors. The modern blast furnace may have a hearth as much as 14 m in diameter, be over 60 m in height, and produce up to 10,000 t/d.

In 1979, there were 168 blast furnaces in the United States, and they produced ca 80,000,000 t of pig iron. Most of the furnaces are located in Pittsburgh and Chicago.

Physical and Chemical Properties

The physical and chemical properties of iron are difficult to define in absolute numbers because even small traces of other elements in iron may have profound affects on the properties. Absolutely pure iron is difficult to obtain but iron of sufficiently high purity has been produced for determination of most of the properties of the metal.

Cooling curves of pure iron are given in idealized form in Figure 1. The points of discontinuity in the cooling curve, or critical points, were termed arrests which has been abbreviated to *A* points; the subscript numerals 1, 2, 3, and 4 are given for purposes of differentiation. The change at 770°C from ferromagnetic, body-centered, alpha iron to paramagnetic, body-centered, alpha iron formerly was designated A_2: this is no longer considered a true allotropic change, as discussed below. The change at 910°C from body-centered, alpha iron to face-centered, gamma iron is A_3. Likewise, the change at 1400°C from the gamma form to body-centered, delta iron is termed A_4. The A_1 point, at 740°C, is not in Figure 1, because it does not appear in carbon-free iron. The A_1 point does occur in cast iron and steel (which are alloys of iron and carbon) at the temperature of formation, 740°C, of the eutectoid (containing 0.83% C) which consists of iron and iron carbide. Iron between the A_2 and A_3 points (the paramagnetic form of alpha iron) was termed beta iron; however, the designation is no longer used, as it has been demonstrated that the change at the A_2 point involves changes in electron distribution within iron atoms rather than an allotropic or crystalline transformation. An accurate knowledge of these transformations is the fundamental basis of the heat treatment of ferrous metals.

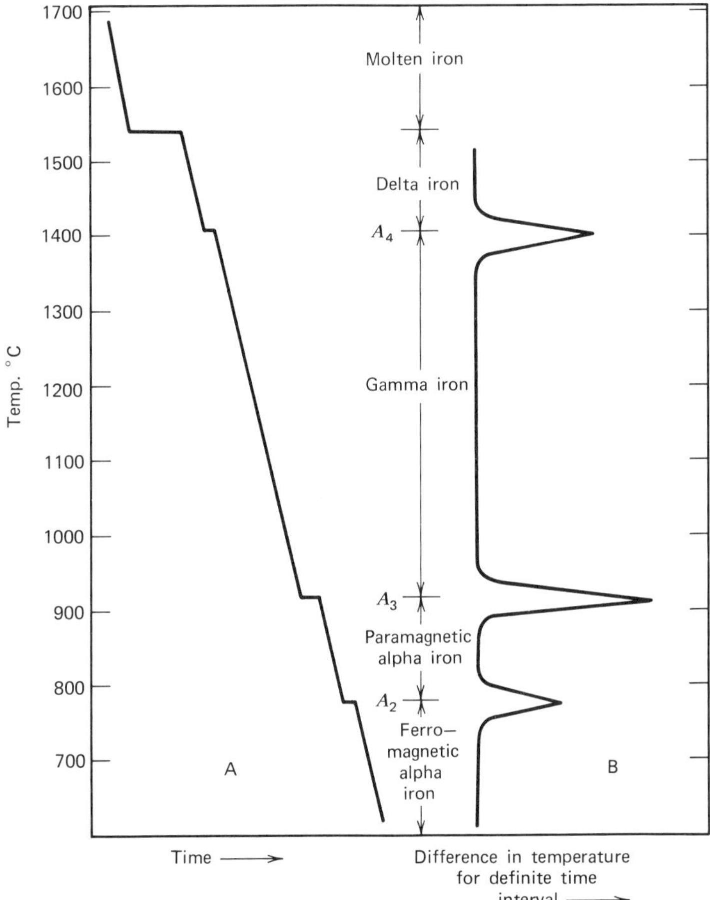

Figure 1. Idealized cooling of pure iron resulting from plotting temperature A against time and B against the difference in temperature between the specimen and a neutral body for a definite time interval (3).

Properties of high purity iron are given in Table 1. Most elements when mixed with iron lower the element's melting and boiling points. The thermal conductivity decreases as the temperature increases. The thermal expansion of iron also increases (almost linearly) with temperature up to 600°C.

However, as iron is heated through the A_3 range, a contraction occurs as alpha iron transforms to denser gamma iron, but the change at A_4 is accompanied by expansion as the structure changes from gamma to delta iron. The electrical properties of iron are significantly affected by traces of impurities. The thermoelectric power of iron as compared to other metals is exploited in temperature measurement; a thermocouple of an iron wire and a constantan (an alloy of copper and nickel) wire is used commonly for temperature measurement (qv). One of the most distinguishing properties of iron is its magnetism (see Magnetic materials, bulk). In its strong magnetism in the presence of a magnetic field or an electric current, iron is superior to all other elements. The magnetic influence produced with high purity iron through in-

Table 1. Properties of Relatively Pure Iron[a]

Property	Value
mp, °C	1532 ± 5°
bp, °C	3000 ± 150°
density (hot rolled), g/cm^3	7.87
thermal conductivity, at 0°C, W/(m·K)	79
electrical resistivity, at 20°C, $\mu\Omega$·cm	9.71
tensile strength, MPa[b]	245–280
yield strength, MPa[b]	70–140
elongation in 5 cm at 20°C, %	40–60
reduction of area, %	65–78
Brinell hardness	82–100
impact strength (Izod notched bar), J/m[c]	
longitudinal	4859
transverse	2990
thermal expansion, (J/m[c])/°C	
at 0–300°C	1.26×10^{-5}
at 0–600°C	1.46×10^{-5}
specific heat, J/g[d]	
at 100°C	0.50
at 500°C	0.67
at 800°C	1.26
hardenability	high-purity iron can be hardened only by unusually heavy cold working

[a] Refs. 2 and 4.
[b] To convert MPa to psi, multiply by 145.
[c] To convert J/m to ft·lbf/in., divide by 53.4 (see ASTM D 256).
[d] To convert J to cal, divide by 4.184.

duction by the presence of a magnetic field of either a permanent magnet or an electric current disappears almost entirely upon removal of the magnet or electric current. The presence of other elements in iron alloys, eg, carbon, cobalt, and nickel, results in substantial increases in magnetic retentivity. Iron when heated loses its magnetism, ie, becomes paramagnetic, at ca 770°C (see Fig. 1).

Manufacturing and Processing

Beneficiation. Beneficiation, as it is applied to iron ores, means to improve the ore's chemical and/or physical properties. When the iron content of the ore is increased, the process generally is referred to as concentration; when the physical structure is improved by making larger particles out of smaller particles, the process is referred to as agglomeration. Some ores can be concentrated simply by crushing, screening, and washing. Other ores must be ground to very small particles before the iron oxides can be separated from the rest of the material. This separation normally is accomplished by magnetic drums (see Magnetic separation) or by flotation (qv). Examples of the types that can be separated magnetically are the magnetite taconites from Minnesota. These ores contain from 25–30% iron as mined. They are ground to 75 μm (200 mesh) or finer and then are concentrated by magnetic separating drums. The final concentrate has ca 60–65% iron and 2–8% silica. Although these concentrates are of acceptable chemical quality, they must be agglomerated into a coarser form before they can be used in furnaces. Agglomeration is one of the principal methods of physical beneficiation of iron ores. Fine particles, whether in natural ores or in concentrates, are undesirable as blast-furnace feed, the most desirable size for blast-furnace feed being from 6–25 mm. Of the numerous methods available, sintering and pelleting are the most important (see Size enlargement).

In the sintering process, a mixture of iron-ore fines or concentrates and coke fines is deposited on a traveling grate. The traveling grate is shaped like an endless loop of conveyor belt forming a shallow trough with small holes in the bottom. The bed of material on the grate first is ignited by passing under an ignition burner that is fired with natural gas and air and then, as the grate moves slowly toward the discharge end, air is pulled down through the bed. As the coke fines burn in the bed, the heat that is generated sinters the small particles. At the discharge end of the machine, the sinter is crushed to remove extra large lumps, then is cooled, and finally screened. In some cases, limestone or dolomite fines are added to the sinter feed to produce a self-fluxing sinter (see Lime and limestone). In a modern integrated steel plant, most of the iron-bearing dusts and slags are recycled by adding them to the sinter-plant feed mix.

In the pelleting process, the ore must be ground to a very fine size, $<75\ \mu m$ (−200 mesh). The ground ore is mixed with the proper amount of water and sometimes with a small amount of bentonite and then is rolled into small balls 10–20 mm in diameter in a balling drum or disk. These green pellets are dried, then are heated to 1200–1370°C to bond the small particles, and finally are cooled. The heating can be done on a traveling grate like that used in sintering, or in a shaft furnace, or by a combination of a traveling grate and a rotary kiln. All three methods are common.

Other potential means of agglomerating are by briquetting or nodulizing. In the briquetting process, ore fines usually are mixed with a binder and are formed into compact masses between two rotating rolls. The rolls exert pressures of 1.5–4.2 t/cm^2 in forming the briquettes. In another briquetting process, the ore is heated to 650–1000°C and then is briquetted while hot. No binder is used in the latter process.

In the nodulizing process, which is relatively uncommon commercially, ore fines are heated in a rotary kiln to a temperature, usually 1250–1370°C, at which the ore begins to melt and bind. The ore balls up in the kiln to form nodules that are discharged and cooled.

The pelletizing process accounted for 65% of the U.S. agglomerate production in 1976. This is understandable in view of the fact that pellets are within the most desirable size range and are desirable chemically (see Table 2).

Blast Furnace. By far the most important method for the manufacture of iron is the blast-furnace process. The blast furnace essentially is a large chemical reactor into which are charged iron ore, coke, and limestone. As the iron ore descends through the

Table 2. Production of Agglomerates in U.S. in 1976, 10^3 t[a]

sinter	32,517
pellets	62,547
briquettes, nodules, and others	454
Total	*95,518*

[a] Ref. 5.

furnace, it is reduced to iron which is melted by a countercurrent flow of hot, reducing gas. The gas is produced by burning coke with preheated air near the bottom of the furnace. The molten iron and slag (formed from the gangue in the ore, the ash in the coke, and the limestone) are removed about every two to four hours from the hearth of the furnace. The raw materials are charged continually into the top of the furnace, thus making the blast furnace a continuous process. The blast furnace is one of the most efficient countercurrent processes. The main product from the blast furnace is hot metal or pig iron. Pig iron generally is refined to make steel or it may be used to make iron castings. The quantities of pig iron produced in the U.S. in 1975–1977 are given in Table 3. Over 90% of the pig iron production (basic and Bessemer iron) is used in steelmaking furnaces.

Raw Materials. The iron-bearing materials (iron ore, sinter, pellets, or briquettes) supply the iron. They usually contain from 50–65% iron in the form of one or more of the oxides of iron, eg, magnetite (Fe_3O_4) or hematite (Fe_2O_3). From 2–20% of the iron-bearing material is gangue (usually silica and alumina) that is separated from the iron in the furnace and is removed as part of the slag. Small quantities of other iron-bearing materials, eg, roll scale, open-hearth slag, and scrap, also are charged to the furnace. Although iron-bearing materials, ranging in size from 150 mm to dust, are used in blast furnaces, the desirable size range is 6–25 mm.

The second principal raw material is coke. The approximate composition of blast-furnace coke is: fixed carbon, 85–90%; ash, 5–13%; moisture, 2–8%; volatile matter, 0.9–1.6%; and sulfur, 0.6–1.3%.

The function of coke is to produce heat and reducing gases when burned with preheated air near the bottom of the furnaces. It helps to support the burden in the furnace and to provide void space for the ascending gases. High mechanical strength of coke is important for smooth operation. Various tests are performed to measure strength. The most desirable size for coke is 15–75 mm.

The third raw material is limestone which is added to adjust the composition of the slag. The proper ratio of basic constituents, CaO and MgO, to acid constituents, SiO_2 and Al_2O_3, must be maintained to obtain a fluid slag with good desulfurizing power. Either limestone or dolomite is used as the blast-furnace flux (see Table 4 for typical compositions). Occasionally gravel, which is principally silica, must be added for slag composition control. The desirable size range for blast-furnace fluxes is from 10–75 mm.

Air also should be considered as a main raw material. Over 1.5 t of air is required to produce one ton of pig iron.

Table 3. Pig Iron Produced From Blast Furnaces in the U.S., 10^3 t

Grade	1975	1976	1977
foundry	1,199	1,351	1,461
basic	68,885	75,221	70,960
Bessemer	986	1,044	230
low phosphorus	127	102	111
malleable	739	726	626
all other	590	386	412
Total	*72,526*	*78,830*	*73,800*

[a] Ref. 5.

Table 4. Typical Compositions of Blast Furnace Flux, %

Constituents	Limestone	Dolomite
$CaCO_3$	95.0	54.8
$MgCO_3$	0.5	39.6
Al_2O_3	0.7	0.4
SiO_2	2.0	1.2
S	0.05	0.02
P	0.02	0.01
H_2O	1.7	4.0

Chemistry. As the coke in the blast-furnace charge descends through the stack, it is preheated to ca 1370°C. This preheated coke burns rapidly in the hot blast air in front of the tuyeres and produces a temperature of ca 1925°C. Since CO_2 is unstable at this temperature in the presence of excess C, the coke burns only to CO. The reaction is

$$C + \tfrac{1}{2}\,O_2 \rightarrow CO \qquad \Delta H = -110\ \text{kJ}\ (-26.3\ \text{kcal}) \tag{1}$$

Water in the air also reacts

$$C + H_2O \rightarrow CO + H_2 \qquad \Delta H = 131\ \text{kJ}\ (31.3\ \text{kcal}) \tag{2}$$

It is important to note that the reaction with H_2O requires heat and, as such, can be used to control the temperature in front of the tuyeres.

In the upper portion of the furnace, where the temperature is below 925°C, the iron oxides are reduced by CO and H_2 as follows:

$$\tfrac{1}{2}\,Fe_2O_3 + \tfrac{3}{2}\,CO \rightarrow Fe + \tfrac{3}{2}\,CO_2 \qquad \Delta H = -13\ \text{kJ}\ (-3.1\ \text{kcal}) \tag{3}$$

$$\tfrac{1}{3}\,Fe_3O_4 + \tfrac{4}{3}\,CO \rightarrow Fe + \tfrac{4}{3}\,CO_2 \qquad \Delta H = -4\ \text{kJ}\ (-0.96\ \text{kcal}) \tag{4}$$

$$FeO + CO \rightarrow Fe + CO_2 \qquad \Delta H = -16\ \text{kJ}\ (-3.8\ \text{kcal}) \tag{5}$$

$$\tfrac{1}{2}\,Fe_2O_3 + \tfrac{3}{2}\,H_2 \rightarrow Fe + \tfrac{3}{2}\,H_2O \qquad \Delta H = 49\ \text{kJ}\ (11.7\ \text{kcal}) \tag{6}$$

$$\tfrac{1}{3}\,Fe_3O_4 + \tfrac{4}{3}\,H_2 \rightarrow Fe + \tfrac{4}{3}\,H_2O \qquad \Delta H = 51\ \text{kJ}\ (12.2\ \text{kcal}) \tag{7}$$

$$FeO + H_2 \rightarrow Fe + H_2O \qquad \Delta H = 25\ \text{kJ}\ (6.0\ \text{kcal}) \tag{8}$$

Although these reactions are shown as going to completion, there are equilibrium ratios of CO_2 to CO and H_2O to H_2 that must be satisfied. In general, 40–50% of the CO is used to reduce the iron oxides and ca 50% of the hydrogen is used. In the upper part of the stack, the limestone is calcined as the temperature reaches 815–870°C

$$CaCO_3 \rightarrow CaO + CO_2 \qquad \Delta H = 179\ \text{kJ}\ (42.8\ \text{kcal}) \tag{9}$$

In the lower part of the stack, where the temperatures are above 925°C, the iron oxides are reduced with CO, but the CO_2 that is formed reacts immediately with C to form CO. The net reaction is

$$FeO + C \rightarrow Fe + CO \qquad \Delta H = 157 \text{ kJ } (37.5 \text{ kcal}) \tag{10}$$

Reaction 10 requires a large amount of heat and sometimes is referred to as direct reduction, whereas reduction by carbon monoxide (as in eq 3–5) is referred to as indirect reduction (see Iron by direct reduction).

The reduction of the other principal metals in pig iron occurs in the hearth of the furnace at temperatures above 1370°C.

$$MnO + C \rightarrow Mn + CO \qquad \Delta H = 272 \text{ kJ } (65.0 \text{ kcal}) \tag{11}$$

$$SiO_2 + 2\,C \rightarrow Si + 2\,CO \qquad \Delta H = 659 \text{ kJ } (158 \text{ kcal}) \tag{12}$$

$$P_2O_5 + 5\,C \rightarrow 2\,P + 5\,CO \qquad \Delta H = 996 \text{ kJ } (238 \text{ kcal}) \tag{13}$$

Most of the sulfur that enters the furnace enters with the coke. It is gasified to H_2S or COS in the tuyere zone and reacts with the CaO and FeO to form CaS and FeS. The FeS is removed from the iron by the following reaction:

$$FeS + CaO + C \rightarrow CaS + Fe + CO \qquad \Delta H = 182 \text{ kJ } (43.5 \text{ kcal}) \tag{14}$$

The extent of sulfur removal depends on the temperature of the slag and on the ratio of CaO and MgO to SiO_2 and Al_2O_3.

Material and energy balances can be used to achieve an overall appraisal of the process. Tables 5 and 6 are balances for a 9-m-diameter hearth furnace operating at 4736 m^3/min (86,000 SCF/min) and at a hot blast temperature of 1093°C. The coke rate used is 436 kg/t of hot metal; fuel oil is injected in the tuyeres at the rate of 83.5 kg/t of hot metal.

Construction. The flow diagram of a blast-furnace plant is shown in Figure 2. Figure 3 is a photograph of a blast-furnace model and shows the principal components. The raw materials, ore, and limestone generally are stockpiled in an ore yard adjacent to the furnace. From the ore yard, the ore and limestone are loaded into bins in the stockhouse by an ore bridge and transfer car for the ore and limestone or by railroad car or conveyor for the coke. In the stockhouse, the materials are withdrawn from the

Table 5. Energy Balance for a 8.5-m Hearth-Diameter Blast Furnace

Energy balance	GJ/t[a] hot metal
energy input	
calorific value of coke	12.6
calorific value of tuyere-injected fuel oil	3.6
Total	*16.2*
energy output	
calorific value of top gas	6.1
top gas consumed in heating blast air	2.1
net energy available for other uses	4.0
heat for chemical reactions, heat loss, and sensible heat of hot metal and slag	12.2
Total	*16.2*

[a] To convert GJ/t to Btu/short ton, multiply by 9.56×10^{11}.

Table 6. Blast Furnace Material Balance

Inputs				Outputs		
	Material	% Fe	kg	Material	m^3	kg
Iron	pellets	61	1008.0	*Dry top gas* (20.0 CO, 18.2 CO_2, 4.3 H_2)	1833	2511
bearing	fluxed sinter	55	515.5			
burden	miscellaneous	60	8.5	*Moisture*	241	85
	steelmaking slag	20	61.5	*Flue dust and sludge*		18
	scrap	85	33.5			
Flux	limestone		17.5			
Fuel	coke		436.0			
	total moisture in charge		54.0			
	Material	m^3	kg	Material		kg
	dry air	1287.0	1665.0	*Slag* (38% SiO_2, 8% Al_2O_3, 40% CaO, 10% MgO) (1.6% S)		320
Blast	enriching oxygen	11.5	16.5			
	moisture	56.0	45.0			
Fuel	oil	87.01	83.5	*Hot metal* (4.1% C, 0.78% Si, 0.60% Mn, 0.026% S) (0.090% P)		1000
				Runner scrap		6

bins into weigh cars or similar weighing devices, and then they are carried to the top of the furnace by a skip car. At the top of the furnace, the charge is admitted through a double-bell system into the furnace. The double-bell system allows the solids to be charged while a gas seal is maintained. The iron is removed from the bottom of the furnace and is conveyed in railroad ladle cars to steelmaking furnaces or the pig-casting machine. The slag also is conveyed from the furnace in railroad ladles.

The blast air is preheated in large, regenerative, refractory stoves. These stoves, generally three or four per furnace, are 8–10 m in diameter and 35–45 m in height. They are steel shells containing refractory checker brick which has multiple flues that are ca 50 mm in diameter. While one stove is heating the blast air, the other stoves are being heated by the burning of the blast-furnace top gas in them.

The top gas leaving the blast furnace passes through a dust catcher where most of the large particles of dust drop out of the gas and are removed from the bottom of the dust catcher. The gas then is washed in a gas washer, eg, a venturi scrubber or a combination of a tower washer and an electrostatic precipitator (see Air pollution control methods; Gas cleaning). If only a venturi scrubber is used, the washing process usually is followed by a gas-cooling tower step. The clean blast-furnace gas is used in the hot blast stoves, in a boiler plant, or in other heating applications within the steel plant. Figure 4 is a photograph of a modern blast furnace at a large Eastern U.S. steel plant. The materials are charged to the top by conveyor.

Direct Reduction. Direct reduction is applied to the reduction of iron ore with a solid or a gaseous reductant, usually without producing a liquid product. Beginning with the ancient ironmaking furnaces, hundreds of direct-reduction processes have been developed. Although these processes do not account for a large percentage of the

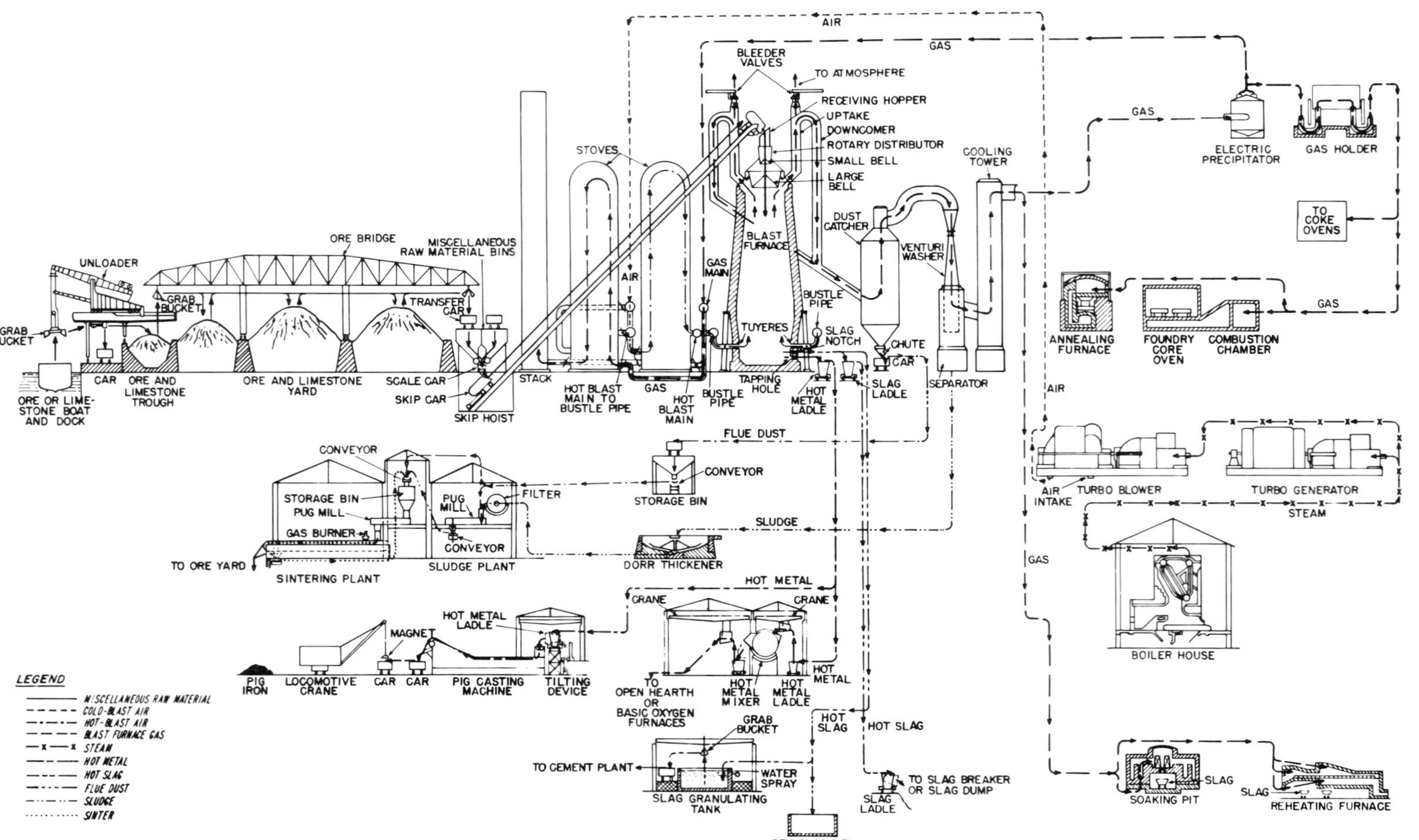

Figure 2. Flow diagram depicting the principal units and auxiliaries in a modern blast-furnace plant, and showing the steps in the manufacture of pig iron from receipt of raw materials to disposal of pig iron and slag, as well as the methods for utilizing the furnace gases.

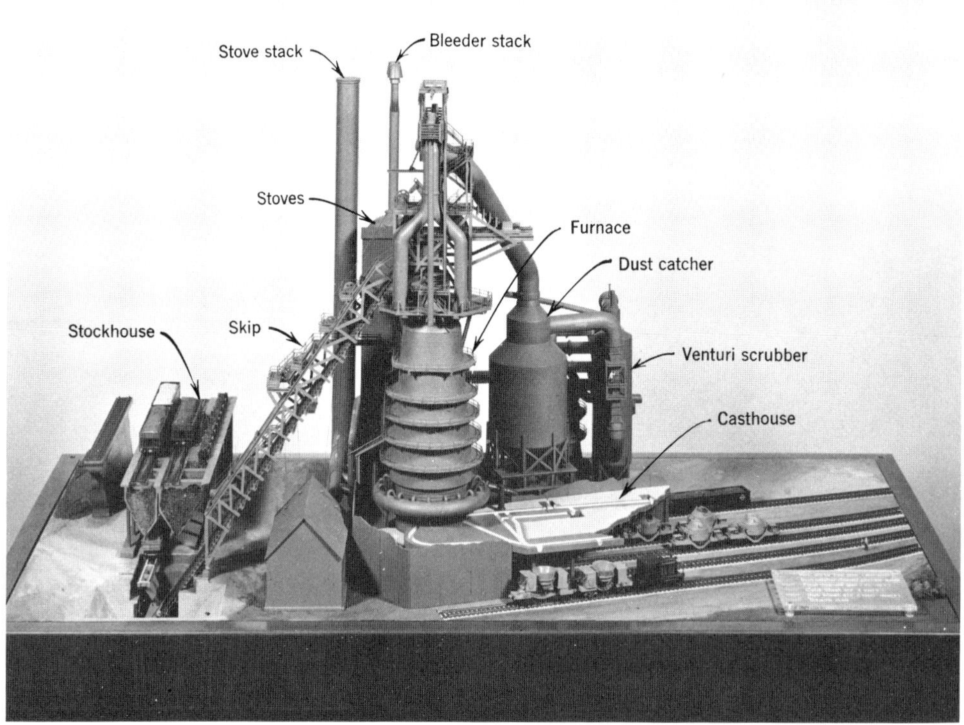

Figure 3. Blast-furnace model showing furnace, stoves, stockhouse, casthouse, dust catcher, venturi scrubber, and skip. Courtesy of U.S. Steel Corporation.

world's production of iron, they are alternatives to the blast-furnace process and production from direct reduction is increasing every year. In 1980, the published annual capacity for direct reduction plants was 20×10^6 t. By 1985 it is projected that there will be a 36.6×10^6 t annual capacity.

It generally is agreed that for a large tonnage ($>10^6$ t/yr) integrated plant, the blast furnace is the most economical process. On the other hand, for a small tonnage plant, a direct reduction process may be used to make a reduced iron product that can be refined and made into steel in a electric furnace (see Furnaces, electric). There are several countries in the world where coking coals are not available but large sources of natural gas are available. For example, Venezuela and Iran are particularly suited for direct-reduction processes: both have plenty of natural gas and no coal. Direct-reduction processes can be divided into two general classifications: processes using solid reductants and processes using gaseous reductions. Table 7 lists some of the major processes in commercial operation as of 1979.

Midrex and HyL are the two leading processes. In the Midrex process (which is a typical shaft-furnace process involving a gaseous reductant) pellets or lump ore are preheated and are reduced as they slowly move down through the furnace. The highest temperature reached in the reduction zone is 800–1000°C. The lower part of the shaft is used as a cooler to lower the pellets or lump ore temperature to ca 110°C. The reducing gas, which is a mixture of carbon monoxide and hydrogen, is made by reforming

Figure 4. Modern blast furnace in Eastern United States.

natural gas with steam in the presence of a catalyst. The top gas from the shaft is recycled through the gas producer.

In the HyL process, pellets or lump ores are reduced in a batch reactor. The charge first is preheated to 800–1000°C, then is reduced with a mixture of carbon monoxide and hydrogen, and finally is cooled. The reducing gas is produced by reforming natural gas.

In the HIB process, the ore is screened and crushed to <3 mm. The ore fines are preheated in a two-stage, fluidized-bed reactor to 850°C and then are reduced in a two-stage, fluidized bed at 750°C. The hot, reduced product is briquetted and the briquettes are cooled in an inert atmosphere. The reduced gas is produced by reforming natural gas.

Table 7. Direct-Reduction Processes in Commercial Operation in 1979

Processes using solid reductants	Processes using gaseous reductants
a. Kiln processes	a. Shaft-furnace processes
SL-RN	Midrex
ACAR	Armco
Kinglor Metor	Purofer
Nippon Steel[a]	Wiberg-Soderfors
Sumitomo[a]	b. Fluidized-bed processes
Kawasaki[a]	HIB
b. Retort processes	FIOR
Höganäs	c. Retort processes
c. Electric-furnace smelting	HyL
Tysland-Hole	

[a] Designed to produce reduced iron pellets from steel-plant waste materials.

Most of the product from the commercial, direct reduction plants is used in an electric furnace as a substitution for steel scrap. These products must be highly reduced (over 90% reduction) and low in gangue content. Some prereduced product can be used in the blast furnace as an enriched feed material; in this case, the product usually is only 75–80% reduced. The use of prereduced ore in a blast furnace results in higher production rates and lower fuel rates. The U.S. Bureau of Mines has proven such a process to be technically and economically feasible.

Grades of Pig Iron. Pig iron, the principal product of the blast furnace, is produced in many grades and is either molten iron or iron that is cast into pigs. The composition ranges for some of the various grades of pig iron are given in Table 8. Most of the iron that is produced in the blast furnace is used directly in the manufacture of steel and is transported while it is liquid to the steelmaking shop. In this form, the iron is referred to as hot metal, and its chemical composition usually is dependent on the type of steelmaking process that is to be used and the kind of steel that is to be made (see Table 9). In most plants in the United States, a basic steelmaking process is used and the hot-metal composition usually is 0.60–1.20% silicon, 0.035% max sulfur, 0.20% max phosphorus, and 0.4–2.0% manganese.

The balance of the blast-furnace production goes into merchant iron or special castings and, in a few instances, into ferroalloys. Recently, the percentage that is used as hot metal has increased, and the amount that is applied to merchant iron and ferroalloys has decreased. The decrease in the demand for merchant iron results in part from the fact that many foundries use a large percentage of steel scrap in both conventional cupolas and in electrical furnaces to make iron for castings.

Production

Minerals. Iron-bearing minerals are numerous and are present in most soils and rocks. However, only a few minerals are important sources of iron. Table 10 shows some of the principal iron-bearing minerals. Commercially, the oxides are the most im-

Table 8. Blast-Furnace Pig Iron[a]

Product	Composition range, %				
	Silicon	Sulfur	Phosphorus	Manganese	Carbon
iron for steelmaking					
basic pig (Northern)	0.50–1.50	0.05 max	0.400 max	1.01–2.00	3.5–4.4
basic pig (Southern)	0.50–1.50	0.05 max	0.700–0.900	0.40–0.75	3.5–4.4
acid pig (Bessemer)	1.00–2.25	0.045 max	0.040–0.135	0.5–1.0	4.15–4.4
oxygen steelmaking pig	0.20–2.00	0.050 max	0.400 max	0.4–2.5	3.5–4.4
merchant iron[b]					
low phosphorus	0.50–3.00	0.035 max	0.035 max	1.25 max	3.0–4.5
intermediate low phosphorus	1.00–3.00	0.050 max	0.036–0.075	1.25 max	3.0–4.5
Bessemer	1.00–3.00	0.050 max	0.076–0.100	1.25 max	3.0–4.5
malleable	0.75–3.50	0.050 max	0.101–0.300	0.50–1.25	3.0–4.5
Northern foundry	3.50 max	0.050 max	0.300–0.700	0.50–1.25	3.0–4.5
Southern foundry	3.50 max	0.050 max	0.700–0.900	0.40–0.75	3.0–4.5

[a] Refs. 3 and 6.

[b] Merchant iron is iron that is sold to foundries and other users.

Table 9. Chief Metallic Products of the Blast Furnace

Product	Composition range				
	Silicon, %	Sulfur, %	Phosphorus, %	Manganese, %	Carbon, %
Iron for steelmaking					
basic pig	1.50	0.05 max	0.400 max	1.01–2.00	3.5–4.40
in steps of	0.25			0.50	
acid pig, open-hearth	0.70–1.50	0.045 max	under 0.05	0.5–2.50	4.15–4.40
oxygen-steelmaking pig	0.20–2.00	0.05 max	0.400 max	0.4–2.50	3.5–4.40
Merchant iron for foundries					
low phosphorus	0.50–3.00	0.035 max	0.035 max	1.25 max	3.0–4.50
intermediate low phosphorus	1.00–3.00	0.050 max	0.036–0.075	1.25 max	3.0–4.50
Bessemer	1.00–3.00	0.050 max	0.076–0.100	1.25 max	3.0–4.50
malleable	0.75–3.50	0.050 max	0.101–0.300	0.50–1.25	3.0–4.50
foundry	3.50 max	0.050 max	0.301–0.700	0.50–1.25	3.0–4.50
all grades available in steps of	0.25			0.25	
Ferroalloys					
spiegel (3 grades)	1.0–4.5	0.05 max	0.14–0.25	16–30	6.5 max
standard ferromanganese (3 grades)	1.2 max	0.05 max	0.035 max	74–82	7.5 max
ferrosilicon, silvery pig	5.00–17.00	0.06 max	0.300 max	1.00–2.00	1.5 max
ferrophosphorus	1.5–1.75	under 0.05	15–24	0.07–0.50	1.10–2.0
Foreign practice					
basic Bessemer (Gilchrist or Thomas)	0.3–1.00	0.20	1.9–2.5	0.7–2.5	3.50–4.0
duplex iron	1.2–1.75	under 0.060	0.7–1.5	0.4–0.90	4.00–4.20

portant class of minerals, followed by carbonates, silicates, and sulfides. Magnetite is strongly magnetic, dark gray to black, and has a specific gravity of 5.16–5.18. Its magnetic properties are very important in that they permit mining exploration and ore beneficiation by magnetic methods. Hematite is steel gray to dull red or is bright red and has a specific gravity of 5.26. Hematite probably is the most plentiful iron mineral being mined. Limonite is the most common of the hydrous iron oxides. It commonly is yellow brown and has a specific gravity of 3.6–4.0. Siderite is white to greenish-gray or is black and has a specific gravity of 3.83–3.88. Siderite is relatively unimportant in U.S. ores but is an important source of iron in England and Europe. Iron silicates generally are not used as an iron ore, because the iron content is too low and the silica content is too high. Iron sulfides generally are used as a source of both iron and sulfur. When pyrite is roasted to recover sulfur in the form of SO_2, iron oxide is formed as a by-product and can be used as an iron ore.

Geography. Iron-ore deposits were formed by many different processes, eg, weathering, sedimentation, hydrothermal, and chemical. Iron ores occur in igneous, metamorphic, and sedimentary deposits. To be classified as an ore, iron-bearing, mineral deposits must be capable of being mined and reduced to iron on a commercial basis. Direct-shipping ores are those that can be shipped as mined, without any beneficiation, and contain from 25–68% iron. Some lower-grade ores are mined but these are beneficiated before being sent to ironmaking furnaces.

United States. The most important source of iron ore for the U.S. is the Lake Superior district. Parts of this district are located in the states of Minnesota, Michigan, and Wisconsin and in Ontario in Canada. Of the 8.18×10^{13} t of ore produced in the

Table 10. Principal Iron-Bearing Minerals

Class and mineralogical name	Chemical compostion of pure mineral	Common designation
oxide		
magnetite	Fe_3O_4	ferrous–ferric oxide
hematite	Fe_2O_3	ferric oxide
ilmenite	$FeTiO_3$	iron–titanium oxide
limonite	$HFeO_2$[a]	
	$FeO(OH)$[b]	hydrous iron oxides
carbonate		
siderite	$FeCO_3$	iron carbonate
silicate		
chamosite	various	
silomelane	and	iron silicates
greenalite	sometimes	
minnesotaite	complex	
grunerite		
sulfide		
pyrite (iron pyrites)	FeS_2	
marcasite (white iron pyrites)	FeS_2	iron sulfides
pyrrhotite (magnetic iron pyrites)	FeS_2	

[a] Goethite.
[b] Lepidocrocite.

Table 11. Shipments of Canadian Iron Ore, 10^3 t

Year	Production
1954–1958 (av)	15,238
1960	19,550
1975	46,866
1976	55,414
1977	51,707

U.S. in 1978, 80% was produced in the U.S. areas of the Lake Superior district. Minnesota accounts for 56% of the total output and Michigan accounts for 22%; the remainder was produced mainly in California, Wyoming, Utah, Missouri, Wisconsin, New York, and Texas.

Iron-ore deposits in the Lake Superior district were formed by sedimentation during the Precambrian era. The original formation consisted of layers of iron oxides, iron carbonates, and iron silicates that were interbedded with layers of chert. Leaching by ground water and oxidation of the iron minerals produced local bodies of enriched oxides, usually soft hematites and limonites, that could be mined commercially without needing major beneficiation. The unaltered iron formation, when containing 20% or more of iron in the form of magnetite, is called taconite; if in the form of hematite, it is referred to as jasper. Taconites and jaspers are of importance because they are readily amenable to beneficiation.

The portion of the Lake Superior district located in the U.S. is comprised of dif-

Table 12. World Production of Iron Ore by Countries, Thousands of Metric Tons[a]

Country	1975	1976	1977
North America			
United States	80,154	80,559	56,660
Canada	46,881	55,432	53,636
Total	*127,035*	*135,991*	*110,296*
Latin America			
Argentina	264	250	250
Brazil	88,519	70,020	67,019
Chile	11,073	10,503	10,203
Colombia	623	601	601
Mexico	4,622	3,501	3,401
Peru	7,756	7,002	7,002
Venezuela	24,110	23,007	22,007
Total	*136,967*	*114,884*	*110,483*
Western Europe (EEC[b])			
Belgium	94	63	47
France	50,156	45,556	36,995
FRG	4,275	3,035	2,870
Denmark			5
Italy	739	643	576
Luxembourg	2,315	2,079	1,538
United Kingdom	4,491	4,585	3,744
Total	*62,070*	*55,961*	*45,775*
Other Western Europe			
Austria	3,834	3,785	3,245
Finland	766	700	1,047
Greece	1,966	2,151	2,111
Norway	4,065	4,292	3,723
Portugal	45	43	50
Spain	8,620	7,702	7,702
Sweden	30,876	30,535	25,423
Turkey	1,991	1,000	1,700
Yugoslavia	5,240	4,267	4,453
Total other Western Europe	*57,403*	*54,475*	*49,454*
Total all Western Europe	*119,473*	*110,436*	*95,229*
Eastern Europe (Communist)			
Bulgaria	2,338	2,301	2,272
Czechoslovakia	1,773	1,851	1,994
GDR	59	50	72
Hungary	386	631	525
Poland	1,192	1,101	660
Romania	3,066	2,301	2,751
U.S.S.R.	232,869	239,067	237,767
Total Eastern Europe	*241,683*	*247,302*	*246,041*
Total Europe	*361,156*	*357,738*	*341,270*
Africa			
Algeria	3,301	3,200	3,001
Angola	3,601	3,301	5,501
Liberia	36,511	35,010	26,508
Mauritania	8,503	8,003	8,302
Morocco	554	350	411
Republic of South Africa	11,194	15,729	15,504
Zimbabwe	601	601	550
Sierra Leone	2,501	2,401	1,500
Swaziland	2,233	1,932	1,901
Tunisia	652	500	370
Total	*69,651*	*71,027*	*63,548*

Table 12 (*continued*)

Country	1975	1976	1977
Middle East			
Iran	650	650	601
Egypt	1,121	1,300	1,200
Total Middle East	*1,771*	*1,950*	*1,801*
Far East			
Burma			
Hong Kong	161	37	31
India	40,282	41,411	41,231
Japan	942	800	684
Malaysia	349	300	326
Philippines	1,353	1,151	1,300
Republic of Korea	525	500	658
Thailand	32	20	19
Total	*43,644*	*44,219*	*44,249*
Far East (Communist)			
Democratic People's Republic of Korea	8,203	6,102	8,102
People's Republic of China	51,013	50,015	50,015
Total	*59,216*	*56,117*	*58,117*
Total Far East	*103,860*	*100,336*	*102,366*
Oceania			
Australia	97,393	92,426	97,527
New Zealand			
Total World	*896,833*	*874,352*	*827,291*

[a] Refs. 2 and 5.
[b] European Economic Community.

ferent iron ranges. The Mesabi range, probably the most important, is located 104 km north of Duluth. In this range, iron formation occurs over a length of 192 km. Since its discovery in 1844, it has produced over 2.3×10^{15} t of ore. Although a large portion of the reserves of direct-shipping ores has been depleted, the reserves of magnetite taconite, from which high-grade agglomerates are being produced, are large.

The Cuyuna range is located in Minnesota, 160 km southwest of Duluth, and has a maximum length of 109 km. Ores produced from the Cuyuna range contain 37–51% iron and 4.5–13.5% manganese, which makes the ores attractive for certain specialized uses.

The Marquette range is located in the upper peninsula of Michigan, and it is ca 48 km long. Most of the direct-shipping ore, in the form of a hard, dense hematite, has been exhausted. However, several large beneficiating plants have been built to process the taconites and jaspers from this range.

The Menominee range is located along the southern boundary of the upper peninsula of Michigan, and the Penokee–Gogebic range is located in northern Wisconsin. Current production from these two ranges is nominal, although they contain extensive iron formations.

The Northeastern district of the U.S. was an important source of iron ore during colonial times, and still supplies ore to several steel mills in the area from underground mines in Pennsylvania, New York, and New Jersey. Most of the mines produce a magnetite ore and beneficiate and agglomerate it before shipment.

Table 13. Identified World Iron Ore Resources, 10^6 Metric Tons[a]

Country	Reserves	Other	*Total*
North America			
Canada	12,000	17,000	*29,000*
United States	4,000	15,600	*19,600*
other	400	200	*600*
South America			
Brazil	18,000	11,000	*29,000*
Venezuela	1,400	2,500	*3,900*
other	1,400	16,700	*18,100*
Europe			
France	1,800	1,800	*3,600*
Sweden	2,200	800	*3,000*
U.S.S.R.	31,000	26,000	*57,000*
other	3,600	1,400	*5,000*
Africa			
Liberia	700	100	*800*
Republic of South Africa	1,200	1,800	*3,000*
other	1,700	2,500	*4,200*
Asia			
India	6,200	2,500	*8,700*
People's Republic of China	3,000	4,100	*7,100*
other	2,000	1,000	*3,000*
Oceania			
Australia	11,800	8,200	*20,000*
other	200	800	*1,000*
World total	*102,600*	*114,000*	*216,600*

[a] Ref. 7.

The Southeastern district of the U.S. is noted for its large deposits of hematite ore (red ore). The ore in this area is of sedimentary origin and is associated with the Clinton formation, which extends the length of the Appalachians. Near Birmingham, Alabama, this formation, the Red Mountain formation, is ca 75 m thick, of which some 2–3 m is of commercial interest. The ore is low in iron content, high in phosphorus content, and largely is self-fluxing as a result of the high lime content. A typical red ore contains 30–35% Fe, 18–29% SiO_2 plus Al_2O_3, 13–15% CaO, and 0.6% P. Up until 1960, this was the chief source of ore for the steel mills in Birmingham. However, because of the increased availability of higher grade, imported ores, most of the ore mines in this district have been closed. If a new or improved ore-beneficiation process is developed to remove the phosphorus and silica from the red ore, the district again could gain commercial importance.

The ore deposits in the Western district of the U.S. are relatively small with the principal ones occurring in Texas, California, Utah, Nevada, Wyoming, and Missouri. Although, originally, many of the ores were mined and shipped directly to the iron blast furnaces, beneficiating and agglomerating plants are the most prominent.

Canada. Until recently, Canada was not a large producer of iron ore. With the advent of beneficiating plants, new deposits of iron ore have been developed. Canadian ore is found principally in four areas: Newfoundland, the Labrador Trough in Quebec, the Northern Shore of the Great Lakes, and the coast of British Columbia. Much of the iron ore is exported from Canada to Europe and to the United States.

The rapid increase in production of ores in Canada from 1954–1958 to 1977 is implied in Table 11.

World. Although it once appeared that the world's iron ore reserves were being depleted rapidly, it appears that the reserves are more than ample for the next 100 yr. Recent discoveries of ore, new technological developments, and expanded production facilities have produced a situation by which there are ample supplies of iron ore available to the world market. Recent discoveries in South America, Africa, Canada, and Australia have provided the latest increase in the world's reserves of iron ore. Table 12 shows the production of iron ore for the main iron-ore-producing countries in the world.

The identifiable world resources of iron ore total more than 800×10^9 t containing over 215×10^9 t of iron. World reserves are estimated at ca 250×10^9 t containing ca 100×10^9 t of recoverable iron. Reserves are those resources that can be profitably mined under present economic conditions. Identified resources include reserves and other iron-bearing materials that may be profitable to mine in the future. These world reserves are adequate to supply demand for over 100 yr. The reserves and resources are shown by country in Table 13.

BIBLIOGRAPHY

"Iron" in *ECT* 1st ed., Vol. 8, pp. 18–45, by Walter Carroll, Republic Steel Corporation; "Iron" in *ECT* 2nd ed., Vol. 12, pp. 1–21, W. A. Knepper, United States Steel Corporation.

1. S. L. Goodale, *Chronology of Iron and Steel,* Penton Publishing Co., Cleveland, Ohio, 1931.
2. H. E. McGannon, *The Making, Shaping, and Treating of Steel,* 8th ed., U.S. Steel Corp., Pittsburgh, Pa., 1964.
3. American Iron and Steel Institute, *Steel Products Manual,* Sec. 1, Washington, D.C., 1951.
4. *Metals Handbook,* 8th ed., American Society of Metals, Metals Park, Ohio, 1961.
5. *Annual Statistical Report,* American Iron and Steel Institute, 1978.
6. Ferrous Metals, *ASTM Standards,* Part I, American Society of Testing and Materials, Philadelphia, Pa., 1961.
7. "Iron Ore," *Minerals Commodity Profiles MPC-13,* U.S. Dept. of Interior, Bureau of Mines, U.S. Govt. Printing Office, Washington, D.C., May 1978.

General References

H. P. Tiemann, *Iron and Steel,* 2nd ed., McGraw-Hill Book Co., Inc., New York, 1919.

C. D. King, *Seventy-Five Years of Progress in Iron and Steel,* American Institute of Mining and Metallurgical Engineers, New York, 1948.

"Pelletizing in North America," *33 Magazine,* 42, 43 (June, 1964).

W. A. Knepper, *Agglomeration,* Interscience Publishers, a division of John Wiley & Sons, Inc., New York, 1962.

D. L. McBride and P. W. Chase, *Blast Furnace Steel Plant 51,* pp. 868–878, Oct. 1963.

R. H. Sweetser, *Blast Furnace Practice,* McGraw-Hill Book Co., Inc., New York, 1938.

W. Gumz, *Gas Producers and Blast Furnaces,* John Wiley & Sons, Inc., New York, 1950.

G. Elliott, *Practical Ironmaking,* Percy Lund, Humphries and Co., London, 1939.

T. M. Rohan, *Iron Age* **196,** 87, 94 (Sept. 16, 1965).

"Iron and Steel," *Bureau of Mines Minerals Yearbook,* U.S. Dept. of Interior, U.S. Govt. Printing Office, Washington, D.C., 1976.

W. A. Knepper
United States Steel Corporation

IRON BY DIRECT REDUCTION

Direct-reduction (DR) processes produce metallic iron from its ores by removing the associated oxygen at temperatures below the melting temperature of any of the materials involved in the process. Products from the DR processes are referred to as direct-reduced iron (DRI).

Before the blast furnace was developed (ca 600 years ago), crude devices were used to reduce iron ores and to produce a solid metal product for making wrought iron and steel (qv) that could be worked into useful objects. These early processes could not be carried out on a large scale and eventually were supplanted by higher productivity, indirect processes that were based on the blast furnace. Blast furnaces reduce iron ore and melt the reduced iron so that it can be withdrawn as a liquid (hot metal) that is refined to steel in a separate process. Hot metal also may be cast into small bars (pigs) for subsequent remelting for foundary castings or for steelmaking. DRI may be refined to steel in any conventional steelmaking process or it may be used in the blast furnace or other smelting process to increase productivity and decrease coke usage rate.

Through the years since the advent of the blast furnace, research has continued in the development of practical technology that would reestablish direct reduction as an alternative or supplement for recovering iron from its ores (1). These attempts have involved practically every known type of apparatus that is suitable for the purpose, eg, pot furnaces, reverberatory furnaces, regenerative furnaces, shaft furnaces, retorts, rotary-hearth furnaces, rotary and stationary kilns, traveling grates, fluidized beds, and various combinations (see Furnaces). Many different reducing agents also have been tried, including coal, coke, char, fuel oil, tar, natural gas, producer gas, coal gas, water gas, carbon monoxide, and hydrogen (see Carbon monoxide; Coal; Fuels, synthetic; Gas, natural). However, despite the intensity of this effort, only 20×10^6 annual metric tons of DRI capacity was installed in noncommunist nations by 1980, which represents less than 5% of the world's total iron-making capacity. The DR processes are expected to remain a minor factor for the foreseeable future except where there are abundant reserves of inexpensive natural gas, noncoking coals, and/or hydroelectric power and where there is access to suitable iron ores or agglomerates. The blast furnace will remain the world's main source of iron for steelmaking on a large scale as long as adequate supplies of metallurgical coke can be provided at competitive cost.

Chemistry

The reduction of iron ore in any DR process is accomplished by the same reactions that occur in the blast furnace stack, including reduction by CO and H_2 and, in some cases, solid carbon, through successive oxidation states to metallic iron, ie, $Fe_2O_3 \rightarrow Fe_3O_4 \rightarrow FeO \rightarrow Fe$ (metallic iron). Where the reduction reactions are carried out below about 1000°C, the reducing agents usually are restricted to CO and H_2 and the DRI that is produced is porous and has essentially the same size and shape as the original iron ore particle or agglomerate. Above about 1000°C, the reduced material begins to sinter and, at ca 1200°C, a pasty, porous mass forms. Above 1200°C, the metallic iron that has been formed absorbs any carbon that is present which results in freezing-point depression (from 1530°C) and subsequent fusing or melting of the solid.

The effectiveness of the commercial processes, eg, fluidized beds (2–3), moving-bed shafts (4–6), fixed-bed retorts (7), and rotary kilns (8–11), depends on sound process design and engineering to ensure that good contact is obtained between reacting gases and solids. Schematic flow sheets of the process concepts are presented in Figure 1. Ultimately, the performance of these processes depends on how closely they can approach the theoretical energy requirements for the reaction at throughputs that are high enough to provide economical plant productivity.

Equilibrium Considerations. The equilibrium for the reactions of CO and H_2 with the oxides of iron are well established (12) and are shown in Figures 2 and 3 for the systems iron–oxygen–carbon and iron–oxygen–hydrogen, respectively, from 500–1200°C. Fe_2O_3 is not shown because there is nearly complete conversion of CO and H_2 for the reaction,

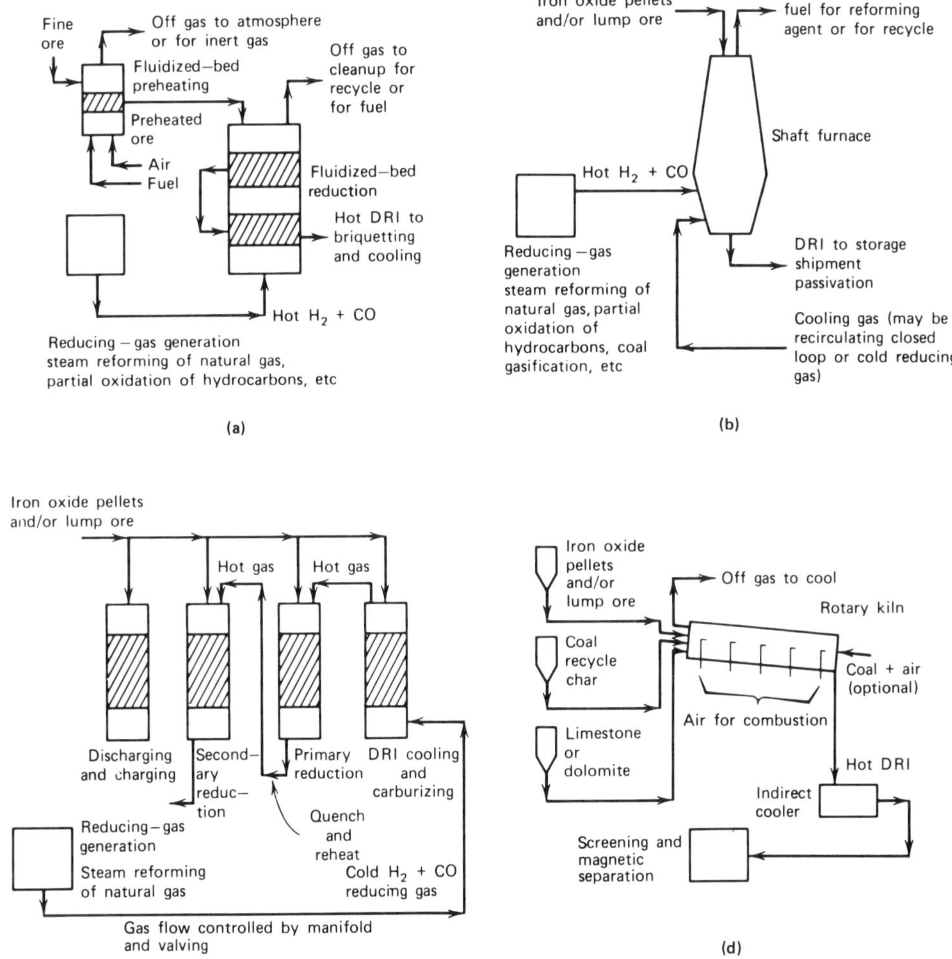

Figure 1. Schematic flow sheets of major DR process concepts. (**a**) Fluidized-bed processes. (**b**) Moving-bed shaft processes. (**c**) Fixed-bed retort processes. (**d**) Rotary-kiln processes. DRI = direct-reduced iron.

$$3\ Fe_2O_3 + \begin{Bmatrix} H_2 \\ CO \end{Bmatrix} \rightarrow 2\ Fe_3O_4 + \begin{Bmatrix} H_2O \\ CO_2 \end{Bmatrix}$$

Below 570°C, Fe_2O_4 is reduced directly to Fe by CO and H_2; above 570°C, Fe_3O_4 first is reduced to FeO which then is reduced to Fe. In the reduction of Fe_3O_4 to FeO, the equilibrium conversions of CO and H_2 increase with increasing reaction temperature. However, in the reduction of FeO to Fe, the equilibrium conversion for H_2 increases with increasing reaction temperature, whereas that for CO decreases. This decrease of equilibrium CO conversion with increasing temperature for the reduction of FeO to Fe is not a limitation on the overall conversion because most DR processes are operated with countercurrent flow of solids and reducing gases. Thus, the spent reducing gas leaves in contact with entering solids which are in their highest oxidation state, and the equilibrium for the reduction of Fe_3O_4 to FeO governs the final gas composition. For DR processes that are based on reduction with mixtures of CO and H_2, the final gas composition usually satisfies the equilibrium for the water–gas reaction at the exit temperature, ie,

$$CO_2 + H_2 \rightarrow CO + H_2O,\ K_{EQ} = \frac{[CO][H_2O]}{[CO_2][H_2]}$$

The preceding comments apply primarily to DR processes that are based on the use of gaseous reductants in fluidized beds, moving-bed shafts, and fixed-bed retorts. In the processes that use solid reductants, eg, coal, the reduction is accomplished to a minor extent first by volatiles and reducing gases that are released as the coal is heated and then by CO that is formed by gasification of coal char with CO_2. The endothermic heat of reduction and sensible energy that is required to heat the reactants in these processes is provided by combustion of volatiles and of CO leaving the bed with air introduced into the free space above the bed. Reduction by solid carbon and coal volatiles in kilns is insignificant. The extent of combustion in the free space and the effectiveness of the transfer to the reduction bed of the energy that is released influence the energy requirements and the productivity per unit of reactor volume. For the DR processes that are based on gaseous reductants in fluidized beds, shafts, and retorts, heat transfer is very efficient as long as uniform gas flow distribution is maintained.

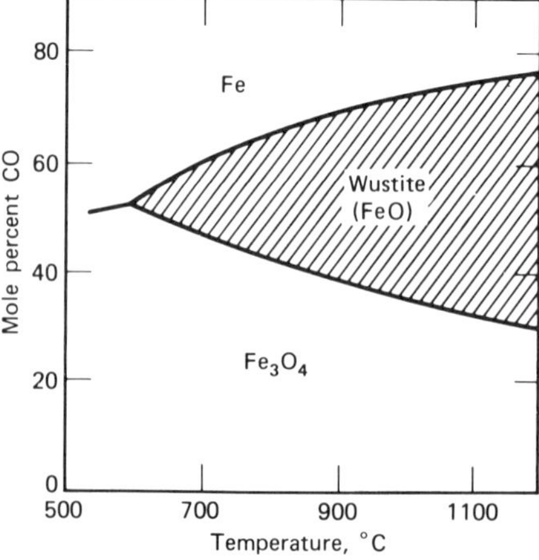

Figure 2. Iron–oxygen–carbon system.

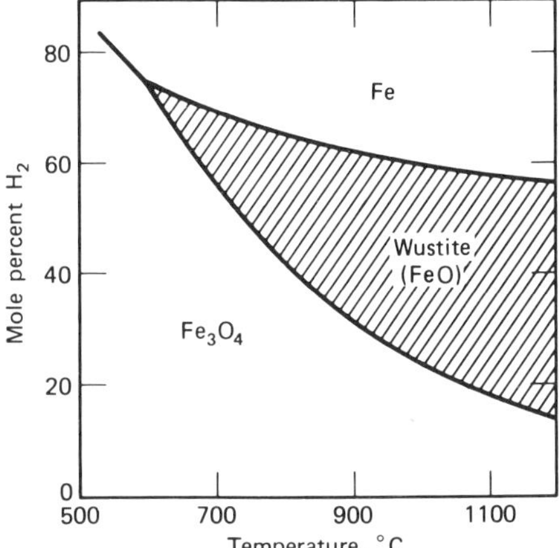

Figure 3. Iron–oxygen–hydrogen system.

Energy Considerations. The energy requirements for DR processes are related directly to the heats of reaction for the reduction reactions over the temperature range of practical interest for DR processes. A summary of heats of reaction for the reduction of Fe_2O_3, Fe_3O_4, and FeO by CO, H_2, and carbon is presented in Table 1 (13).

The reduction of Fe_2O_3 to Fe_3O_4 with H_2 and CO is mildly exothermic and the reduction of FeO to Fe with CO is moderately exothermic. The other reactions are moderately endothermic, with the exception of the reduction of Fe_3O_4 to FeO and of FeO to Fe by carbon; both are highly endothermic. The heat of reduction is not a principal factor in establishing the energy requirements for the gas-based DR processes. Instead, the energy requirements for these processes is comprised mainly of the energy that is used to generate the reducing gas (in most cases, catalytic steam reforming of natural gas) and the chemical energy in the reductants that are consumed in the process. For the DR processes that are conducted in moving-bed shafts and fixed-bed retorts, total energy requirements can be minimized by recovering most of the sensible heat in the DRI and in the spent reducing gases. Such recovery is not practical in the fluidized-bed processes and accounts for the lower total energy requirements that are possible with the moving-bed shafts and fixed-bed retorts, ie, 12 GJ/t (10.5×10^6 Btu/t) compared with ca 18.6 GJ/t (16×10^6 Btu/t) for DR in fluidized beds.

For the DR processes based on the direct use of solid reductants, the highly endothermic reaction of CO_2 with solid carbon is compensated by the highly exothermic combustion reactions in the free space above the reaction bed. There is no recovery of sensible energy in the DRI and in the kiln off gas, and the total energy requirement for the processes based on the direct use of solid reductants also is ca 18.6 GJ/t (16×10^6 Btu/t).

Kinetics. The productivity of the DR processes depends on chemical kinetics and mass- and heat-transport factors that combine to establish the overall rate and extent of reduction of the charged ore. In the case of the fluidized-bed DR processes (Fig. 1), the solids are preheated in a separate fluidized bed by *in situ* combustion of a suitable fuel with air. The reducing gases are preheated in an indirect-fired heat exchanger. The endothermic heat of reduction is introduced as sensible heat in the entering solids

Table 1. **Heats of Reaction of Reduction Reactions in DR Processes**[a,b]**, kJ/kg mol Fe**[c]

Temperature		$Fe_2O_3 \rightarrow Fe_3O_4$			$Fe_3O_4 \rightarrow FeO$			$FeO \rightarrow Fe$		
K	°C	H_2	CO	C	H_2	CO	C	H_2	CO	C
900	627	−733	−6722	21,888	17,573	5606	62,814	16,282	−19,597	152,027
1000	727	−2105	−7920	20,539	17,154	5513	62,430	16,282	−18,631	152,109
1100	827	−2628	−8269	20,027	16,736	5443	62,023	17,119	−16,747	152,981
1200	927	−2791	−8269	19,841	16,608	5652	61,883	17,166	−15,101	152,981
1300	1027	−2803	−8129	19,818	16,375	5722	61,604	16,038	−15,910	151,725

[a] All heats of reaction are based on stoichiometric equation for one mole Fe in reactant oxide.

[b] Negative values are exothermic; positive values are endothermic.

[c] To convert kJ/kg mol Fe to Btu/lb mol Fe, multiply by 0.43.

and gases. The rate of the reduction reactions then are a function of the temperature and pressure in the reduction beds, the porosity and size distribution of the ore, the composition of the reducing gas, the depth of the reduction beds, and the effectiveness of gas/solid contact in the reduction beds. The reduction rate generally increases with increasing temperature and pressure up to about 507 kPa (5 atm). When the reduction occurs below ca 650°C, the DRI from the processes are pyrophoric because the metallic iron that is formed has a very high surface area which is available for reaction with atmospheric oxygen. At higher reduction temperatures, recrystallization and sintering decreases the surface area and the associated reactivity, and the cooled DRI generally is stable when kept dry. Above ca 810°C, there is a tendency for the beds to defluidize as a result of sticking or bogging of the reduced material. The fluidized-bed processes, therefore, are operated from 650–810°C.

In the moving-bed shaft and fixed-bed retort processes, the solids are preheated by countercurrent contact with partially spent reducing gas and the reducing gas is preheated either during generation (in the case of reforming natural gas with stoichiometric steam) or by an indirect-fired heat exchanger when the reforming is achieved with excess steam. With proper design of the shafts and retorts, the rate and extent of reduction generally increases with increasing reducing-gas temperature. The upper limit on temperature is determined by the resistance of the solids to sticking and clustering which interferes with effective solids movement and/or discharge. This upper temperature limit is ca 980°C for the iron-oxide feeds that are most stick resistant. Some materials cannot be reduced at above about 790°C because they tend to stick. The resistance to reoxidation and strength of the DRI from these processes also increases with increasing reduction temperature, which is an important consideration if the DRI is to be shipped to other locations for steelmaking or if it is to be stored in the open before use on site. Because DRI from the fluidized-bed processes is hot briquetted to a specific gravity greater than 5.0, it generally is stronger and more resistant to reoxidation than DRI that is produced in the other gas-based processes and it may be handled as if it were an inert material.

The productivity of the DR processes that are based on solid reductants is influenced by the temperature that is required to achieve the desired high degree of reduction and by the rate at which the required energy can be transferred to the reaction bed. The reduction temperature is a function of the reducibility of the iron ore or agglomerate and of the reactivity of the char that is formed from the coal or other solid reductant. The rate of heat transfer is a function of the reaction bed temperature and the ash-softening temperature of the solid reductant. The ash-softening temperature governs the maximum flame temperature (kiln-lining temperature) that can be used without consequent formation of accretions which would impair the operability of the process.

Suitable coals generally have ash-softening temperatures that are at least 100°C higher than the reduction temperatures required for good productivity with high metallization, are relatively noncaking, and contain less than 1.5% sulfur (for effective steelmaking). Coals that are relatively low in volatile matter also may be preferred because energy losses resulting from incomplete combustion of volatiles that are released close to the feed end of the kiln are minimized. The processes based on the direct use of solid reductants usually operate with reduction temperatures greater than 980°C, so that the DRI that is produced also is relatively strong and resistant to reoxidation.

Production

Annual commercial DRI capacity through 1975 with projected estimates for 1980 and 1985 by region and by type of fuel and reductant is summarized in Table 2. Through 1980, the capacity data are based on projects that were under contract for completion by this date. The 1985 projections include proposed projects that have a good probability of implementation. Most of the production occuring after 1975 was and is slated for South America and the Middle East and will be based on natural gas. In 1975, the installed plants were operated at ca 60% of rated capacity, ie, in many of the plants operating and maintenance problems had to be solved before the plants could achieve the availability and reliability associated with fully established technologies.

The growth of the HyL fixed-bed retort process and the Midrex moving-bed shaft processes is summarized in Table 3. The rapid growth in capacity for both processes after 1975 indicates that these processes have established the reliability and economy of operation that will make it difficult for other processes to compete for future ca-

Table 2. Summary of Installed and Projected Commercial Capacity for DRI (10^6 Metric Tons per Year of DRI Capacity)[a]

Year	North America	South America	Western Europe	Asia	Middle East	Oceania	Africa	*Total*[b]
1960	0	0.4	0.1	0	0	0	0	*0.5*
1965	0	0.4	0.1	0	0	0	0	*0.5*
1970	0.4	0.9	0.1	0.1	0	0.1	0	*1.6*
1975	2.0	2.1	0.1	0.9	0	0.1	0.2	*5.4*
1980	3.0	9.0	1.6	1.7	3.8	0.1	0.5	*19.7*
1985	4.0	16.0	3.1	4.2	5.4	0.2	3.7	*36.6*

Year	*By fuel reductant* Coal	Gas	*Total*[b]
1960	0.1	0.4	*0.5*
1965	0.1	0.4	*0.5*
1970	0.3	1.3	*1.6*
1975	1.9	4.1	*6.0*
1980	2.4	17.3	*19.7*
1985	4.0	32.6	*36.6*

[a] Ref. 14.
[b] Noncommunist countries.

Table 3. Summary of Growth of Installed Capacity for HyL and Midrex DR Processes

	HyL		Midrex	
Year	Capacity, 10^6 t/yr	No. of lines	Capacity, 10^6 t/yr	No. of lines
1960	0.4	2		
1965	0.4	2		
1970	0.9	4	0.3	1
1975	1.6	6	1.5	4
1980	9.9	15	7.3	19

pacity. Processes based on the direct use of coal are not expected to be a major contributor to new capacity after 1980.

The projected total world steel production of ca 10^9 t by 1985 indicates that the rapid growth in DR capacity that is projected during this intervening period still characterizes DR as a relatively minor factor. From the viewpoint of the developing countries where most of this capacity is being installed, however, DR represents the only way that local steel industries can be established with economic justification. In these developing countries, then, DR will make a major contribution to improved living standards and mineral and human resource utilization.

Uses

The main uses for DRI are as prime metallic iron units for electric furnace steelmaking and foundry operations, as a coolant in BOF (Bessemer-oxygen furnace) steelmaking, as a replacement for scrap in the open hearth, and for increasing productivity and decreasing coke ratio in blast furnaces and other smelting processes. Minor uses for DRI are as a chemical reagent for copper cementation and for the manufacture of iron powder for powder metallurgy (qv) applications. Recently, the solid-based DR processes have been adapted to the reduction of agglomerates that are produced from waste iron-bearing materials generated in steel plants primarily for recycling to the blast furnace.

DRI is an excellent electric-furnace charge because it is free of undesirable metallic impurities, has a uniform and precisely known chemical composition, and is easy to transport and handle (15). These properties enable dilution with low-cost scrap without sacrificing quality and permit continuous charging with attendant increased productivity, decreased power consumption, and lower electrode and oxygen consumption. These advantages are offset partially by the presence of gangue in the DRI (DR processes retain all of the original gangue in the charged iron oxide), which imposes an additional energy requirement.

DRI has also been demonstrated to be a very effective metallic charge in the basic BOF and open-hearth steelmaking (16). In the BOF, up to 70% of the scrap was replaced by DRI with no operating difficulties that could not be compensated by adjustments in charging practice. The predicted effect of DRI gangue content on product yield and lime consumption was obtained and, in all cases, the concentration of residual tramp elements in the finished steel decreased with increasing percentage usage of DRI. In the open hearth, DRI serves primarily as a replacement for scrap.

DRI has been used to replace up to 20% of the scrap portion of the charge in foundry cupolas without creating excessive windbox pressures (17). With DRI in the charge, the tramp-element content of the product is decreased; however, additional silicon and manganese generally are required.

The results of tests using DRI as part of the blast-furnace burden show that for every 10% of metallic iron in the burden, ie, for every 100 kg of metallic iron per t of hot metal, the coke ratio decreases by ca 35 kg/t, and the hot-metal production rate increases by 9% (18). Thus, DRI can be used to improve blast-furnace performance during periods of peak demand when metallurgical coke availability is limited and when hot-metal capacity is reduced because blast furnaces are out of operation for extensive servicing. Similar benefits are obtained when DRI is substituted for ore in electric smelting furnaces.

DRI has been used for many years as a chemical reagent for copper extraction. In 1975, a 60,000-t/yr, rotary-kiln, DR plant was built in Arizona to produce DRI specifically for cementation of copper.

The production of iron powder for powder metallurgy is another industry in which DRI is used. Plants in Sweden and New Jersey are based on special fixed-bed retorts, and produced nearly 50% of the total annual production of iron powder (ca 200,000 t) in 1979.

Growth and Future Prospects

Technological improvements in blast-furnace productivity and fuel consumption account for the nearly total lack of implementation of DR at existing integrated steel plants. The ready availability of cheap natural gas in the United States in the early 1950s has given way to scarcity and relatively high cost so that natural gas no longer is considered available for new domestic DR capacity. Almost all DR capacity is being installed in locations where natural gas is abundant and relatively inexpensive, eg, in Venezuela, Iran, Mexico, and the Middle East. These factors account for renewed interest in the United States in basing shaft DR capacity on coal gasification and in improving the efficiency and operating reliability of the kiln processes that directly use noncoking coals. Coal gasification has required too much capital investment to be economical for DR in the United States and the low operating reliability and high energy consumption of the kiln processes (which operate on high-volatile, noncoking coals) have been a barrier to their acceptance for new plants in the United States.

The successful development and demonstration of technology for removing sulfur from coal gasification gas while it is still hot would improve the economics of coal-based DR plants using moving-bed shaft processes. Improvements in the energy efficiency and operating reliability of the kiln processes will require considerable engineering ingenuity in order to provide less expensive and more efficient waste-heat recovery equipment and effective control systems and operating practices that will eliminate or at least minimize the accretion-formation problems that have plagued most of these operations to date. In any event, coal-based DR plants in the United States are likely to be associated with minimill operations or where incremental increases in capacity are needed. The blast furnace probably will continue to be the low-cost route to steel for large integrated steel plants.

BIBLIOGRAPHY

"Iron by Direct Reduction" in *ECT* 2nd ed., Supplement, pp. 535–541, by David L. McBride, United States Steel Corporation.

1. E. P. Barrett, *Sponge Iron and Direct-Reduction Processes,* Bulletin No. 519, United States Bureau of Mines, Washington, D.C., 1954.
2. W. L. Davis, Jr., J. Feinman, and J. H. Gross, "Briquetacion del Mineral de Hierro de Alto Tenor por El Processo de Reducion Directa en Lechos Fluizados," *ILAFA, Memoria Tecnia del Seminario Latinoamericano,* La Reduccion Directa de los Minerales de Fierro y Su Applicacion en America Latino, Ciudad de Mexico, Mexico, Nov. 1973.
3. D. C. Violetta, "FIOR—the Process for Producing Highly Metallized Feed for Steelmaking," in Ref. 2.
4. C. L. Cruse, A. P. Kerschbaum, W. E. Marshall, and R. E. Moon, *J. Met.* **21,** 55 (Oct. 1969).
5. E. R. Morgan, "Reduccion Directa Continua de Los Oxidos de Fierro por el Process Midrex," in Ref. 2.

6. H. D. Pantke and G. H. Lange, *Committee Seminar on Direct Reduction of Iron Ore: Technical and Economic Aspects,* STEEL, Sem. Dir. Red./1-6, Bucharest, Romania, Sept. 1972, pp. 18–23.
7. R. Lawrence, Jr., "The HyL Direct-Reduction Process—Past, Present, and Future," in Ref. 6.
8. G. Meyer and U. Bongers, *Eng. Min. J.* **174**(6), 159 (1973).
9. H. Serbent and W. Janke, "SL/RN Process—Production of Sponge Iron Using Solid Reductants," in Ref. 6.
10. D. A. Bold and N. T. Evans, *Iron Steel Int.* **50,** 145 (June 1977).
11. D. W. Rierson and A. A. Albert, Jr., *AIME Ironmaking Proc.* **36,** 455, 1977.
12. J. F. Elliott and M. Gleiser, *Thermochemistry for Steelmaking,* Vol. 1, Addison-Wesley Publishing Co., Inc., Reading, Pa., 1960.
13. J. P. Coughlin, *United States Bureau of Mines Bulletin No. 542,* 1954.
14. R. Goodman, *Skillings Mining Review* **68**(11), 12 (Mar. 17, 1979).
15. R. M. Smailer, H. B. Jensen, and W. W. Scott, Jr., *AIME Electric Furnace Proc.* **32,** 29 (1974).
16. J. M. Pena and D. Radke, *J. Met.* **23**(8), 27 (1971).
17. R. H. Hafner and S. C. Clow, *Mod. Cast.* **53,** 73 (1968).
18. J. A. Peart and F. J. Pearce, "Prereduced Burdens in the Blast Furnace Process," *Congres International sur la Production et L'Utilization des Minerals Reduits,* Evian, Fr., 1967.

General References

L. von Bogdandy and H. J. Engell, *The Reduction of Iron Ores, Scientific Basis and Technology,* Verlag Stahleisen m.b.H. Dusseldorf, Springer-Verlag, Berlin, FRG, 1971.

R. L. Stephenson, ed., *Direct Reduced Iron—Technology and Economics of Production and Use,* ISS/AIME, Warrendale, Pa., 1980.

H. E. McGannon, ed., *The Making, Shaping, and Treating of Steel,* U.S. Steel Corporation, Pittsburgh, Pa., 1980.

J. FEINMAN
United States Steel Corporation

IRON COMPOUNDS

The usual oxidation states of iron are 2+ (d^6) and 3+ (d^5), the ferrous and ferric states, respectively. The preferred nomenclature indicates oxidation states as iron(II) and iron(III), and this convention is used here. Oxidation states from 4− (as in the phthalocyanine complexes) to 6+ (as in the ferrates) are known, but the 2+ and 3+ oxidation states are of overwhelming importance.

Iron metal dissolves in dilute mineral acids and yields, in the absence of other oxidizing agents (including air), the iron(II) ion. Salts with a tremendous variety of anions may be formed by evaporation of the corresponding aqueous solution. Such salts typically are green or yellow hydrated substances and are subject to air oxidation and either hydration or efflorescence. Aqueous iron(II) salt solutions contain the hexaaquoiron(II) ion which is subject to air oxidation. The oxidation process in basic solution is more rapid than that in acid solution. Upon standing, aqueous iron(II) salt solutions result in precipitated iron oxides or hydroxides. The following potentials (as measured against the hydrogen electrode) are noted:

$$Fe \rightarrow Fe^{2+} + 2\,e- \qquad E° = -0.44\ V$$

$$Fe(H_2O)_6^{2+} \rightarrow Fe(H_2O)_6^{3+} + e^- \qquad E° = +0.77\ V$$

The Russell-Saunders ground-state term for the free ion is ^5D. An octahedral or tetrahedral field splits the configuration into a $^5T_2$ state and a ^5E state. All tetrahedral complexes have magnetic moments of (4.6–4.8) × 10^{-23} J/T (5–5.2 μ_β). Octahedral complexes may be high spin (4.8 × 10^{-23} J/T or 5.2 μ_β). Low-spin octahedral complexes result in strong ligand fields and a fair number of these are known. In a few cases, where strong ligand fields are accompanied by large tetragonal distortion (ie, square planar coordination), the orbital energies increase as d_{yz}, $d_{xz} < d_{z^2} < d_{xy} \ll d_{x^2-y^2}$ and the orbital occupation is $(d_{yz})^2(d_{xz})^2(d_{z^2})^1(d_{xy})^1$; a triplet ground state results.

Iron(III) salts generally are white or pale-colored hydrates. Aqueous solutions of iron(III) salts quickly hydrolyze to form aquo species, eg, $[(H_2O)_4Fe(OH)_2Fe(H_2O)_4]^{4+}$, and a pH of about zero must be maintained to prevent hydrolysis. At a pH of two, more polynuclear species are formed and colloidal gels appear and, at a pH above two, red-brown hydrated ferric oxide precipitates. Iron(III) complexes typically are octahedral but a number of tetrahedral complexes also are known. Oxygen is the favored ligand of iron(III), although strong-field nitrogen ligands also are important. Spectra of iron(III) complexes are dominated by charge-transfer transitions, and the complexes often are very highly colored. The high-spin complexes have magnetic moments of 5.5 × 10^{-23} J/T (5.9 μ_β) (the spin-only value), and the low-spin complexes generally have room temperature magnetic moments of 2.1 × 10^{-23} J/T (2.3 μ_β).

Acetates

If scrap iron is treated with acetic acid, the iron may be solubilized as acetate salts. The initial black liquor solution is concentrated to a 12% solution from which iron acetate [*2140-52-5*] may be obtained. This salt appears to be a mixture of the iron(II) and iron(III) oxidation states of indefinite proportions. It is most safely formulated as $Fe_x(CH_3COO)_y$ and is used as a catalyst of acetylation and carbonylation reactions.

A pure iron(II) acetate [*3094-87-9*], $Fe(C_2H_3O_2)_2$, also may be prepared. It is a colorless compound which may be recrystallized from water to yield hydrated species. Iron(II) acetate is used widely in the preparation of dark shades of inks (qv) and dyes and is used as a mordant in dyeing (see Dyes and dye intermediates). It is available for $60/100 g.

Iron(III) acetate [*1834-30-6*], $Fe(C_2H_3O_2)_3$, is prepared industrially by treatment of scrap iron with acetic acid followed by oxidation of the resulting solution with air. Iron(III) acetate is used as a mordant, as a catalyst in organic oxidation reactions, and as a convenient source of iron in the preparation of other compounds.

Basic iron(III) acetate [*10450-55-2*], $Fe(OH)(C_2H_3O_2)_2$, is a brownish-red compound that can be obtained by boiling a solution of iron(III) acetate and by allowing the hydroxy salt to precipitate. It is insoluble in water but dissolves in alcohols or acids. Basic iron acetate is used extensively as a mordant in dyeing and printing and for the weighting of silk (qv) and felt (see Felts). It may be obtained for $40/100 g (1980).

Carbonates

Iron(II) carbonate [*563-71-3*], $FeCO_3$, appears as a white precipitate when solutions of alkali carbonate salts are added to solutions of iron(II) salts. The compound darkens when subjected to air oxidation. Iron(II) carbonate is used as a flame retardant and as an iron supplement in animal diets (see Flame retardants; Food additives; Pet and other livestock feeds). In solutions of carbon dioxide, iron(II) hydrogen carbonate [*6013-77-0*], $Fe(HCO_3)_2$, forms and easily undergoes air oxidation. Carbon dioxide is evolved during air oxidation of iron(II) hydrogen carbonate, and hydrated iron(III) oxide remains as a precipitate. This process produces water with iron carbonate unacceptable for industrial or domestic use (see Water, water pollution).

Citrates

Iron citrate [*2338-05-8*] is a compound of indefinite ratio of citric acid (qv) and iron in mixed oxidation state. Iron(II) citrate [*23383-11-1*] and iron(III) citrate [*28633-45-6*] are known and also are of indefinite stoichiometry. Iron(III) citrate [*3522-50-7*] (1 Fe:1 citric acid) is available. The iron citrate compounds are slowly soluble in water, more readily soluble in hot water, and exhibit complex solution chemistry wherein a number of monomeric and oligomeric species have been identified. All of these compounds have been used as supplements to animal diets and to soil.

Iron(III) ammonium citrate [*1185-57-5*] also has an indefinite stoichiometry. A brown hydrated compound, iron(III) ammonium citrate [*1332-98-5*], contains 16.5–18.5% iron, ca 9% NH_3, and 65% citric acid; a green hydrated compound, iron(III) ammonium citrate [*1333-00-2*] contains 14.5–16% iron, ca 7.5% NH_3, and 75% citric acid. Iron ammonium citrates are very soluble in water but are insoluble in alcohol. These compounds are used to fortify foods, eg, bread and milk, and are sold for $19.00/500 g or $600/45 kg.

Cyanides

It is possible that the first coordination compound discovered was an iron cyanide, Prussian blue, in 1704. The variety, complexity, and utility of iron cyanide compounds is astounding (1). All of the iron cyanide compounds are complex coordination compounds and, therefore, the names hexakiscyanoferrate(4−) and hexakiscyanoferrate(3−) are preferred to iron(II) cyanide or iron(III) cyanide or ferrocyanide or ferricyanide (see Coordination compounds).

The cyanide ion is very high in the spectrochemical series and produces a very large nephelauxetic effect. The $Fe(CN)_6^{4-}$ ion is diamagnetic. As a monodentate ligand, cyanide very nearly always coordinates through the carbon atom, but cyanide frequently may serve as a bidentate ligand (see below). Bonding to iron by cyanide may best be described by a synergistic mechanism involving σ donation and π acceptance by the ligand.

Hexakiscyanoferrates(4−). Hexakiscyanoferrate(4−) [*13408-63-4*], $[Fe(CN)_6]^{4-}$, may be formed by displacement of water from simple iron(II) salts by aqueous cyanide ion, accompanied by the evolution of a large amount of heat. Electrolytic oxidation results in the formation of hexakiscyanoferrate(3−) [*13408-62-3*], $[Fe(CN)_6]^{3-}$. Alkali or alkaline earth salts of either complex are soluble in water but are insoluble in alcohol. The salts of hexakiscyanoferrate(4−) form yellow crystals and those of hexakiscyanoferrate(3−) form ruby-red crystals. The tremendous variety of iron cyanide complexes arises when one or more cations of the alkali or alkaline earth salts is replaced by a complex cation (eg, ammonium ion), by a representative metal or by a transition metal. Many of the salts have commercial application, although the large majority of industrial production of iron cyanide complexes is of iron blues (eg, Prussian Blue), which are used as pigments (qv). Many of the transition metal salts of $[Fe(CN)_6]^{4-}$ display characteristic colors, and addition of $[Fe(CN)_6]^{4-}$ to an unknown metal salt solution has been used as a qualitative test for those transition metals. All of the salts may be considered salts of ferrocyanic acid, tetrahydrogen hexakiscyanoferrate [*17126-47-5*], $H_4[Fe(CN)_6]$, which is a colorless compound that is subject to air oxidation. It is soluble in water and alcohol but is insoluble in ethyl ether; it may be prepared by addition of sulfuric acid to dibarium [*13821-06-2*] or dilead hexakiscyanoferrate [*14402-61-0*]. Most of the mixed (containing more than one type of cation) iron cyanides may be prepared by the precipitation of the less soluble mixed salt from a solution of a soluble iron cyanide and a soluble salt of another cation, eg,

$$(NH_4)_4[Fe(CN)_6] + MgCl_2 \rightarrow \underset{[66139\text{-}49\text{-}9]}{(NH_4)_2Mg[Fe(CN)_6]} + 2\ NH_4Cl$$

Tetraammonium hexakiscyanoferrate [*14481-29-9*], $(NH_4)_4[Fe(CN)_6]$, is obtained as yellow crystals by the addition of ammonia to $[Fe(CN)_6]^{4-}$ or by the addition of ammonium sulfate to an aqueous solution of the barium or calcium salt of $[Fe(CN)_6]^{4-}$. Tetraammonium hexakiscyanoferrate(II) is soluble in water, is insoluble in alcohol, and is subject to air oxidation. Diammonium barium hexakiscyanoferrate [*60700-20-1*], $(NH_4)_2Ba[Fe(CN)_6]$, and diammonium calcium hexakiscyanoferrate(II) [*60674-40-0*], $(NH_4)_2Ca[Fe(CN)_6]$, salts also are known.

Dibarium hexakiscyanoferrate [*13821-06-2*], $Ba_2[Fe(CN)_6]$, is a light-yellow, slightly water-soluble substance that is prepared by addition of a solution of tetrasodium hexakiscyanoferrate to a concentrated solution of an appropriate barium salt. It is most useful in the preparation of other hexakiscyanoferrate(4−) salts because of the insolubility of barium sulfate. Mixed salts, eg, barium dipotassium hexakiscyanoferrate [*31389-21-6*], $BaK_2[Fe(CN)_6]$, are known.

Dicalcium hexakiscyanoferrate [*13821-08-4*], $Ca_2[Fe(CN)_6]$, is formed as yellow crystals in the reaction of liquid or gaseous HCN with $FeCl_2$ in water having pH ≥ 8 and containing $Ca(OH)_2$ or $CaCO_3$. It has been used to prevent caking and agglomeration of other substances, and it serves as a convenient starting material in the preparation of other hexakiscyanoferrate(II) compounds. Among the mixed salts are calcium dicesium hexakiscyanoferrate [*15415-35-7*], $CaCs_2[Fe(CN)_6]$, and calcium dipotassium hexakiscyanoferrate [*20219-00-5*], $CaK_2[Fe(CN)_6]$.

Dilead hexakiscyanoferrate [*14402-61-0*], $Pb_2[Fe(CN)_6]$, is a white precipitate which forms when lead acetate is added to dicalcium hexakiscyanoferrate. It is insoluble in water or dilute acids but is soluble in hot solutions of ammonium chloride or ammonium succinate. Dilead hexakiscyanoferrate has been used as a qualitative analytical reagent in tests for cadmium and chromate.

Tetrapotassium hexakiscyanoferrate trihydrate [*14459-95-1*], $K_4[Fe(CN)_6].3H_2O$, is a lemon-yellow compound and is known as yellow prussiate of potash. It is slightly efflorescent and the anhydrous material [*13943-58-3*] is obtained at 70°C. Its solubility in water is 0.278 g/cm^3 at 12°C and 0.906 g/cm^3 at 96°C; the solubility of the anhydrous material is 0.145 g/cm^3 at 0°C and 0.74 g/cm^3 at 98°C. The compound also is soluble in acetone but is insoluble in alcohol, ether, or ammonia. The compound undergoes the characteristic reactions of soluble hexakiscyanoferrate(4−) including oxidation to hexakiscyanoferrate(3−) by oxygen in acidic solution or, for example, by O_3, Cl_2, Br_2, H_2O_2, or MnO_4^-. A large number of insoluble or slightly soluble mixed salts of the general formula $K_2M(II)[Fe(CN)_6]$ are known, eg, M = cobalt(II) [*13821-10-8*], M = copper(II) [*14481-39-1*], M = manganese(II) [*15631-19-3*], and M = nickel(II) [*13601-16-6*]. The compounds of formula $K_2M(II)[Fe(CN)_6]$ have face-centered, cubic crystal lattices and the unit cell is ca 110 pm. An exception is the tetragonal unit cell for M = copper(II). Jahn-Teller distortions are significant. Many of the $K_2M(II)$-$[Fe(CN)_6]$ compounds are useful ion-exchange (qv) materials. For example, $K_2Co(II)[Fe(CN)_6]$ absorbs silver(I) ions from wastewater.

Tetrapotassium hexakiscyanoferrate is an important industrial chemical and is prepared from calcium cyanide and iron(II) sulfate at temperatures above 100°C. The soluble dicalcium hexakiscyanoferrate (2:1) is formed first, is separated from insoluble calcium sulfate, is precipitated as $CaK_2[Fe(CN)_6]$ by addition of KCl, and is redissolved as the potassium salt by the addition of potassium carbonate. Calcium carbonate is removed by filtration and tetrapotassium hexakiscyanoferrate is crystallized by rapid cooling. Tetrapotassium hexakiscyanoferrate is used in the synthesis of other hexakiscyanoferrates(4−), in metal coatings, electroplating (qv), dyeing and printing of textiles, engraving and lithography, and in the quantitative titrations of other metal salts, eg, as those of silver, zinc, and the rare earth metals. Prices, in 1980, for tetrapotassium hexakiscyanoferrate trihydrate are $22.00/500 g and $800.00/45 kg.

Tetrasodium hexakiscyanoferrate decahydrate [*14434-22-1*], $Na_4[Fe(CN)_6].10$-H_2O (yellow prussiate of soda), is available as yellow monoclinic crystals. Its solubility in water is 0.318 g/cm^3 at 20°C and 1.56 g/cm^3 at 98°C but it is insoluble in ethanol. It is slightly efflorescent at room temperature and, at 100°C, the anhydrous material, tetrasodium hexakiscyanoferrate [*13601-19-9*], $Na_4[Fe(CN)_6]$, is obtained. The decahydrate is produced from calcium cyanide, iron(II) sulfate, and sodium carbonate in a process which is similar to that for the production of tetrapotassium hexakiscyanoferrate trihydrate.

The crystal structure (2) of the decahydrate reveals octahedra of $[Fe(CN)_6]^{4-}$ that are linked through distorted octahedra built around sodium ions. In each $[Fe(CN)_6]^{4-}$ unit, two of the nitrogen atoms each bridge three sodium ions, two other nitrogens each bridge one sodium ion, and the nitrogens of the remaining two cyanides appear to be involved only in hydrogen bonding with water of hydration. The structural unit is depicted in Figure 1.

Very few slightly soluble salts of the type $Na_2M(II)[Fe(CN)_6]$ have been reported.

Tetrasodium hexakiscyanoferrate is used in the manufacture of trisodium hexakiscyanoferrate (see below), blue and black dyes, as a metal surface coating, and in photographic processing (see Photography). Tetrasodium hexakiscyanoferrate decahydrate is available for $55.00/kg.

Hexakiscyanoferrates(3−). Salts of hexakiscyanoferrates(3−) may be considered salts of ferricyanic acid, trihydrogen hexakiscyanoferrate [*17126-46-4*], $H_3[Fe(CN)_6]$, which may be prepared by adding sulfuric acid to trilead or tribarium bis(hexakiscyanoferrate). Red-brown needles are obtained by evaporation of the aqueous solution. The acid has been used to prevent metal surface corrosion (see Corrosion and corrosion inhibitors; Metal surface treatments—cleaning, pickling, and related processes).

Triammonium hexakiscyanoferrate [*14221-48-8*], $(NH_4)_3[Fe(CN)_6]$, is obtained

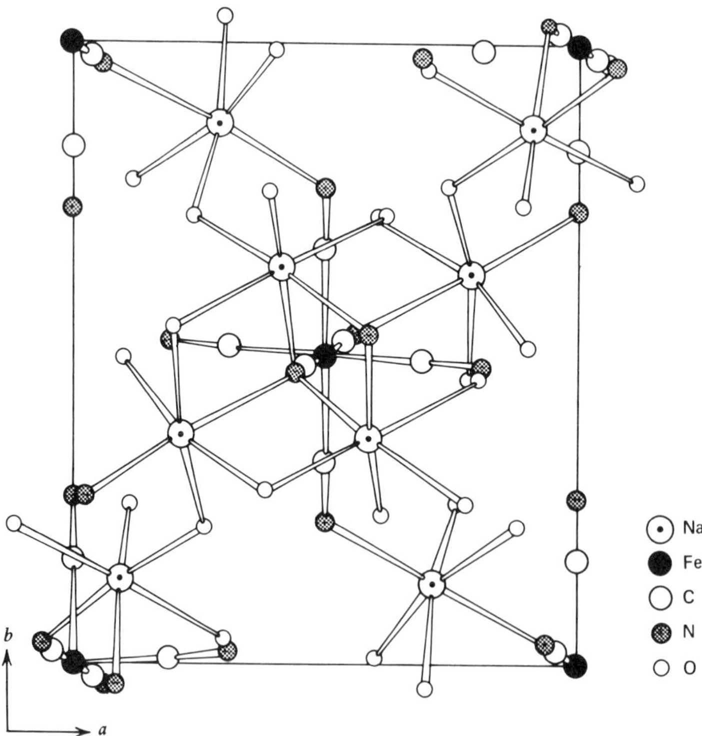

Figure 1. Perspective view of $Na_4(Fe(CN)_6).10H_2O$ along the *c*-direction. Full lines indicate one unit cell. The atoms do not all belong to the same unit cell. Some nitrogen and oxygen atoms belonging to two neighboring unit cells are drawn double with a small separation in the *a*-direction. Courtesy of *Acta Chem. Scand.* (2).

as air-stable, ruby-red crystals which are very soluble in water. Some double salts are known, eg, ammonium disilver(I) hexakiscyanoferrate [*58675-53-9*], $(NH_4)Ag(I)_2$-$[Fe(CN)_6]$.

Tribarium bis(hexacyanoferrate) [*21729-04-4*], $Ba_3[Fe(CN)_6]_2$, forms red crystals which are readily soluble in water but which are insoluble in alcohol.

Tripotassium hexakiscyanoferrate [*13746-66-2*], $K_3[Fe(CN)_6]$, forms anhydrous red crystals, and it must be prepared by chemical or electrolytic oxidation of hexacyanoferrate(4−) (attempts to prepare the compound directly from iron(III) salts and cyanide ion result in significant amounts of the side product, iron(III) hydroxide). The crystalline material is dimorphic. One crystalline form is orthorhombic and the other is monoclinic. The solubility of $K_3[Fe(CN)_6]$ in water is 0.33 g/cm^3 at 4°C or 0.775 g/cm^3 at 100°C. The complex is soluble in acetone but is insoluble in alcohol. The complex is low spin but, at 300 K, the magnetic moment is 2.09×10^{-23} J/T (2.25 μ_β); this high value is attributed to a significant orbital angular momentum contribution. The Curie-Weiss law is obeyed with $\theta = 56$ K. Tripotassium hexakiscyanoferrate is used in the manufacture of pigments, photographic papers, leather (qv), and textiles, and it is used as a catalyst in oxidation and polymerization reactions (see Pigments, inorganic pigments). Its price (1980) is \$28.00/500 g reagent-grade material or \$460/45 kg of technical-grade material.

A number of mixed salts of the general formula, $KM(II)[Fe(CN)_6]$ are known, eg, M = cobalt(II) [*14874-73-8*], M = copper(II) [*53295-15-1*], and M = nickel(II) [*53295-14-0*].

Dipotassium sodium hexakiscyanoferrate [*31940-93-9*], $K_2Na[Fe(CN)_6]$, crystallizes as orange-red monoclinic needles and sometimes is used in place of tripotassium hexakiscyanoferrate. Its solubility in water is 0.50 g/cm^3 at 25°C or 0.80 g/cm^3 at 80°C.

Trisodium hexakiscyanoferrate [*14217-21-1*], $Na_3[Fe(CN)_6]$, forms red deliquescent crystals which are readily soluble in water. The monohydrate, Na_3-$[Fe(CN)_6].H_2O$ [*13755-37-8*], which has a solubility in water of 0.19 g/cm^3 at 0°C and 0.67 g/cm^3 at 100°C, and the dihydrate, $Na_3[Fe(CN)_6].2H_2O$ [*36249-31-7*], are known. Sodium salts of hexakiscyanoferrate are used for many of the same purposes as is the potassium salt.

Prussian Blue Analogues. The literature displays considerable confusion regarding the nomenclature, composition, and structure of Prussian Blue and its analogues, which also may be referred to as mixed oxidation-state complexes. A number of recent publications dealing with the Mössbauer and esr spectra of these complexes are clarifying the situation. Single crystals of Prussian Blue that are suitable for structure determination have been obtained (3–4).

The following aqueous reactions occur:

$$\text{excess Fe(III)} + K_2[Fe(CN)_6] \rightarrow \text{insoluble Prussian Blue}$$

$$\text{excess Fe(II)} + K_3[Fe(CN)_6] \rightarrow \text{insoluble Turnbull's Blue}$$

Insoluble Prussian Blue and insoluble Turnbull's Blue are the same substance which is best formulated as tetrairon(III) tris(hexakiscyanoferrate) [*14038-43-8*], $Fe(III)_4$-$[Fe(CN)_6]_3$. Almost always present as impurities are potassium ions and Fe_2O_3. Additionally, the solid contains a large number of water molecules, some of which are easily removed. If the iron(II) species is assumed to be diamagnetic, the magnetic susceptibility is 5.54×10^{-23} J/T (5.97 μ_{β}) per iron(III).

The structure of the Prussian Blue analogues is understood only in light of the ambidentate nature of the cyanide ligand. Although mononuclear metal–cyanide bonds always occur through carbon, polynuclear species may form by way of the linkage M^B—C—N—M^A (following the notation given in ref. 5). Evidence for the ambidentate cyanide linkage is the fact that, upon its formation, the stretching frequency of the cyanide linkage increases by 30–70 cm^{-1}.

The structure of the tetradecahydrate [*39611-22-8*], $Fe(III)_4[Fe(CN)_6]_3.14H_2O$, reveals that the complex is face-centered cubic. The iron(II) sites are in perfect $[Fe(II)C_6]$ octahedra but are only 75% occupied. The iron(III) sites are fully occupied by either $[Fe(III)N_6]$ octahedra or by sites where one or more nitrogen atoms is replaced by coordinated water. The average structure of the iron(III) sites is $[FeN_{4.5}O_{1.5}]$. Most of the water that is present is not coordinated but is trapped as zeolitic water. A recent neutron-diffraction (4) study confirms the presence of two types of water molecules, eg, interstitial water, which may be removed easily, and coordinated water, which may not be removed without affecting the integrity of the complex. A depiction of the unit cell is shown in Figure 2.

Based on the tetradecahydrate's structure, the structures of a variety of Prussian Blue analogues may be given. The analogues usually are insoluble in water or dilute acids and have served as ion exchangers (especially in radioelement removal from waste water (see Radioisotopes)) and as pigments. The analogues have the general formula, $M_k^A[Fe(CN)_6]_l.xH_2O$ (there also is an extensive series of analogues in which iron is replaced in both sites). The occupancy of the $[Fe(II)C_6]$ site and the average structure

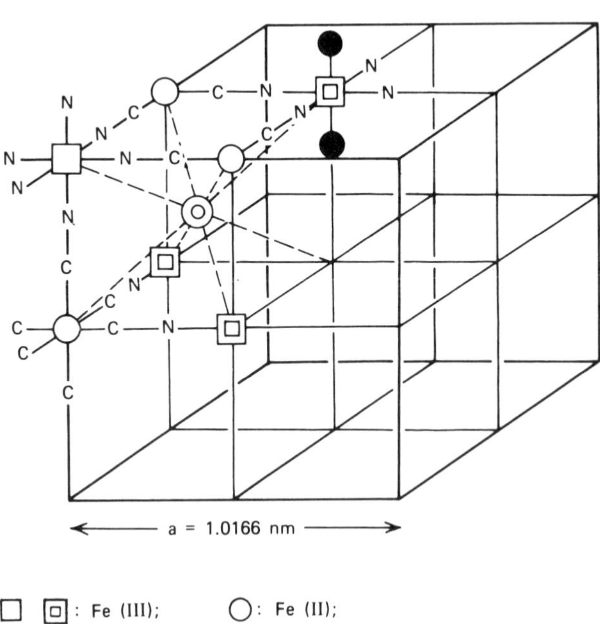

Figure 2. The primitive unit cell of Prussian Blue. Courtesy of *Chimia* (3).

of the M^A site vary with stoichiometry, but the general features of the Prussian Blue structure remain intact. For example, if M^A = cobalt(II) [*15415-49-3*], then $k = 2$ and $l = 1$, the $[Fe(II)C_6]$ sites are one-half occupied and the average Co structure is $[CoN_3O_3]$. If M^A = Cu(II) [*13601-13-3*], Fe(II) [*14460-02-7*], or Mn(II) [*14402-63-2*], the stoichiometry is the same as the cobalt salt. The copper(II) salt is reddish-brown, the iron(II) salt is white but very susceptible to air oxidation to form Prussian Blue, and the manganese(II) salt is greenish-white. If M^A = Ag(I) [*14038-75-6*], then k = 4 and $l = 1$. The strong preference of silver ion for linear coordination is maintained, and linear N–Ag–N sites are present in addition to $[FeC_6]$ octahedra.

In addition to the insoluble Prussian Blues, a series of soluble blues exist and may be prepared by either of the following reactions:

$$FeCl_3 + K_4[Fe(CN)_6] \rightarrow \text{soluble Prussian Blue}$$

$$FeCl_2 + K_3[Fe(CN)_6] \rightarrow \text{soluble Turnbull's Blue}$$

These reaction products are regarded as identical and both have the approximate formulation $KFe[Fe(CN)_6]$ [*25869-98-1*]. This substance is not truly water soluble and is better described as forming colloidal suspensions. Water of hydration and the impurities KCl and $K_4[Fe(CN)_6]$ may be present. Again, low-spin $[Fe(II)C_6]$ units and high-spin $[Fe(III)N_6]$ units are indicated. This compound is used as a pigment, as a cation exchanger, and as a metal coating (see Ion exchange).

Berlin Green may be obtained by the oxidation of Prussian Blue. It is thought that the intense color of this compound results only if some $[Fe(CN)_6]$ octahedra are present as hexakiscyanoferrate(4−). The compound in which only iron(III) is present, $Fe[Fe(CN)_6]$ [*14433-93-3*], is brown and is subject to autoredox processes.

In addition to the variety of compounds outlined above, a vast number of substitution products are known. One (or more) cyano ligands may be replaced by water, ammine, amines, azide, carbonyl, or any other of a series of ligands. Usually substitutions are done on hexakiscyanoferrates(4−) and the corresponding iron(III) salts are obtained by oxidation.

Formates

Iron(II) formate dihydrate [*13266-73-4*], $Fe(HCO_2)_2.2H_2O$, is a green salt which is only slightly soluble in water. It may be prepared from iron(II) sulfate and sodium formate in an inert atmosphere. It is fairly stable to air oxidation and thus is a useful reagent. The anhydrous salt [*3047-59-4*] also is known.

Iron(III) formate [*555-76-0*], $Fe(HCO_2)_3$, exists as red crystals or powder which are soluble in water but only very slightly soluble in alcohol. Aqueous solutions are subject to hydrolysis which results in the formation of basic formates and eventually results in precipitation of iron hydroxide and liberation of formate. The compound is conveniently prepared from iron(III) nitrate [*14104-77-9*] and formic acid in alcohol solution. One or two waters of hydration may be present, in which case the compound appears more yellow.

Fumarates

Iron(II) fumarate [*141-01-5*], $Fe(C_4H_2O_4)$, is the iron salt of 2-butenedioic acid. This red-orange to red-brown powder is favored as a human and veterinary hematinic because it produces few cases of gastrointestinal upset (see Veterinary drugs). Its solubility in water is 0.0014 g/cm^3 at 25°C, but it is more soluble in dilute acid solutions. It is prepared by adding a hot aqueous solution of sodium fumarate to a hot aqueous solution of iron(II) sulfate, followed by filtration of the slightly soluble iron(II) fumarate. It is available under a variety of trade names, and a nonstoichiometric compound [*7705-12-6*] also is available. Iron(III) fumarate [*52118-11-3*], $Fe_2(C_4H_2O_4)_3$, also has been tried as a hematinic in veterinary studies.

Halides

All of the binary halides of iron(II) and iron(III) are known except for the iodide of iron(III), which is stable only in the vapor phase. Complex iron(II) and iron(III) halides have been characterized, again with the exception of complex iron(III) iodides. The number and variety of these are large, and many are described in ref. 6.

Iron(II) fluoride [*7789-28-8*], FeF_2, may be prepared by reaction of iron metal and anhydrous HF at elevated temperatures, or by reaction of $FeCl_2$ and anhydrous HF in a flow system at ca 500°C. Sublimation occurs near 1100°C. Pure FeF_2 is a white crystalline compound with the rutile structure which is axially compressed. A brown color indicates oxide impurity. It is only sparingly soluble in water, slightly soluble in dilute HF, and insoluble in alcohol, ether, and benzene. The Curie-Weiss law is obeyed at temperatures above 100 K with θ = 117 K and magnetic moment = 5.16×10^{-2} J/T (5.56 μ_β). Iron(II) fluoride is used as a catalyst in organic fluorinations, and is available at $50.00/100 g.

Iron(II) fluoride tetrahydrate [*13940-89-1*], $FeF_2.4H_2O$, is obtained by precipitating FeF_2 from dilute HF solution with ethanol. Discrete octahedral units with extensive hydrogen bonding and with fluorines occupying random cis or tris sites characterize this structure. These white rhombohedral crystals are slightly soluble in water, alcohol, and ether; soluble in dilute acid; and begin to decompose at 100°C.

Iron(III) fluoride [*7783-50-8*], FeF_3, is prepared as green rhombohedral crystals from $FeCl_3$ and ClF_3 in a flow system at 500°C or from $FeCl_3$ and anhydrous HF in a flow system at elevated temperatures. Details of the iron(III) fluoride structure are in dispute, although FeF_6 octahedra are recognized. The compound is antiferromagnetic and has a Néel temperature of 394 K. Iron(III) fluoride is slightly soluble in water; very soluble in dilute HF; and insoluble in alcohol, ether, and benzene. Iron(III) fluoride is used as a catalyst of organic reactions, and its price is $140.00/100 g.

Iron(III) fluoride trihydrate [*15469-38-2*], $FeFl_3.3H_2O$, may be crystallized from 40% HF solution in either of two crystalline forms. Low temperatures favor formation of α-$FeF_3.3H_2O$ which is isostructural with α-$AlF_3.3H_2O$; higher temperatures favor β-$FeF_3.3H_2O$ which displays octahedra sharing corners by way of bridging fluorine atoms and one water of hydration per unit cell. The compound, $FeF_3.3H_2O$, may be obtained for $45.00/100 g.

Iron(II) chloride [*7758-94-3*], $FeCl_2$, is prepared from iron and HCl in a flow system at red heat or from iron and HCl or Cl_2 in a flow system at 700°C. A number of preparative methods employ reduction of $FeCl_3$. The compound occurs naturally as the mineral lawrencite. White, very hygroscopic crystals are obtained by sublimation at 700°C in a stream of HCl. Slow cooling of the sublimate results in $FeCl_6$ octahedra that are arranged in cubic close packing, but rapid cooling results in hexagonal close packing. The Curie-Weiss law is obeyed at temperatures >100 K with θ = 48 K and the magnetic moment = 4.99×10^{-23} J/T (5.38 μ_β). The compound is soluble in water, alcohol, and acetone, only slightly soluble in benzene, and insoluble in ether. Iron(II) chloride has many industrial applications as a reducing agent, as a mordant in dyeing, in pharmaceutical preparations, and in metallurgy. Its price is ca $20.00/100 g.

Iron(II) chloride tetrahydrate [*13478-10-9*], $FeCl_2.4H_2O$, is obtained by dissolving iron metal in HCl and allowing the product to crystallize at room temperature. Monomeric *trans*-$FeCl_2(H_2O)_4$ octahedra are formed, but the units are hydrogen bonded. The Curie-Weiss law is obeyed with θ = 11 K and the magnetic moment = 4.82×10^{-23} J/T (5.20 μ_β). The tetrahydrate may be partially dehydrated at 105–115°C to yield iron(II) chloride dihydrate [*16399-77-2*], $FeCl_2.2H_2O$, which is isostructural with manganese dichloride dihydrate. At 150–160°C, the dihydrate yields iron(II) dichloride monohydrate [*23838-02-0*], $FeCl_2.H_2O$, which, at 220°C, loses the last water of hydration. The price of the product, $FeCl_2.xH_2O$, is ca $12.00/100 g.

Iron(III) chloride [*7705-08-0*], $FeCl_3$, may be prepared from Fe and Cl_2 in a flow system at 350°C or from Fe_2O_3 and HCl in a flow system at up to 1000°C. The pure material occurs as hygroscopic, hexagonal, dark crystals which are soluble in water, methyl and ethyl alcohol, and acetone, but which are not soluble in ethyl acetate. Material of high purity is obtained by sublimation in a stream of Cl_2. Iron(III) chloride melts and sublimes at ca 300°C and, at this temperature, thermal decomposition to $FeCl_2$ and Cl_2 is noticeable. In the vapor phase, $FeCl_3$ exists as a dimer which has two iron atoms and two chlorine atoms in a planar ring and the other two pairs of chlorine atoms in a plane that is perpendicular to the Fe_2Cl_2 ring. As a solid, the material is composed of $FeCl_6$ octahedra which share all corners. The Curie-Weiss law is obeyed with the magnetic moment = 5.31×10^{-23} J/T (5.73 μ_β) and θ = 11.5 K. A large number of adducts and substitution products of $FeCl_3$ are known. Iron(III) chloride serves as a convenient starting material in the synthesis of other Fe(III) compounds and salts. It is used as an oxidizing and chlorinating agent and it is used in the manufacture of ink, pigments, and dyes. Extensive use of $FeCl_3$ is made in the treatment of sewage and industrial waste gases. In lots of 100 g, the price of the material is $20, but it may be purchased in lots up to and including tank car quantities.

Iron(III) chloride hexahydrate [*10025-77-1*], $FeCl_3.6H_2O$, is formulated as the complex salt, *trans*-$[FeCl_2(H_2O)_4]Cl.2H_2O$. The octahedra are slightly distorted and extensive hydrogen bonds are indicated. This substance is freely soluble in water, alcohol, ether, and acetone. In most solvents, a series of solvates, eg, $FeCl_3.2C_2H_5OH$ [*74930-86-2*], is formed. In water, hydrolysis and hydroxylation occur to such an extent that the pH of a 5% solution is 1.7 and, upon standing, aqueous solutions of iron(III) chloride hexahydrate precipitate as the hydroxy compounds. When $FeCl_3.6H_2O$ is dried at selected temperatures, a series of hydrated species may be formed. The hexahydrate is available at $4.00/100 g but may be obtained less expensively in bulk quantity.

Iron(II) bromide [*7789-46-0*], $FeBr_2$, may be obtained from the reaction of iron and bromine in a flow system at 200°C, or it may be obtained by reaction in a flow system of HBr with Fe_2O_3 at 200–325°C. The light yellow or brown, hygroscopic material may be sublimed in a vacuum or in a nitrogen stream. It is soluble in water or alcohol. A number of hydrated species are known, including the bluish-green hexahydrate [*13463-12-2*]. The solid material exists as octahedra of $FeBr_6$ which are arranged in hexagonal close-packing. The Curie-Weiss law is obeyed with the magnetic moment = 5.21×10^{-23} J/T (5.62 μ_β) and θ = 6 K. Iron(II) bromide, available at $170.00/100 g (1980), is used as a catalyst of organic brominations and polymerization reactions. An extensive series of adducts of iron(II) bromide is known.

Iron(III) bromide [*10031-26-2*], $FeBr_3$, forms dark red, hygroscopic, rhombic crystals which dissolve easily in water, alcohol, ether, or acetic acid. The material is obtained by reaction of the elements in an inverted V tube. Iron is placed in one arm at 175–200°C and bromine is placed in the other arm at 120°C; sublimation in a bromine atmosphere is the method of purification. The tribromide is not as thermally stable as iron(III) chloride, and it may be completely dissociated to iron(II) bromide and bromine at temperatures slightly above 200°C. The structure of iron(III) bromide is analogous to that of iron(III) chloride. A number of hydrated species are known as are a large variety of adducts. The price of iron(III) bromide is $150.00/100 g. It is used in the catalytic bromination of aromatic compounds.

Iron(II) iodide [*7783-86-0*], FeI_2, is most conveniently prepared by direct reaction of the elements in a sealed tube at ca 500°C. The dark, red-black crystals are soluble in water, alcohol, or ether, and a number of hydrated species are known. A solution of iron(II) iodide in water undergoes oxidation in air. The solid substance has the hexagonal lattice of cadmium(II) iodide and obeys the Curie-Weiss law with a magnetic moment of 5.45×10^{-23} J/T (5.88 μ_β) and θ = 23 K. Many of the adducts known for iron(II) bromide are known for iron(II) iodide. It is available for $100.00/100 g and is used as a catalyst in organic reactions and in veterinary medicine as a source of iron and iodine.

Iron(III) iodide [*15600-49-4*], FeI_3, is known only in the vapor phase.

Gluconates

Iron(II) gluconate dihydrate [*6047-12-7*], $Fe(CH_2OH(CHOH)_2COO)_2.2H_2O$, is a pale, yellow-green powder with an odor of caramel. It is very soluble in water but nearly insoluble in alcohol. It is prepared from barium or calcium gluconate and iron(II) sulfate and is available in isotonic solution. It functions as a hematinic and as a fortifier of foods for both animals and humans, and it is used in the treatment of anemia. The price is $10.00/500 g. An anhydrous salt [*299-29-6*] also is available.

Iron(III) gluconate [*38658-53-6*], $Fe(CH_2OH(CHOH)_4COO)_3$, has been prepared and studied as a nutritional supplement in milk substitutes.

Nitrates

Iron(II) nitrate hexahydrate [*14013-86-6*], $Fe(NO_3)_2.6H_2O$, forms green rhombs which melt at 60.5°C. Its solubility in water is 0.835 g/cm^3 at 20°C or 1.67 g/cm^3 at 61°C. It is prepared by the action of cold nitric acid having a density of less than 1.034 g/cm^3 on iron. Increasing the density of the acid increases the proportion of iron that is oxidized to the iron(III) state. Another method of preparation is the reaction of barium nitrate and iron(II) sulfate. Iron(II) nitrate is used as a catalyst for reduction reactions and as a convenient reagent in the synthesis of other iron compounds.

Iron(III) nitrate nonahydrate [*7782-61-8*], $Fe(NO_3)_3.9H_2O$, forms colorless to pale-violet monoclinic crystals which are somewhat deliquescent. The mp is 47.2°C and decomposition occurs at 125°C. Its solubility in water is ca 0.87 g/cm^3 at 25°C and it also is soluble in alcohol and acetone. Iron(III) nitrate is formed by the action of nitric acid having a density greater than 1.115 g/cm^3 on iron. Acid of too high concentration renders the iron passive. The price of the nonahydrate is $13.00/500 g.

Iron(III) nitrate hexahydrate [*13476-08-9*], $Fe(NO_3)_3.6H_2O$, forms colorless, cubic crystals which melt at 35°C. Its solubility in water at 20°C is 0.835 g/cm^3 and it is infinitely soluble in hot water. Iron(III) nitrate is used as a mordant, in tanning, as a catalyst of oxidation reactions, and as a convenient source of iron in the preparation of other compounds. Nitrate may be conveniently removed from such preparations by reduction to gaseous oxides of nitrogen.

Oxides and Hydroxides

Iron(II) oxide [*1345-25-1*], FeO, is a black substance and forms cubic crystals. Its density is 5.7 g/cm^3 and its mp is 1369 ± 1°C. It is insoluble in water, alcohol, or alkali, but it reacts with acids. It occurs naturally as the mineral wüstite but may be prepared synthetically by thermal decomposition of iron(II) oxalate [*516-03-3*] in a vacuum. Crystalline FeO has the rock salt structure, is easily oxidized in air, is a strong base, and absorbs carbon dioxide. Because the typical true composition is $Fe_{0.90}O$ to $Fe_{0.96}O$ with a slight presence of iron(III), semiconductor properties are displayed. Iron(II) oxide is used in green, heat-absorbing glass, in ceramic mixtures, and in a variety of catalytic preparations, notably those of ammonia synthesis and methanation.

Iron(III) oxide [*1309-37-1*], Fe_2O_3, forms red-brown to black trigonal crystals of density 5.24 g/cm^3. It is insoluble in water but is soluble in hydrochloric or sulfuric acid. It melts and begins to decompose at ca 1565°C. Iron(III) oxide occurs naturally as the mineral hematite, the principal ore of iron, and it may be prepared synthetically by heating brown iron hydroxide oxide, hydrous [*20344-49-4*], Fe(OH)O, at 200°C. Iron(III) oxide has a structure of hexagonal close-packed oxy anions with iron atoms in two thirds of the octahedral holes (the corundum structure). In addition to the hematite structure (α-Fe_2O_3), there exists a structure (γ-Fe_2O_3) of cubic, closely packed oxy anions with iron(III) distributed randomly between the octahedral and tetrahedral holes.

Iron(III) oxide is used in large quantity as a red pigment for paint, rubber, ceramics, and paper, as a coating for steel and other metals, in magnetic recording material, and as a catalyst of oxidation reactions (see Ceramics; Magnetic tape; Rubber, synthetic). Pigments of yellow, orange, or red are obtained depending upon the size and shape of the precipitate, the identity and amount of the impurities that are present, and the amount of water that is present. It may be bought for $9.00/500 g or for $100.00/45 kg.

Triiron tetroxide [*1317-61-9*], Fe_3O_4, forms black, cubic crystals or red-black powder. It has a mp of 1595°C and is insoluble in water, alcohol, or ether, but is soluble in concentrated acids. It may be prepared by heating Fe_2O_3 to above 1400°C. Triiron tetroxide occurs naturally as magnetite and contains both iron(III) and iron(II). The inverse spinel structure is displayed wherein the oxy anions form a cubic closely packed array. All iron(II) ions are in octahedral holes, but the iron(III) ions are equally divided between octahedral and tetrahedral holes. The unit cell length is 839.4 pm. The compound is strongly ferromagnetic and the Curie point is 860 K, at which temperature the effective magnetic moment is 3.9×10^{-23} J/T (4.2 μ_β). Blue steel has a surface coating of triiron tetroxide as a corrosion-resistant film. This compound is used as a pigment for glass, ceramics, and paint; in magnetic recording materials (see Magnetic tape) and many catalytic preparations; and as a polishing compound. Its price is $33.00/2.25 kg.

Iron(II) hydroxide [*18624-44-7*], $Fe(OH)_2$, occurs as pale green, hexagonal crystals or white amorphous powder having a density of 3.4 g/cm^3. It is very slightly soluble in water, insoluble in alkali, soluble in acids, and fairly soluble in ammonium salt solutions. It is oxidized slowly in air and eventually is completely converted to $Fe_2O_3.xH_2O$. It may be prepared by precipitation of an iron(II) salt solution by strong base in the absence of air.

Iron(III) hydroxide [*1309-33-7*], $Fe(OH)_3$ (also known as hydrated iron(III) oxide), forms a red-brown amorphous powder with a density of 3.4–3.9 g/cm^3 (depending on the extent of hydration). It is insoluble in water or alcohol but is soluble in acids. Iron(III) hydroxide usually is obtained as a gelatinous precipitate when strong base is added to an iron(III) salt solution. Salt-free iron(II) hydroxide may be obtained by hydrolysis of iron(III) ethanolate [*5058-42-4*], $Fe(C_2H_5O)_3$. Iron(II) hydroxide is used in abrasives (qv) and as a pharmaceutical carrier.

Oxo Anions

The oxo ligand (O^{2-}) is found in a number of iron anions in which the highest oxidation states are reached. These compounds are studied for their interesting magnetic properties and solid-state phase transitions. The iron(III) oxo compounds (the ferrites, FeO_2^-) of the alkali metals may be obtained by fusing iron(III) oxide with the alkali metal chloride, carbonate, or hydroxide, or by decomposing the alkali iron(VI) oxo compound in boiling water. The sodium salt (sodium ferrite) [*12062-85-0*], $NaFeO_2$, occurs as brown hexagonal plates or needles which are soluble in dilute HCl. Iron(III) oxo compounds of bivalent metal cations, M(II), may be prepared by heating iron(III) oxide with the oxide, MO, or by addition of strong base to a solution of M(II) and iron(III) salts. One of the most studied of these compounds is the magnesium salt, $Mg(FeO_2)_2$ [*12068-86-9*] or magnesium ferrite. It occurs as black cubic crystals having a density of 4.5 g/cm^3. Zinc ferrite [*1317-55-1*], $Zn(FeO_2)_2$, occurs as the mineral franklinite and forms reddish-brown or black cubic crystals. The barium [*12009-00-6*], calcium [*12013-33-1*], and lithium [*12022-46-7*] salts also are known (see Ferrites).

The iron(IV) oxo compounds (the perferrites, FeO_3^{2-}) are prepared by heating a hydroxide, $M(OH)_2$, with iron(III) hydroxide in a stream of oxygen at 600–950°C. The barium salt [*12230-58-9*], $Ba(FeO_3)$, is a black amorphous substance. The calcium salt [*12524-96-8*], $Ca(FeO_3)$, and the strontium salt [*12022-69-4*], $Sr(FeO_3)$, are known.

The iron(VI) oxo compounds (the ferrates, FeO_4^{2-}) are the best characterized iron oxo anions. Ferrate salts of the alkali metals are soluble in water. Potassium ferrate [*13718-66-6*], K_2FeO_4, considered to be a salt of ferric acid (H_2FeO_4), may be obtained by the oxidation of iron(III) hydroxide by potassium hypochlorite. Crystals of potassium ferrate are deep purple and are orthorhombic. The x-ray crystal structure determination of potassium ferrate reveals FeO_4 tetrahedra with an average bond length of 0.1656 nm. The structure is isomorphous with K_2MnO_4 or K_2CrO_4. The FeO_4^{2-}

moiety is even more strongly oxidizing than MnO_4^{2-}. Barium ferrate [*13773-23-4*], $BaFeO_4$, may be precipitated from soluble ferrate salt solution and is the most stable of the various salts. Among other ferrate salts are calcium ferrate [*35764-67-1*], $CaFeO_4$, and sodium ferrate [*13773-03-0*], Na_2FeO_4.

Sulfates

Iron(II) sulfate heptahydrate [*7782-63-0*], $FeSO_4.7H_2O$, forms blue-green, monoclinic crystals having a density 1.898 g/cm^3 and melting at 64°C. It is very soluble in water, soluble in absolute methanol, but only slightly soluble in ethanol. In dry air, the compound is efflorescent and, in moist air, the compound oxidizes to basic iron(III) sulfate. Aqueous solutions also are subject to oxidation, and the rate of oxidation increases with an increase in alkali, temperature, and light. Upon warming to 56°C, the compound loses three waters of hydration to form iron(II) sulfate tetrahydrate [*20908-72-9*], $FeSO_4.4H_2O$, and further warming to 65°C forms iron sulfate monohydrate [*17375-41-6*], $FeSO_4.H_2O$, which is stable to 300°C.

A series of double salts, $FeSO_4.M_2SO_4.6H_2O$, exists where M typically is an alkali metal. If M = NH_4, the compound is iron(II) ammonium sulfate (Mohr's salt) [*7783-85-9*], $FeSO_4.(NH_4)_2SO_4.6H_2O$, which is an accepted primary standard for iron.

Iron(II) sulfate solutions reduce nitrate and nitrite to nitric oxide, whereupon the highly colored $[Fe(H_2O)_5(NO)]^{2+}$ ion is formed. Detection of this ion is the basis of the brown ring test for the qualitative determination of nitrate or nitrite. Most iron(II) sulfate is a by-product of the steel industry. Prior to tinning, galvanizing, electroplating, or enameling, steel surfaces are dipped in sulfuric acid for cleaning (pickling). The resulting pickle liquor contains ca 15% ferrous sulfate and 2–7% free acid. Scrap iron is added to reduce the free acid concentration to ca 0.03%. The solution is filtered and concentrated at 70°C to a specific gravity of ca 1.4 and is allowed to cool to room temperature which induces crystallization of iron(II) sulfate heptahydrate [*7782-63-0*] (commonly called copperas). The iron and steel industries produce 0.5–1.0 $\times 10^6$ t of the heptahydrate. Supply exceeds demand, and the major portion of the pickling liquor presents a serious waste disposal problem. Hydrochloric acid is being substituted for sulfuric acid in the pickling process; the hydrochloric acid may be regenerated with the production of the more useful iron oxide. Of the iron(II) sulfate heptahydrate that is produced, 50% is used in the production of iron oxide pigments and salts, and 35% is used in the production of iron oxo compounds. It has a tremendous variety of uses and is the most important commercial iron compound. The prices (1980) are $11.00/500 g and $245.00/45 kg.

Iron(III) sulfate [*10028-22-5*], $Fe_2(SO_4)_3$, is a yellow substance which is slightly soluble in cold water but which decomposes in hot water. A series of hydrates are known including the monohydrate [*43059-01-4*], the hexahydrate [*13761-89-2*], the heptahydrate [*35139-28-7*], and the nonahydrate [*13520-56-4*]. A series of alum compounds is known, eg, iron(III) potassium alum dodecahydrate [*13463-29-1*], $KFe(SO_4)_2.12H_2O$, and iron(III) ammonium alum dodecahydrate [*7783-83-7*], $(NH_4)Fe(SO_4)_2.12H_2O$, that is important as reagents and as mordants in the textile industry. Iron(III) sulfate may be obtained by oxidation of iron(II) sulfate or by treating iron(III) oxide with sulfuric acid. It is used as a pigment, as a coagulant in water and sewage treatment, and as a mordant (see Water, industrial water treatment; Water, sewage).

Chelate Compounds

A chelate ligand is one which binds at more than one site to a metal atom (see Chelating agents). Chelating ligands may bind at two sites (a bidentate ligand), or they may act as tridentates, tetradentates, etc. A closed-ring structure is formed and, in general, five-membered rings are more stable than six-membered rings. Metal chelates exhibit enhanced stability compared to analogous complexes of unidentate ligands where no rings are formed. This chelate effect probably largely results from a favorable entropic effect.

Bipyridines. The 2,2′-bipyridine ligand is pictured below:

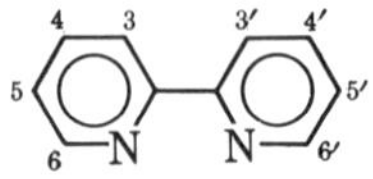

4 3 3′ 4′
5 5′
6 N N 6′

The tris(2,2′-bipyridine)iron(2+) ion [*15025-74-8*] has an intense red color. Formation of this ion serves as a qualitative and quantitative test for iron. The absorption maximum of this complex ion lies at 522 nm and has an absorptivity of 8650 M^{-1} cm^{-1}. The complex is very stable and has an overall formation constant of 10^{17}. A number of complexes with different solubility properties may be synthesized by using 2,2′-bipyridine which has one or more substitutions at different ring positions.

The tris(2,2′-bipyridine)iron(2+) ion, when accompanied by most common counterions, has good solubility in water but may be extracted into organic solvents. The symmetry of the $[FeN_6]$ unit is D_3 with the five-membered chelate ring being coplanar with the remainder of the bipyridine ligand. The complex exists in the low-spin state. Significant bond strength resulting from π donation from metal to ligand is expected. The oxidation potential of the complex is 1.02 V. The complex is used in a variety of electron-transfer reactions. Many stable salts have been isolated, including the dibromide [*15388-40-6*], the dichloride [*14751-83-8*], and the diperchlorate [*15388-48-4*].

The pale-blue tris(2,2′-bipyridine)iron(3+) ion [*18661-69-3*] may be obtained by oxidation of the corresponding iron(II) ion. The complex cannot be prepared directly from iron(III) salts. For example, addition of 2,2′-bipyridine to an aqueous iron(III) chloride solution precipitates species, eg, the doubly hydroxy-bridged $[(bipy)_2Fe(OH)_2Fe(bipy)_2]Cl_4$ [*74930-87-3*]. The tris(2,2′-bipyridine)iron(3+) ion is less stable than the iron(II) ion and has an overall formation constant of 10^{12}. The absorption maximum shifts to 610 nm and the absorptivity is reduced to 330 M^{-1} cm^{-1}. Reduction by water or hydroxide occurs in mildly acidic solution to produce the stable iron(II) ion. The tris(2,2′-bipyridine)iron(3+) ion is used as a probe of electron-transfer mechanisms. The triperchlorate salt [*15388-50-8*] most often is isolated.

Diketones. 2,4-Pentanedione (acetylacetone, acac) may undergo keto–enol tautomerization:

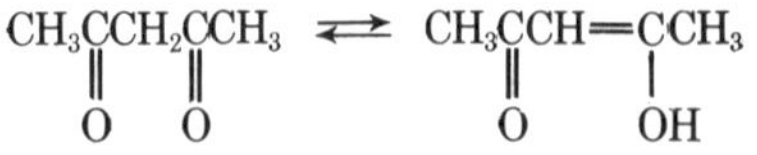

$$CH_3C(=O)CH_2C(=O)CH_3 \rightleftarrows CH_3C(=O)CH{=}C(OH)CH_3$$

Loss of a proton upon metal binding results in the formation of a six-membered chelate ring:

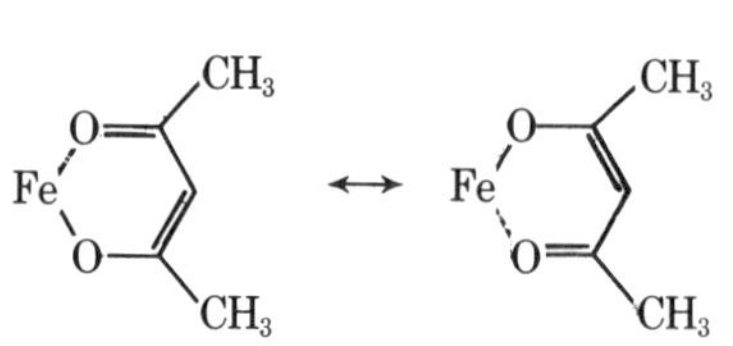

CH₃ CH₃
Fe–O=C(CH₃)–CH=C(CH₃)–O–Fe ↔ Fe–O–C(CH₃)=CH–C(CH₃)=O–Fe

Tris(2,4-pentanedionato)iron(III) [*14024-18-1*], $Fe(C_5H_7O_2)_3$, forms ruby-red rhombs which melt at 184°C. It may be prepared by treatment of iron(III) hydroxide with a slight excess of ligand. This neutral complex is only slightly soluble in hot or cold water but is soluble in alcohol, acetone, chloroform, or benzene. The x-ray structure analysis of this complex reveals an octahedral array of equivalent Fe—O bonds (7). The chelate rings are planar and the C—C distance (139 pm) is the same as that in benzene. The C—O distances are identical (128 pm) and are between those of carbon–oxygen single and double bonds. Generally, it is agreed that the π bonds do not extend to the metal atom. Similar complexes may be formed with other β diketones, or substitution of other β-diketones into tris(2,4-pentanedionato)iron(III) [*14024-18-1*] may be accomplished by heating the complex with the β-diketone in a high boiling solvent. This complex is used extensively as a catalyst in polymerization reactions and in oxidation reactions.

Iron Ethylenediaminetetraacetic Acid. Ethylenediaminetetraacetic acid has six potential chelating atoms: two nitrogen atoms and four oxygen atoms. If the ligand acts as a hexadentate, five five-membered rings are formed and the charge on the complex is four less than that of the metal ion.

Iron(II) ethylenediaminetetraacetic acid [*15651-72-6*], $Fe(EDTA)^{2-}$ or *N,N′*-1,2-ethanediylbis[*N*-(carboxymethyl)glycinato]ferrate(2−), is a colorless, air-sensitive compound. It is a good reducing agent ($E° = -0.117$ V) and has been used as a probe of outer sphere electron-transfer mechanisms. It may be prepared by dissolution of iron wire in hydrochloric acid in an inert atmosphere followed by addition of an equivalent amount of the disodium salt, Na_2H_2EDTA. The diammonium [*56174-59-5*] and disodium [*14729-89-6*] salts have been reported.

Iron(III) ethylenediaminetetraacetic acid [*15275-07-7*], $Fe(EDTA)^{-}$ or *N,N′*-1,2-ethanediylbis(*N*-(carboxymethyl)glycinato)ferrate(1−), is a pale yellow, high-spin complex in which EDTA serves as a hexadentate ligand. However, the complex must be considered seven-coordinate because the x-ray structure analysis reveals a water molecule which is directly bound to iron. The geometry about the iron atom is described as pentagonal bipyramidal and is depicted in Figure 3.

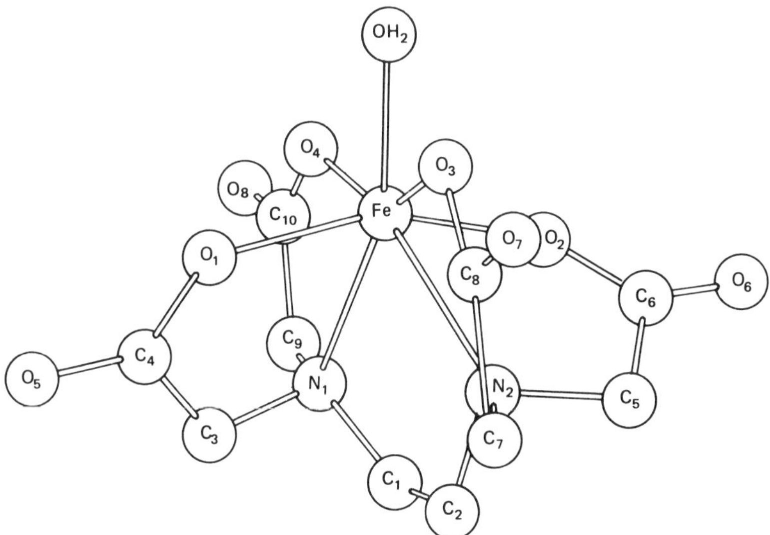

Figure 3. Skeleton model in perspective of $Fe(OH_2)$ (ETDA). Courtesy of *J. Am. Chem. Soc.* (8).

The stability constant for the formation of this complex is 10^{25}. The ammonium [*21265-50-9*] and sodium [*15708-41-5*] salts have been prepared and are used as oxidizing agents, especially in photographic bleaching and fixing preparations.

Oxalates. Iron(II) oxalate [*30948-48-2*], $Fe(ox)_3^{4-}$ or tris(ethanedioato)ferrate(4−), is known but has been studied almost exclusively in solution as an electron-transfer agent.

Iron(III) oxalate [*15321-61-6*], $Fe(ox)_3^{3-}$ or tris(ethanedioato)ferrate(3−), may be prepared by addition of an excess of oxalate to almost any soluble ferric salt. Green crystals of tripotassium tris(oxalato)ferrate trihydrate [*5936-11-8*] (tripotassium tris(ethanedioato)ferrate trihydrate), may be obtained by adding three equivalents of barium oxalate and three equivalents of potassium oxalate to iron(III) sulfate, removing barium sulfate by filtration, and concentrating and cooling to effect crystallization. The crystals are diamagnetic (magnetic moment = 5.33×10^{-23} J/T (5.75 μ_β)). Discrete tris(bidentate) units are present in the structure but the true symmetry is D_3 rather than octahedral. The trihydrate yields the anhydrous tripotassium tris(oxalato)ferrate [*14883-34-2*] at ca 120°C; the salt decomposes at ca 230°C. A number of other salts are known, including triammonium tris(oxalato)ferrate [*14221-47-7*].

Tris(oxalato)ferrate(3−) undergoes photochemical reduction as the first step in the blueprint process. This complex anion is used in other photochemical processes as well as in a variety of redox processes (see Reprography).

Phenanthrolines. The 1,10-phenanthroline ligand and its numbering scheme are shown below:

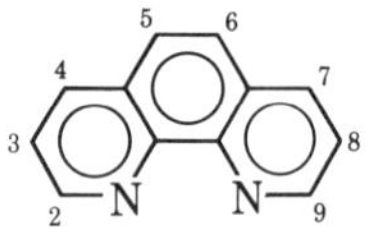

Complexes of 1,10-phenanthroline with iron(II) are intensely colored and 1,10-phenanthroline serves as a reagent for the qualitative and quantitative determination of iron. The orange-red tris(1,10-phenanthroline)iron(2+) ion [*14708-99-7*] has an absorption maximum at 510 nm with an absorptivity of 1.10×10^4 M^1 cm^1. It is an extremely stable iron(II) complex with an overall formation constant of 10^{21}. The complex forms in the pH range 2–9 and forms at higher pH if an effective reducing agent (eg, sodium dithionite) is present. A large variety of complexes may be synthesized from the many substituted phenanthrolines. For example, tris(4,7-diphenyl-1,10-phenanthroline)iron(2+) [*21412-03-3*] sometimes is used to determine iron concentrations because it is a more sensitive reagent (the absorptivity is 2.24×10^4 M^1 cm^1 at 533 nm).

Tris(1,10-phenanthroline)iron(2+), when accompanied by most common counterions, has good solubility in water but may be extracted into nonaqueous solvents. Ease of extraction into nonaqueous solvents increases with the number and size of organic ring substituents. The symmetry about the iron is best described as D_3. Iron(II) exists in the low-spin state and a significant contribution to bonding strength by way of π donation from metal to ligand is expected. The oxidation potential for this complex is 1.06 V. It serves as an electron-transfer mediator in many applications, and it is an excellent indicator in oxidation–reduction titrations. This complex also is a constituent of many oscillating reaction systems. Some of the well known salts of tris(1,10-phenanthroline)iron(2+) include the dichloride [*14978-15-5*], the diiodide [*15553-89-6*], and the diperchlorate [*14586-54-0*].

The blue tris(1,10-phenanthroline)iron(3+) ion [*13479-49-7*] is obtained by oxidation of the corresponding iron(II) ion but is not obtained directly from iron(III) salts. This ion is less stable and less highly colored than the iron(II) ion. The overall formation constant is 10^{14}, and the visible absorption maximum occurs at 590 nm with an absorptivity of 600 M^{1} cm^{1}. In solutions of pH above 4, reduction to the iron(II) complex occurs and, in alkaline solutions, this reduction is instantaneous. At pH values below 2, protons compete with iron for the 1,10-phenanthroline binding site, and coordination is incomplete. The complex is studied most often in solution as an oxidant, but the trichloride [*40273-22-1*] and the triperchlorate monohydrate [*20774-81-6*] salts have been prepared. Crystals of the latter are low-spin, monoclinic and reveal D_3 symmetry about the iron atom.

Terpyridines. The 2,2′:6′,2″-terpyridine ligand is tridentate and is pictured below:

This ligand may form an octahedral complex with iron only as shown in Figure 4.

The bis(2,2′,2″-terpyridine)iron(2+) ion [*17455-70-8*] has an intense purple color which serves as a test for iron(II). The absorption maximum is at 552 nm and the absorptivity is 1.15×10^4 M^{-1} cm^{-1}. The complex has an overall stability constant of 10^{19}, which is less than might be expected for a bis(tridentate) complex. Apparently because the ligand remains planar, strain is introduced into the complex. The fit is not perfect, and N–Fe–N angles are less than 90°. The oxidation potential for the complex is 1.13 V. Addition of perchloric acid to a solution of the ion precipitates the diperchlorate salt [*22079-98-7*].

The tris(2,2′,2″-terpyridine)iron(3+) ion [*47779-99-7*] is a pale bluish-green. The iron(III) salt may be obtained only by oxidation of the corresponding iron(II) ion. It is very unstable with respect to both solvent reduction and ligand dissociation. The perchlorate salt [*21536-42-5*] has been reported.

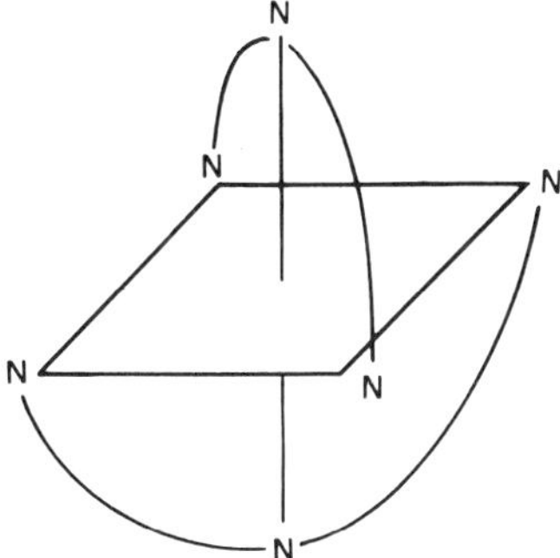

Figure 4. Schematic diagram of 2,2′,2″-terpyridine coordination. Courtesy of *Chem. Rev.* (9).

Macrocycles. An n-dentate, macrocyclic ligand displays enhanced stability over an analogous open-chain, multidentate ligand. The explanation for this observation (which is analogous to the chelate effect) is a more favorable entropic effect upon complex formation; this phenomenon is termed the macrocyclic effect. These large structures usually involve nitrogen as the ligand donor atom but macrocycles with oxygen or sulfur donor atoms or those with mixed donor atoms also are known (for a more complete listing of macrocyclic ligands, see ref.10).

Iron(II) phthalocyanine (phthalocyaninatoiron) [*132-16-1*] is synthesized from phthalocyanine according to

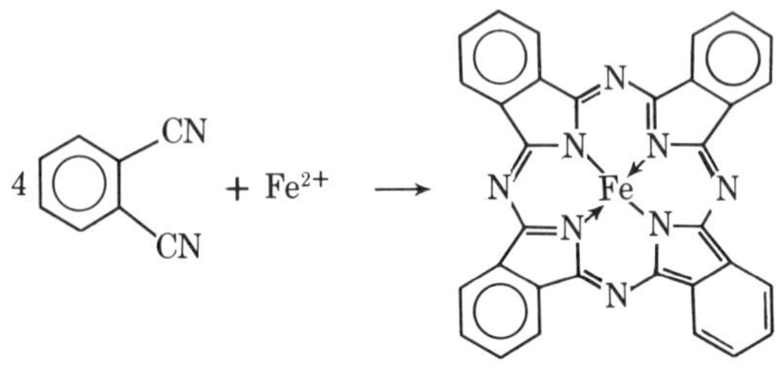

by refluxing in 1-chloronaphthalene and is purified by sublimation at 450°C under partial vacuum (see Dyes, sensitizing). The x-ray crystal structure (11) of this complex reveals square planar geometry about the iron atom, which formally is in the 2+ oxidation state. An interesting feature of this complex is that it exists in an intermediate spin state with S = 1; that is, the average electronic configuration has two unpaired electrons. The Fe—N bond distance of 0.1926 nm is intermediate between that expected for a high-spin complex and that expected for a low-spin complex.

Iron phthalocyanine is a green compound and is insoluble in most solvents. The phthalocyanine ligand stabilizes the low oxidation states of the metal and, through chemical or electrochemical methods, up to four electrons may be added to the complex. Because of the many oxidation states available to this complex, it is used extensively as a catalyst for a variety of chemical and electrochemical redox reactions. It also is an important pigment. A number of adducts are known wherein the iron becomes six coordinate. Often the adducts are nitrogenous bases, eg, phthalocyaninatobis(pyridine)iron [*20219-84-5*]. Soluble iron phthalocyanine complexes may be obtained by sulfonating the phenyl residues to obtain, for example, tetrasodium phthalocyaninetetrasulfonatoferrate [*41867-66-7*].

Iron porphyrins derive most of their recent interest from the fact that they are models for a number of systems found in nature (eg, the oxygen storage and transport proteins, myoglobin and hemoglobin, and the system of electron transport proteins (qv), the cytochromes). Some very elegant models have been prepared (12) (see Oxygen generation systems). Tetraphenylporphinatoiron [*16591-56-3*] is a square-planar complex which exists in the S = 1 ground state, similar to iron phthalocyanine. The x-ray crystal structure (13) supports the intermediate spin-state assignment because the Fe—N bond length is an intermediate distance of 0.1972 nm. The iron atom is in the plane of the porphyrin, as is expected for an iron atom of intermediate spin. The 2-methylimidazole adduct of the tetraphenyl–phorphinatoiron complex may be isolated by precipitation by ethanol from a benzene solution of the components. The crystal structure of (2-methylimidazole)tetraphenylporphinatoiron ethanol solvate [*41754-42-1*] reveals that iron is 0.055 nm above the plane of the porphyrin and that six coordination is sterically impossible. The five-coordinate complex is high spin but remains in the 2+ oxidation state.

Reversible binding of dioxygen was demonstrated in the complex (dioxygen)-(1-methylimidazole) [(21*H*,23*H*-porphine-5,10,15,20-tetracyltetra-2,1-phenylene) tetrabis(2,2-dimethylpropanamidato)]iron [*52629-15-9*], which is one of a series of related models called picket-fence porphyrins. The structure has all four substituents on the same side of the porphyrin plane to form a pocket for the dioxygen molecule which coordinates in end-on fashion. The iron atom is approximately octahedrally coordinated and is in the porphyrin plane. The iron atom remains in the 2+ oxidation state but here is in the low-spin state. The molecule is depicted in Figure 5.

Siderophores

In response to the nearly complete insolubility of iron under physiological conditions, bacteria synthesize powerful iron-binding ligands that are called siderophores (15). Siderophores are excreted into the bacterial medium (eg, the soil, an aqueous environment, or an animal or human host) where they encounter iron, chelate and solubilize it, and transport it to within the bacterial cell.

Siderophores preferentially complex iron(III). The iron is coordinated in octahedral fashion and the complexes are high spin. Many siderophores incorporate three hydroxamic acid moieties

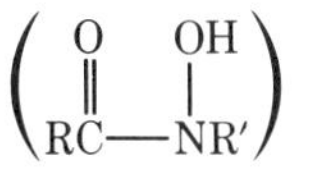

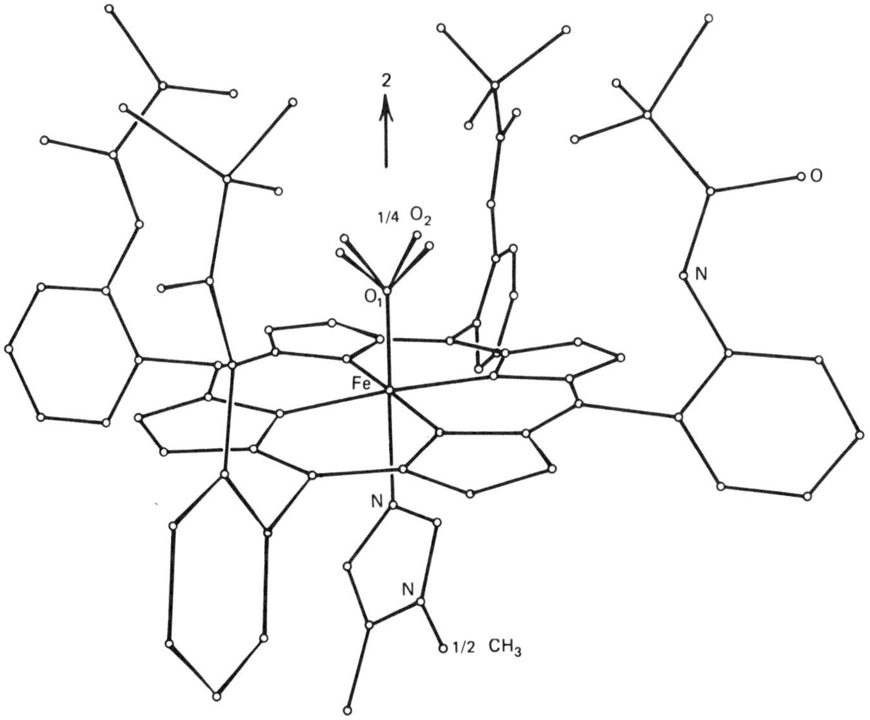

$$\left(\begin{array}{cc} \mathrm{O} & \mathrm{OH} \\ \| & | \\ \mathrm{RC}\!-\!\!-\!\! & \mathrm{NR'} \end{array}\right)$$

2
1/4 O_2
O_1
O
N
Fe
N
N
1/2 CH_3

Figure 5. Perspective view of one molecule of the iron dioxygen complex showing the crystallographic 2-fold axis of symmetry and the four-way statistical disorder of the terminal O_2 oxygen atom of dioxygen. Courtesy of *Proc. Natl. Acad. Sci. USA* (14).

into their structure as the iron-binding residues. The ferrichromes are cyclic polypeptides that have three appended hydroxamic acid side chains, each of which forms a stable five-membered ring upon iron complexation. Their structures are presented in Figure 6. The ferrichromes are extremely stable yellow or orange complexes with overall stability constants typically of ca 10^{30}. Ferrichrome, ferrichrome A, ferrichrome C, ferrichrysin, and ferricrocin are named as the iron(III) complex, and the prefix deferri- is added to signify the free ligand. The x-ray crystal structures of ferrichrome (17) and of ferrichrome A (18) reveal the unusual property that these high-spin iron(III) complexes crystallize exclusively as the Λ-cis isomer. Circular dichroism solution spectra indicate this isomer predominates, perhaps exclusively, in solution as well.

The ferrioxamines also use the hydroxamic acid moiety to coordinate iron but, in the ferrioxamines which may be linear or cyclic as in Figure 7, the hydroxamic acids are an integral part of the aliphatic chain rather than appendages to it. The ferriox-

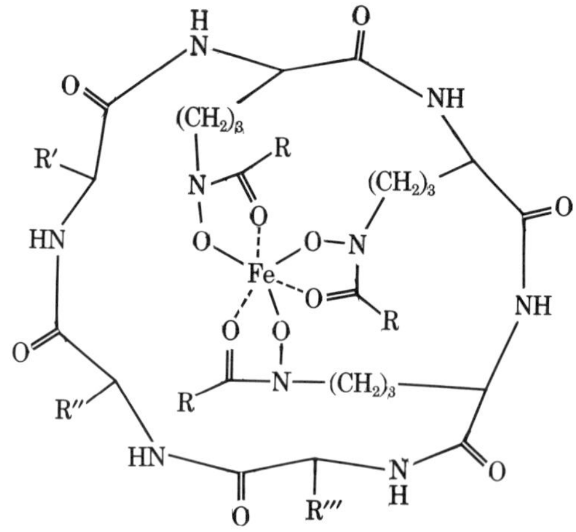

Siderophore	CAS Registry No.	R′	R″	R‴	R	Source
Ferrichrome	*[15630-64-5]*	H	H	H	CH_3	*Aspergillus spp* *Neurospora spp* *Penicillium spp*
Ferrichrome A	*[15258-80-7]*	CH_2OH	CH_2OH	H	$CH{=}C(CH_3)$-CH_2CO_2H (trans)	*Ustilago sphaerogena*
Ferrichrome C	*[56665-78-2]*	H	CH_3	H	CH_3	*Cryptococcus melibiosum*
Ferrichrysin	*[18972-10-6]*	CH_2OH	CH_2OH	H	CH_3	*Aspergillus melleus* *Aspergillus terreus*
Ferricrocin	*[23086-46-6]*	H	CH_2OH	H	CH_3	*Aspergillus fumigatus* *Aspergillus humicola* *Aspergillus versicolor*

Figure 6. Structural diagram of ferrichromes. (**a**) Linear ferrioxamines. (**b**) Cyclic ferrioxamines. Courtesy of *Acc. Chem. Res.* (16).

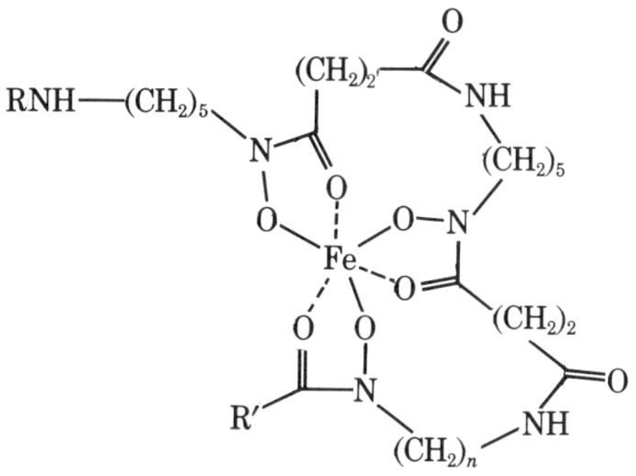

Linear ferrioxamines	CAS Registry No.	R	n	R′	Source
Ferrioxamine B	[*14836-73-8*]	H	5	CH_3	*Nocardia spp* *Streptomyces spp* *Micromonaspora spp*
Ferrioxamine D_1	[*15349-98-1*]	$COCH_3$	5	CH_3	*Streptomyces pilosus*
Ferrioxamine G	[*29825-05-6*]	H	5	$C(CH_2)_2COOH$	*Streptomyces pilosus*
Ferrioxamine A_1	[*14710-31-7*]	H	4	$C(CH_2)_2COOH$	*Streptomyces pilosus*
Ferrimycin A_1	[*15319-50-3*]	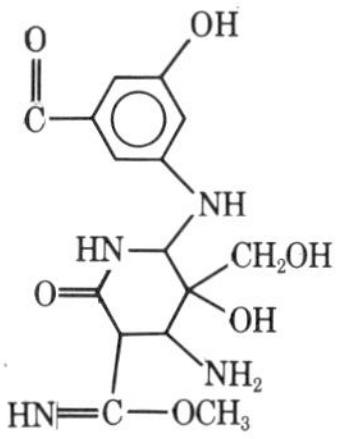	5	CH_3	*Streptomyces galilaeus* *Streptomyces griseoflavus* *Streptomyces lavendulae*

(a)

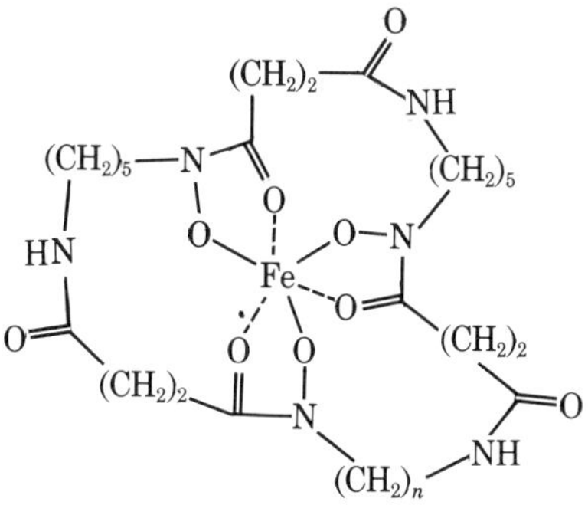

Cyclic ferrioxamines	CAS Registry No.	n	Source
Ferrioxamine E	[*20008-20-2*]	5	*Nocardia spp* *Streptomyces pilosus*
Ferrioxamine D_2	[*29825-04-5*]	4	*Streptomyces pilosus*

(b)

Figure 7. Structural diagram of ferrioxamines. Courtesy of *Acc. Chem. Res.* (16).

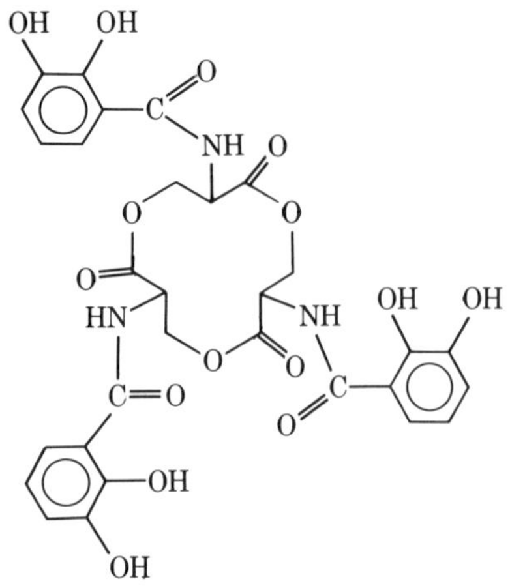

Figure 8. Structural diagram of enterobactin. Courtesy of *Acc. Chem. Res.* (16).

amines form complexes that are as stable as the ferrichromes. Ferrioxamine A_1, ferrioxamine B, ferrioxamine D_1, ferrioxamine D_2, ferrioxamine E, and ferrioxamine G are named as the iron(III) complex and the prefix deferri- is used to indicate the free ligand. Deferriferrioxamine B is marketed as Desferal and is used to treat iron overload that results from accidental poisoning or from chronic transfusions.

Enterobactin (see Fig. 8) is a siderophore which binds iron through three deprotonated catechol

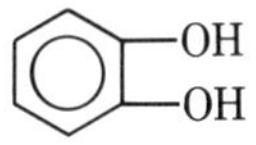

groups. The iron(III) enterobactin complex [*62280-34-6*] exhibits some extraordinary properties, including an overall formation constant of 10^{52} and a reduction potential of −980 mV at pH 10. The absolute conformation of ferric enterobactin has been determined to be Δ-cis—the opposite isomer of the ferrichromes.

In an attempt to develop an effective drug to remove the excess iron in the blood of patients with Cooley's anemia, the iron complex of HBED (Fig. 9) has been studied (19).

Analytical Methods

Gravimetric. Soluble iron samples may be analyzed by precipitation of iron(III) as the hydrated oxide, followed by ignition at 900–1000°C to achieve constant weight of anhydrous Fe_2O_3. The sample must be free of interfering ions, eg, those of aluminum, chromium, titanium, and manganese. The sample first is heated with nitric acid to ensure that all iron is present as iron(III) and then is treated with an excess of ammonia to precipitate hydrated iron(III) oxide as a gelatinous mass. The precipitate is collected on ashless filter paper and is washed with hot 1% ammonium nitrate solution. Paper and precipitate are transferred to a porcelain crucible and ignited.

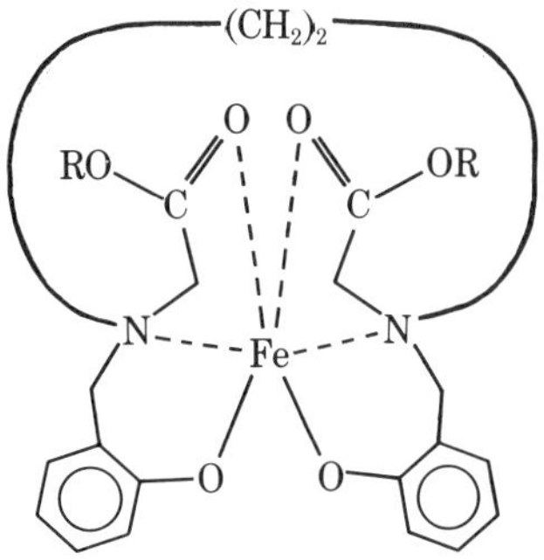

$R = CH_3$ [75535-07-8]; $R = (CH_2)_2CH_3$ [75535-08-9]; $R = (CH_2)_4CH_3$ [75535-09-0]

Figure 9. Iron complexes of esters of HBED. [*N,N′*-bis(2-hydroxybenzyl)ethylenediamine-*N,N′*-diacetic acid].

Volumetric. Iron solutions may be analyzed by first reducing all of the iron that is present to the iron(II) state; this reduction most often is accomplished through the use of a Jones reductor. The iron(II) solution then may be titrated with a standardized solution of potassium permanganate, potassium dichromate, cerium(IV) sulfate, or cerium(IV) perchlorate. Titration with permanganate may be complicated by the tendency of permanganate to oxidize chloride ion, by the temporal instability of the permanganate solution, and by the possibility of uncertain stoichiometry in the redox reaction. Titration with cerium(IV) standard solution eliminates these problems, but cerium(IV) is considerably more expensive, must be used in acidic solution, and requires an indicator (usually 1,10-phenanthroline). Use of potassium dichromate also requires an indicator (usually diphenylaminesulfonic acid). The dichromate is not as strong an oxidant as either permanganate or cerium(IV), and it sometimes may be noticeably slow in reacting. However, potassium dichromate standard solutions may be prepared directly by weight from the primary standard salt.

Colorimetric. A sensitive method for the determination of small concentrations of dissolved iron is the spectrophotometric determination of the orange-red tris(1,10-phenanthroline)iron(II) complex. An even more sensitive reagent is 4,7-diphenyl-1,10-phenanthroline. Because only the iron(II) complex is highly colored, the sample first is treated with an excess of reducing agent (usually hydroxylamine hydrochloride). The complex is stable from pH 2–9, although the analysis preferably is done at ca pH 3.5. Absorbance at 510 nm is compared to a standard curve.

Health and Safety

The only iron compound that is associated with any particular industrial risk is iron(III) oxide, Fe_2O_3, which has an exposure limit of 5 mg/m^3 (20). Chronic inhalation of iron(III) oxide leads to siderosis and adequate ventilation and mechanical filter respirators should be provided to those exposed to the oxide.

Most other iron salts may be safely handled following common principles of laboratory safety. The most serious threat is the ingestion of massive quantities of iron salts which results in diarrhea, hemorrhage, liver damage, heart damage, and shock. A lethal dose is 200–250 mg iron/kg body weight. Most casualties of iron poisoning are children under five years of age.

BIBLIOGRAPHY

"Iron Compounds" in *ECT* 1st ed., Vol. 8, pp. 56–66, by R. S. Casey, W. A. Scheaffer Pen Co., C. S. Grove, Jr., Syracuse University; B. J. Lerner, University of Texas; "Iron Compounds" in *ECT* 2nd ed., Vol. 12, pp. 22–44, by Robert S. Casey, Consultant, and John R. Doyle, State University of Iowa.

1. A. G. Sharpe, *The Chemistry of Cyano Complexes of the Transition Metals,* Academic Press, Inc., New York, 1976.
2. A. Tullberg and N.-G. Vannerburg, *Acta Chem. Scand.* **28A,** 551 (1974).
3. H. J. Buser and A. Ludi, *Chimia* **30,** 99 (1976).
4. F. Herren, P. Fischer, A. Ludi, and W. Halg, *Inorg. Chem.* **19,** 956 (1980).
5. A. Ludi and H. U. Gudel, *Struct. Bonding* **14,** 1 (1973).
6. R. Colton and J. H. Canterford, *Halides of the First Row Transition Metals,* Wiley-Interscience (from Berlin), London, 1969.
7. R. B. Roof, *Acta Cryst.* **9,** 781 (1956).
8. J. L. Hoard, M. Lind, and J. V. Silverton, *J. Am. Chem. Soc.* **83,** 2770 (1961).
9. W. W. Brandt, F. P. Dwyer, and E. C. Gyarfas, *Chem. Rev.* **54,** 959 (1954).
10. F. A. Cotton and G. Wilkinson, *Advanced Inorganic Chemistry,* 4th ed., John Wiley & Sons, Inc., New York, 1980.
11. D. W. Clark and Y. R. Yandle, *Inorg. Chem.* **11,** 1738 (1972).
12. J. P. Collman, *Acc. Chem. Res.* **10,** 265 (1977).
13. J. P. Collman, J. L. Hoard, N. Kim, G. Lang, and C. A. Reed, *J. Am. Chem. Soc.* **97,** 2676 (1975).
14. J. L. Hoard, C. A. Reed, and J. P. Collman, *Proc. Natl. Acad. Sci. (USA)* **71,** 1326 (1974).
15. K. N. Raymond, ed., *Bioinorganic Chemistry II,* American Chemical Society Advances in Chemistry Series Number 162, American Chemical Society, Washington, D.C., 1977.
16. K. N. Raymond and C. J. Carrano, *Acc. Chem. Res.* **12,** 183 (1979).
17. D. van der Helm, J. R. Baker, D. L. Eng-Wilmot, M. B. Hossain, and R. A. Loghry, *J. Am. Chem. Soc.* **102,** 4224 (1980).
18. A. Zalkin, J. D. Forrester, and D. H. Templeton, *J. Am. Chem. Soc.* **88,** 1810 (1966).
19. *Chem. Eng. News* **58,** 42 (Sept. 29, 1980).
20. E. R. Plunkett, *Handbook of Industrial Toxicology,* Chemical Publishing Co., New York, 1976.

General References

Ref. 10 is also a general reference.

J. E. Huheey, *Inorganic Chemistry,* 2nd ed., Harper and Row, New York, 1978.

D. A. Skoog and D. M. West, *Fundamentals of Analytical Chemistry,* 3rd ed., Holt, Rinehart, and Winston, New York, 1976.

F. A. Lowenheim and M. K. Moran, *Industrial Chemicals,* 4th ed., John Wiley & Sons, Inc., New York, 1975.

F. S. Galasso, *Structure and Properties of Inorganic Solids,* Pergamon Press, Oxford, 1970.

J. W. Mellor, *A Comprehensive Treatise on Inorganic and Theoretical Chemistry,* Volume XIII, Longmans, Green, and Co., London, 1934.

F. P. Dwyer and D. P. Mellor, *Chelating Agents and Metal Chelates,* Academic Press, Inc., New York, 1964.

A. A. Schilt, *Analytical Applications of 1,10-Phenanthroline and Related Compounds,* Pergamon Press, Oxford, 1969.

J. M. Arena, *Poisoning,* 4th ed., C. C. Thomas, Springfield, 1979.

JAMES V. MCARDLE
University of Maryland

ISOCYANATES, ORGANIC

Organic isocyanates are compounds in which the isocyanate group, —NCO, is attached to an organic group. They are frequently classified as esters of isocyanic acid, HNCO. The first organic isocyanate was prepared by Wurtz in 1849, but organic isocyanates did not arouse major scientific and commercial interest until after World War II. The reactivity of the highly unsaturated isocyanate group has led to its study and use in a great variety of reactions. Polyfunctional isocyanates, usually with two or three isocyanate groups in the molecule, have been particularly useful for the systematic buildup of polymer molecules having tailored properties; as a result, they have become the cornerstone of a large new branch of the plastics industry. The methods of preparation and the reactions of the isocyanates have been reviewed (1–7) as have the applications of these compounds in the polymer field (2,6,8–9).

Properties

Some of the physical properties of a number of representative mono- and diisocyanates are summarized in Table 1. Physical and chemical properties have been compiled and published (4,10). The isocyanates react readily with a great variety of organic compounds and also may react with themselves. The electronic structure of the isocyanate group indicates that it has the following resonance hybrids:

$$R\ddot{\underset{..}{\bar{N}}}—\overset{+}{C}{=}\ddot{O}: \rightleftharpoons R\ddot{N}{=}C{=}\ddot{O}: \rightleftharpoons R\ddot{N}{=}\overset{+}{C}—\underset{..}{\ddot{O}}:$$

Reactions with Active Hydrogen Compounds. The normal isocyanate reaction ultimately provides addition to the carbon–nitrogen double bond. In reactions involving compounds with an active hydrogen, ie, one that can be replaced by sodium, the hydrogen becomes attached to the nitrogen of the isocyanate, and the remainder of the active hydrogen compound becomes attached to the carbonyl carbon:

$$RN{=}C{=}O + HA \longrightarrow RNH\overset{\overset{\displaystyle O}{\|}}{C}A$$

In most reactions, especially those involving active hydrogen compounds, the aromatic isocyanates are more reactive than the aliphatic isocyanates. Substitution of electronegative groups on the aromatic ring enhances the reactivity, whereas electropositive groups reduce the reactivity of the isocyanates. As would be expected, steric hindrance on either the isocyanate or the active hydrogen compound retards the reaction. All of the reactions are catalyzed by acids and, usually, more strongly by bases. Certain metal compounds are exceptionally powerful catalysts (see Catalysis). Proper selection of a catalyst can assist in the use of isocyanates for the tailoring of polymers through control of the desired reactions (1,9).

The most important reaction of isocyanates is with alcohols. Primary alcohols react at room temperature, and secondary and tertiary alcohols react much more slowly. The usual reaction of an alcohol and an isocyanate leads to a urethane (see Urethane polymers).

Table 1. Physical Properties of Some Isocyanates

Compound	Formula	CAS Registry No.	mp, °C	bp, $°C_{kPa}$ [a]	Density, g/cm^3	Refractive index, n_D^t	Flash point, open cup, °C
methyl isocyanate	CH_3NCO	*[624-83-9]*		38_{101}	0.96_4^{20}	1.36^{20}	−7
n-butyl isocyanate	C_4H_9NCO	*[111-36-4]*	<−70	115_{101}	0.89_4^{20}	1.4064^{20}	ca 20
cyclohexyl isocyanate	$C_6H_{11}NCO$	*[3173-53-3]*	<−80	171_{101}	0.96_4^{20}	1.4557^{20}	54
octadecyl isocyanate	$C_{18}H_{37}NCO$	*[112-96-9]*	21	$170_{0.27}$	0.86_{15}^{25}	1.4468^{30}	184
hexamethylene diisocyanate	$OCN(CH_2)_6NCO$	*[822-06-0]*	−67	$127_{1.3}$	1.05_4^{20}	1.4530^{20}	140
phenyl isocyanate	C_6H_5NCO	*[103-71-9]*	−30	165_{101}	1.10_4^{20}	1.5362^{20}	60
p-chlorophenyl isocyanate	ClC_6H_4NCO	*[104-12-1]*	28	$88_{1.3}$	1.25_4^{40}	1.5543^{40}	110
3,4-dichlorophenyl isocyanate	$Cl_2C_6H_3NCO$	*[102-36-3]*	42	$113_{1.3}$	1.39_4^{50}	1.5710^{50}	142
m-phenylene diisocyanate	$C_6H_4(NCO)_2$	*[123-61-5]*	51	$110_{1.5}$	1.21_4^{60}	1.5573^{50}	107
2,4-toluene diisocyanate	$CH_3C_6H_3(NCO)_2$	*[584-84-9]*	22	$120_{1.3}$	1.22_{15}^{25}	1.5654^{25}	132
80/20-toluene diisocyanate	$CH_3C_6H_3(NCO)_2$	*[26471-62-5]*	11–14[b]	$120_{1.3}$	1.22_{15}^{25}	1.5666^{25}	132
65/35-toluene diisocyanate	$CH_3C_6H_3(NCO)_2$	*[26471-62-5]*	3–5[b]	$120_{1.3}$	1.22_{15}^{25}	1.5663^{25}	132
4,4′-diphenylmethane diisocyanate	$OCN-C_6H_4-CH_2-C_6H_4-NCO$	*[101-68-8]*	38	$196_{0.7}$	1.19_{50}	1.5906^{50}	201
4,4′-diisocyanato-3,3′-dimethoxy-1,1′-biphenyl diisocyanate (dianisidine diisocyanate)	$OCN-C_6H_3(OCH_3)-C_6H_3(OCH_3)-NCO$	*[91-93-0]*	122	$200–210_{0.7}$	1.397_4^{20}		245
4,4′-diisocyanato-3,3′-dimethyl-1,1′-biphenyl diisocyanate (tolidine diisocyanate)	$OCN-C_6H_3(CH_3)-C_6H_3(CH_3)-NCO$	*[91-97-4]*	70	$160–170_{0.5}$	1.156_4^{80}		214

biuret of hexamethylene diisocyanate	$OCN(CH_2)_6N(H)C(O)N[(CH_2)_6NCO]C(O)N(H)(CH_2)_6NCO$	[*4035-89-6*]	−19		1.114_4^{40}		
isophorone diisocyanate	OCN–(3,3,5-trimethylcyclohexyl)–CH_2NCO (ring bearing CH_3, CH_3, CH_3, CH_2NCO, OCN)	[*4098-71-9*]	ca 60	$217_{13.3}$	1.0615_4^{20}	1.4844^{20}	163
polymeric diphenylmethane diisocyanate	OCN–C_6H_4–CH_2–[C_6H_3(NCO)–CH_2]$_n$–C_6H_4–NCO	[*9016-87-9*]			1.236_4^{20}		
4,4′-dicyclohexylmethane diisocyanate	OCN–C_6H_{10}–CH_2–C_6H_{10}–NCO	[*5124-30-1*]	60–71[c]	$245_{7.2}$	1.029_4^{70}	1.4759^{80}	
propyl isocyanate	$CH_3CH_2CH_2NCO$	[*110-78-1*]	<−30	87.8_{101}	0.8948_4^{20}	1.3947^{20}	
isopropyl isocyanate	$(CH_3)_2CHNCO$	[*1795-48-8*]		$70–75_{101}$	0.8669_4^{20}	1.3886^{20}	

[a] To convert kPa to mm Hg, multiply by 7.5.
[b] Freezing point.
[c] Depending on stereoisomer content.

$$RNCO + R'OH \longrightarrow RNH\overset{O}{\overset{\|}{C}}OR'$$

urethane

When this reaction occurs between a diisocyanate and a dihydric (or higher) alcohol, straight-chain (or cross-linked) polymers arise that are used in foams, elastomers (see Elastomers, synthetic), and coatings (qv) (see Foamed plastics).

Most compounds that contain a hydrogen atom bonded to oxygen react with isocyanates under proper conditions. Phenols, being less basic than aliphatic alcohols, react more slowly and require catalysts or higher temperatures (see Phenol; Phenolic resins). The reaction of phenols and isocyanates is reversible at ca 150°C and is used commercially to generate free isocyanates *in situ.* The phenylurethanes also are called blocked isocyanates or splitters. These blocked isocyanates are moisture resistant and are stable to storage (in contrast to the free isocyanates). They are used primarily in baked coating systems, eg, insulating enamel on wire (see Insulation, electric—wire and cable coverings).

Isocyanates are hydrolyzed rapidly by water and give the corresponding disubstituted urea. This reaction is of importance in the preparation of flexible foams (see Toxicity and Handling).

$$RNCO + H_2O \rightarrow RNHCOOH \rightarrow RNH_2 + CO_2$$

$$RNH_2 + RNCO \rightarrow RNHCONHR$$

Halogen acids combine reversibly with isocyanates to give monosubstituted carbamyl halides which, generally, are stable at room temperature but which dissociate at ca 80–100°C.

$$RNCO + HCl \rightleftharpoons RNHCOCl$$

Essentially all compounds containing a hydrogen attached to a nitrogen are reactive.

$$RNCO + HN\begin{smallmatrix} \diagup R' \\ \diagdown R'' \end{smallmatrix} \longrightarrow RNHCON\begin{smallmatrix} \diagup R' \\ \diagdown R'' \end{smallmatrix}$$

The most basic nitrogen compounds usually are the most reactive unless steric hindrance is excessive. Primary amines readily give disubstituted ureas in high yield. Secondary aliphatic amines, as well as primary aromatic amines, react similarly although not as readily.

Amides react slowly at ca 100°C, giving the acylurea (see Amides, fatty acid).

$$RNCO + R'CONH_2 \rightarrow RNHCONHCOR'$$

acylurea

Ureas are moderately reactive at elevated temperatures and form biurets. Desmodur N, a commercial polyisocyanate which is used in high quality coatings, is a biuret mixture from hexamethylene diisocyanate (where R = $(CH_2)_6NCO$):

$$RNCO + RNHCONHR \longrightarrow \begin{array}{l} RNCONHR \\ \;| \\ CONHR \end{array}$$

biuret

Urethanes react in uncatalyzed systems at ca 120–140°C to form allophanates.

$$\mathrm{RNCO + R'NHCOOR'' \rightarrow R'N(CONHR)COOR''}$$

allophanate

Carboxylic acids generally give amides and carbon dioxide by way of the anhydride.

$$\mathrm{RNCO + R'COOH \rightarrow RNHCOOCOR' \rightarrow RNHCOR' + CO_2}$$

Enolizable compounds, eg, malonic esters, react in the presence of basic catalysts to form urethanes which dissociate at relatively low temperatures (11).

$$\mathrm{RNCO + CH_2(COOC_2H_5)_2 \rightarrow RNHCOCH(COOC_2H_5)_2}$$

In general, sulfur compounds react in the same manner as their oxygen analogues but at a much slower rate. The reaction of thioalcohols and thiophenols with isocyanates is slow even in the presence of a mild base, eg, triethylamine, but is rapid in the presence of alkoxides and proceeds to the allophanate stage (12).

Aqueous sodium bisulfite reacts with the isocyanates, particularly the aliphatic derivatives, to give stable, water-soluble derivatives. Dialkyl phosphites and phosphine react similarly.

$$\mathrm{RNCO + NaHSO_3 \rightarrow RNHCOSO_3Na}$$

$$\mathrm{RNCO + HPO(OR')_2 \rightarrow RNHCOPO(OR')_2}$$

$$\mathrm{3\,RNCO + PH_3 \rightarrow (RNHCO)_3P}$$

In addition to the usual reactions with active hydrogen, Friedel-Crafts and Grignard reactions may be performed with isocyanates; both reactions give amides (see Friedel-Crafts reactions; Grignard reaction).

Because isocyanates are highly reactive toward such a large number of active hydrogen compounds, more than one reaction may occur in a system at a given time. This possibility is most pronounced at elevated temperatures and in the presence of catalysts and frequently requires close quality control of reactants. Equilibria may be involved. Trace impurities in resins may act as catalysts. Similarly, the possibility of small amounts of acids in the isocyanate should be considered. Choice of solvents may affect both the rate of an uncatalyzed reaction and the effectiveness of a catalyst.

Catalysis of the isocyanate–hydroxyl reaction is of particular commercial importance. Tertiary amine catalysts are effective in proportion to their base strengths, with the exception of triethylenediamine (bicyclo[2.2.2]-1,4-diazaoctane), which is considerably more powerful than would be predicted, probably because of its configuration. Organotin compounds are another group of catalysts that are used widely in polyurethane chemistry (1–2), especially to control the relative rates of reaction of isocyanates with the secondary hydroxyl of polyethers and water (see Polyethers).

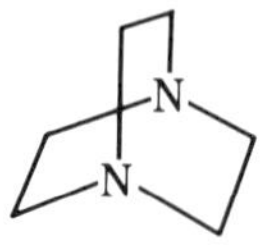

bicyclo[2.2.2]-1,4-diazaoctane

Self-Addition. Isocyanates are capable of reacting with themselves to form dimers, trimers, and polymers. The aromatic isocyanates are readily dimerized by the application of heat; if a catalyst is required, the trialkyl phosphines are very active. The presence of a substituent that is ortho to the NCO group retards the tendency of dimer formation. Perhaps for this reason, toluene diisocyanate [*26471-62-5*] (TDI) is much less prone to dimer formation than either diphenylmethane diisocyanate (MDI) or *para*-phenylene diisocyanate [*104-49-4*]. However, TDI is readily dimerized [*26747-90-0*] by the use of catalysts, and the presence of a trace of dimer can be found in commercial TDI that has been heated or has been stored for an excessive period of time at elevated temperature.

The dimerization of MDI offers more practical problems than that of TDI. Although the reaction is slow, it is rapid enough to become troublesome where storage for weeks is necessary. The dimer [*17589-24-1*] is soluble in liquid MDI at 50°C only to the extent of ca 1% and, therefore, can pose mechanical and chemical difficulties. Although the rate of dimerization increases with temperature below and above the melting point, the slope of the rate curve is higher in the solid. Presumably, this is because the crystal structure is such that two —NCO groups are placed in the correct conformation for formation of the uretidinedione ring. Thus, storage of small quantities of MDI as solid usually is at reduced temperature, and bulk storage and shipment is as a liquid just above the melting point. The dimerization reaction is reversible; dissociation of the dimer becomes complete at above 200°C. At 120°C, the equilibrium mixture contains 6.3% dimer (13).

It generally has been accepted that aliphatic isocyanates do not form dimers (1). However, the dimer [*58247-82-8*] of hexamethylene diisocyanate has been isolated by fractional extraction and has been characterized (14): ir absorption at 1768, 1400, and 780 cm^{-1}, bp 180°C at 0.13 kPa (1 mm Hg).

Both aliphatic and aromatic isocyanates readily form trimers, which have the structure of an isocyanurate ring. The trimerization of aliphatic isocyanates is catalyzed by heat and by trialkyl phosphines. Alkaline salts of carboxylic acids, eg, potassium acetate and many metal compounds that are soluble in organic media, catalyze trimerization of both aliphatic and aromatic isocyanates. Trimerization can occur in polymer reactions, either as an undesirable side reaction, or as a reaction to be promoted to introduce branching and cross-linking. Isocyanurate foams are of commercial interest because of their improved flammability properties. Selected Mannich bases are reported to be useful catalysts in this application.

A number of isocyanate dimers and trimers have been catalogued (10). The polymerization of aromatic isocyanates to nylon-1 also has been reported; molecular weights as high as 300,000–400,000 have been reported for certain of these polymers.

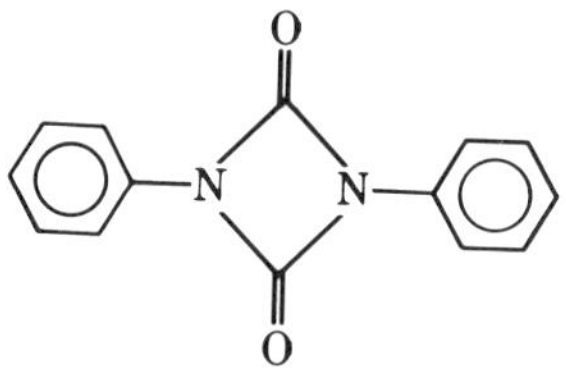

phenyl isocyanate dimer
[*1025-36-1*]

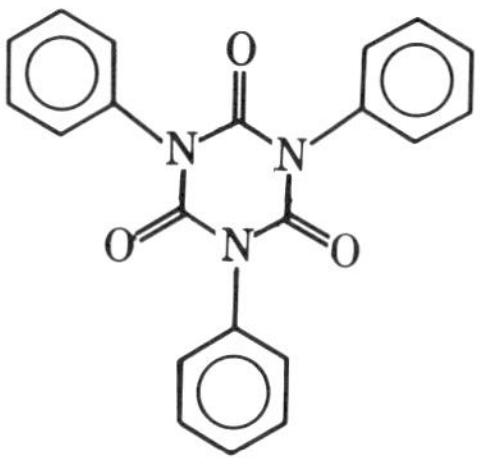

phenyl isocyanate trimer
[*1785-02-0*]

Metallic sodium in dimethylformamide (DMF) was used as catalyst at temperatures well below 0°C to obtain the following product:

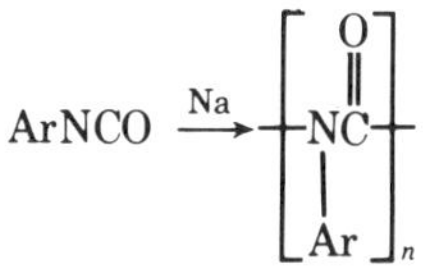

$$ArNCO \xrightarrow{Na} \left[-\underset{\underset{Ar}{|}}{N}-\overset{\overset{O}{\|}}{C}- \right]_n$$

A special case of the polymerization of the isocyanate group is that of vinyl isocyanate (3). The vinyl group of the resulting nylon-1 may be polymerized to produce a polymer bearing a fused ring or ladder structure.

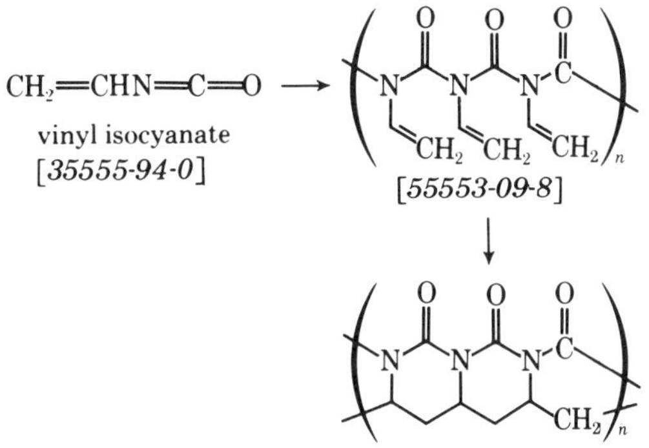

$CH_2{=}CHN{=}C{=}O$ →
vinyl isocyanate
[*35555-94-0*]

[*55553-09-8*]

↓

[*29409-13-0*]

Other Reactions. A number of reactions of the isocyanates (1,11) may be considered to parallel those of the structurally similar ketenes, with the exception that the resulting four-membered ring is unstable and opens with elimination of carbon dioxide, as shown here for the reactions of phenyl isocyanate with *p*-dimethylaminobenzaldehyde, nitrosobenzene, and dimethylformamide (see Ketenes and related substances).

The amidine that is formed by the reaction of isocyanate and DMF may react further with isocyanate to form a six-membered ring adduct which can undergo further addition of isocyanate to form a bicyclic urea (3). On standing, at room temperature, in the presence of DMF, isocyanates may form isocyanurates. These side reactions

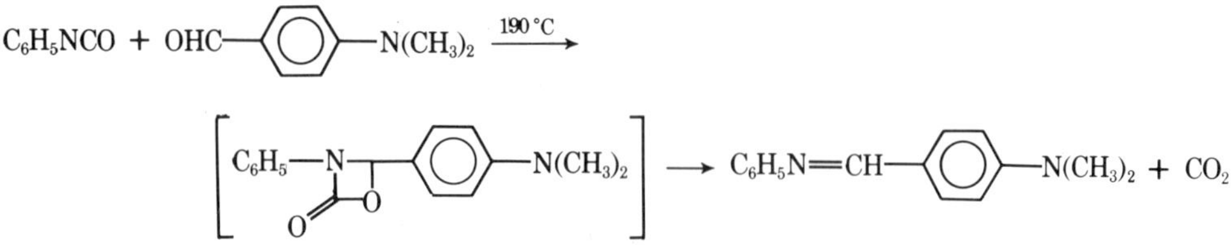

$$C_6H_5NCO + OHC{-}C_6H_4{-}N(CH_3)_2 \xrightarrow{190\,°C} \left[C_6H_5{-}N\text{(4-membered ring with C=O and O)}{-}C_6H_4{-}N(CH_3)_2 \right] \longrightarrow C_6H_5N{=}CH{-}C_6H_4{-}N(CH_3)_2 + CO_2$$

[*74631-09-7*]

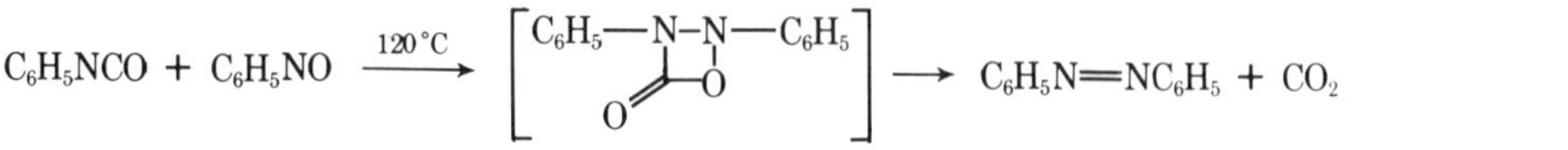

$$C_6H_5NCO + C_6H_5NO \xrightarrow{120\,°C} \left[C_6H_5{-}N{-}N{-}C_6H_5 \text{(4-membered ring with C=O and O)} \right] \longrightarrow C_6H_5N{=}NC_6H_5 + CO_2$$

[*74631-10-0*]

$$C_6H_5NCO + (CH_3)_2NCHO \longrightarrow \left[C_6H_5{-}N\text{(4-membered ring with C=O and O)}{-}N(CH_3)_2 \right] \longrightarrow C_6H_5N{=}CHN(CH_3)_2 + CO_2$$

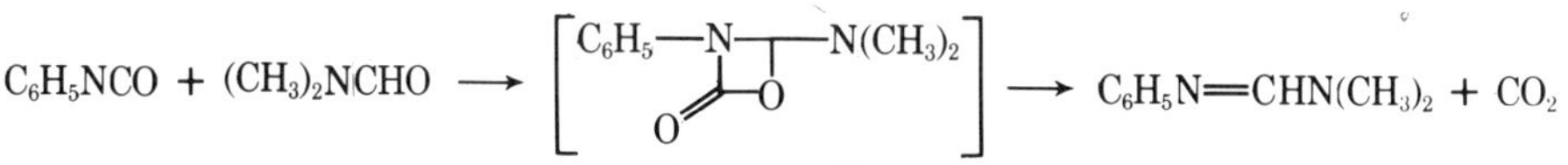

[*74631-11-1*]

and the hazard of CO_2 formation in a closed system indicate that DMF should not be used as a solvent for isocyanate reactions.

1,3-Diphenyluretidone, which is similar in structure to the isocyanate dimer, is formed by reaction with *N*-methyleneaniline.

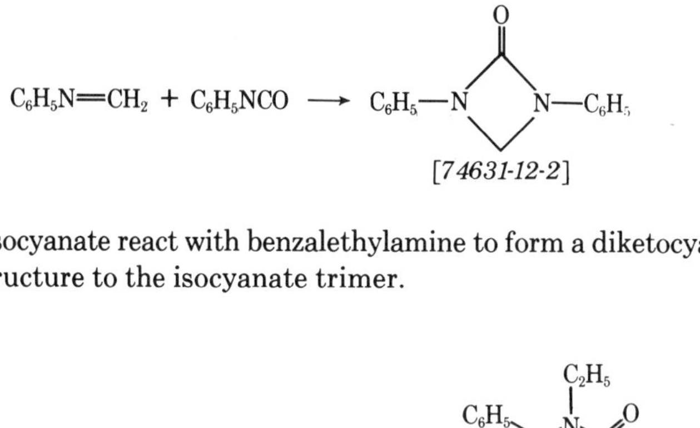

$$C_6H_5N{=}CH_2 + C_6H_5NCO \longrightarrow C_6H_5{-}N\text{(4-membered ring, C=O)}N{-}C_6H_5$$

[*74631-12-2*]

Two moles of isocyanate react with benzalethylamine to form a diketocyanidine, which is similar in structure to the isocyanate trimer.

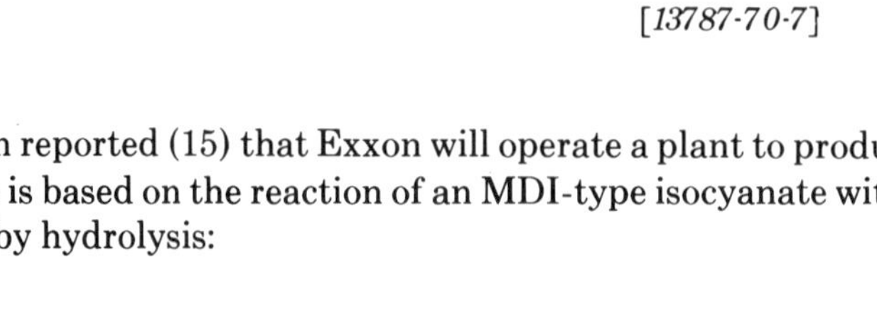

$$C_6H_5CH{=}NC_2H_5 + 2\ C_6H_5NCO \xrightarrow{200\,°C} \text{(six-membered triazine ring: N-}C_2H_5\text{, CH-}C_6H_5\text{, two N-}C_6H_5\text{, two C=O)}$$

[*13787-70-7*]

It has been reported (15) that Exxon will operate a plant to produce a polyparabanic acid that is based on the reaction of an MDI-type isocyanate with hydrocyanic acid, followed by hydrolysis:

$$OCN-C_6H_4-CH_2-C_6H_4-NCO + HCN \longrightarrow \left[-C_6H_4-CH_2-C_6H_4-N\!<\!\text{(ring: C=O, C=NH, C=O)}\!>\!N- \right]_x \xrightarrow{H_2O} \left[-C_6H_4-CH_2-C_6H_4-N\!<\!\text{(ring: C=O, C=O, C=O)}\!>\!N- \right]_x$$

The new materials are said to represent a family of new, high performance polymers that are stable at temperatures up to 200°C and that are 20–40% less expensive than polyimides (qv).

Both alkylene oxides and cyclic carbonate esters add to the isocyanate group with formation of oxazolidin-2-ones:

$$RN{=}C{=}O + R'CH\underset{O}{-}CH_2 \longrightarrow$$

$$RN{=}C{=}O + R'\text{-(cyclic carbonate)} \longrightarrow R'\text{-(oxazolidin-2-one, O, NR, C=O)} + CO_2$$

Extended or excessive heating of isocyanates result in condensation to a carbodiimide with elimination of carbon dioxide. A number of catalysts are effective in accelerating this reaction to the extent of making it a practical synthesis for both symmetrical carbodiimides and, from diisocyanates, polymeric carbodiimides. The phospholine oxides are particularly effective catalysts, although simple trialkylphosphine oxides or even triethyl phosphate may be used. The dimer is not an intermediate in the uncatalyzed reaction; in the presence of a phosphine oxide catalyst, a four-membered-ring intermediate is highly probable.

$$R_3PO + ArNCO \rightleftharpoons \begin{matrix} R_3P-O \\ | \quad\; | \\ ArN-C{=}O \end{matrix} \rightleftharpoons R_3\overset{+}{P}-\overset{-}{N}Ar + CO_2$$

$$\Big\updownarrow ArNCO$$

$$R_3PO + ArN{=}C{=}NAr \rightleftharpoons \begin{matrix} R_3P-NAr \\ | \quad\;\; | \\ O-C{=}NAr \end{matrix}$$

Foams that are based on the formation of carbodiimide have been of recent commercial interest and there are a number of carbodiimide-modified MDI products on the market. The composition of these products is rather complex, in addition to the formation of polycarbodiimides of varying molecular weight, the carbodiimide group equilibrates with free isocyanate to form polyfunctional uretoneimines:

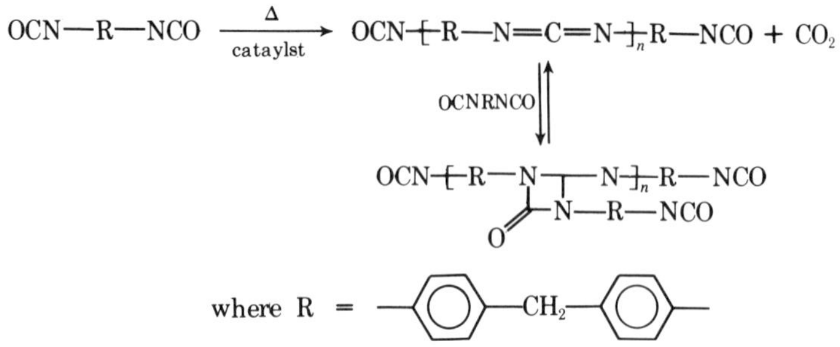

$$OCN-R-NCO \xrightarrow[\text{cataylst}]{\Delta} OCN\text{+}R-N{=}C{=}N\text{+}_n R-NCO + CO_2$$

$$\updownarrow \text{OCNRNCO}$$

$$OCN\text{+}R-N-N\text{+}_n R-NCO \quad (\text{ring: } C(=O)-N-R-NCO)$$

where R = —C₆H₄—CH₂—C₆H₄—

In contrast to the elimination of carbon dioxide in many of the reactions noted above, is the reaction of methyl isocyanate with carbon dioxide to form a six-membered ring.

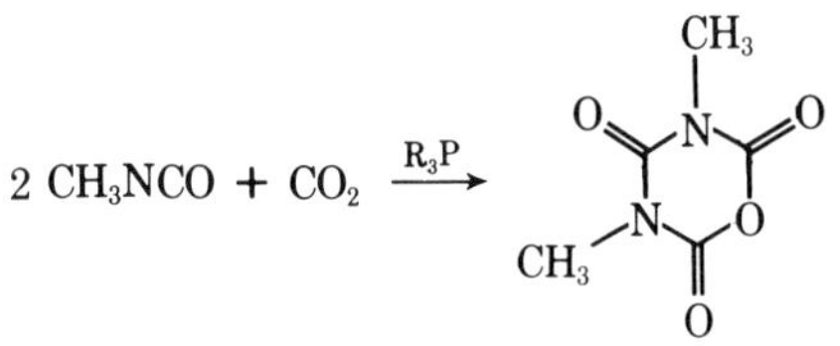

$$2\ CH_3NCO + CO_2 \xrightarrow{R_3P}$$ (six-membered ring with two N–CH_3 groups, one ring O, three C=O)

[36209-52-6]

Carbon disulfide or carbonyl sulfide has not been found to react in this manner.

Phenyl isocyanate is reported to add halogens to give a dihalide, which rearranges to a mixture of *o*- and *p*-halophenyl isocyanates.

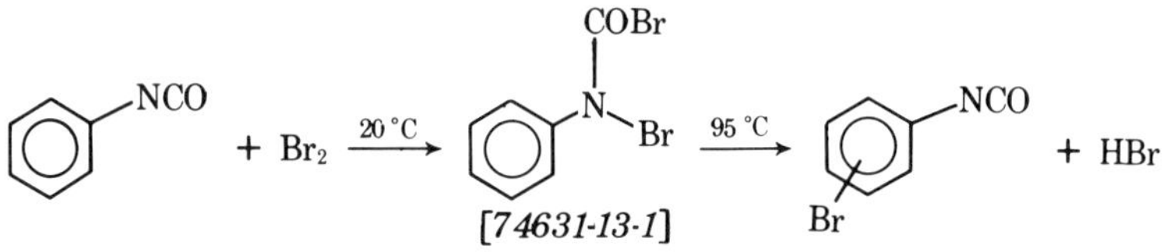

$$C_6H_5NCO + Br_2 \xrightarrow{20\ °C} C_6H_5N(Br)COBr \xrightarrow{95\ °C} BrC_6H_4NCO + HBr$$

[74631-13-1]

Isocyanates may be used as acylating agents in the Friedel-Crafts reaction to obtain amides.

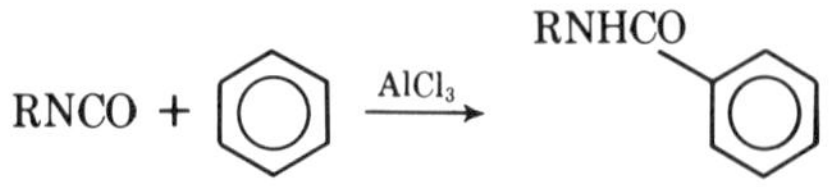

$$RNCO + C_6H_6 \xrightarrow{AlCl_3} RNHCOC_6H_5$$

The conversion of anhydrides to *N*-substituted imides by means of isocyanates is a fairly common reaction.

$$(CH_3CO)_2O + RNCO \rightarrow (CH_3CO)_2NR + CO_2$$

Isocyanates are reducible, but the reaction is not always straightforward. With hydrogen over a nickel catalyst at 190°C, aniline, carbanilide, methane, and carbon dioxide are obtained from phenyl isocyanate. Reduction with lithium aluminum hydride gives fairly good yields of *N*-methylaniline.

Synthesis

Many methods for the preparation of isocyanates have been reported. These can be classified according to the reaction involved, the most common method using the reaction between an amine, or its salt, and phosgene. Curtius, Hofmann, or Lossen rearrangements, and double-decomposition reactions also have been used widely. Many less important miscellaneous methods have been described.

Phosgene Reactions. In 1884, Hentschel obtained an isocyanate from the reaction between phosgene and the salt of a primary amine. Several modifications of this reaction have since been developed. This reaction is especially useful for high boiling isocyanates and diisocyanates, which may be prepared readily and in good yields by the reaction of a slurry of the amine hydrochloride with phosgene in toluene or dichlorobenzene. This method of preparation has been discussed (1,3–4) (see Manufacture).

Curtius, Hofmann, and Lossen Rearrangements. Other than the phosgene reaction, the method most commonly used is the Curtius rearrangement of an acid azide in a neutral solvent. A complete survey of such preparations that have been reported in the literature is given in ref. 16.

The Hofmann rearrangement of amides (17) and the Lossen rearrangement of hydroxamic acids (18) have been used similarly, though not as extensively as Curtius rearrangement.

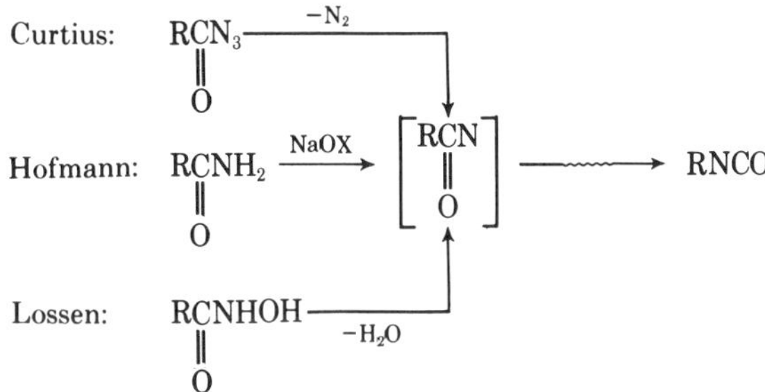

Metathesis. Wurtz used the reaction between organic halides or sulfates and salts of cyanic acid for the first preparation of alkyl isocyanates in 1849. This preparation has received more attention recently (19) because of interest in methoxymethyl isocyanate [*6427-21-0*] as a modifier for extending or cross-linking polymers. By building this group into a polymer, the molecular weight can be increased by heating to effect condensation of the methoxyl group and an active hydrogen:

$$CH_3OCH_2Cl + KCNO \longrightarrow CH_3OCH_2NCO$$

$$\sim\!OC(=O)NHCH_2OCH_3 + \sim\!C(=O)\!-\!N(H)\!\sim \xrightarrow{\Delta} \sim\!C(=O)\!-\!N(CH_2NHC(=O)O\!\sim)\!\sim + CH_3OH$$

From Isocyanate Derivatives. Many alkyl and aryl carbamates and arylureas may be pyrolyzed at temperatures from 135 to 500°C to regenerate the isocyanate from which they might be considered to have been derived (7,11). In some cases, the decomposition is facilitated by the presence of P_2O_5, $COCl_2$, or HCl. As mentioned above, the aryl esters of carbamates are more susceptible to decomposition of exchange reactions than are the alkyl esters.

$$\text{RNHCOOAr} \xrightarrow[\Delta]{} \text{RNCO} + \text{ArOH}$$

$$\text{ArNH}\underset{\underset{\text{O}}{\|}}{\text{C}}\text{NHAr} \xrightarrow[\Delta]{\text{COCl}_2} 2\ \text{ArNCO} + 2\ \text{HCl}$$

$$\text{RNH}\underset{\underset{\text{O}}{\|}}{\text{C}}\text{NHR} \xrightarrow[\Delta]{\text{HCl}} \text{RNCO} + \text{RNH}_2\cdot\text{HCl}$$

From Isocyanic Acid. Formaldehyde and a number of carbonyl compounds bearing electronegative substituents in the α position add to isocyanic acid at temperatures of −70 to 0°C to form α-hydroxy isocyanates (20). At higher temperatures, these isocyanates polymerize and sometimes do so with explosive violence. Many may be used as intermediates to obtain urethanes or urea derivatives. Addition of isocyanic acid to olefins has been accomplished giving fair yields, especially when the olefin is activated, as in the case of isopropenylbenzene or vinyl ethers. Toluenesulfonic acid is used as catalyst.

$$\text{Cl}_3\text{CCHO} + \text{HNCO} \xrightarrow{0\,^\circ\text{C}} \text{Cl}_3\text{C}\underset{\underset{\text{OH}}{|}}{\text{C}}\text{HNCO}$$

$$\text{ROCH{=}CH}_2 + \text{HNCO} \xrightarrow{\text{CH}_3\text{C}_6\text{H}_4\text{SO}_3\text{H}} \text{RO}\overset{\overset{\text{CH}_3}{|}}{\text{C}}\text{HNCO}$$

Other Preparations. Fluorinated alkyl isocyanates have been prepared from nitriles and carbonyl difluoride

$$\text{CF}_3\text{CN} + \text{COF}_2 \xrightarrow[300^\circ\text{C}]{\text{CsF}} \text{CF}_3\text{CF}_2\text{NCO}$$

[356-74-1]

Acyl isocyanates may be prepared by metathesis of the acyl halide and a metal cyanate or, in excellent yield, from the halide and isocyanic acid in the presence of a hydrogen chloride acceptor (21).

$$\text{RCOCl} + \text{HNCO} + \text{C}_6\text{H}_5\text{N} \rightarrow \text{RCONCO} + \text{C}_6\text{H}_5\text{N.HCl}$$

The sulfonyl isocyanates (22) may be prepared by the reaction of the sulfonamide and phosgene under conditions similar to those employed with aromatic amines. The sulfonyl isocyanates react even more rapidly with active hydrogen compounds than the simple aromatic isocyanates. The sulfonamide that is obtained by hydrolysis does not react with more sulfonyl isocyanate to form the substituted urea. Because of the

SO_2NH_2–C$_6$H$_4$–NH_2 $\xrightarrow[150\,°C]{COCl_2}$ SO_2NCO–C$_6$H$_4$–NCO $\xrightarrow{H_2O}$ SO_2NH_2–C$_6$H$_4$–NCO [1773-42-8] $\xrightarrow{H_2O}$ $(H_2NO_2S$–C$_6$H$_4$–NH$)_2$CO

lowered basicity of the nitrogen atom in the sulfonyl isocyanates, they show a lowered tendency to add hydrogen halides. *p*-Isocyanatobenzenesulfonyl isocyanate [*1773-41-7*] is an interesting compound, containing both the simple isocyanate and sulfonyl isocyanate groups (23).

Preparation of isocyanates of aryl esters of phosphorus acids and silicic acids is reviewed in ref. 24. Organophosphorus isocyanates can be prepared from the corresponding phosphorus halide and an alkali metal cyanate (25); alkyl silicon isocyanates can be prepared from the halide and silver cyanate (26).

The trialkylmetallo isocyanates of tin, antimony, and arsenic may be obtained in high yield by fusion of the oxide with urea (27).

$$R_3SnOH + H_2NCONH_2 \xrightarrow{130°C} R_3SnNCO + NH_3 + H_2O$$

$$R_3AsO + 2\,H_2NCONH_2 \rightarrow R_3As(NCO)_2 + 2\,NH_3 + H_2O$$

Ferrocenyl isocyanate [*74930-85-1*] has been prepared by rearrangement of the azide (28).

$$\text{Fc–CON}_3 \xrightarrow{-N_2} \text{Fc–N{=}C{=}O}$$

Manufacture and Processing

World production of isocyanates in 1978 has been estimated as 635,000 metric tons of toluene diisocyanate (TDI) and 454,000 t diphenylmethane-4,4′-diisocyanate (MDI) products. A list of isocyanates that are available commercially in the United States is presented in Table 2. Estimated 1981 capacities of U.S. manufacturers of isocyanates are listed in Table 3.

Phosgenation. The reaction of amines with phosgene (phosgenation) has, for economic reasons, been used almost exclusively for the manufacture of isocyanates. The details of processing may vary with the specific isocyanate and, in particular, for aromatic and aliphatic isocyanates, but the general approach is the same. Because of the several side reactions and associated complications, the development of practical, high yield reaction conditions has been studied extensively. The primary reactions involved in the phosgenation of a simple amine are indicated below (see also Amines, aromatic; Phosgene).

$$RNH_2 + COCl_2 \xrightarrow[-20\text{ to }60\,°C]{\text{solvent}} [RNHCOCl + HCl] \xrightarrow{100\text{–}200\,°C} RNCO + 2\,HCl$$

$$[RNHCOCl + HCl] \xrightarrow{RNH_2} RNH_2HCl \xrightarrow{COCl_2} [RNHCOCl + HCl]$$

$$[RNHCOCl + HCl] \xrightarrow{RNH_2} RNHCONHR; \quad RNCO \xrightarrow{RNH_2} RNHCONHR$$

Table 2. Isocyanate Products Available in the United States

Product composition	Producer[a]							
	Allied	BASF	Dow	Hüls	Mobay	Olin	Rubicon	Upjohn
80/20 mixture of 2,4- and 2,6-toluene diisocyanate (TDI)	Nacconate	TDI	Voranate T-80		Mondur TD-80	TDI-80	Rubinate TDI	
2,4-toluene diisocyanate					Mondur TDS			
65/35 mixture of 2,4- and 2,6-toluene diisocyanate					Mondur TD			
modified toluene diisocyanates	Nacconate 5050				Mondur MT-40			
diphenylmethane-4,4′-diisocyanate (MDI)					Mondur M			Isonate 125M
partially carbodiimidized diphenylmethane-4,4′-diisocyanate					Mondur CD			Isonate 143-L
polymethylene polyphenyl isocyanate (PMPPI)					Mondur MR Mondur MRS		Rubinate M	PAPI; PAPI 20; PAPI 135
4,4′-dicyclohexylmethane diisocyanate					Desmodur W			
tolidine diisocyanate								Isonate 136T
phenyl isocyanate					Mondur P			
o-, *m*-, and *p*-chlorophenyl isocyanates					Mondur PCPI			
3,4-dichlorophenyl isocyanate					[b]			
methyl and ethyl isocyanates					[b]			
n-butyl isocyanate								[b]
cyclohexyl isocyanate					[b]			[b]
octadecyl isocyanate					Mondur O			
hexamethylene diisocyanate					Mondur HX			
biuret of hexamethylene diisocyanate					Desmodur N			
3-isocyanatomethyl-3,5,5-trimethylcyclohexyl isocyanate (IPDI)				[b]				

[a] Allied: Allied Chemical Corporation, Specialty Chemicals Div.; BASF: BASF Wyandotte Corporation, Urethane Division; Dow: The Dow Chemical Co., Organic Chemical Dept.; Hüls: Chemische Werke Hüls A.G.; Mobay: Mobay Chemical Corporation, Polyurethane Div.; Olin: Olin Corporation, Design Products Division; Rubicon: Rubicon Chemicals, Inc.; Upjohn: The Upjohn Company, Polymer Chemicals Division.

[b] Offered by chemical name, no trade names known.

Table 3. Estimated 1981 Nameplate Capacities of U.S. Manufacturers of Isocyanates, 10^3 t

Isocyanate	Allied	Arco	BASF-Wyandotte	Dow	Mobay	Olin	Rubicon	Upjohn	*Total*
TDI	36		57	45	98	64	18		*318*
polymeric		68	68		136		45	127	*444*

Problems increase in the manufacture of a diisocyanate, where the simple by-products may be intramolecular (eg, a mixed carbamyl chloride/amine hydrochloride) and the urea by-product may be polymeric.

It appears that all commercial manufacturing processes for aromatic isocyanates in use to date have the following approach: the solution of an amine in an aromatic solvent, such as xylene, monochlorobenzene, or orthene (*o*-dichlorobenzene) is mixed with a solution of phosgene in the same solvent at a temperature below 60°C; the resulting mixture slurry is digested in one to three stages for several hours at progressively increasing temperatures up to 200°C and is accompanied by the injection of additional phosgene; and the final solution of reaction products is fractionated to recover hydrogen chloride, unreacted phosgene and solvent for recycling, isocyanate product, and distillation residue.

A schematic flow sheet for the production of toluene diisocyanate is presented in Figure 1 (12,29–32). Batch processing, which provides similar conditions of solvent, temperature, and time, may be employed for the production of isocyanates on a scale that is too small to justify continuous operation.

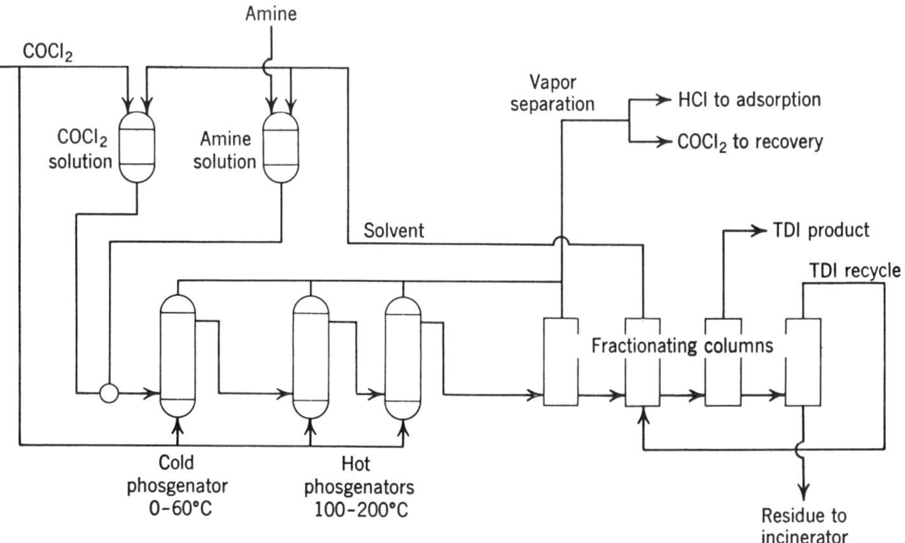

Figure 1. Schematic flow sheet for the production of toluene diisocyanate.

The overall chemistry of an integrated facility for the manufacture of 80/20 TDI is illustrated by the following equations:

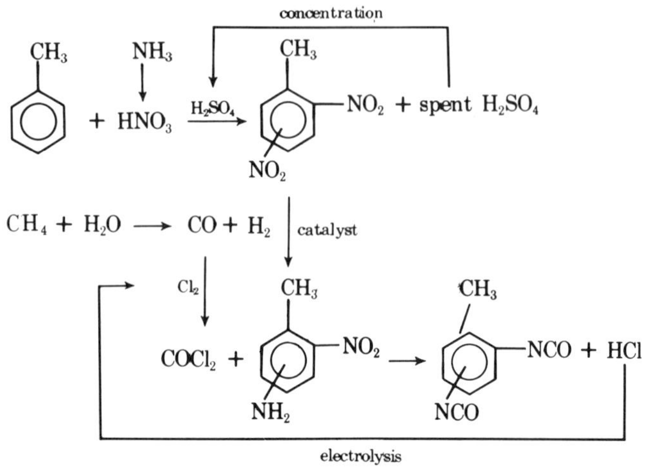

concentration

CH_3 (toluene) + HNO_3 (from NH_3) $\xrightarrow{H_2SO_4}$ CH_3, NO_2, NO_2 (dinitrotoluene) + spent H_2SO_4

$CH_4 + H_2O \longrightarrow CO + H_2$ catalyst

Cl_2

$COCl_2$ + CH_3, NO_2, NH_2 $\longrightarrow$ CH_3, NCO, NCO + HCl

electrolysis

80/20 TDI

Toluene is nitrated with mixed sulfuric and nitric acids by manufacturing procedures that are similar to those employed for the production of TNT (trinitrotoluene) except that less acid and lower temperatures limit the degree of nitration. Both the Biazzi and Meissner nitration processes are used in the United States. The nitration may be effected stepwise to obtain first the mononitrotoluenes, which can be separated by distillation. Further nitration of only the ortho isomer leads to a DNT (dinitrotoluene) mixture of ca 65/35 ratio of 2,4 and 2,6 isomers. The majority of production is effected without separation of the MNT (mononitrotoluene) isomers, leading to the 80/20 ratio of isomers (see Explosives; Nitration).

Carbon monoxide is generated by any one of several approaches, eg, steam reforming of natural gas to produce carbon monoxide and hydrogen. The hydrogen is employed for reduction of the DNT to the diamine. It is probable that both noble metal- and Raney nickel-catalyzed, liquid-phase reduction are employed. The carbon monoxide is combined with chlorine to obtain phosgene. The two gases are metered carefully to provide a slight excess of carbon monoxide beyond an equimolar ratio and are passed through ca 5-cm-diameter iron tubes that are packed with carbon as catalyst; the resulting phosgene vapor is condensed to liquid phosgene or is absorbed in solvent. A solution of the diamine reacts with phosgene to obtain the diisocyanate (see Fig. 1).

Two minor complications of this process scheme deserve mention. The presence of ca 3% 2,3- and 3,4-diaminotoluene leads to significant yield loss as a result of formation of the corresponding benzimidazoles, whose amide hydrogens can react with TDI and reduce the yield. However, if the phosgenation conditions are sufficiently drastic, the ortho diisocyanates may be formed and be present in the product where they polymerize to insoluble Nylon-1-type products and result in a turbid TDI (13).

Disposition of the distillation residue, which has a significant vapor pressure of TDI, requires some care. There are several patents reporting hydrolysis of the residue to recover some value as toluenediamine (TDA) (33), or it may be incinerated while hot and liquid.

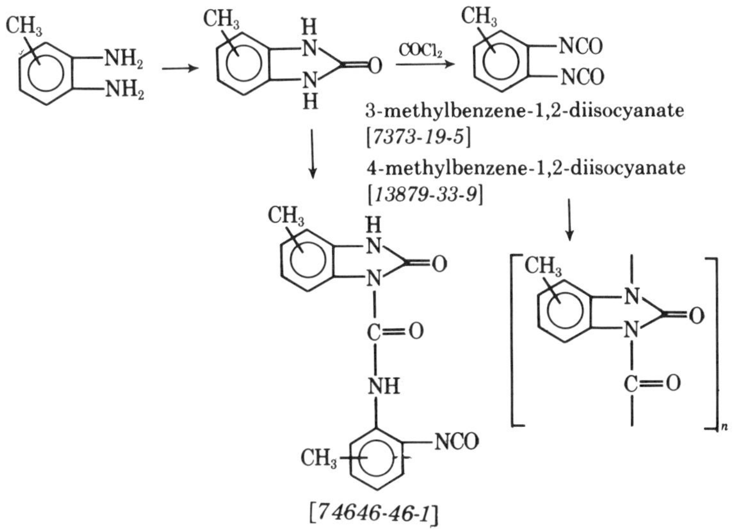

3-methylbenzene-1,2-diisocyanate
[7373-19-5]
4-methylbenzene-1,2-diisocyanate
[13879-33-9]

[74646-46-1]

Although TDI represents over half of aromatic isocyanate production, it is predicted that the growth of rigid foam products will result in the 4,4′-methylenebis-(phenyl isocyanate) (MDI) group overtaking TDI in volume. The large-volume item of the group is the polymeric or polymethylene polyphenyl isocyanates. A variety of this type of product may be obtained by phosgenation of amines that are derived from the condensation of aniline and formaldehyde in varying ratios and under varying conditions. If a large excess of aniline and an equivalent amount of hydrochloric acid is used, the condensation product will contain up to 85–90% of bis(4-aminophenyl)-methane (MDA). Decreasing the excess of aniline results in increasing amounts of polyamines. The primary commercial isocyanates, eg, Mobay's Mondur MR or Upjohn's PAPI are derived from amines of ca 50% diamine and an average composition approaching $n = 1$.

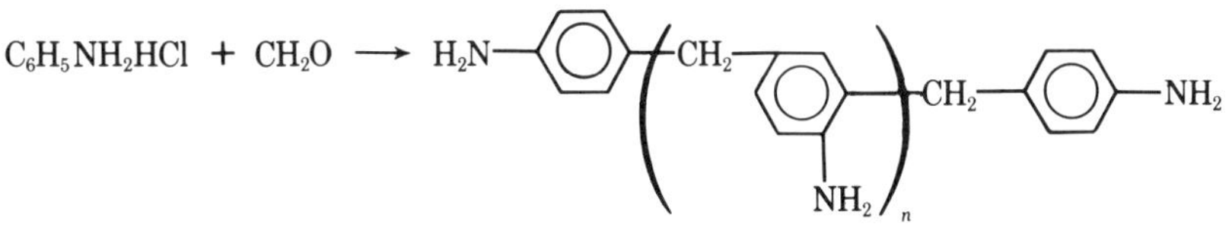

$C_6H_5NH_2HCl + CH_2O \longrightarrow$

The diisocyanate content of such product is over 95% 4,4′ isomer. If a weak acid is used as the catalyst in the aniline–formaldehyde condensation, higher proportions of 2,4′, with a little 2,2′ isomer, result. Thus, in Mondur MRS, the diisocyanate is ca 65% 4,4′ isomer.

The phosgenation conditions for these isocyanates are quite similar to those for TDI, but there is one very significant difference in the process—the product is not distilled. This makes adherence to good process control mandatory if product consistency is to be expected.

Pure MDI enjoys a significant market, if small by comparison to the polymerics. Although it may be produced from a high diamine-content condensation product, it is probable that all producers obtain it by distillation topping of an amine or isocyanate from an amine that has been produced to provide both diisocyanate and polymeric isocyanate. For example:

$$\left.\begin{matrix}\text{60 diamine}\\+\\\text{40 polyamine}\end{matrix}\right\}\rightarrow\begin{matrix}\text{60 diisocyanate}\\+\\\text{40 polyisocyanate}\end{matrix}\rightarrow\begin{matrix}\text{20 diisocyanate}\\+\\\text{40 diisocyanate}\\+\\\text{40 polyisocyanate}\end{matrix}$$

In the case of both polymeric and pure MDI, the process must be such that the product is quench-cooled. Although the mixture of polymeric isocyanates dimerizes less rapidly than MDI alone, slow cooling of either can result in the dimer content exceeding its solubility in the isocyanate (ca 1%). MDI usually is stored and shipped in the molten state for volume applications both for convenience and improved stability.

In the manufacture of aliphatic isocyanates, eg, hexamethylene diisocyanate, the amine frequently is protected by the formation of a salt before phosgenation (34–35). This tends to reduce the formation of urea. Otherwise the process is similar to that employed for aromatic isocyanates.

There are five aliphatic isocyanates of commercial significance, although none approach the volume of TDI or the polymerics. Methyl isocyanate is produced by Union Carbide. Several insecticides and herbicides (qv) are derived from this isocyanate, including the widely used 1-naphthylcarbamate sold under the tradename, Sevin (see Insect control technology). It is estimated that 12,000–14,000 metric tons of methyl isocyanate was produced in the U.S. in 1975 and 24,000 t of Sevin. About half of the Sevin was exported. Growth to 23,000 t of the isocyanate is predicted for 1980. Methyl isocyanate is believed to account for about three-quarters of all monoisocyanates manufactured.

There has been interest in aliphatic or light stable diisocyanates for many years, but price–volume considerations have limited the number of products that are available. Several companies have been interested in producing xylylene diisocyanate, and it has been offered by Takeda in Japan. Interest in the United States has been based upon an ammoxidation route from xylene, but the volume that is necessary to permit low cost has not developed. The same is true of the corresponding bis(isocyanatomethyl)cyclohexane which has been offered for some years by Tennessee Eastman and, more recently, has been offered in sample quantities by Sun Petroleum Products Co. This diisocyanate provides coatings with excellent properties but the volume is not adequate to achieve low cost. Light stability is not required in the large-volume, polyurethane foam application.

$OCNCH_2-C_6H_4-CH_2NCO$

p-xylylene diisocyanate
[*25854-16-4*]

$OCNCH_2-C_6H_{10}-CH_2NCO$

(1,4-bisisocyanatomethyl)cyclohexane
[*10347-54-3*]

Hexamethylene diisocyanate has been available for many years but has not reached large volume usage as such for similar reasons: the advantages gained do not justify the price, and the high vapor pressure poses a handling hazard. However, a low-monomer-content biuret mixture derived from hexamethylene diisocyanate has achieved volume application despite high cost, because it permits the formulation of exceptionally high quality coatings. The biuret may be generated by controlled reaction of the isocyanate with water, a water generator, or an amine:

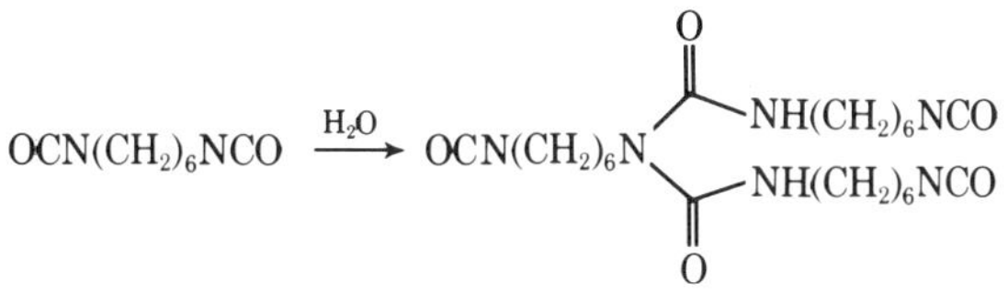

$$OCN(CH_2)_6NCO \xrightarrow{H_2O} OCN(CH_2)_6N\begin{matrix} \overset{O}{\overset{\|}{C}}NH(CH_2)_6NCO \\ \underset{\underset{O}{\|}}{C}NH(CH_2)_6NCO \end{matrix}$$

biuret of hexamethylene diisocyanate

The isocyanate is obtained by phosgenation of the amine salt (probably the CO_2 salt, or polycarbamate),

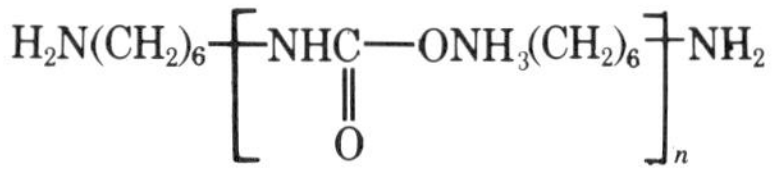

$$H_2N(CH_2)_6\left[NH\underset{\underset{O}{\|}}{C}—ONH_3(CH_2)_6\right]_nNH_2$$

Isophorone diisocyanate, IPDI (5), has been more successful than hexamethylene diisocyanate because it is lower in cost. However, it also has a high vapor pressure with attendant industrial hygiene problems. To overcome this problem, Hüls has introduced a trimerized IPDI as Isocyanurate T 1890. Bayer offers a similar product as Desmodur Z-4370 and a trimerized hexamethylene diisocyanate as experimental product KL5-2444.

The most successful aliphatic diisocyanate to date (1979) has been bis(4-isocyanatocyclohexyl)methane (4,4′-dicyclohexylmethane diisocyanate, Table 1). Several companies were active in developing this product, but DuPont's Hylene W won the market based on economics and the physical state of the product. The amine is obtained by hydrogenation of MDA. The hydrogenated amine is a cis–trans mixture, from which a high trans–trans fraction is separated and used to make the fiber, Qiana. The remaining amine, which has a higher cis content, yields a liquid isocyanate mixture when phosgenated. Thus, Hylene W has the advantages of the economy of being derived from a by-product amine, relatively high molecular weight and low vapor pressure, liquid form, and the ability to confer excellent properties to coatings that are developed from it. With DuPont's decision to curtail production of isocyanates, production of this isocyanate has been taken over by Mobay, and the product is now sold as Desmodur W.

Direct Conversion of Nitro Compounds. Between 1966 and 1979 nearly two hundred patents and literature references appeared which related to preparative schemes for the direct conversion of nitro compounds to isocyanates of urethanes by reaction with carbon monoxide.

One TDI production process offers the advantage that neither chlorine or phosgene are involved, obviating disposal of by-product HCl (36). The disadvantages are the use of noble-metal catalysts, eg, $PdCl_2$, moderate yields, and low conversions. Typical reaction conditions are 7–14,000 kPa (1–2000 psig) of CO_2 at 200°C for 2–4 h.

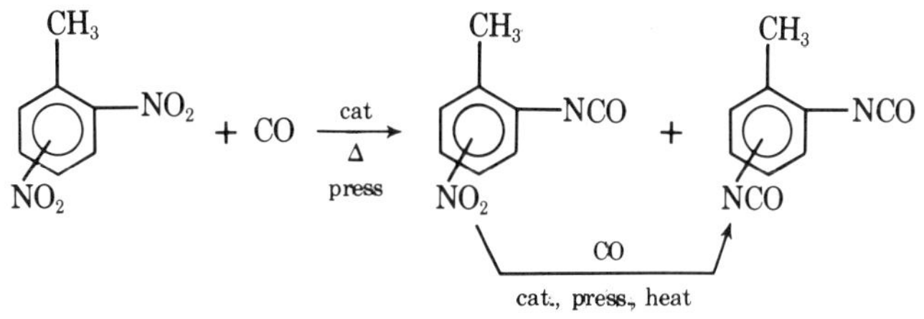

$$CH_3C_6H_3(NO_2)_2 + CO \xrightarrow[\text{press}]{\text{cat, }\Delta} CH_3C_6H_3(NCO)(NO_2) \xrightarrow[\text{cat., press., heat}]{CO} CH_3C_6H_3(NCO)_2$$

The investment for a TDI plant by carbonylation are slightly lower than the phosgenation route but, within margin of error, the net production cost is estimated at \$1.03/kg by each route (32).

The catalyzed reaction of aromatic nitro compounds with carbon monoxide in alcohol solution to form the urethane provides high yields under the proper conditions and the thermal splitting of the urethane to isocyanate also is a high yield reaction (32). Atlantic Richfield (ARCO) has announced plans for commercial production of polymeric isocyanate by this route to be on stream in 1981 (32,37). Although palladium compounds also are effective, selenium appears to be the catalyst of choice.

The advantages of this route are the same as mentioned above. The disadvantages are the problems involved in handling and recovering the toxic selenium catalyst, a comparable degree of processing to that by phosgenation, and the probability that the product will not be identical to and interchangeable with those of the phosgenation producers. Comparative economic estimation (32) of this process route suggests a slightly higher capital investment but a slightly lower net production cost.

$C_6H_5NO_2 \xrightarrow[\text{Se, 690 kPa (1000 psi) 1 hr}]{\text{CO, ROH}} C_6H_5NHCOOR \xrightarrow{CH_2O}$ ROOCHN—C_6H_4—CH_2—C_6H_4—NHCOOR + polyurethanes

$\xrightarrow{\Delta}$ OCN—C_6H_4—CH_2—C_6H_4—NCO + ROH

MDI [*101-68-8*]
+ polyisocyanates

Economic Aspects

It would appear that even with the slowed growth of flexible foam and TDI, U.S. demand must be approaching U.S. capacity. This is reflected in price increases in TDI to a level that may justify the addition of new capacity. The production of TDI in Western Europe is probably 80–90% of that in the United States, that of Japan is ca 10%. On the other hand, the higher growth rate for rigid foam has attracted new entries to the market. The new facilities projected by ARCO and BASF-Wyandotte and expansions by Mobay should satisfy the demand at least into the mid-1980s. The growth rates generally accepted for TDI and polymerics in the late 1970s are ca 5–7% and 12–15%, respectively (38). Bulk prices (\$/kg) of TDI and polymeric isocyanates, respectively, have been reported: 1965, 0.70 and 0.70; 1970, 0.77 and 0.77; 1975, 0.88 and 0.95; and 1980, 1.61 and 1.65.

Specifications and Analytical Methods

Typical analyses for several commercial isocyanates and the ASTM specifications (39) of 80/20 toluene diisocyanate are given in Table 4.

In the last few years, vapor-phase (gas–liquid) chromatography (glc) (41) has

Table 4. Quality of Several Commercial Isocyanates[a]

Property	ASTM Specification for Type II TDI[b]	Typical analyses Mondur TD-80	Mondur M	Mondur MR	Mondur MRS	Mondur HX
NCO, %		48.3	33.6	31.5	31.5	50.0
purity, %, min[c]	99.5	99.7	99.5			99.5
2,4′ isomer, %[c]	80 ± 2	80 ± 1				
2,6 isomer, %[c]	20 ± 2	20 ± 1				
diisocyanate, %[c]		100	100	50	55	
4,4′ isomer, %[c]				95[d]	65[d]	
color (max), APHA	50	10	50	dark brown liquid		15
total chloride (max), %[c]	0.2	0.03	0.05	0.4	0.6	0.1
hydrolyzable chloride, %[c]	0.01 max	0.005[e]	0.002	0.05	0.05	0.02
acidity (as HCl), %[c]		0.003–0.02[e]	0.002	0.05	0.05	0.02
viscosity, 25°C, mPa·s (= CP)				200 ± 50	200 ± 50	
freezing point, °C		14	38	<0	<0	−67
hexane insolubles, %			0.5[f]			

[a] Ref. 42.
[b] Type I and III are identical except for 2,4 isomer contents of 97.5% min and 65 ± 2%, respectively.
[c] Ref. 43.
[d] Refers to the 4,4′-isomer content of the diisocyanate content.
[e] Adjusted to customer requirements.
[f] A measure of urea and dimer content; higher in flake or cast material.

proved to be a most valuable test for product control. In glc application to isocyanates, silicone substrates are stable and generally applicable (42). In applying gas chromatographic methods to isocyanate products containing isocyanate reaction products, an injection block temperature must be chosen such that the monomeric isocyanate is reproducibly volatilized but the reaction products are not decomposed to generate isocyanate. High pressure liquid chromatography and gel-permeation chromatography are useful in separating and analyzing mixtures of isocyanates and isocyanates mixed with their condensation products.

Infrared examination is another useful tool in work with isocyanates and derived products. Table 5 presents the designation of absorption bands used as the reference in one laboratory (43). A tabulation of near infrared absorptions also is available (42). Such reference tables must be used with care and some knowledge of the inferences and shifts that may be encountered.

The rapidly expanding commercial use of isocyanates and tighter governmental regulation of possible health considerations has resulted in increasing attention to the development of simplified, more sensitive methods for determining low levels of isocyanate vapors in air. In the Marcali method (44), detection of TDI to 0.01 ppm is possible. The method is not applicable to aliphatic isocyanates. A thin-layer chromatographic method was developed for analysis of aliphatic and aromatic isocyanates and aerosols of polyisocyanates (45). A lower limit of 0.01 ppm is claimed for TDI based on measurement of a 10-L sample of air. Precision is ca ±20%. This method has been adapted to high speed liquid chromatography (46) using a hexane–ethanol solvent program, a uv detector and commercial silica packing. If the method is applied to a 20-L air sample, TDI levels as low as 0.7 ppb can be detected with a precision of ca

Table 5. Infrared Absorption of Isocyanate Derivatives[a]

Functional group	Most intense bands: Wave number, cm^{-1}	Wavelength, μm
—NCO	2274–2242	4.40–4.46
—N=C=N—	2152–2128	4.64–4.70
—C=O in uretidinedione	1783–1760	5.60–5.68
—C=O in isocyanurate	1713–1685	5.83–5.93
—C=O in urethane	1718–1690	5.82–5.91
	1430–1410	7.0–7.1
—C=O in uretoneimine	1878–1870	5.32–5.35
—C=O in ureas	1685–1625	5.93–6.15
—C=O in allophanate	1731–1718	5.78–5.82
	1692–1683	5.91–5.94
—C=O in biuret	1708–1670	5.85–5.99
	1674–1642	5.97–6.09
	1592–1523	6.28–6.56
—C=O in esters	1750–1715	5.71–5.83
—C=N in carbodiimide dimer	1680	5.95
—CN in carbodiimide trimer	1670–1665	5.98–6.0
	1235–1205	8.1–8.3
amide II band (NH)	1525	6.5
CO aliphatic ether	1110–1075	9.0–9.3
CO, stretch and/or OH deformation	1234	8.10
C=C, skeletal stretch of benzene ring	1613–1600	6.20–6.25
CCH_3, (symmetrical) deformation	1385–1370	7.22–7.30

[a] Ref. 46.

±10%. This detection limit can be lowered by a factor of ca 50 (47) and determination can be made more precise (48). Attention also is being directed toward simplified means of collecting air samples so as to avoid the use of liquid absorbents in impingers (49).

Instruments to monitor the isocyanate level in the work place or to monitor the exposure of personnel are commercially available from MDA Scientific, Inc. Consideration must be given to adsorption of the isocyanates on glass or plastic sampling lines. Such lines should be minimized, especially if accurate results are to be expected for spot or single-breath samples.

Health and Safety

Several useful discussions of isocyanate toxicity and handling are available (50–52) and should be consulted before undertaking work with these chemicals. Detailed discussion of the biologic effects of exposure is given in ref. 51.

All isocyanates are potentially hazardous and require care in handling. They are lacrimators and may have a mild tanning action on the skin, but the primary health effect is respiratory irritation caused by isocyanate vapors. In 2–5% of a population that is exposed to low concentrations of isocyanate vapor, a hypersensitive, asthmalike allergic reaction may result. Skin allergies have been observed but are not common. In general, the lower the vapor pressure of the isocyanate, the fewer the problems in its use, but adequate ventilation should always be employed. Even with low vapor-pressure products, consideration must be given to air-borne aerosols when spray application is made of foams or coatings. Spills of isocyanates should be decontaminated promptly with care, ie, using respiratory protection and rubber or PVC protection of the skin. Both solid absorbents (1,49) and liquid (51) formulations are used to neutralize isocyanate spills.

The level of toluene diisocyanate in air detectable by odor is from 0.1–1 ppm. This is considerably above the current Federal standards, which requires that employees shall not be exposed to more than an 0.02 ppm ceiling limit. NIOSH has recommended a maximum exposure for all diisocyanates of a 10-minute ceiling of 20 ppb and a time-weighted average for a 10-hour day/40-hour week of 5 ppb. The $\mu g/m^3$ equivalents of the latter levels for several isocyanates are shown in Table 6 (52).

The hypersensitive asthmatic reaction of some people has instigated two major studies of the chronic effects and the mechanism of action of exposure to low levels of isocyanate vapors.

A five-year study of workers in a new TDI manufacturing plant has been completed (53). Clinical hypersensitivity was noted in 4.3% of those studied. The longitudinal study showed annual declines in pulmonary function were significantly related to TDI exposure. However, cross-sectional studies showed that annual declines in high exposure categories were not significantly different from those predicted for the general

Table 6. Recommended Time-Weighted Average and Maximum Exposure to Diisocyanates, $\mu g/cm^3$ [a]

Diisocyanate	TWA (5 ppb)	Ceiling (20 ppb)
toluene diisocyanate (TDI)	35	140
diphenylmethane diisocyanate (MDI)	50	200
hexamethylene diisocyanate (HDI)	35	140
isophorone diisocyanate (IPDI)	45	180
dicyclohexylmethane diisocyanate (H-MDI)	55	210

[a] Ref. 55.

Table 7. Selected Animal Toxicity Data [a]

Isocyanate	LD_{50} (oral, rats), g/kg	LD_{50} (skin, rabbits), g/kg	LC_{50} (air, rats) mg/m^3 (h)	Eye injury, rabbits	Subacute oral toxicity, rats: daily dose for 10 d, g/kg	Subacute oral toxicity, rats: deaths/no. fed
phenyl	0.94[b]	3.5–4.4		severe	0.19	1/6
2,4-TDI	4.9–6.7	16	350 (4)	severe	1.5	3/6
MDI	31.6[b]	10[b]	370 (4)	slight		
butyl	0.85[b]	6–8[b]		severe and permanent		
octadecyl	30	12		moderate	6.1	0/6
IPDI	2.5	1.0	123 (4)	slight		
HDI	1.05	1.25	310 (4)	severe		
Desmodur N	19.8	15.8	425[c]	moderate		

[a] Refs. 55, 60.
[b] In corn oil.
[c] Desmodur N-75 solution.

population. It was also noted that study of the worker following removal from the TDI-containing area suggests that TDI reactivity is reversible when isocyanate exposure is stopped. Overall, the study does not show the abnormally high pulmonary decrements reported in several papers (54). These investigators suggest that the most likely mechanism of TDI asthma is pharmacologic.

A series of papers have been published in which evidence is reported that the allergenic reaction to isocyanates is immunological. A tolyl isocyanate antigen was developed and incorporated into a radioimmunoassay for detection of IgE antibodies in workers with TDI hypersensitivities (55). The development of hypersensitivity in guinea pigs exposed to tolyl and hexyl isocyanate-ovalbumin aerosols was demonstrated, and most recently, the sensitization of guinea pigs to TDI itself by detection of tolyl-specific antibodies (56).

That the oral toxicity of the isocyanates is relatively mild may be noted from the data listed in Table 7. No reports are known (52) that indicate production of carcinogenic, teratogenic, or reproductive effects in humans or animals by diisocyanates.

Exposure to moisture leads to formation of carbon dioxide and the development of pressure in closed containers. This can result in rupture of metal and glass containers. Polyethylene or polypropylene containers of TDI may harden and rupture, presumably because of diffusion of moisture through the plastic with deposition of urea in the plastic and development of CO_2 pressure in the container.

Uses

For a detailed discussion of polyurethane products and chemistry, see refs. 2, 6, 8–9, 58–59 (see Urethane polymers; Foamed plastics).

Development by Farbenfabriken Bayer A.G. of a diisocyanate–polyester system led to the production of a foam that is similar to rubber–latex foam and provided the impetus for commercialization of both isocyanates and foams (see Rubber). The foams that contributed significantly to the initial commercialization of isocyanates originally were based on the reaction of a diisocyanate, a hydroxyl-terminated polyester, and water. The carbon dioxide liberated by the reaction of water with isocyanate served to foam the polymerizing mass; the amine, also formed by reaction of isocyanate and water, was incorporated into the polymer by way of urea groups. Toluene diisocyanate was used universally, diols and triols were used for flexible foams, and the more highly functional polyols were used for rigid foams. In 1958–1959, polypropylene glycols largely displaced the polyesters as the backbone of the polymers, thereby decreasing the raw material cost and expanding the volume significantly. Chlorofluoromethanes came into use as inert blowing agents and replaced part or all of the water. Organometallic and amine catalysts and silicone stabilizers were developed for the control of the highly complex foaming–polymerizing system to obtain the desired product. Polyisocyanates, which were made by the phosgenation of polyamines that were derived from the reaction of aniline with formaldehyde at molar ratios of less than 2:1, are of great interest for rigid foams. Flame-retardant additives also are increasing in popularity (see Flame retardants). Thus, the diisocyanate–polyol system lends itself to many broad variations in raw materials and chemistry and results in a great versatility of products which may vary in flexibility from very soft to very rigid, and in density from 16–800 kg/m^3 (1 to 50 lb/ft^3); they also may vary markedly in load-bearing characteristics and other properties.

In the last decade the most significant development in the flexible foam area has been the introduction of molding formulations that require little, if any, oven curing and produce more resilient and higher load-bearing foams. Known as cold cure, ambient cure or high resiliency, among other names, these products are based on highly reactive polyether polyols containing relatively high fractions of primary hydroxyl groups (60–62).

Polyurethane elastomers are essentially linear polymers in contrast to the foams. The most common route to the finished elastomer product, which frequently is a cast item, is by way of a prepolymer. A prepolymer may be purchased or it may be prepared from isocyanate and polyol at the time of use. The prepolymer is then extended to an elastomer polymer by further reaction with such extenders as 1,4-butanediol, the bis(β-hydroxyethyl) ether of hydroquinone, or 3,3′-dichloro-4,4′-diaminodiphenylmethane (MOCA). Some of the variables available to the elastomer designer are choices of polyol, isocyanate, extender, ratio of —NCO to active hydrogen, and the reaction temperature for control of branching and/or cross-linking, etc. It is evident that such a system is extremely versatile chemically and in product properties. The effects of several variables and their interrelationships have been discussed in detail (2,63). Texin, Estane, and Pellethane are examples of thermoplastic polyurethane elastomers that are suitable for extrusion or injection molding (64–65).

One of the most recent advances in the urethane industry has been the development of the reaction injection molding (RIM) process for microcellular elastomeric parts by a very fast one-step process. The liquid ingredients are injected through a mixing device directly into a mold along with a small amount of blowing agent to aid in filling the mold as the reaction proceeds. The process was developed almost exclusively for production of elastomeric bumpers and fascias for the front and rear ends of automobiles (66–67).

Coatings (qv) may utilize a system in which the diisocyanate has been blocked as a phenolic ester. When a blocked isocyanate and polyol mixture are applied and baked, the polyol displaces the phenol and forms a polymer. In the classical polyol–polyisocyanate system, use of primarily difunctional materials results in relatively elastic, abrasion-resistance products. Increased functionality of the components produces not only a harder coating with improved resistance to chemicals, but one that is more brittle and less abrasion resistant.

The most recent significant development in the coatings area has been the introduction of aqueous polyurethane dispersions (68). These materials are more ecologically acceptable than are the conventional solvent-based polyurethanes.

The polyurethane industry in the United States has grown in a period of twenty years from pilot-plant scale to a yearly production of ca 1,100,000 metric tons of products. Worldwide use of polyurethane resins will be just under 3,000,000 t in 1979 and is estimated that it may exceed 3,200,000 t in 1980 (69). The past, present, and future of this industry are illustrated in Table 8.

The nonpolymer applications for isocyanates are largely insecticides, herbicides, and other biologically active products. Commercially available monoisocyanates of these applications include methyl, *n*-butyl, phenyl, and mono- and dichlorophenyl isocyanates.

Pseudo Isocyanates. ***Nitrile Carbonates.*** In 1971–1979 ARCO Chemical (70) offered adiponitrile carbonate [*14642-37-6*] as an exploratory or developmental chemical, with projected pricing of $2.20/kg at a 11,000 t level.

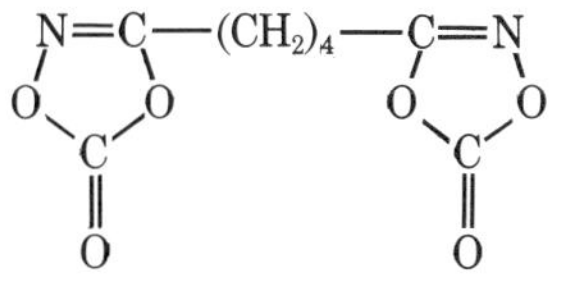

```
N=C—(CH2)4—C=N
/   \       /   \
O     O     O     O
 \   /       \   /
   C           C
   ‖           ‖
   O           O
```

Table 8. Estimated U.S. Consumption of Urethane Products, 10^3 t

Product	1961	1963	1965	1970	1975	1980
flexible foam	41	68	102	159	477	675
rigid foam		14	36	114	155	306
elastomers			16	36	45	91
coatings			9	27	32	41
other			5	12	27	39
Total	*41*	*82*	*168*	*348*	*736*	*1152*

This interesting chemical was obtained from adiponitrile, probably through reaction of the hydroxamic acid with phosgene. At elevated temperature in the presence of a combination of catalysts, such as Dabco and dibutyl tin dilaurate, the carbonate reacts with alcohols and amines to form products equivalent to those that would be obtained from tetramethylene diisocyanate. Using polyester or polyether polyols, it was possible to formulate polyurethane foams, thermoplastics, or coatings. The advantages claimed for the adiponitrile carbonate were stability of the carbonate to storage and moisture, excellent light stability of the resulting urethanes, excellent adhesion of its polyurethane products to metals and glass, and freedom from the vapor hazards attendant with isocyanates. These advantages were evidently insufficient to generate an economic volume–price relationship that justified commercialization.

Aminimides. Aminimides have been rather extensively studied (71–72) as isocyanate precursors for use in a variety of adhesive systems for bonding tire cord and industrial fabrics, for components of solvent-based acrylic surface coatings and as modifiers for acrylic powder coatings. Although they are technically effective for these applications, significant commercial volume has not been achieved.

The structure of the aminimides is similar to the intermediate postulated for the Curtius rearrangement, and the *N-N* cleavage results in an isocyanate.

$$R'(CO\overset{-}{N}\overset{+}{N}R_3)_2 \xrightarrow{\Delta} R'(NCO)_2 + 2\ NR_3$$

Thus a bis(aminimide) may be included in a formulation which is applied, even from aqueous solution, then heated to effect cure through urethane formation.

Derivatives

Modified Isocyanates and Prepolymers. Flexible foams are made almost entirely by the direct reaction of isocyanate and polyol, but prepolymers are used in the manufacture of rigid foams and elastomeric products. The reaction of a diisocyanate with a hydroxyl-terminated polyether or polyester to form an isocyanate-terminated prepolymer may be represented as follows:

$$OCN{-}R{-}NCO + HO{\sim\sim}OH \longrightarrow OCN{-}R{-}NH\overset{\displaystyle O}{\overset{\|}{C}}O{\sim\sim}O\overset{\displaystyle O}{\overset{\|}{C}}NH{-}R{-}NCO$$

The manufacturing process for most "no-cook" type prepolymers is straightforward. The polyol is added with good agitation to a stoichiometric excess of isocyanate in a glass-lined or stainless-steel reactor at 50–100°C for ca 2 h. The mixture is held at this temperature for an additional 2 h to ensure completion of the reaction, and the product is filtered. Many variations are possible in the details (2) of the processing of prepolymers, eg, the NCO/OH ratio, the choice of isocyanate and polyol, the quality of raw material required, the time of reaction, the use or avoidance of catalysts and/or moisture, etc. Multrathane F-242 or E-516, and the Adiprene products are typical elastomer prepolymers. Adiprene polymers are made from TDI and polyethers derived from the polymerization of tetrahydrofuran. Multrathane F-242 is based on MDI and on poly(ethylene adipate); Multrathane E-516 upon MDI and a polyether. Prepolymers of this type are also offered by American Cyanamid (Cyanoprene) and Uniroyal (Vibrathanes).

The modified isocyanates generally are polyfunctional products derived from TDI or MDI, which are difunctional. Thus, reaction of 80/20 TDI at moderate temperature with a polyol, eg, trimethylolpropane (2-ethyl-2-hydroxymethyl-1,3-propanediol) (see Alcohols, polyhydric) results in preferential reaction of the *p*-isocyanate group to produce a urethane triisocyanate that is high in *o*-isocyanate groups. Mondur CB is a product of this type, and is marketed in the United States for use in coatings.

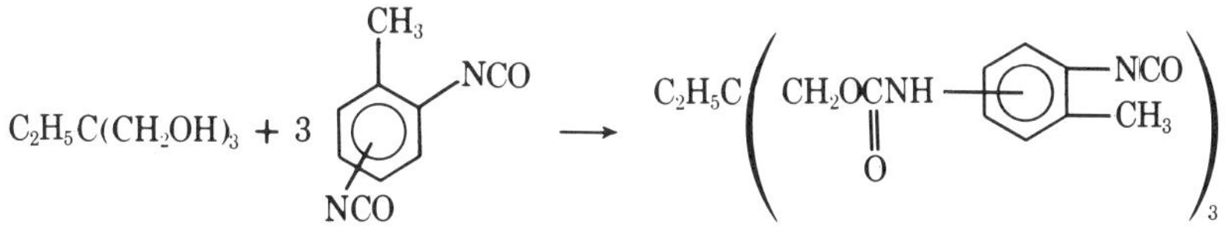

$$C_2H_5C(CH_2OH)_3 + 3\ CH_3C_6H_3(NCO)_2 \longrightarrow C_2H_5C\left(CH_2O\overset{\overset{\displaystyle O}{\|}}{C}NH-C_6H_3(NCO)(CH_3)\right)_3$$

The reaction of a phenol with the remaining free isocyanate groups of a modified isocyanate, eg, a urethane triisocyanate, results in a polyfunctional blocked isocyanate. A material of this type is offered to the wire-coating industry as Mondur S. A related product is Mondur SH, in which TDI that is partially blocked with phenolic groups, is trimerized by heating in the presence of an amine catalyst until condensation has proceeded to the desired degree, as judged by NCO content; the catalyst then is neutralized.

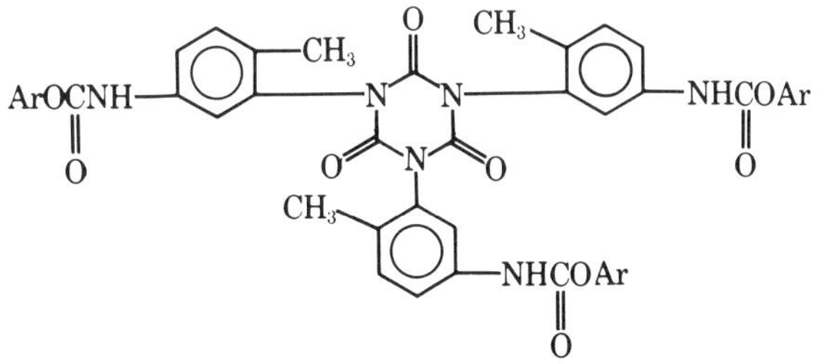

ArOCNH–, CH$_3$, O, CH$_3$, N, N, –NHCOAr, O, O, N, O, O, CH$_3$, –NHCOAr, O

(Ar = C_6H_5, [*28888-79-1*])

Isocyanates may be blocked with many active hydrogen compounds other than phenol; for example, ketoximes and hydroxamic esters. Blocked isocyanates have been reviewed (11).

The modified isocyanates that are based on MDI typically are proprietary products developed for specific applications or even for specific foam producers. Very frequently they are mixtures of MDI or partially carbodiimized MDI with polymeric isocyanates and/or prepolymers. This is especially true for applications in the reaction injection molding (RIM) area (see Plastics processing).

BIBLIOGRAPHY

"Isocyanates, Organic" in *ECT* 2nd ed., Vol. 12, pp. 45–64, by D. H. Chadwick, Mobay Chemical Co., and E. E. Hardy, Monsanto Research Corp.

1. J. H. Saunders and K. C. Frisch, *Polyurethanes: Chemistry and Technology, Part I: Chemistry,* Interscience Publishers, a division of John Wiley & Sons, Inc., New York, 1962.
2. J. H. Saunders and K. C. Frisch, *Polyurethanes: Chemistry and Technology, Part II: Technology,* Interscience Publishers, a division of John Wiley & Sons, Inc., New York, 1964.
3. S. Ozaki, *Chem. Rev.* **72,** 457 (1972).
4. W. Siefken, "Isocyanate" in F. Ullmann, ed., *Enzyklopädie der technischen Chemie,* 3rd ed., Vol. 9, Urban and Schwarzenberg, Munich–Berlin, 1957; *Ann.* **562,** 75 (1949).
5. A. A. R. Sayigh, H. Ulrich, and W. J. Farrissey, Jr., "Diisocyanates" in J. K. Stille and T. W. Campbell, eds., *Condensation Monomers,* Wiley Interscience, New York, 1972.
6. E. Müller, "Polyurethanes" in E. Müller, ed., *Houben-Weyl Methoden der Organischen Chemie,* Vol. 14, Part 2, Thieme-Verlag, Stuttgart, 1963.
7. S. Petersen, *Ann.* **562,** 205 (1949); E. Muller, ed., *Houben-Weyl Methoden der Organischen Chemie,* Vol. 8, Thieme-Verlag, Stuttgart, 1963, p. 119.
8. O. Bayer, *Angew. Chem.* **A59,** 257 (1947).
9. J. W. Britain, "Polyurethanes" in W. M. Smith, ed., *Manufacture of Plastics,* Reinhold Publishing Corp., New York, 1964.
10. D. J. David and H. B. Staley, *Analytical Chemistry of the Polyurethanes,* Wiley-Interscience, New York, 1969.
11. Z. W. Wicks, *Prog. Org. Coat.* **3,** 73 (1975); update with over 200 references accepted for publication, same journal (in press).
12. R. Richter and H. Ulrich, "Syntheses and Preparative Applications of Isocyanates" in S. Patai, ed., *The Chemistry of Cyanates and Their Derivatives,* Part 2, John Wiley & Sons, Inc., New York, 1977.
13. H. J. Twitchett, *Chem. Soc. Rev.* **3,** 226 (1974).
14. Personal communication, R. F. Kress, Mobay Chemical Corporation; Brit. Pat. 1,153,815 (May, 1969), (to Bayer A.G.).
15. *Chem. Eng. News* **58,** 38 (Mar. 10, 1980).
16. P. A. S. Smith, "The Curtius Reaction" in R. Adams and co-eds., *Organic Reactions,* Vol. 3, John Wiley & Sons, Inc., New York, 1946.
17. F. L. Pyman, *J. Chem. Soc.* **103,** 852 (1913).
18. H. L. Yale, *Chem. Rev.* **33,** 209 (1943).
19. Ger. Pat. 1,205,087 (Nov. 18, 1965), K. Zenner, G. Oertel, and H. Holdschmidt (to Bayer A.G.).
20. F. W. Hoover and co-workers, *J. Org. Chem.* **28,** 1825, 2082 (1963); *J. Org. Chem.* **29,** 143 (1964).
21. U.S. Pat. 3,155,700 (Nov. 3, 1964), P. R. Steyermark (to W. R. Grace and Co.).
22. H. Ulrich, *Chem. Rev.* **65,** 369 (1965).
23. J. Smith, Jr., T. K. Brotherton, and J. W. Lynn, *J. Org. Chem.* **30,** 1260 (1965).
24. H. Holtschmidt and G. Oertel, *Angew. Chem. Int. Ed. Engl.* **1,** 617 (1962).
25. U.S. Pat. 2,835,652 (May 20, 1958), A. C. Haven, Jr. (to DuPont).
26. G. S. Forbes and H. H. Anderson, *J. Am. Chem. Soc.* **70,** 1043, 1222 (1948).
27. W. Stamm, *J. Org. Chem.* **30,** 693 (1965).
28. K. Schlögl and H. Seiler, *Naturwissenschaften* **45,** 337 (1958).
29. U.S. Pat. 2,680,127 (June 1, 1954), R. J. Slocombe, H. Flores, and T. H. Cleveland (to Monsanto Chemical Co.).
30. U.S. Pat. 3,188,337 (June 8, 1965), A. M. Gemassmer (to Farbenfabriken Bayer A.G. and Mobay Chemical Co.).
31. *Hydrocarbon Process. Pet. Ref.* **42,** 230 (Nov. 1963).
32. *Process Evaluation, Research Planning, Report 77-5,* Chem. Systems, Inc., New York, 1978.
33. U.S. Pat. 3,128,316 (Apr. 7, 1964), H. J. Koch (to Bayer and Mobay); U.S. Pat. 3,331,867 (July 18, 1967),

I. B. Van Horn and E. L. Powers (to Mobay Chemical Corp); U.S. Pat. 4,091,009 (May 23, 1978), J. R. Cassata (to Olin Corporation); U.S. Pat. 4,137,266 (Jan. 30, 1979), J. R. Cassata (to Olin Corporation).

34. U.S. Pat. 2,953,590 (Sept. 20, 1960), J. Pfirschke (to Mobay Chemical Co., and Farbenfabriken Bayer A.G.).
35. M. W. Farlow, "Hexamethylene Diisocyanate" in N. Rabjohn, ed., *Organic Synthesis,* Coll. Vol. IV, John Wiley & Sons, Inc., New York, 1963, p. 521.
36. *Chem. Week* **120,** (Mar. 9, 1977).
37. W. S. Durnell, J. H. Bateman, and D. S. Meiditch, *Chem. Tech.,* **8,** 383 (June 1978).
38. *Chem. Eng.* **86,** 80 (1979).
39. *Standard Specification for Toluene Diisocyanate,* ANSI/ASTM, D1786, American Society for Testing Materials, Philadelphia, Pa., 1980.
40. *Methods for Testing Urethane Foam Raw Materials,* ASTM Designation: 1638-74, American Society for Testing and Materials, Philadelphia, Pa., 1980.
41. R. L. Sandridge and R. F. Bargiband, "Urethane Polymers" in F. D. Snell and L. S. Ettre, eds., *Encyclopedia of Chemical Analysis,* Vol. 19, John Wiley & Sons, Inc., New York, 1974, p. 262.
42. D. L. David, "Isocyanic Acid Esters" in F. D. Snell and L. S. Ettre, eds., *Encyclopedia of Chemical Analysis,* Vol. 15, John Wiley & Sons, Inc., New York, p. 94.
43. Personal communication, W. E. Kendall and R. L. Sandridge, Mobay Chemical Corporation.
44. K. Marcali, *Anal. Chem.* **29,** 552 (1957).
45. J. Keller, K. L. Dunlap, and R. L. Sandridge, *Anal. Chem.* **46,** 1845 (1974).
46. K. L. Dunlap, R. L. Sandridge and J. Keller, *Anal. Chem.* **48,** 497 (1976).
47. S. P. Levine and co-workers, *Anal. Chem.* **51,** 1106 (1979).
48. Personal communication, R. L. Sandridge, L. Hernandez, and J. Keller, Mobay Chemical Corporation.
49. J. Keller and R. L. Sandridge, *Anal. Chem.* **51,** 1868 (1979).
50. *Properties and Essential Information for Safe Handling and Use of Tolylene Diisocyanate,* Chemical Safety Data Sheet SD-73, Manufacturing Chemists' Association, Washington, D.C., 1959.
51. *Recommendations for the Handling of Aromatic Isocyanates,* International Isocyanate Institute, Inc., New York, 1976.
52. Title 29, Code of Federal Regulations, 1910.1000; Criteria for a Recommended Standard, Occupational Exposure to Diisocyanates, Nat. Inst. for Occ. Safety and Health, U.S. Department of Health, Education, and Welfare, Sept. 1978.
53. H. Weill and co-workers, *Respiratory and Immunologic Evaluation of Isocyanate Exposure in a New Manufacturing Plant,* Final Report, NIOSH Contract No. 210-75-0006, Sept. 30, 1979.
54. D. H. Wegman, J. M. Peters, L. Pagnotto, and L. J. Fine, *Brit. J. Ind. Med.* **34,** 196 (1977).
55. M. Karol and co-workers, *J. Occup. Med.* **20,** 383 (1978); *J. Occup. Med.* **21,** 354 (1979).
56. M. Karol, B. Hauth, and Y. Alarie, *Toxicol. Appl. Pharmacol.* **51,** 73 (1979); M. Karol, C. Dixon, M. Brady, and Y. Alarie, *Toxicol. Appl. Pharmacol.* **53,** 260 (1980).
57. P. D. Ziegler, *Mod. Paint Coat.* **69,** 33 (Feb. 1979).
58. J. K. Backus, "Polyurethanes" in C. E. Schildnecht and I. Skeist, eds., *Polymerization Processes,* High Polymers Series, Vol. 29, John Wiley & Sons, Inc., New York, 1977.
59. J. M. Buist, ed., *Developments in Polyurethanes,* Applied Science Publishers LTD, Essex, UK (1978).
60. D. G. Powell and co-workers, *J. Cell. Plast.* 8(2), 90 (Mar./Apr. 1972).
61. H. J. Fabris, "High Resilience Polyurethane Foams" in K. C. Frisch and S. L. Reegan, eds., *Advances in Urethane Science and Technology,* Vol. 4, Technomic Publishing Co., Westport, Conn., 1976.
62. W. Patten and C. G. Seefried, Jr., *J. Cell. Plast.* **12,** 41 (Jan./Feb. 1976).
63. J. H. Saunders, *Rubber Chem. Technol.* **33,** 1259 (1960).
64. C. S. Schollenberger and K. Dinbergs, "Thermoplastic Polyurethane Molecular Weight–Property Relations" in K. C. Frisch and S. L. Reegan, eds., *Advances in Urethane Science and Technology,* Vol. 3, Technomic Publishing Co., Westport, Conn., 1974.
65. *An Engineering Handbook for TEXIN Urethane Elastoplastic Materials,* Mobay Chemical Corp., 1971.
66. W. E. Becker, ed., *Reaction Injection Molding,* Van Nostrand Reinhold Company, New York, 1979.
67. F. M. Sweeney, *Introduction to Reaction Injection Molding,* Technomic Publishing Co., Westport, Conn., 1979.
68. D. Dietrich and J. N. Rieck, *Adhesives Age* **21,** 24 (Feb. 1978).

69. *Chem. Eng. News* **57,** 12 (Oct. 15, 1979).
70. *Product Bulletin,* Arco Chemical Company, Sept. 1, 1972.
71. W. J. McKillip, in K. C. Frisch and S. L. Reegan, eds. *Advances in Urethane Science and Technology,* Vol. 3, Technomic Publishing Co., Westport, Conn., 1974.
72. *Technical Bulletins 1238, 1247–49,* Ashland Chemicals, 1972–1974.

General Reference

O. Bayer, "The Odyssey of an Invention, Charles Goodyear Medal Address" *Rubber Chem Technol.* **48** (3) G73 (1975).

D. H. CHADWICK
T. H. CLEVELAND
Mobay Chemical Corporation

ISOCYANURIC COMPOUNDS.

See Cyanuric and isocyanuric acids.

ISOPHORONE (3,5,5-TRIMETHYL-2-CYCLOHEXENE-1-ONE).

See Ketones.

ISOPRENE

From the time that Williams isolated isoprene [*78-79-5*] from the pyrolysis products of natural rubber (1), scientific researchers have been attempting to reverse the process, ie, to prepare rubber from isoprene (see Rubber, natural). In 1879, Bouchardat prepared a synthetic rubbery product by treating isoprene with hydrochloric acid (2) but it was not until 1954–1955 that methods were found to prepare a poly(*cis*-1,4-isoprene) which duplicates the structure of natural rubber. In one method (3–4), a Ziegler-type catalyst of trialkyl aluminum and titanium tetrachloride was used to polymerize isoprene in an air-free, moisture-free hydrocarbon solvent to an all *cis*-1,4-polyisoprene. The other catalyst, based on lithium metal, was discovered during a study of structures and molecular weights of polyisoprenes prepared with alkali metals (5). Isoprene polymerization with alkali metals also was investigated in the USSR (6). A polyisoprene with 90% 1,4 units was synthesized with the aid of lithium catalysts as early as 1949.

These discoveries were followed by technical efforts by research and development groups to develop economical processes for large-scale manufacture of isoprene and for polymerization that would enable *cis*-1,4-polyisoprene to compete with natural rubber and with the synthetic rubbers in the world market.

The isoprene unit exists not only in natural rubber but in terpenes, camphors, diterpenes (eg, abietic acid), vitamins A and K, chlorophyll, and other compounds isolated from animal and plant materials (see Terpenoids; Vitamins). The correct structural formula for isoprene was first proposed in 1884 (7):

$$\begin{array}{l} CH_2{=}CCH{=}CH_2 \\ \quad\quad\;\; | \\ \quad\quad\; CH_3 \end{array}$$

The first commercial use of isoprene in the U.S. started in 1940 and then only as a minor comonomer with isobutylene for the preparation of butyl rubber (see Elastomers, synthetic).

Properties

Isoprene, 2-methyl-1,3-butadiene or 2-methyldivinyl or 2-methylerythrene, is a colorless, volatile liquid that is soluble in most common hydrocarbons but is practically insoluble in water. Isoprene forms binary azeotropes with methanol, *n*-pentane, methylamine, methyl formate, ethyl bromide, methyl sulfide, acetone, propylene oxide, isopropyl nitrite, methylal, formaldehyde, ethyl ether, and perfluorotriethylamine. Ternary azeotropes with water–acetone, water–acetonitrile, methyl formate–ethyl bromide also occur (8). Typical properties of isoprene are listed in Table 1.

Many of the common properties of isoprene have been presented graphically (9). These include the vapor pressure from −60 to 210°C, heat of vaporization from −40 to 210°C, the liquid heat capacity from −30 to 70°C, the vapor heat capacity from 0–1000°C, the liquid density from −40 to 210°C, the vapor viscosity from 0–500°C, the liquid viscosity from −30 to 70°C, the surface tension from −30 to 70°C, and the vapor thermal conductivity from −30 to 70°C.

The exact conformation of the isoprene molecule still is in doubt. It generally is accepted that there is no free rotation around the central C—C single bond. Isoprene may be considered as an equilibrium of two conformations, namely a cisoid (*s-cis*) conformation in which both vinyl groups are located on the same side of the C—C bond,

Table 1. Properties of Isoprene

mol wt	68.11
density (of liquid), g/cm^3	
at 0.0°C	0.6984
at 20.0°C	0.6809
at 25.0°C	0.6759
at 30.0°C	0.6707
freezing point, °C	−145.95
bp (at 101.3 kPa = 1 atm), °C	34.067
n_D^{30}	1.41524
flash point, °C	−48
autoignition temperature, °C	220
dipole moment (of liquid), C·m[a]	9.4×10^{-31}
heat of combustion (of gas at constant pressure and at 25°C), kJ/mol[b]	−3012.6
free energy of formation (at 25°C), kJ/mol[b]	146.0
heat of formation (at 25°C), kJ/mol[b]	
liquid	49.40
gas	75.78
coefficient of expansion (−20.6 to 21.1°C)	0.0016

[a] To convert C·m to D, divide by 3.336×10^{-30}.
[b] To convert kJ to kcal, divide by 4.184.

and a transoid (*s-trans*) one with the vinyl groups located on the opposite sides of the bond. The predominance of the trans-planar or nonplanar configuration has been supported by experimental data (10–14).

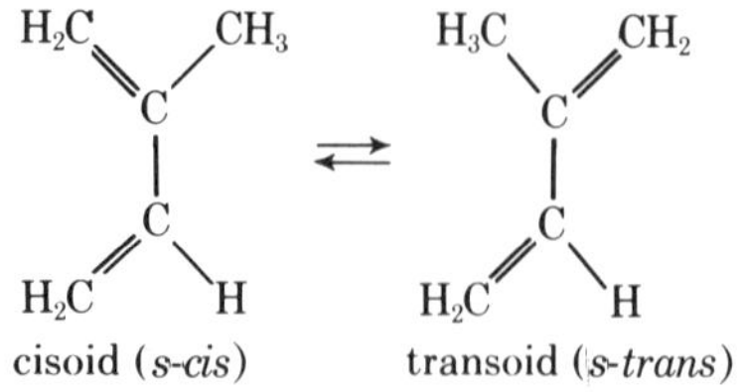

cisoid (*s-cis*) $\rightleftarrows$ transoid (*s-trans*)

The energy required for a trans–cis change in conformation was calculated to be 4.59 kJ/mol (1.10 kcal/mol) (14) and 6.28 kJ/mol (1.50 kcal/mol) (15). This difference in energy is not considered sufficiently large to prevent easy interconversion (16), and it is not great enough to provide a barrier to chemical reactions requiring the cis form, eg, in the Diels-Alder reaction of isoprene with maleic anhydride (14). With butadiene, the transoid geometry is accepted as the more stable one in the ground state but, in the excited state, the cisoid and transoid forms are of nearly equal energy and of similar stability (17). Equal stabilities of the cis and trans forms also were implied for isoprene in the excited state. Conformational considerations are important in the interpretation of structures of stereoregular polyisoprenes that are prepared by various catalysts (see Catalysis). A catalyst in one case need only be stereopreserving and in the other case stereodirecting. Several authors have discussed polymer structures with reference to the conformation of the monomer (18–21).

Reactions. Isoprene is highly reactive both as a diene and through its allylic hydrogens, and its reaction processes are similar to those of butadiene (qv); an extensive review is given in ref. 8.

Apart from polymerization, the most widely investigated isoprene reactions are the formation of six-membered rings by the Diels-Alder reaction:

isoprene + dienophile $\longrightarrow$ adduct

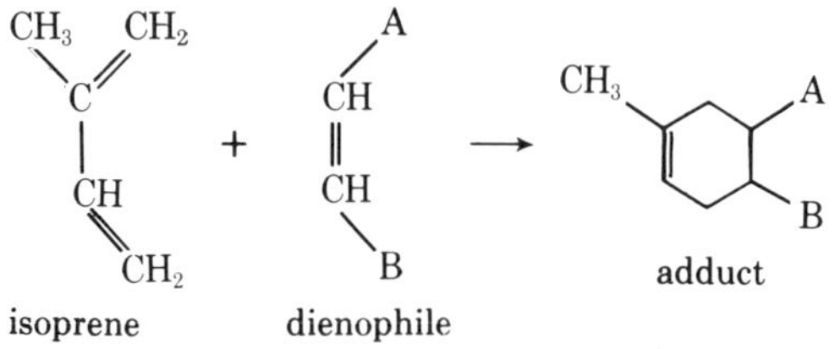

The reaction proceeds readily, depending on the nature of the dienophile and, normally, no catalyst is effective. The substituents, A and B, in the adduct retain their configuration relative to the double bond originally present in the dienophile (22). Many materials having the general structure $CH_2{=}CH{-}X$ have reacted with isoprene to give mixtures of the two isomers shown:

(1) **(2)**

Generally, isomer (**1**), the 1,4 analogue, predominates. Details of the distribution of the 1,4 adduct that is produced from various unsymmetrical dienophiles are given in Table 2. Bulkier substituent groups in the unsymmetrical dienophile favor the production of the 1,4 adduct. This is confirmed by work using dienophiles that contain two substituent groups; eg, $CH_2{=}C[CH(CH_3)_2]COOC_2H_5$, in the reaction at 200°C in the presence of hydroquinone, gives 81% of the 1,4 analogue whereas $CH_2{=}C(CH_3)COOCH_3$ gives only 72% of the 1,4 analogue (25).

Maleic anhydride has been used in many Diels-Alder reactions (29) and the kinetics of its reaction with isoprene have been taken as proof of the essentially transoid structure of isoprene monomer (30). The Diels-Alder reaction of isoprene with chloromaleic anhydride has been analyzed using gas chromatography (31). Reactions with other reactive hydrocarbons have been studied, eg, the reaction with cyclopentadiene yields 2-isopropenylbicyclo[2.2.1]hept-5-ene (32):

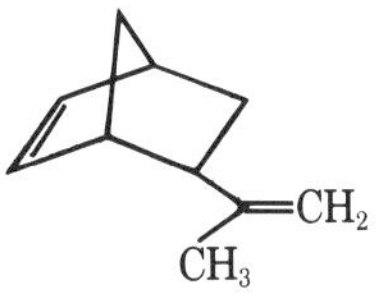

=CH2
CH3

Isoprene may function both as diene and dienophile in Diels-Alder reactions to form dimers.

In the absence of air or peroxides, only cyclic dimers are formed in the thermal dimerization of isoprene (33). Six cyclic dimers are formed in good yields; four substituted cyclohexenes (**3–6**) and two dimethylcyclooctadienes (**7–8**). The latter two are, of course, not Diels-Alder dimers. There is some evidence that the isoprene dimerization mechanism differs from the usual Diels-Alder route.

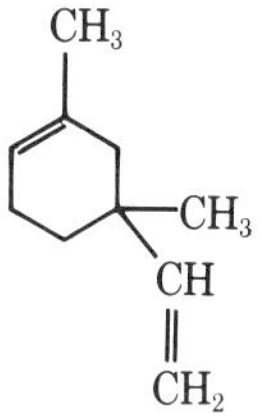

(**3**) *[1611-21-8]* 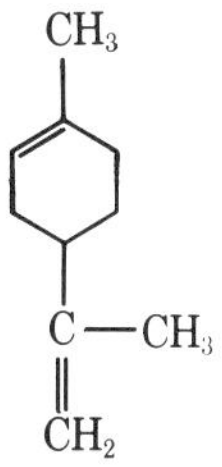 (**4**) *[138-86-3]* 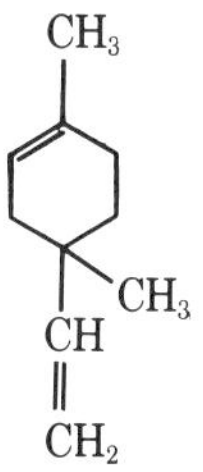 (**5**) *[1743-61-9]* 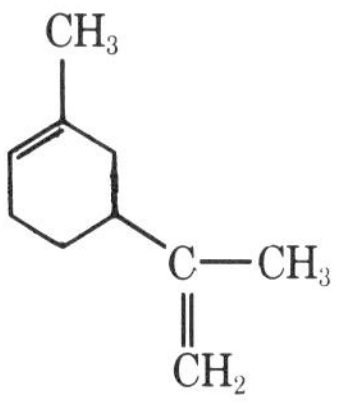 (**6**) *[38738-60-2]* 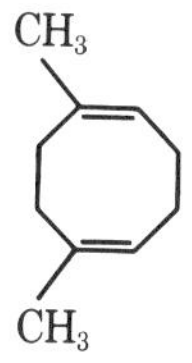 (**7**) *[3760-13-2]* 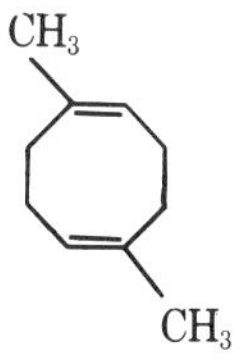 (**8**) *[3760-14-3]*

Table 2. Distribution of Isomers Produced by the Reaction of Isoprene with Unsymmetrical Dienophiles Having the General Structure, $CH_2{=}CHX$

X	Temperature, °C	% 1,4 Analogue	Reference
$COCH_3$	25[a]	93	23
$COCH_3$	120	71	23–24
CHO	25[a]	96	23
CHO	120	59	23
CHO	200[b]	64	25
$COOCH_3$	25–350	70	26
$COOCH_3$	25[b]	85	25
$COOCH_3$	120[b]	80	25
$COOCH_3$	200[b]	67	25
$COOCH_3$	400[b]	58	25
COOH	200[b]	65	25
C_6H_5	200	77	24
NO_2	150	79	24
OC_6H_5	250	75	27–28

[a] In the presence of $SnCl_4.5H_2O$.
[b] In the presence of hydroquinone.

The proportions of the various dimers depend on the reaction conditions. Dimers (**4**) and (**6**) constitute 57–80% of the products that are formed from 100–190°C. At 520°C, compounds (**3**) and (**4**) are the principal reaction products (34). The rate of dimer formation as a function of temperature is listed in Table 3 (35).

The dimerization of isoprene has been accomplished by methods other than heating. Thus, isoprene was dimerized by uv radiation in the presence of photosensitizers to give a complex mixture of cyclobutane, cyclohexene, and cyclooctadiene derivatives (36–37). Sulfuric acid was reported (38) to convert isoprene to linear and cyclic dimers. Ziegler-Natta coordination catalysts containing titanium (39–40), nickel (41), and iron (42–44) dimerize isoprene at low temperatures. In most instances, the proportions of dimers differ from those obtained in thermal dimerizations, and linear trimers and higher homologues also are obtained in the reaction products.

Isoprene can form five-membered hydrocarbon rings (8). A five-membered, sulfur-containing ring is an intermediate in an isoprene purification reaction:

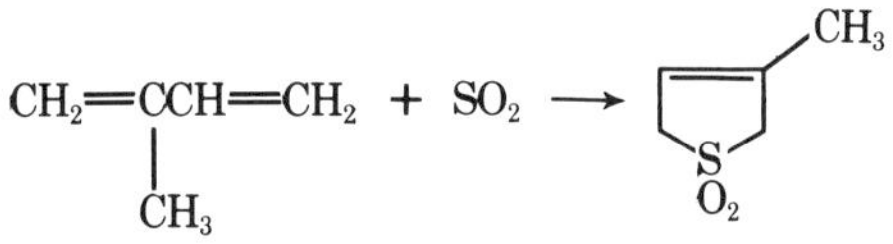

$$CH_2{=}\underset{\displaystyle CH_3}{\underset{|}{C}}CH{=}CH_2 + SO_2 \longrightarrow$$ CH$_3$ S O$_2$

The solid sulfone can be recrystallized and the isoprene can be regenerated by heating the purified sulfone (45–46).

With SO_3–DMF (dimethylformamide) complex, a six-membered ring can form the δ sultone (47):

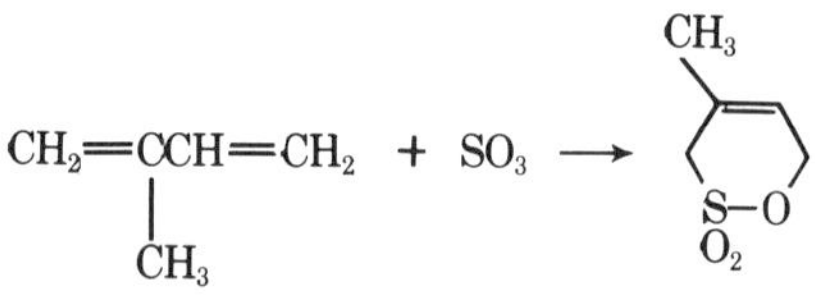

$$CH_2{=}\underset{\displaystyle CH_3}{\underset{|}{C}}CH{=}CH_2 + SO_3 \longrightarrow$$ CH$_3$ S–O O$_2$

Eight-membered rings may be formed thermally or photochemically (see Photochemical technology). Excellent yields can be obtained with ferric acetylacetonate–triethylaluminum–dipyridyl (43):

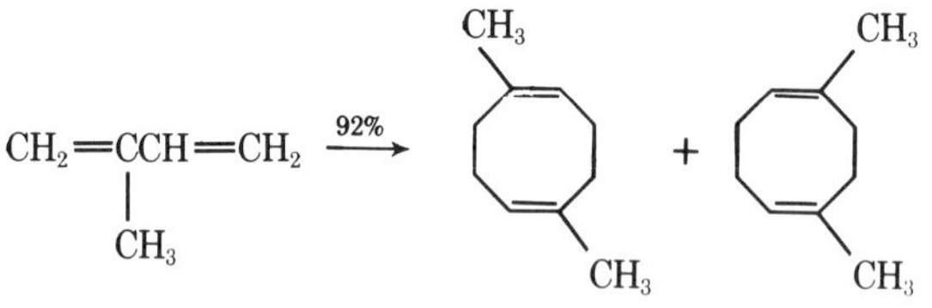

$$CH_2{=}\underset{\displaystyle CH_3}{\underset{|}{C}}CH{=}CH_2 \xrightarrow{92\%}$$ CH$_3$ CH$_3$ + CH$_3$ CH$_3$

Table 3. Effect of Temperature on the Rate of Dimer Formation[a]

Temperature, °C	% Isoprene dimerized per hour
20	0.000017
40	0.00019
60	0.0021
80	0.023
100	0.25

[a] Ref. 35.

Twelve-membered rings have been obtained using coordination catalysts. The *trans,trans,cis*-cyclododecatriene has been prepared with a tetrabutyl titanate–diethylaluminum chloride catalyst (48–49) and with a chromium-based system (50). The trans,trans,trans isomer has been prepared with a nickel system. For the trans,trans,cis:

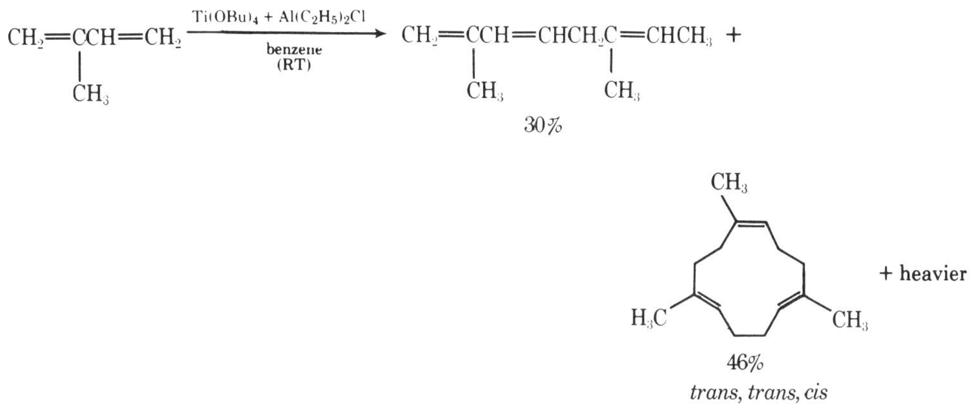

$$CH_2{=}C(CH_3)CH{=}CH_2 \xrightarrow[\text{benzene (RT)}]{Ti(OBu)_4 + Al(C_2H_5)_2Cl} \underset{30\%}{CH_2{=}C(CH_3)CH{=}CHCH_2C(CH_3){=}CHCH_3} +$$

H₃C CH₃ CH₃ (trimethylcyclododecatriene ring) + heavier

46%
trans, trans, cis

Free Radical. Free radicals attack isoprene and two competing mechanisms are postulated:

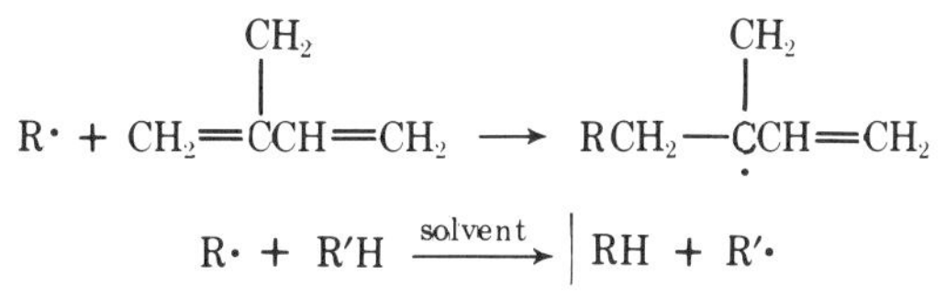

$$R\cdot + CH_2{=}C(CH_2)CH{=}CH_2 \longrightarrow RCH_2{-}\dot{C}(CH_2)CH{=}CH_2$$

$$R\cdot + R'H \xrightarrow{\text{solvent}} RH + R'\cdot$$

The rates of these two reactions have been studied for the attack of trifluoromethyl (51) and methyl radicals (52) in isoprene that has been dissolved in 2,3-dimethylbutane and isooctane, respectively. The rate constants for the reactions with isoprene are much greater than those for the attack on the solvent. The ratio between the two rates for the attack of trifluoromethyl radicals varies from 1090 at 65°C to 233 at 180°C. For the corresponding reaction involving methyl radicals, the ratio is 2090 at 65°C.

The photosensitized dimerization of isoprene in the presence of benzil, $C_6H_5COCOC_6H_5$, has been investigated. Mixtures of substituted cyclobutanes, cyclohexenes, and cyclooctadienes were formed and identified (53). The reaction is believed to proceed by formation of a reactive triplet intermediate.

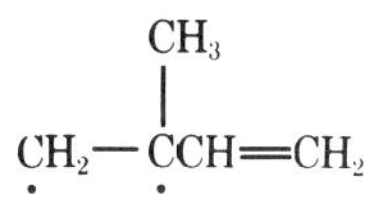

$$\dot{C}H_2{-}\dot{C}(CH_3)CH{=}CH_2$$

The energy for this triplet state presumably is obtained by interaction with the photoexcited benzil species. Under other conditions, photolysis results in the formation of a methylcyclobutene (54–55).

The addition of aromatic and aliphatic thiols, RSH and ArSH, and of thiolacetic acid to isoprene yields mainly the trans-1,4 adduct (56). The aromatic thiyl radicals, ArS·, add almost entirely to the first carbon atom; however, aliphatic thiyl radicals, RS·, also add to the fourth C atom in significant amounts.

Halogens and Halogenated Compounds. The chlorination of isoprene in CCl_4 at −5 to −10°C, using an equimolar ratio of chlorine to isoprene, gives a mixture of 44% of 1,4-dichloro-2-methyl-2-butene and 14% of 3,4-dichloro-2-methyl-1-butene as addition products, along with 42% of the substitution product, 2-chloromethyl-1,3-butadiene (57). For the latter product, the reaction is an electrophilic substitution and the carbon bearing the chlorine is not of the original methyl group in the isoprene molecule (58). At elevated temperatures (850°C) and in the vapor state, chlorine does not add but chlorinates the methyl in a free-radical substitution reaction. Thus, a given reagent appears to react by different mechanisms under different conditions.

Bromination of isoprene using Br_2 at −5°C in chloroform yields only *trans*-1,4-dibromo-2-methyl-2-butene (59). Dry hydrogen chloride reacts with a one-third excess of isoprene at −15°C to form the 1,2 addition product, 2-chloro-2-methyl-3-butene (60). When an equimolar amount of HCl is used, the principal product is the 1,4 addition product, 1-chloro-3-methyl-2-butene (61). It has been proposed that the mechanism of addition is essentially all 1,2 with a subsequent isomerization step which is catalyzed by HCl and which is responsible for the formation of the 1,4 product (60). The 3,4 product, 3-bromo-2-methyl-1-butene, is obtained by the reaction of isoprene with 50% HBr in the presence of cuprous bromide (59). Isoprene reacts with the reactive halogen of 3-chlorocyclopentene (62); 3-chlorocyclopentene first is treated with $SnCl_2$ catalyst in dry acetone, then the isoprene is added slowly to the reaction mixture at 30–35°C. The 1,4 addition product that is obtained after catalyst deactivation with pyridine is 1-chloro-3-methyl-4-(2-cyclopentenyl)-2-butene and is unstable on storage.

The reaction of dihalocarbenes, X_2C, with isoprene yields exclusively the 1,2 (or 3,4) addition product, eg, dichlorocarbene; Cl_2C and isoprene react to give 1,1-dihalo-2-methyl-2-vinylcyclopropane (63). The evidence for the presence of any 1,4 or much 3,4 addition is inconclusive (64). The cycloaddition reaction of 1,1-dichloro-2,2-difluoroethylene to isoprene yields 1,2 and 3,4 cycloaddition products in a ratio of 5.4:1 (65). The main product is 1,1-dichloro-2,2-difluoro-3-isopropenylcyclobutane and the side product is 1,1-dichloro-2,2-difluoro-3-methyl-3-vinylcyclobutane. When the dichlorocarbene is generated from $CHCl_3$ plus aqueous base with a tertiary amine as a phase-transfer catalyst, the addition has a high selectivity that increases (for a series of diolefins) with a decrease in activity (66) (see Phase-transfer catalysis). For isoprene, both mono- (1,2) and diadducts (1,2 and 3,4) could be obtained in various ratios depending on which amine is used.

Isoprene reacts with α-chloroalkyl ethers in the presence of $ZnCl_2$ in diethyl ether from 0–10°C. For example, α-chloromethyl methyl ether at 10°C gives a 6:1 ratio of the 1,4 adduct, (*E*)4-chloro-1-methoxy-2-methyl-2-butene, to the 1,2 adduct, 2-chloro-1-methoxy-2-methyl-3-butene. Other α-chloroalkyl ethers react in a similar manner to give predominantly the 1,4 addition product. A wide variety of allylic chlorides and bromides and α-chloroethers and esters add primarily 1,4 to isoprene in the presence of acid catalysts (8).

A telomerization reaction of isoprene can be carried out by treatment with 2-chloro-3-pentene, prepared by the addition of dry HCl to 1,3-pentadiene (67). An equimolar amount of isoprene in dichloromethane reacts with the 2-chloro-3-pentene at 10°C with stannic chloride as catalyst. 1-Chloro-3,5-dimethyl-2,6-octadiene is obtained in 80% yield by 1,4 addition.

Addition reactions between isoprene and tetrahalomethanes can be induced by peroxides, high energy ionizing radiation, or other radical-generating catalysts (see Initiators). In a radical reaction (68), carbon tetrachloride adds 1,4 to isoprene in 81–86% yield when catalyzed by dichloro(triphenylphosphine)ruthenium(II). In the presence of cuprous chloride, dibromoacetonitrile also adds to isoprene (69). Isoprene reacts with carbon tetrabromide in carbon tetrachloride solvent on ^{60}Co irradiation (200 Gy/min (20,000 rad/min)) (70). Addition is induced using bromotrichloromethane in a 3:1 mol ratio to isoprene by x-irradiation at 60 Gy/min (6000 rad/min) (71). Only small amounts of the 1,2 and 4,3 addition products are obtained. The main reaction yields a mixture of ca 75% 1,4 and 25% 4,1 addition product.

$$CH_2{=}C(CH_3)CH{=}CH_2 + BrCCl_3 \rightarrow \underset{\substack{\text{1,4 addition product}\\ 75\%}}{Cl_3CCH_2C(CH_3){=}CHCH_2Br} + \underset{\substack{\text{4,1 addition product}\\ 25\%}}{BrCH_2C(CH_3){=}CHCH_2CCl_3}$$

Hydrocarbons. The reaction of isoprene with toluene, ethylbenzene, or isopropylbenzene is catalyzed by sodium or potassium (72). The reactions are carried out at 125°C in a pressure autoclave by adding the isoprene slowly to the alkylarene in which the alkali metal is dispersed along with a trace quantity of *o*-chlorotoluene which is used as a chain initiator. The products are chiefly monopentenylated in the side chain and no information can be obtained on whether the addition is 1,4 or 1,2 since, under these conditions, the double bond migrates. The alkene products subsequently are reduced to the alkanes by hydrogenation using 5% palladium on charcoal as catalyst.

```
    R   C                 R C
    |   |                 | |
C6H5CCCCC             C6H5CCCCC
    |                     |
    R'                    R'
  head product          tail product
```

With a sodium catalyst, the head product-to-tail product ratios were 2.77:1, 1.88:1, and 1.98:1 for toluene, ethylbenzene, and isopropylbenzene, respectively; corresponding ratios obtained using potassium were 3.04:1, 2.54:1, and 3.30:1.

Alkylation of cyclohexane with isoprene can be carried out with alkyl radicals formed at 450°C and 20.3 MPa (200 atm) (73). 40% Pentenylcyclohexanes, 20% dipentenes (ie, substances having the general formula $C_{10}H_{16}$), and 40% higher boiling compounds are obtained using a 6.8 molar ratio of cyclohexane to isoprene and a space velocity of 2.5 h^{-1}. Of the pentenylcyclohexanes, the head products (attachment at *C*-4) and tail products (attachment at *C*-1) are in equal amounts. Even stable radicals, eg, triphenylmethyl, add readily to isoprene (74).

Olefins (qv), eg, ethylene, propylene, and styrene, add to isoprene in the presence of coordination catalysts that are based on cobalt, nickel, or iron (8).

Other Compounds. Primary and secondary amines add to isoprene in the 1,4 position (75). For example, dimethylamine in benzene reacts with isoprene in the presence of sodium or potassium to form dimethyl(3-methyl-2-butenyl)amine. Similar results are obtained with diethylamine, pyrrolidine, and piperidine. Under the same conditions, aniline and *N*-methylaniline do not react.

Isoprene reacts with phenol in the presence of aluminum phenoxide (76) or concentrated phosphoric acid (77) to give complex products.

Nitrosobenzene reacts with a solution of isoprene in chloroform to form 2-phenyl-4-methyl-3,6-dihydro-1,2-oxazine (78). Reaction with nitrogen dioxide results in a breakdown of the isoprene structure to give oxidation and nitration products (79).

The reaction with diborane in 1,2-dimethoxyethane forms polymeric organoboranes (80); the heat of reaction is −5.28 MJ/mol (−1262 kcal/mol) of diborane at complete reaction. Mono- and dithioethylene phosphorochloridites react (81) to give 1-β-chloroethylthio-3-methyl-3-phospholene-1-oxide and 1-β-chloroethylthio-3-methyl-3-phospholene-1-sulfide, respectively.

At 165°C and in the presence of chloroplatinic acid as catalyst, isoprene reacts with trichlorosilane, methyldichlorosilane, ethyldichlorosilane, benzyldichlorosilane, and dibenzylchlorosilane (72). The addition is 1,4 with the substituted silane group attaching to the first carbon atom. The reaction is $CH_2{=}CHC(CH_3){=}CH_2 + SiHX_3 \rightarrow CH_3CH{=}C(CH_3)CH_2SiX_3$, where X represents a substituent group in the various silanes. Trimethylsilane does not react under these conditions. However, under similar conditions, heptamethylcyclotetrasiloxane reacts with isoprene by 1,2 addition (82).

Thiophene reacts with isoprene in the presence of phosphoric acid to give mainly 2-(3-methyl-2-butenyl)thiophene and some higher boiling compounds (83). Reactions of isoprene with Grignard reagents have been described (84–85).

Polymerization. Isoprene polymerization can be either by 1,4 or vinyl addition (see Elastomers, synthetic—polyisoprene). 1,4 Addition leads to two possible structures which differ in the configuration of the remaining double bond:

$$\begin{array}{ccc} \sim H_2C & & CH_2\sim \\ & C{=}C & \\ H & & CH_3 \end{array} \qquad \begin{array}{ccc} \sim H_2C & & CH_3 \\ & C{=}C & \\ H & & CH_2\sim \end{array}$$

cis-1,4 *trans*-1,4

Vinyl addition produces two other possible structures, according to whether the 1,2 or the 3,4 double bond takes part in the polymerization reaction:

$$\begin{array}{c} \sim CH_2\overset{*}{C}H\sim \\ | \\ CCH_3 \\ \| \\ CH_2 \end{array} \qquad \begin{array}{c} CH_3 \\ | \\ \sim CH_2C^*\sim \\ | \\ CH \\ \| \\ CH_2 \end{array}$$

3,4 1,2

Any of the four monomer residues can be arranged in a polymer chain in either head-to-head, head-to-tail, or tail-to-tail configurations. Each of the two head-to-tail vinyl forms can exist as syndiotactic or isotactic structures because of the presence of an asymmetric carbon atom (marked with an asterisk) in the monomer unit. Of course, the random mix of syndiotactic and isotactic structures—atactic structures—also exists. Of these possible structures, only three are known. *cis*-1,4-Polyisoprene [*9003-31-0*] is exemplified by natural rubber (hevea, guayule) and synthetic analogues (see Rubber, natural). Balata and gutta percha are natural *trans*-1,4-polyisoprene [*9003-31-0*] and there is a synthetic trans as well. A high 3,4-polyisoprene [*9003-31-0*] has been prepared (86) and is amorphous and, thus, is considered to be atactic. No reports of isotactic or syndiotactic 3,4 material have been found. None of the three possible 1,2-polyisoprenes has been synthesized. Many polymers with mixed cis, trans, or vinyl structures have been prepared. Several reports of mixed head-to-head and head-to-tail chains have appeared (87).

The physical properties of any polyisoprene depend not only on the microstructural features but also on macro features, eg, the molecular weight, the ability of the rubber to crystallize (both rate and extent), the linearity and amount of branching of the polymer chains, and the presence of gel (cross-linked chains).

For a polymer to be capable of crystallization, it must have long sequences where the structure is completely stereoregular. These stereoregular sequences must be linear structures composed exclusively of 1,4, 1,2, or 3,4 isoprene units. If the units are 1,4, then they must be either all cis or all trans. If 1,2 or 3,4 units are involved, they must be either syndiotactic or isotactic. In all cases, the monomer units must be linked in the head-to-tail manner (88). Only two polyisoprene polymers, *cis*- and *trans*-1,4-polyisoprene, exhibit any crystallinity. The only known 3,4-polyisoprene is amorphous and atactic (86).

Al–Ti Catalyst for cis-1,4-Polyisoprene. Of the many catalysts that polymerize isoprene, four have attained commercial importance. One is a coordination catalyst based on an aluminum alkyl and a vanadium salt for making *trans*-1,4-polyisoprene, and the other three form *cis*-1,4-polyisoprene. One of these is a lithium alkyl and the other two are coordination catalysts consisting of a combination of titanium tetrachloride ($TiCl_4$) plus a trialkyl aluminum (R_3Al) (which is designated as Al–Ti) or a combination of $TiCl_4$ with an alane (aluminum hydride derivative) (see Organometallics; Ziegler-Natta catalysts).

The reaction product of an alkyl aluminum with titanium tetrachloride (Ziegler or Ziegler-Natta catalyst) polymerizes isoprene to a high *cis*-1,4-polyisoprene (89–91). Table 4 provides typical specifications of isoprene that are suitable for Al–Ti polymerization (92).

Traditional purification techniques, eg, fractional distillation and extractive distillation have been used to provide an isoprene that is practically free of catalyst poisons. In some cases, however, special procedures are required to remove traces of serious poisons. Cyclopentadiene, probably the single most damaging poison, was reported to completely deactivate the catalyst at a concentration of 1.5 mmol/L. Maleic anhydride, which reacts quantitatively with cyclopentadiene at a very high rate, has been used successfully for removal of traces of cyclopentadiene from isoprene. A few ppm of acetylene poison a polymerization catalyst for isoprene; however, linear C_5 acetylenic compounds have been removed efficiently by use of molecular sieves (qv) (91).

Table 4. Isoprene Specification for the Preparation of High *cis*-Polyisoprene[a]

Reagent	Specification
isoprene	97.0 wt % (min)
isoprene dimer	0.1 wt % (max)
alpha olefins	1.0 wt % (max)
beta olefins	2.8 wt % (max)
acetylenes	50 ppm
allenes	50 ppm
carbonyls	10 ppm
sulfur	5 ppm
piperylene	80 ppm
cyclopentadiene	1 ppm
peroxides	5 ppm
acetonitrile	8 ppm

[a] Ref. 92.

Alkali Metals. The polymerization of isoprene by alkali metals and organoalkali compounds (93–95) (other than organolithium) is a heterogeneous reaction both in bulk and hydrocarbon solvents. A homogeneous reaction will take place only in the presence of highly polar solvents. A comprehensive evaluation of the influence of solvent and positive counterion on polymer structure has been given (96). Only lithium-based initiators in hydrocarbon lead to high *cis*-1,4 structures. A trend of decreasing cis structures is observed with an increase of solvent basicity. Alkali metals, other than lithium, generally yield polymers of mixed structures with little or no *cis*-1,4. The ionic character of the propagating ion pair depends mainly on the metal counterion and the solvent type. In solvents, eg, diethyl ether and tetrahydrofuran, where one might expect large charge separation in the ion pair, all initiators produce similar, though not necessarily identical, polymer structures.

The first successful use of lithium metal for the preparation of a cis-1,4-polyisoprene was announced in 1955 (5); however, lithium metal catalysis was quickly phased out in favor of initiation with organolithium compounds. These initiators provide an easy to handle homogenous system that is not encumbered by poorly reproducible induction periods. Organolithium initiators are used commercially in the production of *cis*-1,4-polyisoprene and other polymers.

The polymerization of isoprene in hydrocarbon solvents with organo (mono) lithium compounds, eg, butyllithium, is a homogenous reaction. In the absence of inhibiting impurities, the reaction starts immediately and proceeds to essentially 100% conversion. At moderate temperatures, in hydrocarbon solvent, and in the absence of compounds having an active hydrogen, there is practically no chain termination or chain transfer. The number-average molecular weight is given by the weight of monomer consumed divided by the moles of initiator (equivalent to the kinetic molecular weight). This polymerization lends itself to the preparation of linear polymers of controllable molecular weight and of very narrow molecular weight distribution. Isoprene and solvents of high purity are essential for good results.

Other Polymerization Systems. Extensive work during the last decade on the polymerization of butadiene in solution with soluble nickel complexes has led to general acceptance of the idea that catalyst sites involve a monometallic π complex with a monomer (97–98). Three 2-alkyl butadienes, including isoprene, have been polymerized with (1,2,3-η)-2-butenyl iodonickel; the results support an intermediate π-complex with a syn–anti equilibrium. Anti species lead to cis-1,4 configurations and syn species lead to trans.

A number of other coordination catalysts may be used to form high or very high *cis*-1,4-polyisoprenes, eg, Zr salts or Mg alkyls (99), a cerium-salt-based catalyst (100–101), and uranium- and thorium-based catalysts (102). Uranium tetraalkoxide–aluminum alkyl–Lewis acid combinations can be used to catalyze the formation of high *cis*-1,4-polybutadiene and *cis*-1,4-polyisoprene (103). Several of the lanthanide and actinide rare earth series make active catalysts and should be of special theoretical interest since they appear to be the only types that produce very high *cis*-1,4 structures from both butadiene and isoprene.

An important group of isoprene polymerization catalysts is based on alanes and $TiCl_4$. In place of alkyl aluminum, derivatives of AlH_3 (alanes) are used and react with $TiCl_4$ to produce an active catalyst for the polymerization of isoprene. These systems are unique because no organometallic compound is involved in producing the active species from $TiCl_4$. The substituted alanes are generally complexed with donor molecules of the Lewis-base type, and they are liquids or solids that are soluble in aromatic solvents. The performance of catalysts prepared from $AlHCl_2.O(C_2H_5)_2$ with $TiCl_4$ has been reported (104).

Among the coordination catalysts, only the vanadium co-catalysts, which can produce high *trans*-1,4-polyisoprene, have reached commercial status. Most other coordination catalysts, including several which produce high *cis*-polybutadienes, result in polyisoprenes with mixed microstructures and mediocre physical properties. In free-radical-initiated, aqueous emulsions (qv), there has been little progress in the last two decades and there are no commercial, radical-catalyzed polyisoprenes. The processes used have been similar to those used for SBR (styrene–butadiene rubber) (see Elastomers, synthetic).

The preparation of high *trans*-1,4-polyisoprene with VCl_3 plus $(C_2H_5)_3Al$ catalyst has been described; it has been concluded that there are several species of active sites all of which give high *trans*-polyisoprene. Similar conclusions have been reached about varying catalyst species with vanadium salts during EPDM (ethylene–propylene–diene rubber) polymerizations. Several vanadium salts (VCl_3, $VOCl_3$, VCl_4) all yield *trans*-polyisoprene catalysts which, perhaps, argues that the active catalyst is a single, lower valent species. *Trans*-polyisoprene also has been made with a η^3-2-propenyl iodonickel catalyst (with no co-catalyst) and with tris(η^3-2-propenyl)chromium that is deposited on aluminosilicate (105).

In aqueous systems, the rhodium catalysts, which give very high *trans*-polybutadiene, yield mixed polyisoprenes (106–107). The unusual potential of such polymerizations is the possibility to vary microstructure which cannot be achieved using the free-radical emulsion initiators. Iridium, platinum, iron, cobalt, and nickel salts also have been tried (108).

Production. The 1976 total noncommunist world production of polyisoprene amounted to 230,000 metric tons. In countries having centrally planned economies, production was 650,000 t (109). Manufacture of synthetic polyisoprene is growing rapidly in the centrally planned economies but declining in the U.S. because of the high cost of raw materials. During the 1960s, three U.S. companies were in commercial production but, in 1979, Goodyear Tire & Rubber Co. was the sole U.S. producer and operated a 60,000-t/y plant in Beaumont, Texas (see also Uses).

Synthesis

Besides the synthetic routes, economical recovery of isoprene as a superfractionated by-product from naphtha cracking recently has gained paramount commercial importance. Most of the isoprene processes and chemistry are covered in great detail in refs. 87, 110–115.

Propylene Dimer. The synthesis of isoprene from propylene (116–117) is a three-step process:

$$2\,CH_2{=}CHCH_3 \xrightarrow{(n\text{-propyl})_3Al} CH_2{=}C(CH_3)CH_2CH_2CH_3$$

$$CH_2{=}C(CH_3)CH_2CH_2CH_3 \xrightarrow[100°C]{\text{silica–alumina}} CH_3C(CH_3){=}CHCH_2CH_3$$

$$CH_3C(CH_3){=}CHCH_2CH_3 \xrightarrow[650\text{–}680°C]{HBr} CH_2{=}C(CH_3)CH{=}CH_2 + CH_4$$

The propylene (qv) is dimerized to 2-methyl-1-pentene. Next, isomerization of 2-methyl-1-pentene to 2-methyl-2-pentene is conducted in the vapor phase over a special silica–alumina catalyst. The last step is the pyrolysis of 2-methyl-2-pentene in a cracking furnace in the presence of HBr. The bromide can be replaced with a catalyst that is formed from the reaction product of H_2S with NH_3 (118–119). Isoprene is formed along with numerous other compounds which are removed in subsequent distillation steps. The isoprene finally is purified by superfractionation. Goodyear produced isoprene with this process beginning in 1962.

Dehydrogenation of Tertiary Amylenes. The starting material is a C_5 fraction which is cut from the catalytic cracking of petroleum. Two of the tertiary amylene isomers, 2-methyl-1-butene and 2-methyl-2-butene, are extracted from the C_5 stream by cold, aqueous sulfuric acid. The amylenes are mixed with steam and are dehydrogenated over a catalyst:

$$CH_2{=}C(CH_3)CH_2CH_3 \text{ or } CH_3C(CH_3){=}CHCH_3 \xrightarrow[600°C]{Fe_2O_3\text{–}K_2CO_3\text{–}Cr_2O_3} CH_2{=}C(CH_3)CH{=}CH_2 + H_2$$

The crude isoprene usually is purified by extractive distillation using acetonitrile (120). This process was utilized by B. F. Goodrich, Texas, and by Shell Nederland Chemie, Pernis, The Netherlands.

Isobutylene–Formaldehyde. Isobutylene is condensed with formaldehyde at 95°C to give the principal product, 4,4-dimethyl-*m*-dioxane. In the second step, the dioxane is decomposed in the presence of an acid catalyst to isoprene, formaldehyde, and water.

$$(CH_3)_2C{=}CH_2 + 2\,CH_2O \longrightarrow \text{4,4-dimethyl-1,3-dioxane}$$

$$\text{4,4-dimethyl-1,3-dioxane} \xrightarrow{H^+} CH_2{=}C(CH_3)CH{=}CH_2 + CH_2O + H_2O$$

Much of the work with regard to this process was done by the French Petroleum Institute (121) and by the Kuraray Co., Japan (115). In the USSR, a similar process which begins with isobutylene in a C_4 fraction was developed (122). A one-step process that goes directly from isobutylene and formaldehyde to isoprene has been reported (123). Other one-step processes that start with isobutylene and methylal or methanol have been disclosed (115,124). Isoprene that is based on isobutylene is, or has been, produced in the USSR and Japan.

Isopentane Dehydrogenation. In isopentane dehydrogenation, which is used industrially in the USSR, isopentane or a C_5 fraction from a catalytic cracker is dehydrogenated to isoprene (6):

$$(CH_3)_2CHCH_2CH_3 \longrightarrow CH_2{=}C(CH_3)CH{=}CH_2 + 2\,H_2$$

Acetone–Acetylene. A process that is based on acetone and acetylene (120,125) first was utilized in Germany during World War I. ANIC of Italy uses this process to produce isoprene.

$$(CH_3)_2C{=}O + HC{\equiv}CH \longrightarrow CH_3C(CH_3)(OH)C{\equiv}CH$$

$$CH_3C(CH_3)(OH)C{\equiv}CH \xrightarrow[Pd]{H_2} CH_3C(CH_3)(OH)CH{=}CH_2$$

Table 5. Energy Requirements for Isoprene Preparation [a]

	Requirements for power, fuel, and steam, MJ/kg isoprene (Btu/lb)
By synthesis	
isobutylene–methanol	51.2 (22,000)
propylene dimerization	67.8 (29,158)
amylene dehydrogenation	104.4 (44,900)
By extraction	
acetonitrile	34.4 (14,795)
N,N-dimethylformamide	28.5 (12,234)
N-methyl-2-pyrrolidone	27.3 (11,736)

[a] Ref. 110.

The ethynylation reaction takes place at 10–40°C and 2 MPa (20 atm) and liquid ammonia is the solvent. The methylbutynol is converted into methylbutenol by selective hydrogenation and then is dehydrated over alumina at 250–300°C. Polymerization-grade isoprene of 98.5% purity is obtained (see Acetylene-derived chemicals).

$$CH_3C(CH_3)(OH)CH{=}CH_2 \xrightarrow[250-300°C]{\text{alumina}} CH_2{=}C(CH_3)CH{=}CH_2 + H_2O$$

Petroleum Cracking. In recent years, the recovery of isoprene from C_5 streams that are obtained in the thermal cracking of naphtha and gas oil has gained commercial importance (111). When naphtha or gas oil is cracked to produce ethylene (qv), a host of by-products is created. Values differ depending on the cracking process but isoprene yields may be from 2–5% of the ethylene yield. Because of the enormous demand for ethylene-based products, this by-product isoprene has become a commercial source.

Isoprene can be recovered from the cracker fractions by extractive distillation with selective solvents or by careful fractional distillation. The commercial value of any process depends on the conversion cost. In addition to fluctuating raw material costs and the value of the nonisoprenic products, the total energy demand has become a significant factor in choosing a process (110). The energy requirements for isoprene production, as shown in Table 5, indicates why by-product isoprene is displacing chemically produced isoprene in the market place. Recovery processes by extractive distillation with acetonitrile and dimethylformamide are used commercially (see Azeotropic and extractive distillation).

Economic Aspects

Prices ($/kg) for isoprene and *cis*-polyisoprene are as follows.

Year	*Isoprene*	*cis-Polyisoprene*
1970	0.20–0.24	0.44–0.46
1975	0.55–0.60	0.77–0.82
1980	0.93–1.54	1.70

These prices are offered by Japanese and/or Italian producers. The 1980 prices are early 1980. The $0.93 price is for a 99% pure isoprene which is not suitable for polymerization to *cis*-polyisoprene without extensive treatment to remove acetylenes and cyclopentadiene. It is suitable for many other uses. A polymerization-grade isoprene might cost ca $1.19/kg in Japan. A very high purity pharmaceutical-grade isoprene at $1.54/kg plus freight is available in Italy. Goodyear is the only U.S. producer of *cis*-polyisoprene and the prices quoted are for Goodyear's Natsyn.

Health and Safety

Isoprene is not known to present serious toxicological hazards in handling (2). A concentration of 2% isoprene in air does not narcotize mice but produces bronchial irritation. However, concentrations of 5% are fatal to mice. In humans, a one-minute inhalation of 0.16 mg isoprene/L air is mildly irritating to the mucous membranes of the eyes, nose, and upper respiratory passages (126). It was proposed that the limit of isoprene concentration on industrial sites be set at 0.04 mg/L air; it also was recommended that the maximum concentration of isoprene in water be set at 0.005 mg/L. The extent of toxic conditions and air pollution by isoprene in the manufacturing areas of synthetic rubber and their vicinity has been dealt with in several USSR publications (127–129).

Isoprene is classified by the ICC as a flammable liquid requiring a red label (130). It forms dangerous peroxides on exposure to air in the absence of inhibitors. A solution having 17 wt % of peroxide does not detonate in a standard drop test. The polymeric peroxide residue that is obtained on evaporation does explode. The reaction of isoprene and oxygen is rapid, with 1% conversion of isoprene in ca 3 h at 50°C; the product (131) is an alternating copolymer of oxygen and isoprene, with the repeating unit being $\leftarrow C_5H_9O_2 \rightarrow$. The peroxide could serve as a source of initiation, which could lead to uncontrolled polymerization either homogeneously or as a popcorn growth. Because of the potential hazards on its exposure to oxygen, isoprene should be stored in an inert atmosphere (nitrogen) in the presence of at least 50 ppm of *t*-butylcatechol or hydroquinone. Since the inhibitor is slowly consumed during storage, it is advisable to analyze the isoprene periodically and to add more inhibitor if needed to prevent the formation of peroxides. Before being used in the laboratory, it should be flash distilled to remove the inhibitor and dimers. In industrial use, it may be more convenient to remove the inhibitor by a caustic water wash. A dangerous reaction of isoprene with ozone has been reported (132): when 1 g of isoprene that was diluted with 50 mL of *n*-heptane was treated with ozone at −78°C, the resulting product exploded shortly after being removed from the cooling bath; however, the product of a similar reaction

Table 6. World Isoprene Demand[a], 10^3 Metric Tons[a]

Use	1975	1980	1985[b]
polyisoprene	660	1040	1305
butyl rubber	20	25	31
copolymers and other	5	25	40
Total	*685*	*1090*	*1376*

[a] Ref. 110.
[b] Estimated.

Table 7. World Polyisoprene Consumption[a], 10^3 Metric Tons

	1965	1970	1975	1980	1985[b]
North America	40	83	125	200	265
Latin America	0	2	4	30	50
Western Europe	2	50	100	210	260
Mid-East and Africa	0	0	1	5	20
Far East and Australia	2	5	30	45	60
Eastern Europe, USSR, and China	16	100	400	550	650
Total	*60*	*240*	*660*	*1040*	*1305*

[a] Ref. 110.
[b] Estimated.

that was carried out at room temperature did not explode. On storage, isoprene forms cyclic dimers at a slow rate which is not affected by the presence of an inhibitor (35). The rate of dimer formation is temperature dependent (as shown in Table 3).

Uses

Almost all isoprene that is produced is used for the preparation of *cis*-1,4-polyisoprene or in small proportions in copolymers with isobutylene for butyl rubber and in thermoplastic elastomeric SIS block polymers. *cis*-Polyisoprene is by far the largest (>95%) product of isoprene. Table 6 indicates the demand for isoprene by use (111). Table 7 indicates synthetic polyisoprene consumption by region, both actual (to 1975) and projected (to 1987) (111).

Acknowledgment

Portions of this article have been taken from *Rubber Chemistry and Technology* (87) with the permission of the Rubber Division of the American Chemical Society.

BIBLIOGRAPHY

"Isoprene" in *ECT* 2nd ed., Vol. 12, pp. 64–83, by Geoffrey Holden and Roger H. Mann, Shell Chemical Company.

1. C. G. Williams, *Proc. Roy. Soc.* **10,** 516 (1860); *Phil. Trans.*, 241 (1860).
2. G. Bouchardat, *Comp. Rend.* **80,** 1117 (1879).
3. U.S. Pat. 3,144,743 (Dec. 17, 1963), S. E. Horne (to Goodrich-Gulf Company).
4. S. E. Horne, J. P. Kiehl, J. J. Shipman, V. L. Folt, C. F. Gibbs, E. A. Wilson, E. B. Newton, and M. A. Reinhart, *Ind. Eng. Chem.* **48,** 784 (1956).
5. F. W. Stavely and co-workers, *Ind. Eng. Chem.* **48,** 778 (1956).
6. I. V. Garmonov, *paper No. 1 presented at International Symposium on Isoprene Rubber,* Moscow, USSR, 1972.
7. W. A. Tilden, *Chem. News* **46,** 120 (1882); W. A. Tilden, *J. Chem. Soc.* **45,** 410 (1884).
8. W. J. Bailey in E. C. Leonard, ed., *Vinyl and Diene Monomers, Part II,* John Wiley & Sons, Inc., New York, 1971, Chapt. 5.
9. R. W. Gallant, *Hydrocarbon Process. Pet. Ref.* **46**(a), 155 (1967); *Physical Properties of Hydrocarbons,* Gulf Publishing Co., Vol. 1, Houston, Tex., 1968, pp. 157–166.
10. R. J. W. Le Feure and K. M. S. Sundaran, *J. Chem. Soc.*, 3547 (1963); R. J. W. Le Feure and K. M. S. Sundaran, *J. Chem. Soc.*, 3518 (1964).
11. L. V. Vilkov and I. N. Sadova, *Zh. Strukt. Khim.* **8,** 398 (1967).
12. G. J. Szasz and N. Sheppard, *Trans. Faraday Soc.* **49,** 358 (1953); M. I. Batuev, A. S. Onischenko, A. D. Matveeva, and N. I. Aronova, *Dokl. Akad. Nauk. SSSR.* **132,** 581 (1960); D. Craig, J. J. Shipman, and R. B. Fowler, *J. Am. Chem. Soc.* **83,** 2885 (1961); H. Dodziuk, *J. Mol. Struct.* **20,** 317 (1974).
13. J. Gresser, A. Rajbenbach, and M. Szwarc, *J. Am. Chem. Soc.* **82,** 5820 (1960).
14. D. A. C. Compton, W. O. George, and W. F. Maddams, *J. Chem. Soc. (Perkins II)* **14,** 1666 (1976).
15. S. Dzhessati, A. R. Kyazimova, V. I. Tyulin, and Yu. A. Pentin, *Vestn. Mosk. Univ. Khim.* **23**(5), 19 (1968).
16. L. Pauling, *The Nature of the Chemical Bond,* 3rd ed., Cornell University Press, Ithaca, N.Y., 1960, p. 291.
17. R. Hoffmann and R. A. Olofson, *J. Am. Chem. Soc.* **88,** 943 (1966).
18. R. S. Stearns and L. E. Forman, *J. Polym. Sci.* **41,** 381 (1959).
19. I. Kuntz and A. Gerber, *J. Polym. Sci.* **42,** 299 (1960).
20. R. J. Orr, *J. Polym. Sci.* **58,** 843 (1962).
21. M. Szwarc, *J. Polym. Sci.* **40,** 583 (1959).
22. R. Adams, ed., *Organic Reactions,* Vol. 4, John Wiley & Sons, Inc., New York, 1948, pp. 60–173.

23. E. F. Lutz and G. M. Bailey, *J. Am. Chem. Soc.* **86,** 3899 (1964).
24. Yu. A. Titov and A. I. Kuznetsova, *Izv. Akad. Nauk SSSR Otd. Khim. Nauk,* 1297 (1960).
25. I. N. Nazarov, Yu. A. Titov, and A. I. Kuznetsova, *Izv. Akad. Nauk SSSR Otd. Khim. Nauk,* 1412 (1959).
26. H. E. Hennis, *J. Org. Chem.* **28,** 2570 (1963).
27. A. N. Volkov, A. V. Bogdanova, and M. F. Shostakovskii, *Izv. Akad. Nauk SSSR Otd. Khim. Nauk,* 1280 (1962).
28. M. F. Shostakovskii, A. V. Bogdanova, and A. N. Volkov, *Izv. Akad. Nauk SSSR Otd. Khim. Nauk,* 1284 (1962).
29. F. Bergmann and H. E. Eschinazi, *J. Am. Chem. Soc.* **65,** 1405 (1943).
30. D. Craig, J. J. Shipman, and R. B. Fowler, *J. Am. Chem. Soc.* **83,** 2885 (1961).
31. E. Gil-Av and Y. Herzberg-Minzly, *Proc. Chem. Soc.,* 316 (1961).
32. A. F. Plate and N. A. Belikova, *Zh. Obshch. Khim.* **30,** 3953 (1960).
33. A. S. Onishchenko, *Diene Synthesis,* Daniel Davey and Co., Inc., New York, 1964, pp. 592–595.
34. Ya. M. Paushkin, A. G. Liakumovich, Yu. I. Michurov, R. B. Valitov, and A. F. Lunin, *Tr. Mosk. Inst. Neftekhim. Gazov. Prom.* (72), 23 (1967).
35. *Storage and Handling of Liquefied Olefins and Diolefins,* Enjay Chemical Co., New York, 1962, p. 19.
36. G. S. Hammond, N. J. Turro, and R. S. H. Liu, *J. Org. Chem.* **28,** 3297 (1963).
37. R. S. H. Liu, N. J. Turro, and G. S. Hammond, *J. Am. Chem. Soc.* **87,** 3406 (1965).
38. B. S. Greensfelder and H. H. Voge, *Ind. Eng. Chem.* **37,** 983 (1945).
39. L. I. Zakharkin, *Dokl. Akad. Nauk SSSR* **131,** 1069 (1960).
40. H. Takahashi and M. Yagamuchi, *Osaka Kogyo Gijutsu Shikensho Kiho* **15,** 271 (1964).
41. L. I. Zakharkin and G. G. Zhigareva, *Izv. Akad. Nauk SSSR Serv. Khim.,* 168 (1964).
42. J. P. Candlin and W. H. Janes, *J. Chem. Soc.* **C,** 1856 (1968).
43. A. Misono, Y. Uchida, M. Hidai, and Y. Ohsawa, *Bull. Chem. Soc. Jpn.* **39,** 3425 (1966).
44. R. F. Heck, *Organotransition Metal Chemistry,* Academic Press, Inc., New York, 1974, p. 162.
45. R. L. Frank and R. P. Seven in E. C. Horning, ed., *Organic Syntheses,* Collective Vol. III, John Wiley & Sons, Inc., New York, 1955, p. 499.
46. R. L. Frank, C. E. Adams, J. B. Blegen, R. Deanin, and P. V. Smith, *Ind. Eng. Chem.* **39,** 887 (1947).
47. T. Akiyama, M. Sugihara, T. Imagawa, and M. Kawanisi, *Bull. Chem. Soc. Jpn.* **51,** 1251 (1978).
48. H. Takahashi and M. Yamaguchi, *Osaka Kozyo Gijutsu Shikensho Kiho* **15,** 271 (1964).
49. Fr. Pat. 1,393,071 (Mar. 19, 1965), A. Carbonaro, A. Bonfardeci, and L. Porri (to Montecatini).
50. U.S. Pat. 3,429,940 (Feb. 25, 1969), F. T. Wadsworth (to Columbian Carbon Co.).
51. J. M. Pearson and M. Szwarc, *Trans. Faraday Soc.* **60,** 553 (1964).
52. A. Rajbenbach and M. Szwarc, *Proc. Roy. Soc. (London)* **A251,** 394 (1959).
53. D. J. Trecker, R. L. Brandon, and J. P. Henry, *Chem. Ind. (London),* 652 (1963).
54. K. J. Crowley, *Proc. Chem. Soc.,* 334 (1962); *Tetrahedron* **21,** 1001 (1965).
55. R. Srinavasan, *J. Am. Chem. Soc.* **84,** 4141 (1962).
56. A. A. Oswald, K. Griesbaum, W. Thaler, and B. E. Hudson, Jr., *Am. Chem. Soc. Div. Petrol. Chem. Prepr.* **7,** 139 (1962).
57. M. J. Ljam, *The Addition of Halogens to Chloroprene and Isoprene,* Ph.D. dissertation, The University of Texas, Austin, Texas, 1961, p. 110.
58. E. G. E. Hawkins and M. D. Philpot, *J. Chem. Soc.,* 3204 (1962).
59. A. A. Petrov, *J. Gen. Chem. USSR* **13,** 741 (1943).
60. A. J. Ultee, *J. Chem. Soc.,* 530 (1948).
61. W. J. Jones and H. W. T. Cleorley, *J. Chem. Soc.,* 832 (1946).
62. V. N. Belov, V. K. Promonenkov, and A. B. Kamenskii, *Zh. Obshch. Khim.* **34,** 3432 (1964).
63. A. Ledwith and R. M. Bell, *Chem. Ind. (London),* 459 (1959).
64. E. C. Herrick, *J. Org. Chem.* **24,** 139 (1959).
65. P. D. Bartlett, *J. Am. Chem. Soc.* **86,** 616 (1964).
66. Y. Kimura, K. Isagawa, and Y. Otsuji, *Chem. Lett.,* (8), 951 (1977).
67. A. A. Petrov, N. A. Razumova, and M. L. Genusov, *Zh. Obshch. Khim.* **28,** 1128 (1958).
68. H. Matsumoto and co-workers, *Chem. Lett.,* (1), 115 (1978).
69. Fr. Pat. 1,409,516 (Aug. 27, 1965), H. W. Moore, F. F. Rust, and H. S. Klein (to Shell).
70. G. Rabilloud and P. Traynard, *Compt. Rend.* **251,** 1505 (1960).
71. C. S. H. Chen and E. F. Hosterman, *J. Org. Chem.* **28,** 1585 (1963).

72. I. Shiihara, W. F. Hoskyns, and H. W. Post, *J. Org. Chem.* **26,** 4000 (1961).
73. L. Kh. Friedlin, N. M. Nazarivam E. F. Litvin, and G. K. Gaivoronskaya, *Neftekhimaya,* **4,** 246 (1964).
74. J. B. Conant and H. W. Scherp, *J. Am. Chem. Soc.* **53,** 1941 (1931).
75. G. T. Martirosyan and E. A. Grigoryan, *Izv. Akad. Nauk Arm. SSR Khim. Nauk* **16,** 31 (1963).
76. K. C. Dewhirst and F. F. Rust, *J. Org. Chem.* **28,** 798 (1963).
77. A. R. Bader and W. C. Bean, *J. Am. Chem. Soc.* **80,** 3073 (1958).
78. Yu. A. Arbuzov and L. K. Lysanchuk, *Dokl. Akad. Nauk SSSR* **145,** 319 (1962).
79. A. P. Altshuller and I. Cohen, *Ind. Eng. Chem.* **51,** 776 (1959).
80. A. E. Pope and H. A. Skinner, *J. Chem. Soc.*, 3704 (1963).
81. N. R. Razumova and A. A. Petrov. *Zh. Obshch. Khim.* **34,** 356 (1964).
82. K. A. Andrianov, V. I. Sidorov, L. M. Khananashvili, G. D. Bagratishvili, G. V. Tsitsishvili, and M. L. Kantariya, *Dokl. Akad. Nauk SSSR* **158,** 133 (1964).
83. H. Pines, B. Kvetinskas, J. A. Vesely, and E. Baclawski, *J. Am. Chem. Soc.* **73,** 5173 (1951).
84. M. S. Kharasch, R. D. Mulley, and W. Nudenberg, *J. Org. Chem.* **19,** 1477 (1954).
85. M. S. Kharasch, P. G. Holten, and W. Nudenberg, *J. Org. Chem.* **19,** 1600 (1954).
86. G. Natta, L. Porri, and A. Carbonaro, *Makromol. Chem.* **77,** 126 (1964).
87. E. Schoenberg, H. A. Marsh, S. J. Walters, and W. M. Saltman, *Rubber Chem. Tech.* **52,** 526 (1979).
88. G. Natta, *Mod. Plast.* **34,** 169 (1956).
89. C. T. Winchester, *Ind. Eng. Chem.* **51,** 19 (1959).
90. W. M. Saltman and E. Schoenberg in J. R. Elliot, ed., *Macromolecular Syntheses,* Vol. 2, John Wiley & Sons, Inc., New York, 1966, p. 50.
91. U.S. Pat. 2,851,505 (Sept. 9, 1958), A. M. Henke and V. N. Hurd (to Gulf Research and Development Co.); Brit. Pat. 837,908 (1962), (to Shell); U.S. Pat. 2,900,430 (Aug. 18, 1959), A. M. Henke and H. C. Stauffer (to Gulf Research and Development Co.).
92. R. Fowler and D. Barker, *Chem. Eng. (London),* 322 (1971).
93. Brit. Pat. 24,790 (Oct. 25, 1911), F. E. Matthews and E. H. Strange.
94. C. H. Harries, *Ann.* **383,** 184 (1911).
95. Ger. Pat. 255,786 (Jan. 27, 1912), and 255,787 (Apr. 9, 1912), I. H. Labhartd (to BASF Badische Anilin).
96. R. V. Tobolsky and C. E. Rogers, *J. Polym. Sci.* **40,** 73 (1959).
97. Ph. Teyssie and F. Dawans in W. M. Saltman, ed., *The Stereo Rubbers,* Wiley-Interscience, New York, 1977, Chapt. 3.
98. B. A. Dolgoplosk, *Kinet Katal* **18,** 1146 (1977).
99. W. Cooper in W. M. Saltman, ed., *The Stereo Rubbers,* Wiley-Interscience, New York, 1977, pp. 50–53.
100. U.S. Pat. 3,297,667 (Jan. 10, 1967), W. C. VonDohlen, T. P. Wilson, and E. G. Caflisch (to Union Carbide).
101. U.S. Pat. 3,541,063 (Nov. 17, 1970), M. C. Throckmorton and W. M. Saltman (to Goodyear Tire & Rubber Co.).
102. U.S. Pat. 3,676,411 (July 11, 1972), M. C. Throckmorton and W. M. Saltman (to Goodyear Tire & Rubber Co.).
103. M. Bruzzone, A. Mazzei, and G. Giuliani, *Rubber Chem. Technol.* **47,** 1175 (1974).
104. W. Marconi, A. Mazzei, S. Cesca, and M. deMalde, *Chim. i ind. (Milan)* **51,** 1084 (1969); W. Marconi, A. Mazzei, S. Cucinella, and M. DeMalde, *Makromol. Chem.* **71,** 118, 134 (1964).
105. I. M. Kosheleva, N. N. Stefanovskya, and V. L. Shmonena, *Kautch. i Rezina.* **6,** 49 (1975).
106. E. Matsui and T. Tsuruta, *Polym. J.* **10**(2), 133 (1978); E. Matsui, T. Tsuruta, A. Yoshioka, T. Sugimura, and M. Takahashi, *J. Macromol. Sci. Chem. A* **11,** 999 (1977).
107. A. A. Entezeami, A. Deluzarche, B. Kaempf, and B. Schue, *Eur. Polym. J.* **13**(3), 203 (1977).
108. D. C. Blackley and R. K. Matthaw, *Brit. Polym. J.* **2**(1–2), 25 (1970).
109. *Eur. Chem. News* **32,** 12 (May 5, 1978).
110. M. J. Rhoad, *Proceedings, 15th Annual Meeting, International Institute of Synthetic Rubber Producers, Inc.*, Kyoto, Japan, 1974; *Rubber Ind. (London)* **9**(12), 68 (1975); *Rubber Plast. News,* 14 (Jan. 23, 1978).
111. S. K. Ogorohnikov and G. S. Idlis, *Isoprene Production,* Izdatelstvo Chimia, Leningrad, USSR, 1973.
112. *Isoprene,* Report No. 28, Stanford Research Insitute, Menlo Park, Calif. 1967.

113. C. Capitani, *Rev. Gen. Caout. Plast.* **51**(4), 195 (1974).
114. A. Mitsutani, *Yuki Gosei Kagaku Kyokai Shi* **32,** 528 (1974).
115. A. Mitsutani and S. Kumano, *Chem. Econ. Eng. Rev.* **3**(2), 35 (1971).
116. V. J. Anhorn, K. J. Frech, J. J. Tazuma, P. H. Wise, and W. E. Morrow, *Chem. Eng. Prog.* **57**(5), 41 (1961).
117. V. J. Anhorn, K. J. Frech, G. S. Schaffel, and D. Brown, *Rubber Plast. Age* **42,** 1212 (1961).
118. H. J. Osterhof, *Rev. Gen. Caout. Plast.* **42,** 529 (1965).
119. U.S. Pat. 3,284,532 (Nov. 8, 1966), K. J. Frech (to Goodyear Tire & Rubber Co.).
120. *Hydrocarbon Process.* **50**(11), 170 (1971).
121. *Hydrocarbon Process. Pet. Ref.* **92**(11), 187 (1963); *Hydrocarbon Process.* **50**(11), 167 (1971).
122. *Hydrocarbon Process.* **50**(11), 168 (1971).
123. *Rubber World* **167**(2), 16 (1972).
124. H. J. Peterson and J. O. Tucker, *Hydrocarbon Process.* **53**(7), 21 (1974).
125. M. DeMalde, *Chim. e Ind.* (*Milan*) **45,** 665 (1963).
126. V. D. Gostinskii, *Gigiena Truda i Prof. Zabolevaniya* **9**(1), 36 (1965).
127. V. I. D'yachkov, V. K. Mishin, L. N. Berezina, and P. A. Podurueva, *Sb. Nauch. Tr. Kuibyschev. Nauch. Issled. Inst. Gig.* (7), 94 (1972).
128. S. A. Pigolev, *Sb. Nauch. Tr. Kuibyschev. Nauch. Issled. Inst. Gig.,* (6), 64 (1971).
129. A. S. Fauston, *Tr. Voronezh. Med. Ist.* (87), 10 (1972).
130. *Isoprene Technical Data Bulletin No. 1,* Goodrich-Gulf Chemicals, Inc., Cleveland, Ohio, 1968.
131. Private communication, D. G. Hendry, Stanford Research Institute, 1972.
132. *Chem. Eng. News* **34,** 292 (1956).

WILLIAM M. SALTMAN
The Goodyear Tire & Rubber Company

ISOPROPANOLAMINES. See Alkanolamines.

ISOTOPES

The term, isotope was coined to describe members of the natural radioactive series that have the same atomic number but different atomic weights. In 1912 Thompson demonstrated (1) that neon has two stable isotopes of mass 20 and 22; a rarer isotope of mass 21 also exists. The modern definition of an isotope, based on the nuclear theory of the atom, derives from the hypothesis that the atomic nucleus is composed of protons and neutrons (2). The number of protons (or atomic number Z which is equal to the number of electrons surrounding the nucleus) is associated with the chemical properties of the element, whereas the nuclear properties are dependent upon both Z and N (the latter being the neutron number). The mass number A (= N + Z) is nearly equal to the atomic mass; the difference (ca <0.1 mass unit) arises mainly from variations in the average nuclear binding energy. (The atomic mass scale is defined by $M(^{12}C) \equiv 12$.)

The environment is composed almost entirely of those isotopes that are stable or that have half-lives of about the age of the earth, ie, 5×10^9 yr or longer, and there are 286 such isotopes of 83 elements (3). In addition, there are small quantities of short-lived isotopes that are produced continuously, mostly as daughters of natural thorium (Z = 90) and uranium (Z = 92) (see Radioisotopes). Many other isotopes are produced in very small quantities by the interaction of cosmic-ray particles on stable matter; at least two of these—^3H and ^{14}C—have important applications. Many radioactive isotopes have been produced artificially; about 1965 isotopes (unique A and Z) of 106 elements are known (3).

The stable and long-lived isotopes are listed in Table 1; many of the elements are polyisotopic. The frequency of occurrence of elements with 0, 1, 2 . . . stable or long-lived isotopes is shown in Figure 1. Most odd-numbered elements consist of one or two isotopes, whereas most even-numbered elements have more than two isotopes and have an average of ca five. Only nine naturally occurring isotopes are odd–odd (ie, odd Z and N, even A). The earth and the rest of the solar system—to the extent that we have been able to observe it by way of moon rocks, meteorites, etc—display a remarkably uniform isotopic composition, which is the basis for accurate gravimetric measurements (see Space chemistry). There are two common exceptions: (*1*) isotope effects produce chemical fractionation of different isotopes of the same element; these effects are pronounced only for the lightest elements; and (*2*) elements possessing stable isotopes with long-lived, natural radioactive parents have varying isotopic compositions, depending on the geologic history of the particular sample. These include argon and calcium (^{40}K decay) strontium (^{87}Rb decay to ^{87}Sr), and lead (natural uranium and thorium decay chains beginning with ^{238}U, ^{235}U, and ^{232}Th and ending with ^{206}Pb, ^{207}Pb, and ^{208}Pb, respectively).

The causes of the variations are much better understood than the reason for the general lack of variations in the natural isotopic abundances. The synthesis of matter is dependent on nuclear characteristics (eg, cross sections and nuclear-level properties) and on ambient conditions (eg, temperature and density). The classic study of nucleosynthesis (5) hypothesizes that most elements are produced continuously in stars; only hydrogen and helium existed in the primordial material of the universe (6). In order to explain the isotopic and elemental abundances, it is necessary to assume the existence of at least eight processes that take place under a variety of very different

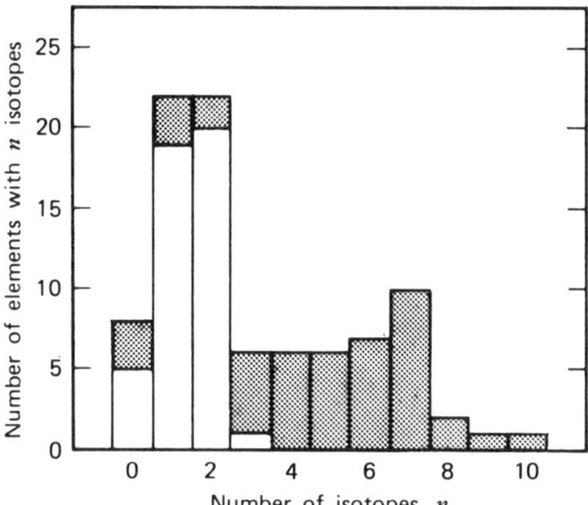

Figure 1. Distribution of the number of stable or long-lived, naturally occurring isotopes per element. Z = 1–92: ▩, even Z; □, odd Z.

conditions which are characteristic of different stages of stellar evolution. The implication of this hypothesis is a universe with nonuniform abundances, since the relative rates at which the processes occur depend on the locally varying populations of stars in different stages. Recent studies (7) have shown the existence of abundance anomalies in meteorites whose explanation, in terms of different nucleosynthetic events, has been debated (8).

Despite the feeble influence of isotopic composition upon ordinary chemical properties (deuterium is an exception; see Isotope Effects), isotopes play a significant role in research and technology, eg, in nuclear energy technology. Uranium, plutonium, and hydrogen isotopes head the list; boron and lithium also are important. Because chemical properties are hardly altered by isotopic substitution, isotopes are excellent labels for atoms within compounds. Not only do they provide a convenient way to assay for an element, they also can act as probes of chemical structure and of reaction mechanisms. The high-energy, ionizing radiations emitted by radioisotopes have important applications, and pose environmental problems in the development of nuclear energy.

Separation

Although stable isotopes can be produced by transmutation, economically feasible processes involve the separation of the isotopes of natural elements. An exception is the production of ^{3}He as the decay product of radioactive ^{3}H by the following reactions (see Fusion energy):

$$^{6}Li + n \rightarrow {}^{4}He + {}^{3}H \qquad (1)$$

$$^{3}H \xrightarrow[t_{1/2} = 12.33\ yr]{\beta^-} {}^{3}He \qquad (2)$$

The most important classes of processes that have been used or proposed for isotope separation are (*1*) processes used for small-scale commercial production of many isotopes: electromagnetic [large (≫1) separation factors], thermal diffusion, distillation (C, N, O), chemical exchange (Li, B), and photochemical or photophysical (laser)

Table 1. Stable and Long-Lived, Naturally Occurring Isotopes[a]

Z	Element	A	CAS Registry No.	Natural abundance, atom %
1	H	1	[*12385-13-6*]	99.985[b]
		2	[*16873-17-9*]	0.015
2	He	3	[*14762-55-1*]	1.38×10^{-4} [b,c]
		4	[*7440-59-7*]	99.99986
3	Li	6	[*14258-72-1*]	7.5[b,d]
		7	[*13982-05-3*]	92.5[b,d]
4	Be	9	[*7440-41-7*]	100
5	B	10	[*14798-12-0*]	19.8[b]
		11	[*14798-13-1*]	80.2[b]
6	C	12	[*7440-44-0*]	98.89[b]
		13	[*14762-74-4*]	1.11[b]
7	N	14	[*17778-88-0*]	99.63
		15	[*14390-96-6*]	0.366[b,c]
8	O	16	[*17778-80-2*]	99.76
		17	[*13968-48-4*]	0.038[b,c]
		18	[*14797-71-8*]	0.204[b,c]
9	F	19	[*14762-94-8*]	100
10	Ne	20	[*13981-34-5*]	90.51[c,d,e]
		21	[*13981-35-6*]	0.27[c,d,e]
		22	[*13886-72-1*]	9.22[c,d,e]
11	Na	23	[*7440-23-5*]	100
12	Mg	24	[*14280-39-8*]	78.99[e]
		25	[*14304-84-8*]	10.00
		26	[*13981-68-5*]	11.01[e]
13	Al	27	[*7429-90-5*]	100
14	Si	28	[*14276-58-4*]	92.23
		29	[*14304-87-1*]	4.67
		30	[*13981-69-6*]	3.10[b]
15	P	31	[*7723-14-0*]	100
16	S	32	[*13981-57-2*]	95.02[b]
		33	[*14257-58-0*]	0.75[b]
		34	[*13965-97-4*]	4.21[b]
		36	[*14682-80-5*]	0.017[b]
17	Cl	35	[*13981-72-1*]	75.77
		37	[*13981-73-2*]	24.23
18	Ar	36	[*13965-95-2*]	0.337[b,c]
		38	[*13994-72-4*]	0.063[b,c]
		40	[*7440-37-1*]	99.60[b,c,f]
19	K	39	[*14092-91-2*]	93.26
		40	[*13966-00-2*]	0.0117[g]
		41	[*13965-93-0*]	6.73
20	Ca	40	[*14092-94-5*]	96.94[f]
		42	[*14333-05-2*]	0.647
		43	[*14333-06-3*]	0.135
		44	[*14255-03-9*]	2.09
		46	[*13981-77-6*]	0.0035
		48	[*13981-76-5*]	0.187
21	Sc	45	[*7440-20-2*]	100
22	Ti	46	[*14304-89-3*]	8.2
		47	[*14304-90-6*]	7.4
		48	[*14304-91-7*]	73.7
		49	[*14304-92-8*]	5.4
		50	[*14304-93-9*]	5.2
23	V	50	[*14391-89-0*]	0.250[e]
		51	[*7440-62-2*]	99.750
24	Cr	50	[*14304-94-0*]	4.35
		52	[*14092-98-9*]	83.79
		53	[*13981-78-7*]	9.50
		54	[*14304-97-3*]	2.36
25	Mn	55	[*7439-95-5*]	100
26	Fe	54	[*13982-24-6*]	5.8
		56	[*14093-02-8*]	91.8
		57	[*14762-69-7*]	2.15
		58	[*13968-47-3*]	0.29
27	Co	59	[*7440-48-4*]	100
28	Ni	58	[*13981-79-8*]	68.3
		60	[*13981-80-1*]	26.1
		61	[*14928-10-0*]	1.13
		62	[*13981-81-2*]	3.59
		64	[*14378-31-5*]	0.91
29	Cu	63	[*14191-84-5*]	69.2[b]
		65	[*14119-06-3*]	30.8[b]
30	Zn	64	[*14378-32-6*]	48.6
		66	[*14378-33-7*]	27.9
		67	[*14378-34-8*]	4.10
		68	[*14378-35-9*]	18.8
		70	[*14378-36-0*]	0.62
31	Ga	69	[*14391-02-7*]	60.1
		71	[*14391-03-8*]	39.9
32	Ge	70	[*14687-55-9*]	20.5
		72	[*13982-21-3*]	27.4
		73	[*15034-58-9*]	7.8
		74	[*15034-59-0*]	36.5
		76	[*14687-41-3*]	7.8
33	As	75	[*7440-38-2*]	100
34	Se	74	[*13981-33-4*]	0.87
		76	[*13981-32-3*]	9.0
		77	[*14681-72-2*]	7.6
		78	[*14833-16-0*]	23.5
		80	[*14681-54-0*]	49.8
		82	[*14687-58-2*]	9.2[h]
35	Br	79	[*14336-94-8*]	50.69
		81	[*14380-59-7*]	49.31
36	Kr	78	[*33580-79-9*]	0.356[d]
		80	[*26110-67-8*]	2.27
		82	[*14191-81-2*]	11.6
		83	[*13905-98-5*]	11.5
		84	[*14993-91-0*]	57.0
		86	[*14191-82-3*]	17.3
37	Rb	85	[*13982-12-2*]	72.17
		87	[*13982-13-3*]	27.83[i]
38	Sr	84	[*15758-49-3*]	0.56
		86	[*13982-14-4*]	9.8
		87	[*13982-64-4*]	7.0[f]
		88	[*14119-10-9*]	82.6
39	Y	89	[*7440-65-5*]	100
40	Zr	90	[*13982-15-5*]	51.5
		91	[*14331-93-2*]	11.2
		92	[*14392-15-5*]	17.1
		94	[*14119-12-1*]	17.4

Table 1 (*continued*)

Z	Element	*A*	CAS Registry No.	Natural abundance, atom %
		96	[*15691-06-2*]	2.80
41	Nb	93	[*7440-03-1*]	100
42	Mo	92	[*14191-67-4*]	14.8
		94	[*14683-00-2*]	9.3
		95	[*14392-17-7*]	15.9
		96	[*14274-76-1*]	16.7
		97	[*14392-19-9*]	9.6
		98	[*14392-20-2*]	24.1
		100	[*14392-21-3*]	9.6
43	Tc	(no stable isotope)		
44	Ru	96	[*15128-32-2*]	5.5
		98	[*18393-13-0*]	1.86
		99	[*15411-62-8*]	12.7
		100	[*14914-60-4*]	12.6
		101	[*14914-61-5*]	17.0
		102	[*14914-62-6*]	31.6
		104	[*15766-01-5*]	18.7
45	Rh	103	[*7440-16-6*]	100
46	Pd	102	[*14833-50-2*]	1.0
		104	[*15128-18-4*]	11.0
		105	[*15749-57-2*]	22.2
		106	[*14914-59-1*]	27.3
		108	[*15749-58-3*]	26.7
		110	[*15749-60-7*]	11.8
47	Ag	107	[*14378-37-1*]	51.83
		109	[*14378-38-2*]	48.17
48	Cd	106	[*14378-39-3*]	1.25
		108	[*14191-62-9*]	0.89
		110	[*14191-63-0*]	12.5
		111	[*14336-64-2*]	12.8
		112	[*14336-65-3*]	24.1
		113	[*14336-66-4*]	12.2[j]
		114	[*14041-58-8*]	28.7
		116	[*14390-59-1*]	7.5
49	In	113	[*14885-78-0*]	4.3
		115	[*14191-71-0*]	95.7[k]
50	Sn	112	[*15125-53-8*]	1.01
		114	[*14998-72-2*]	0.67
		115	[*14191-72-1*]	0.38
		116	[*14191-70-9*]	14.8
		117	[*13981-59-4*]	7.75
		118	[*14914-65-9*]	24.3
		119	[*14314-35-3*]	8.6
		120	[*14119-17-6*]	32.4
		122	[*14119-18-7*]	4.56
		124	[*14392-29-1*]	5.64
51	Sb	121	[*14265-72-6*]	57.3
		123	[*14119-16-5*]	42.7
52	Te	120	[*15125-47-0*]	0.091
		122	[*14390-71-7*]	2.5
		123	[*14304-80-4*]	0.89
		124	[*14390-72-8*]	4.6
		125	[*14390-73-9*]	7.0
		126	[*14390-74-0*]	18.7
		128	[*14390-75-1*]	31.7[l]
		130	[*14390-76-2*]	34.5[m]
53	I	127	[*14362-44-8*]	100
54	Xe	124	[*15687-60-2*]	0.096[d]
		126	[*27982-81-6*]	0.090
		128	[*27818-78-6*]	1.92
		129	[*13965-99-6*]	26.4
		130	[*14808-51-6*]	4.1
		131	[*14683-11-5*]	21.2
		132	[*14155-79-4*]	26.9
		134	[*15751-43-6*]	10.4
		136	[*15751-79-8*]	8.9
55	Cs	133	[*7440-46-2*]	100
56	Ba	130	[*15055-15-9*]	0.106
		132	[*15065-85-7*]	0.101
		134	[*15193-77-8*]	2.42
		135	[*14698-58-9*]	6.59
		136	[*15125-64-1*]	7.85
		137	[*13981-97-0*]	11.2
		138	[*15010-01-2*]	71.7[f]
57	La	138	[*15816-87-2*]	0.089[n]
		139	[*7439-91-0*]	99.911
58	Ce	136	[*15758-67-5*]	0.190
		138	[*15758-26-6*]	0.254[f]
		140	[*14191-73-2*]	88.5
		142	[*14119-20-1*]	11.1
59	Pr	141	[*7440-10-0*]	100
60	Nd	142	[*14336-82-4*]	27.2
		143	[*14336-83-5*]	12.2
		144	[*14834-76-5*]	23.8[o]
		145	[*14336-84-6*]	8.3
		146	[*15411-67-3*]	17.2
		148	[*14280-26-3*]	5.7
		150	[*14683-22-8*]	5.6
61	Pm	(no stable isotope)		
62	Sm	144	[*14981-82-9*]	3.1
		147	[*14392-33-7*]	15.1[p]
		148	[*14913-64-5*]	11.3[q]
		149	[*14392-34-8*]	13.9
		150	[*14907-33-6*]	7.4
		152	[*14280-32-1*]	26.6
		154	[*14833-41-1*]	22.6
63	Eu	151	[*14378-48-4*]	47.9
		153	[*13982-02-0*]	52.1
64	Gd	152	[*14867-54-0*]	0.20[r]
		154	[*14683-24-0*]	2.1
		155	[*14333-34-7*]	14.8
		156	[*14392-07-5*]	20.6
		157	[*14391-32-3*]	15.7
		158	[*15068-71-0*]	24.8
		160	[*14834-81-2*]	21.8
65	Tb	159	[*7440-27-9*]	100
66	Dy	156	[*15720-39-5*]	0.057
		158	[*14913-25-8*]	0.100
		160	[*14683-25-1*]	2.3
		161	[*13967-68-5*]	19.0

Table 1 *(continued)*

Z	Element	A	CAS Registry No.	Natural abundance, atom %	Z	Element	A	CAS Registry No.	Natural abundance, atom %
		162	[*14834-85-6*]	25.5	76	Os	184	[*14922-68-0*]	0.018
		163	[*14391-36-7*]	24.9			186	[*13982-09-7*]	1.6[v]
		164	[*13967-69-6*]	28.1			187	[*15766-52-6*]	1.6[f]
67	Ho	165	[*7440-60-0*]	100			188	[*14274-81-8*]	13.3
68	Er	162	[*15840-05-8*]	0.14			189	[*15761-06-5*]	16.1
		164	[*14900-10-8*]	1.56			190	[*14274-79-4*]	26.4
		166	[*14900-11-9*]	33.4			192	[*15062-08-5*]	41.0
		167	[*14380-60-0*]	22.9	77	Ir	191	[*13967-66-3*]	37.3
		168	[*14833-43-3*]	27.1			193	[*13967-67-4*]	62.7
		170	[*15701-24-3*]	14.9	78	Pt	190	[*15735-68-9*]	0.013[w]
69	Tm	169	[*7440-30-4*]	100			192	[*14913-85-0*]	0.78
70	Yb	168	[*15743-54-1*]	0.135			194	[*14998-96-0*]	32.9
		170	[*13982-08-6*]	3.1			195	[*14191-88-9*]	33.8
		171	[*14041-50-0*]	14.4			196	[*14867-61-9*]	25.3
		172	[*14041-52-2*]	21.9			198	[*15756-63-5*]	7.2
		173	[*14041-51-1*]	16.2	79	Au	197	[*7440-57-5*]	100
		174	[*14683-29-5*]	31.6	80	Hg	196	[*14917-67-0*]	0.15
		176	[*15751-45-8*]	12.6			198	[*13981-21-0*]	10.0[b]
71	Lu	175	[*14391-25-4*]	97.39			199	[*14191-87-8*]	16.8
		176	[*14452-47-2*]	2.61[s]			200	[*15756-10-2*]	23.1
72	Hf	174	[*14922-49-7*]	0.16[t]			201	[*15185-19-0*]	13.2
		176	[*14452-48-3*]	5.2[f]			202	[*14191-86-7*]	29.8[b]
		177	[*14093-09-5*]	18.6			204	[*15756-14-6*]	6.9
		178	[*14265-77-1*]	27.1	81	Tl	203	[*14280-48-9*]	29.5
		179	[*14265-76-0*]	13.7			205	[*14280-49-0*]	70.5
		180	[*14265-78-2*]	35.2	82	Pb	204	[*13966-26-2*]	1.42
73	Ta	180	[*15759-29-2*]	0.0123			206	[*13966-27-3*]	24.1[f]
		181	[*7440-25-7*]	99.9877			207	[*14119-29-0*]	22.1[f]
74	W	180	[*14265-79-3*]	0.13			208	[*13966-28-4*]	52.3[f]
		182	[*14265-80-6*]	26.3	83	Bi	209	[*7440-69-9*]	100
		183	[*14365-81-7*]	14.3	84–89		(no stable isotopes)		
		184	[*14265-82-8*]	30.7	90	Th	232	[*7440-29-1*]	100[x]
		186	[*14265-83-9*]	28.6	91	Pa	(no stable isotope)		
75	Re	185	[*14391-28-7*]	37.40	92	U	235	[*15117-96-1*]	0.720[d,e,y]
		187	[*14391-29-8*]	62.60[u]			238	[*24678-82-8*]	99.275[d,e,z]

[a] Data taken from refs. 3–4. Natural abundances are given in atom %; the uncertainty in the last place is ≲5 units. (Note that, because of rounding off, the abundances for an element do not always add to 100%.)

[b] Natural isotopic composition of normal terrestrial matter varies.

[c] Atmospheric sources.

[d] Isotopic composition of some commercially available sources has been altered by processing.

[e] Natural isotopic composition can vary for some anomalous sources.

[f] Isotopic composition can vary as a result of the existence of a radioactive parent.

[g] Radioactive: $t_{1/2} = 1.28 \times 10^9$ yr.

[h] Slightly radioactive: $t_{1/2} = 1.4 \times 10^{20}$ yr.

[i] Radioactive: $t_{1/2} = 4.8 \times 10^{10}$ yr.

[j] Slightly radioactive: $t_{1/2} = 9 \times 10^{15}$ yr.

[k] Slightly radioactive: $t_{1/2} = 5.1 \times 10^{14}$ yr.

[l] Slightly radioactive: $t_{1/2} = 1.5 \times 10^{24}$ yr.

[m] Slightly radioactive: $t_{1/2} = 2 \times 10^{21}$ yr.

[n] Radioactive: $t_{1/2} = 1.1 \times 10^{11}$ yr.

[o] Slightly radioactive: $t_{1/2} = 2.1 \times 10^{15}$ yr.

[p] Radioactive: $t_{1/2} = 1.06 \times 10^{11}$ yr.

[q] Slightly radioactive: $t_{1/2} = 8 \times 10^{15}$ yr.

[r] Slightly radioactive: $t_{1/2} = 1.1 \times 10^{14}$ yr.

[s] Radioactive: $t_{1/2} = 3.6 \times 10^{10}$ yr.

[t] Slightly radioactive: $t_{1/2} = 2.0 \times 10^{15}$ yr.

[u] Radioactive: $t_{1/2} = 4 \times 10^{10}$ yr.

[v] Slightly radioactive: $t_{1/2} = 2 \times 10^{15}$ yr.

[w] Radioactive: $t_{1/2} = 6 \times 10^{11}$ yr.

[x] Radioactive: $t_{1/2} = 1.41 \times 10^{10}$ yr.

[y] Radioactive: $t_{1/2} = 7.038 \times 10^8$ yr.

[z] Radioactive: $t_{1/2} = 4.468 \times 10^9$ yr.

processes [large (≫1) separation factors, pilot plants planned or under construction]; (*2*) processes used for large-scale commercial production: chemical exchange (H), distillation (H), electrolysis (H), gaseous diffusion (U), gravitational (gas centrifuge) (U, commercial plants planned), aerodynamic (separation nozzle) (U, commercial plants planned), and photochemical or photophysical (laser) processes (U, H) [large (≫1) separation factors, pilot plants planned or under construction]; (*3*) other processes: chromatographic [ion exchange (qv)], mass and sweep diffusion, adsorption, electromigration, and biological processes. Because most of these processes have small intrinsic separation capabilities, many separations in series are required in order to achieve a product of high isotopic purity. The use of repeated separations requires the application of cascade theory in most separation processes; cascade theory and all isotope separation processes that involve diffusion are discussed in detail in the article Diffusion separation methods (9).

The effectiveness of the separation is measured by the enhancement of the desired isotope and is defined by the separation factor:

$$\alpha = \frac{y/(1-y)}{x/(1-x)} \tag{3}$$

where y = the composition of the heads stream and x = the composition of the tails stream. α is nearly independent of the compositions for many separation processes, whereas the heads separation factor,

$$\beta = \frac{y/(1-y)}{z/(1-z)} \tag{4}$$

where z is the composition of the feed stream, depends on the cut, or ratio of the heads flow rate to the feed flow rate. (x, y, and z are expressed as mol fractions of the desired isotope.)

A single unit of a cascade is referred to as a separating unit. It is the smallest unit capable of performing some separation, eg, a single diffusion barrier, a gas centrifuge, or one plate of a distillation (qv) column. An ensemble of separating units in parallel constitutes one stage of the cascade. The number of separating units in a given stage and the manner in which the output (heads and tails) from one stage are connected to the feed of other stages can be varied. The theory of separation cascades deals with the manner in which the individual separation units are interconnected and operated (ie, the value of β).

The ideal cascade is approximated by all large-scale plants based on processes for which the separation factor is close to unity. It has the following properties: the heads separation factor is constant; there is no mixing of streams of different composition (an ideal cascade is sometimes referred to as a no-mixing cascade); and the total interstage flow, and thus the cost and energy, that is required to produce a product and waste with specified compositions is minimum.

The factors which dictate the choice of a process for use in the laboratory or in a small-scale plant—where flexibility and low capital cost are important—are very different from the factors governing industrial-scale plants, where low energy and materials use, ease of operation, and high reliability tend to be more important. Some generalizations (10–11) are: (*1*) the most versatile means for the production of research quantities of isotopes is the electromagnetic method; (*2*) the simplest and most inexpensive means for small-scale separation of many isotopes is the Clusius thermal-diffusion column; (*3*) distillation and chemical exchange are the most economical methods for the large-scale separation of the lighter elements; (*4*) gaseous diffusion and the gas centrifuge are most economical for the large-scale separation of the heaviest elements.

Electromagnetic. Electromagnetic separators are high current, mass spectrometers in which the ion-current detector is replaced by a collector. In 1943 a very large facility, known as Y-12, was constructed at Oak Ridge to separate ^{235}U. When the process was abandoned in 1946 in favor of gaseous diffusion, some of the more than 1100 units were saved and used in a program to produce stable isotopes for research (12). The separators at Y-12, known as calutrons (13), are 180° magnetic focusing devices having radii of 61 or 122 cm. Figure 2 is a partial cutaway view of a calutron and shows the ion orbits. According to electromagnetic theory, the orbital radius is:

$$r = \sqrt{2\,mV/e}/B \tag{5}$$

where m and e are the mass and charge of the ion, V is the accelerating potential, and B is the magnetic field strength. The spatial separation between two isotopes at the

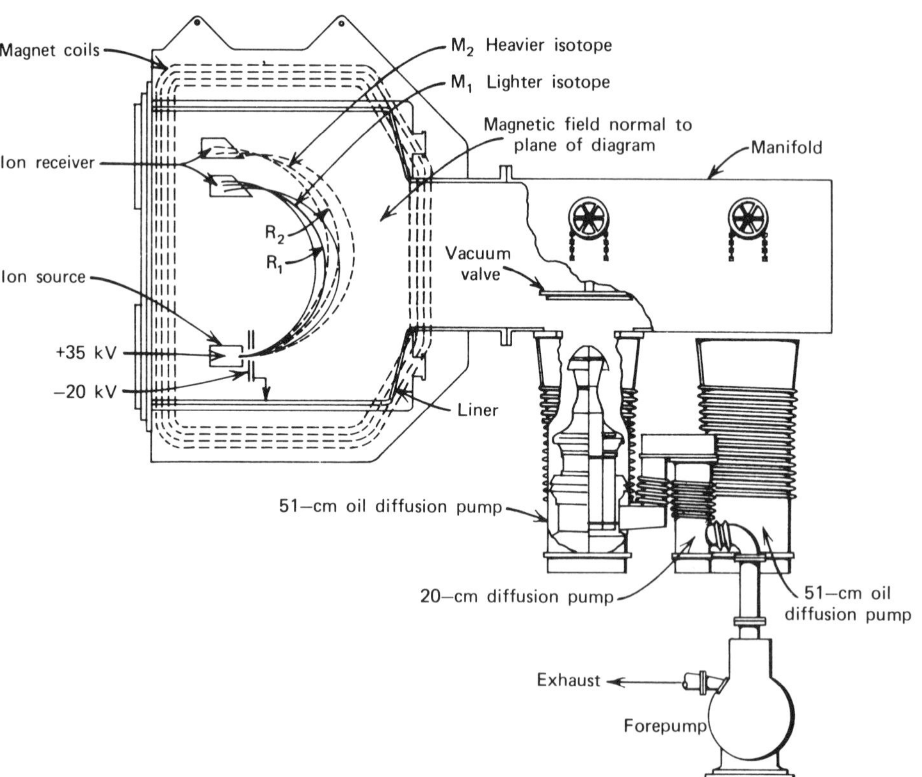

Figure 2. Diagram of a 61-cm calutron. The magnetic field is perpendicular to the plane of the page (12).

focal plane is:

$$2\Delta r = r\Delta m/m \quad (6)$$

Values for $r = 61$ cm range from 2.5 mm for adjacent isotopes of plutonium to 94 mm for lithium isotopes. Since the beam spread is quite small, the electromagnetic method (almost uniquely) provides a good separation in a single stage; in the calutrons, the enrichment factor (β) ranges from 30 to as high as 80,000 (12). A limitation is the restriction of beam current that is necessary to prevent the loss of beam quality or stability. The maximum current of the calutron beams, about 0.01–0.1 A (depending on the element), is equivalent to 0.009–0.09 mol/d for a singly ionized beam. The corresponding separative capacity—assuming a separation factor β of 100—is about 0.04–0.4 mol SWU/d [SWU = separative work unit (see Diffusion separation methods (9))]. Since the throughput limitation applies to the total beam current, the production rate for a pure isotope is inversely proportional to its abundance in the feed material. The beam intensity is high, however, in terms of its energy dissipation (3.5 kW for a 0.1-A beam), which necessitates careful collector design and cooling in order to avoid the loss of the product by vaporization or sputtering. Special collector design and, in some cases, special ion source design are required for each element. The electrical power consumption of the calutrons is roughly 0.1 MW per tank, so that roughly 10^5–10^6 MJ (1 MJ = 239 kcal or 949 Btu) is required to separate one mol of material in a single pass, or 2–20 × 10^4 MJ/mol SWU.

The elements that have been separated from 1966 through 1979 and which include all natural polyisotopic elements except hydrogen and the noble gases are listed in Table 2. Separated isotopes are sold or loaned from a pool; sales for the period 1966–1972 were close to \$1 × 10^6/yr, with the equivalent value of new loan materials in the range \$(3–7) × 10^6/yr (information on prices and availability is given in ref. 15). At current, reduced production rates, the inventories accumulated in previous years are decreasing. A program of electromagnetic separation was begun by the USSR around 1968. Pricing is competitive with Oak Ridge and USSR sales to European customers in the last few years have been significant. An electromagnetic separator at Harwell (UK), Hermes, has been used to separate plutonium and other radioactive or toxic substances (16).

Thermal Diffusion. The thermal diffusion effect, the tendency of lighter components of a fluid subject to a temperature gradient to concentrate in the hotter region, was observed in 1916 (17); practical applications followed from the invention of the countercurrent thermal diffusion column in 1938 (18). The theory of thermal diffusion columns (19–20) is in satisfactory agreement with the measured performance of precisely constructed columns (20). Mound Laboratory (Miamisburg, Ohio) uses this method to produce isotopes of carbon and the noble gases. Up to 19 columns are interconnected in parallel series to form an appropriate squared-off cascade (9). Figure 3 shows several arrangements used to separate krypton isotopes. Based on the data presented (21), the single-column separation factor for ^{84}Kr–^{86}Kr is about 2.6, and the separative capacity of one column is about 1 g Kr SWU/d. Commercial distribution of separated isotopes by Mound was begun in 1960; a summary of 1978 sales is given in Table 3.

Energy consumption by thermal diffusion columns is high. For example, the columns used to separate UF_6, which had a separative capacity of 2.46 kg U SWU/yr,

Table 2. Elements Separated Electromagnetically at Oak Ridge, 1966–1979[a]

Element	Total weight separated, g	Element	Total weight separated, g	Element	Total weight separated, g
H		As	[c]	Tb	2.1[b,c]
He		Se	1,191	Dy	1,178
Li	2,985	Br	572	Ho	[c]
Be	0.003[b,c]	Kr		Er	1,664
B	65	Rb	1,586	Tm	[c]
C	150	Sr	13,710	Yb	3,584
N	41	Y	[c]	Lu	376
O	0.022	Zr	3,134	Hf	1,084
F	[c]	Nb	[c]	Ta	1,427
Ne		Mo	12,385	W	9,576
Na	[c]	Tc	[d]	Re	370
Mg	2,889	Ru	114	Os	334
Al	[c]	Rh	[c]	Ir	195
Si	5,757	Pd	385	Pt	294
P	[c]	Ag	801	Au	[c]
S	3,447	Cd	4,312	Hg	852
Cl	752	In	531	Tl	6,974
Ar		Sn	6,123	Pb	5,280
K	2,200	Sb	384	Bi	51[b,c]
Ca	52,460	Te	7,410	Po	[d]
Sc	[c]	I	[c]	At	[d]
Ti	2,561	Xe		Rn	[d]
V	186	Cs	[c]	Fr	[d]
Cr	4,959	Ba	1,963	Ra	[d]
Mn	[c]	La	294	Ac	[d]
Fe	32,657	Ce	2,123	Th	35[b]
Co	[c]	Pr	[c]	Pa	[d]
Ni	15,401	Nd	2,571	U	6,653[b]
Cu	3,517	Pm	[d]	Np	[d]
Zn	4,552	Sm	2,666	Pu	1,506[b,d]
Ga	713	Eu	627	Am	2[b,d]
Ge	1,984	Gd	1,911	Cm	0.5[b,d]

[a] Data from refs. 12 and 14. The weight includes all production in the years indicated, with exception of a few "waste" isotopes that were discarded.
[b] Includes artificially produced isotopes.
[c] Monoisotopic element.
[d] No stable isotopes.

consumed about 93 kW of power in the form of steam at 559 K (23). The power consumption was about 10^6 MJ/kg U SWU.

Distillation. In the distillation of a mixture of compounds that differ by isotopic substitution, the lighter components tend to concentrate in the vapor phase. For a compound containing n equivalent atoms of the element in question, the separation factor is closely approximated by:

$$\alpha = \sqrt[n]{\pi(\mathrm{XA}_n)/\pi(\mathrm{XB}_n)} \tag{7}$$

where $\pi(\mathrm{XA}_n)$ is the vapor pressure of the isotopically pure compound XA_n, and the mol fractions in the definition of α (eq. 3) refer to the total concentration of the isotope A or B, irrespective of the chemical form, ie:

$$x(\mathrm{A}) = nx(\mathrm{XA}_n) + (n-1)x(\mathrm{XA}_{n-1}\mathrm{B}) + \ldots \tag{8}$$

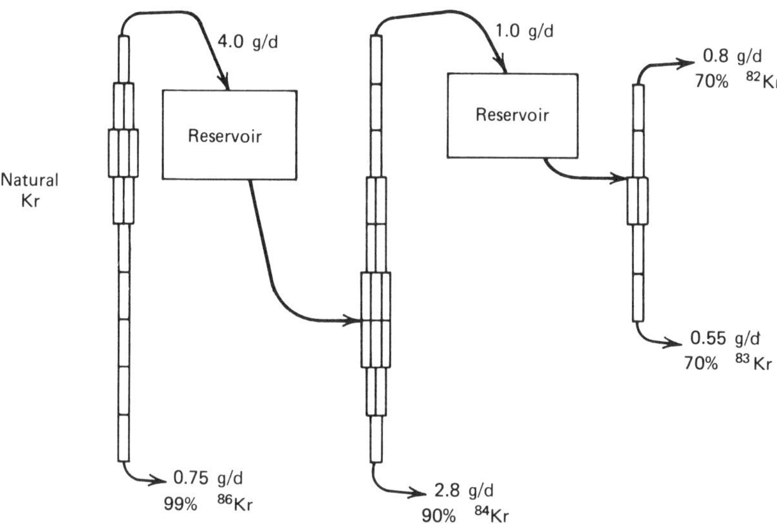

Figure 3. Cascade arrangement of thermal diffusion columns used at Mound Laboratory to separate krypton isotopes (21).

Deviations from this Raoult's law value are important only for ^3He–^4He and H_2–HD–D_2 mixtures; with these two exceptions, the value of ln α predicted by equation 7 is accurate to better than 10% (23). Some examples of separation factors that are estimated using equation 7 are listed in Table 4. The separation factor decreases with increasing temperature so that low temperatures generally are advantageous.

Because of the rapid decrease of the separation factor with increasing mass, distillation is a suitable method only for separation of light elements. The most important application today is as the final step in the separation of deuterium, where distillation of water is the preferred method (23). The use of water distillation for low enrichment portions of the process generally is unfavorable because of the large columns required (24). However, primary enrichment of deuterium using low-grade waste heat has been proposed (25) (see Deuterium and tritium).

Hydrogen (H_2) distillation (23) has the advantage of a large separation factor (see Table 4) but poses problems because of the very low temperatures required. Like all deuterium plants whose feed is elemental hydrogen, hydrogen distillation plants are parasitic to synthetic ammonia production, which sets a limit on deuterium production rates. Ammonia distillation often is preferred as the final stage of enrichment of deuterium plants whose primary enrichment is based on ammonia–hydrogen exchange (23).

Cryogenic distillation of carbon monoxide is the preferred method for separation of ^{13}C in quantities greater than 500 g/yr (see Cryogenics) (21). Distillation of CO and the distillation of nitric oxide to separate isotopes of both oxygen and nitrogen is carried out at Los Alamos Scientific Laboratory (21,26).

Chemical Exchange. The exchange of two isotopes (^1A,^2A) between two compounds (AX, AY) is a chemical reaction:

$$^1AX + {^2AY} \rightleftarrows {^2AX} + {^1AY} \tag{9}$$

whose equilibrium constant is:

Table 3. Stable Isotope Sales by the Mound Facility in 1978[a]

Isotope	Enrichment, atom %	Total sales, g[b]	Average price, $/g[c]
^{3}He	99.9–99.9995	963	804
^{12}C	99.9, 99.95	327	8.24
^{13}C	20–99	2,861	69
^{14}N	99.99	689,818	0.156
^{15}N	40, 99	1,353	95
^{16}O	99.98	9,093	1.41
^{17}O	20–55	73.7	281
^{18}O	20–99	867	90
^{20}Ne	99.95	113	92
^{21}Ne	90	0.20	16,732
^{22}Ne	70–99.9	40	225
^{34}S	90	17.7	313
^{35}Cl	90	12.0	278
^{37}Cl	90	16.6	480
^{36}Ar	99.5	2.98	2,506
^{38}Ar	25, 95	0.80	2,654
^{40}Ar	99.95	253	8.33
^{78}Kr	8–99	17.2	688
^{80}Kr	70, 90	0.46	7,400
^{82}Kr	70, 90	38.4	2,048
^{83}Kr	70	2.89	2,293
^{84}Kr	90	24.3	240
^{86}Kr	99	47.0	168
^{124}Xe	1–40	3.18	15,924
^{126}Xe	2		
^{129}Xe	60	1.15	760
^{131}Xe	60	0.12	2,096
^{134}Xe	50		
^{136}Xe	80–99	61.2	1,282

[a] Ref. 22. Nitrogen, oxygen, and (in part) carbon isotopes were enriched at Los Alamos Scientific Laboratory by distillation; all other isotopes were enriched at Mound Laboratory by thermal diffusion.

[b] Quantities stated in liters (22) were converted to grams, assuming STP (273 K, 101.3 kPa = 1 atm).

[c] Rounded to the nearest dollar for values over $50.

$$K = \frac{a(^{2}AX)\,a(^{1}AY)}{a(^{1}AX)\,a(^{2}AY)} \tag{10}$$

To the approximation that the activities $[a(AX), a(AY)]$ of the substances are proportional to the concentrations ([AX], [AY]), the separation factor is equal to K. For compounds containing more than one atom of element A, the concentrations of ^{1}A and ^{2}A entering into the definition of α (eq. 3) are given by equation 8, and the equivalence of α and K is no longer valid. In the limit of low abundance for the isotope of interest, say ^{1}A, the separation factor for exchange between compounds A_nX and A_mY—having n and m exchangeable atoms of element A, respectively—is given by:

$$\alpha = \frac{n}{m} K \tag{11}$$

Table 4. Separation Factors for Distillation[a]

Compounds	Triple point		Normal bp	
	T, °C	α	T, °C	α
ortho H_2–HD	−259.4	3.61	−252.9	1.81
NH_3–ND_3	−77.7	1.080	−33.6	1.036
H_2O–D_2O	0.0	1.120	100	1.026
^{12}CO–^{13}CO	−205.7	1.0113	−191.3	1.0068
^{14}NO–^{15}NO	−163.6	1.033	−151.8	1.027
$N^{16}O$–$N^{18}O$	−163.6	1.046	−151.8	1.037
$H_2{}^{16}O$–$H_2{}^{18}O$	0	1.010	100	1.0046
^{20}Ne–^{22}Ne	−248.6	1.046	−245.9	1.038
^{36}Ar–^{40}Ar	−189.4	1.006	−185.7	
^{128}Xe–^{136}Xe	−111.8	1.000	−109.1	

[a] Estimated (eq. 7) from vapor pressure ratios; ref. 23.

Equation 11 is a reasonable approximation to the true separation factor for most systems of interest.

Separation factors for some reactions are listed in Table 5; α decreases rapidly with increasing atomic number. Although chemical exchange methods have been studied for many elements (including carbon, nitrogen, sulfur, calcium, and uranium), they are the methods of choice only for hydrogen, lithium, and boron. Of the hundreds of methods that have been studied for heavy-water production (24), most involve chemical exchange.

Bithermal Exchange Processes: the GS Process. One of the factors that favor the GS (Girdler sulfide) process is the use of a bithermal exchange mechanism to avoid the once-through use of costly reagents. The effectiveness of the mechanism is illustrated by contrast with a monothermal system (23) in which water is enriched by exchange with hydrogen sulfide, which then is restored to its original deuterium concentration:

$$H_2S + \tfrac{1}{2}\,O_2 \rightarrow H_2O + S \tag{31}$$

$$S + \tfrac{2}{3}\,Al \rightarrow \tfrac{1}{3}\,Al_2S_3 \tag{32}$$

$$\tfrac{1}{3}\,Al_2S_3 + H_2O \rightarrow \tfrac{1}{3}\,Al_2O_3 + H_2S \tag{33}$$

In this hypothetical process, the net penalty for the phase conversion is the degradation of aluminum to aluminum oxide (less a credit for the heat produced by the partial oxidation of H_2S). Since reflux ratios for the production of D_2O are large—because of its low natural abundance (0.015%)—the penalty for phase conversion is large. For the system described, the ratio is at least:

$$\frac{1}{z_F} \times \frac{\alpha}{\alpha - 1} = \frac{1}{0.00015} \times \frac{2.32}{1.32} = 11{,}700 \tag{34}$$

So that at least 11,700 × ⅔ = 7800 mol of aluminum are consumed per mol of D_2O produced.

The phase-conversion step, which is analogous to reboiling in distillation and is unavoidable in a monothermal process, usually is costly in materials and/or energy.

Table 5. Separation Factors for Some Isotopic Exchange Reactions[a]

Equation no.	Reaction	α (at 25°C)
	Hydrogen	
(12)	$H_2O(l) + HD \rightleftarrows HDO(l) + H_2$	3.81
(13)	$H_2O(l) + HDS \rightleftarrows HDO(l) + H_2S$	2.37
(14)	$NH_3(l) + HD \rightleftarrows NH_2D(l) + H_2$	3.62
(15)	$CH_3NH_2(l) + HD \rightleftarrows CH_3NHD(l) + H_2$	4.0
(16)	$H_2O(l) + NH_2D \rightleftarrows HDO(l) + NH_3$	1.00
(17)	$H_2O(l) + PH_2D \rightleftarrows HDO(l) + PH_3$	2.44
(18)	$H_2O(l) + DCl \rightleftarrows HDO(l) + HCl$	2.51
	Lithium	
(19)	^6LiX (in $(CH_3)_2NOCH$) + $^7Li(Hg) \rightleftarrows {}^7LiX$ (in $(CH_3)_2NOCH$) + $^6LiX(Hg)$ [X = Cl, Br]	1.05
(20)	$^6LiOH\ (aq) + {}^7Li(Hg) \rightleftarrows {}^7LiOH\ (aq) + {}^6Li(Hg)$	1.07
	Boron	
(21)	$^{10}BF_3 + (CH_3)_2O.{}^{11}BF_3(l) \rightleftarrows {}^{11}BF_3 + (CH_3)_2O.{}^{10}BF_3(l)$	1.027[b]
(22)	$^{10}BF_3 + (CH_3)_2S.{}^{11}BF_3(l) \rightleftarrows {}^{11}BF_3 + (CH_3)_2S.{}^{10}BF_3(l)$	1.037[b]
	Carbon	
(23)	$^{13}CO_2 + H^{12}CO_3^- \rightleftarrows {}^{12}CO_2 + H^{13}CO_3^-$	1.012
(24)	$H^{12}CN + {}^{13}CN^- \rightleftarrows H^{13}CN + {}^{12}CN^-$	1.013
	Nitrogen	
(25)	$^{15}NO + H^{14}NO_3\ (aq) \rightleftarrows {}^{14}NO = H^{15}NO_3\ (aq)$	1.055
(26)	$^{14}NH_3 + {}^{15}NH_4^+ \rightleftarrows {}^{15}NH_3 + {}^{14}NH_4^+$	1.034
	Sulfur	
(27)	$^{34}SO_2 + H^{32}SO_3^- \rightleftarrows {}^{32}SO_2 + H^{34}SO_3^-$	1.019
	Calcium	
(28)	$^{40}Ca^{2+}(aq) + {}^{44}Ca^{2+}(C_{20}H_{36}O_6) \rightleftarrows {}^{44}Ca^{2+}(aq) + {}^{40}Ca^{2+}(C_{20}H_{36}O_6)$	1.004[c]
(29)	$^{235}UF_6 + {}^{238}UF_5NOF \rightleftarrows {}^{238}UF_6 + {}^{235}UF_6NOF$	1.0016
(30)	^{235}U(IV, Dowex 50) + $^{238}UO_2^{2+} \rightleftarrows {}^{238}U$(IV, Dowex 50) + $^{235}UO_2^{2+}$	1.0005[d]

[a] Data from ref. 23 and references contained therein, unless otherwise noted.
[b] Ref. 27.
[c] Ref. 21. The ligand is a crown polyether and is dissolved in an organic solvent.
[d] Ref. 29.

Dual-temperature processes substitute thermal reflux for phase conversion by making use of the decrease of the separation factor with increasing temperature. The effective separation factor in a dual-temperature process, in which one substance is separated into enriched and depleted fractions and the other substance is recycled continuously, is given by:

$$\alpha = \frac{\alpha_c}{\alpha_h} \tag{35}$$

where α_c and α_h are the separation factors at the lower and higher temperatures, respectively.

The temperature dependence of the separation factor, upon which any bithermal process relies, is illustrated in Figure 4 for some exchange reactions of hydrogen. The data are fitted to curves of the form:

$$\ln \alpha \simeq \ln K = \frac{b}{T} + a \tag{36}$$

where:

$$b = \frac{\delta \ln K}{\delta(1/T)} = -\frac{\Delta H}{R} \tag{37}$$

$$a = \ln K - \frac{b}{T} = \frac{\Delta F}{RT} + \frac{\Delta H}{RT} = \frac{\Delta S}{R} \tag{38}$$

For comparison with the principal hydrogen-exchange reactions, a boron-exchange reaction also is shown in Figure 4. The small temperature dependence of the separation factor precludes the use of bithermal processes for elements other than hydrogen.

Only one of the two large GS plants built in the United States is still operational. The Savannah River plant (23) uses two stages (each comprised of multistage towers) to enrich the D_2O content of water from 0.0147% to 15%. Further concentration to reactor-grade (99.75%) D_2O is accomplished by water distillation and electrolysis. Production of the Savannah River plant has been cut back from 480 to 69 t D_2O/yr; a similar plant at Dana was shut down in 1958. Canadian plants are described in refs. 24 and 28. Four large, dual-temperature, Canadian plants that are operational are Port Hawkesbury and Glace Bay in Nova Scotia, each having a capacity of 400 t/yr, and Bruce A and B in Ontario, each with a capacity of 800 t/yr (23–24). Two more plants are planned; the addition of these two 800-t/yr plants is expected to meet demand until about 1990 (24).

Ammonia–Hydrogen Exchange. Ammonia–hydrogen exchange has both a larger separation factor (Table 5) and a larger temperature dependence (Fig. 4) than the H_2O–H_2S exchange reaction. Thus, processes based on the reaction are favored by columns with fewer theoretical stages, lower reflux rates, and higher deuterium re-

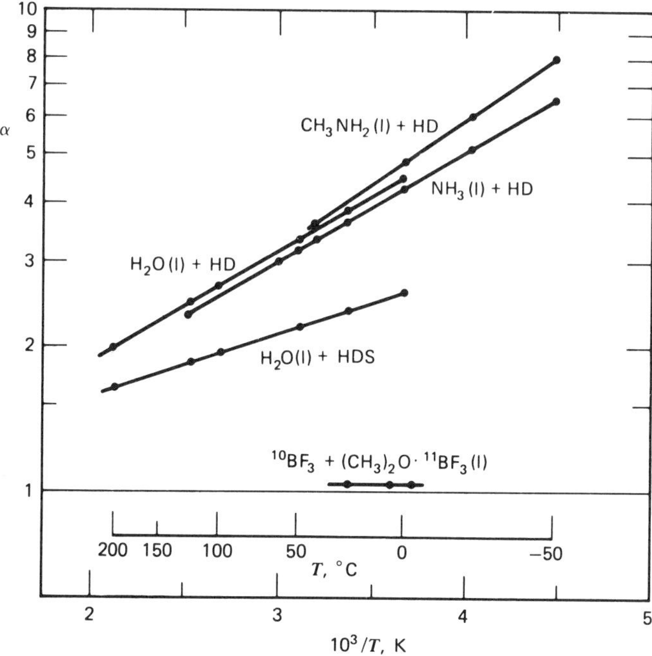

Figure 4. Temperature dependence of the separation factor for some chemical-exchange reactions (23,27).

covery from the source. There are, however, several disadvantages. The reaction proceeds at a sufficient rate only in the presence of a catalyst of 1–2 mol % potassium amide (KNH_2). Even with a catalyst, the reaction is slow, and specially designed gas–liquid contactors, high pressures, and large tower heights are required. The catalyst is dissolved in the liquid (ammonia) fraction with which it is in isotopic equilibrium, necessitating the separation of catalyst from the product and, in the case of the monothermal process, a further stripping operation to recover deuterium before recycling of the catalyst. The catalyst is very reactive, combining with CO_2, CO, H_2O, and O_2 to form solid impurities and with O_2 to form KN_3 which is explosive; concentrations of these impurities in the feed gas must be reduced to less than a part per million. The process is fed by ammonia synthesis gas ($N_2 + 3\ H_2$). In the monothermal process, chemical reflux (phase conversion) is accomplished by dissociation of ammonia that is rich in deuterium and by recycling part of the ammonia made from synthesis gas that is depleted in D_2. Several monothermal plants have been planned, and one has been operated (23).

A bithermal process (30) eliminates the dissociation of ammonia, the synthesis of additional ammonia for reflux, and the catalyst stripping step. However, it requires far more internal stages, larger columns, and more reflux. A flow sheet, based on a partial description of the process (31), is given in ref. 22. Friedrich UHDE GmBH is constructing such a plant at Talcher, India (30).

Amine–Hydrogen Exchange. Potential advantages of amine–hydrogen exchange over ammonia–hydrogen exchange include higher separation factors (and larger temperature dependence) and a faster reaction rate which permits the use of a lower temperature. The reaction, which is catalyzed by potassium methylamide (CH_3NHK), has an interesting process chemistry involving side reactions that were unexpected on the basis of experience with ammonia–hydrogen exchange (32). Sulzer Brothers Ltd. have generated a complete design package for a plant to be built in Alberta (33).

H_2O–H_2 Exchange, Electrolysis, and the CECE Process. The large separation factors that can be obtained by the exchange of hydrogen with steam or water (Table 5 and Fig. 4) are, in part, offset by the slow reaction rate. Additional difficulties are the high chemical reflux energy required to decompose water for a monothermal system and the absence (until recently) of a catalyst whose activity is not destroyed by liquid water. However, the combination of H_2O–H_2 exchange with water electrolysis offers gains over either process by itself. Separation factors for electrolysis are about 5–13, depending on the temperature, the electrolyte, the cathode material, and the cell construction. (Electrolysis has the highest separation factor of any commercial process and requires the largest energy input.) Hydrogen from some of the early stages of an electrolytic cascade, whose deuterium concentration is in excess of natural abundance but is not of sufficient value to be worth recycling, can be exchanged with H_2O (steam) to recover most of the deuterium without sacrificing hydrogen gas production. Norsk Hydro electrolysis plants at Rjukan and Glomfjord, Norway, were converted to the combined process after World War II (23). A similar plant was built by the Manhattan District at Trail, B.C., Canada in 1944 (34).

Reduced cost and energy use are attainable by use of a two-phase system, which is made feasible by a catalyst that is not wetted by, and that retains its activity in, the presence of liquid water (35). Electrolysis with heterogeneous-phase, water–hydrogen exchange—or CECE (combined electrolysis and chemical exchange) process—would appear to be the best method to produce heavy water as a by-product of hydrogen, provided a catalyst of sufficient activity can be prepared (36). A heavy-water pilot plant is planned at Chalk River National Laboratory (Canada). Another application of the process is the recovery of tritium from reactor moderator water (either heavy or light water); a pilot plant is planned at Mound Laboratory (37). A combination of vapor-phase, catalytic exchange and hydrogen distillation has been used at Grenoble, France, to purify heavy-water moderator from a research reactor of both tritium and light water (38) (see Nuclear reactors).

The use of a bithermal water–hydrogen exchange process involving a hydrophobic catalyst has been suggested (39). Because of its larger effective separator factor, which results in lower gas and liquid flow rates, this process might compete favorably with the GS process if a catalyst of sufficient activity could be produced at a sufficiently low cost. H_2O–H_2 exchange also might provide a transfer mechanism between water as feed and H_2-based processes.

Comparison of Processes for Production of Heavy Water. The high cost of heavy water and its importance as a reactor moderator have generated an active competition for the best production methods; some of the factors that are relevant to this competition are listed in Table 6. An economic process must have a separation factor of at least 1.1–1.2 and must use less than about 40–50 GJ/kg D_2O (24).

The cumulative world heavy-water production and the breakdown by process to May, 1977, is shown in Figure 5. Production, by country, is: United States, 59%; Canada, 35%; Norway, 4%; and India and France, 1% each (24). The world capacity (2500 t/yr) shows an even greater dependence on the GS process (98%) but most of the heavy water (95%) is produced in Canada. By 1982 the world capacity is projected to expand to nearly 4400 t/yr, of which 94.7% will be by the GS process. 91% of this capacity will be located in Canada (24); India, which also employs heavy-water reactors, will be the next largest producer. The almost complete dominance of the GS process largely results from the fact that it is the only practical process that is fed by water as the deuterium source. The only other source of sufficiently large amounts of deuterium is synthesis gas, produced from natural gas or petroleum (30). Although the total world production of synthesis gas could support over ten times the present demand for heavy water as a by-product, the largest single plant would yield only 100 t/yr of D_2O. The use of electrolytic processes (including CECE) is prohibitively expensive in terms of energy consumption if undertaken for deuterium production alone; in locations where excess hydroelectric capacity makes it profitable to produce electrolytic hydrogen, deuterium can be obtained as a by-product at almost no additional cost.

The price charged for heavy water from Savannah River by the U.S. DOE, $215/kg in 1977 (40), reflected a diseconomy of scale for this plant, which partially has been shut down; the DOE was willing to negotiate a lower price for quantities large enough to require operation at a higher capacity (23). The Canadian price in 1977 was around $150/kg (41), of which 60% was allocated to return on capital, 25% to energy, and 15% to other operating expenses (39). Energy consumption in the Canadian plants is about 6800 kW·h thermal, plus 700 kW·h electric, per kg of D_2O.

Chemical Methods for Other Elements. Lithium isotopes are separated by exchange between lithium and amalgam and aqueous lithium hydroxide (23,42). A plant that utilizes this reaction to produce 1000 kg/yr of 99.99% ^7Li at a price of about $3/g is being built at Quapaw, Oklahoma (23). (The amount of ^6Li used for nuclear weapons

Table 6. Comparison of Some Methods for the Production of Heavy Water[a]

Process	Effective separation factor[b]	Energy input, GJ/kg D_2O[c,d]	Deuterium recovery[e], %	Operating conditions[e]			Reflux ratio[e,g] (10^3 moles per mole of D_2O)	
				T or T_c, °C	T_h, °C	Pressure, MPa[f]	Gas	Liquid
GS	1.29	30	14	32	138	2.0	73	37
monothermal NH_3–H_2	5.2	20	85	−25		35	16	4
bithermal NH_3–H_2	1.7	7	85	−25	60	35	37	4.3
bithermal CH_3NH_2–H_2	2.2	11	80	−50	40	5	14	2.2
bithermal H_2O–H_2	1.6	19	30	50	170	7	59	22
H_2 distillation	1.5	21	90	−249		0.25	20	
CECE	≃4	500 (2)[h]	66	60[i]		7	10	10
electrolysis[j]	5–10	700	24	20		0.10	28	28
H_2O distillation[j]	1.04	70	2	82		0.05	200	

[a] Ref. 24.
[b] The listed separation factor for a bithermal process is α_c/α_h.
[c] Thermal energy plus electrical energy divided by 0.4.
[d] To convert GJ/kg to Btu/lb, multiply by 430,200.
[e] Typical values for actual plants or estimates for proposed processes.
[f] To convert MPa to psi, multiply by 145.
[g] Average value. Actual flows may vary in different parts of the system (eg, hot or cold towers).
[h] The lower energy for parasitic CECE represents the additional energy needed to produce D_2O as a by-product of hydrogen.
[i] Temperature of the H_2O–H_2 reactor.
[j] Economically unfavorable.

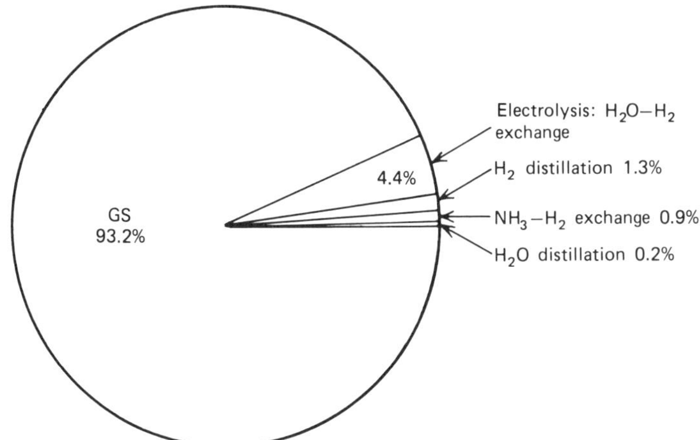

Figure 5. Cumulative world production of heavy water to May, 1977 (24). Total: 11,000 t.

must be much larger than this figure suggests; authoritative figures are unavailable.) Other reactions for separation of ^6Li–^7Li are discussed in refs. 43 and 44. Boron isotopes are separated by exchange distillation (reaction 21 in Table 5) (27). A plant at Quapaw is being expanded to a capacity of 1000 kg/yr of ^{10}B, which is to be sold at \$5–15/g (23). Processes for separation of ^{14}N–^{15}N based on reaction 25 of Table 5 have been proposed (23,43,45–46).

The enrichment of carbon isotopes by CO_2–carbamate exchange has been studied and appears to be uneconomic compared to CO distillation for quantities above 0.5 kg/yr (21). Mound Laboratory has studied (32) the enrichment of sulfur isotopes by SO_2–$NaHSO_3$ exchange and the enrichment of Ca and other metals by extraction with organic solvents containing crown ethers with which they form a complex (reaction 28 in Table 5). Uranium separation by reactions 29 and 30 of Table 5 has been studied (23,29); the low separation factors make it highly unlikely that any chemical process will compete with the industrial-scale methods described below.

Gaseous Diffusion. Since the first gaseous diffusion plant, known as K-25, began operation at Oak Ridge in 1944, almost all of the world's enriched ^{235}U has been produced by this method. Although other processes are expected to compete in the future, over 98% of the present uranium separation capacity is embodied in diffusion plants.

In 1972 the combined uranium separation capacity of the three United States plants was 17.23×10^6 kg U SWU/yr (47). At full capacity, these plants consumed about 6100 MW of electric power. Their efficiency (power usage per unit of separative capacity or energy per unit of separative work) was 0.352 kW/(kg U SWU/yr) (11 GJ/kg U SWU). The 1980 price of separative work in these plants is about \$100/kg U SWU (23). Improvements that are being implemented are expected to add another 10.5×10^6 kg U SWU/yr to the capacity and to reduce energy consumption to about 8.4 GJ/kg U SWU. In 1978 the energy consumption of United States gaseous diffusion plants was 9.0 GJ/kg U SWU (48). Additional diffusion plants are operating in the USSR ($7–10 \times 10^6$ kg U SWU/yr), in France and the UK (about 0.5×10^6 kg U SWU/yr each), and in China (capacity unknown). Eurodif, a European consortium, is constructing a plant in France that has an initial capacity of 10.8×10^6 kg U SWU/yr and plans 50% expansion by the late 1980s.

The DOE enriches uranium for the nuclear power industry on a toll basis. The cost of uranium of a given enrichment (around 3% ^{235}U for light-water reactors) depends on the cost of natural uranium feed, the cost of a unit of separative work, and the fraction of ^{235}U remaining in the tails x_W. For a lower tails assay, more separative work but less uranium is required; x_W can be adjusted to minimize the total cost. Recent figures are about \$89/kg of natural uranium, \$100/kg SWU, which yields a minimum cost of about \$897/kg of 3% ^{235}U when x_W = 0.24% (23). With uranium prices rising faster than the cost of separative work, a lower tails assay should be optimal in the future, and it will be economic to reuse the tails from previous separation as feedstock.

Gas Centrifuge. URENCO-Centec, a combination of British, Dutch, and German interests, operates three pilot plants that have a total capacity of 0.12×10^6 kg SWU/yr and plans expansion (some of it already under construction) to about 10×10^6 kg SWU/yr. The capacity of the URENCO machines is estimated as 2–20 (49) and Ca 5 (50) kg SWU/yr. Japan is completing a pilot plant and is considering a 6×10^6 kg SWU/yr plant to be built in 1985. The United States is committed to an expansion of uranium separation facilities; after completion of improvements in the three gaseous diffusion plants, further additions to capacity will be in the form of centrifuges. A test facility is operating at Oak Ridge National Laboratory, and a plant with a capacity of 8.8×10^6 kg SWU/yr is to be built at Portsmouth, Ohio, by 1988. United States centrifuges have been estimated to have ten times the capacity of the URENCO machines (49).

The capital plus operating cost associated with a centrifuge plant is thought to be comparable with or lower than that for a gaseous diffusion plant of the same capacity (51). The advantage of the centrifuge is its lower energy consumption; the energy per separative work unit has been estimated as about 4% of that required for gaseous diffusion (23,52) or about 0.4 GJ/kg U SWU. The centrifuge process has an estimated unit capital cost of about \$500/(kg SWU/yr) (53).

Laser Isotope Separation. The conditions for laser separation of isotopes are (51): (*1*) there must be at least one absorption line of the isotope being separated that does not significantly overlap any absorption lines of other isotopes in the mixture; (*2*) radiation at the chosen frequency must be available with the characteristic power, duration, divergence, and monochromaticity that is necessary for the separation method chosen; (*3*) there must be a primary photophysical or photochemical process that permits easy separation of the excited species from the mixture; (*4*) the selectivity for the desired isotope must be maintained against all competing photophysical or photochemical processes (see Lasers).

Laser separation is applicable to atoms or molecules. Either may be excited to a state that undergoes a photochemical reaction with another atom or molecule, that can be deflected by an external field or by the pressure of the laser beam, or that can be ionized by a second photon. Additionally, a molecule can be excited to a state that isomerizes to a different compound or that dissociates either spontaneously or upon absorption of a second photon. Dissociation upon absorption of a single photon can be isotopically selective only if the excited state is quasi-stable, ie, with a lifetime long enough to yield a sufficiently narrow linewidth (predissociation).

The existence of well-resolved absorption lines, even for the heaviest elements, gives laser-based processes the potential for very high separation factors. In conjunction with the high efficiencies that are attainable in the production and use of photons, high separation factors result in costs and energy efficiencies that are very attractive compared to the present methods for research-scale isotope separation and that are promising competitors with industrial-scale uranium- and deuterium-separation methods. An example of a process that has been studied extensively is the photochemical dissociation of formaldehyde which, under appropriate conditions, yields stable products with a quantum efficiency that is close to unity.

$$H_2CO \xrightarrow{h\nu\ (340\text{–}350\ \text{nm})} H_2 + CO \tag{39}$$

The reaction allows separation of ^{1}H–^{2}H–^{3}H, ^{12}C–^{13}C–^{14}C, and ^{16}O–^{17}O–^{18}O. The highest enrichment factors obtained experimentally are 180 for hydrogen–deuterium (54), 80 for ^{12}C–^{13}C (55), 9 for ^{16}O–^{17}O, and 17 for ^{16}O–^{18}O (54). By use of this process, ^{13}C of high (>90%) enrichment could be prepared in two stages at a cost of about 12¢/g to power the laser and amortize its capital cost, compared to an average price of around $69/g in 1978 for ^{13}C that is enriched by CO distillation (22). Although the overall costs are much higher than those associated with the laser, it appears that separation of carbon and perhaps oxygen isotopes by this method already has a strong economic advantage over existing small-scale commercial processes. The separation of deuterium by predissociation of formaldehyde must compete with low cost, energy-efficient, industrial-scale processes. A research effort aimed at a 1-t/yr pilot plant in 1983 is underway (56); isotopic selectivities of 10^3–10^4 and laser efficiencies approaching 2–3% are needed to make the process competitive.

Several schemes for separation of uranium isotopes are being investigated. One involves the selective ionization of ^{235}U atoms in a beam of uranium metal vapor and removal of the ions in an electric field (57). Product concentrations of 6% ^{235}U have been obtained from natural uranium in a single stage with nearly complete stripping of the tails. The process has the capability of economically producing enriched uranium for light-water reactors ($\simeq$3% ^{235}U) from the large stockpiles of tails from gaseous diffusion plants. Jersey-Nuclear-Avco-Isotopes, Inc. (JNAI) is constructing a pilot separation plant based on this method. The economics are not well established (23); however, one estimate is that a full-scale plant would cost about $195/(kg SWU/yr) and use about 0.6 GJ/kg SWU (57). The capital cost is much lower than for gaseous diffusion or the gas centrifuge, and the energy consumption is comparable to that expected for the centrifuge.

Another type of process of recent interest is the use of isotope-selective, vibrational excitation of molecules (51,58). The multiple-photon dissociation of SF_6 has been studied (51,59–61) using wavelengths around 10,600 nm that are obtained from a CO_2 laser. The dissociation yields lower fluorides, which are less volatile and thus easily separated from unreacted SF_6. The USSR is producing small quantities of the rare (0.017%) isotope ^{36}S at enrichments of 70% and at about $180,000/g and, presumably, by this method (62); the USSR is producing also several grams per day of 95% ^{13}C by multiple-photon dissociation of CF_3I (63). A similar method is under consideration for separation of uranium by dissociation of UF_6 by multiple vibrational excitation or by vibrational excitation plus electronic excitation at another wavelength (23). In either case, the absorption bands are too broad to permit good selectivity, and it is necessary to cool the molecules by expansion through a nozzle in order to obtain a high separation factor (64).

The main technical problem facing large-scale laser separation of uranium and deuterium is the need for lasers of higher power (51). The power requirements are particularly severe for economic separation of deuterium.

Aerodynamic (Nozzle) Process. Only one other method, the aerodynamic or nozzle process, has been considered for commercial-scale enrichment. This process, which depends on the centrifugal acceleration of a high speed stream of gas molecules, is discussed in references 9 and 23 and references contained therein. The process involves costs and energy efficiencies that are comparable to those for gaseous diffusion.

Analytical and Test Methods

The primary instrument for detection and measurement of isotopes is the mass spectrometer, in which a substance is vaporized and ionized, accelerated by an electric field, and the resulting ion beam analyzed by a magnetic field or a combination of electric and magnetic fields. Mass spectroscopy has been reviewed in refs. 65 and 66. The resolution requirements for isotopic analysis, $\Delta M/M \ll 1/A$, are not stringent. The measurement of accurate isotope ratios encounters a number of problems including partial fractionation of isotopes in the ion source, which may change the apparent ratios by as much as 1% per mass unit. In order to circumvent such problems, absolute or calibrated measurements—comparison to known standards prepared by mixing separated isotopes—are required. Isotopic ratios have been measured with mass spectrometers with reproducibilities as high as 0.002% and absolute accuracies as high as 0.05% (67). Other methods of assaying stable isotopes include nmr, radioassay of transmutation products that are induced by bombardment with neutrons or charged particles (activation analysis), and atomic absorption spectrometry, particularly in the ir region (see Analytical methods).

Applications of Stable Isotopes

Nuclear Power. Nuclear electric power generation involves a number of isotope applications and accounts for most of the separated isotope production (see Nuclear reactors). Materials whose nuclear properties are important in a reactor include the fuel, moderator, coolant, neutron absorbers, and other materials (eg, structural) that are present within the nuclear reaction zone. Separated ^{10}B is sometimes used in control elements when high neutron absorption per unit volume is important. $^{7}LiOH$ is used to control the acidity of the coolant/moderator, because it does not absorb neutrons strongly or yield radioactive neutron-capture products. However, the predominant isotope applications are the use of ^{235}U as the fuel in light-water reactors and D_2O as the moderator in heavy-water reactors.

Most of the enriched uranium required in the next 30 yr will serve as fuel in light-water reactors utilizing 2–4% enriched ^{235}U. Enrichment services of ca 0.5×10^6 kg U SWU for the initial core load plus 0.10×10^6 kg U SWU/yr are required to support one 1000 MWe (MW of energy produced) reactor. When the enrichment is performed in a gaseous diffusion plant, the energy required is equivalent to 0.25 yr of the reactor's power production for the initial core load plus 5% of its continuing output. In view of the declining projections for nuclear power, the additional separative capacity planned (gas centrifuges) or potential (lasers), and the probable limitations imposed by the size of the uranium resource base (53,68–69), it is unlikely that a serious shortage of separative capacity will become a problem in the United States or even worldwide.

The total heavy-water inventory of a representative 600 MWe CANDU reactor is about 770 t per 1000 MWe (70). Yearly make-up requirements are only 6–10 t per 1000 MWe (30), so that the heavy-water requirement is nearly proportional to the amount of new generating capacity being built. The present world heavy-water capacity of about 2500 t/yr could support a construction rate of about 3200 MWe/yr; by 1982, if planned plants were to be completed on schedule, the world capacity would support a growth rate of 5700 MWe/yr. The amount of deuterium required to fuel a fusion reactor is insignificant by comparison.

Tracers. The use of isotopes as tracers is a powerful method for following the path of an element through a chemical, biological, or physical system. Radioisotopes are most commonly employed as tracers because they are easily detected, can be used in small quantities, and often can be detected remotely through intervening material as a result of the emission of penetrating gamma rays of characteristic energy (see Radioisotopes). On the other hand, there are a number of reasons for using stable isotopes as tracers: there may be no radioisotope of the desired element with a convenient half-life or with easily detected radiations; there may be constraints on the use of radioactivity because of damage to the chemical or biological system under study, to a patient, to the experimenter, or to the environment. These reasons apply especially to medical, biological, and agricultural applications, which account for much of the ^{13}C and ^{15}N used, and which form a large potential market for inexpensive, laser-enriched isotopes. These isotopes also are convenient tracers for use in chemistry laboratories where ir or nmr spectrometers often are available as detectors.

Production of Radioisotopes. Radioisotopes are produced by bombardment of a stable isotope with neutrons or charged particles. Enriched isotopic targets, although not essential, may be used because other isotopes of the target element yield radioisotopes that interfere with the application or emit excessive or long-lasting radiation, or because the target isotope is a rare one, and high yield, high specific activity, or both, are desired. Both reasons are often important for nuclear physics experiments, in which high radiochemical purity is usually important, and high yield or specific activity ("massless" sources) may be necessary to obtain the desired sensitivity or resolution. In medical applications, the use of separated isotopes as targets may lower the dose to the patient from long-lived impurities.

Nuclear Physics Research. Nuclei can be studied via radioactive decay or via nuclear reactions observed in-beam. Separated isotopes are used routinely as targets for in-beam studies. Although most accelerators and/or associated beam transport systems provide good mass discrimination, separated isotopes whose natural abundances are low are used as beam particles in order to obtain a high beam intensity. Of particular importance is ^{3}He; other separated isotopes that are used as beam particles including ^{13}C, ^{15}N, $^{17,18}O$, ^{48}Ca, and ^{136}Xe.

Structural Chemistry. The usefulness of nmr to determine chemical structure is enhanced by the substitution of isotopes that have easily observed resonances for ones that do not. There are three reasons for isotopic substitution in nmr experiments: (*1*) The abundant isotope has no magnetic moment, ie, its spin is zero. Three of the 83 naturally occurring elements have no nmr-detectable isotopes—argon, cerium, and thorium. For the remaining 40 elements with even Z, detectable (odd-mass) isotopes tend to have low abundances; 11 elements have no detectable isotopes of abundance greater than 5%. (*2*) The abundant isotope has a quadrupole moment ($J > 1/2$), which both splits and broadens the nmr resonances, making it difficult to detect them and to interpret their structure. (*3*) Selective isotopic substitution at specific sites of a molecule permits examination of the electron-nucleus interaction at those sites without interference from the same element at other sites.

Two of the most important isotopic substitutions used in nmr studies are ^{13}C ($J = 1/2$) for ^{12}C ($J = 0$) and ^{15}N ($J = 1/2$) for ^{14}N ($J = 1$). Isotopic substitution can be used similarly in esr measurements to produce a fine structure that is identifiable with a particular site in a molecule (see Analytical Methods.)

Isotope Effects. Isotope effects include the exchange of isotopes between two compounds or phases and the change in properties of a chemical compound or reaction upon substitution of one isotope for another. They may be classified as thermodynamic effects (the equilibrium isotope effect) or as rate changes (the kinetic isotope effect). The former are of particular interest for isotope separation. The latter provide a useful tool for the study of reaction mechanisms, and play an important role in isotope biology.

Isotope effects are well understood (71). They are vibrational in nature, and depend only on the isotopic masses and the force constants for bonds to the isotopic atom; isotope effects can be predicted quantitatively from reduced masses and either the calculated bond force constants or the measured vibrational frequencies. Examples of the size of equilibrium effects for distillation and exchange reactions are given in Tables 4 and 5.

Kinetic isotope effects are classed as primary if the rate-determining step involves the bond to the isotopically substituted atom. Secondary effects are at most about 10% as large as primary effects; observation of a change in the rate constant by more than this amount identifies the reaction site with a bond to the substituted atom (72).

Isotope Biology. Isotope substitution in living systems is of interest both for its effect on the organism and for the production of isotopically labeled organic compounds. The field has been reviewed in refs. 73 and 74 (see Deuterium and tritium).

Isotope Geology. The many variations in the isotopic composition of natural materials are the basis for a number of applications to the geosciences. The isotopes ^{40}K, ^{87}Rb, ^{176}Lu, ^{187}Re, ^{232}Th, ^{235}U, and ^{238}U, have half-lives that are within a factor of 10 of the age of the earth and, therefore, are useful as geologic clocks. The somewhat longer-lived isotopes, ^{138}La, ^{144}Sm, and possibly, ^{190}Pt, also might serve as clocks. More recent events can be measured by shorter lived isotopes that occur in trace quantities as the result of cosmic ray bombardment or as daughters of long-lived isotopes. (Refs. 75–80 include discussions of some isotope dating applications.)

Small variations in the stable isotope abundance ratios R have been exploited to yield information about underlying geological, hydrological, or astrophysical causes. These variations are expressed conventionally by the quantity (81):

$$\delta(‰) = \frac{R_{sample} - R_{standard}}{R_{standard}} \times 1000 \qquad (40)$$

Water is of particular interest because it is ubiquitous. The abundance variations of hydrogen (including tritium), oxygen, and carbon (dissolved as CO_2–bicarbonate) in groundwater are related to the origin, sources of recharge, and temperature of the water being studied (82–86). Meteoric water is characterized by an oxygen isotope ratio that is dependent linearly on precipitation temperature (87) and related linearly to the hydrogen ratio (88). Both correlations are the result of differential separation that occurs when water condenses from vapor with which it is in isotopic equilibrium and is removed continuously as raindrops (Rayleigh condensation). Geothermal waters generally have a D/H ratio characteristic of local precipitation, which is their original source, but elevated $^{18}O/^{16}O$ ratios, because of the exchange of oxygen with deep-lying rock. The temperature dependence of the equilibrium constant for rock–water exchange reactions (either isotopic or chemical) is the basis for several geothermometers, from which temperatures at depth can be estimated from the analysis of the composition of surface hot springs (84).

$^3He/^4He$ abundances can vary over 8 orders of magnitude (89–90); the main cause of the variation is the admixture of primordial helium, containing 0.01–0.1% ^3He, with pure ^4He from the decay of natural alpha emitters. Variations in stable isotope abundances have many causes in addition to those described above, including fractionation in biologic systems. Refs. 81 and 91–92 offer general treatments of this extensive subject.

A striking recent discovery is the onetime existence, about 1.8×10^9 years ago (93), of a natural nuclear reactor in Gabon, Africa. At that time, the ^{235}U abundance would have been about 3.1%, sufficient to sustain a chain reaction with water as the moderator. Evidence for the reactor includes anomalously low ^{235}U abundances and certain elements, such as neodymium, whose isotopic compositions are characteristic of fission products rather than terrestrial material (94–95).

Nomenclature

A	= mass number
CANDU	= Canadian Deuterium Uranium Reactor
CECE	= combined electrolysis and chemical exchange
D	= deuterium, ^2_1H
d	= deuteron, $^2_1H^+$
DOE	= United States Department of Energy and its forerunners, ERDA and the AEC
GS	= Girdler sulfide
$h\nu$	= photon
J	= nuclear spin
M	= atomic mass, scale defined by $M(^{12}C) = 12$
N	= neutron number
n	= neutron, ^1_0n
SWU	= separative work units
t	= metric ton, 10^3 kg
t	= triton, $^3_1H^+$
x	= mole fraction of the desired isotope in the tails stream
x_W	= mole fraction of the desired isotope in the waste stream from a plant
y	= mole fraction of the desired isotope in the heads stream
Z	= atomic number
z	= mole fraction of the desired isotope in the feed stream
z_F	= mole fraction of the desired isotope in the input stream to a plant
α	= alpha particle, $^4_2He^{2+}$, or separation factor
α_c	= separation factor at lower (cold) temperature
α_h	= separation factor at upper (hot) temperature
β	= heads separation factor
π	= vapor pressure
‰	= per mil (thousandths, analogous to % = hundreths)

BIBLIOGRAPHY

1. J. J. Thompson, *Rays of Positive Electricity and their Application to Chemical Analyses,* Longmans, Green and Co., Ltd., London, Eng., 1913.
2. J.Chadwick, *Proc. R. Soc. London Ser. A* **136,** 692 (1932).
3. C. M. Lederer and V. S. Shirley, eds., *Table of Isotopes,* 7th ed., John Wiley & Sons, Inc., New York, 1978.
4. N. E. Holden, *Brookhaven National Laboratory Report No. BNL-NCS-50605,* 1977.
5. E. M. Burbidge, G. R. Burbidge, W. A. Fowler, and F. Hoyle, *Rev. Mod. Phys.* **29,** 547 (1957).
6. A. A. Penzias, *Science* **205,** 549 (1979).
7. H. R. Heydegger, J. J. Foster, and W. Compston, *Nature* **278,** 704 (1979).
8. R. V. Ballad, L. L. Oliver, R. G. Downing, and O. K. Manuel, *Nature* **277,** 615 (1979).
9. R. L. Hoglund, J. Shacter, and E. Von Halle, "Diffusion Separation Methods" in M. Grayson and D. Eckroth, eds., *Encyclopedia of Chemical Technology,* 3rd ed., Vol. 7, John Wiley & Sons, Inc., New York, 1979, pp. 639–723.
10. M. Benedict and T. H. Pigford, *Nuclear Chemical Engineering,* 1st ed., McGraw-Hill Book Co., New York, 1957.
11. W. Spindel in P. A. Rock, ed., *Isotopes and Chemical Principles, ACS Symposium Series No. 11,* American Chemical Society, Washington, D.C., Chapt. 5, 1975.
12. L. O. Love, *Science* **182,** 343 (1973).
13. H. Childs, *E. O. Lawrence, An American Genius,* E. P. Dutton and Co., Inc., New York, 1968, Chapt. XIII.
14. E. Newman, Oak Ridge National Laboratory, private communication, July, 1980.
15. *Research Material, Separated Isotopes and Radioisotopes, and Special Preparations,* Oak Ridge National Laboratory Isotope Sales.
16. S. Villani, *Isotope Separation,* American Nuclear Society, Washington, D.C., 1976.
17. S. Chapman, *Philos. Trans. R. Soc. London Ser. A* **216,** 279 (1916); **217,** 115 (1917).
18. K. Clusius and G. Dickel, *Naturwissenschaften* **26,** 546 (1938).
19. W. H. Furry, R. C. Jones, and L. Onsager, *Phys. Rev.* **55,** 1083 (1939); R. C. Jones and W. H. Furry, *Rev. Mod. Phys.* **18,** 151 (1946).
20. W. M. Rutherford, *Sep. Purif. Methods* **4,** 305 (1975).
21. R. A. Schwind and W. M. Rutherford in *Radiopharmaceuticals and Labeled Compounds, International Atomic Energy Agency (Vienna) Report IAEA-SM-171/47,* Vol. II, 1973, p. 255.
22. A. H. Ruwe, Jr., *Mound Laboratory Report MLM-2594,* Mar. 1979.
23. M. Benedict, T. H. Pigford, and H. W. Levi, *Nuclear Chemical Engineering,* 2nd ed., McGraw-Hill Book Co., New York, 1980.
24. H. K. Rae in H. K. Rae, ed., *Separation of Hydrogen Isotopes, ACS Symposium Series No. 68,* American Chemical Society, Washington, D.C., 1978, Chapt. 1.
25. G. M. Keyser, D. B. McConnell, N. Anyas-Weiss, and P. Kirkby in ref. 24, Chapt. 9.
26. R. E. Schreiber, *ICONS at LASL, Los Alamos Scientific Laboratory Report LA-4759-MS,* 1971.
27. A. A. Palko and J. S. Drury in R. F. Gould, ed., *Isotope Effects in Chemical Processes, ACS Advances in Chemistry Series No. 89,* American Chemical Society, Washington, D.C., 1969, Chapt. 3.
28. G. D. Davidson in ref. 24, Chapt. 2.
29. J. Shimokawa, G. Nishio, and F. Kobayashi in K. Ogata and T. Hayakawa, eds., *Recent Developments in Mass Spectrometry,* University Park Press, Baltimore, Md., 1969, p. 397.
30. E. Nitschke, H. Ilgner, and S. Walter in ref. 24, Chapt. 6.
31. E. Nitschke, *Atmowirtschaft,* 274 (June 1973).
32. W. J. Holtslander and W. E. Lockerby in ref. 24, Chapt. 3.
33. N. P. Wynn in ref. 24, Chapt. 4.
34. G. M. Murphy, H. C. Urey, and I. Kirshenbaum, *Production of Heavy Water,* McGraw-Hill Book Co., New York, 1955.
35. J. P. Butler, J. H. Rolston, and W. H. Stevens in ref. 24, Chapt. 7.
36. M. Hammerli, W. H. Stevens, and J. P. Butler in ref. 24, Chapt. 8.
37. M. L. Rogers, P. H. Lamberger, R. E. Ellis, and T. K. Mills in ref. 24, Chapt. 13.
38. P. H. Pautrot and M. Damiani in ref. 24, Chapt. 12.
39. A. I. Miller and H. K. Rae, *Chem. Can.* **27,** 25 (1975).

40. *Fed. Reg.* **42,** 14768 (Mar. 16, 1977).
41. J. B. Marling, J. R. Simpson, and M. M. Miller in ref. 24, Chapt. 10.
42. Brit. Pat. 902,755 (Jan. 19, 1960), E. Saito and G. Dirian.
43. T. I. Taylor and W. Spindel in *Proc. Intl. Symp. on Isotope Separation,* Wiley-Interscience, New York, 1958, p. 158; W. Spindel and T. I. Taylor, *J. Chem. Phys.* **24,** 626 (1956).
44. D. A. Lee in ref. 27, Chapt. 4.
45. G. A. Garrett and J. Shacter in *Proc. Intl. Symp. on Isotope Separation,* Wiley-Interscience, New York, 1958, p. 17.
46. M. Jeevandam and T. I. Taylor in ref. 27, Chapt. 8.
47. *Report ORO-684,* U.S. Atomic Energy Commission, Washington, D.C., Jan. 1972.
48. D. R. Olander, *Adv. Nucl. Sci. Tech.* **6,** 105 (1972).
49. D. R. Olander, *Sci. Am.* **239**(2), 37 (1978).
50. *Nucl. Fuel,* 15 (Oct. 31, 1977).
51. V. S. Letokhov and C. B. Moore in C. B. Moore, ed., *Chemical and Biochemical Applications of Lasers,* Vol. III, Academic Press, Inc., New York, 1977, Chapt. 1.
52. *United States Gas Centrifuge Program for Uranium Enrichment,* U.S. Department of Energy, Washington, D.C., 1978.
53. *Final Report of the Committee on Nuclear and Alternative Energy Systems,* National Academy of Sciences–National Research Council, Washington, D.C., 1979.
54. J. Marling, *J. Chem. Phys.* **66,** 4200 (1977).
55. J. H. Clark, Y. Haas, P. L. Houston, and C. B. Moore, *Chem. Phys. Lett.* **35,** 82 (1975).
56. J. C. Vanderleeden, *Laser Focus* **13**(6), 51 (1977).
57. G. S. Janes, H. K. Forsan, and R. H. Levy, *Am. Inst. Chem. Eng. Symp. Ser. No. 169,* **73,** 62 (1977).
58. R. V. Ambartzumian, N. P. Furzikov, Yu. A. Gorokhov, V. S. Letokhov, G. N. Makarov, and A. A. Puretzkii, *Opt. Commun.* **18,** 517 (1976).
59. R. V. Ambartzumian, Yu. A. Gorokhov, V. S. Letokhov, and G. N. Makarov, *ZhETF Pis'ma Red.* **21,** 375 (1975); *JETP Lett.* **21,** 171 (1975).
60. R. V. Ambartzumian, Yu. A. Gorokhov, V. S. Letokhov, and G. N. Makarov, *Zh. Eksp. Teor. Fiz.* **69,** 1956 (1975); *Sov. Phys. JETP* **42,** 993 (1975).
61. R. V. Ambartzumian, N. V. Chekalin, Yu. A. Gorokhov, V. S. Letokhov, G. N. Makarov, and E. A. Ryabov, *Kvantovaya Elektron. (Moscow)* **3,** 802 (1976); *Sov. J. Quantum Electron.* **6,** 437 (1976).
62. S. Raman, Oak Ridge National Laboratory, private communication, Nov. 1979.
63. C. B. Moore, University of California (Berkeley), private communication, Dec. 1979.
64. R. J. Jensen, J. G. Marinuzzi, C. P. Robinson, and S. D. Rockwood, *Laser Focus* **12**(5), 51 (1976).
65. G. P. Barnard, *Modern Mass Spectrometry,* The Institute of Physics, London, Eng., 1953; R. W. Kiser, *Introduction to Mass Spectrometry and its Applications,* Prentice-Hall, Inc., Englewood Cliffs, N.J., 1965.
66. K. Ogata and T. Hayakawa, eds., *Recent Developments in Mass Spectrometry,* University Park Press, Baltimore, Md., 1969; N. R. Daly, ed., *Advances in Mass Spectrometry,* Vols. 7a and 7b, Heyden and Sons, Ltd., London, Eng., 1978; *Proc. 8th Intl. Mass Spectrometry Conf., Oslo, Summer 1979.*
67. P. de Bievre in N. R. Daly, ed., *Advances in Mass Spectrometry,* Vols. 7a and 7b, Heyden and Sons, Ltd., London, Eng., 1978, p. 395.
68. H. Brooks and J. M. Hollander, *Annu. Rev. Energy* (4), 1 (1979).
69. C. L. Wilson, Director, *Report of the Workshop on Alternative Energy Strategies,* McGraw-Hill Book Co., New York, 1977.
70. A. V. Nero, Jr., *A Guidebook to Nuclear Reactors,* University of California Press, Berkeley, Calif., 1979.
71. J. Bigeleisen in ref. 11, Chapt. 1; J. Bigeleisen, M. W. Lee, and F. Mandel, *Ann. Rev. Phys. Chem.* **24,** 407 (1973).
72. E. K. Thornton and E. R. Thornton in C. J. Collins and N. S. Bowman, eds., *Isotope Effects in Chemical Reactions, ACS Monograph No. 167,* Van Nostrand Reinhold Co., New York, 1970, Chapt. 4.
73. J. J. Katz, R. A. Uphaus, H. L. Crespi, and M. I. Blake in ref. 11, Chapt. 9.
74. J. J. Katz and H. L. Crespi in ref. 72, Chapt. 5.
75. *Proc. Symp. on Radioactive Dating, Athens, Nov. 19–23, 1962,* International Atomic Energy Agency, Vienna, 1963.
76. W. F. Libby, *Radiocarbon Dating,* 2nd ed., University of Chicago Press, Chicago, Ill., 1955.
77. S. Moorbath, *Sci. Am.* **236**(3), 92 (1977).
78. M. A. Lanphere, G. J. F. Wasserburg, A. L. Albee, and G. R. Tilton in H. Craig, S. L. Miller, and G. J.

Wasserburg, eds., *Isotopic and Cosmic Chemistry,* North-Holland Publ. Co., Amsterdam, The Netherlands, 1964, p. 269.
79. C. Patterson in ref. 78, p. 244.
80. R. D. Dallmeyer, *Lectures in Isotope Geology (Course),* Dept. of Geology, University of Georgia, Springer Verlag, Berlin, FRG, 1979, p. 77.
81. J. Hoefs, *Stable Isotope Geochemistry,* Springer-Verlag, Berlin, FRG, 1973.
82. D. E. White, *Econ. Geol.* **69,** 954 (1974).
83. D. E. White in *Proc. 1st U.S. Symposium on the Development and Utilization of Geothermal Resources, Pisa, Italy, Fall 1970; Geothermics* **1**(Special Issue 2), 55 (1970).
84. A. H. Truesdell in *Proc. 2nd U.N. Symposium on the Development and Utilization of Geothermal Resources, San Francisco, Calif., May 1975,* U.S. Government Printing Office, Washington, D.C., 1976, p. liii.
85. *Ibid.,* Section III, p. 687 ff.
86. W. G. Mook, *Geol. Mijnbouw* **51,** 131 (1972).
87. W. Dansgaard, *Tellus* **16,** 436 (1964).
88. H. Craig, *Science* **133,** 1702 (1961).
89. V. I. Kononov and B. G. Polak in ref. 84, p. 767.
90. L. K. Gutsalo in ref. 84, p. 745.
91. J. Pilot, *Les Isotopes en Geologie (Methodes et Applications),* Doin, Paris, 1974.
92. *Lectures in Isotope Geology (Course),* Dept. of Geology, University of Georgia, Springer Verlag, Berlin, FRG, 1979.
93. G. A. Cowan, E. A. Bryant, W. R. Daniels, and W. J. Maeck in *The Oklo Phenomenon,* IAEA-SM-204/35, International Atomic Energy Agency, Vienna, 1975, p. 341.
94. G. A. Cowan, *Sci. Am.* **235**(1), 36 (1976).
95. *The Oklo Phenomenon,* IAEA-SM-204/35, International Atomic Energy Agency, Vienna, 1975.

General References

References 3, 5, 9, 23, 51, and 81 are general references.

H. K. Rae, ed., *Separation of Hydrogen Isotopes, ACS Symposium Series No. 68,* American Chemical Society, Washington, D.C., 1978.

R. F. Gould, ed., *Isotope Effects in Chemical Processes, ACS Advances in Chemistry Series No. 89,* American Chemical Society, Washington, D.C., 1969.

Proc. Intl. Conf. on Uranium Isotope Separation, London, Eng., Mar. 5–7, 1975, British Nuclear Engineering Society, 1975.

M. Kamen, *Isotopic Tracers in Biology,* 3rd ed., Academic Press, Inc., New York, 1957.

P. D. Klein and S. V. Peterson, eds., *Proc. First Intl. Conf. on Stable Isotopes in Chemistry, Biology, and Medicine, May 9–11,* Argonne National Laboratory, Argonne, Ill., 1973.

J. Bigeleisen, M. W. Lee, and F. Mandel, *Ann. Rev. Phys. Chem.* **24,** 407 (1973).

C. J. Collins and N. S. Bowman, eds., *Isotope Effects in Chemical Reactions, ACS Monograph No. 167,* Van Nostrand Reinhold Co., New York, 1970.

C. MICHAEL LEDERER
Lawrence Berkeley Laboratory
University of California

ISOTOPES, RADIOACTIVE. See Radioisotopes.

ITACONIC ACID AND DERIVATIVES

Itaconic acid [*97-65-4*] (methylenebutanedioic acid, methylenesuccinic acid), $CH_2{=}C(COOH)CH_2COOH$, is a crystalline, high melting acid produced commercially by fermentation. Isolated from the pyrolysis products of citric acid in 1836, this α-substituted acrylic acid received its name by rearrangement of aconitic, the acid from which it is formed by decarboxylation.

Itaconic acid is a constituent of various synthetic resins, particularly styrene–butadiene and acrylic latexes for textile, paper, and paint applications. Incorporation of the acid improves stability of the emulsion and performance characteristics of the resin such as adhesion of coatings to substrates (see also Citric acid).

Physical Properties

The physical properties of itaconic acid are given in Table 1, and solubilities in some common organic solvents and water are shown in Table 2 and Figure 1, respectively.

Table 1. Physical Properties of Itaconic Acid

Property	Value	References
mol wt	130.10	
mp, °C[a]	167–168	
crystal density, g/cm^3	1.49 ± 0.01	1, 2
heat of combustion, MJ/mol[b]	1.98	3
heat of formation, kJ/mol[b]	840	3
dissociation constant in H_2O at 25°C		4
K_1	1.40×10^{-4}	
K_2	3.56×10^{-6}	
soly, g/100 mL H_2O at 25°C[c]	9.5	5

[a] Of octahedra formed on crystallization from water; unit cell is orthorhombic.
[b] To convert J to cal, divide by 4.184.
[c] Increases with temperature and pH (see Fig. 1).

Table 2. Solubility of Itaconic Acid in Organic Solvents at 25°C[a]

Solvent	Solubility, g/100 mL solvent	Solvent	Solubility, g/100 mL solvent
acetic acid	5.8	ethyl ether	1.4
acetone	8.7	ethylene glycol	25.2
benzene	<0.01	methanol	35.8
1-butanol	6.8	methyl ethyl ketone	3.7
n-butyl acetate	1.6	methyl isobutyl ketone	1.1
chloroform	<0.01	1-propanol	10.7
cyclohexane	<0.01	2-propanol	12.5
dioxane	14.1	propylene glycol	7.2
ethanol	19.8	tetrahydrofuran	23.6

[a] Ref. 6.

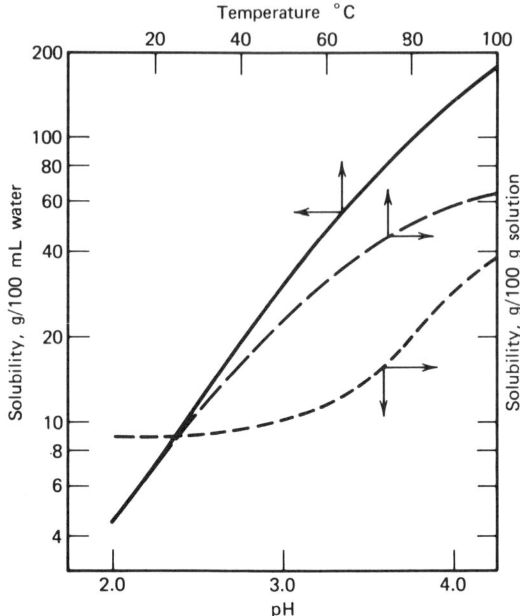

Figure 1. Solubility of itaconic acid in water: (—) g/100 mL water vs temp; (– –) g/100 g solution vs temp; (- - -), g/100 g solution vs pH.

Chemical Properties

Itaconic acid (**1**) is isomeric with citraconic (**2**) and mesaconic (**3**) acids. Under acidic, neutral, or mildly basic conditions and at moderate temperatures, itaconic acid is stable. At elevated temperatures or under strongly basic conditions, the isomers are interconvertible, probably via the species (**A**) where Y = H^+ or a cation (7). In aqueous systems, small amounts of the hydroxy acids resulting from hydrations of the double bonds also may be formed. Esters undergo isomerization, especially in presence of base (8–9).

Itaconic anhydride is readily formed by treatment of the acid with acetic anhydride (10) or acetyl chloride (11). Both the acid and anhydride are converted to itaconyl chloride by phosphorus pentachloride (12). Standard esterification methods provide dialkyl esters in good yields. Esterification proceeds almost entirely through the 4-alkyl ester (**4**), which is also the main product obtained from itaconic anhydride (13). The other monoalkyl ester, 1-monoalkyl 2-methylenebutanedioate (see ref. 13 for nomenclature) has been reported, but is less readily available. Elevated temperatures, commonly applied to the manufacture of unsaturated polyesters from unsaturated dicarboxylic acids and glycols, when used with itaconic acid, result in the presence of citraconate and mesaconate in addition to itaconate units in the polyester (14).

Nucleophilic reagents such as alkoxides (8,15), thiols (16), and amines (17) add to the itaconic double bond. The reaction product of dimethyl itaconate and ammonia, incorrectly formerly assigned the diamide structure, is the pyrrolidinone (**5**) which results from such an addition and subsequent cyclization (18).

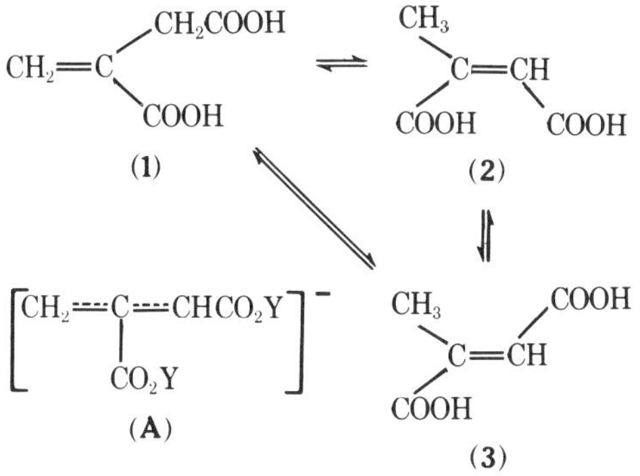

$$CH_2{=}C(CH_2COOH)COOH \ (1) \rightleftharpoons CH_3C(COOH){=}CHCOOH \ (2)$$

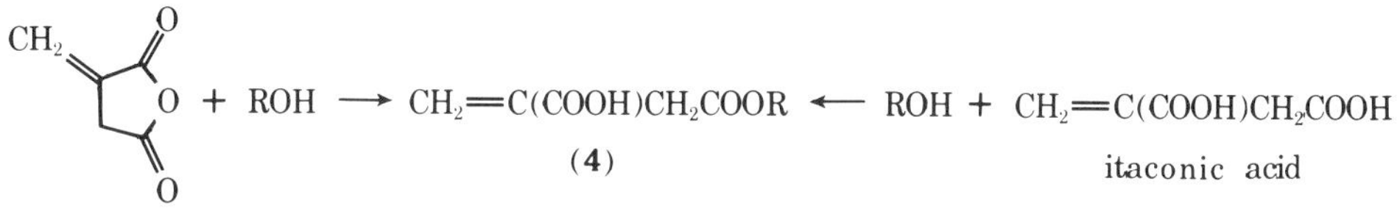

$$(1) \rightleftharpoons (3); \quad (2) \rightleftharpoons (3) \ CH_3C(COOH){=}CHCOOH$$

$$\left[CH_2{\overset{...}{=}}C(CO_2Y){\overset{...}{=}}CHCO_2Y\right]^- \ (\mathbf{A})$$

$$\text{itaconic anhydride} + ROH \longrightarrow CH_2{=}C(COOH)CH_2COOR \ (4) \longleftarrow ROH + CH_2{=}C(COOH)CH_2COOH \ \text{(itaconic acid)}$$

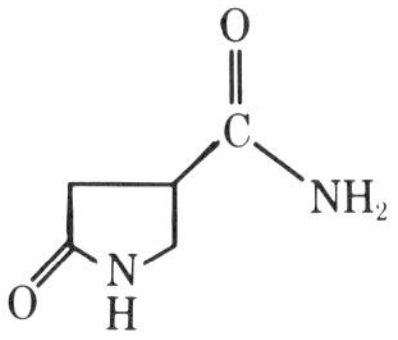

(5)

Itaconic acid, anhydride, and mono- and diesters undergo vinyl polymerization. Rates of polymerization and intrinsic viscosities of the resulting homopolymers are lower than those of the related acrylates (19–20). These characteristics have been attributed to chain transfer to the itaconic monomer with formation of a stabilized allylic radical (**B**) incapable of further polymerization (21). Since normal polymerization kinetics are observed (22–23), degradative chain transfer does not occur and the radical (**B**), if formed, must initiate further polymerization.

$$R\cdot + CH_2{=}C(COOH)CH_2COOH \rightarrow CH_2{=}C(COOH)\dot{C}HCOOH + RH \quad (\mathbf{B})$$

Molecular weights as determined by light scattering for a series of homologous poly(dialkyl itaconates) are reported to be typically in the range of $(0.2 \text{ to } 0.5) \times 10^6$ but extending up to 2.5×10^6. The values are not as low as the intrinsic viscosities would suggest (20,24). Decreased chain-propagation rates originating in steric crowding, increased rates of termination because of the availability of hydrogens α to both the radical carbon and a carboxyl group (25), and branching (26) may contribute to the observed lowered polymerization rates and intrinsic viscosities.

Poly(itaconic acid) [*25119-64-6*] has been reported not to have the repeating itaconic acid structure (27–28). Other evidence, however, strongly suggests that the repeating itaconic unit is present in the initially formed polymer which can subsequently decarboxylate, most likely via a β-keto acid formed by condensation of pairs of adjacent itaconic units (29–31):

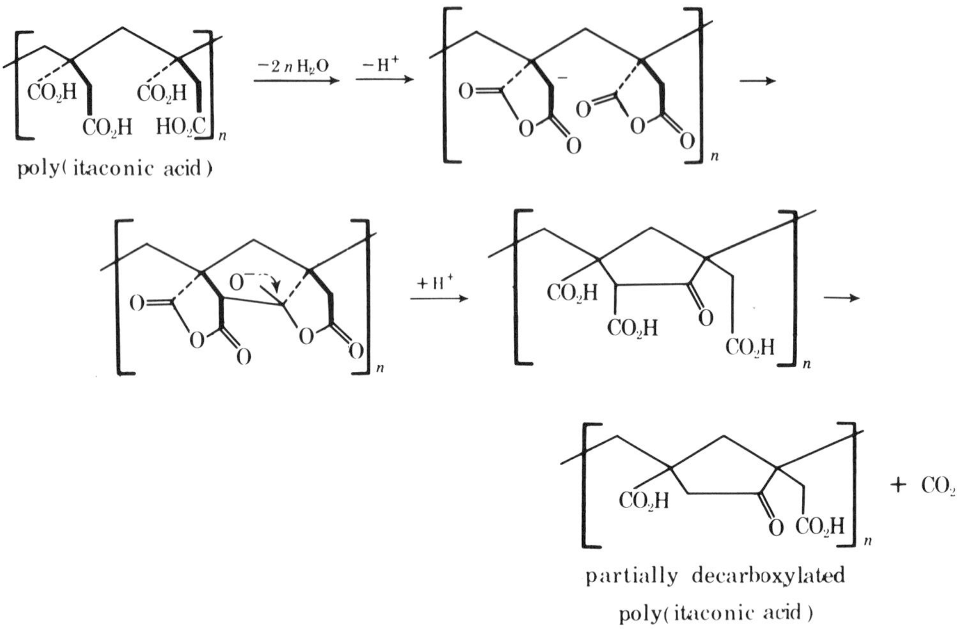

CO_2H CO_2H CO_2H HO_2C n $\xrightarrow{-2n\,H_2O}$ $\xrightarrow{-H^+}$ → $\xrightarrow{+H^+}$ CO_2H CO_2H CO_2H → CO_2H CO_2H + CO_2

poly(itaconic acid)

partially decarboxylated poly(itaconic acid)

Itaconic acid and its mono- and dialkyl esters readily undergo copolymerization with conjugated vinyl monomers such as the acrylates and styrene. Copolymerization with unconjugated monomers, eg, vinyl acetate, is also possible, but monomer reactivity ratios are less favorable. Rates of polymerization and intrinsic viscosities of the polymers are usually higher for copolymerization.

Manufacture and Processing

Itaconic acid is manufactured by fermentation of carbohydrates (32–35). Large-scale production utilizes molasses as the raw material because of low cost, but glucose, cane, or corn sugar are acceptable substrates. Initial carbohydrate concentration in the fermenter charge may be of the order of 15–25%. Smaller amounts of other ingredients such as ammonium salts to supply nitrogen, and essential metals, eg, magnesium and zinc, are also introduced. After the mixture is sterilized and inoculated with a selected strain of *Aspergillus terreus*, the broth is agitated and aerated at 35–40°C for 3–7 days. Itaconic acid is produced in the broth from citric acid **(6)**

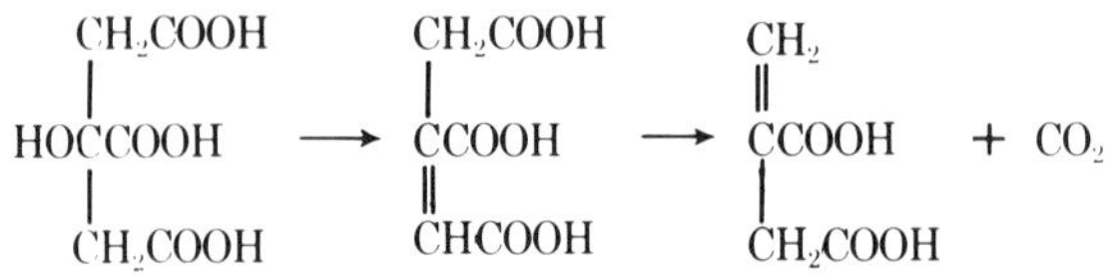

$$\underset{\textbf{(6)}}{\begin{array}{c}CH_2COOH\\ |\\ HOCCOOH\\ |\\ CH_2COOH\end{array}} \longrightarrow \underset{\textbf{(7)}}{\begin{array}{c}CH_2COOH\\ |\\ CCOOH\\ \|\\ CHCOOH\end{array}} \longrightarrow \underset{\textbf{(1)}}{\begin{array}{c}CH_2\\ \|\\ CCOOH\\ |\\ CH_2COOH\end{array}} + CO_2$$

generated in the tricarboxylic acid cycle. Citric acid is dehydrated by the enzyme, aconitate hydratase, to *cis*-aconitic acid (**7**). Subsequent decarboxylation of (**7**) by aconitate decarboxylase results in itaconic acid. As acid is formed, the pH is controlled by addition of a base, usually lime. Concentration of itaconic acid is 5–10% at the end of the fermentation; stoichiometric yields of over 80% have been reported. The product must be separated from broth which contains mycelium, residual sugar, fermentation products including other organic acids, and inorganic salts. Acidification, clarification, and other purification treatments are followed by concentration, crystallization, and collection of itaconic acid. Solvent extraction of the broth may also be used to recover product. By-products of manufacture are biodegradable organic materials and inorganic salts which constitute no environmental hazard.

Numerous chemical syntheses have been reported but none are practiced commercially. Some of the more feasible routes include: thermolysis of citric (36) or aconitic acids (37), oxidation of isoprene (38) or mesityl oxide (39–40) to citraconic acid with subsequent isomerization, carbonylation of acetylenic derivatives such as propargyl chloride (41) or ethyl 3-butynoate (42), and condensation of formaldehyde with succinic derivatives (43).

Economic Aspects

In 1979, Pfizer was the only U.S. manufacturer of itaconic acid, although there are at least four companies that produce and sell the material abroad. Esters were offered in the past, but are no longer sold. No published information is available on itaconic production or sales volume. In 1979 the U.S. price for itaconic acid was \$1.93–2.04/kg, in 22.5-kg multiwall bags.

Standards

The acid is stable in storage, and no polymerization inhibitor is present in commercially supplied material. The crystalline product is not hygroscopic and does not cake even at high relative humidities. The quality of acid produced in the U.S. is indicated by the specifications provided by the manufacturer (see Table 3).

Table 3. Specifications for Refined Itaconic Acid[a]

Property	Value
appearance	free-flowing, white to off-white crystals or powder
odor	slight, typical
assay[b], min %	99.0
loss on drying, 2 h, 100°C, max %	0.5
residue on ignition[c], max %	0.5
heavy metals, max ppm	50
%	0.005
sulfate, max ppm	300
%	0.03

[a] Pfizer Inc.
[b] By bromination.
[c] Sulfated.

Analytical Methods

Chromatographic techniques are usually applied for the detection and determination of the itaconic acid derivatives and their isomers. Paper (44–47), thin-layer (47–49), or ion exchange (50–51) chromatography are particularly suitable for detection of the acid and its isomers. Gas chromatography is applicable to both detection and quantitation of the esters, and after esterification, the acid (52). In the absence of other acids, itaconic acid can be quantitatively titrated to a phenolphthalein end point, or in the presence of other saturated organic acids by quantitative addition of bromine to the double bond (52).

Health and Safety Factors

Itaconic acid is a high melting solid exhibiting essentially no volatility at room temperatures. Accepted practices for handling organic acids should be followed.

Daily application of a 10% aqueous solution of itaconic acid (pH, 1.9) to the skin of rabbits caused significant irritation after four days. Instillation of the solution into the rabbit eye produced moderate erythema and swelling. Acute toxicities in the rat are: oral LD_{50}, 4000 mg/kg; intraperitoneal LD_{50}, 450 mg/kg (53). Toxic reactions were observed in cats when disodium itaconate [*5363-69-9*] was given orally at 500 mg/kg or intravenously at 150 mg/kg. However, daily oral or intravenous administration of 100 mg/kg of itaconic acid or equivalent doses of the disodium salt for fourteen weeks produced no indications of systemic toxicity. The toxic effects noted for the higher sodium itaconate doses were similar to those seen with sodium carbonate and were attributed to the high base content of the salt (54).

Itaconic acid and di-*n*-butyl itaconate [*2155-60-4*] have been approved for use in food-packaging materials when incorporated into acrylic, styrene–butadiene, vinylidene chloride and other resins (55).

Uses

Itaconic acid is a specialty monomer that affords performance advantages to certain polymeric coatings when relatively low levels of the acid are incorporated. Styrene-butadiene latexes containing under 10%, typically 1–5% of itaconic acid, are widely used in carpet backings (56–58) and paper coatings (59–60) (see Coatings; Paper). Emulsion stability, flow properties of the formulated coating, and adhesion to substrates are improved by the acid. Resistance to picking of coatings from paper in high speed printing with tacky inks is improved, for example. The highly polar itaconic monomer tends to react primarily at the particle surface where the functional effect of the resulting charge is high (61). Itaconic acid also undergoes little homopolymerization in the aqueous phase at the low concentrations commonly present. Consequently, the aqueous phase is not excessively thickened as may occur when acidic acrylic monomers are utilized.

Acrylic latexes containing itaconic acid have applications as nonwoven fabric binders. Vinylidene chloride coatings containing the monomer exhibit improved adhesion to paper, cellophane, and poly(ethylene terephthalate) films. Polyacrylonitrile fibers incorporating low levels of the acid exhibit improved dye receptivity which allows more efficient dyeing to deeper shades (62–63).

Dental cements composed of itaconic–acrylic acid copolymers that are cured with polyvalent metal compounds such as aluminosilicates or oxides of zinc and magnesium possess good compressive and adhesive strength and physiological compatibility (64).

Derivatives

Properties of the isomers and of commonly encountered itaconic derivatives are presented in Table 4.

Table 4. Physical Properties of Common Itaconic Derivatives

Derivative	CAS Registry No.	Mp, °C	Bp, °C (kPa[a])	n_D	Reference
Isomers					
citraconic acid	[*498-23-7*]	93			65
mesaconic acid	[*498-24-8*]	204			66
Anhydrides					
itaconic anhydride	[*2170-03-8*]	68			67
citraconic anhydride	[*616-02-4*]	8	108 (2.9)		65
Monoalkyl esters					
4-methyl 2-methylenebutanedioate	[*7338-27-4*]	68			13, 68–69
4-*n*-butyl 2-methylenebutanedioate	[*6439-57-2*]	40			13, 69
Dialkyl esters					
dimethyl itaconate	[*617-52-7*]	36	91 (1.3)	1.441^{20}	69
diethyl itaconate	[*2409-52-1*]		81 (0.3)	1.4355^{30}	69
di-*n*-butyl itaconate	[*2155-60-4*]		145 (1.3)	1.442^{25}	69

[a] To convert kPa to mm Hg, multiply by 7.5.

BIBLIOGRAPHY

"Itaconic Acid" in *ECT* 1st ed., Vol. 8, pp. 105–107 by C. J. Knuth, Chas. Pfizer & Co., Inc.; "Itaconic Acid" in *ECT* 2nd ed., Vol. 12, pp. 83–86, by C. J. Knuth, Chas. Pfizer & Co., Inc.

1. A. Goldstein, G. Mandel, and D. Pindzola, *Acta Crystallogr.* **14,** 94 (1961).
2. R. L. Harlow and C. E. Pfluger, *Acta Crystallogr. Sect. B* **29,** 2965 (1973) (in English).
3. R. C. Wilhoit and I. Lei, *J. Chem. Eng. Data* **10**(2), 166 (1965).
4. H. W. Ashton and J. R. Partington, *Trans. Faraday Soc.* **30,** 598 (1934).
5. B. E. Tate in E. C. Leonard, ed., *Vinyl and Diene Monomers,* Wiley-Interscience, New York, 1970, Pt. 1, p. 212.
6. *Ibid.*, p. 215.
7. M. Sakai, *Bull. Chem. Soc. Jpn.* **49**(1), 219 (1976).
8. E. H. Coulson and G. A. R. Kon, *J. Chem. Soc.*, 2568 (1932).
9. M. Sakai, *Bull. Chem. Soc. Jpn.* **50**(5), 1232 (1977).
10. R. Fittig and K. Bock, *Ann. Chem.* **331,** 172 (1904).
11. R. Anschütz and W. Petri, *Chem. Ber.* **13,** 1539 (1880).
12. N. Rabjohn, *Organic Syntheses,* Vol. IV, John Wiley & Sons, Inc., New York, 1963, p. 554.
13. Ref. 5, p. 227. *Chem. Abstr.*, 8th Coll. Index, uses methylenesuccinic acid for itaconic acid, the 9th, methylenebutanedioic acid.
14. W. Bruegel and K. Demmler, *J. Polym. Sci. Part C* **22**(2), 1117 (1967).
15. C. K. Ingold, C. W. Shoppee, and J. F. Thorpe, *J. Chem. Soc.*, 1477 (1926).
16. C. Knuth, A. Bavley, and W. A. Lazier, *J. Org. Chem.* **19,** 845 (1954).

17. P. L. Paytash, M. J. Thompson, and W. B. Clarke, *J. Am. Chem. Soc.* **76,** 3500 (1954).
18. H. Akashi, *Kogyo Kagaku Zasshi* **69,** 1416 (1966).
19. B. E. Tate, *Adv. Polymer Sci.* **5,** 215 (1967).
20. J. Velickovic and S. Vasovic, *Makromol. Chem.* **153,** 207 (1972).
21. C. S. Marvel and T. H. Shepherd, *J. Org. Chem.* **24,** 599 (1959).
22. S. Nagai and K. Yoshida, *Kobunshi Kagaku* **17,** 79 (1960).
23. M. G. Baldwin and S. F. Reed, *J. Polym. Sci. A-1,* 1919 (1963).
24. J. M. G. Cowie, S. A. E. Henshall, I. J. McEwen, and J. Velickovic, *Polymer* **18**(6), 612 (1977).
25. Ref. 5, p. 246.
26. E. A. Bekturov and L. A. Bimendina, *Izv. Akad. Nauk Kaz. SSR, Ser. Khim.* **23**(5), 79 (1973).
27. D. Braun and I. A. Aziz El Sayed, *Makromol. Chem.* **96,** 100 (1966).
28. B. L. Gafurov, L. N. Semenova, and M. A. Askarov, *Uzb. Khim. Zh.* **15**(5), 71 (1971).
29. B. E. Tate, *Makromol. Chem.* **109,** 176 (1967).
30. S. Nagai and F. Fujiwara, *J. Polym. Sci. Part B* **7,** 177 (1969).
31. K. Yokota, T. Hirabayashi, and T. Takashima, *Makromol. Chem.* **176,** 1197 (1975).
32. U.S. Pat. 2,385,283 (Sept. 18, 1945), J. H. Kane, A. C. Finlay, and P. F. Amann (to Chas. Pfizer & Co., Inc.).
33. V. F. Pfeifer, C. Vojnovich, and E. N. Heger, *Ind. Eng. Chem.* **44,** 2975 (1952).
34. U.S. Pat. 3,162,582 (Dec. 22, 1964), M. Batti (to Miles Laboratories, Inc.).
35. U.S. Pat. 3,044,941 (July 17, 1962), R. C. Nubel and E. J. Ratajak (to Chas. Pfizer & Co., Inc.).
36. A. H. Blatt, *Organic Syntheses,* Vol. II, John Wiley & Sons, Inc., New York, 1943, p. 368.
37. U.S. Pat. 2,448,831 (Sept. 7, 1948), E. J. Roberts, J. A. Ambler, and A. L. Curl (to the United States of America as represented by the Secretary of Agriculture).
38. H. Pichler, F. Obenaus, and G. Franz, *Erdoel Kohle Erdgas Petrochem.* **20**(3), 188 (1967).
39. U.S. Pat. 3,963,759 (June 15, 1976), E. J. Strojny (to The Dow Chemical Co.).
40. U.S. Pat. 4,100,179 (July 11, 1978), R. G. Berg and D. S. Hetzel (to Pfizer Inc.).
41. G. P. Chiusoli, *Angew. Chem.* **72**(2), 74 (1960).
42. E. R. H. Jones, G. H. Whitham, and M. C. Whiting, *J. Chem. Soc.,* 1865 (1954).
43. U.S. Pat. 3,835,162 (Sept. 10, 74), B. E. Tate and R. G. Berg (to Pfizer Inc.).
44. J. R. Howe, *J. Chromatog.* **3,** 389 (1960).
45. R. D. Hartley and G. J. Lawson, *J. Chromatogr.* **7,** 69 (1962).
46. V. Sanda, Z. Prochazka, and H. Le Moal, *Chem. Listy* **52,** 1546 (1958).
47. Ref. 5, p. 221.
48. G. Pastuska and H.-J. Petrowitz, *J. Chromatogr.* **10,** 517 (1963).
49. E. Knappe and I. Rohdewald, *Z. Anal. Chem.* **210**(3), 183 (1965).
50. C. Davies, R. D. Hartley, and G. J. Lawson, *J. Chromatogr.* **18,** 47 (1965).
51. S. Egashira, H. Nakasuka, and Y. Takeyama, *Bunseki Kagaku* **14,** 636 (1965).
52. Ref. 5, p. 222.
53. Ref. 5, pp. 218–220.
54. M. Finkelstein, H. Gold, and C. A. Paterno, *J. Am. Pharm. Assoc.* **36**(6), 173 (1947).
55. *Code of Federal Regulations, Title 21, Food and Drugs,* Parts 175–177, 179 and 181, U.S. Government Printing Office, Washington, D.C., 1978.
56. Neth. Pat. Appl. 6,405,106 (Nov. 11, 1964), (to Dunlop Rubber Co. Ltd.).
57. U.S. Pat. 3,928,695 (Dec. 23, 1975), D. H. Philp and N. L. Madison (to The Dow Chemical Co.).
58. U.S. Pat. 3,759,860 (Sept. 18, 1973), C. R. Peaker (to Uniroyal, Inc.).
59. U.S. Pat. 3,575,913 (Apr. 20, 1971), G. L. Meier (to The Dow Chemical Co.).
60. U.S. Pat. 3,793,244 (Feb. 19, 1974), J. F. Megee and R. G. Nickerson.
61. G. W. Ceska, *J. Appl. Polym. Sci.* **18,** 427 (1974).
62. A. A. Kharkharov and G. S. Saribekov, *Tekstil. Prom.* **26**(5), 74 (1966).
63. Ital. Pat. 572,367 (Jan. 25, 1958), E. Condorelli and co-workers (to Laboratori Italiani di Ricerca Chimica (LIRC)).
64. Ger. Offen. 2,439,882 (Mar. 20, 1975), S. Crisp and A. D. Wilson (to National Research Development Corp.).
65. Ref. 36, p. 140.
66. Ref. 36, p. 382.
67. S. Nagai, *Bull. Chem. Soc. Jpn.* **37,** 369 (1964).
68. B. R. Baker, R. E. Schaub, and J. H. Williams, *J. Org. Chem.* **17,** 116 (1952).
69. Ref. 5, p. 216.

General References

L. S. Luskin in R. H. Yocum and E. B. Nyquist, eds., *Functional Monomers,* Marcel Dekker, Inc., New York, 1974, Vol. 2, pp. 465–501.
B. E. Tate in E. C. Leonard, ed., *Vinyl and Diene Monomers,* Wiley-Interscience, New York, 1970, Pt. 1, pp. 205–261.
B. E. Tate, *Adv. Polymer Sci.* **5,** 214 (1967).
R. H. Billington, *Chem. Process. (London)* **15**(9), 8 (1969).
H. Warson, *Polym. Paint Colour J.* **161** (3809), 3337 (1972).
H. Akashi, *Kagaku (Kyoto)* **21,** 70 (1966) (in Japanese).
Ibid., 172 (66).

BRYCE E. TATE
Pfizer Inc.

J

J ACID, $H_2N(OH)C_{10}H_5SO_3H$. See Naphthalene derivatives.

JASMIN, JASMINE. See Oils, essential; Perfumes.

JUNIPER. See Oils, essential.

JUTE. See Fibers, vegetable.

K

KETENES AND RELATED SUBSTANCES

Ketenes and analogous compounds are characterized by a heterocumulene structure, $R_2C{=}C{=}X$, where the R substituents may be any combination of hydrogen, alkyl, aryl, acyl, halogen, and various functional groups. If X is O, S, or NR, the compounds are ketenes, thioketenes, or ketenimines, respectively. The parent compound ketene, $CH_2{=}C{=}O$, is the only compound of this type that is manufactured commercially, but ketenes play an important role in many organic reactions and processes. In general, ketenes and their analogues are highly reactive; however, the electrophilic center (the central carbon on the heterocumulene triad) is strongly affected by the nature of the substituents. Although a few ketenes are moderately stable, most of these compounds are transient species, not isolable as such. Ketenes were discovered in 1905 by Staudinger (1) who studied them intensively for twenty years (2).

The principal application of ketene is in the manufacture of acetic anhydride. A dimer of ketene, commonly called diketene, is becoming an increasingly important organic intermediate, and is now the main source of acetoacetic esters, acetoacetamides, and acetoacetanilides. The latter are used in manufacture of fine chemicals, drugs, pesticides, and pigments.

Apart from the commercial development of ketene and diketene, relatively little work was done until the late 1950s, when interest was revived in a wide range of ketenes and like compounds.

Monomeric Ketenes

Physical Properties. Ketenes range in properties from colorless gases such as ketene and carbon suboxide to highly colored liquids such as diphenylketene and carbon subsulfide. Many ketenes are unstable and cannot be isolated. The ketenes listed in Table 1 are the best known examples in the chemical literature.

Table 1. Monomeric Ketenes

Name	CAS Registry No.	Properties		Reference
		Physical state	mp or bp[a], °C	
ketene	[463-51-4]	colorless gas	bp −49.8	3
methylketene	[6004-44-0]	colorless gas		4
dimethylketene	[598-26-5]	yellow liq	bp 34	5
diphenylketene	[525-06-4]	orange liq	bp 118–120 (0.1 kPa)	6
pentamethyleneketene	[22589-13-5]	yellow liq	bp 40–41 (0.4 kPa)	7
dichloroketene	[4591-28-0]	not isolable		8
bis(trifluoromethyl)ketene	[684-22-0]	colorless gas	bp 5	9–10
tert-butylcyanoketene (**36**)	[29342-22-1]	not isolable		11
trimethylsilylketene	[4071-85-6]	colorless liq	bp 81–82	12
triphenylphosphoranylidene-ketene	[15596-07-3]	white cryst	mp 172–174	13
carbon suboxide (**28**)	[504-64-3]	colorless gas	bp 6.8	14
di-*tert*-butylthioketene	[16797-75-4]	purple liq	bp 59–60 (0.8 kPa)	15
bis(trifluoromethyl)thioketene (**41**)	[7445-60-5]	red-orange liq	bp 52–53	16
carbon subsulfide	[627-34-9]	dark red liq	bp 60–70 (1.6 kPa)	17
triphenylketenimine	[14181-84-1]	yellow cryst	mp 54–56	18–19

[a] At 101.3 kPa, unless otherwise stated; to convert kPa to mm Hg, multiply by 7.5.

Chemical Properties. The chemistry of ketenes is dominated by addition reactions. Reagents with labile hydrogen atoms (HX, where X is hydroxyl, amino, halogen, acyl, and the like) form the corresponding carboxylic acid derivatives (**1**) (20):

$$R_2C{=}C{=}O + HX \rightarrow \underset{(\mathbf{1})}{R_2CHCOX} \qquad (1)$$

The most important application of this reaction is the manufacture of acetic anhydride (**2**) from ketene and acetic acid (21) (see also Acetic acid and derivatives).

$$CH_2{=}C{=}O + CH_3CO_2H \rightarrow \underset{(\mathbf{2})}{(CH_3CO)_2O} \qquad (2)$$

Carbonyl compounds with active hydrogen atoms undergo enol acetylation when treated with ketene in the presence of a strong acid catalyst (22):

$$CH_2{=}C{=}O + CH_3COCH_3 \longrightarrow CH_2{=}\overset{\overset{\displaystyle CH_3}{|}}{C}{-}O{-}\overset{\overset{\displaystyle O}{\|}}{C}{-}CH_3 \qquad (3)$$

Many other reagents add to ketenes, eg, halogens (23) and nitrosyl chloride (24):

$$ClCH_2COCl \xleftarrow{Cl_2} CH_2{=}C{=}O \xrightarrow{NOCl} [ONCH_2COCl] \qquad (4)$$

The latter reaction is a key step in the synthesis of caprolactam (**3**) from toluene; in practice, the reagent is nitrosylsulfuric acid, and the intermediate nitroso compound (**4**) is formed, decarboxylated to cyclohexanone oxime (**5**), and rearranged to caprolactam in one step (25).

$$C_6H_5CH_3 \rightarrow C_6H_5CO_2H \rightarrow C_6H_{11}CO_2H \rightarrow$$

$$C_6H_{10}{=}C{=}O \xrightarrow{NOHSO_4} C_6H_{10}(NO)(CO_2SO_3H)\ (\mathbf{4}) \xrightarrow[-CO_2]{H_2O} C_6H_{10}{=}NOH\ (\mathbf{5}) \rightarrow \text{caprolactam}\ (\mathbf{3}) \qquad (5)$$

Acid halides, acetals, and ortho esters add by a ketene insertion (26). Starting with phosgene, this reaction is the first step in a process for synthesis of citric acid (qv) (**6**) (27):

$$COCl_2 + CH_2{=}C{=}O \longrightarrow CO(CH_2COCl)_2 \xrightarrow{ROH}$$

$$CO(CH_2CO_2R)_2 \xrightarrow{HCN} (NC)(HO)C(CH_2CO_2R)_2 \xrightarrow[H^+]{H_2O} (HOOC)(HO)C(CH_2CO_2H)_2\ (\mathbf{6}) \qquad (6)$$

The reaction of methylal (**7**) with ketene provides methyl 3-methoxypropionate (**8**) (28), an intermediate for acrylic esters that avoids the use of propiolactone:

$$\underset{(7)}{CH_2(OCH_3)_2} + CH_2{=}C{=}O \rightarrow \underset{(8)}{CH_3OCH_2CH_2CO_2CH_3} \qquad (7)$$

Ketenes characteristically undergo cycloaddition (29–32) with many compounds containing nucleophilic olefinic, carbonyl, azomethine, or azo linkages. These reactions generally result in 1,2-cycloadditions, and are favored by electrophilic ketenes and nucleophilic ketenophiles.

Addition to olefinic compounds takes place readily with enamines (**9**, X = NR_2) (33), less readily with vinyl ethers (**9**, X = OR) (34), and least with simple olefins (**9**, X = R) (35).

$$R_2C{=}C{=}O + \underset{(9)}{RCH{=}CHX} \longrightarrow \text{cyclobutanone } (R_2, O, R, X) \qquad (8)$$

The cycloaddition is 1,2 even with conjugated systems, as illustrated by the dichloroketene–cyclopentadiene adduct [*5307-99-3*] (**10**), which is a useful intermediate for synthesis of tropolone (**11**) (36):

$$C_5H_6 + Cl_2C{=}C{=}O \rightarrow (\mathbf{10}) \xrightarrow{H_2O} (\mathbf{11}) \qquad (9)$$

Addition of ketenes to carbonyl bonds requires an acidic catalyst. This reaction was formerly employed in the production of propiolactone (**12**), an intermediate for acrylic esters (**13**) (37) (see also Acrylic acid and derivatives), and is now used in the manufacture of sorbic acid (qv) (**14**) (38–39). In these processes, ketene adds to formaldehyde and to crotonaldehyde (**15**) with subsequent conversion of the β-lactone to the unsaturated acid or ester.

$$CH_2{=}C{=}O + CH_2O \xrightarrow{H^+} \underset{(\mathbf{12})}{\underbrace{OCH_2CH_2CO}} \xrightarrow{ROH} \underset{(\mathbf{13})}{CH_2{=}CHCO_2R} \qquad (10)$$

$$CH_2{=}C{=}O + \underset{(\mathbf{15})}{CH_3CH{=}CHCHO} \longrightarrow CH_3CH{=}CH\underbrace{CHCH_2COO} \longrightarrow \underset{(\mathbf{14})}{CH_3CH{=}CHCH{=}CHCO_2H} \qquad (11)$$

The reaction of ketenes and azomethines occurs not only by the 1:1 addition to form β-lactams (**16**) (40), but also by addition of two moles of the ketene to one of the ketenophile (41).

Ketenes add readily to cis-azo compounds (**17**), but not to the trans isomers (**18**) (42).

There are a number of exceptions to the general 1,2-cycloaddition of ketenes; examples are addition across the 1,4-systems in aminovinyl ketones (**19**) (43) and *o*-quinones (**20**) (44), and of 1:1 and 2:1 additions to single carbon atoms by reaction with diazomethane (**21**) (45) and with isonitriles (**22**) (46).

In the absence of active ketenophiles, most ketenes undergo self-addition to form dimers, trimers, and polymers. Exceptions are very electrophilic or sterically hindered ketenes, such as bis(trifluoromethyl)ketene, diphenylketene, and di-*tert*-butylketene

$$R_2C{=}C{=}O + ArN{=}CHAr \longrightarrow \text{(16)} \qquad (12)$$

(**16**) [β-lactam ring: C(R)(R)–C(=O)–N(Ar)–CH(Ar)]

$$2\,R_2C{=}C{=}O + RN{=}CHAr \longrightarrow \qquad (13)$$

[six-membered ring: C(R)(R)–C(=O)–O–C(=CR_2)–N(Ar)–CH(Ar)]

$$\text{no reaction} \xleftarrow{R_2C{=}C{=}O} \underset{(\mathbf{18})}{\text{trans-}ArN{=}NAr} \xrightarrow{h\nu} \underset{(\mathbf{17})}{\text{cis-}ArN{=}NAr} \xrightarrow{R_2C{=}C{=}O} ArN{-}N(Ar)CR_2CO \qquad (14)$$

$$R_2C{=}C{=}O + \underset{(\mathbf{19})}{R_2NCH{=}CHCOR} \longrightarrow \text{[6-R-3,3-R}_2\text{-4-NR}_2\text{-3,4-dihydro-2-pyranone]} \qquad (15)$$

$$R_2C{=}C{=}O + \underset{(\mathbf{20})}{\text{[tetrachloro-}o\text{-benzoquinone]}} \longrightarrow \text{[5,6,7,8-tetrachloro-3,3-R}_2\text{-1,4-benzodioxan-2-one]} \qquad (16)$$

$$CH_2{=}C{=}O + \underset{(\mathbf{21})}{CH_2N_2} \xrightarrow{-N_2} \text{cyclopropanone} \qquad (17)$$

$$2\ (C_6H_5)_2C{=}C{=}O + \underset{(\mathbf{22})}{RN{\equiv}C} \longrightarrow \text{[RN=, }(C_6H_5)_2C{=}\text{ and }{=}C(C_6H_5)_2\text{ substituted dioxolane ring]} \qquad (18)$$

[*59578-99-3*]. Ketenes dimerize either symmetrically to form 1,3-cyclobutanediones, or unsymmetrically to β-lactones. Ketene itself forms primarily the lactone dimer (**23**); less than 5% is converted to a symmetrical dimer (**24**), which in turn is converted to a trimer (**25**) by enol acetylation (47). Higher ketenes dimerize symmetrically unless acidic or basic catalysts are present (48–49).

$$\underset{(\mathbf{24})}{\underline{CH_2COCH_2CO}} \longleftarrow CH_2{=}C{=}O \longrightarrow \underset{(\mathbf{23})}{CH_2{=}\underline{CCH_2COO}} \qquad (19)$$

$$(\mathbf{24}) \xrightarrow{CH_2{=}C{=}O} \underset{(\mathbf{25})}{\underline{CH_2COCH{=}C}OCOCH_3}$$

$$\underline{R_2CCOCR_2CO} \longleftarrow R_2C{=}C{=}O \longrightarrow R_2C{=}\underline{CCR_2COO} \qquad (20)$$

Thioketenes form another type of symmetrical dimer (**26**) (16):

$$(CF_3)_2C{=}C{=}S \xrightarrow{R_3N} \underset{(\mathbf{26})}{(CF_3)_2C{=}\text{[1,3-dithietane]}{=}C(CF_3)_2} \qquad (21)$$

No ketenes show this behavior, but similar linkages through the heteroatoms are formed in polymerization; depending on the catalyst, dimethylketene is polymerized in this (50) and two other ways (51–52).

The polymerization of ketenes has been reviewed (53).

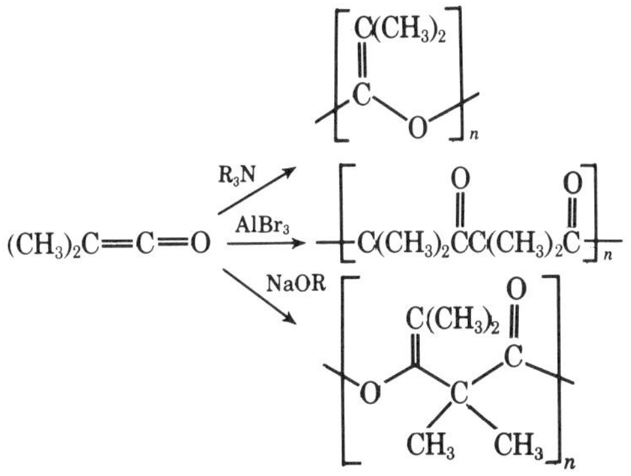

$$(CH_3)_2C{=}C{=}O \xrightarrow{R_3N} \left[-C(=C(CH_3)_2)-O- \right]_n \quad (22)$$

$$(CH_3)_2C{=}C{=}O \xrightarrow{AlBr_3} \left[-C(CH_3)_2\overset{O}{\overset{\|}{C}}C(CH_3)_2\overset{O}{\overset{\|}{C}}- \right]_n \quad (23)$$

$$(CH_3)_2C{=}C{=}O \xrightarrow{NaOR} \left[-O-C(=C(CH_3)_2)-C(CH_3)_2-\overset{O}{\overset{\|}{C}}- \right]_n \quad (24)$$

Adducts of acylketenes, including dimers, have structures corresponding to 1,4-addition (54). The dimer of diketene, dehydroacetic acid (**27**) has this structural relationship.

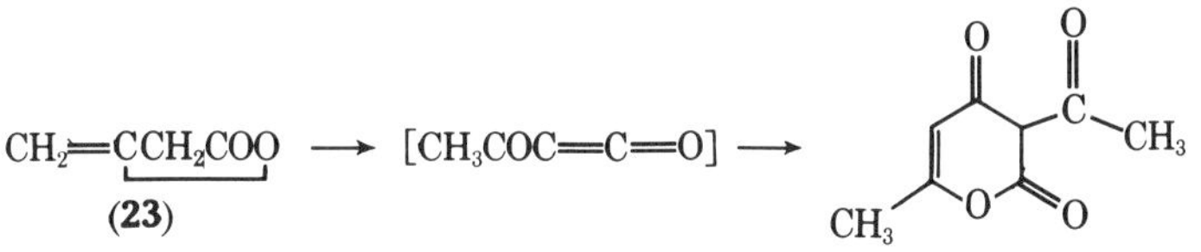

$$CH_2{=}CCH_2COO\ (\mathbf{23}) \longrightarrow [CH_3COC{=}C{=}O] \longrightarrow (\mathbf{27})\,[520\text{-}45\text{-}6] \quad (25)$$

The polymer of carbon suboxide (**28**), so-called red carbon, also contains pyrone units (**29**) (55–56).

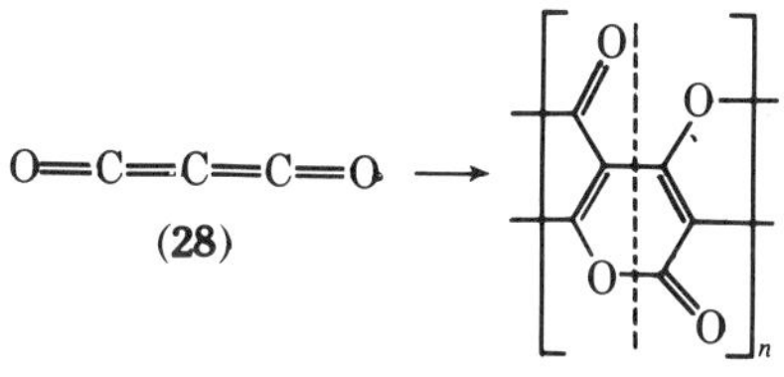

$$O{=}C{=}C{=}C{=}O\ (\mathbf{28}) \longrightarrow (\mathbf{29})_n \quad (26)$$

Preparation. Ketenes are the product of abstraction of one mole of water per mole of corresponding carboxylic acid, but few ketenes are prepared directly. The parent compound ketene is manufactured by pyrolysis of acetic acid at 700–800°C under reduced pressure (10–50 kPa, or 0.1–0.5 atm). A phosphate ester is injected to provide an acidic catalyst. After removal of water and unconverted acetic acid, the gaseous ketene is absorbed immediately in an appropriate reaction medium; eg, acetic anhydride is prepared by passage into a mixture of acetic acid and anhydride. The direct pyrolysis of acetic acid to ketone is quite efficient; the reaction is less effective with higher acids; better results are obtained by cracking the corresponding carboxylic anhydrides (57–58). Very pure ketene is best obtained by pyrolysis of acetic anhydride (59).

The use of acidic or basic dehydrating reagents is limited to very stable ketenes; eg, bis(trifluoromethyl)ketene is produced by heating the corresponding acid and phosphorus pentoxide (9).

Diphenylketene is best made by abstraction of hydrogen chloride from diphenylacetyl chloride with triethylamine (6). If the latter method is used to generate ketenes that dimerize easily, only dimers are produced; however, the technique is useful for generation of ketenes *in situ* for capture by an active ketenophile (8).

Dialkylketenes are obtained by the pyrolysis of dialkylmalonic anhydrides. If Dimethylmalonic acid (**30**) is heated with acetic anhydride, with removal of acetic acid at reduced pressure, the intermediate cyclic acylal of dimethylketene and dimethylmalonic acid (**31**) can be isolated (60). This compound generates dimethylketene by pyrolysis at 130°C.

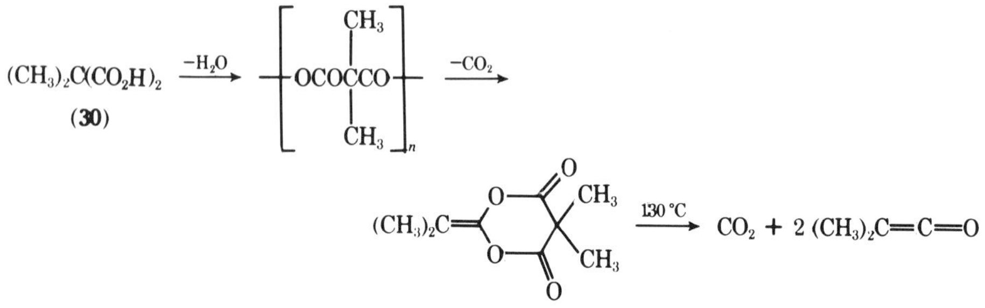

$$(CH_3)_2C(CO_2H)_2 \xrightarrow{-H_2O} \left[-OCOC(CH_3)_2CO- \right]_n \xrightarrow{-CO_2}$$

(**30**)

$$(\mathbf{31}) \xrightarrow{130\,°C} CO_2 + 2\,(CH_3)_2C{=}C{=}O \qquad (27)$$

(**31**)

In the preparation of acid chlorides of monosubstituted malonic acids (**32**), ketenes (**33**) are formed by loss of hydrogen chloride, even at moderate temperatures (ca 100°C) (61).

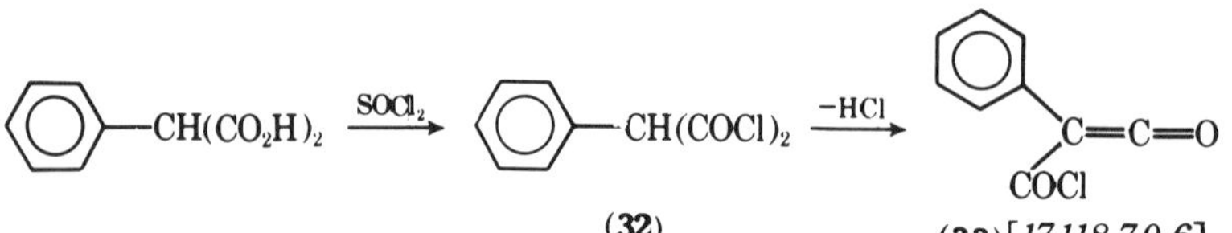

$$C_6H_5CH(CO_2H)_2 \xrightarrow{SOCl_2} C_6H_5CH(COCl)_2 \xrightarrow{-HCl} C_6H_5C(COCl){=}C{=}O \qquad (28)$$

(**32**) (**33**)[*17118-70-6*]

The decomposition of a diazoketone (**34**), either thermally or photochemically, gives a ketocarbene which usually rearranges to a ketene. The reaction, known as the Wolff rearrangement, is the key step of the Arndt-Eistert synthesis, a method for converting an acid to the next higher homologue (62):

$$RCO_2H \xrightarrow{SOCl_2} RCOCl \xrightarrow{CH_2N_2} RCOCHN_2 \xrightarrow{-N_2} [RCH{=}C{=}O] \xrightarrow{H_2O} RCH_2CO_2H \qquad (29)$$

(34)

The corresponding decomposition of a cyclic diazoketone (**35**) results in a ring

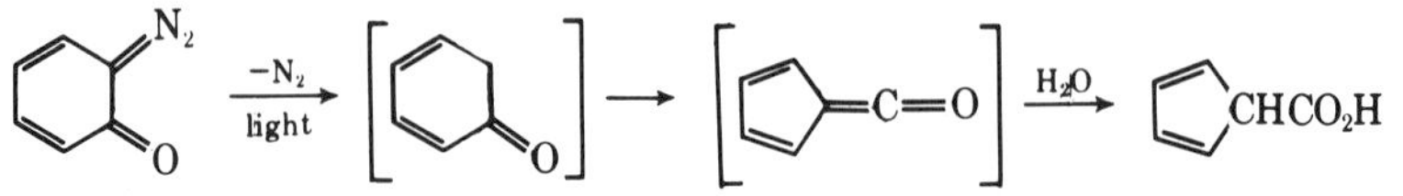

$$(\mathbf{35}) \xrightarrow[\text{light}]{-N_2} [\text{ketocarbene}] \longrightarrow [C_5H_4{=}C{=}O] \xrightarrow{H_2O} C_5H_5CHCO_2H \qquad (30)$$

(**35**) [*4727-22-4*]

contraction. This process is used in the diazotype process, where the image arises from reaction of the diazoketone and its decomposition product to form a dye (63). Relatively stable diazoketones are used in a photomechanical process for printing plates (64) (see Printing processes). After exposure of the light-sensitive layer containing the diazoketone, the plate is developed with alkali to remove the carboxylic acid. The diazo compound left in nonexposed areas picks up ink for printing. The process thus is a positive-to-positive process.

Although diazoketones are rarely used to prepare isolable ketenes, many unusual ketenes can be prepared *in situ*. Acylketenes are obtained from 2-diazo-1,3-diketones (65). A similar decomposition occurs in the generation of *tert*-butylcyanoketene (**36**) from the benzoquinone (**37**) (11):

$$\text{2,5-diazido-3,6-di-}tert\text{-butyl-1,4-benzoquinone } (\mathbf{37}) \longrightarrow \left[\text{2-azido-3-}tert\text{-butyl-5-}tert\text{-butyl-5-cyanocyclopent-2-ene-1,4-dione}\right] \longrightarrow \underset{(\mathbf{36})}{(CH_3)_3C(NC)C{=}C{=}O} \quad (31)$$

Some monomeric ketenes are best generated by thermolysis of their dimers, eg, ketene from diketene (66). Dimethylthioketene [*44158-23-8*] (**38**) is obtained by the cracking of tetramethyl-1,3-cyclobutanedithione (**39**) (67), prepared from the corresponding dione (**40**) (68). Bis(trifluoromethyl)thioketene (**41**) is prepared from its dimer (**42**), which is synthesized from malonic ester, thiophosgene, and sulfur tetrafluoride (16).

$$(CH_3)_2C{=}C{=}O \longleftarrow \underset{(\mathbf{40})}{\text{tetramethyl-1,3-cyclobutanedione}} \xrightarrow{P_2S_5} \underset{(\mathbf{39})}{\text{tetramethyl-1,3-cyclobutanedithione}} \longrightarrow \underset{(\mathbf{38})}{(CH_3)_2C{=}C{=}S} \quad (32)$$

$$CH_2(CO_2R)_2 + CSCl_2 \longrightarrow (RO_2C)_2C{=}\langle S_2 \rangle{=}C(CO_2R)_2 \xrightarrow[HF]{SF_4} \underset{(\mathbf{42})}{(CF_3)_2C{=}\langle S_2 \rangle{=}C(CF_3)_2} \xrightarrow{750^\circ C} \underset{(\mathbf{41})}{(CF_3)_2C{=}C{=}S} \quad (33)$$

Acetylenic ethers (**43–44**) are enol ethers of ketenes and give ketenes by pyrolysis (12,69).

$$(CH_3)_3SiCl + LiC{\equiv}COC_2H_5 \rightarrow \underset{(\mathbf{43})}{(CH_3)_3SiC{\equiv}COC_2H_5} \rightarrow (CH_3)_3SiCH{=}C{=}O \quad (34)$$

$$CH_2{=}CH{-}C{\equiv}COC_2H_5 \rightarrow \underset{[50888\text{-}73\text{-}8]}{CH_2{=}CH{-}CH{=}C{=}O} \quad \text{(44)} \tag{35}$$

Carbon suboxide is prepared by an unusual thermolysis of diacetyltartaric anhydride (**45**) (14).

$$\underset{(45)}{\text{diacetyltartaric anhydride } (CH_3CO_2)_2} \rightarrow O{=}C{=}C{=}C{=}O \tag{36}$$

Carbon subsulfide results from the passage of carbon disulfide through an electric arc between graphite electrodes (17).

Ketenimines (**46**) are prepared by methods analogous to those used for ketenes (ie, dehydrohalogenation of imino chlorides, dehydration of amides), or from ketenes by exchange with iminophosphoranes (**47**) (19).

$$R_2C{=}C{=}O + \underset{(47)}{(C_6H_5)_3P{=}NR} \rightarrow \underset{(46)}{R_2C{=}C{=}NR} + (C_6H_5)_3PO \tag{37}$$

Certain reactions of isoxazolium salts (**48**) with bases proceed through intermediate acylketenimines (**49**).

$$\underset{(48)}{\text{isoxazolium salt}} \xrightarrow{B^-} \underset{(49)}{RC(COR){=}C{=}NR} \xrightarrow{R'COOH} \underset{(50)}{R(OCOR')C{=}C(R)CONHR} \tag{38}$$

These so-called Woodward reagents are used to convert carboxylic acids to enol esters (**50**), which are valuable acylating agents in peptide syntheses (70).

Health and Safety Factors; Analytical Methods. Ketenes are sensitive to moisture and many are sensitive to air. The lower dialkylketenes react instantly with oxygen to form dangerously explosive polymeric peroxides (71). In the case of dimethylketene, peroxides can be avoided by handling the material above 70°C (72). Diarylketenes, although very sensitive to oxygen, do not form hazardous peroxides.

In view of their reactivity, volatile ketenes are hazardous compounds. Ketene itself ranks with phosgene in toxicity with a TLV of 0.5 ppm (73). Inhalation leads to lung edema, and in case of excessive exposure, administration of oxygen is indicated.

Ketenes are quantitatively determined by preparation of an anilide or an ester. A volatile derivative is determined conveniently by gas chromatographic techniques (74).

Ketene Dimers

Physical Properties. Several better known ketene dimers are listed in Table 2. Dimers of relatively simple ketenes are 1,3-cyclobutanediones (**51**) and unsaturated β-lactones (**52**). The former are usually crystalline solids and the latter, high boiling

Table 2. Ketene Dimers

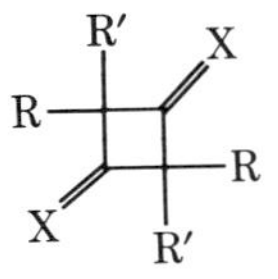

(51) 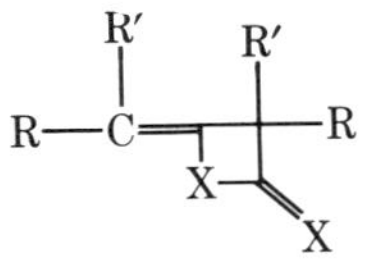 (52) 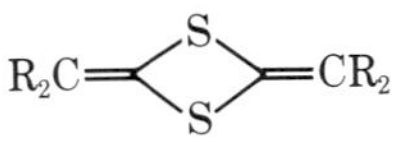 (53)

Type	R	R′	X	CAS Registry No.	mp or bp[a], °C	Reference
(51)	H	H	O	[*15506-53-3*]	mp 119	75
(52)	H	H	O	[*674-82-8*]	mp −6.5; bp 127	66
(51)	CH_3	H	O	[*20117-50-4*]	mp 141–143	4
(52)	CH_3	H	O	[*5659-14-3*]	mp −49; bp 48 (1.7 kPa)	76
(51)	C_2H_5	H	O	[*75379-05-4*]	mp 87–89	4
(52)	C_2H_5	H	O	[*5659-15-4*]	bp 92–95 (4.1 kPa)	77
(51)	CH_3	CH_3	O	[*933-52-8*]	mp 115–116; bp 159–161	78
(51)	CH_3	CH_3	S	[*10181-56-3*]	mp 124–125	68
(52)	CH_3	CH_3	O	[*3173-79-3*]	mp −18; bp 83–85 (5.2 kPa)	52
(52)	CH_3	CH_3	S	[*10181-61-0*]	bp 112–116 (2.6 kPa)	68
(53)	CH_3			[*10181-64-3*]	mp 170–172	68
(51)	C_6H_5	C_6H_5	O	[*3469-15-6*]	mp 247–248	79
(52)	C_6H_5	C_6H_5	O	[*6925-23-1*]	mp 148	48
(52)	H	$C_{16}H_{33}$	O	[*10126-68-8*]	mp 64	80
(52)	CF_3	CF_3	O	[*10436-05-2*]	bp 93	9
(53)	CF_3			[*7445-61-6*]	mp 84; bp 170	16

[a] At 101.3 kPa, unless otherwise stated; to convert kPa to mm Hg, multiply by 7.5.

liquids. Thioketenes also form dithietane-type dimers (**53**), which are usually crystalline solids.

Chemical Properties. The most important ketene dimer is the lactone dimer of ketene itself, commonly called diketene (**23**). In most of its reactions, diketene forms the products that would be expected from the hypothetical acetylketene (eq. 25): alcohols form acetoacetic esters, ammonia and alkylamines readily form acetoacetamides, whereas arylamines react more slowly to give acetoacetanilides. Arylhydrazines produce 3-methyl-1-aryl-5-pyrazolones (**54**).

$$CH_2{=}\underline{CCH_2COO}\ (\mathbf{23}) \xrightarrow{ArNH_2} CH_3COCH_2CONHAr;\quad \xrightarrow{ROH} CH_3COCH_2CO_2R;\quad \xrightarrow{R_2NH} CH_3COCH_2CONR_2;\quad \xrightarrow{ArNHNH_2} C_6H_5\underline{NN{=}C(CH_3)CH_2C{=}O}\ (\mathbf{54}) \qquad (39)$$

Diketene is converted to the ketene tetramer, dehydroacetic acid (**27**), when heated with a basic catalyst (tertiary amine) (81). In the presence of water, 1,3,5-heptanetrione (**55**) and 2,6-dimethyl-4-pyrone (**56**) are the main products (82).

$$CH_2{=}\underline{CCH_2COO}\ (\mathbf{23}) \xrightarrow{R_3N} (\mathbf{27}) \qquad (40)$$

$$CH_2{=}\underline{CCH_2COO}\ (\mathbf{23}) \xrightarrow{R_3N,\ H_2O} CH_3COCH_2COCH_2COCH_3\ (\mathbf{55}) + \text{2,6-dimethyl-4-pyrone}\ (\mathbf{56}) \qquad (41)$$

Diketene reacts with urea, thiourea, amidines, and guanidine to give pyrimidine derivatives; the formation of 6-methyluracil (**57**) from urea is illustrative (83):

$$CH_2{=}\underline{CCH_2COO}\ (\mathbf{23}) + NH_2CONH_2 \longrightarrow \text{6-methyluracil}\ (\mathbf{57}) \qquad (42)$$

In a related reaction, diketene is used to prepare a hydroxypyrimidine (**58**) which is the intermediate for the pesticide dimpylate (84).

Acetoacetylation of fluorosulfonamide with diketene is the first step in synthesis of the sweetening agent (**59**) (85).

Diketene reacts with ketones in the presence of an acid catalyst to form dioxinones (**60**). Thermolysis of these adducts has been used to force reactions of diketene, presumably by generation of acetylketene (54).

Addition of hydrogen chloride to diketene produces the unstable acetoacetyl chloride. Acetoacetyl fluoride (**61**), from the addition of hydrogen fluoride, is more stable. Addition of chlorine forms γ-chloroacetoacetyl chloride (**62**) (86).

$$CH_2{=}CCH_2COO + (CH_3)_2CHCONH_2 \longrightarrow (CH_3)_2CHCONHCOCH_2COCH_3 \xrightarrow{NH_3} \textbf{(58)} \quad (43)$$

$$CH_2{=}CCH_2COO\ \textbf{(23)} + H_2NSO_2F \longrightarrow CH_3COCH_2CONHSO_2F \longrightarrow \textbf{(59)} \quad (44)$$

$$CH_2{=}CCH_2COO\ \textbf{(23)} + CH_3COCH_3 \xrightarrow{H^+} \textbf{(60)} \longrightarrow [CH_3COCH{=}C{=}O] \xrightarrow{ArCN} \quad (45)$$

$$\underset{\textbf{(61)}}{CH_3COCH_2COF} \xleftarrow{HF} CH_2{=}CCH_2COO \xrightarrow{Cl_2} \underset{\textbf{(62)}}{ClCH_2COCH_2COCl} \quad (46)$$

Like diketene, other lactone dimers react with hydroxylic and amino compounds to form β-keto esters and amides. Lactone dimers of long-chain monoalkylketenes are valuable sizing agents in the paper industry (see Papermaking additives). These materials (Aquapels) can be applied under alkaline conditions and provide a permanent finish by reacting with the hydroxyl groups of the cellulose (87).

The symmetrical ketene dimers are strained cyclic 1,3-diketones and can be cleaved readily by alkaline reagents. In the presence of alcohols or amines, β-keto esters or amides are formed, identical with those obtained from the isomeric lactone dimers (88–89).

$$R_2CCOCR_2CO \xrightarrow[NaOR]{ROH} R_2CHCOCR_2CO_2R \xleftarrow{ROH} R_2C{=}CCR_2COO \quad (47)$$

If active hydrogen compounds are absent, dione dimers may isomerize to lactone dimers (52) or disproportionate to trimers (**63**) (90), depending on the catalyst.

Dione dimers undergo typical carbonyl reactions, such as formation of anils (**64**) or oximes (89), and reduction to alcohol or methylene groups (88).

$$R_2C{=}\underline{CCR_2COO} \xleftarrow{AlCl_3} \underline{R_2CCOCR_2CO} \xrightarrow{NaOR} \text{(63)} \quad (48)$$

(63)

$$\text{(64)} \xleftarrow[H^+]{ArNH_2} \underline{R_2CCOCR_2CO} \xrightarrow[Ru]{H_2} \quad (49)$$

(64)

Treatment with phosphorus pentasulfide converts dione dimers to dithiones, the dimers of the corresponding thioketenes (68). Phosphorus pentachloride converts the carbonyl groups to dichloromethylenes (91).

$$\underline{R_2CCSCR_2CS} \xleftarrow{P_2S_5} \underline{R_2CCOCR_2CO} \xrightarrow{PCl_5} \quad (50)$$

Photolysis of dione dimers gives ring cleavage and elimination of carbon monoxide. This reaction has provided a convenient route to tetramethylcyclopropanone (**65**) (92).

$$\underline{(CH_3)_2CCOC(CH_3)_2CO} \xrightarrow[-CO]{h\nu} \text{(65)} \xrightarrow[-CO]{h\nu} (CH_3)_2C{=}C(CH_3)_2 \quad (51)$$

(65)

Preparation. Ketene dimers are usually produced by spontaneous dimerization. Diketene is manufactured by absorption of gaseous ketene in diketene followed by distillation (93). Removal of last traces of acetic anhydride is difficult, but very pure diketene can be prepared by crystallization (94).

Other easily dimerized ketenes form cyclobutanedione dimers, but many give lactone dimers in the presence of acidic or basic reagents. Lactone dimers of monoalkylketenes are prepared easily by treatment of acid chlorides with triethylamine. This process is used commercially to prepare the long-chain ketene dimer sizing agents.

Ketene dimers can also be prepared by indirect routes (see preparation of thioketene dimers, eq. 50). The dione dimer of ketene was first prepared by addition of ketene to ethoxyacetylene, followed by selective hydrolysis of the dimer enol ether (75).

$$CH_2{=}C{=}O + HC{\equiv}COC_2H_5 \longrightarrow \underline{CH_2COCH{=}COC_2H_5} \longrightarrow \underline{CH_2COCH{=}COH} \quad (52)$$

With the exception of the tetramer, dehydroacetic acid, the higher oligomers and polymers of ketenes have no significant uses.

Health and Safety Factors; Shipping. Diketene is shipped under nitrogen in refrigerated tank trucks. It must be kept cold (0–5°C) and free of acidic or basic contaminants that catalyze exothermic polymerization. Diketene is classified as a flammable liquid, and must be handled with adequate ventilation and protective equipment. The vapor is very irritating; the liquid can cause severe eye injury and skin burns (95). There is no evidence that diketene is mutagenic.

Other lower lactone dimers have similar properties and require careful handling. Higher dimers of all types can be handled without special precautions. The ketene tetramer, dehydroacetic acid, is also a low hazard material.

Acetoacetic Acid Derivatives

Physical Properties. Acetoacetic esters are high boiling liquids with pleasant odors. Acetoacetamide is a low melting solid, whereas its lower *N*-alkyl derivatives are liquids; all are soluble in water. The acetacetanilides are moderately high melting solids. Properties of some commercially available derivatives are listed in Table 3.

Chemical Properties. Acetoacetic esters have long been used in organic syntheses, owing to the ease of substitution at the active methylene group and decarboxylation or deacetylation of the product.

$$CH_3COCH_2CO_2R \xrightarrow[NaOH]{RX} CH_3COCHRCO_2R \begin{cases} \xrightarrow{\text{dil NaOH}} CH_3COCH_2R' \\ \xrightarrow{\text{conc NaOH}} R'CH_2CO_2H \end{cases} \quad (53)$$

$$CH_3COCH_2CO_2R + CH_2{=}CHC(CH_3)_2(OH) \xrightarrow{-ROH} CH_3COCH_2CO_2C(CH_3)_2CH{=}CH_2 \xrightarrow{-CO_2} (CH_3)_2C{=}CHCH_2CH_2COCH_3 \ (\mathbf{66}) \quad (54)$$

Table 3. Acetoacetic Acid Derivatives

	CH_3COCH_2COX		
Substituent X	CAS Registry No.	mp or bp[a], °C	Reference
OCH_3	*[105-45-3]*	bp 169–170, 73–74 (1.6 kPa)	96
OC_2H_5	*[141-97-9]*	bp 181, 74 (1.9 kPa)	96
NH_2	*[5977-14-0]*	mp 54	96
$NHCH_3$	*[20306-75-6]*	bp 100–102 (0.1 kPa)	97
$N(CH_3)_2$	*[2044-64-6]*	bp 75–77 (0.2 kPa)	97
C_6H_5NH	*[102-01-2]*	mp 84–85	98
o-$CH_3C_6H_4NH$	*[93-68-5]*	mp 104	96
2,4-$(CH_3)_2C_6H_3NH$	*[97-36-9]*	mp 89	96
o-$CH_3OC_6H_4NH$	*[92-15-9]*	mp 87	96
o-ClC_6H_4NH	*[93-70-9]*	mp 105	96

[a] At 101.3 kPa, unless otherwise stated; to convert kPa to mm Hg, multiply by 7.5.

Allylic alcohols can be converted to acetoacetates by ester exchange; during the reaction or subsequent thermolysis, a rearrangement provides a route to unsaturated ketones, such as methylheptenone (**66**), an intermediate for vitamin A (99).

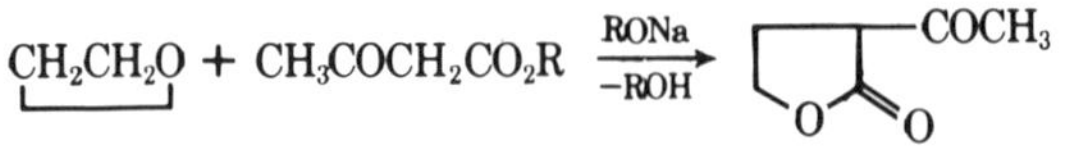

$$\underbrace{CH_2CH_2O} + CH_3COCH_2CO_2R \xrightarrow[-ROH]{RONa} \textbf{(67)} \quad (55)$$

(67)

Alkylation with ethylene oxide produces acetylbutyrolactone (**67**), an intermediate for thiamine (100).

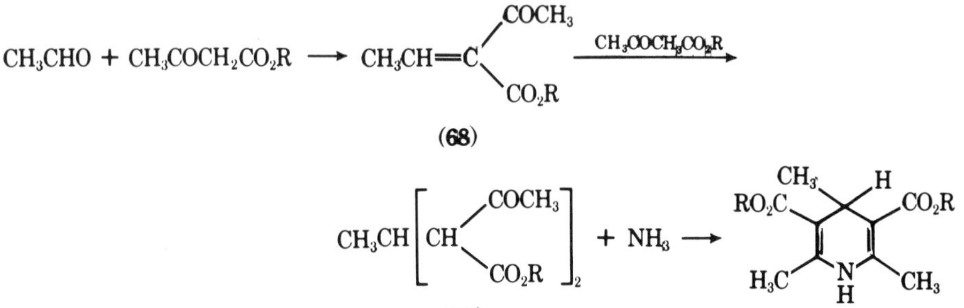

$$CH_3CHO + CH_3COCH_2CO_2R \longrightarrow CH_3CH{=}C(COCH_3)(CO_2R) \xrightarrow{CH_3COCH_2CO_2R}$$

(68)

$$CH_3CH[CH(COCH_3)(CO_2R)]_2 + NH_3 \longrightarrow \quad (56)$$

(69) **(70)**

Carbonyl compounds react at the active methylene center of acetoacetic esters; acetoacetic ester can then add to the unsaturated carbonyl system (**68**). In the Hantzsch pyridine synthesis, the resulting 1,5-diketone (**69**) is converted with ammonia to a dihydropyridine (**70**), which is easily oxidized to the corresponding pyridine (101).

Compounds with 1,2- and 1,3-systems reactive with carbonyl groups form 5- and 6-membered rings with acetoacetic esters. Ureas, for example, yield 6-methyluracils; halogenation gives herbicides, such as bromacil (**71**, R = *sec*-butyl, X = Br) and terbacil (**71**, R = *tert*-butyl, X = Cl) (102).

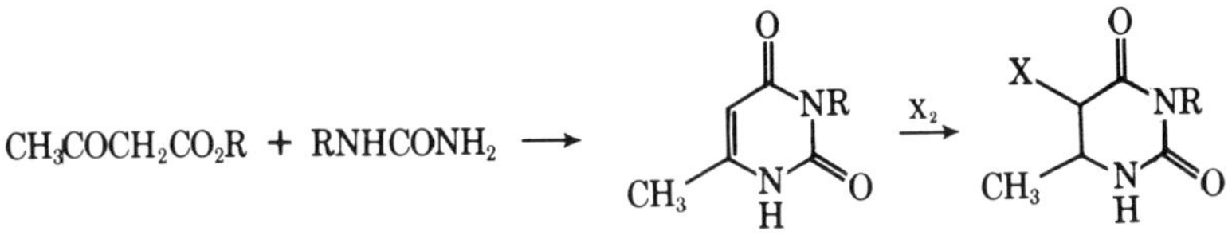

$$CH_3COCH_2CO_2R + RNHCONH_2 \longrightarrow \xrightarrow{X_2} \quad (57)$$

(71)

Acetoacetamide results from treatment of diketene with ammonia; addition of

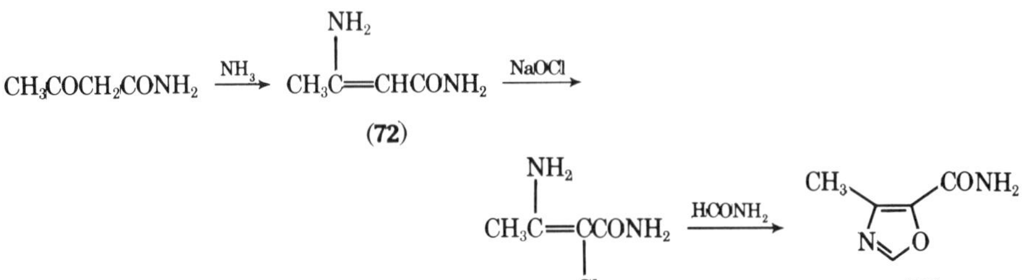

$$CH_3COCH_2CONH_2 \xrightarrow{NH_3} CH_3C(NH_2){=}CHCONH_2 \xrightarrow{NaOCl}$$

(72)

$$CH_3C(NH_2){=}C(Cl)CONH_2 \xrightarrow{HCONH_2} \quad (58)$$

(73)

more ammonia forms the enamine, 3-aminocrotonamide (**72**). Chlorination by hypochlorite, followed by ring closure with formamide yield the oxazole (**73**), an intermediate for pyridoxine (103).

Similar chemistry is involved in a synthesis of leucine (**74**) (104–105).

$$CH_3COCH_2CONH_2 \xrightarrow{\text{i-BuBr}} (CH_3)_2CHCH_2CH(COCH_3)CONH_2 \xrightarrow{\text{KOBr}} (CH_3)_2CHCH_2C(COCH_3)(Br)CONH_2 \xrightarrow[\text{HCl}]{\text{KOH}} \underset{(\mathbf{74})}{(CH_3)_2CHCH_2CH(NH_2)CO_2H} \quad (59)$$

$$CH_3COCH_2CON(CH_3)_2 \xrightarrow[\text{HCl}]{\text{NaNO}_2} CH_3COC(=NOH)CON(CH_3)_2 \xrightarrow{\text{Cl}_2}$$

$$(CH_3)_2NCOC(=NOH)Cl \xrightarrow[\text{NaOH}]{\text{CH}_3\text{SH}} (CH_3)_2NCOC(=NOH)SCH_3 \xrightarrow{\text{CH}_3\text{N=C=O}} \underset{(\mathbf{75})}{(CH_3)_2NCOC(=NOCONHCH_3)SCH_3} \quad (60)$$

N,N-Dimethylacetoacetamide is used to prepare the pesticide, oxamyl (**75**) (106).

$$CH_3COCH_2CONRCH_3 \xrightarrow{\text{Cl}_2} CH_3COCHClCONRCH_3 \xrightarrow{\text{P(OCH}_3)_3} \underset{(\mathbf{76})}{CH_3C(OP(O)(OCH_3)_2){=}CHCONRCH_3} \quad (61)$$

$$ClCH_2COCH_2COCl + NaNO_2 \longrightarrow ClCH_2COC(=NOH)COCl \xrightarrow{\text{H}_2\text{O}} \underset{(\mathbf{77})}{ClCH_2COCH{=}NOH} \quad (62)$$

N-Methyl- and *N,N*-dimethylacetoacetamide are also used for synthesis of

$$ClCH_2COCH_2COCl \longrightarrow ClCH_2COCH_2CO_2R \longrightarrow$$

$$\underset{(\mathbf{79})}{\text{2,5-dioxocyclohexane-1,4-}(CO_2R)_2} \xrightarrow{\text{C}_6\text{H}_5\text{NH}_2} \underset{(\mathbf{78})}{\text{quinacridone}} \quad (63)$$

$CH_3COCH_2CO_2R + C_6H_5NH_2$

H^+, 25 °C → CO_2R, N, CH_3 → 250 °C → OH, N, CH_3 **(80)** (64)

conc. H_2SO_4, 100 °C → CH_3, O, N, H, O → CH_3, N, OH **(81)** (65)

pesticides of much different structure, such as monocrotophos (**76,** R = H) and dicrotophos (**76,** R = CH_3) (107).

γ-Chloroacetoacetyl chloride is an intermediate of growing interest (108); eg, it is used to prepare β-chloropyruvaldoxime (**77**), an intermediate in pteridine syntheses (109).

Converted to an ester and self-condensed under proper conditions, γ-chloroacetoacetyl chloride has provided a new route to succinylsuccinic esters (**78**), which are intermediates for quinacridone (**79**) pigments (see also Cyanine dyes) (110).

Acetoacetic esters undergo two modes of condensation with aromatic amines. The Conrad-Limpach reaction gives 4-hydroxyquinolines (**80**) and the Knorr synthesis produces 2-hydroxyquinolines (**81**) (111).

Preparation. Acetoacetic esters are prepared by heating diketene with alcohols in the presence of acidic (112) or basic (113) catalysts. The process may be conducted continuously in a distillation column (114).

Acetoacetamides are prepared by addition of diketene to ammonia or alkylamines. *N*-Methyl- and *N,N*-dimethylacetoacetamides are usually sold as concentrated (60–80%) aqueous solutions (115).

Acetoacetanilides are also prepared by addition of diketene to an aqueous suspension of the arylamine, or simultaneous addition of diketene and the amine to the aqueous reaction mixture (116). These materials are used mainly to manufacture pigments (qv).

BIBLIOGRAPHY

"Ketenes" in *ECT* 1st ed., Vol. 8, pp. 109–113, by John R. Caldwell, Tennessee Eastman Company; "Ketenes and Related Substances" in *ECT* 2nd ed., Vol. 12, pp. 87–100, by Robert H. Hasek, Tennessee Eastman Company.

1. H. Staudinger, *Ber.* **38,** 1735 (1905).
2. H. Staudinger, *Das Wissenschaftliche Werk von Hermann Staudinger, Bd. 6: Die Ketene,* Hüthig & Wepf, Basel, Switz., 1976.
3. B. G. Reuben, *J. Chem. Eng. Data* **14,** 235 (1969).
4. C. C. McCarney and R. S. Ward, *J. Chem. Soc. Perkin Trans. 1,* 1600 (1975).
5. U.S. Pat. 3,201,474 (Aug. 17, 1965), R. H. Hasek and E. U. Elam (to Eastman Kodak Co.).
6. E. C. Taylor, A. McKillop, and G. H. Hawks, *Org. Syn.* **52,** 36 (1972).
7. G. Sioli, R. Mattone, L. Giuffre, R. Trotta, and E. Tempesti, *Chim. Ind. (Milan)* **53,** 133 (1971).
8. W. T. Brady, *Synthesis,* 415 (1971).
9. D. C. England and C. G. Krespan, *J. Am. Chem. Soc.* **88,** 5582 (1966).

10. Yu. A. Cheburkov and I. I. Knunyants, *Fluorine Chem. Rev.* **1,** 107 (1967).
11. W. Weyler, Jr., W. G. Duncan, M. B. Liewen, and H. W. Moore, *Org. Syn.* **55,** 32 (1976).
12. R. A. Ruden, *J. Org. Chem.* **39,** 3607 (1974).
13. C. N. Matthews and G. H. Birum, *Tetrahedron Lett.*, 5707 (1966).
14. T. Kappe and E. Ziegler, *Angew. Chem. Int. Ed. Engl.* **13,** 491 (1974).
15. E. U. Elam, F. H. Rash, J. T. Dougherty, V. W. Goodlett, and K. C. Brannock, *J. Org. Chem.* **33,** 2738 (1968).
16. M. S. Raasch, *J. Org. Chem.* **35,** 3470 (1970).
17. W. Stadlbauer and T. Kappe, *Chem. Ztg.* **101,** 137 (1977).
18. H. Staudinger and E. Hauser, *Helv. Chim. Acta* **4,** 887 (1921).
19. G. R. Krow, *Angew. Chem. Int. Ed. Engl.* **10,** 435 (1971).
20. H. Eck, *Chem. Ztg.* **97,** 62 (1973).
21. H. Eck, "Acetic Anhydride and Other Fatty Acid Anhydrides" in E. Bartholome, E. Biekert and H. Hellman, eds., *Ullmanns Encyklopadie der Technische Chemie,* 4th ed., Vol. 11, Verlag Chemie, Weinheim, FRG, pp. 75–87.
22. H. Spes, *Chem. Ing. Tech.* **38,** 961 (1966).
23. U.S. Pats. 3,758,569 and 3,758,571 (Sept. 11, 1973), D. E. Bissing and V. W. Gash (to Monsanto Co.).
24. U.S. Pat. 3,686,280 (Aug. 22, 1972), T. W. Rave (to Hercules, Inc.).
25. G. Sioli and L. Giuffre, *Hydrocarbon Process.* **53,** 124 (1974).
26. H. Eck and H. Prigge, *Justus Liebigs Ann. Chem.* **755,** 177 (1972).
27. U.S. Pat. 3,963,775 (June 15, 1976), N. Heyboer (to Akzona, Inc.).
28. H. Spes, *Chem. Ing. Tech.* **38,** 957 (1966).
29. H. Ulrich, *Cycloaddition Reactions of Heterocumulenes,* Academic Press, Inc., New York, 1967, pp. 38–109.
30. F. I. Luknitskii and B. A. Vovsi, *Usp. Khim.* **38,** 1072 (1969).
31. R. W. Holder, *J. Chem. Educ.* **53,** 81 (1976).
32. L. Ghosez and M. J. O'Donnell in A. P. Marchand and R. E. Lehr, eds., *Pericyclic Reactions,* Vol. 2, Academic Press, Inc., New York, 1977, pp. 79–109.
33. R. H. Hasek and J. C. Martin, *J. Org. Chem.* **28,** 1468 (1963).
34. R. H. Hasek, P. G. Gott, and J. C. Martin, *J. Org. Chem.* **29,** 1239 (1964).
35. J. C. Martin, P. G. Gott, V. W. Goodlett, and R. H. Hasek, *J. Org. Chem.* **30,** 4175 (1965).
36. R. A. Minns, *Org. Syn.* **57,** 117 (1977).
37. G. Kijima, *Jpn. Chem. Quart.* **5,** 12 (1969).
38. U.S. Pat. 3,759,988 (Sept. 18, 1973), G. Künstle and H. Spes (to Wacker-Chemie Gmbh).
39. U.S. Pat. 3,574,728 (Apr. 13, 1971), I. Takasu, M. Higuchi, and Y. Hijioka (to Daicell, Ltd.).
40. J. C. Sheehan and E. J. Corey in R. Adams, ed., *Organic Reactions,* Vol. 9, John Wiley & Sons, Inc., New York, 1957, pp. 395–399.
41. J. C. Martin, V. A. Hoyle, Jr., and K. C. Brannock, *Tetrahedron Lett.* 3589 (1965).
42. R. C. Kerber and T. J. Ryan, *Tetrahedron Lett.*, 703 (1970).
43. J. C. Martin, K. R. Barton, P. G. Gott, and R. H. Meen, *J. Org. Chem.* **31,** 943 (1966).
44. W. Reid and R. Kraemer, *Ann.* **681,** 52 (1965).
45. N. J. Turro, *Acc. Chem. Res.* **2,** 25 (1969).
46. T. El Gomati, J. Firl, and I. Ugi, *Chem. Ber.* **110,** 1603 (1977).
47. L. Tenud, M. Weilenmann, and E. Dallwigk, *Helv. Chim. Acta* **60,** 975 (1977).
48. R. Anet, *Chem. Ind. (London),* 1313 (1961).
49. D. G. Farnum, J. R. Johnson, R. E. Hess, T. B. Marshall, and B. Webster, *J. Am. Chem. Soc.* **87,** 5191 (1965).
50. G. F. Pregaglia, M. Binaghi, and M. Cambini, *Makromol. Chem.* **67,** 10 (1963).
51. G. Natta, G. Mazzanti, G. F. Pregaglia, M. Binaghi, and M. Peraldo, *J. Am. Chem. Soc.* **82,** 4742 (1960).
52. R. H. Hasek, R. D. Clark, E. U. Elam, and J. C. Martin, *J. Org. Chem.* **27,** 60 (1962).
53. G. F. Pregaglia and M. Binaghi, "Ketene Polymers" in N. M. Bikales, ed., *Encyclopedia of Polymer Science and Technology,* Vol. 8, Interscience Publishers, New York, 1968, pp. 45–57.
54. G. Jaeger and J. Wenzelburger, *Justus Liebigs Ann. Chem.*, 1689 (1976).
55. E. Ziegler, H. Junek, and H. Biemann, *Monatsh. Chem.* **92,** 927 (1961).
56. H. Sterk, P. Tritthart, and E. Ziegler, *Monatsh. Chem.* **101,** 1851 (1970).
57. G. Pregaglia and M. Binaghi, *Macromol. Syn.* **3,** 152 (1968).

58. G. Sioli, R. Mattone, L. Giuffre, R. Trotta, and E. Tempesti, *Chim. Ind. (Milan)* **53,** 133 (1971).
59. S. Andreades and H. D. Carlson, *Org. Syn. Coll. Vol.* **5,** 683 (1973).
60. H. Bestian and D. Günther, *Angew. Chem.* **75,** 841 (1963).
61. S. Nakanishi and K. Butler, *Org. Prep. Proced. Int.* **7,** 155 (1975).
62. W. E. Bachmann and W. S. Struve in R. Adams, eds., *Organic Reactions,* Vol. 1, John Wiley & Sons, Inc., New York, 1942, p. 38.
63. O. Süs, *Ann.* **556,** 65 (1944).
64. O. Süs, *Ann.* **579,** 139 (1953).
65. L. Capuano, W. Fischer, H. Scheidt, and M. Schneider, *Chem. Ber.* **111,** 2497 (1978).
66. S. Andreades and H. D. Carlson, *Org. Syn. Coll. Vol.* **5,** 679 (1973).
67. G. Seybold, *Tetrahedron Lett.,* 555 (1974).
68. E. U. Elam and H. E. Davis, *J. Org. Chem.* **32,** 1562 (1967).
69. J. K. Terlouw, P. C. Burgers, and J. L. Holmes, *J. Am. Chem. Soc.* **101,** 225 (1979).
70. R. B. Woodward, R. A. Olofson, and H. Mayer, *Tetrahedron* **22**(Suppl. 8), 321 (1966).
71. N. Turro, M.-F. Chow, and Y. Ito in B. Ranby and J. F. Rabek, eds., *EUCHEM Conf., 1976,* John Wiley & Sons, Inc., Chichester, Eng., 1978, pp. 174–181.
72. Fr. Pat. 1,381,831 (Dec. 11, 1964), T. E. Stanin and V. L. Brown, Jr. (to Eastman Kodak Co.).
73. *Threshold Limit Values for Chemical Substances,* American Conference of Governmental Industrial Hygienists, Cincinnati, Oh., 1978.
74. G. M. Breuer, F. J. Grieman, and E. K. C. Lee, *J. Phys. Chem.* **79,** 542 (1975).
75. H. H. Wasserman and E. V. Dehmlow, *J. Am. Chem. Soc.* **84,** 3786 (1962).
76. J. R. Johnson and V. J. Shiner, *J. Am. Chem. Soc.* **75,** 1350 (1953).
77. R. L. Wear, *J. Am. Chem. Soc.* **73,** 2390 (1951).
78. J. L. E. Erickson and G. C. Kitchens, *J. Am. Chem. Soc.* **68,** 492 (1946).
79. H. Das and E. C. Kooyman, *Rec. Trav. Chim.* **84,** 965 (1965).
80. E. S. Rothman, *J. Org. Chem.* **32,** 1683 (1967).
81. U.S. Pat. 3,483,224 (Dec. 9, 1969), J. T. Fitzpatrick and W. Van der Hoeven (to Union Carbide Corp.).
82. E. Marcus, J. K. Chan, and C. B. Strow, *J. Org. Chem.* **31,** 1369 (1966).
83. Ger. Pat. 2,138,802 (Feb. 22, 1973), H. Eck and H. Spes (to Wacker-Chemie Gmbh).
84. U.S. Pat. 4,018,771 (Apr. 19, 1977), J. T. Gupton, A. M. Jelenevsky, T. U. Miyazaki, and H. E. Petree (to Ciba-Geigy Corp.).
85. U.S. Pat. 4,052,453 (Oct. 4, 1977), H. Pietsch, K. Clauss, H. Jensen, and E. Schmidt (to Hoechst A.-G.).
86. K.-J. Boosen, *Chimia* **27,** 541 (1973).
87. W. O. Kincannon, Jr. and S. H. Watkins, *TAPPI Monograph Ser.* **33,** 157 (1971).
88. R. H. Hasek, E. U. Elam, J. C. Martin, and R. G. Nations, *J. Org. Chem.* **26,** 700 (1961).
89. R. H. Hasek, E. U. Elam, and J. C. Martin, *J. Org. Chem.* **26,** 4340 (1961).
90. R. H. Hasek, R. D. Clark, E. U. Elam, and R. G. Nations, *J. Org. Chem.* **27,** 3106 (1962).
91. U.S. Pat. 3,345,420 (Oct. 3, 1967), H. G. Gilch (to Union Carbide Corp.).
92. N. J. Turro, W. B. Hammond, and P. A. Leermakers, *J. Am. Chem. Soc.* **87,** 2774 (1965).
93. U.S. Pat. 3,865,846 (Feb. 11, 1975), G. Schulz, G. Matthias, W. Kasper, and E. Haarer (to BASF A.-G.).
94. Swiss Pat. 423,754 (May 13, 1967), H. Keller (to Lonza, Ltd.).
95. *Bulk Storage and Handling of Eastman Diketene,* Publication No. GN-301A, Eastman Chemical Products, Inc., Kingsport, Tenn., 1978.
96. *Dictionary of Organic Compounds,* 4th ed., Oxford University Press, New York, 1965.
97. U.S. Pat. 2,908,605 (Oct. 13, 1959), E. Beriger and R. Sallmann (to Ciba Ltd.).
98. J. W. Williams and J. A. Krynitsky, *Org. Syn. Coll. Vol.* **3,** 10 (1955).
99. U.S. Pat. 3,975,446 (Aug. 17, 1976), T. Kitagaki, A. Yamamoto, and T. Ishihara (to Shinetsu Chemical Co.).
100. P. I. Pollak, "Thiamine" in A. Standen, ed., *Kirk-Othmer Encyclopedia of Chemical Technology,* 2nd ed., Vol. 20, Interscience Publishers, New York, 1969, pp. 178–179.
101. H. S. Mosher in R. C. Elderfield, ed., *Heterocyclic Compounds,* Vol. 1, John Wiley & Sons, Inc., New York, 1950, p. 462.
102. U.S. Pat. 3,235,357 (Feb. 15, 1966), H. M. Loux (to E. I. du Pont de Nemours & Co.).
103. U.S. Pat. 4,093,654 (June 6, 1978), D. L. Coffen (to Hoffmann-LaRoche).
104. M. Yamoto and K. Oshima, *Yakugaku Zasshi* **87,** 943 (1965).

105. M. Yamoto and M. Fukuyama, *Yakugaku Zasshi* **87,** 1431 (1967).
106. U.S. Pat. 3,557,089 (Jan. 19, 1971), E. W. Raleigh (to E. I. du Pont de Nemours & Co.).
107. U.S. Pat. 2,802,855 (Aug. 13, 1957), R. R. Whetstone and A. R. Stiles (to Shell Development Co.).
108. K.-J. Boosen, *Helv. Chim. Acta* **60,** 1256 (1977).
109. E. C. Taylor and R. C. Portnoy, *J. Org. Chem.* **38,** 806 (1973).
110. U.S. Pat. 3,803,209 (Apr. 9, 1974), E. Greth (to Lonza, Ltd.); P. Pollak, *Chimia* **30,** 357 (1976).
111. R. C. Elderfield in R. C. Elderfield, ed., *Heterocyclic Compounds,* Vol. 4, John Wiley & Sons, Inc., New York, 1952, p. 32.
112. U.S. Pat. 3,651,130 (July 25, 1968), O. Marti and W. Zimmerli (to Lonza, Ltd.).
113. U.S. Pat. 3,542,855 (Nov. 24, 1970), A. Moschel (to Farbwerke Hoechst A.-G.).
114. Ger. Pat. 1,912,406 (Oct. 1, 1970), W. Kasper, G. Matthias, and G. Schulz (to BASF A.-G.).
115. *Mono-N-methylacetoacetamide,* Publication No. GN-323; *N,N-Dimethylacetoacetamide,* Publication No. GN-308A, Eastman Chemical Products, Inc., Kingsport, Tenn., 1973.
116. U.S. Pat. 3,304,328 (Feb. 14, 1967), R. L. Pelley (to FMC Corp.).

General References

D. Borrmann "Methoden zur Herstellung and Umwandlung von Ketenen, dimeren Ketenen, . . ." in E. Müller, ed., *Methoden der Organischen Chemie (Houben-Weyl),* Vol. 7, Part 4 (Sauerstoff Verbindungen II), Georg Thieme Verlag, Stuttgart, FRG, 1968, pp. 53–339.

Y. Etienne and N. Fischer "β-Lactones" in A. Weissberger, ed., *Compounds with Three- and Four-Membered Rings, Heterocyclic Compounds,* Vol. 19, Interscience Publishers, New York, 1964, Part 2, pp. 729–884.

R. N. Lacey "Ketene in Organic Synthesis" in R. A. Raphael, E. C. Taylor, and H. Wynberg, eds., *Advances in Organic Chemistry: Methods and Results,* Vol. 2, Interscience Publishers, Inc., New York, 1960, pp. 213–263.

R. N. Lacey "Ketenes" in S. Patai, ed., *The Chemistry of the Alkenes,* Interscience Publishers, New York, 1964, pp. 1161–1227.

G. Quadbeck "Ketene in Preparative Organic Chemistry" in W. Foerst, ed., *Newer Methods of Preparative Organic Chemistry,* Vol. 2, Academic Press, New York, 1963, pp. 133–161.

The Chemistry of Ketenes, Allenes and Related Compounds, Parts 1 and 2, S. Patai, ed., John Wiley-Interscience, New York, 1980.

ROBERT H. HASEK
Tennessee Eastman Company

KETONES

Ketones are organic compounds containing one or more carbonyl groups bound to two carbon atoms and are represented by the general formula

$$\overset{O}{\overset{\|}{RCR'}}$$

and

$$R\overset{O}{\overset{\|}{C}}-R''-\overset{O}{\overset{\|}{C}}R'''$$

Depending on the nature of the hydrocarbon groups (R, R′, R″, and R‴) attached to the carbonyl group ketones are classified as symmetric or simple aliphatic, such as acetone, CH_3COCH_3, symmetric aromatic, such as benzophenone, $C_6H_5COC_6H_5$, unsymmetric or mixed aliphatic–aromatic, such as methyl ethyl ketone, $CH_3COC_2H_5$, and acetophenone, $CH_3COC_6H_5$, alicyclic, such as cyclopentanone, or heterocyclic. Although the carbonyl group of ketones may contain heteroatoms in place of the oxygen, as in the case of thioketones, this review deals only with ketones containing carbon, hydrogen, and oxygen. The most important ketone is acetone (see Acetone).

Ketones may be named by adding the word ketone to the names of the two hydrocarbon groups bonded to the carbonyl carbon (common name); dropping the e and adding the suffix one to the parent hydrocarbon; and using the prefix keto or oxo (or other prefixes such as acetyl, acetonyl, benzoyl, etc) for the carbonyl group. Some mixed ketones are commonly named by adding the suffix phenone or naphthone to the stem of the aliphatic acid, eg,

$$CH_3\overset{O}{\overset{\|}{C}}C_6H_5,\quad C_6H_5\overset{O}{\overset{\|}{C}}C_6H_5,\quad \text{and}\quad CH_3CH_2\overset{O}{\overset{\|}{C}}C_{10}H_7$$

are named acetophenone, benzophenone, and propionaphthone (1- [*33744-50-2*] and 2- [*21567-68-0*]), respectively.

Physical Properties

Simple and mixed low molecular weight aliphatic and cycloaliphatic ketones are stable, colorless liquids and generally have a pleasant, slightly aromatic odor. They are relatively volatile with boiling points slightly above those of corresponding paraffins. The members of the series up to C_5 are soluble in water and are excellent solvents for nitrocellulose, vinyl resin lacquers, cellulose ethers and esters, and various natural and synthetic gums and resins. In contrast, even the simplest aromatic ketone ($CH_3COC_6H_5$) is a high-boiling colorless liquid with a fragrant odor. Aromatic ketones are useful as intermediates in chemical manufacture.

Functionalized and cyclic ketones also have excellent solvent properties but can be colored (eg, 1,2-diketones and some quinone derivatives), have characteristic odors (cyclic ketones), and form metal complexes (1,3-diketones). They also are used as intermediates in chemical manufacture. Ring size and the type and location of functional groups affect odor, color, and reactivity of cyclic ketones. Physical properties of some common ketones are listed in Tables 1 and 2. Most ketones are commercially available.

Pure ketones are not associated liquids and tend to form azeotropes with water and/or a variety of organic compounds (see Table 3). This property is used in industrial extractive distillation processes (see Azeotropic and extractive distillation).

Chemical Properties

Ketones undergo addition, redox, and condensation reactions forming alcohols, ketals, acids, and amines (17). The reactivity of ketones is a function of the polarity and electrophilic nature of the carbonyl group and its influence on the activity of nearby functional groups (eg, hydrogens alpha to the carbonyl group). In diketones, cyclic and unsaturated ketones, such as 2,4-pentanedione, cyclohexanone, and mesityl oxide, respectively, these chemical properties are enhanced, increasing their usefulness as chemical intermediates.

Reduction. Most ketones are reduced readily to the corresponding alcohols by a variety of hydrogenation processes. For example, 4-methyl-2-pentanol (methyl isobutyl carbinol) is commercially produced by catalytic reduction of methyl isobutyl ketone in the vapor phase with a nickel catalyst (18–19):

$$CH_3COCH_2CH(CH_3)_2 \xrightarrow[\text{Ni–Cr}]{H_2} CH_3CH(OH)CH_2CH(CH_3)_2$$

Oxidation. Ketones are oxidized readily with air or oxygen to produce peroxides; oxidation at about 100°C, with powerful oxidizing agents such as chromic or nitric acid, causes carbon—carbon bond cleavage and gives carboxylic acids. Ketone peroxides may explode in concentrated form (>30%) but dilute solutions are useful in curing unsaturated polyester resin mixtures (see Peroxides, organic, initiators).

Condensations. ***Base Catalyzed.*** Depending on the nature of the hydrocarbon groups attached to the carbonyl, ketones can either undergo self-condensation or condense with other activated reagents in the presence of base. Name reactions which describe these condensations include the Aldol reaction, the Darzens-Claisen condensation, the Claisen-Schmidt condensation, and the Michael reaction. Production of diacetone alcohol and 4-ethoxy-4-methyl-2-pentanone [*27921-36-4*] are Aldol (20) and Michael (21) reactions, respectively:

$$2\,CH_3COCH_3 \xrightleftharpoons{Ba(OH)_2} CH_3COCH_2C(OH)(CH_3)_2$$

$$CH_3COCH{=}C(CH_3)_2 \xrightarrow[\text{anion-exchange resin}]{C_2H_5OH} \underset{82\%}{CH_3COCH_2C(OC_2H_5)(CH_3)_2}$$

Acid Catalyzed. Although ketonic carbonyl groups are less reactive than aldehydic carbonyls in the presence of basic catalysts, this is not the case with acid catalysts. Thus, acetone undergoes aldol condensation in the presence of sulfuric acid to give

Table 1. Physical Properties of Ketones[a]

CAS name	Common name	CAS Registry No.	Mol wt	Freezing point, °C	Melting point, °C	Boiling point at 101.3 kPa[b], °C	Refractive index, n_D^{20}	Apparent specific gravity, 20/20°C	Viscosity at 20°C, mPa·s (= cP)	Vapor pressure at 20°C, kPa[b]	Heat of vaporization at 101.3 kPa[b], kJ/mol[c]	Flash point, °C Closed cup	Flash point, °C Open cup[d]	Solubility at 20°C, wt % In water	Solubility at 20°C, wt % Water in
Simple and mixed acyclic ketones															
acetone, 2-propanone	acetone, dimethyl ketone	*[67-64-1]*	58.08	−94.7		56.1	1.3590	0.7905	0.33	186	29.53	−18[e]	−16	complete	complete
acetophenone	acetophenone, phenyl methyl ketone	*[98-86-2]*	120.15	19.7	19–20	201.7	1.5342	1.0296	0.93	0.3	45.69	82[f]	93[g]	0.55	1.65
benzophenone	benzophenone, diphenyl ketone	*[119-61-9]*	182.22		48–49.5	305									
1,3-dihydroxy-2-propanone	dihydroxyacetone, DHA	*[96-26-4]*	90.08		75–80										
3,4-dihydro-1(2*H*)-naphthalenone	1-tetralone	*[529-34-0]*	146.19			258		1.097					130[g]		
2-heptanone	methyl *n*-amyl ketone	*[110-43-0]*	114.19	−35		151.5	1.4087	0.8166	0.77	2	43.51	49[e]	47	0.43	1.45
3-heptanone	ethyl butyl ketone	*[106-35-4]*	114.19	−39	−39	147.3	1.4088	0.8197	0.84	4	36.61		41	0.43	0.78
2-hexanone	methyl *n*-butyl ketone	*[591-78-6]*	100.16	−55.8		127.5	1.4007	0.8125	0.62	10	36.07		28	1.75	3.7
3-hexanone	ethyl propyl ketone	*[589-38-8]*	100.16			123.2	1.4003	0.8174		12	35.69	35[f]		1.57	
3-methyl-2-butanone	methyl isopropyl ketone	*[563-80-4]*	86.13	−92		94.2		0.8044		40	30.63			6.53	
5-methyl-2-hexanone	methyl isoamyl ketone	*[110-12-3]*	114.12	−73.9		144.9	1.4069	0.8127	0.77	4	35.02		41	0.54	1.28
2-nonanone	methyl heptyl ketone	*[821-55-6]*	142.24		−21	192 at 96.6 kPa[b]	1.4210	0.8320				64[f]			

2-octanone	methyl hexyl ketone	[*111-13-7*]	128.22	−20.5	−16	173.3	1.4153	0.8197		<1	40.88	62[f]			
3-octanone	ethyl amyl ketone	[*106-68-3*]	128.22	−46		167–168	1.4150	0.8220				46[f]			
2-pentanone	methyl propyl ketone	[*107-87-9*]	86.13	−77.8	−78	102.4	1.3902	0.8076	0.51[d]	27	33.39	7[f]		4.3	3.3
3-pentanone	diethyl ketone	[*96-22-0*]	86.3	−39.4		101.8		0.8155		27		13[e]		3.4	2.6
propiophenone	propiophenone, phenyl ethyl ketone	[*93-55-0*]	134.17	18.2		218.0	1.5265	1.012		0.2	45.48	85[e]	96	0.2	1.0
2,6,8-trimethyl-4-nonanone	isobutyl heptyl ketone	[*123-18-2*]	184.32	−75[h]		218.2	1.4257	0.8180	1.9	<0.1		88[f]	90[g]	<0.01	0.2
Unsaturated ketones															
3-butene-2-one	methyl vinyl ketone	[*78-94-4*]	70.09	−6		79–80 at 98.1 kPa[b]	1.4130								
4-methyl-4-penten-2-one	isomesityl oxide	[*3744-02-3*]	98.15	−52.9		121.5	1.4458	0.8548	0.64						
Substituted ketones															
2,2-diethoxy-acetophenone		[*6175-45-7*]					1.513	1.0398	7			110[e]	128		
2,4-dihydroxy-benzophenone		[*131-56-6*]	214.2	[i]	142	194 at 0.13 kPa[b]									
2-hydroxy-4-methoxybenzo-phenone		[*131-51-7*]	228.24	<1	63–64.5	150–160		1.324[j]							
Cyclic ketones															
cyclohexanone	ketohexamethylene; pimelic ketone	[*108-94-1*]	98.15	−31.2	−40.5	155.8	1.4502	0.9482	2.2	2	37.66	43[f]	46	2.5	8.0

[a] Refs. 1–14.
[b] Unless otherwise stated; to convert kPa to mm Hg, multiply by 7.5.
[c] To convert J to cal, divide by 4.184.
[d] Determined by ASTM Method 1310, using the Tag open cup, unless otherwise stated.
[e] Determined by ASTM Method D 56, using the Tag closed cup.
[f] Determined by closed cup in accordance with DOT regulations.
[g] Determined by ASTM Method D 92, using the Cleveland open cup.
[h] Sets to glass below this temperature.
[i] Sets to glass below −80°C.
[j] 25/25°C.

Table 2. Physical Properties of Other Ketones[a]

Property	MEK *[78-93-3]*[a]	MIBK *[108-10-1]*[b]	DAA *[123-42-2]*[b]	DIBK *[108-83-8]*[c]	MSO[d] *[141-79-7]*[b]	Isophorone *[78-59-1]*[c]	2,4-Pentane-dione *[123-54-6]*[c]	2,5-Hexane-dione *[110-13-4]*[c]	Biacetyl *[431-03-8]*[c]
molecular weight	72.10	100.16	116.16	142.24	98.15	138.21	100.12	114.15	86.09
boiling point at 101.3 kPa[e], °C	79.57	116.2	169.2	169.4	129.8	215.3	140.4	192.3	90.2
freezing point, °C	−85.9	−84.0	−44.2	−41.5	−46.4	−8.1	−23.5	−5.4	−2.5
refractive index, n_D^{20}	1.3780	1.3957	1.4226	1.4127	1.4434	1.478	1.4510	1.4260	1.3938
density at 20°C, g/L	804.5	800.8	938.7	806.4	858.4	921.4	973.6	971.7	982.6
specific gravity, 20/20°C	0.8062	0.8020	0.9406	0.8076	0.8598	0.9229	0.9753	0.9734	0.9843
surface tension at 20°C, mN/m (= dyn/cm)	24.6	23.6	31	22.2	28.4	32	31		
dielectric constant at 20°C	15.45								
dipole moment	3.18								
specific heat									
vapor at 137°C, J/(kg·K)[f]	1732								
liquid at 20°C, J/(kg·K)[f]	2084	1920	1883		2176	1799	1983	1958	
heat of fusion, kJ/(kg·K)[f]	103.3								
heat of combustion at 25°C, kJ/mol[f]	2435	3078	−418	−5687	−3543	−5165	−2687	−3327	−2065
heat of formation at constant pressure, $-\Delta H_f^\circ$, kJ/mol[f]	279.5								

latent heat of vaporization at 101.3 kPa[e], kJ/mol[f]	32.8	35.6	41.6	39.3	43.1	43.4	35.4	36.6	34.3
critical temperature, °C	260	298	332	344	327	442	328	390	267
critical pressure, MPa[g]	4.4	3.3	3.5	2.5	3.4	3.3	3.9	3.4	4.5
viscosity, mPa·s (= cP)									
0°C	0.54	0.80	5.34	1.48	0.80	4.2	0.77		
20°C	0.41	0.61	3.20	1.02	0.60	2.6	0.58	1.6	
40°C	0.34	0.48	2.05	0.74	0.47	1.71	0.45		
flash point, °C									
closed cup	−6	23	47	49	31				
open cup	−6	23	61	49	29				
solubility at 20°C, wt %									
in H_2O	26.8	1.9	complete	0.05	3.1				
H_2O in	11.8	1.6	complete	0.75	3.4				

[a] Refs. 7 and 15.
[b] Refs. 7 and 16.
[c] Ref. 7.
[d] Contains usually 1–9% isomesityl oxide.
[e] To convert kPa to mm Hg, multiply by 7.5.
[f] To convert J to cal, divide by 4.184.
[g] To convert MPa to atm, divide by 0.101.

Table 3. Azeotropes of Ketones [a]

Components	Component bp at 101.3 kPa[b], °C	Azeotrope bp at 101.3 kPa[b], °C	Azeotrope composition at 20°C, wt %
acetone	56.1		59
hexane	68.7	49.8	41
acetone	56.1		56.5
isopropyl ether	68.5	53.3	43.5
acetone	56.1		48
methyl acetate	57	55.6	52
acetone	84[c]		98.7
water	126[c]	81.4[c]	1.3
acetophenone	201.7		18.5
water	100	99.1	81.5
diacetone alcohol	169.2		13.0
water	100	99.6	87.0
diisobutyl ketone	169.4		48.1
water	100	97.0	51.9
isobutyl heptyl ketone	218.2		16
water	100	99	84
isophorone	215.3		16.1
water	100	99.5	83.9
mesityl oxide	129.8		65.3
water	100	91.8	34.7
methyl ethyl ketone	79.6		37.5
benzene	80.1	78.4	62.5
methyl ethyl ketone	79.6		66
ethanol	78.3	74.8	34
methyl ethyl ketone	79.6		70
isopropanol	82.3	77.3	30
methyl ethyl ketone	79.6		88
water	100	73.4	12
methyl isobutyl ketone	116.2		76
water	100	87.9	24
2,4-pentanedione	140.4		59
water	100	94.4	41

[a] Ref. 1.
[b] To convert kPa to mm Hg, multiply by 7.5.
[c] At 137.9 kPa.

$$2\ CH_3COCH_3 \xrightarrow[-H_2O]{H_2SO_4} CH_3COCH{=}C(CH_3)_2 \xrightarrow{CH_3COCH_3} [(CH_3)_2C{=}CH]_2CO + C_6H_3(CH_3)_3 + 2\ H_2O$$

mesityl oxide which then condenses with a third molecule of acetone to give a mixture of phorone [*504-20-1*] (2,6-dimethyl-2,6-heptadien-4-one) and mesitylene. This reaction undoubtedly proceeds through the enol form of acetone.

Ketones also condense with activated aromatic compounds in the presence of sulfuric acid to give coupled aromatic products. For example, acetone and phenol condense to 4,4′-isopropylidenediphenol, bisphenol A [*80-05-7*] which is used to prepare epoxy resins, modified phenolic resins, polycarbonates, aromatic polyesters, and polysulfones (22).

Preparation of Amines. Amines are prepared by heating aliphatic, aromatic, or cyclic ketones with ammonium formate, formamide, or an *N*-substituted ammonium formate at 165–190°C (Leuckart reaction). Formic acid is believed to be the reducing agent. Thus, α-methylbenzylamine is prepared in 60–66% yield by the reaction of acetophenone with ammonium formate (23):

$$C_6H_5COCH_3 + HCO_2NH_4 \longrightarrow C_6H_5CH(NH_2)CH_3 + CO_2 + H_2O$$

Thermal Stability. The C_4–C_{12} ketones are thermally stable up to pyrolysis temperatures (500–700°C). At these high temperatures, decomposition of ketones can be controlled to produce useful ketene derivatives (24–25). Ketene itself is produced commercially by pyrolysis of acetone at temperatures just below 550°C (26) (see Ketenes).

Miscellaneous Reactions. Ketones react quantitatively under mild conditions with hydroxylamine, phenyl hydrazine, hydrazine, or semicarbazide. The resultant solid derivatives are useful in analytical identification of ketones.

Health and Safety Factors

As a general class of chemicals, low molecular weight (C_3–C_{12}) saturated aliphatic ketones present a low toxicity hazard. However, the toxicity of diketones and α,β-unsaturated ketones is significantly greater. The primary health hazard of ketones is owing to inhalation of vapors, especially those of low molecular weight derivatives, which cause central nervous system depression. For example, rats died after a 4-h exposure to 2000–4000 ppm of ketone vapors. Exposure at lower concentrations (about 500 ppm) for a few hours typically produces irritation of the mucous membranes of the eyes, nose, and respiratory tract, usually providing adequate warning of overexposure.

Saturated aliphatic ketones have low chemical reactivity under normal conditions, and thus the liquids are not irritating to the skin. However, prolonged or repeated contact with the skin should be avoided because ketones are powerful degreasers and may lead to dermatitis. Some ketones cause moderate to severe eye irritation on contact.

Ketones should not be used for internal consumption; accidental swallowing of small amounts may result in minor injury but not death.

Odors of ketone vapors generally can be detected at the 5–25 ppm level (27) and provides adequate warning of the presence of hazardous concentrations of these materials.

Table 4 shows the toxicological properties of some ketones.

Environmental Aspects

Ketones were widely used as solvents in coatings for many years (see Coatings). During the late 1960s and early 1970s, various state and local air pollution regulatory agencies placed restrictions on the amount of photochemically reactive solvents, in-

Table 4. Toxicity of Ketones[a]

Ketone	Ingestion LD_{50} rats, mg/kg	Skin penetration, LD_{50} rabbits, g/kg	Inhalation LC_{Lo}[b] rats, ppm/4 h	Eye injury[c] rabbits	TLV air[d], ppm
acetone	9,750	20	64,000	minor	1000
methyl ethyl ketone	3,400	13	2,000	moderate	200
methyl isobutyl ketone	2,080	more than 20 mL/kg	4,000[e]	minor	100
diacetone alcohol	4,000	14.5[f]		moderate	50
methyl n-amyl ketone	1,670	13	4,000		100
diisobutyl ketone	1,416[g]	20	2,000	trace	25
ethyl butyl ketone	2,760		2,000		50
2,4-pentanedione	1,000	5.0	1,000	minor	
mesityl oxide	1,120	5.99	1,000	moderate	25
isophorone	2,330	1.5	1,840	moderate	5
cyclohexanone	1,620	1.0	2,000	moderate	50
acetophenone	900	15.9[f]		severe	

[a] Refs. 1, 28–29.
[b] The lowest concentration in air that caused death in 4 hours.
[c] Refers to surface damage.
[d] Refers to time-weighted concentrations for 7 or 8-hour workday and 40-hour workweek.
[e] ppm/15 min.
[f] mL/kg.
[g] Mouse.

cluding certain ketone derivatives and aromatic hydrocarbons, that could be emitted into the atmosphere. Most regulatory agencies adopted Rule 66-type solvent emission regulations. According to this standard, the amounts of olefinic and branched ketones used in a coating solvent system are restricted to 5 and 20 vol % max, respectively (30). In addition, all photochemically reactive solvents cannot exceed 20 vol % if the total solvent system is to be classified as photochemically unreactive or exempt. All ketones not falling under the above classifications, ie, straight-chain derivatives and cyclohexanone are considered photochemically unreactive. Table 5 shows the status of some ketones under Rule 66 (31).

Currently, 44 states are in the process of developing and adopting even more stringent air pollution control regulations to comply with the Clean Air Act Amendments of 1977. Since essentially all solvents, including nonketonic derivatives, show some photochemical reactivity under proper conditions, the new regulations are designed to reduce overall solvent emissions from coatings applications and not just restrict the use of highly photoreactive solvents. These new regulations, which replace Rule 66-type regulations, generally follow the EPA control technique guidelines developed to provide assistance to states (see Air pollution; Air pollution control methods).

Uses

The lower aliphatic ketones are excellent solvents for nitrocellulose, vinyl resin lacquers, cellulose ethers and esters, and various gums and resins. Because of their ability to form azeotropes with water and organic liquids, they are employed in ex-

Table 5. Status of Ketones as Solvents for Coatings Under Rule 66[a]

Ketone	Distillation range, °C	Status under Rule 66[b]
acetone	56–57	NPCR
cyclohexanone	154–158	NPCR
diacetone alcohol	145–172	PCR-20%
diisobutyl ketone	163–173	PCR-20%
ethyl *n*-amyl ketone[c]	156–162	PCR-20%
ethyl butyl ketone	143–151	NPCR
isophorone	210–218	PCR-5%
mesityl oxide	125–132	PCR-5%
methyl *n*-amyl ketone	147–154	NPCR
methyl *n*-butyl ketone	124–130	NPCR
methyl ethyl ketone	78–81	NPCR
methyl isoamyl ketone	141–148	PCR-20%
methyl isobutyl ketone	114–117	PCR-20%
methyl propyl ketone	97–106	NPCR

[a] Ref. 31.
[b] NPCR = Nonphotochemically reactive; PCR-20% and PCR-5% = photochemically reactive, volume % without requiring emission control.
[c] Flash point, 57°C determined by ASTM-D 1310 (Tag open cup); for flash points of the other compounds, see Tables 1 and 2.

tractive distillation. Aromatic ketones are useful intermediates for the manufacture of pharmaceuticals, fragrances, resins, corrosion inhibitors, and dyestuffs. Functionalized and cyclic ketones form metal complexes and are used as solvents and synthetic intermediates.

Aliphatic Ketones

Acetone, cyclohexanone, methyl ethyl ketone, and methyl isobutyl ketone are among the 100 largest volume organic chemicals (in order of decreasing volume). The three aliphatic ketones are used as solvents, whereas cyclohexanone is primarily used in polymer manufacture. Next in importance are aliphatic ketones such as diacetone alcohol, mesityl oxide, and diisobutyl ketone, which are primarily used as solvents. Several ketones produced by the aldol process represent a significant outlet for acetone. Most raw materials are inexpensive. Pricing and environmental concerns play key roles in determining the future of industrially important ketones and their manufacturing processes.

Methyl Ethyl Ketone. ***Physical Properties.*** Methyl ethyl ketone, MEK, 2-butanone, is a colorless stable liquid with the characteristic acetone-type odor of low molecular weight aliphatic ketones. It is highly miscible with water and many conventional organic solvents and forms azeotropes with a number of organic liquids (see Table 3). Methyl ethyl ketone is distinguished by its exceptional solvency for a wide variety of resinous materials. Some physical constants of methyl ethyl ketone are listed in Table 2. Vapor pressures at various temperatures are shown in Figure 1.

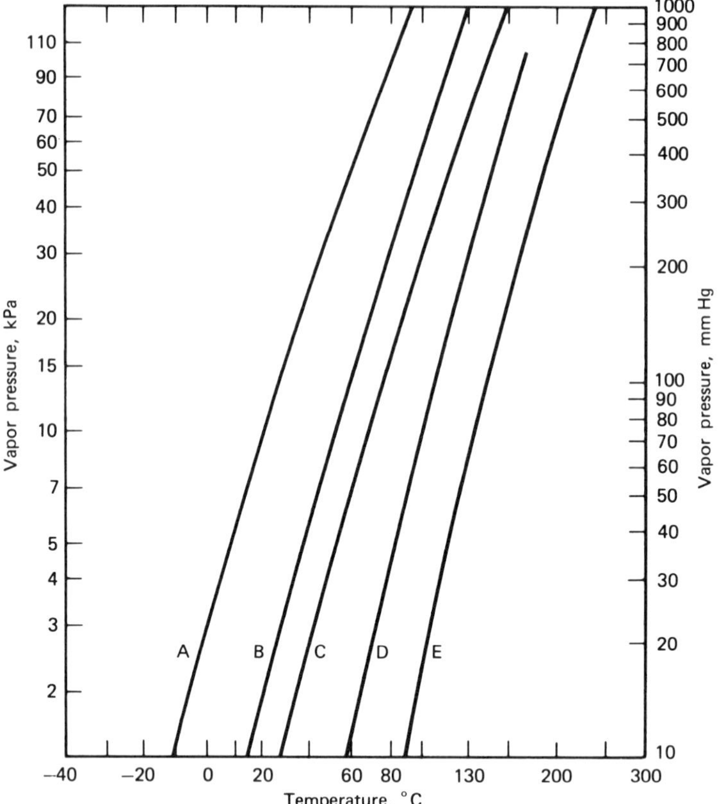

Figure 1. Vapor pressure of ketones (1). A, methyl ethyl ketone; B, methyl isobutyl ketone; C, 2,4-pentanedione; D, diacetone alcohol; E, isophorone.

Chemical Properties. Methyl ethyl ketone undergoes reactions typical of carbonyl groups with activated hydrogen atoms on adjacent carbon atoms, and condenses with a variety of reagents. Manufacture of methyl isopropenyl ketone [*814-78-8*] (3-methyl-3-buten-2-one) and *sec*-butylamine by condensation with formaldehyde and ammonia–hydrogen, respectively, illustrate this reactivity:

$$CH_3\overset{O}{\overset{\|}{C}}C_2H_5 + HCHO \xrightarrow{NaOH} CH_3\overset{O}{\overset{\|}{C}}\underset{CH_3}{\underset{|}{C}H}CH_2OH \xrightarrow{-H_2O} CH_3\overset{O}{\overset{\|}{C}}\underset{CH_3}{\underset{|}{C}}{=}CH_2$$

$$CH_3\overset{O}{\overset{\|}{C}}C_2H_5 + NH_3 \xrightarrow[cat.]{Ni} CH_3\underset{NH_2}{\underset{|}{C}H}CH_2CH_3 + H_2O$$

Condensation with aliphatic esters under strongly alkaline conditions produces 1,3-diketones.

$$CH_3\overset{O}{\overset{\|}{C}}C_2H_5 + CH_3\overset{O}{\overset{\|}{C}}OC_2H_5 \xrightarrow{base} C_2H_5\overset{O}{\overset{\|}{C}}CH_2\overset{O}{\overset{\|}{C}}CH_3 + C_2H_5OH$$

Direct oxidation (32) yields 2,3-butanedione (biacetyl), a valuable flavorant (15):

$$CH_3\overset{\overset{O}{\|}}{C}C_2H_5 + O_2 \xrightarrow[275°C]{Cu_2O} CH_3\overset{\overset{O}{\|}}{C}\overset{\overset{O}{\|}}{C}CH_3 + H_2O$$

Or methyl ethyl ketone peroxide:

$$HO\overset{\overset{CH_3}{|}}{\underset{\underset{C_2H_5}{|}}{C}}OO\overset{\overset{CH_3}{|}}{\underset{\underset{C_2H_5}{|}}{C}}OH$$

This is an important initiator for polyester production (see Initiators).

Manufacture. Methyl ethyl ketone is usually made by the vapor-phase dehydrogenation of 2-butanol, like the production of acetone from isopropanol. A two-step process from butenes, which are first hydrated to give 2-butanol, is involved. The dehydrogenation is catalyzed by zinc- or copper-based catalysts at high temperatures and low pressures (<345 kPa or 3.4 atm) (33–34). The process gives high conversion of 2-butanol and high selectivity to MEK, about 90–95 mol % for each:

$$CH_3CH{=}CHCH_3 \xrightarrow[H_2SO_4]{\text{aqueous}} CH_3\overset{\overset{OH}{|}}{C}HCH_2CH_3$$

$$CH_3\overset{\overset{OH}{|}}{C}HCH_2CH_3 \xrightarrow[400\text{–}500°C]{\text{ZnO or pumice}} CH_3\overset{\overset{O}{\|}}{C}CH_2CH_3 + H_2$$

Short catalyst life and high reaction temperatures limit its usefulness.

In Europe, liquid-phase dehydrogenation of 2-butanol is practiced at lower temperatures (35–36). The reaction is conducted over Raney nickel or copper chromite at 150°C and MEK and hydrogen are separated in the vapor state as they form. Other advantages of this process include a better yield (about 3%) longer catalyst life, simple product separation, and lower energy consumption.

Methyl ethyl ketone is also commercially available in significant quantities as a by-product from liquid-phase oxidation of butane to acetic acid (37–38). For example, in 1976, Union Carbide made 36,000 metric tons t MEK as by-product from a 226,000 t/yr acetic acid process (39). Depending on the demand for acetic acid, methyl ethyl ketone can be marketed or recycled. This process may have a slight economic advantage over dehydrogenation of 2-butanol, but the key factor is the availability and price of butane vs butenes.

A seemingly attractive commercial route to methyl ethyl ketone via direct oxidation of n-butenes in aqueous solutions of palladium and cupric chlorides has been patented (40–43) but its commercial status in the United States is unknown.

$$n\text{-}C_4H_8 + 1/2\ O_2 \xrightarrow[\substack{120°C,\ 1\text{–}2\ MPa \\ (10\text{–}20\ atm)}]{PdCl_2\text{–}CuCl_2} CH_3\overset{\overset{O}{\|}}{C}CH_2CH_3$$

This process, like the Wacker process for the oxidation of ethylene to acetaldehyde (44), uses a unique catalyst regenerating system (see Acetaldehyde). Product yields are excellent (about 85%), but titanium equipment is necessary because of the corrosive nature of the aqueous chloride salts used. Other disadvantages include formation of chloroketone by-products and problems associated with recovering the product from dilute aqueous solutions.

A recent USSR patent claims that the low conversions (about 30% per pass) of the butene oxidation process can be increased two to three times by employing an aliphatic alcohol as a coreactant and solvent (45).

Several other interesting methods for producing methyl ethyl ketone are described in the patent literature:

Free-radical addition of acetaldehyde to ethylene (46):

$$CH_3CHO + CH_2{=}CH_2 \xrightarrow[\text{initiator}]{\text{free-radical}} CH_3\overset{\overset{\displaystyle O}{\|}}{C}CH_2CH_3$$

Isomerization of butylene oxide (47–48):

$$CH_3CH_2\underset{\diagdown O \diagup}{CHCH_2} \xrightarrow[65°C]{[Co(CO)_4]_2} CH_3CH_2\overset{\overset{\displaystyle O}{\|}}{C}CH_3 \text{ (80\% yield)}$$

This method can be applied equally well for 1,2- and 2,3-epoxide derivatives.

Isomerization of isobutyraldehyde (49–50):

$$(CH_3)_2CHCHO \xrightarrow[400\text{–}500°C]{\text{acidic catalyst}} CH_3CH_2\overset{\overset{\displaystyle O}{\|}}{C}CH_3$$

Economic Aspects. The price of MEK was stable from 1955 to 1974 when it sharply increased 20–30¢/kg because of diversion of raw material to acetic acid manufacture (51). Since then the price has risen about 10%/yr to 77¢/kg in 1980. The tight supply was alleviated with expansions by Exxon and Shell in 1977 (52).

Demand for MEK increased from 1964 to 1974 at the rate of 6.5%/yr and is expected to continue to grow at about this rate through 1980 (53–54). Apparently the long-expected decline in demand owing to the increase in solventless coatings has not materialized (51). Recently, Japan has also experienced MEK shortages. European producers appear to have kept pace with demand and even export to the United States.

Total 1977 United States nameplate capacity for methyl ethyl ketone production was 418,000 t (see Table 6); this figure includes the 273,000 t expansions by Exxon and Shell.

Health and Safety Factors. Methyl ethyl ketone is neither highly toxic nor does it exhibit cumulative toxicological properties (it is slightly more toxic than acetone, see Table 4). Inhalation of the vapors over prolonged periods may irritate the mucous membranes and cause nausea, and eventually loss of consciousness. Methyl ethyl ketone exhibits a strong degreasing action and may result in dermatitis. The OSHA

Table 6. Methyl Ethyl Ketone Capacities and Processes[a]

Producer	Plant location	1977 Capacity, 10^3 t	Process
Arco	Channelview, Tex.	29.5	butanol dehydrogenation
Celanese	Pampa, Tex.	40.8	by-product from butane oxidation
Dixie Chemical	Houston, Tex.	1.4	butadiene by-product
Exxon	Bayway, N.J.	136.0	butanol dehydrogenation
Shell	Houston, Tex.	45.4	butanol dehydrogenation
Shell	Norco, La.	127.0	butanol dehydrogenation
Union Carbide	Brownsville, Tex.	38.6	by-product from butane oxidation
		Total *418.7*	

[a] Refs. 52, 55–56.

standard time weighted average in air is 200 ppm. Methyl ethyl ketone is highly flammable and should be used with caution.

Uses. Methyl ethyl ketone exhibits outstanding solvent properties and is one of the lowest priced solvents in its boiling range. It is competitive on a cost–performance basis with ethyl acetate as a solvent for nitrocellulose and is widely used as a solvent in a great variety of coating systems. As a solvent for lacquers, methyl ethyl ketone is particularly advantageous because it provides low viscosity solutions at high solids content without affecting film properties.

Methyl ethyl ketone is a dewaxing agent in the refining of lubricating oils and is a solvent for adhesives, rubber cement, printing inks, paint removers, and cleaning solutions. It is used in vegetable-oil extraction processes and in azeotropic separation schemes in refineries.

Other applications involve use as an intermediate for the preparation of catalysts, flavors, antioxidants and perfumes and in the manufacture of ethyl *n*-amyl ketone and methyl ethyl ketone peroxide. The latter is widely used as a hardening agent for the production of reinforced polyester fiberglass parts for cars, sailboats, recreational vehicles, and chemical storage tanks. Table 7 lists the main uses of methyl ethyl ketone for 1977.

Methyl Isobutyl Ketone. Methyl isobutyl ketone, MIBK, 4-methyl-2-pentanone, is a water-white liquid that boils higher than either acetone or MEK and lower than mesityl oxide and diacetone alcohol. Some physical constants are shown in Table 2 and a vapor pressure curve is given in Figure 1.

Table 7. Methyl Ethyl Ketone Uses[a]

Use	Percentage
vinyl coatings	34
nitrocellulose coatings	14
adhesives	14
acrylic coatings	12
miscellaneous coatings	7
lube-oil dewaxing	7
miscellaneous and export	12

[a] Ref. 57.

Chemical Properties. Condensation of methyl isobutyl ketone with another methyl ketone can produce ketones containing 9–15 carbons. For example, condensation with acetone affords diisobutyl ketone, whereas reaction with another mol MIBK produces a 12-carbon ketone–alcohol intermediate for synthesis of 2,6,8-trimethyl-4-nonanone. Condensation with 2-ethylhexanal gives 1-tetradecanol (7-ethyl-2-methyl-4-undecanol), a valuable surfactant intermediate (58).

Hydrogenation gives methyl isobutyl carbinol. Methyl isobutyl carbinol is made in conjunction with methyl isobutyl ketone by hydrogenation of mesityl oxide. Typically the reduction is conducted over a nickel catalyst at 100–140°C and 1–1.5 MPa (10–15 atm) (59–60):

$$(CH_3)_2CHCH_2COCH_3 + H_2 \xrightarrow{Ni} (CH_3)_2CHCH_2\overset{\overset{\displaystyle OH}{|}}{C}HCH_3$$

High hydrogen concentrations are required and the heat of reaction is about 69.56 MJ/(kg·mol) [16,625 kcal/(kg·mol)].

Methyl isobutyl ketone reacts with methanol over copper–alumina catalyst at 275–550°C, 101–345 kPa (1–3.4 atm) and a liquid hourly space velocity (LHSV) of 1–5, to give ethyl isobutyl ketone [*565-69-5*] in high yield (61). Oxidation of MIBK gives methylisopropylglyoxal (62–63).

Manufacture. Methyl isobutyl ketone is produced commercially in three steps from acetone:

Liquid-phase condensation:

$$2\ CH_3COCH_3 \xrightarrow[\text{fixed bed}]{\text{alkali}} (CH_3)_2\overset{\overset{\displaystyle OH}{|}}{C}CH_2\overset{\overset{\displaystyle O}{\|}}{C}CH_3$$

diacetone alcohol

Acid-catalyzed dehydration:

$$(CH_3)_2\overset{\overset{\displaystyle OH}{|}}{C}CH_2\overset{\overset{\displaystyle O}{\|}}{C}CH_3 \xrightarrow[100°C]{\text{acid}} (CH_3)_2C{=}CH\overset{\overset{\displaystyle O}{\|}}{C}CH_3 + H_2O$$

mesityl oxide

Selective hydrogenation:

$$(CH_3)_2C{=}CH\overset{\overset{\displaystyle O}{\|}}{C}CH_3 + H_2 \xrightarrow[120\text{–}165°C]{\text{Ni or Cu}} (CH_3)_2CHCH_2\overset{\overset{\displaystyle O}{\|}}{C}CH_3$$

MIBK

The first and second steps are equilibrium-limited and involve corrosive acids, respectively. The third step can be carried out in the presence of nickel (59–60,64–72), palladium (73–75), copper (76–77), or nickel–chromium (72) catalyst in either the liquid or vapor phase with yields of >93%. Vapor-phase hydrogenation is typically conducted at 150–200°C, whereas liquid-phase reduction is run at 100–140°C under 1–1.5 MPa (10–15 atm) pressure. The hydrogenation is highly exothermic, yielding about 146 kJ/mol (34.9 kcal/mol) (78).

In the Hibernia Scholven process (79) acetone and dilute sodium hydroxide are fed continuously to a reactor (see Fig. 2). After the reaction is completed (about 1 h), the product is stabilized with phosphoric acid and stripped free of acetone, leaving an 80–90% diacetone alcohol mixture. Crude diacetone alcohol is mixed with more phosphoric acid and fed to the dehydration column where mesityl oxide is recovered overhead. The heads are redistilled to separate residual acetone, which is recycled, whereas 98–99% pure mesityl oxide is taken at the bottom of the column. Mesityl oxide is then subjected to distillative hydrogenation which comprises simultaneous hydrogenation and rectification. Methyl isobutyl ketone is obtained in >96% yield by hydrogenation with a palladium catalyst at 110°C and atmospheric pressure.

Owing to recent catalyst developments, a long-known one-step hydrogenation of acetone has become commercially feasible:

$$2\,CH_3COCH_3 + H_2 \xrightarrow{\text{catalyst}} (CH_3)_2CHCH_2COCH_3 + H_2O$$

Mixed catalysts such as KOH–alumina–Pd (80), Mg–silica–Pd (81), cation-exchange resin–Pd (82–85), zirconium phosphate–Pd (86), and Cu–Cr (87) reportedly effect acid aldolization, dehydration, and hydrogenation in a single stage. Hydrogen and acetone in a molar ratio between 0.2 and 2.0 are preheated and fed to a fixed-bed reactor operating at 80–150°C and 5–10 MPa (50–100 atm). The LHSV of acetone usually ranges from about 1 to 10 liquid volumes of acetone per hour per volume of catalyst. Acetone conversion is 30–40% and selectivity of acetone to methyl isobutyl ketone is >90%.

In spite of the good selectivities obtained with the newly developed catalysts,

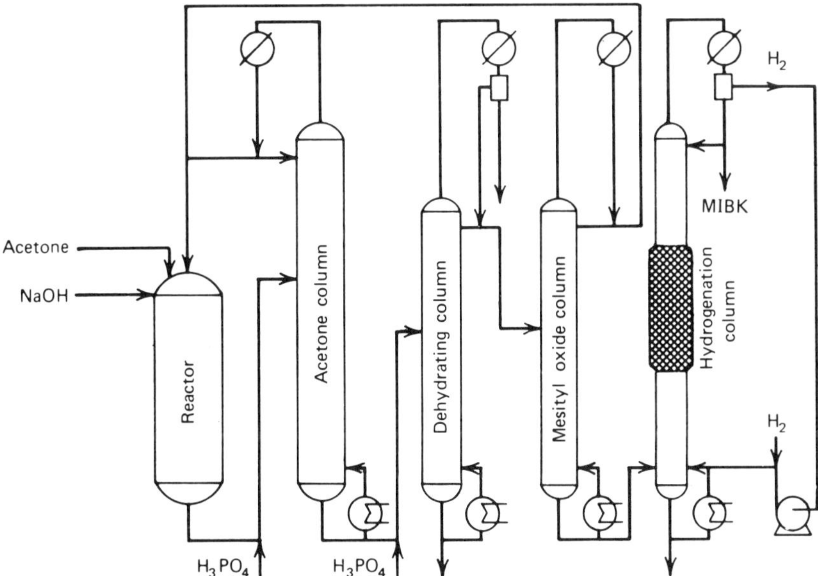

Figure 2. Three-step process for methyl isobutyl ketone manufacture. Courtesy of Chemische Industrie, Dusseldorf, FRG (79).

formation of various by-products continues to be a problem. By-product diisobutyl ketone is usually marketed.

Relatively high operating pressures are another disadvantage of the single-step process. A recent patent (88) claims the use of a TiO_2–SnO_2–Ni catalyst system which permits use of atmospheric pressure at temperatures between 155 and 270°C and a LHSV of about 0.5. Isopropanol, diisobutyl ketone, and mesityl oxide were by-products.

Veba-Chemie, in Germany, was the first to come on stream (1968) with a plant to manufacture methyl isobutyl ketone by the one-step process (83,89). A Japanese one-step MIBK process, the Tokuyama Soda process, has also been reported (86). A flow scheme of the Veba-Chemie process (83) is shown in Figure 3.

In this process, hydrogen and acetone (mol ratio = 1.7) are preheated and react at 80°C under a total pressure of 6.1 MPa (60 atm) in a fixed-bed reactor. The reactor output temperature is maintained at 135°C by circulating hydrogen. The LHSV of acetone is 1.0. The product mixture and circulating gas are condensed and separated at 50–60°C and the methyl pentane fraction is removed in the first distillation column as an azeotrope with acetone (bp 44°C). Unreacted acetone is stripped from the product mixture and isopropanol and water are removed in the third column of the distillation train. Finally, pure MIBK is taken overhead; the column tails fraction contains higher condensed reaction products (diisobutyl ketone, etc). Under these operating conditions, acetone conversion is 34.4% per pass; product distribution is given in Table 8.

Other noncommercial routes include vapor-phase condensation of isopropanol

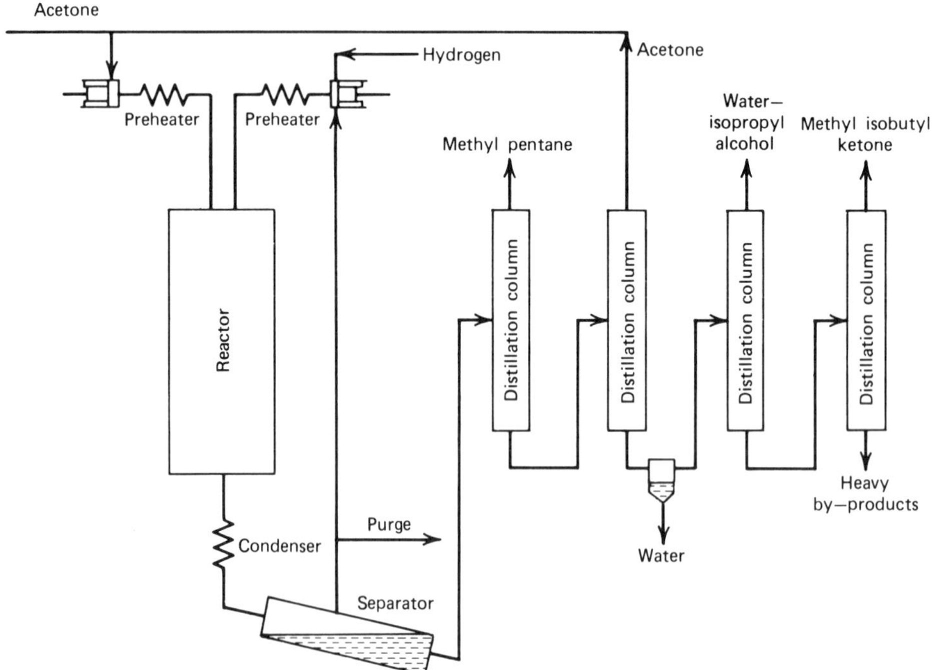

Figure 3. Veba-Chemie process for methyl isobutyl ketone manufacture (83,89).

Table 8. Product Distribution in the Veba-Chemie Process[a]

Compound	Mol %
isopropanol	0.08
2-methyl pentane	0.45
2-methyl-2-pentene	0.01
methyl isobutyl ketone	96.50
mesityl oxide	0.01
methyl isobutyl carbinol	0.03
diisobutyl ketone	2.20
4,4-dimethylheptanone	0.55
residue	0.17

[a] Ref. 83.

(90–94), condensation of isopropanol and acetone (95–96), and oxidation of 4-methyl-1-pentene (97).

Economic Aspects. Because of its nonexempt status under Rule 66, the price of MIBK has remained stable, ranging about 55–66¢/kg from 1975 to 1979.

Reported capacities for methyl isobutyl ketone are shown in Table 9. Generally, facilities manufacturing MIBK also make other acetone derivatives.

Methyl isobutyl ketone is the most important product derived from acetone (see Table 10). As solventless coating systems are developed to meet future pollution requirements, demand for methyl isobutyl ketone is expected to stabilize at about 91,000 t/yr resulting in some overcapacity (100).

Health and Safety Factors. Like other low molecular weight ketones, methyl isobutyl ketone is an anesthetic chemical with no highly significant cumulative toxicological effects (101). Inhalation of vapors causes irritation of eyes and nasal mucosa; continuous exposure in high concentrations could anesthetize.

Methyl isobutyl ketone is nonexempt under Rule 66 because of its branched ketonic nature. Its use in solvents is limited to a maximum of 20 vol % under this regulation. Pending new regulations under the Clean Air Act Amendments of 1977 should temporarily relieve the stress on Rule 66 nonexempt materials until low solvent coatings are developed.

Table 9. Methyl Isobutyl Ketone Production[a]

Producer	Plant location	U.S. 1977 capacity, 10^3 t/yr
Eastman	Kingsport, Tenn.	11.3
Exxon	Linden, N.J.	18.1
Shell	Dominguez, Calif.	7.3
Shell	Houston, Tex.	36.3
Union Carbide	Institute, W.Va.	34.0
	Total	*107.0*

[a] Ref. 98.

Table 10. United States Acetone Derivatives Production, 1973[a]

Derivative	Production, thousand metric tons per year
methyl isobutyl ketone	119.5
diacetone alcohol	21.8
hexylene glycol	22.7
methyl isobutyl carbinol	20.0
isophorone	15.9
mesityl oxide	12.3
diisobutyl ketone	[b]
Total	*212.2*

[a] Ref. 99.

[b] Data not available but estimated to be about 10–20% of methyl isobutyl ketone production.

Uses. Like acetone and MEK, methyl isobutyl ketone is highly compatible with a variety of organic reagents and is a good solvent for a wide range of industrial materials. Its principal uses are in coating solvents (70%) and for rare-metal extraction (5%), export (11%), and miscellaneous solvent and denaturant uses (14%) account for the remaining production (102). As a solvent for cellulose-based (eg, nitrocellulose and cellulose acetate butyrate) and resin-based (eg, acrylic, alkyd and vinyl) coating systems, MIBK is unsurpassed. Attempts to replace it with straight-chain solvents which are exempt from Rule 66 have met with great difficulty. However, MIBK and other branched ketones will eventually be removed from most coating systems to comply with government regulations (100,102).

Methyl isobutyl ketone is an unusually effective separating agent for certain metals from solutions of their salts, such as plutonium from uranium, niobium from tantalum, and zirconium from hafnium (103) (see Extraction).

Other applications include as a solvent for DDT, pyrethrins, adhesives, rubber cements, aircraft and model dopes, and drugs (eg, tetracycline). Its dewaxing and separating properties are used in the purification of pharmaceuticals, mineral oils, tall oil, stearic acid, and butanol. It is also used to denature ethyl alcohol solvent formulations.

As a nonexempt solvent, methyl isobutyl ketone growth has been minimal during the past 10 years. There are no new uses, export opportunities are declining, and demand for coatings has stabilized. However, projected decline has not materialized (104) because coating formulators cannot find a suitable replacement.

Diacetone Alcohol. Diacetone alcohol, DAA, 4-hydroxy-4-methyl-2-pentanone, is a colorless liquid with a mild odor and unusual solvent and wetting properties. It is completely miscible with water and polar organic solvents. It has special utility in the coatings industry owing to its capability for dissolving cellulose acetate to give solutions with high tolerance for water (105). Some physical constants are given in Table 2 and a vapor pressure curve is given in Figure 1. The toxicity of diacetone alcohol is similar to that of MEK (see Table 4).

As a compound containing both alcohol and ketone groups, diacetone alcohol undergoes reactions characteristic of these functionalities. Reactions include its oxidation to 1-acetyl-2-methyl-1,2-epoxypropane [*4478-63-1*] (106–107), complex formation with copper salts (108), and reaction with hexylene glycol to give 2-(2-methyl-1-propenyl)-2,4,6-trimethyl-4,5-dihydropyran (109). Hexylene glycol is produced by hydrogenation of diacetone alcohol at moderate temperatures (79):

$$(CH_3)_2C(OH)CH_2COCH_3 + H_2 \xrightarrow[\substack{100°C,\\ 30\ MPa\\ (300\ atm)}]{Ni} (CH_3)_2C(OH)CH_2CH(OH)CH_3$$

Because of its high boiling point, low freezing point, low vapor pressure, good chemical stability, and excellent solvent power, it is widely used in hydraulic brake fluids. It is also used in many industries as a coupling agent and stabilizer in lubricating and cutting oils, as a wetting and dispersing agent in polishes and soaps for textiles, printing inks, paper coatings, and leather goods, as a coalescing agent in latex paints, and as a solvent for the wood preservative, pentachlorophenol (110).

Manufacture. Diacetone alcohol is manufactured by the liquid-phase self-condensation of acetone in the presence of an alkaline catalyst:

$$2\ CH_3COCH_3 \underset{5\text{–}20°C}{\overset{alkali}{\rightleftharpoons}} (CH_3)_2C(OH)CH_2COCH_3$$

The reaction is equilibrium controlled and low temperatures favor diacetone alcohol. However, even under these conditions conversions are low and yields are limited. The following conversions are obtained as a function of temperature (111–112):

Temperature, °C	−20	−10	0	10	20	30
Conversion, %	27.3	25.6	23.1	16.9	12.1	9.1

Acetone condensation at 0°C is very slow. It is exothermic and the heat of reaction is reported to be 14.6 kJ/mol (3.49 kcal/mol).

In commercial practice acetone is passed at about 5–20°C through a bed of solid alkaline catalyst contained in tank reactors. Suitable catalysts include the hydroxides of sodium (113), potassium (114–115), calcium (116–119), and barium (120–124) with residence times ranging from 20 to 60 minutes. Ferric oxide (125–126) and anion-exchange-resin (127–129) catalysts have been patented but are not believed to be used on an industrial scale. Catalysts are generally pure solids but may be supported, eg, on alumina. Low temperatures may be achieved by cascading reactors; the first reactor is near 25°C, followed by reactors at successively lower temperatures. The last reactor is held at about 0°C (130). The crude, slightly alkaline acetone solution is neutralized with, eg, phosphoric acid (131), and stripped free of excess acetone. The acetone is recycled to the converters but it must first be dried (<1 wt % H_2O) to minimize mesityl oxide formation (131–133). Diacetone alcohol of >99% purity is recovered by vacuum distillation. It tends to revert to acetone on distillation at atmospheric pressure, and therefore, recoveries are only about 60–80%.

Higher recovery of refined diacetone alcohol is hampered by its ease of reversing to acetone and mesityl oxide. For example, relative reaction rates in the presence of 0.1 *N* sodium hydroxide show that the reverse reaction to acetone is greater than nine thousand times faster than the forward reaction, and surprisingly, diacetone alcohol is dehydrated to mesityl oxide even under alkaline conditions (134):

$$2\ CH_3COCH_3 \underset{9140}{\overset{rates,\ 1.0}{\rightleftharpoons}} \underset{DAA}{(CH_3)_2C(OH)CH_2COCH_3} \underset{284}{\overset{rates,\ 17}{\rightleftharpoons}} \underset{\text{mesityl oxide}}{(CH_3)_2C{=}CHCOCH_3} + H_2O$$

Moreover, the tendency of diacetone alcohol to undergo condensations with itself, acetone, or mesityl oxide during refining to give a complex array of higher molecular weight products further complicates its recovery in high purity.

Another by-product (and precursor for some impurities) is a stable acetone trimer, *sym*-triacetone dialcohol [*3682-91-5*], formed by condensation of acetone with diacetone alcohol. The proportion of trimer increases with decreasing condensation temperature (111). Dehydration of *sym*-triacetone dialcohol yields semiphorone [*5857-71-6*] (6-hydroxy-2,6-dimethyl-2-hepten-4-one), 2,2,6,6-tetramethyl-γ-pyrone [*1197-66-6*], or phorone [*504-20-1*] (135).

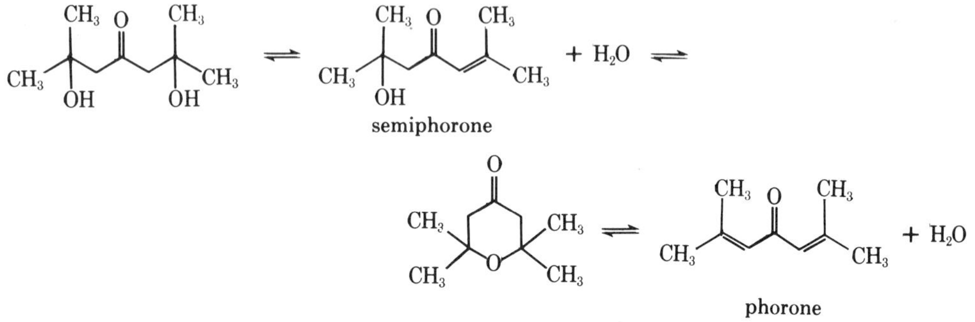

$\rightleftharpoons$ semiphorone $+ H_2O \rightleftharpoons$

$\rightleftharpoons$ phorone $+ H_2O$

Similarly, an unsymmetrical trimer can also be formed which readily dehydrates to 2,4-dimethyl-2,4-heptadiene-6-one, a compound that discolors diacetone alcohol (111):

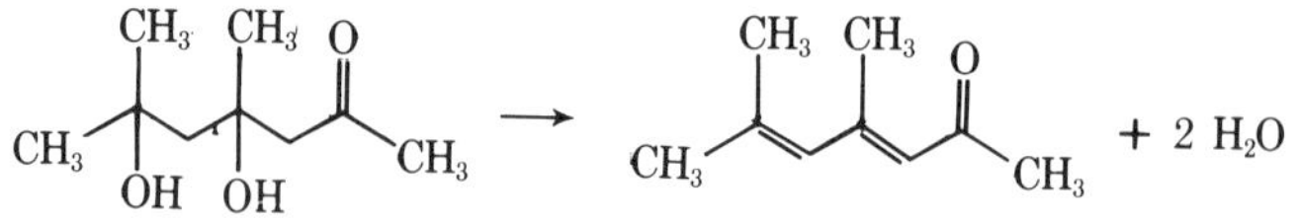

[*74631-04-2*] $\longrightarrow$ [*29179-02-0*] $+ 2\ H_2O$

Generally, catalyst life is about 1 year; it can be reactivated by successive washing with hot water and acetone (136). Highly condensed impurities affect catalyst activity (137).

Recent Japanese patents (138–139) claim that neutralization of crude, alkaline diacetone alcohol with dibasic acids before distillation inhibits both retroaldol condensation and dehydration to mesityl oxide; eg, less than 200 ppm of phthalic anhydride is effective. Stabilization against discoloration and odor formation is accomplished by hydrogenation with a Ni–Al alloy catalyst at 77°C and 862 kPa (8.5 atm) followed by two-step distillation (140).

Economic Aspects. Demand for diacetone alcohol remains strong in spite of its nonexempt Rule 66 status which limits its use to 20 vol % in coatings applications. As raw material for hexylene glycol and mesityl oxide, diacetone alcohol represents a significant portion of acetone derivatives production (Table 10). It is currently (October, 1980) available in tank car quantities for 92¢/kg.

Uses. Diacetone alcohol is used as a lubricant and viscosity regulator in hydraulic brake fluids (see Hydraulic fluids), and as a solvent in the coatings industry. It is generally used in brake fluids with castor oil, in which it is miscible over a wide range of temperatures. However, in this application it is being replaced by hexylene glycol in the United States (105). As a coatings solvent, diacetone alcohol is most useful in hot lacquers which require high boiling components and in brushing lacquers where its mild odor, good blush resistance, flow, and gloss are desired. Diacetone alcohol is also a solvent for nitrocellulose, cellulose acetate, vinyl chloride–vinyl acetate, and epoxy resins (58). For solvents with comparable evaporation rates (31), it has one of the highest aromatic hydrocarbon dilution ratios of solutions of nitrocellulose. Diacetone alcohol is a useful solvent for dyestuffs, in textile printing, and in removing dyestuff from cellulose-based films and fabrics, and ink from printing rollers. It is also used in fuel additives to prevent freezing and remove carbon deposits.

Diisobutyl Ketone. Diisobutyl ketone, DIBK, 2,6-dimethyl-4-heptanone, is a colorless stable liquid with a peppermint odor. Some physical properties are given in Table 2 and vapor pressure data are shown in Figure 4.

Diisobutyl ketone is a by-product in the manufacture of methyl isobutyl ketone. It is produced by the hydrogenation of phorone which, in turn, is prepared by the acid-catalyzed aldol condensation of acetone. Diisobutyl ketone has also been prepared by the decomposition of isovaleric acid in the presence of heavy metal salts.

Union Carbide Corporation is the main United States producer of diisobutyl ketone. It is available at a minimum purity of about 98% in tank cars for 95¢/kg (October, 1980 price).

Diisobutyl ketone is an excellent dispersant for organosols and aids in controlling

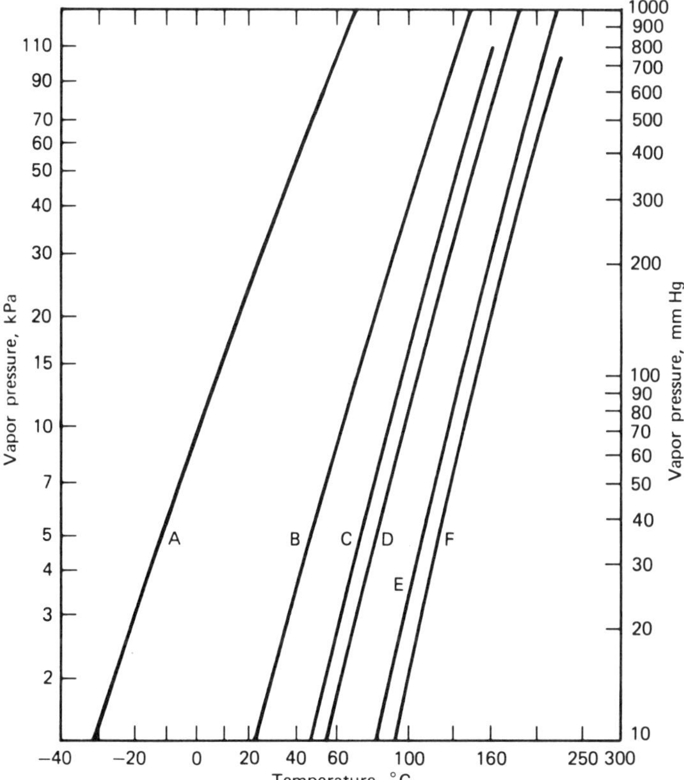

Figure 4. Vapor pressure of ketones (1). A, acetone; B, mesityl oxide; C, cyclohexanone; D, diisobutyl ketone; E, acetophenone; F, isobutyl heptyl ketone.

the evaporation rate of lacquers based on nitrocellulose. It is also used in spray lacquers, as a coupler for soap, in solvent systems, as a dewaxing agent for lubricating oils, and for the separation of tantalum and niobium (141).

Catalytic hydrogenation of diisobutyl ketone affords the diisobutyl carbinol (2,6-dimethyl-4-heptanol).

Unsaturated Ketones

Mesityl oxide and isophorone, the most important members of this group, have three reactive sites: a carbonyl group, a double bond, and an activated α-methyl group. This polyfunctionality enables these compounds to be important synthetic intermediates in the preparation of a broad range of chemicals and polymers.

Mesityl Oxide. Mesityl oxide, MSO, 4-methyl-4-penten-2-one, is a colorless oily liquid with a honeylike odor; it exhibits the versatility and unusual reactivity associated with conjugated α,β-unsaturated carbonyl compounds (142).

Physical properties and vapor-pressure curves are given in Table 2 and Figure 4, respectively.

On standing in air, mesityl oxide slowly forms bis(3,5,5-trimethyl-1,2-dioxolanyl)-3-peroxide (143). This peroxide can be produced in 45% yield from a mixture of sulfuric acid, hydrogen peroxide, and mesityl oxide:

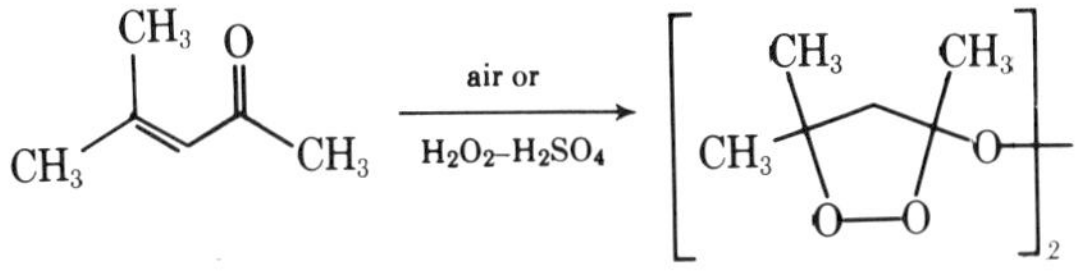

$CH_3C(CH_3){=}CHC(O)CH_3 \xrightarrow[H_2O_2\text{–}H_2SO_4]{\text{air or}} [(CH_3)_2C_3H_2(CH_3)(O{-}O)\text{–}O\text{–}]_2$

After prolonged exposure to air, peroxides may explode during distillation.

Prolonged contact with air promotes discoloration of mesityl oxide. Stabilizers such as tetramethylhexahydro-1,3-diazine (144) or diisopropylamine (145) have been suggested.

Owing to its conjugated α,β-unsaturated structure, mesityl oxide exhibits unusual spectroscopic behavior, especially in the ultraviolet region. The strong absorption (extinction coefficient $\epsilon = 12{,}600$) maximum in the ultraviolet at 237 nm can be used to detect its presence at extremely low concentrations (<5 ppm) (146).

Mesityl oxide undergoes simple additions; eg, with methanol high yields of 4-methoxy-4-methyl-2-pentanone [*107-70-0*] are obtained:

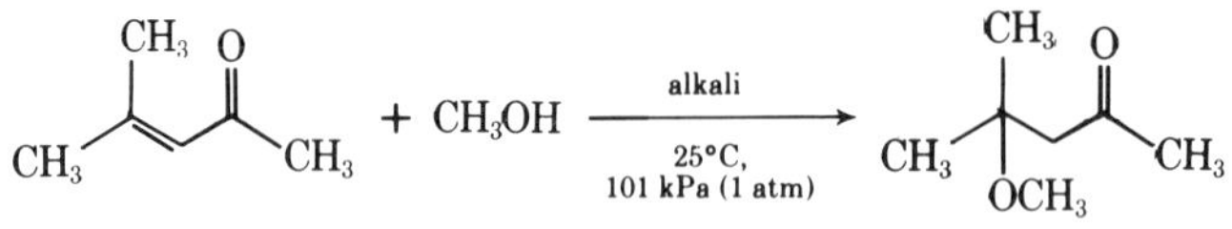

$$CH_3C(CH_3){=}CHC(O)CH_3 + CH_3OH \xrightarrow[\text{25°C, 101 kPa (1 atm)}]{\text{alkali}} CH_3C(CH_3)(OCH_3)CH_2C(O)CH_3$$

Usually, 1,4-addition precedes 1,2-addition and frequently both products are obtained. Thus, reduction gives a mixture of methyl isobutyl ketone and methyl isobutyl carbinol:

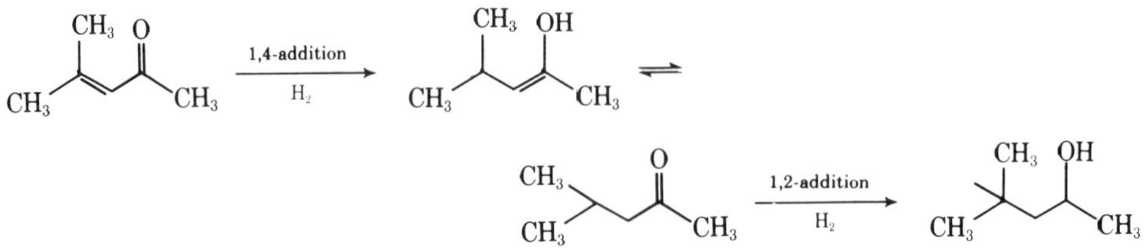

$$CH_3C(CH_3){=}CHC(O)CH_3 \xrightarrow[H_2]{\text{1,4-addition}} CH_3CH(CH_3)CH{=}C(OH)CH_3 \rightleftharpoons$$

$$(CH_3)_2CHCH_2C(O)CH_3 \xrightarrow[H_2]{\text{1,2-addition}} CH_3CH(CH_3)CH_2CH(OH)CH_3$$

A variety of condensation reactions lead to carbocyclic and heterocyclic compounds. For example, self-condensation to 1-acetyl-2,4,6,6-tetramethylcyclohexa-1,3-diene [*13834-80-5*] (30%) and isophorone (10%) can be effected with barium or calcium oxide catalysts. Mesityl oxide reacts with diethyl malonate under alkaline conditions to form methone [*126-81-8*] (5,5-dimethyl-1,3-cyclohexanedione), a product used as an analytical reagent for aldehydes:

$$(CH_3)_2C{=}CHCOCH_3 + CH_2(CO_2C_2H_5)_2 \xrightarrow{NaOC_2H_5} \text{[4029-25-8]} \longrightarrow \text{methone (70–85\%)} + CO_2$$

Unlike simple ketones, the carbonyl group of mesityl oxide is not as reactive toward 1,2-additions. Very strong reactants such as Grignard reagents are required.

Mesityl oxide undergoes aldol reactions with aldehydes but condensation occurs at the methyl of the CH_3CO-group. For example reaction with benzaldehyde provides $(CH_3)_2C{=}CHCOCH{=}CHC_6H_5$ [*55901-61-6*].

Cycloaddition of mesityl oxide with dienes in Diels-Alder reactions is slow. However, mesityl oxide reacts with certain dienes, eg, 2,3-dimethylbutadiene gives 1,2,4,4-tetramethyl-5-acetylcyclohexene [*74631-05-3*].

Commercial mesityl oxide can contain 0.5–20% of the β,γ-unconjugated isomer, isomesityl oxide [*141-79-7*] (4-methyl-4-pentene-2-one); typically about 1–3% is present (8). At equilibrium, the mixture contains 91% of α,β-unsaturated mesityl oxide and 9% of the β,γ-isomer (147–149). Equilibrium is catalyzed by either acid or alkali.

$$(CH_3)_2C{=}CHCOCH_3 \rightleftharpoons CH_2{=}C(CH_3)CH_2COCH_3$$

mesityl oxide isomesityl oxide

Isomesityl oxide has been isolated in 96% purity (149). Some physical properties of isomesityl oxide are given in Table 1 and reported in ref. 8.

Manufacture. Mesityl oxide is produced by liquid-phase dehydration of diacetone alcohol in the presence of acidic catalysts at 100–120°C and 101 kPa (1 atm). Preferred catalysts are dilute aqueous phosphoric acid (79,150–151) and sulfuric acid (134,152–153). Yields of about 95% with conversions of greater than 90% are obtained.

Dehydration is generally conducted batch-wise in the base of a distillation column. Dry acetone is removed overhead and mesityl oxide recovered from a sidestream as an azeotropic mixture with water; mesityl oxide so produced is saturated with water (3.4 wt %). Distillation gives an anhydrous product with 98–99.5% purity.

Mesityl oxide can also be prepared by direct one-step condensation from acetone; however, this is not practiced on a commercial scale. A variety of catalysts, eg, metal phosphates (154), phosphoric acid on silica gel (155), cation-exchange resins (156), aluminum oxides (157), zinc oxide (158) and zinc oxide/zirconium oxide (159) have been described but conversions and selectivities are unacceptably low. By-products such as phorone, semiphorone, mesitylene, isophorone, 2,2,4,4-tetramethyl-γ-pyrone and xylitone isomers (160) have been reported. These impurities are generally formed by additional condensations with acetone (157,160–163), followed by dehydration:

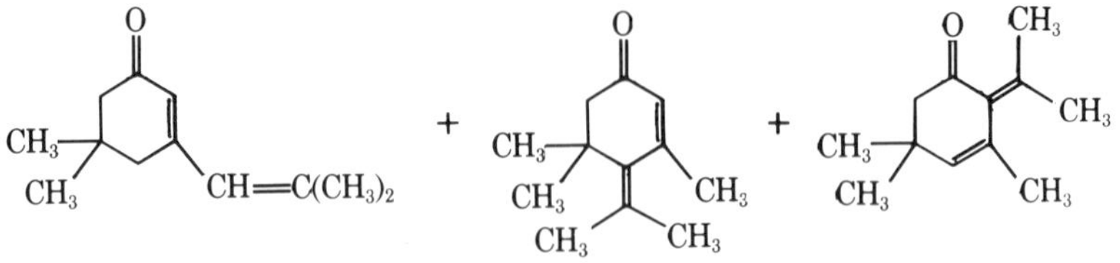

xylitone isomers

Strong acids favor trimerization of acetone. Thus, by-products are also obtained from the acid-catalyzed dehydration of diacetone alcohol.

Economic Aspects. Mesityl oxide and its derivatives (methyl isobutyl ketone and methyl isobutyl carbinol) are important industrial chemicals as judged by the amount of acetone consumed in their production (Table 10). Combined, these products accounted for about 20% of the total United States consumption of acetone in 1974 (164). Mesityl oxide is currently (October, 1980) selling for \$1.01/kg in tank car quantities.

Health and Safety Factors. Mesityl oxide is more toxic than saturated ketones but does not present a serious health hazard from exposure to normal industrial handling (see Table 4).

High concentrations of mesityl oxide vapor have a narcotic effect but odor and vapors cause marked irritation and discomfort to the eyes, nose, and throat long before maximum allowable concentration in air (25 ppm) is reached. However, frequent exposure to vapor concentrations may cause anemia and leukopenia (165–166). In animal tests, exposure to 5000 parts per million was dangerous to life in 30–60 min (27).

Minor, if any, injury may occur due to skin penetration from occasional short contact with liquid mesityl oxide. Sustained or repeated contact may cause dermatitis (27,166).

The odor detection level is 12–25 ppm (27).

Uses. Mesityl oxide is a chemical intermediate in the preparation of drugs, insecticides, solvents, drying oils and plasticizers for poly(vinyl acetal) resins (167–170). It is an excellent solvent for nitrocellulose and vinyl chloride–vinyl acetate resins, paint and varnish removers, carburetor cleaners, and stain removers. Mesityl oxide is also employed as an extractant for actinide elements such as thorium and uranium and in ore flotation (171).

Isophorone. Isophorone, 3,3,5-trimethyl-2-cyclohexen-1-one, is a cyclic α,β-unsaturated ketone derived from the condensation of three molecules of acetone:

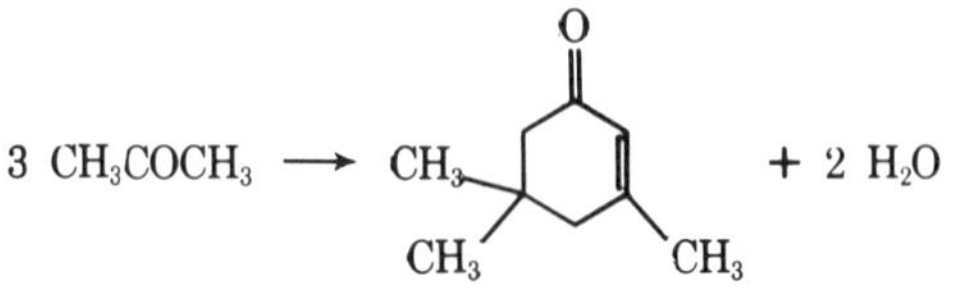

$$3\ CH_3COCH_3 \longrightarrow \text{isophorone} + 2\ H_2O$$

Physical constants and vapor pressure curves are given in Table 2 and Figure 1, respectively. The ultraviolet spectrum of isophorone exhibits a maximum at 235 nm with an extinction coefficient ϵ of 13,300 (172).

In the presence of *p*-toluenesulfonic (173) or adipic acid (174), the α,β-unsaturated system of isophorone tends to isomerize to the unconjugated β,γ-system giving 3,5,5-trimethyl-3-cyclohexen-1-one [*471-01-2*] (173,175–176).

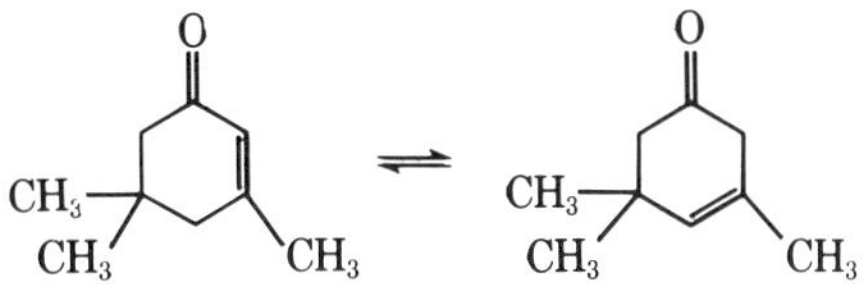

The term α-isophorone is sometimes used in referring to the α,β-unsaturated system whereas β-isophorone connotes the unconjugated derivative. Some physical property data for the latter are available in the literature (173,177).

Isophorone reacts readily with hydrogen peroxide to give 2,3-epoxy-3,5,5-trimethylcyclohexanone [*10276-21-8*] (172), which has the potential of forming hazardous peroxides after prolonged contact with air.

A large number of isophorone derivatives are available from Veba-Chemie (178), including alcohols, amines, glycols, diamines, diisocyanates, diesters, amino alcohols, and dicarboxylic acids. The diamines and diisocyanates are available in both cyclic and acyclic form.

A special distillative hydrogenation process designed to reduce only the carbon–carbon unsaturation (79) gives 3,3,5-trimethylcyclohexanone [*873-94-9*] in 97% yield. Complete hydrogenation gives the cyclohexanol derivative in 97% yield:

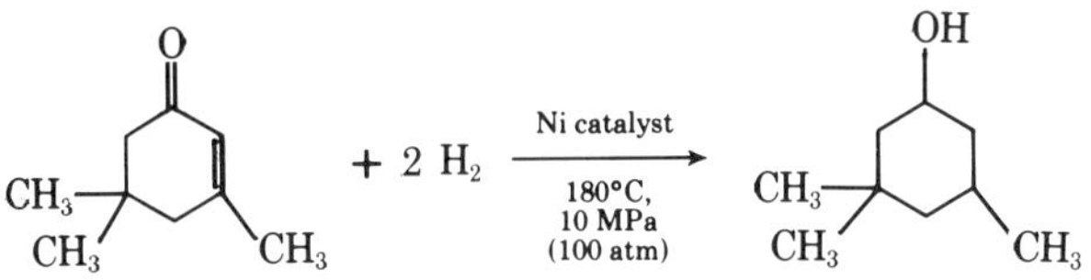

$$+ 2\ H_2 \xrightarrow[\substack{180°C,\\ 10\ MPa\\ (100\ atm)}]{\text{Ni catalyst}}$$

The saturated cyclic derivatives, such as 3,3,5-trimethylcyclohexanol, exist in two stereoisomer forms which can be produced in various ratios depending upon the reaction conditions.

Reductive amination of isophorone by a continuous-film hydrogenation process provides 3,3,5-trimethylcyclohexylamine in 90% yield (79):

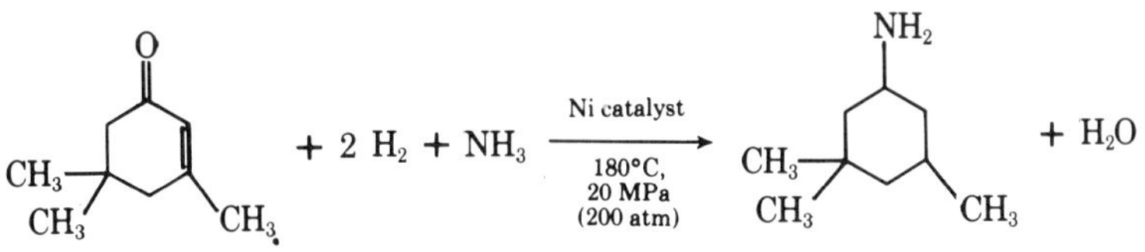

$$+ 2\ H_2 + NH_3 \xrightarrow[\substack{180°C,\\ 20\ MPa\\ (200\ atm)}]{\text{Ni catalyst}} \quad + H_2O$$

Cycloaliphatic diamines and diisocyanates are prepared by 1,4-addition of hydrogen cyanide to isophorone followed by reductive amination (79).

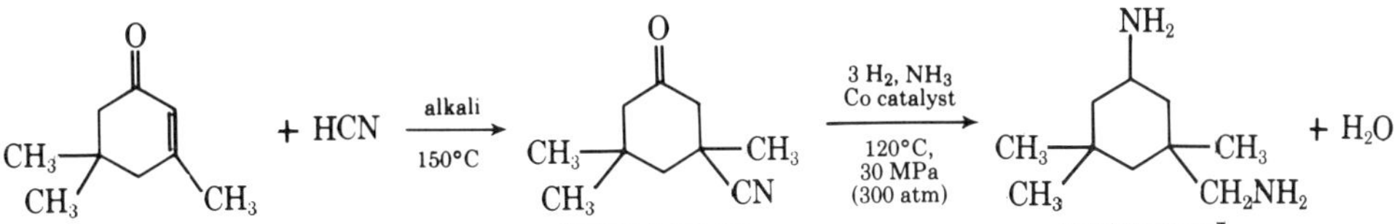

$$+ HCN \xrightarrow[150°C]{\text{alkali}} \quad \xrightarrow[\substack{120°C,\\ 30\ MPa\\ (300\ atm)}]{\substack{3\ H_2,\ NH_3\\ \text{Co catalyst}}} \quad + H_2O$$

[*58114-70-8*] [*2855-13-2*]

The resultant diamine is treated with phosgene to give the diisocyanate derivatives. These difunctional derivatives are excellent epoxy hardeners and polyurethane intermediates (see Isocyanates; Urethane polymers).

Other reactions further demonstrate the versatility of isophorone. For example, heating with catalytic amounts of methyl iodide at 500–600°C produces 3,5-xylenol (179–180). Conversions of 100% and selectivities of >90% to 3,5-xylenol were obtained:

$$\text{isophorone} \xrightarrow[570°C]{CH_3I,\ 1.5\%} \text{3,5-xylenol} + CH_4 + H_2$$

A cyclic dione, 2,2,6-trimethylcyclohex-5-ene-1,4-dione [*1125-21-9*], is prepared by oxidation of the unconjugated isophorone isomer (181). This compound is useful in perfumery and fragrance industries and is a starting material for certain carotenoids. Air oxidation of α-isophorone at 100°C in the presence of phosphomolybdic acid catalyst also yields the 1,4-dione (182).

$$\beta\text{-isophorone} + O_2 \xrightarrow[70°C]{\text{vanadium acetylacetonate}} \text{2,2,6-trimethylcyclohex-5-ene-1,4-dione}$$

91% yield

Manufacture. Like most industrial ketones, isophorone is produced under alkaline aldol conditions but more drastic reaction conditions are required for condensation of the third molecule of acetone. Consequently, a complex mixture of by-products contaminate the product and low yields are obtained. A continuous vapor-phase or liquid-phase mode is employed. In the older vapor-phase process, acetone vapor is passed over a catalyst bed of magnesium aluminate (183), zinc oxide–bismuth oxide (184), or calcium oxide (185) at 300–400°C and about 0.5 LHSV. The yield of isophorone is about 50%.

The liquid-phase process consists in feeding a mixture of acetone, water (as much as 30%), and potassium hydroxide (about 0.1%), to a pressure column operated at about 200°C and about 3.5 MPa (35 atm) (79,186–189). The column has an upper condensation section and a lower hydrolysis section, where isophorone flows downward and unreacted acetone passes upward. The presence of water, which serves both as a solvent and reactant, and alkali in the lower part of the column, causes hydrolysis of many by-products to isophorone and also produces recyclable acetone. The conversion is 6–10% and about 70% yield of isophorone is obtained.

Isophorone and acetone are also obtained by hydrolysis of isophorone residues containing mesityl oxide, xylitone isomers, higher C_{15}, C_{18}, and C_{24} ketones, and small amounts of phorone and mesitylene. An alumina or sodium hydroxide catalyst at 175°C and 930 kPa (9.2 atm) pressure provided 20% conversion to isophorone (190).

Veba-Chemie (79,186) (and probably BP Chemicals (187–188)) in Europe employ the liquid process. It offers improved energy utilization, higher isophorone yields, and a product of superior quality.

Both methods produce substantial amounts of by-products that not only consume about 30% of the acetone feed but also cause refining difficulties. The refining operation can require as many as three distillation columns and acidic discoloration treatment (188).

The residues consist primarily of mesitylene, phorone, and xylitone isomers (191). Substantial amounts of 3,3,6,8-tetramethyl-1-tetralone [*5409-55-2*] are also formed, particularly in the vapor-phase process (192). This tetralone has been synthesized from isophorone and mesityl oxide and it can thus be assumed to be a product of these two materials in the isophorone process (193–194):

$$\text{isophorone} + CH_3C(CH_3){=}CHCOCH_3 \xrightarrow[300^\circ C]{MgO} \text{3,3,6,8-tetramethyl-1-tetralone} + CH_4 + H_2O$$

Small amounts of isophorone isomer, phorone isomers, and isophorone dimers and trimer are also present (173,195–197). The liquid-phase process yields dimerized products whereas the more drastic conditions employed in the vapor-phase process favor the trimer (197).

Isophorone tends to discolor more than does mesityl oxide. High-purity (about 99%) material generally has a platinum–cobalt color value of about 30–50 (APHA); further yellowing can be expected on prolonged storage. Stabilization against color formation is provided by treatment with *p*-toluenesulfonic acid (188,198), acidified fuller's earth (199), diazines (144), or diisopropylamine (145). Commercial isophorone usually contains some unconjugated isomer (up to 5%) and small amounts (<1%) of xylitone.

Economic Aspects. Isophorone is produced in the United States by Union Carbide Corporation and Exxon Corporation. About 16,000 t were produced in 1973 (Table 10). In Europe, production is higher. Isophorone is currently (October, 1980) available in drums and tank cars for 1.34 and 1.21 \$/kg, respectively.

Uses. Isophorone is a powerful solvent for a large number of natural and synthetic polymers, resins, waxes, fats, and oils. However, it does not dissolve polyethylene, polypropylene, polyamides, polyureas, polyurethane, polycarbonates, polyacrylonitrile, or cellulose triacetate (178).

The commercial utility of isophorone stems from its unique chemical structure and excellent solvation power. It is used as a solvent for formulating highly concentrated vinyl chloride–acetate-based coating systems for metal cans. It provides systems with good storage stability and enhances flow, flexibility, and adhesion. Isophorone is used as the high boiling solvent in nitrocellulose finishes. It allows highly concentrated solutions to be made with increased blush and blistering resistance and improved gloss. In the paint industry, it is used as a leveling aid to prevent blistering and promote flow for metal paints from polyacrylates, alkyds, and epoxy and phenol–formaldehyde resins (178).

In printing inks for plastics, isophorone provides low viscosity and highly concentrated formulations and imparts good adhesion and gloss. It is an outstanding solvent for pesticide and herbicide concentrates. When used as a solvent for adhesives for plastics, it causes the plastics to swell. It is exceptionally effective for bonding articles of poly(vinyl chloride) and polystyrene.

Health and Safety Factors. Isophorone vapors are more toxic than lower saturated aliphatic ketones and mesityl oxide (see Table 4). However, because of its high vapor pressure, isophorone concentration in air can be kept within the maximum allowable limit with adequate ventilation (178).

Cyclic Ketones

Cycloaliphatic ketones are colorless liquids with boiling points that increase regularly with increasing molecular weight. Virtually all members of the series have characteristic odors depending on ring size. Neither ring substitution nor nature of the ring atoms (eg, lactones) have much effect on odor. Relationship of ring size and odor is given in Table 11 and physical properties of some compounds are listed in Table 12.

Many cyclic ketones occur in natural oils:

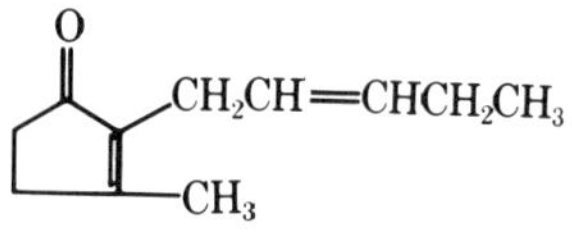

jasmone [*488-10-8*] (jasmine fragrance)

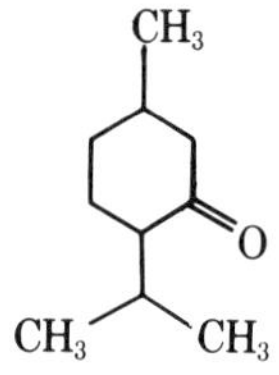

l-menthone [*14073-97-3*] (peppermint oils)

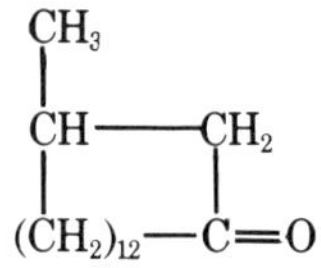

muscone [*541-91-3*] (musk oil)

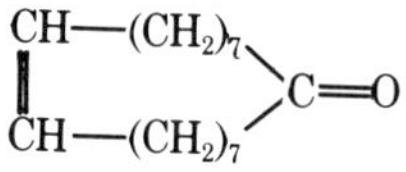

civetone [*542-46-1*] (civet cat odor)

Jasmone (3-methyl-2-(2-pentenyl)-2-cyclopenten-1-one) is an odoriferous component (5%) of the oil obtained from jasmine flowers. The synthetic material is a highly valued perfume base (205). Other large-ring ketones, such as cyclopentadecanone and muscone, are valued for their musklike odor. Cyclopentadecanone is sold as a perfume under the name exaltone (see Oils, essential; Perfumes).

The chemical properties of cyclic ketones also vary with ring size (Table 12). The carbonyl groups of lower members of cyclic ketones are significantly more reactive, eg, toward addition reactions, than corresponding acyclic ketones. The C_8–C_{12} cyclic ketones are unreactive, reflecting the strain and high enol content of medium-size ring systems (Table 12). Lactones are prepared from cyclic ketones by the Bayer-Villiger oxidation reaction with organic peracids. Caprolactone is manufactured from cyclohexanone by this process:

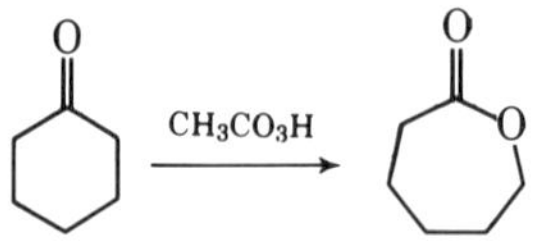

$$\xrightarrow{CH_3CO_3H}$$

Table 11. Relationship of Cyclic Ketone Ring Size to Odors[a]

Ring size	Odor
5	bitter almonds
6	peppermint
7–9	transition to camphorlike
10–12	camphorlike
13	cedarwoodlike
14–18	musklike

[a] Ref. 200.

Table 12. Properties of Cyclic Ketones, $\overline{(CH_2)_x C}{=}O$ [a]

CAS name	CAS Registry No.	Formula, x =	Bp, °C (kPa[b])	Melting point, °C	Refractive index, n_D^{20}	Density at 20°C[c], g/L	Ir C=O stretching frequency, max, cm^{-1}	Enol, %
cyclopropanone	[5009-27-8]	2	[d]					
cyclobutanone	[1191-95-3]	3	100–102		1.4195	938	1788	0.55
cyclopentanone	[120-92-3]	4	130		1.4359	951	1746	0.09
cycloheptanone	[502-42-1]	6	179–181		1.4611	951	1703	0.56
cyclooctanone	[502-49-8]	7	74 (1.6)	42		958	1703	9.3
cyclononanone	[3350-30-9]	8	93–95 (1.6)	34	1.4770	959	1702	4.0
cyclodecanone	[1502-06-3]	9	107 (1.7)	29	1.4820	958		6.1
cycloundecanone	[878-13-7]	10	108 (1.6)	10	1.4804			
cyclododecanone	[830-13-7]	11	125 (1.6)	61		906		
cyclotridecanone	[832-10-0]	12	138 (1.6)	32	1.4790	927		
cyclotetradecanone	[3603-99-4]	13	155 (1.6)	53				
cyclopentadecanone	[502-72-7]	14	120 (0.04)	63		897		
cyclohexadecanone	[2550-52-9]	15	138 (0.04)	60				
cycloheptadecanone	[3661-77-6]	16	141 (0.13)	63				
cyclooctadecanone	[6907-37-5]	17	158 (0.04)	72				
cyclononadecanone	[6907-38-6]	18	160 (0.04)	72				
cyclocosanone	[6907-39-7]	19	171 (0.04)	59				

[a] Refs. 201–204.
[b] To convert kPa to mm Hg, multiply by 7.50.
[c] For solids, the density is given for the liquid at melting point temperature.
[d] Rapidly polymerizes at room temperature. A stable hydrate, mp 71–72°C, is formed in water.

Reference 206 reviews various methods of synthesizing cyclic ketones.

Limited toxicological data are available on cyclic ketones; cyclohexanone is not considered to be highly toxic (13). As with most ketones, breathing concentrated vapors may be harmful but humans cannot tolerate high concentrations long enough to become ill. Cyclic ketones are not particularly harmful to the skin but eye contact is irritating and causes an effect similar to that of strong soap. Table 13 summarizes toxicological data for some cyclic ketones. Comparison with MEK and mesityl oxide shows that cyclic ketones are among the least toxic commonly used ketones; toxicity increases with increasing ring size. Interestingly, this is in reverse order of their reactivity.

Table 13. Toxicological Properties of Cyclic Ketones Compared with Some Aliphatic Ketones[a]

Compound	Administered into the peritoneal[b] cavity LD_{50}, mouse, mg/kg	Administered under the skin LD_{lo}[c], mouse, mg/kg
cyclopentanone	1950	2600
cyclohexanone	1350	1300
cycloheptanone	750	930
cyclooctanone	740	
acetone	1297	
mesityl oxide	268	
methyl ethyl ketone	616	

[a] Ref. 28.
[b] Membrane that lines the cavity of the abdomen.
[c] The lowest lethal dose.

Cyclohexanone. Cyclohexanone is by far the most important cyclic ketone (see Cyclohexanol and cyclohexanone). In 1973, seven United States companies (Allied, Dow Badische, Celanese, DuPont, Monsanto, Nipro, and Union Carbide) produced about 500,000 t (207–208). Current United States nameplate capacity is estimated in the range of 900,000–1,200,000 t/yr (207–208). Most cyclohexanone is produced as a mixture with cyclohexanol (KA oil) by air oxidation of cyclohexane. Some is also produced by hydrogenation of phenol, followed by dehydrogenation of the resulting cyclohexanol. The principal use of pure cyclohexanone is in manufacture of γ-caprolactam (by rearrangement of its oxime) for nylon-6. Crude cyclohexanone–cyclohexanol is used to make adipic acid for nylon-6,6 (209) (see Polyamides).

Cyclohexanone is a solvent for vinyl resins, polystyrene, ethyl cellulose, acrylic resins, dyes, insecticides, and lubricating oil sludges (13), and especially poly(vinyl chloride) and its copolymers (210). Physical property data and vapor pressure curves are given in Table 1 and Figure 4, respectively; the ir C=O stretching frequency is 1715 cm^{-1} max; % enol = 1.18.

Diketones

Diketones are generally used as specialty chemical intermediates for the pharmaceutical, flavor, fragrance, and dye industries.

1,2-Diketones. Aliphatic 1,2-diketones are liquids with boiling points that increase with increasing molecular weight. The boiling point is lowered with branching. 1,2-Diketones are yellow and absorb in the visible region at λ max 370–440 nm (211). Biacetyl and acetyl propionyl are soluble in water; increasing molecular weight decreases water solubility rapidly. Aliphatic 1,2-diketones have sharp and penetrating odors in high concentrations but diluted possess a sweet, aromatic odor. Most of the cyclic and aromatic 1,2-diketones are yellow solids, and possess less penetrating odors than aliphatic derivatives. Like cyclic monoketones, cyclic 1,2-diketones demonstrate enolic tautomerism with solvent polarity affecting tautomeric equilibrium:

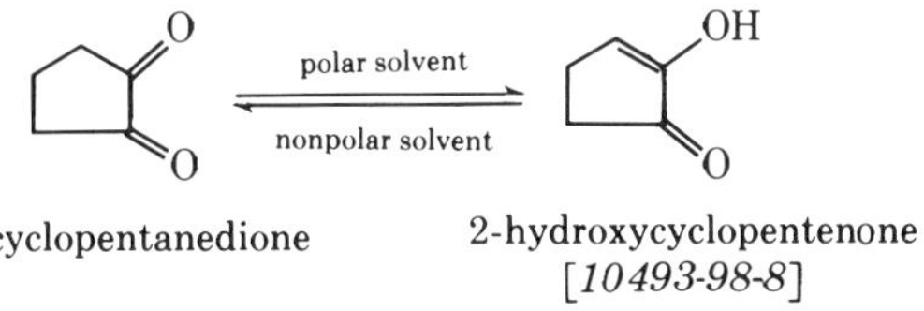

polar solvent / nonpolar solvent

1,2-cyclopentanedione ⇌ 2-hydroxycyclopentenone [*10493-98-8*]

Diosphenol [*490-03-9*], the main constituent of oil of buchu leaves, presents an example of such tautomerism (212).

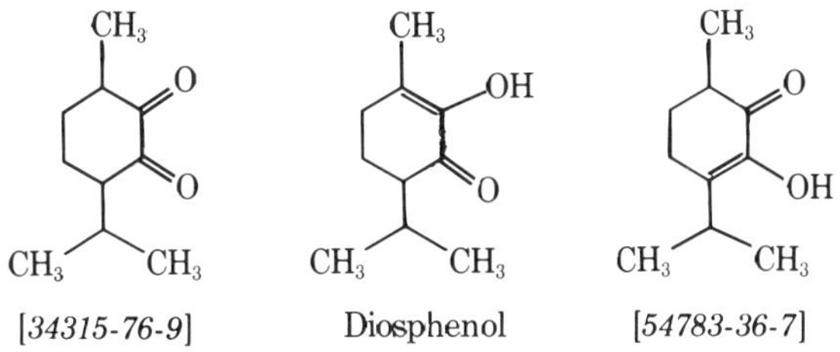

[*34315-76-9*] Diosphenol [*54783-36-7*]

Many 1,2-diketones exist in nature, eg, biacetyl is present in butter and natural oils (lemongrass oil), and five cyclic 1,2-diketones contribute to the aroma and flavor of coffee (qv) (213–214). Several 2,3-hexanediones and 3,4-heptanediones have also been identified in a coffee concentrate (215).

Physical properties of representative 1,2-diketones are given in Table 14.

Compounds of this series are generally prepared by oxidation of the corresponding monoketone (211) or α-hydroxyketone (216).

1,2-Diketones are used extensively as chemical intermediates for the preparation of pharmaceuticals, flavors, and fragrances. They remove undesirable odors from dairy products (217) and certain compounds are used to determine ureas, guanidines, nitrates, and resorcinol (218) and as polymerization catalysts (219).

The toxicity of 1,2-diketones is on the order of that of acetone and methyl ethyl ketone. Toxicity data for biacetyl and benzil and various diketones are given in Table 15.

Biacetyl and Benzil. Biacetyl, 2,3-butanedione, is made commercially by passing vinylacetylene into a solution of mercuric sulfate in sulfuric acid and decomposing the insoluble product with dilute hydrochloric acid (220). It can also be prepared by oxidation of methyl ethyl ketone over a copper oxide catalyst (32), by oxidation of 2,3-butanediol (221) or from the reaction of acetal with formaldehyde (222).

Flavor-grade biacetyl is available in bottles at \$24/kg (October, 1980). It is used as an odorant for tobacco and food products (see Food additives). Biacetyl is a better preservative against various yeasts than benzoic acid or sodium benzoate (223) and has been suggested as a hardening agent for gelatin in photography and adhesive compositions, and as a component in insecticide formulations (224).

Table 14. Physical Properties of 1,2-Diketones[a]

CAS name	Synonyms	CAS Registry No.	Bp, °C (kPa[b])	Melting point, °C	Refractive index, n_D^t	Density, g/L, d_4^t	Color	Ultraviolet spectrum λ_{max}, nm
Aliphatic								
2,3-pentanedione	acetylpropionyl	*[600-14-6]*	108		1.4014^{19}	956.5^{19}	yellow	419
2,3-hexanedione	acetylbutyryl	*[3848-24-6]*	128			934.0^{19}		
3,4-hexanedione	bipropionyl	*[4437-51-8]*	130		1.4130^{21}	941.0^{21}		435
4-methyl-2,3-pentadione	acetylisobutyrl	*[7493-58-5]*	115–116			921.5^{11}	yellow	429
3,4-heptanedione	propionylbutyrl	*[13706-89-3]*	147 (98)			885.0^{0}		
5-methyl-2,3-hexanedione	acetylisovaleryl	*[13706-86-0]*	138		1.4119^{20}	908.0^{22}		432
2,3-octanedione	acetylcaproyl	*[585-25-1]*	172–173 (98)					
4,5-octanedione	bibutyryl	*[5455-24-3]*	168			934.0^{0}	yellow	435
2,5-dimethyl-3,4-hexanedione	biisobutyryl	*[4388-87-8]*	144–145		1.4206^{20}	923.2^{20}		436
5-methyl-3,4-heptanedione		*[13678-56-3]*	63–67 (5.3)					
6-methyl-3,4-heptanedione		*[3131-90-6]*	53–54 (2.0)		1.4151^{20}	901.9^{20}		
Cyclic								
1,2-cyclopentanedione		*[3008-40-0]*	105 (2.7)	55–56				
1,2-cyclohexanedione		*[765-87-7]*	193–195	35–38	1.4995^{20}			
Aromatic								
benzil	bibenzoyl	*[134-81-6]*	346–348 188 (1.6)	94–95			yellow	
1-phenyl-1,2-propanedione	acetylbenzoyl	*[579-07-7]*	125 (3.1)			1101^{20}		
1,2-naphthalenedione	1,2-naphthoquinone	*[524-42-5]*		145–147 (dec)			golden yellow	

[a] Refs. 177, 203, 211, 215.
[b] To convert kPa to mm Hg, multiply by 7.50.

Table 15. Toxicological Properties of Diketones[a]

Ketone	Ingestion, LD_{50}, rats, mg/kg	Inhalation, LC_{Lo}[b], rats, ppm/4 h	Injection into the peritoneal[c] cavity LD_{50}, mg/kg	Injection into the peritoneal[c] cavity LD_{Lo}[d], mg/kg
1,2-Diketones				
benzil	2710			
biacetyl	1580		400[e]	
1,3-Diketones				
2,4-pentanedione	1000	1000	750[f]	
1,3-cyclohexanedione				64[f]
1,4-Diketones				
2,5-hexanedione	2700	200		
1,4-cyclohexanedione				100[e]

[a] Ref. 28.
[b] The lowest concentration in air which caused death in 4 hours.
[c] Membrane lining the cavity of the abdomen.
[d] The lowest dose resulting in death.
[e] Rats.
[f] Mouse.

Benzil is prepared commercially by oxidation of benzoin (225–226). Nearly quantitative yields are obtained by oxidation of benzyl phenyl ketone [*451-40-1*] (227). Benzil is used in the cosmetics industry and has been suggested as a chigger repellant (228) (see Repellants).

Physical property data for biacetyl and benzil are given in Tables 2 and 14, respectively. Biacetyl is greenish-yellow and shows a uv absorption at 421 nm.

Biacetyl and benzil have low acute oral toxicity (see Table 15). Likewise, skin irritation and inhalation hazards are low as demonstrated by use as flavorants and in cosmetics (see Flavors and spices).

1,3-Diketones. Aliphatic and lower aliphatic–aromatic 1,3-diketones are colorless liquids with boiling points that increase with increasing molecular weight. Cyclic 1,3-diketones are colorless solids. The lowest member of the aliphatic series (2,4-pentanedione) has a typical ketonic odor, whereas higher molecular weight derivatives possess lasting sweet, esterlike odors: 3-methyl-2,4-heptanedione [*13152-54-0*] has a fruity aroma, tetramethylcyclobutanedione [*29714-52-3*] a minty, camphorlike odor, and benzoylacetone [*93-91-4*] a balsamlike odor. All blend well in perfume formulations (229) (see Perfumes). 1,3-Diketones are highly miscible with most organic solvents and a few low molecular weight derivatives have appreciable water solubility, eg, a 16.6% solution of 2,4-pentanedione in water can be prepared. The physical properties of various 1,3-diketones are listed in Table 16.

1,3-Diketones are characterized by a highly reactive methylene group between two activating carbonyls. Consequently, 1,3-diketone derivatives exist as an equilibrium mixture with enolic forms. The equilibrium distribution varies with structure and solvent (229,232):

1,3-Diketone	*Enol, %*
2,4-pentanedione	76.4
2,4-hexanedione	80.2
2,4-heptanedione	83.6
1-phenyl-1,3-butanedione	99
1,3-cyclohexanedione	100

Table 16. Physical Properties of 1,3-Diketones[a]

CAS name	Common name	CAS Registry No.	Bp, °C (kPa[b])	Melting point, °C	Refractive index, n_D^{20}	Density, g/L, d_4^{20}	Copper chelate Color	Copper chelate Mp, °C
Aliphatic								
2,4-hexanedione	propionylacetone	[*3002-24-2*]	158		1.4516	959	green	198
3,5-heptanedione	dipropionylmethane	[*7424-54-6*]	47 (0.8)			944.5		209–214
2,4-heptanedione	butyrylacetone	[*7307-02-0*]	174–175			941.1[c]	blue	161
3,5-octanedione	butyrylpropionylmethane	[*6320-18-9*]	189–190				blue	158
5-methyl-2,4-hexanedione	isobutyrylacetone	[*7307-03-1*]	168					171
2,6-dimethyl-3,5-heptanedione	diisopropionylmethane	[*18362-64-6*]	66 (1.1)					
2,4-octanedione	valerylacetone	[*14090-87-0*]	79–83 (2.7)		1.4559	923.3		
5,5-dimethyl-2,4-hexanedione	pivaloylacetone	[*29284-62-6*]	164–167 (99)					
6-methyl-2,4-heptanedione	isovalerylacetone	[*3002-23-1*]	73 (2.4)					
Cyclic								
1,3-cyclopentanedione		[*3859-41-4*]		151–153				
1,3-cyclohexanedione	dihydroresorcinol	[*504-02-9*]		103–105				
5,5-dimethyl-1,3-cyclohexanedione	dimedone, methone	[*126-81-8*]		149–151				
Aromatic								
1-phenyl-1,3-butanedione	benzoylacetone	[*93-91-4*]	98–100 (0.3)	58–60		1090	sea green	199
1-phenyl-1,3-pentanedione	benzoylpropionylmethane	[*5331-64-6*]	92–94 (0.13)		1.5731[d]		sea green	135
1,3-diphenyl-1,3-propanedione	dibenzoylmethane	[*120-46-7*]	219–221	77.5–79			green	296–300
1-phenyl-2,4-pentanedione		[*3318-61-4*]	144		1.5837			

[a] Refs. 177, 211, 230–231.
[b] To convert kPa to mm Hg, multiply by 7.50.
[c] At 15°C.
[d] At 25°C.

The enol forms are cyclic and acidic and form colored, solid chelates with metals:

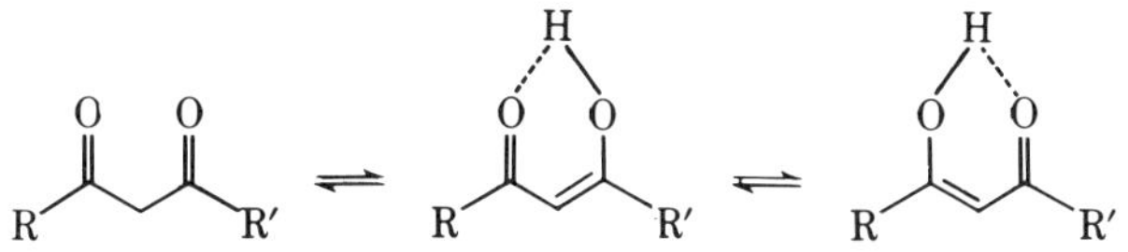

Thus, ferric chloride produces red complexes whereas copper chelates are blue or green (see Table 16). The chelates are covalent compounds, insoluble in water and soluble in benzene (see Chelating agents). For bivalent metals, such as copper, the chelate structure of 2,4-pentanedione is as follows:

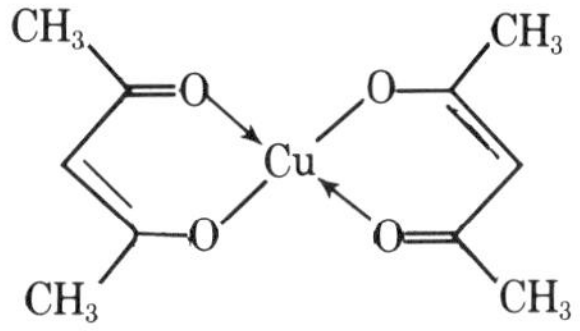

1,3-Diketones undergo typical ketone reactions. Notable is the relative ease of basic cleavage to a ketone and acid:

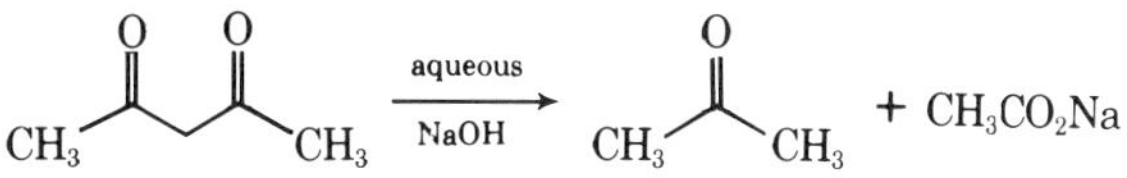

$$CH_3COCH_2COCH_3 \xrightarrow[\text{NaOH}]{\text{aqueous}} CH_3COCH_3 + CH_3CO_2Na$$

Reaction is more facile than with analogous β-keto-esters.

1,3-Diketones can be obtained by a variety of synthetic routes (233); eg, by reaction of a ketone with a carboxylic acid ester in the presence of strong base. Highest yields are obtained with sodium amide or sodium hydride catalysts:

$$RCO_2C_2H_5 + CH_3\overset{O}{\overset{\|}{C}}R' \xrightarrow{\text{alkali}} R\overset{O}{\overset{\|}{C}}CH_2\overset{O}{\overset{\|}{C}}R' + C_2H_5OH$$

Five- and six-membered cyclic 1,3-diketones can be prepared by either treating corresponding γ-keto-acids with concentrated sulfuric acid, or keto esters with base. Thus, the ethyl ester of levulinic acid [*123-76-2*] produces 1,3-cyclopentanedione.

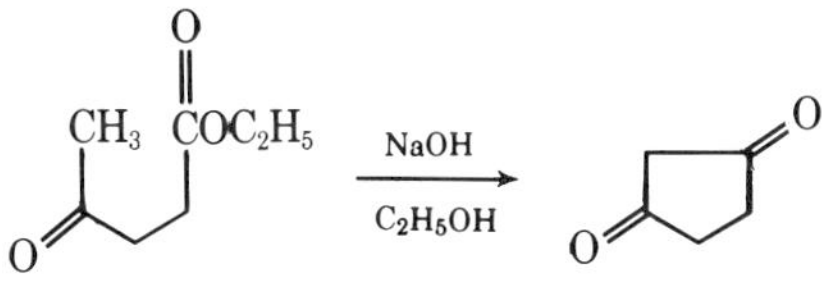

$$CH_3COCH_2CH_2COC_2H_5 \xrightarrow[C_2H_5OH]{\text{NaOH}} \text{1,3-cyclopentanedione}$$

This reaction resembles Dieckmann cyclizations.

1,3-Diketones have a variety of uses depending on the specific derivative. Generally, they are used for extraction and identification of metals and as raw materials for synthesis of heterocyclic compounds. Other applications take advantage of their enolizability; eg, acetylbenzoylmethane [*93-91-4*] and dibenzoylmethane are used for determination of boron (234) and uranium (235), as light stabilizers for ABS polymers (236), as catalyst for cross-linking polyesters (237), for extending the pot life of polyurethane propellants (238), for improved spinning properties of spandex-fiber spinning dopes (239), as copper chelates for increasing storage stability of unsaturated polyester resins (240), and as copper and nickel chelates for increasing ultraviolet light resistance of spandex polymers (241) (see Uv stabilizers).

1,3-Cyclopentanedione is an intermediate in the synthesis of 8-azasteroids and the sesquiterpenes, cedrone [*3477-88-3*] and (±)-laurene (242).

1,3-Diketones are more toxic than 1,2- and 1,4-diketones, probably because of their ability to tie-up essential metal ions in animal systems (Table 15). Skin irritation and eye damage are less severe (9).

2,4-Pentanedione. 2,4-Pentanedione, acetylacetone, is a colorless liquid with a mild ketonelike odor. The enol form predominates (70–90%) in the pure liquid state, the vapor state, and when dissolved in most organic solvents. In water only 12% of the enol is present at equilibrium. It is acidic with a pK_a of 8.93 at 25°C (243). Physical properties are given in Table 2.

Several attractive production methods are described:

Thermal and metal catalyzed isomerization of isopropenyl acetate (244–246):

$$CH_3COOC(CH_3){=}CH_2 \xrightarrow[500\text{–}600°C]{\text{Mo cat.}} CH_3COCH_2COCH_3$$

Condensation of acetone with ethyl acetate (247–251):

$$CH_3COCH_3 + CH_3COOC_2H_5 \xrightarrow[\text{or } BF_3]{\text{alkali}} CH_3COCH_2COCH_3 + C_2H_5OH$$

Condensation of ethyl acetoacetate and ketene (252–255).

$$CH_3COCH_2COOC_2H_5 + CH_2{=}C{=}O \longrightarrow$$

$$CH_3COCH(COCH_3)COOC_2H_5 \xrightarrow[\text{or boric acid}]{CH_3COOH} CH_3COCH_2COCH_3 + C_2H_5OH + CO_2$$

The efficiency of the 1,3-dione structure of 2,4-pentanedione in forming metal chelates is used in extraction processes. It is also used as an intermediate for manufacture of pharmaceuticals and dyes, and in metal plating and resin modification (9).

2,4-Pentanedione is appreciably more toxic by oral ingestion and vapor inhalation than either 1,2- or 1,4-diketones and saturated monoketones (Tables 4 and 15). Its acute oral toxicity is high and internal consumption should be avoided. It is comparable in this respect to mesityl oxide. Breathing 2,4-pentanedione vapors may cause dizziness, headache, nausea, vomiting, and loss of consciousness (256). Skin irritation appears less hazardous. However, eye burns may result from a large application, similar to soap.

1,4-Diketones. With exception of 2,5-hexanedione which is a high boiling liquid, 1,4-diones are low melting white solids with only faint odors. Lower members are highly soluble in organic solvents and water. Since the carbonyl groups are separated by two methylene groups, 1,4-diketones do not form salts with metals and are insoluble in alkali. Properties of representative 1,4-diketones are shown in Table 17.

1,4-Diketones are intermediates for synthesis of natural products and several preparative methods have been developed (258); in the simplest preparative methods, ketone enolates are oxidatively dimerized (259):

$$2\ C_6H_5\text{—}\overset{\displaystyle OLi}{\overset{|}{C}}\text{=}CH_2 + 2\ CuCl_2 \xrightarrow[-78^\circ C]{DMF} C_6H_5\text{—}\overset{\displaystyle O}{\overset{\|}{C}}CH_2CH_2\overset{\displaystyle O}{\overset{\|}{C}}\text{—}C_6H_5 + 2\ CuCl + 2\ LiCl$$

95% yield

1,4-Diketones are readily transformed to cyclic derivatives, such as cyclopentenones and furans. Thus, they are used to prepare cis-jasmone (260). 2,5-Heptanedione and 6-methyl-2,5-heptanedione improve the taste and aroma of tobacco (261).

2,5-Hexanedione. 2,5-Hexanedione is the only commercially available member of this series (Fritzsche Dodge & Olcott Inc.). It is a colorless high boiling liquid and is prepared by hydrolysis of 2,5-dimethylfuran (262). Properties of 2,5-hexanedione are listed in Table 2. Its main use is in solvent systems and as a raw material for chemical synthesis.

2,5-Hexanedione is reportedly not highly toxic (263). It exhibits a narcotic effect about comparable to methyl ethyl ketone and its vapor causes some irritation of mucous membranes (Table 15). Owing to its low volatility, danger by inhalation has been difficult to estimate but it seems to be slightly toxic. No toxic effects from exposure have been recorded (263).

Table 17. Physical Properties of 1,4-Diketones[a]

CAS name	Common name	CAS Registry No.	Bp, °C (kPa[b])	Melting point, °C	Refractive index, n_D^{20}
3,4-dimethyl-2,5-hexanedione		*[25234-79-1]*	92 (4.0)		1.4330
3,3,4,4-tetramethyl-2,5-hexanedione		*[25328-38-3]*	40 (0.7)		1.4522
2,5-heptanedione	acetylpropionylethane	*[1703-51-1]*	90 (2.8)		
3,6-octanedione	dipropionylethane	*[2955-65-9]*	98 (1.9)	34–35	
6-methyl-2,5-heptanedione		*[13901-85-4]*	91 (1.6)		
2,5-decanedione		*[41368-32-5]*	132 (2.3)		
2,5-dodecanedione		*[32781-66-1]*	148 (1.6)	40.5	
1,4-cyclohexanedione		*[637-88-7]*		77–78.5	
1,4-diphenyl-1,4-butanedione	1,2-dibenzoylethane	*[495-71-6]*		145–147	

[a] Refs. 177, 257.
[b] To convert kPa to mm Hg, multiply by 7.50.

Other Diketones. Several higher diketones are reported in the literature but have little commercial value (264). The properties of several 1,5-diketones and their use in cyclocondensations to produce cyclohexenone derivatives has been recently reported (265).

Cyclodihydroaromatic 1,2- and 1,4-diketones, such as benzoquinone, naphthoquinone, and anthraquinone are used in the preparation of dyes. A few occur in nature in molds, fungi, and mushrooms and some possess biological activity (see Quinones).

Polyketones

Unsaturated ketones such as methyl vinyl ketone [*78-94-4*] and methyl isopropenyl ketone [*814-78-8*] polymerize readily with peroxide initiators. The products are structurally similar to poly(methyl acrylate) and poly(methyl methacrylate), respectively, but inferior physical properties have precluded commercialization:

$$\text{-}[\text{CH}_2\underset{\text{COCH}_3}{\underset{|}{\text{CH}}}]_n\text{-}$$

poly(methyl vinyl ketone)
[*25038-87-3*]

$$\text{-}[\text{CH}_2\overset{\text{CH}_3}{\overset{|}{\underset{\text{COCH}_3}{\underset{|}{\text{C}}}}}]_n\text{-}$$

poly(methyl isopropenyl ketone)
[*25988-32-3*]

Polyketones are also produced by copolymerization of ethylene with either carbon monoxide or aliphatic aldehydes at high temperatures under pressure (266–267):

$$\text{CH}_2\text{=CH}_2 + \text{CO} \xrightarrow[\substack{140°\text{C} \\ 10\text{–}100\ \text{MPa} \\ (100\text{–}1000\ \text{atm})}]{\text{peroxide}} (\text{CH}_2\text{CH}_2)_x\ (\text{CH}_2\text{CH}_2\text{CO})_y$$

$$\text{CH}_2\text{=CH}_2 + \text{RCHO} \xrightarrow[\text{pressure}]{>140°\text{C}} \text{RCO(CH}_2)_x\text{H}$$

However, only very limited success has been achieved in producing polymers with acceptable mechanical properties.

Styrene–methyl vinyl ketone copolymers [*100-42-5*] or styrene–phenyl vinyl ketone copolymers [*27340-61-0*] are useful as photodegradable polymers in packaging applications (268).

Miscellaneous Ketones

Several commercially significant aliphatic, and aliphatic–aromatic ketones are produced in low volumes by one of several processes:

The *Aldol reaction* of methyl ketones with aldehydes gives eg, methyl isoamyl ketone from isobutyraldehyde and acetone.

$$(\text{CH}_3)_2\text{CHCHO} + \text{CH}_3\text{COCH}_3 \xrightarrow[\substack{(2)\ -\text{H}_2\text{O} \\ (3)\ \text{H}_2}]{(1)\ \text{alkali}} (\text{CH}_3)_2\text{CHCH}_2\text{CH}_2\text{COCH}_3$$

Aldol condensations of this type, where both reactants are capable of aldolization, usually give poor yields and separation and purification are expensive. Special catalysts have been developed which favor reaction on the methyl site of the ketone (269).

Dehydrogenation of the corresponding alcohol, eg, preparation of methyl *n*-hexyl ketone from 2-octanol is generally preferred where the alcohol is readily available.

Vapor-phase decarboxylation of carboxylic acids, eg, the preparation of aceto-

$$CH_3CH(OH)(CH_2)_5CH_3 \xrightarrow[\text{vapor phase, } 200\text{–}300^\circ C]{\text{Cu Catalyst}} CH_3CO(CH_2)_5CH_3$$

phenone from benzoic acid and acetic acid, is suitable for preparing mixed as well as simple ketones:

$$C_6H_5COOH + CH_3COOH \xrightarrow[300\text{–}450^\circ C]{CaO} C_6H_5COCH_3 + CO_2 + H_2O$$

However, numerous by-products are obtained at high temperatures (270–271). Used for cross-decarboxylation of two different acids, one product can be favored by adjustment of the reaction stoichiometry.

Friedel-Crafts reaction between benzene and carboxylic acids (or their acid chloride) is limited to the preparation of mixed aliphatic–aromatic and simple aromatic ketones, eg, propiophenone from benzene and propionic acid (or propionyl chloride):

$$C_6H_6 + CH_3CH_2COCl \xrightarrow[85^\circ C]{AlCl_3} C_6H_5COCH_2CH_3 + HCl$$

Reaction requires one mol of anhydrous aluminum chloride per mol of acid chloride (272). Excellent yields of ketones (>85%) are obtained but difficulties associated with handling anhydrous aluminum chloride and disposal of large amounts of hydrogen chloride are involved (see Friedel-Crafts reaction).

Oxidation of α-olefins to methyl ketones, eg, 1-hexene to 2-hexanone (273–274) has been recently patented.

$$CH_2{=}CH(CH_2)_3CH_3 \xrightarrow[\substack{550\text{ kPa }(5.4\text{ atm})\\105^\circ C}]{\substack{O_2\\ \text{Pd–Cu–alkali}}} CH_3C(=O)(CH_2)_3CH_3$$

Hexene conversions of greater than 75% with quantitative selectivity to hexanones are obtained when oxidation is conducted in a multiphase diluent system containing a quaternary ammonium halide surfactant (274).

Producers and uses of low volume ketones are shown in Table 18. In general, they are used for either solvents or as intermediates for pharmaceuticals and dyes. Biacetyl is used for its odor and 2,6,8-trimethyl-4-nonanone [*123-17-1*] for its dispersant qualities.

Several classes of ketones have gained much attention in recent years but are not yet commercially available. For example, interest in bis(1,3-diketones), ie, tetraketones, arose from the synthesis of thermally stable polymers containing heterocyclic groups, such as pyrazoles and quinoxalines, in the main polymer chain and from preparing coordination polymers with metals (see Metal-containing polymers). A review of the preparation, reactions, and properties of these ketone derivatives has been published (275).

Table 18. Suppliers and Uses of Low Sales Volume Ketones

Ketone	Supplier–producer	Uses
acetophenone	Union Carbide	perfume base for bath soaps; chemical intermediate for resins, pharmaceuticals, corrosion inhibitors, and dyestuffs; solvent for gums, resin dyestuffs, and high melting aromatic chemicals
benzophenone	BASF Wyandotte, Upjohn	fixative for perfumes in soaps; raw material for the manufacture of insecticides, hypnotics, and antihistamines
diethyl ketone	Union Carbide	chemical intermediate for the production of herbicides and dyestuffs; solvent in resin formulation
dihydroxyacetone	G. B. Fermentation Industries	tanning agent; browning agent in bakery products, an aroma enhancer, and a meat flavorer[a]
ethyl amyl ketone	Shell Chemical	alarm pheromone for myrmicine ants
ethyl n-butyl ketone	Eastman Chemical Products	high boiling solvent for air-dried and baked finishes based on nitrocellulose and vinyl resins; organosols for stable, low viscosity dispersions
ethyl n-propyl ketone	Eastman Chemical Products	solvent
isobutyl heptyl ketone	Union Carbide	dispersant for plastisols and organosols with low viscosities; solvent for organic stabilizers; chemical intermediate for the preparation (by hydrogenation) of the corresponding alcohol derivative, 2,6,8-trimethyl-4-nonanol
methyl n-amyl ketone	Eastman Chemical Products	solvent for synthetic resin finishes for metal roll coatings (imparts good blush resistance in lacquers)
methyl n-butyl ketone		medium-evaporating solvent for nitrocellulose, acrylates, vinyl, and alkyd coatings
methyl heptyl ketone	Eastman Chemical Products	solvent
methyl hexyl ketone	Union Camp	solvent
methyl isoamyl ketone	Eastman Chemical Products	solvent
methyl isopropyl ketone		solvent
methyl n-propyl ketone	Eastman Chemical Products	solvent
propiophenone	Union Carbide	chemical intermediate for fragrances, such as phenylpropanolamine hydrochloride, and propoxyphene, and pharmaceuticals, including antiallergy–decongestant–cold preparations, aspirin substitutes, and diet control formulations

[a] Ref. 275.

BIBLIOGRAPHY

"Ketones" in *ECT* 1st ed., Vol. 8, pp. 113–151, by H. J. Hagemeyer, Jr., Tennessee Eastman Co., P. R. Rector, Carbide and Carbon Chemicals Co., B. O. Blackburn, Shell Development Co., and G. A. Reynolds, Meyer

Ream, and J. A. Van Allan, Eastman Kodak Co.; "Ketones" in *ECT* 2nd ed., Vol. 12, pp. 101–169, by Arnold P. Lurie, Eastman Kodak Co.

1. *Ketones, Brochure F-41971A,* Union Carbide Corporation, New York, Sept. 1975.
2. *2,2-Diethoxyacetophenone, Product Bulletin F-44713B,* Union Carbide Corporation, New York, Dec. 1975.
3. *Eastman Inhibitor DHBP, Technical Data Sheet No. X-127,* Eastman Chemical Products, Inc., Kingsport, Tenn.
4. *Uvinu, Technical Bulletin 7543-036,* General Aniline and Film Corporation, New York.
5. *Cyansorb® UV 9 Light Absorber,* technical brochure, American Cyanamid Company, Intermediates Department, Bound Brook, N.J.
6. *1-Tetralone, Product Data Sheet F-41354,* Union Carbide Corporation, New York, Apr. 1966.
7. Unpublished physical property data, Union Carbide Corporation, Chemicals and Plastics Division, New York.
8. F. H. Stross, J. M. Monger, and H. de V. Finch, *J. Am. Chem. Soc.* **69,** 1627 (1947).
9. *2,4-Pentanedione, Product Information Brochure F-40313D,* Union Carbide Corporation, New York, Mar. 1979.
10. *Propiophenone, Product Information Brochure F-44467A,* Union Carbide Corporation, New York, Mar. 1979.
11. *Diethyl Ketone, Product Information Brochure F-47214,* Union Carbide Corporation, New York, 1979.
12. *Acetophenone, Product Information Brochure F-41646A,* Union Carbide Corporation, New York, Feb. 1968.
13. *Cyclohexanone, Product Information Brochure F-41750,* Union Carbide Corporation, New York, Dec. 1967.
14. *Methyl n-Butyl Ketone, Product Information Brochure F-43977,* Union Carbide Corporation, New York, Apr. 1972.
15. *Methyl Ethyl Ketone, Technical Publication SC:50-2,* Shell Chemical Company, New York, 1950.
16. *Organic Chemicals Brochure, Technical Publication SC:52-10,* Shell Chemical Corporation, New York, 1952.
17. S. Patai, ed., *The Chemistry of Functional Groups,* Vol. 2, Interscience Publishers, a division of John Wiley & Sons, Inc., New York, 1966.
18. A. A. Grigor'ev and co-workers, *Nauchno Issled. Inst. Sint. Spirtov Org. Prod.* **6,** 65 (1974); *Chem. Abstr.* **85,** 159318 (1976); K. Hoshiai, *Kogyo Kagaku Zasshi* **60,** 1150 (1957).
19. F. Gajewski, J. Ogonowski, and E. Roth, *Czas. Tech. M* (3), 34 (1974); Neth. Appl. 6,609,628 (Jan. 10, 1967), (to Esso Research and Engineering Co.).
20. J. B. Conant and N. Tuttle in H. Gilman, ed., *Organic Synthesis,* Collective Vol. I, John Wiley & Sons, Inc., New York, 1941, p. 199.
21. N. B. Lorette, *J. Org. Chem.* **23,** 973 (1958).
22. W. R. Sorenson and T. W. Campbell, *Preparative Methods of Polymer Chemistry,* Interscience Publishers, a division of John Wiley & Sons, Inc., New York, 1968.
23. E. E. Royals, *Advanced Organic Chemistry,* Prentice-Hall, Inc., New York, 1954, p. 650.
24. F. C. Whitmore, *Organic Reactions,* D. Van Nostrand Co., New York, p. 252.
25. *Methyl Ethyl Ketone, Technical Publication SC:50-2,* Shell Chemical Company, New York, 1950.
26. Ref. 22, p. 588.
27. F. A. Patty, *Industrial Hygiene and Toxicology,* 2nd ed., Vol. II, Interscience Publishers, a division of John Wiley & Sons, Inc., New York, 1962, pp. 1725–1749.
28. E. J. Fairchild, ed., *Registry of Toxic Effects of Chemical Substances,* Vol. II, National Institute for Occupational Safety and Health (NIOSH), U.S. Department of Health, Education and Welfare, Cincinnati, Ohio, Sept. 1977.
29. *Isophorone, Material Safety Data Sheet F-43065A,* Union Carbide Corporation, New York, July 1976.
30. Headquarters bulletin, *Air Pollution Regulations and Their Influence on Solvent Utilization in Coatings Formulations,* Union Carbide Corporation, Coatings Intermediates, New York.
31. *Solvent Selector, Bulletin F-7465T,* Union Carbide Corporation, New York, June 1973 (also *Bulletin F-7465S,* Aug. 1976).
32. Br. Pat. 586,754 (Mar. 31, 1947), G. W. Hearne, M. L. Adams, and V. W. Buls (to Shell Development Co.).

33. Brit. Pat. 665,376 (Jan. 23, 1952), (to Standard Oil Development).
34. Jpn. Pat. 43,3163 (Feb. 5, 1968), T. Miyota and co-workers (to Toyo Rayon).
35. *Chem. Eng.,* 63 (Feb. 8, 1960).
36. U.S. Pat. 2,829,165 (Apr. 1, 1958), F. Coussemant.
37. U.S. Pat. 3,196,182 (July 20, 1965), N. R. Cox (to Union Carbide Corporation).
38. U.S. Pat. 2,704,294 (Mar. 15, 1955), C. S. Morgan and co-workers (to Celanese Corporation).
39. J. B. Saunby and B. W. Kiff, *Hydrocarbon Process.* **55,** 247 (Nov. 1976).
40. U.S. Pat. 3,236,897 (Feb. 22, 1966), L. Hoernig and co-workers (to Hoechst).
41. U.S. Pat. 3,215,743 (Nov. 2, 1965), W. Riemenschneider (to Hoechst).
42. U.S. Pat. 3,247,084 (Apr. 19, 1966), C. H. Worsham (to Esso Research and Engineering).
43. Neth. Pat. 68 17,336 (June 8, 1970), (to Maruzen Oil).
44. R. P. Lowry and A. Aguilo, *Hydrocarbon Process.* **44,** 103 (Nov. 1973).
45. USSR Pat. 472,928 (June 5, 1975), A. Grigorev.
46. C. Walling and E. S. Huyser in A. C. Cope, ed., *Organic Reactions,* Vol. 13, John Wiley & Sons, Inc., New York, 1963, Chapt. 3, p. 91.
47. U.S. Pat. 3,151,167 (Sept. 29, 1964), J. L. Eisenmann and co-workers (to Diamond Alkali).
48. J. L. Eisenmann, *J. Org. Chem.* **27,** 2706 (1962).
49. U.S. Pat. 3,384,668 (May 21, 1968), F. C. Canter (to Eastman Kodak).
50. Brit. Pat. 1,049,990 (Nov. 30, 1966), F. C. Canter (to Eastman Kodak).
51. *Chem. Eng. News,* 10 (Aug. 8, 1976).
52. *Chem. Mark. Rep.,* 16 (Oct. 3, 1977).
53. *Chem. Mark. Rep.,* 9 (Jan. 14, 1974).
54. *Chem. Mark. Rep.,* 9 (Dec. 27, 1976).
55. *Chem. Mark. Rep.,* 3 (July 4, 1977).
56. *Chem. Mark, Rep.,* 3, 36 (Aug. 28, 1978).
57. *Chem. Mark. Rep.,* (Apr. 1, 1977).
58. P. W. Sherwood, *Erdol Kohle* **8**(12), 884 (1955).
59. Neth. Appl. 6,503,470 (Sept. 19, 1966), (to Shell Internationale Research Maatschappij N.V.).
60. Ger. Pat. 1,243,665 (July 6, 1967), A. Heykoop and F. A. Van Dijk (to Shell Internationale Research Maatschappij N.V.).
61. U.S. Pat. 2,701,264 (Oct. 1, 1951), T. J. Deahl (to Shell Chemical Company).
62. U.S. Pat. 2,686,813 (May 8, 1952), L. M. Peters (to Shell Chemical Company).
63. U.S. Pat. 2,393,532 (Nov. 27, 1942), G. W. Hearne (to Shell Chemical Company).
64. Brit. Pat. 574,446 (Jan. 7, 1946), C. Weizmann.
65. K. Hoshiai, *Kogyo Kagaku Zasshi* **60,** 1150 (1957).
66. K. P. Grinevich and V. A. Zaitsev, *Khim. Prom.,* 276 (1958).
67. Neth. Appl. 6,403,176 (Oct. 28, 1964), (to Scholven Chemie A.G.).
68. Fr. Pat. 1,478,704 (Apr. 28, 1967), B. W. Turnquest and O. H. Thomas (to Sinclair Research, Inc.).
69. Brit. Pat. 1,116,037 (June 6, 1968), L. J. Sirois, F. J. Herrmann, and M. E. Oldweiler (to Esso Research and Engineering Co.).
70. Neth. Appl. 6,609,628 (Jan. 10, 1967), (to Esso Research and Engineering Co.).
71. F. Gajewski, J. Ogonowski, and E. Roth, *Czas. Tech. M* (3), 34 (1974).
72. A. A. Grigor'ev and co-workers, *Nauchno-Issled. Inst. Sint. Spirtov Org. Prod.* **6,** 65 (1974); *Chem. Abstr.* **85,** 159318 (1976).
73. Brit. Pat. 574,446 (Jan. 7, 1946), R. N. Lacey.
74. Jpn. Kokai 72 15,810 (May 11, 1972), K. Takagi and K. Manabe (to Sumitomo Chemical Co., Ltd.).
75. Jpn. Kokai 72 15,809 (May 11, 1972), K. Takagi and K. Manabe (to Sumitomo Chemical Co., Ltd.).
76. A. Pirvulescu, M. Panaitescu, and T. Bota, *Rev. Chim.* **12,** 377 (1961).
77. Jpn. Kokai 72 15,808 (May 11, 1972), K. Tayagi and K. Manabe (to Sumitomo Chemical Co., Ltd.).
78. U.S. Pat. 3,361,822 (Jan. 2, 1968), K. Schmitt, J. Disteldorf, H. Schnurbusch, and W. Hilt (to Scholven-Chemie A.G.).
79. K. Schmitt, *Chem. Ind. (Duesseldorf)* **18**(4), 204 (1966).
80. U.S. Pat. 2,499,172 (Feb. 28, 1950), E. F. Smith (to Commercial Solvents).
81. Brit. Pat. 1,015,003 (Dec. 31, 1965), (to Distillers).
82. Jpn. Pat. 46-2009 (Jan. 19, 1971), (to Showa Denko).

83. U.S. Pat. 3,953,517 (Apr. 27, 1976), K. Schmitt, J. Disteldorf, W. Flakus, and W. Hubel (to Veba-Chemie A.G.).
84. Ger. Pat. 1,238,453 (Apr. 13, 1967), H. Giehring (to Rheinpreussen A.G.).
85. Ger. Pat. 1,260,454 (Feb. 8, 1968), (to Rheinpressen A.G.).
86. Y. Onoue, Y. Mizutani, S. Akiyama, Y. Izumi, and Y. Watanabe, *Chem. Tech.*, 36 (Jan. 1977).
87. USSR Pat. 445,263 (Mar. 25, 1978), T. M. Khanonov, A. G. Akhmetov, R. M. Masagutor, and G. N. Kivichenko.
88. Jpn. Pat. 52-35646 (Sept. 10, 1977), T. Kiyoura, T. Takahashi, and Y. Hayashi (to Mitsui-Toatsu Chemical Company).
89. *Chem. Eng.*, 108 (May 6, 1968).
90. U.S. Pat. 3,047,630 (July 7, 1958), L. E. Addy (to British Hydrocarbon Chemicals Ltd.).
91. Jpn. Pat. 34-517 (Feb. 9, 1959), R. Fujii and co-workers (to Kyowa Hakko Kogyo).
92. U.S. Pat. 2,891,095 (Feb. 25, 1954), W. Opitz (to Knapsack-Griescheim A.G.).
93. Brit. Pat. 868,024 (July 7, 1958), L. E. Addy (to British Hydrocarbon Chemicals Ltd.).
94. Can. Pat. 646,643 (Aug. 8 and Dec. 18, 1957), A. McLean (to British Hydrocarbon Chemicals Ltd.).
95. T. Kanamori, *Kogo Kagaku Zasshi* **61,** 1576 (1958).
96. *Ibid.*, p. 1578.
97. Brit. Pat. 1,024,684 (Mar. 30, 1966),(to Distillers).
98. *Chem. Mark. Rep.*, 9 (Dec. 12, 1977).
99. *Chem. News.*, 1 (Feb. 1975).
100. *Chem. Mark. Rep.*, 26 (Aug. 21, 1978).
101. *Industrial Hygiene Bulletin, Safety and Toxicity Data Sheet IC:69-10,* Shell Chemical Company, New York, Mar. 1969.
102. *Chem. Mark. Rep.*, (Apr. 1, 1978).
103. G. T. Austin, *Chem. Eng.*, 149 (June 24, 1974).
104. *Chem. Mark. Rep.*, (July 1, 1975).
105. S. K. Datta, *Indian Chem. J.* **9**(8), 535 (1975).
106. Ger. Pat. 2,223,299 (Nov. 23, 1971), C. Neri and E. Perrotti.
107. R. Bianchi, C. Neri, and E. Perrotti, *Ann. Chim.* **65**(1–2), 45 (1975).
108. M. V. Artemenko and K. F. Slyusarenko, *Ukr. Khim. Zh.* **41,** 806 (1975).
109. J. Kulesza and co-workers, *Int. Congr. Essent. Oils, 6th,* 71 (1974); *Chem. Abstr.* **84,** 135827e (1976).
110. *Glycols, Brochure F-41515A,* Union Carbide Corporation, New York, July 1971.
111. E. C. Craven, *J. Appl. Chem.* **13,** 71 (1963).
112. M. Bourdiol and co-workers, *Bull. Soc. Chim. Fr.* **8,** 375 (1941).
113. U.S. Pat. 2,889,369 (June 6, 1959), P. Godet (to Usines de Melle).
114. Ger. Pat. 1,052,970 (Mar. 19, 1959), K. Schmitt and J. Disteldorf (to Bergwerksgesellschaft Hibernia A.G.).
115. A. A. Grigor'ev and co-workers, *Nauchno-Issled Inst. Sint. Spirtov Org. Prod.* **6,** 65 (1974).
116. Jpn Kokai 58 456 (Jan. 30, 1958), S. Yamada, S. Noguchi, and S. Tago (to Nippon Oil Company).
117. J. Herscovici, T. Bota, and D. Sireteanu, *Rev. Chim. (Bucharest)* **15**(12), 736 (1964).
118. U.S. Pat. 1,550,792 (Aug. 25, 1926), W. J. Edmonds.
119. Brit. Pat. 504,337 (Apr. 24, 1939), S. H. McAllister and E. F. Bullard (to N. V. de Bataafsche Petroleum Maatschappi (Shell)).
120. B. P. Ershov and V. L. Pridoragin, *J. Appl. Chem.* (USSR) **19,** 38 (1946).
121. Ger. Pat. 800,661 (Nov. 27, 1950), W. Schlenk (to Badische Anilin- und Soda-Fabrik).
122. I. Ichikizaki, N. Kawamura, and T. Hoshino, *Chem. High Polym.* **5,** 75 (1948).
123. Jpn. Kokai 72 18,731 (Nov. 5, 1968), S. Hattori and co-workers (to Mitsubishi Chem. Ind. Company, Ltd.).
124. Jpn. Kokai 72 18,730 (Nov. 5, 1968), S. Hattori and co-workers (to Mitsubishi Chem. Ind. Company, Ltd.).
125. USSR Pat. 110,060 (June 25, 1958), S. A. Levina and N. F. Ermolenko.
126. S. A. Levina and N. F. Ermolenko, *Zh. Obshch. Khim.* **29,** 1920 (1959); *Chem. Abstr.* **54,** 8251e (1960).
127. C. J. Schmidle and R. C. Mansfield, *J. Polym. Sci.* **8,** 313 (1952).
128. Fr. Pat. 1,023,805 (Mar. 24, 1953), G. V. Austerweil (to Campagnie de Produits Chimques).
129. Z. N. Verkhovskaya and co-workers, *Khim. Prom.* **43**(7), 500 (1967).

130. Brit. Pat. 1,527,033 (Oct. 4, 1978), C. Hawkins and B. Yeomans (to B. P. Chemicals, Ltd.).
131. J. Przondo and co-workers, *Przem. Chem.* **51**(8), 504 (1972).
132. U.S. Pat. 1,714,378 (May 21, 1929), A. Knorr and A. Weissenborn (to Winthrop Chem. Company).
133. Brit. Pat. 881,918 (Nov. 8, 1961), H. D. Burgess (to Union Carbide Corporation).
134. D. S. Noyce and W. L. Reed, *J. Am. Chem. Soc.* **81,** 624 (1959).
135. E. E. Connolly, *J. Chem. Soc.,* 338 (1944).
136. P. W. Sherwood, *Pet. Ref.* **33**(12), 144 (1954).
137. S. Kudo and co-workers, *Chem. Eng. Jpn.* **30,** 1124 (1966).
138. Jpn. Pat. 73 72,110 (Sept. 29, 1973), S. Kuwata and T. Uchida (to Mitsubishi Chemical Industries Co., Ltd.).
139. Jpn. Pat. 73 52,715 (July 24, 1973), Y. Moriyasu and co-workers (to Mitsubishi Chemical Industries Co., Ltd.).
140. U.S. Pat. 3,137,732 (June 16, 1964), D. G. Kuper (to Shell Oil Company).
141. *Ketones,* brochure, Union Carbide Chemical Co., New York, 1959.
142. M. Hauser, *Chem. Rev.* **63,** 311 (1963).
143. A. Rieche, E. Schmitz, and E. Grunderman, *Ber.* **93,** 2443 (1960).
144. U.S. Pat. 2,566,792 (May 25, 1948), H. Dannenberg (to Shell).
145. U.S. Pat. 2,444,006 (Mar. 13, 1945), H. Dannenberg (to Shell).
146. H. N. A. Al-Jallo and E. S. Waight, *J. Chem. Soc. B,* 75 (1966).
147. C. R. Noller, *Chemistry of Organic Compounds,* 3rd ed., W. B. Saunders Company, Philadelphia, Pa., 1965.
148. J. C. Aumiller and J. A. Whittle, *J. Org. Chem.* **41,** 2959 (1976).
149. D. S. Noyce and M. Evett, *J. Org. Chem.* **37,** 397 (1972).
150. J. Przondo and E. Bielous, *Przem. Chem.* **53**(9), 541 (1974).
151. U.S. Pat. 2,827,490 (Mar. 18, 1958), A. W. Martin (to Celanese).
152. Romanian Pat. 44,576 (Dec. 13, 1966), J. Herscovici and co-workers (to Romania Ministry of the Chemical Industry).
153. U.S. Pat. 1,913,159 (Dec. 23, 1929), H. Guinot (to Usiner de Melle).
154. U.S. Pat. 3,946,079 (Mar. 23, 1976), Y. Mitzutani, Y. Izumi, and Y. Watanabe (to Tokuyama Soda).
155. A. A. Shaumarova and co-workers, *Deposited Doc., VINITI,* 2370 (1976); *Chem. Abstr.* **89,** 108004f (1978).
156. F. G. Klein and J. T. Banchero, *Ind. Eng. Chem.* **48,** 1278 (1956).
157. H. J. Seebald and W. Schumack, *Arch. Pharm.* (*Weinheim, Ger.*) **305**(6), 406 (1972).
158. Brit. Pat. 1,116,037 (June 6, 1968), L. J. Sirois (to Esso Research and Engineering).
159. U.S. Pat. 3,153,068 (Oct. 13, 1964), W. J. Porter and co-workers (to Esso Research and Engineering).
160. J. Wiemann, B. Furth, and G. Dana, *Compt. Rend.* **250,** 3674 (1960).
161. U.S. Pat. 3,155,730 (Nov. 3, 1964), J. Eng (to Esso Research and Engineering).
162. U.S. Pat. 2,451,350 (Oct. 12, 1948), H. O. Mottern and co-workers (to Standard Oil Development Company).
163. Can. Pat. 769,084 (Oct. 10, 1967), G. Kohan and co-workers (to Shawinigan Chemicals).
164. *Chem. Mark. Rep.,* 1 (Apr. 1, 1975).
165. S. Ito, *Yokohama Igaku* **20,** 253 (1969).
166. *Mesityl Oxide Industrial Hygiene Bulletin, Safety Data Sheet No. SC:57-105R,* Shell Chemical Company, New York.
167. U.S. Pat. 2,317,663 (Feb. 18, 1939), C. C. Allen (to Shell).
168. U.S. Pat. 2,288,589 (Dec. 4, 1939), F. A. Bent (to Shell).
169. U.S. Pat. 2,395,453 (Feb. 26, 1944), H. A. Bruson (to Rohm and Haas Company).
170. U.S. Pat. 2,309,650 (Feb. 12, 1940), S. H. McAllister (to Shell).
171. S. M. Khopkar, *J. Sci. Ind. Res.* **24,** 242 (May 1965).
172. R. L. Wasson and H. O. House, *Org. Syn.* **37,** 58 (1957).
173. E. C. Craven, *J. Appl. Chem.* **12,** 120 (1962).
174. U.S. Pat. 4,005,145 (Jan. 25, 1977), E. Widmer (to Hoffmann-La Roche, Inc.).
175. M. S. Kharasch and P. O. Tawney, *J. Am. Chem. Soc.* **67,** 128 (1945).
176. M. S. Kharasch and P. O. Tawney, *J. Am. Chem. Soc.* **63,** 2308 (1941).
177. J. R. A. Pollock and R. Stevens, *Dictionary of Organic Compounds,* Vol. 5, Oxford University Press, New York, 1965, p. 3168.

178. *Isophorone Technical Information Bulletin,* Veba-Chemie A.G., Department Di 127, P.O. Box 45, D4660 Gelsenkirchen-Buer, FRG; U.S. distributor is Thorson Chemical Corporation, 299 Park Avenue, New York.
179. U.S. Pat. 4,086,282 (Apr. 25, 1978), F. Wattimena (to Shell Oil Company).
180. W. Ger. Pat. 2,804,114 (Aug. 3, 1978), L. Huizer and co-workers (to Shell Internationale Research Maatschappij B.V.).
181. U.S. Pat. 4,046, 813 (Sept. 6, 1977), W. Brenner (to Hoffmann-La Roche Inc.).
182. U.S. Pat. 3,960,966 (June 1, 1976), (to Hoffmann-La Roche Inc.).
183. U.S. Pat. 2,373,510 (June 23, 1941), W. A. Bailey, Jr. (to Shell).
184. U.S. Pat. 2,419,142 (June 29, 1944), V. N. Ipatieff (to Universal Oil Products Company).
185. U.S. Pat. 2,183,127 (Mar. 19, 1938), T. H. Vaughn (to Union Carbide Corporation).
186. U.S. Pat. 3,337,633 (Aug. 22, 1967), K. Schmitt, J. Disteldorf, and W. Baron (to Hibernia Chemie G.m.b.H.).
187. U.S. Pat. 3,981,918 (Sept. 21, 1976), J. R. Walton and B. Yeomans (to B. P. Chemicals International, Ltd.).
188. U.S. Pat. 4,059,632 (Nov. 22, 1977), C. Cane and B. Yeomans (to B. P. Chemicals International, Ltd.).
189. U.S. Pat. 3,337,423 (Aug. 22, 1967), K. Schmitt, J. Disteldorf, and W. Baron (to Hibernia Chemie G.m.b.H.).
190. U.S. Pat. 2,419,051 (Dec. 5, 1944), S. A. Ballard (to Shell).
191. E. C. Craven and W. R. Ward, *J. Appl. Chem.* **10,** 18 (1960).
192. H.-G. Franck and co-workers, *Liebigs Ann. Chem.* **724,** 94 (1969).
193. A. Roger, J. J. Godfroid, and J. Wiemann, *Bull. Soc. Chim. Fr.* **6,** 2590 (1968).
194. E. Cyrot, *Ann. Chim.* **6,** 413 (1971).
195. W. T. Reichle and co-workers, *J. Org. Chem.* **42,** 1600 (1977).
196. Z. W. Wolkowski, *Tetrahedron Lett.* (12), 821 (1971).
197. E. Cyrot, N. Thoai, and J. Wiemann, *Tetrahedron Lett.* (1), 83 (1970).
198. U.S. Pat. 2,968,677 (Nov. 12, 1957), M. W. Fewlass (to Distillers Co., Ltd.).
199. Brit. Pat. 832,124 (Apr. 6, 1960), E. C. Craven and M. W. Fewlass (to Distillers Co., Ltd.).
200. D. Lloyd, *Alicylic Compounds,* American Publishing Company, Inc., New York, 1963, p. 68.
201. S. Coffey, ed., *Chemistry of Carbon Compounds,* 2nd ed., Vol. I, Part B, Elsevier Publishing Company, New York, 1968, p. 409.
202. G. H. Whitham, *Alicyclic Chemistry,* Oldbourne Press, London, Eng., 1963, p. 7.
203. *Catalog Handbook of Fine Chemicals,* Catalog 20, Aldrich Chemical Company, Inc., Milwaukee, Wisc., 1981–1982.
204. A. Gero, *J. Org. Chem.* **26,** 3156 (1961).
205. T. Fiyita, S. Watanabe, K. Suga, and T. Inaba, *J. Chem. Tech. Biotechnol.* **29,** 100 (1979).
206. Ref. 201, p. 406.
207. *Chem. Mark. Rep.,* 11 (May 21, 1973).
208. *Chem. Week,* 19 (Apr. 24, 1974).
209. *Hydrocarbon Process.* **52,** 118 (Nov. 1973).
210. J. V. Koleske and L. H. Wartman, *Poly(Vinyl Chloride),* Gordon and Breach Science Publishers, New York, 1969, pp. 24–25.
211. Y. Ogata and M. Yamashita, *Tetrahedron* **27,** 3395 (1971).
212. K. Kulka, *Am. Perfum.* **76**(11), 23 (1961).
213. M. A. Gianturco and P. Friedel, *Tetrahedron* **19,** 2039 (1963).
214. M. A. Gianturco, A. S. Giammarino, and R. G. Pitcher, *Tetrahedron* **19,** 2051 (1963).
215. M. Stoll, M. Winter, F. Gautschi, I. Flament, and B. Willhalm, *Helv. Chim. Acta* **50,** 628 (1967).
216. S. B. Bowlus and J. A. Katzenellenbogen, *J. Org. Chem.* **39,** 3309 (1974).
217. U.S. Pat. 2,400,454 (May 14, 1946), N. P. Christensen and R. E. Sturhan.
218. A. Steigman, *J. Soc. Chem. Ind.* **65,** 233 (1940).
219. S. G. Cohen, B. E. Ostberg, D. B. Sparrow, and E. R. Blout, *J. Polym. Sci.* **3,** 264 (1948); U.S. Pat. 2,413,973 (Jan. 7, 1947) B. Howk and R. Jacobson (to E. I. du Pont de Nemours & Co., Inc.).
220. U.S. Pat. 2,062,263 (Nov. 24, 1937), E. Eberhardt (to I. G. Farbenindustrie).
221. U.S. Pat. 2,455,631 (Dec. 7, 1948), O. Weinkauff (to Monsanto Chemical Co.).
222. U.S. Pat. 1,899,094 (Feb. 28, 1933), A. J. Kluyver and M. A. Scheffer (to Thomas H. Verhane); H. Soumalainen and L. Jannes, *Nature* **157,** 335 (1946).
223. H. Lagoni, *Zentralbl. Bakteriol. Parasitenkol. Infektionskr. Hyg. Abt. 2* **103,** 225 (1941).

224. Indian Pat. 110,625 (July 12, 1969), R. Banerjee.
225. U.S. Pat. 2,377,749 (June 5, 1945), C. A. Bordner (to E. I. du Pont de Nemours & Co., Inc.).
226. U.S. Pat. 2,658,920 (Nov. 10, 1953), D. X. Klein, T. A. Girard, and H. R. Johnson (to Heyden Chemical Corp.).
227. H. H. Hatt, A. Pilgrim, and W. J. Curran, *J. Chem. Soc.*, 93 (1936).
228. F. M. Snyder and F. A. Morton, *J. Econ. Entomol.* **39,** 385 (1946).
229. K. Kulka, *Am. Perfum.* **78**(9), 27 (1963).
230. W. M. Muir, P. D. Ritchie, and D. J. Lyman, *J. Org. Chem.* **31,** 3790 (1966).
231. D. C. Nonhebel and H. D. Murdoch, *J. Chem. Soc.*, 2153 (1962).
232. C. Weygand and H. Baumgartel, *Ber.* **62B,** 574 (1929).
233. Ref. 203, Vol. I, Part D, p. 68.
234. M. Marcantonator, G. Camba, and D. Mannier, *Anal. Chim. Acta* **67,** 220 (1973).
235. J. M. Haigh and D. A. Thornton, *J. Inorg. Nucl. Chem.* **33,** 1787 (1971).
236. Neth. Appl. 6,606,707 (Nov. 22, 1966), (to Carlisle Chemical Works).
237. U.S. Pat. 3,581,076 (June 8, 1971), E. Chetakian (to Norac Co., Inc.).
238. U.S. Pat. 3,314,834 (Apr. 18, 1967), E. J. Walden and J. K. West (to United Aircraft Corp.).
239. U.S. Pat. 3,506,620 (Apr. 14, 1970), B. Davis and D. L. Nealy (to Eastman Kodak Co.).
240. Jpn. Pat. 69 31,832 (Dec. 18, 1969), T. Sarai, H. Miyashita, and Y. Sano (to Takeda Chemical Industries, Ltd.).
241. U.S. Pat. 3,499,869 (Mar. 10, 1970), G. R. Lappin and G. C. Newland (to Eastman Kodak Co.).
242. *Aldrichimica Acta,* **10**(1), 19 (1977).
243. M. L. Eidinoff, *J. Am. Chem. Soc.* **67,** 2072 (1945).
244. U.S. Pat. 2,395,800 (Mar. 5, 1946), A. B. Boese, Jr., and F. G. Young, Jr. (to Union Carbide Corp.).
245. U.S. Pat. 3,794,686 (1974), H. Spes and G. Kunstle (to Wacker-Chemie).
246. U.S. Pat. 2,787,642 (April 2, 1957), E. Enk and F. Buttner (to Wacker-Chemie).
247. K. K. Georgieff, *Ind. Eng. Chem.* **49,** 1067 (1957).
248. Brit. Pat. 681,696 (1952), R. Decker and H. Holz.
249. Brit. Pat. 741,563 (1955), (to Wacker-Chemie).
250. Ger. Pat. 1,068,236 (1959), A. Stocker (to Lonza).
251. C. E. Denoon, *Org. Syn.* **20,** 6 (1940).
252. U.S. Pat. 2,395,012 (Feb. 19, 1946), W. H. Reeder and G. A. Lescisin (to Union Carbide Corp.).
253. Brit. Pat. 1,213,255 (Nov. 25, 1970), (to Wacker-Chemie).
254. Brit. Pat. 1,213,253 (Nov. 25, 1970), (to Wacker-Chemie).
255. U.S. Pat. 2,432,499 (Dec. 16, 1947), A. B. Boese, Jr. (to Union Carbide Corp.).
256. *2,4-Pentanedione, Material Safety Data Sheet F-43569, 2/72-5M,* Union Carbide Corporation, New York, Feb. 1972.
257. M. S. Kharasch, H. C. McBay, and W. H. Urry, *J. Am. Chem. Soc.* **70,** 1269 (1948).
258. Y. Kobayashi, T. Taguchi, and E. Tokuno, *Tetrahedron Lett.* (42), 3741 (1977).
259. Y. Ito, T. Konoike, T. Harada, and T. Saegusa, *J. Am. Chem. Soc.* **99,** 1487 (1977).
260. J. E. McMurry and T. E. Glass, *Tetrahedron Lett.* (27), 2575 (1971).
261. U.S. Pat. 3,381,691 (May 7, 1968), J. N. Schumacher and D. L. Roberts (to R. J. Reynolds Tobacco Co.).
262. F. B. Wells and C. F. H. Allen, *Org. Syn.* **Coll. Vol. II,** 219 (1943).
263. E. Browning, *Toxicity and Metabolism of Industrial Solvents,* Elsevier Publishing Co., New York, 1965, p. 446.
264. Ref. 201, p. 73.
265. M. Larcheveque, G. Valette, and T. Cuvigny, *Synthesis* (6), 424 (1977).
266. U.S. Pat. 2,495,286 (Jan. 24, 1950), M. M. Brubaker (to E. I. du Pont de Nemours & Co., Inc.).
267. U.S. Pat. 2,517,684 (Aug. 8, 1950), E. C. Ladd (to U.S. Rubber Co.).
268. J. E. Guillet, *Polymers and Ecological Problems,* Plenum Publishing Corporation, New York, 1973.
269. I. Kuwajima, T. Sato, M. Arai, and N. Minami, *Tetrahedron Lett.* (21), 1817 (1976).
270. E. E. Royals, *Advanced Organic Chemistry,* Prentice-Hall, Inc., New York, 1954, pp. 573–574.
271. U.S. Pat. 2,661,771 (Jan. 19, 1954), A. C. Zettlemoyer and W. C. Walker (to Food Machinery and Chemical Corp.).
272. I. Kh. Fel'dman, N. N. Bel'tsova, and A. A. Ginesina, *Zh. Priklad. Khim.* **35,** 1364 (1962).
273. U.S. Pat. 4,022,837 (July 24, 1970), R. A. Schneider (to Chevron Research Co.).
274. U.S. Pat. 4,152,354 (May 1, 1979), P. R. Stapp (to Phillips Petroleum Company).

275. E. S. Krogauz and A. M. Berlin, *Russ. Chem. Rev.* **38**, 655 (1969).

ANTHONY J. PAPA
PAUL D. SHERMAN, JR.
Union Carbide Corporation

KRYPTON. See Helium-group gases.

L

LAMINATED AND REINFORCED METALS

The desire to satisfy diverse and competing design requirements in a cost-effective manner has been coupled with the drive for energy conservation as a result of the energy shortage and increasing energy costs, and the need to develop cost-effective alternatives for critical materials.

Metallic laminates and fiber-reinforced metals are serious contenders for structural applications. In this article, these laminates are called metallic-matrix laminates (MMLs). MMLs are relatively expensive at this time (1980) because fibers, eg, boron fibers, are expensive and the manufacturing processes are time consuming and costly. They have been used mainly in aerospace structures. However, MMLs will find more extensive use as energy and other design considerations, including scarcity of critical materials, override the material costs.

A large body of information has been generated about MMLs over the last fifteen years. Significant developments of MMLs are reported in the proceedings of the Society for the Advancement of Materials and Processing Engineering (SAMPE). These proceedings include papers that are presented at the two SAMPE annual meetings (spring and fall). Three recently compiled bibliographies with abstracts (1–3) cover technical articles and government reports that have been published since about 1965. An extensive review of metal-matrix composites covering developments up to 1972 is in ref. 4. Boron-fiber reinforced aluminum (boron–aluminum composites) and graphite-fiber reinforced aluminum (graphite–aluminum composites) are reviewed in ref. 5. The present article is a selective review of the state-of-the-art of MMLs, it describes only material that is not directly defense-related. The emphasis of the review is on design–analysis procedures for structural components that have been made from MMLs (see Composite Materials).

Definitions of Selected Laminates

The types of MMLs that are reviewed herein include those made from fiber-reinforced metals, superhybrids, and those made from layers of different metals. Fiber-reinforced metals consist of unidirectional-fiber-composite (UFC) laminates, as depicted schematically in Figure 1**a**, and angleplied (APL) laminates, Figure 1**b.** In both UFC and APL laminates, metallic foils may be used between plies to enhance certain mechanical properties as is discussed later. Superhybrid composites (SHC) consist of outer metallic foils, boron–aluminum plies (B–Al), graphite-fiber–resin (UFC) inner or core plies, and adhesive film between these as shown in the photomicrograph in Figure 1**c.** Metallic laminates consist usually of alternate layers from two or more metals as depicted schematically in Figure 1**d.** The various procedures that are used to fabricate these laminates are described below. The combinations of materials that are used to make these laminates also are described below.

The basic unit used to study, design and fabricate UFC laminates is the single layer (ply, monolayer, lamina) which consists of stiff, strong fibers embedded in a metal matrix. The fibers and the matrix are generally called the constituent materials, or constituents, of the laminate (composite). Various constituents that have been used to make UFC are summarized alphabetically under the heading fiber-reinforced-metal laminates (first two columns, Table 1). A large number of constituent materials are used for both fibers and matrices. The materials for fibers range from alumina to whiskers. Those for matrices range from aluminum to superalloy. The constituents used thus far for SHC are those summarized under superhybrids in Table 1. The constituents that have been used for metallic laminates are summarized also in Table 1. An extensive list of constituent combinations for metallic laminates is tabulated in ref. 6, and constituent combinations for metallic laminates made by explosive bonding are tabulated in ref. 7.

Mechanical and physical properties of constituent fiber reinforcements for MMLs are summarized in Table 2. Corresponding properties for metal matrices and metallic constituents for MMLs are summarized in Table 3.

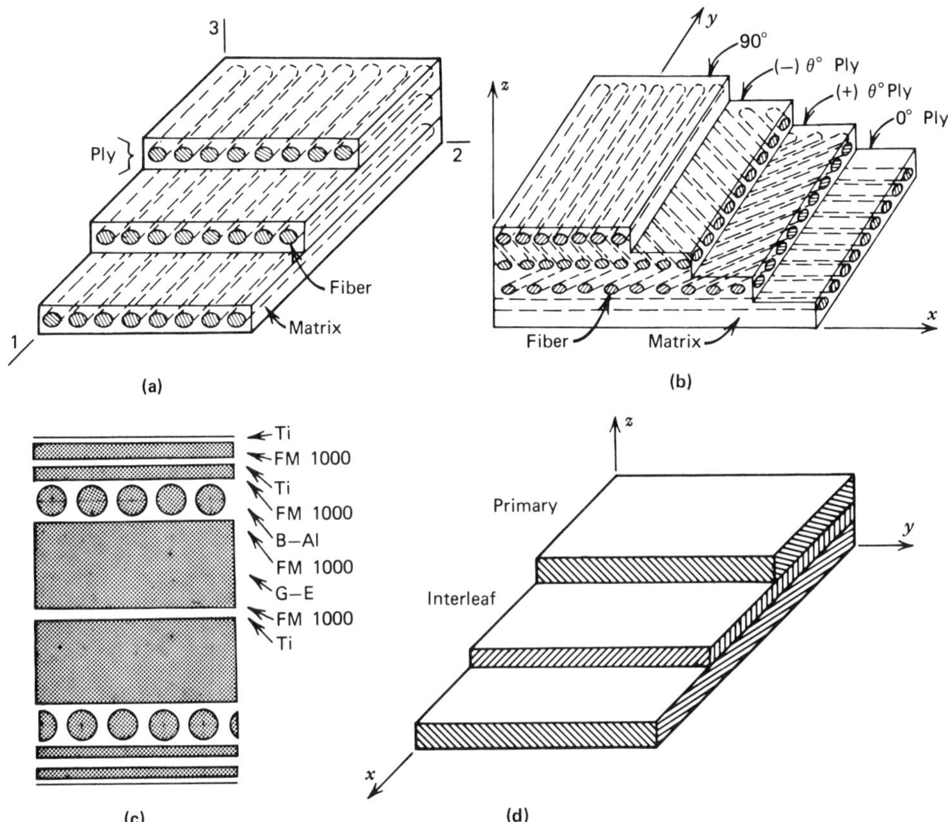

Figure 1. General types of metal-matrix laminates: (**a**) unidirectional fiber composite (UFC); (**b**) angleplied laminate (APL); (**c**) superhybrid composite (SCH); (**d**) metal–metal laminate (MML).

Hybrids

A general definition for a hybrid composite is a composite that combines two or more different types of fibers in the same matrix, or one fiber type in two different matrices or combinations of these (ICCM, International Conference on Composite Materials) (10). Superhybrids (Fig. 1**c**) are a generic class of composites that combine appropriate properties of fiber–metal-matrix composites, fiber–resin-matrix composites and/or metallic plies in a predetermined manner in order to meet competing and diverse design requirements (11–12). Tiber hybrids are trivial names for titanium–beryllium adhesively bonded metallic laminates (13).

Boron-fiber-reinforced 1100, 2024, 5052, or 6061 aluminum alloys have been investigated for use in fan blades for aircraft engines. Different diameter boron fibers (142 and 203 μm) may be included in the same hybrid. The high impact resistance of the 203-μm dia boron fiber in the 1100 aluminum alloy matrix (Fig. 2) is combined with the high transverse and shear properties of the 142- or 203-μm dia boron fiber in either a 2024, 5052, or 6061 aluminum alloy matrix. Fan blades made from some of these hybrids and subjected to the impact of a small bird (85 g) are shown in Figure 3. The advantages and disadvantages of these types of hybrids are described in ref. 14, and their use for fan blades is described in ref. 15.

Table 1. Summary of Constituent Materials for Metallic Laminates

Fiber-reinforced metal laminates		Superhybrids			
		Metal-matrix composite		Metal–metal laminates	
Fiber	Metal	Fiber	Matrix	Primary	Interleaf
alumina (FP)	aluminum	boron	aluminum	aluminum	aluminum
	lead				
	magnesium			beryllium	titanium
beryllium	titanium	Resin-matrix composite			
boron	aluminum	Fiber	Matrix	steel	steel
	magnesium	graphite	epoxy		
	titanium		polyimide	titanium	aluminum
					titanium
borsic	aluminum	kevlar	epoxy		
	titanium			tungsten	copper
		S-glass	epoxy		superalloy
graphite	aluminum				
	niobium			tungsten	tantalum
	copper				
	lead	Metal foil			
	magnesium	titanium			
	nickel				
	tin				
	zinc				
molybdenum	superalloy				
silicon carbide	aluminum				
	superalloy				
	titanium				
steel	aluminum				
	nickel				
tantalum	superalloy				
tungsten	niobium				
	superalloy				

Superhybrids have been developed primarily for use in fan blades of high-bypass-ratio turbojet engines. These types of superhybrids generally have: longitudinal strengths and stiffness comparable to advanced fiber composites; transverse flexural strength comparable to titanium; impact resistance comparable to aluminum; transverse and shear stiffness comparable to aluminum; and density comparable to glass-fiber–resin composites (11). In addition, superhybrids are notch insensitive and do not degrade when subjected to thermal fatigue (12). Impact-resistance data of superhybrids, other fiber composites, and some metals are summarized in Table 4. The high velocity-impact resistance of superhybrid wedge-type cantilever specimens relative to other composites and titanium is shown in Figure 4 (15). Large fan blades made by using superhybrids-shell–titanium-spar (either leading edge or center) have been described (16).

Experimental data generated at the NASA Lewis Research Center showed that tiber hybrids can be made which have: moduli equal to that of steel; tensile fracture stress comparable to the yield strength of titanium; flexural fracture stress comparable to the ultimate strength of titanium; and density comparable to aluminum (13). The relatively high stiffness of tiber hybrids and their relatively low density compared to

Table 2. Typical Properties of Constituent Fiber Reinforcements for Metal–Metal Laminates (Along the Fiber)[a]

Fiber	Density, g/cm^3 [b]	Mp, °C	Heat capacity, kJ/(kg·K)[c]	Thermal cond., W/(m·K)[d]	Coeff. of thermal exp., $10^{-6}/°C$	Tensile strength, MPa[e]	Modulus, GPa[e]	Dia, μm	Remarks
boron on tungsten	2.49	2100	1.3	38	5.0	3,620	400	102–203	monofilament
borsic	2.71	2100	1.3	38	5.0	3,100	400	102–203	monofilament
boron on carbon	2.21	2100	1.3	38	5.0	3,450	360	102–203	monofilament
graphite									
PAN HM	1.86	3650	0.7	1003	−1.1	2,210	380	7	10,000 filaments per ton
PAN HTS (T300)	1.74	3650	0.7	1003	−1.1	2,340	210	8	3000 filaments per yarn
rayon (T50)	1.66	3650	0.7	1003	−1.1	2,170	390	6	1440 filaments per 2 ply yarn
Thornel 75 (T75)	1.83	3650	0.7	1003	−1.1	2,660	520	5	1440 filaments per 2 ply yarn
pitch (type P)	1.99	3650	0.7	1003	−1.1	1,380	340	5–10	2000 filaments per yarn
pitch UHM	2.05	3650	0.7	1003	−1.1	2,410	690	11	2000 filaments per yarn
silicon carbide on tungsten	3.32	2690	'1.2	16	4.3	3,100	430	102–203	monofilament
silicon carbide on carbon	3.04	2690	1.2	16	4.3	3,450	400	102	monofilament
beryllium	1.86	1280	1.9	150	11.5	970	290	127	monofilament
alumina (FP)	3.96	2040			8.3	1,520	380	20	210 filaments per yarn
glass-S	2.49	840	0.7	13	5.0	4,140	80	9	1000 filament per strand
-E	2.49	840	0.7	13	5.0	2,700	70	9	1000 filaments per strand
molybdenum	1.02	2620	0.3	145	4.9	660	320	127	monofilament
steel	7.75	1400	0.5	29	13.3	2,070	210	127	monofilament
tantalum	16.88	3000	0.2	55	6.5	1,520	190	508	monofilament
tungsten	19.38	3400	0.1	168	4.5	3,170	390	381	monofilament
whisker ceramic[f] Al_2O_3	3.96	2040	0.6	24	7.7	42,760	450	10–25	
metallic (Fe)	7.75	1540	0.5	29	13.3	13,100	200	127	

[a] Adapted from ref. 8.
[b] To convert g/cm^3 to $lb/in.^3$, divide by 27.68.
[c] To convert kJ/(kg·K) to Btu/(lb·°F), divide by 4.184.
[d] To convert W/(m·K) to Btu·ft/(ft^2·h·°F), divide by 1.729.
[e] To convert MPa to psi, multiply by 145; to convert GPa to psi, multiply by 145,000.
[f] Ref. 9.

Table 3. Typical Properties of Metal Matrices and Metallic Constituents for Metal–Metal Laminates

Metal	Density, g/cm^3 [a]	Mp, °C	Heat capacity, kJ/(kg·K) [b]	Thermal cond., W/(m·K) [c]	Thermal exp. coeff., 10^{-6}/°C	Tensile strength, MPa [d]	Modulus, GPa [d]	Remarks
aluminum	2.8	580	0.96	171	23.4	310	70	6061 (T6)
beryllium	1.9	1280	1.88	150	11.5	620	290	annealed
copper	8.9	1080	0.38	391	17.6	340	120	oxygen-free hardened
lead	11.3	320	0.13	33	28.8	20	10	1% Sb
magnesium	1.7	570	1.00	76	25.2	280	40	AZ31B-H24
nickel	8.9	1440	0.46	62	13.3	760	210	nickel 200 hardened
niobium	8.6	2470	0.25	55	6.8	280	100	
steel	7.8	1460	0.46	29	13.3	2070	210	ultra-high strength (MOD.H-11)
superalloy	8.3	1390	0.42	19	16.7	1100	210	Inconel X-750
tantalum	16.6	2990	0.17	55	6.5	410	190	
tin	7.2	230	0.21	64	23.4	10	40	
titanium	4.4	1650	0.59	7	9.5	1170	110	Ti–6 Al–4 V
tungsten	19.4	3410	0.13	168	4.5	1520	410	
zinc	6.6	390	0.42	112	27.4	280	70	alloy agada

[a] To convert g/cm^3 to $lb/in.^3$, divide by 27.68.
[b] To convert kJ/(kg·K) to Btu/(lb·°F), divide by 4.184.
[c] To convert W/(m·K) to Btu·ft/(ft^2·h·°F), divide by 1.729.
[d] To convert MPa to psi, multiply by 145; to convert GPa to psi, multiply by 145,000.

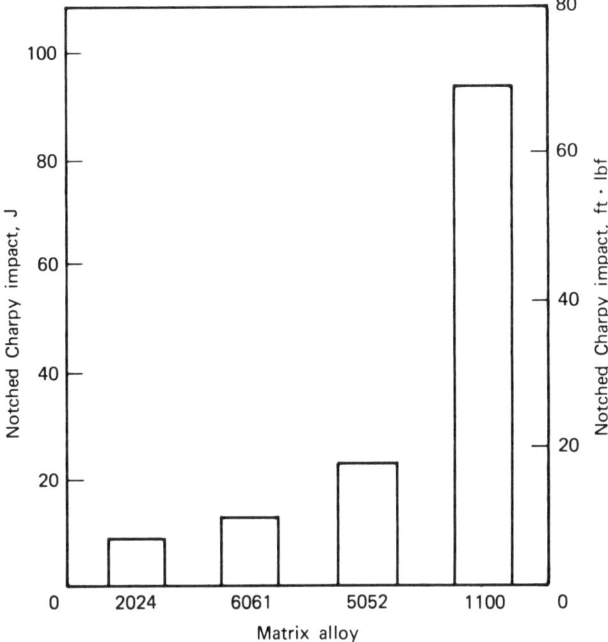

Figure 2. Impact resistance of boron–aluminum unidirectional metallic-matrix laminates (203-μm dia fiber, 0.50 fiber–volume ratio).

conventional metals make them good candidates for compression members in aircraft and space structures. Buckling stresses for plates and shells from tiber hybrids are compared with those from advanced unidirectional composites and from conventional metals in Table 5. Tiber hybrids have superior buckling resistance compared with either advanced composites or conventional metals. Another potential used for tiber hybrids is high-tip-speed fan blades (Fig. 5) for turbojet engines. Finite element analysis results showed that tiber-hybrid fan blades would have higher frequencies and lower tip distortions compared to those made from graphite fiber–resin composites (17). The comparisons for the first five frequencies are summarized in Table 6.

Properties

Physical, thermal, and mechanical properties of fibrous MMLs that have been made are summarized in Table 7. The last three entries in this table are superhybrid composites (Fig. 1**c**) where the core is made from graphite-fiber composite (see Carbon, carbon and artificial graphite), S-glass-fiber composite (see Glass), and Kevlar-fiber composite (see Aramid fibers).

Fibrous MMLs have relatively low transverse *T* strengths ranging from about 35–520 MPa (5,000–75,000 psi).

The transverse moduli range from 13.8 GPa (2×10^6 psi) for superhybrids to 207 GPa (30×10^6 psi) for SiC–Ti. The transverse properties for several fibrous MMLs are not available. These properties as well as those of metallic laminates (Fig. 1**d**), can be predicted approximately by the methods described below.

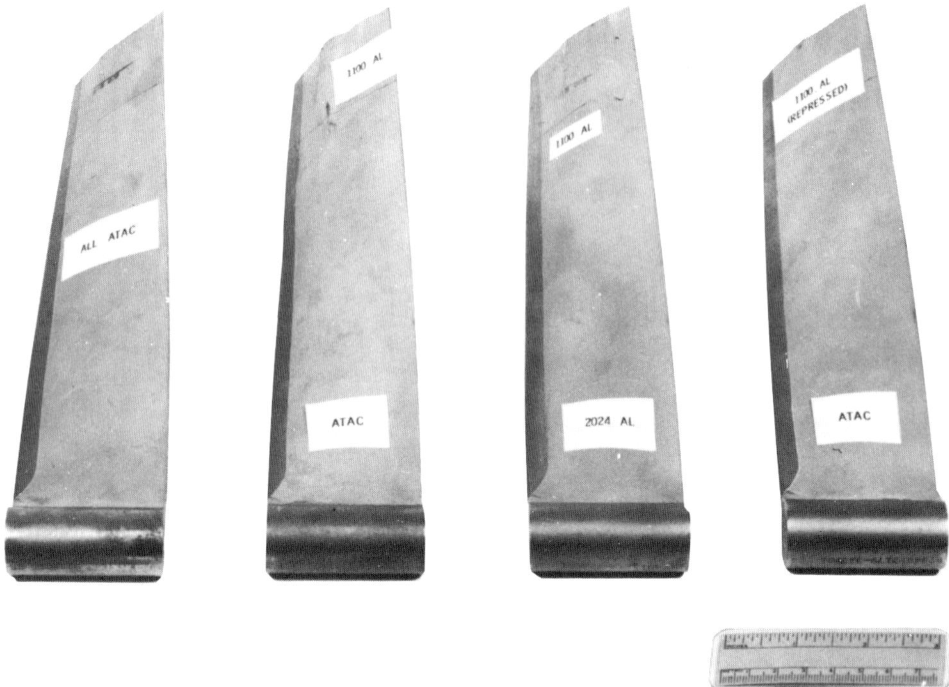

Figure 3. Boron–aluminum metallic-matrix fan blades after impact by a small bird (85 g).

Mechanical thermal and other physical properties are used in selecting fibrous MMLs for possible use in structural components during the preliminary design phases. These properties are then verified by selective testing and are subsequently used in the detailed analysis and the final design phases.

Fabrication Procedures

Laminates or composites from MMLs are fabricated using diffusion-bonding, roll-bonding, coextrusion, explosive bonding, and brazing in general.

Several other methods are used depending on the type of constituent metals to be used in the laminate. These methods include vacuum-infiltration casting, high energy forming, flow-molding, plasma-spraying, hot-pressing, continuous infiltration, power-metallurgy (qv) methods for discontinuous-fiber composites, explosive welding, and superplastic forming (4–5).

In diffusion-bonding, the filament or the interleaf layer (plies) are hot-pressed between layers of the matrix material. The pressure usually is 6.9–20.7 MPa (1000–3000 psi) and the temperature is 455–540°C. In roll-bonding the layers of metal–metal laminates are bonded by mill-rolling under specified heat and pressure. For fiber-reinforced metals, the ply (monolayer) is formed by diffusion-bonding, or any of the other methods, then the laminate of the specified number of plies is made by roll-bonding. In coextrusion, the constituents are assembled as in a billet and are extruded through a given die at specified temperatures and pressures depending on the con-

Table 4. Summary of Thin-Specimen Izod Impact-Strength Results and Comparisons with Some Other Materials (Specimen Nominal Dimensions: 1.27-cm Wide by 1.52-mm Thick)

Laminate type	Constituents	Test direction	Izod impact strength, kJ/m[a] Low	High	Number of specimens
I	Gr–Ep	longitudinal	1.45	1.59	4
		transverse	0.19	0.21	2
II	B–Al[b]	longitudinal	1.23	1.27	2
		transverse	1.01	1.10	2
III	B–Al[c]	longitudinal	0.68	0.96	2
		transverse	0.43	0.71	4
IV	Ti, B–Al	longitudinal	1.09	1.13	2
		transverse	0.79	1.00	4
V	superhybrid (Ti, B–Al, Gr–Ep)	longitudinal	2.82	3.20	2
		transverse	0.82	0.90	2
Other materials					
HT-S–PMR-PI[d]		longitudinal	0.91	0.92	2
		transverse	0.17	0.19	2
glass-fabric–epoxy			1.11	1.13	3
102 μm-dia B–6064 Al		longitudinal	1.13	1.21	2
aluminum 0.6061			3.36	4.07	2
titanium (6 Al–4 V)			11.23	11.38	2

[a] To convert kJ/m to ft·lbf/in., divide by 53.4×10^{-3} (see ASTM D 256).
[b] Diffusion-bonded.
[c] Adhesive-bonded.
[d] PMR = polymerization of monomeric reactants; PI = polyimide.

stituents used. The primary bonding mechanism in the coextrusion process is diffusion-bonding. Coextrusion is particularly suited for round and rectangular bar stocks. In explosive-bonding the constituent metal plies are bonded into a laminate by the high pressure generated through explosive means (see Metallic coatings, explosively clad metals). The amount of charge used is determined by the metallurgical bond required between the plies. This method is especially suitable for fabricating MMLs from metal plies with widely different melting temperatures. In brazing, bonding of the constituent metal plies into a laminate is accomplished by a third metal (brazing foil) which acts as a wetting liquid-metal phase and which has a lower melting temperature than either of the constituent metals (see Solders and brazing alloys). The plies to be bonded are stacked into a laminate with brazing foils between them. Then the temperature is raised between the melting temperature of the brazing foils and the constituent plies and appropriate pressure is applied. Upon solidification, the brazing foil bonds the adjacent constituent plies together into a laminate. Boron–aluminum plies are fabricated at about 600°C and <1.4 MPa (200 psi).

Superhybrids are fabricated by adhesively bonding titanium outer plies over boron–aluminum plies and graphite-fiber–epoxy inner plies (core) with a titanium ply at the center (11–12). The adhesive bond between the metallic plies and between the metallic and composite plies is provided by use of FM 1000 adhesive film approximately 25-μm thick. The bonding process is accomplished under specified pressure and temperature normally used for epoxy-matrix composites. This process consists of a 3-h cure at 150°C and <700 kPa (100 psi). The same bonding process (fabrication procedure) is used to fabricate the nongraphitic superhybrids, the tiber (titanium–beryllium) hybrids and the adhesively bonded metallic-ply laminates.

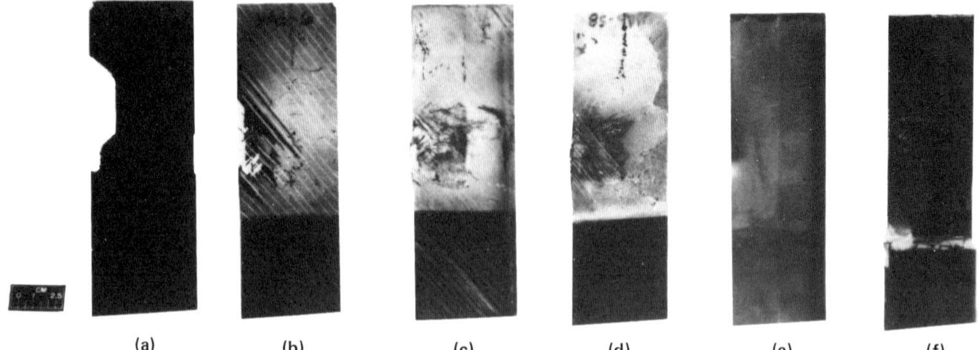

Figure 4. Relative high velocity impact assessment of superhybrids. Projectile was 2.5-cm dia gelatin sphere; specimens (**a**) through (**e**) oriented at 30° incidence angle, specimen (**f**) at 19° incidence angle. To convert kJ/m to ft·lbf/in., divide by 0.0534 (see ASTM D 256).

Specimen	Construction	Leading edge thickness, μm	Midchord thickness, mm	Projectile velocity, m/s	Slice, %	Kinetic energy per thickness, kJ/m
(**a**)	graphite–epoxy	737	3.81	252	55	43.45
(**b**)	graphite–glass–epoxy	711	3.73	284	50	55.94
(**c**)	boron–glass–polysulfone–graphite–epoxy	737	4.09	285	40	42.76
(**d**)	boron–glass–epoxy	762	4.14	269	40	37.04
(**e**)	titanium–boron–aluminum–graphite–epoxy superhybrid	635	3.96	281	50	63.63
(**f**)	solid titanium (6 Al–4 V)	356	3.89	222	50	23.65

Design–Analysis Procedures

Design requirements for structures may be: maximum strength with light weight, long life service with minimum strength degradation, notch or other defect insensitivity with high stiffness, impact resistance with high stiffness, damage tolerance with high stiffness and low cost. Many recent designs for aircraft, spacecraft, automobiles, complex machine parts, and electronic computers have design requirements that include several of the above as well as ease of fabrication.

Design of structures with metal-matrix laminates means cost-effective use of MMLs in structures and structural parts, ie, use of MMLs collectively satisfy all of the design requirements better than other materials for a given structure or structural part.

Alternative design concepts cannot be evaluated by trial and error in the building or fabricating of complex structures. Therefore, alternative design concepts for a specific case are evaluated on paper. The formal way for evaluating structural concepts with respect to given design requirements is the use of structural analysis. Structural analysis consists of a collection of mathematical models (equations). These equations describe the response of the structure when the structure is subjected to the anticipated loads which the structure will have to resist safely during its life span.

Table 5. Assessment of Tiber Hybrids for Possible Aerospace Structural Applications

Material	Specific buckling stress, $\sigma_{CR}/t_c\rho$	
	Plate	Cylindrical shell
tiber hybrids		
70% Ti–30% Be	402×10^3	4160
50% Ti–50% Be	532	6230
30% Ti–70% Be	818	9150
other composites		
B–Al (0.50 FVR)	883×10^3	5120
B–E (0.50 FVR)	464	1310
T75–E (0.60 FVR)	ca 238	1180
AS–E (0.60 FVR)	ca 233	1150
metals		
steel	331×10^3	2442
aluminum	307	2263
titanium	307	2263

Table 6. Comparisons of Frequencies of High-Tip-Speed Composite Blade

Mode, frequency	HTS–K 601 (±40°, ±20°)[a]	HTS–PMR (±40°, ±20°)[a]	Tiber hybrid (titanium–beryllium) (30%–70%)[b]
1, Hz	361	400	662
2, Hz	939	960	1608
3, Hz	1178	1418	2108
4, Hz	1485	1658	2333
5, Hz		2427	3253
density, g/cm³ [c]	ca 1.38	ca 1.52	ca 2.35

[a] 40°, 20° denotes angle measurement.
[b] Lamina thickness: titanium, 127 μm; beryllium, 254 μm.
[c] To convert g/cm³ to lb/in.³, divide by 27.68.

A general structural analysis model in equation form is given by:

$$M\ddot{u} + C\dot{u} + Ku = F \tag{1}$$

Equation 1 describes the structural response at any point in the structure in terms of acceleration $\ddot{u}$, velocity $\dot{u}$, and displacement u for a given mechanical and/or thermal load condition F. The structure geometric configuration and material are represented in equation 1 in terms of mass M, damping C, and stiffness K. Equation 1 can be applied to simple and complex structures, and to structures made from any material. In order to use equation 1 for a structure or structural part made from a given material, the values in M, C, and K for this material must be known.

Procedures for using equation 1 for the analysis and/or design of structures made from composite laminates are extensively discussed in refs. 18–19. Herein, the MML properties for M, C, and K are determined, as well as the strength and thermal properties which are needed to evaluate and/or select MMLs for specified design requirements.

If the MML behaves like a general orthotropic solid (20), then physical, thermal and mechanical properties that are needed for structural analysis of MMLs (Fig. 1**b**

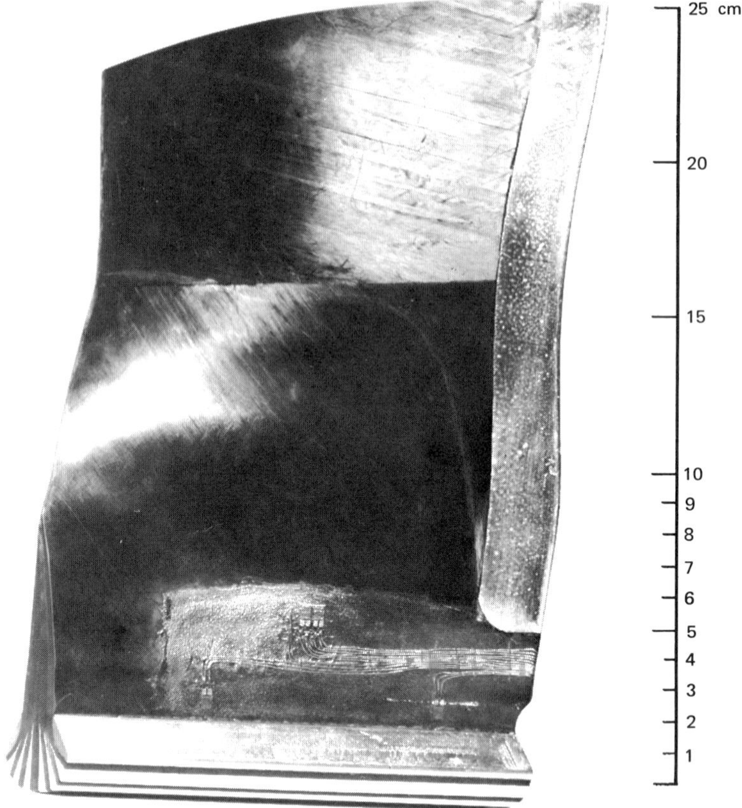

Figure 5. High-tip-speed composite fan blade.

and **d**) include: density ρ, heat capacity H_c, three thermal heat conductivities K, three thermal expansion coefficients α, three normal (Young's) moduli E, three shear moduli G, three Poisson's ratios ν, and nine strengths s. Except for ρ and H_c, the other properties are given with respect to three mutually orthogonal directions which are taken to coincide with the planes of elastic symmetry of the MML. These directions are either 1, 2, and 3 in Figure 1**a,** called the material axes of the single layer (ply), or x, y, and z in Figure 1**b** and **d,** called structural or load axes of the laminate (composite). It is customary to use subscripts to denote the directions along which the properties are given. The subscript l with combinations of 1, 2, or 3 subscripts are used to denote ply material-axis properties whereas the subscript c with combinations of x, y, or z subscripts are used to denote composite structural-axis properties. For example: E_{l11} denotes the modulus of elasticity (normal modulus) along the 1-direction and G_{l12} for the shear modulus in the 1–2 plane (Fig. 1**a**). The corresponding moduli along the structural axes of the MML for Figures 1**b** and **d** are E_{cxx} and G_{cxy}. These properties are summarized in symbolic form in Table 8 for MMLs made from plies with three types of symmetry. The material-axis properties are for the single layer (ply), and the structural-axis properties are for the laminate (composite).

Composite-mechanic theories based on constituent properties have been developed for predicting material-axes and/or structural-axis properties. Composite me-

Table 7. Typical Mechanical Properties of Metal-Matrix Composites

Fiber	Matrix	Reinforcement, vol %	Density, g/cm^3 [a]	Longitudinal tensile strength, MPa[b]	Longitudinal modulus, GPa[b]	Transverse tensile strength, MPa[b]	Transverse modulus, GPa[b]
G T 50	201 Al	30	2.380	620	170	50	30
G T 50	201 Al	49		1120	160		
G GY 70	201 Al	34	2.380	660	210	30	30
G GY 70	201 A1	30	2.436	550	160	70	40
G HM pitch	6061 Al	41	2.436	620	320		
G HM pitch	AZ31 Mg	38	1.827	510	300		
B on W, 142 μm fiber	6061 Al	50	2.491	1380	230	140	160
borsic	Ti	45	3.681	1270	220	460	190
G T 75	Pb	41	7.474	720	200		
G T 75	Cu	39	6.090	290	240		
FP	201 Al	50	3.598	1170	210	(140)	140
SiC	6061 Al	50	2.934	1480	230	(140)	140
SiC	Ti	35	3.931	1210	260	520	210
SiC whisker	Al	20	2.796	340	100	340	100
B_4C on B	Ti	38	3.737	1480	230	>340	>140
G T 75	Mg	42	1.799	450	190		
G HM	Pb	35	7.750	500	120		
G T 75	Al–7% Zn	38	2.408	870	190		
G T 75	zinc	35	5.287	770	120		
G T 50	nickel	50	5.295	790	240		
G T 75	nickel	50	5.342	828	310	30	40
G (81.3 μm)	2024 Al	50	2.436	760	140		
G (142 μm)	2024 Al	60	2.436	1100	180		
superhybrid	grafitic	60	2.048	860	120	220	60
superhybrid	S-glass	60	2.159	740	80	190	30
superhybrid	Kevlar	60	1.799	700	80	190	10

[a] To convert g/cm^3 to $lb/in.^3$, divide by 27.68.
[b] To convert MPa to psi, multiply by 145; to convert GPa to psi, multiply by 145,000.

Table 8. Summary of Physical, Thermal, and Mechanical Properties Needed for Structural Analysis and Design for Metal-Matrix Laminates Made from Plies with Three Types of Symmetry in Symbolic Form [a]

	Composite with type of symmetry of:					
	Generally orthotropic		Transverse isotropic		Isotropic	
Property	Ply	Composite	Ply	Composite	Ply	Composite
density, g/cm^3	ρ_l	ρ_c	ρ_l	ρ_c	ρ_l	ρ_c
heat capacity	H_{cl}	H_{cc}	H_{cl}	H_{cc}	H_{cl}	H_{cc}
thermal heat	K_{l11}	K_{cxx}	K_{l11}	K_{cxx}	K_l	K_{cxx}
conductivity	K_{l22}	K_{cyy}	K_{l22}	K_{cyy}	K_l	$K_{cyy} = K_{cxx}$
	K_{l33}	K_{czz}	$K_{l33} = K_{l22}$	K_{czz}	K_l	K_{czz}
thermal expansion	α_{l11}	α_{cxx}	α_{l11}	α_{cxx}	α_l	α_{cxx}
coefficient	α_{l22}	α_{cyy}	α_{l22}	α_{cyy}	α_l	$\alpha_{cyy} = \alpha_{cxx}$
	α_{l33}	α_{czz}	$\alpha_{l33} = \alpha_{l22}$	α_{czz}	α_l	α_{czz}
elastic and shear	E_{l11}	E_{cxx}	E_{l11}	E_{cxx}	E_l	E_{cxx}
moduli	E_{l22}	E_{cyy}	E_{l22}	E_{cyy}	E_l	$E_{cyy} = E_{cxx}$
	E_{l33}	E_{czz}	$E_{l33} = E_{l22}$	E_{czz}	E_l	E_{czz}
	G_{l12}	G_{cxy}	G_{l12}	G_{cxy}	$G_l = \dfrac{E_l}{2(1 + \nu_l)}$	$G_{cxy} = \dfrac{E_{cxx}}{2(1 + \nu_{czy})}$
	G_{l23}	G_{cyz}	$G_{l23} = \dfrac{E_{l22}}{2(1 + \nu_{l23})}$	G_{cyz}	$G_l = \dfrac{E_l}{2(1 + \nu_l)}$	G_{cyz}
	G_{l13}	G_{cxz}	$G_{l13} = G_{l12}$	G_{cxz}	$G_l = \dfrac{E_l}{2(1 + \nu_l)}$	$G_{cxz} = G_{cyz}$
Poisson ratios	ν_{l12}	ν_{cxy}	ν_{l11}	ν_{cxy}	ν_l	ν_{cxy}
	ν_{l23}	ν_{cyz}	ν_{l23}	ν_{cyz}	ν_l	ν_{cyz}
	ν_{l13}	ν_{cxz}	$\nu_{l13} = \nu_{l12}$	ν_{cxz}	ν_l	$\nu_{cxz} = \nu_{cyz}$
strengths	$S_{l11T,C}$	$S_{cxxT,C}$	$S_{l11T,C}$	$S_{cxxT,C}$	$S_{lT,C}$	$S_{cxxT,C}$
(T = tension;	$S_{l22T,C}$	$S_{cyyT,C}$	$S_{l22T,C}$	$S_{cyyT,C}$	$S_{lT,C}$	$S_{cyyT,C} = S_{cxxT,C}$
C = com-	$S_{l33T,C}$	$S_{czzT,C}$	$S_{l33T,C} = S_{l22T,C}$	$S_{czzT,C}$	$S_{lT,C}$	$S_{czzT,C}$
pression;	S_{l12S}	S_{cxyS}	S_{12S}	S_{cxy}	S_{lS}	S_{cxyS}
S = shear)	S_{l23S}	S_{cyzS}	S_{l23S}	S_{cyz}	S_{lS}	S_{cyzS}
	S_{l13S}	S_{cxzS}	$S_{l13S} = S_{l12S}$	S_{cxz}	S_{lS}	$S_{cxzS} = S_{cyzS}$

[a] Subscripts refer to directions shown in Figure 1.

chanics is subdivided into micromechanics, macromechanics, and laminate theory. Micromechanics embodies the various theories which are used to predict material-axis properties of unidirectional fiber composites (plies) using constituent fiber and matrix properties. Typical results predicted for boron–aluminum plies using composite micromechanics are shown in Figure 6. Macromechanics usually consist of transformation equations which are used to transform material axes properties along any other axes. Macromechanics also includes failure theories and criteria for plies subjected to combined stress states. Typical results predicted for boron–aluminum MMLs using composite macromechanics are shown in Figure 7 for thermal and elastic properties and in Figure 8 for strengths. Laminate theory embodies the equation and procedures that are used to predict the laminate properties using ply properties. Laminate theory is also used to generate the properties required to form M, C, and K (17,23–24). In addition, laminate theory is used to predict the residual stresses of the lamination in

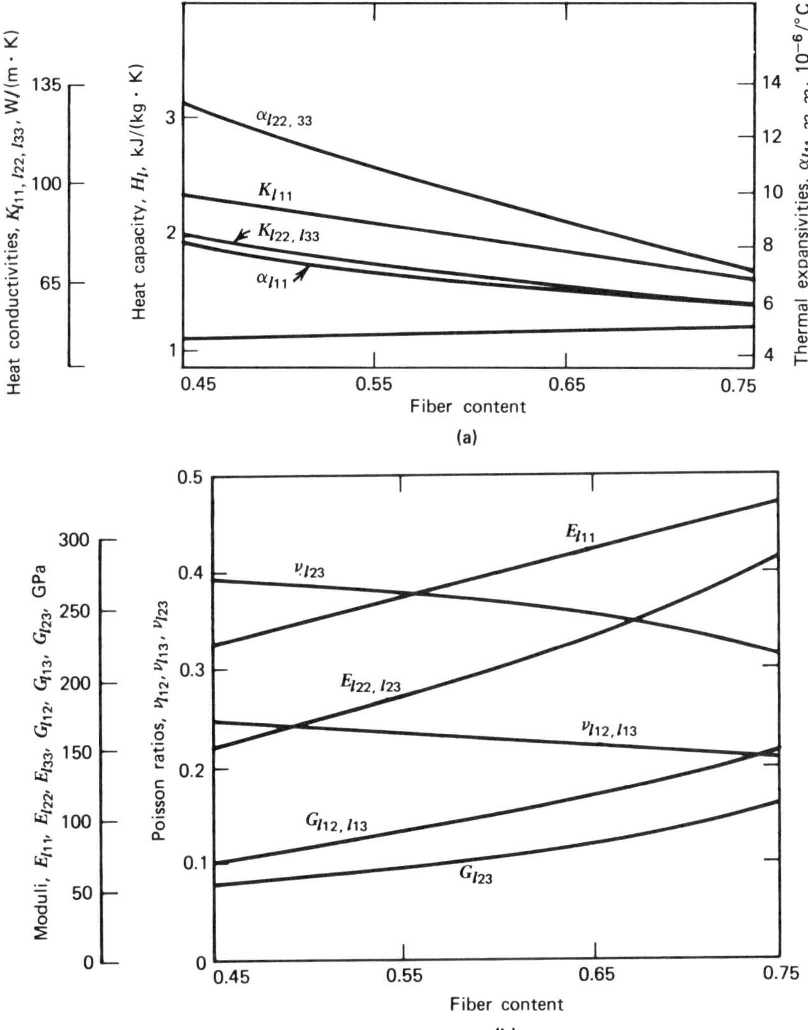

Figure 6. Typical unidirectional composite (ply) properties predicted using composite micromechanics. (**a**) Boron–aluminum unidirectional thermal properties. To convert W/(m·K) to Btu·ft/(ft^2·h·°F), divide by 1.729; to convert kJ/(kg·K) to Btu/(lb·°F), divide by 4.184. (**b**) Boron–aluminum unidirectional elastic properties. To convert GPa to psi, multiply by 145,000. (**c**) Boron–aluminum uniaxial strengths; ○ measured data. To convert MPa to psi, multiply by 145.

the plies. These residual stresses result from the difference in the processing and use temperature as well as the difference in the thermal expansion coefficients (25–26). Typical results predicted for boron–aluminum MMLs using laminate theory are shown in Figure 9 for thermal expansion coefficients and elastic properties, and in Figure 10 for lamination residual stresses. Lamination residual stresses (strains) affect significantly the laminate mechanical behavior of boron–aluminum MMLs (26). Different types of heat treatment also affect the mechanical behavior of MMLs, especially the transverse properties (27).

MMLs are made from isotropic plies (Fig. 1**d**). The analysis is considerably simpler

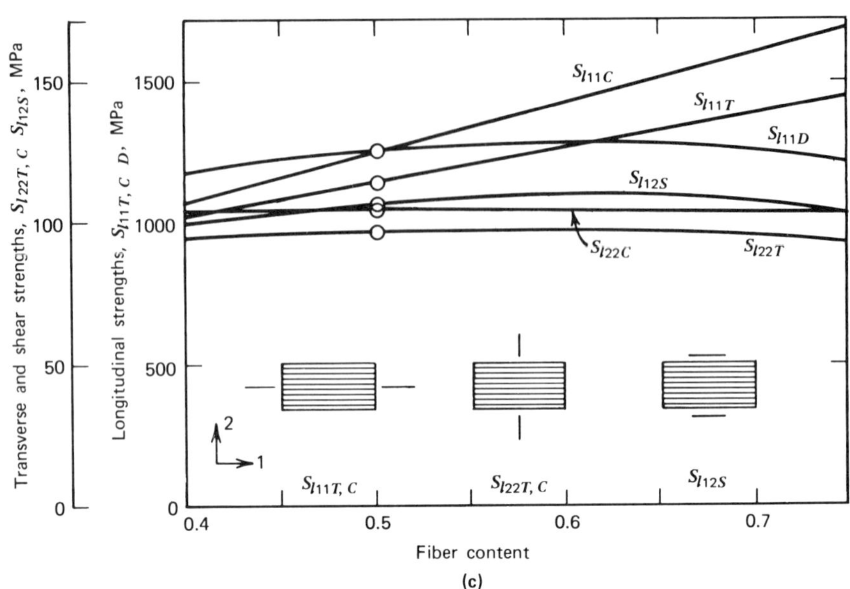

(c)

Figure 6. *(continued)*

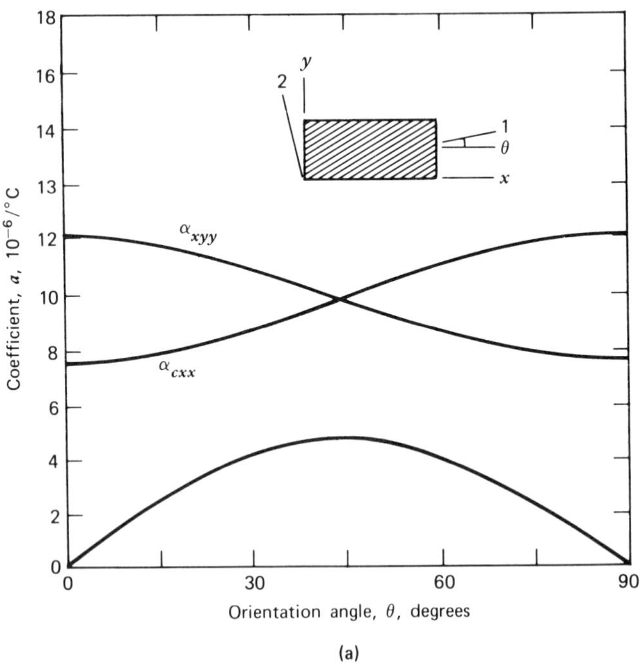

(a)

Figure 7. Typical thermal and elastic properties of metal-matrix composites predicted using composite macromechanics (about 0.5 fiber–volume ratio). (**a**) Thermal coefficients of expansion for off-axis unidirectional boron–aluminum composites. (**b**) Moduli and Poisson ratios for off-axis unidirectional boron–aluminum composites. To convert GPa to psi, multiply by 145,000. To convert MPa to psi, multiply by 145.

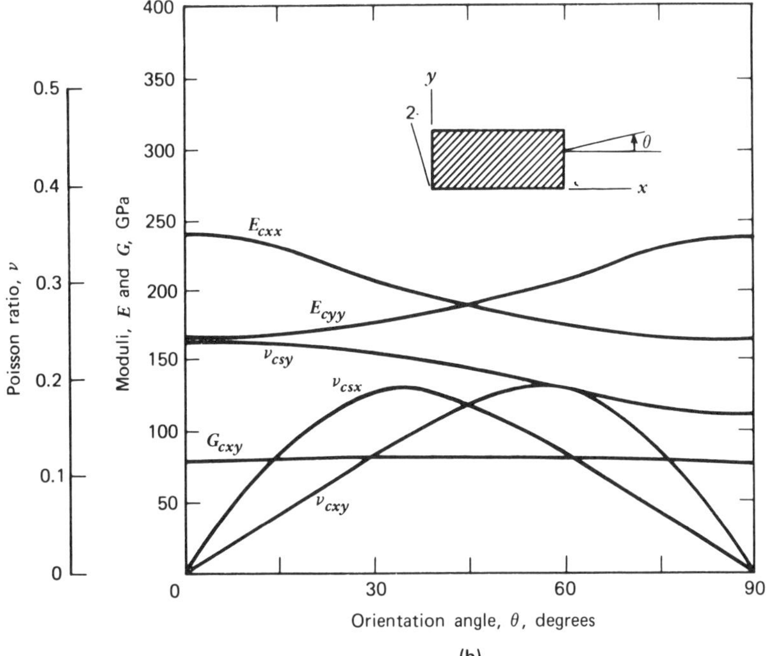

(b)

Figure 7. *(continued)*

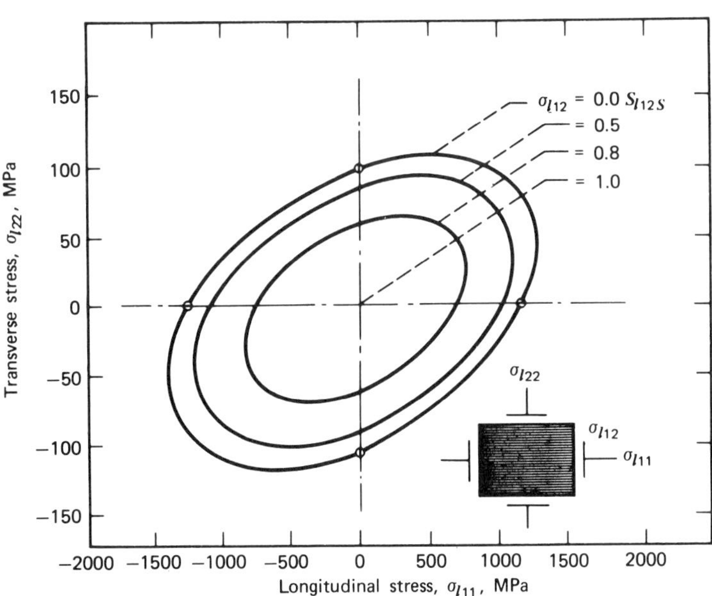

Figure 8. Typical strengths of metal matrix composites predicted using composite macromechanics (about 0.5 fiber–volume ration). (**a**) Failure envelopes under combined stress for boron–aluminum. ○ experimental data from ref. 22. (**b**) Boron–aluminum off-axis normal load failure envelopes. ○ experimental data. (**c**) Boron–aluminum off-axis shear load failure envelopes. ○ experimental data.

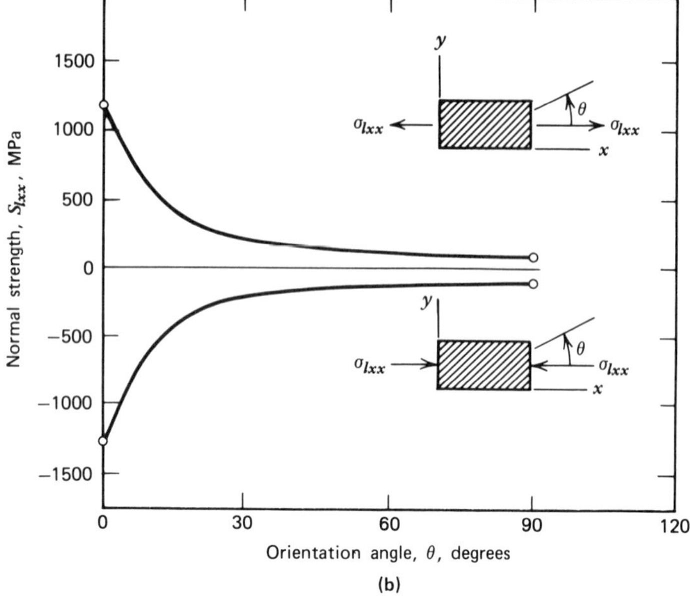

(b)

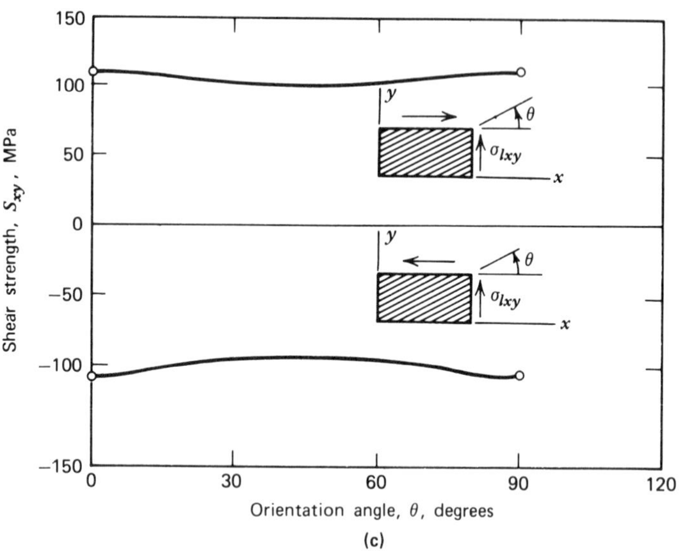

(c)

Figure 8. *(continued)*

as compared to that used for fiber-composite plies. One such analysis is described in detail in ref. 13. Typical results obtained using this analysis are summarized in Table 9 for titanium–beryllium (tiber) hybrid MMLs.

The thermal and mechanical properties of MMLs described above constitute only a minimum of those usually required to assess the suitability of a relatively new ma-

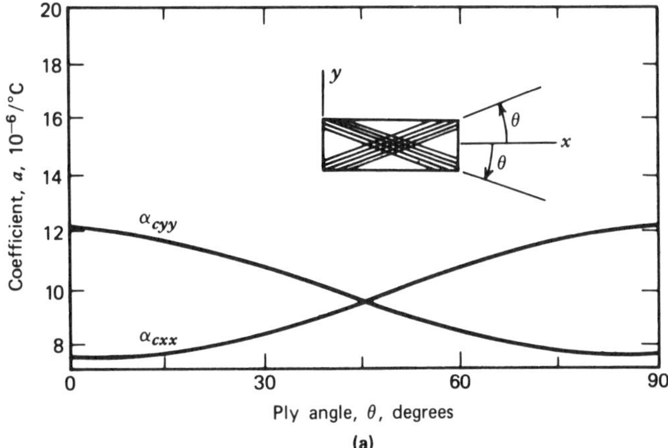

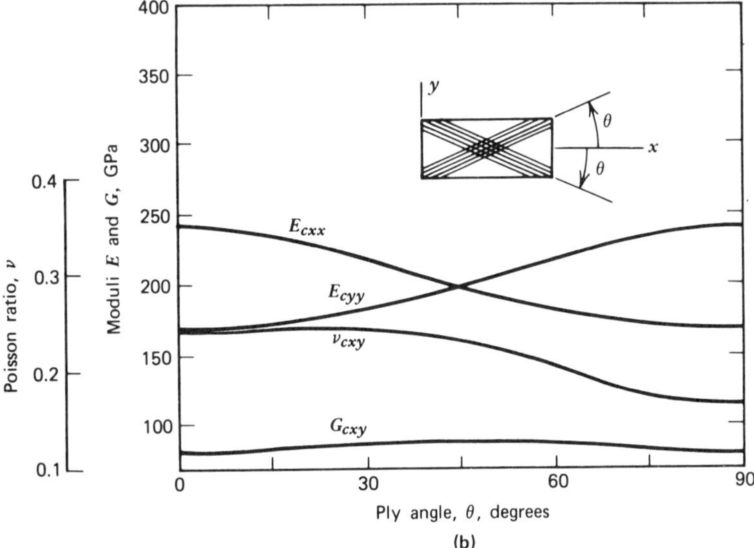

Figure 9. Typical thermal (**a**) and elastic (**b**) properties for metal matrix laminates (boron–aluminum angle-ply composites) predicted using laminate theory (about 0.5 fiber–volume ratio). To convert GPa to psi, multiply by 145,000.

terial at the preliminary design stages. Several other important factors need be considered simultaneously with the thermal and mechanical properties, eg, fatigue resistance, creep, impact resistance, erosion and corrosion resistance, service environment effects, notch insensitivity and fracture toughness, damage tolerance and repairability, fabrication and quality control, reliability and durability, inspectability and maintainability, design data development costs and reproducibility, design–analysis experience of the staff, and acceptance of the public agency which sets and administers structural integrity–safety requirements.

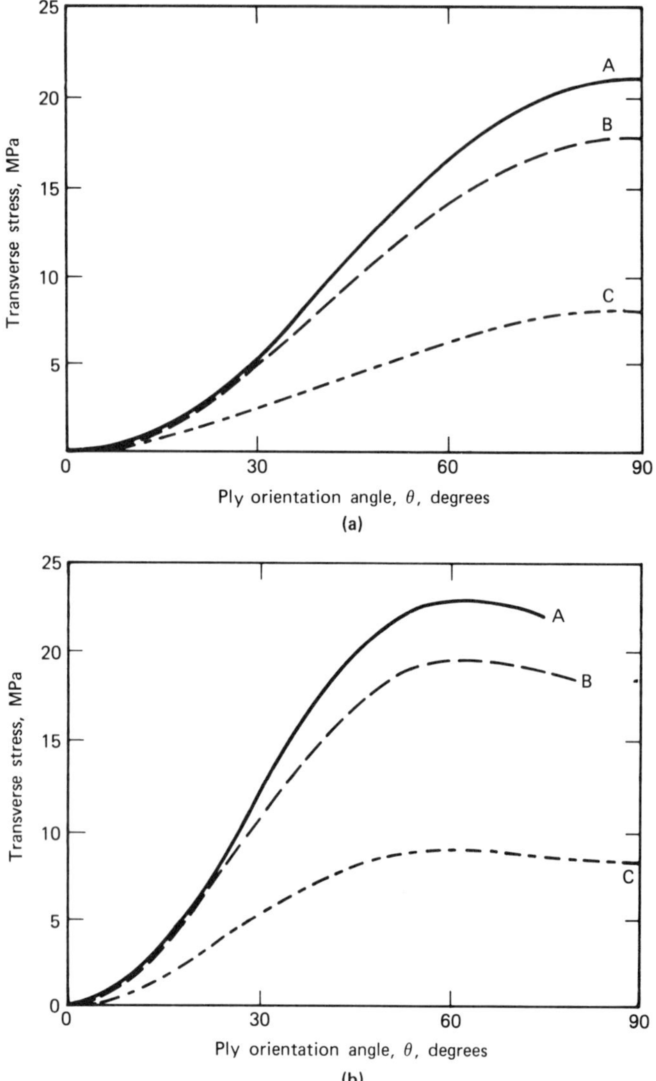

Figure 10. Lamination residual stresses in metal-matrix laminates predicted using laminate theory (boron fiber–6061 aluminum matrix, 482°C temperature difference). Fiber–volume ratio: A, 0.35; B, 0.55; C, 0.75. (**a**) Stress in the 0° ply. (**b**) Stress in the $\pm\theta$ ply. (**c**) Stress in the $+\theta$ ply. To convert MPa to psi, multiply by 145.

Special Types of Metal–Metal Laminates

The primary reason for making and investigating these types of MMLs is their potential for fracture control and damage tolerance. Fracture control is usually assessed by using a material property called fracture toughness. Special types of metal/metal MMLs that have been investigated (shown in Table 1) include different kinds of steels such as mild, high strength, and maraging (alloyed steel); aluminum–aluminum; titanium–titanium and titanium–aluminum; tungsten–superalloy and tungsten–tan-

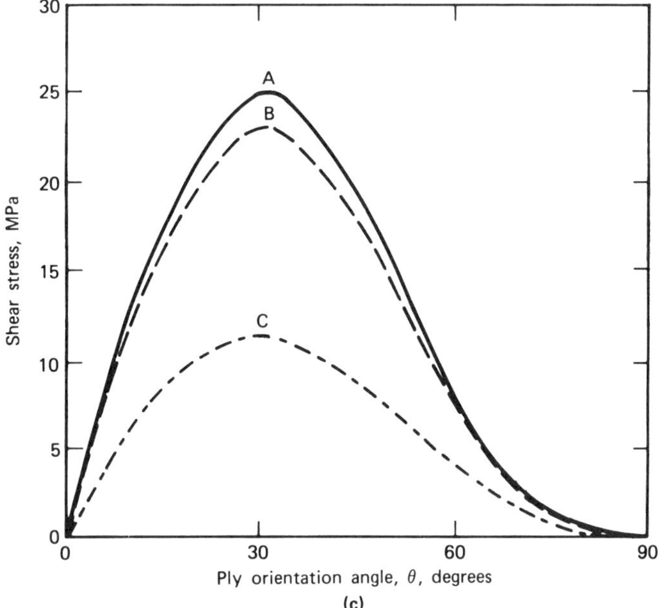

Figure 10. (*continued*)

talum; and titanium–beryllium. The fracture toughness of a plate-form material, with a cracklike defect, is the stress that this material can resist just prior to onset of rapid crack propagation. Fracture toughness varies for different materials. It also can vary for the same material but with different alloying elements, thickness and heat treatment. In addition fracture toughness depends on the use temperature. In principle, by interleafing materials with different fracture toughness, the fracture toughness of metal–metal MMLs can be altered.

For analysis–design purposes, fracture toughness is used to determine the level of stress that a structural member with an assumed crack length can safely support. This level of stress is determined using linear elastic fracture mechanics (LEFM) in general. Equation 2 is an elementary equation from LEFM:

$$\sigma = \frac{K_c}{\sqrt{a}} F \tag{2}$$

where σ is the average or gross stress (stress without the crack), K_c is the material fracture toughness parameter corresponding to primary loading conditions and anticipated crack propagation depicted schematically in Figure 11, a is the crack length, and F represents the stress state at the crack tip and depends on: material, geometry, and loading condition. Values for K_c for different materials are found in reports published by the Metals and Ceramics Information Center, Columbus, Ohio, as well as various handbooks dealing with aerospace structures, and pressure-vessel materials and design.

The designer can use MMLs to control fracture and, therefore, provide damage tolerance either by using plies of materials with different fracture toughness to divide the fracture-driving stress (crack divider), or by using plies with higher fracture

Table 9. Comparison of Measured and Predicted Properties of Tiber Laminates

	Tiber laminate and direction					
	I—40% Ti–58% Be		II—55% Ti–36% Be		III—63% Ti–31% Be	
Property identification	RD	TD	RD	TD	RD	TD
modulus, GPa[a]						
measured	203.5	206.9	165.5	165.5	175.9	169.0
predicted	211.0	212.4	163.5	165.5	157.9	160.0
difference, %	3.7	2.7	−1.2	0	−10.2	−5.1
Poisson ratio						
measured	0.20	0.25	0.26	0.27	0.26	0.28
predicted	0.27	0.27	0.28	0.29	0.29	0.29
difference, %	35.0	8.0	7.7	7.4	11.5	3.6
fracture stress, MPa[b]						
measured	646.9	466.2	723.5	713.8	787.0	716.6
predicted	677.3	642.8	713.8	694.5	767.0	749.0
difference, %	4.7	37.9	−1.4	−2.7	−2.5	−4.7
density, g/cm³ [c]						
measured	2.85		3.24		3.40	
predicted	2.82		3.21		3.46	
difference, %	0.9		−0.8		1.6	

[a] To convert GPa to psi, multiply by 145,000.
[b] To convert MPa to psi, multiply by 145.
[c] To convert g/m³ to lb/in.³, divide by 27.68.

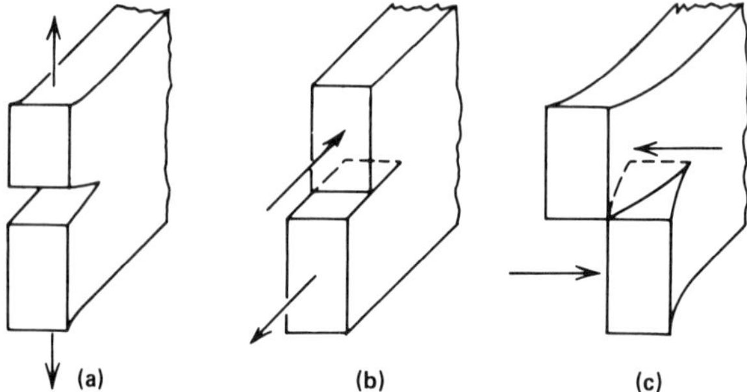

Figure 11. Primary load conditions and corresponding fracture modes used in fracture mechanics: (**a**) mode I loading; (**b**) mode II loading; (**c**) mode III loading.

toughness to arrest the fracture-driving stress (crack arrest). Both of these methods are illustrated schematically in Figure 12.

In order for either concept to work, the type of bond selected must meet three general criteria; it must be strong enough to constrain the laminate to respond structurally (with respect to displacement, buckling and frequency) like a homogenous material; weak enough to permit each ply to fracture independently of its neighbors; and brittle enough to fail by local delamination in the vicinity of the advancing crack front. Examples of the fracture toughness of aluminum–aluminum MMLs by diffusion, roll, or explosive bonding are described in ref. 28, and those made by adhesive bonding in ref. 29. Photomicrographs depicting arrested cracks in actual samples are shown in ref. 22; a concise treatment of fracture analysis for aerospace metals is given in ref. 21; and ref. 30 provides a comparable treatment for fatigue (see Ablative materials).

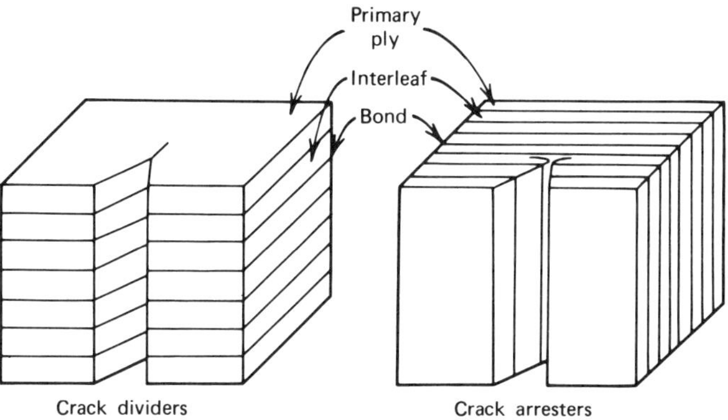

Figure 12. Metal laminate concepts for fracture control.

The root portion of helicopter blades is an example showing where MMLs are used for fracture control. This part of the blade may have cracklike defects because of the joint, and it is subjected to high cycle fatigue. Wings of military aircraft and helicopter booms are potential applications of MMLs in order to provide damage tolerance to projectile impact.

In addition to fracture control or damage tolerance, MMLs also are used in applications where the interleaf may be the stronger, stiffer material whereas the primary material provides erosion, corrosion, oxidation, or other service–environmental resistance. Examples of such an application are tantalum/tungsten MMLs which have been investigated for possible use in aircraft engine turbine blades. Tantalum coated with suitable coatings is used to resist the corrosive environment of the burning fuel, and tungsten is used for strength and stiffness in order to meet mechanical-design requirements.

Special Types of Fiber-Reinforced Metals

Boron-fiber–aluminum-matrix (B–Al) MMLs have been made and investigated more extensively than any other fiber-reinforced metal. These types of laminates combine several of the desirable features of aluminum and in addition provide about a threefold improvement in modulus and about a threefold improvement in strength over that of aluminum. The key disadvantage limiting its application is the high cost of the boron fiber; the main part of this boron-fiber cost is the tungsten substrate. In order to reduce the fiber cost, carbon fibers are used for the substrate, as well as boron fibers of larger diameter.

Boron-fiber–aluminum-matrix MMLs have excellent fatigue-, creep-, corrosion-, and erosion resistance. Galvanic action may degrade the interfacial bond, depending on the surface coating of the fiber. These MMLs have good temperature resistance up to about 150°C. They may be used with relatively small property-loss penalty up to 315°C in stiffness-controlled designs such as dimensional stability, and buckling and vibration frequencies. B–Al MMLs have improved fracture toughness compared to the aluminum matrix, and they are notch-insensitive. However, B–Al MMLs have about one half the impact resistance of aluminum. An extensive discussion on various mechanical properties of B–Al MMLs is given in ref. 31. The elevated temperature effects are discussed in ref. 32. The cryogenic temperature conditions have negligible effect on the tensile properties of B–Al MMLs from the limited available data (33).

Angleplied MMLs (Fig. 1**b**) from B–Al undergo inelastic deformations at relatively small load (about 10–20% of the fracture load) (34). In a cyclic-load condition, this inelastic deformation may progressively improve, degrade or have no effect on the mechanical properties of the laminate (35). Significant parameters affecting joints and joint design are discussed in ref. 36.

Boron–aluminum MMLs have been made for: aircraft fuselage skin and stringers, aircraft wing skins, aircraft-wing boxes, aircraft-engine fan and compressor blades, propeller shells, landing gear struts, thrust support structure for the space shuttle, shafts for torque transmission, and rocket-motor cases. An extensive discussion of the application of B–Al MMLs for aerospace structures is given in ref. 37. Because of the high cost [about $550/kg (1980 dollars)] B–Al MMLs have not been considered seriously for use outside the aerospace industry.

Graphite-fiber–aluminum-matrix (Gr–Al) MMLs are investigated mainly for B–Al MMLs replacement because of their low cost (about 25–50%) potential compared to B–Al MMLs. In addition, Gr–Al MMLs have excellent thermal-dimension stability. They also are good contenders in friction-wear applications because of the inherent lubricating properties of the graphite fibers. However, Gr–Al MMLs are susceptible to galvanic corrosion as are B–Al MMLs. Special surface treatment of the fibers is required to minimize this galvanic action. Gr–Al MMLs exhibit rule-of-mixtures properties along the fiber direction (Table 7). However, the transverse properties are relatively poor. Alternatives such as heat-treating and metal-foil-interleaving are used to improve the transverse properties. These alternatives are selected with considerable caution since they tend to reduce the longitudinal properties. Gr–Al MMLs have good fracture toughness and damage tolerance. They also have excellent mechanical- and thermal-fatigue resistance. Their corrosion and erosion resistance can be comparable to that of B–Al. An extensive discussion and a good review of several important aspects of Gr–Al MMLs up to 1976 is given in ref. 38.

Borsic-fibers–titanium and borsic–aluminum matrix MMLs have been investigated primarily for possible use in aircraft-engine fan blades. Borsic/titanium MMLs have about twice the stiffness and about 80% the density of titanium (39–40). The combination of these two properties is generally sufficient to eliminate the midspan shrouds which are presently used to meet vibration and flutter design requirements. This diminution of shrouds also is apparent with B–Al.

Tungsten-fiber–superalloy (TF–SA) MMLs are investigated for their potential use in aircraft-turbine blades. The excellent mechanical properties retention of the tungsten fibers at high temperatures (about 1100°C) is the key feature for investigating these types of laminates. However, TF–SA have two main disadvantages: high density, and thermal-fatigue degradation. The high density disadvantage may be circumvented to some extent by appropriate structural-design configurations, such as hollow blades. On the other hand, the thermal-fatigue degradation only can be minimized by metallurgical considerations. Most of the research to date for TF–SA MMLs was conducted at the NASA Lewis laboratory (41–42). Limited research recently has been initiated to make turbine blades from these laminates (43–44). Whisker-reinforced metals and ceramic laminates also are investigated for possible use in internal combustion engines and other high temperature applications in general. The major disadvantages of whisker-reinforced MMLs are their poor fracture toughness and poor impact resistance characteristics. These poor characteristics may be improved by designing the laminate to operate in preferential compression.

Some of the other fiber-reinforced MMLs listed in Table 7 are investigated for specific applications. For example, graphite-fiber–magnesium-matrix MMLs are investigated for space antennas because of their desirable thermal distortion and density properties, and graphite-fiber–lead-matrix MMLs are investigated for use in batteries where weight is an important design consideration. Some other MMLs listed in Table 7 are investigated for metallurgical considerations at the fiber–matrix interface (borsic–aluminum, silicon carbide–aluminum).

Nomenclature

a	= crack length
B–Al	= boron fiber–aluminum matrix
C	= damping
E	= modulus of elasticity
F	= stress state at crack tip
F	= mechanical and/or thermal load condition
G	= shear modulus
Gr–A	= graphite fiber–aluminum matrix
H_c	= heat capacity
ICCM	= International Conference on Composite Materials
K	= stiffness
$K_{subscript}$	= heat conductivity
K_c	= fracture toughness parameter
LEFM	= linear elastic fracture mechanics
M	= mass
MML	= metal matrix laminate
u	= displacement
$\dot{u}$	= velocity
$\ddot{u}$	= acceleration
SHC	= superhybrid composite
TF–SA	= tungsten-fiber–superalloy
UFC	= graphite-fiber–resin
α	= thermal expansion core
σ	= average or gross stress
ρ	= density

BIBLIOGRAPHY

"Ceramic Composite Armor" in *ECT* 2nd ed., Suppl. Vol., pp. 138–157, by George Rugger, Dept. of the Army, and John R. Fenter, Dept. of the Air Force.

1. M. F. Smith, *Metal Matrix Composites, NTIS/PS-78/0684, N79-10155,* Vol. 2, National Technical Information Service, Springfield, Va., 1978.
2. D. M. Cavagnaro, *Boron-Reinforced Composites, NTIS/PS-79/0476, N78-27189,* Vol. 2, NTIS, Springfield, Va., 1979.
3. D. M. Cavagnaro, *Boron-Reinforced Composites, NTIS/PS-78/0357, N78-27189,* Vol. 2, NTIS, Springfield, Va., 1978.
4. K. G. Kgeider, ed., *Composite Materials,* Vol. 4 of L. Broutman and R. Krock, eds., *Metallic-Matrix Composites,* Academic Press, Inc., New York, 1974.
5. W. J. Renton, ed., *Hybrid and Select Metal-Matrix Composites: A State-of-the-Art Review,* America Institute of Aeronautic and Astronautics, New York, 1977.
6. Ref. 4, p. 40.
7. Ref. 4, p. 49.
8. L. Rubin, *SAMPE J.* 4 (July–Aug. 1979).

9. L. R. McCreight, H. W. Rauch, and W. H. Sutton, *Ceramic and Graphite Fibers and Whiskers*, Academic Press, Inc., New York, 1965.
10. B. R. Noton and co-eds., *ICCM2, Proceedings of the 1978 International Conference on Composite Materials,* AIME, Warrendale, Pa., 1978, pp. 1337–1339.
11. C. C. Chamis, R. F. Lark, and T. L. Sullivan, *Boron/Aluminum-Graphite/Resin Advanced Composite Hybrids, NASA TND-7879,* NASA, 1975.
12. C. C. Chamis, R. F. Lark, and T. L. Sullivan, *Super-Hybrid Composites—an Emerging Structural Material, NASA TMX-71836,* NASA, 1975.
13. C. C. Chamis and R. F. Lark, *Titanium/Beryllium Laminates: Fabrication Mechanical Properties and Potential Aerospace Applications, NASA TM-73891,* NASA, 1978.
14. D. A. McDanels and R. A. Signorelli, *Effect of Fiber Diameter and Matrix Alloys on Impact Resistant Boron/Aluminum Composites, NASA TND-8204,* NASA, 1976.
15. J. W. Brantly and R. G. Stabrylla, *Fabrication of J79 Boron/Aluminum Blades, Final Report, NASA CR-159566,* NASA, 1979.
16. C. T. Salemmee and G. C. Murphy, *Metal Spar/Superhybrid Shell Composite Fan Blades, NASA CR-159594,* NASA, 1979.
17. C. C. Chamis, *J. Aircraft* **14**(7), 644 (1977).
18. C. C. Chamis, ed., *Structural Design and Analysis,* Part I, Vol. 7, Academic Press, Inc., New York, 1974.
19. C. C. Chamis, ed., *Structural Design and Analysis,* Part II, Vol. 8, Academic Press, Inc., New York, 1975.
20. Ref. 18, p. 8.
21. H. J. Oberson, Jr., *SAMPE J.* 4 (Nov.–Dec. 1977).
22. S. T. Mileiko and V. M. Anishenkov, *Sci. Adv. Mater. Process Eng. Ser.* **24**(Book 1), 799 (1979).
23. Ref. 19, p. 231.
24. C. C. Chamis, *Comput. Struct.* **3,** 467 (1973).
25. C. C. Chamis, *Lamination Residual Stresses in Multilayered Fiber Composites, NASA TND-6146,* NASA, 1971.
26. C. C. Chamis, *Residual Stresses in Angleplied Laminates and Their Effects on Laminate Behavior, NASA TM-78835,* NASA, 1978.
27. Ref. 5, p. 72.
28. R. D. Goolsby, "Fracture of Crack Divider Al/Al Laminates" in *Proceedings of the 1978 International Conference on Composite Materials, Metallurgical Society of AIME, Toronto, Canada, April 16–20, 1978,* pp. 941–960.
29. G. H. Koch, *SAMPE Q.* **11**(1), 7 (1979).
30. K. H. Miska, *Mater. Eng.,* 31 (June 1978).
31. Ref. 5, pp. 67–97.
32. P. G. Sullivan, *Elevated Temperature Properties of Boron/Aluminum Composites, NASA CR-159445,* NASA, 1978.
33. R. E. Schramm and M. B. Kasen, *Adv. Cryog. Eng.* **22,** 205 (1977).
34. C. C. Chamis and T. L. Sullivan, *A Computational Procedure to Analyze Metal Matrix Laminates with Nonlinear Lamination Residual Strains, NASA TMX-71543,* NASA, 1974.
35. C. C. Chamis and T. L. Sullivan, *Nonlinear Response of Boron/Aluminum Angleplied Laminates, Under Cyclic Tensile Loading: Contributing Mechanisms and Their Effects, NASA TMX-71490,* NASA, 1973.
36. S. Janes, *Application of Conventional Joining Techniques to Boron Fiber and Carbon Fiber Reinforced Aluminum, Report D2532 H1,* Battelle Institute, Frankfurt, FRG, 1976.
37. Ref. 5, pp. 99–157.
38. Ref. 5, pp. 159–255.
39. Ref. 4, pp. 269–318.
40. B. R. Collins, W. D. Brentnall, and I. J. Toth, *Properties and Fracture Modes of Borsic Titanium, AFML-TR-73-43,* AFB, Ohio, 1972.
41. R. A. Signorelli, *Review of Status and Potential of Tungsten-Wire–Superalloy Composites for Advanced Gas Turbine Engine Blades, NASA TMX-2599,* NASA, 1972.
42. Ref. 4, pp. 229–267.
43. W. D. Brentnall, *FRS Composites for Advanced Gas Turbine Engine Components, Final Report TRW-ER-7887-F, AD-AC50 59518ST,* Cleveland, Ohio, 1977.

44. D. W. Petrasek, E. A. Winsa, L. J. Westfall, and R. A. Signorelli, *Tungsten Fiber Reinforced Fe-CrAlY—A First Generation Composite Turbine Blade Material, NASA TM-79094,* NASA, 1979.

General References

Reference 4 is a general reference.

C. C. Chamis, "*Characterization and Design Mechanics for Fiber-Reinforced Metals, NASA TND-5784,* NASA, 1970.

K. M. Prewo, *Fabrication and Evaluation of Low Cost Alumina Fiber-Reinforced Metal Matrices, Report UTRC (United Technologies Research Center)/R77-912547-13, Contract F33615-76-C5199,* June 15, 1977.

A. K. Dhingra and W. H. Krueger, "Magnesium Castings—Reinforced with Du Pont Continuous Alumina Fiber FP, *paper presented at the 36th World Conference on Magnesium, Oslo, Norway, June 25–26, 1979.*

A. K. Dhingra, *Phil. Trans. R. Soc. London Ser. A* **294,** 559 (1980).

C. C. CHAMIS
National Aeronautics and Space Administration

LAC. See Shellac.

LACQUERS. See Coatings, industrial; Paint.

LACRIMATORS. See Chemical warfare.

LACTIC ACID. See Hydroxy carboxylic acids.

LACTONITRILE, $CH_3CHOHCN$. See Cyanohydrins.

LACTOSE, $C_{11}H_{22}O_{11}$. See Carbohydrates; Milk products; Sugars.

LAMINATED AND REINFORCED PLASTICS

Reinforced plastics are combinations of fibers and polymeric binders or matrices that form composite materials (qv). The strongest geometry in which any solid can exist is as a wire, a fiber, or a whisker, ie, the strength of any solid is determined by the defects it contains—voids, cracks, discontinuities, etc—and the magnitude of their weakening influence depends upon their absolute size. Thus, if a fine-diameter fiber can be drawn or grown from a material, any defect it contains must be very small: the material will be stronger than in the bulk form (1). But fibers, even if they are strong when pulled, have very limited structural utility by themselves. They bend easily, especially if they are small in diameter, and when pushed axially, they buckle under very low forces. To remedy these inadequacies, a supporting medium is required, surrounding each fiber, separating it from its neighbors, and stabilizing it against bending and buckling. These are the functions of the matrix and they are best fulfilled when good adhesion exists between the two. If the fiber–matrix interaction is only a mechanical fit, without adhesion, the combined function is impaired and both components are underutilized.

Adhesion requires intimate contact, on the scale of a few hundred picometers (10^{-8} cm), over large surface areas. The best practical way to achieve this is by rendering one component a liquid, having it completely wet the surface of the other, and then solidifying it after the contact has been established (see Adhesives). Polymers readily make the liquid–solid phase change, and do not require much energy or elaborate processing to do so. Many thermoplastic polymers melt in the range of 150–250°C without large inputs of heat, and they readily solidify upon cooling. During the molten stage, even though their viscosity may be high, they are liquids and they possess the ability to wet most surfaces so their adhesion potential is readily available. The thermosetting polymers pass through a liquid phase just once during their life, while they are being polymerized and cross-linked into heat-infusible forms. During this liquid phase, the viscosity is very low, however, and they can easily infiltrate fiber bundles and fabrics, encapsulating every fiber; once this is done, solidification by catalysis or mild heating is quite straightforward.

The combination of strong fibers and synthetic polymers to form reinforced plastics and laminates derives from several basic considerations of materials science: the inherent strength of fine fibers, the wetting requirement for adhesion, and the ease of the liquid–solid phase change by synthetic polymers. These and other factors are responsible for the continuing rapid growth in the production of reinforced plastics, which has averaged about 10–12%/yr for many years and shows no signs of lessening, despite energy and raw materials availability problems.

At all levels of technological sophistication, the use of reinforced plastics continues to expand. Aircraft, marine, automobile and chemical markets hold additional opportunities, but one important fact must be recognized: reinforced plastics are not cheap, either by volume or by weight. At present, their cost ranges from $2.75 to as much as $165 per kilogram, in finished form. To compete with traditional materials—metals, wood, ceramics, concrete, and glass—their unique characteristics must be fully utilized in any application. The more this criterion can be fulfilled, the greater is the success of the application. Filament-wound pipe for corrosive service in oil fields is an almost perfect example (see Piping systems; Plastic building products). It is resistant to acids, water, oil, and decay. It is light, easy to transport, handle, and place. On-site joining with adhesives requires simple equipment, materials, and procedures; highly skilled labor is not necessary. Conventional tools and practices require only modest changes to be appropriate. Automated production promotes consistent, high quality units within an established context of industry standards, at tolerable costs. The cost in-place is absolutely competitive with metal pipe; the cost over the service life dramatically favors reinforced plastic.

Properties

The specific gravity of reinforced plastics is low, in the range of 1.5–2.25, compared to 3 for aluminum, 7.9 for steel, 2.5 for concrete, and 2.7 for natural granites and marbles. Only wood, at 0.5, is lower among the structural materials. This low density, and the ease of forming into intricately curved or corrugated forms, makes possible very high stiffness-to-weight structural elements and shapes. If strong, continuous fiber reinforcements are used, comparably high strength-to-weight ratios also can be obtained. These two factors, the high specific stiffness and specific strength as they are called, are the reasons why both commercial and military aircraft use fiber-reinforced plastics so widely. Weight can be kept to a minimum, extra fibers can be placed and oriented only where needed, section thickness changes easily are made, and joints, fasteners, and discontinuities are eliminated. Obviously, special design procedures and criteria are required for the materials, since their properties and behavior are much different from conventional, isotropic metals (2–4).

In contrast to most metals, polymers do not oxidize or corrode in normal moist air. Even in more severe environments—water, acids, bases, common solvents—specific polymers can be formulated to exhibit unusual resistance, often bordering on inertness (see also Coatings, resistant). Thus, reinforced plastics are used very widely in chemical applications, such as pipes, tubes, tanks, hoods, ducts, flues, fittings, and general hardware. The success of one of the most familiar products, boats, depends directly upon the resistance to water-based degradation, both biological and chemical, and the ease of curved, monocoque-shell construction without any joints.

The same general characteristics make reinforced plastics attractive to the automobile designer, now under such pressure to reduce vehicle weight, and this market is growing rapidly. Reinforced plastics offer another advantage: complex subassemblies involving welds, rivets, screws, and bolts executed in metals often can be molded in one piece, eliminating substantial labor costs even though the plastics may be more expensive in terms of materials costs. A classic example of this is the front-end grille assembly which also supports the many lights used in current automobiles. In metals, several dozen different parts require manufacture and assembly; in reinforced plastics, the whole unit is made in one molding operation, and it weighs 50% less. The cost in place is substantially reduced.

Most polymers are poor conductors of heat and electricity and if the fibers added to them have similar properties, strong, stiff insulating composites result. This was the motivation for the development of one of the earliest laminates, which is still very popular. As the supply of natural mica declined, its use to insulate motors, generators, and transformers expanded, so the need for a substitute became urgent. Thus, sheets of Kraft paper impregnated with phenol–formaldehyde polymer [*9003-35-4*] were squeezed together under heat and pressure to produce a replacement "for mica," a trade name familiar to all, although the current product is used more for decorative and protective purposes than for electrical insulation.

Fibers

Table 1 presents data about most of the fibers used in composites and laminates. The two predominant fibers are glass and cellulose.

Glass. Fibrous glass comprises well over 90% of the fibers used in reinforced plastics because it is inexpensive to produce and possesses high strength, high stiffness, low specific gravity, chemical resistance, and good insulating characteristics. Originally, the composition was developed for standoff insulators for electrical wiring, in which resistance to surface adsorption of water is important to reduce arcing. Only later was it discovered to be an excellent fiber-former: the operating window of temperature and rates within which fibers can be drawn from the melt is unusually large, and full-scale continuous production of such E-glass (Electrical grade) fibers became practical. Based on this ease of production, and the attractive properties of the fibers, uses were sought. Reinforced plastics and fireproof textiles have emerged as the principal markets.

In reinforced plastics, the glass is used in various forms. When chopped into short lengths (6–76 mm) and gathered into a felt or matte, it is easy to handle and low in cost. The discontinuities, however, penalize the strength properties of the final composite; as a result, a large variety of woven textile forms, or fabrics, also are available. These textiles use continuous fibers, gathered into yarns of low twist, and woven with essentially conventional textile processes. The fabrics cost more but offer better properties, although some sacrifices of formability into complex curved shapes may be incurred unless the design of the weave anticipates this. Contrary to what might be expected, sometimes the dry fabrics are not easily handled because wracking, unravelling, and other problems can arise as they are cut, tailored, and put into place in molds or on forms.

The best properties are achieved with nonwoven fabrics in which all the fibers are straight, continuous, and aligned parallel in a single direction (see Nonwoven textiles). Layers of these can be stacked atop each other, with each layer oriented in a specific direction, thus providing any option between high directionality, useful in a rod or a pole for example, to approximately nondirectional properties, referred to

Table 1. Fiber Properties

Material	Specific gravity	Tensile strength, MPa[a]	Tensile modulus, GPa[a]	Cost range, $/kg
E-glass	2.6	3570	85	2–5
carbon[b]	1.6	2035	357	30–150
carbon[b]	1.9	1790	430	30–150
Kevlar 29[c,d]	1.44	2860	64	5–10
Kevlar 49[c,d]	1.44	3750	135	5–10
bulk spruce wood	0.46	104	10	0.60–1.00

[a] To convert MPa to psi, multiply by 145; to convert GPa to psi, multiply by 145,000.
[b] Properties depend upon carbon/graphite ratio.
[c] E. I. du Pont de Nemours & Co., Inc.
[d] See Aramid fibers.

as quasi-isotropic, which would be suitable for structural plates or sheets similar to metals. Obviously, such nonwoven fabrics are not easy to handle unless they are impregnated with a partially cured matrix (prepregs); if not, most of their use is in automated machine processes which operate continuously producing pipes, cylindrical tanks, rods, and other similar shapes generated by revolving forms or molds.

Although compounded to be resistant, E-glass is not impervious to water and, like all glasses, its surface also is sensitive to abrasion damage. To protect the surface, and to promote adhesion to the polymeric matrix, coupling agents are applied to the fibers as soon after forming as possible. These are complex molecules, one end of which is intended to react with the glass and the other with the polymer as the latter solidifies. Thus, a primary valence-bond bridge joins the reinforcement to the matrix. This straightforward idea continues to be the subject of much technical controversy and dispute but from an empirical viewpoint there is no question: a glass fiber-reinforced composite lacking a coupling agent will lose half its strength after a month under water, whereas one with a proper coupling agent shows no strength loss at all. Consequently, only rarely are coupling agents not used irrespective of the type of reinforcing fiber employed (5).

There is one other characteristic of glass that is not widely recognized. When subjected to tensile loads–pulls for prolonged periods of time, glass breaks at stress levels much below those measured in short-time (2–5 min) laboratory tests (6). This behavior, known as static fatigue, effectively reduces the useful strength of glass if it is intended to sustain such loads for months or years in service. The reduction can be as much as 70–80%, depending upon the load duration, temperature, moisture conditions, and other factors, and if fracture does occur, it gives little or no prior warning, since glass is a brittle material. Conservative design practices and periodic inspections are recommended preventive measures in such applications where failure could have critical consequences (see Glass).

Cellulose. Cellulose fibers come from many sources and are used in many forms. Cotton (qv), jute, hemp, sisal, and bagasse (qv) from sugar cane provide economical comparatively long fibers with attractive properties of stiffness, strength, low specific gravity, and resistance to handling damage (see Fibers, vegetable). Often these are spun into yarns, then woven into fabrics such as burlap or canvas (see Textiles). Disadvantages include water sensitivity, lack of flame resistance, and susceptibility to biodegradation. The common source of short cellulose fibers, of course, is wood and the conventional use form is paper. As mentioned above, Kraft paper is very commonly used in laminates, where it is impregnated with phenolic resin and then fused under high pressure and moderate temperature to form electrical insulation stock. A refined wood product, alpha cellulose, also is used as a reinforcing fiber in plastic molding compounds to reduce curing shrinkage and brittleness and to improve impact resistance (see Cellulose).

Graphite. The stiffest fibers known are composed of graphite, which theoretically can be almost five times more rigid than steel. Practically, those now produced commercially are 1.5–2 times stiffer and laboratory specimens a bit more than 3 times stiffer have been made. The fibers themselves are composites: only part of the carbon present has been converted to graphite, in tiny crystalline platelets specially oriented with respect to the fiber axis. The higher the graphite content, the stiffer becomes the fiber; unfortunately, stiffness and strength are inversely related (7) (see Carbon).

Despite much work over many years by numerous technical organizations, the cost of graphite fibers remains high: the cheapest and lowest quality still are 15–20 times more expensive than glass. Barring unusual breakthroughs, this disparity is likely to persist since the starting materials are expensive and the processes of carbonization and graphitization consume much time, energy, and materials, and require close control throughout. As a result, their use in composites is limited to applications that place a premium on saving weight: aircraft, missiles (see Ablative materials), some sports equipment, and certain specialized hardware. For a time it was thought that automotive applications would be forthcoming, but better use of conventional materials and public acceptance of smaller automobiles have delayed such developments. Further substantial cost reductions will be necessary to penetrate this market.

Aramids. Aramid fibers (qv) are relatively recent industrial products and can be considered as an evolutionary product following nylon (7). In the stiffness range between glass and steel, they are lighter than glass, comparably strong, and much tougher and absorb considerable energy before breaking, even under impact conditions. Aramid fibers are two to five times the cost of glass, and they are used in composites in sports and transportation equipment, and in protective systems where ballistic stopping exploits their superior impact resistance. These fibers are highly crystalline and directional in character. Their transverse strength is very low, so that scuff resistance is poor and fibrillation can become excessive when the fibers are abraded or worked mechanically.

Polymeric Binders

In principle, any polymeric resin that can be liquefied and thereby used to wet the reinforcing fibers can be used as a matrix for a composite material. In practice, there are examples of almost any resin in some kind of composite formulation. Realistically, however, the bulk of reinforced plastics produced is based on polyester, epoxy, or a few thermoplastic matrix materials (see Elastomers, synthetic—thermoplastic; Epoxy resins; Polyesters). Only these are discussed below.

Polyesters. A polyester is the reaction product of a diol and a dicarboxylic acid. About ten different diacids and seven or eight different diols are in common use, from which different polyesters are available. A polyester is a low viscosity, clear liquid, composed of linear molecules with the potential for further chemical combinations. The latter are effected by adding another low molecular weight liquid, a cross-linking agent, which forms bridges or cross-links between the polyester chains when catalyzed by chemical additives or by heat, irradiation, or other energy inputs. The mixture thickens, releases heat, solidifies, and shrinks. Again, seven or eight different monomers are commonly used, and a very wide variety of polyester matrix resins are commercially available; most are comparatively inexpensive and easy to apply. The flexibility and variety of composition also facilitates the tailoring of properties for specific applications. This art has reached a fairly high level of sophistication so that electrical, chemical, or structural grades (among many) have been developed for optimum performance in such service (8). The volumetric shrinkage upon hardening (curing) is approximately 8% and since any reinforcement present does not shrink, this can cause internal stresses and cracking as well as dimensional, inaccuracy, surface roughness, and instability. Much of the art of molding is concerned with minimizing and preventing such actions, which greatly affect the quality of the final composite.

Epoxy Resins. Epoxy resins shrink less than polyesters (4%), adhere to most surfaces better, are affected less by water and heat, cure in a more controllable manner, and are more expensive. The basic material is a diglycidyl ether, a low viscosity liquid of moderate molecular weight, formed by reaction of epichlorohydrin and Bisphenol A (see Chlorohydrins; Alkylphenols). This resin can be cross-linked or solidified by a number of different amines, anhydrides, and acids, and it is capable of reacting with many other polymers to form copolymer substances. Therefore, the epoxy family has composition and property versatility comparable to the polyesters (9). Materials used for the purpose include phenolics, melamines (see Amino resins), polyamides (qv), esters, and many elastomers. The result is a very large number of commercially available epoxies formulated to optimize certain characteristics sought in specific applications: metal coatings, electrical insulation, chemical resistance, structural strength, adhesives, etc. Despite their superior inherent adhesion, however, when epoxies are reinforced by fibers it is necessary to use coupling agents to retain the strength of the composite if exposed to water.

In practice, both polyester and epoxy resins contain a large number of additional substances. Since both are adversely affected by exposure to sunlight, uv absorbers (qv) often are added. Stabilizers are used to prolong storage life and to prevent gradual gelation. Viscosity modifiers, fire retardants, mold-release agents (see Abherents), additives to improve molded surface smoothness, impact improvers, chemical accelerators, and catalysts are among the materials commonly used in such resins. The result is a very complex system which is heterogeneous on a microscale.

Thermoplastic Materials. As mentioned above, a number of thermoplastic materials also are used as matrix resins. These include nylon, polystyrene, polyethylene, polypropylene, styrene/acrylonitrile, polycarbonate, and polysulfone (see Polyamides; Styrene plastics; Olefin polymers; Acrylonitrile polymers; Polycarbonates; Polymers containing sulfur). Usually the reinforcement is glass, although the use of graphite fibers is growing. All of these resins melt reversibly when heated; as a result, high production processes such as injection molding and extrusion are used to fabricate parts. Short (≤6.4 mm) chopped fibers encapsulated in polymer enter such machines but the rigorous mixing and high shear flow patterns encountered cause much damage to them. In the finished part, the fiber length has been reduced by 10–100 times, so that its principal contribution is stiffening rather than strengthening. This can be especially important at elevated use temperatures where the fibers reduce creep and enhance dimensional stability; the burning characteristics also can be changed favorably by their presence. Impact resistance often is improved and, on balance, fiber reinforcement of many thermoplastics is attractive enough to ensure a continued, rapid growth of the field (10).

Fillers

Fillers differ from fibers in that they are small particles of very low cost materials; they are used extensively in reinforced plastics and laminates. Typical fillers include clay, silica, calcium carbonate, diatomaceous earth, alumina, calcium silicate, carbon black, and titanium dioxide. Sometimes they comprise as much as half the volume of a composite. They provide bulk at low cost, and they confer other valuable properties such as hardness, stiffness, abrasion resistance, color, reduced molding shrinkage, reduced thermal expansion, flame resistance, chemical resistance, and a sink for the heat evolved during curing. If coupling agents are used on their surfaces, many of them improve the impact strength of the composite. The original motive for using fillers was to replace part of the expensive resin phase, and this skill is important, but now their technology is becoming more complex and their use more versatile. Effects of particle size, size distribution, shape, surface treatment, and blends of particle types on the flow behavior and final properties of the composite are better understood, ensuring that their use will continue to expand (7) (see Fillers).

Products and Processes

Statistics in this area are imprecise, but as shown in Tables 2 and 3, a reasonable estimate of use categories is: marine (boats, decks, shields, etc)—20%; transportation (autos, trucks, trailers, body components)—24%; construction (corrugated sheet, space dividers, showers, tubs, lightweight control panels)—21%; chemical (pipes, ducts, hoods, tanks)—12%; electrical (printed circuits, insulation panels, switchgear)—9%; appliances, aircraft, recreational (housings, partitions, luggage racks, floor pan-

Table 2. Reinforced Plastics Markets[a]

	1979 (1000 metric tons)
Reinforced polyesters	
marine	159
transportation	191
construction, other (eg, farm equipment)	171
chemical	98
electrical	73
appliance, aerospace, consumer	106
Total	*798*
Epoxies	
electrical laminates	10
filament winding	9
other	5
Total	*24*

[a] Ref. 11.

Table 3. Reinforced Plastics Materials[a]

	1979 (1000 metric tons)
Reinforced thermoplastics	
nylon	20
polycarbonate	5
polyester	22
polypropylene	34
styrenics	15
other	9
Total	*105*
Decorative and industrial laminates	
Kraft phenolic	181 (est)
melamine phenolic	45 (est)

[a] Ref. 11.

els)—13%. In 1979, approximately one million (10^6) metric tons of reinforced plastics were used in the United States, the three principal resins being polyesters, thermoplastics, and epoxies. As mentioned above, over 90% of the fiber reinforcement was glass.

Hand Lay-Up/Spray-Up. The construction of boat hulls and comparably large articles is done by hand lay-up methods: one or two layers of dry fabric or matte are placed in molds or on forms and then liquid polyester is poured on them. The resin is distributed throughout the fibers by hand-roller action, and after several cycles, the desired thickness is achieved. Curing takes place by catalysis, either at room temperature or by means of heat lamps or warm-air ovens. Cycles are lengthy (8–24 h), production rates are low, molds are not expensive, and the quality of the product can vary widely. However, hulls as long as 49 m have been produced by this method. In the U.S., more than 90% of all boats less than 15 m are built by this procedure. (A variation of the technique, spray-up, uses a hand-held gun which propels chopped fiber and liquid resin against a mold surface until the desired thickness has been attained. This is used commonly for smaller-size boats).

Die-Molding. Most automotive parts and general hardware are made in heated matched metal dies or molds mounted in hydraulic presses operating semi-automatically. A charge of SMC (sheet-molding compound) or BMC (bulk-molding compound) is placed in the mold, then cured under moderate heat and pressure (120–175°C, 3.5–13.8 MPa (500–2000 psi)) for cycles ranging from 15–90 s. Typical recipes are 100 parts resin, 150 parts filler, and 30 parts chopped glass fiber. The molded piece is removed hot and allowed to air-cool before undergoing further operations such as trimming, assembly or painting. The quality of the product is fairly consistent, much subassembly work normal to metal practice can be avoided, and die costs are much lower than for sheet-metal stamping or metal die casting. Thus, shorter production runs are economical and more design flexibility and variety can be achieved.

Filament Winding. Pipes, tubes, and cylindrical tanks are made by a process known as filament winding. A mandrel or form is rotated about one axis as continuous yarn or roving which passes through a bath of liquid resin is wet-wound on it, usually in complex patterns controlled by automated machinery. Once the required wall thickness is reached, heat curing in autoclaves, ovens, or by heating the mandrel is begun. Because the reinforcement is continuous and its orientation subject to close control, very efficient structures can be produced consistently by this method, and in the case of smooth-bore pipes, the process can be continuous. Once cured, the pipe is removed from the metal mandrel and cut into proper lengths; mandrels for tanks and pressure bottles often are inflatable, soluble, or collapsible, so they can be removed through end ports. Pipe production for oil fields, the chemical processing industry, water distribution, and sewage disposal has reached very large volumes, in diameters from several centimeters to as much as 6 m; fittings also can be filament wound or die-molded. The attributes of light weight, ease of assembly (by field-applied adhesives), corrosion resistance, and economical production have combined to make this a very attractive business.

Flat or corrugated sheet stock normally is made in simple, multi-opening presses, often with prepreg (preimpregnated fabric or matte). Generally, quality and consistency are related directly to pressure and temperature, ie, high performance composites usually are made in this way. High fiber contents, good resin penetration, and full, controlled cures are easiest to achieve here and partially automated production methods keep costs moderate. Most of the electrical insulating board and printed circuit panels are high pressure laminates, using woven glass fabric or cellulose paper, with epoxy or phenolic resins.

Injection Molding. Injection-molded, fiber-reinforced thermoplastics compete directly with die-cast metals in many hardware and automotive parts. Higher production rates, lighter weight, and assembly simplification are their principal advantages. The fibers stiffen and allow higher operating temperatures. Often, however, the flow patterns of the molten material into the mold cavity cause significant orientation of the fibers and the localized anisotropy which results can adversely affect strength and thermal-expansion properties. Considerable skills are required of the mold designer and the machine operator. Also, the reuse of fiber-reinforced thermoplastic scrap material from runners, sprues, gates, and flash presents problems that can impair the overall efficiency of material utilization. Despite these difficulties, however, this is the fastest growing segment of the entire composites family (see Plastics processing).

Miscellaneous. The four processing techniques that have been mentioned are the dominant ones in use today: hand lay-up/spray-up; matched metal die molding (including flat-press laminating); filament winding; and injection molding. However, there are many others that may be variations of one of the above or are distinctive in themselves. For example, vacuum-bag techniques can be used to help remove air from a hand lay-up and to provide modest positive pressure on large, low cost molds too bulky for mechanical pressing. Better quality composites result from the reduction in void content and from the superior impregnation of fiber bundles by the liquid-matrix resin.

Pultrusion, in which continuous filaments are drawn through an orifice, which also meters out the encapsulating resin, can produce a variety of profiles containing high volume fractions (50–70%) of reinforcement. Very strong and stiff rods, tubes, and structural shapes result, but the properties are highly directional; transverse behavior depends almost entirely on the matrix which is at least an order of magnitude weaker and more flexible.

Transfer molding, borrowed from rubber technology, employs a piston to force compound to flow from a cool cylindrical reservoir into a hot mold cavity where it cures. Less fiber damage and fewer orientation effects are encountered, compared to injection molding, but cycles are longer and the process is discontinuous. This process has improved quality, permits more intricate detail and thicker sections, but it also has higher costs.

A variation of filament winding is centrifugal casting and some pipe is produced in this manner. Prewoven braids or sleeves of reinforcement can be impregnated quickly and effectively by the liquid resin and it is easy to maintain pressure on the assembly during curing. A very good quality product results.

Certain thermoplastic sheet materials such as polypropylene, nylon, polycarbonate, and others can be formed at temperatures well below their melting points. Even if these contain fibers or fabrics, some such materials still possess cold formability and they can be stamped, roll-formed, or bent into final shape simply by mechanical action. A substitute for sheet metal results that is light, stiff, strong, corrosion resistant, and has insulating properties.

Reaction injection molding (RIM) processes pump two or more liquid streams under high pressure into an impingement chamber where they are mixed intimately and then are forced into a mold cavity at much lower pressures. By heat, catalysis, and the mixing action, rapid polymerization occurs from reactions among the components in the cavity, which can tolerate no gaseous by-products. Urethanes are the principal polymers used in this process which emphasizes speed, large parts, and automation (see Urethane polymers). Reinforced RIM is being developed (RRIM) containing short chopped glass fibers or particulate fillers for stiffness, reduced shrinkage, and higher use temperatures. By this route, the "forgiving fender" for automobiles is predicted.

The area of fiber-reinforced plastics is complex and dynamic. A review of this type can describe only basic ideas and a few highlights, but comprehensive treatments are available (12–14) and several technical journals report new developments on a continuing basis (15–18).

Perhaps one of the most exciting developments in fibrous reinforcement is now emerging: molecular composites. Using special extrusion and drawing techniques, synthetic polymer fibers with ultrahigh orientations can be produced, and they exhibit strength and stiffness characteristics approaching theoretical limits, directly comparable to the strongest metals available (19). If the technology of these can be manipulated so they can be used in and reinforce matrices of the same or similar compositions, nearly ideal composite materials can be postulated (20).

BIBLIOGRAPHY

"Lamination and Laminated Products" in *ECT* 1st ed., Vol. 8, pp. 185–192, by H. W. Narigan and G. E. Vybiral, Panelyte Division, St. Regis Paper Co.; "Laminated and Reinforced Plastics" in *ECT* 2nd ed., Vol. 12, pp. 188–197, by C. S. Grove, Jr., Syracuse University, and D. V. Rosato, Consultant, Plastics World.

1. A. Kelly, *Strong Solids,* Clarendon Press, Oxford University Press, London, 1973.
2. R. M. Jones, *Mechanics of Composite Materials,* Scripta Book Co., Washington, D.C., 1975.
3. *Advanced Composites Design Guide,* 3rd ed., Vols. I–V, Air Force Materials Laboratory, Wright-Patterson Air Force Base, Oh., 1973 to date, with revisions.
4. L. R. Calcote, *Analysis of Laminated Composite Structures,* Van Nostrand Reinhold Co., New York, 1969.
5. "Composite Materials" in E. P. Plueddemann, ed., *Interfaces in Polymeric Matrix Composites,* Vol. 6, Academic Press, Inc., New York, 1974.
6. E. B. Shand, *Glass Engineering Handbook,* McGraw-Hill, New York, 1958.
7. H. S. Katz and J. V. Milewski, eds., *Handbook of Fillers and Reinforcements for Plastics,* Van Nostrand Reinhold Co., New York, 1978.
8. P. F. Bruins, ed., *Unsaturated Polyester Technology,* Gordon and Breach Science Publishers, New York, 1976.
9. H. Lee and K. Neville, *Handbook of Epoxy Resins,* McGraw-Hill, New York, 1967.
10. J. Agranoff, *Modern Plastics Encyclopedia, 1979–1980,* Vol. 56, No. 10A, McGraw-Hill, New York, 1979.
11. *Mod. Plast.* **57**(2), (Jan. 1980).
12. S. S. Oleesky and J. G. Mohr, *Handbook of Reinforced Plastics,* Reinhold Publishing Corp., New York, 1964.
13. G. Lubin, ed., *Handbook of Fiberglass and Advanced Plastics Composites,* Van Nostrand Reinhold Co., New York, 1969.
14. L. J. Broutman and R. H. Krock, eds., *Composite Materials,* Vols. 1–8, Academic Press, Inc., New York, 1974.
15. *Annual Proceedings—Reinforced Plastics/Composites Institute,* Society of the Plastics Industry, New York.
16. *J. Compos. Mater.,* Technomic Publishing Co., Westport, Conn.

17. *J. Sci. Technol. Reinforced Mater.*, I.P.C. Science and Technology Press Ltd., Bury St., Guilford, Surrey, England, GU25BH.
18. *Polymer Composites,* Society of Plastics Engineers, Inc., Brookfield Center, Conn.
19. A. Ciferri and I. M. Wards, eds., *Ultra-High Modulus Polymers,* Applied Science Publishers, London, 1979.
20. A. Ciferri, *Polym. Eng. Sci.* **15**(3), (Mar. 1975).

FREDERICK J. MCGARRY
Massachusetts Institute of Technology

LAMINATED MATERIALS, GLASS

A laminate is an orderly layering and bonding of relatively thin materials. A commonly laminated material is glass. Usually, two pieces of float or sheet glass are bonded with poly(vinyl butyral) (PVB) (see Vinyl polymers, poly(vinyl acetals)) to produce a highly transparent safety glass, eg, an automotive windshield. This combining of transparent abrasion-resistant glass and resilient plastic achieves the durability and safety demanded of such products. Other materials that may be incorporated in laminated glass are colorants, electrically conducting films or wires, and rigid plastics. The value of the laminate is the utilization of the desirable properties from each of the constituents. In the case of laminated glass, the excellent weathering properties of the glass protect the impact-energy-absorbing plastic interlayer from deterioration, abrasion, and soiling.

Benedictus, a French chemist who accidentally broke a flask that contained dried-on cellulose nitrate, is credited with founding the laminated-glass industry (1). The first patent was issued in 1906 (2).

The growth of the laminated-glass market was slow until automobile numbers and automotive speeds increased to the point that glass-caused injury was of concern. By the late 1920s, laminated windshields were standard in United States automobiles. The most common construction was two pieces of plate glass bonded with cellulose nitrate. However, the plastic interlayer introduced problems of haze, discoloration, and loss of strength, and it was replaced by cellulose acetate in 1933. Cellulose acetate demonstrated improved stability to sunlight but lacked strength over a broad temperature range and produced haze. The advent of the poly(vinyl butyral) resins in 1933 permitted the development of the modern interlayers that are used to make the majority of laminated safety glass in use today; the resins were adopted for all automotive laminates by 1939.

Laminated glass is not a true composite material (see Composite materials). The glass needs the safety-net effect of the interlayer if impacted, and the interlayer needs the durability and rigidity of the glass for useful service other than during impacts. Exceptions where laminated glass more truly fits the definition of a composite are when it is used for noise attenuation (see Insulation, acoustic) or bullet resistance. In these applications, the alternate layering of rigid and soft materials achieves results beyond those produced by either alone.

Properties

Laminated materials frequently have limits on properties below those found in one of the components. Laminated glass, with a PVB interlayer, has a service temperature not exceeding 70°C (conservative), far below that of solid glass. The strength of laminated glass is dependent upon the number, thickness, and strength of the individual glass plies and upon the characteristics of the particular interlayers used. For the majority of laminates consisting of two plies of annealed glass and one PVB interlayer, the bending strength is about 0.6 of that for an equal thickness of solid glass.

Laminated glass becomes more rigid with a decrease in temperature and, below −7°C, it approaches the performance of solid glass. At temperatures above 38°C, it responds more nearly like two glass plies separated by a fluid. Some applications utilize heat-strengthened or tempered glass for additional strength. Figure 1 is an example of a wind-load chart for the combination of heat-strengthened and laminated glass (3). Wind-load information is used jointly by the architect, glazing contractor, and glass manufacturer to determine the permissible glazing area and glass thickness that is required to meet the design wind load.

Most laminated glass applications are concerned with impact strength, and minimum performance levels are required by specification. The impact strength of two plies of laminated annealed glass and various PVB thicknesses are reported in ref. 4. Aircraft laminates may utilize electrical resistance heating to improve laminate impact resistance (see Aircraft Windshields).

Automotive and architectural laminates of PVB develop maximum impact strength near 16°C, as shown in Figure 2. This balance is obtained by the plasticizer-to-resin ratio and the molecular weight of the resins. It has been adjusted to this optimum temperature based on environmental conditions and auto population at various ambient temperatures. The frequency and severity of vehicle occupant injuries vs temperature ranges at the accident location has been studied (5), and the results

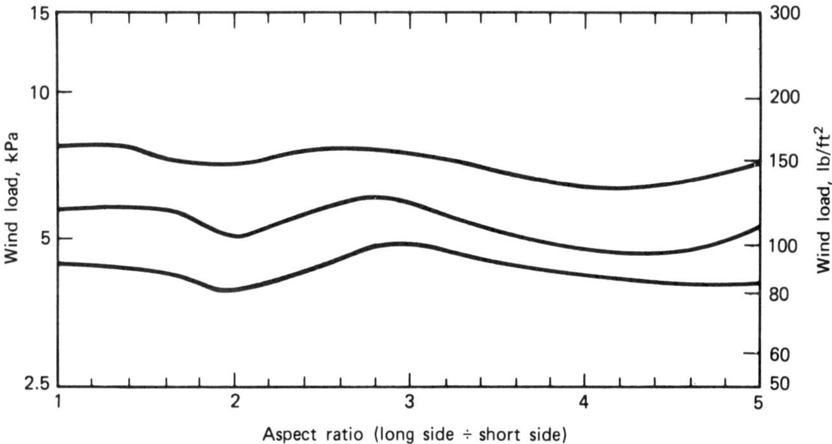

Figure 1. Wind-load data for heat-strengthened and laminated 3.2-mm glass. Architect's specified probability of breakage is 8/1000 laminates for a 1-min uniform wind-load duration. Four sides supported in weathertight rabbet. Courtesy of PPG Industries, Inc.

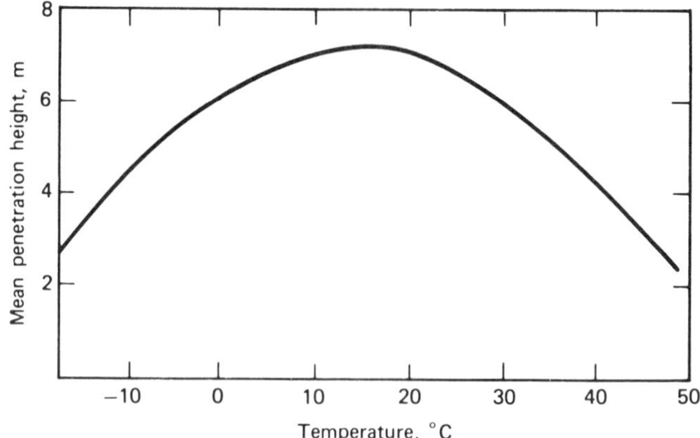

Figure 2. Typical penetration resistance vs temperature data from laboratory procedure (305 × 305 mm laminates, 0.76 mm PVB; 2.27-kg ball impact). Courtesy of Monsanto Co.

confirm the selection of the maximum performance temperature and the decreasing penetration resistance at temperature extremes.

The optical properties of laminated glass are required to be equal to solid glass, since most applications are in vision areas. Light scattering by the interlayer essentially is nonexistent if PVB is used. Clean-room practices can reduce the dust and lint that is attracted to the surfaces (see Sterile techniques). Visible-light transmittance of a typical automotive laminate (2.1 mm glass–0.76 mm PVB–2.1 mm glass) is nearly equal to solid glass (Fig. 3), and noticeable color usually is absent. Visible-light transmittance is about 88% for clear glass laminates and 78% for tinted laminates. All light is absorbed below 370 nm and several discrete absorption bands are in the infrared beyond 1100 nm. The uv absorption may be enhanced when additional protection of color dyes is required, eg, gradient shade bands in automotive windshields or merchandise in

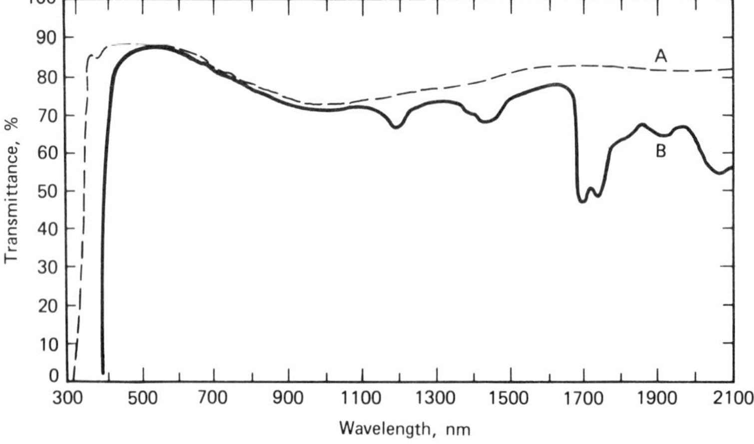

Figure 3. Visible-light transmittance of automotive laminate: A, 4.75-mm monolithic glass; B, 5-mm laminated glass. Courtesy of Ford Motor Co.

window displays (see Uv absorbers). Generally, the solar uv internal transmittance (neglecting losses resulting from surface reflections) is on the order of 30–35%, and the infrared is about 97% for 0.76-mm thick PVB.

The index of refraction of PVB at 1.48 is close enough to glass (which has a refractive index of 1.520) to couple the two glass plies, with a loss by reflectance of 0.02%:

$$R = \frac{(n_2 - n_1)^2}{(n_2 + n_1)^2} \tag{1}$$

where R is the reflectance at the interface and n_1 and n_2 are the different refractive indexes. The absorption coefficient for visible light (400–700 nm) generally is −0.25 to −0.45 for PVB. This produces a transmission loss within the PVB of 0.7–1.3%, which is mostly in the blue portion of the spectrum.

Subsequent to the lamination process, some defects may appear that were not visible previously. One of these phenomena is called a bull's eye when found in windshields. These are small depressions that are formed in the glass pair during bending by the presence of glass chips or other debris between the plies. Upon lamination, the pockets fill with PVB and become convex lenses. Conversely, shallow ridges on an internal glass surface may be absorbed in the PVB and the optics are improved. Various other optical distortions may be caused by nonparallel plies of glass or of PVB or nonuniform fabrication pressures.

Manufacture

Practically all laminated glass utilizes plasticized poly(vinyl butyral) (PVB) as the interlayer. Curved, laminated windshields are by far the principal product; silicone and cast-in-place urethane resins are used seldom. Laminators purchase PVB in rolls up to 500 m long, up to 270 cm wide, and from 0.38–1.52 mm thick. There are several plasticizers that are used and at different ratios of plasticizer-to-resin content, depending on the product being manufactured. Typically, Flexol 3 GH (bis(2-ethyl butanoic acid), triethyleneglycol ester) manufactured by Union Carbide, is utilized at about 44 parts per 100 parts of resin. Other plasticizers used are di-*n*-hexyl adipate and di-butyl sebacate. There are at least five companies offering these products with manufacturing facilities in the United States, Japan, Belgium, Germany, and Mexico. Since the plastic is an adhesive material, it is shipped either with a dusting of parting agent on the surface or is refrigerated so it does not cohere (see Abherents). The refrigerated material is clean, moisture adjusted, and ready to laminate. The dusted material requires washing and moisture conditioning. Removal of the parting agent by warm water, followed by a chilled water rinse, adds about 0.2% water content to the plastic which must be compensated either by overdrying before washing or drying after washing so as to achieve the desired 0.3–0.5% H_2O content. These steps are done more efficiently on the continuous roll before cutting.

The drying stage is carefully controlled to relax the sheeting of physical stresses as well as to adjust the moisture content. It consists of draping the sheeting over slat conveyors or driven rolls in a temperature- and humidity-controlled oven. The gradient-band sunshade that appears in many windshields is printed continuously on the interlayer roll at the PVB-manufacturing plant. To permit a more pleasing conformance to the curved glass, the interlayer may be preshaped which causes the extremities of the band to more nearly parallel the horizon in the installed windshield. The shaping of the interlayer may be carried out on the continuous roll using a vinyl expander prior to cutting the blanks (6). The radius of curvature is preset, depending on the particular windshield being manufactured. The interlayer then is cut to approximate laminate size and is accumulated in low stacks (150 mm) ready for assembly. Another method for shaping the interlayer involves warping in special ovens after the blanks are cut. The interlayer stacks must be stored in cooled, moisture-controlled rooms to prevent water absorption and blocking.

The glass for laminating may be annealed, heat-strengthened, tempered, flat or curved, clear or colored. Thicknesses of 1.5–12 mm or more are used. For flat laminates, the glass is cut to size, edged and treated, if specified, washed, and delivered to the clean room by conveyor. The washing process, in addition to cleaning, can affect the interlayer bond. Common water hardness residues at invisible levels can decrease the adhesion to the interlayer. The desired level is achieved by controlling the hardness of the final rinse water and by removing the water by air stripping as opposed to evaporative drying. The glass is cooled during drying to prevent premature grabbing when the interlayer is placed on the glass, thereby permitting easier positioning of the components.

In order to manufacture curved laminates, the glass is preshaped before laminating. This is achieved by simultaneously bending a pair of glass templates which are usually cut to the shape of the finished windshield and are separated by an inert powder to prevent fusing of the plies. The bending process occurs on a peripheral support metal fixture or mold; the pair slowly travels through a lehr so that the glass sags to shape by gravity. Glass temperatures of 600°C are required to achieve the shape, and the shaping is followed by annealing. Banded windshields usually are constructed with one or more pieces of tinted, heat-absorbing glass to enhance occupant comfort and to reduce air-conditioning load.

The clean room is operated at 18°C and 26% rh, which is an equilibrium condition for interlayer moisture control. The interlayer is placed on one piece of glass (Fig. 4) with the gradient band, if present, carefully positioned above the designated eye position. The adjacent piece is superimposed, excess interlayer is trimmed, and the sandwich is conveyed from the room through a series of heaters and rolls that press the assembly together while expelling air. Temperature is increased stepwise to 90°C and pressures of 170–480 kPa (25–70 psi) are applied. Solid rubber rolls usually are used with flat laminates, and curved glass requires segmented rolls on a swivel frame (Fig. 5) (7–8). For more complex shapes, peripheral gaskets may be applied and the assembly may be evacuated (9), or the entire assembly may be placed in a bag and evacuated. The bag may or may not be removed prior to autoclaving but, when using an oil autoclave where the oil would damage one of the components, an oil-resistant bag of poly(vinyl alcohol) is used (10). The tacked assembly is loaded onto racks for autoclaving, which may be either in an air or oil vessel that is capable of pressing the sandwiches at 1.38–1.72 MPa (200–250 psi) and 100–135°C for 30–45 min. Curved laminates and multiple laminates may require longer cycles.

In the autoclave cycle, the pressure is increased more rapidly than the temperature and then it is maintained towards the end of the cycle as the temperature is lowered, to reduce any chance of delamination. The exit temperature must be near 50°C to avoid thermal breakage. During the cycle, residual air is absorbed by the interlayer, and the embossed surface of the interlayer flows and wets the glass surface, thereby producing

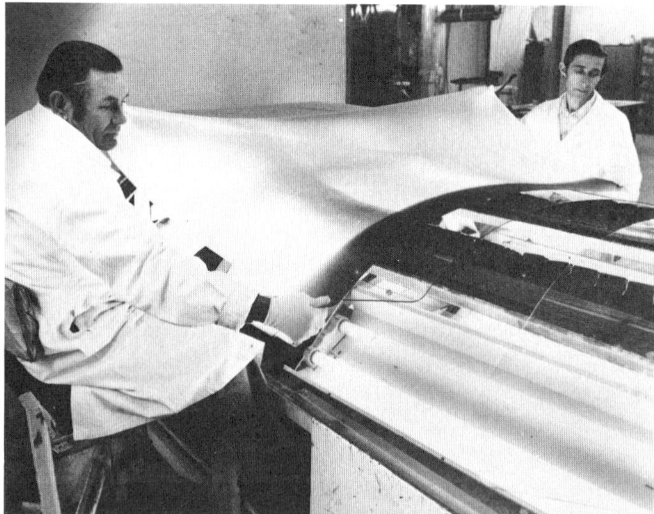

Figure 4. The vinyl interlayer is inserted into the matched set of bent glass in a cold room where the humidity and temperature are carefully controlled. Courtesy of Ford Motor Co.

Figure 5. The glass and vinyl sandwich is fed through a special deair machine to remove trapped air and increase the adhesion of the vinyl to the glass. Courtesy of Ford Motor Co.

the clear laminate. Occasional small residual bubbles of trapped air can be removed by an additional autoclaving cycle. The air–autoclave process is becoming more popular because it eliminates the subsequent washing of oily residues that are produced in oil autoclaving. The elimination of the process oil and subsequent wash-water waste products also are environmental improvements. In some cases, additional trimming of the interlayer or finishing of the glass edge is required. The protruding interlayer may be removed by wire brushing, but the glass edge is finished by abrasive belt-seaming or diamond-wheel edging (see Abrasives). The labeling of safety glass is done by grit-blasting through a mask or by enameling (see Specifications). A final step in windshield manufacture is the bonding of a small metal plate to the windshield. This plate, which is used to support the rearview mirror, is laminated to the glass using poly(vinyl butyral).

Radio antennas have been incorporated in some windshield models by inclusion of a very fine copper wire placed across the top of the windshield and vertically at the center. The wire is tacked to the interlayer prior to assembly and is embedded into the interlayer during autoclaving. An electrical connector that is soldered to the antenna wire is bonded to the bottom edge of the windshield for ease of connection.

Production and Shipment

Chemical attack, particularly from moisture and alkaline conditions, is prevented by use of acidic packing materials and open, ventilated packages. Good crate design and proper handling throughout shipment avoids mechanical damage. Glass-to-glass contact is never permitted. Long-term storage must be in well-ventilated areas, never in sealed containers.

Flat laminates are separated only by newsprint or plastic beads and are bound into a block to prevent movement between the laminates. Curved laminates are spaced to prevent abrasion and require supporting dunnage at several points to prevent breakage by excessive flexing during shipment. Banding and blocking are designed to add compressive forces only. Staining is not a problem in the uncovered areas, but the supporting members are specified to have an acidic content.

Laminated-glass products are considered noncombustible and are shipped without DOT hazardous-warning labels. Flat-laminate packs, because of their high density, do not fill the car or trailer and require sturdy bracing to prevent shifting. Glass products are shipped and stored in a vertical plane, and during transportation, they are placed so that each plate has an edge in the direction of travel.

Economic Aspects

The growth of laminated glass closely followed the growth of motor vehicles from the late 1920s to the 1960s. Windshields and side glasses of all domestic vehicles were laminated during this period. In the early 1960s, the flat laminated side glass was almost entirely replaced by curved tempered glass. The curved windshield laminated-glass market continued to follow automotive trends, but the flat-glass products redeveloped around architectural uses. Architectural products represent 5–10% of the laminated-glass volume. Increased consumer-safety awareness and security needs are expanding the flat-laminate market. Safety codes (eg, 16 CFR 1201 and ANSI Z97.1) specify laminated glass as one means of meeting their requirements (11–12).

Flat laminated glass and windshield-replacement production has been estimated to 1982 using trend lines of historical data (13). A similar technique has been used to determine original equipment manufacture (OEM) of windshields for worldwide use and the data are compiled in Table 1 (14). Laminated windshields, as opposed to tempered-glass windshields, are gaining in the market share outside of North America. From 37% of the non-North American market of 1976, they are estimated to have

Table 1. Laminated Glass and Windshield Production

Production	1976	1979	1982[a]
U.S. laminated flat glass, 10^6 m^2 [b]	2.7	3.5	4.6
wholesale values[a], 10^6 \$ U.S. (1979)	87	114	150
worldwide laminated windshield, 10^6 units			
U.S. and Canadian OEM[c]	13.1	14.7	17
U.S. replacements[b]	4.6	4.8	5.4
foreign	9.7	15.3	24
Total	*27.4*	*34.8*	*46.4*
wholesale value[a], 10^6 \$ U.S. (1979)	919	1137	1471

[a] Estimate.
[b] Courtesy of U.S. Glass Publications, Inc. (13).
[c] In part from ref. 14.

reached 55% by 1978 and are projected to obtain 75% by 1982. In addition to North America: Belgium, Italy, and the Scandinavian countries permit only laminated windshields, and other nations are increasing use by customer option.

Specifications

Practically all laminated glass is made, tested, and certified to comply with certain safety performance standards. In the United States, there are two types of standards: automotive and architectural. For the former, *ANSI Z26.1-1966* is used and is incorporated in the Federal Motor Vehicle Safety Standard 205 (15). It specifies safety performance, durability, and optical quality. Specific tests are required depending on the location in the vehicle where the glazing is to be used. Item 1, the most difficult to meet, may be used in any location in the vehicle. It requires, in addition to other tests, support of a 2.3-kg ball dropped from 3.7 m onto a 305-mm square of laminate. Item 2 (safety glazing for any location except windshields) may be met by laminated glass using thinner PVB because it does not require the 2.3-kg-ball test nor the distortion tests. Laminated glass also may be used in locations specifying item 3 (no visible transmittance requirement) or item 11A (bullet-resistant glass; also requires appropriate tests, eg, ballistic tests). Automotive safety glass is required to be labeled as to the manufacturer, code, item, and model number that will identify the type of construction.

Conformance to the standard is achieved by submitting samples to an approved laboratory for evaluation and submitting the laboratory report to the American Association of Motor Vehicle Administrators (AAMVA). The approved certificate is sent to the manufacturers with copies to the state and provincial jurisdictions for which the AAMVA serves as approvals agent (16).

Laminated glazing materials that are used in building locations specified by federal regulations are certified to comply to *Federal Standard 16 CFR 1201* by the Safety Glazing Certification Council (SGCC) (17). Other locations requiring safety glazing specified by state or local code may use *ANSI Z97.1-1975* (12). Glass complying to these standards is labeled permanently as to the standard (or standards) that it meets, including thickness, identification of the manufacturer, and plant. Also, it usually contains a date of manufacture. In situations where the large laminated sheets are cut into smaller pieces by the local distributor or installer, each piece is permanently labeled to indicate that it was cut from glass meeting the standard.

Certification to these standards is obtained by submitting a test report from an approved laboratory to the SGCC. Once certified, the product is assigned an SGCC certification number to identify it and the factory at which it was made. Subsequently, samples are selected randomly by the administrator at least twice a year to ensure continued adherence to the standard. Based on these reevaluation reports, SGCC authorizes continued use of the certification label and the product listing published in its directory. The building standards are concerned mainly with body impact, and they require testing by impact on the glazing with a 45-kg bag. Detailed testing procedures and interpretations are given in refs. 11–12.

Bullet-resistant glass products are tested according to *UL 752* (18). The test specifies that three shots are fired from 4.6 m and impacting within 100 mm of each other in a triangle, and that there is no penetration of the projectile nor any glass imbedded in the corrugated board. The level of approval is determined by the velocity and energy level of the bullet at the muzzle of the firearm. Additional tests required include impacts 38 mm apart and tests over temperature ranges of 13–35°C for indoor use and −31.7 to 49°C for outdoor use.

The above-mentioned codes contain requirements for accelerated durability tests. In addition, interlayer manufacturers and laminators expose test samples for several years under extreme weather conditions, eg, the Florida coast and Arizona desert. The laminated products weather extremely well, with no change in the plastic interlayer. Occasionally, clouding is noted around the edges when exposed to high humidity for long periods, but this is reversible.

Analytical and Test Methods

Interlayer moisture is one of the important controls for adhesion to glass. The moisture content equilibrates with the relative humidity to which the interlayer is exposed and, thus, is variable. Prior to lamination, interlayer moisture content is measured by one of three methods. The most rapid is by ir absorption using a spectrophotometric technique to determine a ratio of the 1.925 nm to the 1.705 nm wavelength peak (Fig. 3). A slower but less expensive method is weighing the interlayer before and after vacuum desiccation. The third and classical method is by Karl Fisher reagent. The infrared method, in addition to being the most rapid, permits measurement of the interlayer moisture content while the interlayer is in the laminate.

Interlayer bond strength is determined by either pummeling the laminate at −18°C to break away the glass and to determine the amount of adhering glass particles or by compressively shearing the laminate sample in a universal test machine. The optimum pummel range is three to six units on an arbitrary scale established by the industry. The relationship of pummel and compressive shear data to water content of the interlayer is given in Figure 6 and that of pummel to mean penetration height is given in Figure 7. These data are influenced also by residual hardness of the water used to wash the glass.

Subsequent to processing, an inspection is made for incomplete bonding, inside dirt, and glass quality. In the case of windshields, rigid optical standards are required, and these must be evaluated for the completed windshield. Extensive test requirements are described in the appropriate codes (11–12,15,18–24), and they include light stability, humidity, boil test, abrasion, and assorted impact tests.

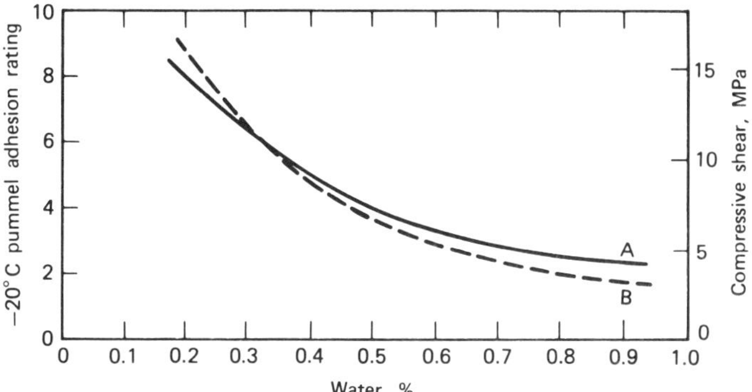

Figure 6. Typical effect of moisture on PVB adhesion. A, Pummel data from Monsanto Co.; B, compressive shear data from DuPont Co. To convert MPa to psi, multiply by 145.

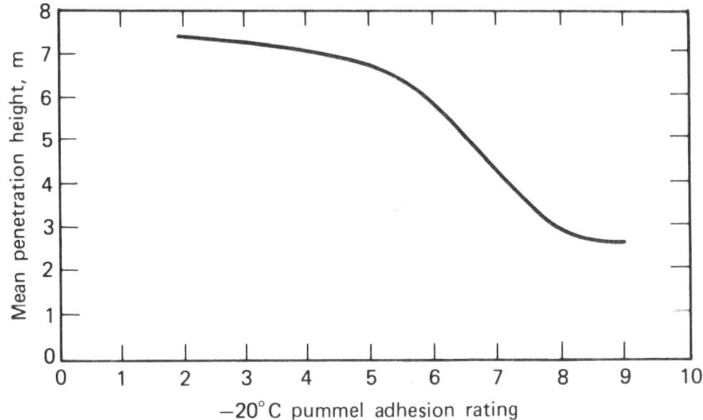

Figure 7. Typical variation of mean penetration height with adhesion (2.27-kg ball, impact at 20°C on 305 × 305-mm laminates; 0.40% water, 0.76 mm PVB). Courtesy of Monsanto Co.

Uses

Impact-Resistant Windshields. Performance difference between windshields manufactured in the FRG and the United States was reported in the early 1960s (25–26). Three variables contribute to the greater safety of the German windshields against impact: thinner glass (especially the inboard member), thicker plastic interlayer, and higher moisture content of the interlayer. The latter acts as a plasticizer and adhesion control (see Plasticizers). By reducing the adhesion of the interlayer to the glass, more interlayer area can be released and stretched during impact. Also, the thicker interlayer, in addition to having more inherent strength, causes more fracturing of the glass during impact than in the older model of windshield. This, in turn, increases the amount of released interlayer for impact-energy absorption. Upon review by the SAE Glazing Committee, it was agreed that the improvement was desirable, if it could be accomplished without taking the risk of increased water content. The U.S. PVB manufacturers subsequently developed controlled-adhesion interlayers without increasing moisture above the previous standard content, and the glass fabricators utilized this material to produce laminated windshields with more than twice the impact resistance of the pre-1966 windshields. This product was introduced in limited production in 1965 and was used in all United States car lines for the 1966 models (27).

The ASA (now ANSI) performance code for Safety Glazing Materials (*Z26.1*) was revised in 1966 to incorporate these improvements in windshield construction. The addition of test no. 26 requiring support of a 2.3-kg ball dropped from 3.7 m defined this level of improvement. It was based on a correlation established between 10-kg, instrumented, head-form impacts on windshields, on 0.6 × 0.9 m flat laminates and the standard 0.3 × 0.3 m laminate with the 2.3-kg ball (28). Crash cases involving the two windshield interlayer types were matched for car impact speeds and were compared (29). The improved design produced fewer, less extensive, and less severe facial lacerations than those produced in pre-1966 models.

Additional improvements have been incorporated since 1966 with the availability of thinner float glass. Glass thickness and interlayer thickness have been studied to optimize the product for occupant retention, occupant injury, and damage to the windshield from external sources (30–31). The thinner float-glass windshields are more resistant to stone impacts than plate-glass windshields. The majority of laminated windshields are made of two pieces of 2–2.5 mm of annealed glass and 0.76 mm of controlled-adhesion interlayer.

Special Laminated Windshields. Combinations of strengthened glass and interlayer offer advantages of lessened weight, higher impact resistance, lowered laceration potential, and resistance to bending stresses. These may be needed in high speed aircraft, helicopters, and motor vehicles. The additional strengthening is achieved by chemical or thermal processes. The chemical process utilizes a special glass of high alumina content, which, after bending to shape, is processed by ion exchange (qv) in molten potassium salts to produce a highly compressed skin and a center tension with a bending stress up to 276 MPa (40,000 psi). The product, offering reduced laceration potential, was used briefly on some United States-produced cars (32–33). A thermal process capable of inducing high stress into thin float glass and incorporating a precise forming capability is known as Triplex Ten-Twenty and is a patented process of the Triplex Safety Glass Company Limited (UK) (34–35). For automobile windshields, this process is used to induce a high stress into the inner glass whereas the outer glass is strengthened partially but to a level that is much less than that which would cause dicing by stone impact. Combined with 0.76-mm automotive PVB, these glass plies give a windshield that causes significantly less facial soft tissue damage upon impact than windshields made with annealed glass. This product is available commercially and is used in several European cars.

Another variation of special construction windshields is the Securiflex windshield made by St. Gobain (Fr.) (36). A fourth ply, a plastic film on the innermost surface, is formed and bonded onto a thin conventional windshield. It essentially prevents the occupant from scraping across the broken glass at impact.

A deicing–defogging windshield and back window were produced from 1974 to 1976 by Ford, and it is comprised of a Sierracin conductive film that is laminated within the interlayer rather than being coated onto the glass (37). This requires 1–2 kW for rapid deicing, necessitating a special alternator in the vehicle. The product has 70–75% visible transmittance and good ir reflectance which improves driving comfort as well as efficiently deicing the glazing.

Another type of deicing windshield and back window is made with resistance wires embedded in the plastic interlayer. Light diffraction from the wires and visible distortion from the heated interlayer have been recognized and modifications have been propcsed (38–40) to undulate the wires randomly, and to vary the plasticizer content of the interlayer containing the wires (41). The process has been used in back windows and is in limited use in windshields (particularly for public service vehicles) in Europe.

Other automotive uses of laminated glass include colored glass and decorated glass. The privacy glass used in the side and rear glazing of vans frequently is laminated from a brown PVB and clear glass. Opera windows containing metallic ornaments and sufficient plastic interlayer to accomodate their thickness are popular. Laminated roof glazing usually is a combination of metallized glass and a colored PVB.

Aircraft Windshields. Aircraft windshields have extreme requirements in service temperature and pressurization and they must be resilient against high velocity bird impact. In addition, they must offer excellent visibility, both from optics and deicing capability, and an aerodynamic design. These highly specialized windshields are produced in low volume and are made by few companies (Sierracin, PPG Industries, Triplex Safety Glass Company). Construction varies with the need and service potential of the aircraft. Small planes of limited altitude and speed usually have acrylic monolithic windshields. Slower commercial aircraft use flat laminated glass made with aircraft-grade PVB (Monsanto Saflex PT). These aircraft require deicing capability which may be given by a conductive film that is pyrolytically or vacuum deposited on a glass surface or a conductive plastic film that is laminated in the sandwich. The third general class of aircraft windshield is for the modern, commercial, wide-body aircraft. These windshields become extremely complex, large in size, and expensive. A fourth type is for high speed, low flying military aircraft where birds, high skin temperature, and gunfire warrant extremely complex construction. The third and fourth types are multilayer constructions; typical examples are shown in Figures 8 and 9 (42–43).

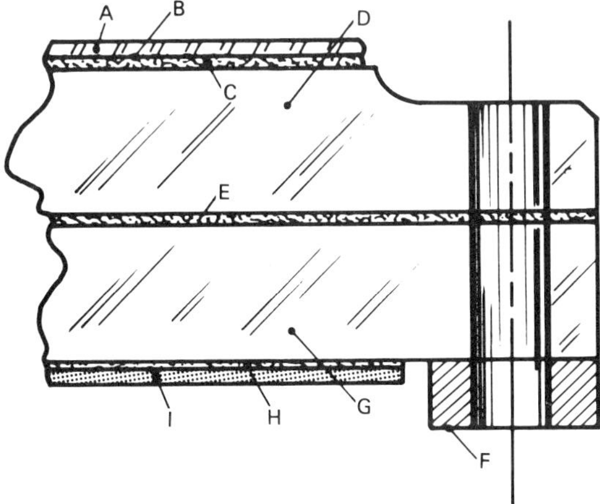

Figure 8. Cross section of Sierracin windshield used on Boeing 747 (42). A, 2.2-mm chemically strengthened glass; B, Sierracote 3 conductive coating; C, 1.9-mm PVB; D, 23-mm stretched acrylic; E, 1.3-mm PVB; F, laminated cloth spacer ring; G, 23-mm stretched acrylic; H, 0.6-mm PVB; I, 3.0-mm Sierracin 900.

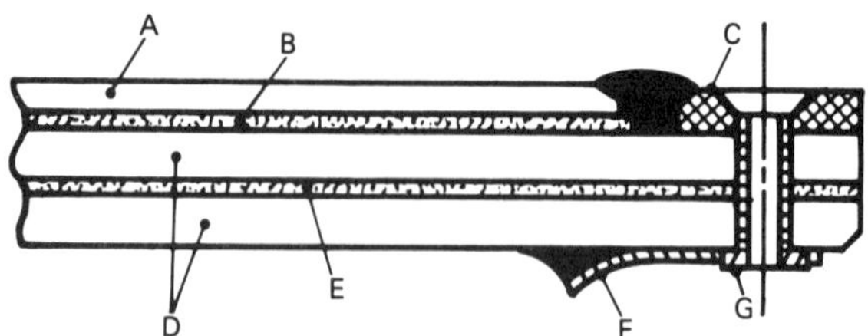

Figure 9. Sierracin lightweight, birdproof F-111 windshield cross section (43). A, 3.0-mm as-cast acrylic face ply; B, S-100 silicone interlayer; C, fiberglass retainer; D, 6.4-mm polycarbonate structural ply; E, S-120 polyurethane interlayer; F, stainless steel bearing strip; G, stainless steel bushing.

The Boeing 747 windshield (Fig. 8) is about 1 × 1.1 m and is curved to increase the pilot viewing area and to reduce air drag and air noise. Composed of seven plies, it weighs about 64 kg (42). The outer strengthened glass skin and the inner plastic shield may be replaced when damaged. Triplex Safety Glass Company also makes wide-body aircraft windshields, flat and curved, for Boeing and others. For the Boeing 747, two precurved, 12-mm plies of Ten-Twenty glass are laminated with PVB and covered with a 3-mm ply of Ten-Twenty glass bent to conform to the curved windshield. An electroconductive coating, Hyviz, is applied to the inner surface of the outer ply and then is laminated to the 12-mm Ten-Twenty ply (44).

The construction of the F-111 windshield shown in Figure 9 replaced a glass–silicone laminate previously used. The all-plastic windshield has improved impact resistance so that it is birdproof to 250 m/s (43). In this instance, the scratch resistance of glass was waived to obtain the impact performance at the allowed weight.

Architectural Products. Many specialized laminated glasses are made for architectural needs such as safety, sound attenuation, solar control, and security. These products may be further enhanced with colors and patterns for decorative effects. Safety glasses are specified in potentially higher risk breakage areas and overhead or sloped glazing (defined as more than 15° from vertical). Overhead glazing materials have varied in the past but more localities are accepting laminates. Sloped and overhead glazing frequently have heat-strengthened or tempered glass used in the construction of the laminate (45). Vertical passageway glazing usually is a 0.76-mm interlayer and sloped glazing is constructed with a 1.52-mm interlayer to accommodate the waviness of heat-treated glasses when they are used.

Noise attenuation is achieved effectively with laminated glass by the combination of the vibration damping effect of the plastic interlayer and usually an unbalanced glass thickness. Typical construction is 0.76–1.5-mm interlayer that is laminated with 3–10-mm glass. The type of glass or strength is not a factor in noise attenuation. A Sound Transmission Class Index of 34–41 is achieved with single laminate glazing and can be improved if combined with double glazing that has large air spaces. Mounting of the glass in an air-tight but flexible gasket reduces sound transmission (46). Airports, hotels, factory offices, and control rooms benefit from laminated acoustical glazing (47) (see Insulation, acoustic; Noise pollution).

Laminated glass is used for solar control, particularly where a highly reflective surface is not desired and where the laminate contributes other benefits. In these applications, a uniformly pigmented interlayer is obtained from the manufacturer and the laminate can be prepared by the conventional process. A broad range of colors and transmission levels is available with shading coefficients as low as 0.41. Pigmented interlayer is considered to be more color stable than dyed interlayer. Browns, blues, greens, pink, white, and clear plastic containing uv absorbers are readily available. Body-colored glasses may be used also, usually with clear interlayer. In these cases, the laminate is dependent only upon the solar properties of the glass.

All laminated glass increases the level of security to some extent. However, depending on the application, security glass is constructed of multiple layers of glass and PVB and, in some instances, additional rigid plastic is included. Laminated glass permits the same visual observation as normal glass but prevents or delays entry (or exit) until the attempt is detected. It complies with test UL 972 (48).

Bullet-resistant glass is constructed of many layers of glass and aircraft-type PVB depending on the level of resistance desired. Typical products are 38–50 mm thick and weigh 90–130 kg/m^2.

A third type of security glass is installed in modern penal institutions. This product is utilized for prisoner detention and obviates iron bars and their demeaning aspect. Typical construction utilizes three or more layers with at least one ply of thick PVB. Strengthened glass and electrically conductive circuits for alarms may be included. Large, heavy sections of similar construction have been used for underwater windows for boats, submarines, and aquariums. Four plies of fully tempered, 10-mm glass plus three plies of 1.9-mm PVB totaling 44.5 mm in thickness has a modulus of rupture of 172 MPa (25,000 psi) (49).

Glazing of laminated architectural glass requires additional care in the selection of sealants (qv) and drainage design. Sealants must be free of solvents (particularly aromatics) and mineral or vegetable oils (3) and must not provide pockets that would trap water at the glass–PVB edge. Similarly, the glazing detail must be designed with proper drainage (45). Generally, the practice is similar to that of glazing organically sealed insulating units (50–51).

BIBLIOGRAPHY

1. A. F. Randolph, *Mod. Plast.* **18**(10), 31, 98 (1941).
2. U.S. Pat. 830,398 (Sept. 4, 1906), J. C. Wood.
3. *PPG Glass Thickness Recommendations to Meet Architects' Specified 1-Minute Wind Load,* PPG Industries, Pittsburgh, Pa., 1979.
4. R. G. Rieser and G. E. Michaels, "Factors in the Development and Evaluation of Safer Glazing," *Proceedings of the Ninth Stapp Car Crash Conference,* University of Minnesota, 1965, pp. 181–203.
5. R. L. Morrison, "Influence of Ambient Temperature on Impact Performance of HPR Windshields," *paper presented at Fifteenth Stapp Car Crash Conference, SAE, 1971,* pp. 603–612.
6. U.S. Pat. 3,885,899 (May 27, 1975), D. J. Gurta and G. A. Koss (to Ford Motor Company).
7. U.S. Pat. 2,983,635 (May 9, 1961), R. E. Richardson (to Pittsburgh Plate Glass Company).
8. U.S. Pat. 3,009,850 (Nov. 21, 1961), J. P. Kopski and L. H. Schmidt (to Ford Motor Company).
9. U.S. Pat. 2,994,629 (Aug. 1, 1961), R. E. Richardson (to Pittsburgh Plate Glass Company).
10. U.S. Pat. 2,374,040 (Apr. 17, 1945), J. D. Ryan (to Libbey-Owens-Ford Glass Company).
11. *Standard 16 CFR 1201,* Consumer Products Safety Commission, Bethesda, Md.
12. *Safety Performance Specifications and Methods of Test for Safety Glazing Material Used in Buildings, ANSI Z97.1-1975,* American National Standards Institute, New York, 1975.
13. R. C. Cunningham, *U.S. Glass Metal and Glazing,* U.S. Glass Publications, Memphis, Tenn., Jan. 1979, p. 28.
14. *Ward's Automotive Yearbook,* 39th and 41st ed., Ward's Communications, Inc., Detroit, Mich., 1977 and 1979.
15. *Safety Code for Safety Glazing Materials for Glazing Motor Vehicles Operating on Land Highways, Z26.1-1966,* American National Standards Institute, New York.

16. *Manufacturer's Guide for Safety Equipment Services,* American Association of Motor Vehicle Administrators, Washington, D.C., 1979.
17. *CPSC Certified Products Directory,* Safety Glazing Certification Council, Hialeah, Fla., 1980.
18. *Standard for Bullet Resisting Equipment UL 752, ANSI SE 4.6-1973,* Underwriters' Laboratories, Inc., Melville, L.I., N.Y., 1973.
19. *ASS AS-R1-1968,* Standards Association of Australia, North Sydney, Australia, 1968.
20. *Brazilian Contran Resolution,* 483/74, Federal Official Gazette, Brazilia, Brazil, 1974.
21. *BS 5282-1975,* British Standards Institute, London, Eng., 1975.
22. *Specifications Relating to Safety Glass Requirements for Land Vehicles and Their Trailers,* Ministere De L'Equipement, Paris, Fr., 1975.
23. *Requirements on Safety Glass for Automotive Glazing,* Bundesministerim Ür Verkehr, Godesberg, FRG, 1973.
24. *A Tutte Gly Impettorati-Compartmentali Della Motorizzazione-Civille E Dei Transporti N Concessione E Sezioni,* Ministero Dei Transporti, Rome, Italy, Articles 218 and 297–302, 1959.
25. G. Rodloff, *Automobiltech. Z. (ATZ)* **64**(6), 1979 (1962); *English Translation 62-18916,* National Translation Center, Chicago, Ill.
26. G. Rodloff, *Automobiltech. Z. (ATZ)* **66**(12), 353 (1964); *English Translation 65-11982,* National Translation Center, Chicago, Ill.
27. J. C. Widman, *Recent Developments in Penetration Resistance of Windshield Glass, SAE 650474,* SAE, 1965.
28. E. R. Smith, "SAE Test Procedure for Quality Control of Windshields," *paper presented at Ninth Stapp Conference,* SAE, 1965, pp. 277–281.
29. D. F. Huelke, W. G. Grabb, and R. O. Dingman, *Automobile Occupant Injuries from Striking the Windshield, Report No. Bio-5,* Highway Safety Research Institute, Ann Arbor, Mich., 1967.
30. R. G. Rieser and J. Chabel, *Safety Performance of Laminated Glass Structures, SAE 700481,* SAE, 1970.
31. H. M. Alexander, P. T. Mattimoe, and J. J. Hofmann, *An Improved Windshield, SAE 700482,* SAE, 1970.
32. J. R. Blizard and J. S. Howitt, *Development of a Safer Nonlacerating Automobile Windshield, SAE 690474,* SAE, 1969.
33. L. M. Patrick, K. R. Trosien, and F. T. DuPont, *Safety Performance of a Chemically Strengthened Windshield, SAE 690485,* SAE, 1969.
34. S. E. Kay, J. Pickard, and L. M. Patrick, "Improved Laminated Windshield with Reduced Laceration Properties," *paper presented at Seventeenth Stapp Car Crash Conference,* SAE, 1973, pp. 127–169.
35. S. E. Kay, V. J. Osola, J. Pickard, and P. A. Brereton, *ATZ* **79**(9), 389 (1977).
36. U.S. Pat. 3,979,548 (Sept. 7, 1976), W. Schäfer and H. Raedisch (to St. Gobain Industries).
37. B. P. Levin, *Development of the Sierracin Electrically Heatable Safety Glass Interlayer, SAE 740156,* SAE, 1974.
38. U.S. Pat. 3,522,651 (Aug. 4, 1970), J. E. Powell and M. W. Lacey (to Triplex Safety Glass Company).
39. U.S. Pat. 3,895,433 (July 2, 1975), G. A. Gruss (to General Electric Company).
40. U.S. Pat. 3,954,547 (May 4, 1976), W. Genther (to St. Gobain Industries).
41. U.S. Pat. 3,903,396 (Sept. 2, 1975), P. T. Boaz and J. S. Maluchnik (to Ford Motor Company).
42. G. L. Wiser, "Sierracin® Glass/Plastic Composite Windshields," *paper presented at Conference on Transparent Materials for Aerospace Enclosures, U.S. Air Force and University of Dayton, June 25, 1969.*
43. J. B. Olson, "Design, Development and Testing of a Lightweight Bird-Proof Cockpit Enclosure for the F-111," *paper presented at the Conference on Aerospace Transparent Materials and Enclosures, Long Beach, Calif., Apr. 24–28, 1977.*
44. R. W. Wright, "High Strength Glass in Service—A Status Report," *paper presented at the Conference on Aerospace Transport Materials and Enclosures, Tech. Report AFML-TR-76-54, Atlanta, Ga., 1975.*
45. *Archit. Rec.* (6), 143 (1979).
46. *Architectural Saflex® for Sound Control, Tech. Bulletin No. 6295,* Monsanto Polymers and Petrochemicals, St. Louis, Missouri, 1972.
47. J. M. Clinch, *Study of Reduction of Glare, Reflection Heat and Noise Transfer in Air Traffic Control Tower Cab Glass, FAA-RD-72-65, AD747069,* NTIS, Springfield, Va., 1972.

48. *Burglary-Resisting Glazing UL 972,* Underwriters' Laboratories, Inc., Melville, L.I., N.Y., 1978.
49. *The New Look—Prisons Without Bars, Sierracin Field Report,* Sierracin Corporation, Sylmar, Calif., 1972.
50. *Alum. Curtain Walls* **6**, 24 (Sept. 1972) (Architectural Aluminum Manufacturers Association, Chicago, Ill.).
51. *FGMA Glazing Manual,* Flat Glass Marketing Association, Topeka, Kansas, 1974.

General References

R. N. Pierce and W. R. Blackstone, *Impact Capability of Safety Glazing Materials, PB195040,* Southwest Research Institute, San Antonio, Tex., 1970; contains detailed descriptions of test equipment, methods, and results for all types of glazings.

SAE Transactions (annual), *SAE Handbook* (annual), Society of Automotive Engineers, Warrendale, Pa.

Stapp Car Crash Conference series (annual, 1956 and continuing), Society of Automotive Engineers, Warrendale, Pa.; for safety and construction of automotive glass.

ROBERT M. SOWERS
Ford Motor Company